Ion Exchange
and
Solvent Extraction
of Metal Complexes

Ion Exchange
and
Solvent Extraction
of Metal Complexes

Y. Marcus
and
A. S. Kertes

Department of Inorganic Chemistry,
The Hebrew University of Jerusalem
Israel

WILEY—INTERSCIENCE

A Division of John Wiley and Sons Ltd

London   New York   Sydney   Toronto

Library of Congress catalog card No. 69–19061
SBN 471 56858 9

Composed by Jerusalem Academic Press, Jerusalem, Israel
Printed in Great Britain by J. W. Arrowsmith Ltd., Bristol

to

Charles D. Coryell

Preface

Some six years ago it appeared to us that a book on the ion-exchange and solvent extraction of metal complexes would be useful to the scientific community in the fields of solution chemistry, coordination chemistry and the various aspects of separation chemistry—analytical as well as industrial. The aim was to provide a physico-chemical monograph in which the theoretical framework of the title subject is elaborated, general relationships derived and experimental material presented in a systematic way. Our thesis, in this book, is that ion exchange and solvent extraction are complementary techniques, and that they can be applied intelligently only if the behaviour of electrolytes and molecules in both aqueous solutions and in organic solvents are clearly understood. We therefore organized this monograph according to the following lines.

The material is presented essentially in three sections: one, common to both methods, is devoted to the theory of electrolyte solutions, including the chemistry of complex formation, the second section to ion exchange and the third to solvent extraction chemistry. This is an obvious division, one which permits an emphasis on fundamental similarities for the two methods.

In the first section an effort is made to give a condensed picture of the chemistry of electrolytes in aqueous and organic solutions. Though similar material is partly presented in greater detail in a number of more specific monographs, this approach permits the quantitative discussion of distribution equilibria used in subsequent chapters to be traced back to the basic principles of ionic equilibria so that anyone unfamiliar with the whole field can work his way into the main topic. The equilibria of formation of complexes in solution are also considered.

The second section contains chapters devoted to a detailed discussion of ion exchangers, their swelling and exchange characteristics. Particular attention is paid in these chapters to the methods of calculation of the stability of metal complexes as derived from distribution data using ion exchangers. A command of the method implies above all an understanding of its rationale and a power of adapting its numerous modifications to particular requirements.

The chapters dealing with the chemistry of extraction are designed to afford as complete a picture as possible of the basic chemistry of the process. An initial chapter devoted to the quantitative aspects in such heterogeneous systems is followed by chapters dealing with the various types of solvent extraction systems. This sets the stage for the final chapter on the applications of distribution methods, where the ion-exchange and solvent extraction methods are compared, and their application to metal complex formation studies is illustrated by several examples. Our original intention to give a full account of the application of these methods to complex formation, systematically ligand by ligand and metal by metal, proved much overambitious.

We have tried to give in our treatment an integrated view of the state of each subject as it is today, and a fair presentation of controversial points of view, in the many cases where a definite treatment of a problem is as yet impossible. Certain distribution equilibria could be discussed in terms of mass-action-law equations and of their deviations from ideal behaviour. For others, on the other hand, the fundamental knowledge of solution behaviour available permitted a discussion only in general, qualitative terms. The importance of a topic is reflected, in a way, by the volume of the literature concerning it, and this, in turn, is also a factor in the depth as well as the breadth of the treatment afforded it in this book. Since many of the subjects dealt with are developing rather rapidly, it has been impossible to keep the discussion completely up to date. These factors may account for its shortcomings, some of which the authors are well aware, and of which they would certainly like to learn, in order to improve possible future editions.

The book is addressed to practising chemists, both those who want to study complex formation in solution, and those who employ distribution methods in analysis, radiochemistry and preparative work. The book is directed also to graduate students, who can find here a convenient introduction to solution chemistry, and a fundamental discussion of separation methods. Inorganic chemists, analytical chemists, physical chemists and radiochemists will find important aspects of their disciplines covered in this book.

The authors finally wish to thank all those who helped them in making this book. Y. M. is indebted to the Israel Atomic Energy Commission, Soreq Nuclear Research Centre, which permitted him to spend much of his time on this work, while he was in a responsible position with it. A.S.K. acknowledges a partial support by the USAEC (Contract AT–30–1–905) while he was with the Department of Chemistry and the Laboratory for Nuclear Science, Massachusetts Institute of Technology, where a part of the book was written. Both are obliged to Miss Miriam Balaban, who encouraged us in the conception of this book and helped throughout the process of its publication. We are grateful to the many girls who patiently typed pages and pages of drafts and the final manuscript, and in particular to Mrs. Malka Shavit, who carried the main burden of this job. We thank our many friends all over the world, who supplied us with reprints of their work, and discussed their ideas with us at visits and conferences, and appreciate their patience, waiting for this book, which took a long time to complete. Our wives and children, who have been the main victims of our ambition, are thus finally thanked for their forbearance.

Y. MARCUS
A. S. KERTES

Jerusalem, June 1968

Contents

1

Aqueous Solutions
of Electrolytes

The ion exchange and solvent extraction methods discussed in this book generally deal with two-phase systems. One of these phases is usually an aqueous solution of electrolytes. Much of the information we have on the ion exchange and solvent extraction processes comes from thermodynamic considerations. Hence a good understanding of the thermodynamics of aqueous solutions of electrolytes is necessary, and is a good introduction to the contents of this book. Structural considerations, which play a major role in the understanding of ion exchange and solvent extraction processes, will also be discussed briefly. Since many excellent books on thermodynamics in general, and on solutions of electrolytes in particular are available, the introduction, section 1.A, only recapitulates the most important concepts and considerations, which are of direct use for the rest of this book. It gives the formal thermodynamics of electrolyte solutions, and discusses briefly the structure of water.

Ionic hydration is of prime importance in determining the extractability and ion exchange properties of ions, and is discussed in detail. The thermodynamic functions of the hydration process are related, as far as possible, to the properties of single ions, as are also ionic radii and volumes and hydration numbers. The effect of hydration on the structure of the electrolyte solutions is thus related to an acceptable model of hydration (section 1.B).

The interactions in aqueous solutions of electrolytes are discussed in turn: from the ion–solvent interactions (section 1.B), which occur even at infinite dilution, through ion–ion interactions occurring in dilute solutions (section 1.C), where the Debye–Hückel treatment is successful, and pH measurements are significant, to concentrated solutions (section 1.D). Theories relating activity coefficients to appropriate models of the solutions are discussed, including alternative distribution functions, statistical entropy terms as well as empirical equations. Two concepts which have been found fruitful in concentrated solutions: the Hammett acidity function and the effective ligand activity, are then developed.

Most extraction and ion exchange processes deal with mixtures of electrolytes, which are discussed in section 1.E, mainly in terms of the Guggenheim and Harned treatments. The alternative way of looking at mixtures, using excess functions, is also briefly presented. The information available on electrolyte mixtures is made use of in a discussion of the constant ionic medium method for studying ionic interactions in aqueous solutions.

A final section (1.F) deals with the behaviour of non-electrolytes in electrolyte solutions, i.e. with the problems of salting-in and salting-out. The mechanical, electrostatic and hydration theories of salting are reviewed, and their relevance to extraction systems are discussed.

A. INTRODUCTION

An (aqueous) electrolyte solution is a homogeneous mixture of two components, i.e. substances which may be added independently. One of them is the solvent S (water W), usually present in a large excess, the other is an electrolyte G, which is considered to be anhydrous unless otherwise stated. The composition is thus defined by stating the number of moles: N_W of water and N_G of anhydrous solute G, contained in the mixture. The electrolyte is capable of dissociating into v ions, of which v_+ are cations, each bearing a charge $z_+ e$, and v_- are anions, each bearing a charge $z_- e$, which is to be taken algebraically, because the electrolyte is electrically neutral.

$$v_+ z_+ + v_- z_- = 0 \qquad v_+ + v_- = v \tag{1A1}$$

When the charge z is not specified as pertaining to cations or anions, it is to be taken algebraically.

a. Formal Thermodynamic Treatment

For our purposes, the most important thermodynamic quantity associated with an electrolyte solution is its free energy G

$$G = N_W G_W + N_G G_G + G^M \tag{1A2}$$

It is the sum of the contributions of the free energies of the water and the electrolyte, and the free energy of mixing, G^M. At a given temperature and pressure, the molar free energy of the solution, G, depends only on the composition, and in the following, the conditions of constant temperature and pressure are assumed. The chemical potential μ of a component in the mixture is the partial free energy, i.e. the free energy change when a mole of the component is added to an infinite amount of the solution.

The chemical potential is related to the composition through the concepts of the standard state and the activity. For the solvent, these quantities are defined by

$$\mu_W = (\partial G / \partial N_W)_{N_G} = \mu_W^0 + R T \ln a_W \tag{1A3}$$

Since the solution is in equilibrium with solvent vapour

$$\mu_W^0 \text{ (solution)} + R T \ln a_W = \mu_W \text{ (solution)} = \mu_W \text{ (vapour)} \tag{1A4}$$

$$= \mu_W^0 \text{ (vapour)} + R T \ln p/p^0$$

assuming the ratio of vapour pressures p/p^0 to be the same as the ratio of the fugacities. This approximation is valid for aqueous solutions, since low pressures and small differences are involved. The standard state for the vapour is pure water at 1 atm pressure at the given temperature, having the vapour pressure p^0, and the same standard state is used for the solution, so that μ_W^0 (vapour) $= \mu_W^0$ (solution). This defines then the water activity in the solution according to (1A3) and (1A4) as

$$a_W = p/p^0 \tag{1A5}$$

Since electrolytes are usually only partially soluble in water, it is not convenient to use the pure solid electrolyte as its standard state. The formal definitions

$$\mu_G = (\partial G / \partial N_G)_{N_W} = \mu_G^0 + R T \ln a_G \tag{1A6}$$

$$\mu_G = \mu_+ + \mu_- = \mu_G^0 + v_+ R T \ln a_+ + v_- R T \ln a_- \tag{1A7}$$

again hold, but the standard state is the hypothetical state of a solution at unit electrolyte activity, which has the same properties as that of a reference state, which is the ideal solution at infinite dilution.

Ideal solutions may be nearly realized at very high dilutions, and are considered as the limit at infinite dilution. An ideal solution is defined by

$$\Delta H_{id}^M = 0 \tag{1A8}$$

$$\Delta S_{id}^M = - R(x_w \ln x_w + x_G \ln x_G) \tag{1A9}$$

where x is the mole fraction, which for the case of the solute is defined by

$$x_G = \frac{N_G}{v N_G + N_w} = \frac{N_G}{v_+ N_+ + v_- N_- + N_w} \tag{1A10}$$

(it is conventional to count the cations and anions separately). Use of the relationship $\Delta G_{id}^M = \Delta H_{id}^M - T \Delta S_{id}^M$ and equations (1A2) and (1A6) yields Henry's law, as a practical characterization of the ideal solution

$$a_{G id}^{1/v}/x_G = constant \tag{1A11}$$

Setting the constant equal to unity defines $\mu_{G(x)}^0$ and $a_{G(x)}$ for the mole fraction concentration scale.

There are two other practical concentration scales in use. The molality m_G of the electrolyte is the number of moles of electrolyte dissolved in 1000 g (or $1000/M_w = 55.51$ moles) of water. It is related to x_G as

$$x_G = m_G/(v m_G + 1000/M_w) = 1/(v + 55.51/m_G) \tag{1A12}$$

The molarity c_G of the electrolyte is the number of moles of electrolyte dissolved in one litre of solution. The density of the solution being ρ, one litre contains $1000 \rho - c_G M_G$ g water, hence

$$c_G = m_G \rho/(1 + 0.001 m_G M_G) \tag{1A13}$$

$$m_G = c_G/(\rho - 0.001 c_G M_G) \tag{1A14}$$

Henry's law may be expressed in terms of molarity, and if, again, the constant is set equal to unity, activities on the molal scale $a_{G(m)}$ and molar scale $a_{G(c)}$ are defined.

Differentiation of (1A2) for an ideal solution with respect to the solvent, or using the Gibbs–Duhem relationship to an ideal solution which obeys Henry's law, leads to Raoult's law

$$a_{w id}/x_w = constant = a_w^0 \tag{1A15}$$

which is also a characterization of an ideal solution. To be consistent with equations (1A5) and (1A11), the standard-state activity a_w^0 must be equal to unity, so that for an ideal liquid $x_w = p/p^0$.

Except at the reference state, electrolyte solutions are non-ideal and the activity of the electrolyte is related to its concentration through the activity coefficient. For the rational (mole fraction), molal and molar concentration scales, respectively, the stoichiometric activity coefficients and the mean ionic activity coefficients are given by

$$f_G = a_{G(x)}^{1/\nu}/x_G = (\nu_+^{\nu_+} \nu_-^{\nu_-})^{1/\nu} f_\pm = (\nu_+ f_+)^{\nu_+/\nu}(\nu_- f_-)^{\nu_-/\nu} \quad (1A16a)$$

$$\gamma_G = a_{G(m)}^{1/\nu}/m_G = (\nu_+^{\nu_+} \nu_-^{\nu_-})^{1/\nu} \gamma_\pm = (\nu_+ \gamma_+)^{\nu_+/\nu}(\nu_- \gamma_-)^{\nu_-/\nu} \quad (1A16b)$$

$$y_G = a_{G(c)}^{1/\nu}/c_G = (\nu_+^{\nu_+} \nu_-^{\nu_-})^{1/\nu} y_\pm = (\nu_+ y_+)^{\nu_+/\nu}(\nu_- y_-)^{\nu_-/\nu} \quad (1A16c)$$

From the definition of the standard state for the electrolyte it is evident that as the solutions approach infinite dilution

$$\lim_{x_G \to 0} f_G = 1 \qquad \lim_{m_G \to 0} \gamma_G = 1 \qquad \lim_{c_G \to 0} y_G = 1 \quad (1A17)$$

The activity coefficients on the different scales are related

$$\ln \gamma_\pm = \ln c_G/m_G + \ln y_\pm + (\mu_{G(c)}^0 - \mu_{G(m)}^0)/\nu RT \quad (1A18)$$

Since at infinite dilutions $c_G/m_G = \rho_0$, the density of water (1A13), and $\ln \gamma_\pm = \ln y_\pm = 0$, the last term in (1A18) is equal to $-\rho_0$. Hence

$$m_G \gamma_\pm = c_G y_\pm/\rho^0 = 1.003 c_G y_\pm \text{ (at } 25°) \simeq c_G y_\pm \quad (1A19)$$

The rational and molal activity coefficients are related by

$$f_\pm = \gamma_\pm (1 + \nu M_w m_G/1000) = \gamma_\pm(1 + 0.018\nu m_G) \quad (1A20)$$

The deviations of the solvent from ideality are expressed through the osmotic coefficient. The molal osmotic coefficient is defined by

$$\phi = -(1000/\nu m_G M_w)\ln a_w = -(55.51/\nu m_G)\ln a_w \quad (1A21)$$

Using the Gibbs–Duhem relationship in the form $N_W d\mu_W = - N_G d\mu_G$ and equations (1A3), (1A6), (1A16b) and (1A21) leads to the relationship between the molal activity coefficient and the molal osmotic coefficient

$$\ln \gamma_\pm = (\phi - 1) + \int_0^{m_G} (\phi - 1) \, d\ln m_G \qquad (1A22)$$

This relationship is important for the experimental determination of the activity coefficient.

b. Structural Models

For the purpose of further elaborations, and a theory for the change of free energy in electrolyte solutions with concentrations, it is necessary to consider a molecular model. The most naive model, which nevertheless is sufficient for many purposes, is one where the solvent is a continuous structureless fluid of constant dielectric constant, and the ions are point charges of magnitude $z_+ e$ or $z_- e$ dispersed in this fluid. This model is useful only for extremely dilute solutions, since for less dilute solutions it is necessary to ascribe to the ions a finite size. Ions are then considered as spheres with an average diameter a. In these dilute solutions only long-range forces are important, i.e. electrostatic forces between ions. In more concentrated solutions it is necessary to consider also the structure of the solvent, i.e. the size of its molecule, the change of dielectric constant with field strength, and ion–dipole interactions between ions and solvent.

A currently widely accepted model of pure liquid water is the 'flickering cluster' model[1]. Water is considered to consist of a mixture of hydrogen-bonded clusters of molecules and of single non-hydrogen-bonded molecules. The clusters are short-lived (some 100–1000 times the period of a molecular vibration) and contain water molecules in an 'ice–like' state, with as many molecules as possible being in the interior rather than the surface, and being quadruply hydrogen bonded to neighbours. The single, monomeric water molecules in rapid equilibrium with the clusters, are considered to be in a 'gas-like' state although their free rotation may be hindered by the strong polar fields of the neighbouring molecules.

Detailed statistical–mechanical calculations on this model have been made[2,3], the results differing somewhat, depending on the assumptions made. These consist, principally, of recognizing molecules in five different states (those with zero, one, two, three or four hydrogen bonds)[2], or in only two states (those in clusters and the rotating monomers)[3], and of considering the freedom to rotate of the non-bonded and singly bonded molecules. The partition functions obtained permit the calculation of the free energy, enthalpy, entropy, vapour pressure, molar volume and heat capacity, as well as the size of the clusters (about 57 water molecules at 20°)[2], the fraction of water in the unbonded state (30% at 20°)[2] and in the states with one to four bonds, and the fraction of hydrogen bonds relative to ice remaining unbroken (46% at 20°)[2], all as functions of the temperature. The agreement with the experimental values is fairly good (within 10% for most quantities). The main qualitative result is that the cluster size, the fraction of water in the ice-like (cluster) form, and the fraction of hydrogen bonding retained, decrease as the temperature increases, while the fraction of water as unbonded molecules increases. These results of the statistical–thermodynamic calculations can be checked by spectroscopic results, although the latter have not been interpreted in an unequivocal manner. Raman spectral[4], n.m.r.[5] and near-infrared spectral data[6] have been interpreted in terms of fractions of water molecules donating zero, one or two hydrogen bonds. The fraction of unbonded molecules is near 30% around 30°C [6], in good agreement with the calculated values[2]. These results have been challenged both as regards the assignment and intensity of the infrared peaks[7], and as regards the fractions of unbonded water, which has to be much lower (< 1%) to fit some other data[8], although the latter may be interpreted in an alternative manner, which permits a fraction near 30% of unbonded water[8]. Several other types of measurements, such as x-ray diffraction, neutron scattering, solubility of noble gases, etc., are in qualitative agreement with the two-state structure of water, namely ice-like and gas-like molecules, which are in equilibrium with each other. These however cannot be interpreted independently in a unique manner in terms of cluster size and fraction of monomeric water. Thus despite the impressive success of the statistical – thermodynamic calculations based on the flickering cluster model, to explain the thermodynamic and spectroscopic data on water, the problem of its structure is still open, as shown in a recent review[9].

Some more information on the structure of water and of aqueous solutions is obtained from a study of the solutions, and of the interactions of ions of electrolytes, and of non-electrolyte solutes, with water. These topics are discussed in sections 1.B and 2.D below.

The properties of water as a solvent are summarized in Table 1A1.

TABLE 1A1. Properties of liquid water.

Temperature (°C)	Density (g/ml)	Vapour pressure (mm Hg)	Viscosity (centipois)	Dielectric constant	Ionisation constant, pK_W
0	0.99987	4.579	1.792	88.90	14.943
10	0.99973	9.209	1.308	83.95	14.534
20	0.99823	17.535	1.005	80.18	14.166
25	0.99707	23.756	0.895	78.36	13.996
30	0.99568	31.824	0.801	76.58	13.833
40	0.9922	55.324	0.656	73.15	13.535
60	0.9832	149.38	0.468	66.76	13.017
80	0.9718	355.1	0.356	61.22	
100	0.9583	760.0	0.284	55.90	

Values at 25°C

ρ = density 0.997075 g/ml = 0.997047 g/cm^3

p = vapour pressure 23.756 mm Hg = 0.03220 bar

η = viscosity 0.008949 poise

ε = dielectric constant 78.358

α = thermal expansion coefficient 2.570×10^{-4}cm^3/deg

β = compressibility coefficient 4.565×10^{-5} bar^{-1}

V = molar volume 18.06 ml

M = formula weight 18.015 g

$(\partial \ln \varepsilon / \partial T)_p = -4.588 \times 10^{-3}$deg^{-1}

$(\partial \ln \varepsilon / \partial P)_T = 4.710 \times 10^{-5}$bar^{-1}

n_D = refractive index 1.33251

c_W = molar concentration of water 55.3 M

References pp. 85–92

B. IONIC HYDRATION

a. Introduction

When a crystalline salt dissolves in water, a certain amount of energy, the lattice energy U, is lost, i.e. work must be done in order to separate the ions to distant places from their sites in the lattice. The ions, however, exert a high electrical potential $z_i e/r_c$, which acts on the dipoles of the water molecules. If the charge number z_i is sufficiently large or the radius r_c sufficiently small for the ion–dipole interaction to be so strong as to break the tetrahedral hydrogen-bonded quasi-structure of the water, the water molecules will rearrange themselves around the ions. In certain cases, the ion can fit itself into the quasi-structure, and little structural energy of the water is lost. However if the ion is very large, or if its field is very strong, the ions will show structure-modifying behaviour. Usually, the ion–dipole interactions are so strong as to overcome the losses in lattice energy and water-structure energy, and a net hydration energy results.

For certain ions of the transition metal series, further energy is gained if the hydration water arranges itself in an ordered geometrical co-ordination sphere around the ion, e.g. $Nd(OH_2)_9^{3+}$, $Cr(OH_2)_6^{3+}$, $Cu(OH_2)_4^{2+}$. More water molecules, influenced by the central field, may be 'bound' in a second layer, around the primary coordination layer. For other ions, the first layer may not conform to a rigid spatial arrangement, but will be bound no less tightly. In these cases, however, it is impossible to ascertain how many water molecules are in the first hydration layer, and how many in the second and further layers, except by arbitrary operational definitions.

Let us first examine the thermodynamics of the hydration process. These considerations pertain to the total observed effect of hydration, and do not give information on the manner in which the water molecules are arranged around the ion. This is obtained from structural studies and considerations. It is also convenient to describe the interactions in terms of a hydration number, this number describing the total interaction, but being defined only operationally. This quantity is also connected with the radius of a hydrated ion, and with the molar volume of the hydrated electrolyte. These subjects are treated in the following parts of this section.

b. Thermodynamic Functions of Hydration

Thermodynamic functions of hydration, i.e. the free energy, enthalpy and entropy of hydration, of electrolytes can be obtained experimentally. Thus entropies of hydration can be obtained from the temperature coefficient of the solubility, or of the e. m. f. of suitable cells. Enthalpies of hydration can be obtained from the heat of solution, corrected for the lattice energy and the heat of dilution to the standard state. The generalized molar thermodynamic function Y is additive for the ions

$$Y = \sum v_i Y_i = v_+ Y_+ + v_- Y_- \tag{1B1}$$

Of interest are the individual ionic values of Y, which must be obtained by non-thermodynamic procedures. One of these is to define conventional values, based on the following conventions[10]:

a) For the formation of any species in its standard state from the elements in their standard state at a given temperature, ΔH_f^0 and ΔG_f^0 are the standard molal enthalpy and free energy changes.

b) The molar entropy of a species in the standard state is S^0. For any reaction $\Delta S^0 = (\Delta H^0 - \Delta G^0)/T$ must hold.

c) Aqueous hydrogen ion in its standard state at any temperature has zero values for ΔH_f^0 and ΔG_f^0, and the value of S^0 is half the molar entropy of hydrogen gas at that temperature and a fugacity of 1 atm, namely 15.606 cal/deg mole.

d) The electron in a 'half-reaction' involving aqueous ions has zero values for ΔH_f^0, ΔG_f^0 and S^0.

Since for any reaction $\Delta Y^0 = \sum \Delta Y_f^0$ (products) $- \sum \Delta Y_f^0$ (reactants), it follows from these conventions that for the half-reaction

$$\tfrac{1}{2} H_2\,(g) \rightarrow H^+\,(aq) + e^- \tag{1B2}$$

$\Delta Y_{H(con)}^0 = \Delta G_H^0,\ \Delta H_H^0,\ \Delta S_H^0 = 0$, where the subscript (con) refers to the conventions (c) and (d). Furthermore, the conventional change in thermodynamic function for the half-reaction for the formation of any aqueous ion Z^z (aq)

$$Z\,(ss) \rightarrow Z^z\,(aq) + z\,e^- \tag{1B3}$$

where (ss) designates the standard state, and z is taken algebraically, is equivalent to that for the reaction

$$Z(\text{ss}) + z\,H^+\,(\text{aq}) \rightarrow Z^z(\text{aq}) + \tfrac{1}{2}z\,H_2(g) \tag{1B4}$$

and will be designated as ΔY^0_{aq}. Conventional values ΔY^0_{ion} can be assigned to the ionization reaction

$$Z(\text{ss}) \rightarrow Z^z(g) + z\,\bar{e}\,(g) \tag{1B5}$$

if values are assigned to the electron $\bar{e}(g)$: $\Delta H^0_f = (5/2)RT = 1481$ cal/mole and $\Delta S^0 = 4.994$ cal/mole deg. Values for the functions ΔG^0, ΔH^0 and ΔS^0 for reactions (1B3) (or 1B4) and (1B5) have been tabulated[11].

It is now possible to define the hydration process

$$Z^z(g) \rightarrow Z^z(\text{aq}) \tag{1B6}$$

and the change in the thermodynamic functions involved

$$\Delta Y^0_{hyd} = \Delta Y^0_{aq} - \Delta Y^0_{ion} \tag{1B7}$$

which can be calculated in a straightforward manner from the tabulated data, and some of the values are reported in Table 1B1[12]. It must be remembered that these values of ΔY^0_{hyd} are conventional, since ΔY^0_{aq} values are conventional.

The conventional standard molal entropies for aqueous ions, $\Delta S^0_{i\,aq}$ have been related to the charge z and the crystalline radius r_{ic} of the ions by various empirical equations:

$$\Delta S^0_{i\,aq} = \tfrac{3}{2}R \ln M_i + 37 - 270\,z_i/r^2_{eff\,i} \tag{1B8}$$

where $r_{eff\,i}$, the effective radius to be used in this expression[13], is $r_{eff+} = r_{c+} + 2.0$ Å and $r_{eff-} = r_{c-} + 1.0$ Å, for monatomic ions, or alternatively[14]

$$\Delta S^0_{i\,aq} = \tfrac{3}{2}R \ln M_i + 10.2 - 11.6\,z_i/r^2_{ci} \tag{1B9}$$

TABLE 1B1. Thermodynamic functions of hydration, ΔY^0_{hyd}, at 25°C.

Ion	$-\Delta G^0_{con}$ (kcal/mole) (ref. 12)*	$-\Delta H^0_{con}$	$-\Delta S^0_{con}$ (e.u.)	$-\Delta G^0_{el}$ (kcal/mole) (ref. 29)	$-\Delta H^0_{el}$	$-\Delta S^0_e$ (e.u.)
H^+	364.0	360.9	10.4	260.5	267.4	23.0
Li^+	227.0	223.2	12.8	122.1	129.7	25.4
Na^+	208.7	207.1	5.3	98.4	103.6	17.5
K^+	175.9	176.9	−3.2	80.6	83.4	9.4
Rb^+	169.1	170.9	−6.1	75.5	77.5	6.6
Cs^+	161.1	163.1	−6.8	67.8	69.6	5.9
NH_4^+	177	176	2.3			
Ag^+	215.8	213.8	6.7	114.5	120.2	19.3
Tl^+	176.8	178.1	−4.2	82.0	84.5	8.4
Be^{2+}	791*	780*	55*	582.3	608	86
Mg^{2+}	669.6	659.8	32.5	454.2	473.3	63.9
Ca^{2+}	586.7	581.0	19.0	379.5	394.5	50.4
Sr^{2+}	555.8	545.7	34.0*	339.7	359.2	65.4
Ba^{2+}	514.1	512.0	7.0*	314.0	375.5	38.4
Mn^{2+}	650.4	641.4	30.3	436.4	454.9	61.9
Fe^{2+}	670.9	659.3	38.9*	451.8	472.8	70.3
Co^{2+}	706.2	691.4	49.6*	481	504.9	81.0
Ni^{2+}	718.7	703.7	50.3*	492.8	517.2	81.7
Cu^{2+}	712.8	702.3	35.3	496.2	515.8	65.8
Zn^{2+}	698.7	689.0	32.7	483.3	502.4	64.1
Cd^{2+}	639.0	632.0	23.4	429.1	445.5	54.9
Hg^{2+}	643.2*	637.0*	15.9*	434.9	449.1	47.3
Pb^{2+}	555.8	554.1	5.6	356.5	367.6	37.0
Al^{3+}	1433.3	1414.3	63.9	1100.5	1134.7	114.8
Sc^{3+}	1251.6*	1247.1	47*	938.5	967.5	97
Y^{3+}	1172.1*	1165.8	41*	859.0	886.2	91
La^{3+}	1091.9*	1085.2	39.9	778.8	805.7	90.3
Ce^{3+}	1162	1150	41		870	
Cr^{3+}	1350*	1329*	70*	1037	1073	120
Fe^{3+}	1368	1347	64.8	1033	1067	114
Ga^{3+}	1442.7	1420.3	75.1	1103.4	1140.8	125.5
In^{3+}	1299.1	1282.6	55.3	979.9	1011.4	105.7
Tl^{3+}	1289*	1300.6	101.3*	975.9	1021.1	151.7
Ce^{+4}		1950			1579	
Th^{+4}		1800				
F^-	−10.3	5.4	52.7	89.5	97.8	27.8
Cl^-	−23.7	−12.1	39.1	76.1	80.3	14.0
Br^-	−30.7	−20.2	35.4	69.2	72.2	10.0
I^-	−39.5	−30.6	29.9	60.3	61.8	5.0
OH^-		−15.2	−24.1**			
CN^-			− 0.6			
SCN^-			29**			
NO_3^-			27.8**			
ClO_3^-		−33.3	28**			
ClO_4^-			23.9**			
ReO_4^-			26.6**			
SO_3^{2-}			85**			
SO_4^{2-}			66**			
CO_3^{2-}			87**			
PO_4^{3-}			140**			

* Those marked with asterisk: ref. 29.   ** A. P. Altshuler, *J. Chem. Phys.*, **28**, 1254 (1959).

For oxyanions a variety of formulae exist[14-16] which depend on the central element–oxygen bond distance, the number of such bonds, and structural factors.

The entropy of hydration can also be calculated from a consideration of the compression of the water in the field of the ion, or from a consideration of the restriction of the degrees of freedom of hydration water compared to bulk water[17], adding a term which is the derivative of the Born energy (see below) with respect to temperature.

Conventional thermodynamic functions of hydration of individual ions cannot be compared directly with the functions for non-ionic solutes, nor with predictions from theory. For this purpose absolute values are necessary, which are obtained from the experimental values for electrolytes by division among the ions, based on non-thermodynamic considerations. Bernal and Fowler[18] regard the heat of hydration to be divided equally between cation and anion, in the case where they have the same crystalline radius, except for a small corre⟨ on. The correction arises from the preference of the cation for the oxygen side, and of the anion for the hydrogen side of the water dipole, and from the non-central position of the latter. Taking potassium and fluoride ions to have the same radii, heats of hydration of these two ions were obtained, and from them those of all other ions could be calculated. Alternatively, if differences in the heats of hydration of lithium fluoride and those of the heavier halides of lithium, or those of the fluorides of the heavier alkali metals, are plotted against the reciprocal of the crystalline radii, the intercepts at infinite radius correspond, according to Verwey[19], to the individual heats of hydration of lithium and of fluoride ions. The results obtained by the two methods disagree, and the premises of the allocation of the individual hydration heats have been criticized[20].

The absolute value of $\Delta Y^0_{hyd\,abs}$ is related to the conventional value, obtained from equation (1B7), $\Delta Y^0_{hyd\,con}$, by the expression

$$\Delta Y^0_{hyd\,abs} = \Delta Y^0_{hyd\,con} + z\,\Delta Y_{hyd\,H}$$

$$= \Delta Y^0_{aq} - \Delta Y^0_{ion} + z\,\Delta Y_{hyd\,H}$$

(1B10)

The conventional ionic entropies of hydration could therefore be converted to absolute ionic partial molal standard entropies, if the absolute value for the hydrogen ion, or another ion, were known. Various estimates have been proposed for $\Delta S^0_{H^+\,aq}$, e.g. -5.4 e.u. by Lee and Tai[21], (from the temperature coefficient of a cell involving an acid at 'unit hydrogen ion activity' and mercury at its electrocapillary maximum), -5.5 e.u. by Gurney[22] (the value that makes plots of the B coefficients of the viscosities against the ionic entropies for cations and anions coincide) and -2.1 e.u. by Eastman[23] (from thermo-cells, but cf. refs. 20, pp. 57–60, and 24).

The free energy change of process (1B6) can be written according to the usual Born treatment[25] as

$$\Delta G_{\text{Born}}^0 = G_{\text{el aq}}^0 - G_{\text{el vac}}^0 + \Delta G_{\text{neut}}^0 \qquad (1B11)$$

where the last term refers to non-electrostatic hydration effects which are small compared to the total effect. The electrostatic terms can be written in terms of the radii and dielectric constant

$$\Delta G_{\text{el}}^0 = \tfrac{1}{2} N z^2 e^2 \left(\frac{1}{r_{i\,\text{aq}} \varepsilon_{\text{aq}}} - \frac{1}{r_{i\,\text{vac}} \varepsilon_{\text{vac}}} \right) \qquad (1B12)$$

Where ε_{vac} is of course unity. Use of the crystalline radius r_c for both $r_{i\,\text{aq}}$ and $r_{i\,\text{vac}}$ and of the bulk dielectric constant of water, ε_{w}, for ε_{aq} leads to values of ΔG_{Born}^0 in great disagreement with the estimated values of $\Delta G_{\text{hyd abs}}^0$, and more seriously, the value calculated from (1B12) for an electrolyte, as the sum of terms for the ions, disagrees with the experimental value. Several approaches have been tried to overcome this difficulty.

The *effective aqueous radius* approach has been used by the earlier workers[26]. They set $\varepsilon_{\text{aq}} = \varepsilon_{\text{w}}$ and $r_{i\,\text{aq}} = r_{i\,\text{vac}} = r_{\text{eff}}$, an effective radius, where $r_{\text{eff}+} = r_{c+} + 0.8$ Å and $r_{\text{eff}-} = r_{c-} + 0.1$ Å, and obtained good agreement between the Born free energy and the experimental values for electrolytes. The effective radii are interpreted as the radii of cavities in the dielectric medium, in which the ions are situated. The empirical correction of $r_{i\,\text{vac}}$ is, however, unjustified, and this approach has been criticized[20]. New values for the crystal ionic radius, calculated as the distance to the electron-density minimum between the two ions in the crystal, change this picture. If these radii are used for both $r_{i\,\text{vac}}$ and $r_{i\,\text{aq}}$, better correlation of (1B12) with experimental data is obtained[27]. The use of the crystalline radius for the solution has been justified, but for the ions in vacuum another approach has been suggested[28].

The *dielectric saturation* approach[29] employs the crystalline radius for both states of the ions: $r_{i\,\text{aq}} = r_{i\,\text{vac}} = r_{ci}$, but uses an effective dielectric constant[30] $\varepsilon_{aq} = \varepsilon_{eff}$, much lower than the bulk value because of dielectric saturation in the high field near the ions. From a fit of the Born free energies to the experimental values, ε_{eff} has been calculated for many ions, and it was found empirically that

$$\varepsilon_{eff} = 1.000 + 1.376(r_c - 0.054) \tag{1B13}$$

A modified form of this approach is to consider a layer of thickness Δ around the ions with unit dielectric constant, and use the bulk value beyond that[31]. This in a sense is a combination of the effective radius and effective dielectric constant approaches. The thickness $\Delta = 0.55$ Å was found to give agreement with experiment of the theoretical equation

$$\Delta G^0 = -\tfrac{1}{2}Nz^2e^2[(n_W^2 - 1)n_W^{-2}/(r_{ci} + \Delta) + (3\varepsilon_W/\beta(T)\,ez)^{1/2}(n_W^{-2} - \varepsilon_W^{-1})$$

$$(\tfrac{2}{3} - \tfrac{1}{2}R + \tfrac{1}{12}R^3)] \tag{1B14}$$

where$_\urcorner$ n_W is the refractive index of water, $\beta(T)$ describes the field-strength dependence of the dielectric constant $\beta = 0.116/T$ cm (e.s.u.)$^{-1}$, and $R = (r_{ci} + \Delta)/(9\varepsilon_W/\beta(T)ez)^{1/2}$.

Finally, estimated *radii* r_v for the *ions in vacuum*, effectively van der Waals radii, together with an effective dielectric constant $\varepsilon_{aq} = \varepsilon_{eff} = 9$ for a layer of thickness Δ ($= 2.8$ Å for monovalent cations, the diameter of a water molecule) and bulk dielectric constant beyond this, and $r_{i\,aq} = r_{ci} + \Delta$, have been used to calculate free energies of hydration in good agreement with experiment[28].

The term ΔG^0_{neut} in (1B11) must also be considered, although it is small. Stokes[28] followed Noyes[29], who suggested two alternative assumptions. According to the zero energy assumption, only enthalpy and entropy changes accompanying the compression of Z^z(g) (molar volume 24.47 litres at 25°) to Z^z(aq) (molar volume in standard state 1 litre) are considered, hence $\Delta S^0_{neut} = -R\ln(24.47/1) = -6.350$ cal/deg mole, $\Delta H^0_{neut} = -(24.47 - 1)$ litre atm/mole $= -56.8$ cal/mole and $\Delta G^0_{neut} = 1325$ cal/mole. According to the inert gas assumption, ΔG^0_{neut} is the same as the experimental value for an inert gas atom of the same size, and the data for these can be summarized as $\Delta G^0_{neut} = 160 + 6760/r$ cal/mole and $\Delta S^0_{neut} = -36.86 + 24.70/r$ cal/deg mole. The former assumption gives better agreement with experiment, in particular if a term in r^2, accounting for surface effects, is added[29].

The 'Born' term of the entropy is obtained by differentiation of (1B11) with respect to temperature

$$\Delta S^0{}_{\text{Born}} = -(\partial G^0{}_{\text{Born}}/\partial T)_p = (\tfrac{1}{2}Nz^2e^2/r_{i\,\text{aq}}\varepsilon_{\text{aq}})(\partial \ln \varepsilon_{\text{aq}}/\partial T)_p + \Delta S^0{}_{\text{neut}}$$

$$(1B15)$$

Again the effective dielectric constant ε_{eff} can be used for ε_{aq} instead of the bulk ε_{w}. As regards the enthalpy of hydration, Bernal and Fowler[18] showed that the heat of hydration of an ion i is made up from three terms

$$\Delta H^0{}_{i\,\text{hyd}} = -\frac{Nz^2e^2}{2(r_{ci}+2r_{\text{w}})}(1-1/\varepsilon_{\text{w}}) - \frac{h_i\,z\,e\,\mu_{\text{w}}}{(r_{ci}+r_{\text{w}})^2} + U_{\text{w}} \qquad (1B16)$$

The first is a Born term describing the transfer of charge from vacuum to a medium with dilectric constant ε_{w}. The second is an ion–dipole interaction term, where μ_{w} is the dipole moment of water and h_i the number of water molecules interacting in this manner. The last term is the electrostatic energy of the one molecule of water in its water medium, substituted by the ion, i.e. the energy for cavity formation, which represents $\Delta H^0{}_{\text{neut}}$. Refinements in this theory were introduced by Eley and Evans[17] and by Verwey[19], who considered orientation differences of μ_{w} for cations and anions, and dipole–dipole interactions.

Azzam[32] developed for the heat of hydration a treatment which is not a modification of the Born approach, as far as the molecules of water nearest the ions are concerned. According to him, the heat of hydration is given by

$$\Delta H^0{}_{\text{hyd}} = 4\pi n_i \int_{a_i}^{r_p} u_r \exp(-(W_r - W_r)/kT)\,r^2 dr$$

$$(1B17)$$

$$+ 4\pi n_0 \int_{r_p}^{r_s} u_r \exp(-(u_r - W_0)/kT)\,r^2\,dr$$

$$-\tfrac{1}{2}(Nz^2e^2/r_s)[(1-1/\varepsilon_{\text{w}}) - T\varepsilon_{\text{w}}^{-2}(\partial\varepsilon_{\text{w}}/\partial T)_P]$$

where the first term pertains to the region from the ion surface $(r = a_i)$ to the limit of the primary hydration $(r = r_p)$, the second term from there to the limit of the secondary hydration $(r = r_s)$, and the third term, the Born transfer term, pertains to the region beyond this (using $\Delta H = \Delta G - T(\partial \Delta G / \partial T)_P$). In the first two terms n_i is the number of ions per cm^3, u_r is the ion–dipole interaction energy, W_r is the potential energy for the interactions of the ion with the water dipole, of two water molecules in the second layer, and of the water dipoles in the hydration layers with those in the bulk, W_0 is the potential in the bulk. The value of r_s is the same for all monovalent cations (2.4 Å) and anions (2.8 Å) hence the third term, which makes the major contribution, is also the same (68.2 kcal/mole for cations). The first term, which would be thought to make an important contribution from the interaction of the water with the nearest water molecules, does not (4.5 kcal/mole for K^+).

This treatment, as well most of the others, has been criticized recently by Halliwell and Nyburg[33] and by Conway[34], who recommends these authors'[33] final selection of values. They base their treatment on the consideration of the experimental enthalpies of hydration for electrolytes, and calculate absolute values not from a division of the measured data between cation and anion, but from the trend of the difference between the conventional enthalpies for monovalent cations and anions with the ionic radii according to

$$\Delta H^0_{\text{hyd con (cations)}} - \Delta H^0_{\text{hyd con (anions)}} = \Delta(\Delta H^0_{\text{hyd abs}}(r)) - 2\Delta H^0_{\text{hyd H}}$$

$$(1B18)$$

As the function of the radius, they show that $C(r_{ci} + r_W)^{-3}$, where C is a constant, is the most suitable and a plot of the left-hand side of (1B18) (obtained from smoothed plots of each term against r or a function of it, since there are no cations and anions of the same r), against this function of r gives as intercept $\Delta H_{\text{hyd H}} = -260.7 \pm 2.5$ kcal/mole. This value is, however, what other authors designate as $\Delta H^0_{\text{hyd abs}}$ for the hydrogen ion, and the constant factor $\Delta H^0_{\text{hyd H}}$, which is the difference between the conventional and absolute enthalpies of hydration for all monovalent ions, is -367.1 kcal/mole $(= \Delta G^0_{\text{con}} + T\Delta S^0_{\text{con}}) + 260.7$ kcal/mole $= 106.4$ kcal/mole. This value is somewhat high compared to values obtained by other authors.

Summarizing, it is possible to calculate $\Delta Y^0_{\text{hyd abs}}$ values from (1B12), (1B14) and (1B15) or their appropriate modifications, as a function of r_i for ions, for example those of Noyes[29] (Table 1B1, shown as ΔY^0_{el}), subtract these values from $\Delta Y^0_{\text{hyd con}}$ for these ions, and the difference according to (1B10) should be $z \Delta Y^0_{\text{hyd H}}$. Somewhat different estimates have been obtained by different workers for this quantity, ranging for $\Delta G^0_{\text{hyd H}}$ from $100.0\,[28]$ through $100.3\,[35]$, $103.3 - 104.8\,[27]$ to $123\,[36]$ kcal/mole. Similarly, for $\Delta S^0_{\text{hyd H}}$ the values range from -16.8 (or $-18.9)\,[27]$ to $-29.2\,[35]$ cal/deg mole. (cf. the values on p. 13 of $\Delta S^0_{\text{H}^+\text{aq}}$ which convert to $\Delta S^0_{\text{hyd H}}$ on substraction of 15.6 cal/deg mole). If any of these values are accepted, e.g. $\Delta G^0_{\text{hyd H}} = 104$ kcal/mole and $\Delta S^0_{\text{hyd H}} = -18$ cal/deg mole (hence $\Delta H^0_{\text{hyd H}} = 99$ kcal/mole), absolute functions of hydration can be calculated for any ion from tabulated conventional values of ΔY^0_{hyd} (Table 1B1) and equation (1B10).

c. Structural Considerations

Structural information on solutions may be obtained from neutron and x-ray diffraction studies. An increase of the concentration of ions has the same effect on the quasi-structure of water as an increase of temperature, so that electrolyte solutions can be characterized by their 'structural temperature'[18]. In concentrated solutions, the water molecules were found to be more tightly packed than in their ordinary tetrahedral quasi-structure, but different ions break the open structure to different degrees. Thus the lithium ion fits into the structure without breaking it appreciably[37], while potassium ions exert a structure-breaking influence[22]. Some ions (e.g. fluoride), on the contrary, have a structure-making effect.

These effects are shown best by $\Delta S_w = S_w - S^0_w$ the difference in the partial molal entropy of water in the solution and in pure water. Frank and Robinson[38] showed that at low concentration the ordering of the water molecules in the hydration shell makes ΔS_w negative, but for ions which are structure breaking, ΔS_w becomes increasingly positive as the concentration increases. Structure is enhanced by ions (F^-, OH^-, OH_3^+, NH_4^+) which can enter the water structure by hydrogen bonding, while structure is broken more, the larger the ions[22].

There are however two different modes of structure making: one shown by small ions such as Li^+, Na^+, F^- and OH^- (and perhaps also H_3O^+), the other shown by large ions with centrally localized charge, such as $(C_2H_5)_4N^+$, $(C_3H_7)_4N^+$, etc. (Large ions in which the charge can be smeared out at the surface, such as BF_4^-, ClO_4^-, $FeCl_4^-$, etc., are structure breakers, not makers). In the former group of structure makers, the small ion fits more or less into the existing structure of water, and enhances its stability through its charge. For the latter group of structure makers, 'hydrophobic bonding' occurs, in which the inert hydrocarbon material can be envisaged as guarding the water structure near it from disruption by thermal agitation of other water molecules[39].

If an electrolyte is made up of ions of widely differing structural properties (e.g. one structure making, the other breaking, such as LiI or CsOH, or both structure making but of different mode, such as $(C_3H_7)_4NF$), its chemical potential is strongly increased, and conversely, if they have similar structural properties (e.g. LiF or CsI), its chemical potential is decreased[22]. Extreme examples are the larger tetraalkylammonium fluorides[40]. They have the highest mean molal activity coefficient of all 1:1 electrolytes at any given molality. The same trend is observed for the tetraalkylammonium chlorides, although to a much smaller extent[41], while for the bromide and iodide, where both ions of the electrolyte are structure breakers, activity coefficients are low[41]. This can be ascribed to water-structure enforced ion pairing[42], since when the ions are near each other, a part of the surrounding water sheath is common to both ions, and less structure is broken, as pointed out already by Gurney[22]. Measurements of the enthalpy and entropy of dilution for aqueous solutions of these salts do not however, completely support this explanation. These quantities are much larger than the change in free energy, but show very little dependence on the size of the anion, an effect which comes out only in the free energy, i.e. in their difference[41]. Better criteria for the water-structure modifying effects of the tetraalkylammonium ions are their effects on the viscosity of the solutions and their mobilities. According to these, tetramethylammonium ions are structure-breaking, tetraethyl- and tetraethanol ammonium ions are indifferent, and tetrapropyl- and tetrabutylammonium ions are hydrophobic-bond type structure makers.[43]

A somewhat different way to look at the problem of the effects of ions on the structure of water is to consider their effect on the size of the water cluster (section 1.A.b). Evidence from infrared spectroscopy[44] and from the solubility of argon[45] in aqueous electrolyte solutions shows these effects: the ions La^{3+}, Ca^{2+}, Mg^{2+}, H^+, OH^- and F^- produce 'order', i.e. increase cluster size, while the ions K^+, Na^+, Li^+, Cs^+, Ag^+, ClO_4^-, I^-, Br^-, NO_3^-, Cl^- and SCN^- destroy order, i.e.

decrease the cluster size. This sense of 'order' is not exactly the same as that discussed above.

The competing tendencies of structure making (either the ordered structure of water in the hydration shell or the ice-like structure of water near large hydrophobic ions) and structure breaking are best discussed quantitatively in terms of the concentric-shells model of Frank[46]. This idea has been developed by Azzan[47], who has also pointed out that the dielectric constant near the ion is a function of the hydration. Indeed, the change of the dielectric constant with the distance from the ion, as calculated theoretically[30,48-50], can form the basis for the assignment of the water molecules to the various concentric shells (Figure 1B1). Below a certain radius r_p, dielectric saturation is practically complete and ε is near unity (or near n_W^2, depending whether voids near the ion are recognized or not). Above a certain radius r_s, the dielectric constant is near that of pure water. The values $r_p = 1.8 z^{1/2}$ and $r_s = 8.0 z^{1/2}$ Å have been proposed[49], but others have been suggested[47,50]. The effective dielectric constant ε_{eff} discussed above in connection with the Born equation can now be understood as a mean between the local value near the ion and the bulk value[28,29].

Consider now an ion of type A (Figure 1B1), a small ion to which an effective radius r_e can be assigned so that $r_c < r_e < r_p$. The size of r_e can be determined by either adding a constant distance Δ to r_c[28,31], by using a constant ratio r_e/r_c (e.g. 1.25[48]) or by treating r_e as a parameter to be determined from a fit to experimental data (e.g. simultaneously with data for electrostriction (cf. below), entropy of hydration and free energy of hydration[49]). (Table 1B2). Some water of hydration may be included in the interval from r_c to r_e, and this may be 'permanent hydration' as for H_3O^+, or perhaps also for $Al(OH_2)_6^{3+}$, and the entity of an ion with permanent hydration may be called 'cationium'[47] (anions are too large to be thus hydrated). Since $r_e < r_p$, the 'cationium' is further hydrated, and all the water up to r_p may be called 'primary hydration'. Primary hydration water is highly structured and has a much lowered entropy. Water in the interval from r_p to r_s may be called 'secondary hydration' (this includes also a part of 'hydrodynamic hydration'[47]), and can be considered as 'thawed', i.e. its structure is broken and its entropy increased. For sufficiently small ions, the increased structure up to r_p more than compensates for the decreased structure between r_p and r_s, so the ion is a net structure maker.

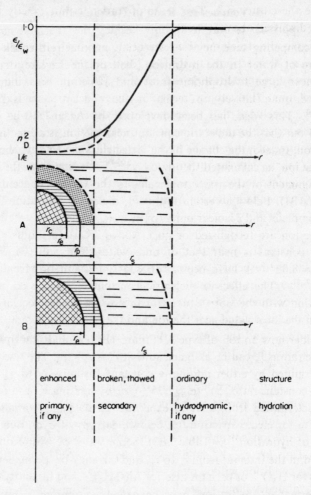

Figure 1B1.   The concentric shell in hydration model. Upper part: the variation of the dielectric constant in water as a function of the distance from the ion. Lower part: the regions bound by spheres having the crystalline radius r_c, the effective radius r_e, the primary hydration radius r_p, and the secondary hydration radius r_s for a small ion A and a large ion B.

Consider now an ion of type B (Figure 1B1), a large ion, for which $r_e > r_p$ (r_e assigned as above). It has no primary, only secondary hydration, and there is nothing to compensate the structure breaking in this region, so the large ion is a net structure breaker.

d. Volumes and Radii of Hydrated Ions

A third approach to ion hydration besides the thermodynamic and structural, is the stoichiometric–volumetric approach, in which a hydration number h_i is assigned to the ion. The problem of hydration numbers is intimately connected with that of the radius and the volume of the hydrated ions. This subject has been reviewed by Stern and Amis[51], who have concluded that these quantities depend at least as much on the methods used for measurement, as on the system studied.

The volume of water in the hydration shell is the total volume of the hydrated ion v_{hi} less the intrinsic volume of the bare ion v_{ci}. The ratio of this difference to the volume of a molecule of water in the hydration shell, v_{hW}, gives the hydration number h_i

$$h_i = (v_{hi} - v_{ci})/v_{hW} = (4\pi/3 \times 0.58)(r_{hi}^3 - r_{ci}^3)/(v_W^0 - \delta v_W) \qquad (1B19)$$

In the last term, the hydrated radius r_{hi} and intrinsic radius r_{ci} are defined in terms of the respective volumes, the factor 0.58 arising from the void volume inherent in the closest packing of spheres[52] (thus the numerical coefficient becomes 4.35 rather than 2.51), and the molecular electrostriction δv_W is defined as the difference between the volume of water in the bulk v_W^0 and the water in the hydration shell v_{hW}. Instead of the molecular volumes, $1/N$ times the respective molar volumes V_{hi}, V_{ci}, V_{hW}, V_W^0 and δV_W can of course be used. The problem in the application of (1B19) is that none of the quantities h_i, r_{hi} (or V_{hi}), r_{ci} (or V_{ci}) and v_{hW} (or δV_W) are known with good confidence. In the following, the condition of infinite dilution will be assumed in order to be concerned only with the ion–solvent interactions, and not with the ion–ion interactions. Thus h_i^0, v_{hi}^0, V_{hi}^0 and r_{hi} will be used to designate this state. Later this condition will be removed.

The hydrated volume of an ion v_{hi}^0 can be estimated from the conventional partial molal volume of this ion $V_i'^0$ as follows. The partial molal volume of the electrolyte at infinite dilution is

$$V_G'^0 = \lim_{c \to 0} \varnothing_V = \lim_{c \to 0}(M_G/\rho_W - 1000(\rho - \rho_W)/\rho_W c) \qquad (1B20)$$

where $\varnothing_V$ is the apparent molar volume and must be divided between the cation and the anion. Conventionally[10], this is done by assigning the value $V_{H^+}'^0 = 0$ (Table 1B2). An absolute scale can be based on the equation[53][55]

$$V_i'^0 = A r_{ci}^3 - B z_i^2/r_{ci} = V_{ic} - \Delta V_{Wi} \qquad (1B21)$$

The first term gives the intrinsic volume, which according to (1B19) is $(4\pi/3 \times 0.58)\,r_{ci}^3$, so that $A = 4.35$. The second term arises from electro-striction of the solvent[29,55], and may be obtained from the Born equation (1B12) by differentiation with respect to pressure

$$\Delta V_{el}'^0 = (\partial \Delta G_{el}^0/\partial P)_T = - [(e^2/2\varepsilon_{aq})(\partial \ln \varepsilon_{aq}/\partial P)_T]z_i^2/r_{aq} \qquad (1B22)$$

If the value ε_W is used for ε_{aq}, the coefficient in the square bracket becomes $B = 4.18$ ml Å/mole (or 6.0^{55}). Application of (1B21) and (1B22), yields $V_{H^+}'^0 = -5.5$ ml/mole.

Mukerjee[54] has shown that the empirical equation with $A = 4.5$ and $B = 8.0$, leading to $V_{H^+}'^0 = -4.5$ ml/mole and thus to equal volumes of monatomic cations and anions of the same crystalline radius, agrees with the data within 2 ml/mole. Instead of taking into account the packing ratio of 0.58 of the volume, Mukerjee uses an effective radius $r_{eff} = 1.213 r_{ci}$ (compare the ratio 1.25 used by Laidler and Pegis[48], and $1.91^{1/3} = 1.24$ proposed by Benson and Copeland[55]). Using the same assumption of independence of the volume from the sign of the charge, Couture and Laidler[56] have previously proposed another empirical equation, as successful as Mukerjee's based on $V_{H^+}'^0 = -6.0$ ml/mole. They set $A = 4.9$ and instead of the second term they used a term $16 - 26|z_i|$.

The above assumption is in conflict with the fact that for cations the negative side of the water dipole points towards the ion, while for anions it is the positive side, i.e. the hydrogen atoms, which may hydrogen bond to the anion. Differences between the volumes of hydrated cations and anions are thus expected[57]. Indeed, Hepler[53] has proposed equation (1B21) with $A = 5.3$ and $B = 4.7$ for cations, and $A = 4.6$ and $B = 19$ for anions, as an empirical relation, with a value of $V_{H^+}'^0 = 0.1$ ml/mole, but this appears to be too large (algebraically). A better basis for the division of V_G^0 between the ions may be the consideration of ions so large that there is no electrostriction of the water, and extrapolation of a series of such ions to zero size of one ion[57]. For example if $\emptyset_V^0$ for the series $(CH_3)_4NI, (C_2H_5)_4NI, (C_3H_7)_4NI$ and $(C_4H_9)_4NI$ is plotted against $M_{R_4N^+}$, a value of 42.3 ± 0.2 ml/mole is obtained for the iodide ion, which corresponds to -5.2 ± 1.0 ml/mole for $V_{H^+}'^0$, in agreement with Mukerjee's basis for absolute ionic partial molal volumes.

The hydrated volume of ions can be calculated from the partial molal ionic volume from the equation[57-59]

$$V_{hi}^0 = V_i'^0 + V_w'^0(V_{ci} - V_i'^0)/\delta V_w = V_i'^0 + h_i^0\, V_w^0$$

$$\text{(1B23)}$$

$$= V_{ci} + h_i^0 V_w^0 - \Delta V_w$$

but at this stage, neither h_i^0 nor V_{ci} and δV_w are known.

The most direct way to measure V_{hi}^0 is from the ionic mobilities, using Stokes' law

$$r_{hi} = 0.820\, z_i/\lambda_i \eta \qquad \text{(1B24)}$$

where the coefficient 0.820 contains the constant $1/6\pi$, conversion from cm to Å, and from volts to c.g.s. units. However the Stokes radii $r_{hi(Stokes)}$ calculated from (1B24) are of reasonable magnitude ($r_{hi\,(Stokes)} > r_{ci}$), only for very small or very large ions, while for ions of intermediate size (such as K^+ and Cl^-), they are too small ($r_{hi(Stokes)} < r_{ci}$). This is explained by noting that Stokes' law applies only to spheres which are much larger than the particles of the medium, i.e. the water molecules, whereas hydrated potassium or chloride ions are of approximately the same size. Robinson and Stokes[60] and Nightingale[61] proposed to use $r_{hi\,(Stokes)}$ only for the large ions such as the tetraalkylammonium ions (except $N(CH_3)_4^+$), and by plotting $r_{hi\,(Stokes)}$ for these ions against r_{ci} (obtained from bond lengths, Van der Waals radii and molar volumes), they obtained a curve relating the two quantities. Nightingale proposed to equate r_{hi} of any ion to the value read from this curve, corresponding to its experimental $r_{hi\,(Stokes)}$ from (1B24), and his values are shown in Table 1B2. In order to obtain hydrated ionic volumes V_{hi}^0 from the radii r_{hi}, it is necessary to take into account also the interstitial volumes in the hydrated ion. The packing factor $1/0.58$ (cf. equation 1B19) can be used, or else a term proportional to the surface area of the ion added [62,63]

$$V_{hi}^0 = 2.51\, r_{hi}^3 + 3.15\, r_{hi}^2 \qquad \text{(1B25)}$$

Another way to arrive at hydrated ionic volumes and hydration numbers is through the use of compressibilities. If it is assumed that the hydrated ion, in which the water is already compressed by electrostriction, is not compressible by external pressure[64], then the incompressible volume fraction of the solution is given by $(1 - \beta/\beta_w)$, where β is the compressibility of the solution, $-(\partial \ln V/\partial P)_T$. The hydrated volume of the electrolyte can therefore, be defined operationally, as[58]

$$V_{hG} = 1000(\beta_w - \beta)/c\, \beta_w \qquad \text{(1B26)}$$

If an apparent compressibility $\emptyset_K$ is defined as

$$\emptyset_K = - (\partial \emptyset_V / \partial P)_T = \emptyset_K^0 + S_K c^{1/2} \qquad (1B27)$$

and compared with Masson's equation for the apparent molar volume

$$\emptyset_V = \emptyset_V^0 + S_V c^{1/2} \qquad (1B28)$$

calculation shows that

$$\beta = \beta_W + (\emptyset_K^0 - \beta_W \emptyset_V^0) c / 1000 + (S_K - \beta_W S_V) c^{3/2} / 1000 \qquad (1B29)$$

in complete analogy with Root's equation for the density ρ. These relationships lead to the equations

$$V_{hG}^0 = \emptyset_V^0 - \emptyset_K^0 / \beta_W \qquad (1B30)$$

$$\delta V_W = V_W^0 S_V \beta_W / S_K \qquad (1B31)$$

The theoretical value for δV_W is[58] 2.1 ml/mole, obtained from (1B31)[65], so that hydration numbers for electrolytes can be calculated from the experimental values of $V_{hG}^{\prime 0}$ and $V_G^{\prime 0}$ from (1B23). Individual ionic values can be obtained if h can be appropriately aportioned into its h_i components. Procedures essentially similar to that of Padova [58] have been used by others[66, 67], and have been applied to new ultrasonic data on compressibilities[67].

The use of the theoretical value of $\delta V_W = 2.1$ ml/mole, which is independent of the ion charge and radius, has been criticized[57], and a varying value, dependent on the field, is preferable. The empirical values of S_K and S_V for each electrolyte can be used with (1B31) to obtain δV for this electrolyte. The theory of electrostriction[59], in which an effective pressure is calculated as a function of the field, which would give the observed volume change, leads to a numerical solution, relating $\Delta V_{Wi} = h_i^0 \, \delta V_W$ to the field $E = z e / \varepsilon_{eff} r_{eff}^2$ as follows

r_{ci}(Å) (for $z = 1$)	5.3	3.5	1.76	0.43		
$E(10^5$ e.s.u.)	0.23	0.60	1.10	2.7	8.2	10.2
ΔV_{Wi} (ml/mole)	0.54	1.00	1.36	2.36	5.08	5.71

For the calculation of E, $\varepsilon_{eff} = n_D^2 = 1.78$ and $r_{eff} = r_{ci} + r_W = r_{ci} + 1.38$ Å.

A still different way of calculating the molar volumes of ions in solution uses the picture of structured water around the ions instead of a continuum, i.e. considers the voids occuring round the ions (cf. ref. 52), using an equation similar to (1B21), where the first term refers to V_{ci}, the intrinsic volume of the ion, and the second to the total electrostriction around the ion ΔV_{W_i}[62].

The intrinsic volume is obtained from the consideration that an ion replaces a water molecule[55], so if the void space is equated to a hollow spherical shell of thickness Δ, then $(r_W + \Delta)^3 = V_W^0/(4\pi N/3)$, which gives with $r_W = 1.38$ Å the value $\Delta = 0.55$ Å, exactly as in the treatment of the free energy and entropy of hydration[31] (section 1.B.b.). Thus for any ion $V_{ci}^0 = (4\pi N/3)(r_{ci} + \Delta)^3 = 2.51(r_{ci} + \Delta)^3$. This differs from the first term of (1B21) in having terms also in r_{ci}^2, r_{ci} and a constant. Non-monatomic ions can also be described by a suitable equation: $V_{ci}^0 = 2.51(r_O + r_W)^3$, where r_O is the distance from the centre to the centre of the oxygen atom in a tetrahedral oxyanion (such as ClO_4^-), while the added value r_W equals the radius of the oxygen atom. The electrostriction term utilizes an empirical constant, and has the form $(B \pm C)z^{3/2}(z^{1/2}/r_{ci} + r_W \pm \delta)$, with $B = 33.7$, $C = 5.2$, $\delta = 0.2$ Å describes the displacement of the dipole center of the water from the molecule center, + refers to cations and − to anions, and the charge dependence is split into two parts, to show the dependence of the electrostriction on $z^{1/2}/\bar{r}$, where $\bar{r}$ is the effective radius, $r_{ci} + r_W \pm \delta$.

Glueckauf[62] used an indefinite form of the dependence of the electrostriction on the field, but for numerical calculations used a modified Born term (1B22), equating the change of pressure in the hydration shell with field strength to a term in E^2 with an invariant dielectric constant. Padova[49] considered the change of ε with the field strength (section 1.B.c), obtaining a rather complicated implicit equation $\Delta V_w = F(r_e)$, which is introduced in (1B32) as

$$V_i^{\prime 0} = 2.51 r_e^3 - F(r_e) \qquad (1B32)$$

so that the effective radius of the ion, r_e can be calculated from the experimental $V_i^{\prime 0}$ (using $V_{I-}^{\prime 0} = 37.1$ ml/mole for the assumed unhydrated iodide ion). As stated in section 1.B.c the same r_e gives good agreement with experimental free energies and entropies of hydration too. The values of r_e and r_{hi} calculated from (1B23) are tabulated in Table 1B2. This treatment differs from that of Desnoyers and coworkers[59] reported above, mainly in evaluating the field strength E from the effective radius. It is also shown[49] that the assumption that primary hydration water (i.e. water between r_e and r_p) is incompressible compared with bulk water[64] (section 1.B.d) is indeed valid.

TABLE 1B2. Radii and molar volumes of ions.

Ion	V'^{0}_{con} (ml)	r_{ci} (Å)	r_{e}^{49} (Å)	r_{hi}^{61} (Å)	r_{hi}^{58} (Å)	r_{hi}^{71} (Å)
H^+	(0.00)		1.27	2.82		2.44
Li^+	−1.20	0.60	1.23	3.82	2.00	2.50
Na^+	−1.60	0.95	1.21	3.58	2.28	2.17
K^+	8.60	1.33	1.64	3.31	2.17	1.75
Rb^+	13.77	1.48	1.84	3.29	2.24	1.53
Cs^+	20.95	1.69	2.08	3.29	2.07	1.47
NH_4^+	17.80	1.48	1.98	3.31	1.59	1.88
$(CH_3)_4N^+$		3.47		3.67		
$(C_2H_5)_4N^+$		4.00		(4.00)		
$n-(C_3H_7)_4N^+$		4.52		(4.52)		
$n-(C_4H_9)_4N^+$		4.94		(4.94)		
Ag^+	−1.57	1.26	1.22	3.41	1.97	
Tl^+	10.60	1.44	1.73	3.30	2.18	
Be^{2+}		0.31		4.59		
Mg^{2+}	−21.10	0.65	1.44	4.28	3.09	2.96
Ca^{2+}	−18.15	0.99	1.46	4.12	3.10	2.72
Sr^{2+}	−18.50	1.13	1.46	4.12		2.74
Ba^{2+}	−12.80	1.35	1.57	4.04	3.13	2.48
Mn^{2+}	−19.10	0.80	1.47	4.38		
Fe^{2+}	−26.10	0.75	1.38	4.28		
Co^{2+}	−25.40	0.72	1.38	4.23		
Ni^{2+}	−25.40	0.70	1.38	4.04		
Cu^{2+}		0.70		4.19		
Zn^{2+}	−22.90	0.74	1.42	4.30		3.07
Cd^{2+}	−13.16	0.97	1.55	4.26	2.64	
Hg^{2+}		1.10				
Pb^{2+}	−16.10	1.32	1.51	4.01		
Al^{3+}	−43.20	0.50	1.77	4.75		
Sc^{3+}		0.81				
Y^{3+}		0.93				
La^{3+}	−38.65	1.15	1.81	4.52	3.32	3.19
Nd^{3+}	−43.47	1.08	1.77		3.41	3.15
Yb^{3+}	−44.70	0.94	1.76	4.65	3.52	
Fe^{3+}	−44.70	0.60	1.76	4.57		
OH^-	−5.70	(1.38)	1.11		3.18	2.46
F^-	−2.00	1.36	1.23	3.52	3.06	
Cl^-	18.20	1.81	2.02	3.32	2.55	1.81
Br^-	25.28	1.95	2.21	3.30	2.42	1.92
I^-	36.70	2.16	2.48	3.31	2.04	2.16
NO_3^-	29.60	1.76	2.34	3.35	2.23	2.03
CNS^-	40.90		2.57		2.12	
$CH_3CO_2^-$	35.5		2.57		2.96	
BF_4^-	44.0					
MnO_4^-	43.40		2.60			
ReO_4^-	48.7	3.3				
ClO_4^-	44.20	2.98	2.64	3.38	2.43	2.31
ClO_3^-	36.40		2.48		2.56	
BrO_3^-	33.70		2.44		2.72	
IO_3^-	25.70		2.23		3.27	
SO_4^{2-}	15.07	2.90	2.18	3.79	3.64	
CrO_4^{2-}	19.90	3.0	2.32		3.88	
CO_3^{2-}	−4.10	1.48	1.74		3.94	

e. Hydration Numbers

Hydration numbers can be defined only operationally, i.e. according to the method used for their evaluation. If the concentric shell model[45-47] is accepted it is possible to count the number of water molecules belonging to the permanent, primary, secondary and hydrodynamic categories. However different methods put different boundaries to the regions around an ion designated by these terms.

Some of the theoretical treatments used for dividing the volume around the ion into the respective regions have been discussed above. If the mean volume of a water molecule in each region is known, hydration numbers can be calculated from the appropriate equations, (1B19) or (1B23).

An attempt at the direct theoretical evaluation of hydration numbers[68] uses the equation

$$h_i^0 = \int_{r_{cl}}^{r_s} 4\pi r^2 n_W \exp(-W_r/kT)\,dr \qquad (1B33)$$

where n_W is the number of water molecules per unit volume in the bulk, and W_r is the total interaction energy (including terms for Born transfer energy, ion–dipole interaction, and mutual interaction of water dipoles) to calculate the hydration number up to and including secondary hydration. Calling the 'permanent' hydration number n, the primary p, and the secondary s, the hydration of the hydrogen ion is $[H(H_2O)_{n=1}(H_2O)_{p=3}(H_2O)_{s=9}]^+$, of the hydroxide ion $[OH(H_2O)_{p=4.7}(H_2O)_{s=17}]^-$ and of divalent ions $[M(H_2O)_{n+ip=8}(H_2O)_{op+s}]^{2+}$, where ip designates inner primary and op outer primary, whatever their significance[47]. The detailed theory of Azzam[68] has not been accepted generally, since the expressions used for the change of dielectric constant with field and the interaction potentials calculated do not withstand criticism.

Practical calculations of hydration numbers from experimental data have been made according to the following methods.

a) Partial molar volumes and electrostriction, using (1B19) and (1B23). The data of Padova[49] are shown in Table 1B3.

b) Compressibility. The equations (1B23) and (1B30) lead to the expression

$$h_G^0 = -\emptyset_K^0/\beta_W V_W^0 \qquad (1B34)$$

and it is necessary to aportion h_G^0 for the electrolyte between the ions. This can be done if certain ions, such as NH_4^+, I^- and SCN^- are considered to be unhydrated[58,69], and the values thus obtained are shown in Table 1B3. Values obtained from new (ultrasonic sound speed) compressibility data[67] are also shown there. These are based on the assignment of $h_{NO_3^-}^0 = 2$, for splitting h_G^0.

c) Stokes' radii. Hydration numbers have been calculated by dividing the volume of the hydration shell calculated from the Stokes' radius r_{hi}, (equation 1B24), by the volume of a molecule of water. The criticism raised against the use of Stokes' radii (section 1.B.d), and the uncertainty of the volume v_{hw} to be used, invalidates this method.

d) Entropy of aqueous ions. Ulich[70] considered that the entropy change per molecule of water in the hydration shell equals the average amount per molecule of water of crystallization in solids, -6 e.u., so that $h_i^0 = \Delta S_{i\,aq}^0 / (-6)$. The oversimplifications in this consideration are obvious.

e) Activity coefficients. The change of free energy with concentration contains a statistical term, depending on the number of particles, or the fractional volume they occupy. These quantities are modified by hydration, compared with the stastics applying to non-hydrated ions, used as the basis for stoichiometric activity coefficients. Their dependence on the hydration number h_G^0 will be discussed in detail in section 1.D, and for the sake of comparison, the values derived by Glueckauf[71] and by Monk[72] are shown in Table 1B3.

f) Dielectric constant. The change of the dielectric constant near an ion can be ascribed to hydration[50], and calculated according to

$$h_i = \sum_j Z_j (\varepsilon_W - \varepsilon_j)/(\varepsilon_W - n_W^2) \qquad (1B35)$$

where Z_j is the coordination number for the jth coordination shell, and ε_j the mean dielectric constant there. Estimates based on experimental data of the dielectric constants and dielectric loss in electrolyte solutions have also been made[73].

These and other methods give the value of $h^0 = n + p + s$, where different parts of the secondary hydration, s, are taken into account, hence the different values obtained. Methods based on ion mobility measure, also, a part of the hydrodynamic hydration beyond the secondary layer. Methods purported to measure only the (permanent and) primary hydration are those based on the transport of water with the ions[20], and certain spectroscopic methods.

g) Hittorf transference numbers. Very early attempts to estimate hydration numbers used the quantity of water transported with the ions in Hittorf transference experiments, referred to a non-electrolyte such as sugar[74]. Since however some transport of the non-electrolyte also occurs[75], this method is unreliable. The Recent measurements of the amount of D_2O transported per ion into a H_2O solution[76] may be more reliable if isotope effects are indeed negligible.

h) *Nuclear magnetic resonance.* The evaluation of hydration numbers from n.m.r. data has recently been made possible by the availability of water enriched in ^{17}O. Early attempts to use proton magnetic resonance, although yielding reasonable ionic molar chemical shifts[77], could not be unequivocally used for calculation of hydration numbers[78]. The presence of a paramagnetic ion as a probe can however lead to definite hydration numbers[79], especially if more highly enriched ^{17}O water is used (from relaxation times or hyperfine interaction)[80] or from proton magnetic resonance (from chemical shifts)[81]. In the case that the exchange period of the water is longer than ca. 10^{-3} sec, two peaks in the n.m.r. spectrum should be observable, one for nuclei in the bulk water, the other for nuclei in the solvation sphere. Exchange times are as a rule shorter (except for Cr^{3+}), even for ions such as Ni^{2+}, Be^{2+} or Al^{3+}, which have relatively long relaxation times. Using the paramagnetic probe ion, however, chemical shifts are observed when diamagnetic ions are added, since water can now occur in a new environment. The observed shift σ_i for a given cation at concentration c_i (at fixed anion concentration and fixed paramagnetic ion concentration) is

$$\sigma_i = \sigma_p \frac{h_i c_i}{55.3 - h_i c_i} \tag{1B36}$$

where σ_p is the (unknown) shift caused by the paramagnetic ion, and the numerator and denominator are respectively the molar concentrations of the bound and free water. The quantity σ_p can be calibrated by ions of known hydration (e.g. $Al(OH_2)_6^{3+}$ or $Be(OH_2)_4^{2+}$, or secondary standards calibrated with them, e.g. H^+ or NH_4^+, which for this purpose are unhydrated[81], so that h_i can be calculated for other ions from observed shifts. Some primary hydration numbers obtained from n.m.r. are shown in Table 1B3.

It is impossible here to review many other methods used for estimating ion hydration. These have been reviewed, among others, by Conway and Bockris[20], Robinson and Stokes[60], Monk[72] and Conway[34, 82].

Most of the discussion above was concerned with infinitely dilute solutions, for which the values of h_i^0 (as shown in Table 1B3) are valid. Most of the methods are also capable of yielding the values of h_i at any given concentration of electrolyte in the solution. For example, for the compressibility method, equations (1B23) (1B27–30) and (1B34) yield the following expression for the hydration of the electrolyte

$$h = h^0(1 - (S_K/(- \varnothing_K^0))c^{1/2}) \qquad (1B37)$$

but at each concentration, it must be decided afresh how to divide h among the cation and anion. Since $\varnothing_K^0$ is a negative quantity it is evident that $h < h^0$, so that the ions are dehydrated as the concentration of the electrolyte increases,

Few studies indeed have investigated the change of the hydration number (or the hydrated ion volume or radius) with the concentration. Similarly, no detailed information is available for the hydration of ions in mixtures at finite concentration, where one ion affects the hydration of another.

In concentrated solutions, such as 16 m hydrochloric acid or 20 m lithium chloride (near saturation), there is certainly not enough water present (altogether 55.51 m) to permit complete solvation of each ions with h_i^0 moles of water. In fact, the solubility limit is partly dependent on the minimal amount of hydration water required for keeping the ions apart. Even long before saturation, drastic changes in the hydration of the ions occur. The non-availability of complete hydration sheaths of water causes a) enhanced interaction, due to the decrease of the minimal interionic distance of approach as discussed in section 1.D; b) association of the ions to ion pairs (see section 3.A) or higher associations, if removal of the water occurs because of replacement by non-solvating solvents (section 2.D); and c) preference of another phase, e.g. an ion exchanger or an organic solvent, if through appropriate interactions a lower free energy is attained (chapters 4–10).

If the temperature is sufficiently high, so that water can be removed while still leaving the system in a liquid state (a possibility in a few systems, such as ammonium nitrate or some tetraalkylammonium or related salts), an anhydrous molten salt is attained at the end of this process. The interesting interactions occuring in such systems, permitting also some ion exchange and liquid-distribution equilibria to occur, are outside the scope of this book[83, 84].

TABLE 1B3. Hydration numbers of ions, h_i^o.

Method	Electrostriction	Compressibility		Activity		N.m.r.
Reference	49	69	67	71	72	81
H^+	3			3.9	3.0	(0)
Li^+	3	2	3.6	3.4	4.3	
Na^+	3	3	4.8	2.0	2.9	
K^+	1	2	4.1	0.6	1.2	
Rb^+	1	2		0	0.5	
Cs^+	1	2		0	0	
NH_4^+	1	(0)	2.0	0.2	1.1	(0)
$(CH_3)_4N^+$			5.0			
$(C_2H_5)_4N^+$			8.7			
Ag^+	3	2	4.0			
Tl^+	1	2				
Be^{2+}						(4)
Mg^{2+}	14	8	7.9	5.1	8.3	3.8
Ca^{2+}	14	8		4.3	7.2	4.3
Sr^{2+}	14			3.7	6.7	5.0
Ba^{2+}	11	8		3.0	5.2	5.7
Zn^{2+}	17			5.3	9.3	3.9
Cd^{2+}	11	5				4.6
Hg^{2+}						4.9
Pb^{2+}	12					5.7
Al^{3+}	29					(6)
La^{3+}	26	11		7.5	11.0	
Nd^{3+}	29	12			11.2	
Yb^{3+}	29	13				
Fe^{3+}	29					
F^-	3	7				
Cl^-	1	3	2.3	0.9	0.1	
Br^-	1	2	1.7	0.9	(0)	
I^-	1	(0)	1.2	0.9	(0)	
OH^-	4	8	5.2	4.0	3.5	
NO_3^-	1	1	(2.0)	(0)	(0)	
ClO_3^-	1	2				
BrO_3^-	1	3				
IO_3^-	1	7				
ClO_4^-	1			0.3	0.2	
SCN^-	1	(0)		0.3		
$CH_3CO_2^-$	1	4	2.0	2.6		
SO_4^{2-}	4	11	7.9			
CrO_4^{2-}	4	13				
CO_3^{2-}	8	15				

( ) designates standard used for division of h_G^o between cation and anion.

References pp. 85–92

C. DILUTE SOLUTIONS

a. The Debye-Hückel Treatment

The well-known theory of Debye and Hückel[85] relates the long-range electrostatic forces in extremely dilute solutions to the electrical contribution to the free energy of the solution. Consider ions with charge $z_i e$ interacting in a medium of dielectric constant ε_0 (the field independence of the dielectric constant is an approximation, cf. section 1.C.b). Poisson's equation relates the electrostatic potential ψ at any point of the medium with the local charge density ρ at distance r from a given ion considered as the origin of coordinates

$$\frac{1}{r^2} \frac{d}{dr} \left(r^2 \frac{d\psi}{dr} \right) = - \frac{4\pi\rho}{\varepsilon_0} \tag{1C1}$$

where spherical polar coordinates are used because of spherical symmetry. (At higher concentrations spherical symmetry may not be justified because of the particulate nature of the ions, cf. sections 1.C.b and 1.D.a). The local charge density is also given by the charge of the ions times their number concentration, n_i particles per unit volume (cm^3)

$$\rho = \sum_i n_i z_i e \tag{1C2}$$

At zero potential this quantity equals zero, because of the electroneutrality of the solution as a whole

$$\rho_0 = \sum_i n_{i0} z_i e = 0 \tag{1C3}$$

but at a point where the potential is ψ there may be an excess of a certain type of ions, and the number density is given by the Boltzmann distribution

$$n_i = n_{i0} \exp(- z_i e\psi / kT) \tag{1C4}$$

The theory of Debye and Hückel now introduces the approximation of expanding the exponent (1C4), and retaining only the first two terms in the expanded form obtained from (1C2) and (1C4)

$$\rho = \sum_i n_{i0} z_i e - \sum_i \frac{n_{i0} z_i^2 e^2 \psi}{kT} + \sum_i \frac{n_{i0} z_i^3 e^3 \psi^2}{2k^2 T^2} + \cdots \tag{1C5}$$

This linearization is also required by the principle of linear superposition of fields, and the third and higher terms must then be zero, as must also be the first term, from (1C3). If this principle is disregarded, the third.

fifth, etc. terms will still be zero for binary (1:1, 2:2 etc) electrolytes. From these considerations and (1C1) the expression

$$\frac{1}{r^2} \frac{d}{dr} \left(r^2 \frac{d\psi}{dr} \right) = \frac{4\pi e^2}{\varepsilon_0 kT} \sum_i n_{i0} z_i^2 \psi = \kappa^2 \psi \qquad (1C6)$$

is obtained. The quantity κ has the dimensions of reciprocal length, and is termed the average reciprocal diameter of the 'ionic atmosphere' which may be considered to surround a given ion.

Equation (1C6) may be solved to give the charge density of ions of type k as function of distance r from a given ion j, taking into consideration the boundary condition that no other ions are present within a sphere of diameter a, which may be regarded as the average diameter of the ions, or as the distance of closest approach of the centres of two ions. The Debye–Hückel treatment considers a to be the same for all pairs of ions in solution, and because of its independence of concentration this may cause deviations from expected behaviour even at infinite dilution. The solution for the charge density is

$$\rho_k = - \frac{\kappa^2 z_k e}{4\pi r} \frac{e^{\kappa a} e^{-\kappa r}}{1 + \kappa a} \qquad (1C7)$$

A combination of (1C1), (1C6) and (1C7) yields the expression for the potential acting on these ions

$$\psi_j = \frac{z_j e}{\varepsilon_0} \frac{e^{\kappa a}}{1 + \kappa a} \frac{e^{-\kappa r}}{r} \qquad (1C8)$$

Equations (1C7) and (1C8) are Debye and Hückel's fundamental expression for the distribution and the potential. Other expressions have been proposed[86,87] but need not be discussed here, since the Debye–Hückel expression has been found to be adequate in the region where long-range forces are predominant. The potential ψ_j may be considered to be made up of the potential due to the ionic atmosphere ψ_{ia} and that due to the ion $z_j e/\varepsilon_0 r$, hence

$$\psi_{ia} = \frac{z_j e}{\varepsilon_0} \left(\frac{e^{\kappa a}}{1 + \kappa a} \frac{e^{-\kappa r}}{r} - \frac{1}{r} \right) \qquad (1C9)$$

but the continuity of the potential when r varies from $r < a$ to $r > a$ requires for $r = a$

$$\psi_{ia} = - \frac{z_j e}{\varepsilon_0} \frac{\kappa}{1 + \kappa a} \qquad (1C10)$$

b. Free Energy and Activity Coefficients

The electrical contribution to the free energy per mole of ions j, or the electrical contribution to the chemical potential of j, $\mu_{j(el)}$, is N times the contribution for the addition of one ion j to a solution in which each ion has its ionic atmosphere characterized by κ and ψ_{ia}. This charging process, proposed by Güntelberg[88], is carried out in two steps. In the first, the ion is added without its charge, causing a negligible contribution to the free energy from the volume change, and in the second step the charge is raised from zero to $z_j e$. The electrical contribution to the chemical potential therefore is

$$\mu_{j(el)} = \frac{-Ne^2}{\varepsilon_0} \int_0^{z_j} \frac{\kappa z_j}{1 + \kappa a} dz_j \tag{1C11}$$

Fowler and Guggenheim[89] showed that the total electrical free energy of the system is

$$G_{(el)} = V \sum n_i \mu_{i(el)} = \frac{e^2 V \sum n_i z_i^2}{3\varepsilon_0} \kappa \tau(\kappa a) \tag{1C12}$$

where V is the total volume of the system, $\mu_{i(el)}$ refers to one ion of type i, and $\tau(\kappa a)$ is defined by

$$\tau(\kappa a) = \frac{3}{(\kappa a)^3} (\ln(1 + \kappa a) - \kappa a + \tfrac{1}{2}(\kappa a)^2)$$
$$\simeq 1 - (3/4)\kappa a + (3/5)(\kappa a)^2 - \cdots \text{ for } \kappa a \lesssim 1 \tag{1C13}$$

Differentiation with respect to $N_j = n_j V/N$, the total number of moles of j in the system, yields the chemical potential

$$\mu_{j(el)} = -\frac{Ne^2 z_j^2}{2\varepsilon_0} \frac{\kappa}{1 + \kappa a} + \frac{V_i' kT}{24\pi} \kappa^3 \sigma(\kappa a) \tag{1C14}$$

where V_i' is the partial molar volume $(\partial V/\partial N_i)_{N_{j \neq i}}$ and $\sigma(\kappa a)$ is defined by

$$\sigma(\kappa a) = \frac{3}{(\kappa a)^3} \left[1 + \kappa a - \frac{1}{1 + \kappa a} - 2\ln(1 + \kappa a) \right]$$
$$\simeq 1 - (6/4)\kappa a + (9/5)(\kappa a)^2 - \cdots \text{ for } \kappa a \lesssim 1 \tag{1C15}$$

Values of $\sigma(\kappa a)$ are tabulated[90]. The last term in equation (1C14) is numerically small, and is ordinarily neglected.

Differentiation of (1C12) with respect to the number of moles of solvent in the system leads to

$$\mu_{W(el)} = \frac{V_W' kT}{24\pi} \kappa^3 \sigma(\kappa a) \tag{1C16}$$

from which by (1A2) the ideal term $\ln x_W$, the activity of water in the electrolyte solution

$$\ln a_W = -\ln(1 + 0.018 v m_G) + \frac{V'_W}{24\pi N} \kappa^3 \sigma(\kappa a) \qquad (1C17)$$

and the osmotic coefficient $\phi = -(55.5/vm_G) \ln a_W$ are obtained. Since the volume of a mole of solution is $x_W V'_W + \Sigma v_i x_i V'_i$, it can be shown that (1C17) may be obtained by applying the Gibbs–Duhem relation to (1C14).

The activity coefficient of the electrolyte may be obtained from (1C14) as follows, ignoring the (negligible) last term. Ideal solutions have been defined by their obeying Henry's law (1A11) in terms of molalities (section 1.A.*a*). Therefore deviations from ideality expressed by the molal activity coefficient may be ascribed to the contribution to the chemical potential of the electrical interactions characteristic of the long-range forces:

$$\ln \gamma_j = \frac{\mu_{j(el)}}{RT} = -\frac{z_j^2 e^2}{2k\varepsilon_0 T} \frac{\kappa}{1 + \kappa a} \qquad (1C18)$$

Let us introduce the concept of ionic strength I, defined as[91]

$$I = \tfrac{1}{2} \sum_i c_i z_i^2 = \frac{\rho}{2} \sum_i \frac{m_i z_i}{1 + m_i M_i / 1000} \qquad (1C19)$$

In dilute solution the denominator of the right-hand side in equation (1C19) is very close to unity, and so is ρ in aqueous solutions, so that I is often[88] set equal to $\tfrac{1}{2} \Sigma m_i z_i^2$. In nonaqueous solutions and at higher concentrations this approximation is less satisfactory. Since $c_i = 1000 n_{io}/N$, the value of κ^2 defined in (1C6) may by expressed as

$$\kappa^2 = \frac{4\pi e^2}{\varepsilon_0 k T} \sum_i n_{io} z_i^2 = \frac{8\pi N e^2}{1000 \varepsilon_0 k T} I \qquad (1C20)$$

Introducing into (1C18) the ionic strength I from (1C20) and the parameters

$$A = \left(\frac{8\pi N e^2}{1000\,k}\right)^{\frac{1}{2}} \frac{e^2}{2 \times 2.303\,k} (\varepsilon_0 T)^{-3/2} = 1.825 \times 10^6 (\varepsilon_0 T)^{-3/2} \qquad (1C21)$$

$$= 0.511 \quad \text{for water at } 25°C$$

$$B' = \left(\frac{8\pi N e^2}{1000\,k}\right)^{\frac{1}{2}} (\varepsilon_0 T)^{-\frac{1}{2}} = 5.03 \times 10^9 (\varepsilon_0 T)^{-\frac{1}{2}} \qquad (1C22)$$

$$= 3.29 \times 10^7 \text{ for water at } 25°C$$

the expression

$$\log \gamma_j = \frac{-z_j^2 A I^{\frac{1}{2}}}{1 + B' a I^{\frac{1}{2}}} = \frac{-z_j^2 A I^{\frac{1}{2}}}{1 + B \mathring{a} I^{\frac{1}{2}}} \qquad (1C23)$$

is obtained, where on the right-hand side $B = 10^{-8} \times B' = 0.329$ for water at 25°, and $\mathring{a}$ is the value of a in Ångstroms. Since activity coefficients of single ions, j, are not measurable experimentally (vide infra, section 1.C.*b*), expression (1C23) may be converted to the familiar Debye–Hückel expression for the activity coefficient using $\log \gamma_{\pm} = 1/v \sum_i z_i \log \gamma_i$

$$\log \gamma_{\pm} (\text{D.H.}) = \frac{|z_+ z_-| A I^{\frac{1}{2}}}{1 + B \mathring{a} I^{\frac{1}{2}}} \qquad (1C24)$$

It must be remembered that the parameters A and B are functions of temperature and of dielectric constant, as seen in (1C21) and (1C22) A compilation of numerical values of (1C24) for 0–250°C and ionic strengths of 10^{-5} to 20 in water, with $\mathring{a}$ values between 0.25 and 3.25, may be found useful[92].

Let us summarize the main assumptions leading to (1C24):

a) Same average diameter applies for all ions.

b) Solvent is structureless and bulk dielectric constant applies throughout.

c) Spherical symmetry of the ionic atmosphere occurs.

d) Linear approximation to the Boltzmann distribution holds.

e) Long-range electrostatic interaction energy describes molal activity coefficient.

Assumptions (a) and (b) are independent of concentration. At low concentrations the ions are sufficiently wide apart for each to be affected only by the bulk of the solvent, so that changes of dielectric constant with r are unimportant at large r values. Assumption (b) breaks down for ions of higher charges and at higher temperatures[93]. Assumption (c) breaks down at concentrations where r is too small for symmetrical packing of the ions[94,95], i.e. at concentrations where nearest-neighbour interactions become important compared to long-range interactions. Assumption (d) is valid at such low concentrations that ψ (depending on κ, and hence on concentration according to (1C10) and (1C20)) is sufficiently small for only the first two terms in the expansion of (1C4) to be necessary. Since κ contains also the dielectric constant ε_0, the temperature T and the charge z_i, the approximation improves at a given concentration the higher ε_0 and T, and the lower z_i are. Higher terms in the expansion could be added, but this would conflict with the principle of superposition of fields (section 1.C.a). Generally, other assumptions break down before (d) does, so that there is no justification for including the higher terms in the expansion. Finally, if we accept that the long-range electrostatic interactions are sufficient to describe all deviations from ideality, there still remains the question as to which concentration scale is chosen for describing Henry's law. At very low concentrations assumption (e) is valid, since differences between activity coefficients on the molal and rational scale are slight. Some authors prefer to ascribe the D.H. term to the rational activity coefficient, i.e. replace $\log \gamma_{\pm}$(D.H.) in (1C24) by $\log f_{\pm}$ (D.H.).

From the above considerations it follows that the Debye–Hückel treatment is valid only for very dilute solutions ($\kappa a \leq 0.2$), and may be considered as a limiting law, useful for extrapolations. Approaching infinite dilution, the denominator in (1C24) approaches unity and the simple square-root law

$$\log \gamma_{\pm}(\text{D.H., limiting}) = - \left| z_+ z_- \right| A I^{\frac{1}{2}} \qquad (1C25)$$

results. Actual dilutions where (1C25) holds are only seldom reached. Since the parameter $\mathring{a}$ is arbitrary, values can usually be chosen to make (1C24) describe measured activity coefficients adequately up to $I = 0.1$ M. It should be remembered that in spite of this agreement, the theoretical basis of (1C24) breaks down already below $I = 0.001$ M. Nevertheless, many authors use (1C24) to account for the contribution of the long-range forces to the non-ideal free energy at even much higher ionic strengths than $I = 0.1$ M.

c. Individual Ionic Activity Coefficients

The theory described above deals with the contribution of the interaction energy of given ionic species j to the free energy of the solution. Thermodynamics, contrary to the molecular approach, is interested only in the properties of components, i.e. electrically neutral entities. It is however possible to select reasonable values for $\mathring{a}$ in (1C23), the diameter of an ion in solution (e.g. $2r_h$, twice the hydrated radius, cf. section 1.B.d) and calculate[96] values of individual ionic activity coefficients. Their value is mainly that suitable combinations of them by multiplication or division, resulting in activity coefficients of neutral entities, do have thermodynamic significance. Thus values of $\frac{1}{2}$ ($\log \gamma_{Na+} + \log \gamma_{Cl-}$), calculated using $\overset{\circ}{a} = 4.5$ Å for Na^+ and $\overset{\circ}{a} = 3$Å for Cl^-, agree with experimental log $\gamma_{\pm NaCl}$ up to $I = 0.1$ M.

The small thermodynamic significance of individual-ion activity coefficients is intimately connected with the practical impossibility of measuring them. A detailed discussion of this was given by Bjerrum[97] or more recently by Bates[98]. The problem boils down to the impossibility of isolating in a solution a measurable quantity of an ionic species, because of the huge space charge created thereby, and to the impossibility of measuring the activity of an ion in solution by a single electrode[99], i.e. without either the presence of another reversible electrode, or a liquid junction. The former gives only mean ionic activity coefficients $\gamma_\pm$, the latter requires knowledge of liquid-junction potentials, which depend on the individual-ion activity to be measured. It is nevertheless useful to employ in intermediate steps in thermodynamic calculations individual ionic activity coefficients[100], and to assign to them conventional values. Two conventions have been proposed, and both have found use in electrochemical work in dilute solutions. McInnes[101] has introduced the assumption of the equality of the activity coefficients of potassium and chloride ions in pure aqueous potassium chloride of any concentration. The individual activity coefficients are thus assumed to be equal to the mean ionic activity coefficient in these solutions, and also to have the same value in other solutions of the same ionic strength and the same potassium or chloride ion concentration. For example (using molal instead of the original rational activity coefficients)

$$\gamma_-(0.1 \text{ M KCl}) = \gamma_\pm(0.1 \text{ M KCl}) = \gamma_-(0.1 \text{ M HCl})$$

$$= \gamma_-(0.01 \text{ M HCl} + 0.09 \text{ M NaCl}) = \gamma_+(0.1 \text{ M KCl}) \quad (1C26)$$

$$= \gamma_+(0.1 \text{ M KBr}) = \text{etc.}$$

Guggenheim[102] introduced the more general assumption that in dilute solutions the activity coefficients of the ions of binary electrolytes CA are equal

$$\gamma_{+(CA)} = \gamma_{-(CA)} = \gamma_{\pm(CA)} \quad (1C27)$$

For the region where equation (1C24) is valid, this requires for $z_+ = z_-$ that $\overset{\circ}{a}_+ = \overset{\circ}{a}_- = \overset{\circ}{a}_\pm$. For other charge types this becomes

$$\log\gamma_i = \frac{z_i^2}{v_+ z_+^2 + v_- z_-^2} \log\gamma_{\pm(C_{v_+}A_{v_-})} \quad (1C28)$$

Other assumptions, e.g. derived from conventional pH measurements, are also possible[100].

d. The pH of Solutions

The application of these conventions to mixtures of electrolytes with no common ions is not straightforward, but this question is of fundamental importance in attempts to define a reasonable pH scale[98]. The common method is to assign to a given buffer solution (or to more than one) a conventional pH value. The procedure by which to arrive at such a value[98] involves the following steps. A determination of

$$- \log m_{H^+} \gamma_{H^+} \gamma_{Cl^-}$$

in buffer solutions containing chloride ions is made from measurements on cells without liquid junctions, such as Pt; H_2, NaA, HA, NaCl, AgCl; Ag (*vide infra*, end of section 1.C.d)

$$\mathrm{pwH} = - \log m_{H^+} \gamma_{H^+} \gamma_{Cl^-} = \frac{(E - E^0)F}{2.303RT} + \log m_{Cl^-} \qquad (1C29)$$

Values of pwH are then extrapolated to zero chloride concentration to obtain $\mathrm{pwH^0}$. Finally, conventional values are assigned to γ_{Cl^-}, and pH values of the standard buffer calculated

$$\mathrm{pH_{std}} = - \log m_{H^+} \gamma_{H^+} = \mathrm{pwH^0} + \log \gamma_{Cl^-} \text{(conventional)} \qquad (1C30)$$

The United States National Bureau of Standards selected the value $\overset{\circ}{a}_{Cl} = 4.4$ Å and equation (1C23) with $B\overset{\circ}{a} = 1.5$ at 25°C, or equivalently the assumption (1C28), to assign to the recommended buffer, 0.025 M with respect to both potassium dihydrogen phosphate and disodium hydrogen phosphate, a value $\mathrm{pH_{std}} = 6.86 \pm 0.01$ at 25°. The assigned limits of uncertainty indicate the variation introduced by using other reasonable values for $\overset{\circ}{a}_{Cl}$ and other reasonable assumptions regarding the individual activity coefficient of the chloride ion[103]. Other buffers are recommended[98] for other regions of the pH scale and have been assigned $\mathrm{pH_{std}}$ values. The pH of any solution is defined as

$$\mathrm{pH} = \mathrm{pH_{std}} + \frac{(E - E_{std})}{2.303\,RT} \simeq \log c_{H^+} y_{H^+} \qquad (1C31)$$

where E is the e.m.f. measured by a suitable cell with the test solution, and E_{std} is the e.m.f. measured in the same cell with the standard buffer solution. Thus 0.1 M hydrochloric acid, which may be used as a standard for low pH values, was assigned the value 1.10 at 25° using the Guggenheim convention (1C27), but use of the McInnes convention (1C26) instead, would have resulted in the value 1.085. This illustates the uncertainty of 0.01 units associated with the assigned values.

The practical pH defined by (1C31), is neither a measure of hydrogen ion concentration (which is a perfectly defined quantity), nor of the hydrogen ion activity (which is not thermodynamically defined), but a useful and conventional measure of the practical effect of hydrogen ions in a given solution. Differences in pH, measured using the same standard, are significant for any solution and may attain precisions of ± 0.001 or better. Absolute pH values on the other hand, can be based only on conventional values assigned to standards, and these have been restricted to dilute aqueous solutions, where errors not exceeding ± 0.01 pH unit can be reasonably expected from disagreement on the conventional individual activity coefficients. In more concentrated aqueous solutions or in nonaqueous or mixed solvents, the basis for selection of conventional individual activity coefficients is less firm, but acceptable pH standards have now become available[104]. Hence absolute pH values in such solutions, based on comparison with dilute aqueous buffers, have little significance.

Often, values of pH, measured according to (1C31), are used in mass-action law expressions (chapter 3) in place of either $-\log c_{H^+}$ or $-\log c_{H^+} y_{\pm HA}$ or similar terms. The limitations of the precision or the definiteness of the pH values are then transferred to the equilibrium constants calculated from them.

e. Measurement of Activity Coefficients

It is appropriate to recount here briefly the methods found most useful for the measurement of activity coefficients. These methods have been discussed in details in standard works, and only an outline is given here. The methods are classified into those where the activity of the solvent is measured, and that of the solute is obtained from the Gibbs–Duhem relation, and those where the activity coefficient of the electrolyte is measured directly.

To the first group belong methods of measuring the vapour pressure of the solvent, which may be made statically (the vapour pressures p^0 of solvent and p of solution are compared) or dynamically (gas is blown through solvent and solution, and the amounts evaporated from each, compared). The bithermal equilibration method measures the temperature diffence of solvent and solution which are brought adiabatically to vapour equilibrium[105], or the electrolyte concentration is noted, which gives the same vapour pressure at a given temperature as pure water at a known lower temperature. The most convenient and accurate (above about 0.1 M) method seems to be the isopiestic equilibration method[106]. Solutions are brought to vapour equilibrium isothermally inside an evacuated desiccator, and the concentration of the test solution G is noted, which has the same vapour pressure as a solution of a reference salt R. The solutions have then the same solvent activity a_S, and from (1A21)

$$\phi_G = \frac{v_R m_R}{v_G m_G} \phi_R \qquad (1C32)$$

Using (1A22) this leads to

$$- \ln \gamma_G = (1 - \phi_G) + \int_0^{m_G} (1 - \phi_G) d \ln m_G = - \ln \gamma_R - \ln \frac{v_R m_R}{v_G m_G}$$
$$+ \int_0^{m_R} \frac{v_G m_G - v_R m_R}{v_G m_G} d \ln (m_R \gamma_R) \qquad (1C33)$$

or when $v_G = v_R$

$$- \ln \gamma_G = \ln m_G - \ln (m_R \gamma_R) + \int_0^{m_R} (1 - m_R/m_G) d \ln (m_R \gamma_R) \qquad (1C34)$$

Improved graphical and numerical methods for the integration in (1C34) have been recently proposed[107,108].

Many of the activity coefficients found in compilations were obtained by this method, using as references sodium chloride, potassium chloride, calcium chloride or sulphuric acid. Activity coefficients for these electrolytes must, of course, be obtained by different methods, as must also values for concentrations below about 0.1 M, where the isopiestic method becomes unreliable. Solutions in other solvents besides water may be treated similarly, provided data for a reference salt are known.

Freezing point depression measurements give in principle the same information as vapour-pressure measurements but the computations for obtaining activity coefficients are much more involved, and in order to obtain data at ordinary temperatures, rather than at the freezing temperature, further data (heat capacity) and calculations are necessary. The same may be said of methods based on the elevation of the boiling point of solutions, their osmotic pressure, etc.

Of the methods in which the solute activity is measured, the measurements of e.m.f. of concentration cells are the most important. In cells without transport (such as Pt; H_2, HCl(m), AgCl; Ag), the activity coefficient of the (binary) electrolyte G may be calculated from

$$- \log \gamma_G = \frac{F}{v 2.303 RT}(E - E^0) + \log m_G \qquad (1C35)$$

The value of E^0 is obtained by extrapolating a suitable deviation function, such as

$$E' = E + \frac{v\, 2.303\, RT}{F} (\log m_G - A m_G^{1/2} / 1 + B \overset{\circ}{a} m_G^{1/2}) \qquad (1C36)$$

to $m_G = 0$, where $E' = E^0$. For non-binary electrolytes a numerical term in v_i etc. is added.

Concentration cells with transference, such as Ag; AgCl, NaCl(m_1); NaCl(m_2), AgCl; Ag are simpler experimentally, but require the knowledge of the variation of the transport number of the cation t_+, with concentration. The activity coefficient is given by

$$\log \gamma_2 = \log \gamma_1 + \log(m_1 / m_2) - \frac{F}{2 \times 2.303\, RT} \times \frac{E}{t_{+(1)}}$$

$$- \frac{E}{2 \times 2.303\, RT} \int_{m_1}^{m_2} \left(\frac{1}{t_{+(2)}} - \frac{1}{t_{+(1)}}\right) dE \qquad (1C37)$$

where the solution at concentration m_1 is considered as reference. The last term, involving the integral, is small compared with the rest, and its graphical or tabular evaluation does not lead to loss of accuracy.

Most other useful methods for measuring activity coefficients involve the distribution of the solute between two phases. In solubility measurements, the activity of the solute in solution is constant, because of the equality of the chemical potential in the two phases at equilibrium, and the constancy of the activity of the pure solid. The variation of the activity coefficients of the solute under the influence of added foreign elctrolytes can thus be determined by measuring the changes in the solubility, i.e. the concentration of the sparingly soluble solute. The distribution of a solute between aqueous solutions and ion exchanger or solvent phases, which is the subject of the bulk of this book, is also a useful method for determining activity coefficients, and will be discussed in detail in later chapters.

The activity coefficients determined by the methods indicated above are stoichiometric mean activity coefficients, i.e. the ratio of the measured thermodynamic activity of the solute and the stoichiometric concentration (on the appropriate scale) of the (anhydrous) solute. Effects of association or other interactions, such as with the solvent, are not taken into account. Sometimes it is useful to have values for an electrolyte as if it were a fully dissociated strong electrolyte. These may be obtained from the stoichiometric values if the fraction ionized is known. This point will be elaborated on in chapter 3.

A tabulation of activity and osmotic coefficients for a number of common electrolytes will be found in the Appendix. The table is based on the best available values, but small shifts may be indicated from time to time, when better values are obtained for reference substances, to which most data are referred[109].

D. CONCENTRATED SOLUTIONS

a. Introduction

As an illustration of the problems in concentrated solutions, the activity coefficients of calcium and cadmium chlorides are plotted as functions of their molalities in Figure 1D1. The cations of these salts have approximately the same crystalline radius, and may be expected to have the same hydrated radius too. Activity coefficients calculated from the Debye–Hückel equation (1C24) with an ion-size parameter making $B\overset{\circ}{a} = 1$ are shown as a broken curve. The figure shows deviations from the Debye–Hückel curve for both salts, but in opposite directions. The activity coefficients for cadmium chloride fall below the expected values, and this effect is explained by association, as discussed in chapter 3. The data for calcium chloride, however, lie above the Debye–Hückel curve, and at higher concentrations show an upward trend of the activity coefficient, rather than the steady fall expected from long-range forces alone. This deviation is partly explained by ion–solvent interaction (section 1.B), and is the subject of the present section. As will be noted, there exists a variety of treatements of the problem, with little agreement. The most important approaches are described briefly, with some discussion on their relative merits. A successful comprehensive theory which is able to describe all the facts has, unfortunately, not yet been proposed and accepted.

The general expression for the mean molal activity coefficient in solutions more concentrated than those very dilute ones to which the Debye–Hückel derivation of section 1.C.b applies, will be

$$\ln \gamma_{\pm} = \ln f_{el} + \ln f_{stat} + \ln f_A + \ln f_B + \cdots \qquad (1D1)$$

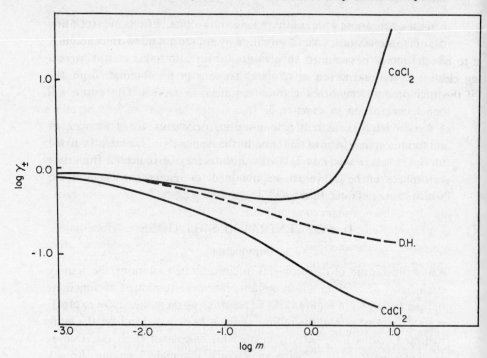

Figure 1D1. The mean ionic molal activity coefficients of calcium chloride and cadmium chloride in aqueous solutions, and the activity coefficient curve according to the Debye and Hückel theory, extrapolated to higher concentratuions (— — — D.H.).

In this expression additivity of effects is assumed, and care must be taken not to make the same correction twice. The expression is written in terms of rational activity coefficient, and the term f_{stat} takes care of conversion between concentration scales, as well as effects of hydration, etc. The term f_{el} takes care of the long-range electrostatic effects. The terms f_A, f_B, etc., take care of any additional effects deemed relevant, such as mutual salting-out of ions, etc.

b. Alternative Distribution Functions

Let us examine again the assumptions on which the Debye–Hückel treatment is based (section 1.C.b). It has been shown that the assumption of spherical symmetry, on which the distribution laws (1C1) or (1C6) are based, is the first to break down as the concentration of electrolyte increases. Frank and Thompson[110] have shown that even below 0.001 M, half of the electrostatic effect is produced by not more than two neighbours, which cannot be assumed to present a spherically symmetrical aspect either to the central ion or to the other ions in the ionic atmosphere, which produce the other half of the electrostatic effect. At these or higher concentrations short-range forces between pairs of ions become of similar importance to the long-range forces between a central ion and its atmosphere. Above this boundary concentration of about 0.001 M (for a 1:1 salt in water at 25°), $1/\kappa$, the 'average diameter of the ionic atmosphere' of the Debye–Hückel treatment, is smaller than the average distance between ions. Frank and Thompson propose to use a cube-root law above this boundary concentration. The theoretical justification[60,111] is an extension of earlier views[112,113] involving a diffuse lattice ionic atmosphere, characterized by an appropriate Madelung constant A_M, and an inverse dependence on a 'lattice-length' λ, the cube root of the volume available per ion

$$\lambda = (V/\nu N_G N)^{1/3} = (\nu N c_G/1000)^{-1/3} \tag{1D2}$$

where V is the volume of solution in cm^3 containing N_G moles of electrolyte. The free energy of electrostatic interaction for a litre of this solution is[60]

$$\Delta G \text{ (el., dif. lattice)} = \frac{- N N_G A_M F_T e^2 \sum z_i^2}{\varepsilon_0 \lambda}$$

$$= \frac{- N A_M F_T e^2 \sum z_i^2}{\varepsilon_0} \times \left(\frac{\nu N}{1000}\right)^{1/3} c_G^{4/3} \tag{1D3}$$

where F_T is a function of $(T_0/T)^{1/2}$ (T_0 being a reference temperature) taking account of the perturbation by the Brownian movement of the long-range lattice forces, described by the Madelung constant A_M. The activity coefficient is obtained from $\ln y_\pm = (d \sum \nu_i G_i/d c_G)/\nu RT$:

$$\log y_\pm = \frac{- 4/3\, N^{1/3} e^2 A_M F_T \sum (\nu_i z_i)^2}{2.303 \times 10\, \nu \varepsilon_0 k\, T} c_G^{1/3} = -0.293\, \mathring{a}\, \frac{\sum (\nu_i z_i)^2}{\nu} c_G^{1/3} \tag{1D4}$$

where $\mathring{a}$ is an 'exclusion distance' (in Ångstrom), describing the volume unavailable for ions around a central ion. This distance may be regarded as a free parameter, specific for each electrolyte, permitting the fitting of (1D4) to experimental data.

Glueckauf[114] presented a similar treatment, based on Bjerrum's[113] ideas, ending up with the expression $\log y_{\pm}$ (el., dif. lattice) $= -0.385\ c_G^{1/3}$, which differs from (1D4) because a different value for A_M was selected and the function F_T was neglected. Both treatments permit a better fit to the experimental data than the Debye–Hückel expression in the range 0.001 to 0.1 M (1D4) or in conjuction with non-electrostatic terms up to concentrations of a few molar[111,114].

Another way to deal with the 'exclusion distance' is to equate its effect to that of a van der Waals covolume. Van Rysselberghe and Eisenberg[115] derived the following equation in terms of $\overset{\circ}{a}$, Debye-Hückel's distance of closest approach

$$\ln f \text{ (covolume effects)} = \nu N_G \frac{4\pi}{3} a^3 + \frac{15}{64}\left(\nu N_G \frac{4\pi}{3} a^3\right)^2$$

$$= 2.206 \times 10^{-3}\, \overset{\circ}{a}{}^3 \nu c_G + 2.627 \times 10^{-6}\overset{\circ}{a}{}^6 \nu^2 c_G^2 \qquad (1D5)$$

The introduction of a linear and a square term permits the fitting of activity coefficients of hydrochloric acid and the alkali halides up to 4 M, with but a single parameter, $\overset{\circ}{a}$.

A different approach to the problems of the exclusion volume and to the non-conformation to spherical symmetry is to consider the ions as occupying a limited number of 'sites' around the central ion. Wicke and Eigen[116,117] introduced the 'occupation number' N_i, which is the reciprocal of the effective volume v_{hi} occupied by an (hydrated) ion i, and use a modified distribution function instead of (1C5), the linearized form of the Boltzmann distribution

$$n_i' = n_{i0} - \frac{n_{i0}z_i e}{kT}\left(1 - \frac{n_{i0}}{N_i}\right)\psi = n_{i0}\left(1 - \frac{z_i e}{kT}(1 - n_{i0}v_{hi})\psi\right) \qquad (1D6)$$

This leads to a modified diameter of the ionic atmosphere

$$\kappa'^2 = \frac{4\pi e^2}{\varepsilon_0 kT}\, \Sigma n_{i0} z_i^2 (1 - n_{i0}v_{hi}) \qquad (1D7)$$

The validity of this modification of the distribution law has been criticized[119]. The electrostatic contribution of the activity coeffiients takes the form (*vide supra* section 1.C.a)

$$\ln f_{el} = \frac{-\left|z_+ z_-\right|e^2}{2\varepsilon_0 kT}\, \frac{\kappa'}{1 + \kappa'a} \qquad (1D8)$$

instead of (1C18), where the term $(1 - n_{i0}v_{hi})$ in (1D7) has the effect of causing an increase in $\ln\gamma_{\pm}$ with an increase in concentration $c_G = 1000 \times n_{i0}/N$ after a minimum is reached. The volume v_{hi} is related to a, the distance of closest approach, so that equation (1D8) is a one-parameter expression. It has been shown by Eigen and Wicke[116] to reproduce experimental activity coefficients for salts such as calcium chloride (Figure 1D1), including also the sharp increase at higher concentrations.

c. Statistical Entropy Terms

The general trend in dealing with moderately concentrated solutions of electrolytes, however, is to assume the validity of the Debye–Hückel treatments of the electrostatic (long-range) forces term even at concentrations higher than those where its underlying assumptions are justified, and to apply correction terms describing the short-range forces. Thus, equating f_{el} with $f_{\pm}$ (D.H.),

$$\log\gamma_{\pm} \text{ (conc. solns.)} = \frac{-A|z_+ z_-|I^{1/2}}{1 + B\mathring{a}I^{1/2}} + \text{Correction terms}$$

$$(1D9)$$

One term is based on the argument that the ideal concentration scale leading from (1C14) to (1C18) should be the rational, rather than the molal (1A11), so that the Debye–Hückel term in (1C24) and (1D9) is equal to $\log f_{\pm}$ (D.H.), and by (1A20)

$$\log f_{\text{stat}} = \log\gamma_{\pm} - \log f_{\pm} \text{ (D.H.)} = -\log(1 + 0.018\,v\,m_G) \quad (1D10)$$

For a 1:1 electrolyte this introduces a correction of up to 7% in the activity coefficient. This may be considered a statistical entropy term (cf. equation 1A9) in the expression for the free energy. As an alternative to the mole-fraction statistics, on which (1D10) is based, volume statistics can be used[71, 119], the ideal entropy of mixing being

$$\Delta S_{\text{id}}^M = -R(x_W \ln\varphi_W + v x_G \ln v\varphi_G) \quad (1D11)$$

where φ is the fraction of the total volume contributed by each component, i.e. $\varphi_G = N_G V_G/(N_G V_G + N_W V_W)$, and similarly for φ_W. Setting $r_v = V_G/V_W$ for the ratio of the molar volumes and noting that $N_G/N_W = 0.018\,m_G$, leads to

$$\log f_{\text{stat}} = -\log(1 + 0.018\,m_G r_v) + \frac{0.018\,m_G r_v(r_v - v)}{2.303\,v(1 + 0.018\,m_G r_v)} \quad (1D12)$$

This term is important only for large ions $(r_v \gg 1)$[119], or when short-range ion–solvent interactions are considered[71], as discussed below.

Bjerrum was the first to point out that hydration of the ions introduces a term in the expression for the activity coefficient, because of the virtual increase in concentration, when counting hydrated ions in the remaining free water[120].

Robinson and Stokes[52,60,121], Glueckauf[71,114], Miller[122], Ikeda[123] and others developed this theory for the effect of ion–solvent interactions, or hydration, on the activity coefficients and other thermodynamic quantities in electrolyte solution at moderate concentration (from about 0.1 M, the limit of the Debye–Hückel treatment, to a few molar). They consider the ions hydrated by a fixed number of water molecules, h_+ per cation, h_- per anion, $h = \sum v_i h_i$ for the electrolyte irrespective of concentration. The amount of *free* water is therefore less than if no hydration occurred, and the mole fraction (or volume fraction) of the electrolyte, in its hydrated form, is larger. Moreover, the electrostatic term f_{el} pertains to the actual ions interacting in the solution, whereas the molal activity coefficient $\gamma_\pm$ formally pertains to the anhydrous electrolyte. This necessitates the introduction of terms containing h to the statistical part of the activity coefficient in equation (1D9). One advantage of this treatment is that the hydrated ions may be considered as interacting through a medium of the bulk dielectric constant, a supposition which is nearer the truth than the case of non-hydrated ions. Since deviations from ideality occur also with non-electrolytes (*vide infra*), the extra deviations introduced by electrolytes may be confined to long-range electrostatic interactions and to ion–solvent interactions, to a good approximation. These may be completely taken care of by the f_{el} term and the terms involving h. Depending on whether one retains the term for the contribution of the solvent to the electrical free energy of the solution, i.e. the last term in equation (1C14), or not, one may include a term $h \log a_W$ in the equations[52,71,121].

In the simplest derivation[122], neglecting changes in the activity of the water, the fact that the number of moles of free water per kg of solvent decreases by $h m_G$ because of hydration leads directly to the expression

$$\log f_{stat} = -\log(1 - 0.018(h - v)m_G) \qquad (1D13)$$

instead of (1D10). This relation was used by Miller[122] to fit the activity coefficient of 48 electrolytes, and similar equations were used by Ikeda[123] for 30, and by Stokes and Robinson for 36 electrolytes. In most cases the fit in the range $I = 0.1$ to approximately $I = 4$ M is remarkably good. The treatment involves two adjustable parameters, $\overset{\circ}{a}$ and h, and has the drawback that the values of h found for the best fit are not additive in terms of h_i for the ions. Furthermore, these values of h are twice to three times the values of reasonable hydration numbers (Table 1B3).

The use of volume statistics[52,71,114,119] (cf. section 1.D.c) improves the situation. The free energy of the solution is according to (1A2) and (1D11)

$$G = N_W \mu_W^0 + N_G \mu_G^0 + G_{el}$$

$$+ (RT/1 + N_G/N_W)\left[\ln \frac{1}{1 + r_v N_G/N_W} + (v N_G/N_W)\ln \frac{v r_v N_G/N_W}{1 + r_v N_G/N_W}\right]$$

$$(1D14)$$

Differentiating (1D14) at constant N_G with respect to the number of moles of water (either N_W, the total water present, or $N_{Wf} = N_W - h N_G$, the free water, which is the same operation at constant N_G and h), and setting $N_G/N_W = 0.018\, m_G$, yields

$$\log a_W = \frac{\mu_W^{(el)}}{2.303 R T} + \log \frac{1 - 0.018\, h\, m_G}{1 + 0.018\, r_v\, m_G} + \frac{0.018\,(r_v + h - v)m_G}{2.303\,(1 + 0.018\, r_v\, m_G)}$$

$$(1D15)$$

for the activity of water. Differentiation with respect to N_G at constant N_W gives for the unhydrated ion[52,71]

$$\log \gamma_\pm = \frac{\mu_G^{(el)}}{2.303 R T} - \frac{h}{v} \log \frac{1 - 0.018\, h\, m_G}{1 + 0.018\, r_v\, m_G} - \log(1 + 0.018\, r_v\, m_G)$$

$$+ \frac{r_v}{v} \times \frac{0.018\,(r_v + h - v)\,m_G}{2.303\,(1 + 0.018\, r_v\, m_G)} \tag{1D16}$$

while for the hydrated electrolyte, differentiation with respect to N_G at constant N_{Wf} gives

$$\log \gamma_\pm \text{ (hydrated)} = \frac{\mu_G^{(el\ hyd)}}{2.303\, R\, T} + \log \frac{1 - 0.018\ h\, m_G}{1 + 0.018\, r_v\, m_G}$$

$$+ \frac{(r_v + h)}{v} \times \frac{0.018\,(r_v + h - v)\,m_G}{2.303\,(1 + 0.018\, r_v\, m_G)}$$

$$(1D17)$$

The last expression differs from (1D16) in having the term $\mu_G^{(el\ hyd)}$ instead of $\mu_G^{(el)}$, in having the coefficient of the second term unity instead of h/v, in not having a second logarithmic term and in having $(r_v + h)$ in the numerator of the last term instead of r_v. Miller's expression[122] for the activity coefficient of the hydrated electrolyte, based on volume-fraction statistics, is similar to (1D17), except that he identifies $\mu_G^{(el\ hyd)} + \log(1 - 0.018\,h\,m_G)$ with $\log f_{\pm}(D.H.)$ as in (1D10), and his symbol for the volume fraction has to be identified with $r_v + h$, i.e. the ratio of the molar volume of the hydrated electrolyte to that of water. Glueckauf[71] and Stokes and Robinson[52] differ in their interpretation of the terms $\mu_G^{(el)}$ and $\mu_G^{(el\ hyd)}$ in equations (1D16) and (1D17) to a minor extent. It seems best[52] to write[89]

$$\frac{\mu_G^{(el)}}{2.303\,R\,T} = \frac{-A\,|z_+\,z_-|\,I^{1/2}}{1 + B\,\mathring{a}\,I^{1/2}} + \frac{2\,A\,I^{3/2}V_G}{3000}\,\sigma(\kappa a) \qquad (1D18)$$

where the last term, which was introduced above (section 1.C.b), is is equal to $r_v\,\mu_W^{(el)}/2.303\,R\,T$, and is numerically small. On this basis $\mu_G^{(el\ hyd)}/2.303\,R\,T$ would be equal to the Debye–Hückel term in (1D18), as already proposed (section 1.D.b).

The advantage of the treatment involving volume-fraction statistics is that the hydration numbers h required to fit equation (1D16) to the data are found[71] to be additive, $h = v_+\,h_+ + v_-\,h_-$, and of a reasonable magnitude. Stokes and Robinson[52] propose to relate h and $\mathring{a}$ by

$$\mathring{a} = r_{c+}\,(1 + 0.58\,h_+/r_{vc+})^{1/3} + r_{c-}\,(1 + 0.58\,h_-/r_{vc-})^{1/3} \qquad (1D19)$$

where the crystalline radii r_{ci} are in Ångstroms, and r_{vci} are the ratios of the molar volumes of water and anhydrous ion, and the factor 0.58 is the fraction of the total volume occupied by randomly close-packed spheres of different sizes (cf. section 1.B.d). As discussed above (section 1.B) hydration numbers h can be determined independently, and need not be taken as free parameters. An attempt may thus be made to fit the activity coefficients with only universal constants, the temperature, the bulk dielectric constant, crystal radii of the ions and independent, experimental hydration numbers.

The above treatment does not take into account the observed variation of the hydration numbers h, and the hydrated molar volumes (or ratios r_{vh}) with concentration. Glueckauf[71] showed that if h is treated as constant, but r_{vh} is allowed to vary, an approximate term

$$\frac{-3 S_r (r_{vh} + h - v) m^{1/2}}{2.303 \times 2v(r_{vh} + h)}$$

has to be added to (1D16), where S_r is the coefficient of the square-root term of the variation of r_{vh} with concentration. This addition of a square-root term to $\log \gamma_{\pm}$ has the effect that $\overset{\circ}{a}$ has to be increased to obtain a fit, and this may improve the rather too low values found for large ions. The non-constancy of h at high concentrations actually results in serious deviations of measured activity cofficients from (1D16), for highly hydrated ions. This occurs usually when about 15% of the water molecules are bound, or at $h m_G > 8$ m.

Still another approach to the problem of concentrated solutions considers the ion–solvent interactions not in terms of a hydration number h, but in terms of an electrostatic salting-out theory, essentially that of Debye and McAulay[124]. Scatchard[125] proposed to use for a 1:1 electrolyte ($v = 2$)

$$\log \gamma_{\pm} = -\log(1 + 0.036\, m_G) + \frac{b'_H m_G (2 + V_G m_G)}{2.303\, R T (1 + V_G m_G)^2} - \frac{e^2}{2.303 \times 2kT\varepsilon_o}$$

$$\times \left[\frac{\kappa''}{1 + \kappa'' a}(1 + 2 V_G m_G) - \kappa'' V_G m_G \sigma(\kappa'' a) - 2\left(\frac{v_+}{a_-} + \frac{v_-}{a_+}\right) m_G \right] \tag{1D20}$$

where the first term corrects for statistical entropy on mole-fraction scale (1D10) and the second term accounts for non-electrolyte interactions (see section 1.D.c). The electrical term contains the Debye–Hückel term for long-range forces, modified for the volume occupied by the ions and corrected for the solvent contribution, and a final term which describes the salting-out effect of the ions on the solvent, a_+ and a_- being modified radii of the ions. The variation of the dielectric constant of the bulk is taken care of by assuming it to be proportional to the volume fraction of water, and defining $(\kappa'')^2$ in terms of $m_G \rho_o / \varepsilon_o$, instead of using κ^2 proportional to c_G/ε, with the variable dielectric constant. This approach is a development of earlier ideas of Scatchard[126] and Hückel[127], the latter introducing instead of the last term on the right, an empirical term $b\, m_G$, to take account of the salting-out effect.

The second term in (1D20) may be rewritten as

$$\frac{1}{2.303\, R T}\left(\frac{b'_H}{V_G}\right)\left(\frac{1}{(1 + V_G m_G)^2} - 1\right)$$

and is characteristic of regular solution behaviour, on the basis of volume fractions, as derived by Scatchard[128]. Using mole fractions however, the term obtained is rather

$$\log f(\text{regular}) = \frac{b_H}{2.303\,RT}\left(\frac{1}{(1+0.018vm_G)^2} - 1\right) \simeq \frac{-2 \times 0.018vb_H m_G}{2.303\,RT}$$

$$(1D21)$$

the last expression being valid for dilute solution. This contribution to non-ideality involves a non-zero heat of mixing, which is related to the constant b_H. Such a term would describe the non-ideality of the electrolyte solution if the ions were uncharged.

d. Empirical Equations

It would not be correct to sum all the contributions to the free energy of the solution listed above, in order to calculate the activity coefficients according to (1D1), since certain interactions and effects will undoubtedly be counted twice. One might however regard the statistical entropy (including the size effect), the ion–solvent interactions, and the non-electrolyte-type interactions (1D12, 1D13 and 1D21) as relatively independent. At concentrations of about 1 m the term $0.018\,m_G$ is small compared to unity, and $\log(1 + 0.018\,m_G)$ may be approximated by $0.0078\,m_G$ so that the above equations become

$$\log f\,(\text{size effect}) = 0.0078\left(\frac{r_v}{v} - 2\right)r_v m_G \qquad (1D22)$$

$$\log f\,(\text{hydration}) = 0.0078\,(h - v)\,m_G \qquad (1D23)$$

$$\log f\,(\text{heat of mixing}) = -0.0078\,\frac{2v}{RT}b_H m_G \qquad (1D24)$$

i.e. all terms linear in m_G. Setting

$$0.0078\left[\left(\frac{r_v}{v} - 2\right)r_v + (h - v) - \left(\frac{2v}{RT}b_H\right)\right] = b,$$

one may set $b\,m_G$ equal to the 'correction terms' in (1D8) and write

$$\log \gamma_\pm\,(\text{intermediate concentration}) = \frac{-A\left|z_+ z_-\right|I^{1/2}}{1 + B\mathring{a}I^{1/2}} + b\,m_G$$

$$(1D25)$$

It was shown empirically[129] that for the alkali halides b is related to the difference in standard molal partial entropies of the cation and anion, and at 25^0, $b = 0.059 - 0.00525 \Delta S^0_{i\,aq}$.

It must be remembered that the quasi-theoretical derivation of (1D25) is valid only for the range where the logarithms in (1D12) and (1D13) may be approximated by the first linear term. Using higher terms in these exponents and in the ratio (1D21) will yield terms in m_G^2 etc., similar to (1D5), which fit experimental data to relatively high concentrations. The use of the full expressions (1D16) and (1D21) will describe the activity coefficients with four parameters ($\mathring{a}, h, r_v$ and b_H) but the variation of h and of r_v with concentration cannot, with the presently available knowledge, be taken into account.

Various empirical relationships have been proposed to describe the activity coefficients in concentrated solutions. The simplest is Güntelberg's[130], who dispensed with an adjustable parameter, and wrote $B\mathring{a} = 1$ (or $\mathring{a} = 3.04$ Å), so that (1C24) becomes

$$\log \gamma_{\pm} = \frac{-A|z_+ z_-| I^{1/2}}{1 + I^{1/2}} \tag{1D26}$$

which fits a few electrolytes up to 0.1 M. Scatchard[125] proposed $B\mathring{a} = 1.5$ ($\mathring{a} = 4.57$ Å) to give a better fit for more electrolytes. Hückel[127] proposed (1D25) with $\mathring{a}$ and b as adjustable parameters, and this expression, indeed, is successful for many electrolytes, over a wide concentration range. Since $\mathring{a}$ is now a free parameter, rather than a physically meaningful 'distance of closest approach' (this may be brought about by the arbitrary device of mutual partial penetration of the hydration sheaths of a pair of oppositely charged ions), its function may be completely taken over by an adjusted b parameter. Hence Guggenheim[131] showed that using a constant value of $B\mathring{a} = 1$, the expression

$$\log \gamma_{\pm} = \frac{-A|z_+ z_-| I^{1/2}}{1 + I^{1/2}} + bI \tag{1D27}$$

adequately expresses the activity coefficients of all electolytes, with a single adjustable, specific parameter b up to a few tenths molal. Davies[132] showed that up to $I = 0.2$, equation (1D27) may be used with the constant value $b = 0.15|z_+ z_-|$ for all electrolytes. Guggenheim's equation (1D27) has been used by Pitzer and Brewer[133] for representing activity coefficients relative to that of potassium chloride, noting that for the ratio $\log(\gamma_{\pm G}/\gamma_{\pm KCl})$ the square-root term cancels, and only the term

$(b_G - b_{KCl})\ I$ remains. Activity coefficients for potassium chloride have been obtained with high precision, and many activity coefficient determinations are in any case made isopiestically relative to this salt. The tabulations by Pitzer and Brewer show that $(b_G - b_{KCl})$ is not constant as proposed by Guggenheim, and by Åkerlof and Thomas[134], although it is a rather slowly varying function of I, particularly at high I.

The activity of water in the solutions may be obtained from (1D15), using (1C16) for the first term. If instead of (1D15) with the parameters h and r_v, the empirical expression (1D25) is used, the Gibbs–Duhem relation yields

$$\log a_w = -\log(1 + 0.018\,v m_G) + 0.018\,v A(B\mathring{a})^{-3}\sigma(B\mathring{a}I^{1/2}) - \tfrac{1}{2}(0.018\,v b\,m_G^2)$$

$$(1D28)$$

ignoring terms involving $\partial I/\partial m_G$. The osmotic coefficients, which are a more sensitive expression for the non-ideality of the solutions, are obtained as usual from

$$\phi = -(2.303 \times 55.5/v\,m_G)\log a_w$$

e. The Acidity Function

The proton plays an important role in solutions, and especially in aqueous solutions. As has been seen above (section 1.C.d), the concept of pH has little significance in concentrated solutions. Hammett[135] has introduced in its place the concept of 'acidity function', a quantity measurable by the use of indicators

$$H_i = pK_I - \log\frac{(H_{i+1}I^{i+1})}{(H_iI^i)} = -\log a_{H+} - \log\frac{y_{H_iI^i}}{y_{H_{i+1}I^{i+1}}}$$

$$= -\log a_{H+} - \log y_I' \qquad (1D29)$$

where i may be -1, zero or $+1$, depending on the indicator H_iI^i used. The ratio of concentrations $(H_{i+1}I^{i+1})/(H_iI^i)$ can be measured, e.g. colorimetrically, and knowing pK_I, H_i can be calculated. Since a small amount of indicator with $i = 0$, for example, will not change the acidity, or H_0, of a given solution, two indicators I_1 and I_2 in two portions of the solution will be related by

$$pK_{I_1} - \log(HI_{1+})/(I_1) = H_0 = pK_{I_2} - \log(HI_{2+})/(I_2) \qquad (1D30)$$

Hence, knowing pK_{I_1}, and measuring the concentration ratios in the two solutions, pK_{I_2} can be calculated. The dissociation constants of indicators may thus be related one to another, until an indicator is reached which is sufficiently basic to add a proton even in dilute solutions, where $\log y'_I + \log a_{H+}$ can be evaluated by reference to the ordinary pH scale. By these sequential relations, pK_I values may therefore be calculated even for very feebly basic indicators, so that H_0 (or H_+ or H_-, depending on indicator) values can be measured also in very highly acidic solution.

The importance of the H_i concept is that Hammett[135] and others[136], have shown that $\log y_I'$ is the same function of a_{H+} for all indicators, for a given i, within about 0.1 units or better, so that H_i is a unique measure of $-\log a_{H+}$. The limit of $\log y'_I$ at low concentrations is zero, so that H_i approaches pH as the concentration of acid (and electrolytes) is decreased.

Another class of indicators, triarylcarbinols ROH, ionize by dissociation of the hydroxyl group in strong acids, forming a carbonium ion and water[137]. The acidity function can be defined as

$$H_R = -pK_R - \log(R^+)/(ROH) = -\log a_{H+} + \log a_{H_2O} - \log y_{ROH}/y_{R+}$$

$$(1D31)$$

Since the charge of the triarylcarbonium ion is delocalized, it is expected that the last term is very small. The quantity $H_R' = H_R - \log a_{H_2O}$ should then be a direct measure of the activity of the hydrogen ions in the solution[138]. The two acidity functions can be related by

$$H_R' - H_0 + \log y'_I = -\log y_{ROH}/y_{R+} = -\log y'_R \qquad (1D32)$$

The quantity y_I' can be estimated by measuring the solubility of the indicator I (yielding y_I), and of a suitable salt of the protonated indicator, relative to the same salt of tetraethylammonium (yielding y_{IH+}/y_{Et_4N+}), and assuming[139] that the coefficient for the latter cation would not show specific effects. Inserting the known values of H_R' and H_0, and the estimated y_I' into (1D32) indeed yields values y_R' almost invariant with sulphuric acid concentration up to 70%[138]. However although H_R' is a unique function of the acidity, and independent of the indicator ROH at any given acidity, as Hammett indicators are required to be, it is not a unique function of the water activity[140], a feature that distinguishes the H_0 Hammet indicators.

The H_0 function shows the same dependence on $\log a_w$ for several acids (perchloric, nitric, sulphuric, hydrochloric, hydrobromic)[141,142] and at a given molality, i.e. ratio of water to H^+ ions, H_0 is independent of the acid[143]. This throws some light on the hydration of the species participating in the indicator protonation equilibrium. If a total of h moles of water are released by the protonation of I by $H_3O(H_2O)_p^+$ the same h applies for any of the acids and

$$H_0 = -\log c_{H+} + h \log a_w - \log y_I'' \qquad (1D33)$$

In the last term of (1D33), the activity coefficient of the indicator I is measurable from solubilities[139, 144], and can be introduced explicitly into the equation, leaving the ratio $y_{H+(hydrated)}/y_{HI+(hydrated)}$, which could be very near unity[144]. The water activity in (1D33) must be expressed on the same concentration scale as the other terms, i.e. the molar scale, and can be written $a_w = c_{H_2O(free)} y_w$, where $c_{H_2O(free)} = 55.3 c_{H+}$ $(1 - 0.018\, h\, m_{H+})/m_{H+}$. The acidity function may then be written as

$$H_0 = (h-1)\log c_{H+} + h(1.754 + \log(1-0.018\ h\ m_{H+})/m_{H+}) - \log y'' \qquad (1D34)$$

which permits the evaluation of $y'' = y_{H_3O(H_2O)_p^+}\ y_{I(H_2O)_q}/y_{HI(H_2O)_r^+} y_w$, where $p + q - r = h - 1$. This quantity has been shown to be quite small, since the first three terms on the right of (1D34) were found to describe H_0 very well (with $h = 4$), up to 8 m concentration for several acids[143]. The quantity $h = 4$ was used, assuming the large indicator ions to be non-hydrated ($q = r = 0$), and the proton tetrahydrated ($p = 3$ for $O(HOH_2)_3^+$ or $H(OH_2)_4^+$), as in dilute solutions. These assumptions are not necessarily true, although $h = 4$ is experimentally established. The use of a_w on a mol-fractional base (as ordinarily tabulated), although leading to reasonable (and reasonably constant) values of $h \sim 2$ and of . $= \log y'/c_{H+}$, as excepted from salting-out considerations[145], is not warranted. The H_- acidity function, which has found much use in strongly basic solutions, has also been related to the activity of the water[146-149]. The equation used [149]

$$H_- = pK_w + \log c_{OH-} - (h+1)\log c_{H_2O(free)}/55.3 \qquad (1D35)$$

with $h = 3$ (assigned to the hydroxide ion, neglecting hydration of the indicator species), and neglecting the activity coefficient term, is analogous to (1D33) and (1D34).

For acid solutions, the most reliable acidity function is still H_0. Although various new indicators have been proposed, e.g. amides[150], the original choice of substituted anilines (mainly chloro- and nitro-)

has proven to be fortunate. Recent revisions of the H_0 scale for sulphuric acid[151] and perchloric acid[152] have shown good consistency, and agreement with the best available previous estimates for sulphuric[153], perchloric[136], hydrochloric[136,153], hydrobromic[136] and nitric acid[154]. The data for these acids are shown in Table 1D1.

TABLE 1D1. The Hammett acidity function, $-H_0$.

M*	HNO_3^{154}	$H_2SO_4^{153}$	$HCl^{136,153}$	HBr^{136}	$HClO_4^{152}$
.5	−0.08	−0.13		−0.20	−0.02
1	0.30	0.26	0.20	0.20	0.33
2	0.78	0.84	0.69	0.71	0.87
3	1.08	1.38	1.05	1.11	1.32
4	1.40	1.85	1.40	1.50	1.82
5	1.68	2.28	1.76	1.93	2.33
6	1.92	2.76	2.12	2.38	2.94
7	2.15	3.32	2.56	2.85	3.50
8	2.37	3.87	2.86	3.34	4.22
9	2.58	4.40	3.22	3.89	5.21
10	2.79	4.89	3.59	4.44	6.20
11	2.98	5.41	3.99		7.08
12	3.19	5.59	4.41		8.16
13	3.40	6.44	4.82		
14	3.61	6.96			
15	3.83	7.47			

* At a given m, H_0 is independent of the acid [143].

The acidity function is useful not only in concentrated solutions of acids, but also in concentrated salt solutions, containing some acid. It has been found that concentrated salt solutions (lithium and calcium chlorides, sodium nitrate and perchlorate) obey the rule

$$pH_{G.E.} - H_0 = f(c_{salt}) \neq f(c_{acid}) \qquad (1D36)$$

provided the acid concentration is low[155]. The values of $pH_{G.E.}$, measured by a glass electrode, correlate well also with activities of the acid as measured with a quinhydrone electrode, at least in the case of nitric acid in lithium nitrate solutions[156]. Although changes in y_I, for the un-uncharged indicator, play an important role in the the the salt effect on acidity[144,157], and can be separately measured and corrected for[144]. Another major effect is the dehydration of the hydronium ion[145], probably from $H_9O_4^+$ to H_3O^+, possibly in a number of steps. The hydration of acids, and changes in hydration in concentrated solutions, have been studied by Högfeldt[158].

Since $y_I' y_{H+}$ has been shown to be the same for all acids at a given H_0, it may be calculated from the values of H_0 in strong acids, where $(H^+) = c_{HA}$. It is then possible[159] to calculate (H^+) for weak acids, from the measured H_0, and the degree of dissociation $\alpha = (H^+)/c_{HA\,(weak)}$.

It is difficult to judge where the above interesting correlations are going to lead, but probably many observations in concentrated acid solutions may be best interpreted in terms of acidity function data.

f. Effective Ligand Activity

The most important point in the above discussion is that y_I', a ratio of activity coefficients, was found to be independent of the indicator I and of its concentration, and to depend only on the acid concentration. For a formally similar problem, Bjerrum[160] has shown that for a complex-formation reaction $MA_{n-1}^{m+1-n} + A^- \rightleftharpoons MA_n^{m-n}$, omitting charges,

$$\log y_M' = \log y_{MA_n}/y_{MA_{n-1}} y_A = \log K_n^0 - \log(MA_n)/(MA_{n-1})(A)$$

$$\text{(1D37)}$$

$$= A' + b' c_A$$

with A' and b' constants independent of c_A and n. He showed that for $M = Cu^{2+}$ and $A = Cl^-$, b' was a function of the bulk cation, having the values -0.20 in HCl, -0.23 in LiCl, and -0.18 in MgCl$_2$. For comparison, in the bulk electrolyte $\log y_\pm = A + b c_A$, and the values for b are 0.18 in HCl and 0.20 in CaCl$_2$. Gamlen and Jordan[161] used the same concept for $M = Fe^{3+}$, while Haigt[162] used it for $M = Sn^{2+}$. The former authors assume in addition that $y_{Cl^-} = y_{\pm HCl}$ and $\log y_{MA_n}/y_{MA_{n-1}}$ $= 0$, while the latter assumes that $a_{Cl^-} = c_{Cl^-} \cdot y_{\pm HCl}$ and that $\log (y_{MA_n}/y_{MA_{n-1}}) = A' + A + (b' + b)c_{Cl^-}$ is constant, independent of c_{Cl^-}, but not necessarily zero, i.e. $b' = -b$. The validity of the assumption concerning the b coefficients is made plausible by the values given above for Cu^{2+} in halide solutions. Similar assumptions have also been used by many other authors[163]. Marcus and Coryell[164] and Marcus and Maydan[165] have introduced the term 'effective ligand activity' a

$$a = (A^-)y_{\pm CA} \text{ (for 1:1 electrolytes)} \tag{1D38}$$

as the quantity relating the stepwise thermodynamic formation constant K_n^T to the metal complex ratio

$$\log K_n^T = \log(MA_n^{m-n})/(MA_{n-1}^{m+1-n}) - \log a + \log F_n \tag{1D39}$$

Marcus[163] has shown that the function F_n, the effective activity coefficient quotient is given by

$$F_n = y_{\pm(MA_n)}^{n+1-m}/y_{\pm(MA_{n-1})}^{n-m}y_{\pm CA} \qquad \text{(for } n > m\text{)}$$
$$= y_{\pm(MA_n)}^{m+1-n}y_{\pm CA}/y_{\pm(MA_{n-1})}^{m+2-n} \qquad \text{(for } n \leq m\text{)} \tag{1D40}$$

and is a very useful quantity in discussing equilibria in concentrated solutions. It can be evaluated from the following expression, based on the empirical equation (1D25), found to hold well in concentrated electrolyte solutions

$$\log F_n \approx -Ac_{CA}^{1/2}(1 + B\mathring{a}_{CA}c_{CA}^{1/2})^{-1}\Delta z_i + (\sum z_i b_i)c_{CA} \tag{1D41}$$

where $\Delta z_i = 2(n - m) - 1$, and $\sum z_i b_i = [(|m - n| + 1) b_{MA_n} - (|m - n| + 1 \pm 1)b_{MA_{n-1}} + b_{CA}]$, the plus sign applying for $n < m$, and the minus for $n > m$. The last term in (1D41) is equivalent to $(b' + b)c_{CA}$ in Bjerrum's treatment[160], the coefficient having in some cases a very small value. The variation of $\log F_n$ with c_{CA} for various values of $\sum z_i b_i$ and of Δz_i is shown in Figure 1D2. Between 2 and 12 M CA, the slopes are seen to be very small, and for shorter regions, F_n may be assumed to be independent of c_{CA}. For hydrochloric acid $\log a = 0.3$ at 2 M and $\log a = 2.7$ at 12 M so that the effective ligand activity a may be changed by a factor of 250, with only a slight change in F_n.

This treatment of metal complexes in concentrated ligand salt solutions is seen to be similar to that of the acidity function (section 1.D.e). The metal M takes the place of the indicator I, the ligand number n that of the proton number i, the ligand A^- that of the proton H^+ [163]. It is therefore not surprising that functions such as y_M', F_n and a can be as useful as y_I' and H_i, or that assumptions of the constancy of certain

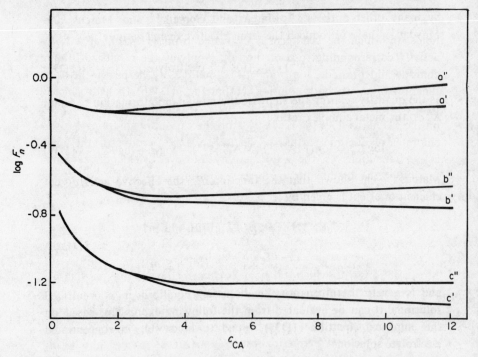

Figure 1D2. Plots of the function $\log F_n$ (equation 1D40) against electrolyte concentration, with Δz_i: $a = 1$, $b = 3$, $c = 5$ and $\sum z_i b_i$: prime $= 0.01$ and double prime $= 0.02$ (equation 1D41). (By permission from *Record Chem.Progr.* **27**, 112 (1966)).

functions are approximately true. It may be expected that correlations such as those noted above for H_0 will also be useful for the metal–ligand systems. Further discussion of the use of effective ligand activities in complex-formation equilibria is given in section 3.B.

A different approach has been suggested by Högfeldt and Leifer[166], who propose the use of ion activity functions, defined for acids as

$$\log \varphi_0 \, a_{H_3O^+} = \log(y_I/y_{HI^+}) a_{H_3O^+} = -H_0 + \log a_w \qquad (1D42a)$$

$$\log \varphi_0^{-1} a_{A^-} = \log(y_{HI^+}/y_I) a_{A^-} = H_0 + 2\log a \qquad (1D42b)$$

for the cation and anion respectively. Using the observation[141-143] that H_0 is the same for several acids at a given water activity, they assume that the anion activity function ($\log \varphi_0^{-1} a_{A^-}$) is the same for both salt and (strong) acid, at the same water activity. Thus for a cation of a salt, the expression

$$\log \varphi_0 \, a_{C+} = 2 \log a - \log \varphi_0^{-1} \, a_A \qquad (1D43)$$

holds, in which the last term is evaluated from (1D42b), at a water activity corresponding to a of the salt. Applying the results to lithium chloride and bromide, $\log \varphi_0 \, a_{Li+}$ was found[166] to be nearly the same function of a_W in both solutions. Applying (1D42b) to hydrochloric, hydrobromic and perchloric acids, $\log \varphi_0^{-1} \, a_{A-}$ was found to depend on the anion, as does $\log \varphi_0^{-1} \, a_{ClO_4-}$ on the cation, in solutions of perchlorates of alkaline earths and lead. The slope of plots of the ionic activity functions vs. the water activity, however, yields direct information on the ionic hydration in these solutions,

E. MIXED ELECTROLYTES

a. Introduction

Most solutions with which this book is concerned contain more than one electrolyte, for example they may contain a metal salt together with a ligand salt or acid. The thermodynamic relationships of mixed electrolyte solutions need therefore to be understood before proceeding to ion exchange or solvent extraction processes. Our knowledge about concentrated mixed electrolyte solutions is, unfortunately, even smaller than about concentrated solutions of single electrolytes (section 1.D) and the systematizations and correlations possible at present are much fewer. The main thermodynamic relationship that has to be considered is the cross-differentiation expression

$$\frac{\partial \mu_G}{\partial N_J} = \frac{\partial^2 G}{\partial N_G \partial N_J} = \frac{\partial \mu_J}{\partial N_G} \qquad (1E1)$$

For a mixture of two $1:1$ electrolytes this becomes

$$\left(\frac{\partial \log \gamma_G}{\partial m_J} \right)_{m_G} = \left(\frac{\partial \log \gamma_J}{\partial m_G} \right)_{m_J} \qquad (1E1a)$$

and for other types this is modified by terms involving ν_G and ν_J. These relationships can be used as a check on the self-consistency of any treatment of mixed electrolytes.

The osmotic coefficients of a solution of mixed electrolytes may be regarded as made up of additive terms for the contribution of each electrolyte (i.e. using $\kappa = B I^{1/2}$ in (1C16)) with possibly a correction

for the mutual interactions. In fact, it has been found[167] that to a good
aproximation, within about 2%, the water activity is given by the simple
additive relationship

$$1 - a_{\mathrm{w}} = \frac{\Delta p_{\mathrm{w}}}{p_{\mathrm{w}}^0} \approx \frac{I_{\mathrm{G}}}{I_{\mathrm{G}} + I_{\mathrm{J}}}\left(\frac{\Delta p_{\mathrm{w}}}{p_{\mathrm{w}}^0}\right)_{\mathrm{G}} + \frac{I_{\mathrm{J}}}{I_{\mathrm{G}} + I_{\mathrm{J}}}\left(\frac{\Delta p_{\mathrm{w}}}{p_{\mathrm{w}}^0}\right)_{\mathrm{J}} \tag{1E2}$$

where I_{G} is the contribution of electrolyte G to the total ionic strength,
and $(\Delta p_{\mathrm{w}}/p_{\mathrm{w}}^0)_{\mathrm{G}}$ is the relative lowering of the water vapour pressure in
solutions of G at total ionic strength $I = I_{\mathrm{G}} + I_{\mathrm{J}}$. Deviations therefrom,
however, have been noted[168].

b. Guggenheim's and Harned's Treatments.

A corresponding approximation does not however, hold for the activities
of the individual electrolytes. Instead, recourse can be made to Brönsted's[169] theory of specific interactions, which states that ions are
uniformly influenced by ions of their own sign and specifically influenced
only by ions of the opposite sign. Guggenheim[131] proposed to use this
theory for expressing the activity coefficient of G in a solution of total
molality m and total ionic strength I (assuming for simplicity 1:1 electrolytes, see section 1.D.d)

$$\log \gamma_{\pm \mathrm{G}} = \frac{-AI^{1/2}}{1 + I^{1/2}} + [2x_{\mathrm{G}}b_{\mathrm{G}\pm} + (1 - x_{\mathrm{G}})(b_{\mathrm{G}_+\mathrm{J}_-} + b_{\mathrm{G}_-\mathrm{J}_+})]m \tag{1E3}$$

where $x_{\mathrm{G}}m$ is the molality of G in the solution, and the b's are inter-
action coefficients. According to Brönsted's theory $b_{\mathrm{G}_+\mathrm{J}_+} = b_{\mathrm{G}_-\mathrm{J}_-} = 0$.
The first term is independent of the specific electrolytes present, according
to Guggenheim's formulation of the activity coefficients of a single
electrolyte (1D27), and by this device the difficult problem of assigning
a value to $\overset{\circ}{a}$ in mixed electrolyte solutions is avoided. Guggenheim's
equation (1E3) leads at the limit for $x_{\mathrm{G}} = 0$ to

$$\log \gamma_{\pm \mathrm{G}(0)} = \frac{-AI^{1/2}}{1 + I^{1/2}} + (b_{\mathrm{G}_+\mathrm{J}_-} + b_{\mathrm{G}_-\mathrm{J}_+})m \tag{1E4}$$

and for $x_{\mathrm{G}} = 1$ to

$$\log \gamma_{\pm \mathrm{G}(1)} = \frac{-AI^{1/2}}{1 + I^{1/2}} + 2b_{\mathrm{G}\pm}m \tag{1E5}$$

so that

$$\log \gamma_{\pm G} = x_G \log \gamma_{G(1)} + (1 - x_G) \log \gamma_{G(0)}$$

$$\text{(1E6a)}$$

$$= \log \gamma_{G(1)} + (1 - x_G)(\log \gamma_{G(0)} - \log \gamma_{G(1)})$$

$$\log \gamma_{\pm J} = (1 - x_G) \log \gamma_{J(1)} + x_G \log \gamma_{J(0)}$$

$$\text{(1E6b)}$$

$$= \log \gamma_{J(1)} + x_G(\log \gamma_{J(0)} - \log \gamma_{J(1)})$$

and also $\log \gamma_{J(0)} = \log \gamma_{G(0)}$. Plots of $\log \gamma_G$ and of $\log \gamma_J$ against x_G are straight lines having different slopes unless the electrolytes have a common ion, i.e. $J_+ = G_+$ or $J_- = G_-$. In this case the activity coefficient of trace electrolyte G or J in the other, becomes the geometric mean of the activity coefficients of the two electrolytes. Extension of (1E6a) to cases where there are more than two electrolytes is possible. The above equations yield by the Gibbs–Duhem relation the expression for the osmotic coefficient

$$\phi_{x_G} = (\text{term in } I, \text{ independent of } x_G) + [b_{G\pm} x_G^2 + b_{J\pm}(1 - x_G)^2$$

$$\text{(1E7)}$$

$$+ (b_{G_+ J_-} + b_{G_- J_+}) x_G \times (1 - x_G)] m$$

which is not linear in x_G.

Writing for the difference between (1E4) and (1E5)

$$\log \gamma_{\pm G(0)} - \log \gamma_{\pm G(1)} = (b_{G_+ J_-} + b_{G_- J_+} - 2b_{G\pm}) m = -\alpha_G m \quad \text{(1E8)}$$

leads from the right-hand side of (1E6a) to the expression

$$\log \gamma_{\pm G} = \log \gamma_{G(1)} - (1 - x_G) \alpha_G m = \log \gamma_{G(1)} - \alpha_G m_J \quad \text{(1E9)}$$

Equation (1E9) is known as Harned's rule[170], but according to Harned, α_G, although independent of x_G, may still depend on m, contrary to Guggenheim's treatment of the b coefficients. Equation (1E9) adequately expresses the measured activity coefficients of many electrolytes at constant molality and varying mole fraction. In some cases deviations occur, and a square term is needed

$$\log \gamma_G = \log \gamma_{G(1)} - \alpha_G m_J - \beta_G m_J^2 \quad \text{(1E10)}$$

The cross-differentiation relationship (1E1) then requires that $(\beta_G + \beta_J)$ $= const. - (\alpha_G + \alpha_J)/2\,m$. In many cases $(\alpha_G + \alpha_J)$ is independent of m, so that $(\beta_G + \beta_J)$ is either inversely proportional to m or zero, and in the latter case most probably $\beta_G \simeq \beta_J \simeq 0$. For the lithium–sodium chloride system, for instance, $(\alpha_{LiCl} + \alpha_{NaCl}) = -\,0.013 + 0.004\;m$ so that $\beta_{LiCl} \simeq \beta_{NaCl} \simeq -\,0.001$ and similarly for lithium–potassium chloride and for lithium chloride–nitrate mixtures. In some other cases, as with the lithium–caesium chloride system, $\beta_G \neq \beta_J$, and $(\alpha_{LiCl} + \alpha_{CsCl})$ $= 0.082 + 0.009\,m$, $\beta_{LiCl} = -\,0.005$, $\beta_{CsCl} = 0.001$. The difference between the α's for the two electrolytes can be shown to be related to the osmotic coefficients of the individual solutions

$$\alpha_G - \alpha_J = \frac{2(\phi_G - \phi_J)}{2.303\,m} \tag{1E11}$$

where ϕ_G is the osmotic coefficient in a pure solution of G at molality m. For electrolyte mixtures with a common ion Guggenheim's treatment gives $\alpha_G + \alpha_J = 0$ or $\alpha_J = -\,\alpha_G$, hence with the aid of (1E11) both α's may be calculated individually.

From the relation between ϕ and γ the expression

$$(\alpha_G - \alpha_J) = \frac{2}{m^2} \int_0^m m\, d\ln(\gamma_{J(1)}/\gamma_{G(1)}) \tag{1E12}$$

is obtained. According to Åkerlof and Thomas[134] $\log\gamma_{J(1)}/\gamma_{G(1)} = b_{JG}\,m$, where b_{JG} is independent of m (see section 1.D.d) so that (1E4) leads simply to $(\alpha_G - \alpha_J) = b_{JG}$, and is independent of m.

c. Excess Functions of Mixing

Another way to express the deviation of a mixture from ideality is through the excess functions. The excess molar free energy of a mixture is defined as

$$\Delta G^E = \Delta G^M - \Delta G_{id}^M \tag{1E13}$$

where ΔG^M is the total observed free energy change on mixing, and ΔG_{id}^M is the ideal contribution to it. The activity coefficients are then given, for component G, for example, by

$$\ln \gamma_G = (1/R\,T)(\partial \Delta G^E/\partial x_G)_m \qquad (1E14)$$

for mixing two solutions at constant m, where $x_G = m_G/m$ in the mixture, and symmetrical electrolytes are discussed for simplicity. The other excess functions are obtained from (1E13) by appropriate differentiation: $\Delta H^E = - T^2 \partial(\Delta G^E/T)/\partial T$, $\Delta S^E = - \partial \Delta G^E/\partial T$, $\Delta V^E = (\partial \Delta G^E/\partial P)_T$, etc., noting, that $\Delta H_{id}^M = \Delta V_{id}^M = 0$.

For an electrolyte that obeys Harned's rule

$$\Delta G^E = - 2.303\,R\,T\,x_G(1 - x_G)\,m\,(\alpha_G + \alpha_J) \qquad (1E15)$$

but in the general case, where a square term (equation 1E10)[171,172] or even higher terms are required for expressing activity coefficients, a much more complicated expression results[173]. It contains a number of inter-action parameters, which are not generally known, hence the treatment cannot be applied directly to many practical problems.

Systems that obey Brönsted's principle (section 1.E.b) and have a common ion, have zero excess free energy (from (1E15), since $\alpha_G + \alpha_J = 0$) hence also, zero excess heats and volume changes of mixing. Most systems indeed show small $\Delta H^{E\,172,\,174,\,175}$ and $\Delta V^{E\,176,\,177}$, and very careful measurements are required in order to obtain good results. Excess heats are only a few to a few tens of calories per mole, as obtained both from the temperature dependence of the Harned coefficients and from direct microcalorimetry[175]. The excess volume change has been found[177] to be so small, that the apparent molar volume deviates at most by 0.7 ml from the additivity rule[174] for ternary solutions

$$\emptyset_V = (V - 55.5\,V_W^0)/(m_G + m_J)$$
$$(1E16)$$
$$= (m_G \emptyset_{VG}^0 + m_J \emptyset_{VJ}^0)/(m_G + m_J)$$

where $\emptyset_{VG}^0$ and $\emptyset_{VJ}^0$ are the respective apparent volumes measured in binary solutions at the same total molality $m = m_G + m_J$, and V is the volume of the solution containing 1000 g water.

Instead of mixing binary solutions to form a ternary solution at constant electrolyte molality, it is useful to mix such solutions at constant solvent activity, i.e. solutions which are in isopiestic equilibrium (section 1.C.e). When no chemical interactions occur, such mixtures often obey Zdanovskii's rule[178], which for a mixture of two electrolyte solutions is

$$m_G/m_G^0 + m_J/m_J^0 = 1 \qquad (1E17).$$

where m_G^0 and m_J^0 pertain to the binary solutions of equal water activity, and m_G and m_J to the mixture, which, if (1E17) holds, will also be in isopiestic equilibrium with them. Such mixtures can be termed 'simple' mixtures[179], and resemble ideal solutions by being standards against which to measure deviations. The activity coefficient of a component of a simple mixture is given by

$$\gamma_G = \nu_G \, m_G^0 \, \gamma_G^0 / (\nu_G m_G + \nu_J m_J) \qquad (1E18)$$

For its calculation we need to know m_G^0 and γ_G^0 for the binary solution which is isopiestic with the mixture. The osmotic coefficient or water activity is obtained from tables, for increasing values of m_G^0, and from them the values of m_J^0 are read. The values of m_G and m_J for the mixture are used with (1E17) to calculate $(m_G + m_G^0 m_J/m_J^0)$ as a function of m_G^0, to give the m_G^0 value of the mixture studied, and finally γ_G in the mixture. Mixtures of the alkali metal halides, alkaline earth–alkali metal chlorides or hydrochloric acid (except with potassium), and uranyl nitrate—and other nitrates were found[179] to be simple. Mixtures involving ions with very different water structure-breaking or -making properties (cf. 1.B.c) such as KCl–$MgCl_2$, are not simple, and do not obey the above considerations.

A simple mixture can also be shown[179] to obey additivity rules concerning heats and volumes of mixture. The molar enthalpy is

$$H = (m_G/m_G^0)H_G^0 + (m_J/m_J^0)H_J^0 \qquad (1E19)$$

and correspondingly for the volume.

d. Experimental Methods and Data

The activity coefficients in mixed solutions and the Harned α and β coefficients may be determined by a variety of methods. According to Harned's treatment, the osmotic coefficient of the mixed solution at a mole fraction x_G is given by

$$\phi_{x_G} = \phi_J + \frac{2.303}{2} m((\alpha_G + \alpha_J) x_G^2 - 2\alpha_J x_G) \qquad (1E20)$$

This expression may be transformed into

$$\frac{\phi_J - \phi_{x_G}}{2.303\,m_G} = \alpha_J - \tfrac{1}{2}(\alpha_G + \alpha_J)\,x_G \qquad (1E21)$$

and a plot of the left hand side against x_G yields the values of α_G and α_J individually (at constant total molality m).

If however the β's are not zero, and water activities are measured instead of osmotic coefficients, then (still at constant m)

$$\frac{-55.5}{2\,x_G\,m^2} \log \frac{a_{W(x)}}{a_{W(J)}} = \tfrac{2}{3}\,x_G{}^2\,m(\beta_J - \beta_G)$$

$$+ x_G\,(\tfrac{1}{2}(\alpha_G + \alpha_J) - (\beta_J - \beta_G)\,m) - \alpha_J \qquad (1E22)$$

It is usually more convenient to work at constant a_W than at constant m, for example when isopiestic measurements are used, as discussed above. Following McKay and Perring[180], one may write for a ternary mixture of $1:1$ electrolytes (cf. ref.[179] for generalization to multicomponent systems)

$$0.036 \ln \frac{m\gamma_G}{m^0\gamma_G^0} = -\int_0^{\ln a_W} [m^{-2}(\partial m/\partial \ln x_J)_{a_W} + (m^0 - m)/m\,m^0]\,d\ln a_W \qquad (1E23)$$

where m^0 and γ_G^0 pertain to the binary solution of pure G having the same solvent activity a_W as the mixed solution. This equation has been successfully applied by McKay and Perring to published data of Owen and Cooke[181], and more recently by McCoy and Wallace[182] and by Robinson[183] for LiCl–KCl, KCl–KBr and NaCl–KCl respectively.

When one of the electrolytes is volatile, the measurements of partial vapour pressures permits the determination of the activity coefficients of all components, using the Gibbs–Duhem equation. The experiments are most conveniently carried out at a constant molality of hydrochloric acid, and varying salt molalities (i.e. $N_G/N_{H_2O} \propto N_G/N_{HCl}$). This approach was used by Moore[184] for hydrochloric acid–transition metal chloride solutions. The activities of the salts were found to increase enormously in the presence of hydrochloric acid, and much of this effect can be ascribed to association of the metal cations with the chloride anion.

In addition to vapour pressure and isopiestic measurements, some other methods have proved to be of value for the determination of

activity coefficients in mixed electrolyte solutions. The activity coefficient of hydrochloric acid was obtained potentiometrically in many
salt solutions, (see ref. 185 and recent determinations by Harned[186] and
others[187]) and where suitable electrodes are available activity coefficients
in mixed salt solutions can be determined. Activity coefficients of both
electrolytes can be measured with suitable cells, using electrodes reversible
for either the two cations when the anion is common (cf. measurements
on mixtures of cadmium chloride and hydrochloric acid[188]), or vice versa.

The activity coefficient of an electrolyte G in a mixture can often be
determined by measuring its solubility[189,190], its extractability[184,191] or
its ion exchange distribution[192,193] in a solution of the other, J. In these
cases its concentration is often small compared with that of J, so that
values of $\gamma_{\pm G(0)}$ are obtained, although, except for solubility measurements, this is not a necessary restriction. Some published values of α are
shown in Table 1E1.

e. The Constant Ionic Medium Method

The activity coefficient of trace G in bulk J will be obtained from equation
(1E9) as

$$\log\gamma_{\pm G(0)} = \log\gamma_{\pm G(1)} - \alpha_G m \qquad (1E24)$$

i.e. dependent only on the concentration of J, and not on its own concentration. The activity coefficient of J will be simply $\log\gamma_{\pm J(1)}$ at the
concentration $m = m_J$, and will, of course, not be influenced by the
admixture of a trace amount of G. McTigue[194] has shown that the
hydration theory of activity coefficients (section 1.D.c) can be applied
to this problem.

These facts are utilized by the 'constant ionic medium' method. Biedermann and Sillèn[195] reviewed the history of this method. The principle
employed is the use of a bulk inert electrolyte at a constant, high concentration, with the purpose of keeping constant the activity coefficients
of reacting ions, the concentration of which is small and variable. According to Brönsted's principle of specific interaction, when the re-
acting ions, are cations only (e.g. $M(OH_2)_h^{m+} \leftrightarrows M(OH_2)_{h-1}OH^{(m-1)+}$
$+ H_{aq}^+$) it is essential to keep the anion concentration of the bulk electrolyte (e.g. sodium perchlorate) constant, and let the cation concentration
vary slightly, in order to balance small variations in metal and hydrogen
ion concentrations. Conversely, if the reacting ions are anions only
(e.g. $MCl_N^{(N-m)-} \leftrightarrows M(H_2O)Cl_{N-1}^{(N-m-1)-} + Cl_{aq}^-$) it is essential to keep the

TABLE 1E1. Some Harned's rule coefficients at 25°.

G	J	m**	$100\alpha_G$	$-100\alpha_J$	β_G or β_J^*	Reference
HCl	LiCl	1	0.5	1.2	0	a
		3	0.4	1.3	0	a
		6	−0.9			a
	NaCl	1	3.2	5.8	0	a
		3	3.1	5.8	0	a
		6	2.9		0	a
	KCl	1	5.6	7.2		a
		3	6.2	5.4		a
	NH$_4$Cl	7.3		3.5		c
	CsCl	1	10.0	6.0		a
		3	9.8	4.1		a
HBr	LiBr	2	0.35	3.0		a
	NaBr	2	3.7	6.3		a
	KBr	2	7.9	6.4		a
LiCl	NaCl	3	3.5	3.5	0	b
		6	4.2	3.3	0	b
	KCl	3	7.5	2.5	0	b,d
		6	8.4	2.4	0	b
	CsCl	3	11.4	−0.4	0	b
		6	12.8	0.8		b
	LiNO$_3$	3	1.0	2.0	0	b
		6	2.7	2.6	0	b
		10	5.0	2.3	0	b
NaCl	KCl	3	2.4	0.8	0	b
	CsCl	3	4.3	0.5	0	b
KCl	CsCl	3	1.1	0.5	0	b
	KBr	3	2.8	2.7		d
	CaCl$_2$	1.5**	5.5	4.5		i
	BaCl$_2$	1.5**	5.8	8.1		i
HCl	SrCl$_2$	1–5**	5.8	7.4		i
	BaCl$_2$	1–3**	6.7	8.7		i
	ZnCl$_2$	1**	6.9	6.8		i
	CdCl$_2$	1**	18.0			e
	NiCl$_2$	7–13**	13.0			f
	AlCl$_3$	1**	6.1	6.4		i
		5**	6.3	11.8		i
	CeCl$_3$	1**	8.7	1.8		i
		3**	9.2	5.1		i
UO$_2$(NO$_3$)$_2$	HNO$_3$	12**	−2.0			g
	LiNO$_3$	4**	−2.8			g
	NaNO$_3$	4**	6.0	6.2		g,h
		12**	3.2			g,h
	KNO$_3$	12**	7.4			g
	NH$_4$NO$_3$	4**	8.6			g
		12**	6.3			g
	Mg(NO$_3$)$_2$	5**	2.3			g
	Ca(NO$_3$)$_2$	4**	.2.6	5.7		g
		12**	1.5			g
	Cu(NO$_3$)$_2$	4**	1.5	2.5		g
		12**	0.0			g
	Al(NO$_3$)$_3$	12**	0.7			g

* Since values of α_G or α_J are known to ±0.001, the effect of a term $|\beta| \leqq 0.0025$ is within the experimental error, and designated as 0 in the Table. Where there are no entries, either $|\beta|$ is larger, or is unknown[c].

** At constant ionic strength, I_J replacing m_J in (1E9).

— a. Ref. 185, p. 607–632. — b. R. A. Robinson, *Trans. Faraday Soc.*, 49, 1144, 1147 (1953); *J. Am. Chem. Soc.*, 74, 6035 (1952); *J. Phys. Chem.*, 65, 662 (1961). — c. Ref. 189. — d. Ref. 182. — e. Ref. 88. — f. Ref. 184. — g. L. Jenkins and H. A. C. McKay, *Trans. Faraday Soc.*, 50, 107 (1954). — h. Ref. 191. — i. Ref. 187.

cation concentration of the bulk electrolyte constant. When the reacting ions are both cations and anions, as with many ordinary complex-formation reactions, the position is not clear. It is customary to keep either the concentration of the bulk electrolyte constant, or to keep the total ionic strength of the solution constant. When the concentrations of the reacting ions are very small, as they should be for the constant ionic medium method to operate, the difference is immaterial. At somewhat higher concentrations, serious difficulties may result, as shown by Leden[196]. A small replacement of 3M sodium perchlorate by sulphate at constant ionic strength permitted the determination of the first complex formation constant (cf. chapter 3) for cadmium sulphate with both cadmium amalgam and silver (metal-displacement method) electrodes. At higher concentrations (of the order of a few tenths molar), divergent results were obtained by the two methods. Errors may thus result from careless application of the constant-medium method.

It is of interest to see how the activity coefficient of a trace electrolyte M will vary in a constant medium of electrolyte J, when part of the J ions are replaced by G ions (with a common ion of opposite sign). McKay[197] has shown that at constant total m (medium concentration)

$$\Delta \log \gamma_{M(0)} = \tfrac{1}{2} m_G (\log \gamma_{G(1)} - \log \gamma_{J(1)})/m \qquad (1E25)$$

but the change in $\log \gamma_{M(0)}$ is proportional to m_G rather than as stated by Rossotti[198], to m_G/m, since the difference in activity coefficients $(\log \gamma_{G(1)} - \log \gamma_{J(1)})$ itself is proportional[134] to m (cf. section 1.E.b), hence

$$\Delta \log \gamma_{M(0)} = \tfrac{1}{2}(b_G - b_J) m_G = \tfrac{1}{2} b_{GJ} m_G = \tfrac{1}{2}(\alpha_G - \alpha_J) m_G \qquad (1E26)$$

In order to make the variation in $\log \gamma_{M(0)}$ as small as possible it is therefore best to use as small as possible absolute concentrations of G replacing J, and it is of no special advantage to use a high value of m, the constant concentration[171]. Thus a complete replacement of 0.1 m G for a 0.1 m J solution would produce the same change in the activity coefficient of tracer M, as a 3.3% replacement of G for J in 3 m J. A complete replacement of 3 m G for 3 m J, of course, will produce a much larger $\Delta \log \gamma_{M(0)}$, so that when it is necessary to use high concentrations of G, the constant-medium method using J as bulk electrolyte is of no use (cf. section 3.B.b), unless J can be selected so that $\alpha_G - \alpha_J$ is very small[171].

The problem of the constancy of the activity coefficients, when a part of an ionic medium of $m = 3$ m sodium perchlorate is replaced by perchloric acid, has been studied experimentally by Biedermann and Sillèn[195]. They measured the e.m.f. of suitable cells, and found the deviation E'_j, of the measured e.m.f., from that calculated by the Nernst equation using concentrations, to vary linearly with (H^+). Writing $E'_j = E_j - F(\gamma)$, i.e. a liquid-junction term and an activity coefficient term, one obtains

$$E_j = \frac{-RT}{nF}\ln\left(1 + \frac{\lambda(H^+)}{m}\right) \approx -\frac{RT\lambda}{nFm}(H^+) \qquad (1E27)$$

$$F(\gamma) = \Delta b(H^+) \qquad (1E28)$$

where λ is a function of the mobilities of the ions, and n the number of electrons involved in the electrode reaction. Both terms are seen to be proportional to (H^+) to a good approximation. Biedermann and Sillèn chose to assume that $\Delta b \approx 0$ and $E'_j \approx E_j$, for reactions involving only cations, while for reactions involving anions they assumed the same E_j and from that obtained a finite $\Delta b = (E'_j - E_j)/(H^+)$. When m was permitted to vary, they found E'_j to vary linearly with $(H^+)/m$, independently of m, up to about 10% replacement. This would result from, (1E27) if $F(\gamma) = 0$, and it may be approximately true up to 10% replacement if Δb is small, or inversely proportional to m. When however, halide or hydroxide (A) were substituted for perchlorate, it was found[199] that E'_j remained constant, independent of replacement, so that $\Delta b_{(A)} = -RT\lambda_A/nFm$. The ratio of the activity coefficients for various anions was also constant, independent of replacement.

Zielen and Sullivan[200] found that $E' = A + B(H^+)$, where $E' = E^0$ - (activity coefficient term) + (hydrolysis term), when replacing sodium or lithium perchlorate by perchloric acid in certain cells. The parameter B was approximately zero for Fe^{2+}/Fe^{3+} and Hg^0/Hg_2^{2+} couples and had the values 2.5 $\mu V/M$ for Hg_2^{2+}/Hg^+, $-1.5\mu V/M$ in $LiClO_4$ and $-4.8\mu V/M$ in $NaClO_4$ for NpO_2^+/NpO_2^{2+}. The hydrolysis term for Hg_2^{2+} is about 30 μV, so an error in it is larger than $B(H^+)$, even for full substitution of 2M perchloric acid for 2M lithium or sodium perchlorate media. The activity coefficient term seems thus to be very small, except for neptunium where no hydrolysis correction was used by the authors. In similar measurements, Rabideau[201] found that the potential changed by 1 mV for the Pu^{3+}/Pu^{4+} couple measured in a 0.1M $HClO_4$ + 1.9M $LiClO_4$ medium, if the lithium ions are exchanged for sodium ions.

Dilution of the metal ion (cadmium) concentration in a constant medium (3.00 M NaClO$_4$) caused no deviations from Nernst's law expressed in terms of concentrations[202].

In conclusion it is seen that the constant ionic medium method is useful if an electrolyte J can be found that a) provides cations and anions not interacting to an appreciable extent with those whose reactions are to be studied, b) minimizes b_{GJ} in equation (1E26), G being the electrolyte providing the reacting ions (e.g. hydrogen ions or ligand ions). A further requirement for the method is that the concentration of G used is not 'too high' for the permitted magnitude of the quantity given by (1E26). If these conditions are fulfilled, it may be expected that $\Delta \log \gamma_{M(0)}$ terms can be neglected in equilibrium calculations involving activities, and that concentrations of reacting species can be used instead of their activities, provided that equilibrium constants are adjusted to include (constant) activity coefficient terms. These adjusted constants will then pertain to the specified medium used. Tests of whether these expectations are true have been devized[197].

F. SALTING-IN AND -OUT

a. Introduction

The coulomb forces and short-range interactions have been seen (sections 1.C and 1.D) to affect the activity coefficients of the electrolyte and of the water in solution. In the case where another electrolyte is present, its activity is affected too (section 1.E), and so is that of a non-electrolyte N, in a solution containing an electrolyte. In most cases it has been observed experimentally that salting-out occurs, i.e. that the activity coefficient of the non-electrolyte is larger in the electrolyte solution than in pure water. In order to attain a given activity in the aqueous electrolyte solution, as in equilibrium with another phase of constant activity (e.g. pure N), the concentration of the non-electrolyte will have to be smaller than for a pure aqueous solution. In some cases salting-in, i.e. a smaller activity coefficient of N relative to that in water, has been observed. The non-dissociated part of weak electrolytes is influenced by the same factors as are non-electrolytes, and may be treated similarly.

The effects which were seen to be responsible for the increase of the activity coefficient of electrolytes, over the Debye–Hückel term (section 1.D), should also be responsible for salting-out. In a naive way, the relative decrease of the amount of 'free' water in a highly hydrated electrolyte solution (section 1.D.c), should cause an increase in the concentration of non-electrolyte over its stoichiometric value, i.e. salting-out. The phenomenon of salting-in is harder to explain, and is much rarer. It occurs for certain combinations of electrolyte and non-electrolyte, whereas with other combinations salting-out occurs. It is therefore impossible to assign to a given electrolyte or non-electrolyte universal exclusive salting-in or -out properties. A more direct interaction between ions and non-electrolyte is sometimes responsible for anomalous salting behaviour, which is not accounted for by the non-specific interactions which cause the ordinarily observed effects. These 'chemical' interactions are discussed in the next chapter, while the present section deals with the non-specific electrostatic and mechanical interactions.

The best measure for the salting-out or -in behaviour of a non-electrolyte is its activity coefficient in the electrolyte solution, relative to that in pure water[203]. It is customary to extrapolate the activity coefficient to zero non-electrolyte concentration for both salt solution and water in order to avoid complications arising from the self-interaction of non-electrolyte molecules. It is the relative activity coefficient $f_N = f_{N(G)}/f_{N(W)}$ which is determined, usually setting $f_{N(W)} = 1$ and $f_{N(G)} = y_{N(G)}$. Hence for constant activity (e.g. in solubility experiments, or in extractions normalized to constant activity in the organic phase) the expression

$$f_N := y_{N(G)} = c_{N(G)}/c_{N(W)} \tag{1F1}$$

will be approximately valid. It is the dependence of f_N on c_G which is the subject of this section.

Empirically, the Setchenov equation (1F2) has been found to express the variation of f_N with c_G adequately, at least for low values of c_G (a few tenths molar):

$$\log f_N = k_s c_G \tag{1F2}$$

The proportionality constant k_s is called the salting-out (-in, if negative) constant. Sometimes a better representation of the data is obtained using

$$f_N = 1/(1 - k'_s c_G) \tag{1F3}$$

which reduces to (1F2) at low c_G with $k'_s = 2.303 \, k_s$. At higher concentrations of electrolyte, deviations from the linear relation (1F2) occur, and two-parameter equations have been proposed[204, 205], such as the following[204]

$$\log f_N = k''_s c_G{}^{k_p} \tag{1F4}$$

They have been justified on a semi-theoretical basis and have their value in summarizing experimental data, but usually have little advantage over the simple Setchenov equation (1F2). Most theoretical treatments of the problem deal mainly with a way to calculate k_s from known properties of the components and their concentrations. It is then usually the practice to define k_s as

$$k_s = \lim_{c_G \to 0} \mathrm{d}\log f_N/\mathrm{d}\,c_G \tag{1F5}$$

The temperature coefficient of the salting constant is obtainable from direct measurement, as well as from calorimetric enthalpy-of-transfer measurements[205].

b. The McDevit and Long Treatment

Of the thermodynamic theories of the salting effect, the best known is McDevit and Long's[206]. The theory considers the excess work required in order to introduce an element of volume of the non-electrolyte, $\mathrm{d}V_{(N)}$, into the solution instead of into water, in terms of the relative compressibilities of solution and water. This quantity may be calculated from the change in volume on mixing the electrolyte with water, per mole of electrolyte, divided by the compressibility of water, β_w. The change in volume on mixing is given by the difference of the intrinsic volume of the electrolyte $V_{G(in)}$ (i.e. the molar volume of the hypothetical supercooled liquid salt) and the partial molar volume in solution $V_G^{\prime 0}$. At infinite dilution the expression (see section 1.B.d)

$$\lim_{c_G \to 0} \frac{\mathrm{d}^2 G}{\mathrm{d}c_G\,\mathrm{d}V_{(N)}} = \frac{V_{G(in)} - V_G^{\prime 0}}{\beta_w^0} = \frac{h^0(V_w^0 - V_{hw})}{\beta_w^0} = \frac{h^0\delta V_w}{\beta_w^0} \tag{1F6}$$

is obtained[206], where the last two terms on the right express the volume change per mole of electrolyte as the product of the hydration number at infinite dilution of the electrolyte h_G^0 and the difference between the molar volume of 'free' water V_w^0, and that of 'electrostricted' water in the hydration shell[207]. Using the relationships

$$k_s = 0.4343\,(1/R\,T) \times \lim_{c_G,\,c_N \to 0} \mathrm{d}\mu_N/\mathrm{d}\,c_G$$

and

$$\mu_N = (\mathrm{d}G/\mathrm{d}\,V_N)\,(\mathrm{d}\,V_N/\mathrm{d}\,N_N) = (\mathrm{d}\,G/\mathrm{d}V_N)\,V_N^0$$

one obtains from (1F6)

$$k_s = \frac{V_N^0(V_{G(in)} - V_G^{\prime 0})}{2.303\,\beta_w\,R\,T} = \frac{V_N^0 h_G^0\;\delta V_w}{2.303\,\beta_w^0\,R\,T} \tag{1F7}$$

Theoretically (ref. 185, p. 173) V_W should be independent of the electrolyte, so that the electrostriction per water molecule is $\delta V_W = 2.1$ml. The
salting constant therefore depends on the electrolyte only through h_G^0,
and on the non-electrolyte through V_N^0. The expression

$$k_s(\text{theor.}) = 0.00081 \, V_N^0 h_G^0 \qquad (1F8)$$

then results. Actually however δV_W was found to depend on the electrolyte. For some electrolytes $(V_{G(in)} - V_G^{\prime 0}) = h_G^0 \delta V_W$ was claimed to be
negative[206, 208], and in these cases salting-in was found. Exclusively
salting-out behaviour would be predicted by this theory for all electrolytes,
since $h_G^0 \delta V_W$ is indeed universally positive. Exceptions to this were
however found with some non-electrolytes, which show both salting-in
and salting-out behaviour for electrolytes, all of which have a positive
$h_G^0 \delta V_W$ (ref. 185, p. 736, 737). In general, values of k_s predicted from
(1F7) are about three times larger than the observed values[208].

c. Electrostatic Theories

The molecular theories of the salting effect are mainly electrostatic in
nature, and relate the interaction of non-electrolyte and ions with the
dielectric constant of the non-electrolyte solution. An early attempt to
explain the salting-out phenomena is that of Debye and McAulay[124],
who calculated the difference in electrical work $\Delta G_{(el)}$ required for
charging the ions in water and in the non-electrolyte solutions. For a
solution containing c_j moles per litre of ion j

$$\Delta G_{(el)} = \frac{N e^2}{2000} \left(\frac{1}{\varepsilon} - \frac{1}{\varepsilon_W} \right) \sum_j \frac{c_j z_j^2}{r_j} \qquad (1F9)$$

Writing for a solution which contains both electrolyte and non-electrolyte

$$\varepsilon = \varepsilon_W - \delta_G c_G - \delta_N c_N \qquad (1F10)$$

and neglecting terms containing powers of c_G and c_N higher than the first,
yields $\delta_G c_G + \delta_N c_N$ for the factor $(1/\varepsilon) - (1/\varepsilon_W)$. Using this and
$c_j = v_j c_G$ in (1F9)

$$k_s = 0.4343 \lim_{c_G, c_N \to 0} \frac{\partial^2 \Delta G_{(el)}}{\partial c_G \partial c_N} = \frac{N e^2 \delta_N}{4606 \, R T \varepsilon_W^2} \sum \frac{v_j z_j^2}{r_j} \qquad (1F11)$$

which is the equation of Debye and McAulay. Differentiating (1F9) with
respect to c_G at $c_N = 0$ gives the self-salting-out of the ions

$$\log f_{\pm} \text{(self-salting)} = \frac{N e^2 \delta_G c_G}{2.303 \, R T \varepsilon_w^2} \sum \frac{v_j z_j^2}{r_j} \qquad (1F12)$$

Since $\frac{1}{2} c_G \sum v_j z_j^2 = I$, expression (1F12) is of the form bI ($b = $ constant), and is the salting-out term in Hückel's equation for the activity coefficient of an electrolyte (section 1.D.c). Sandved[209] has written for (1F11) $\log f_N = (\text{const. } (d\varepsilon/d c_N) \tilde{z}^2/\tilde{r}) c_G$, where $\tilde{z}$ and $\tilde{r}$ are the average charge and radius of the ions, which is of the form of the Setchenov equation (1F2). The Debye–McAulay treatment relates the salting constant to the properties of the non-electrolyte through δ_N and to those of the ions through their charges and radii. Salting-in is predicted for those non-electrolytes having negative δ_N values, i.e. that are more polar than water.

A more sophisticated theory was proposed a little later by Debye[210]. It considers the distribution of non-electrolyte molecules in the field of the ions. Debye's main equation for the distribution as a function of distance r from an ion j is

$$\ln \frac{x_N}{x_N^0} = \frac{-z_j^2 e^2}{8 \pi k T v_w \varepsilon^2} \left(v_N \frac{\partial \varepsilon}{\partial n_w} - v_w \frac{\partial \varepsilon}{\partial n_N} \right) \frac{1}{r^4} = -\left(\frac{\bar{R}_j}{r} \right)^4 \qquad (1F13)$$

where x_N is the mole fraction of the non-electrolyte N (x_N^0 at infinite distance from ions, $r = \infty$), v_w and v_N are the molar volume of water and N divided by N, n_w and n_N are the number of molecules of water and N per cm^3, and $\bar{R}_j$ is a characteristic length. Equating f_N with x_N^0/x_N, and integrating over all ions and the total available volume, leads to

$$1/f_N = 1 - 4\pi \sum n_j \int_{b_j}^{\infty} (1 - \exp(-(\bar{R}_j/r)^4)) r^2 dr \qquad (1F14)$$

where n_j is the number of ions of kind j per cm^3 and b_j their closest distance of approach to the non-electrolyte. Debye[210] and Harned and Owen (ref. 185, p. 87) express the dielectric constant as (cf. section 2.B.g)

$$\varepsilon = \varepsilon_w x_w + \varepsilon_N x_N \qquad (1F15)$$

i.e. as a linear function of $x_N \approx c_N/V_w$, which identifies δ_N/ε_w with $V_w (\varepsilon_w - \varepsilon_N)/\varepsilon_w$ so that

$$\bar{R}_j^4 = \frac{1000 \, e^2 z_j^2 \delta_N}{8 \pi R T \varepsilon_w^2} \qquad (1F16)$$

Inserting this in (1F14), defining $J_j = \int_{b_j}^{\infty} (1 - \exp -(\bar{R}_j/r)^4) r^2 dr$, and using (1F3) and $n_j = N c_G v_j/1000$ yields

$$k_s = \frac{4\pi N}{2303} \sum v_j J_j \tag{1F17}$$

which is the Debye (Harned and Owen) expression for the salting constant. Values of J_j are given by the series[185]

$$J_j = \bar{R}_j^3 (1.21 - 0.33(\bar{R}_j/b_j)^{-3} + \cdots) \text{ for } \bar{R}_j > b_j \tag{1F18a}$$

$$J_j = \bar{R}_j^3 ((\bar{R}_j/b_j) - 0.10(\bar{R}_j/b_j)^5 + \cdots) \text{ for } \bar{R}_j < b_j \tag{1F18b}$$

As with the Debye–McAulay treatment, the salting constant is related to the properties of the non-electrolyte through δ_N, and with those of the electrolyte through v_j, z_j and b_j, which is to be taken as the sum of the radii of the ion j and the non-electrolyte.

Modifications of Debye's theory have been proposed. Albright[211] used an expression for the dielectric constant different from (1F15), which has the effect of multiplying $\bar{R}_j^4$ in (1F16) by the factor $(V_W + x_N V_N)/(V_W - x_N \delta_N/\varepsilon_W)$, which makes $\bar{R}_j$ concentration dependent and has no advantage. Another modification[212] makes use of equation (1F10) directly in (1F13) by writing $v_N(\partial\varepsilon/\partial n_W) = (V_N/N)(\partial\varepsilon/\partial c_G)$ $(\partial c_G/\partial n_W)$, $(\partial\varepsilon/\partial c_G) = -\delta_G$ and $(\partial c_G/\partial n_W) = -1000 v_w/(\varnothing_{VG} + h V_W)$ where $\varnothing_{VG}$ is the apparent molar volume of the electrolyte in solution, and only 'free' water molecules are counted in n_W, and $(\partial\varepsilon/\partial n_N) = -1000\delta_N/N$. Introducing these values in (1F13) yields for $\bar{R}_j^4$

$$\bar{R}_j^4 = \frac{1000 e^2 z_j^2}{8\pi R T \varepsilon^2} \left(\frac{V_N}{\varnothing_{VG} + h V_W} \delta G + \partial N \right)$$

$$= \frac{1000 e^2 z_j^2 V_N}{8\pi RT \varepsilon^2} \left(\frac{\delta_G}{\varnothing_{VG} + h V_W} + \frac{\varepsilon_W - \varepsilon_N}{1000} \right) \tag{1F19}$$

as according to Butler[213], $\delta_N = V_N(\varepsilon_W - \varepsilon_N)/1000$. Equations (1F18) may be rewritten as $J_j = \bar{R}_j^3 \ f(b_j) \approx \bar{R}_j^3$ since it has been found[213] that for most cases $\bar{R}_j$ is somewhat greater than b_j. When this is introduced into (1F17), together with (1F19), noting that at the limit $c_G \to 0$ $\varnothing_{VG} + h V_W = V_G'^0 + h^0 V_W$ one finally obtains[212]

$$k_s = \frac{4\pi N}{2.303} \left(\frac{1000 e^2}{8\pi R T \varepsilon_W^2} \right)^{3/4} V_N^{3/4} \left(\frac{\delta_G}{V_G'^0 + h^0 V_W} + \frac{\varepsilon_W - \varepsilon_N}{1000} \right)^{3/4}$$

$$\times \sum v_j z_j^{3/2} f(b_j) \tag{1F20a}$$

or at 25°

$$k_s \approx 0.025 V_N^{3/4} \left(\frac{\delta_G}{V_G'^0 + h^0 V_W} + \frac{\varepsilon_W - \varepsilon_N}{1000} \right)^{3/4} \sum v_j z_j^{3/2} \tag{1F20b}$$

This equation would permit salting-in for non-hydrated electrolytes with negative partial molar volumes (an unlikely condition, see above) or for very large ions, if $\varepsilon_N > \varepsilon_W$, so that the second term in the parenthesis in (1F20b) is negative, and its absolute value larger than the first term. The salting constant depends on the non-electrolyte through V_N (as in McDevit and Long's theory (1F7)) and ε_N (as in Debye's theory, through δ_N(1F16)), and on the electrolyte through v_j and z_j (as in Debye's theory (1F17)), $V_G'^0$ and h^0 (as in McDevit and Long's theory (1F7)), and also through δ_G, the molar dielectric decrement of the electrolyte[212].

Another way to take account of the relative polarizabilities of water and non-electrolyte was proposed by Butler[213]. Instead of using the dielectric constants, he uses the polarizabilities. Butler's equation may be written as

$$k_s = \frac{2 \pi N e^2 (\alpha_W - \alpha_N)}{2.303 \, R T \varepsilon_W^2 \, V_N} \sum \frac{v_j z_j^2}{b_j} \qquad (1F21)$$

where α_W and α_N are the molar polarizabilities of water and non-electrolyte. It is almost equivalent to Debye and McAulay's equation (1F11), remembering that $\alpha_i = V_i(\varepsilon_i - 1)/4 \pi N$. The dipole moment of the non-electrolyte was taken into account explicitly in a theory proposed by Kirkwood[214], which yields values for the salting constant very similar[203] to those obtained from Debye's theory.

The above equations, however, cannot account for all the phenomena observed. In mixed aqueous–organic solvents, for instance, the trend of the salting constant with the dielectric constant of the solvent is not that expected from the theories[215]. This has been explained by the preferential solvation of the ions by water. Further cases of 'anomalous salting-in' are known, usually for large alkylammonium ions in water, which are not explained by the coulombic forces described above, but rather by the structure-making (hydrophobic bonding) properties of these ions (section 1.B.c). It has been proposed to invoke the effect of dispersion forces which, for large and polarizable ions and non-electrolyte molecules should be important, and these are the species involved in the 'anomalous salting-in'[216,217]. A term

$$\frac{4\pi}{2.303 \, R T} \int_{b_j}^{\infty} \left(\frac{\lambda_W(V_N/V_W) - \lambda_N}{r^6} \right) r^2 \mathrm{d}r$$

$$= \frac{4\pi}{2.303 \times 3 \, R T}(\lambda_W(V_N/V_W) - \lambda_N) \times \frac{1}{b^{3\prime}} \qquad (1F22)$$

is added to (1F21) to give the total (coulomb and dispersion) salting effect[216]. This term is negative for large ions and non-electrolytes, in the case $\lambda_N \gg \lambda_W$, the λ_i are functions of the electronic polarizabilies (i.e. proportional to b^3), the characteristic frequencies and the electronic dielectric constant. Values calculated for the salting effect of tetraalkyl-ammonium iodides on benzoic acid[216] show that the dispersion term is negative and absolutely larger than the coulombic term, so that salting-in is predicted. Experimentally salting-in was indeed observed, but the measured values of k_s did not agree with the calculated ones.

The idea of adding a dispersion term to the coulombic term[216] has recently been further developed by Mikhailov[218] who defined a 'salting-out sphere' around each ion, i.e. the average volume of solution available per ion, with radius $r_n = (3000/4\pi N)^{1/3} (\sum v_j c_G)^{-1/3}$. The integration over the volume then proceeds for each ion from the distance of closest approach b_j to r_n. Writing $E_j = \bar{R}_j{}^4$ in (1F13) and D_j for the corresponding dispersion term (see 1F22) leads to the following equation instead of (1F14)

$$
1/f_N = \frac{4\pi N}{1000} c_G \left[\frac{1}{3} \sum v_j \int_{b_j}^{r_n} \exp\left(-\frac{E_j}{r^4} - \frac{D_j}{r^6} \right) r^2 \, dr \right]
$$

$$
= \frac{4\pi N}{1000} c_G \left[\left(\frac{1}{3} \sum v_j b_j{}^3 + \sum v_j \int_{b_j}^{\infty} \left(1 - \exp\left(-\frac{E_j}{r^4} - \frac{D_j}{r^6} \right) \right) r^2 \, dr \right. \right.
$$

$$
\left. \left. - \sum v_j \int_{r_n}^{\infty} \left(1 - \exp\left(-\frac{E_j}{r^4} - \frac{D_j}{r^6} \right) \right) r^2 \, dr \right) \right]
$$

(1F23)

The first two terms in the square bracket in (1F23) are concentration independent, and will be designated by Q, while the last term is concentration dependent through r_n. Hence, approximating the exponent by the first linear term, the salting constant will be given by

$$
2.303 \, k_s = \frac{4\pi N}{1000} Q - \left[27 \left(\frac{4\pi N}{3000} \right)^{4/3} \sum v_j E_j \left(\sum v_j \right)^{1/3} \right] c_G^{1/3}
$$

(1F24)

$$
- \left[\left(\frac{4\pi N}{3000} \right)^2 \sum v_j D_j \left(\sum v_j \right) \right] c_G
$$

Experimentally[218], the last term involving the dispersion forces, was found to be small for the salting-out of benzene with sodium chloride or of tributyl phosphate with various salts, since a plot of k_s against $c_G^{1/3}$ was linear. The slope of the deviation from the Setchenov equation calculated from (1F24) however was found to disagree with the experimental results.

d. Hydration Theories

Many authors consider in one way or another the decrease in the number of moles 'free' water molecules, in order to calculate the increased mole fraction of the non-electrolyte. The simplest treatment is to apply the statistical considerations used in discussing activity coefficients of electrolytes, and consider the amount of free water available to dissolve the non-electrolyte. In molal units this gives $f_N = 55.5/(55.5 - h_G m_G)$[219,220] and if the volume occupied by the hydrated ions is considered, this becomes $f_N = 1000/(1000 - c_G V_{hG})$[221]. The salting coefficient is the limit of $(\log f_N)/c_G$ at infinite dilution, and becomes in the absence of electrostatic effects

$$k_s = (V_w^0/2303)(h_G - v) = 0.0078(h_G^0 - v) \qquad (1F25)$$

setting the intrinsic volume of an ion equal to that of a molecule of water. For non-hydrated ions ($h_G < v$) this predicts salting-in, and only for hydrated ions ($h_G > v$) is salting-out predicted from this source, the more the higher the hydration. Eucken and Hertzberg[219] considered the changes in the structure of water near the ions caused by the presence of non-electrolyte molecules but obtained unreasonable hydration numbers (e.g. 20 for sodium chloride) from their treatment.

A combination of hydration and electrostatic considerations is therefore more likely to be able to account for the observed salting behaviour. The modification[212] to Debye's treatment given in equation (1F20) takes hydration into account by the term in $\delta_G/(V_G'^0 + h^0 V_w^0) = \delta_G/V_{hG}^0$, and also by identifying the radius of closest approach of ion to non-electrolyte b_j, with the hydrated radius r_h of the ion. This is equivalent to excluding the non-electrolyte completely from a volume given by $(4\pi N/3)c_G r_h^3$ per litre of solution, i.e. a term of the form of (1F25) being contributed to k_s. The same treatment in terms of an exclusion volume, has also been used by others[220,221], in addition to an electrostatic term. The use of a radius r_n to describe the volume available per ion[218-220], is immaterial for calculating the salting coefficient k_s, which pertains to $c_G \rightarrow 0$, hence $r_n \rightarrow \infty$, and the term involving the integral from r_n to infinity (e.g. the last term in (1F23), or corresponding terms in refs.

215 and 221) tends to zero. Thus the treatments of Conway and co-workers[220] and of Ruetschi and Amlie[221] do not differ in these respects from that of Givon and coworkers[212]. The contention that non-polar solutes do not show a dielectric decrement, hence their salting behaviour has no electrostatic term[221], is not correct, since these solutes too occupy volume, otherwise occupied by water. The use of the molar polarization[220] instead of $V_N \varepsilon_N$ in (1F16) or (1F19), or the polarizability in (1F21), makes little difference, and it is impossible to say which approach approximates best the effect of replacing a volume in the ion field occupied by water in the binary solution, with a volume of non-electrolyte.

Another approach to the correction of non-electrolyte activity for ionic hydration has been proposed by Baranowski and Sarnowski[222] who differentiated (1F9) with respect to volume to obtain the salting contribution to the osmotic pressure π of the solution[223]. Neglecting the term in δ_G, they obtained

$$\pi_{\text{salting}} = \frac{N e^2}{2000 \, \varepsilon_W^2} \, c_G \, c_N' \, \delta_N \, \sum \frac{\nu_j z_j^2}{r_j} \qquad (1F26)$$

Here c_N' is the concentration of non-electrolyte corrected for hydration, involving the interaction energies W_N and W_W of nonelectrolyte and water with the ions, and the relative solvation parameter h_{NW} (which is the relative number of non-electrolyte and water molecules solvating the ions)

$$c_N' = \frac{0.018 \, h_{NW} \, c_N^2 \exp(W_N - W_W)/RT}{1 + 0.018 \, c_N \exp(W_N - W_W)/RT} \qquad (1F27)$$

Much of the work on the salting effect was done in connection with the extraction of uranyl nitrate from solutions[224]. As discussed in later chapters, this reaction involves a complicated mechanism of which salting-out is one step, However one may consider non-dissociated uranyl nitrate, once formed, as a non-electrolyte, and then what was discussed above applies to it. Among the treatments of the salting-out of uranyl nitrate is the hydration approach of Adamskii[225]. He considered the number of moles of water bound to the (cation of the) salting agent G and to the uranyl ion U as $h_G c_G$ and $h_U c_U$ per litre, equating the hydration numbers with those at saturation, which are taken as the ratio $55.5/m$(sat.). This volume of one litre contains approximately $(1000 - c_U V_U)/V_W$ moles of water and $3c_U$ moles of ions in a solution containing no salting agent. The relative concentration of combined water is then $S_U = h_U c_U/(3c_U + (1000 - c_U V_U)/V_W)$, and $S_{UG} = (h_U c_U + h_G c_G)/(3c_U + (v_+ + v_-)c_G + (1000 - c_U V_U - c_G V_G)/V_W)$, in solutions without and with salting agent. Adamskii [225] showed that the concentration of uranyl nitrate in the organic phase is a unique function of S_U for each solvent, and it may be calculated for given c_G and c_U from this function by using the calculated S_{UG} instead. This is a completely empirical approach, and little can be said on the significance of the S functions, the relative concentration of bound water except that they predict increased salting-out with increasing h_G and V_G (using the hydrated volume), cf. section 1.F.d.

It is possible also to relate the salting effect to a property introduced by Samoilov [226], the surface density of water in the first coordination sphere defined by

$$\rho' = N/4\pi(r_c + 1.38)^2 \qquad (1F28)$$

where N is the coordination number of the cation of the salting agent, and 1.38 is the diameter of a water molecule, r_c being in Å. The hydration number is here equated with the coordination number of the ion. Solovkin [227] uses this to define an effective salting parameter $\rho''_{ef} = m_G z_G \rho'$, but inconsistently with published information on the salting effect, and with what is stated in the same article, he relates the logarithm of the activity coefficient not to ρ'_{ef} but to $3 \log \rho'_{ef}$, showing the latter quantity to describe his experimental results. But he also showed that the activity coefficient conforms to a Harned-law-type equation (cf. 1E9) $\log y_N = \log y_N^0 - \alpha_G I_G$, with $\alpha_G = 0.424 - 6.9\rho'$. This behaviour is more reasonable than the logarithmic dependence mentioned above.

Summarizing, the logarithm of the activity coefficient of solutes in electrolyte solutions was found to contain a term linear in the electrolyte concentration, as a first approximation. For a solution of a single electrolyte G, this is the term bI_G or bm_G, which adds to the electrical (Debye–Hückel) term (cf. equation 1C24). For an electrolyte J, in the presence of another electrolyte G it is the Harned-law parameter αm_G (cf. equation 1E9), which adds to the activity coefficient of J at the total concentration m. For a non-electrolyte N it is the Setchenov equation $k_s c_G$ or $k_s' m_G$. This comparable behaviour is partly due to the statistical entropy effect, which introduces a term for converting from the rational to the molal activity coefficient, partly to the hydration of ions, which has the effect of repulsing other molecules from the immediate vicinity of the ions, with corresponding entropy and energy terms, and partly to electrical interactions of the ions of G with the ions of G or of J, or the dipoles (permanent or induced) of N. The linear term has found some success in correlating data and, as was seen above, has a theoretical basis. However, many of the details of the interaction involved are not at all clear, so it is impossible with present knowledge to predict accurate values of the coefficients of the linear terms from independent information, or the expected deviations from the linear term over a wide concentration range.

G. REFERENCES

1. H. S. Frank and W.-Y. Wen, *Discussions Faraday Soc.*, **24**, 133 (1957); H. S. Frank, *Proc. Roy. Soc. (London)*, **A247**, 481 (1958).
2. G. Nemethy and H.A. Scheraga, *J. Chem. Phys.*, **36**, 3382 (1962).
3. R. P. Marchi and H. Eyring, *J. Phys. Chem.*, **68**, 221 (1964).
4. G. E. Walrafen, *J. Chem. Phys.*, **40**, 3249 (1964).
5. J. C. Hindman, *J. Chem. Phys.* **44**, 4582 (1966).
6. K. Buijs and G. R. Choppin, *J. Chem. Phys.*, **39**, 2035 (1963); M. R. Thomas, H. A. Scheraga and E. E. Schrier, *J. Phys. Chem.*, **69**, 3722 (1965).
7. D. F. Hornig, *J. Chem. Phys.*, **41**, 3129 (1964); K. Buijs and G.R. Choppin, *J. Chem. Phys.*, **41**, 3130 (1964); G. Boeltger, H. Harders and W. A. P. Luck, *J. Phys. Chem.* **71**, 459 (1967).
8. D. P. Stevenson, *J. Phys. Chem.*, **69**, 2145 (1965).
9. J. L. Kavanau, *Water and Solute–Water Interactions*, Holden-Day, Inc., San Francisco, 1964.
10. R. M. Noyes, *J. Chem. Educ.*, **40**, 2, 116 (1963).
11. F. Rossini and coworkers., *Nat. Bur. Std. (U.S.)*, No. 500, 1952; W. M. Latimer, *Oxidation Potentials*, 2nd. ed.. Prentice Hall, Englewood Cliffs, N.J., 1952.
12. L. Benjamin and W. Gold, *Trans. Faraday Soc.*, **50**, 757 (1954); cf. also ref. 19.
13. R. E. Powell and W. M. Latimer, *J. Chem. Phys.*, **19**, 1139 (1951).
14. K. J. Laidler, *Can. J. Chem.*, **34**, 1197 (1956); A. M. Couture and K. J. Laidler, *Can. J. Chem.*, **35**, 202 (1957).
15. R. E. Connick and R. E. Powell, *J. Chem. Phys.*, **21**, 2206 (1953).
16. J. W. Cobble, *J. Chem. Phys.*, **21**, 1443, 1446 (1953).
17. D. D. Eley and M. G. Evans, *Trans. Faraday Soc.*, **34**, 1093 (1938).
18. J. D. Bernal and R. H. Fowler, *J. Chem. Phys.*, **1**, 515 (1933).
19. E. J. W. Verwey, *Rec. Trav. Chim.*, **61**, 127 (1942).
20. B. E. Conway and J. O. M. Bockris, 'Ionic Solvation', in *Modern Aspects of Electrochemistry* (Eds. Bockris and Conway), Vol. 1, Butterworths Sci. Publ., London, 1954, p. 51.
21. Lee and Tai, *J. Chinese Chem. Soc.*, **8**, 60 (1941); cf. ref. 20, p. 56.
22. R. W. Gurney, *Ionic Processes in Solution*, McGraw Hill Book Co. Inc. New York, 1953, p. 175; W. G. Breck and J. Lin, *Trans. Faraday Soc.*, **61**, 2223 (1965).

23. E. D. Eastman, reported by J. C. Goodrich and coworkers, *J. Am. Chem. Soc.*, **72**, 4411 (1950).

24. E. D. Eastman, *J. Am. Chem. Soc.*, **50**, 283, 292 (1928).

25. M. Born, *Z. Physik.*, **1**, 45 (1920).

26. W. M. Latimer, K. S. Pitzer and C. M. Slansky, *J. Chem. Phys.*, **7**, 108 (1939).

27. M. J. Blandamer and M. C. R. Symons, *J. Phys. Chem.*, **67**, 1304 (1963).

28. R. H. Stokes, *J. Am. Chem. Soc.*, **86**, 979 (1964).

29. R. M. Noyes, *J. Am. Chem. Soc.*, **84**, 573 (1962); **86**, 971 (1964).

30. J. Webb, *J. Am. Chem. Soc.*, **48**, 2589 (1926).

31. E. Glueckauf, *Trans. Faraday Soc.*, **60**, 572 (1964).

32. A. M. Azzam, *Can. J. Chem.*, **38**, 2203 (1960).

33. H. F. Halliwell and S. C. Nyburg, *Trans. Faraday Soc.*, **59**, 1126 (1963).

34. B. E. Conway in *Modern Aspects of Electrochemistry* (Eds. J. O'M. Bockris and B. E. Conway) Vol. 3, Butterworth Sci. Publ., London, 1964, p. 56.

35. G. Kortüm and J. O'M. Bockris, *Textbook of Electrochemistry*, Elsevier Publ. Co., Amsterdam, 1951.

36. E. J. W. Vervey, *Chem. Weekblad*, **37**, 530 (1940).

37. F. Vaslow, *J. Phys. Chem.*, **67**, 2773 (1963).

38. H. S. Frank and R. A. Robinson, *J. Chem. Phys.*, **8**, 933 (1942).

39. A. Ben-Naim, *J. Phys. Chem.*, **69**, 1922 (1965).

40. W. -J. Wen, S. Saito and C. Lee, *J. Phys. Chem.*, **70**, 1244 (1966).

41. S. Lindenbaum and G. E. Boyd, *J. Phys. Chem.*, **68**, 911 (1964); S. Lindenbaum, *J. Phys. Chem.*, **70**, 814 (1966).

42. R. M. Diamond, *J. Phys. Chem.*, **67**, 2513 (1963).

43. R. L. Kay and D. F. Evans, *J. Phys. Chem.*, **70**, 2325 (1966); R. L. Kay T.Vituccio, C. Zawoyski and D. F. Evans, *J. Phys. Chem.*, **70**, 2336 (1966); D. F. Evans, G. P. Cunningham and R. L. Kay, *J. Phys. Chem.*, **70**, 2974 (1966).

44. G. E. Choppin and K. Buijs, *J. Chem. Phys.*, **39**, 2042 (1963).

45. A. Ben-Naim and M. Egel Thal, *J. Phys. Chem.*, **69**, 3250 (1965).

46. H. S. Frank and M. W. Evans, *J. Chem. Phys.* 3, 507 (1945); H. S. Frank and W.-Y. Wen. *Discussions Faraday Soc.*, **24**, 133 (1957).

47. A. M. Azzan, *Can. J. Chem.*, **38**, 993 (1960); *Z. Physik. Chem.*, (*Frankfurt*) 32, 309 (1962).

48. K. J. Laidler and C. Pegis, *Proc. Roy. Soc.*, (*London*) **A241**, 80 (1957).

49. J. Padova, *J. Chem. Phys.*, **39**, 1552 (1963).

50. E. Glueckauf, *Trans. Faraday Soc.*, **60**, 1637 (1964).

51. K. H. Stern and E. S. Amis, *Chem. Rev.*, **59**, 1 (1959).

52. R. H. Stokes and R. A. Robinson, *Trans. Faraday Soc.*, **53**, 301 (1957).

53. L. G. Hepler, *J. Phys. Chem.*, **61**, 1426 (1957).

54. P. Mukerjee, *J. Phys. Chem.*, **65**, 740 (1961); **70**, 2708 (1966).

55. S. W. Benson and C. S. Copeland, *J. Phys. Chem.*, **67**, 1194 (1963).

56. A. M. Couture and K. J. Laidler, *Can. J. Chem.*, **34**, 1209 (1956).

57. B. E. Conway, R. E. Verrall and J. E. Desnoyers, Z. *Physik. Chem.*, (*Leipzig*) **230**, 157 (1965); but see also B. E. Conway, and R. E. Verrall, *J. Phys. Chem.*, **70**, 3952, 3961 (1966).
58. J. Padova, *J. Chem. Phys.*, **39**, 2599 (1963); **40**, 691 (1964).
59. J. E. Desnoyers, R. E. Verrall and B. E. Conway, *J. Chem. Phys.*, **43**, 243 (1965).
60. R. A. Robinson and R. H. Stokes, *Electrolyte Solutions*, 2nd ed., Butterworths Sci. Publ., London, 1959.
61. R. E. Nightingale, *J. Phys. Chem.*, **63**, 1381 (1959).
62. E. Glueckauf, *Trans. Faraday Soc.*, **61**, 914 (1965).
63. J. E. Desnoyers and B. E. Conway, *J. Phys. Chem.*, **70**, 3017 (1966).
64. A. Passinsky, *Acta Physicochim. URSS*, **8**, 835 (1938).
65. H. S. Harned and B. B. Owen, *Physical Chemistry of Electrolyte Solutions-* 3rd ed., Reinhold Publ. Co., New York, 1958, p. 173.
66. K. Tamura and T. Sasaki, *Bull. Chem. Soc. Japan*, **36**, 975 (1963).
67. D. S. Allam and W. H. Lee, *J. Chem. Soc.*, **5**, 426 (1966).
68. A. M. Azzam, *Z. Electrochem.*, **58**, 889 (1954).
69. J. Padova, *Bull. Res. Council Israel*, **10A**, 63 (1961).
70. H. Ulich, *Z. Electrochem.*, **36**, 497 (1930); *Z. Physik. Chem. (Leipzig)*, **168**, 141 (1934).
71. E. Glueckauf, *Trans. Faraday Soc.*, **51**, 1235 (1955).
72. C. B. Monk, *Electrolyte Dissociation*, Acadamic Press, London, 1961, p. 270.
73. J. B. Hasted, D. M. Ritson and C. H. Collie, *J. Chem. Phys.*, **16**, 1 (1948); G. H. Haggis, J. B. Hasted and T. Buchanan, *J. Chem. Phys.*, **20**, 1452 (1952); J. B. Hasted and G. W. Roderick, *J. Chem. Phys.*, **29**, 17 (1958); J. A. Lane and J. A. Saxton, *Proc. Roy. Soc. (London)*, **A214**, 531 (1952).
74. E. W. Washburn and E. B. Millard, *J. Am. Chem. Soc.*, **31**, 322 (1909); **37**, 694 (1915).
75. L. G. Longworth, *J. Am. Chem. Soc.*, **69**, 1288 (1947).
76. A. J. Rutgers and Y. Hendrikx, *Trans. Faraday Soc.*, **58**, 2184 (1962).
77. C. J. Hindman, *J. Chem. Phys.*, **36**, 1000 (1962).
78. H. G. Herz, *Ber. Bunsen Ges.*, **67**, 311 (1933).
79. J. A. Jackson, J. F. Lemons and H. Taube, *J. Chem. Phys.*, **32**, 553 (1960).
80. T. J. Swift and R. E. Connick, *J. Chem. Phys.*, **37**, 307 (1962); R. E. Connick and D. Fiat, *J. Chem. Phys.*, **44**, 4103 (1966).
81. T. J. Swift and W. G. Sayre, *J. Chem. Phys.*, **44**, 3567 (1966).
82. B. E. Conway in *Ann. Rev. of Phys. Chem.*, **17**, 481 (1966).
83. Y. Marcus, 'Solvent Extraction from Molten Salts', in *Solvent Extraction Chemistry*, North Holland Publ. Co., Amsterdam, 1967, p. 555.
84. E. C. Freiling, 'Cation Exchange Equilibria with Fused Salts', in *Thermodynamics*, Vol. 1, Intl. Atomic Energy Agency, Vienna, 1966, p. 435; in *Advances in Ion Exchange* (Ed. J. A. Marinsky), Vol. 2, M. Dekker Inc., New York, 1968.
85. P. Debye and E. Hückel, *Z. Physik.* **24**, 185, 334 (1923); **25**, 97 (1924).

86. T. H. Gronwall, V. K. LaMer and K. Sandred, *Z. Physik.*, **29,** 358 (1928).
87. E. A. Guggenheim, *Trans. Faraday Soc.*, **55,** 1714 (1959).
88. E. Güntelberg, quoted by N. Bjerrum, *Z. Physik. Chem. (Leipzig)*, **119,** 155 (1926).
89. R. H. Fowler and E. A. Guggenheim, *Statistical Thermodynamics*, Cambridge Univ. Press, 1949, Ch. 9. P. Debye and E. Hückel, *Z. Physik*, **24,** 185 (1923).
90. Cf. ref. 65, p. 176.
91. G. N. Lewis and M. Randall, *J. Am. Chem. Soc.*, **43,** 1112 (1921).
92. M. N. Lietzke, *USAEC Rept.* ORNL–2628 (1958).
93. E. Glueckauf, *Trans. Faraday Soc.*, **60,** 776 (1964).
94. J. H. De Lap, *Ph. D. Thesis*, Duke Univ., 1960.
95. H. S. Frank and P. T. Thompson in *Structure of Electrolyte Solutions* (Ed., W. J. Hamer), J. Wiley & Sons, Inc., New York, 1959.
96. J. Kielland, *J. Am. Chem. Soc.*, **59,** 1675 (1937).
97. N. Bjerrum, *Acta Chem. Scand.*, **13,** 945 (1958).
98. R. G. Bates, *Determination of pH, Theory and Practice*, J. Wiley & Sons, I c., New York, 1964.
99. I. Oppenheim, *J. Phys. Chem.*, **68,** 2959 (1964).
100. H. S. Frank, *J. Phys. Chem.*, **67,** 1554 (1963).
101. D. A. McInnes, *J. Am. Chem. Soc.*, **41,** 1086 (1919).
102. E. A. Guggenheim, *J. Phys. Chem.*, **34,** 1758 (1930).
103. R. G. Bates and S. F. Acree, *J. Res. Nat. Bur. Std.*, **34,** 873 (1945).
104. M. Paabo, R. A. Robinson and R. G. Bates, *J. Am. Chem. Soc.*, **87,** 415 (1965); cf. also R. G. Bates *Nat. Bur. Std.*, *U. S.*, Tech. Notes 271 (1965) and 400 (1966).
105. R. H. Stokes, *J. Am. Chem. Soc.*, **69,** 1291 (1947).
106. D. A. Sinclair, *J. Phys. Chem.*, **37,** 495 (1933); R. A. Robinson and D. A. Sinclair, *J. Am. Chem. Soc.*, **56,** 1830 (1934).
107. M. L. Lakhanpal and B. E. Conway, *Can. J. Chem.*, **38,** 199 (1960).
108. M. H. Lietzke and R. W. Stoughton, *J. Phys. Chem.*, **66,** 508 (1962).
109. E. A. Guggenheim and R. H. Stokes, *Trans. Faraday Soc.*, **54,** 1646 (1958).
110. H. S. Frank and P. T. Thompson, *J. Chem. Phys.*, **31,** 1086 (1959); cf. ref. 95.
111. J. E. Desnoyers and B. E. Conway, *J. Phys. Chem.*, **68,** 2305 (1964).
112. J. C. Ghosh, *J. Chem. Soc.*, **113,** 449, 707 (1918).
113. N. Bjerrum, *Z. Elektrochem.*, **24,** 321 (1918); *Medd. Vetensk. Akad. Nobelinst.* 5, No. 16 (1919); *Z. Anorg. Allgem. Chem.*, **109,** 275 (1920).
114. E. Glueckauf in *Structure of Electrolyte Solutions* (Ed. W. J. Hamer), J. Wiley & Sons, Inc., New York, 1959, p. 106.
115. P. Van Rysselberghe and S. Eisenberg, *J. Am. Chem. Soc.*, **61,** 303 (1939); **62,** 451 (1940).
116. E. Wicke and M. Eigen, *Naturwiss.*, **38,** 453 (1951); **39,** 545 (1952); *Z. Elektrochem.*, **56,** 551 (1952); **57,** 319 (1953); *Z. Naturforsch.*, **8A,** 161 (1953).
117. M. Eigen and E. Wicke, *J. Phys. Chem.*, **58,** 702 (1954).

118. G. Scatchard, *J. Phys. Chem.*, **58**, 712 (1954); E. Hückel and G. Krafft, *Z. Physik. Chem.* (*Frankfurt*), **3**, 135 (1955); ref. 60. p. 82, 86.
119. B. E. Conway and R. E. Verrall, *J. Phys. Chem.*, **70**, 1473 (1966).
120. N. Bjerrum, *Z. Anorg. Allgem. Chem.*, **109**, 275 (1920).
121. R. H. Stokes and R. A. Robinson, *J. Am. Chem. Soc.*, **70.**, 1870 (1948).
122. D. G. Miller, *J. Phys. Chem.*, **60**, 1296 (1956).
123. T. Ikeda, *Rept. Liberal Arts Fac.*, *Shizuoka Univ. Japan*, No. 1, p. 25, 1960.
124. P. Debye and J. McAulay, *Z. Physik.*, **26**, 22 (1925).
125. G. Scatchard, in *Structure of Electrolyte Solutions* (Ed. W. J. Hamer), J. Wiley & Sons, Inc. New York, 1959, Chap. 1.
126. G. Scatchard, *Z. Physik.*, **33**, 22 (1922); *Chem. Rev.*, **19**, 309 (1936).
127. E. Hückel, *Z. Physik*, **26**, 93 (1925).
128. G. Scatchard, *Chem. Rev.*, **8**, 321 (1931).
129. D. T. Burns, *Electrochim. Acta*, **11**, 1545 (1964).
130. E. Güntelberg, *Z. Physik. Chem.* (*Leipzig*), **123**, 199 (1926).
131. E. A. Guggenheim, *Phil. Mag.*, **19**, 588 (1935).
132. C. W. Davies, *Ion Association*, Butterworths Sci. Publ, London, 1962, Chap. 3. This supercedes the earlier value $b_. = 0.10 \ |z_+ z_-|$ proposed by him, *J. Chem. Soc.*, 2093 (1938).
133. K. S. Pitzer and L. Brewer, *Thermodynamics* (*Revision of 'Lewis and Randall'*), 2nd ed. McGraw-Hill Book Co. New York, 1961.
134. G. Åkerlof and H. C. Thomas, *J. Am. Chem. Soc.*, **56**, 593 (1934).
135. L. P. Hammett, *Physical Organic Chemistry*, McGraw-Hill Book Co., New York, 1940, p. 267; L. P. Hammett and A. J. Deyrup, *J. Am. Chem. Soc.*, **54**, 2721 (1932).
136. M. A. Paul and F. A. Long., *Chem. Rev.*, **57**, 1 (1957).
137. N. C. Deno, J. J. Jaruszelski and A. Schriescheim, *J. Am. Chem. Soc.*, **77**, 3044 (1955); N. C. Deno and C. Perizzolo, *J. Am. Chem. Soc.*, **79**, 1345 (1957); N. C. Deno, P. Groves and G. Saires, *J. Am. Chem. Soc.*, **81**, 5790 (1959); N.C. Deno, P. Groves, J. J. Jaruszelski and M. Lugasch, *J. Am. Chem. Soc.*, **82**, 4719 (1960).
138. K. Yates, *Can. J. Chem.*, **42**, 1239 (1964).
139. R. H. Boyd, *J. Am. Chem. Soc.*, **85**, 1555 (1963).
140. E. Högfeldt, *Acta Chem. Scand.*, **16**, 1054, (1962).
141. P. A. H. Wyatt, *Discussions Faraday Soc.*, **24**, 162 (1957).
142. E. Högfeldt, *Acta Chem. Scand.*, **14**, 1627 (1960).
143. K. N. Bascombe and R. P. Bell, *Discussions Faraday Soc.*, **24**, 158 (1957).
144. M. Ojeda and P. A. H. Wyatt, *J. Phys. Chem.*, **68**, 1857 (1964).
145. D. Rosenthal J. S. Dwyer, *Can. J. Chem.*, **41**, 80 (1963).
146. R. Stewart and J. P O'Donnell, *Can. J. Chem.*, **42**, 1681 (1964).
147. C. H. Rochester, *Trans. Faraday Soc.*, **59**, 2820 (1963).
148. J. T. Edward and I. C. Wang, *Can. J. Chem.*, **40**, 399 (1962).
149. G. Yagil and M. Anbar, *J. Am. Chem. Soc.*, **85**, 2376 (1963).
150. K. Yates, J. B. Stevens and A. R. Katritzki, *Can. J. Chem.*, **42**, 1957 (1964).

151. M. J. Jorgenson and D. R. Hartter, *J. Am. Chem. Soc.*, **85**, 879 (1963).
152. K. Yates and H. Wai, *J. Am. Chem. Soc.*, **86**, 5409 (1964).
153. E. Högfeldt and J. Bigeleisen, *J. Am. Chem. Soc.*, **82**, 15 (1960).
154. J. G. Dawber and P. A. H. Wyatt, *J. Chem. Soc.*, 3588 (1960).
155. D. Rosenthal and J. S. Dwyer, *J. Phys. Chem.*, **66**, 2687 (1962).
156. Y. Marcus and M. Givon, *J. Phys. Chem.*, **68**, 2230 (1964).
157. D. G. Lee and R. Stewart, *Can. J. Chem.*, **42**, 486 (1964).
158. E. Högfeldt, *Svensk Kem. Tidskr.*, **75**, 63 (1963).
159. E. Högfeldt, *J. Inorg. Nucl. Chem.*, **17**, 302 (1961).
160. J. Bjerrum, *Kgl. Danske Videnskab. Selskab, Mat.-Fys. Medd.*, **22**, No. 18 (1946).
161. G. A. Gamlen and D. O. Jordan, *J. Chem. Soc.*, 1435 (1953).
162. G. P. Haigt, J. Zoltewicz and W. Evans, *Acta Chem. Scand.*, **16**, 311 (1962).
163. See Y. Marcus, *Record Chem. Progr.* (*Kresge-Hooker Sci. Lib.*), **27**, 105 (1966).
164. Y. Marcus and C. D. Coryell, *Bull. Res. Council Israel*, **8A**, 1 (1959).
165. Y. Marcus and D. Maydan, *J. Phys. Chem.*, **67**, 979 (1963).
166. E. Högfeldt and L. Leifer, *Acta Chem. Scand.*, **17**, 338 (1963); L. Leifer and E. Högfeldt, 1st. *Australian Conf. Electrochemistry*, Sydney, 1963.
167. R. A. Robinson and R. H. Stokes, *Trans. Faraday Soc.*, **41**, 752 (1945); *Electrolyte Solutions*, Butterworths Sci. Publ. London, 2nd ed., 1959, p. 445; J. D'Ans and H. Tollert, *Z. Elektrochem.*, **43**, 81 (1937); R. A. Robinson and V. E. Bower, *J. Res. Natl. Bur. Stda.*, **69A**, 365 (1965).
168. M. S. Stakhanova and V. A. Vasilev, *Zh. Neorg. Khim.*, **6**, 1240 (1961).
169. J. N. Brönsted, *J. Am. Chem. Soc.*, **44**, 877 (1922).
170. H. S. Harned, *J. Am. Chem. Soc.*, **51**, 1865 (1935).
171. Cf. ref. 133, p. 572.
172. J. H. Stern and C. W. Anderson, *J. Phys. Chem.*, **68**, 2528 (1964).
173. G. Scatchard, *J. Am. Chem. Soc.*, **83**, 2636 (1961).
174. T. F. Young and M. B. Smith, *J. Phys. Chem.*, **58**, 716 (1954).
175. H. S. Harned, *J. Phys. Chem.*, **63**, 1299 (1959); **64**, 112 (1960); **67**, 1739 (1963); J. H. Stern and A. A. Paschier, *J. Phys. Chem.*, **67**, 2420 (1963); Y. C. Wu, M. B. Smith and T. F. Young, *J. Phys. Chem.*, **69**, 1868, 1873 (1965).
176. R. M. Rush and G. Scatchard, *J. Phys. Chem.*, **65**, 2240 (1961).
177. H. E. Wirth, R. E. Lindstrom and J. N. Johnson, *J. Phys. Chem.*, **67**, 2339 (1962).
178. A. B. Zdanovskii, *Tr. Solyanoi Lab. Akad. Nauk SSSR*, **6**, 1 (1936).
179. V. M. Vdovenko and N. A. Ryazanov, *Radiokhimiya*, **7**, 39, 442 1965).
180. H. A. C. McKay and J. K. Perring, *Trans. Faraday Soc.*, **49**, 163 (1952).
181. B. B. Owen and T. F. Cooke, *J. Am. Chem. Soc.*, **59**, 2273 (1937).
182. W. H. McCoy and W. E. Wallace, *J. Am. Chem. Soc.*, **78**, 1830 (1956).
183. R. A. Robinson, *J. Phys. Chem.*, **65**, 662 (1961).
184. T. E. Moore, E. A. Gootman and P. C. Yates, *J. Am. Chem. Soc.*, **77**, 298 (1955); F. Dyer, E. H. Gilmore and T. E. Moore, *J. Am. Chem. Soc.*, **77**, 4223 (1955); T. E. Moore, F. W. Burtch and C. E. Miller,

J. Phys. Chem., **64**, 1454 (1960); *U. S. Rept.* AD–268721 from Oklahoma State Univ., 1961.

185. H. S. Harned and B. B. Owen, *Physical Chemistry of Electrolyte Solutions*, 3rd ed., Reinho d Publ. Co., New York, 1958.

186. H. S. Harned, *J. Phys. Chem.*, **58**, 683 (1954); **63**, 1299 (1959); H. S. Harned and A. B. Gancy, *J. Am. Chem. Soc.*, **76**, 5924 (1954); **77**, 1995, 4695 (1955); *J. Phys. Chem.* **62**, 627 (1958); **63**, 2079 (1959); H. S. Harned and R. Gary, *J. Phys. Chem.*, **63**, 2086 (1959).

187. W. J. Argersinger, Jr. and D. M. Mohilner, *J. Phys. Chem.*, **61**, 99 (1957); N. J. Meyer, W. J. Argersinger, Jr. and A. W. Davidson, *J. Am. Chem., Soc.*, **79**, 1024 (1957); H. Schonhorn and H. P. Gregor, *J. Am. Chem. Soc.*, **83**, 3576 (1961).

188. L. Leifer, W. J. Argersinger, Jr. and A. W. Davidson, *J. Phys. Chem.*, **66**, 1321 (1962).

189. H. A. C. McKay, *Trans. Faraday Soc.*, **51**, 903 (1955).

190. J. N. Brönsted and V. K. LaMer, *J. Am. Chem. Soc.*, **46**, 555 (1924).

191. E. Glueckauf, H. A. C. McKay, and A. R. Mathieson, *J. Chem. Soc.*, Suppl. 2, S299 (1949).

192. R. H. Betts and A. N. MacKenzie, *Can. J. Chem.*, **30**, 146 (1952).

193. S. W. Mayer and S. D. Schwartz, *J. Am. Chem. Soc.*, **72**, 5106 (1950); **73**, 222 (1951).

194. P. T. McTigue, *Trans. Faraday Soc.*, **60**, 127 (1964).

195. G. Biedermann and L. G. Sillèn, *Arkiv Kemi*, **5**, 425 (1953).

196. I. Leden, *Acta Chem. Scand.*, **6**, 971 (1952).

197. H. A. C. McKay, *Proc. Intern. Conf. Coord. Comp.*, Amsterdam, 1955, p. 188.

198. F. J. C. Rossotti and H. S. Rossotti, *Determination of Stability Constants*, McGraw Hill Book Co., New York, 1961, p. 26.

199. B. Carell and A. Olin, *Acta Chem. Scand.*, **15**, 727 (1961).

200. A. J. Zielen and J. C. Sullivan, *J. Phys. Chem.*, **66**, 1065 (1962).

201. S. W. Rabideau, *J. Am. Chem. Soc.*, **79**, 3675 (1957).

202. J. F. Tate and M. M. Jones, *J. Inorg. Nucl. Chem.*, **24**, 1010 (1962).

203. F. A. Long and W. F. McDevit, *Chem. Rev.*, **51**, 119 (1952).

204. T. J. Morisson, *Trans. Faraday Soc.*, **48**, 43 (1944); M. Nakajima, *J. Electrochem. Soc. Japan.*, **21**, 166 (1953).

205. J. H. Stern and A. Hermann, *J. Phys. Chem.* **71**, 306, 309 (1967).

206. W. F. McDevit and F. A. Long, *J. Am. Chem, Soc.*, **74**, 1773 (1952).

207. J. Padova, *J. Chem. Phys.*, **40**, 691 (1964).

208. N. C. Deno and C. H. Spink, *J. Phys. Chem.*, **67**, 1347 (1963).

209. K. Sandved, *Kgl. Norske Videnskab Selskabs Forh.*, **6**, 90, (1933).

210. P. Debye, *Z. Physik. Chem. (Lepzig)*, **130**, 56 (1927).

211. P. S. Albright, *J. Am. Chem. Soc.*, **59**, 2098 (1937); P. S. Albright and Williams, *Trans. Faraday Soc.*, **33**, 247 (1937).

212. M. Givon, Y. Marcus and M. Shiloh, *J. Phys. Chem.*, **67**, 2495 (1963).

213. J. A. V. Butler, *J. Phys. Chem.*, **33**, 1015 (1929).

214. J. G. Kirkwood, *Chem. Rev.*, **24**, 233 (1939)., but see A. G. Leiga and J. N. Sarmonsäkis, *J. Phys. Chem.*, **70**, 3544 (1966).

215. J. O'M. Bockris and H. Egan, *Trans. Faraday Soc.*, **44,** 151 (1948).
216. J. O'M. Bockris, J. Bowler-Reed and J. A. Kitchener, *Trans. Faraday Soc.*, **47** 184 (1951).
217. W. F. McDevit and F. A. Long, *18th Mtg. Am. Chem. Soc.*, Chicago, Ill., Sept. 1950.
218. V. A. Mikhailov, *Zh. Fiz. Khim.*, **36,** 306 (1962).
219. A. Eucken and G. Hertzberg, *Z. Physik. Chem.* (*Leipzig*), **195,** 1 (1950).
220. B. E. Conway, J. E. Desnoyers and A. C. Smith, *Phil. Trans. Royal Soc. London*, **256A,** 389 (1964); J. E. Desnoyers and B. E. Conway, *J. Phys. Chem.*, **70,** 3017 (1966).
221. P. Ruetschi and R. F. Amlie, *J. Phys. Chem.*, **70,** 718 (1966).
222. B. Baranowski and M. Sarnowski, *Roczniki Chem.*, **32,** 135 (1958).
223. H. Schmutzer, *Z. Physik. Chem.* (*Leipzig*), **204,** 131 (1955).
224. I. L. Jenkins and H. A. C. McKay, *Trans. Faraday Soc.*, **50,** 107 (1954).
225. N. M. Adamskii, *Radiokhimiya*, **2,** 653 (1960).
226. O. Sa. Samoilov, *Structure of Aqueous Electrolyte Solutions and Hydration of Ions*, (in Russian), State Publ. House Moscow, 1957.
227. A. S. Solovkin, *Zh. Neorg. Khim.*, **5,** 2119 (1960).

2

Nonaqueous Solutions of Electrolytes

Solvent extraction deals with two liquid phases, and the distribution of solutes, which usually are electrolytes, between them. One of the phases is generally an aqueous solution of electrolytes, discussed at length in Chapter 1. The other phase is a nonaqueous, usually organic, solvent, and it is therefore appropriate here to deal with nonaqueous solutions of electrolytes and discuss their properties. Although the solvents used in extraction are necessarily practically immiscible with water, it is of importance to deal also with solutions in water-miscible solvents. On the one hand, much of our knowledge of the behaviour of electrolytes in organic solvents pertains to this class of solvents, on the other, these solvents as well as their mixtures with water, find increasing utilization in ion exchange processes and studies. Up till now, many workers using organic solvents with ion exchangers, have not considered in sufficient details the intricacies of their behaviour, as discussed in this chapter.

A tabulation of the properties of the most widely used solvents is found in appendix C, but the general properties of the solvents as liquids are discussed in section 2.A. The distribution and potential functions of liquids are defined, as well as the concept of corresponding states. The most important physical properties, such as thermodynamic functions, dielectric constant and dipole moments, and chemical properties such as autoionization, basicity and capability of hydrogen bonding, are discussed.

The behaviour of mixtures of solvents, both aqueous and nonaqueous, is of major importance in this monograph. In section 2.B are therefore discussed the general aspects of the thermodynamics of mixtures of liquids. Particular cases, such as athermal and regular solutions, the concept of solubility parameters arising from them, and the average-potential (corresponding solutions) model, valid mainly for mixtures of non-polar molecules, are then discussed. This leads to a view of the behaviour of mixtures involving polar solvents, and to mixed aqueous–organic solvents. The problems of mutual miscibility and of the dielectric constant in solvent mixtures are particularly stressed.

Once the properties of organic solvents and their mixtures are known, it is possible to understand the behaviour of electrolytes in organic solvents, treated in sections 2.C and 2.D. In the former, the problems of solubility and solvation are treated, and inert and active solvents are discussed in turn. The subject of this section is of course closely related to that of chapter 9, where the relationship of these problems to that of extraction is treated. The role of water, and the problem of preferential hydration in solvating solvents are the subject of the rest of this section.

Section 2.D then follows with a discussion of electrolytic dissociation in organic solvents. This aspect of solutions in organic solvents is of great importance in extraction systems, such as of acido-complexes (section 9.C) and of long-chain ammonium salts (section 10.B). Since ionic dissociation can be studied most directly by conductivity measurements, these are discussed briefly, as well as the ionic size and mobility in organic solvents. In solvents of high dielectric constant, where dissociation is appreciable, electrolytes form non-ideal solutions, and the, unfortunately scant, information on the activity coefficients is reviewed. In solvents of low dielectric constants, ions aggregate to ensembles larger than ion pairs. Aggregation culminates in micelle formation, on which most information is available for aqueous, rather than organic solutions. Although ionic detergents in water are not completely similar to aggregated electrolytes in organic solvents, many of the concepts used, such as the critical micelle concentration (*cmc*), and the factors determining micelle stability, are similar for aqueous micelles and the inverted micelles found in organic solvents. A discussion of the similarities, and of the important differences, concludes this chapter.

A. PROPERTIES OF SOLVENTS

a. Introduction

For the purpose of this book, substances which are liquid at least in a part of the temperature range of the existence of liquid water, and which are chemically stable in contact with water, are termed solvents. Some of these substances show only poor solvent power towards electrolytes, while others may have a high solvent power towards them. The substances which have found actual application in solvent extraction and ion exchange studies are all organic liquids, either hydrocarbons, or substituted hydrocarbons with one or more functional group.

In chapter 7 solvents are classified according to the extraction mechanism and the types of solvent (section 7.A). For the purpose of extraction a solvent must be reasonably immiscible with water. In this chapter the properties of the solvents themselves and their mixtures are considered, so that also water-miscible solvents are included. These may play a role, for example in ion exchange processes.

Solvents may be classified according to their physical and chemical properties. For many purposes, it is a bulk physical property, such as the bulk dielectric constant, which is of prime importance. For other purposes a molecular physical property, such as the dipole moment, determines the suitability of a solvent. The most straightforward classification is according to the chemical constitution and structure, the nature, number and placement of the functional groups.

In appendix C solvents are listed according to this classification, with the appropriate values of their physical properties, where known. The standard nomenclature is used, and for complicated cases, or commercial solvents, skeletal formulae are given. The numerical values are taken from standard reference sources or commercial literature, if necessary.

b. Solvents as Liquids

The properties of solvents all devolve on the fact that they are liquids at the temperatures of application. Liquids as a class are much less well understood than are solids or gases, but there are empirical means to describe their properties, and also theories based on reasonable models, of considerable predictive value. Only a few of the concepts and the relationships in the presently accepted theories of the liquid state will be here introduced.

The liquid state differs from the solid crystalline state by the disorder of the molecules, although they are fairly closely packed. The disorder and packing are measurable by the radial distribution function, obtained from x-ray scattering. The probability of finding a particular molecule in a volume element dV at a distance r from a reference molecule is $g(r)dV/V$, where $g(r)$ is the radial distribution function. The number of molecules in a spherical shell of thickness dr is

$$n_r(\mathrm{d}r) = n\,g(r)\,4\pi r^2\,\mathrm{d}r \qquad (2\mathrm{A}1)$$

where n is the number of molecules per unit volume in the bulk. If r_i is the radius of a molecule, $g(r < r_i) = 0$ and $g(r = 2r_i) > 1$, i.e. the nearest neighbours are at an inter-centre distance of a molecular diameter. The function $g(r)$ fluctuates between values smaller and larger than unity at low r, but approaches unity at high r.

The potential function $U(r)$ describes the potential energy between two molecules of the liquid at a distance r apart. The number of pairs of molecules at a distance r apart in a mole of liquid is $\frac{1}{2}NV^{-1}g(r)4\pi r^2\,\mathrm{d}r$. The potential energy per mole of liquid is therefore[1]

$$E = 2\pi N^2 V^{-1} \int U(r)\,g(r)\,r^2\,\mathrm{d}r \qquad (2\mathrm{A}2)$$

The molar energy of vaporization E^V should be equal to $-E$. The potential $U(r)$ is the sum of attractive and repulsive potentials. As attractive forces, London dispersion forces can be cited, as well as dipole and induced-dipole interactions. For spherical non-polar molecules the Lennard–Jones '6–12' law is a good approximation[2,3]

$$U(r) = -2\,U_{ii}^{*}(r_{ii}/r)^6 + U_{ii}^{*}(r_{ii}/r)^{12} \qquad (2\mathrm{A}3)$$

where U_{ii}^{*} and r_{ii} are the coordinates of the minimum in the $U(r)$ curve. For other types of molecules a modified potential function must be used. Dipole interactions introduce another term in r^{-6}[4]

$$U(r)\ (\text{dipole}) = -(2/3)\,\mu^4/k\,T\,r^6 \qquad (2\mathrm{A}4)$$

where μ is the dipole moment of the molecule. The pair potential energy given by (2A3) is the same for the gaseous and the liquid states. Other interactions would be considerably modified by the presence of numerous other molecules in the vicinity of the pair. Thus only for spherical, non-polar molecules can the parameters U_{ii}^{*} and r_{ii} be obtained from the bulk properties of the liquids or gases. The second virial coefficient, $B(T)$, in the expression

$$PV/RT = 1 + B(T)/V + C(T)/V^2 + \cdots \qquad (2\mathrm{A}5)$$

is given in terms of the pair potential as

$$B(T) = 2\pi N \int (1 - \exp(U(r)/k\,T)\,r^2\,\mathrm{d}r \qquad (2\mathrm{A}6)$$

Values of U_{ii}^{*} and of r_{ii} obtained in this way have been tabulated[3].

An important attribute of liquids is their having a 'free volume', that is, the molar volume of the liquid is larger than the 'covolume', or the actual volume of the molecules

$$V_i^f = V_1 - (4\pi/3) N r_i^3 \tag{2A7}$$

The radius r_i is, of course, a parameter not directly obtained from bulk properties of the liquid. A good approximation for the free volume is based on a cage model[5]

$$V_i^f = (4\pi/3) N (\gamma V_i/N)^{1/3} - r_{ii})^3 \tag{2A8}$$

where r_{ii} is the internuclear distance of minimum energy used above, and γ is a packing factor, which can be obtained from the radial distribution function, but can be well approximated by the value 1.3. It can be shown that (2A8) is equivalent to an expression using the macro properties of the liquid: its coefficient of thermal expansion α and its coefficient of isothermal compressibility β

$$V_i^f = (4\pi/3)\gamma V^{-2} (R \beta/\alpha)^3 \tag{2A9}$$

A final general concept which has been found of value is that of corresponding states. Perfect liquids [6] are those for which the following assumptions hold. The translational degrees of freedom should be classical, i.e. the spacing between the translational quantum levels should be small compared to kT. The internal degrees of freedom in the liquid should be the same as in the gas. The intermolecular potential energy $U(r)$ should be a function of the distances r only, excluding thus polar liquids or hydrogen-bonded liquids, allowing only for London dispersion forces. Finally, the function $U(r)$ is a universal function (such as the Lennard–Jones, or any similar law). Very few liquids indeed exhibit this perfect behaviour (Ar, Kr, Xe and to a lesser degree CH_4, N_2, O_2 and CO, but not H_2 or He). For a perfect liquid, the ratio of the critical constants is a universal constant

$$P_c V_c/T_c = 0.292 R \tag{2A10}$$

Other properties can be expressed in terms of the reduced functions $P_r = P/P_c$, $V_r = V/V_c$ and $T_r = T/T_c$. The reduced vapour pressure of a perfect gas is a universal function of the reduced temperature.

The entropies of vaporization of perfect liquids are the same at corresponding states, i.e. at the same reduced pressures, volumes or temperatures

$$\Delta S^V = f(P_r) = g(V_r) = h(T_r) \qquad (2A11)$$

where f, g and h are universal functions. If all liquids had the same critical pressure P_c, the normal boiling points of the liquids (at 1 atm pressure) would be corresponding states. All liquids should then have had the same entropies of vaporization at their normal boiling points, and Trouton's rule

$$\Delta H^V / T_b = const. = 21 \, cal/mole/deg \qquad (2A12)$$

would hold vigorously. Similarly, if all liquids had the same critical volume, then Hildebrand's rule[7] of equal entropies of vaporization at equal vapour volumes, would be valid. Both rules apply strictly only to perfect liquids, and are approximations for many actual, imperfect liquids.

c. Physical Properties of Solvents

Some of the properties of solvents have a great practical importance, while others are important for understanding their behaviour. The melting and boiling points, and the vapour pressure at room temperature define the temperature range of applicability of a given solvent. For many solvents the melting point is so low that it is of little interest, but it is important to note that such solvents can be used below the freezing point of water, in order to study phenomena which are too fast at higher (i.e. room) temperature, or for other purposes. Some solvents have a freezing point at a convenient temperature, and can be used for cryoscopic studies, e.g. benzene (m.p. = 5.533°). Too high a vapour pressure at room temperature is inconvenient for many purposes, and often, when other things are equal, a solvent boiling at 100–120° (e.g. toluene, b.p. = 110.6°) is preferable to one boiling at 60–80° (e.g. benzene, b.p. 80.1°).

The density, viscosity and surface tension of a solvent play a role in the practical applications of extraction, since they govern the rate and efficiency of phase separation. These properties have few direct effects on the thermodynamics of the distribution process.

The thermochemical properties, such as the heat of vaporization ΔH^V and the heat of freezing ΔH^F are of importance in several respects. The heat of freezing is connected with the molar freezing point depression, the cryoscopic constant

$$K_f = \lim_{m \to 0} \Delta T_f / m = M_S R T_f^{0\,2} / 1000 \Delta H^F \qquad (2A13)$$

where T_f^0 is the freezing point of the pure solvent of molecular weight M_S. The ebullioscopic constant is connected by a similar equation to the heat of vaporization ΔH^V. This quantity, in turn, can be calculated approximately by Trouton's rule (2A12). It can be equated with the molar energy of vaporization E^V (section 2.A.b) and can, in principle, be calculated from the distribution and potential functions according to (2A2).

The molar volume of a solvent is an important quantity, which directly affects the miscibility of solvents, the solubility of solutes and the activity of the solvent in a solution

$$V_S = M_S/\rho_S \tag{2A14}$$

Whereas the molecular weight varies from 50 to 500 for many solvents, the density varies only from 0.7 to 1.3 for these solvents, so that the molar volume is more sensitive to the former quantity.

The dielectric constant ε of a solvent, and the polarity and polarizability of solvent molecules are properties which affect strongly their interactions with themselves and with solutes, hence their solvent power. In a gas, Debye's theory [8] gives the relationship for the molar polarization

$$P = \frac{\varepsilon - 1}{\varepsilon + 2} V_S = \frac{4\pi}{3} N(\alpha + \mu^2/3kT) \tag{2A15}$$

where α is the polarizability and μ the dipole moment. For a liquid, α is the sum of the atomic and the electronic polarizabilities, which is given to a good approximation by

$$\alpha = (3V_S/4\pi N)(n_D^2 - 1)/(n_D^2 + 2) \tag{2A16}$$

where n_D is the refractive index for visible light. In this approximation the atomic polarization is neglected, but this is compensated by a slight over-estimation of the electronic polarization by the use of n_D. Equations (2A15) and (2A16) should, thus, yield a value for the dipole moment μ from measurements of the dielectric constant ε and the refractive index n, but the values thus obtained for liquids do not agree with the values accepted for the free solvent molecules from measurements on gases [9]. The discrepancy is expressed by a coefficient g, in a modified expression derived by Onsager [10] and Kirkwood [11]

$$(\varepsilon - 1)(2\varepsilon + 1)/9\varepsilon = \frac{4\pi}{3} N(\alpha + g\mu^2/3kT) \tag{2A17}$$

The value of g can be obtained from the slope of the left-hand side of the equation plotted against $1/T$, and values of μ obtained experimentally for the vapour of the solvent (from 2A15 and 2A16) or by calculation from the bond moments and the geometry. The parameter g represents the hindrance of free rotation in the liquid

$$g = 1 + Z \int \cos\gamma \exp(-W/kT)\,d\gamma \qquad (2A18)$$

where Z is the number of nearest neighbours in the liquids, obtainable from the integral of the $g(r)$ function up to the first minimum in the curve, being about 8–10, γ is the angle between the dipoles of representative molecules, and W the potential energy of torque hindering the rotation. This parameter g is near unity for 'normal' liquids, but has values as high as 3 or 4 for associated liquids such as water (2.7), ethanol (3.0) or hydrogen cyanide (4.1).

Summarizing this part of the discussion, it is seen that solvents can be classified as non-polar ($\mu = 0$) or polar ($\mu > 0$), and slightly polarizable ($\alpha \lesssim 2\,\text{Å}^3$) or highly polarizable ($\alpha \gtrsim 10\,\text{Å}^3$). Among the non-polar are included symmetrical molecules such as cyclohexane, p-xylene and carbon tetrachloride. Other molecules, of lower symmetry, show a small dipole moment, e.g. toluene (0.39 Debye) or p-dioxane (0.45 Debye), while very unsymmetrical molecules, with functional groups such as $C=O$, $-NO_2$ or $-CN$ may show high dipole moments, e.g. acetone (2.72 Debye), nitrobenzene (3.99 Debye) and benzonitrile (4.05 Debye). Multiple bonds, and bonds between carbon and the heavier halogen and sulphur atoms contribute highly to the polarizability of organic molecules, whereas single $C-C$ or $C-O$ bonds, $C-H$ and $C-F$ bonds contribute much less.

d. Chemical Properties of Solvents

The basicity (in the Lewis sense) of solvent molecules is one of the most important chemical properties. Molecules having functional groups with oxygen or nitrogen atoms with unshared electron pairs are basic, the more so the higher the electron density on these atoms. Groups that withdraw or provide electrons diminish or enhance the basicity respectively. Lewis bases are also bases in the Brönsted sense, since they will accept protons. It is rare to find among organic solvents Lewis acids, but protic solvents, i.e. those capable of yielding a proton to a base stronger than their own conjugate base, are Brönsted acids. The lower the electron density on the atom to which the proton is bound, the higher is the acid strength of the solvent.

In a pure solvent, autoprotolysis or self-dissociation can occur according to the scheme

$$HS + HS \rightleftharpoons H_2S^+ + S^- \tag{2A19}$$

Since most of the solvents have a low dielectric constant, the equilibrium (2A19) will lie far to the left, unless the solvent molecules are sufficiently polar or polarizable, to provide strong solvation of the ions formed through ion–dipole interactions. There are very few quantitative data concerning the equilibrium constants of equation (2A19) for organic solvents. Most of the data are shown in Table 2A1.

TABLE 2A1. Autoprotolysis products for several protic solvents at 25°.

Solvent	$-\log(SH_2^+)$ (S^-)	pK_s	Reference
H_2O	14.0	15.8	
CH_3OH	16.7		b
		16.7	a
C_2H_5OH	19.1		b
		19.1	a
$n-C_3H_7OH$	19.4		b
$i-C_3H_7OH$	20.8		b
HOC_2H_4OH	~11		b
$HCOOH$	6.2		b
CH_3COOH	>13		a,b
	14.5		b
$HCONH_2$	16.8		b
		16.8	a
CH_3CONH_2	10.5		b
CH_3CN^*	19.5		b

* Dissociates to $CH_2CN^- + CH_3CNH^+$.
a. R. P. Bell, *The Proton in Chemistry*, Cornell Univ. Press, Ithaca, New York, 1959, pp. 36–47.
b. G. Charlot and B. Tremillon, *Les Réactions Chimiques dans les Solvants et les Sels Fondus*, Gauthier Villars, Paris, 1963.

Even if no appreciable self-dissociation occurs, solvents often provide dissociating media for stronger acids or bases than themselves, according to the schemes

$$SH + B \rightleftharpoons BH^+ + S^- \tag{2A20}$$

$$S(H) + HA \rightleftharpoons SH(H)^+ + A^- \tag{2A21}$$

For permitting dissociation of acids according to (2A21) the solvents can be either protic, i.e. of the type SH, capable of reacting according to (2A20) with sufficiently strong bases B, or aprotic, i.e. of the type S, not capable of reacting thus (e.g. trichloritrifluoroacetone). The base B or the acid HA can be neutral or charged, as the case may be.

If a solvent SH is itself a sufficiently strong acid, the equilibrium (2A20) will be far to the right even for relatively weak bases B. The solvent thus has a levelling effect for bases. On the other hand, such solvents are often reluctant to add on another proton, so that the equilibrium (2A21) will be far to the left, except for strong acids HA. The solvent thus has a differentiating effect for acids. The converse arguments apply for a solvent SH which is itself a strong base[12,13]. Amphoteric solvents or solvents which are neither acidic nor basic, will permit a range of dissociation constants for acids and bases, centred roughly around the neutral point.

A final property to be discussed is the capability of hydrogen-bond formation. If the atom to which a hydrogen atom is bound is sufficiently electronegative, and if an electronegative atom with a free electron pair is available in the same or another molecule, the hydrogen atom can form a hydrogen bond between the two electronegative atoms. Fluorine, oxygen, chlorine and nitrogen atoms are considered sufficiently electronegative in this sense, as well as certain groups, such as Cl_3C— in chloroform. Intramolecular hydrogen bonds in a solvent affect some of its properties, but of more far-reaching consequences are intermolecular hydrogen bonds since their formation causes an association of the solvent to dimers or larger aggregates, making the solvent an 'abnormal' liquid as far as the vapour pressure, boiling point, heat of vaporization, dielectric constant and other properties are concerned. Although the energy of the hydrogen bond (3–6 kcal/mole) is small compared with that of ordinary covalent bonds, it is large compared with that of other intermolecular interactions or the kinetic energy of molecules at room temperature ($\sim$ 0.6 kcal/mole). Teh formation of hydrogen bonds is observable in the infrared and proton magnetic resonance spectra of the solvents for solutions) involved. It is important to distinguish between solvents forming hydrogen-bonded dimers (such as acetic acid or dibutylphosphoric acid) and those forming hydrogen-bonded chains or larger aggregates (such as alcohols or water). Obviously, the latter class shows the greater deviations from 'normal' liquid behaviour.

Dipole moment measurements have contributed to the assessment of the degree of aggregation (cf. section 2.A.c). Cyclic aggregates have a low dipole moment, while chain polymers have a large one[14], the size and degree of branching of the R groups in alcohols ROH affecting the configuration and the resultant moments[15, 16]. This and other physical properties and thermodynamic information has provided indirect evidence for the hydrogen-bonded association of alcohols and similar substances.

The infrared absorption spectrum has provided more direct evidence for hydrogen-bond formation and aggregation, particularly of alcohols. Smith and Creitz[17] attributed the O—H stretching frequency at 2.74 μ to the monomers, that at 2.86 and 2.76 μ to the terminal groups R—O—H···O— and R—O(H)···H— of aggregates, and that at 2.96 μ to interior $(ROH)_n$ groups. The latter band is broad, and unaffected by the nature of the R groups. The position of the other bands is affected by the R groups, and by dilution[18]. In the n.m.r. spectrum there are large chemical shifts due to even weak hydrogen bonds[19,20]. A problem in the interpretation of the infrared band intensities and the n.m.r. chemical shifts arises from the fact that the measurements are made on dilute systems, using mainly carbon tetrachloride or benzene as diluents. The diluents were considered to be 'inert', their only function being to increase the mutual average distance between alcohol molecules, causing, thereby, hydrogen bonds to be broken[21-23]. This is, however, not completely true, since the appreciable polarizability of these diluents permits specific interactions[24], e.g. with phenolic OH groups[25] or carboxylic COOH groups[26], reducing thereby the degree of aggregation more than by mere dilution. With undiluted liquids, however, this objection does not arise, and the temperature effect on the infrared band intensities can be correlated with the aggregation equilibria[21, 23]. In certain cases the equilibrium constants for the monomer–dimer–polymer aggregation steps could be calculated[23].

In conclusion, it is the chemical properties (acid–base strength and hydrogen-bonding capacity) and the physical properties (polarity, polarizability, molar volume, heat of vaporization, etc.) which determine the behaviour of an organic liquid as a solvent, towards other organic liquids, towards non-polar solutes, towards water and towards electrolytes.

B. SOLVENT MIXTURES

a. Thermodynamics of Mixtures

Section 1.A discussed briefly some of the thermodynamic relationships pertaining to mixtures. There, however, the interest was in the solutions of a solute G, present at relatively low concentrations, in a solvent S, present in a large excess. Furthermore, the solute G was an electrolyte, dissociating into v ions. For the present section, interest lies in two solvents 1 and 2, present in any relative amounts. If the free energy per mole of pure solvent i is G_i^0, the free energy of mixing of N_1 moles of solvent 1 with N_2 moles of solvent 2 is

$$G^M = G - N_1 G_1^0 - N_2 G_2^0 \tag{2B1}$$

where G is the free energy of the mixture. The molar free energy of mixing will then be

$$G_{1,2}^M = G_{1,2} - x_1 G_1^0 - x_2 G_2^0 \tag{2B2}$$

Similar relations hold also for the molar entropy of mixing $S_{1,2}^M$ and molar volume (change) of mixing $V_{1,2}^M$, as well as for the other thermodynamic functions, substituting S, V, etc. for the symbol G in (2B1).

It is useful to consider two types of solutions: *perfect solutions* and *ideal solutions*, to which actual solutions may approach under certain circumstances. For a perfect solution

$$G_{1,2}^M = RT(x_1 \ln x_1 + x_2 \ln x_2), \text{ so that} \tag{2B3}$$

$$S_{1,2}^M = -R(x_1 \ln x_1 + x_2 \ln x_2) > 0 \text{ and} \tag{2B4}$$

$$H_{1,2}^M = 0 \quad V_{1,2}^M = 0 \quad \text{etc.} \tag{2B5}$$

It should be noted that like the enthalpy change (heat) of mixing and the volume change of mixing, all thermodynamic quantities of mixing which do not include the entropy are zero. The entropy of mixing for a perfect solution may be related to the statistical thermodynamic increase in randomness made possible in the mixing process of two molecular kinds which may occupy completely equivalent positions. In reality, only mixtures of very similar molecules, such as isotopes or optical isomers, exhibit this property.

From (2B3) the chemical potential of a component of a perfect mixture is obtained by partial differentiation

$$\mu_i = \mu_i^0 + RT \ln x_i \tag{2B6}$$

An ideal solution obeys relationship (2B6) only over a certain range of concentrations, but not up to $x_i = 1$, hence the standard state is not that of the pure liquid i, but an extrapolated one, for which μ_i^{0*} is the standard chemical potential (cf. section 1.A.a). For an ideal solution, then, the change of chemical potential on mixing differs from that for a perfect solution by $\mu_i^{0*} - \mu_i^0$ (cf. section 7.B.b).

The excess functions may now be considered. The *excess free energy* for a mole of the mixture is defined as (cf. section 1.E.c)

$$G_{1,2}^E = G_{1,2}^M - RT(x_1 \ln x_1 + x_2 \ln x_2) \tag{2B7}$$

and it is obvious that for perfect solutions $G_{1,2}^E = 0$. The excess free energy can also be expressed in terms of activity coefficients

$$G_{1,2}^E = RT(x_1 \ln f_1 + x_2 \ln f_2) \tag{2B8}$$

The excess free energy of mixing can also be written as

$$G_{1,2}^E = bx_1 x_2 = bx_1(1 - x_1) \tag{2B9}$$

where b in the general case is dependent on x_1, but for many systems is independent of it. (Such systems are said to form regular solutions, as discussed further below). In the latter case, the thermodynamic excess functions are symmetrical with respect to the composition. Since, for perfect solutions, the enthalpy and volume change of mixing are zero, the functions for actual mixtures are thus the appropriate excess functions

$$H_{1,2}^E = H_{1,2}^M = H_{1,2} - x_1 H_1^0 - x_2 H_2^0 \tag{2B10}$$

$$V_{1,2}^E = V_{1,2}^M = V_{1,2} - x_1 V_1^0 - x_2 V_2^0 \tag{2B11}$$

The excess entropy is given by an equation analogous to (2B7), with S replacing G, and T omitted in the last term.

A most direct evaluation of the deviations from ideality are the deviations from Raoult's law for both components (cf. section 1.A.a). Thus the per cent deviation of the total vapour pressure of the mixture of solvents from its additive value

$$\% \,\Delta p = 100(p - x_1 p_1^0 - x_2 p_2^0)/p \tag{2B12}$$

is an obvious measure of the deviations. The excess free energy can be computed from total vapour pressure data, with or without vapour composition data[27, 28].

Table 2B1 is a collection of values of a number of excess functions at 25^0 for $x_1 = x_2 = 0.5$ for some solvent mixtures. Most of the data were taken from Rowlinson's[29] book.

The above relations help to describe the behaviour of two solvents on mixing. If $G_{1,2}^M$ is plotted against x_1, the curve, for a perfect mixture ($G_{1,2}^E = 0$) is concave upwards. Actual mixtures ($G_{1,2}^E \lesssim 0$) may show a region where the curve is concave downwards. The curve then has two inflexion points, where

$$\left(\frac{\partial^2 G_{1,2}^M}{\partial x_1^2}\right)_{T,P} = 0 \qquad (2B13)$$

A mixture exhibiting this behaviour is unstable as a single phase, and will separate into two liquid phases. With a change in the temperature, the two values of x_1, where (2B13) holds, may approach each other, and at a *critical consolute temperature* (T_c) these points will meet at x_c, where

$$\left(\frac{\partial^2 G_{1,2}^M}{\partial x_c^2}\right)_{T_c,P} = \left(\frac{\partial^3 G_{1,2}^M}{\partial x_c^3}\right)_{T_c,P} = 0 \qquad (2B14)$$

Beyond the temperature T_c, the two solvents are miscible at all proportions. The temperature T_c may be an upper or a lower critical consolute temperature, depending upon whether phase separation occurs below or above T_c. For symmetrical excess functions, $x_c = 0.5$ at the critical consolute temperature, and from (2B9) and (2B13)

$$T_c = 0.5 \, b/R \qquad (2B15)$$

Thus, b can be calculated for a mixture which obeys the relationship (2B9) from the measured T_c. Conversely, a mixture which has an excess free energy above ca. RT will separate into two immiscible liquid phases, if the corresponding temperature is in the liquid range.

b. Athermal and Regular Solutions

It is useful now to introduce two further types of solutions. *Athermal solutions* are characterized by $H^E = H^M = 0$, as for perfect solutions, but unlike them, $S^E \neq 0$[30, 31]. The entropy of mixing is given by[32] (cf. section 1.D.c)

TABLE 2B1. Excess functions for some equimolar mixtures.
(Curves highly asymmetric in $H^E_{1,2}$ with respect to x_1 are marked*.)

Components	Temperature (°C)	$G^E_{1,2}$ (cal/mole)	$H^E_{1,2}$ (cal/mole)	$V^E_{1,2}$ (ml)	UCCT (°C)
Cyclohexane +					
neo-Pentane	0	44		−1.1	
n-Hexane	20	17	51	0.1	
n-Heptane	20	~0	64	0.27	
i-Octane	30			<0.04	
Benzene	40	72	194		
Toluene	25	95	142	0.59	
CCl$_4$	25	17	35	0.17	
Aniline					29.6
Nitrobenzene	80	395			−4
Methanol			≫0	>0	45
Ethanol	25	315		0.55	
Cyclo-C$_6$F$_{12}$	25	322	494		43
Benzene +					
n-Pentane	60	48			
neo-Pentane	0	136		−0.5	
Cyclopentane				0.32	
n-Hexane	20	77	205		
Cyclohexane	40	72	194		
n-Heptane	60	72	180	0.62	
Toluene	20	~0	16	<0.05	
CCl$_4$	25	20	26	<0.02	
Chlorobenzene	24		−2		
Acetonitrile	45	160	120		
Acetone	45	70	38		
Aniline	119	104			
CF$_3$C$_6$F$_{11}$	25	295			86
Methanol*	35	316	163	−0.016	
Ethanol*	45	267	286		
CCl$_4$ +					
neo-Pentane	0	76	74	−0.5	
Cyclohexane	25	17	35	0.17	
Benzene	25	20	26	0.02	
Chloroform	25	26	55	0.17	
Ethylene dichloride		70			
SiCl$_4$	25	20	31	0.015	
SnCl$_4$	25	~0		0.44	
Acetonitrile*	45	285	222		
Acetone*	45	130	62		
Nitromethane	45	320	340	0.17	
Methanol*	35	335	67		
Ethanol*	45	284			
n-C$_7$F$_{16}$	25	265			59
CF$_3$C$_6$F$_{11}$	25	249		3.4	27
Methanol +					
n-Pentane					14.8
n-Hexane					45
n-Heptane					51.5
Cyclohexane			≫0	>0	45
i-Octane					42.5
Benzene*	35	316	163	0.016	
CCl$_4$*	35	335	67		
C$_2$Cl$_4$					−10
CS$_2$					36.2

$$S_{1,2}^M \text{ (athermal)} = -R(x_1 \ln \varphi_1 + x_2 \ln \varphi_2) \qquad (2B16)$$

where φ_i is the volume fraction of component i, defined as

$$\varphi_i = N_i V_i / (N_1 V_1 + N_2 V_2) \qquad (2B17)$$

In this definition, the volume change on mixing is ignored, i.e. $V^M = 0$ and $V = N_1 V_1 + N_2 V_2$, V_i being the molar volume of the pure component. Also, the variation of V_2/V_1 with temperature is ignored, and its value at some reference temperature, say 25°C, is used.

An alternative definition of the volume fraction (and of athermal solutions[31]) depends on considering component 1 to occupy one site in the liquid quasi-lattice, and component 2 to occupy r_v sites. The second component can thus be considered as an r_v-mer of the first, and indeed this derivation has been developed for solutions of polymers or of chain compounds (e.g. hexadecane in hexane), or even of dimers, trimers, etc.[33] (e.g. diphenyl in benzene). It is thus assumed that $r_v = V_2/V_1$, so that the relations between volume fraction and mole fractions are

$$\varphi_1 = x_1/(x_1 + r_v x_2) \quad x_1 = r_v \varphi_1 / (1 - \varphi_1) + r_v \varphi_1 \qquad (2B18)$$

Thus the entropy of mixing of athermal solutions can be written as

$$S_{1,2}^M = -R[x_1 \ln x_1/(x_1 + r_v x_2) + x_2 \ln r_v x_2 (x_1 + r_v x_2)] \qquad (2B19)$$

Another way of dealing with solutions, the components of which have very different volumes, is to consider r_v not as the ratio of molar volumes of the components, but as the ratio of free volumes, as discussed in section 2.A.a. Then $r_v^f = V_2^f/V_1^f$ should replace r_v in (2B19)[30].

The relationships shown for athermal solutions are useful mainly for mixtures of molecules differing considerably in size (large r_v). Even if the heat of mixing is not zero, the entropy of mixing given here should be used instead of the usual formulation in terms of mole fractions (equation 2B4).

Regular solutions have been introduced by Hildebrand, and can be defined[30, 34] by

$$S_V^E \text{ (regular)} = S_{1,2}^E - (\alpha/\beta)V_{1,2}^E + \cdots = 0 \qquad (2B20)$$

where S_V^E is the excess entropy at constant volume, and the ratio of thermal expansion to compressibility coefficients (α/β) pertains to the mixture. In fact, the term in $V^E_{1,2}$ is usually small and can be neglected as a good approximation. Regular solutions can therefore be termed those solutions which have zero excess entropy of mixing. The heat of mixing does not vanish for regular solutions but is practically equal to the excess free energy.

$$H^M_{1,2} \text{ (regular)} = bx_1x_2 \tag{2B21}$$

Guggenheim[35] has taken (2B21) to be a definition of regular solutions, with the further restriction that b is not only independent of the composition, but also of temperature. This requirement is rather restrictive, and, according to Hildebrand and Scott[34], unnecessary. The parameter b will thus be considered as independent of the composition only, as far as regular solutions are concerned.

Regular solutions are therefore symmetrical with respect to composition, and the activity coefficients are given by the simple relation

$$\ln f_1 = (b/R\,T)(1 - x_1)^2 = (b/R\,T)x_2^2 \tag{2B22}$$

and similarly for component 2.

Regular behaviour is approximated by many liquid mixtures, and can be used as a reference to which mixtures which do not conform can be compared.

c. Solubility Parameters

The parameter b in equation (2B9), or in a similar equation based on volume fractions, with φ_i replacing x_i, which is closely connected with the heat of mixing in regular solutions (2B21), is an important quantity, which can be related to the molecular properties of the solvents, and in particular to their pair potentials, equation (2A3).

For regular solutions, which obey the assumptions of zero volume of mixing, of random orientation and distribution of the molecules, and of having mutual pair-potential energies independent of surrounding molecules and temperature, Scatchard[36] has shown that the energy of mixing is given by

$$E^M = (x_1V_1 + x_2V_2)(C_{11} + C_{22} - 2C_{12})\,\varphi_1\varphi_2 \tag{2B23}$$

where $C_{ii} = \Delta E_i^V/V_i$ is the 'cohesive energy density', ΔE_i^V being the energy of vaporization. If the geometric-mean assumption $C_{12}^2 = C_{11}C_{22}$ is accepted, and the square root of the cohesive energy density is designated by δ, the solubility parameter[1],

$$E^M = (x_1V_1 + x_2V_2)(\delta_1 - \delta_2)^2\varphi_1\varphi_2 \qquad (2B24)$$

Equation (2B23) is related to the molecular parameters of the components and the mixture in the following way. The potential energy of the pure liquid i is given by equation (2A2) as a function of the pair potential and the radial distribution. The energy of one mole of mixture, of volume $V = x_1V_1 + x_2V_2$ is

$$E \text{ (mixture)} = (2\pi N^2/(x_1V_1 + x_2V_2))\left[x_1\left(x_1\int U_{11}g_{11}r^2dr + \right.\right.$$
$$\left.\left. x_2\int U_{21}g_{21}r^2dr\right) + x_2\left(x_1\int U_{12}g_{12}r^2dr + x_2\int U_{22}g_{22}r^2dr\right)\right] \qquad (2B25)$$

This equation can be simplified considerably. According to the assumptions made above, universal potential and distribution functions apply: $U_{ij} = U_{ij}^*f(r/r_{ij})$ and $g_{ij} = g(r/r_{ij})$, furthermore, $U_{12} = U_{21}$ and $g_{12} = g_{21}$. Setting $y = (r/r_{ij})$, and subtracting the energies of the pure components, leads to the energy of mixing

$$E^M = (x_1V_1 + x_2V_2)\varphi_1\varphi_2[(2U_{12}^*r_{12}^3/V_1V_2) - (U_{11}^*r_{11}^3/V_1^2)$$
$$- (U_{22}^*r_{22}^3/V_2^2)]2\pi N^2\int f(y)g(y)y^2dy \qquad (2B26)$$

The quantity C_{ij} in Scatchard's equation (2B23) is thus identified as

$$C_{ij} = U_{ij}^*r_{ij}^3V_i^{-2}2\pi N^2\int_0^\infty f(y)g(y)y^2dy \qquad (2B27)$$

The application of the geometric-mean assumption to the C's is therefore tantamount to its application to $U_{ij}^*r_{ij}^3$. The further assumptions of $r_{11} = r_{22} = r_{12}$ and $V_1 = V_2$ lead to the simple equation $E^M = bx_1x_2$ with b a constant, independent of temperature and composition (cf. section 2.B.a), but these assumptions are too restrictive to be valuable.

The solubility parameter $\delta_i = C_i^{1/2}$ of a solvent i can be calculated from equation (2B27) from published values of the parameters U_{ii}^* and r_{ii} and the universal functions f(r) and g(r), or more directly from the energy of vaporization ΔE_i^V and the molar volume V_i. The energy of vaporization is related to the calorimetrically available heat of vaporization at low pressures: $\Delta E_i^V = \Delta H_i^V - RT$, and a negligible error will be made if ΔH_i^V at the normal boiling point is used instead.

If these quantities, ΔE_i^V or ΔH_i^V are unavailable, δ can be estimated from other data: from the temperature dependence of the vapour pressure: $\Delta H_i^V = pT(\mathrm{dln}\,p/\mathrm{d}T)\,(V^g - V)$, from Hildebrand's rule (section 2.A.c), from the solubility of suitable solutes (see below), and from less suitable methods[30,34], such as the internal pressure $(\delta^2 \approx (\partial E/\partial V)_T \approx T(\partial P/\partial T)_V \approx T\alpha/\beta)$, the critical constants $(\delta \approx 1.25\,P_c^{1/2})$, etc. A list of solubility parameters[30] is shown in Table 2B2.

TABLE 2B2. Solubility parameters δ (cal$^{1/2}$/ml$^{1/2}$) at 25°.

Solvent	δ	Solvent	δ
Perfluoro-n-hexane	5.9	Benzene	9.2
Perfluorotributylamine	5.9	Chloroform	9.2
Perfluorocyclohexane	6.1	Methylene chloride	9.8
2,2-Dimethylpropane	6.2	1,2-Dichloroethane	9.9
2-Methylbutane	6.8	Carbon disulphide	10.0
Trichlorotrifluoroethane	7.3	1,2-Dibromoethane	10.2
1-Hexene	7.3	Bromoform	10.5
n-Hexane	7.3	Methylene iodide	11.8
n-Heptane	7.4		
Diethyl ether	7.4		
n-Octane	7.5	Solute	δ
Silicon tetrachloride	7.6		
Methylcyclohexane	7.8	Germanium tetrachloride	8.1
n-Hexadecane	8.0	Tin tetrachloride	8.7
Cyclohexane	8.2	Naphthalene	9.9
Ethyl chloride	8.3	Anthracene	9.9
Carbon tetrachloride	8.6	Bromine	11.5
m- or p-Xylene	8.8	Tin tetraiodide	11.7
Ethylbenzene	8.8	Osmium tetroxide	12.6
Mesitylene	8.8	Sulfphur	12.7
Toluene	8.9	Phosphorus	13.1
Ethyl bromide	8.9	Iodine	14.1

It should be noted that the mixing process leading to (2B23) is done at constant volume. The Gibbs free energy change $G_{1,2}^M$ is obtained at constant pressure, but it differs from the Helmholtz free energy $F_{1,2}^M$ for constant volume only by the insignificant terms $(1/2V\beta)(V_{1,2}^E)^2 + \cdots$ (cf. section 2.B.b, equation 2B20), where for the excess entropy a term in the first power of the excess volume was considered negligible). Since $S_{1,2}^E = 0$ for regular solutions

$$G_{1,2}^E \approx F_{1,2}^E = E_{1,2}^E - TS_{1,2}^E = E_{1,2}^E = E^M \qquad (2B28)$$

$$G_{1,2}^E = (x_1 V_1 + x_2 V_2)\varphi_1\varphi_2(\delta_1 - \delta_2)^2 \qquad (2B29)$$

$$\ln f_2 = V_2\varphi_1^2(\delta_1 - \delta_2)^2/RT \qquad (2B30)$$

These equations combine the regular-solution and solubility-parameter concepts. It should be noted that whereas (2B29) is a good approximation to the excess free energy (considering the approximations made in its derivation), similar expressions for the excess entropy and volume or the heat of mixing show much poorer agreement with experiment. This is so, since in the derivation, zero volume change and complete randomness have been assumed, which are inconsistent with finite values for the excess entropy or volume.

The solubility of a solid, 2, in a liquid solvent, 1, can be obtained from (2B30), equating the activities of 2 in both phases

$$\ln x_2 = \ln a_2 \text{ (solid)} - V_2\varphi_1^2(\delta_1 - \delta_2)^2/RT \qquad (2B31)$$

The activity of the solid in (2B31) is referred to the pure liquid 2 as a reference, and hence must be calculated, e.g. from

$$\ln a \text{ (solid)} \approx (\Delta S_F/R) \ln (T/T_F) \qquad (2B32)$$

where ΔS_F is the entropy of melting at the freezing point. The same approach applies to the solubility of a liquid 2, but here often it is not the pure liquid which is at equilibrium with the solution. If the two liquid phases are designated as A and B, the resulting expression is

$$RT[(1/V_1) \ln x_{1B}/x_{1A} - (1/V_2) \ln x_{2B}/x_{2A}]$$
$$= 2 (\varphi_{1B} - \varphi_{1A})(\delta_1 - \delta_2)^2 \qquad (2B33)$$

The critical consolute temperature for two partly immiscible liquids is obtained from regular solution–solubility-parameter theory as

$$T_c = 2(\delta_1 - \delta_2)^2 x_1 x_2 V_1^2 V_2^2 / (x_1 V_1 + x_2 V_2)^3 R \qquad (2B34)$$

or if the volume statistical ideal entropy is used (equation 1A17)

$$T_c = 2(\delta_1 - \delta_2)^2 V_1 V_2 / (V_1^{\frac{1}{3}} + V_2^{\frac{1}{3}})^2 R \qquad (2B35)$$

These equations should be compared with equation (2B16) and the discussion concerning b.

The main criticism on the use of solubility parameters devolves on the inherent uncertainties in their determination (different methods give values in a range of ca. $\pm$ 0.3 units around the accepted value) on the one hand, and the limited total range of their values for practical solvents and compounds on the other. Values of δ range from about 6.25 for isobutane or *neo*-pentane to 12.6 for osmium tetroxide or the *polar* nitromethane, most non-polar liquids to which the theory applies having values in the range of 7.3 to 10.5. Since the square of the difference of the solubility parameters is the important quantity, it is obvious that very often differences between near numbers, of uncertainties comparable to the difference, have to be used. Thus gross errors in calculated solubilities, activity coefficients, etc. and inversions of the order of δ values should be expected, and often occur.

Hildebrand and Scott[34] themselves have pointed out that the use of solubility parameters for polar substances is questionable. Unfortunately, solubility parameters have been used for polar solutes, polar solvents and even hydrogen-bonded solvents such as water, with little discrimination. If some empirical success has been obtained with slightly polar substances such as chloroform or dioxane, use for highly polar systems is not recommended.

d. The Average-Potential Model

An alternative treatment of the intermolecular forces, using a different model, which leads to results not confined to regular solutions, is that of Prigogine[37]. It is based on the average-potential model and on the corresponding-solutions principle. The latter holds for *conformal substances*, which are substances, the potentials of which are related to each and to a reference substance (subscript zero) by

$$U_{ii}(r) = (U_{ii}^*/U_{00}^*)U^0(r\, r_{00}/r_{ii}) \qquad (2B36)$$

where $U^0(r)$ is a universal potential function, e.g. the Lennard–Jones '6–12' law. If the equation of state of the reference substance is $F^0(P, V, T) = 0$, then that of conformal substance 1 is

$$F^0[P(r_{ii}/r_{00})^3(U_{00}^*/U_{ii}^*),\, V(r_{00}/r_{ii})^3,\, T(U_{00}^*/U_{ii}^*)]$$

$$= F^1(P, V, T) = 0 \qquad (2B37)$$

Equation (2B37) is a statement of the *corresponding-states principle*. It should be noted that (U_{ii}^*/U_{00}^*) and (r_{ii}/r_{00}) are scale factors independent of pressure and temperature, so that (2B37) holds also for the critical point, and

$$(P_c V_c/T_c)_1 = (P_c V_c/T_c)_0 \qquad (2B38)$$

The rare gases, having monatomic spherical molecules, are conformal substances. As a good approximation, all non-polar spherical molecules, which have central potentials, will be considered conformal.

For mixtures of solvents which obey the above relationships, it is useful to define normalizing parameters γ, ρ and θ, as follows

$$\gamma = (U_{22}^*/U_{11}^*) - 1,\, \rho = (r_{22}/r_{11}) - 1,\, \theta = (U_{12}^*/U_{11}^*) - \tfrac{1}{2}(\gamma + 2) \qquad (2B39)$$

using solvent 1 as a reference solvent. It has been shown[37] that the excess free energy of the mixture is related to the molar enthalpy of vaporization of the reference solvent H_1^V and to its heat capacity C_{p1}, through the normalizing parameters and equation (2B9) as follows

$$b = H_1^V(2\theta - 18\rho^2) + \tfrac{1}{2}\, T\, C_{p1}(\gamma^2 - 4\theta\gamma(1 - x_1) - 4\theta^2 x_1(1 - x_1))$$

$$- 3RT(19\rho/4 + \theta(2x_1 - 1) + \tfrac{1}{2}\gamma) \qquad (2B40)$$

If $\theta = \gamma = \rho = 0$ the intermolecular forces are identical for the two solvents and the mixture, so that $b = G_{1,2}^E = 0$ and a perfect solution results. In general, however, these factors do not vanish, and five cases may be distinguished:

a) If $\theta = 0$, the interaction energy of the two solvents is the arithmetical mean of the energies U_{11}^* and U_{22}^* of the two pure solvents.

In that case b becomes independent of x_1 (2A40) and the excess function curves are symmetrical.

b) More common is the case where U_{12}^* is the geometrical mean of U_{11}^* and U_{22}^*, where to a good approximation $\theta = -\gamma^2/8$, b depends on x_1 and $G_{1,2}^E$ is unsymmetrical with respect to composition. It also follows from (2B40) that $G_{1,2}^E > 0$, $H_{1,2}^E > 0$, $S_{1,2}^E < 0$ and $V_{1,2}^E < 0$ if the molecules are of the same size ($\rho = 0$).

c) If there is a very strong interaction between two unlike solvent molecules, an addition compound may result, so that $\theta > 0$ and $\gamma \sim 0$. Then $G_{1,2}^E < 0$, $H_{1,2}^E < 0$, $S_{1,2}^E < 0$ and $V_{1,2}^E < 0$.

d) If one of the solvents shows much stronger self-interaction than both the other and the pair, e.g. $U_{11}^* \sim U_{12}^* \ll U_{22}^*$, then $\theta \sim -\frac{1}{2}|\gamma|$.

e) Conversely, if one of the solvents shows a much weaker self-interaction than the other and the pair, e.g. $U_{11}^* \sim U_{12}^* \gg U_{22}^*$, then $\theta \sim +\frac{1}{2}|\gamma|$.

In case (e) as in case (c), all the excess functions are negative, while in case (d) they are all positive. These signs, again, hold if $\rho = 0$. If, however, ρ is permitted to increase, $S_{1,2}^E$ and $V_{1,2}^E$ will change signs at some value, while $G_{1,2}^E$ and $H_{1,2}^E$ will retain their sign (section 2.B.e).

If the solvents consist of polar, polarizable and non-spherical molecules, these relationships must be somewhat modified.

In order to apply equation (2B40), values of θ, γ and ρ are required, as well as of H_1^V and C_{p1} for the reference component 1. The combining rules of $U_{12}^{*2} = U_{11}^* U_{22}^*$ (i.e. $\theta = -\gamma^2/8$) and $r_{12} = \frac{1}{2}(r_{11} + r_{22})$ will be used, so that only ρ and γ have to be determined. They can be obtained from data for the pure solvents 1 and 2: from the second virial coefficient (2A6) (cf. section 2.A.b), the viscosity in the gas phase, the critical temperature, pressure and volume, the heat of vaporization and the temperature dependence of the density. For solvents of molecular weight of above 50, the two former methods do not produce accurate results[37]. The critical data are useful (equations 2B37 and 2B38)

$$1 + \gamma = T_{c2}/T_{c1} = P_{c2}V_{c2}/P_{c1}V_{c1} \tag{2B41}$$

$$(1 + \rho)^3 = V_{c2}/V_{c1} = P_{c1}T_{c2}/P_{c2}T_{c1} \tag{2B42}$$

Any other corresponding temperatures may be used, for example those yielding the same entropy of vaporization or molar volume (cf. section 2.A.b). Table 2B3 gives some values obtained for $1 + \gamma$ and $1 + \rho$ for mixtures of solvents with cyclohexane and with carbon tetrachloride.

Note that all the solvents used are highly symmetrical. Less agreement is expected for unsymmetrical and polar solvents.

TABLE 2B3. Values for the normalizing factors $1 + \gamma$ and $1 + \rho$ for mixtures with $x_1 = 0.5$ for cyclohexane and carbon tetrachloride[29].

Solvent 2	Solvent 1	$1 + \gamma$	$1 + \rho$	$G^E_{1,2}$ (expt.) (cal/mole)	$G^E_{1,2}$(calc.) (cal/mole)
Carbon tetrachloride	Cyclohexane	1.00	0.96	17	24
neo-Pentane	Cyclohexane	0.77	1.00	44	49
Silicon tetrachloride	Cyclohexane	1.07	1.00		5
Benzene	Cyclohexane	1.02	0.92	75	41
Carbon disulphide	Cyclohexane	0.97	0.83		168
Cyclopentane	Cyclohexane	0.92	0.95		46
neo-Pentane	Carbon tetrachloride	0.77	1.04	76	75
Benzene	Carbon tetrachloride	1.02	0.96	21	23
Silicon tetrachloride	Carbon tetrachloride	1.07	1.04	19	20
Cyclopentane	Carbon tetrachloride	0.91	0.99		9

From the value of $G^E_{1,2}$ or of b the activity coefficients of the solvents can be obtained. Thus, whether $\theta = 0$ or not, differentiation of (2B9) and application of the Gibbs–Duhem relation yields (2B22), but if $\theta \neq 0$, b is dependent on the composition.

e. Mixtures Involving Polar Solvents

Mixtures of polar solvents, or polar and polarizable solvents, can be discussed in terms of the average-potential model[37], introducing perturbations to the intermolecular potentials due to dipole interactions. This can be done, provided the dipole interactions are not too strong, i.e. that randomness of mixing can still be assumed. The geometric mean of the pair potentials will be assumed for U^*_{12}, so that $\theta = -\gamma^2/8$ (p. 116).

Consider first a case of non-polarizable solvents ($\alpha_1 = \alpha_2 = 0$), the solvent 2 having a permanent dipole $\mu = \mu_2$, while solvent 1 is non-polar. A perturbation parameter can be defined (cf. equation 2A4) as

$$\Gamma = \mu_2^4/r_{22}^6 U^*_{22} k T \tag{2B43}$$

This leads to an expression for the excess free energy of mixing, or to b of equation (2B9)

$$b = \tfrac{2}{3}\Gamma\left[-H_1^V(1 + \tfrac{1}{2}\gamma - 3x_2\rho) - \tfrac{3}{2}C_{p1}T\gamma - x_1 RT(1 - \tfrac{1}{2}\gamma - 3\rho)\right] \tag{2B44}$$

which can be compared with (2B40) for the unperturbed system. For the special case of $\gamma = \rho = 0$, there are only dipole effects (otherwise perfect-solution behaviour and $b = 0$ would be exhibited), and

$$b = -\tfrac{2}{3}\Gamma(H_1^V + x_1 R T) \tag{2B45}$$

If the two solvents are polarizable, and only solvent 2 has also a permanent dipole μ_2, there are two perturbation parameters

$$\Gamma_{12} = \alpha_1\mu_2^2/r_{12}^6 U_{12}^* \tag{2B46a}$$

$$\Gamma_{22} = \alpha_2\mu_2^2/r_{22}^6 U_{22}^* \tag{2B46b}$$

which, when applied to the free energy equation, lead to

$$b = (2H_1^V + \tfrac{1}{2}R T)[\Gamma_{12} - \Gamma_{22} + (\tfrac{1}{2}\gamma + 3\rho)(\tfrac{1}{2}x_1\Gamma_{12} - \tfrac{1}{2}x_2(\Gamma_{12} - 2\Gamma_{22}))]$$

$$+ (H_1^V + TC_{p1})\gamma(\tfrac{1}{2}\Gamma_{12} - \Gamma_{22}(1 + x_1)) \tag{2B47}$$

If $\gamma = \rho = 0$ and $H_1^V \gg R T$, the ratio of the two effects (pure dipole and dipole with induction)is

$$b \text{ (dipole)}/b \text{ (dipole and induction)} = \mu_2^2/3k T(\alpha_1 - \alpha_2) \tag{2B48}$$

so that the relative importance of the various contributions can be estimated.

If the solubility-parameter theory is extended[38] to include also polarity effects, the C_i terms in equation (2B23) are no longer equal to δ_i^2 but to $\delta_i^2 + \omega_i^2$, where ω_i^2 might be equated[30] with $8\pi N\mu_i^4/9r_i^6 k TV_i'$. If again the geometric mean assumption $\omega_{ij}^2 = \omega_i\omega_j$ is used, inclusion of the dipole interactions in (2B23) and (2B29) yields

$$G_{1,2}^E = (x_1V_1 + x_2V_2)\varphi_1\varphi_2[(\delta_1 - \delta_2)^2 + (\omega_1 - \omega_2)^2] \tag{2B49}$$

As an example, the case of mixtures of chloroform ($\mu_2 = 1.05\,\text{D}$, $\alpha_2 = 8.23 \times 10^{-24}\,\text{e.s.u}$, $r_{22} = 6.10\,\text{Å}$, $U_{22}^* = 4.38 \times 10^{-14}\,\text{erg}$) and carbontetrachloride ($\mu_1 = 0$, $\alpha_1 = 10.5 \times 10^{-24}\,\text{e.s.u}$ $r_{11} = 6.62\,\text{Å}$, $U_{11}^* = 4.38 \times 10^{-14}\,\text{erg}$) can be taken. The experimental excess free energy at $25°$ and $x_1 = x_2 = 0.5$ is $G_{1,2}^E = 25.6\,\text{cal/mole}$. According to the average-potential treatment[37], $\gamma = 0$, $\rho = -0.08$ and $\Gamma = 0.013$, hence the value calculated from (2B44) is $G_{1,2}^E = 17\,\text{cal/mole}$. According to the solubility-parameter treatment $\delta_1 = 8.6$, $\delta_2 = 9.3$, $\omega_1 = 0$, $\omega_2 = 0.52$, and the value of $G_{1,2}^E$ calculated from (2B49) is also $17\,\text{cal/mole}$. Both values can be considered to be in good agreement with experiment.

If one of the solvents in the mixture is capable of hydrogen-bond formation, its self-association will be affected by the admixtures of the second component (cf. section 2.A.d). Whether a solvent associates with itself only, or also with the other solvent, it can be shown that the equilibrium condition requires that the chemical potential of each solvent will equal that of its monomeric unassociated form

$$\mu_i = \mu_{i1} \text{ and } \mu_j = \mu_{j1} \tag{2B50}$$

If solvent i is capable of hydrogen bonding and solvent j not, then in a dilute solution of i in j

$$f_i/f_j = x_{i1}/x_i \tag{2B51}$$

The ratio of the activity coefficients is a thermodynamic property, obtainable from the vapour pressure, for example, while the ratio of monomeric to total concentration of associating solvent i can be obtained spectroscopically (cf. section 2.A.d). The validity of equation (2B51) has been demonstrated for a number of systems: mixtures of carbon tetrachloride with methanol, ethanol or benzyl alcohol, and of *t*-butanol with cyclohexane. In other systems there are deviations, such as in ethanol–carbon disulphide and ethanol–dioxane.

Table 2B1 contains some data for the excess thermodynamic functions for equimolar mixtures. Whenever $G^E_{1,2}$ considerably exceeds $\frac{1}{2}RT$ between the melting point of the solvent mixture and its boiling point (or gas–liquid critical point, if pressure is applied), separation into two immiscible phases occurs. The solvents listed have upper critical consolute temperatures (UCCT). Lower critical consolute temperatures, or closed solubility loops are much rarer, even in highly polar liquid mixtures, except when strong hydrogen bonding is involved, as discussed below for aqueous solutions. A general observation is that the UCCT's of many mixtures of polar solvents, such as nitromethane or acetonitrile, with aliphatic hydrocarbons, are above room temperatures, while those of mixtures with aromatic hydrocarbons are below it. This can be explained according to solubility-parameter theory (extended to polar substances), noting that the former hydrocarbons have δ's in the range 7–8, while the latter have δ's in the range 9–10. The polar solvents would then have δ's above 10, consistent with their relatively high heats of evaporation.

Another generalization that can be made from data such as given in Table 2B1 is that in mixtures of an associated liquid with a non-polar liquid, the excess function curves are highly asymmetric. The excess heat of mixing is always positive, with a maximum at low fractions of the polar component. The excess entropy is always negative, with the minimum at high fractions of the polar component. Consequently, the excess free energy is large and positive, and nearly symmetrical. The heat of mixing measures essentially the number of broken hydrogen bonds, the entropy loss measures the number of hydrogen bonds remaining or being formed in the mixture.

f. Mixed Aqueous–Organic Solvents

What has been said above about mixtures of polar or associated liquids with non-polar liquids is generally true, *a fortiori*, about mixed aqueous–organic solutions. A collection of thermodynamic excess functions is shown in Table 2B4. The following generalizations can be made. Positive excess free energies are the rule, the only exceptions are polyalcohols and polyamines, such as ethylene glycol and ethylenediamine and homologues, which can participate in tridimensional hydrogen bonding. These systems are very stable thermodynamically and never separate into two immiscible phases. For many other systems $G^E_{1,2}$ exceeds $\frac{1}{2}RT$ in the liquid range, so that an UCCT is observable. All the systems also have negative excess entropies of mixing. If these are large enough, a LCCT results (amines) or even a closed solubility loop (cyclic amines, alkoxyl alcohols). The excess heats of mixing are small for nitriles, ketones, ethers and alcohols, either positive or negative. The large negative excess entropies of the amines are accompanied by large negative excess heats. Except for the nitriles and ketones, there are negative excess volumes, i.e. the solvents contract on mixing.

The lower homologues of polar organic solvents are miscible with water at all proportions at all temperatures where the liquid mixtures exist Appendix D. The list includes alcohols, aldehydes, ketones, carboxylic acids, amides and substituted amides, amines, ethers, nitriles and other substances, with up to 4–6 carbon atoms altogether, depending on the functional group and structure. Molecules with more than one functional group permit more carbon atoms in the completely miscible solvent (e.g. polyalcohols, polyamines, polyethers, alcoholethers, etc.), but with too many functional groups, compounds which are solid at room temperature result, and these are not classified as solvents.

TABLE 2B4. Excess functions in equimolar aqueous–organic solutions.

Organic solvent	Temperature (°C)	$G_{1,2}^E$	$-TS_{1,2}^E$ (cal/mole)	UCCT	LCCT (°C)
Methanol	25	90	240	Miscible	
Ethanol	25	180	330	Miscible	
n-Butanol	25	305		125	
i-Butanol				133	
t-Butanol	25	250	245	Miscible	
Diethylamine	49	200	1320		140
Di-n-butylamine					< −2
Methyldiethylamine	47	360	720		49.4
Triethylamine	16	330	880		18.4
Pyridine	80	230	420	Miscible	
3-Methylpyridine	80	246	346		
4-Methylpyridine	80	233	406		
2,4-Dimethylpyridine	80	302	615	189	23
2,5-Dimethylpyridine				207	13
2,6-Dimethylpyridine				231	34
N-Methylpiperidine					48
N-Methylpiperidine					7
2-Methylpiperidine				227	80
3-Methylpiperidine				235	57
Dioxane	25	240	220	Miscible	
Ethyl methyl ketone		$\gg 0$		151	
β-Butoxyethanol				128	49
β-Isobutoxyethanol				150	24.5
Acetonitrile	35	252	52	Miscible	
Propionitrile		> 250		104	

Since solubilities depend on $G_{1,2}^E$, and this in turn on $S_{1,2}^E$ and $H_{1,2}^E$, some of the classifications as completely or partly miscible are marginal, in the sense that a slightly different balance of a negative $H_{1,2}^E$ and a negative $S_{1,2}^E$ would lead to completely different solubilities. Thus the p-dioxane–water system has $G_{1,2}^E$, too small by only 10 % for the formation of a closed solubility loop over a 125° range[39], while butanol and ethyl methyl ketone have a closed solubility loop at high pressures (100 atm)[40], and diethyl ether would show a LCCT if crystallization did not occur before a sufficiently low temperature is reached[41]. The solubility of ether in water in weight per cent increases as the temperature t is lowered, as $11.64 - 0.290\,t + 0.00264\,t^2$. That the pressure affects the mutual solubility of water and an organic solvent strongly is evident also from the effect of electrolytes on the solubility. The salting-out effect has been interpreted in terms of the internal pressure generated[42] (cf. also section 1.F.b). Thus $x = 0.005$ only of hydrogen chloride is sufficient to cause phase separation in dioxane–water mixtures[39], 1.2 % by weight of potassium chloride for *t*-butanol–water and 6.6 % for n-propanol–water at 60°C, and 9.3 % for acetone–water at 37° or above[42].

The mutual solubilities of water and some solvents which are not completely miscible are also listed in Appendix D. Although water is polar, and strongly associated, it is possible to assign to it a solubility parameter, even though the solubility-parameter theory does not apply to such solvents. (See also comments below.) From the solubility of water in paraffins, aromatic hydrocarbons and their halogen-substituted compounds[30, 43, 44], a value $\delta_{H_2O} = 24.5$ has been deduced. For this the solubility in ml of water per ml of solution, s_W, is convenient, since from (2B17) and (2B33), the simple relationship

$$\ln s_W + 1 - V_W/V_S + V_W(\delta_S - \delta_W)^2/R\,T = 0 \qquad (2B52)$$

follows, where S denotes the organic solvent.

The value $\delta_{H_2O} = 24.5$ (or 24.0[44]) agrees well, but fortuitously[44] with the value calculated from the heat of vaporization, 23.8. It is, however, not at all compatible with the complete miscibility of water and ethanol on the one hand, and of ethanol and benzene on the other. Thus a solubility parameter for hydrogen-bonded solvents such as water or ethanol seems to be of very little value. The solubility of water in the organic solvent is also a measure of the interaction of water with the solvent. In solvents where the solubility is small, so that no interaction with the solvent is expected, self-association of the water is still possible. However, this does not usually occur, and in benzene, where the solubility of water is 0.0349 M at 25⁰, it has been shown that the concentration of even dimers must be very small[45]. The solubility of water has been found to be linear with its activity, in an isopiestic experiment, so that monomers should predominate. The opposite solubility, of benzene in water, has also been measured recently with high precision over a broad temperature range[46], and as usual, is smaller than that of water in benzene. Solubilities of many oxygenated solvents are discussed in section 9.A.

The thermodynamic functions and the mutual solubilities of water and organic solvents can be interpreted in terms of hydrogen bonding and basicities[47]. In the organic-solvent-rich phase, or part of the diagram, the hydrogen bonds between dissolving water molecules will break, causing a positive enthalpy and a positive entropy change. If, however, the solvent is sufficiently basic to interact with the water, association between the two by formation of new hydrogen bonds occurs, and the enthalpy change is less positive, and may become negative if sufficiently strong bonds are formed. Ketones and nitriles give positive H^E, alcohols and amines negative H^E in this region. On the other hand, in the water-rich portion or phase, the addition of solvent breaks few hydrogen bonds, since it can be accommodated interstitially unless the molecules are too large. Thus S^E is always negative in this region. Negative H^E is also observed in this region if hydrogen bonds between water and the solvent are formed, but not for weak proton acceptors, such as ketones or nitriles, where H^E is still positive.

The fraction of hydrogen bonds broken or being formed is obtained generally from spectrophotometry at the infrared stretching frequencies of the O—H$\cdots$O bonds[48,49], or from n.m.r. chemical shifts[49]. Because of the strong infrared absorption by water, aqueous mixtures have not been studied very widely by this technique.

Two further effects besides the dissociation of the hydrogen-bonded water structure and the increase in the fraction of unbonded water must be considered. One is the hydrophobic bonding effect, which has been discussed by Nemethy and Scheraga[50, 51]. Consider the first water layer around a hydrocarbon molecule: it contains $Y^C = 12 + 2n_C$ water molecules, where n_C is the number of carbon molecules in an alkane chain. The energy of these water molecules is raised, since instead of being near bondable unbound water molecules, they are near the hydrophobic hydrocarbon. When two hydrocarbon chains meet, the water molecules in the space between them are pushed aside and returned to the bulk water, the free energy contribution to the hydrophobic bond formed being $Y^C(G_W^0 - G_W^C)$. The total excess free energy of the solution is

$$G^E = (G_{HR} - G_{HR}^0) + Y^C(G_W^C - G_W^0) \qquad (2B53)$$

where G_{HR} for the hydrocarbon in the solution includes terms for the self-interaction energy, the configurational entropy and the dispersion interaction with first water layer, while the second term pertains to the first water layer. The excess volume can also be calculated according to

$$V^E = -(43/24)\, Y^C x_4^C \qquad (2B54)$$

where x_4^C is the fraction in the water layer of the water molecules which are bonded to four neighbours. Not only hydrocarbons should follow this behaviour, but also hydrocarbon chains to which functional groups are attached which fit into the water structure, such as alcohols. Indeed, the (negative) excess volumes for several alcohols in dilute aqueous solutions have been measured[52], and compared with the prediction from (2B54), using known x_4^C values for hydrocarbons[53]. Generally good agreement has been obtained for the difference of V^E between two alcohols, which may be ascribed to the effect of the hydrophobic CH_2 groups, the interaction of the hydroxyl groups with water being assumed to be equal and to cancel out the difference.

The total magnitude of the effects, however, depends on another factor: the direct associative interaction of the solvent and the water. This effect is studied best in ternary systems, i.e. in systems where the water and the reacting solvent are diluted by a relatively inert diluent. Thus the solubility of water in mixtures of alcohols with benzene has been measured[54]. The solubilities conform to the expression $\log x_W = A - B/T$, where A and B depend on the nature of the alcohol and its mole fraction. The data can be interpreted by assuming the water to dissolve both in a 'free' state, according to the mole fractions of the two solvents, and as a 'complex' with the reactive solvent[55]

$$x_W = (1 + K_{AW} x_A)(s_A x_A + s_B x_B) \qquad (2B55)$$

where subscripts W, A and B refer to water, alcohol and benzene, K_{AW} is the association constant of the 1:1 water-alcohol complex and s_A and s_B are the solubilities of the 'free' water in the respective solvents. The quantity s_B is obtained from independent data, and K_{AW} and s_A can be calculated from the data, which for methanol at 25°, for example, conform to (2B55) within about 6%. The data for the solubility of water in ether solutions in benzene[55] can be interpreted similarly, with subscript E replacing A in the equation. The constant K_{EW} can be obtained independently from measurements of dipole moments. If $\mu_W = 1.51$ D is the dipole moment of water, $\mu_{W(E)} = 1.99$ D the effective moment of water measured in ether solution, and $\mu_{EW} = 2.44$ D that of the complex, calculated from the bond moments and the geometry, it was shown[55] that $K_{EW} = (\mu_{W(E)}^2 - \mu_W^2)/(\mu_{EW}^2 - \mu^2{}_{W(E)}) = 0.84$.

Ternary systems involving water and two organic solvents have received a great deal of attention, and a large body of experimental information is scattered in the literature which cannot be reveiwed here. Only one further system will be discussed, illustrating the capacity of a component, miscible separately with the other two, to bring about their mutual miscibility. In the water–ethyl methyl ketone system[56] there are two phases, the aqueous phase containing 12.6% wt. of ketone and the organic phase containing 24.8% wt. water. Addition of 12.6% wt. of acetone causes complete miscibility of the two phases.

There is as yet no fundamental comprehensive theory that relates the properties of aqueous–organic solvent mixtures to the molecular properties of the components and of the hydrogen-bonded associate between them. The hydrophobic-bonding concept is a step in this direction, but does not as yet permit the calculation of excess thermodynamic functions from the properties of the solvents and a general combining law, such as the geometric-mean law, as in the average-potential model. As far as this model is concerned, $\theta \geqslant 0$, and must be introduced as a separate parameter, obtainable perhaps from the dipole moments, as briefly mentioned above. Equations such as (2B44), (2B47) or (2B49) cannot be used, since they pertain only to solvents with moderate dipole moments, and do not take into account the hydrogen bonding.

Thus, properties of these mixtures such as density surface tension viscosity, refractive index. etc., although relatable by appropriate expressions to thermodynamic quantities such as excess free energies, and relatable through the expressions derived above (sections 2.B.c, d and e) to the molecular properties of the pure components and to the composition, cannot generally be calculated *a priori*. Except for the density (through the excess volume) and perhaps the dielectric constant, very little progress has been made in this direction. Again, there exists a wealth of emprical information scattered in the literature, which cannot be reviewed here. Often it is a good approximation to consider the property to be linear with the composition, expressed by the mole fractions, or if the molar volumes of the two components are very disparate, by the volume fractions (equation 2B17). In the following paragraphs, only one property will be treated further: the dielectric constant.

g. The Dielectric Constant in Mixed Solvents

The dielectric constant of dilute solutions of a polar liquid in a non-polar liquid has been the subject of many studies, in connection with the estimation of the dipole moment of the polar liquid. Generally, the

Debye equation or a modification of it is employed, although it is not strictly correct, neglecting the internal field[57]. The dielectric constant of the mixture, ε, is related to the polarizabilities α_i and dipole moments μ_i of the components according to

$$(\varepsilon - 1)/(\varepsilon + 2) = (4\pi/3) \sum n_i(\alpha_i + \mu_i^2/3\,k\,T) \qquad (2B56)$$

where n_i is the number of i molecules per unit volume. Using the relationship $n_i = N(\rho/\tilde{M})x_i$, where ρ is the density of the solution and $\tilde{M}$ the mean molecular weight, and the Mossotti relationship for the molar polarization, P, then for a mixture of a polar component 2 in a non-polar one 1 ($\mu_1 = 0$),

$$P = x_1 P_1 + x_2 P_2 = V(\varepsilon - 1)/(\varepsilon + 2) \qquad (2B57)$$

$$P = (4\pi/3)N[(1 - x_2)\alpha_1 + x_2(\alpha_2 + \mu_2/3kT)] \qquad (2B58)$$

For the non-polar component, $P_1 = V_1(\varepsilon_1 - 1)/(\varepsilon_1 + 2) = (4\pi/3)N\alpha_1$, and it is possible to define an ideal molar polarization for the polar component

$$P_2^0 = (P - (1 - x_2)P_1)/x_2 = (4\pi/3)N(\alpha_2 + \mu_2^2/3k\,T) \qquad (2B59)$$

from which the dipole moment μ_2 can be calculated, for example by extrapolating x_2 to zero, according to Debye, or from Guggenheim's treatment[58]

$$P_{2(A+O)}^0 = (3M/x_2\rho)[(\varepsilon - n_\infty^2)/(\varepsilon + 2)(n_\infty^2 + 2) -$$
$$(\varepsilon_1 - n_{1\infty}^2)/(\varepsilon_1 + 2)(n_{1\infty}^2 + 2)] + (V_2/V_1)P_{1(A)} \qquad (2B60)$$

where $P_{2(A+O)}^0$ and $P_{1(A)}$ are the atomic and orientational, and the atomic components of the polarizations, and n_∞ is the refractive index extrapolated to infinite wavelength. If the internal fields are taken into account, then Böttcher's formula[57]

$$\frac{(\varepsilon - 1)(2\varepsilon + 1)}{9\varepsilon} = \frac{4\pi}{3} \sum \frac{n_i}{1 - f_i\alpha_i}(\alpha_i + \mu_i^2/3k\,T(1 - f_i\alpha_i))$$
$$\qquad (2B61)$$

where $f_i = (1/\alpha_i)^3(2\varepsilon - 2)/(2\varepsilon + 1)$, results.

It is always possible to reverse the procedures, use dipole moments determined in the gas phase, and employ (2B56) or (2B61) for mixtures of polar molecules, or (2B58) for a mixture of a polar and a non-polar solvent, to calculate the dielectric constant of the mixed Solution. The Debye treatment implies the volume-fraction additivity rule for the dielectric constant expressions

$$(\varepsilon - 1)/(\varepsilon + 2) = \varphi_1(\varepsilon_1 - 1)/(\varepsilon_1 + 2) + \varphi_2(\varepsilon_2 - 1)/(\varepsilon_2 + 2) \quad (2B62)$$

or the mole-fraction additivity of the molar polarization, equation (2B57), provided $V^E = 0^{58}$. Empirically, various additivity rules for the dielectric constants themselves have been proposed[59],

$$\varepsilon = \varphi_1\varepsilon_1 + \varphi_2\varepsilon_2 \qquad (2B63a)$$

$$= x_1\varepsilon_1 + x_2\varepsilon_2 \qquad (2B63b)$$

$$= w_1\varepsilon_1 + w_2\varepsilon_2 \qquad (2B63c)$$

Akerlöf[60] has determined the dielectric constant of a number of aqueous-organic solvent mixtures (with methanol, ethanol, the propanols, t-butanol, glycol, glycerol, acetone and dioxane); those for mixed organic solvents (e.g. of various pairs among methanol, ethanol, i-propanol t-butanol, acetone and dioxane[61]) have also been measured[59] (Table 2B5). For the aqueous-methanol mixture, for instance, (2B62) or the similar harmonic approximation $1/\varepsilon = \varphi_1/\varepsilon_1 + \varphi_2/\varepsilon_2$, give values approximately 20% too low, (2B63b) gives values about 12% too high, (2B63a) gives values about 4% too low, but the weight-fraction equation (2B63c) gives values deviating only about + 0.7% from the experimental data. Similarly, for dioxane–water mixtures below 50% wt. dioxane, ε follows (2B63c) strictly, and for most other systems[60, 61] approximately, over a wide concentration range.

Mixtures of such polar solvents as alcohols, water, acetone, etc., are indeed not expected to follow Debye's equation. Onsager[10] proposed that $(\varepsilon - 1)/(\varepsilon + 2)$ in Debye's equation should be replaced by $(\varepsilon - n_\infty^2)(2\varepsilon + n_\infty^2)/\varepsilon(n_\infty^2 + 2)$, while Kirkwood[11] showed that n_∞^2 in Onsager's equation can be safely replaced by 1 for polar liquids, yielding the expression $(\varepsilon - 1)(2\varepsilon + 1)/9\varepsilon$. Including this expression in (2B62) and rewriting gives

$$\varepsilon = \varepsilon_1 - ((\varepsilon_1 - 1) - (\varepsilon_2 - 1)(2\varepsilon_2 + 1)/2\varepsilon_2)\varphi_2 = \varepsilon_1 - \delta_2c_2 \quad (2B64)$$

TABLE 2B5. Dielectric constants of some solvent mixtures.

Solvent A	% Solvent B				Solvent B	References
	0	30	70	100		
p-Dioxane	2.2	7.6	15.0	20.7	Acetone	a
		6.6	16.1	24.3	Ethanol	a
		9.2	23.0	32.6	Methanol	a
		17.7	51.8	78.4	Water	a
Benzene	2.3	6.7	14.3	21.1	Acetone	b
		9.5	23.3	33.6	Methanol	b
		9.5	22.5	35.8	Nitrobenzene	b
		11.3	25.3	36.8	Acetonitrile	b
Chloroform	4.8	3.9	2.9	2.2	CCl$_4$	b
		11.2	17.2	21.1	Acetone	b
t-Butanol	12.5	13.2	16.7	20.7	Acetone	a
		17.6	21.1	24.3	Ethanol	a
		20.0	27.2	32.6	Methanol	a
		21.4	52.6	78.4	Water	a
Acetone	21.1	16.0	10.1	6.2	Ethyl acetate	b
		20.4	21.9	24.3	Ethanol	a
		24.2	29.4	32.6	Methanol	a
		25.1	31.0	35.8	Nitrobenzene	b
		35.7	61.0	78.4	Water	a
Ethanol	24.3	26.6	30.1	32.6	Methanol	a
		38.0	61.1	78.4	Water	a
Benzonitrile	25.7	18.8	9.5	2.6	Triethylamine	b
		29.0	33.5	36.8	Acetonitrile	b
Methanol	32.6	33.1	33.6	35.8	Nitrobenzene	b
		45.0	64.3	78.4	Water	a
Water	80.4	69.1	50.0	36.8	Acetonitrile	b

a. % weight at 25°C, ref. 60,61.
b. % volume at 20°C, ref. 59.

where δ_2 is the molar dielectric decrement for dilute solutions of the less polar component 2 in the highly polar component 1, e. g. water[62]. In the derivation, 1 has been neglected compared with 2ε or $2\varepsilon_1$, and c_2 has been substituted for $1000 \, \varphi_2/V_2 = 1000 \, x_2/V$. If also $2\varepsilon_2$ is large compared with unity, Butler's relation[63] for δ_2 is obtained

$$\delta_2 = \lim_{c_2 \to 0} (\mathrm{d}\varepsilon/\mathrm{d}c_2) = V_2(\varepsilon_1 - \varepsilon_2)/1000 \qquad (2B65)$$

Values of δ_2 calculated for aqueous methanol (1.8), ethanol (3.1), n-propanol (4.2) and acetone (4.1) can be compared with the observed values 1.4, 2.6, 4.0 and 3.2 respectively[62].

Self-association in highly polar solvents has been taken into account through the parameter g (cf. section 2.A.b), and this has been extended to mixtures[64], using the expression

$$(\varepsilon - 1)(2\varepsilon + 1)V/9\varepsilon = (4\pi N/3)(\alpha_1 + \alpha_2 + (g_1\mu_1^2 + g_2\mu_2^2)/3kT) \quad (2\text{B}66)$$

where, however, no cross term in $g_{12}\mu_1\mu_2$, is used, contrary to the expectation for highly mutually associated liquids. Progress in the description of the dielectric constant of such mixtures may perhaps be made when the g factors of the components and the mixture are considered in further detail. For pure alcohols, good agreement with experimental values was obtained[65] by considering them to be mixtures of finite-length orientable chains of n members. If a common association constant K per link exists, and if α is the fraction of the monomer $\alpha = c_1/c$, it was shown that the mean value of g is given by $\tilde{g} = 2.39 - 2.00 \, \alpha^{1/2} + 0.61 \, \alpha$. If the chains were considered infinite, then the value $g = 1 + 2f\cot^2(\phi/2)$ is obtained[66], where $f = (\mu_{OH} + \mu_{OR}) (\mu_{OH} + \mu_{OR} \cos \theta)/(\mu_{OH}^2 + \mu_{OR}^2 + 2\mu_{OH}\mu_{OR} \cos \theta)$ and θ is the angle between the two dipoles along the bonds O—H and O—R. For water $f = 1$ and $g = 1 + Z \cos^2(\theta/2)$, where Z is the average number of nearest neighbours[66]. Fairly good agreement with the experimental values of g is obtained, but the consideration of a finite number of members in the hydrogen-bonded alcohol chain is a definite improvement. It is, however, not obvious how a value of g is to be obtained for, say, alcohol–water mixtures, where both tridimensional and one-dimensional association occurs, and how to apply an average g to the other quantities in the formula relating dipole moments and polarizabilities to the dielectric constant, equation (2B66).

Finally, the polarization properties of a mixture have been related to their thermodynamic properties[67, 68]. In a mixture of a polar component 2 with a non-polar component 1, it was shown that the empirical relation

$$R \, T \ln f_2 = - N\mu_2^2 d^{-3}[(\varepsilon - 1)/(2\varepsilon + 1) - (\varepsilon_2 - 1)/(2\varepsilon_2 + 1)] \quad (2\text{B}67)$$

holds, where d is disposable length parameter. It has been applied[69] to aqueous t-butanol solutions, and found to hold over a certain composition range, provided the alcohol–water association products are short-lived compared with the orienting units in the more polar component, water, itself. As long as the relationship (2B67) holds, ε of the mixture can be estimated from the activity coefficient f, if d is obtained from the value at one composition.

C. ELECTROLYTE SOLUTIONS IN ORGANIC SOLVENTS

a. Introduction

The previous sections dealt with solutions of one non-electrolytic substance — mainly an organic solvent — in another organic solvent. Whether the first substance has the attributes of a solvent, or is a solid (such as iodine), a liquid (such as silicon tetrachloride) or a gaseous non-electrolyte (such as methane), is largely immaterial, since the general relationships are obeyed in every case, for example according to the solubility-parameter or average-potential concepts. Polar non-electrolytic solutes show deviations from these relationships, and electrolytes are expected to show quite different behaviour, due to ionic dissociation.

Ionic dissociation, leading to electrical conductivity, is the most important distinguishing feature of electrolyte solutions, and this makes the dielectric constant of the solvent a property of prime importance. The dielectric constants of solvents are listed in appendix C, and those of some mixtures in Table 2B5. Ionic dissociation and conductance are further discussed in sections 2.D and 3.A.

Electrolytes, being extreme cases of polar substances, are expected to show low solubilities in non-polar organic liquids according to the trends discussed in section 2.B, and may show low solubilities even in polar solvents. This statement refers mainly to such electrolytes as are made up from rare-gas-like ions, typified by sodium chloride. Electrolytes made up from strongly coordinating cations and preferably large anions, typified by cobalt thiocyanate, may be much more soluble in polar organic solvents, and those made up of at least one large organic ion (such as tetrabutylammonium bromide or potassium tetraphenylboride) will be appreciably soluble even in solvents of relatively low dielectric constant and polarity. Before discussing in detail the properties of organic electrolyte solutions, it is therefore necessary to ascertain the solubility relations of electrolytes of different types in the various kinds of organic solvents.

b. Solubility in Inert Solvents

Inert solvents for the present purpose are solvents which do not possess donor atoms capable of coordination to metal ions, i.e. solvents which are not basic in the Lewis sense. They may be slightly polar (such as chloroform or toluene), and may interact with solutes through π-electron systems. Thus aliphatic and aromatic hydrocarbons and their halogen-substituted homologues are included in this category.

The literature contains some data on the solubility of salts in such solvents, but not very many. The data available for benzene and chloroform are summarized in Table 2C1. The solubilities are expressed in

TABLE 2C1. Molal solubilities of salts in benzene and chloroform.

Salt	10^3s Benzene	Temp. (°C)	10^3s Chloroform	Temp. (°C)	Reference
$AlCl_3$	9.0	17	3.6	25	a
$AlBr_3$	3260.	20			a
$FeCl_3$					
$CoCl_2$	1.6	25			b
$NiCl_2$	2.3	25			b
CdI_2	2.6	35			a
$InCl_3$			7.0	20	b
$InBr_3$	1.4	20	9.0	20	b
InI_3	13.5	20	20.1	20	b
SnI_4	298.	25	143.	28	a
SbF_3	.035	25			a
$SbCl_3$	1920.	20			a
$SbBr_3$	825.	35			a
$TaCl_5$			107.	20	a
$HgCl_2$	22.	25	3.7	25	c,d
$HgBr_2$	21.	25	4.4	19	c,d
HgI_2	6.1	25	3.5	19	c,d
$AgClO_4$	251.	25			b
$AgNO_3$	1.3	30			a
$Th(NO_3)_4$	0.4	25	0.2	25	b
HCl	524.	20	389.	20	d
NaN_3	15.4	80			a
KN_3	18.5	80			a
$(CH_3)_4NN_3$	34.5	20			a
$(C_4H_9)_4NBr$	>1400.	25			e
$(C_4H_9)_4NSCN$	>5080.	25			e
NH_4IBr_2			5.6	25	a
$NH_3(CH_3)IBr_2$			0.9	25	a
$NH_2(CH_3)_2IBr_2$			0.026	25	a
$NH(CH_3)_3IBr_2$			1.7	25	a
$N(CH_3)_4IBr_2$			0.026	25	a
$N(C_2H_5)_4Cl$			465.	25	a
$N(C_2H_5)_4Br$			1140.	25	a
$N(C_2H_5)_4I$			60.	25	a
			980.	25	a
$N(C_3H_7)_4I$			1730.	25	a
$(C_6H_5)_4AsClO_4$			1.04	30	e
$(C_6H_5)_4AsMnO_4$			1.06	30	e
$(C_6H_5)_4AsReO_4$			1.60	30	e

a. H. Stephen and T. Stephen (Eds.), *Solubilities*, Pergamon Press London, 1963.

b. A. Seidell and W. F. Linke, *Solubilities*, 3rd ed-, Van Nostrand, New York, 1952.

c. I. Eliezer, *J. Chem. Phys.*, 42, 3625 (1965).

d. R. P. Bell, *J. Chem. Soc.*, 1371 (1931).

e. M. Stiller, *M. Sc. Thesis*, Hebrew Univ. of Jerusalem, 1964.

TABLE 2C2. Solubility of hydrogen chloride in inert solvents.

Solvent	m_{HCl}	c_{HCl}	$100x_{HCl}$	$100\varphi_{HCl}$
n-Hexane	0.233	0.152	1.97	0.72
n-Octane	0.267	0.188	2.96	0.81
n-Dodecane	0.190	0.143	3.14	0.58
Cyclohexane	0.218	0.143	1.80	0.63
Benzene	0.524	0.460	4.25	1.52
Toluene	0.580	0.495	5.04	1.50
Chloroform	0.389	0.575	4.44	1.76
Carbon tetrachloride	0.120	0.189	1.81	0.59
Chlorobenzene	0.431	0.477	4.65	1.45
Bromoethane	0.300	0.430	3.19	1.31

Calculated from data of R. P. Bell, *J. Chem. Soc.* 1371 (1931).

terms of molalities, i.e. moles per kg of solvent (12.8 moles of benzene and 8.37 moles of chloroform), so that mole fractions in the saturated solutions can be readily calculated. The mole-fraction scale is required for the comparison of the solubility of a solute in several solvents. Since however the molecular weights of the common inert solvents vary only from 78 for benzene to 170 for dodecane, molal solubilities give a sufficiently clear view of solubility relationships in a great variety of solvents, at least for moderately soluble solutes.

The solubility of hydrogen chloride in a variety of solvents[70] in four different concentration scales (m, c, x and φ) is shown in Table 2C2, for comparing the significance of variables such as the molecular weight and volume of the solvent and solute. The molar volume of hydrogen chloride at 20°C is taken as 30.8 ml, the value at the boiling point. In the case of hydrogen chloride, the difference, between various inert solvents are small. The differences are larger in the case of mercury(II) chloride[71], where there is a definite preference (a hundred fold) for aromatic or unsaturated solvents over saturated aliphatic solvents. In some cases, the effect of even the slight change from benzene to toluene is very great. The solubility of silver perchlorate is 0.251 m in benzene and 4.88 in toluene at 25°C[72], and a somewhat smaller effect is shown by antimony(III) bromide (0.825 and 6.28 m respectively). Other highly soluble salts are antimony(III) chloride, aluminium bromide, tetraalkylammonium bromides, iodides and thiocyanates, etc.

The large solubilities observed in some cases are explained by direct interactions of the metal cation with the solvent, usually an aromatic hydrocarbon, as discussed in detail e.g. for silver perchlorate[73, 74] and the mercury halides[71]. In other cases, the salts have a large degree of covalency, and can be treated as non-electrolytes, e.g. according to the solubility-parameter concept (germaniun and tin tetrahalides, cf. section 2.B.c). The solubility relationships among the more electrolyte-like (i.e. ionically dissociating) salts, such as alkylammonium or similar salts, are more difficult to explain. Solvation in the ordinary sense is of course absent in the inert solvents (by definition, so to say), but enthalpy can still be released by dipole interactions (ΔH_{dip}) — some of the solvents being highly polarizable — and by hydrogen bonding, as for example with chloroform or trialkylammonium salts ($\Delta H_{\text{H bond}}$). On the other hand, work must be done against the cohesive energy of the solvent in order to introduce the solute (ΔH_{hole}) and against the lattice energy of the solute in order to disperse it (ΔH_{sub})

$$\Delta G \text{ (solution)} = -T\Delta S - \Delta H_{dip} - \Delta H_{\text{H bond}} + \Delta H_{sub} + \Delta H_{hole} \quad (2C1)$$

The hole-formation energy can be estimated from the macro surface tension of the solvent, σ, and the hole surface area, i.e. $4\pi r^2$, where r is the radius of a molecule of solute: $\Delta H_{hole} = 4\pi N\sigma r^2$ per mole of solute. Thus the smaller the solute, the easier will be its dissolution in a structured inert solvent[76]. The lattice energy for a solute dissolving in a low-dielectric-constant solvent, where no ionic dissociation occurs, is represented by the sublimation energy of the salt, ΔH_{subl}[76]. Solubility of salts in mixtures of solvents will be governed by the enthalpy terms ΔH_{dip}, $\Delta H_{\text{H bond}}$ and ΔH_{hole}, the first of which is expected to be linear with the composition, and by any losses in translational entropy occuring because of interaction with the solvent.

Some of the solutes discussed here are soluble in water or in aqueous acid solutions with little change, except solvation and perhaps some ionic dissociation. Since the distribution coefficient of a distribuend is closely related to the ratio of its solubilities in the two phases (section 7.B.b), the relative solubilities in different organic solvents can be directly related to the relative values of the distribution coefficients for these solvents and a given aqueous solution. These quantities can thus be calculated from each other. The solubilities of nitric acid, obtained from data at 14 M HNO_3 relative to benzene = 1, are toluene = 1.17, dodecane = 0.37, carbon tetrachloride = 0.068, kerosene and cyclohexane = 0.017[77], and those of iron(III) chloride (or of $HFeCl_4$?), tobained from data at 4.856 M HCl and trace iron, relative to benzene=1, are carbon tetrachloride = 0.73, carbon disulphide = 1.2, chloroform = 4.0

and 1,2-dichloroethane = 147[78]. The distribution of solutes between aqueous solutions and inert solvents is discussed more extensively in section 7.B.

The solubility of metal chelates in inert organic solvents is the basis of their extractive chemistry, as discussed in detail in section 8.A. Although the chelates are not counted ordinarily as electrolytes, and their solubility relationships could perhaps best be described in terms of the concepts developed in section 2.B, solubility parameters usually are unknown. A more qualitative discussion in terms of equation (2C1) is therefore helpful. The same holds for the important groups of very polar ion pairs made up of a large organic ion and an inorganic one[79], either simple or complex, e.g. trilaurylammonium tetrachloroferrate(III). The application of the solubility-parameter concept to such systems[80], as well as to chelates[81] has been attempted, but must be considered to be empirical. Unfortunately, there is too little information available concerning most solute–solvent combinations, to make an evaluation of the terms in (2C1) possible.

c. Solubility in Active Solvents

Active solvents differ from inert solvents by being so polar that they permit strong ion–dipole interactions, having in many cases atoms with donor properties, so that they can coordinate and solvate metal ions. Oxygen and nitrogen atoms with unshared electron pairs are the most important donor atoms. The solvation reaction can be envisaged as a generalized acid–base (in the Lewis sense) reaction. Solvation is thus one of the main features of electrolyte solutions in active solvents.

Active solvents, being usually highly polar, have often appreciable dielectric constants, so that some ionic dissociation occurs (e.g. methanol, $\varepsilon = 32.6$, acetone, $\varepsilon = 20.7$, acetonitrile, $\varepsilon = 36.2$). On the other hand, active solvents may have quite low dielectric constants (e.g. dioxane $\varepsilon = 2.2$, 2-methylbutanol $\varepsilon = 5.8$, triethylamine $\varepsilon = 2.4$), so that electrolytes dissolved in them would be only slightly dissociated. The consequences of electrolytic dissociation, which is an important but not general feature of solutions in active solvents, are discussed in section 2.D.

An early discussion of the solubility of electrolytes in non-aqueous solvents has been given by Walden[82], who also compiled many of the data then available. These data, with a few further ones obtained from standard compilations[83] are shown for some representative salts in Table 2C3 for methanol, ethanol, acetone, diethyl ether, acetonitrile, pyridine and *N*-methylacetamide[84]. Some more recent solubility data in methanol[85, 86], acetonitrile[85] and ethylene glycol, diaminoethylene and ethanolamine[87] should be mentioned.

Katzin[76] discusses three characteristic features in which organic solutions of electrolytes differ from aqueous solutions:

a) In organic solutions, cations and anions are often covalently associated to give a neutral molecular species, while in aqueous solutions hydrated cations and anions interact electrostatically, mainly. The covalent interaction is exhibited for instance in mixtures of cobalt(II) perchlorate and lithium bromide solutions in isopropanol, which have absorption bands in the region 210–350 mμ, absent in the separate solutions of the components.

b) There is a definite coordination of anion ligands, solvent molecules, and, if present, water molecules, around the metal cation in organic solutions. The coordination of the Lewis bases, whether anionic or neutral, is essentially covalent.

c) The solubilities of different coordination configurations (e.g. octahedral and tetrahedral) with a given ligand anion and solvent can be quite different in organic solvents.

The dissolution of a salt in an organic solvent is furthered by a favourable entropy and heat of solution. The former is, of course, always positive, the latter is within the range of $\pm$ 10 kcal/mole. The heat of solution depends strongly on whether the salt is anhydrous, a hydrate or a solvate with the solvent employed, as does the solubility itself. Thus the solubility of anhydrous lithium perchlorate in diethyl ether is 53.21 weight % (0.44 mole fraction), while that of the trihydrate is only 0.196 weight %[88]. On the other hand, the solubility of anhydrous thorium nitrate[89] or cobalt chloride[90] in organic solvents such as diethyl ether, acetone or dimethyl formamide is slight, compared with the appreciable solubilities of the hydrates in these or other solvents[90, 91]. Data for the heats of solution of anhydrous salts in organic solvents are very scarce, but those for several hydrates of cobalt chloride[90], and for cobalt[92], thorium[93] and uranyl[94] nitrates, and potassium, tetraethyl- and tetrapropylammonium iodides[95], have been reported. The results for the heavy-metal salts were summarized and compared by Katzin[76], who concluded that there is a correlation of ΔH^0_{soln} with the base strength of the solvents, although this correlation in far from perfect, and the order of solvents is not the same for the four salts.

The heat of solvation of the ions by the solvent, which must over-come the lattice energy of the salt, is included as an important term in the heat of solution. With organic solvents, where solvation is less strong than with water (they are weaker Lewis bases), the extra energy is given by the cation–anion bond formation. Thus whereas lattice energies of 140–240 kcal/mole are involved in the dissolution of the alkali halides in water, only sublimation energies of 45–70 kcal/mole are involved in their dissolution in non-dissociating organic solvents. The differences are much higher with the more highly charged salts. The process can thus be envisaged as

$$C^+A^- \text{ (cryst)} \xrightarrow{-\Delta H_{subl}} C^+A^- \text{ (vap)} \xrightarrow{\Delta H_{solv}} C^+A^- \text{ (soln)} \quad (2C2)$$

compared with the process for aqueous solutions (or solvents of high dielectric constant)

$$C^+A^- \text{ (cryst)} \xrightarrow{-\Delta H_{latt}} C^+A^- \text{ (gas)} \xrightarrow{\Delta H_{hydr}} C^+ \text{ (aq)} + A^- \text{(aq)} \quad (2C3)$$

Only very strong Lewis base solvents can supply the heats of solvation balancing ΔH_{subl} of at least 50 kcal/mole of metal fluorides, so these salts are insoluble in most organic solvents. The same is true for the other alkali metal, alkaline earth and divalent transition metal halides (except for beryllium), which have $\Delta H_{subl} = 45$ kcal/mole. Other salts, such as zinc(II), mercury(II), tin(II), aluminium(III), indium(III), antimony(III), titanium(IV) and zirconium(IV) chlorides, bromides and iodides, as well as cadmium(II) and gallium(III) iodides, have $\Delta H_{subl} < 35$ kcal/mole, and are indeed soluble in solvating organic solvents (see Table 2C3). The heat of solvation in the process of (2C2) is however not the same for all salts in a given solvent, since it depends directly on the number of solvent molecules coordinated. The higher this number, the higher the expected solubility.

Heats of sublimation of hydrated salts are unknown, so that the considerations applied above to the anhydrous salts cannot be used. In a qualitative manner, however, lower hydrates may be expected[76] to show higher solubilities than both higher hydrates or anhydrous salts, because of favourable coordination conditions, i.e. covalent bonding of

TABLE 2C3. Several molal solubilities in organic solvents at room temperature.

Salt	Metha-nol	Etha-nol	Acetone	Ethyl ether	Aceto-nitrile	Pyri-dine	NMA**
LiCl	10.0	6.1	0.97		0.033	1.84	0.41
LiI	24.8	18.9	3.20		11.4		
LiClO$_4$	17.1	14.2	12.7	10.7			
NaCl	0.24	0.012	0.0006		0.00004		0.34
NaClO$_4$	4.16	1.19	4.20				5.00
KCl	0.067	0.0046			0.0003		0.12
KBr	0.127	0.011	0.0019	0.0017	0.0020		0.43
KI	1.00	0.106	0.100	0.016	0.124	0.016	1.36
KSCN			2.15		1.17	0.64	2.17
Ca(NO$_3$)$_2$	7.75	3.12	1.04				
Ca(ClO$_4$)$_2$	10.3	6.9	2.58	0.0108			
FeCl$_3$	8.8	12.2	3.88				
CoCl$_2$	3.63	4.33	0.66	0.0016	0.315	0.044	
CuCl$_2$	5.05	4.00	0.216	0.0082	0.12	0.021	
CuCl$_2$. 2H$_2$O		1.38*	0.53	0.065			
ZnCl$_2$		9.3	3.18			0.21	
CdCl$_2$	0.093	0.083				0.038	
CdBr$_2$	0.675	1.06	0.055	0.015			
CdI$_2$	6.00	3.05	0.67	0.75		0.0012*	
HgCl$_2$	2.47	1.83	2.22	0.24		0.71	
AgNO$_3$	0.22	0.018	0.021		17.1	2.72	
Th(NO$_3$)$_4$		2.65	3.03	1.56			
UO$_2$(NO$_3$)$_2$. 6H$_2$O		6.65	0.032	0.125			

* Molarity.
** *N*-Methylacetamide, molarity (L. R. Dawson and coworkers, *J. Phys. Chem.*, 67, 281 (1963)).

cation and anion, and not so tight arrangement as in anhydrous salts. They should therefore exhibit relatively low heats of sublimation. This prediction has not yet been tested experimentally in a quantitative manner.

The phase diagram in the ternary system cobalt chloride, water and acetone, Figure 2C1, shows the very complicated solubility relationships that result from the above considerations[96-98]. At 25°, above about 30% acetone, there is a two-liquid-phase region, where the solution splits into a light-blue acetone-rich phase containing tetrahedral CoCl$_2$(CH$_3$COCH$_3$, H$_2$O)$_4$, and a magenta-coloured water-rich phase, containing more cobalt, as the octahedral Co(H$_2$O)$_6^{2+}$ mixed with CoCl$_2$ (CH$_3$COCH$_3$, H$_2$O)$_4$. Above about 86% acetone in the mixture, the two phases coalesce again, and a minimum solubility is observed at ca.

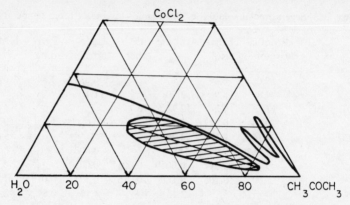

Figure 2C1. Ternary phase diagram of the system $CoCl_2$—H_2O—CH_3COCH_3. Compositions in the shaded area split into two liquid phases.

90% acetone. At a still higher acetone concentration, however, there is a sharp upturn in the solubility, and cobalt chloride tetra- and trihydrates are at equilibrium with above 20% cobalt chloride in solution. When more acetone is added the solubility again drops, but rises at still higher acetone concentrations again, to yield a 25% cobalt chloride solution, containing only 3% water, at equilibrium with cobalt chloride dihydrate and various acetonate crystals. The solubility drops to rather low values for the anhydrous salt in pure acetone.

It has often been stated that the most important correlation of the solubility of an electrolyte in a series of solvents can be made with the dielectric constant of the solvents. Further, this correlation has been often shown to hold for mixtures of solvents too. Walden[82] has shown that indeed in many cases there is a direct proportionality between the mole fraction of the solute i in the saturated solution and the cube of the dielectric constant of the solvent j

$$x_{i,j} \text{ (satd)} = K_i\, \varepsilon_j^3 \qquad (2C4)$$

This empirical rule was confirmed for tetraethylammonium iodide, for example, in such diverse solvents as water, acetonitrile, ethanol, acetone, methyl formate, bromobenzene and others ($K_{Et_4NI} = 9.1 \times 10^{-6}$), but not for lithium chloride (K_{LiCl} varies from 5.5×10^{-5} for water to 1.27×10^{-3} for ethanol to 3.6×10^{-3} for isopentanol). Equation (2C4) has been shown[99, 100] to be followed also by solvent mixtures, such as water–ethanol and even water–dioxane, down to $\varepsilon = 34$ (50 % dioxane). It has been shown[99] that equation (2C4) is derived from the Debye–Hückel limiting law for electrolytes which are considerably dissociated.

Another relationship is derived from the Born equation and relates the solubilities in two solvents j and k

$$\log (s_j/s_k) = 0.4343 \ (Ne^2/RT)(r_+ + r_-)^{-1} (1/\varepsilon_j - 1/\varepsilon_k) \qquad (2C5)$$

where it is assumed that the sum of the radii of the cation and anion of the salt is independent of the solvent. This relationship has been confirmed by data for lead[101], silver[102], thallium(I)[101] and caesium[102] chlorides and caesium sulphate[103] in aqueous alcohols, dioxane and acetone, down to $\varepsilon = 15$. In other cases, equation (2C5) is not obeyed, as found for example for silver bromate[104], acetate[105, 106] and sulphate[106] in aqueous alcohols or dioxane. Thus the dielectric constant is not the sole factor determining solubilities, both in pure solvents and in solvent mixtures. Specific interactions, such as solvation, play also a major, and independent role.

d. Solvation in Active Solvents

Solvation of the solute, in low-dielectric constant solvents, or of its ions in high-dielectric-constant solvents, is determined by the polarity and donor properties of the solvent molecules. The solvating power, as measured for example by ΔH_{soln} for a given solute with a given ΔH_{subl} or ΔH_{latt} (cf. equations 2C2 and 2C3), correlates well with the dipole moment of the solvent molecule. There are however several factors which superimpose their effects on that of the total dipole moment. In large molecules it is the moment of the polar bond, rather than the total moment, which determines the solvating power. When the polar bond is sterically hindered, the solvating power suffers. Solvents capable of hydrogen bonding can solvate anions by forming such bonds. Finally, the nature of the donor atom, and the electron density on it, are important in the solvation of metal ions, in as much as direct coordination by covalent bond formation between metal atom and donor atom in the solvent occurs.

Since it is the bond dipole moment, and not the total dipole moment, which is important, then even molecules with two functional groups which cancel each other's contribution to the total moment, would be good solvating solvents. Thus all the active solvents are in principle dipolar. They can be dipolar–aprotic, i.e. not have a dissociable proton available, or protic. Even aprotic solvents may in some cases lose a proton, after molecular tautomerism, e.g. in acetonitrile to $H_2C = C = N^- + H^+$). Ordinarily, however, the dipolar aprotic solvents are basic, as are many of the protic ones, except, of course, acids such as formic acid or phenol. There is thus a good correlation between the heats of solution in a series of solvents and their base strengths, as deduced independently, e.g. from infrared spectroscopy[107]. Among the aprotic dipolar solvents the order of decreasing solvating power is[108]

$$(CH_3)_2SO, (CH_3)_2NCOCH_3 > (CH_3)_2NCOH, H_2O > (CH_3)_2CO,$$

$$(-CH_2-)_4=SO_2 \text{ (sulpholane)} > CH_3OH \gg CH_3CN, CH_3NO_2 >$$

$$C_6H_5CN, C_6H_5NO_2$$

with the two protic solvents, water and methanol, included for comparison. Parker[109] has reviewed the solvating characteristics of the dipolar aprotic solvents, in particular with respect to anion solvation.

If the dissolved electrolyte is not dissociated appreciably in the organic solvent, it may be solvated as a whole, or the two ions may share solvent molecules. Thus in carbon tetrachloride solutions, a methanol molecule is shared betwen a tetrabutylammonium and aniodide ion, which are therefore no longer in contact when methanol is added to the solution[110]. A similar effect occurs when octanol is added to solutions of long-chain alkylammonium salts in inert diluents[111] (cf. section 10.C). Complexes involving a larger number of solute ion pairs and solvent molecules are formed in solutions such as those of lithium perchlorate in ether, where solvated aggregates $(LiClO_4 \cdot x\,(C_2H_5)_2O)_q$ $(1 < x < 2, q \gg 1)$ are observed[112]. The aggregation of solutes in organic solutions will be discussed in the next section.

The most important aspect of solvation, as far as this book is concerned, is the solvation of metal ions. Solvation by water-immiscible solvents is treated in detail in section 9.C in connection with the extraction of the metal ions from aqueous solutions. The same considerations apply also to water-miscible solvents, and therefore need not be discussed here extensively. Complex-formation reactions in solvents and molten salts have received considerable attention, and have been recently reviewed[76, 113]. The solvation of the metal ions may lead to definite coordination[111], and the main point concerning solvation is that solvents compete with anionic or neutral ligands for coordination sites at the metal ions, in addition to the general medium (low-dielectric-constant) effects (section 2.C.c). Complex formation, and in particular solvation, stabilize oxidation states of certain ions which are unstable in aqueous solutions. Copper(I) becomes stable, even without anionic ligands such as chloride or cyanide, when solvated in n-propanol, i-propanol, acetone or nitromethane[114]. In these solvents, it is the relatively low solvation energy of copper(II) which contributes most to the stabilization of copper(I). In nitrile solvents[115], the high solvation energy of copper(I) makes the major contribution to its stabilization. Some reactions are affected not because of the solvation of the metal ions but because of the solvation of anions. The solubility of silver halides in excess halide solutions is lower in methanol than in the aprotic solvents acetonitrile, dimethyl sulphoxide, nitroethane or acetone[116]. This is explained by the ability of methanol to hydrogen-bond to the halide anion, solvate it, thereby decrease its relative activity in the solution, and its reaction with the silver halide to form the soluble complex AgA_2^-.

Summarizing, it is seen that both the cation and the anion can be solvated, or both can share a solvent molecule. Solvation depends on the dipole moment of the polar bond of the solvent, or its base strength, and its specific donor properties, including hydrogen bonding. Because of these specific factors, no complete correlation between solvating power (i.e. heats of solution) and any general factor can be made.

e. Solvation and Hydration in Mixed Solvents

Mixed solvents, for the present purpose, are defined as mixtures of water and an organic solvent. The properties of electrolyte solutes in such mixtures differ from those in pure aqueous solutions largely because of the lowered dielectric constant. Some water-miscible solvents have very low dielectric constants (e.g. dioxane, $\varepsilon = 2.2$), so that even in mixed solvents rather low values of this parameter can be attained. As a zeroeth approximation, then, the equations given for electrolyte behaviour in Chapters 1 and 3, for aqueous solutions, should be obeyed in mixed solvents, with the appropriate value of the dielectric constant inserted.

Many systems show considerable deviations from this simple dependence on the dielectric constant, as discussed above for the solubilities. A first approximation can be reached when changes in the ionic radii, because of changes in the solvating medium, are considered. This point is of prime importance in explaining the effect of the solvent on the association and conductance of electrolytes in mixed solvents.

Another refinement leading to a better aproximation to the observed behaviour is the inclusion of specific solvation in the model, i.e. competition between water and the organic solvent for coordination sites around the ions. The preferential solvation can be treated in a formal thermodynamic manner as follows. From the thermodynamics of fluids in an electrical field[117] follows the relationship[118]

$$\ln\left(x_W/x_W^0\right)/(x_S/x_S^0) = \tfrac{1}{2}\int_{r_e}^{\infty}\int_{0}^{E_r}(\partial\varepsilon/\partial x_W)_{P,E}(\partial\mu_W/\partial\ln x_W)_{P,E}^{-1}\,d(E^2)r^2dr \tag{2C6}$$

where W and S refer to water and the organic solvent, superscript 0 to the bulk of the solution, while absence of superscript refers to the vicinity of an ion, E is the electrical field at distance r from the ion, E_r being that at the effective limit, r_e, of the dielectrically saturated region around the ion. For an ideal solvent mixture and at infinite dilution, equation

(2C6) reduces to the simple form relating the relative concentration of the solvents around the ion to that in the bulk, and the partial free energies of solvation of the ion in the two solvents G'_W and G'_S

$$\alpha = \log\left(x_W/x_S\right)/(x_W^0/x_S^0) = (0.4343/2\,RT)\,(G'_W - G'_S) \tag{2C7}$$

The solvent sorting coefficient α can be calculated from published ionic free energies[119], and some representative values are shown in Table 2C4. They are all positive, and recalling that component 1 is water, this means

TABLE 2C4. Some ionic solvent sorting coefficients (equation 2C7).

	H^+	Li^+	Na^+	Cs^+	Zn^{2+}	Cd^{2+}	Cl^-	Br^-	I^-
CH_3OH	2.9	1.5	1.5	2.6	9.9	9.5	1.1	1.1	0.0
C_2H_5OH	3.7	2.9	2.2	3.7	14.7	10.6	1.5	1.5	1.5
$(CH_3)_2CO$	2.9	1.5	10.3	4.4			3.3	0.9	2.6

that water is preferred, often to several orders of magnitude, around these ions over the solvents indicated. With other solvents and ions, such as silver and amine or acid solvents, negative values of α are obtained. It should however be stressed that (2C7) is valid only for ideal solvent mixtures and a linear variation of dielectric constant with composition.

A different approach, starting from the Born equation and minimizing the electrostatic free energy with composition of the solvent around the ions, leads to the relationship[120, 121]

$$dG'/R\,T = - (h_W^0 \, d\ln a_W + h_S^0 \, d\ln a_S)$$

$$- (Ne^2/2RT\,\varepsilon)(1/r_+ + 1/r_-)\,d\ln \varepsilon$$

(2C8)

where h_i^0 is the solvation number in the pure solvent, and a_i refers to the solvent in the vicinity of the ions, while the radii refer to the solvated ions. The change in partial free energy of the salt with solvent composition has been obtained from vapour pressure measurements, and ranges from $dG'/dx_W = -17.6$ for sodium hydroxide (which is strongly hydrated) to $+15.4$ for sodium tetraphenylboride (which is solvated by dioxane). Assuming that the values of dG'/dx_W are the same for the two large ions tetraphenylphosphonium and tetraphenylboride, and that it differs by the electrostatic term $-(Ne^2/2\,r\varepsilon)\,(d\ln \varepsilon/dx_W)$ from that of the large non-electrolyte tetraphenylmethane (which agrees with the former assumption within 2 ± 1.5 kcal/mole), Grunwald[121] could calculate single ionic values for dG'/dx_W. They range from 0.7 kcal/mole for K^+ to 6.6 kcal/mole for H^+, and from -14.5 kcal/mole for Cl^- to -4.7 kcal/mole for ClO_4^-. The positive values for the small cations can be explained only if the solvated 'uncharged' analogue has a positive value of dG^0/dx_W, and it was found that solvation of cations by dioxane in the mixed solvent (at 50% weight) is not negligible, in spite of the low polarity and low dielectric constant of this solvent. It was estimated that the small inorganic cations are solvated by $h_S^0 = 2$ moles of dioxane, in addition to an undetermined number h_W^0 of water molecules, that the hydrogen ion forms the solvated ions $O(H \cdots OC_4H_8O)_3^+$, that small anions are not dioxanated, but perhaps hydrated, and that around large organic ions the solvents are distributed as in the bulk[121].

The thermodynamic functions of transfer of electrolytes from aqueous solutions to solutions in mixed solvents can be calculated from solubilities, the e.m.f. of cells without liquid junctions and direct calorimetrical measurements. The data for electrolytes can often be split into ionic contributions, and these, in turn, can be interpreted in terms of ionic solvation. The free energy of transfer can be used to calculate ionic distribution coefficients $\Delta G^0_{W,S} = -RT \ln D_{W,S}$, which are a measure of the preference of an ion for a given solvent. Univalent cations prefer water strongly over ethanol (log $D_{W,S}$ ranges from 2.1 for Ag^+ to 2.7 for $(C_2H_5)_4N^+$ to 4.1 for K^+) as do anions (log $D_{W,S}$ ranges from 0.7 for ClO_4^- to 2.5 for Cl^-), and as certainly do more highly charged ions, but, as expected, long-chain carboxylate ions prefer ethanol over water (log $D_{W,S} = -2.4$ for $C_{11}H_{23}COO^-$)[122]. These data with the appropriate Born equation can be used to calculate the ionic radii in the mixed solvent[123]. The corresponding values for methanol solutions have been obtained from e.m.f. data for aqueous methanol[124], and $\Delta S^0_{W,S}$ and $\Delta H^0_{W,S}$ values were obtained from them and published heats of solution. Distribution coefficients calculated from these values are log $D_{W,S} = 0.86$ for K^+, 2.3 for Li^+ and -5.8 for Cl^-. These values do not correspond with those given above for ethanol, and the discrepancy may be due to the manner in which $\Delta G^0_{W,S}$ for the electrolyte has been split into terms for the constituent ions, and to the failure of the Born equation, on which the data for ethanol are based, to even nearly approximate the observed behaviour[124].

With solvents which are not completely miscible with water the transfer of the electrolyte can be considered to occur in stages

$$C^+A^-(H_2O) \rightarrow C^+A^-(H_2O \text{ satd. with S}) \qquad \Delta G^0_{W,W(S)} \quad (2C9a)$$

$$C^+A^-(H_2O \text{ satd. with S}) \rightarrow C^+A^- (S \text{ satd. with } H_2O)\ \Delta G^0_{W(S),S(W)} \quad (2C9b)$$

$$C^+A^- (S \text{ satd. with } H_2O) \rightarrow C^+A^-(S) \qquad \Delta G^0_{S(W),S} \quad (2C9c)$$

with $\Delta G^0_{W,S} = \Delta G^0_{W,W(S)} + \Delta G^0_{W(S),S(W)} + \Delta G^0_{S(W),S}$. The free energies of transfer for steps (2C9a) and (2C9c) for certain salts and S = nitromethane were obtained from solubilities, those for step (2C9b) from partition data[125]. Ionic values of the free energies of transfer could however not be calculated, since the solubilities of the salts were not of suitable magnitude. Enthalpies of transfer of ions from water to propylene carbonate ($\varepsilon = 65$) have been determined calorimetrically, by measuring heats of solution[126]. Values of $\Delta H^0_{w,s}$ relative to that of Na^+ are in kcal/mole: Li^+ 3.17, K^+ $-$ 2.80, Rb^+ $-$ 3.43, Cs^+ $-$ 3.96, $(CH_3)_4N^+$ $-$ 1.45, $(C_2H_5)_4N^+$ 2.65, Cl^- 3.83, Br^- 0.85, I^- $-$ 3.22, ClO_4^- $-$ 6.37. The value for sodium should be near zero, so that these values are near the absolute single-ion values. The enthalpies, as found for the free energies, are inconsistent with the predictions from the Born equation, for any reasonable molecular model of the solvation[126].

The relative binding energies of water and the organic solvent in mixed solvates has been obtained from measurements of the heat of solution of metal salt hydrates in the organic solvents and in their mixtures with water[90-94]. Octahedral cobalt(II) is capable of binding two solvent molecules, in addition to two water molecules and two nitrate ions in organic solution, so the heats of solution of $Co(NO_3)_2 \cdot 2H_2O$ relate directly to the binding of the organic solvent, ranging from $\Delta H_{soln} = -17.76$ kcal/mole for dimethylformamide to -1.65 kcal/mole for acetone. The difference between the heats of solution of the tetrahydrate and the dihydrate, -8.78 kcal/mole, is the corresponding heat of adding two water molecules. This value is reasonable with respect to the relative base strengths of the solvents compared. Addition of a small amount of water to an acetone solution of the nitrate dihydrate gives essentially the same value.

In addition to information deduced from solubility and heat of solution measurements, the relative binding of water and solvent has been measured by a large variety of methods. Data obtained from vapour pressure[121] and e.m.f. measurements[124] have been mentioned above. Results from crysoscopic measurements in dioxane indicate that acids associate strongly with water, perchloric acid forming $HClO_4 \cdot 2H_2O$, hydrochloric acid forming the dimer $(HCl \cdot H_2O)_2$ and sulphuric acid forming $H_2SO_4 \cdot H_2O$, all solvated to an unknown extent with dioxane[127]. The number of moles of solvent transported with a mole of calcium chloride electroosmotically has been measured in aqueous methanol[128]. At a mole fraction of 0.7 of methanol there are 7 moles of each solvent transported with the calcium ions, 2 moles of each with the chloride ions. The hydration of closed-shell ions such as the alkalies, the alkaline earths, zinc, cadmium, lanthanum and thorium in 50% by volume of aqueous dioxane has been estimated from the chemical shifts of n.m.r. signals[129]. The shift is negligible for dioxane, and marked for the water, in the presence of the salts, showing preferential hydration over dioxanation, although not excluding some of the latter. More explicit information has been obtained for paramagnetic ions, such as cobalt(II) and nickel(II)[130] by the n.m.r. technique. The hexasolvated $Co(CH_3OH)_6^{2+}$ in pure methanol exchanges methanol and forms the monohydrate and the dihydrate as water is gradually added. With robust solvates such as those of chromium(III) it is possible to quench the equilibrium mixture, and separate by cation exchange various mixed hydrates-solvates[131]. For the reaction

$$Cr(H_2O)_n(CH_3OH)_{6-n}^{3+} + H_2O \rightleftarrows Cr(H_2O)_{n+1}(CH_2OH)_{5-n}^{3+} + CH_3OH$$

$$(2C10)$$

at 30 or 60°C, the equilibrium constants are 50, 25, 26 and 1.73 for $n=2,3,4$ and 5 respectively. This means that water is preferred over methanol, the more methanol there is already bound to the chromium ion. Similar results were obtained polarographically for copper(II) in acetone, where the constants are 56, 32, 10 and 5.6 for $n = 1, 2, 3$ and 4 respectively[132] The relative solvating power of ethanol, acetone and nitromethane for copper(II) decreases in this order, as is seen from the finding that in the presence of 0.01 M water, only 1% of the copper ions are hydrated to some extent in ethanol, 43% are hydrated in acetone and as much as 94% are hydrated in nitromethane solution. When however ethanol or acetone are added to the copper(II) hexahydrated ion in aqueous solution, these solvents can displace up to two water molecules, as seen from the change in the absorption spectrum[133]. This result is in agreement with the relatively low value of the equilibrium constant for displacement of the fourth acetone molecule by water given above.

The relative solvating power of water, formamide, methanol and ethanol for zinc, cadmium and lead has also been measured polarographically[134]. Thallium(I) prefers methanol to water even up to 5 M of the latter solvent, while lead forms hydrates up to $Pb(OH_2)_2^{2+}$ and zinc and cadmium up to the trihydrate. Formamide cannot displace the solvent from the coordination shell in aqueous solutions, but it does in methanol, and even more so in ethanol. Water can displace all, and methanol most, of the solvent around cadmium ions in acetone, but ethanol cannot, while it can displace some of the solvent from lead ions. Zinc is most hydrated, and least solvated by formamide, while lead shows the opposite behaviour.

Spectrophotometric measurements have also been employed to study the hydration of ions in alcohols. Half hydration, i.e. formation of the species $M(H_2O)_3 (C_2H_5OH)_3^{m+}$ occurs[135] at 0.93 M H_2O for Co^{2+}, 1.2 M for Ni^{2+}, 4 M for Cr^{3+} and 5.3 M for Cu^{2+}, while if the neodymium ion is also considered to be hexacoordinated, half hydration occurs[136] at 3 M H_2O. Addition of an inert solvent such as carbon tetrachloride was found not to affect the relative affinities of the solvents for the ions. However, association with the anion is superimposed on the exchange of solvating solvent[137], and the water concentration required to reach half hydration for cobalt(II) *p*-toluenesulphonate is 2.2 M, compared with 0.93 M for the nitrate. It was concluded that water binds about ten times more strongly than ethanol to the ions[125], but a large excess of water is required to remove all the solvating ethanol. The same is also obtained for neodymium[138], where a thirty-fold excess was found to be necessary.

Finally, solvent extraction shows many metal ions to be accompanied by water of hydration when extracted into solvents such as alcohols, ketones, esters or ethers. This is discussed in detail in section 9.B. Here it will suffice to cite three examples only. Uranyl nitrate is accompanied with about four molecules of water when extracted into divers solvent-vents[139, 140]. Zinc, lithium and calcium chlorides are accompanied by 2–3, cobalt and nickel perchlorates by 10–15 molecules of water when extracted by 2-octanol[141]. Thorium nitrate is accompanied by 6–7 molecules of water when extracted into several oxygenated solvents[91]. These figures are the slopes of plots of the water content, corrected for that extracted in the absence of the metal salt, against metal salt content in the solvent. It is seen that the water occupies all, or most, of the coordination sites on the metal.

Summarizing, it is seen that water is strongly preferred over other solvents by most ions, but that some solvation by active solvents does occur in mixed solvents. For 'hard' ions in the Pearson sense[142], the order of preference is approximately

$$H_2O > HCONH_2 > CH_3OH > C_2H_5OH > (CH_3)_2CO > CH_3NO_2 > \text{dioxane}$$

'Soft' ions may show deviations from this order, and specific interactions should always be considered when estimating the relative solvation of ions by the components of mixed solvents.

D. DISSOCIATION AND AGGREGATION

a. Electrolytic Dissociation

Electrolytes dissolved in organic solvents undergo divers reactions, the one of prime importance being that of ionic dissociation, accompanied by solvation of the ions. The main factor which determines the degree of dissociation is the dielectric constant of the solvent. In solvents of high dielectric constants (e.g. water, $\varepsilon_{25°} = 78.5$) ionic dissociation is virtually complete, and the equilibrium

$$C^+A^- + (h_c + h_a)S \rightleftarrows CS_{h_c}^+ + AS_{h_a}^- \tag{2D1}$$

lies far to the right. For salts of higher charges, and at higher concentrations, reversal of the equilibrium occurs even in aqueous solutions. It is then convenient to speak of the association of the ions in certain cases, rather than of the dissociation of the electrolyte. In solvents of low dielectric constants (e.g. dioxane, $\varepsilon_{25°} = 2.2$), on the other hand, ionic dissociation is so slight, that it is best to discuss the extent of reaction (2D1) going as written from left to right. The nomenclature employed by workers in the field of electrolyte solutions thus differs, K_{ass} being used for high-dielectric-constant solvents, and pK_{diss} for low-dielectric-constant solvents.

The theory of ionic association is discussed in detail in chapter 3. The qualitative observation that $pK = pK_{diss}$ decreases (K_{ass} decreases) as the dielectric constant of the solvent increases, implicit in the above introduction, is given there a theoretical and quantitative expression. For the present purposes it is sufficient to employ the thermodynamically derived expression of Denison and Ramsay[143], based on a Born cycle

$$pK = (e^2/2.303\ k\ T)\ (1/r_C + 1/r_A) \times 1/\varepsilon = A \times 100/\varepsilon \tag{2D2}$$

Provided that the contact distance of the two ions does not vary with the composition and nature of the solvents in a series of solvents, i.e. $r_C + r_A$ = const., then for a given salt and a given temperature, A is a constant, and a plot of pK vs. $100/\varepsilon$ is strictly linear.

Behaviour according to equation (2D2) has indeed been observed for many electrolytes, both for series of pure solvents of different ε, and for mixtures of solvents, mainly in the interval $\varepsilon = 10$ to $\varepsilon = 40$. Instances which may be cited are lanthanum ferricyanide and zinc and copper malonates and sulphates[144] in aqueous acetone, glycol, ethanol and dioxane, aqueous ethanolic solutions of cobalt and magnesium sulphates, potassium bromate, silver nitrate, zinc perchlorate[145] and alkaline earth thiosulphates[146], among inorganic electrolytes, and tetraalkylammonium halides, nitrates, picrates and tetraphenylborides in acetonitrile, carbon tetrachloride[147], aqueous dioxane[148], and the series nitrobenzene, acetone, pyridine, 1,1-dichloroethane, chlorobenzene and *meta*-dichlorobenzene[149], among the organic electrolytes.

There are however many cases where (2D2) is not obeyed, i.e. that A is not a constant for a given salt for a series of solvents. For series of pure solvents, this is mainly because of failure of the assumption of a constant value of $r_C + r_A$, which implies constant solvation. Better adherence to (2D2) is expected for electrolytes having both a large cation and a large anion, neither of which is solvated (e.g. tetrabutylammonium tetraphenylboride). Even large ions, if they can partake in hydrogen bonding, may be solvated, and to different degrees with different solvents. Thus the picrate anion, although large, hydrogen-bonds to water and alcohols, and is solvated by them to a different degree than by other solvents.

Even for large, non-solvated ions, deviation from (2D2) occurs, in particular for solvent mixtures, or in solvents consisting of a mixture of conformers, such as in 1,2-dichloroethane. Although this liquid has a dielectric constant of $\varepsilon = 10.23$, it permits more dissociation than do other solvents of nearly the same dielectric constant, such as 1,1-dichloroethane ($\varepsilon = 9.90$) or *o*-dichlorobenzene ($\varepsilon = 10.1$), and behaves as if it had $\varepsilon = 11.8$ to 12.4[149, 150]. The explanation for this lies probably in a shift of the equilibrium between the polar, *gauche*, conformation of 1,2-dichloroethane and the non-polar, *trans* conformer. There is spectroscopic evidence that such a shift in the equilibrium indeed occurs under the influence of ionic fields[151], and that the ratio of the polar to the non-polar forms increases from 1.3 in the absence of ions to 1.9 in the presence of 1.2 M ($x \sim 0.1$) of tetrabutylammonium perchlorate. A similar explanation has been offered[152] for the deviations observed in dioxane–water solutions where a shift in the equilibrium between the polar boat and non-polar chair conformers of dioxane is involved.

The effect of a polar constituent in a mixture is apparent even more when different solvents are mixed, at a given macro dielectric constant. Whereas the pure solvents, methanol ($\varepsilon = 32.6$), acetonitrile ($\varepsilon = 36.2$) nitromethane ($\varepsilon = 35.9$) and nitrobenzene ($\varepsilon = 34.8$), have sufficiently close dielectric constants, to cause similar degrees of association in many electrolytes, there are gross differences in the solvating power, which, decreases strongly in this series[153]. Differences in solvation are enhanced in mixtures[154, 155], as shown, for example, by a series of experiments at $\varepsilon = 13.2$ attained by mixing dioxane with water (22.6% wt), methanol (42.4% wt), acetonitrile (29.6% wt) or *p*-nitroaniline (19.9% wt)[158]. Mixtures containing the last compound lead to negligible association of such solutes as tetrabutylammonium picrate, bromide, or tetraphenylboride. For the former two electrolytes, moderate association ($pK = 2.6 - 3.1$) is obtained with acetonitrile, and for all three electrolytes with the other solvents[155]. This is cited as evidence for the solvation of even large anions by compounds of very high dipole moment (for *p*-nitroaniline, $\mu = 6.32$ D, for acetonitrile, $\mu = 3.51$ D). This is in line with the assignment of a local ε, higher than the bulk dielectric constant, thus leading to increasing dissociation (smaller pK), in mixtures of polar and non-polar solvents, causing pK values to fall below the line of equation (2D2). It is important, however, to use care in discussing pK values below about 1.5, derived from conductivity data, since apparent discrepancies[152] may be due to incorrect pK values, rather than to real effects[156].

b. Conductivity of Electrolyte Solutions

The dissociation constants discussed above have been derived mainly from conductivity data. The conductance theory of Onsager–Fuoss[157] has shown that for the dielectric constant interval $15 < \varepsilon < 100$ and for concentrations $c < 2 \times 10^{-4} \varepsilon$ (corresponding to $\kappa \mathring{a} < 0.2$ at room temperature and $\mathring{a}$ about 5 Å, in the ion atmosphere term, section 1.C.a), the following equations account for the equivalent conductivity for a 1:1 electrolyte in terms of all the known effects

$$\Lambda (1 + F C) = \alpha[\Lambda_o - S (\alpha c)^{1/2} + E (\alpha c) \log (\alpha c) + J\alpha c] \quad (2D3)$$

$$K_{\text{diss}} = \alpha^2 cy^2 (1 - \alpha)^{-1} = \alpha^2 c (1 - \alpha)^{-1}$$

$$\times \exp [- 4.606 A (\alpha c)^{1/2}/(1 + B\mathring{a} (\alpha c)^{1/2})] \quad (2D4)$$

where α is the degree of dissociation and y the mean ionic activity coefficient. There are four unknown parameters in these equations: the limiting conductivity at infinite dilution Λ_o, the centre-to-centre contact distance of the two ions $\mathring{a}$, the dissociation constant of the electrolyte K_{diss}, and an occupied-volume correction to the viscosity F. There are also the constants S, E, J, A and B, which include universal constants and depend in a known way on the dielectric constant and the temperature of the solvent, and in the cases of S, E and J also on the limiting conductivity Λ_o and the viscosity of the solvent η, and in the case of J also on $\mathring{a}$. The values of A and B have been given in section 1.C.b ($A = 1.825 \times 10^6 \ (\varepsilon T)^{-3/2}$ and $B = 50.3 \ (\varepsilon T)^{-1/2}$ for $\mathring{a}$ in Å). The values of the other constants are

$$S = 8.204 \times 10^5 \, (\varepsilon T)^{-3/2} \, \Lambda_0 + 62.50 \, \eta_0^{-1} \, (\varepsilon T)^{-1/2} \qquad (2D5)$$

$$E = 6.775 \times 10^{12} \, (\varepsilon T)^{-3} \, \Lambda_0 - 9.977 \times 10^7 \, \eta_0^{-1} \, (\varepsilon T)^{-2} \qquad (2D6)$$

$$J = 5.884 \times 10^{12} \, (\varepsilon T)^{-3} \, [(2b^2 + 2b - 1)/b^3 + 1.646$$
$$+ \ 0.4343 \log \mathring{a} \, (\varepsilon T)^{-1/2}] \, \Lambda_0 + 3.804 \times 10^3 \, \mathring{a} \, \eta_0^{-1} (\varepsilon T)^{-1}$$
$$- \ 8.666 \times 10^9 \, \eta_0^{-1} \, (\varepsilon T)^{-2} [0.957 + 0.4343 \log \mathring{a} \, (\varepsilon T)^{-1/2}]$$
$$(2D7)$$

where $\mathring{a}$ is expressed in Å and η in poise, and b is the characteristic Bjerrum ratio (section 3.A.b) $e^2/a\varepsilon k T = 1.671 \times 10^5/\mathring{a} \, (\varepsilon T)$. Using the approximation $(1 + Fc)^{-1} \approx 1 - Fc$, neglecting terms in c higher than first order, and substituting α from (2D4) into (2D3) converts the latter equation into

$$\Lambda = \Lambda_0 - S \, (\alpha c)^{1/2} + E \, (\alpha c) \log (\alpha c) + J \, (\alpha c)$$
$$- \ F\Lambda_0 c - (\alpha c) \, y^2 \Lambda / K_{\text{diss}} \qquad (2D8)$$

The viscosity correction term F can be obtained directly from viscosity measurements[158] on the solution

$$\eta = \eta_0 (1 + 1.031 \, (\eta_0 \Lambda_0)^{-1} \, (\varepsilon T)^{-1/2} c^{1/2} + F c) \qquad (2D9)$$

where η_0 is the viscosity of the pure solvent, the term in $c^{1/2}$ has the Falkenhagen coefficient, which describes effects due to velocity gradients in the solution, and must be slightly modified if the ions are of very unequal size, and the term in c is the Einstein volume effect, where

$F = \pi N R^3/300$, R being the (unknown) hydrodynamic radius of the ions, see below. The viscosity correction is thus important only for large ions. Where no viscosity data are available, the viscosity term can be combined with the other term linear in c, giving a new coefficient $J' = J - F\Lambda_0$[157]. The quantity $(1 - \alpha)\Lambda_0 F$ is thereby ignored in the Fouss treatment.

Besides the effects of ionic association, described in the last term of (2D8), the term with the coefficient S in (2D3), the Onsager relaxation term, is the most important correction to Λ_0. It is usually much larger than the other two terms combined, since $E(\alpha c) \log(\alpha c)$ is negative ($\alpha c < 1$), while $J\alpha c$ is positive and of similar magnitude. For example, for aqueous 0.007066 M KCl (where $\alpha = 1$), $- S c^{1/2} = - 7.949$, $Ec \log c = - 0.893$ and $Jc = 1.377$, the total correction amounting to $- 7.465$ relative to $\Lambda_0 = 149.955$. Since the terms in E and J are higher order in c than is the term in S, they become important only at higher concentrations, although still below the limit of the theory at $\kappa \mathring{a} = 0.2$. It has been shown that inclusion of all the terms is necessary to describe conductance data in the range of validity of the Onsager–Fuoss conductance theory[159]. Higher terms, in $c^{3/2}$ or above, have been shown to be unnecessary within this concentration range.

In order to find the dissociation constant from conductivity measurements, $\Lambda = f(c)$, one may proceed as follows. If viscosity data are available, Λ is first corrected to give $\Lambda_\eta = \Lambda(1 + Fc)$, and if not, the F term is included in the J term. In the following $\Lambda = \Lambda_\eta$ will be used. A zeroth approximation to the degree of dissociation is made according to the Arrhenius assumption

$$\alpha_0 = \Lambda/\Lambda_0 \qquad (2D10)$$

The initial value of Λ_0 for this is obtained from a) the ionic conductivities of fully dissociated electrolytes containing the ions in question, if available, or b) from the limit of Λ as $c \to 0$, graphically or otherwise, if the function is not too steep, or, as a last resort c) from Walden's rule

$$\Lambda_0 \eta_0 = const. \sim 0.60 \text{ poise/ohm } M \qquad (2D11)$$

Using (2D10) in the Onsager term leads to the first approximation for the degree of ionization

$$\alpha_1 = \Lambda/(\Lambda_0 - S\Lambda_0^{-1/2}(c\Lambda)^{1/2}) \qquad (2D12)$$

The conductance equation (2D8) is then rewritten in the form

$$[\Lambda - \Lambda_0 + S(\alpha_1 C)^{1/2} - E(\alpha_1 c) \log(\alpha_1 c)]/(\alpha_1 c) = J - y^2 \Lambda / K_{\text{diss}} \quad (2D13)$$

The left-hand side is plotted against $y^2\Lambda$, and if a straight line results, the correct value of Λ_0 has been selected. Refined values of α are then calculated from K_{diss} using (2D4) and used to iterate (2D13). At low dielectric constants (where pK_{diss} would be > 2), (2D13) can be used directly with data of high precision (0.05 % or better). At higher dielectric constants K_{diss} becomes larger, and J becomes the dominant term. The first approximation then may yield $\alpha_1 > 1$, which is physically absurd. Then $\alpha \approx 1$ is used to transform (2D8) into

$$\Lambda - Sc^{1/2} + E c \log c = \Lambda_0 + (J - y^2 \Lambda / K_{\text{diss}}) c \quad (2D14)$$

and a plot of the left-hand side of (2D14) against c yields Λ_0 and the combined term in J and K_{diss} from which K_{diss} can be obtained, if large, only when $\mathring{a}$ can be independently estimated.

Instead of the approximation (2D12) Shedlovsky[160] used the approximation

$$\alpha_1 = (\Lambda/\Lambda_0)(1 + S(\alpha_1 c)^{1/2}\Lambda_0{}^{-1}) \approx (\Lambda/\Lambda_0)(1 + S\Lambda^{1/2}c^{1/2}\Lambda_0{}^{-3/2})$$

$$(2D15)$$

while the Davies school[161] has employed (2D12), and both ignored the terms in E and J. This is permissible only when $pK_{\text{diss}} > 2$, as discussed above, since in this range of dielectric constant the two terms cancel each other approximately. In the range $20 < \varepsilon < 60$, however, the full equation (2D12) must be used[157].

A final test of the J and K_{diss} values obtained from (2D13) may be made, in that they must yield $\mathring{a}$ values consistent with each other. The connection of J and $\mathring{a}$ is given by (2D7). The value of $\mathring{a}$ appears again in the Debye–Hückel expression of y, (equation 2D4). Finally, K_{diss} is connected with $\mathring{a}$ through the Denison and Ramsay[143] expression (2D2), which may be rewritten in the form of the Gilkerson[162] equation

$$K_{\text{diss}} = K_{\text{diss}}^0 e^{-b} = K_{\text{diss}}^0 \exp(-e^2/\mathring{a}\,\varepsilon kT) \quad (2D16)$$

in which $K_{\text{diss}}^0 = \exp(-E_s/k\,T)$ takes care of ion–solvent interactions, i.e. ion–dipole and polarization effects. The same value of $\mathring{a}$, of physically reasonable magnitude, must be used in the three expressions.

c. Ion Size and Mobility in Organic Solvents

The mobility u of an ion is its velocity in a unit field, and the equivalent conductance is given by its product with the Faraday

$$\Lambda_0 = \Lambda_{0+} + \Lambda_{0-} = F(u_+ + u_-) \qquad (2D17)$$

A unit field, of 1 volt/cm $= 1/300$ e.s.u./cm, produces a force of $e/300$ dynes on a charge $+e$ of a cation, and causes it to move in a viscous medium according to Stokes' law at a velocity

$$v_+ = u_+ = (e/300)/6\,\pi\,\eta_0\,R_+ \qquad (2D18)$$

and similarly for the anion. Inserting this into (2D17) yields

$$\Lambda_0 = (Fe/1800\,\pi)\,\eta_0^{-1}\,(1/R_+ + 1/R_-) \qquad (2D19)$$

and this, in turn, yields Walden's rule[163], expressed above in equation (2D11), where $const. = (Fe/1800\,\pi)\,(1/R_+ + 1/R_-)$. Provided the sum of the reciprocal hydrodynamic radii for a given electrolyte is constant, Walden's rule should be obeyed in a series of solvents or mixtures, independent of temperature and dielectric constant.

The deviations from Walden's rule often observed, point to the breakdown of the assumption on which (2D11) is based. The most obvious possibility is that the hydrodynamic radii of the ions vary with the solvent, in the case where the ions are solvated. Where transference data and ionic conductivities are known, the ionic radius R_i can be calculated[164], otherwise the mean radius $R = 2R_+ R_-/(R_+ + R_-)$ can be used as a measure of the solvation[165], for each electrolyte–solvent combination,

$$R = Fe/900\,\pi\,\Lambda_0\,\eta_0 \qquad (2D20a)$$

$$R_i = Fe/900\,\pi\,\Lambda_i\,\eta_0 \qquad (2D20b)$$

R_i can be related, in the same manner as in aqueous solution (section 1.B.b) to the solvation number[164]. The larger the ions intrinscially, the less is the interference from solvation, and the more nearly correct is the description of the motion in terms of spheres in a continuous viscous medium, to which Stokes' law applies.

Hydrogen-bonded solvents show large deviations (although aqueous solutions often behave according to the rule). A more subtle effect is that in polar solvents the solvent molecules must orient before, and relax behind, the moving ion, so that in a viscous medium the non-instantaneous process leads to an inhomogeneous electric field, which retards the motion of the ion[166]. This effect is proportional to the dipole moment of the solvent molecules, hence is related directly to the dielectric constant of the solvent. It has been observed[167] that a function $\Lambda_0\,\eta_0\,\varepsilon_0^{-n}$ is more nearly constant than Walden's $\Lambda_0\eta_0$, although n is certainly not as large as unity. There is however no quantitative treatment that can show which relationship of $\Lambda_0\,\eta_0$ and ε_0 will produce a constant of theoretical significance[168].

Diffusion coefficients have a bearing on the same problem, and they can also relate data on non-electrolytes to those for electrolytes. For univalent ions, the self-diffusion coefficient at infinite dilution is

$$D_{i0} = R\,T\Lambda_{i0}/F^2 = (k\,T/1800\,\pi)\,\eta_0^{-1}\,R_i^{-1} \tag{2D21}$$

so that if Stokes' law is obeyed, $D_{i0}\,\eta_0/T$ should be a constant. For the sodium ion in aqueous methanol and $1-$propanol this product varies only slowly with the composition[169], as does the corresponding product for non-electrolytes such as urea or sucrose[169, 170]. The near but not exact constancy has been attributed to the structure-making or -breaking effects of the solutes[168].

The parameters $\mathring{a}$ and R both indicate ionic sizes, the former in so far as contacts of ions of opposite charge are concerned, the latter with respect to the entity which moves in concentration or electric field gradients. It is expected that $r_{c+} + r_{c-} \lesssim \mathring{a} < 2R \sim R_+ + R_-$, and that the differences between the various radii are related to the solvation of the ions. It is also expected, from equations (2D2) or (2D16), that pair-wise association increases, the smaller the intrinsic size (i.e. the crystalline radii r_c) of the ions. These expectations are not always borne out by experiment, in particular in structured (hydrogen-bonded) solvents and for non-spherical ions, or those which are neither much smaller nor much larger than the solvent molecules. Thus for ions such as K^+, Cs^+, $(CH_3)_4N^+$, Br^-, I^- or NO_3^-, it is not possible to generalize as to whether association should increase with the crystalline radius or decrease with it. Data for the alkali halides in the lower alcohols[171] show that association increases with increasing r_{c+}, but decreases with increasing r_{c-}. Picrate showed decreasing association with increasing r_{c+}, as expected, while nitrate behaved as the halides[171, 172]. It is the competition of the interaction of the two ions with each other, and of each with the solvent, that determines whether the association should increase with the ion size or decrease, and no completely valid generalization can at present be made.

d. Activity Coefficients

Activity coefficients of ions in organic and mixed solvents can be calculated from the Debye–Hückel theory (section 1.C) in the range of its validity

$$\ln \gamma_j = - (z_j^2 e^2 / 2 k \varepsilon_o T) \kappa / (1 + \kappa a) \qquad (2D22)$$

where κ is the reciprocal radius of the ionic atmosphere

$$\kappa = (4 \pi N e^2 / 1000 \varepsilon_o k T)^{1/2} \left(\sum c_i z_i^2 \right)^{1/2} \qquad (2D23)$$

The range of validity of this treatment, however, is smaller than in aqueous solutions, because of the lower dielectric constant. The limitations of the theory have been discussed in section 1.C.b, and the practical limit $\kappa a < 0.2$ is often used (cf. section 2.D.a), which means $c < 2 \times 10^{-4} \varepsilon_o$. The assumption that the dielectric medium is a continuum with a dielectric constant ε_o breaks down completely for a mixed solvent, where selective solvation occurs. Association, furthermore, decreases the concentration of ions participating in the ionic atmosphere, so that $\sum \alpha_i c_i z_i^2$ must be used in (2D23) (cf. section 3.A.d). Thus the use of the simple Debye–Hückel expression

$$\log \gamma_\pm = - 1.825 \times 10^6 (\varepsilon_o T)^{-3/2} \left| z_+ z_- \right| I^{1/2} / (1 + 50.3 (\varepsilon_o T)^{-1/2} \mathring{a} I^{1/2})$$

$$(2D24)$$

with the values for the bulk dielectric constant ε_o of the solvent and the stoichiometric ionic strength I is severely limited.

One way around the difficulty is to ascribe all departures from ideality, except those described formally by (2D24), to association.

This approach is discussed in detail in section 3.A. In the very dilute region, $c < 2 \times 10^{-4} \, \varepsilon_o$, this has met with considerable success, as evidenced from the discussion on conductivity and association above.

A refinement of this treatment takes into account the solvation of the ions, and adds terms linear in the concentration. This treatment is formally valid also for mixed solvents[173] (ignoring the problem of the validity of the bulk dielectric constant in 2D24). In the same manner as equation (1D10) has been derived for aqueous solutions, it is possible to add a term $- \ln [1 + 0.001 \, M_W (x_W + (1 - x_W) \, M_W/M_S)^{-1}(v - h_W)m]$ to (2D24), assuming the electrolyte to be solvated by h_W moles of the component W of the solvent mixture, but not at all by the component S. Thus the electrostatic term (2D24) tends to lower the activity coefficient, usually below the values for aqueous solutions (Table 2D1), while at higher concentrations the solvation term causes them to increase again. Although this increase is observed, and is probably due to solvation, the quantitative justification for this term has not been proved experimentally. It is therefore concluded, that at present there is no valid theoretical treatment of deviations from ideality, for electrolytes at higher concentrations, in pure and mixed solvents of low dielectric constants.

In the absence of a valid theory, stoichiometric activity coefficients must be obtained by direct measurements. Unfortunately, only few electrolytes and solvents have been examined experimentally with high precision. Most of the information has been obtained for solutions of hydrogen chloride, from the e.m.f. of cells without liquid junction[174-185]. Data for alkali metal chlorides have been obtained similarly from amalgam cells[186-189], and recently also by the use of metal-cation-sensitive glass electrodes[190-192]. The data available for hydrochloric acid and sodium chloride at one concentration (0.1 m) and one temperature (25°) for several solvents and aqueous solvent mixtures are shown in Table 2D1. Data for other temperatures and concentrations, as well as for a few other electrolytes (lithium, potassium, caesium, lead and thallium chlorides, sulphuric acid) have been compiled[193-195]. The total amount of information available is however very scant, and only for hydrochloric acid is the information in any manner complete.

In this respect it is fortunate that a certain empirical relationship, called Harned's second rule, has been shown to hold[196]. In its most generalized form it says that the ratio of the mean molal activity coefficients of

TABLE 2D1. Activity coefficients of HCl and NaCl at 0.1 m and 25°C.

Solvent	Wt. %	x	ε	$\gamma_{\pm HCl}$	$\gamma_{\pm NaCl}$	Reference
N-Methylformamide	100	1.00	182.4		0.885	189
					0.940	190
N-Methylacetamide	100	1.00	165.5	0.947		179
Formamide	100	1.00	109.3	0.894		185
Water	100	1.00	78.4	0.796	0.778	
Methanol	10.0	0.059	74.1	0.780	0.762	181
	20.0	0.123	69.2		0.745	181, 186
	43.3	0.300	58.0	0.702		194
	50.0		54.9		0.600	195
	64.0	0.500	47.9	0.653	0.630	187, 194
	70.0		45.0		0.504	195
	84.2	0.750	38.2	0.563		194
	90.0		35.7		0.350	195
	100	1.00	31.5	0.453		194
				0.438		182
Ethanol	10.0	0.042	72.8	0.778		176
	20.0	0.089	67.0	0.755		176
	71.8	0.500	31.0	0.521		175,184
	100	1.00	24.3	0.232		185
				0.352		195
Isopropanol	10.0	0.032	71.4	0.773		183
Ethylene glycol	10.0	0.031	75.6	0.770		195
	20.0	0.068	72.8	0.764		195
	40.0	0.162	66.6	0.750		195
	60.0	0.303	59.4	0.717		195
Dioxane	20.0	0.049	60.8	0.720		176, 177
	50.0	0.170	34.3	0.382		180
	70.0	0.323	17.7	0.212		176, 177
	82.0	0.480	9.7	0.043		176, 177

two electrolytes at a given concentration and temperature is independent of the composition of the solvent. This broad statement has not
been tested experimentally, but there exists experimental justification
for the more restricted statement

$$\gamma_{\pm \text{ MX(S)}} = \gamma_{\pm \text{ HX(S)}} \gamma_{\pm \text{ MX(W)}}/\gamma_{\pm \text{ HX(W)}} \qquad (2D25)$$

where X is chloride and M lithium, sodium or potassium, and S is 10 or
20% wt. methanol in water[196, 197] or even higher concentrations of
aqueous methanol, ethylene glycol, dioxane, dimethylformamide and
several other solvents[191]. It is therefore possible to calculate approximate
values for $\gamma_{\pm \text{MCl(S)}}$ from the more complete data of $\gamma_{\pm \text{HCl(S)}}$ and the
extensive data for aqueous solutions.

Mixtures of electrolytes in organic or mixed solvents have also received
some, although limited attention. Harned's (first) rule (cf. section 1.E.b)
seems to be obeyed by hydrochloric acid–sodium chloride mixtures in
up to 60 % wt. of methanol in water[197], as for aqueous solutions[197], and
the coefficient α is not only independent of m_{NaCl} but also of the solvent
composition, at 1 m total concentration. Similarly, when small concentrations of hydrogen ions are exchanged for lithium ions in 0.5 M
perchlorate and nitrate solutions in 50% vol. of aqueous methanol,
ethanol or dioxane, the activity coefficients of hydrochloric acid do not
vary[198]. The same holds for replacement by sodium ions in aqueous
methanol, but in the other two solvents, $\log \gamma_{\pm \text{HCl}}$ is linear with the
sodium ion concentration, according to Harned's (first) rule, in the
range $0.1 < m_{\text{H}^+} < 0.5$[198].

The activity coefficients discussed above are defined with the usual
reference state of infinite dilution in the given solvent. For mixed solvents,
in particular, it is of interest also to estimate the activity coefficient
referred to infinite dilution in water. This introduces the (primary) medi-
um effect $_\text{M}\gamma_\text{G}$, defined as the ratio of the activity coefficient of the electrolyte G in water to that in the solvent at the given concentration

$$_\text{M}\gamma_\text{G} = {_\text{W}}\gamma_\text{G}/{_\text{S}}\gamma_\text{G} \qquad (2D26)$$

This medium effect does not vanish as the electrolyte concentration tends
towards zero, but is related to the difference in the standard chemical
potentials of the of the electrolyte in the two solvents, or to the standard
free energy of transfer of the electrolyte from water to the solvent

$$R\, T \ln \,_M \gamma_G = \,_W \mu_G^0 - \,_S \mu_G^0 = \Delta_W G_G^0 - \Delta_S G_G^0 \qquad (2D27)$$

In cells without liquid junction the primary medium effect is also related to the difference in the standard e.m.f. of the cell in water and in the solvent

$$_S E^0 = \,_W E^0 - (n\, F)^{-1}\, R\, T \ln \,_M \gamma_G \qquad (2D28)$$

The medium effect for electrolytes is thus an experimental quantity. In pH measurements, however, effects for single ions are required, which are, in organic or mixed solvents, as little defined thermodynamically as in aqueous solutions. One way to circumvent this difficulty is to use conventional values, as is done for aqueous solutions (cf. section 1.C.d). This approach has been tested experimentally for aqueous alcoholic solutions of several buffers, and found to lead to reasonable results[199], i.e. that the expression

$$pH - p(a_H \gamma_{Cl}) - \log \,_S \gamma_{Cl} = (0.4343\, F/R\, T)\, E_j + \log \,_M \gamma_{Cl} \qquad (2D29)$$

$$= \delta + \log \,_M \gamma^2_{HCl}$$

is constant. The first two terms on the left are experimental quantities $pH = pH_{std} + (0.4343\, F/R\, T)\,(E - E_{std})$ for the cell (I)

Pt, H_2/soln. of buffer or standard in solvent/reference electrode, and $p(a_H \gamma_{Cl}) = \log m_{Cl} + (0.4343\, F/R\, T)\,(E - \,_W E^0)$ for the cell (II)

Pt, H_2/buffer in solvent, Cl^-/AgCl, Ag, while the third is calculated conventionally from (2D24) with $\mathring{a} = 4.5$. E_j is the (appreciable) liquid junction in the first cell. It is thus possible to define and measure the quantity pH*, referred to the standard state in the solvent

$$pH^* = pa_H^* = pH - \delta = - (\log m_H \,_S \gamma_H) \qquad (2D30)$$

If a pH^*_{std} is assigned to a given standard buffer, then pH* values are obtained directly from cell (I) above. For each aqueous solvent a different pH^*_{std} value must however be conventionally assigned to any given standard buffer.

Other extrathermodynamic approaches have also been more or less successful. An empirical analysis of the solvent effect on the pK_a's of acids led to values of $_M \gamma_H$ in aqueous ethanol[200]. Redox couples, consisting of large, presumably unsolvated, ions, such as iron or osmium ferrocene, dipyridyl or phenanthroline[201] complexes, have been proposed for an evaluation of the liquid-junction potentials. Since absence of solvation for these highly charged, though large, ions is not assured, a different approach must be sought. Recently, the splitting of $_M\gamma_G$ into its constituent ionic parts has been proposed by Popovych[202], on the basis of the very large, and nearly equal sizes of the monovalent ions triisoamylbutyl-ammonium and tetraphenylboride

$$\tfrac{1}{2} \log {}_M\gamma_{i-Am_3BuNBPh_4} \simeq \log {}_M\gamma_{i-Am_3BuN^+} \simeq \log {}_M\gamma_{BPh_4^-} \quad (2D31)$$

The same electrolyte was used for evaluating single-ion conductivities[153] (section 2.D.b). The values obtained for other ions, through $_M\gamma_G$ of electrolytes containing the tetraalkylammonium or tetraphenylboride ions obtained from solubility data[203] (for sparingly soluble 1:1 salts $_s\gamma_G \simeq 1$, and $_M\gamma_G \simeq {}_w s/_s s$) were shown to be reasonable, in qualitative agreement with those estimated from various assumptions concerning the solvation of the ions[204]. Thus for methanol, $\log {}_M\gamma$ for hydrogen, potassium and chloride ions is near 2.0, for silver and rubidium ions near 1.0 and for the picrate ion near -1.0. This, or another convention can be used to split the medium effect for electrolytes into those of the ions, if strict additivity is exhibited. Values of the medium effect, $_M\gamma_G$, can be obtained from the standard e.m.f. of suitable cells (equation 2D28), from solubility[205], or from calorimetric enthalpies of transfer[126], with estimated entropy values. Enthalpies of transfer from one solvent to another for single ions, have been estimated, just as the free energies of transfer discussed above, e.g. for propylene carbonate, $\varepsilon = 65.1$[126].

e. Ionic Aggregation in Organic Solvents

As the concentration of an electrolyte in an organic solvent is increased from extreme dilution, deviations from ideal behaviour, characterized by complete dissociation and limiting conductivity Λ_0, arise. At extremely low concentrations, the limiting laws $\Lambda = \Lambda_0 - S c^{1/2}$ and $\log y_{\pm} = -A c^{1/2}$ hold, and up to $c \sim 2 \times 10^{-4} \varepsilon_0$ the more complicated laws (2D8) for the conductivity, and (2D24) for the activity coefficient hold, recognizing the partial association of the electrolyte to ion pairs, the expression (2D4) connecting the degree of dissociation α with the dissociation constant K_{diss}. The available theories cannot deal with the deviations occuring at higher concentrations with any degree of confidence. Unmixing into two organic solutions of the electrolyte might occur, caused by interactions of the dipole of the ion pairs with the polar, or polarized, solvent[206]. This has been observed for silver perchlorate, tetraisoannylammonium picrate and gallium(I) tetrachlorogallate(III) in benzene, ammonium tetrachloroferrate in ether[206], cobalt chloride in aqueous acetone (section 2.C.c) and many long-chain ammonium salts in several solvents (Chapter 10), among other cases.

There are two effects, which contribute to deviations at higher concentrations, which must now be taken into account, although at present there is no quantitative theory that can do this. One effect is the ion–solvent interaction, which among other consequences reduces the activity of the free solvent. Some aspects of this effect have been discussed above (section 2.D.b[161], section 2.D.c[173]). Another effect is the association of ions beyond the formation of ion pairs, i.e. the interaction of the ion pairs with ions to form ion triplets, and further aggregation of ions and ion pairs to even higher aggregates, ultimately to micelles.

Much of the information concerning the initial stages of the aggregation process is, again, due to conductance measurements. It is observed that at $\varepsilon < 20$ there appears a minimum in the $\Lambda = f(c)$ curve, which is not explicable in terms of the Fuoss–Onsager theory. Empirically[207], it was found that $c_{min} \simeq 3 \times 10^{-5} \varepsilon^3$. Since $\Lambda = f(c)$ is a decreasing function in the range where ionic atmosphere and ion-pair association effects predominate, the increase in Λ beyond the minimum must be explained. It has been ascribed to the formation of triple ions[208], which, for a symmetrical electrolyte (1:1, etc.) would be charged, contrary to the uncharged ion pair. Considerations similar to those involved in the theories of ion pairing of Denison and Ramsay[143], of Bjerrum[208] or of Fuoss[157] lead to an expression for the equilibrium constant for the formation of the triple ions (cf. section 3.A)

$$C^+ A^- + C^+ \rightleftarrows C^+ A^- C^+ \qquad (2D32a)$$

$$A^- + C^+ A^- \rightleftarrows A^- C^+ A^- \qquad (2D32b)$$

If the ions are of approximately the same size, it is expected that the constants for (2D32a) and (2D32b) are the same, K_{trip}. This constant has the form

$$K_{trip} = (\pi \, N \, a_3^3/1000) \, I \, (b_3) \qquad (2D33)$$

where a_3 is the exclusion distance for the formation of the triple ion, b_3 is $e^2/a_3 \, \varepsilon \, k \, T$ and $I \, (b_3)$ is a function, which according to the Bjerrum theory developed by Fuoss and Kraus[208] is a complicated integral, or according to a new theory of Fuoss[157] has the simpler form $\exp \, (- 3 \, R/2 + b_3/2 - E_S/k \, T)$, the last term accounting for solvent-solute interactions[162] (section 2.D.b). The exclusion distance can be envisaged as being given by the dimension of a cylinder made up of the ion pair, and a cation on the anionic side of it, and an anion on the cationic side. This cylinder has the diameter of the ions, $å$, and a length of 4 $å$, hence a volume of $\pi \, å^3$. Thus according to the Fuoss theory a_3 has the same magnitude as $å$, the diameter of the ions. According to the Bjerrum–Fuoss–Kraus theory, however, a_3 sets the limit of the dielectric constant, above which triple ions do not form (potential energy larger than $k \, T$): $\varepsilon_{limit} = (3 \, e^2/8 \, k \, T)/a_3$, or the distance of approach beyond which an ion in the vicinity of an ion pair is not considered to form a triple ion with it.

In both theories the minimum in the conductance curve is accounted for, and the minimum conductivity occurs at a concentration

$$c_{min} = (\Lambda_{0(3)}/\Lambda_0)/K_{trip} \qquad (2D34)$$

Using Walden's empirical rule $c_{min} = 3 \times 10^{-5} \, \varepsilon^3$, and values of K_{trip} calculated from (2D33) for ε around 3, leads to $\Lambda_{0(3)}/\Lambda_0 \simeq 0.5$, i.e. the reasonable conclusion that the equivalent conductivity of the hypothetical triple-ion electrolyte $(C_2A)^+ \, (CA_2)^-$ is about half that of the electrolyte $C^+ \, A^-$. The conductivity equation for the case where single ions, ion pairs and triple ions exist in solution is

$$\Lambda = \Lambda_0 K_{diss}^{1/2} \, c^{-1/2} + \Lambda_{0(3)} K_{trip} K_{diss}^{1/2} \, c^{1/2} \qquad (2D35)$$

neglecting activity coefficients and interionic effects on mobility, which is valid at $\varepsilon < 4$, since then, extremely small concentrations must be used. At $4 < \varepsilon < 10$ the interionic forces must be taken into account, which can be done in the form

$$\Lambda \, c^{1/2} \, g \, (c) = \Lambda_0 \, K_{diss}^{1/2} + \Lambda_{0(3)} \, K_{trip} K_{diss}^{1/2} (1 - \Lambda/\Lambda_0) \, c \qquad (2D36)$$

where

$\log g(c) = 0.4343 \, [S \, c^{1/2} \, \Lambda^{1/2} \, \Lambda_0^{-3/2} - 2.303 \, A \, c^{1/2} \, \Lambda^{1/2} \, \Lambda_0^{-1/2} + 0.5 \, \Lambda \Lambda_0^{-1}],$

S and A being the Onsager and Debye–Hückel limiting slopes, respectively (section 2.D.b). The left-hand side of (2D36) being experimentally known (with an estimated value of Λ_0), it can be plotted against $(1 - \Lambda/\Lambda_0)\,c$ to yield linear plots, from which K_{diss}, K_{trip} and $\Lambda_{0(3)}/\Lambda_0$ can be estimated. Experimental data for tetraisoamylammonium nitrate in dioxane–water solutions $(2.4 < \varepsilon < 6)$ confirm this treatment[208].

At very low dielectric constants (dioxane, benzene) and relatively high concentrations, the conductivity data[209], as well as freezing point depression data[210], show quadrupole formation, i.e. association of a triple ion with the oppositely charged single ion, or dipole–dipole association of two ion pairs. A theory has also been proposed for the equilibrium constant of this reaction[211], requiring the dipole moments of the electrolytes in the solution[212]. These can be estimated from the measured dielectric constants[213] (cf. section 2.A), yielding values of the order of 7–20 D, a little too small for the expected separation between the charges.

It is reasonable to assume that such quadrupoles are formed, and at higher concentrations also aggregates of five, six, $\cdots$ ions, in conformation with the damped sine-curve form of the conductance curve $\Lambda = f(c)$. However the theories[211] need so many free parameters as to be useless for giving an acceptable description of the species existing, their relative stabilities, and their contribution to such properties as conductivity, osmotic pressure, etc. It is therefore preferable to use a different approach to the problem of higher aggregation.

f. Higher Aggregation — Micelle Formation

Electrolytes are usually not sufficiently soluble in organic solvents, to show the effects of aggregation higher than to ion pairs or triplets. An important exception is the class of electrolytes which have long aliphatic chains (tails) and ionic heads, *amphipatic* electrolytes, or *ionic detergents*. These may dissolve to a considerable extent in solvents even of low dielectric constant, and form so-called 'inverted' micelles[214]. In order to understand the behaviour of the inverted micelles, in which the polar ionic heads of a number (5–50) of ionic detergent molecules are concentrated in an inner region, and the hydrocarbon chains are located in an outer region, facing the organic solvent, it is important also to discuss the behaviour of such detergents in aqueous and other polar solvents.

Typical ionic detergents have a straight-chain aliphatic hydrocarbon tail, with ca. 8 to 20 methylene units, bound covalently to an ionic head, which may be cationic (e.g. trimethylammonium or pyridinium) or anionic (e.g. sulphonate or sulphate), the charge of which is neutralized by an appropriate counter ion. Non-ionic detergents which have polar but not ionically dissociable heads (such as $(-OCH_2CH_2-)_n$ groups, ester groups or amine oxide groups) often behave quite similarly to the ionic detergents, and can sometimes be made ionic at appropriate pH values (carboxylic acids at high pH, amine oxides at low pH).

At low (often very low, $\lesssim 10^{-6}$M) concentrations of the ionic detergent in water, it exists as an ordinary electrolyte, dissociated into the amphipatic (long-chain) ion and the counter ion. The same behaviour is also shown by it in an organic solvent if sufficiently polar, otherwise it is associated to ion pairs[215,216]. When the concentration is increased, a point is reached, the *critical micelle concentration* (*cmc*), above which there is an abrupt change of many of the properties of the solution, such as conductivity, surface tension, etc. The *cmc* can be defined operationally in terms of a given property, or a set of properties that yield mutually consistent *cmc*'s. The experimentally most justified definition is the intersection of the straight lines resulting when plots of the property of the solution above and below the *cmc* are extrapolated towards the *cmc*[217]. Consistent results are then obtained for the *cmc* from a number of experimental methods, such as conductivity, light scattering or dye solubilization[217].

Above the *cmc* the amphipatic electrolyte is present mainly as *micelles*, which are responsible for the changed properties of the solutions. In aqueous solutions the micelles consist of a core containing the hydrocarbon tails, a surface at which the ionic heads extend from the core, but having also areas which are hydrocarbon surfaces, and a double layer. The ionic heads with some of the counter ions reach from the inner side, and the rest of the counter ions and the solvent form the outer side of this double layer, while hydrating water molecules constitute the intermediate material[218]. (Figure 2D1a). The structure of the micelles in organic solvents as mentioned above, is inverted[214], the core consisting of the polar part, the ionic heads, possibly hydrated, while the tails surround this core and constitute the surface of the micelle[219] (Figure 2D1b). The main properties of the micellar solutions, in which they differ from ordinary electrolyte solutions, is their solubilizing action, e.g. on water-insoluble dye molecules. These are incorporated into the micelles, and can therefore no longer be separated by filtration or centrifugation[220], although the micelles (which incorporate the dye) are retained by the membrane in a dialysis experiment[221].

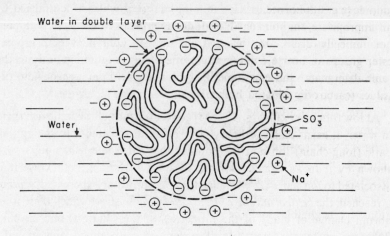

a) Micelle of sodium dodecyl sulphonate

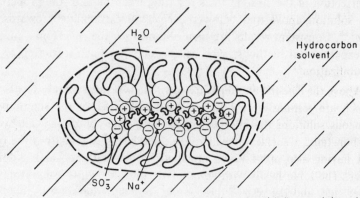

b) Inverted micelle of hydrated sodium dinonylnaphtalene sulphonate

Figure 2D1. Schematic representations of micelles a) in aqueous solutions, and b) inverted, in organic solvents.

In a system detergent–water the following equilibrium is set up

Detergent ions ⇄ Monomeric detergent ⇄ Micelles

(2D37)

Solid detergent ⇄ Hydrated solid detergent

Below the *cmc*, the concentration of micelles is negligible. At a sufficiently low temperature, the solid (hydrated) detergent is at equilibrium with the monomeric detergent solution, and an ordinary solubility–temperature curve results. At a certain temperature, the *Krafft point*, micelles are also at equilibrium with the excess solid. Above the Krafft point the solubility rises very sharply, and the *cmc*–temperature curve continues[222], until at a certain temperature the micelles begin to dissociate again to simple monomers[223]. Below the Krafft point, micelles are not found at any detergent concentration below the solubility limit.

The *cmc* is a fundamental property of the process of micellization. Properties of micellar solutions are often extrapolated down to the *cmc*, as the limit of zero micelle concentration, where inter micellar interactions are negligible, and the activity coefficients of the micelles are unity. This procedure is valid, although the assumption of zero micelle concentration at the *cmc* is not strictly true, and the effect of electrolyte present (both detergent monomers and added salt) will cause activity coefficients to deviate from unity. An added electrolyte G affects the *cmc*, it usually lowers it (section 10.A). Over certain concentration ranges a logarithmic law is obeyed[224, 225]

$$\log cmc = A - B \log c_G \qquad (2D38)$$

whereas a linear dependence of log *cmc* on c_G is more characteristic of non-ionic detergent behaviour, which conforms to the ordinary salting-out relationship[226] (section 1.F.a). The counter ion plays the major role in determining the salt effect, co-ions having very little influence. The effect of the ions is roughly according to the lyotropic series, the degree of hydration, or the disturbance of the water structure by the ions (section 1.B. c), both for cationic[227] and anionic[225] detergents. Similar effects, only in the opposite direction, are produced by urea: the *cmc* in its presence increases, roughly linearly with its concentration. This again is explained by the water structure-breaking effect (cf. discussion below on micelle formation and water structure)[225, 228].

The shape and size of a micelle are important properties that can be measured. Mean molecular weights can be estimated from light-scattering, ultracentrifugation, self-diffusion, conductivity, viscosity and x-ray diffraction measurements. The former two methods have been discussed recently[229], and shown to yield consistent results for detergents such as sodium lauryl sulphate or lauryltrimethylammonium bromide. The aggregation numbers range from about 60 in the absence of added electrolyte to

140, respectively 90, in the presence of 0.5 M salt. These results are in approximate agreement (within 10%) with results obtained from the reliable methods listed above. The hydrodynamic methods[230] yield also information on the shape of the micelles (they are spherical for sodium laury sulphate), their size (the radius of the hydrated micelle is 24 Å, that of the unhydrated micelle 16 Å), and the degree of hydration (about 33 % by volume). For non-spherical micelles (plate- or rod-like shapes) the simple relationship[231] between intrinsic viscosity and micelle volume v, $\eta_{intr} = 2.5\ Nv/M$ no longer holds, and an empirical relationship $\eta_{intr} = Kv^{1/2}/M$ holds instead. Absolute intensities of low-angle x-ray scattering also yield definite information concerning micelle size and shape[231]. At low concentrations the micelles are spherical, but become rod-like at higher concentrations, for detergents such as sodium lauryl sulphate, sodium laurate or cetyltrimethylammonium bromide. The radius and aggregation number obtained with this technique for spherical sodium laurylsulfate are 24.0 Å and 67, in good agreement with the data reported above.

The following generalizations can be made about the size and shape of the micelles. In dilute aqueous solutions, micelles are usually spherical, consisting of about 50–100 monomer units for ionic detergents, and somewhat more for non-ionic ones, and having a radius of ca. 20 Å. Micelles in organic solvents, which are inverted, are usually smaller (10–20 monomer units) and plate-like. As the concentration of detergent increases, or the electrolyte concentration in aqueous solutions increases, the micelles grow slowly, until at a certain concentration they change their shape from spherical to rod-like, and at a still higher concentration the solution separates into two distinct phases, the detergent-rich phase being a 'liquid crystal' or 'middle phase' with order detected by x-ray diffraction, or a true liquid phase, or a hydrated solid.

g. Micelle Stability

Two approaches have been used to explain the phenomenon of micelle formation and the properties of micelles. The most important property in this discussion is the fact that near the *cmc* the micelle has a definite size (the most probable aggregation number N). According to the mass-action approach[224, 232], the addition of a detergent monomer D to a smaller micelle to form the aggregate D_n is considered[233]

$$D_{n-1} + D \rightleftharpoons D_n \qquad (2D39)$$

The standard free energy change for this reaction (i.e. per monomer added) is $\Delta G_n^0 = -R\,T\ln(c_n/c_{n-1}c_1)$, where the subscripts to the concentration symbols denote the aggregation number. Since D_N is the most probable micelle, $c_N \simeq c_{N-1}$ and near the *cmc*, $c_1 = cmc$, hence

$$R\,T\ln\,cmc = \Delta\,G_N^0 \qquad (2D40)$$

If the micelles show ionic dissociation, with charge p per monomer unit, it would be better to write reaction (2D39) in term of ions, e.g. for an anionic detergent salt $Na^+ D^{-}$[234]

$$n\,D^- + n\,(1-p)\,Na^+ \rightleftharpoons D_n^{np-} \qquad (2D41)$$

where D_n^{np-} is the micelle incorporating sodium ions, so that

$$\Delta\,G_N^0/N\,R\,T = -(1/N)\ln c_N + \ln\,cmc + (1-p)\ln c_{Na^+} \quad (2D42)$$

Since N is large (50–100), the first term on the right is negligible, so that (2D42) can be rearranged to yield the empirical equation (2D38), where A is identified with $\Delta G_N^0/NRT$ and B with $2.303\,(1-p)$. Because of the large size of N, the activity of the monomers and of the micelles increases only slowly above the *cmc*, but N is not sufficiently large to prevent changes in the activity, as dialysis experiments[221] have shown. Although the micelles themselves (of sodium lauryl sulphate) are not transported across the membrane, as shown by dye tagging, the monomers are transported between two solutions, both above the *cmc* but at non-equal concentrations, i.e. the monomer activities are not the same.

There is, however, conflicting evidence that the activity of the species does remain essentially constant, at least in certain cases. Thus for solutions of calcium dinonylnaphthalenesulphonate in decane, the law of mass action requires for the $N = 12$ found, a considerable change of the surface tension in the range $cmc = 10^{-6}$ M to 10^3 *cmc*, whereas none has been found[219]. For systems with a constant activity above the *cmc* the phase-separation approach applies[222, 235, 236], according to which the micelles are a separate phase of constant activity. The *cmc* is thus simply the solubility limit of the detergent, and the state of aggregation of the separating micelles is the liquid one. According to this model the standard free energy of micelle formation is

$$\Delta G^0_{M(N)} = 2 R T \ln cmc \qquad (2D43)$$

in the absence of electrolyte, similar to the solubility product. In the presence of common ions from added electrolyte, the multiplier on the right-hand side is modified to $(1 + p)$, and it is clear that $\Delta G^0_M = N \Delta G^0_N$ in agreement with both (2D40) and (2D41), provided that in the latter, the term $(1/N) \ln c_N$ is neglected. This means that for sufficiently large micelles the two models lead to the same thermodynamic expressions[222], but that for small micelles $(N < 20)$ there is a difference, and the choice must be based on experimental evidence, as to whether activities remain constant or not. The derived thermodynamic expressions, such as enthalpy, entropy, volume change, etc. of micellization can be obtained from (2D43) by differentiation, e.g.

$$\Delta H^0_M = - R T^2 (1 + p) (\partial \ln cmc / \partial T)_p \qquad (2D44)$$

It now remains to show from molecular energetical arguments why micelles are at all stable, and why they form above the *cmc*, instead of a precipitate of a macro phase (liquid or solid). It should be remembered that the latter reaction is the rule for ordinary solutes, and that formation of micelles is the exception occuring in the case of the long-chain detergent. It is thus necessary to show that there is a minimum in the free energy of the aggregation process, occuring at a certain number N of monomers per aggregate (or a narrow range of values around the most probable value). There must be a special feature in the detergent molecules producing this minimum, and it is clear that a minimum would result from the balance of two opposing tendencies: of aggregation on the one hand and of repulsion on the other. It is necessary now to treat differently the micelles formed in aqueous solutions from the inverted micelles formed in organic solvents of low polarity.

In aqueous solutions, having a pronounced hydrogen-bonded structure and a highly polar nature, the following interactions and effects should lead to aggregation:

a) London dispersion forces between the hydrocarbon tails;
b) increased freedom of tails to find advantageous conformations;
c) increased randomness of ionic heads at the not completely covered surface;
d) hydrophobic bonding of the hydrocarbon parts exposed at the surface to the aqueous solvent;

e) disruption of the water structure by the ionic heads;

while the following interactions and effects would lead to dissociation:

f) loss of translational and rotational entropy of monomer;

g) repulsion of ionic heads on the surface of the micelle;

h) hydration of the ionic heads;

i) crowding of hydrocarbon tails in the micelle core.

In organic solutions, on the other hand, other factors lead to aggregation, rather than to complete phase separation or monomeric solution:

k) interaction of the tails with the solvent;

l) dipole attraction of ionic-head–counter-ion ion-pairs;

m) hydrogen-bonded structure and ionic hydration of solubilized water;

while the following factors hinder micelle formation:

f) loss of translational and rotational entropy of monomer (as above);

n) repulsion of ionic heads in the interior of the micelle;

o) crowding of branched hydrocarbon groups near the ionic heads in the micelle interior.

A quantitative theory has as yet been developed only for micelles in aqueous solutions, the statistical–mechanical treatment of Poland and Scheraga[218] being the most comprehensive, taking most of the factors (a–i) above into account. The free energy change per mole of N-sized micelles formed can be written as

$$\Delta G_{M(N)} = \Delta G_{trans} + \Delta G_{rot} + \Delta G_{int} + \Delta G_{solv} \qquad (2D45)$$

i.e. as a sum of terms corresponding to the translational, rotational, internal-disorder and solvent-interaction partition functions. In these terms, the contributions of the factors listed above is taken into account, in terms of the following variables: the aggregation number of the micelle N (which in turn determines its mass, volume, radius and surface area), the length of the monomer tail, i.e. the number u of —CH_2— links (which determines the mass, etc. of the monomer, as well as the maximal radius of the micelle, the extended length of the monomer), the temperature T and the concentration of detergent c_D, as well as numerical and physical constants. The four free energy terms have then the form

$$\Delta G_{trans}/R\,T = N^{-1} \ln (A N^{5/2} T^{3/2} u^{3/2} c_D^{-1}) \qquad (2D46a)$$

$$\Delta G_{rot}/R\,T = N^{-1} \ln (B N^{5/2} T^{3/2} u^{5/2}) \qquad (2D46b)$$

$$\Delta G_{int}/R\,T = \ln\left(C(N^{-1/3} - N^{-4/3})\,T^{3/2}\,u^{5/2}\right) \qquad (2D46c)$$

$$\Delta G_{solv}/R\,T = D\,u\,(1 - N^{-1/3}) \qquad (2D46d)$$

In these expressions, A and B are known constants, C is an unknown constant, and D is $1/R\,T$ times the free energy of formation of hydrophobic bonds per —CH_2— link, in the hydrocarbon chain in water, i.e., also a constant.

The conditions for micelle stability at any given temperature and detergent concentration are then

$$(\partial\,\Delta G_{M(N)}/\partial N)_{T,u,c_D} = 0 \qquad (2D47a)$$

$$(\partial^2\,\Delta G_{M(N)}/\partial N^2)_{T,u,c_D} > 0 \qquad (2D47b)$$

$$-\,\Delta G_{M(1)} < -\,\Delta G_{M(N)} \qquad (2D47c)$$

i.e. that the free energy have an extremum at the value N, that the extremum be a minimum, and that there will be a driving force for aggregation starting from the monomer. At the *cmc* the inequality (2D47c) becomes an equality, since the free energies of the micelles and monomer, are equal at that point*. It is not possible to solve equations (2D45), (2D46) and (2D47) explicitly, but a graphical representation of $\Delta G_{M(N)}$ = f (N, c_D) for given values of u and T yields values of the constant C and of *cmc* for the formation of stable micelles. A consequence of this analysis is to show that, for example, the *cmc* is related in a known way to the other variables

$$\ln cmc \simeq u\,D + 4\ln u + 3\ln T + 1/3\ln N + const. \qquad (2D48)$$

The hydrophobic-bonding contribution to micelle stability can be estimated from the solubility of hydrocarbon gases in aqueous micellar solutions[237], yielding a value of -1.27 for the constant D, compared with -1.52 for the hydrophobic-bond free energy for solubility in water.

---

* It is assumed that there are no intermediate species, i.e. oligomers. Such species have, however, been found in some systems where inverted micelles are probably formed (see below), and a different treatment is then required.

An alternative evaluation of $\Delta G_{M(N)}$ can be made in terms of ΔG_{el} + ΔG_{hc}, i.e. electrostatic repulsion and hydrocarbon attraction terms[233]. Now $\Delta G_{el} = Ne\psi_{(N)}$, where $\psi_{(N)}$ is the electrostatic potential at the surface of the micelle, and can be computed numerically from the Poisson–Boltzmann equation[233, 238], using experimental values of N and the radius of the micelle, obtained from x-ray data[231]. If this approach is correct, then for detergents having the same tail (e.g. sodium lauryl-sulphate and lauryltrimethylammonium bromide), ΔG_{hc}, calculated as

$$\Delta G_{hc} = (1 + p) R T \ln cmc - Ne\psi_{(N)} \qquad (2D48)$$

should be the same, irrespective of the nature of the ionic head, of N, or of the added-salt concentration. This was approximately confirmed[233].

An important parameter in evaluating the thermodynamic functions of micellization from experimental data, is the fractional charge on the micelle, p, i.e. its 'degree of ionization'. If in the above analysis of micellization ΔG_{hc} is estimated from independent data[233], values of p can be estimated from (2D48). For the two detergents studied $p = 0.45$ is obtained. More direct information on this, and lower values ($p \sim 0.26$) are obtained from conductivity data[239]. The total micellar charge Np varies with micelle size, as does the zeta potential, but most of the counter ions are closely associated with the ionic heads, and only 26 % of them (for long-chain ammonium chlorides) are in the diffuse double layer around the micelle, constituting the ionic atmosphere. Fractional charges, of similar magnitude (p ranging from 0.15 at no added bromide, to 0.25 at 0.5 M added NaBr, for lauryl trimethylammonium bromide, and from 0.15 to 0.22 for 0.1 − 0.4 M added NaCl, for sodium lauryl sulphate) have also been derived from light-scattering and ultracentrifugation measurements[229].

The question now arises as to why there is such a large degree of association for the micelles, while the same detergent electrolytes are completely dissociated below the cmc[216]. Is the association to the micelle specific, or does it involve only the strong electrostatic field caused by the ionic heads, concentrated in the vicinity of each other on the micelle surface, as in a polyelectrolyte? Most of the experimental evidence points to non-specific, electrostatic association[240]. The low charge obtained experimentally ($p = 0.15 - 0.25$) compared with that calculated from equation (2D48), based on a smooth spherical model ($p \geq 0.45$), and the discrepancy between the zeta potential derived from electrophoretic data, and the charge on the double layer, have led to the view that the micelle surface is 'rough'[241], i. e. that the ionic heads reach out from the micelle core into the double-layer region, and that some of the counter ions are situated in the valleys created between the ionic heads at the surface.

There are, however, two models of interaction through which counter ions can affect micelles specifically, beyond the general electrostatic association. One model operates in the case where charge transfer between ionic head and counter ion becomes possible, as with the alkyl-pyridinium iodides[242, 243]. Whereas in solvents such as chloroform or ethanol, laurylpyridinium iodide forms simple ion pairs, with a characteristic charge-transfer bond[242], in aqueous solutions micelles are formed[242, 243], and those counter ions which are associated with the micelle, again show charge transfer[243]. In the case of the iodide, but not the bromide or chloride where there is no charge transfer, about 70 % of the counter ions are estimated to be intimately bound to the micelles[243]. The charge-transfer band also yields information concerning the effective dielectric constant of the solution near to the micelle surface. Dielectric saturation causes an effective value $\varepsilon \sim 40$ in a layer about 4–5 Å from the mean micelle surface, in agreement with the value $\varepsilon = 36$, resulting from the comparison of the charge-transfer peak for the micelles with those of ion pairs in solvents of varying polarity[243]. This reduced dielectric constant is also in qualitative agreement with the increased value of acid association constants of indicator dyes, adsorbed on the surface of the micelles, according to the considerations discussed in section 2.D.a. Thus the apparent pH at the surface, is higher than in the solution, when measured with indicators[244].

Another contribution to the specificity that counter ions can exert on micelle formation arises from the terms in free energy of micelle formation that depend on the structure of the water. Although the bulk concentration of the counter ions and of added electrolyte can be small, the ions are concentrated in the double layer, forming a 3–6 M solution, in which considerable disturbance of the water structure does occur. That the structure of the water is an important condition for micellization through the gain in entropy made possible, when the structures formed around hydrocarbon chains are lost by micellization (hydrophobic-bonding term), can be seen in the fact that the structure-modifying urea, although increasing the polarity of the aqueous solution (increasing dielectric constant), reduces micelle formation (i.e. increases the

$cmc)^{225, 228}$, whereas in anhydrous sulphuric acid, which is also more polar than water but is very strongly structured, micelle formation (of protonated long-chain carboxylic acids) is very pronounced[245].

h. Inverted Micelles

The discussion above shows that much progress has been made in understanding micelle formation of ionic detergents in aqueous solutions. Much less is known about the inverted micelles[214] formed by ionic micellar compounds in organic solvents. As mentioned above (section 2.D.f), inverted micelles consist of aggregates of 5–50 monomers, in which the ionic heads are located in the core of the micelle, and the tails point outward, towards the the non-polar solvent. These micelles are smaller than those usually formed in aqueous solutions by the same ionic detergents, i.e. have lower values of N. The conditions for their formation have been briefly summarized in section 2.D.g. The inter-action of the tails with the solvent, dipole attraction of the ionic head–counter-ion pairs and accomodation of water in the core, favour micellization, while loss of monomer entropy, repulsion of ionized ionic heads and crowding of large organic groups in the core tend to limit the aggregation process. The inverted micelles can be envisaged as the end members of a dipole–dipole aggregation process, such as discussed in section 2.D.e. This does not, therefore, exclude intermediate species between the monomer and the micelle (section 10.A).

The polarity of the solvent has a major effect on the aggregation of the detergent. A light-scattering study has shown[246] that whereas in methanol and ethanol, sodium (bisdioctyl)sulphosuccinate forms only trimers and heptamers, respectively, it forms micelles with $N = 42$ in benzene, and 100 in cyclohexane. Other kinds of measurements, such as conductivity, have shown that micelles are not formed in solutions of long-chain alkylammonium chlorides in methanol and ethanol, of octadecanoate salts in nitrobenzene and dichloroethylene, and of some dinonyl naphthalenesulphonates in methanol, ethanol, butanol or acetone[216, 247] Laurylpyridinium iodide forms only ion pairs in ethanol or chloroform[242], as does also zinc laurate in isobutanol and pyridine[248]. However, the latter salt does form large aggregates in toluene[248], and non-polar aliphatic and aromatic hydrocarbons are the media in which inverted micelle formation has been generally studied. Details of similar obser-vations on long-chain ammonium salt solutions are given in section 10.A.

The size of the aggregates formed by divalent metal carboxylates with $u = 7$ to 17 methylene links is so as to make constant the product uN, (about 55 for the zinc salts in toluene), but it is larger, the larger the dipole moment of the metal-anion ion pair[248]. Phenyl stearates show also great sensitivity to the nature of the counter ion[249]. The viscosity of anhydrous benzene solutions of alkali metal and alkaline earth metal salts is very large, and is proportional to r_{+}^{-1}. As polar substances are added to the benzene solution, a breakdown of the micelles occurs, the effectiveness of substances in this respect decreasing along the series phenylstearic acid, water, phenol and ethanol. When up to one mole of water or ethyleneglycol is added per equivalent of detergent, small compact inverted micelles are stable, with the water 'solubilized' into the centre of the micelles, having a higher total concentration in the detergent solution than in pure benzene. When, however, more than one mole of water per equivalent of detergent is added, long rod-like micelles result, the water forming bridges between the micelles, until finally, when two moles of water per equivalent of alkaline earth, or four to five moles for alkali metal salts, are added, precipitation occurs. The rheological properties of the solutions lead to molar volumes of $\sim 15{,}000$ ml for the micelles, i.e. $N \sim 40$.

A considerable body of information exists concerning the behaviour of dinonylnaphthalenesulphonates (DNNS) in organic solvents. In contrast to the phenylstearates, the aggregation of DNNS salts of lithium, sodium, ammonium, caesium, magnesium, calcium, barium, zinc and aluminium is independent of the water content, amounting to $N = 9$–14, as found from several independent methods[250]. The volume of the micelle seems to depend largely on the coordination capability of the metal ions in the core of the micelle. The radius of such micelles is about 13 Å[250], although more recent information suggests that the micelles are rod-like rather than spherical. In the case of calcium DNNS, the rheological properties suggest an axial ratio of about two for the micelle, which consists of 12 monomers, the smaller radius being 14 Å. The nonyl groups (1, 2, 5, 5-tetramethylpentyl groups) are coiled considerably in this structure. Solubilized water may increase the size of the micelles somewhat. Not only water, but also other small, polar molecules can be solubilized by DNNS micelles, such as diethyl ether, acetone, n-propylamine, acetic acid, methyl acetate and methanol[251]. Both solvation of the cations, and accomodation in the micelle core beyond that, are involved here. Large polar molecules, such as nitronaphthalene, are not solubilized by DNNS micelles[223]. The detergents were found to be completely miscible with such solvents as aliphatic and aromatic hydrocarbons, chloroform or carbon tetrachloride, showing extremely low conductivities, and forming undissociated micelles. Osmotic coefficients, measured by vapour-pressure lowering, shows $N \sim 7$ in these solutions, depending somewhat on the cation[223]. In more polar solvents, such as ethyl acetate, isobutyl methyl letone, acetone and ethanol, solubility is again unlimited, but aggregation is concentration dependent, contrary to the former class of solvent. Conductivities here are appreciable, increasing with the dielectric constant of the solvent, aggregation and ion pairing decreasing with it. The solubility of DNNS detergents is however limited in another class of solvents, which includes siloxane oligomers, perfluorinated esters and alcohols, nitroparaffins and alkyl cyanides, while very low solubility is found in fluorocarbons[223].

These observations have been explained as follows[209], in particular for solvents of low polarity. The dissolving detergent is envisaged as a solvent-swollen phase, essentially a liquid, since the solubility conforms to the relations of binary liquid systems (section 2.B). The hydrocarbon part of the micelle, forming the outer mantel of the micelle, determines the solubility, hence the great similarity to the behaviour of the hydrocarbon, dinonylnaphthalene. Therefore the solubility parameters of the solute and solvent, rather than the polarity, determine the solubility. As the solubility parameter of the solvent increases, the micelles become smaller, untill they are best treated as dimers to tetramers, or finally as monomers, rather than as micelles. The alkali metal DNNS micelles behave as does dinonylnaphthalene, which has $\delta = 7.5$, and is therefore miscible at all proportions with solvents with $5 < \delta < 10$. A difference $\Delta\delta = 3.5$ between solute and solvent is required for limited solubility (cf. section 2.B.c). The aggregation number N of the micelles decreases as δ of the solvent increases, in the range 6.5 to 10.0, the decreased micelle size being tantamount to an increase of δ of the solute, i.e. there is a tendency for matching between the solubility parameters of solute and solvent. In the other direction, 15–20 DNNS units is the limit to which the micelles can grow as δ of the solvent decreases, because of crowding of the aromatic groups in the core. The extrapolation of N with δ leads to a value of $\delta = 10.5$ for the DNNS salt monomer, which is reasonable, in view of the heats of evaporation of related compounds. In solvents of high solubility parameters, $\delta > 10$, the DNNS salts should be soluble in all proportions as the monomer. Conductivity measurements do not confirm this, however[216]. Here, solvation and ionic dissociation play a more important role than does the solubility parameter, i.e. London forces.

Assuming that the solubility parameters and polarities of solvents increase parallel to are another, the behaviour of ionic detergents can be summarized as follows. In solvents of very low δ and polarity (fluorocarbons), only very slight solubility of the detergent is found, the unswollen solid is stable in contact with the solvent. As δ increases (aliphatic hydrocarbons) large inverted micelles are formed, which decrease in size as δ and solvent polarity increase (aromatic hydrocarbons, ketones, etc.), until dimers or monomers, partly dissociated to solvated ions, result in solvents such as methanol or nitrobenzene. As the dielectric constant increases, and solvent structure through hydrogen bonds becomes important, as in water, micelles reform, this time with the ionic heads pointing outward, the tails inward. It would be of interest to see whether such micelles are the rule also for other highly hydrogen-bonded, structured, high-polarity solvents, such as formamide, glycerol or hydrogen fluoride. The data for sulhpuric acid[245] point that way.

E. REFERENCES

1. J. H. Hildebrand and S. E. Wood, *J. Chem. Phys.*, **1**, 817 (1933).
2. J. E. Lennard-Jones, *Proc. Roy. Soc. (London)*, **A112**, 214 (1926).
3. J. O. Hirschfelder, C. F. Curtiss and R. B. Bird, *Molecular Theory of Gases and Liquids*, J. Wiley & Sons, New York, 1954.
4. W. H. Keesom, *Z. Physik.*, **22**, 126, 643 (1921); **23**, 225 (1922).
5. J. E. Lennard-Jones and P. E. Devonshire, *Proc. Roy. Soc. (London)*, **A163**, 59 (1937).
6. K. S. Pitzer, *J. Chem. Phys.*, **7**, 583 (1939).
7. J. H. Hildebrand, *J. Am. Chem. Soc.*, **37**, 970 (1915); **40**, 45 (1918).
8. P. Debye, *Polar Molecules*, Chemical Catalog Co. Inc., New York, 1929
9. J. R. Weaver and R. W. Parry, *Inorg. Chem.*, **5**, 703 (1966).
10. L. Onsager, *J. Am. Chem. Soc.*, **58**, 1486 (1936).
11. J. G. Kirkwood, *J. Chem. Phys.*, **7**, 911 (1939); *Trans. Faraday Soc.*, **42A**, 7 (1946).
12. P. Walden, *Salts, Acids and Bases*, McGraw Hill Book Co. Inc., New York, N.Y. 1929.
13. G. J. Janz and S. S. Danyluk, *Chem. Rev.*, **60**, 209 (1960).
14. I. N. Wilson, *Chem. Rev.*, **25**, 377 (1939).
15. R. J. W. Lefevre and A. J. Williams, *J. Chem. Soc.*, **108**, 115 (1960).
16. J. S. Dryden and R. J. Meakins, *Rev. Pure Appl. Chem.*, **7**, 15 (1957).
17. I. H. Smith and E. C. Creitz, *J. Res. Natl. Bur. Std.*, **46**, 145 (1951).

18. L. J. Bellamy, *The Infrared Spectra of Complex Molecules*, Methuen and Co., London, 1958.
19. L. M. Jackman, *Nuclear Magnetic Resonance Spectroscopy*, Pergamon Press, London, 1959.
20. W. G. Schneider, *J. Chem. Phys.*, **28**, 601 (1958).
21. R. Mecke, *Discussions Faraday Soc.*, **9**, 161 (1950).
22. W. C. Coburn and E. Grunwald, *J. Am. Chem. Soc.*, **80**, 1318 (1958).
23. G. C. Pimentel and A. L. McClellan, *The Hydrogen Bond*, W. H. Freeman and Co., San Francisco, California, 1960.
24. M. M. Davis, *J. Am. Chem. Soc.*, **84**, 3623 (1962).
25. L. J. Bellamy and R. L. Williams, *Proc. Roy. Soc. (London)*, **A254**, 119 (1960).
26. H. A. Pohl, M. E. Hobbs and P. M. Gross, *J. Chem. Phys.*, **9**, 408 (1941).
27. J. A. Barker, *Australian J. Chem.*, **6**, 207 (1953).
28. A. G. Williamson and B. L. Scott, *J. Phys. Chem.*, **65**, 275 (1961).
29. J. S. Rowlinson, *Liquids and Liquid Mixtures*, Butterworths Sci. Publ., London, 1959.
30. J. H. Hildebrand and R. L. Scott, *Solubility of Non-electrolytes*, Reinhold Publ. Co., New York., 3rd ed., 1950.
31. E. A. Guggenheim, *Mixtures*, Oxford Univ. Press, 1952, p. 184.
32. P. J. Flory, *J. Chem. Phys.*, **9**, 660 (1941); **10**, 51 (1942).
33. M. L. Huggins, *Ann. N.Y. Acad. Sci.*, **43**, 9 (1942).
34. J. H. Hildebrand and R. L. Scott, *Regular Solutions*, Prentice-Hall, Englewood Cliffs, N.J., 1962.
35. E. A. Guggenheim, *Thermodynamics*, North Holland Publ. Co., Amsterdam, 1949.
36. G. Scatchard, *Chem. Rev.*, **8**, 321 (1931); *J. Am. Chem. Soc.*, **56**, 995 (1934); *Trans. Faraday Soc.*, **33**, 160 (1937).
37. I. Prigogine, *Molecular Theory of Solutions*, North Holland Publ. Co., Amsterdam, 1957.
38. A. E. van Arkel, *Trans. Faraday Soc.*, **42B**, 81 (1946).
39. G. N. Malcolm and J. S. Rowlinson, *Trans. Faraday Soc.*, **53**, 921 (1957).
40. F. Zernike, *Chemical Phase Theory*, Kluwer, Antwerp, 1956, pp. 108–115.
41. A. E. Hill, *J. Am. Chem. Soc.*, **45**, 1143 (1923).
42. J. Timmermans and J. Lewin, *Discussions Faraday Soc.*, **15**, 188 (1953).
43. J. R. Jones and C. B. Monk, *J. Chem. Soc.*, 2633 (1963).
44. J. H. Hildebrand, *J. Chem. Phys.*, **17**, 1346 (1949).
45. S. D. Christian, H. E. Affsprung and J. R. Johnson, *J. Chem. Soc.*, 1896 (1963).; J. R. Johnson, S. D. Christian and H. E. Affsprung, *J. Chem. Soc.*, 77 (1966); W. L. Masterton and M. C. Gendrans, *J. Phys. Chem.*, **70**, 2895 (1966).
46. F. Franks, M. Gent and H. H. Johnson, *J. Chem. Soc.*, 2716 (1963).
47. J. S. Rowlinson in *Hydrogen Bonding* (Ed. D. Hadzi), Pergamon Press, London, 1959, p. 423; also ref. 29, p. 183.
48. e. g. J. Errera, *Helv. Chim. Acta*, **20**, 1373 (1937); J. Errera, R. Gasport and H. Sack, *J. Chem. Phys.*, **8**, 63 (1940); J. Errera and H. Sack,

Trans. Faraday Soc., **34**, 728 (1938); E. Greinacher, W. Lüttke and R. Mecke, *Z. Elektrochem.*, **59**, 23 (1955).

49. J. R. Holmes, D. Kirelson and W. C. Drinkard, *J. Am. Chem. Soc.*, **84**, 4677 (1962); A. D. Cohen and C. Reid, *J. Chem. Phys.*, **25**, 790 (1956); S. Weinberg and J. R. Zimmerman, *J. Chem. Phys.*, **23**, 748 (1955).

50. G. Nemethy and H. A. Scheraga, *J. Chem. Phys.*, **36**, 3382, 3417 (1962).

51. G. Nemethy and H. A. Scheraga, *J. Phys. Chem.*, **66**, 1773 (1962); **67**, 2888 (1963).

52. M. E. Friedman and H. A. Scheraga, *J. Phys. Chem.*, **69**, 3795 (1965).

53. G. Nemethy and H. A. Scheraga, *J. Chem. Phys.*, **41**, 680 (1964).

54. L. A. K. Stavely, R. J. S. Jones and B. L. Moore, *J. Chem. Soc.*, 2516 (1951).

55. L. Ehrenberg and I. Fischer, *Acta Chem. Scand.*, **2**, 657 (1948).

56. D. F. Othmer, M. M. Chudgar and S. L. Levy, *Ind. Eng. Chem.*, **44**, 1872 (1952).

57. C. J. F. Böttcher, *Theory of Electrical Polarization*, Elsevier Publ. Co., Amsterdam, 1952; R. Fowler and A. E. Guggenheim, *Statistical Thermodynamics*, Cambridge Univ. Press, 1952; J. N. Wilson, *Chem. Rev.*, **25**, 377 (1939).

58. A. E. Guggenheim, *Trans. Faraday Soc.*, **45**, 714 (1949).

59. D. Decocq, *Bull. Soc. Chim. France*, 127 (1964).

60. G. Akerlöf, *J. Am. Chem. Soc.*, **54**, 4125 (1932); G. Akerlöf and O. A. Short, *J. Am. Chem. Soc.*, **58**, 1241 (1936).

61. A. V. Celiano, P. S. Gentile and M. Cefola, *J. Chem. Eng. Data*, **7**, 391 (1962).

62. G. Oster, *J. Am. Chem. Soc.*, **68**, 2036 (1946).

63. J. A. V. Butler, *J. Phys. Chem.*, **33**, 1015 (1929).

64. J. G. Kirkwood in *Proteins ,Amino Acids and Peptides* (Eds. E. J. Cohn and J. T. Edsall), Reinhold Publ. Co., New York, 1943.

65. W. Danhauser and R. H. Cole, *J. Chem. Phys.*, **23**, 1762 (1955).

66. G. Oster and J. G. Kirkwood, *J. Chem. Phys.*, **11**, 175 (1943).

67. R. P. Bell, *Trans. Faraday Soc.*, **27**, 797 (1931); **31**, 1557 (1935).

68. J. G. Kirkwood, *J. Chem. Phys.*, **2**, 351 (1934).

69. A. C. Brown and D. G. J. Ives, *J. Chem. Soc.*, 1608 (1962).

70. R. P. Bell, *J. Chem. Soc.*, 1371 (1931).

71. I. Eliezer, *J. Chem. Phys.*, **42**, 3625 (1965).

72. A. E. Hill, *J. Am. Chem. Soc.*, **50**, 2678 (1928).

73. R. S. Mulliken, *J. Am. Chem. Soc.*, **74**, 811 (1952); *J. Chim. Phys.*, **61**, 20 (1964).

74. D. R. R. Gilson and C. A. McDowell, *J. Chem. Phys.*, **39**, 1825 (1963); **40**, 2413 (1964).

75. M. Taube, *J. Inorg. Nucl. Chem.*, **12**, 174 (1959); cf. Y. Marcus, *Chem. Rev.*, **63**, 161 (1963).

76. L. I. Katzin in *Advances in Transition Metal Chemistry* (Ed. R. L. Carlin), M. Dekker, Inc., Vol. 3, New York, 1967.

77. C. J. Hardy, B. F. Greenfield and D. Scargill, *J. Chem. Soc.*, 90 (1961).

78. M. Mori and R. Tsuchiya, *Nippon Kagaku Zasshi*, **77**, 1525 (1956).

79. H. L. Friedman, *J. Phys. Chem.*, **66**, 1595 (1962).
80. A. S. Kertes, *J. Inorg. Nucl. Chem.*, **27**, 209 (1965).
81. T. Wakahayashi, S. Oki, T. Omori and N. Suzuki, *J. Inorg. Nucl. Chem.*, **26**, 2255 (1964); T. Omori, T. Wakahayashi, S. Oki and N. Suzuki, *J. Inorg. Nucl. Chem.*, **26**, 2265 (1964); S. Oki, T. Omori, T. Wakahayashi and N. Suzuki, *J. Inorg. Nucl. Chem.*, **27**, 1141 (1965).
82. P. Walden, *Electrochemie Nichtwässeriger Lösungen*, Barth, Leipzig, 1924.
83. Cf. refs. a and b at bottom of Table 2.C.1.
84. L. R. Dawson and J. W. Vaughn, *J. Phys. Chem.*, **67**, 281 (1963).
85. T. Pavlopoulos and H. Strehlow, *Z. Physik. Chem. (Leipzig)*, **202**, 474 (1953).
86. R. E. Hamer, J. B. Syndor and E. S. Gilreath, *J. Chem. Eng. Data*, **8**, 411 (1963).
87. H. S. Isbin and K. A. Kobe, *J. Am. Chem. Soc.*, **67**, 464 (1945).
88. H. H. Willard and G. F. Smith, *J. Am. Chem. Soc.*, **45**, 286 (1923).
89. J. R. Ferraro, L. I. Katzin and G. Gibson, *J. Am. Chem. Soc.*, **77**, 327 (1955).
90. L. I. Katzin and J. R. Ferraro, *J. Am. Chem. Soc.*, **75**, 3821 (1953).
91. L. I. Katzin, J. R. Ferraro, W. W. Wendlandt and R. L. McBeth, *J. Am. Chem. Soc.*, **78**, 5139 (1956).
92. L. I. Katzin and J. R. Ferraro, *J. Am. Chem. Soc.*, **74**, 6040 (1952).
93. L. I. Katzin, D. M. Simon and J. R. Ferraro, *J. Am. Chem. Soc.*, **74**, 1191 (1952).
94. J. R. Ferraro, L. I. Katzin and G. Gibson, *J. Inorg. Nucl. Chem.*, **2**, 118 (1956).
95. P. Walden, *Z. Physik. Chem. (Leipzig)*, **58**, 479 (1907); **59**, 192 (1907).
96. L. I. Katzin and J. R. Ferraro, *J. Am. Chem. Soc.*, **74**, 2752 (1952).
97. J. N. Murrell, L. I. Katzin and B. L. Davies, *Nature*, **183**, 459 (1959).
98. L. I. Katzin, *J. Inorg. Nucl. Chem.*, **4**, 187 (1957).
99. J. E. Ricci and T. W. Davies, *J. Am. Chem. Soc.*, **62**, 407 (1940); J. E. Ricci and A. R. Lea, *J. Phys. Chem.*, **45**, 1096 (1941); J. E. Ricci and G. J. Nerse, *J. Am. Chem. Soc.*, **64**, 2305 (1942).
100. J. Kratohvil and B. Težak, *Arhiv Kemiju*, **26**, 243 (1954).
101. E. Hodge and A. B. Garrett, *J. Am. Chem. Soc.*, **63**, 1089 (1941); O. D. Black and A. B. Garrett, *J. Am. Chem. Soc.*, **65**, 862 (1943); M. V. Noble and A. B. Garrett, *J. Am. Chem. Soc.*, **66**, 231 (1944).
102. N. A. Izmailov and V. S. Chernyi, *Zh. Fiz. Khim.*, **34**, 127 (1960).
103. S. A. Voznesenskii and R. S. Biktimirov, *Zh. Neorg. Khim.*, **4**, 623 (1959).
104. B. B. Owen, *J. Am. Chem. Soc.*, **55**, 1922 (1933).
105. F. H. MacDougall and C. E. Bartsch, *J. Phys. Chem.*, **40**, 649 (1936); F. H. MacDougall and W. D. Larson, *J. Phys. Chem.*, **41**, 417 (1937).
106. T. W. Davis, J. E. Ricci and C. G. Santer, *J. Am. Chem. Soc.*, **61**, 3274 (1939).
107. W. Gordy and S. C. Stanford, *J. Chem. Phys.*, **8**, 170 (1940).
108. J. Miller and A. J. Parker, *J. Am. Chem. Soc.*, **83**, 117 (1961).
109. A. J. Parker, *Quart. Rev. (London)*, **16**, 163 (1962); *J. Chem. Soc.*, 220 (1966).

110. F. S. Larkin, *Trans. Faraday Soc.*, **59**, 403 (1963); T. R. Griffiths and M.C.R. Symons, *Mol. Phys.*, **3**, 90 (1960).
111. I. I. Antipova-Karataeva and E. E. Vainshtein, *Zh.Neorg.Khim.*, **6**, 1115 (1961); A. S. Kertes in *Solvent Extraction Chemistry of Metals*, (Ed. H.A.C. McKay), McMillan, London, 1966, p. 377.
112. K. Ekelin and L. G. Sillèn, *Acta Chem. Scand.*, **7**, 987 (1953).
113. G. Charlot and B. Tremillon, *Les Reactions Chimiques dans les Solvants et les Sels Fondus*, Gauthier-Villars, Paris, 1963.
114. I. V. Nelson, R. C. Larson and R. T. Iwamoto, *J. Inorg. Nucl. Chem.*, **22**, 279 (1961).
115. I. M. Kolthoff and J. F. Coetzee, *J. Am. Chem. Soc.*, **79**, 870, 1825 (1957); R. C. Larson and R. T. Iwamoto, *J. Am. Chem. Soc.*, **82**, 3239, 3526 (1960).
116. D. C. Luehrs, R. T. Iwamoto and J. Kleinberg, *Inorg. Chem.*, **5**, 201 (1966).
117. H. S. Frank, *J. Chem. Phys.*, **23**, 2023 (1955).
118. J. Padova, *J. Chem. Phys.*, **39**, 1552 (1963); *Israel J. Chem.*, **1**, 258 (1963); Israel A. E. C. Semi Annual Reports IA–900 and IA–920 (1963).
119. N. Izmailov, *Dokl. Akad. Nauk SSSR*, **148**, 1364 (1963).
120. E. Grunwald in *Electrolytes* (Ed. B. Pesce), Pergamon Press, Oxford, 1962, p. 62.
121. E. Grunwald, G. Baughman and G. Kohnstam, *J. Am. Chem. Soc.*, **82**, 5801 (1960).
122. N. Bjerrum and E. Larrson, *Z. Physik. Chem.* (*Leipzig*), **127**, 358 (1927).
123. H. Strehlow, *Z. Electrochem.*, **56**, 827 (1952).
124. D. Feakins and P. Watson, *J. Chem. Soc.*, 4734 (1963).
125. H. L. Friedman and G. R. Haugen, *J. Am. Chem. Soc.*, **76**, 2060 (1954); G. R. Haugen and H. L. Friedman, *J. Phys. Chem.*, **60**, 1363 (1956); **67**, 1757 (1963).
126. Y.-C. Wu and H. L. Friedman, *J. Phys. Chem.*, **70**, 501, 2020 (1966).
127. J. Koskikallio and S. Syrjapalo, *Acta Chem. Scand.*, **19**, 425 (1965).
128. H. Schneider and H. Strehlow, *Z. Electrochem.*, **66**, 309 (1962).
129. A. Fratiello and D. C. Douglass, *J. Chem. Phys.*, **39**, 2017 (1963).
130. Z. Luz and S. Meiboom, *J. Chem. Phys.*, **40**, 1056 (1964).
131. J. C. Jayne and E. C. King, *J. Am. Chem. Soc.*, **86**, 3989 (1964).
132. I. V. Nelson and R. T. Iwamoto, *Inorg. Chem.*, **3**, 661 (1964).
133. N. J. Friedman and R. A. Plane, *Inorg. Chem.*, **2**, 11 (1963).
134. P. K. Migal and N. Kh. Grinberg, *Zh. Neorg. Khim.*, **6**, 727 (1961); **7**, 527, 531, 1309 (1962).
135. C. K. Jørgensen, *Acta Chem. Scand.*, **8**, 175 (1954).
136. J. Bjerrum and C. K. Jørgensen, *Acta Chem. Scand.*, **7**, 951 (1953).
137. L. I. Katzin and E. Gebert, *Nature*, **175**, 425 (1955); C. K. Jørgensen and J. Bjerrum, *Nature*, **175**, 425 (1955).
138. I. S. Pominov, *Zh. Fiz. Khim.*, **31**, 1926 (1957).
139. H. A. C. McKay and A. R. Mathieson, *Trans. Faraday Soc.*, **47**, 428 (1951); E. Glueckauf, H. A. C. McKay and A. R. Mathieson, *Trans. Faraday Soc.*, **47**, 437 (1951).

140. L. I. Katzin and J. C. Sullivan, *J. Phys. Chem.*, **55**, 346 (1951).
141. P. C. Yates, R. Laran, R. E. Williams and T. E. Moore, *J. Am. Chem. Soc.*, **75**, 2212 (1953).
142. R. G. Pearson, *J. Am. Chem. Soc.*, **85**, 3533 (1963).
143. J. T. Denison and J. B. Ramsay, *J. Am. Chem. Soc.*, **77**, 2615 (1955).
144. J. C. James, *J. Am. Chem. Soc.*, **71**, 3248 (1949); *J. Chem. Soc.*, 1094 (1950); 153 (1951).
145. C. W. Davies and G. O. Thomas, *J. Chem. Soc.*, 3660 (1958).
146. J. R. Bevan and C. B. Monk, *J. Chem. Soc.*, 1392 (1956).
147. F. Accascina, S. Petrocci and R. M. Fuoss, *J. Am. Chem. Soc.*, **81**, 1301 (1959).
148. R. M. Fuoss, *J. Am. Chem. Soc.*, **80**, 5059 (1958).
149. Y. H. Inami, H. K. Bodenseh and J. B. Ramsay, *J. Am. Chem. Soc.*, **83**, 4745 (1961).
150. J. B. Ramsey and E. L. Colichman, *J. Am. Chem. Soc.*, **69**, 3041 (1947); F. Healy and A. E. Martell, *J. Am. Chem. Soc.*, **73**, 3296 (1951); J. T. Denison and J. B. Ramsay, *J. Chem. Phys.*, **18**, 770 (1950); K. H. Stern, F. Healy and A. E. Martell, *J. Chem. Phys.*, **19**, 1114 (1951).
151. Y. H. Inami and J. B. Ramsay, *J. Chem. Phys.*, **31**, 1297 (1959).
152. J. B. Hyne, *J. Am. Chem. Soc.*, **85**, 304 (1963).
153. M. A. Coplan and R. M. Fuoss, *J. Phys. Chem.*, **68**, 1181 (1964); D. F. Evans and C. Zawoyski, *J. Phys. Chem.*, **69**, 3878 (1965).
154. G. Atkinson and S. Petrucci, *J. Am. Chem. Soc.*, **86**, 7 (1964); *J. Phys. Chem.*, **70**, 2050 (1966).
155. A. D'Aprano and R. M. Fuoss, *J. Phys. Chem.*, **67**, 1704, 1722 (1963).
156. E. Hirsch and R. M. Fuoss, *J. Am. Chem, Soc.* **82**, 1018 (1960); R. L. Kay and D. F. Evans, *J. Am. Chem. Soc.*, **86**, 2748 (1964).
157. R. M. Fuoss and F. Accascina, *Electrolytic Conductance*, Interscience, New York, 1959, p. 225.
158. G. Jones and M. Dole, *J. Am. Chem. Soc.*, **51**, 2950 (1929); H. Falkenhagen and M. Dole, *Z. Physik. Chem. (Frankfurt)*, **6**, 159 (1929).
159. R. M. Fuoss and L. Onsager, *J. Phys. Chem.*, **61**, 668 (1957); **62**, 1339 (1958).
160. T. Shedlovsky, *J. Franklin Inst.*, **225**, 739 (1938).
161. C. W. Davies, *Ion Association*, Butterworths Sci. Publ., London, 1962, p. 9.
162. W. R. Gilkerson, *J. Chem. Phys.*, **25**, 1199 (1956).
163. P. Walden, *Z. Physik. Chem. (Leipzig)*, **55**, 207 (1906).
164. J. M. Natley and M. Spiro, *J. Phys. Chem.*, **70**, 1502 (1966).
165. E. S. Amis, *J. Phys. Chem.*, **60**, 428 (1956).
166. R. Zwanzig, *J. Chem. Phys.*, **38**, 1603 (1963).
167. N. G. Forster and E. S. Amis, *Z. Physik. Chem. (Frankfurt)*, **3**, 365 (1955).
168. R. L. Kay and D. F. Evans, *J. Phys. Chem.*, **70**, 2325 (1966).
169. A. E. Marchinkowsky, H. O. Phillips and K. A. Kraus, *J. Phys. Chem.*, 3968 (1965).

170. L. G. Longsworth, *J. Am. Chem. Soc.*, **75**, 5705 (1953); *J. Phys. Chem.*, **67**, 689 (1963).

171. R. L. Kay, *J. Am. Chem. Soc.*, **82**, 2099 (1960); J. L. Hawes and R. L. Kay, *J. Phys. Chem.*, **69**, 2420 (1965); R. L. Kay and J. L. Hawes, *J. Phys. Chem.*, **69**, 2787 (1965).

172. G. D. Parfitt and A. L. Smith, *Trans. Faraday Soc.*, **59**, 257 (1963).

173. R. A. Robinson and R. H. Stokes, *Electrolyte Solutions*, Butterworths Sci. Publ., London, 2nd ed., 1959, pp. 251–253.

174. G. Scatchard, *J. Am. Chem. Soc.*, **47**, 2098 (1925).

175. Z. Z. Lucasse, *Z. Physik. Chem. (Leipzig)*, **121**, 254 (1926).

176. H. S. Harned and coworkers., *J. Am. Chem. Soc.*, **55**, 2179 (1933); **60**, 339, 2128, 2133 (1938).

177. H. S. Harned, J. D. Morrisson, F. Walker, J. G. Donelson and C. Calmon, *J. Am. Chem. Soc.*, **61**, 48 (1939).

178. H. Tanaguchi and G. J. Janz, *J. Phys. Chem.*, **61**, 688 (1957).

179. L. R. Dawson, R. C. Sheridan and H. C. Eckstrom, *J. Phys. Chem.*, **65**, 1829 (1961).

180. E. Grunwald and A. L. Bacarella, *J. Am. Chem. Soc.*, **80**, 3840 (1958).

181. H. S. Harned and H. C. Thomas, *J. Am. Chem. Soc.*, **58**, 761 (1936).

182. G. Nonhebel and H. Hartley, *Phil. Mag.*, **50**, [6], 298, 729 (1923).

183. H. S. Harned and C. Calmon, *J. Am. Chem. Soc.*, **61**, 1491 (1939).

184. H. S. Harned and M. H. Fleysher, *J. Am. Chem. Soc.*, **47**, 82 (1925).

185. R. K. Aggarwal and B. Nayak, *J. Phys. Chem.*, **70**, 2568 (1966).

186. G. Akerlöf, *J. Am. Chem. Soc.*, **52**, 2353 (1930).

187. J. P. Butler and A. R. Gordon, *J. Am. Chem. Soc.*, **70**, 2276 (1948).

188. M. M. Shultz and A. E. Parfenow, *Vestn. Leningr. Univ. Ser. Fiz. i. Khim.*, **16**, (3), 118 (1958).

189. E. Luksha and C. M. Criss, *J. Phys. Chem.*, **70**, 1496 (1966).

190. Y. M. Povarov, P. I. Gorbanev, K. M. Kessler and I. V. Safonova, *Dokl. Akad. Nauk. SSSR*, **142**, 1128 (1962); **155**, 1411 (1964).

191. R. O. Lanter, *J. Phys. Chem.*, **69**, 2697 (1965).

192. G. A. Rechnitz and S. B. Zamochnik, *Talanta*, **11**, 979 (1964).

193. H. S. Harned and B. B. Owen, *The Physical Chemistry of Electrolytic, Solutions*, Reinhold Publ. Co., New York 3rd. ed., 1958, pp. 717–729.

194. J. B. Conway, *Electrochemical Data*, Elsevier Publ. Co., Amsterdam, 1952.

195. R. Parsons. *Handbook of Electrochemical Constants*, Butterworths Sci. Publ., London, 1959 (compiled from undisclosed sources).

196. H. S. Harned, *J. Phys. Chem.*, **66**, 589 (1962).

197. G. Akerlöf, J. W. Tear and H. Turk, *J. Am. Chem. Soc.*, **59**, 1916 (1937).

198. F. J. C. Rossotti and H. S. Rossotti, *J. Phys. Chem.*, **68**, 3773 (1964).

199. R. G. Bates, M. Paabo and R. A. Robinson, *J. Phys. Chem.*, **67**, 1833 (1963).

200. E. Grunwald and B. J. Berkowitz, *J. Am. Chem. Soc.*, **73**, 4939 (1951). B. Gutbezahl and E. Grunwald, *J. Am. Chem. Soc.*, **75**, 559, 565 (1953).

201. H. M. Koepp, H. Wend and H. Strehlow, *Z. Elektrochem.*, **64**, 483 (1960); W. Ward, *Ph. D. thesis*, Univ. of Iowa, 1958; I. V. Nelson and R. T. Iwamoto, *Anal. Chem.*, **33**, 1795 (1961); **35**, 867 (1963); I. M. Kolthoff and F. G. Thomas, *J. Phys. Chem.*, **69**, 3049 (1965).
202. O. Popovych, *Anal. Chem.*, **38**, 558 (1966).
203. O. Popovych and R. M. Friedman, *J. Phys. Chem.*, **70**, 1671 (1966).
204. N. A. Izmailov, *Dokl. Akad. Nauk SSSR*, **126**, 1033 (1959); **127**, 104 (1959); **149**, 884, 1364 (1963).
205. C. F. Coleman, *J. Phys. Chem.*, **69**, 1377 (1965).
206. H. L. Friedman, *J. Phys. Chem.*, **66**, 1595 (1962).
207. P. Walden, *Z. Physik. Chem. (Leipzig)*, **147A**, 1 (1930).
208. R. M. Fuoss and C. A. Kraus, *J. Am. Chem. Soc.*, **55**, 2387 (1933).
209. R. M. Fuoss and C. A. Kraus, *J. Am. Chem. Soc.*, **55**, 21, 3614 (1933); W. F. Luder and C. A. Kraus, *J. Am. Chem. Soc.*, **58**, 255 (1936).
210. F. M. Baston and C. A. Kraus, *J. Am. Chem. Soc.*, **56**, 2017 (1934); C. A. Kraus and R. A. Vingee, *J. Am. Chem. Soc.*, **56**, 511 (1934).
211. R. M. Fuoss and C. A. Kraus, *J. Am. Chem. Soc.*, **57**, 1 (1935).
212. R. M. Fuoss, *J. Am. Chem. Soc.*, **56**, 1031 (1934).
213. J. A. Geddes and C. A. Kraus, *Trans. Faraday Soc.*, **32**, 585 (1936); G. S. Hooper and C. A. Kraus, *J. Am. Chem. Soc.*, **56**, 2265 (1934).
214. C. R. Singleterry, *J. Am. Oil Chem. Soc.*, **32**, 446 (1955).
215. G. D. Parfitt and A. L. Smith, *J. Phys. Chem.*, **66**, 942 (1962).
216. R. C. Little and C. R. Singleterry, *J. Phys. Chem.*, **68**, 2709 (1964).
217. R. J. Williams, J. N. Phillips and K. J. Mysels, *Trans. Faraday Soc.*, **51**, 728 (1955); H. Schott, *J. Phys. Chem.* **70**, 2296 (1966).
218. D. C. Poland and H. A. Scheraga, *J. Phys. Chem.*, **69**, 2431 (1965).
219. F. M. Fowkes, *J. Phys. Chem.*, **66**, 1843 (1962).
220. P. Mukerjee and K. J. Mysels *J. Am. Chem. Soc.*, **77**, 2937 (1955).
221. K. J. Mysels, P. Mukerjee and M. Abu-Hamdiyya, *J. Phys. Chem.*, **67**, 1943 (1963); M. Abu-Hamdiyya and K. J. Mysels, *J. Phys. Chem.* **71**, 418 (1967).
222. K. Shinoda and E. Hutchinson, *J. Phys. Chem.*, **66**, 577 (1962).
223. R. C. Little and C. R. Singleterry, *J. Phys. Chem.*, **68**, 3453 (1964).
224. M. L. Corrin and W. D. Harkins, *J. Am. Chem. Soc.*, **69**, 683 (1947).
225. M. J. Schick, *J. Phys. Chem.*, **68**, 3585 (1964).
226. M. J. Schick, S. M. Atlas and F. R. Eirich, *J. Phys. Chem.*, **66**, 1326 (1962); M. J. Schick, *J. Colloid Sci.*, **17**, 810 (1962).
227. E. W. Anacker and H. M. Ghose, *J. Phys. Chem.*, **67**, 1713 (1963).
228. P. Mukerjee and A. Ray, *J. Phys. Chem.*, **67**, 190 (1963).
229. E. W. Anacker, R. M. Rush and J. S. Johnson, *J. Phys. Chem.*, **68**, 81 (1964).
230. W. L. Courchene, *J. Phys. Chem.*, **68**, 1870 (1964).
231. F. Reiss-Husson and V. Luzzati, *J. Phys. Chem.*, **68**, 3504 (1964).
232. R. C. Murray, *Trans. Faraday Soc.*, **31**, 207 (1935); J. N. Phillips, *Trans. Faraday Soc.*, **51**, 561 (1955); K. J. Mysels, *J. Colloid Sci.*, **10**, 507 (1955).
233. M. F. Emerson and A. Holtzer, *J. Phys. Chem.*, **69**, 3718 (1965).

234. P. Mukerjee, *J. Phys. Chem.*, **66**, 1375 (1962).
235. A. E. Alexander, *Trans. Faraday Soc.*, **38**, 54 (1942), G. Stainsby and A. E. Alexander, *Trans. Faraday Soc.*, **45**, 585 (1949); **46**, 587 (1950).
236. E. Hutchison, A. Inaba and L. G. Bailey, *Z. Physik. Chem. (Frankfurt)*, **5**, 344 (1955).
237. A. Wishnia, *J. Phys. Chem.*, **67**, 2079 (1963).
238. M. Nagasawa and A. Holtzer, *J. Am. Chem. Soc.*, **86**, 531 (1964).
239. H. W. Hoyer and A. Marmo, *J. Phys. Chem.*, **65**, 1807 (1961).
240. S. Lapanje and S. A. Rice, *J. Am. Chem. Soc.*, **83**, 496 (1961); L. Kotin and M. Nagasawa, *J. Am. Chem. Soc.*, **83**, 1026 (1961); P. Mukerjee, *J. Phys. Chem.*, **66**, 943 (1962).
241. D. Stigter and K. J. Mysels, *J. Phys. Chem.*, **59**, 45 (1955).
242. W. D. Harkins, H. Krizek and M. L. Corrin, *J. Colloid Sci.*, **6**, 576 (1951).
243. P. Mukerjee and A. Ray, *J. Phys. Chem.*, **70**, 2138, 2145, 2150 (1966).
244. P. Mukerjee and K. Banerjee, *J. Phys. Chem.*, **68**, 3569 (1964).
245. J. Steigman and N. Shane, *J. Phys. Chem.*, **69**, 968 (1965).
246. A. Kitahara, T. Kobayashi and T. Tachibana, *J. Phys. Chem.*, **66**, 363 (1962).
247. A. W. Ralston and C. W. Hoerr, *J. Am. Chem. Soc.*, **68**, 851, 2460 (1946); H. E. Weaver and C. A. Kraus, *J. Am. Chem. Soc.*, **70**, 1707 (1948); W. J. McDowell and C. A. Kraus, *J. Am. Chem. Soc.*, **73**, 2173 (1957).
248. S. M. Nelson and R. C. Pink, *J. Chem. Soc.*, 1804 (1951); 1744 (1952).
249. J. G. Honig and C. R. Singleterry, *J. Phys. Chem.*, **58**, 201 (1954); **60**, 1108 (1956).
250. S. Kaufman and C. R. Singleterry, *J. Colloid Sci.*, **10**, 139 (1957); **12**, 465 (1959); T. F. Ford, S. Kaufman and O. D. Nichols, *J. Phys. Chem.*, **70**, 3726 (1966).
251. S. Kaufman, *J. Phys. Chem.*, **68**, 2814 (1964).

3

Complex Formation
in Solution

The background material to a discussion of the solvent extraction and ion exchange of metal complexes requires an account of the formation of complexes in solution. The methods that have been proposed to study complex formation have been critically reviewed in a number of comprehensive monographs. It is therefore necessary here, to discuss only those aspects of this field which bear directly on extraction or exchange problems.

The discussion of ionic association in section 2.D is mainly qualitative, so the quantitative treatments, such as those of Bjerrum and of Fuoss, concerning ion-pair formation in solution, are dealt with in section 3.A. Since there are controversies concerning the validity of the ion-pairing concept, in particular when the ions are not in contact, this problem is discussed in relation to the methods used to detect and study ion pairing.

Complex formation in solution involves interactions stronger than the electrostatic ones leading to ion pairing, and directed valence bonds are involved. A discussion of these is outside the scope of this book, and only the thermodynamics of complex formation is briefly discussed in section 3.B. The determination of the complex formation constants is, however, an important aspect of solvent extraction and ion exchange, as are the ways of describing the equilibrium mixtures of species in a complex formation system. Space is therefore devoted to the mathematical treatment of equilibrium data, to graphical and numerical methods of solving the equilibrium equations, and to appropriate methods of presenting the results. The concepts and equations are frequently used in Chapters 5, 6 and 7. The best mathematical methods, however, cannot remove the problem of the reality of apparent species, which may be thought to be present. Section 3.B is therefore concluded with a discussion of the reality of complexes, and an attempt to disentangle medium effects from the formation of chemical species.

This monograph cannot review, or even list, the available material concerning association and complex formation in aqueous solutions. Compilations of such data, although not critical reviews, have been published. Still, some of the results, particularly those concerning the mineral acids, from which most of the extraction and ion exchange processes take place, ought to be included for ready reference. This has been done in section 3.C, where the dissociation of strong acids, the hydrolysis of metal cations, and the association of salts are briefly treated, with results shown in tables and graphs.

A. ION-PAIR FORMATION

a. Introduction

The interactions of ions in solution discussed in the previous sections included hydration (1.B), solvation (2.C), long-range coulombic interactions (1.C), interactions with molecules of non-electrolytes (1.F) and various short-range interactions in water (1.D) or solvents in general (2.D). Among the interactions was mentioned briefly ionic association to ion-pairs. One effect of this interaction is to decrease the number of (free) ions in solution, and therewith the activity coefficient (Figure 1D1), this being the most important effect as regards the subject of this book. Other effects include decreased equivalent conductivity, certain changes in the ultraviolet (electron transfer to solvent) spectra, etc. In very dilute solutions, where best agreement between theory and experiment is to be expected, conductivity measurements are the most sensitive test for association theories, and therefore the phenomenon called 'ion-pair formation' is operationally best defined in terms of the observed effects on the conductivity. The 'ion-pair' concept may, however, be different operationally when other methods are used to measure it, as discussed below.

The main feature that characterizes an ion-pair is the electrostatic nature of the interaction*, so that there is no ionic specificity, apart from size effect, and no fixed geometrical arrangement or coordination. For a symmetrical electrolyte, association essentially stops with the formation of an ion-pair. Higher aggregation will occur only in low dielectric constant solvents, where a few ions may be associated to an aggregate (cf. section 2.D). The electrostatic forces between ions and multipolar aggregates are however not saturable. Although some treatments require the partners of an ion-pair to be in contact, this contact may still be between solvated ions. The primary solvation need not be disrupted, and one (or at most two) solvent molecules may be located between the two ions, which may for certain purposes still constitute an ion-pair. In these respects ion-pair formation differs from both incomplete ionization of weak electrolytes, such as acetic acid, and complexation to form a coordinated complex ion, as the cupric tetramine cation. The latter type of association will be discussed more fully in section 3.B.

A simple model for ion-pair formation considers implicitly only two

---

* For a discussion of the non-electrostatic water-structure-enforced ion pairing, see section 1.B.

oppositely charged ions in contact as constituting a pair, while all other ions are considered free. All other interactions between the ions, and with the solvent, are neglected. Other approaches use a more sophisticated way to deal with the potential energy between ions, and more complicated distribution functions. Other effects, as interactions with the solvent, are also taken into account. The theories are judged by their success to account for the change of the association with the concentration, the dielectric constant of the solvent, (the temperature) and the known size of ions. The latter quantity is connected with the parameter a, the distance of closest approach, for which different values are obtained by the various theories, some of which may conflict with known data on ionic sizes. These points have been reviewed by Kraus[1] and are discussed more fully below.

b. Bjerrum's Theory

Let us consider the distribution of ions i around a central ion j. In a solution containing n_{i0} ions i per unit volume, their average number in a spherical shell of thickness dr at distance r from ion j will be

$$n_{ij} = n_{i0}[\exp(U_{ij}/kT)]4\pi r^2 dr \qquad (3A1)$$

At short distances, the screening by other ions will be negligible. Using the notation

$$u = |z_i z_j|e^2/\varepsilon kT \qquad (3A2)$$

the mutual potential energy of the ions will be $U_{ij} = -ukT/r$ if the ions i and j have the same sign, and $U_{ij} = ukT/r$ if they have opposite signs. In the first case, the number of ions i in the shell is negligible at small r, and increases continuously as r increases. In the second case, there will be a distance q, where this number for a given shell thickness dr is a minimum, as can be seen by differentiation. This distance is

$$q = u/2 \qquad (3A3)$$

Bjerrum[2] proposed to treat the regions $r > q$ and $r < q$ differently. In the former region the ions are considered free, and the usual Debye–Hückel treatment applies to them. Since these ions can approach ion j only to a distance q by Bjerrum's hypothesis, this is the value to be used for the a parameter in Debye–Hückel's equation (1C7 and 1C24). On the other hand, ions in the region $r < q$ are considered associated with the ion j. In solutions with not too low dielectric constant, the number of such ions, calculated from (3A1), will be negligible for ions of the same sign as j, and will not exceed unity for ions of opposite sign, so that one can talk of an (associated) ion-pair. The exact meaning of the term 'associated' is not specified in this treatment. Bjerrum's ion-pair constitutes a dipole, which rotates in an external electrical field but does not migrate, and does therefore not count as two free ions in conductivity measurements. The mutual interaction energy of the ions in the pair is larger than $2kT$ (or equal to it at the limit $r = q$). The ions in the pair cannot, however, approach each other closer than a distance a, which is at least the sum of the crystal radii and usually contains also some of the thickness of the hydration shells. At low ε or high z's, the value of q is sufficiently large for the inclusion of a considerable number of solvent molecules between the ions j and i. The association is therefore more comprehensive than that based on effects depending on the proximity of the ions, such as some optical phenomena. For ions of low charges and at a high dielectric constant, q is smaller than a, so that there is no region where ions associate to pairs. In water at 25° $q/z_i z_j = 3.57\,\text{Å}$, so that for 1:1 electrolytes q is smaller than the usual a values (a around 4.5 Å), and no ion-pair formation is expected. On the other hand, for higher-valency-type electrolytes, and in low dielectric constant solvents, considerable ion-pair formation should occur.

Consider the average volume available in the solution for a j ion. The fraction of associated ions it contains is equal to the probability of an i ion being within the critical volume from a to q, given by

$$4\pi n_0 \int_a^q \exp(u/r) r^2 \, \mathrm{d}r \tag{3A4}$$

Considering the whole solution, the fraction of associated ions is $(1 - \alpha)$, and the ion concentration is $Nc/1000$, so that

$$(1 - \alpha) = \frac{4\pi Nc}{1000} \int_a^q \exp(u/r) r^2 \mathrm{d}r \tag{3A5}$$

Setting $t = u/r$ and $b = u/a$ the above expression is transformed into

$$(1 - \alpha) = \frac{4\pi N c u^3}{1000} \int_2^b t^{-4} \exp(t) \, dt = \frac{4\pi N c u^3}{1000} Q(b) \qquad (3A6)$$

Values of the integral $Q(b)$ were calculated by Bjerrum[2] and others are shown in Table 3A1.

TABLE 3A1. Values of the integral $Q(b)$, equation (3A6).

b	$-\log Q(b)$	b	$\log Q(b)$	b	$\log Q(b)$
2.0	∞	6	0.016	30	7.19
2.1	1.358	7	0.150	35	9.08
2.2	1.074	8	0.300	40	11.01
2.3	(0.912)	9	0.470	45	12.99
2.4	0.808	10	0.655	50	14.96
2.5	0.726	11	0.885	55	16.95
2.6	0.662	12	1.125	60	18.98
2.7	(0.610)	13	(1.40)	65	21.02
2.8	0.562	14	1.68	70	23.05
2.9	(0.524)	15	1.96	75	25.01
3.0	0.490	16	2.28	80	27.15
3.5	0.356	17	2.59		
4.0	0.260	18	2.92		
4.5	(0.192)	19	(3.24)		
5.0	0.124	20	3.59		
5.5	0.048	25	5.35		
6.0	-0.016	30	7.19		

H. S. Harned and B. B. Owen, *Physical Chemistry of Electrolyte Solutions*, 3rd ed., Reinhold Publ. Co. New York, 1958; R. A. Robinson and R. H. Stokes, *Electrolyte Solutions*, 2nd ed. Butterworths., Sci. Publ., London, 1959; E. A. Guggenheim, *Discussion Faraday Soc.*, 24, 59 (1957).
( ) —interpolated values.

The ion-pair formation can be treated as an equilibrium reaction, the association constant of which is given by $K_{ass} = (1 - \alpha) y_u / \alpha^2 y_\pm^2 c$, where y_u is the activity coefficient of the (uncharged, for symmetrical electrolytes, but dipolar) ion-pair, and $y_\pm$ is the mean activity coefficient of the free ions, calculated using q as the distance of closest approach. In very dilute solutions $\alpha \simeq y_u \simeq y_\pm \simeq 1$, so that $(1-\alpha)/c = K_{ass}$, which from (3A6) is at all concentrations

$$K_{ass} = \frac{4\pi N}{1000}(u^3)Q(b) = \frac{4\pi N}{1000}\left(\frac{z^2 e^2}{\varepsilon k T}\right)^2 Q(b) \tag{3A7}$$

and at 25°

$$\log b = \log|z_i z_j| + 2.746 - \log\varepsilon - \log a \ (a \text{ in Å}, \ b \text{ in Å}^{-1}) \tag{3A7a}$$

$$\log K_{ass} = 3\log|z_i z_j| + 6.120 - 3\log\varepsilon + \log Q(b) \tag{3A7b}$$

Questions have been raised about the arbitrariness of the choice of the upper integration limit q. The exact value of q (i.e. whether $q = u/4$, $u/2$ or u) was however shown[3] not to affect K_{ass} to an appreciable extent. The value of K_{ass}, found experimentally, leads through equations (3A7) to a value for a (see for example, ref. 4.). It has been found that these a values depend on the dielectric constant of the solvent. For example, in a water–dioxane mixed solvent, if a were the distance of nearest approach for ions preferentially solvated by water, no changes of a with solvent composition should occur.

c. Other Treatments

Some authors have objected to the notion implied in Bjerrum's treatment, that ions can be considered to be paired, without being in actual contact. For higher-valency-type electrolytes and at low dielectric constants, q is sufficiently large for a considerable number of solvent molecules to be found between the two partners. On the one hand this strengthens Bjerrum's use of the macroscopic dielectric constant in his expression for the potential (3A2). On the other hand, this corresponds to a less probable state than if the partners are in contact, as can be seen from the distribution function. Since some (e.g. optical) methods count as pairs only ions in contact, the molecular theory, it is argued, should use this concept in its model. This was done by Denison and Ramsay[5], and more recently, in a more detailed manner, by Fuoss[6].

Denison and Ramsay[5] used in their simple thermodynamic approach to ion association, a Born[7] cycle to calculate the electrical work required to separate a pair of ions from contact (distance between centres a) to an infinitely large distance, in a medium of dielectric constant ε. Equating the free energy of association to N times this work, they obtained

$$K_{ass} = \exp(-\Delta G^0_{ass}/RT) = \exp(b) \tag{3A8}$$

$$b = e^2/akT\varepsilon \tag{3A9}$$

for monovalent ions, K_{ass} being the association equilibrium constant.

Fuoss considered in his model the cations as conducting spheres with a radius a, and the anions as spheres with a negligibly small radius. The potential within the cation sphere is the same as at its surface ($r = a$), and is, from (1C8)

$$\psi_+ = z_+ e/a\varepsilon(1 + \kappa a) \qquad (3A10)$$

An anion is considered paired with a cation when it is situated just inside it, the distance between their centres being a, which corresponds electrostatically to contact between the partners, whatever the model. When an increment of electrolyte is added to a solution containing N_+ moles of cations and N_- moles of anions, the probability of the anions remaining free is proportional to the volume of solution V (less the volume of the cations $NN_+ (4\pi a^3/3)$ which is negligible at high dilutions), while the probability of their being paired is proportional to the volume occupied by the free cations, modified by a Boltzmann distribution term. This quantity is $NN_{+f}(4\pi a^3/3)\exp(z_- e\psi_+/kT)$, where N_{+f} is the number of moles of free cations. Since the number of free anions added dN_{-f}, is equal to that of free cations added dN_{+f}, we have for the increment in the number of paired anion

$$dN_{-p} = 2N(4\pi a^3/3)V^{-1}\exp(z_+ z_- e^2/a(1 + \kappa a)\varepsilon kT)N_{+f}dN_{+f} \quad (3A11)$$

assuming the same proportionality constants for the probability, and remembering that for each of the added anions which is paired, an added cation is also paired. The exponent in (3A11) may be written as $\exp(u/a)\exp(u\kappa/(1 + \kappa a))$; the first exponent equals e^b (3A9), while the second exponent equals $\gamma_{\pm(f)}^2$ (1C18), the square of the activity coefficient of the free ions. Converting to practical concentrations ($c = 1000 N/V$, $(1 - \alpha) = c_p/c$, the fraction of paired ions, and $\alpha = c_f/c$, the fraction of free ions), and integrating (3A11), leads to

$$(1 - \alpha) = (4\pi Na^3/3000)e^b\gamma_{\pm(f)}^2\alpha^2 c \qquad (3A12)$$

Since $K_{ass} = (c_p/c_f\gamma_{\pm(f)})^2 = (1 - \alpha)/\alpha^2\gamma_{\pm(f)}^2 c$, neglecting the activity coefficient of the ion-pair, which is very near to unity, one finally obtains

$$K_{ass} = 4\pi Na^3 e^b/3000 \qquad (3A13)$$

as Fuoss' expression for the association constant. It contains the factor $e^b/3$ instead of $b^3 Q(b)$ in Bjerrum's theory, which at high b values is approximately e^b/b.

Since it has been found[8] that the solvent dependence is not restricted to the dependence on ε (different K_{ass} being found for solvents with virtually the same ε, section 2.D.a), while changes of a for a given electrolyte with solvents are not expected for non-solvating solvents, some other solvent effect must be taken into account. Gilkerson[8] used Kirkwood's[9] zeroth approximation for the partition function of a particle in solution. He obtained for the entropy of association to ion-pairs a term involving the masses, free volumes and rotational and vibrational contributions of the ions and the ion-pair, and a term which is the difference U_S between the interaction energies of the ion-pair with the solvent, and the ions with the solvents. The association constant is

$$K_{ass} = A(T)\exp(U_S/RT)\exp(b) = K_{ass}^0\exp(b) \qquad (3A14)$$

where $A(T)$ is independent of the solvent, but depends on temperature as $T^{-3/2}$, and is specific for the electrolyte, and the second exponent is the contribution of the electrical work for separating the partners of the ion-pair, i.e. equation (3A8). The energy difference U_S was shown to be independent of ε, but to be proportional to the dipole moment of the solvent, and the K_{ass} values for three solvents of the same ε were found to be interrelated as expected from equation (3A14), using known values of the dipole moments[8]. An expression generally similar to (3A14) has been derived by Magnusson[10], for correlating the free energy change of association with the radii of the reacting ions.

Still different approaches to the problem of ion-pair formation use other distribution functions. Fuoss[6] considered assigning to each ion one, and only one, partner, which is the nearest ion not yet a partner of another ion. De Lap[11] developed this idea, obtaining expressions for the distribution different from those used by Bjerrum or Fuoss, hence also different association constants. He extended his treatment to cases where the ions are of different size for a given electrolyte, to non-symmetrical electrolytes, and to electrolyte mixtures with a common ion. Guggen-heim[12] showed that for 2:2 electrolytes the Debye–Hückel linear approximation $ze\psi \ll kT$ is not valid, and that the solution proposed by Müller[13] should be used instead, leading to a new distribution function. He also showed that instead of using Bjerrum's $q = 14$ Å (for water), any value between 10 and 14 Å could be used, with no appreciable change of the calculated osmotic or activity coefficients (although K_{ass} is changed). In fact, the calculated values agree well with experimental values for 2:2 electrolytes. Thus although correct osmotic and activity coefficients can be obtained from this treatment, the values of K_{ass} suffer from the arbitrariness by which q is fixed.

More recently, Guggenheim[14] proposed to use the non-arbitrarily defined terms sociation and supersociation, instead of the association derived above. For solutions of non-electrolytes and for gases a 'sociation constant' $b_s = 4\pi \int_0^\infty (\exp(-W/kT) - 1) r^2 \, dr$ is proposed, W being the mutual interaction energy. For electrolyte solutions the 'degree of super-sociation' is defined as

$$4\pi n_0 \int_0^\infty (\exp(-W_r/kT) - \exp(-W_\infty/kT)) r^2 dr \qquad (3A15)$$

where n_0 is the stoichiometric number of ions per cm^3, and W_r the mutual interaction energy[15] (W_∞ as r tends to infinity). This definition contains no arbitrary parameter (such as q in Bjerrum's treatment) but it lacks the intuitive picture of ions associated to ion pairs. The 'degree of supersociation' equals $(1 - \alpha)$ in (3A5) when association is definite.

d. Activity Coefficients

The participation of an ion in an ion-pair removes it from the ionic atmosphere of other ions. If only a fraction α of the electrolyte exists as free ions in solutions, the parameter of the ionic atmosphere will be $\alpha\kappa^2$, and the ionic strength will be αI. The activity coefficient of the free ions will therefore be different from the value it would have if no association occurred, and so will the stoichiometric activity coefficient, i.e. the ratio of the measured activity to the stoichiometric concentration. For a binary electrolyte

$$\gamma_\pm(\text{stoichiometric}) = \alpha\gamma_\pm(\text{hypothetical, free}) \qquad (3A16)$$

while for other valency types, terms in v_i have to be included (ref. 16, p. 36). This decrease in the stoichiometric activity coefficient is illustrated for cadmium chloride in Figure 1D1. If values of $\gamma_\pm$ (hypothetical, free) can be estimated on theoretical grounds, for example by comparison with ions of similar sizes and hydration tendencies, the degree of association $(1 - \alpha)$ can be calculated from (3A16) and from measured activity coefficients, hence the association constant is also known. The activity coefficient of the free ions can be calculated from the now known α, for example by using the Debye-Hückel equation (1C24)

$$\log \gamma_\pm (\text{D.H., free}) = \frac{-\left| z_+ z_- \right| A(\alpha I)^{1/2}}{1 + B\mathring{a}(\alpha I)^{1/2}} \qquad (3A17)$$

If the Debye–Hückel equation is valid in the range of measurement, the values of $\gamma_\pm$ (D.H., free) should agree with $\gamma_\pm$ (hypothetical, free). Since the osmotic coefficients show less individuality for completely dissociated electrolytes up to 0.3 m, compared with the activity coefficients, those for associated electrolytes can be used with advantage for estimating association constants[17]. The constants thus obtained agree well with published values.

In the discussions above, the activity coefficient of the ion pair was neglected, i.e. it was set equal to unity. For non-binary electrolytes, the ion pair will, of course, be charged, and influenced by the ionic atmosphere. For binary electrolytes the ion pair is uncharged, but it is a dipole, that is influenced by the potential other ions exert. Although it will not migrate under the influence of an external field, it will show non-ideal thermodynamic properties[18].

e. Interpretation of the Association

The various modes of considering association, discussed above, lead also to different interpretations of the properties of ion-pairs. An important question is whether the ions lose their solvation shells, partly or completely, on association. It is reasonable to believe that even if the associated partners are in contact, the resulting dipole is still sufficiently strong to orient solvent molecules around the pair. This, however, does not answer the question of whether there are solvent molecules between the partners. Bjerrum's treatment permits the occurrence of solvent molecules between the partners. It was found experimentally that no water of hydration is lost on pairing of bivalent metal sulphates[19] This was shown by those electrolytes not obeying the law $\Delta S_{\text{ass}}^0 = const. - \Delta S_{\text{h}}^0$, obeyed by those systems where there is other evidence for strong interaction. In this expression the entropy of association ΔS_{ass}^0 for a series of electrolytes with a common ion differs by a constant from the entropy of hydration ΔS_{h} of the other ion, which is interpreted as loss of hydration-water on association.

Another property which could give an indication of the presence of solvent molecules between the partners is the light-absorption behaviour. Light absorption can arise from different kinds of electronic transitions. Thus, forbidden *d–d* or *f–f* transitions characterize the bands in the visible region of transition metal and lanthanide ions respectively. These orbitals are screened from interaction by distant ions, and only strong interaction, i.e., complex formation with directed bonds, can effect these transitions[20]. Ion-pair formation, proven to occur from thermodynamic properties or conductance measurements, and expected from calculation by Bjerrum's or some other method, cannot therefore cause appreciable changes in these spectra[20], or those of tetraphenylarsonium salts[21], etc. Other transitions, however, such as charge transfer to solvent, which are responsible for bands in the ultraviolet, may be affected by more distant ions. In particular ions influencing the solvent molecules in a solvation shell, even causing them to be removed, can be expected to affect these bands strongly[20]. The observation of spectral changes in these cases, therefore, need not imply cation–anion contact. One or two solvent molecules[20,22,23], or even ligands in a first coordination sphere, as with the association of sulphate with aquopentamine cobaltic ions[24], may be situated between the ions in such an ion pair.

It has often been observed that association constants calculated from the light-absorption properties are lower then those calculated from conductivity or thermodynamic properties (activity coefficients, solubility, etc.). This has been explained[25] by the inclusion of distant ion pairs in the concept of 'association', and in its calculation, in the latter methods, and their exclusion by the spectrophotometric method. For high q values of Bjerrum's theory, there can be many more solvent molecules between the partners than the one or two considered above as still permitting optical effects. If ion-pairs are detected optically only up to a distance p, instead of q, then the optically measured association constant will be[25]

$$K_{ass}(\text{optical}) = \frac{4\pi N}{1000} u^3 Q(b') \qquad (3A18)$$

$$Q(b') = Q(b) - Q(b''); \quad b'' = u/p$$

with the same symbols as before (section 3.A.b). Since $p < q$ one obtains $K_{ass}(\text{optical}) < K_{ass}$. One must however be careful in the calculation of the experimental value of $K_{ass}(\text{optical})$ to specify correctly the molar absorbance of the ion-pair. If the optical properties of ions further apart than p are those of free ions, and for all ions which are nearer there is some average molar absorbance, characteristic for the ion-pair, then Cohen's treatment[25] holds. If, on the contrary, the decrease of the optical effects with distance is gradual, without a sharp boundary at distance p, then an average molar absorbance of the ion-pair, which takes into account all optical effects, applies. In this case, however, the optical association constant may be higher than the thermodynamic or theoretical values, since interactions which do not lead to 'association' are also counted. Thus the dehydrating effects of perchlorates, or of a concentrated ionic medium, can change the molar absorbancy of a well-defined species, such as a hydrated metallic ion, as well as that of an ion-pair, and produce spectral changes which are erroneously counted as association[26,27]. The spectra of anions can be affected similarly by cations, without there being any association detectable by other methods. This is illustrated by the sensitivity of the u.v. band of iodide ions to polarization of the water molecules around them by cations[28,29]. On the other hand, a high concentration of an inert ionic medium will assure that only near reactive ions will interact to form an ion pair, with a definite molar absorbance, while distant ions will interact rather with the abundant 'inert' ions, without causing optical effects[30]. If this is true, then the association constant can be calculated with no arbitrariness introduced by the choice of a variable molar absorbance. The association constant will still depend on the distance q and different constants will be obtained by different methods, unless the same q applies.

In conclusion, it has been seen that on theoretical grounds interaction between near ions is expected to lead to 'association', which may express itself by the following effects. The associated ion-pair, if uncharged, does not migrate in an external field, and it does not contribute to the ionic atmosphere of other ions. There may be spectral effects, particularly if the spectral band is due to charge transfer, to solvent transition, and if the paired ions are very near. It is possible to define the concept of 'association' in a number of ways, and different theoretical treatments lead to different association constants, partly because different distribution functions are used, partly because different integration limits are applied. From the effect on thermodynamic properties such as solubility or activity coefficients, there is no justification to the limitation of the ions considered as paired, to ions in actual contact, and even if ions in contact are emphasized, they need not be bare, i.e. unsolvated in the region between the partners.

B. COMPLEX FORMATION

a. Introduction

Up to now, all the interactions which have been discussed were of a simple electrostatic nature without any changes in the electronic configuration of the ions or molecules, except polarization effects. The only exception may be some of the interactions involved in hydration, or solvation in general, which require some sort of bonding beyond coulombic interaction of charges or dipoles. There is, however, an important class of species formed in solution from interactions between ions or molecules capable of independent existence, which do involve chemical bonding: the complexes. These may be cationic, neutral or anionic, mononuclear or polynuclear. Ususally they involve a central metal cation, which may be monatomic (e.g. Co^{2+}), polyatomic (e.g. UO_2^{2+}) or an inner-sphere complex (e.g. $Co(NH_3)_6^{3+}$), and coordinated ligands, arranged in a definite geometrical arrangement (most common are octahedral and tetrahedral arrangements) around the central cation. The ligands may be anions (e.g. Cl^-) or neutral molecules (e.g. NH_3), and they are bound to the central cation through a donor atom (chlorine and nitrogen in the above examples) for monodentate ligands, or through more than one for polydentate ligands, which form chelated complexes. The class of complexes formed in solution is, however, much wider than the examples above suggest, and includes all interactions involving electron acceptors and electron donors. The following discussion will be mostly limited to metal cations M^{m+} as central groups, forming mononuclear complexes (i.e. those containing only one central group) with monodentate ligands L^{l-}. For the sake of clarity, charges will usually be omitted from M, L and the complexes.

An important attribute of the central cation is its coordination number N, i.e. the number of donor atoms that occupy sites in its coordination sphere. This number depends on the relative sizes of the central cation and the donor atoms, and on the bond-hybridization preferences of the central cation. Some cations show a maximal coordination number only under certain conditions, and a typical coordination number under milder conditions, an example being Hg^{2+}, forming complexes of the type ML_2 easily, and of the type ML_4 more reluctantly. It has been emphasized by the Bjerrums[31,32] that complex formation usually occurs stepwise, that is, that intermediate complexes $ML, ML_2, \cdots, ML_{N-1}$ occur at certain concentrations, besides the coordinatively saturated complex ML_N.

Since the cation M is solvated in solution, the formation of a complex with L usually involves substitution of a solvent molecule: $MS_N + L \rightleftharpoons MS_{N-1}L + S$, rather than simple addition, so that a fixed total co-ordination number N is retained. In this respect, complexes may be said to differ from ion-pairs. Since usually the solvent is present in a large excess over both M and L, its concentration is scarcely changed by the complex-formation reaction, and so its participation is ignored, and a general reaction for n ligands may be written as

$$M + nL \rightleftharpoons ML_n \qquad (3B1)$$

b. Thermodynamics of Complex Formation

The change of free energy of reaction (3B1), ΔG_n, can be written as

$$\Delta G_n = \Delta G_n^0 + RT \ln a_{ML_n}/a_M a_L^n \qquad (3B2)$$

and since at equilibrium $\Delta G = 0$, the equilibrium constant K_n for this reaction is obtained

$$-\Delta G_n^0 = RT \ln K_n = RT \ln(a_{ML_n}/a_M a_L^n)_{equil} \qquad (3B3)$$

For the rest of this section equilibrium will be assumed, so that the suffix 'equil' will be omitted. The numerical value of K_n depends on the concentration scale used in defining ΔG_n^0 (or the activities), and in the following, the molar concentration scale will be used for solutes. Conversion factors to other scales have been discussed elsewhere[33]. It should be emphasized that the equilibrium constant and the standard free energy change are relative to the binding of a given solvent by the central cation (and the ligand), so that it is necessary to modify (3B1–3) by including the solvent explicitly when complex formation in different solvents is to be compared.

The enthalpy and entropy of complex formation (e.g. reaction 3B1) can be obtained from (3B3). Assuming constant ΔH_n^0 and ΔS_n^0 over a narrow temperature range they can be obtained from

$$-\Delta H_n^0 = RT^2 \, d\ln K_n/dT = -R \, d\ln K_n/d(T^{-1}) \qquad (3B4a)$$

$$\Delta S_n^0 = R\ln K_n + \Delta H_n^0/T = R\ln K_n + RT \, d\ln K_n/dT \qquad (3B4b)$$

These thermodynamic quantities may usually be obtained more accurately from direct calorimetric measurements. Exceptions are systems where slow reactions occur, or where low solubilities or non-availability of material (as with radioactive elements) require work with trace quantities, and use of (3B4), since equilibrium constants may be determined under these conditions also. The enthalpy of complex formation can usually be interpreted as the strength (energy) of the bonds between central cation and ligand, relative to those between central cation and solvent molecules. The entropy of complex formation includes the difference between the translational (and rotational, for polyatomic ligands) entropy lost by the ligand and the corresponding entropy gained by the displaced solvent molecules. There is in addition to a statistical term, describing the number of ways n ligands may be coordinated in N coordination sites, and the number of ways they may be dissociated off again

$$\Delta S_n^0 (\text{statistical}) = \mathbf{R} \ln (N!/(N-n)! \, n!) \qquad (3B5)$$

Strong complex formation, expressed by a high value of K_n, requires a large negative enthalpy and a large positive entropy of complex formation. A large negative enthalpy change can compensate small or negative entropy changes, and vice versa. The thermodynamics of complex formation have been discussed extensively and interpreted in a number of publications[33-35]. In spite of the scarcity of reliable data, there is good correlation between these quantities and the properties of the central cation and the ligand.

The thermodynamic functions discussed above are related to the thermodynamic equilibrium constant of the complex formation reaction, i.e. the thermodynamic stability constant of the complex. Consider now the stepwise reaction

$$ML_{n-1} + L \rightleftharpoons ML_n \qquad (3B6)$$

to which may be assigned a thermodynamic stepwise formation constant k_n^T defined as

$$k_n^T = a_{ML_n}/a_{ML_{n-1}}a_L = (ML_n)y_{(n)}/(ML_{n-1})y_{(n-1)}(L)y_L \qquad (3B7)$$

For the sake of definiteness, the molar concentration scale has been chosen, and the molar ionic activity coefficients $y_{(n)}$, $y_{(n-1)}$ and y_L apply to the species designated. The constant k_n^T can be regarded as a product of the equilibrium concentration quotient Q_n and an activity coefficient quotient Y_n

$$k_n^T = Q_n Y_n = [(ML_n)/(ML_{n-1})(L)] \, [y_{(n)}/y_{(n-1)}y_L] \qquad (3B8)$$

If conditions are chosen so that Y_n is a constant, then the concentration quotient $Q_n = k_n^T/Y_n$ is also a constant, and can serve as the stepwise formation constant of the complex. Such conditions can be provided in a constant ionic medium, as discussed in section 1.E.

Most experimental methods used to study complex formation yield finally a value of Q_n for each complex species present. If the reference state of activities is chosen as infinite dilute solutions of all active species $(ML_n(n = 0, 1 \cdots)$ and L) in the constant medium, then $Y_n = 1$ at the reference state with the medium considered as solvent. Q_n is then a thermodynamic equilibrium constant in this solvent, and is as 'valid' as other constants, determined in other solvents. Of course there is a medium effect if comparison with any given solvent (such as pure water) is desired (cf. section 2.D). Since k_n^T is the equilibrium constant for the solvent water, Y_n is the medium effect. For aqueous ionic media Y_n can be estimated, e.g. using Guggenheim's expression for the activity coefficients (equation 1D26)[36]

$$\log Y_n = (-A\sqrt{I}/(1 + \sqrt{I}))\sum z^2 + I\sum b \qquad (3B9)$$

where $\sum z^2 = z_{ML_n}^2 - z_{ML_{n-1}}^2 - z_L^2$ and $\sum b = b_{ML_n} - b_{ML_{n-1}} - b_L$. Guggenheim's expression is convenient, since it contains a 'universal' Debye–Hückel term, dependent only on the ionic strength of the medium, and not on the ions, except through their known charges. Expression (3B9), or an equivalent expression involving mean ionic activity coefficients rather than individual ones[37], agrees well with experimental data. If it is desired to obtain k_n^T for the solvent water from a series of measurements of $Q_n(I)$, the linear extrapolation to $I = 0$ of the expression

$$\log k_n^T - (\sum b)I = \log Q_n(I) - (A\sqrt{I}/(1 + \sqrt{I}))\sum z^2 \qquad (3B10)$$

of which the right-hand side is known, yields, $\log k_n^T$ as intercept, and the unknown $\sum b$ as slope.

In the case where one of the participants in a complex-formation equilibrium is the hydrogen ion, the conventionally defined and measured pH (equation 1C31) is often used in place of either $-\log(H^+)$ in determining Q, or in place of $-\log(H^+)y_{H^+}$ or $-\log(H^+)y_{\pm HL}$ in determining k_n^T. Because of the conventions involved in determining pH, the resulting constants are conventional 'mixed' equilibrium constants, and can be related to the quotient Q_n or the constant k_n^T only through conventional assignments of individual ionic activity coefficients.

In some cases, neither sufficiently dilute solutions, where (3B9) and (3B10) apply, nor a constant medium, where $Y_n = const.$, can be used, especially for studying weak complexes. For such cases an effective activity coefficient quotient[37] $F_n = Y_n y_{\pm CA}$ can be used, in particular when studying complexes of metal ions present as traces in a 1:1 electrolyte CA, providing the ligand A^-. This function has been defined and discussed in section 1.D.f (equation 1D39), and was shown to be practically independent of the ligand concentration, c_{CA}. The effective ligand activity $a = c_{CA} y_{\pm CA}$ was also introduced there. An effective stepwise formation constant can now be defined

$$k_n^* = k_n^T / F_n = Q_n / y_{\pm CA} = (MA_n)/(MA_{n-1})a \qquad (3B11)$$

It can be expressed in terms of measurable quantities, and is independent of the ligand concentration to a good approximation, at least in the region of concentrations where the transition $MA_{n-1} \to MA_n$ takes place. Indeed, for a hypothetical case, where $k_1^* = 10 M^{-1}$ and $k_2^* = 0.1 M^{-1}$, with the typical dependencies $F_1 = -0.02 c_{CA}$ (Figure 1D2) and $\log a = 0.22 c_{CA}$, the expression $k_1^{*'} = k_1^* a^{-0.09}$ results, where k_1' is the apparent constant, which now somewhat depends on the ligand concentration. Still, the deviations of the fractions α_n (the ratio of the metal in the form of the nth complex to the total metal present, see below section 3.B.c), from the values calculated assuming no dependence, are only a few percent, and negligibly low where α_n is large, and the nth complex is important, Figure 3B1.

c. Mathematical Treatment of Data

The terms that have been found useful to describe complex formation in solution have been established by Bjerrum[32]. The complex formation reaction (3B1) is characterized by the overall stability constant β_n

$$(ML_n) = \beta_n(M)(L)^n \qquad (3B12)$$

When the parenthesis denote activities, β_n^T is the thermodynamic equilibrium constant for the reaction, if they denote concentrations β_n is the equilibrium quotient. Considering one step of ligand addition in a series of (mononuclear) complexes, equation (3B6), the stepwise complex-formation quotient k_n (written Q_n above) is defined by

$$(ML_n) = k_n(ML_{n-1})(L) \qquad (3B13)$$

with significance as above for β_n. A combination of (3B12) and (3B13) yields

$$\beta_0 = k_0 = 1, \quad \beta_1 = k_1, \quad \beta_n = k_1 k_2 \cdots k_n \qquad (3B14)$$

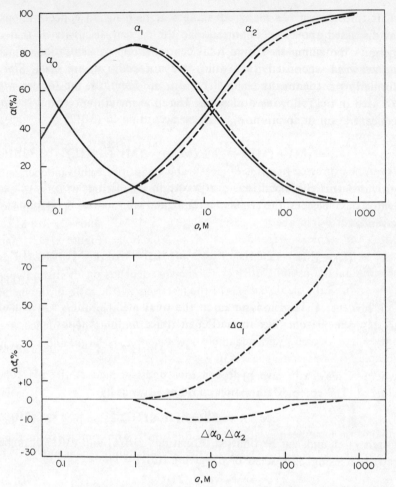

Figure 3B1. Deviations in the fractions a_n for non-constant k_n^*. Upper part: —————— calculated for $k_n^* = 10\mathrm{M}^{-1}$ and $k_2^* = 0.1\,\mathrm{M}^{-1}$; ————— calculated with $k_1^{*'} = k_1^* a^{-0.09}$ and same k_2^*. Lower part: curves ————— for same constants, relative deviations. (By permission from *Record of Chem. Progr.*, **27**, 114 (1966)).

In the following it will be assumed that reactions (3B1) and (3B6) follow the mass-action law in terms of concentrations (L) (e.g. by use of a constant ionic medium). If the concept of effective ligand activity a is used, one obtains from (3B11) and (3B14)

$$(\mathrm{ML}_n) = \beta_n^*(\mathrm{M})\, a^n \tag{3B15}$$

This formulation has many advantages at high ligand concentrations, as discussed above. Of importance to the present discussion is that it permits the summation over total central group concentration, which is awkward when activities, which are not additive, are used. Since formally the treatments using (L) and a are identical, the former will be used in the following discussion. The total metal ion (central-group) concentration in solution, c_M, will be according to (3B12)

$$c_M = (M) + (ML) + \cdots + (ML_n) = (M) \sum_{n=0}^{N} \beta_n(L)^n \qquad (3B16)$$

provided that the metal ions participate in no other reaction (such as hydrolysis) besides (3B1). The total ligand concentration is likewise given as

$$c_L = (L) + (ML) + 2(ML_2) + \cdots + n(ML_n) = (L) + (M) \sum_{n=1}^{N} n\beta_n(L)^n$$
$$(3B17)$$

If however, the ligand is an anion of a weak acid H_jX, it is convenient to describe the complex formation in terms of the reaction

$$M + nH_jX \rightleftharpoons MX_n + njH^+ \qquad (3B18)$$

where X may still have hydrogen ions undissociated at the pH considered. The complex formation constant is given by

$$(MX_n) = \beta_{jn}^H(M)(H_jX)^n(H)^{-nj} \qquad (3B19)$$

Often $j = 1$ and can be omitted. Equations (3B16) and (3B17) can be rewritten using β_n^H instead of β_n, giving (for $j = 1$)

$$c_M = (M) \sum_{n=0}^{N} \beta_n^H(HX)^n(H)^{-n} \qquad (3B20)$$

$$c_X = (X) + (HX) + (M) \sum_{n=1}^{N} n\beta_n^H(HX)^n(H)^{-n} \qquad (3B21)$$

At low pH and c_M, c_X can replace (HX) in the equations. Otherwise, using K_X as the acid dissociation constant for HX according to $HX \rightleftharpoons H^+ + X^-$, (HX) is replaced by $(X)(H)/K_X$, and (X) is obtained by successive approximations from $(X) = (c_X - \tilde{n}c_M)/(1 + (H)/K_X)$. The stability constants β_n^H are related to β_n in terms of the anions X^- as

$$\beta_n^H = \beta_n K_X^n \qquad (3B22)$$

For a metal cation M^{m+} and monovalent ligands A^- (or ligands B^{b-} where the ration m/b is an integer) it is sometimes convenient[37,38] to consider the complex formation reaction as pertaining to the neutral complex MA_m

$$MA_m \rightleftharpoons MA_{m-i} + iA \qquad (3B23)$$

where i may be both positive or negative. In the former case cationic complexes are formed, in the latter anionic complexes, in both cases complexes with charge i. (When the ligand is B^{b-}, m is replaced by m/b, and the charge of the complex will be ib). The index i is called the 'charge number' of the complex MA_{m-i}. Overall complex formation constants on this plan will be primed

$$(MA_{m-i}) = \beta_i'(MA_m)(A)^{-i} \qquad (3B24)$$

$$c_M = (MA_m)\sum_{i=m}^{m-N} \beta_i'(A)^{-i} \qquad (3B25)$$

Comparison of (3B23) and (3B1) yields as the relation between the two indexes

$$i = m - n \qquad (3B26)$$

and between the complex formation constants

$$\beta_i' = \beta_n/\beta_m \qquad (3B27)$$

It must be stressed that for this formal treatment involving the neutral species MA_m, the question as to whether the electrolyte MA_m is completely dissociated in solution or not is irrelevant. The species MA_m may often be stable only in the presence of a large excess of anions A.

A useful function for relating experimental data is the 'complexity function' X, introduced by Fronaeus[39], which is the ratio of total central-group concentration to its free concentration, and from (3B16)

$$X = c_M/(M) = 1 + \beta_1(L) + \cdots + \beta_N(L)^N = \sum_{n=0}^{N} \beta_n(L)^n \qquad (3B28)$$

The function X may be determined experimentally by a number of methods: e.g. in a concentration cell with electrodes reversible to the metal M, and with equal c_M in both half-cells, but with a free-ligand concentration equalling (L) in one and zero in the other, the e.m.f. will be (for a z-electron reaction)

$$E = \frac{RT}{zF}\ln X \qquad (3B29)$$

Another useful function is the fraction α_n of central group in the form of a particular complex ML_n

$$\alpha_n = (ML_n)/c_M = \beta_n(L)^n/X \tag{3B30}$$

The fraction α_n is also called the degree of formation of the nth complex. The fraction of M in the form of free cation, α_0, is the reciprocal of X. Of course, $\sum_0^N \alpha_n = 1$, if only complexes with L are formed (i.e. no hydrolysis etc.). Some properties of the solution are proportional to α_n, such as the molar absorbance of the solution, if only the complex ML_n absorbs at the wavelength λ_n considered

$$\varepsilon_M(\lambda_n) = \sum \varepsilon_n(\lambda_n)\alpha_n \tag{3B31}$$

The distribution coefficient D in some extraction systems is proportional to the concentration of the neutral complex MA_m in a system

$$D = P_m\alpha_m \tag{3B32}$$

where P_m is the distribution constant of the neutral species (section 7.B).

Finally, the function $\tilde{n}$, coined the (average) ligand number of the system, should be mentioned[32]; it has been widely used to describe the degree of complexation of a system under given conditions. The ligand number is defined as the ratio of the concentration of bound ligand (i.e. total ligand minus free ligand) to the total concentration of metal

$$\tilde{n} = (c_L - (L))/c_M \tag{3B33}$$

Using (3B16), (3B17) and (3B28) this becomes

$$\tilde{n} = \frac{\sum_1^N n\beta_n(L)^n}{\sum_0^N \beta_n(L)^n} = \frac{d\log X}{d\log(L)} \tag{3B34}$$

For very strong complexes, if $c_L < Nc_M$, $\tilde{n} \simeq c_L/c_M$. Two solutions with different c_L and c_M may still have the same (L), and therefore (from 3B34) the same $\tilde{n}$. These solutions will be called 'corresponding'[40] and for them

$$(L) = \frac{c_{M(2)}c_{L(1)} - c_{M(1)}c_{L(2)}}{c_{M(2)} - c_{M(1)}} \tag{3B35a}$$

$$\tilde{n} = \frac{c_{L(2)} - c_{L(1)}}{c_{M(2)} - c_{M(1)}} \tag{3B35b}$$

Some properties of a solution depend on (L) only, e.g. the molar absorbance, hence solutions with the same molar absorbance ε_M are corresponding, and no detailed knowledge of the dependence of ε_M on (L) is needed in order to determine (L) and $\tilde{n}$ for these solutions.

Similarly to $\tilde{n}$, an average charge number $\bar{i}$ can be defined for a system

$$\bar{i} = \frac{d\log\alpha_m}{d\log(A)} = m - \tilde{n} \tag{3B36}$$

The slope of a logarithmic plot of the distribution coefficient against free-ligand concentrations for systems obeying (3B32) yields $\bar{i}$ directly.

d. Determination of Stability Constants

The stability constants β_n and k_n (or β_i' or β_n^* etc.) are determined (for mononuclear-complex systems) from the relationships $\alpha_n = F(L)$, $X = F(L)$ and $\tilde{n} = F(L)$.

In order to use the relations involving the stability constants it is necessary to know the free-ligand concentration, (L). In case of tracer metal concentrations $c_L \gg c_M$ and $(L) \simeq c_L$. In all cases where $c_L > Nc_M$, the value $(L) \sim c_L$ can be used as a first approximation, but it must be subsequently refined, using $(L) = c_L - \tilde{n}c_M$, with $\tilde{n}$ calculated from the first approximation. An approximate value of $\tilde{n}$ can also be obtained from (3B30) or (3B32) using c_L in the derivative. If many accurate data of X as function of c_M and c_L are known (e.g. obtained potentiometrically), the Hedström–McKay[41] equation can be used to obtain (L)

$$\log(L)/c_L = \left[\int_0^{c_M} \left(\frac{\partial \log 1/X}{\partial c_L} \right)_{c_M} dc_M \right]_{c_L} \tag{3B37}$$

The most reliable method is, of course, to determine (L) experimentally, e.g. potentiometrically.

When $\tilde{n}$, X or α_n have been determined as functions of (L) by one of the methods referred to above, the problem of calculating the various β_n or k_n values remains. This is essentially a mathematical, rather than a chemical problem. The only chemical consideration that enters the calculation is the selection of the correct value of N, and a final check on the significance of the values obtained. Negative values of stability constants, which may be obtained by calculation, have of course no physical significance, and are a sure sign for a systematic error in the data (or an error in the calculation). It must be remembered that (for mononuclear complexes) the variables $\tilde{n}$, X or α_n are functions of (L) and the parameters β_n (or k_n), only. In principle, therefore, a set of N pairs of values (variable, (L)) could be solved for N parameters β_n. It is preferable, of course, to have many more experimental points, and the methods employed should use all of them, weighted according to their precision.

The usefulness of various methods depends on N, or rather the number of parameters β_n necessary to describe the data adequately in the range of measurements, and on the spacing of the individual stability constants. This latter property has been discussed in detail by Bjerrum[32]. The ratio of consecutive stepwise stability constants $k_n/k_{n+1} = \beta_n^2/\beta_{n-1}\beta_{n+1}$ is of prime importance, and four cases may be discussed.

a) If $k_n/k_{n+1} \gtrsim 10^4$ the steps of the successive equilibria are so widely spaced that each may be treated individually, at a given range of (free-) ligand concentration, as if only one complex is formed. If $\tilde{n}$ values have been obtained, the stepwise constant k_n may be obtained from

$$\log k_n = -\log(L)_{\text{at }\tilde{n}=n-0.5} \tag{3B38}$$

This method employs, however, only a single experimental point. In order to use all the points one of the methods discussed below for the case $N = 1$ should be used.

b) When $10^4 \gtrsim k_n/k_{n+1} \gtrsim 10^2$ the consecutive equilibrium constants are nearer together, and at a given (free-) ligand concentration more than two complexes are present simultaneously. However, only two will be important at any concentration, the others giving only a small contribution. Therefore equation (3B38) may be used as a first approximation, and the value of k_n can be refined by using approximate values of k_{n-1} and k_{n+1} to correct for the other species. Again, full utilization of all experimental points involves the methods discussed below for $N \geq 2$.

c) When $10^2 \gtrsim k_n/k_{n+1} \gtrsim 1$ there are simultaneous equilibria involving at least three species ML_{n-1}, ML_n and ML_{n+1} of similar relative importance. Equation (3B38) is now rather useless, and great care must be taken to disentangle the equilibrium steps from one another, by methods discussed below.

d) Finally, if $k_n/k_{n+1} \lesssim 1$, the actual existence of the species ML_n may be suspect. Calculations should be made to see if the data cannot be equally well described by an expression in which $k_n = 0$, the values of the other parameters being adjusted accordingly. If this is the case, it must be admitted that the data, at their given precision, do not show positive evidence for the formation of ML_n, although minor amounts may be present.

The methods used to determine the constants may conveniently be discussed under the headings of graphical and numerical methods. A thorough discussion of them has been given by Rossotti and Rossotti[33] and other reviews[42-45]

e. Graphical Methods

Graphical methods are of three main types: extrapolation methods, permitting the determination of one parameter as the intercept, and an approximation to another as the slope; linear plots, yielding two parameters as intercept and slope; and curve fitting, allowing the simultaneous determination of two parameters and an approximation of a third.

The extrapolation method is best applied to data $X = F(L)$[39,46]. Data $\tilde{n} = F(L)$ can be converted using the relation

$$X = \exp \int_0^{(L)} (\tilde{n}/(L)) \, d(L) \tag{3B39}$$

while data α_n can be converted to an expression similar to X

$$\frac{(L)^n}{\alpha_n} = X/\beta_n = \frac{1}{\beta_n} + \frac{\beta_1}{\beta_n}(L) + \cdots \tag{3B40}$$

An extrapolation of $(X - 1)/(L)$ to $(L) = 0$ will give β_1 as intercept and β_2 as the initial slope

$$\frac{X - 1}{(L)} = \beta_1 + \beta_2(L) + \beta_3(L)^2 + \cdots \tag{3B41}$$

If there are more than two complexes, the plot will curve upwards, and values of β_2 will not be very precise; if $\beta_{n>2} = 0$ a straight line is obtained and β_2 will have good precision. Where $\beta_{n>2} \neq 0$ the procedure may, in principle, be repeated, to obtain in a general way

$$\frac{X - \sum_0^{t-1} \beta_n(L)^n}{(L)^t} = \beta_t + \beta_{t+1}(L) + \sum_{t+2}^{N} \beta_n(L)^{n-t} \tag{3B42}$$

However, any errors in the early members of the series will accumulate in the later ones, and their precision will decrease. In principle, again, the procedure will yield a straight line of intercept β_{N-1} and slope β_N in the last step. A much better procedure for obtaining the higher members of the series is to extrapolate a plot $X(L)^{-N}$ against $1/(L)$, to $1/(L) = 0$, giving β_N as intercept and β_{N-1} as slope, and iterate between both ends of the series in order to refine the constants by successive approximations. A system with $N = 4$ can thus be evaluated rather well, one with $N = 6$ moderately so, provided that the ratios k_n/k_{n+1} are not too small, and a sufficient range of (L) is covered.

When $N = 2$ the above method gives a linear plot, as mentioned. Linear plots can also be obtained from other types of data. When $N = 1$ the functions $(1 - \alpha_0)/\alpha_0$ or $\alpha_1/(1 - \alpha_1) = \tilde{n}/(1 - \tilde{n})$ plotted against (L) will be straight lines of slope $k_1 = \beta_1$ and zero intercept. When $N = 2$ various transformations of the functions $\alpha_n = F(L)$ and $\tilde{n} = F(L)$ can be made in order to yield linear plots, e.g.

$$\frac{(1 - \alpha_2)(L)^2}{\alpha_2} = \frac{1}{\beta_2} + \frac{\beta_1}{\beta_2}(L) \tag{3B43}$$

$$\frac{\tilde{n}}{(1 - \tilde{n})(L)} = \beta_1 + \beta_2 \frac{(\tilde{n} - 2)(L)}{(1 - \tilde{n})} \tag{3B44}$$

Plots according to the second example[47] become very imprecise at values of $\tilde{n}$ near zero, one and two. Any part of an extended series of complexes may be considered, as a first approximation, to involve only three species ML_{n-1}, ML_n and ML_{n+1} in equilibrium, over a limited range of (free-) ligand concentration, and the parameters $k_n = \beta_n/\beta_{n-1}$ and $k_n k_{n+1} = \beta_{n+1}/\beta_{n-1}$ can be obtained from linear plots in this range, as first approximations.

Curve-fitting methods[48] utilize calculated curves, which involve normalized variables, and which are compared with the experimental data plotted to the same scale. For $N = 1$ the variable $\log u = \log(L) + \log \beta_1$ is used as abcissa and α_0 or $\alpha_1 = \tilde{n}$ as ordinate, and the function $\alpha_0 = 1/(1 + u)$ or $\alpha_1 = u/(1 + u)$ is plotted for arbitrary values of the variable u, from zero to infinity. These curves are compared with plots of the experimental points $\alpha_0 = F(\log(L))$ or $\alpha_1 = F(\log(L))$. Moving the plots along the abcissa to coincidence gives $\log \beta_1 = -\log(L)_{\log u = 0}$. The advantage of the method is the utilization of the directly measured variables $\tilde{n}$ or α_0 and $\log(L)$, the logarithmic plot of ligand concentration being preferred because of the wide range of concentration (a few powers of ten) which should be used. Thus experimental errors can be easily discerned and corrected for, since the curve has a unique form.

For $N = 2$ and data $\alpha_n = F(L)$ or $X = F(L)$ two variables can be normalized. For instance, setting $u = (X - 1)/\beta_1(L)$ and $v = \beta_2(L)/\beta_1$ yields $u = 1 + v$, so that the calculated curve becomes a plot of $\log u$ against $\log v$, while the experimental curve is a plot of $\log(X - 1)/(L)$ against $\log(L)$, as shown in Figure 3B2. One plot is moved over the other, with parallel axis, until the curves coincide, when $\log \beta_1 = -\log(X - 1)/(L)$ at $\log u = 0$, and $\log \beta_2/\beta_1 = -\log(L)$ at $\log v = 0$, i.e. the position of the origin of the calculated plot on the coordinate space of the experimental plot gives the values of the constants. If the data are in the form $\tilde{n} = F(L)$ the case is more complicated. Setting $u = \beta_2^{1/2}(L)$ and $p = \beta_1\beta_2^{-1/2}$ gives $\tilde{n} = (pu + 2u^2)/(1 + pu + u^2)$, which may be plotted against $\log u$ to obtain a family of curves with p as parameter, which are compared with the experimental plot of $\tilde{n} = F(\log(L))$. The disadvantage of this method, compared to the one above, is that the plot is not unique, so that experimental errors cannot be detected so easily, and the value of p is obtained with only poor precision.

When it is necessary to determine three parameters, the same technique can be used, as will be illustrated for data $X = F(L)$. The normalized variables are $u = (X - 1)/\beta_1(L)$, $v = \beta_2(L)/\beta_1$ and $p = \beta_1\beta_3/\beta_2^2$, so that $u = 1 + v + pv^2$, and the calculated family of curves have the form $\log u = F(\log v)_p$, with which are compared the experimental data $\log(X - 1)/(L) = F(\log(L))$, Figure 3B3. From the location of the origin of the $\log u = F(\log v)$ plots on the experimental plot are obtained $\log \beta_1 = -\log(X - 1)/(L)$ at $\log u = 0$, $\log \beta_2/\beta_1 = -\log(L)$ at $\log v = 0$, and β_3 is obtained from the value of p for that curve among the family which gives the best fit. However, p is not very precise, so that another transformation of the data, e.g. $u = 1 + qt + t^2$ with the same u, but with $q = p^{-1/2}$ and $t = (L)\sqrt{\beta_3/\beta_1}$, will give $\log \beta_3/\beta_1 = -2\log(L)$ at $\log t = 0$ and β_2 is obtained from the parameter q.

In some cases the data are of the form $D = K\alpha_p = F(L)$ with experimental values of D, proportional to the degree of formation of the pth complex. Curve fitting can then conveniently be used to obtain K and two complexity constants, e.g. with $p = 0$, the constants β_1 and β_2, by using the normalized variables $\log u = \log K - \log D = \log(1 + v + qv^2)$ with $v = \beta_1(L)$ and $q = \beta_2/\beta_1^2$. Many other variations are possible[49].

In cases there are more than three parameters to be determined, graphical methods can only be used in conjunction with successive

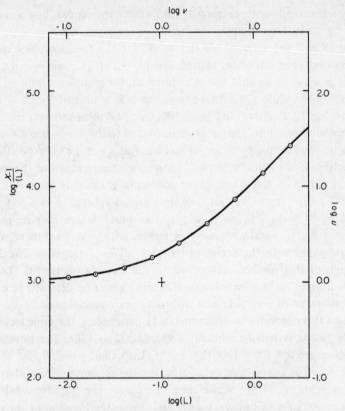

Figure 3B2.   Curve fitting for determining two equilibrium constants. Points are hypothetical experimental data, pertain to left-hand ordinate and lower abscissa; curve is a plot of $\log u = \log (1 + v)$ against $\log v$, on the right-hand ordinate and upper abscissa respectively, on a separate sheet of paper, which is transparent. When the curve and points are brought to coincide, as in the Figure, the two parameters $\log \beta_1$ and $\log \beta_2$ are obtained from the position of the $(\log u = 0, \log v = 0)$ origin, $+$, in the $\log (X - 1)/(L) - \log (L)$ space.

approximations, provided the range of data can be divided into regions where at most four species coexist (with three interconversion equilibria). Otherwise an approximation, the 'two-parameter' method, proposed by Dyrssen and Sillèn[50], can be used. In practice only seldom are data precise enough to warrant a more detailed treatment, if many species coexist over a narrow range of ligand concentrations. The two parameters can be chosen, e.g., as β_N and R, the latter being defined as an average spreading factor, the parameters giving approximate values of $\log \beta_n = (n/N)\log \beta_N + n(N + n)\log R$. Both $\tilde{n}$ and α_n can be expressed

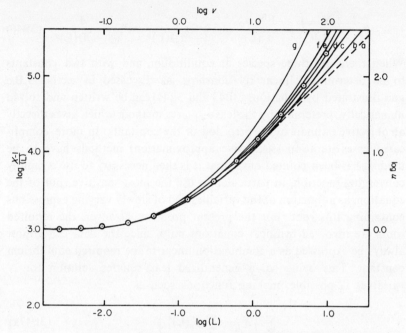

Figure 3B3. Curve fitting for determining three equilibrium constants. Points are hypothetical experimental data, pertaining to the left-hand ordinate and lower abscissa; curves a—g are the functions $\log u = \log(1 + v + pv)$ against $\log v$, pertaining to the right-hand ordinate and upper abscissa, for different values of p, on a separate sheet of paper, which is transparent. As in Figure 2B2, the equilibrium constants are obtained when the points are brought to coincide with one of the curves, from the position of the origin, $+$, and the p value of the coinciding curve.

as functions of β_N, R, n and (L) only, using, for example, a normalized variable $u = \beta_N^{1/N}(L)$ for the expression

$$X = \sum_0^N R^{n(N-n)} u^n \qquad (3B45)$$

Plotting $\log X$ against $\log(L)$ and comparing this with normalized curves $\log X = F(\log u)_R$ will give the values of β_N and of R.

f. Numerical Methods

In very simple cases, straightforward numerical calculation of the constants is possible. When only one complex is formed, ML_p, beside the free metal cation M (or any lower complex, with appropriate changes in the definitions), then

$$\beta_p = \frac{X - 1}{(L)^p} = \frac{1 - \alpha_0}{\alpha_0 (L)^p} = \frac{\alpha_p}{(1 - \alpha_p)(L)^p} = \frac{\tilde{n}}{(p - \tilde{n})(L)^p} \quad (3B46)$$

When there are three species at equilibrium, and with two constants to be determined, linear relationships, as discussed in section 3.B.e and illustrated by equations (3B43 and 3B44) can be written and solved numerically, preferably by the least-squares method, which gives directly an objective estimate of the precision of the constants. In more complicated cases iteration (successive approximation) methods have to be used. It has been pointed out[51] that it is then necessary to use a rapidly converging function, in particular to use the most sensitive root of the equation, as a function of the variables and of slowly varying expressions containing this root. For the present problem, however, the required roots are involved in linear equations only, since the data can almost always be expressed as a combination linear in the required equilibrium constant. This being so, a generalized least-squares solution for N variables is possible, making functions such as

$$\sum_j \left[X_j - \sum_0^N \beta_n (L)_j^n \right]^2 \quad (3B47a)$$

$$\sum_j \left[\sum_n (\tilde{n}_j - n)\beta_n (L)_j^n \right]^2 \quad (3B47b)$$

minimal, by correct choice of the parameter β_n. This process is greatly aided by the use of high-speed digital computers, which have lately become available for this kind of work.

One advantage of the use of computors for the least-squares method is the ease of weighting the variables with their appropriate statistical weights. It is usual to neglect experimental errors in (L), in particular when trace metal concentrations are used, and to assign the standard experimental error to the quantity depending on (L), e.g. X or α_n. The calculation then yields a set of parameters, each with its standard error. Programmes have been devized which automatically eliminate negative parameters, or parameters closer to zero than one standard deviation, and make a new calculation in terms of one fewer parameter[52]. Some programmes also apply a significance test, to show whether the experimental data arc better expressed in terms of N or $N - 1$ parameters. The method of 'pit-mapping' has been found useful when it is possible to estimate approximate values connected implicitly, though not necessarily linearly, with the 'best set' of parameters[53]. The method is said to increase the speed of calculation and to ensure rapid converging of iterations.

Programmes are now available for a variety of electronic computers[53-57] with a variety of input data: pH titration[56], metal-electrode potentiometry[57], solvent extraction[52], cation exchange[58], etc. The programmes are not difficult to translate in order to be applicable to the computer that happens to be at hand.

It has, however, been shown, that results obtained by the 'precomputer era' methods seldom yield results not compatible with those obtained by the computers. Most often almost identical results are obtained, i.e. virtually the same stability constants. Since very often a large measure of arbitrariness is involved in the process of weighting the individual experimental points, the contention that the computer method, contrary to graphical methods, for example, gives completely objective estimates to the limits of error of the computed parameters, may be an overstatement. In many cases, a preliminary graphical solution of the problem is required in order to give the computer the necessary input data. As will be discussed below (section 3.B.h), we only know that a species, when present in solution at a considerable relative concentration, will make a set of data produce on computation a constant which corresponds to the formation of this species. The reverse statement may sometimes be erroneous.

g. Presentation of Results

It is customary to present the results of the calculations as a set of stability constants, e.g. $\beta_1, \beta_2 \cdots \beta_N$, or the part of the complete set involved in the range of ligand concentrations used. A visually more descriptive way is to calculate values of α_n, the degrees of formation of the individual complexes, and to plot them as function of the (free-) ligand concentration (Figures 3B1 and 3B4a). Alternatively, $\sum_0^t \alpha_n$ is plotted for values of t from unity to $N-1$, (Figure 3B4b). These plots may be the only way available to represent the results of the complex formation, since at high ligand concentrations, lack of knowledge of activity coefficients may preclude the calculation of meaningful stability constants. In some cases, a plot of $\bar{n}$ against (free) ligand concentration (Figure 3B4c) may also yield important information, such as the spacing of the individual k_n values, section 3.B.d, and the maximal number of ligands, whether N or less.

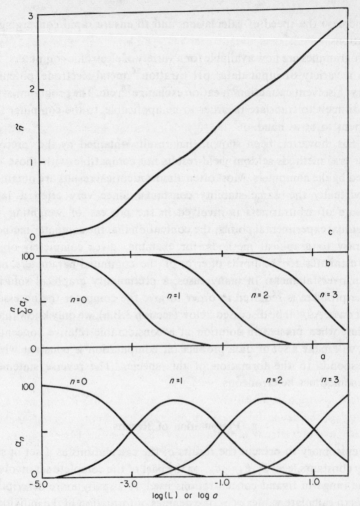

Figure 3B4.   Graphical representation of complex formation data with $\beta_1 = 10^3 \mathrm{M}^{-1}$ $\beta_2 = 10^4 \mathrm{M}^{-2}$, $\beta_3 = 10^3 \mathrm{M}^{-3}$, $\beta_{>3} = 0$. Lower portion (a) shows a_n; middle portion (b) shows comulative values $\sum_0^n a_i$; upper portion shows $\bar{n}$; all as functions of the logarithm of ligand concentration or, effective activity.

When it is necessary to plot the formation of complexes as a function of two variables (such as total ligand concentration c_L and pH, or two independent ligands L_1 and L_2) a device such as predominance area diagrams[59] may prove useful. Here the boundaries of domains, in each a given species has the highest concentration, are plotted in a coordination space determined by the two independent concentration variables.

h. The Reality of Complexes

A few comments have already been made on how 'real' a complex species is, if it is detected in solution by a given method. The case of ion-pairs has been discussed in detail above (section 3.A.d), and the same question arises also with weak complexes (distinguished from ion pairs by there being more than one ligand per central cation, and a definite geometrical arrangement of the ligands). Young and Jones[60] strongly criticized investigations of the formation of weak complexes, concluding that 'doubtless many complexes reported will be found not to exist in appreciable amounts'. Similarly, Wormser[61] emphasized that species, for which the only evidence of their formation is a good parametric fit of an expression with data, need not exist at all. Indeed, many methods for studying complex formation have proved to be unreliable, and in many investigations not sufficient precautions have been taken[33] to obtain significant data and analyse them in the most reliable way. Thus Young and Jones' criticism serves to force investigators to design their experiments better, and treat experimental errors, random as well as systematic, and shortcomings in the theory, less optimistically. However it cannot be concluded that complex formation constants of the order of ten or less have no significance, as has been done by some critics, provided that the experimental and mathematical methods used are adequate[33].

Methods become unreliable, if they conclude that a complex exists, because of the deviation of a set of measurements, only indirectly related to this complex, from a theoretically expected function. A notorious example is the conductivity method, which gives, indirectly, the concentrations of free ions in solution. If these are low compared to the stoichiometric concentration, the presence of a complex is strongly indicated. If, however, the free concentration approaches the stoichiometric one, then the uncertainties of parameters and functions of the conductance theory, together with variations in activity coefficients, and even only small experimental error, can lead to errors in the association constant of orders of magnitude.

Again, sometimes the sole evidence for the formation of a complex is the deviation of data, forming a power series, from another power series, in which the last term, pertaining to this complex, is omitted. In such cases then this species can be rather suspect, particularly if it is the third or subsequent species in a series. Errors in data accumulate in the extrapolation procedures often employed (section 3.B.e), so that the measurement of free central cation concentration (e.g. potentiometrically, or by cation exchange), is not a very reliable method for studying the higher complexes.

Only methods that directly measure a property of a given complex can be said to establish the existence of this complex. Thus finding the metal migrating to the anode in a well-designed electromigration experiment can be taken as irrefutable evidence for the formation of an anionic complex. The extraction of a metal into an inert solvent establishes the presence of a neutral complex *in this solvent*. The formation of a new absorption band, or the appearance of a new Raman line establish the formation of a new species, a complex, the composition and stability of which is to be determined. The mere shift of a band, or a change in its intensity may or may not be due to the formation of a complex.

These considerations will be illustrated with two examples: the case of bivalent metal sulphates, and the case of perchlorate complexing. In the first case, association constants of about 100–300 were determined by a variety of methods (conductivity, cryoscopy, potentiometry, spectrophotometry) for the association of sulphate anions with bivalent metal ions such as magnesium, zinc, copper, etc. There is, however, no standard of behaviour of non-associated 2:2 electrolytes, by which to gauge the association, since the theoretical equations (of activity coefficients, conductivity, etc.) break down for highly charged electrolytes at such low concentrations where association is not yet significant[62]. Matters become worse if it is attempted to measure complex formation beyond the first stage, i.e. the stability of $M(SO_4)_2^{2-}$. Leden[63] obtained discordant results from potentiometric measurements, using a cadmium-amalgam electrode, which indicated formation of $Cd(SO_4)_2^{2-}$ and even $Cd(SO_4)_3^{4-}$, and using a competitive method involving silver ions and a silver electrode, where no complex higher than $CdSO_4$ could be found. Evidence from anion-exchange measurements is conflicting too: Leden[64] finding no complexes, Fronaeus[65] finding some, though weak, formation of an anionic complex. Similarly, Ahrland[66] found potentiometrically that $UO_2(SO_4)_3^{4-}$ is formed, although spectrophotometric evidence hardly confirmed this, and other workers[67] could not confirm the formation of this species. In all these cases, the weakness of the complexes formed necessiated a high ligand concentration, producing consequently changes in activity coefficients, in spite of a nominally constant medium (or rather, ionic strength) (section 1.E).

The case for the formation of perchlorate complexes in aqueous solutions is similarly obscure. The problem is to disentangle stoichiometric from medium effects. Most of the positive evidence on perchlorate complexes comes from spectrophotometry. Usually, addition of a perchlorate to a solution containing a metal cation produces little, if any changes in the light absorption. Occasionally, changes are observed, and then it is a question whether they are due to a medium effect (since they were obtained also at a constant, low, ionic strength[68]) or more specifically to dehydration of the cation, permitting stronger interaction with anions and changes in the spectrum[27,69], or to formation of perchlorate complexes[70,71]. Since it is uncertain that molar absorbancies are independent of temperature and of ionic strength, it is not possible to depend on spectrophotometric measurements alone, in deciding that addition of perchlorate ion leads to complex formation, rather than to a general, non-stoichiometric medium effect, such as dehydration or changes in the ionic atmosphere (activity coefficients). The view that the second explanation is more nearly correct is strengthened by the enhancement of the apparent stability of the complex between iron(III) and chloride ions on the addition of perchlorate[27,69], rather than the opposite effect, expected from competition if a perchlorate complex were formed.

Thus although there is no justification for the conclusion that perchlorate complexes are not formed in aqueous solutions, there is also no firm evidence for their formation[72,73]. In the case where a constant ionic medium is employed with a large excess of perchlorate ions, their concentration remains virtually unchanged during complex formation, so that they may be considered to belong to the solvent, and ignored (section 3.B.a). It is sometimes useful to remember that the generalized species M in a medium of aqueous $C^+ClO_4^-$ is rather $\sum_i \sum_j \sum_k C_i M(H_2O)_j (ClO_4)_k$. Thus complex formation constants are relative to possible substitution of water and perchlorate in this medium, which might produce changes in the entropy of complex formation compared to simple addition. Perchlorate is found, at least, to produce complexes generally less stable than those formed with any other anion, and is probably the most suitable to serve as an inert anion in a constant medium. In special cases, where oxidation by perchlorate is feared (as for ruthenium(II)), tetrafluoroborate and p-toluenesulphate may also serve well[74].

The choice of the cation is governed by the need to have acid present, to repress hydrolysis of highly charged cations. Too much acid may sometimes produce special 'acid effects' (cf. section 6.C for anion exchange), and in any case will interfere with the ionization of weak acids to provide adequate ligand concentration. The requirement of keeping the activity coefficients as constant as possible leads to the selection of electrolytes for the ionic medium having small Harned's-rule values in the mixture (section 1.E).

Even at a constant ionic strength, the ionic medium must be kept essentially constant to obtain meaningful results, rather than imagined species. Thus, neither the cations of the medium may be exchanged to a large extent, nor, of course, the inert anion for the ligand. Early work of Mironov and coworkers[75] suggested the existence of species such as $PbCl_5^{3-}$, $PbCl_6^{4-}$, $PbBr_5^{3-}$, $CdCl_5^{3-}$, $CdCl_6^{4-}$, etc., all in cases where a large fraction of the medium was exchanged. Later work retracted this interpretation of the data in favour of cation effects[76], i.e. association of anionic complexes with cations to form ion-pairs. This interpretation, too, need not withstand further tests, since non-specific activity effects could be responsible for the observed deviations, rather than specific associations.

C. ELECTROLYTE ASSOCIATION AND HYDROLYSIS

a. Introduction

Having examined the theories of ion-pair formation, and the possibilities for formation of complexes, and the experimental and mathematical methods for their evaluation, it will now be convenient to consider the association of a number of specific electrolytes. In addition, the hydrolytic properties of metal ions in aqueous solution will be briefly considered. Information on the formation of complexes between specific metallic cations and anionic or other ligands is so abundant on the one hand, but controversial and inconsistent on the other[77], that it cannot be reviewed here. The following discussion will therefore be confined to a few chosen cases, on which there is some agreement. It is now generally accepted that strong electrolytes are completely ionized in solution, but the ions may interact with one another electrostatically to produce an associated ion pair. In weak electrolytes, on the other hand, some kind of chemical bond exists in the unionized form. Ionization may occur in two stages, the first being the production of an associated ion-pair, the second its dissociation. These semantic differences are not very important, except that it is now customary to talk of ion association, rather than of electrolyte dissociation, and of association constants, rather than of dissociation constants, which are their reciprocals (cf. section 2.D) (except for acids, where the term 'dissociation' is still common). This change in convention parallels that involved in complex stability, where it is now customary to talk of formation and stability constants, rather than of their reciprocals, instability constants, still favoured by Russian workers (section 3.B).

In any case, the association reaction obeys the mass-action law, and is described by its equilibrium constant, the association constant.

$$\log K_{ass} = pK_{diss} = \log a_{CA}/a_{C^+}a_{A^-} = \log(CA)/(C^+)(A^-) - \log y_u/y_\pm^2$$

$$= \log k_{ass} + \log y_u/y_\pm^2 \qquad (3C1)$$

In many cases of strong electrolytes, K_{ass} is so small that a high concentration of the anion A^-, for example, is required to obtain a measurable ratio of $(CA)/(C^+)$. At these high concentrations the estimates of y_u and of $y_\pm$ (free) are rather uncertain, since we measure only the stoichiometric activity coefficients, not those of the species. Therefore the estimate of K_{ass} becomes rather uncertain.

b. Association of Strong Acids

The hydrogen ion is well known to associate strongly with many Brönsted bases, particularly with anions, to form acids. Extensive data for the more highly associated acids, the weak acids, have been published[77]. Information available on the relatively strong acids is less certain, and most quantitative information is available for water as solvent. Some information on solutions in organic solvents is presented in Chapter 2 and the following is confined to aqueous solutions.

The association constant for perchloric acid has been determined by the nuclear magnetic resonance (n.m.r.) method[78], the value found being $\log K_{ass} = -1.58$ at 25°. This rather high value of the association constant has been confirmed recently[79], but the degrees of dissocation leading to this value disagree with early Raman spectroscopic data[80], and also with a comprehensive recent Raman study[81]. The temperature coefficient[79] leads to $\Delta H° = -1800$ cal/mole and $\Delta S° = +1.20$ e.u. The degree of dissociation is shown in Table 3C1 and Figure 3C1 as a function of concentration, as obtained from Raman data[81] as well as from n.m.r.[78-80]. Using the lower degrees of association obtained from the Raman data[81], it is concluded that $\log K_{ass} \ll -1.58$, but a definite value cannot be given, owing to the long extrapolation required.

No reliable information is found concerning the association of the hydrohalic acids in water. In nonaqueous solvents it is usually observed that the order of strength is $HClO_4 > HI > HBr > HHSO_4 > HCl > HNO_3$[82], and if this holds for aqueous solutions too, then $\log K_{ass}$ must lie in the narrow range between -1.58, the value for perchloric acid, and -1.38, the value for nitric acid. Indeed, Posner[83] calculated from literature data on activity coefficients for hydrochloric acid, $\log K_{ass} = -1.51$. The vapour pressure of hydrogen chloride in equilibrium with concentrated hydrochloric acid does indicate the presence of undissociated HCl. Undissociated HCl molecules have been detected by Raman spectroscopy above $6M$[84]. The degree of dissociation of the hydrohalic acids in concentrated solutions has been estimated from H_0 data[85], as shown in Figure 3C1.

The association of nitric acid has received much attention. Both Raman spectroscopy and n.m.r. yield reliable results, leading to to $\log K_{ass} = -1.37$ (Raman[86]) to -1.44 (n.m.r.[79]) at $25°$, with $\Delta H° = -1970$ cal/mole and $\Delta S° = -4.46$ e.u. Earlier results[87] agree with these values as well as some more recent estimates[88], but a new estimate of -1.28 ± 0.06 has been based on newly determined activity coefficients, and should replace older values[89]. The degree of dissociation is shown in Table 3C1 and Figure 3C1. McKay[90] has used the degree-of-dissociation data of Krawetz[86] to calculate the activity coefficient of the ionized part of nitric acid γ_h, taking into account the effect of the unionized part. His equation is

$$\log \gamma_h = \log(\gamma_\pm/\alpha) - \tfrac{1}{2}k_s(1-\alpha)m \qquad (3C2)$$

where m and $\gamma_\pm$ are the stoichiometric molality and mean ionic activity coefficient, and k_s is the salting constant for undissociated nitric acid by the ions, found to be 0.048 ± 0.005. The value of $\log K_{ass} = -1.37$, which fits this treatment, is consistent with the values reported above, but not with the more recent activity coefficients reported[89].

The n.m.r. method has also been used for determining the degree of dissociation of nitric acid in aluminium nitrate solutions[91]. The common-ion effect, of course, causes a decrease in the dissociation. Another important effect is the dehydration of the ions by competition with the highly hydrated aluminium ion, which results in small distances of closest approach, hence more extensive association.

The association of the bisulphate ion with protons is very slight even up to fairly high concentrations. Raman spectroscopic results[92] and

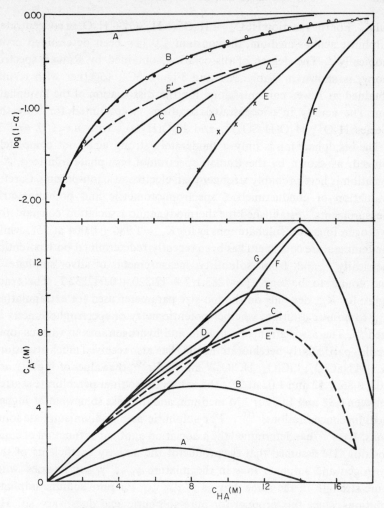

Figure 3C1. Acid dissociation as a function of the stoichiometric acid concentration. Upper portion: the fraction of acid in the associated form, $\log (1 - a)$; lower portion: the concentration of the anion of the acid. The acids are A = HSO_4^- B = HNO_3, C = HCl, D = HBr, E = $HClO_4$ (from Raman data, primed from p.m.r. data), F = H_2SO_4. Curve G = A + F. (In curve B: ● n.m.r. data, ○ Raman; in curve E': △ from Cl—n.m.r. data).

n.m.r. data[93] agree that up to 14 M sulphuric acid the concentration of H_2SO_4 is very low. No association constant could be established for this species, based on a dilute solution in water as a standard state. An estimate $\log K_{ass} = -8.57$ has been made from spectrophotometric data[94]. For the reaction $HSO_4^- + H_3O^+ \rightleftharpoons H_2SO_4 + H_2O$, in concentrated sulphuric acid as medium, the constant 1.0 has been determined cryoscopically[95]. The degree of dissociation, obtained by Raman spectroscopy, is shown in Table 3C1 and Figure 3C1, together with results obtained at lower concentrations for the dissociation of the bisulphate ion. The acidity in concentrated solutions is accounted for[96] by the species H_3O^+, $H_3O(H_2SO_4)^+$, and $H_3O(H_2O)_n^+$ with $n = 1$, 2 and 3.

The bisulphate ion is only a moderately strong acid, not completely ionized, as shown by the Raman spectrum of bisulphate solutions. Association is here probably stronger than electrostatic ion-pairing. Careful evaluation of conductimetric, spectrophotometric and potentiometric measurements[97] established the thermodynamic association constant for hydrogen ions and sulphate ions as $\log K_{ass} = 1.986 \pm 0.004$ at $25°$, while the temperature coefficient has been recently redetermined both potentiometrically[98] and from solubility measurements of silver sulphate[99], and found to be $\log K_{ass} = 1283.1/T - 12.320 + 0.042232\,T$. The exact figure for K_{ass} depends on the ion-size parameter used for extrapolating K_{ass} to infinite dilution, whether potentiometry or spectrophotometry is used[100]. The association of sulphate and hydrogen ions in various ionic media, particularly perchlorate media, has also received much attention. In 1 M $NaClO_4$, $LiClO_4$, $HClO_4^{101}$ and $NaBr^{102}$, the values of $\log K_{ass}$ are 1.02, 0.86, 0.52 and 1.02 at $25°$; the values for sodium perchlorate scatter between 0.98 and 1.05 for 1 M medium, and increase somewhat at higher medium concentrations[77,101]. For sulphuric acid–sodium sulphate solutions, Baes[103] has determined the association quotient as function of composition. He assumed that the product of the activity coefficients of the hydrogen and sulphate ions in the mixture $\gamma_{h(H^+)}^2\gamma_{h(SO_4^{2-})}$ varies with ionic strength in the same way as $\gamma^3_{\pm(Na_2SO_4)}$ for pure sodium sulphate solutions, since this product for pure sulphuric acid does vary so. He used the values obtained from

$$\gamma_{h(H^+)}^2\,\gamma_{h(SO_4^{2-})} = 4\,m_{H_2SO_4}^3\gamma_{\pm(H_2SO_4)}^3/(H^+)^2(SO_4^{2-}) \qquad (3C3)$$

with Raman spectroscopic[92] values of (H^+) and (SO_4^{2-}). The association quotient fitting these results is

$$\log K_{ass} = 1.991 - 2.036\,I^{1/2}/(1 + 0.4\,I^{1/2}) \qquad (3C4)$$

The association quotients and species concentrations in sodium bisulphate solutions or mixtures of sulphuric acid and sodium sulphate were calculated

TABLE 3C1. Dissociation of the strong acids.

c	HNO$_3$(B)* 1—a	(HNO$_3$)	(NO$_3^-$)	H$_2$SO$_4$(F)** 1—a	(H$_2$SO$_4$)	(HSO$_4^-$)	HSO$_4^-$ (A)* 1—a	(HSO$_4^-$)	(SO$_4^{2-}$)
1	0.015	0.015	0.985						
2	0.042	0.084	0.916				0.66	1.32	0.68
3	0.071	0.213	1.79				0.66	2.01	0.99
4	0.120	0.48	3.51				0.68	2.73	1.27
5	0.163	0.81	4.19				0.68	3.4	1.6
6	0.225	1.35	4.65				0.70	4.2	1.8
7	0.282	1.98	5.02				0.74	5.2	1.8
8	0.368	2.95	5.05				0.78	6.2	1.8
9	0.425	3.85	5.15				0.81	7.3	1.7
10	0.490	4.90	5.10				0.85	8.5	1.5
11	0.559	6.15	4.84				0.90	9.9	1.1
12	0.614	7.35	4.65				0.93	11.2	0.75
13	0.68	8.8	4.2				0.96	12.5	0.45
14	0.73	10.1	3.8	0.012	0.17	13.6	0.97	13.6	0.25
15	0.79	11.9	3.1	0.16	2.4	12.6			
16	0.85	13.6	1.4	0.32	5.1	10.9			
17	0.89	15.1	1.9	0.53	9.0	8.0			
18				0.80	14.4	3.6			

c	(E)* 1—a	HClO$_4$(E')* 1—a	(HClO$_4$)	(ClO$_4^-$)	HCl(C)* 1—a	(Cl$^-$)	HBr(D)* 1—a	(Br$^-$)
1		0.015	0.015	0.985	0.00	1.00	0.00	1.00
2		0.036	0.072	1.928	0.00	2.00	0.00	2.00
3		0.056	0.168	2.83	0.00	3.00	0.00	3.00
4		0.078	0.31	3.69	0.04	3.85	0.00	4.00
5		0.098	0.49	4.51	0.09	4.55	0.00	5.00
6	0.000	0.120	0.72	5.28	0.11	5.35	0.04	5.75
7	0.003	0.143	1.00	6.00	0.11	6.25	0.07	6.50
8	0.009	0.173	1.39	6.61	0.15	6.80	0.09	7.30
9	0.027	0.205	1.85	7.15	0.17	7.45	0.13	7.85
10	0.056	0.235	2.35	7.65	0.24	7.60		
11	0.090	0.270	2.95	8.05	0.26	8.15		
12	0.146	0.315	3.75	8.25	0.29	8.50		
13	0.23	0.37	4.8	8.2	0.31	9.00		
14	0.35	0.45	6.3	7.7				
15		0.54	8.1	6.9				
16		0.68	10.9	5.1				
17		0.89	15.1	1.9				

* Curve in Figure 3C1. Curve E from Raman data, E' from n.m.r.
** Curve G designates sum A + F.

from these equations, using successive approximations for the ionic strength, and were tabulated[103].

Among the strong acids must be counted also the monobasic tetrahedral acids of the general formula HMO_4. Spectrophotometric results[104] led to $\log K_{ass}$ of -2.3 and -1.3 for permanganate and perrhenate respectively, the value for pertechnetate being expected to be intermediate, but, surprisingly, is apparently much higher, $\log K_{ass} = 0.3$[105]. Although the acids HMO_4^- are only moderately strong (sulphuric acid) or weak (chromic acid), they associate with a hydrogen ion only reluctantly, thus formation H_2CrO_4 proceeds[104] with $\log K_{ass} = -1.0$, and formation of H_2SeO_4 is expected to be as difficult as that of H_2SO_4. Association of the last hydrogen ion to form H_3MO_4 is comparatively easy ($\log K_{ass}$ for H_3PO_4 being 2.15). Sulphonic acids[106], as well as chloric acid, tetrafluoroboric acid, and some others are considered to be strong, but for the latter acids no reliable association constants are known. The moderately strong acids include thiocyanic acid ($\log K_{ass} = 0.85$[107], or 0.70[108]), iodic acid($\log K_{ass} = 0.77$ at 25°)[77], and some others[109]. Dissociation constants for weak acids have been tabulated in many publications [77]. Values for the stronger acids are summarized in Table 3C2.

TABLE 3C2. Dissociation constants of strong and medium acids at 25°.

Acid	Log K_{diss}	Reference*	Remarks
$HHSO_4$	$\gg 2(8.57)$	(94)	Figure 3C1
$HHSeO_4$	$\gg 2$	104	
$HClO_4$	$>4(1.58)$	81(78)	Raman (n.m.r.)
$HMnO_4$	2.25	104	Spectrophotometry
HBr	>2.15	85	H_0, tentative value
HCl	>1.05	85	H_0, tentative value
CH_3SO_3H	1.87	106	Raman
HNO_3	1.37(1.44)	86(79)	Raman, n.m.r.
	1.28	89	With new activity values
$HReO_4$	1.25	104	Spectrophotometry
$p\text{-}CH_3C_6H_4SO_3H$	1.06	106	Raman
$HHCrO_4$	0.98	104	Spectrophotometry
$HClO_3$	Strong	77	Calculated hypothetically
$HTcO_4$	-0.3	105	
HIO_3	-0.77	77	Conductivity, 25°
$HSCN$	$-0.85(-0.70)$	107(108)	Potentiometry, 25°
HSO_4^-	-1.98	97	Potentiometry, 25°
H_3PO_4	-2.15	77	Potentiometry, 25°

* Reference in parenthesis refers to values in parenthesis.

c. Association of Bases and Hydrolysis of Metal Cations

The hydroxides of monovalent cations with rare-gas electronic configurations are relatively strong electrolytes. The hydroxide anion can, however, approach cations much more closely than most anions, since the arrangement M(OH)HOH competes with the arrangement M(OHH)OH. Davies[110,111] has shown that the association constants vary with charge and crystalline radius of the cation M^{z+}, with a noble-gas core, as

$$\log K_{ass} = -1.150 + 0.607 \, z/(r_c + 0.74) \tag{3C5}$$

with r_c in Å, and 0.74 being the radius of the hydroxide ion.

Recent determinations of the hydrolysis in 3 M perchlorate media gave lower values for $\log k_{ass}$ for the alkaline earths[112], 0.640, 0.225 and 0.005, compared to $\log K_{ass}$ 1.40, 0.85 and 0.64 for Ca, Sr and Ba, all[77] at 25°, and a value for La[112], $\log k_{ass} = 4.1$, which is higher than the value calculated from (3C5). Ions with a high field, show deviations from Davies' relationship, and a correction $-0.05 z/r_c$ has been suggested[111] for the denominator in (3C5). Indeed, hydrolysis is more extended than the association of the metal cation with one hydroxide anion, and polynuclear hydrolytic complexes are formed rather easily, even at low concentrations[113]. An initial step in many cases is the acid dissociation, illustrated by hydrolysis of iron(III)[114]

$$Fe(H_2O)_6^{3+} + H_2O \rightleftharpoons Fe(H_2O)_5OH^{2+} + H_3O^+ \tag{3C6}$$

This remains the only step in a few cases (mercury(I)), or the only step at low metal concentrations, where polymerization is negligible (cadmium, iron(II), manganese, zinc, lanthanum). In a few other cases, a second step, forming $M(OH)_2$, is the only other hydrolytic reaction that need be taken into account; this occurs with thallium(III) and with mercury(II) and silver ions which form linear complexes. A common mode of hydrolysis is to form oligonuclear species, i.e. hydrolytic complexes containing a few metal cations and hydroxide anions. Dimers $M_2(OH)_2$ are formed by iron(III), VO^{2+}, copper, tin(II), scandium, indium and perhaps some others; a trimer $M_3(OH)_3$ is formed by beryllium, and a tetramer by lead(II). At high concentrations, many metallic cations form polynuclear hydrolysis products, usually with formulae more complicated than the simple oligomers considered above. Thus uranium-(VI), scandium, indium, thorium, uranium(IV), bismuth, aluminium, among others, form species with a large number of metal cations per complex, hydrolysed, ion in certain cases an indefinite number. Some data are summarized in Table 3C3.

Knowledge of the hydrolysis constants permits the calculation of the relative amount of cation in the form of each species at given pH and metal concentration. Obviously, if only mononuclear species are formed, the distribution of the cation among the species is independent of its concentration. If complications due to hydrolysis are to be avoided, it is necessary to work at a pH so low that all hydrolysis products can be neglected in comparison with the free metal cation. A limit of the metal concentration may also be imposed, in some cases, because of the in-

TABLE 3C3. Hydrolysis of metal ions—fraction a_0 of non-hydrolysed trace metal ions at several pH.

Metal ion	pH=1	pH=3	pH=7	pH=10	Ref.	Medium
Li^+–Cs^+	1.00	1.00	1.00	1.00		
Be^{2+}*	1.00	1.00	0.05	$10^{-7.3}$	a	0.1 (NaClO$_4$)
Mg^{2+}–Ba^{2+}	1.00	1.00	1.00	1.00		
Sc^{3+}*	1.00	1.00	$10^{-3.8}$		b	1 (NaClO$_4$)
Y^{3+}*	1.00	1.00	1.00	0.11	c	3 (Li) ClO$_4$
La^{3+}*	1.00	1.00	1.00	0.56	d	3 (Li) ClO$_4$
Nd^{3+}	1.00	1.00	0.97	0.03	e	3 (Na) ClO$_4$
Lu^{3+}	1.00	1.00	0.29	$10^{-3.4}$	f	Dilute
TiO^{2+}	0.11	$10^{-4.9}$			g	2 NaClO$_4$
Zr^{4+}, Hf^{4+}*	≪1.00					
Th^{4+}*	1.00	0.93	$10^{-6.3}$		h	1 NaClO$_4$
VO^{2+}*	1.00	1.00	0.09		i	3 (Na)ClO$_4$
Pa^{5+}*	≪1.00					
U^{4+}*	1.00	0.045			j	2 (NaClO$_4$)
UO_2^{2+}*	1.00	0.98	$10^{-4.5}$		k	0.1 (NaClO$_4$)
Mn^{2+}	1.00	1.00	1.00	0.70	l	0 corr
Fe^{2+}	1.00	1.00	1.00	0.25	m	1 (Na)ClO$_4$
Fe^{3+}*	1.00	0.42	$10^{-7.7}$		n	3 (Na)ClO$_4$
Cu^{2+}*	1.00	1.00	0.68	0.002	p	0 corr
Ag^+	1.00	1.00	1.00	1.00	q	3 Na(ClO$_4$)
Zn^{2+}*	1.00	1.00	1.00	0.18	p	0 corr
Cd^{2+}*	1.00	1.00	1.00	0.31	r	3 (Na)ClO$_4$
Hg^{2+}	1.00	0.52	$10^{-7.8}$		s	3 (Na)ClO$_4$
Al^{3+}*	1.00	0.99	0.01		t	0 corr

TABLE 3C3. Hydrolysis of metal ions—fraction a_0 of non-hydrolysed trace metal ions at several pH (cont.)

Metal ion	pH$=$1	pH$=$3	pH$=$7	pH$=$10	Ref.	Medium
Ga^{3+}	1.00	0.45	10^{-4}		u	0.5 (Na)ClO_4
In^{3+}*	1.00	0.96	$10^{-5.7}$		v	3 (Na)ClO_4
Tl^+	1.00	1.00	1.00	1.00		
Tl^{3+}	0.51	$10^{-3.4}$			w	3 (Na)ClO_4
Sn^{2+}*	1.00	0.88	$10^{-3.1}$		x	3 (Na)ClO_4
Pb^{2+}*	1.00	1.00	0.88	$10^{-2.66}$	y	3 (Na)ClO_4
Sb^{3+}*	$\ll 1.00$					
Bi^{3+}*	0.80	0.04			z	3 (Na)ClO_4
Po^{4+}*	$\ll 1.00$					

* Definite evidence for polynuclear hydroxo complexes at macro concentrations. For references, those recent data were selected, that give information on mononuclear species.

a. G. Schwarzenbach, *Pure Appl. Chem.*, 5, 377 (1962).
b. G. Biedermann and coworkers *Acta Chem. Scand.*, 10, 1327 (1956).
c. G. Biedermann, *Proc. 7 Intl. Conf. Coordination Chem.* 1962, Abstr. p. 159.
d. G. Biedermann and L. Ciavatta, *Acta Chem. Scand.*, 15, 1347 (1961).
e. R. S. Tobias and A. B. Garrett, *J. Am. Chem. Soc.*, 80, 3532 (1958).
f. E. J. Wheelwright and coworkers., *J. Am. Chem. Soc.*, 75, 4196 (1953).
g. C. J. Garrignes, *Publ. Sci. Tech. Min. Air (Paris)*, No. NT93 (1960).
h. K. A. Kraus and R. W. Holmberg, *J. Phys. Chem.*, 58, 325 (1954).
i. F. J. C. Rossotti and H. S. Rossotti, *Acta Chem. Scand.*, 9, 1177 (1955).
j. J. C. Sullivan and J. C. Hindman, *J. Phys. Chem.*, 63, 1332 (1959).
k. J. Stary, *Collection Czech. Chem. Commun.*, 25, 890 (1960).
l. D. D. Perrin, *J. Chem. Soc.*, 2197 (1962).
m. B. O. A. Hedstrom, *Arkiv Kemi*, 5, 457 (1953).
n. K. Schlyter, *Trans. Roy. Inst. Techn. Stockholm*, No. 196 (1962).
p. F. Achenza, *Ann. Chim. (Rome)*, 48, 565 (1950).
q. P. J. Antikainen and D. Dyrssen, *Acta Chem. Scand.*, 16, 1785 (1962).
s. I. Ahlberg, *Acta Chem. Scand.*, 16, 887 (1962).
t. C. R. Frink, *Thesis*, Cornell Univ., 1960.
u. A. S. Wilson and H. Taube, *J. Am. Chem. Soc.*, 74, 3509 (1952).
v. G. Biedermann, *Arkiv Kemi*, 9 277 (1956).
w. G. Biedermann, *Arkiv Kemi*, 5, 441 (1953).
x. R. S. Tobias, *Acta Chem. Scand.*, 12, 198 (1958).
y. B. Carrell and Å. Olin, *Acta Chem. Scand.*, 14, 1999 (1960).
z. Å. Olin, *Acta Chem. Scand.*, 11, 1445 (1957).

solubility of the hydrolysis products. Hydrolysis equilibria are notoriously slow in some cases, and local precipitation or hydrolysis, due to inadequate diffusion and stirring, can lead to non-reproducible results. This phenomenon leads to the need to dilute concentrated solutions of some metals (e.g. zirconium, polonium, protactinium, plutonium(IV)) with dilute acid rather than with water. In general, work with tetravalent or more highly charged metals at acidities below 0.1 M should be avoided, unless the metals are very strongly complexed.

References pp. 234–238

d. Association of Salts

According to Bjerrum's theory on ion association (section 3.A.b), 1:1 electrolytes should be completely non-associated in water, since $a > q = 3.57 Å$ at 25°. If, however, ions at contact are considered associated, according to Fuoss and others (section 3.A.c), then association occurs with all electrolytes, only its extent is minimal for low charges and large diameters (small b) in a given solvent. It may, however, be argued that the association measured by such methods as conductivity or calculated from activity coefficient does not require the removal of the solvation shells of the associating ions. In this case it is expected that the more highly hydrated ions, which usually show a larger hydrated radius (Table 1B2), would show less association than only slightly hydrated ions, unless very large. These considerations can help us to understand some features of the observed association behaviour (Table 3C4). A further aspect of this problem is the concept of localized hydrolysis[115]. Cations with a high field strength polarize the water of hydration so strongly that a proton is dissociated and hydrolysis occurs, as described above. Cations with a somewhat smaller field cause the hydrogens of the water molecules to become more available to basic anions, so that a hydrogen bond ensues between the hydrated cation and the anion. This bond leads to a decrease of the freedom of the ions, to association, and to decreasing activity coefficients. This was found with hydroxides, fluorides, acetates etc. of the alkali metals. Diamond[116] suggested that the same mechanism is responsible also for the observed behaviour of the alkali halides. This effect should be important with the high-field lithium cation and decrease with sodium, becoming unimportant with rubidium. Since the chloride ion has a higher local charge density on the oxygen of the water of its hydration shell than the less polarizing and less hydrated iodide ion, it should show the stronger interaction with the cation. According to these concepts rubidium, caesium and iodide salts should show little 'association-through-local-hydrolysis', while lithium, sodium and chloride salts should show this effect most strongly, as indeed is borne out by the experimental activity coefficients. Association through hydrogen bonding thus shows the opposite trend with ion size to that shown by simple electrostatic association discussed previously. It is difficult to place the activity coefficients of the perchlorate ion in the sequence of the halides, and rather *ad-hoc* arguments are necessary to fit them in (perhaps using the water-structure-enforced ion-pairing concept[117]). Those of the nitrates, on the other hand, show

TABLE 3C4. Association constants of salts, entries are log K_{ass} values.

	Li^+	Na^+	K^+	Rb^+	Cs^+	Ag^+	Mg^{2+}	Ca^{2+}	Sr^{2+}	Ba^{2+}
OH^-	−0.08	−0 7	—	—	—	c	2.58	1.30	0.96	0.64
F^-						0.4	1.82			
Cl^-	—	—	—	−0.8	−0.4	c	—	—	—	−0.13
Br^-			−0.40		0.05	c	—			
NO_3^-	−1.0	−0.6	−0.22		0.01	−0.2	—	0.28	0.82	0.92
ClO_3^-		−0.5	−0.1			0.2				0.7
BrO_3^-		−0.5	−0.4							0.85
IO_3^-		−0·47	−0.24			0.6	0.72	0.89	1.0	1.1
ClO_4^-		−0.48								
ReO_4^-		−0.72								
SO_4^{2-}	0.64	0.72	0.96			1.3	2.23	2.28		
$S_2O_3^{2-}$		0.68	0.92			c	1.83	1.95	2.04	2.28
$P_2O_7^{4-}$	3.1	2.4	2.3				c	c	c	c
$CH_3CO_2^-$						0.73	0.78	0.77	0.44	0.41

	Mn^{2+}	Co^{2+}	Ni^{2+}	Cu^{2+}	Zn^{2+}	UO_2^{2+}	Fe^{3+}	La^{3+}	Ce^{3+}	Th^{3+}
F^-				1.23	1.26	c	c	c	c	c
Cl^-	0.0	−0.18[b]	−0.24[b]	0.09[b]	0.2[c]	0.3	1.5	−0.12	0.1	1.38
Br^-		−0.12[c]	−0.11[i]	−0.02[b]	−0.5[a]	−0.2	0.60		0.4	
NO_3^-						0.5	1.0	0.26[j]	0.13[e]	−0.80
SO_4^{2-}	2.28	2.36	2.32	2.36	2.31	2.96	4.2	3.62	3.4	c
$S_2O_3^{2-}$	1.95	2.05	2.06		2.40		3.25			
$CH_3CO_2^-$		0.22[d]	0.42[f]			c	3.2[g]	c	c	
CNS	0.66[h]	0.95[h]	1.18[h]		0.91[h]	c	c			1.08

c. Complexation rather than ion-pair formation.
— No detectable association, log $K_{ass} < -1$.
a. W. Yellin and R. A. Plane, *J. Am. Chem. Soc.*, **83**, 2448 (1960).
b. M. W. Lister and P. Rosenblum, *Can. J. Chem.*, **38**, 1827 (1960).
c. S. Tribalat and C. Dutheil, *Bull. Soc. Chim. France*, 160 (1960).
d. N. Tanaka, M. Kamada, H. Osawa and G. Sato, *Bull. Chem. Soc. Japan*, **33**, 1412 (1960).
e. L. H. Sutcliffe and J. R. Weber, *Trans. Faraday Soc.*, **15**, 1892 (1959).
f. N. Tanaka and K. Kato, *Bull. Chem. Soc. Japan.*, **32**, 516 (1959).
g. D. D. Perrin, *J. Chem. Soc.*, 1710 (1959).
h. S. Fronaeus and R. Larson, *Acta Chem. Scand.*, **16**, 1433 (1962).
i. M. W. Lister and D. W. Wilson, *Can J. Chem.*, **39**, 2606 (1961).
j. E. E. Kriss and Z. A. Sheka, *Radiokhimiya*, **4**, 312 (1962).
Other data are from
C.W. Davies, *Ion Association*, Butterworths Sci. Publ., London, 1962.
C.B. Monk, *Electrolyte Dissociation*, Academic Press, London, 1961.
J. Bjerrum, G. Schwarzenbach and L. G. Sillèn, *Stability Constants*, Chem. Soc. Spec. Publ., **7**, London 1958. L. G. Sillèn and A. E. Martell, *Stability Constants*, 2nd ed., Chem Soc. Spec. Publ., London 17, 1964.

the expected behaviour of an anion somewhat more basic than chloride, but still not sufficiently so for the localized hydrolysis effect to become more important than the hydration affect, so that the activity coefficients decrease from Li to Cs and from Mg to Ba as for the heavier halides, rather than in the opposite direction as for the acetates.

The experimental methods usually used to determine ion association of strong electrolytes measure the concentrations of the ions, rather than those of the pairs, and their presently attainable precision preclude the reliable detection of less than about 1% pairing. These methods also require a non-associated standard of behaviour, so they are effective only in dilute solutions, where theoretical models, such as that of Debye and Hückel are operative, say below 0.1 M. This sets a lower limit for the association constant which has any significance, $\log K_{ass} \geq -1$. The values of association constants found for a number of common strong electrolytes are summarized in Table 3C4, using published data[77,111]. Davies[111] has recently presented an extensive discussion of this problem. In addition to the values in the Table it should be noted that information available at present indicates that the perchlorates of divalent cations are usually completely dissociated, as are also nitrates, bromides and iodides of first-row transition metal cations, and little association of the chlorides of these cations is found. Second- and third-row transition metal, and post-transition metal cations (such as Tl^+ and Pb^{2+}) show more appreciable association with chloride and nitrate, with some indication of 'complex formation' rather than 'ion-pairing'. The sulphates of bivalent and trivalent metals are associated to an appreciable extent to ion-pairs, and more strongly basic anions again show 'complex formation' rather than 'ion-pairing'. The distinctions between these concepts has been discussed in sections 3.A.e and 3.B.a.

D. REFERENCES

1. C. A. Kraus, *J. Phys. Chem.*, **60**, 129 (1956).
2. N. Bjerrum, *Kgl. Danske Videnskab. Selskab.*, **7**, No. 9 (1926).
3. R. M. Fuoss, *Trans. Faraday Soc.*, **30**, 967 (1934).
4. e.g. recently: H. S. Dunsmore, T. R. Kelly and G. H. Nancollas, *Trans. Faraday Soc.*, **59**, 2606 (1963).
5. J. T. Denison and J. B. Ramsay, *J. Am. Chem. Soc.*, **77**, 2615 (1955).
6. R. M. Fuoss, *J. Am. Chem. Soc.*, **80**, 5059 (1958)
7. M. Born, *Z. Physik.*, **1**, 45 (1920).
8. W. R. Gilkerson, *J. Chem. Phys.*, **25**, 1199 (1965).

9. J. G. Kirkwood, *J. Chem. Phys.*, **18**, 380 (1950).

10. L. B. Magnusson, *J. Chem. Phys.*, **39**, 1953 (1963).

11. J. H. De Lap, *Thesis*, Duke Univ., 1960.

12. E. A. Guggenheim, *Discussions Faraday Soc.*, **24**, 53 (1957).

13. H. Müller, *Z. Physik.*, **28**, 324 (1929).

14. E. A. Guggenheim, *Trans. Faraday Soc.*, **56**, 1159 (1960).

15. E. A. Guggenheim, *Trans. Faraday Soc.*, **55**, 1714 (1959).

16. R. A. Robinson and R. H. Stokes, *Electrolyte Solutions*, 2nd ed., Butterworth Sci. Publ. London, 1959.

17. W. L. Masterton and L. H. Berka, *J. Phys. Chem.*, **70**, 1924 (1966).

18. D. R. Rosseinsky, *J. Chem. Soc.*, 785 (1962).

19. J. H. B. George, *J. Am. Chem. Soc.*, **81**, 5530 (1959).

20. J. M. Smithson and R. J. P. Williams, *J. Chem. Soc.*, 457 (1958); M. C. R. Symons and P. A. Rrevalion, *J. Chem. Soc.*, 3503 (1962).

21. A. I. Popov and R. E. Humphrey, *J. Am. Chem. Soc.*, **81**, 2043 (1959).

22. D. H. Richards and K. W. Sykes, *J. Chem. Soc.*, 3632 (1960).

23. P. Chang, R. V. Slates and M. Sczwarc, *J. Phys. Chem.*, **70**, 3180 (1966).

24. H. Taube and F. Posey, *J. Am. Chem. Soc.*, **78**, 15 (1956).

25. S. R. Cohen, *J. Phys. Chem.*, **61**, 1670 (1957).

26. T. W. Newton and G. M. Arcand, *J. Am. Chem. Soc.*, **75**, 2449 (1953).

27. H. Coll, R. V. Naumann and P. W. West, *J. Am. Chem. Soc.*, **81**, 1284 (1959).

28. M. Smith and M. C. R. Symons, *J. Chem. Phys.*, **25**, 1074 (1956).

29. D. Meyerstein and A. Treinin, *J. Phys. Chem.*, **66**, 446 (1962).

30. D. A. L. Hope, R. J. Otter and J. E. Prue, *J. Chem. Soc.*, 5226 (1960).

31. N. Bjerrum, *Kgl. Danske Videnskab. Selskabs Skrifter*, (7) **12**, No. 4 (1915).

32. J. Bjerrum, *Metal Amine Formation in Aqueous Solution*, P. Haase and Son, Copenhagen, 1941.

33. F. J. C. Rossotti and H. S. Rossotti, *Determination of Stability Constants*, McGraw Hill Book Co. Inc., New York, 1961.

34. G. H. Nancollas, *Discussions Faraday Soc.*, **24**, 108 (1957); *Quart. Rev. (London)*, **14**, 402 (1960).

35. K. B. Yatsimirski, *Z. Neorg. Khim.*, **2**, 491 (1957); F. J. C. Rossotti in *Modern Coordination Chemistry* (Eds. J. Lewis and R. G. Wilkins), Interscience Publ. Inc., New York, 1960.

36. E. Rabinowitch and W. H. Stockmayer, *J. Am. Chem. Soc.*, **64**, 335 (1942); R. Näsänen ,*Acta. Chem. Scand.*, **1**, 204 (1947); **3**, 179 (1949); **4**, 140, 816 (1950); **5**, 1293 (1951); **6**, 1384 (1952); **7**, 1261 (1953); **11**, 1308 (1957); R. M. Milburn and W. C. Vosburgh, *J. Am. Chem. Soc.*, **77**, 1362 (1955).

37. Y. Marcus, *Record Chem. Progr.* (*Kregse-Hooker Sci. Lib.*), **27**, 105 (1966).

38. Y. Marcus and C. D. Coryell, *Bull. Res. Council Israel.*, **8A**, 1 (1959).

39. S. Fronaeus, *Thesis*, Lund, 1948.

40. J. Bjerrum, *Kgl. Danske Videnskab. Selskab. Mat.-Fys. Medd.*, **21**, No. 4(1944).

41. B. O. A. Hedström, *Acta Chem. Scand.*, **9**, 613 (1955); H. A. C. McKay, *Trans. Faraday Soc.*, **49**, 237 (1953).
42. H. L. Schläfer, *Komplexbildung in Lösung*, Springer Verlag, Berlin, 1961.
43. J. C. Sullivan and J. C. Hindman, *J. Am. Chem. Soc.*, **74**, 6091 (1952).
44. H. Irving and F. J. C. Rossotti, *J. Chem. Soc.*, 3397 (1953).
45. J. Z. Hearon and J. B. Gilbert, *J. Am. Chem. Soc.*, **77**, 2594 (1955).
46. I. Leden, *Thesis*, Lund, 1943.
47. F. J. C. Rossotti and H. S. Rossotti, *Acta Chem. Scand.*, **9**, 1166 (1955).
48. L. G. Sillèn, *Acta Chem. Scand.*, **10**, 186, 803 (1956).
49. Y. Marcus, *Acta Chem. Scand.*, **11**, 599 (1957).
50. D. Dyrssen and L. G. Sillèn, *Acta Chem. Scand.*, **7**, 663 (1953).
51. A. E. Lansner, *Acta Chem. Scand.*, **12**, 1136 (1958).
52. J. Rydberg and J. C. Sullivan, *Acta Chem. Scand.*, **13**, 186, 2057 (1959), J. C. Sullivan, J. Rydberg and W. F. Miller, *Acta Chem. Scand.*, **13**; 2023 (1959).
53. L. G. Sillèn, *Acta Chem. Scand.*, **16**, 159 (1962); N. Ingri and L. G. Sillèn, *Acta Chem. Scand.*, **16**, 173 (1962).
54. J. A. Chopoorian, G. R. Choppin, R. C. Grifith and R. Chandler *J. Inorg. Nucl. Chem.*, **21**, 21 (1961).
55. J. Rydberg, *Acta Chem. Scand.*, **14**, 157 (1960); **15**, 1723 (1961).
56. G. Anderegg, *Helv. Chim. Acta*, **44**, 1673 (1961).
57. S. W. Rabideau and R. H. Moore, *J. Phys. Chem.*, **65**, 371 (1961).
58. A. Zielen, *J. Am. Chem. Soc.*, **81**, 5022 (1959).
59. Y. Marcus, *Acta Chem. Scand.*, **11**, 811 (1957); *Second U.N. Conference Peaceful Uses Atomic Energy*, Geneva, **3**, 465 (1958).
60. T. F. Young and A. G. Jones, *Ann. Rev. Phys. Chem.*, **3**, 385 (1952).
61. Y. Wormser, *Bull. Soc. Chim. France*, 387 (1954).
62. E. A. Guggenheim, *Trans. Faraday Soc.*, **56**, 1150 (1960).
63. I. Leden, *Acta Chem. Scand.*, **6**, 971 (1952).
64. I. Leden, *Svensk Kem. Tidskr.*, **64**, 145 (1952).
65. S. Fronaeus *Acta Chem. Scand.*, **8**, 1174 (1954).
66. S. Ahrland, *Acta Chem. Scand.*, **5**, 1151 (1951).
67. K. A. Allen, *J. Am. Chem. Soc.*, **80**, 4133 (1950); W. J. McDowell and C. F. Baes, Jr., *J. Phys. Chem.*, **62**, 777 (1958).
68. A. R. Olson and T. R. Simonson, *J. Chem. Phys.*, **17**, 1322 (1949).
69. R. N. Heistand and A. Clearfield, *J. Am. Chem. Soc.*, **85**, 2566 (1963).
70. K. W. Sykes, *J. Chem. Soc.*, 2473 (1959); *Chem. Soc. Spec. Publ.*, **1**, 64, (1954).
71. C. H. Sutcliff and J. R. Weber, *Trans. Faraday Soc.*, **52**, 1225 W. D. Bale, E. W. Davies, D. B. Morgans and C. B. Monk, *Discussions Faraday Soc.*, **24**, 97 (1957); T. E. Rogers and G. M. Waind, *Trans. Faraday Soc.*, **57**, 1360 (1961); K. M. Jones and J. Bjerrum, *Acta Chem. Scand.*, **19**, 974 (1965).
72. M. M. Jones, E. A. Jones, D. F. Harmon and R. T. Semmes, *J. Am. Chem. Soc.*, **83**, 2038 (1961).
73. F. Klanberg, J. P. Hunt and H. W. Dodgen, *Inorg. Chem.*, **2**, 139 (1963).
74. E. E. Mercer and R. R. Buckley, *Inorg. Chem.*, **4**, 1692 (1965).

75. V. E. Mironov, *Zh. Neorg. Khim.*, **6**, 405, 897 (1961); F. Ya. Kulba, V. E. Mironov, G. S. Troitskaya and N. G. Maximova, *Zh. Neorg. Khim.*, **6**, 1865 (1961).

76. V. E. Mironov, *Zh. Neorg. Khim.*, **8**, 764 (1963); V. E. Mironov, F. Ya. Kulba and V. A. Nazarov, *Zh. Neorg. Khim.*, **8**, 916 (1963).

77. L. G. Sillèn and A. E. Martell, *Stability Constants*, 2nd ed. *Chem. Soc. Spec. Publ.* No. 17, Chemical Society, London, 1964.

78. G. C. Hood, O. Redlich and C. A. Reilly, *J. Chem. Phys.*, **22**, 2067 (1954); O. Redlich, *Monatsh. Chem.*, **86**, 329 (1955); Y. Masuda and T. Kanda, *J. Phys. Soc. Japan.*, **9**, 82 (1954).

79. G. C. Hood and C. A. Reilly, *J. Chem. Phys.*, **32**, 127 (1960).

80. O. Redlich, E. K. Holt and J. Bigeleisen, *J. Am. Chem. Soc.*, **66**, 13 (1944).

81. A. K. Covington, M. J. Tait and Lord Wynne–Jones, *Proc.Roy. Soc. London*, **A286**, 235 (1965); cf. also K. Heinzinger and R. E. Weston, Jr., *U.S. At. Energy Comm. Rept.* BNL 8393 (1964); T. F. Young, private communication to G. C. Hood, ref. 79.

82. S. Bruckenstein and I. M. Kolthoff, *J. Am. Chem. Soc.*, **78**, 2974 (1956), G. Swarzebach and P. Stensby, *Helv. Chim. Acta.*, **42**, 2342(1959), I. M. Kolthoff, S. Bruckenstein and M. K. Chantooni, *J. Am. Chem. Soc.*, **83**, 3927 (1961).

83. A. M. Posner, *Nature*, **171**, 519 (1953).

84. G. S. Karetnikov, *J. Phys. Chem. Moscow.*, **28**, 1331 (1954).

85. E. Högfeldt, *J. Inorg. Nucl. Chem.*, **17**, 302 (1961).

86. A. A. Krawetz, *Thesis*, Chicago, 1955; T. F. Young and A. A. Krawetz, *Abstr.* 130*th Mtg. Am. Chem. Soc.*, Miami, Florida, 18R (1957); T. F. Young, L. F. Maranville and H. M. Smith in *Structure of Electrolyte Solutions* (Ed.W. J. Hamer), J. Wiley & Sons, New York, 1959.

87. J. Chedin, *Ann. Chim.*, (II) **8**, 243 (1937). O. Redlich, *Z. Physik. Chem. (Leipzig)*, **182A**, 42 (1938); O. Redlich and J. Bigeleisen, *J. Am. Chem. Soc.*, **65**, 1883 (1943).

88. E. Högfeldt, *Acta Chem. Scand*, **17**, 785 (1963); E. Högfeldlt and B. Bolander, *Arkiv. Kemi*, **21**, 161 (1963).

89. W. Davis, Jr., and H. J. DeBruin, *J. Inorg. Nucl. Chem.*, **26**, 1069 (1964).

90. H. A. C. McKay, *Trans. Faraday Soc.*, **52**, 1568 (1956).

91. R. C. Axtman, W. E. Shuler and B. B. Murray, *J. Phys. Chem.*, **64**, 57 (1960).

92. T. F. Young, *Record. Chem. Progr.* (*Kresge-Hooker Sci. Lib.*), **12**, 81 (1951).

93. G. C. Hood and C. A. Reilly, *J. Chem. Phys.*, **27**, 1126 (1957).

94. A. I. Gel'bstein, *Dokl. Akad. Nauk SSSR*, **107**, 108 (1956); V. A. Pal'm, *Zh. Fiz. Khim;* **32**, 380 (1958).

95. R. H. Flowers, R. J. Gillespie and E. A. Robinson, *J. Chem. Soc.*, 821 (1960).

96. E. B. Robertson and H. B. Dunford, *J. Am. Chem. Soc.*, **86**, 5080 (1964).

97. H. S. Dunsmore and G. H. Nancollas, *J. Phys. Chem.*, **68**, 1579 (1964).

98. V. S. K. Nair and G. H. Nancollas, *J. Chem. Soc.*, 4147 (1958).

99. M. H. Lietzke, R. W. Stoughton and T. F. Young, *J. Phys. Chem.*, **65**, 2247 (1961).

101. A. K. Covington, J. V. Dobson and Lord Wynne-Jones, *Trans. Faraday Soc.*, **61**, 2057 (1965).

101. R. W. Ramette and R. F. Stewart, *J. Phys. Chem.*, **65**, 243 (1961).

102. A. N. Fletcher, *J. Inorg. Nucl. Chem.*, **26**, 955 (1964).

103. C. F. Baes, Jr., *J. Am. Chem. Soc.*, **79**, 5611 (1957).

104. N. Bailey, A. Carrington, K.A.K. Lott and M.C.R. Symons, *J. Chem. Soc.*, 290 (1960).

105. C. L. Rulfs, R. F. Hirsch and R. A. Pacer, *Nature*, **199**, 66 (1963).

106. O. D. Bonner and A. L. Torres, *J. Phys. Chem.*, **69**, 4109 (1965); J.H.R. Clarke and L. A. Woodward, *Trans. Farady Soc.* **62**, 2226 (1966).

107. T. Suzuki and H. Hagizawa, *Bull. Inst. Phys. Chem. Res. Tokyo*, **21**, 601 (1942).

108. S. Tribalat and J. M. Caldero, *Bull. Soc. Chim. France*, 114 (1966).

109. J. H. Boughton and R. N. Keller *J. Inorg. Nucl. Chem.* **28**, 2851 (1966).

110. C. W. Davies, *J. Chem. Soc.*, 1256 (1951).

111. C. W. Davies, *Ion Association*, Butterworths Sci. Publ. London, 1962.

112. B. Carrell and Å. Olin, *Acta Chem. Scand.*, **15**, 727 (1961); G. Biedermann and L. Ciavatta, *Acta, Chem. Scand.*, **115**, 1347 (1961).

113. L. G. Sillèn *Quart. Rev. (London)*, **13**, 146 (1959).

114. A. Werner, *Chem. Ber.*, **40**, 272 (1907).

115. R. A. Robinson and H. S. Harned, *Chem. Rev.*, **28**, 419 (1941).

116. R. M. Diamond, *J. Am. Chem. Soc.*, **80**, 4808 (1958).

117. R. M. Diamond, *J. Phys. Chem.*, **67**, 2513 (1963).

4.

Ion Exchangers

The ion exchange of metal complexes takes place between an aqueous solution (or possibly a mixed or nonaqueous solution) containing the metal ions and suitable ligands and inert electrolytes, and a second phase, the ion exchanger. The exchanger is usually an organic polymer, a resin, with functional groups responsible for the exchange. It is necessary to become familiar with the properties of ion exchangers, and with the exchange characteristics of simple ions, before the exchange of complexes can be dealt with.

Section 4.A describes the structure of ion exchange resins, mainly of the polystyrene–divinylbenzene type, but also some other types, and exchangers with functional groups other than the common sulphonate (for cation exchangers) or trimethylmethyleneammonium (for anion exchangers), as well as inorganic exchangers. The main characteristic of an ion exchanger is its capacity, which is defined and discussed. A short discussion of the rate of ion exchange processes concludes this section.

Ion exchangers do not operate, unless swollen by a solvent. Swelling equilibria are therefore dealt with in detail, in section 4.B. The probable cause of swelling, solvation of the ions in the exchanger, is discussed, as well as the sorption of non-electrolytes, besides the solvent. This leads to a thorough treatment of Donnan equilibria and electrolyte invasion, assuming several alternative models and formal thermodynamic treatments.

The methods developed for the cases where only swelling by the solvent, or only invasion by one electrolyte are considered, are utilized in discussing the selectivity of the exchanger towards two different ions, in section 4.C. The numerical magnitude of selectivity depends on its definition, which is considered carefully. The generalization that selectivity increases as swelling decreases is examined, and the behaviour of many ions is reviewed. Thermodynamic theories of selectivity can now be discussed, in view of the many factors which contribute towards ion exchange selectivity.

The theoretical treatments are valid usually for exchangers equilibrated with dilute aqueous solutions of electrolytes. Applications of ion exchangers to problems of metal complex formation or to separation processes, often deal with concentrated aqueous or with nonaqueous or mixed solutions. The little that is known about ion exchange in these media is discussed in section 4.D. Cation exchange selectivity shows inversions at high concentrations, and anionic species are sorbed on these exchangers. Anion exchangers are invaded by electrolytes, a fact that effects strongly the distribution of anions. Organic solvents too, cause inversions in ion selectivities. These points are discussed, though, because of a lack of valid theories, only along qualitative lines.

A. STRUCTURE AND PROPERTIES OF ION EXCHANGERS

a. Introduction

Ion exchangers are materials which exchange reversibly ions which they contain, for ions in the solution, while being insoluble in the solvents in which they are applied. Their two main characteristics are a given *capacity*, i.e. the amount of ions that are exchangeable per unit quantity of exchanger, and a *selectivity*, i.e. the restriction that only ions of a certain type, e.g. cations, are thus exchanged. The exchange proceeds on an *equivalent* basis. The ions that can be exchanged for those in solution are called *counter* (*gegen*) *ions*. In order that electroneutrality be retained, the exchanger must contain an equivalent amount of charge fixed to be insoluble material, the *skeleton* of the exchanger, in the form of functional groups carrying the *fixed ions*. The most important exchangers are those having negative fixed ions, which are then *cation exchangers*, capable of exchanging positive counter ions, and those having positive fixed ions, exchanging negative counter ions, i.e. *anion exchangers*. (Terms such as cationic exchanger or resin should be avoided since it is unclear whether the fixed or the counter ions are the cations). Exchangers of less general importance are those specific for a certain type of ions (such as chelating exchangers) or so called electron exchangers, which change the oxidation number of the exchanged ions. Materials which do not ionize, but interact coordinatively with ions or molecules in the solution, and do this on an equivalent basis rather than adsorptively, are akin to ion exchangers, and will be briefly treated below. Ion exchangers have been made from a variety of materials, both natural and synthetic. Synthetic materials are usually superior, because their properties can be better controlled. *Synthetic inorganic exchangers* are treated further below, while the synthetic organic exchangers, *the ion exchange resins*, which are far more widely used, will now be discussed.

b. Structure of Polystyrene–DVB-type Ion Exchangers

Ion exchange resins were introduced as practical exchangers by Adams and Holmes[1], with the polymeric condensation of formaldehyde with suitable phenols or arylamines. The aromatic rings carry functional groups such as sulphonate for cation exchan ers, or alkylamino for anion exchangers. The polymeric chains are highly crosslinked through attachment of more than two methylene bridges to the same aromatic ring. Better resins were later introduced by D'Alelio[2], based on the copolymerization of styrene with the crosslinking agent divinylbenzene (DVB). The resulting crosslinked polystyrene can be sulphonated[2] to produce a cation exchanger, or chloromethylated and treated with a tertiary amine, to produce an anion exchanger[3]. These exchangers have been produced commercially for some twenty years, and have reached high reproducibility of properties, besides being stable and of fairly high capacity. Since they are typical and highly popular, and have been used for most of the studies considered in this book, they will be described in greater detail.

The skeleton of the resins is a copolymer of styrene and *p*-divinylbenzene, DVB, prepared by suspension polymerization. The monomers are heated in a well-stirred aqueous suspension at 80°C or above. The success of the polymerization depends among other factors on the removal of inhibitors, presence of a catalyst (such as 1% benzoyl peroxide) and of suspension stabilizers. The resulting polymer (1) is in the form of spherical beads. The crosslinked polymer is insoluble, hydrophobic and does not swell with water.

The sulphonate cation exchanger (2) is prepared by sulphonation of the beads, preswollen by a solvent such as toluene, to avoid cracking. The reaction can be performed with concentrated sulphuric acid or with chlorosulphonic acid at 100°C, 1% silver sulphate being a suitable catalyst. Practically every benzene ring, including those belonging to the bridges, is sulphonated, preferentially in the *para* position, but double sulphonation of a single ring is very rare. The resulting exchanger (Figure 4A1) is therefore monofunctional, and should have an equivalent weight close to the theoretical for $=(CH_2CH)_{1+X/100}(C_6H_{4-X/100}) SO_3H$, i.e. $184 + 0.26X$, where X is the percentage of DVB in the polymer. Commercial DVB contains, however, only about half *p*-divinylbenzene, the rest being ethylstyrene. Therefore, an X–10 resin, with nominal DVB content of $X = 10$ mole percent pure DVB, will have an equivalentweight of about

(1)   (2)

$184 + 0.5X = 189$, and one kg of dry material will have a capacity of 5.30 equivalents. This figure is typical of ordinary commercial resins.

The quaternary ammonium anion exchanger is prepared by chloromethylation of the polystyrene–DVB beads with chloromethyl ether, which is also useful for preswelling, using, for example, zinc chloride as catalyst. The intermediate (3) is then reacted with trimethylamine to produce the exchanger (4) (Figure 4A2).

Again by taking suitable precautions, all aromatic rings carry one functional group, and the exchanger has the formula

$$=(CH_2CH)_{1+X/100}(C_6H_{4-X/100})CH_2N(CH_3)_3Cl,$$

with equivalent weight $216 + 0.5X$. The ordinary commercial $X - 10$ resin has therefore an equivalent weight of 221, and a dry capacity of 4.52.

Uncrosslinked ionized polyelectrolytes are known[4] to stretch in solution so as to maximize the distances between the like charges of the fixed ions. The ion exchangers differ from polyelectrolytes in their being crosslinked. This restrains the swelling which occurs on contact of the resins with water, or solvents in general. The *skeleton* of the exchanger, which consists of the aliphatic $-CH_2CH-$ chains (with attached benzene rings) and the DVB bridges, is fairly elastic, and considerable changes of volume are possible. Let us consider a fully swollen cation- or anion-exchanging X–10 resin. The aromatic rings are connected to the skeleton in repeating units, about 8 Å apart, the rings striving for a parallel position in order to maximize mutual interactions. At the ends of the benzene residues are located the functional groups, the *exchange sites*. Crosslinking has the effect of producing cavities of average diameter 50 Å, which are lined by the exchange sites. These cavities are filled by water. A schematic representation of a fragment of a cation exchanger is re-

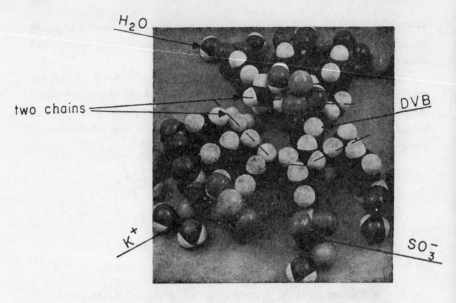

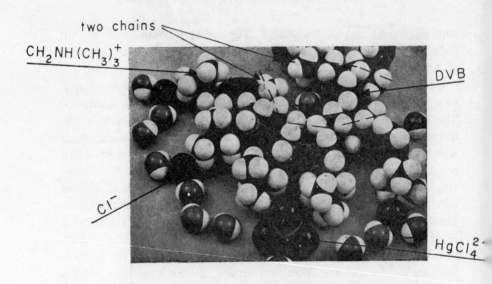

Figures 4A1 and 2. Photographs of three-dimensional models of polystyrene–DVB-type ion exchangers, nominally 12% crosslinked, showing the relative sizes of functional groups, ions, hydrocarbon skeleton and "free" space.

produced in Figure 4A3. The two-dimensional projection of the model appears somewhat more crowded than the model actually is.

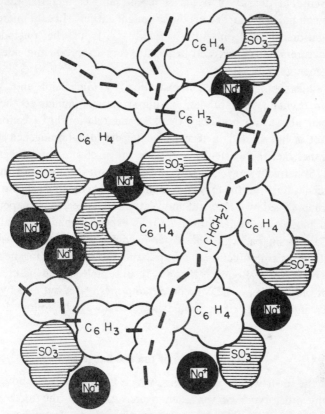

Figure 4A3. Two-dimensional projection of a model of a cation exchanger, nominally 16% crosslinked. (By permission from *Molecular Science and Engineering*, ed. A. R. von Hippel, The M.I.T. Press, Cambridge, 1959.)

An important feature of ion exchange resins is the randomness of the crosslinks, which causes a completely disordered structure of the exchanger. There is a wide range of distances between sites, and the environment of a site may vary from preponderately organic material (for sites near branching points) to freely available space, occupied by mobile solvent. The flexibility of the chains causes, of course, changes of the environment with time also. Early commercial resins showed considerable variation among individual batches, and even among beads within a batch[5, 6], that led to differing capacities and swelling and hence selectivities[7]. A later examination of a number of batches of a commercial cation exchanger[8] showed much less variation in capacity (up to 3%) but some variation in water uptake, although the uncertainty limits were not given. The density of beads of recent samples of resin was found[9] to be much more homogeneous than thought earlier[6]. Recent microscopic investigations showed only little variability[10]. It may be concluded that manufacturers have succeeded in recent years in producing adequately homogeneous material.

More disturbing is the demonstration by Glueckauf and Watts[11] that ion exchange membranes, having structures similar to the resins described above, do not obey the Donnan relationship (section 4.B). This fact is interpreted in terms of extensive non-homogeneity of the membranes, i.e. the existence of a wide range of sites of different properties. Spherical ion exchange resin beads, however, need not show the same deviation from the expected Donnan-law behaviour, shown by the membranes tested by Glueckauf and Watts, and indeed recent experiments showed much smaller deviations[10]. It is expected that the smaller the size of the resin particles, the easier becomes the regular crosslinking and attachment of functional groups, and the more homogeneous the resin. The advice of Bonner[8], that any resin sample should be characterized not only by nominal DVB content and mesh size, but also by actual capacity and swelling data, should however always be followed.

c. Other Types of Ion Exchangers

Besides the most widely used resins, which were discussed above, there are many other types of ion exchangers. A list of the most important commercially available exchangers is given in Appendix E. Even polystyrene–DVB resins have been varied. It is possible to produce a highly porous resin, without varying the crosslinking or the DVB content. Such a resin (e.g. Amberlyst, macroreticular resin[12]) although having large diameter particles, still permits rapid exchange without excessive swelling. The capacity of a resin can be varied by desulphonation[13], or by special synthesis for both cation and anion exchangers[14]. Resins of various particle sizes can be produced by careful control of the copolymerization stage.

Different functional groups confer on the resin different properties as regards ionization, i.e. resins can be strongly acid (or basic) or weakly acid (or basic) or even amphoteric. The resins discussed in section 4.A.b. are strong electrolytes, and are fully dissociated over the entire pH range. Carboxylate or tertiary-amine resins, on the other hand, are weak electrolytes, and do not 'ionize' at low or high pH values, respectively.

It is not too difficult to attach different functional groups to the polystyrene–DVB copolymer. Instead of sulphonation, treatment with phosphorus trichloride, with or without oxidation, and subsequent hydro-

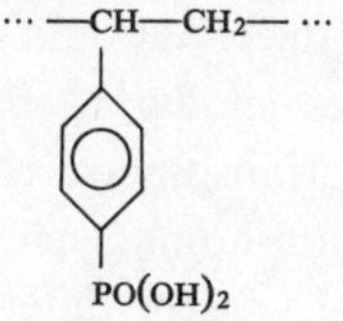

$$\cdots -CH-CH_2- \cdots \qquad \cdots -CH-CH_2 \qquad \cdots -C-CH_2-CH-CH_2- \cdots$$

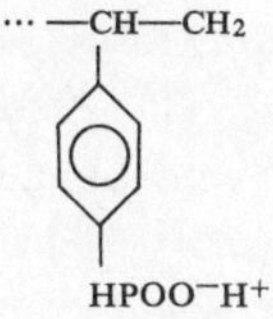

PO(OH)₂ — (5) HPOO⁻H⁺ — (6) COOH / CH₃ ... —CH—CH₂— ... (7)

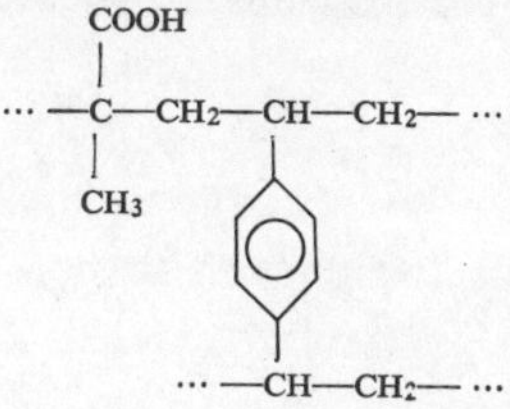

lysis, produces phosphonic (5) or phosphinic (6) acid (e.g. Duolite C–63 or C–62) cation exchangers[15]. Carboxylic exchangers are produced on copolymerization of DVB and methacrylic acid[16] (7) (e.g. Amberlite IRC–50).

For anion exchangers, quaternization of (3) with dimethylethanolamine produces a variant (8) (e.g. Dowex–2 or Amberlite–410) with higher affinity for hydroxide ions than has resin quaternized with trimethylamine, (4) (e.g. Dowex–1 or Amberlite–400). Use of secondary or primary

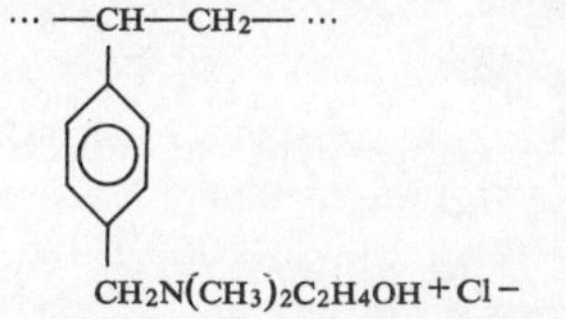

$$\cdots -CH-CH_2- \cdots$$

$CH_2N(CH_3)_2C_2H_4OH + Cl -$

(8)

amines or ammonia leads to tertiary or polyfunctional weak-base anion exchangers[17], (e.g. Amberlite IR–45). Dimethyl sulphide produces with (3) a tertiary-sulphonium strong-base anion exchanger[18] (9). A copolymer of DVB with 4–vinylpyridine produces on peroxidation with ammonium persulphate an intermediate-strength anion exchanger (10), with pyridinium-*N*-hydroxide groups[19].

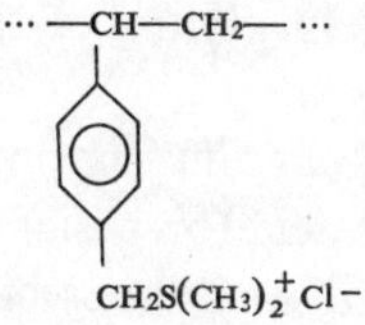

$CH_2S(CH_3)_2^+ Cl -$

(9)

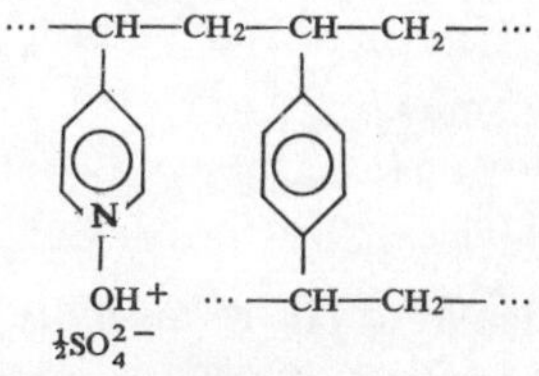

$OH +$
$\tfrac{1}{2}SO_4^{2-}$

(10)

Completely different synthetic methods are used to produce resins not based on a polystyrene–DVB skeleton. Earlier preparations usually employed condensation polymerization of formaldehyde with suitable aromatic materials producing a skeleton of benzene rings bridged by methylene groups *meta* to each other. Since hydroxyl or amine groups on the benzene rings are necessary for condensation, *polyfunctional*

(11)   (12)

resins are produced, when the product is sulphonated or converted to quaternary ammonium exchangers. Polyfunctional monomers, such as phenol sulphonic acid, salicylic acid (11) or phenylenediamine (12) can be used in the polymerization for direct introduction of the desired functional groups[20]. Inumerable variations are possible, and some of them were exploited commercially, although most of these resins are now obsolete. It is possible to produce completely aliphatic ion exchangers, e.g. based on the sulphate ester[21] of polyvinyl alcohol, which is an intermediate-strength cation exchanger (13), or on polyacrylonitrile, which can be hydrogenated and quaternized[22] (14). Polymeric diallyl phosphate[23] (15) is an example of an aliphatic resin with phosphate instead of sulphonate groups.

(13)   (14)   (15)

For particular applications more specialized resins may be appropriate. A resin which has gained some popularity is the chelating resin, which

incorporates imidodiacetic acid groups into a polystyrene–DVB framework (16) (e.g. Dowex–A1).

(16)   (17)

It can be prepared[24] by treating the chloromethylated intermediate (3)
with ammonia and chloroacetic acid. However, not every aromatic ring
carries an imidodiacetic acid group[25], and this resin seems to be even
less homogeneous than ordinary commercial ion exchangers. The
earliest example of a specific resin is that prepared by Skogseid[26]
for binding potassium ions. Since dipicrylamine is a soluble reagent
specific for K^+, a resin incorporating this group (17) should show
specificity for K^+, as it indeed does. High selectivity is shown also by
ionic materials containing functional groups showing specificity towards
certain ions. Thus, the phosphate group has a high affinity towards the
uranyl ion, and polymeric crosslinked triallyl phosphate or phosphonate[27]
or diethyl (polystyrenemethylene)phosphonate (18)[28] are examples of
highly specific materials, where the phosphate or phosphonate group
coordinates with the uranyl ion through the $P \rightarrow O$ oxygen, without
there occuring any ion exchange. These resins do not swell in water,
and are reactive only in organic solvents, this behaviour being a con-
consequence of their being non-ionic c.f. section 5.C.f.

(18)   (19)

Another group of non-ionic materials are the so-called 'electron exchangers', i.e. resins capable of reversibly reducing or oxidizing substances in solution. They are characterized by a given *redox capacity*, and by the *potential* of their couples. Most of these electron exchangers or redox resins are based on hydroquinone, or analogues, incorporated into a resin network. Sometimes, in order to increase their hydrophillic nature, and thereby their swelling and rate of reaction, they are made into cation exchangers, e.g. by sulphonation[29, 30]. Polymerization of hydroquinone, phenol and formaldehyde[31] produces a condensation-type resin, and the relatively more stable addition polymers are obtained by polymerization of vinyl hydroquinone dibenzoate and DVB with subsequent hydrolysis of the esters[32] (**19**). Other redox resins can be prepared which do not involve hydroquinone, but rather methylene blue[33] or sulphydryl groups[34], etc., covering a wide range of redox potentials.

d. Inorganic Ion Exchangers

Organic ion exchangers, such as the resins discussed above, are a later development of a subject that began more than a century ago, with the realization of the ion exchange properties of soils[35], later found to be mainly due to their clay content. In particular, natural aluminosilicate minerals, such as zeolites and montmorillonites, are cation exchangers of fairly high capacity. The former group of minerals, of which chabazite, (Ca, Na) $(Si_4Al_2O_{12} . 6H_2O)$, is an example, has a structure containing many cavities interconnected by channels, formed by a lattice of SiO_4 and AlO_4 tetrahedra, sharing oxygen atoms. The second class of minerals, of composition $Na(Al_2Si_4O_{10} . nH_2O)$, has a layer structure, capable of swelling in a direction perpendicular to the layers. The cations are in all cases more or less mobile, and can be exchanged for others[36,37]. The natural exchangers, however, are relatively unstable, are decomposed by acids, and do not have a high capacity.

Synthetic zeolites are cation exchangers with a regular crystal lattice, produced by crystallization from solutions containing alkali silicate and aluminate at high temperatures[37, 38]. Their advantage is the regularity of their pore sizes, which is regulated by the diameters of the channels between the cavities, which may be 4 Å (in type A4 zeolite) or 10 and 13 Å (in type X zeolite). This property led to the name 'molecular sieve', since the materials exclude molecules or ions which are too large.

More recently, relatively acid-insoluble exchangers have been introduced, which are more useful, in that they are useful also for ions other than those of the alkali metals and alkaline earths. Ammonium salts of heteropoly acids can be used[39], the phosphotungstate[40] being even more stable to acid than the phosphomolybdate. Still more stable are the materials produced by coprecipitation of highly charged cations, such as zirconium or antimony, with anions such as phosphate, tungstate, etc. Such materials are cation exchangers, and have received much attention[41, 42], because of their thermal and radiation stability, in connection with processing of irradiated nuclear fuels and wastes. Of special importance is zirconium phosphate, which is obtained by precipitation from zirconyl chloride or nitrate solutions with phosphoric acid, with careful drying of the resulting gel. Overheating during the drying process drives away the water irreversibly[43, 44], so that the exchanger does not swell and the ion exchange capacity decreases. To some extent an increase in drying temperature increases selectivity, in a way corresponding to an increase in crosslinking of an organic exchanger[44]. Some improvement in the properties, in particular an increase in grain size, is obtained when citrate is coprecipitated with the phosphate[45]. The structure of zirconium phosphate exchangers has recently been determined[46] on crystalline material of stoichiometric ratio $P:Zr = 5:3$. A chain unit of this exchanger has a formula weight of 878, and structure (20) has been suggested.

$$
\begin{array}{ccccccc}
OH & & OPO(OH)_2 & OH & & OH & \\
| & & | & | & & | & \\
-Zr-O- & & Zr-O- & Zr-O-P-O- & \cdots \\
\cdots \ \ | & & | & | & & | & \\
OPO(OH)_2 & OPO(OH)_2 & OPO(OH)_2 & O &
\end{array}
$$

(20)

Inorganic anion exchangers have also been prepared. These include mainly insoluble oxide of multivalent cations, such as Sn^{IV}, Zr, Th, etc.[47–49]. Although amphoteric to some degree, they show anion exchange properties in acid solutions, in which they are fairly insoluble. Coprecipitation of two hydroxides of different valency can also produce materials having excess positive fixed charges, capable of exchanging anions[49, 50].

e. Capacity

An important characteristic of ion exchangers is the property of equivalent exchange. In dilute solutions the counter ions neutralize exactly the fixed ions in the resin, without there being any additional counter ions together with their *co-ions*, i.e. ions of the same charge as the fixed ions, sorbed in the exchanger as invading electrolyte. In an exchange process under these conditions, for every amount of ions leaving the exchanger, there is an equivalent amount (on charge basis) entering it, to preserve charge balance. The total amount of exchangeable counter ions is, therefore, constant, equal to the amount of fixed ions bound as functional groups to the resin skeleton. This amount is called the *capacity* of the resin.

The dry capacity $\bar{C}_d$ of the resin is simply the reciprocal of the equivalent weight, as discussed in section 4.A.b. For 1–16X resins it decreases only slightly with increasing crosslinking (because of the higher equivalent weight of DVB compared to styrene), but at higher crosslinking the introduction of functional groups becomes progressively more difficult, and not every aromatic ring contains a fixed ion in the resin, so that the dry capacity is lower. It is usually given in units of milliequivalents per gram of dry H^+ (for cation exchangers) or Cl^- (for anion exchangers) form resin. It is virtually impossible, however, to completely dry the resin, because of its high hydration affinities (see next section), although it is possible to approach complete dryness by using anhydrone ('anhydrous' magnesium perchlorate) for extended periods in an evacuated desiccator, or for shorter periods at 60^0, or a 'drying pistol' with phosphoric anhydride at 100^0. It is then still very difficult to handle the dry resin while weighing, and it is seldom practical to determine directly the dry capacity.

Capacities are often reported on an 'air-dry' basis, for resin from which most of the water been removed by drawing the laboratory air through it, until it is free flowing. This air-dry, free-flowing, resin can be stored in bottles and handled in the laboratory, with only slight changes in weight (and water content) on contact with laboratory air. Determination of the *air-dry capacity* $\bar{C}_{ad}$ of this resin, and the water content ($\bar{W}_{ad}$ in weight percent) of a separate aliquot of the homogeneous sample, is equivalent to the determination of the dry capacity (4A1)

$$\bar{C}_d = \bar{C}_{ad} \times \frac{100}{100 - \bar{W}_{ad}} \qquad (4A1)$$

The water content of the air-dry resin depends on the crosslinking and the ionic form of the resin, and is of the order of 30%, so that the air-dry capacities of typical (10X) cation and anion exchangers are 70% of the dry capacities, or 3.7 and 3.0 meq/g air-dry resin, respectively. Because of the uncertainty of the water content, a determination of capacity is seldom much more precise than $\pm 1\%$.

Another useful measure is the *swollen capacity* of the resins, $\bar{C}_s$ meq/g swollen resin, i.e. its capacity when completely swollen in pure water. Because swelling is strongly dependent on crosslinking, so is the wet capacity, varying approximately from one quarter to one half of the dry capacity as crosslinking varies from $2X$ to $16X$. The more important connected property is the volume swollen capacity, $\bar{C}_{vs}$ meq/ml, or its reciprocal, the *equivalent swollen volume* of the resin, $\bar{V}_s$ litre/equiv. The densities of swollen resins $\bar{\rho}_s$ increase from about 1.1 for $2X$ to 1.3 for $16X$ in hydrogen or chloride form, so that the equivalent swollen volume can be calculated from the dry capacity and the water content of the swollen resin $\bar{W}_s$

$$\bar{V}_s = (\bar{\rho}_s \bar{C}_s)^{-1} = (\bar{\rho}_s \bar{C}_d)^{-1} \times \frac{100}{100 - \bar{W}_s} = (\bar{\rho}_s \bar{C}_{ad})^{-1} \times \frac{100 - \bar{W}_{ad}}{100 - \bar{W}_s} \quad (4A2)$$

The volume swollen capacity, $\bar{C}_{vs} = (1/\bar{V}_s)$, is the normality of ions in the exchanger, and for resins swollen in pure water is in the range of 1 to 2, as crosslinking changes from $2X$ to $16X$. Of greater importance than the normality of ions is their molality, i.e. their molar concentration per kg of water in the swollen resin. This quantity, $\bar{m}^0$, is given by

$$\bar{m}^0 = \bar{C}_s \times \frac{100}{\bar{W}_s} = \bar{C}_d \times \frac{100 - \bar{W}_s}{\bar{W}_s} = \bar{C}_{ad} \times \frac{100(100 - \bar{W}_s)}{\bar{W}_s(100 - \bar{W}_{ad})} \quad (4A3)$$

The water content of a typical anion exchanger, changes from $\bar{W}_s = 78\%$ for $2X$ to 48% for $16X$[51], while the dry capacity changes from 4.60 to 4.46, hence the molality of the fixed and the counter ions in the swollen exchanger varies from $\bar{m}^0 = 5.9$ to 9.3 m.

The final quantity, related to capacity, of interest for this discussion, is the *bed capacity* of an ion exchange column. Optimum packing of spheres of unequal size (as ion exchanger beads are apt to be) allows them about 58% of the total space, with 42% as the void fraction[52]. Often, packing in the resin bed does not attain maximal density, and the void fraction is larger. The bed density $\bar{\rho}_b$ is the weight of wet resin and interstitial water per unit volume of resin bed, and with the water density ρ_0 and the void volume fraction, i, the bed capacity $\bar{C}_b$ meq/ml resin bed is

$$\bar{C}_b = (\bar{\rho}_b - i\rho_o)\,\bar{C}_s = (\bar{\rho}_b - i\rho_o) \times \frac{100 - \bar{W}_s}{100 - \bar{W}_{ad}} \times \bar{C}_{ad} \qquad (4A4)$$

For a bed density $\bar{\rho}_b = 1.10$ g/ml, a void fraction $i = 0.45$, a swollen water content of $\bar{W}_s = 55\%$, and an air-dry water content $\bar{W}_{ad} = 30\%$ the ratio $\bar{C}_{ad}/\bar{C}_b$ becomes 2.4 ml/g. This figure converts volume distribution coefficients obtained from tracer elution in columns to (weight) distribution coefficients, obtained from batch experiments (cf. Chapters 5 and 6), for the resin having these characteristics.

Experimental methods for determining the quantities dealt with above, such as capacity $\bar{C}$, density $\bar{\rho}$, or water content $\bar{W}$, have been discussed in other monographs[53].

f. The Rate of Ion Exchange Processes

It is outside the scope of this book to treat rate processes in great detail, but a short description of the factors which govern the rate of ion exchange is in place here. A thorough and comprehensive discussion of the kinetics of ion exchange has recently been published (chapter 6 of ref. 53).

Ion exchange has been found[54] to be a diffusion-controlled process. It is generally accepted that cases where the chemical exchange reaction is rate controlling are extremely rare[55], if occuring at all[56, 57]. The rate-controlling mechanism can be *film diffusion*, if the slow step is diffusion across the Nernst, or hydrodynamic, film that surrounds the resin particles, or *particle diffusion*, if the slow step is diffusion inside the resin beads themselves. In the first case, a concentration gradient is set up within the liquid film, whereas inside the resin a uniform concentration of ions prevails. In the sscond case the concentration gradient occurs within the resin, while the film has a uniform composition. What are, then, the factors that determine which step is rate controlling?

Film diffusion may become the slow step when the agitation of the solution is insufficient, so that the film thickness is large, and when the solution is very dilute. Furthermore, conditions which increase the rate of particle diffusion (see below), such as high capacity, low crosslinking (i.e. high swelling), low affinity of ions to the exchange sites, and small particle size, combined with a low rate of film diffusion, can lead to film-diffusion control. Helfferich derived the criterion (ref. 53, pp. 255, 277)

$$\frac{\bar{C}_{vs}\bar{D}\delta}{cD\bar{r}}(5 + 2K_B^A) \lessgtr 1 \qquad (4A5)$$

For values of this quantity much larger than unity, film diffusion is rate controlling, while for values much smaller than unity particle diffusion determines the rate. In this expression $\bar{C}_{vs}$ is the volume swollen resin capacity, c the solution concentration, $\bar{D}$ and D the diffusion coefficients in the two phases repsectively, δ the film thickness (of the order of 10^{-3}cm), $\bar{r}$ the resin-bead radius, and K_B^A the selectivity coefficient for the exchange (section 4.C). For a typical mesh size of resin particles of 100–200 ($\bar{r} = 4 - 8 \times 10^{-3}$cm), and a capacity $\bar{C}_{vs} = 2$ M, the main factors become the solution concentration c, and the ratio of diffusion coefficients $\bar{D}/D$. At solution concentrations above 0.1 M the latter quantity, which is usually $<10^{-2}$, causes the process to be particle-diffusion controlled. Studies of the kinetics of ion exchange are, therefore, often oriented towards the investigation of diffusion coefficients in the resin, $\bar{D}$.

It has been found[53, 58] that in a process where an ion A exchanges for an ion B, there occurs an electric coupling of the opposite fluxes of the two ions, so that the rate of interdiffusion is dependent on the state of conversion of the resin from form B to form A, and diffusion coefficients in the resin are not constant with time. This makes the expressions describing the rate of the process rather complicated. Earlier theories neglected this effect, and this is permissible when the ions have the same mobility. However, an approximate indication of the rate may be obtained if an average, constant, mobility is assumed. In the case the of isotope exchange this assumption is of course very nearly correct, and the rates of isotope exchange can rather easily be related to the mobilities of the ions. In a relatively simple experiment, the rate of effusion of a radioactively tagged ion from resin particles, which are in equilibrium with a solution, into an 'infinite' volume of this solution, is followed. The fractional attainment of equilibrium $F(t) = (\bar{Q}(0) - \bar{Q}(t))/(\bar{Q}(0) - \bar{Q}(\infty))$, where $\bar{Q}(t)$ is the amount of tagged isotope in the exchanger at time $t = 0$ (initial stage), t, or ∞ (equilibrium state), is related to the diffusion coefficient $\bar{D}$ and resin-particle radius $\bar{r}$ by the expression (valid for particle-diffusion control)

$$F(t) = 1 - 6\pi^{-2} \sum_{n=1}^{\infty} n^{-2} \exp(-\pi^2 n^2 D t \bar{r}^{-2})$$ (4A6)

or to a good approximation by

$$F(t) \simeq 1 - \exp(-\pi^2 \bar{D} t \bar{r}^{-2})^{1/2}$$ (4A7)

from which the diffusion coefficient is obtained as (approximately)

$$\bar{D} = -0.233\ \bar{r}^2 t^{-1} \log(1 - F^2(t)) \tag{4A8}$$

and the half-time for exchange as

$$t_{\frac{1}{2}} = 0.03\bar{r}^2/\bar{D} \tag{4A9}$$

The rate for particle diffusion-controlled isotope exchange is thus seen to be inversely proportional to the square of the particle radius. The appropriate equation for the case where film diffusion is rate controlling (for isotope exchange and an 'infinite' solution volume) is

$$F(t) = 1 - \exp\left(-3\frac{cDt}{\bar{C}_{vs}\bar{r}\delta}\right) \tag{4A10}$$

from which the half-time is

$$t_{1/2} = 0.23\ \frac{\bar{C}_{vs}}{c} \times \bar{r}\delta/D \tag{4A11}$$

the difference from the corresponding equations (4A7 and 9) for particle-diffusion control is obvious, the rate in the present case being inversely proportional to the first power of the particle radius.

It remains to discuss the factors determining the magnitude of $\bar{D}$, the (self-) diffusion coefficient in the resin phase. In the above short discussion, the resin is considered to be a homogeneous phase, in which the mobility of the ions is described by the diffusion coefficients $\bar{D}_i$. An alternative model, treating the resin as a porous medium, with movement of ions restricted to the pores, is also possible, though less instructive. In any case, $\bar{D}_i$ can be looked upon as the effective diffusion coefficient of ion i.

Both the properties of the diffusing ion and of the resin determine the magnitude of $\bar{D}_i$. The diffusing ion carries with it its hydration shell, and the larger its hydrated radius r_h (cf. section 1.B), the smaller $\bar{D}_i$, just as in aqueous solutions[59]. Under dehydrating conditions, the series of relative $\bar{D}_i$'s may be reversed. The higher the affinity of the ion i to the resin (section 4.C), the lower $\bar{D}_i$. Co-ions have as a rule higher diffusion coefficients in a given resin environment than counter ions[59]. Since affinity usually increases with the charge on the ions, $\bar{D}_i$ decreases with it, even for different tracers moving in a resin of the same ionic form[60]. The diffusion of ions in anion exchangers is slower than in cation exchangers for monovalent ions[61]. The predominant resin factor which governs diffusion is its swelling: the higher the swelling the higher generally the diffusion coefficients, approaching at maximum swelling those in aqueous solutions. As discussed in detail in the next section, swelling decreases with increasing crosslinking, with increasing valency of the counter ions, and with decreasing hydration tendency of these ions. It decreases also in weak-acid (or -base) resins with decreasing degree of ionization. With strong-acid (or -base) exchangers, the capacity plays only a minor role in determining the diffusion rate. The effect of nonaqueous solvents on the rate of ion exchange processes is in line with the above. With solvents where swelling is appreciable, so is the rate of exchange, although it is often smaller than in water-swollen resins. However, there exist hardly any quantitative data on diffusion coefficients of solvents or ions in resins swollen by nonaqueous solvents. Table 4A1 is a general survey of diffusion coefficients in the resin, $\bar{D}_i$, using data of Boyd and Soldano[59-61].

TABLE 4A1. Self-diffusion coefficients in ion exchangers at 25° (10^{-8}cm^2/sec)
a) As a function of ionic charge[a,b].

Dowex 50[a]	2 X (5.6 meg/dry g)	8.6 X (5.24 meg/dry g)	16 X (5.10 meg/dry g)
Ion			
Na$^+$		94.4	11.0
K$^+$		134.	
Cs$^+$		137.	
Ag$^+$		64.2	27.5
Zn^{2+}	73.7	6.3	1.16
Sr^{2+}		3.2*	0.30
Y^{3+}	11.4	0.31*	
La^{3+}	13.0	0.92	0.051
Th^{4+}		0.022*	

* Crosslinking is ca. 12 *X* (Dowex 50) rather than 8.6 *X*.

Dowex-2[b]	2 X (3.39 meg/dry g)	6 X (3.27 meg/dry g)	16 X (2.66 meg/dry g)
Cl$^-$		35.4	
Br$^-$	64.	38.7	25.7
I$^-$		13.3	
BrO$_3^-$		45.5	
WO$_4^{2-}$		18.0	
PO$_4^{3-}$		5.7	

258   *Ion Exchange and Solvent Extraction of Metal Complexes*

b) As a function of crosslinking.

X	Na^+	Ag^+	Zn^{2+}	La^{3+}	Br^-*	$H_2O(H^+$-form resin)[c]	Br^-**
		(Cation exchanger)[a]					
Water	1330					3010	2080
1			101.	18.8			91.2
2			73.7	13.0			64.
3							45.2
4	141.			6.9	320	914	
6							38.7
8			6.3	0.92	111	540	20.0
8.6	94.4	64.2					
12***	28.8		2.9				
16	24.0	28.5	0.54	0.051	24	220	25.7
24	10.0	11.3	0.26				

* At 3.4°C.
** In Dowex-2 anion exchanger
*** Approximately, Dowex-50 resin.

c) As a function of the ionic fraction in the resin[d].

Ion	8.6 X %-Other ion		10^8D	Ion	16 X %-Other ion		10^8D
Na^+	0		94.4	Na^+	0		24.0
	69	H^+	93.6		91.9	H^+	39.0
	93	H^+	88.2		96.6	K^+	36.3
	57	Rb^+	112.		98.2	Cs^+	26.2
	99.4	Cs^+	98.0		23.7	Zn^{2+}	18.1
					80.0	Zn^{2+}	12.0
Rb^+	0		138.		99.3	Zn^{2+}	8.6
	43	Na^+	132.		90	La^{3+}	4.1
				Zn^{2+}	0		1.16
					20.0	Na^+	1.43
					67.3	Na^+	1.81
					95.8	Na^+	2.21

d) As a function of the capacity (in desulfonated Dowex– 50)[e].

Capacity meq/g	12 X (? 'Dowex– 50') Na^+ 10^8D	Zn^{2+}	Y^{3+}	Capacity meq/g	16 X Na^+ 10^8D	Ag^+	Zn^{2+}	La^{3+}
5.14	28.8	2.9	0.32	5.10	24.0	27.5	1.16	0.049
4.36	65.6	10.5	1.27	4.29	38.6	33.2		0.078
3.74		7.3		3.19	39.6	19.0		0.076
0.98		5.9		2.52	30.1	7.0	1.20	0.060

References
a. G. E. Boyd and B. A. Soldano, *J. Am. Chem. Soc.*, 75, 6091 (1953).
b. B. A. Soldano and G. E. Boyd, *J. Am. Chem. Soc.*, 75, 6099 (1953).
c. G. E. Boyd and B. A. Soldano, *J. Am. Chem. Soc.*, 75, 6105 (1953).
d. B. A. Soldano and G. E. Boyd, *J. Am. Chem. Soc.*, 75, 6107 (1953).
e. G. E. Boyd, B. A. Soldano and O. D. Bonner, *J. Phys. Chem.*, 58, 456 (1954).

The expected effect of temperature, of increasing diffusion coefficients in the resin, is indeed found. Energies of activation of diffusion in the resin are usually somewhat higher than in aqueous solutions, probably because of the additional effect of higher flexibility of the resin skeleton. The rate of exchange in film-diffusion and particle-diffusion-controlled processes increases by 3 to 5% and 4 to 8% per degree, respectively, corresponding to 4–10 kcal/mole in activation energy[53].

B. SWELLING AND SORPTION EQUILIBRIA

a. Introduction

The process of ion exchange occurs between an exchanger and a solution, therefore, an important aspect of ion exchange is the interaction between exchanger and solvent. Usually, a dry exchanger swells when brought into contact with solvent, even in the vapour phase, until a state of equilibrium is reached, due to the limited stretching allowed by the cross-linking of the exchanger. Knowledge of the changes of volume, the free energy associated with the process, the molecular state of the solvent inside the exchanger, and their dependence on the structure and ionic form of the exchanger and the nature of the solvent, is important in understanding ion exchange processes. Thermodynamics is a potent tool for studying these effects, and has been widely employed for this purpose. Recently, some other methods, such as nuclear magnetic resonance and infrared spectroscopy, have added to our understanding of the phenomenon of swelling.

Since ion exchangers carry non-diffusible fixed charges, the diffusion of counter ions to the solvent outside of the exchanger soon sets up a *Donnan potential*, preventing further counter ions from diffusing out, or co-ions in the outer solution from diffusing in. The furthest limit to which the exchanger can swell acts like a semi-permeable membrane, which does not permit the fixed ions to diffuse through. The equilibrium system ion exchanger – electrolyte solution with common counter ion is described in terms of the *Donnan membrane equilibrium*[62] below. The more complicated case, where the counter ion is not common to the two phases, permitting ion exchange to occur, will be treated in the next section. For a single counter ion, the Donnan equilibrium is expressed as an exclusion of co-ions from the exchanger, and as a concentration of counter ions much higher inside the exchanger than in the solution. Exclusion by the Donnan mechanism is, however, not absolute, and even in dilute solutions some *invasion* of the electrolyte (i.e. of co-ions together with counter ions) occurs. This phenomenon becomes much more important in concentrated solutions.

The discussion which follows will be mainly in terms of the poly-styrene–DVB crosslinked strong-acid and strong-base ion exchangers, described in detail in section 4.A.b, of water as a solvent, and of solutions containing uni-univalent electrolytes. Other systems behave similarly, with the necessary quantitative differences, which are not difficult to derive.

b. Swelling Equilibria

The swelling of the resin will be described in terms of the osmotic pro-perties of the 'internal' solution of the resin. The resin model employed is a concentrated solution of fixed ions and counter ions in the imbibed water, confined in the pores surrounded by the resin skeleton or matrix. At equilibrium with an infinite amount of pure water, the free energy of the system would decrease with imbibement of water by dilution of the internal solution, liberating free energy of mixing, and decreasing electrostatic repulsion between neighbouring fixed ions. This process is limited by the finite elasticity of the resin, imposed by the crosslinking. A stage is reached, at which the osmotic pressure $\bar{\pi}$, caused by difference in water activity between internal concentrated solution, $\bar{a}_w$, and external pure water, $a_w = 1$, is equal to the opposite, contracting pressure exerted by the resin matrix. This pressure is called the *swelling pressure* of the resin. At equilibrium the chemical potential of water is equal in the two phases. The external water is at its standard state, i.e. pure water at 1 atm pressure, but the internal water is under pressure $\bar{P}$, so

$$\bar{\mu}_w(\bar{P}) = \bar{\mu}_w^0(\bar{P}) + RT\ln \bar{a}_w$$

$$= \bar{\mu}_w^0 + (\bar{P} - 1)\bar{V}_w' + RT\ln \bar{a}_w \qquad (4B1)$$

$$= \mu_w = \mu_w^0$$

The second term in the second row is the partial free energy correcting from a standard state at pressure $\bar{P}$, to one at 1 atm, assuming the partial molar volume of water in the resin $\bar{V}_w'$ to be independent of pressure. The last equality expresses the equilibrium with pure water. Calling the pressure difference $\bar{P} - 1 = \bar{\pi}$, the relation

$$\ln \bar{a}_W = - \bar{\pi}\bar{V}'_W/RT \qquad (4B2)$$

valid for equilibrium with pure water, defines $\bar{a}_W$, by using the same standard state in the two phases, $\bar{\mu}^0_W = \mu^0_W$, both at 1 atm.

The activity of water in the resin can be varied directly, by changing the vapour pressure above the resin, cf. equation (1A5), which is done most easily isopiestically[63] (cf. section 1.C.e). Equation (4B2) must then be modified by inclusion of the term $+ \ln p/p^0$ on the right side. From equation (4B1) it is clear that $\bar{a}_W$ is independent of the pressure, hence two resins of the same structure, differing only in crosslinking, will have the same $\bar{a}_W$ when they have the same water content at given pressure p^0, but different vapour pressures p. Hence

$$\ln \bar{a}_W = - \bar{\pi}_1 \bar{V}'_W/RT + \ln p_1/p^0$$

$$= - \bar{\pi}_2 \bar{V}'_W/RT + \ln p_2/p^0 \qquad (4B3)$$

or $\qquad RT \ln (p_1/p_2) = (\bar{\pi}_1 - \bar{\pi}_2)\bar{V}'_W$

Assigning to a 'strainless', i.e. a very weakly crosslinked resin ($X = 0.5$), a value $\bar{\pi}_2 = 0$, permits calculation of $\bar{\pi}_1$ for other resins from relative vapour-pressure measurements at given water contents. It has been found[63] that swelling pressures vary linearly with the equivalent volume of the resin, independent of ionic form, and since the latter varies approximately linearly with crosslinking, so does the swelling pressure ('Hookes law' behaviour). For two typical resins, Dowex– 50 $X8$ and Dowex– 1 $X6$ the relationships $\bar{\pi} = 1.24 \, \bar{V}_s - 238$ and $\bar{\pi} = 0.854 \, \bar{V}_s - 245$, respectively, were found to hold[63].

Osmotic coefficients ϕ^0 can be calculated (1A21), knowing the solvent activity $\bar{a}_W$, the molality $\bar{m}^0$ of the resin (i.e. concentration of fixed, or counter ions, per kg of imbibed water), and the number of ions into which the resin dissociates, $\bar{v}$. It is reasonable to assume that $\bar{v} = 1$,*

---

* The fixed ions are bound to a polymeric skeleton of (relatively) infinite molecular weight, which makes the product $\bar{v}_f \bar{m}_f = 0$ for the fixed ions (i.e. one polyelectrolyte ion for the total amount of water in a resin bead) and taking $\bar{m}_f = \bar{m}_c$ (for the counter ions) $= \bar{m}^0$, makes $\bar{v}_f = 0$ and $\bar{v}_c = 1$.

although this point has not yet been settled. It is always possible to leave this question open, and to define the product

$$\phi^0 \bar{v} = (- 55.51 \ln \bar{a}_W)/\bar{m}^0 \qquad (4B4)$$

The Gibbs–Duhem equation can be used (1A22) to calculate the activity coefficients of the resin counter ion electrolyte, if data on ϕ^0 (obtained from isopiestic water equilibrations) as a function of $\bar{m}^0$ are available[64]. Values of $\bar{a}_W$ may be estimated from (4B4) if it is assumed that $(\phi^0 \bar{v})$ for the resin has the same value as ϕv for the monomeric form of the resin in the same ionic form at the same molality. Indeed, the curves of ϕv from many alkylbenzene sulphonic acids as function of molality do behave like $\phi^0 \bar{v}$ for the corresponding sulphonated polystyrene–DVB cation exchanger[65].

None of the above equations, however, permits a prediction of the amount of swelling, i.e. the amount of water taken up by unit amount of exchanger or the equivalent water uptake $\bar{n}_W$, moles per one equivalent of resin. This quantity must be determined experimentally, since there is no direct way of measuring $\bar{\pi}$ or $\bar{a}_W$. Swelling can be stated both in terms of moles of water uptake $\bar{n}_W$, and in terms of change in volume. The partial molar volume of water taken up in the resin at high water contents is the same as in pure water, but this is not so at low water contents, where electrostriction occurs[66]. An average value of $\bar{V}'_W = 17.2$ ml/mole has been proposed[67]. Since our knowledge of electrostriction even in aqueous solutions is far from perfect (section 1.B), it is impossible to predict the volume changes in the resin on swelling, from independent data. The equivalent volume of the resin $\bar{V}_s$ has been related empirically[63] to the osmotic pressure $\bar{\pi}$ (earlier this section). It can be obtained either from density values, or microscopically[68, 69]. Since the resin beads are nearly perfectly spherical, precise diameter measurements of beads at equilibrium with a given solvent can be related to equivalent volume values with a precision of $\pm 0.3\%$[69]. The equivalent volume can be stated both in terms of the solvated, and also in terms of the unsolvated ions

$$\bar{V}_s = \bar{V}_{hR} + \bar{V}_{hi}/|z_i| + \bar{n}_{fW}\bar{V}_{fw} = \bar{V}_R + \bar{V}_i/|z_i| + \bar{n}_W\bar{V}_W \qquad (4B5)$$

where $\bar{V}_R$ is the equivalent volume of the resin matrix with non-hydrated fixed ions, $\bar{V}_i/|z_i|$ the equivalent volume of the non-hydrated counter ion

TABLE 4B1. Equivalent swollen volume, V_s (ml/equiv) and water uptake $\bar{n}_w = 55.5/\bar{m}^0$ (mole/equiv) of resins.

		$\bar{n}_w$ for cation exchangers Dowex-50						
Ion	X	1	2	4	8*	12	16	24
H+ a			52.5	23.2	12.2	8.1	7.1	5.3
b		63.4	34.6	18.8	12.4	9.1	8.3	
c					10.8 (326)			
d					10.3			
Li+ e			34.7	19.8	10.9	7.9	6.6	4.5
c					9.8 (313)			
Na+ a			28.5	16.8	9.6	6.4	5.5	4.0
b				15.8	10.6	7.8	7.2	
c					8.6 (292)			
d					9.0 (234)			
NH+ c					8.0 (285)			
K+ a			27.8	16.3	9.3	6.2	5.3	3.8
c					7.8 (272)			
Cs+ a			19.2	13.0	8.0	5.5	4.8	3.3
b					7.3 (281)			
(CH$_3$) N+ d					6.8 (276)			
Ag+ d					6.7			
Mg^{2+} d					10.0 (308g)			
Ca^{2+} d					8.4			
Ba^{2+} d					7.1 (262g)			

		$\bar{n}_w$ for anion exchangers							
Ion		Dowex-1		Dowex-2^f					
	X	6c*	8e	1	2	4	6	8	16
OH$^-$		10.4 (403)							
F$^-$		12.3 (431)	13.4	23.5	19.2	16.1		10.3	7.1
Cl$^-$		9.6 (400)	9.7	18.5	14.7	12.9	9.8	7.7	6.3
Br$^-$		7.6 (373)	8.0	11.9	10.0	8.9	6.6	5.5	4.8
I$^-$			5.9	4.8	4.7	4.6		3.7	3.6
IO$_3^-$			11.7						
BrO$_3^-$			9.5						
ClO$_3^-$			7.6						
NO$_3^-$			7.9						
IO$_4^-$			4.1						
ClO$_4^-$		2.4 (300)	3.5						

* Values of $\bar{v}_s$ in parenthesis.
a. G. E. Myers and G. E. Boyd, *J. Phys. Chem.*, **60**, 521 (1956).
b. G. Scatchard, *M.I.T. Lab. Nucl. Sci. Ann. Rep.*, May 1963.
c. G. E. Boyd and B. A. Soldano, *Z. Electrochem.*, **57**, 162 (1953).
d. H. P. Gregor, B. R. Sundheim, K. M. Held and M. Z. Waxman, *J. Colloid. Sci.*, **7**, 511 (1952).
e. B. Kahn, *Ph. D. Thesis*, M.I.T., 1960.
f. B. A. Soldano and D. Chestnut, *J. Am. Chem. Soc.*, **77**, 1334 (1955).
g. D. S. Flett and P. Mears, *J. Phys. Chem.*, **70**, 1841 (1966).

of charge z_i, suffix h pertains to the values for the hydrated ions, and f to free water. The sum $(\bar{V}_R + \bar{V}_i/|z_i|)$ can be obtained from measurements on the dry resin, hence the average values of the molar volume of water is $[\bar{V}_s - (\bar{V}_R + \bar{V}_i/|z_i|)]/\bar{n}_W$. The difference $\bar{V}_i/|z_i| - \bar{V}_j/|z_j|$ for two ions i and j can be obtained assuming $\bar{V}_R$ not to vary. Assuming a value for $\bar{n}_{fW}$ (see below) and using $\bar{V}_{fW} = 18$ ml/mole, permits the volumes for the hydrated ions in the resin to be calculated.

Representative values of water uptake, $\bar{n}_W$, and equivalent volume, $\bar{V}_s$, are shown in Table 4B1. As expected, swelling decreases strongly with increasing crosslinking of the resin, as well as with the charge of the counter ion. However, whereas $\bar{n}_W$ diminishes in a series of similar ions with decreasing crystalline and increasing hydrated volumes, the equivalent volumes $\bar{V}_s$ become larger[70].

c. Solvation in the Resin

Crosslinked polystyrene polymers swell in contact with organic solvents such as benzene, and even kerosene[71]. However, with ion exchange resins carrying ionic groups, the interaction by London forces responsible for swelling by non-polar organic solvents is much smaller than the interaction between ions and solvent. This latter interaction is mainly in the form of solvation, in the case of polar solvents, and salting out, in the case of non-polar solvents. (Large organic ionic groups, such as those of anion exchangers (section 4.A.b) also show London-type interactions, and these resins do swell in benzene or kerosene[71]).

Solvation has been considered as the primary cause of swelling[70]. Usually there is some 'free' solvent in the resin, side-by-side with bound solvent. Only little work has been done with organic solvents, but there is a fair amount of information on hydration. The strength of the interaction can be shown by the inability to determine by the usual Karl–Fisher method the last mole of water present in a cation exchanger[72].

Evidence on hydration comes from a variety of sources. As discussed previously (section 1.B), hydration may be expressed in terms of hydration numbers. Methods involving the mobility of ions (giving so-called primary hydration numbers) have been of little use for ion exchangers, but thermodynamic and spectroscopic data proved to be useful. It has been observed[66] that on gradual addition of a few moles of water per equivalent of dry resin the total volume of the resin plus water system first decreases, then remains unchanged. The decrease is ascribed to the electrostriction of the water in the (first) hydration shell of the ions (section 1.B.d), whereas the free water has virtually the same molar volume as in pure water, $\bar{V}_{fw} = 18.07$ ml/mole. The limiting number of moles of water per equivalent of resin, beyond which no decrease of volume takes place, is taken as the hydration number.

Another approach uses isotherms for water uptake from the vapour phase at different temperatures to calculate free energies of swelling, and hydration numbers[73, 74]. It has been shown, at least for cation exchangers, that up to a few moles of water per equivalent (about four[74] or less, depending on the ion[73]) there is a large heat effect on swelling of an initially dry resin, with enthalpies of above 4 kcal/mole. Beyond this there is little heat effect. This is interpreted as showing that the first few moles of water associate strongly with the ions, i.e. hydrate them. Indeed, the heat effect of the first stage is independent of crosslinking, but strongly dependent on the nature of the ion, in the direction expected from hydration effects (section 1.B). After the stage of hydration is complete, further water uptake depends on crosslinking, indicative of non-hydrative volume swelling of the resin. The total water uptake, through the vapour phase, from mixed water–ethanol solutions, extrapolated to zero nonaqueous content, has been taken as the hydration of the resin ions[75]. These numbers, however, certainly pertain both to hydration and volume swelling, and are much higher than expected for hydration only.

A thermodynamic method, complementary to studying the heat effect of swelling, is studying the heat effect of freezing that portion of the total water in the resin, which is not bound to the ions[76]. Heat capacity measurements in the region $-100°C$ to $25°C$ showed that about six moles of water per equivalent remained unfrozen in cation exchangers in the hydrogen and sodium forms, and about 7 moles in anion exchangers in the chloride form. The number for the anion exchanger seems unexpectedly high, if it should represent the hydration number, in view of the relatively low hydration of the chloride anion and quaternary ammonium cation (section 1.B).

Spectroscopic studies of ion exchangers have been made recently. Some information on the nature of water[77, 78] and some other solvents[77] inside the resin has been obtained by proton magnetic resonance studies. The main conclusion is that the solvent itself inside the resin does not exchange rapidly with the solvent outside it (even on the surface), and that the interior solvent is under the influence of the ions and skeleton of the resin. The effect of the ions is, indeed, similar to that in homogeneous aqueous solutions. However, protons in hydration water exchange too rapidly with protons in free water—and with protons of the sulphonic acid group, where the resin is in this form—for differentiation of the various kinds of water to be made by means of the p.m.r. method. Studies by infrared spectroscopy[77, 79] could differentiate between unionized sulphonic acid in the dry resin and the dissociated form in the hydrated resin. Furthermore, working with thin $(5\mu$ thick) foils[79], it was possible to show that water hydrates the cations, rather than the sulphonic group, even in the acid form, where the hydrated form is $-SO_3^-H_3O^+$ at low, and $-SO_3^-H_7O_3^+$ at high, water activities.

It must be recognized that in highly crosslinked resins some counter ions cannot be fully hydrated, since the total equivalent water uptake $\bar{n}_w$ may be less than the regular hydration number of the counter ion, h (cf. Tables 4B1 and 1B3). The limited stretching permitted by the skeleton imposes spatial restrictions on the hydration. The partially dehydrated counter ion may be thought of as associating with the fixed ions under these conditions (cf. section 4.D).

d. Sorption of Non-electrolytes

Ion exchangers in the dry state can sorb a great variety of substances besides water. This can occur both from the gas phase (e.g. ethanol and methanol vapour[80]) and from a liquid phase or solution[65,68,81]. A prerequisite for a large amount of sorption is a strong interaction with the constituents of the resin: the hydrocarbon skeleton, the ions, and any solvent already present. Interaction with the skeleton occurs through London forces, as discussed above. Interaction with the ions may occur through solvation. This, however, is not the only mechanism possible, because solutes may be sorbed also from aqueous solution, where hydration is preferred over solvation by the solute. Hydrogen bonding of some types of solutes, such as alcohols, with the hydrate water, may then be a major factor. The nature of the ions present is still of importance, determining the relative amounts of water and nonaqueous solvent sorbed in the resin[82, 83]. Up to fairly high mole fractions of solvent in the solvent–water mixture, there is a strong preference for water in the resin. In some cases, however, swelling may become higher in anhydrous solvents than in pure water, as with some anion exchangers[71, 84]. With aqueous alcohols, swelling may become higher than with either pure water or pure alcohol[85]. Increasing crosslinking usually decreases sorption, there being less water for hydrogen bonding, and higher concentrations of water-preferring ions[86]. The more swollen the resin, the higher becomes its sorptive power for polar solutes.

The nature of the non-electrolyte solute plays, of course, a major role in determining its sorption. A series of aliphatic alcohols or acids of increasing chain length shows increasing sorption, up to a point, because of increasing interaction of the alkyl radicals with the skeleton[87]. A counteracting tendency of the increase in size $\bar{V}_N$ of the solute N is to increase the pressure–volume term, at a given swelling pressure, thereby decreasing its activity in the resin

$$\ln \bar{a}_N = \ln a_N - \bar{\pi}\bar{V}_N/RT \tag{4B6}$$

Solutes which can interact by complex formation with the counter ions are sorbed very strongly. Cation exchangers with transition metals as counter ions sorb strongly ammonia or amines[88, 89], for example, and anion exchangers in bromide or iodide form sorb strongly bromine and iodine, forming probably Br_7^- and I_7^{-}[90, 91]. In these cases sorption is a stoichiometric process, and beyond a certain concentration of solute the resin becomes saturated with it, the sorption isotherm flattening out.

Many electrolytes are only slightly dissociated in organic solvent solutions, and may be sorbed as neutral compounds by ion exchange resins. Lithium chloride, for example, can be sorbed on an anion exchanger from acetone solution[92]. The mechanism of its absorption has not been established: it is improbable that it undergoes interaction with chloride present as counter ions, but it may be a simple distribution of a solute between two liquids: the acetone in the pores of the resin, and the outer solution. With a solute such as cobalt chloride, complex formation with the counter ions is likely[92]. This point is discussed thoroughly in Chapter 6. In the case of mixed solvents, the mole fraction of water in the resin is usually much higher than in solution, and then many electrolytes prefer the more aqueous environment in the resin[82].

e. Donnan Equilibria, Electrolyte Invasion

The equilibria of the resin with the solvent and with non-electrolytes have been dealt with above; this section discusses the equilibrium of the resin containing the fixed ion R and the counter ion X with an (aqueous) solution of an electrolyte with the same ion X, and co-ion Y. For the sake of simplicity the main treatment will be in terms of the symmetrical electrolyte XY, and will be common to cation and anion exchangers, i.e. charges will be omitted. The system can be best treated in terms of the *Donnan membrane equilibrium*[62]. Diffusion of the least quantity of counter ions X (which cannot be accompanied by the fixed ions R) outside of the resin, sets up a double-layer electrical potential, the *Donnan potential*, which tends to pull in ions X and repel ions Y. At equilibrium the electrochemical potential of each ionic species must be the same in both phases, and electroneutrality must prevail everywhere. The partial free energy of ion i (of charge z_i) will be made up of two terms

$$\Delta G_i' = \mu_i + z_i F \psi \tag{4B7}$$

the second, electrical, term depending on the potential ψ. The Donnan potential E_{Donnan} is the potential difference $\bar{\psi} - \psi$ between the two phases

$$E_{Donnan} = \bar{\psi} - \psi = (\mu_i - \bar{\mu}_i)/z_i F \tag{4B8}$$

The chemical potentials of the individual ions X and Y conform to (4B8), but since $z_X = - z_Y$, the chemical potentials of the Gibbsian component XY are seen to be the same in both phases

$$\mu_{XY} = \bar{\mu}_{XY} \tag{4B9}$$

A number of models have been proposed to relate equations (4B8) and (4B9) to measurable quantities, and three of them will now be discussed.

An important model, which has been proposed quite early by Gregor[70], involves consideration of the hydrated ions R, X and Y, and the free water. The chemical potential has a composition term and a pressure–volume term

$$\bar{\mu}_{Wf} = RT \ln \bar{a}_{Wf} + \bar{\pi} \bar{V}_{Wf} = \mu_W \tag{4B10}$$

$$\bar{\mu}_{XY_h} = \bar{\mu}_{XY_h}^0 + RT \ln \bar{a}_{X_h} \bar{a}_{Y_h} + \bar{\pi} \bar{V}_{XY_h} = \mu_{XY_h} \tag{4B11}$$

the last equation on the right expressing condition (4B9), that of (4B11) being correct only if the hydration numbers are the same in both phases, i.e. the components XY_h and $XY_{\bar{h}}$ are identical. Otherwise, as pointed out by Kakihana[93], among others, it must be replaced by (cf. section 1.D.c):

$$\bar{\mu}_{XY_{\bar{h}}} = \mu_{XY_h} + (\bar{h} - h)\mu_W \tag{4B12}$$

$$\ln\left(\bar{a}_{X_{\bar{h}}}\bar{a}_{Y_{\bar{h}}}/a_{X_h}a_{Y_{\bar{h}}}\right) = (\bar{h} - h)\ln a_W - \bar{\pi}\bar{V}_{XY_{\bar{h}}}/RT$$
$$= (\bar{h} - h)\ln a_W + (\bar{V}_{XY_{\bar{h}}}/\bar{V}_{Wf})\ln \bar{a}_W/a_W \tag{4B13}$$

h being the hydration numbers, and using the same standard state for both phases. This derivation by Kakihana[93] is, however, based on setting $\bar{\mu}^0_{XY_{\bar{h}}} = \bar{\mu}^0_{XY_h}$, which may not be valid when $\bar{h} \neq h$. In dilute solutions $\ln a_W$ is nearly zero, $\bar{\pi}$ is of the order of three hundred atm, $\bar{V}_{XY_{\bar{h}}}$ may be estimated from the hydrated radii in Table 1B2, allowing for reduced hydration numbers in highly crosslinked resins (section 4.B.c) and may be of the order of 80 cm³, hence with $R = 82$ cm³ atm/deg, the value of the left-hand term in (4B13) comes out to be of the order of unity. Writing activities in terms of molalities and activity coefficients, and setting $\gamma^2_{\pm XY_h} \approx 1$ in dilute solution (where $\bar{m}_Y \ll \bar{m}_R$) yields

$$\bar{m}_Y \simeq m^2_{XY}\bar{m}_R^{-1}\bar{\gamma}^{-2}_{\pm XY_{\bar{h}}}(\bar{a}_W/a_W)^{\bar{V}_{XY_{\bar{h}}}/\bar{V}_W} \tag{4B14}$$

$$\ln \bar{\gamma}_{\pm XY_{\bar{h}}} \simeq \ln m_{XY} - \tfrac{1}{2}\ln \bar{m}_R\bar{m}_Y - \bar{\pi}\bar{V}_{XY_{\bar{h}}}/2RT \tag{4B15}$$

the last term contributing a factor of 0.6 to $\bar{\gamma}_\pm$. In more concentrated solutions the term $(\bar{h} - h)\ln a_W$ becomes important, but unfortunately we do not know the variation with concentration of either h or $\bar{h}$, or $\bar{V}_{XY_{\bar{h}}}$. Estimates of these quantities might, nevertheless, help us to understand the contribution of ion dehydration to phenomena at high concentrations (section 4.D).

Lack of knowledge of the parameters pertaining to hydration favours using another treatment, proposed among others by Bauman[94], Boyd[95] and Glueckauf[96], involving the non-hydrated ionic species and the total water. Starting again from (4B9) one obtains

$$\bar{\mu}_W = RT\ln \bar{a}_W + \bar{\pi}\bar{V}_W = \mu_W \tag{4B16}$$

$$\bar{\mu}_{XY} = \bar{\mu}^0_{XY} + RT\ln \bar{a}_X\bar{a}_Y + \bar{\pi}\bar{V}_{XY} = \mu_{XY} \tag{4B17}$$

$$\ln(\bar{a}_X \bar{a}_Y / a_X a_Y) = -\bar{\pi} \bar{V}_{XY} / RT = \frac{\bar{V}_{XY}}{\bar{V}_W} \ln(\bar{a}_W / a_W) \qquad (4B18)$$

The last equation may be generalized for an electroyle G dissociating into v ions by writing

$$\left(\frac{\bar{a}_\pm}{a_\pm}\right)^v = \left(\frac{\bar{a}_W}{a_W}\right)^{\bar{V}_G / \bar{V}_W} \qquad (4B19)$$

as a general statement of the Donnan equilibrium. In dilute solution, again

$$\bar{m}_Y \simeq m_{XY}{}^2 \, \bar{m}_R{}^{-1} \, \bar{\gamma}_{\pm XY}{}^{-2} (\bar{a}_W / a_W)^{\bar{V}_{XY} / \bar{V}_W} \qquad (4B20)$$

which must of course be equivalent to (4B14) through different $\bar{\gamma}_\pm$ and $\bar{V}_{XY}$

$$\ln \bar{\gamma}_{\pm XY} \simeq \ln m_{XY} - \tfrac{1}{2} \ln \bar{m}_R \bar{m}_Y - \bar{\pi} \bar{V}_{XY} / 2RT \qquad (4B21)$$

Since the partial molal volume of (unhydrated) electrolytes are many-fold less than the hydrated volumes, the ratio

$$\bar{\gamma}_{\pm XY} / \bar{\gamma}_{\pm XY\bar{h}} = \exp\left(\frac{\bar{\pi}}{2RT}(\bar{V}_{XY\bar{h}} - \bar{V}_{XY})\right)$$

can be a factor of 1.5.

A third approach[97] neglects the pressure–volume term, or rather includes it implicitly in the activity coefficients. The activities of the ions in the resin are defined by

$$\bar{\mu}_{XY} = \bar{\mu}_{XY}^{0\prime} + RT \ln \bar{a}_X' \bar{a}_Y' \qquad (4B22)$$

and the reference states are no longer infinite dilution as in the previous treatments, but are such, that $\bar{\mu}_{XY}' = \bar{\mu}_{XY} = \mu_{XY}$ and hence $\bar{\gamma}_{\pm XY}' \neq \bar{\gamma}_{\pm XY}$. This formalism leads to the simple equation

$$\bar{a}_\pm' = (\bar{a}_X' \, \bar{a}_Y')^{\frac{1}{2}} = a_\pm \qquad (4B23)$$

and in dilute solutions

$$\bar{m}_Y \simeq m_{XY}^2 \, \bar{m}_R^{-1} \, \bar{\gamma}_{\pm XY}'^{-2} \qquad (4B24)$$

This is, again, equivalent to (4B14) or (4B20) through the different $\bar{\gamma}'_{\pm}$

$$\ln \bar{\gamma}'_{\pm XY} = \ln m_{XY} - \tfrac{1}{2}\ln \bar{m}_R \bar{m}_Y \qquad (4B25)$$

$$= \ln \bar{\gamma}_{\pm XY} + \frac{\bar{\pi}\bar{V}_{XY}}{2RT} \qquad (4B25a)$$

For not too high crosslinking and not too large electrolytes (e.g. X–8 with $\bar{\pi} = 120$ atm and NaCl with $\bar{V}_{XY} = 16$ cm^3) the last term on the right in (4B25a) is small (0.04 in the above example, i.e. 1.7% difference in $\bar{\gamma}_{\pm}$) and may be neglected.

Experimentally, the invasion of electrolyte, i.e. the concentration of co-ion in the resin $\bar{m}_Y$, is measured as a function of the external electrolyte concentration. The expression obtained for the activity coefficient has practically always been in terms of the third approach, equation (4B24) in dilute solutions and (4B23) in more concentrated solutions. The results of a number of such investigations are summarized in Table 4B2. Over considerable ranges of concentration it is seen that the water content of the resin changes as invasion ($\bar{m}_Y$) proceeds with increasing m_{XY} in such a manner as to make

$$\bar{m}_X = \bar{m}_R + \bar{m}_Y = A + B\,\bar{m}_Y \qquad (4B26)$$

The constant A is equal to $\bar{m}_R$ when there is no invasion ($\bar{m}_Y = 0$), but $\bar{m}_R$ increases with increasing invasion because of the decrease in water content.

Further, at sufficiently high invasion it is found that the ratio $\Gamma = \bar{\gamma}'_{\pm}/\gamma_{\pm}$ approaches a constant value, not far from unity (e.g. 0.60 for HCl[98], 0.82 for LiCl[98], 0.70 for LiNO$_3$[99], for a representative anion exchanger, Dowex–1). This requires a linear relationship between $\bar{m}_Y$ and m_{XY}, and this, coupled with the linear relationship between $\log \gamma_{\pm XY}$ and m_{XY} at high XY concentrations (cf. section 1.D.e) leads to the empirical relationships

$$\log \bar{\gamma}'_{\pm XY} = A' + B'\bar{m}_Y \qquad (4B27a)$$

$$= A^* + B^*\bar{m}_X \qquad (4B27b)$$

Equation (4B27a) is to be preferred at low concentration, while (4B27b) holds (Table 4B2) between rather wide limits at higher concentrations.

TABLE 4B2. Activity coefficients of electrolytes in ion exchangers

$$\log \bar{\gamma}'_{\pm XY} = A^* + B^* \bar{m}_X (X = \text{counter ion}) \quad (\text{equation 4B27b})$$

Electrolyte	Exchanger	$-A^*$	B^*	$\bar{m}_{XY}$ range	Reference
HCl	Dowex–1 X 10	1.06	0.104	0.1–16	a
HCl	Dowex–1 X 10	0.99	0.091	0.1–10	b
HCl	Dowex–50 X 5.5	0.53	0.135	0.1– 4.4	c
HCl, HClO$_4$	Dowex–50 X 4	−0.18	−0.003	3 –15	d
LiCl	Dowex–1 X 10	0.96	0.103	2 –13	b
LiCl	Amberlite–410 X 3–6	0.84	0.108	0.3–15	e
NaCl	Dowex–1 X 10	0.31	0.032	0.2– 6	b
KCl	Dowex–1 X 7.5	0.21	0.00	1 – 4	f
KCl	Dowex–50 X 5.5	0.36	0.014	0.2– 2	c
HNO$_3$	Dowex–1 X 8	0.79	0.025	1 – 3	g
LiNO$_3$	Dowex–1 X 8	0.49	0.036	3 –13	g
NH$_4$NO$_3$	Dowex–1 X 8	0.56	0.007	1 –22	g
H$_2$SO$_4$	Dowex–1 X 10	2.64	0.114	0 – 4	b
(NH$_4$)$_2$SO$_4$	Dowex–1 X 10	0.35	−0.044	3 – 5	b
Na$_2$S$_2$O$_3$	Dowex–1 X 10	1.5	0.0	0.1– 2	h

$$\log \bar{\gamma}'_{\pm XY} = A' + B' \, \bar{m}_Y (Y = \text{co-ion}) \quad (\text{equation 4B27a})$$

Electrolyte	Exchanger	$-A'$	B'	$\bar{m}_{XY}$ range	Reference
KCl	Dowex–50X2	0.64	−0.5	0.016 –0.26	i
MgCl$_2$	Dowex–50X2	0.31	0.3	0.0025–0.13	i
MgCl$_2$	Dowex–1X2	0.33	2.0	0.011 –0.18	i
HCl	Amberlite XE100 (X4.8)	0.09	0.45	0.1 –1.0	j
NaCl	Amberlite XE100 (X4.8)	0.21	0.11	0.1 –3.0	j
NaI	Amberlite XE100 (X4.8)	0.18	0.16	0.1 –3.4	j
NaH$_2$PO$_4$	Amberlite XE100 (X4.8)	0.29	0.0	0.02 –3.7	j
CaCl$_2$	Amberlite XE100 (X4.8)	0.27	0.25	0.1 –1.6	j
LaCl$_3$	Amberlite XE100 (X4.8)	0.43	0.0	0.02 –0.25	j
NaCl	Amberlite 1R120 (X8)	0.19	0.55	0.2 –1.0	j
NaH$_2$PO$_4$	Amberlite 1R120 (X8)	0.20	0.0	0.1 –1.1	j

a. K. A. Kraus and G. E. Moore, *J. Am. Chem. Soc.*, 75, 1457 (1953).
b. K. A. Kraus and F. Nelson, *J. Am. Chem. Soc.*, 80, 4154 (1958).
c. C. W. Davies and G. D. Yeoman, *Trans. Faraday Soc.*, 49, 968 (1953).
d. R. H. Dinius and G. R. Choppin, *J. Phys. Chem.*, 68, 425 (1964).
e. R. Barbieri, *J. Inorg. Nucl. Chem.*, 26, 845 (1964).
f. D. H. Freeman, *J. Phys. Chem.*, 64, 1048 (1960).
g. J. Danon, *J. Phys. Chem.*, 65, 2039 (1961).
h. Y. Marcus, *Acta Chem. Scand.*, 11, 619 (1957); molar scale.
i. D. H. Freeman, V. C. Patel and T. M. Buchanan, *J. Phys. Chem.*, 69, 1477 (1965).
j. R. L. Gustafson, *J. Phys. Chem.*, 70, 957 (1966).

Arguments have been presented[98, 100] that a model in which the invaded resin is pictured as a concentrated mixture of aqueous electrolytes RX and XY, should lead to equation (4B27) in the case where Harned's rule is valid, as shown by Scatchard[101] (cf. section 1.E.b). This implies that the constant A' in (4B27a) should be equal to log $\gamma_{\pm XY} - B'\bar{m}_X$ at a concentration $m = \bar{m}_X$. This concentration is usually much higher than accessible in aqueous XY solutions, but can be estimated by extrapolation. For HCl on Dowex-1 of various crosslinkings invaded by HCl of different concentrations, Kraus and Nelson[98] have indeed confirmed the identity of A' with the above expression at each constant total $\bar{m}_{Cl}$, with B' values, corresponding to Harned's rule α_{XY}, varying with total concentration as $B' = 0.21 - 0.0032\bar{m}_X$. Treatment of the invaded resin as a mixed concentrated electrolyte should lead to the possibility of using osmotic-coefficient data[102] to evaluate the activity coefficients of both imbibed electrolyte and resinate (cf. section 1.E.c), knowing the osmotic coefficients of the pure water-swollen resinate RX and of the electrolyte XY as functions of their respective molalities. A detailed development of this idea has not yet been made.

A useful concept which may be introduced here is the 'effective counter-ion activity' in the resin, $\bar{a}_X$, defined in the same manner as in solutions (cf. section 1.D.f)[103], although here the molal scale is more convenient

$$\bar{a}_X = \bar{m}_X \bar{\gamma}'_{\pm XY} \qquad (4B28)$$

Whereas in an aqueous 1:1 electrolyte $a = a_{\pm}$, this is not the case in the resin, since $\bar{m}_X > \bar{m}_Y$, and here the quantity $\bar{a}_X$ should not be equated with $\bar{a}_X$ appearing in (4B23). The effective counter-ion activity can be calculated from

$$\log \bar{a}_X = \log a + \tfrac{1}{2}(\log \bar{m}_X - \log \bar{m}_Y) \qquad (4B29)$$

where $a = m_{XY}\gamma_{\pm XY}$. At high concentrations an empirical relationship $\log \bar{a}_X = A'' + B'' \log a$ may be used to relate the resin and aqueous effective activities. The constant A'' depends slightly on the electrolyte, but more so on crosslinking, while B'' depends mainly on the electrolyte, as found for various chlorides and Dowex–1 anion exchanger[104,105]. It will be shown below (section 4.D.d) that $\bar{a}_X$ can also be determined to a good approximation by relating the distribution ratio D_B of a tracer counter ion B^{-b} to the effective activity in solution a

$$\log \bar{a}_X = \frac{1}{b}(\log D_B - \log D_B^0) + \log a \qquad (4B30)$$

where D_B^0 is a normalizing quantity, which makes $\bar{a}_X = A''$ if it is taken as D_B at $a = 1$[106].

Returning to the region of low invasion, in dilute electrolyte solutions, equations (4B14), (4B20) and (4B24) lead to the expectation that $\bar{m}_Y$ varies as the square of m_{XY}, since $\bar{\gamma}'_{\pm XY}$ should tend to a constant value as the ionic strength in the resin tends to that of non-invaded resin in equilibrium with pure water. This, however, is not generally the case[100], as $\bar{\gamma}'_{\pm XY}$, calculated from (4B25), tends to very low values as m_{XY} is decreased, and so does the ratio Γ, since $\gamma_{\pm XY}$ tends towards unity. This is equivalent to a change of m_Y with m_{XY} at a lower power than second. Various reasons have been proposed for this effect, and some may operate at the same time. The solution retained on the surface of the beads and in the interstices between them in the usual centrifugation method[66] has the same molality as the bulk of the solution, m_{XY}, which can introduce serious errors when $\bar{m}_Y \ll \bar{m}_{XY}$, the so-called retention error. Impurities in the resin may interact with ions Y, and bind them even when present at low concentrations. Calling the measured invasion $\bar{m}_Y^*$, the true invasion $\bar{m}_Y$, the fractional retention $q = $ 'retained water'/ 'imbibed water', and the molality of impurities in the resin $\bar{m}_Q$, Freeman[100] obtained the relationship

$$\bar{m}_Y^* - \bar{m}_Y = q(m_{XY} - \bar{m}_Y^*) + \bar{m}_Q \qquad (4B31)$$

enabling him to calculate the constants q and $\bar{m}_Q$ by assuming (4B27) to hold, down to zero concentration, and $\bar{m}_Y$ to be given by (4B24). Conversely, $\bar{m}_Q$ may be reduced to zero by proper choice of the resin, and q may be estimated independently from results on glass beads[107], surface-sulphonated polystyrene[66], or special centrifugation procedures[108], to permit calculation of $\bar{m}_Y$ values which do follow the square relationship (4B24).

f. Alternative Treatments

The activity coefficients used in equations (4B15), (4B21) or (4B25) are stoichiometric coefficients, hence can be interpreted as being caused by several different interactions. Treating the resin as a concentrated electrolyte solution makes it reasonable to obtain Γ approximately, constant at high concentrations. As invasion decreases with decreasing external solution concentration, it is again expected that Γ approximates a (different) constant, as $\gamma_{\pm XY}$ approaches unity, and $\bar{\gamma}'_{\pm XY}$ a constant value characteristic of the non-invaded resin, having a given, high, ionic strength. Here it is assumed that the deviations introduced by retention and impurities have been corrected for, i.e. that $\bar{m}_Y$ is used instead of the observed $\bar{m}_Y{}^*$ (equation 4B31). For many systems (Table 4B2) Γ has a value near unity, as expected for species which are the same in both phases, only subjected to different ionic media. For some systems, however, Γ decreases to low values, which may be taken as indicating the presence of a species in the resin different from that in solution. Thus for an anion exchanger and the acids hydrochloric[98], sulphuric[98], nitric[99, 109] either alone or in their corresponding lithium salt solutions[98, 109], and the salt cobalt chloride[98], low values of Γ are obtained, which are interpreted as indicating ion-pairing or some sort of association of the electrolyte in the resin phase. For the acids, the series

$$\Gamma_{HCl} > \Gamma_{HHSO_4} > \Gamma_{HNO_3}$$

has been found[99], indicative of the increasing tendency of the acids to associate. No quantitative correlations of Γ (or $\bar{\gamma}'_{\pm XY}$) with the properties of the electrolytes have, however, appeared as yet.

Another interpretation of Γ or $\bar{\gamma}'_{\pm XY}$ decreasing with dilution, besides the effects of retention or impurities shown by Freeman[100], has been proposed by Tye[110]. He considers the resin to be inhomogeneous (cf. section 4.A.b), which can be expressed in a formal manner by the occurence of a constant Donnan potential over the whole volume, except a fraction β, where it is zero. This is a simplified expression of the existence of regions in the polymer which have not been sulphonated orchloromethylated, leading mathematically to the observed effects.

Glueckauf[111] broadened the above concept to include not only a region with zero capacity and another with a fixed capacity, but a whole range of regions, with local capacities $\bar{m}_R$ over a fraction φ of the whole resin, leading to an average capacity

$$\overset{\approx}{m}_R = \int_{\varphi=0}^{1} \bar{m}_R \, d\varphi \quad \text{and} \quad \int_{\bar{m}_R=0}^{\bar{m}_R=B} d\varphi = 1 \qquad (4B32)$$

If this is the only cause for the observed deviations, then the ratio Γ can

be treated as constant, and if (4B18) is used to define the activity co-
efficients, the coefficient $\alpha = \Gamma^{-1}\exp(-\bar{\pi}\bar{V}_{XY}/2RT)$ takes care of them.
The distribution function in (4B32) is of the form $d\varphi/d\bar{m}_R = k_R\bar{m}_R^{-x}$,
and the maximal capacity is B, which can be related to the mean molal
capacity $\tilde{\bar{m}}_R$ by (4B32) as

$$B \simeq \tilde{\bar{m}}_R(2 - x)/(1 - x) \text{ and } k_0 \simeq (1 - x)^{2-x}/(2 - x)^{1-x}\tilde{\bar{m}}_R^{1-x}$$

From the invasion equation $\bar{m}_Y(\bar{m}_Y + \bar{m}_R) = (\alpha m_{XY})^2$ and the above
distribution function, Glueckauf finally obtained for the invasion

$$\tilde{\bar{m}}_Y = k_0(\alpha m)^{2-x} \times I(\bar{m}_R/\alpha m_{XY}, B) \tag{4B33}$$

where $I(\bar{m}_R/\alpha m_{XY}, B)$ is an integral, of which the values have been
tabulated[111]. In the range of m_{XY} from 0.001 m to 1.0 m the integral
depends only slightly on m_{XY}, and since α is roughly independent of
m_{XY}, one finally obtains

$$\tilde{\bar{m}}_Y \simeq const.\, m_{XY}^{2-x} \tag{4B34}$$

The experimental values that Glueckauf finds[11] for ion exchange
membranes (where there is no retention error) correspond to $x = 0.63$
for the cation exchanger and $x = 0.743$ for the anion exchanger. Invasion
data in the literature (uncorrected for impurities or retention, and with
constant Γ) were interpreted by Glueckauf to indicate x values between
0.55 and 0.86. As shown above, the Donnan treatment with constant
capacity corresponds to $x = 0$.

It is difficult to tell which treatment is more advantageous. The reality
of retention, impurity and association effects cannot be doubted, but
the parameters which are introduced to account for them are arbitrary
or approximate. It is possible to 'correct' most experimental results so
that they conform to Donnan's square law (i.e. $x=0$). Furthermore, it may
be argued that the membranes examined experimentally by Glueckauf[11]
are indeed inhomogeneous, not representative of carefully manufactured
spherical ion exchange beads. For materials which are known to be
inhomogeneous, on the other hand, Glueckaf's treatment may have
much merit.

C. ION EXCHANGE SELECTIVITY

a. Introduction

Previous sections dealt with equilibria of the ion exchanger with the solvent and with an electrolyte solution having one ion the same as the counter ion of the exchanger. The main usefulness of ion exchangers is, of course, under conditions where there are at least two different counter ions, with a concentration ratio in the resin differing from that in the solution. This difference makes the exchanger a useful tool for changing concentrations and studying the behaviour of ions in solution, and is based on the preference of the exchanger for one ion over the other (of the same sign), i.e. the *selectivity* of the exchanger. This property is expressed quantitatively in terms of various selectivity coefficients and constants.

In order to avoid confusion it is necessary to note that two concentration scales for the resin phase have been widely used in the literature: the rational and the molal scales. The former relates all quantities to one equivalent of resin, and uses mole fractions $\bar{x}$ and rational activity coefficients $\bar{f}$, all quantities referring to this scale will have the suffix r (activity $\bar{a}_r$, selectivity coefficient Q_r, thermodynamic selectivity constant K_r, etc.). The latter concentration scale relates all quantities to 1000 g solvent (water) imbibed in the exchanger, using molalities $\bar{m}$, molal activity coefficients $\bar{\gamma}$, and other quantities with suffix m ($\bar{a}_m$, Q_m, K_m, etc.). For the aqueous phase molalities are usually used, sometimes also molarities.

Consider an ion exchange reaction

$$z_B A^{z_A} + z_A \overline{B}^{z_B} \rightleftharpoons z_B \overline{A}^{z_A} + z_A B^{z_B} \qquad (4C1)$$

in which ions A^{z_A} replace ions B^{z_B} on the exchange sites. (Accompanying changes in the water and invaded electrolyte concentrations are ignored at this stage, and will be dealt with further below). This reaction can be rewritten to pertain to one equivalent of resin sites, by dividing throughout by $z_A z_B$

$$1/z_A \times A^{z_A} + 1/z_B \times \overline{B}R_{z_B} \rightleftharpoons 1/z_A \times \overline{A}R_{z_A} + 1/z_B \times B^{z_B} \qquad (4C2)$$

At equilibrium both phases will contain appropriate concentrations of both ions A and B. The (molal) *selectivity coefficient* Q_{mB}^A will be defined as

$$Q_{mB}^A = Q_m = \bar{m}_A^{z_B} m_B^{z_A} / m_A^{z_B} \bar{m}_B^{z_A} \qquad (4C3)$$

where the symbol Q_{mB}^A with the two suffixes pertains to a reaction where
the ion in the upper suffix, A, displaces from the resin the ion in the
lower suffix, B. In the following, the ions A and B are understood, and
the suffixes will be omitted. Similarly, the rational selectivity coefficient
Q_{rB}^A is obtained from

$$(Q_r)^{z_A z_B} = \bar{x}_A^{z_B} m_B^{z_A} / m_A^{z_B} \bar{x}_B^{z_A} \tag{4C4}$$

When $z_A = z_B = z$, $\bar{m}_A/\bar{m}_B = \bar{x}_A/\bar{x}_B$, and $Q_m = Q_r^{z^2}$, and when $z = 1$,
$Q_m = Q_r$.

Another useful quantity is the *distribution coefficient* D, where for
the displacing ion A

$$D_{mA} = \bar{m}_A/m_A \tag{4C5}$$

Distribution coefficients are often defined as

$$D_A = \frac{\text{quantity of A per unit weight of (dry } H^+\text{-or } Cl^-\text{-form) resin}}{\text{quantity of A per unit volume of solution}} \tag{4C6}$$

The numerator is proportional to $\bar{x}_A$, the denominator to c_A, so that
$D_A = k\bar{x}_A/c_A$, the constant k being independent of the ion, the resin
composition and crosslinking, and the solution concentration. The
separation factor S_B^A between the ions is given by

$$S_B^A = D_A/D_B \qquad S_{mB}^A = D_{mA}/D_{mB} \tag{4C7}$$

When $z_A = z_B$, $S_B^A = S_{mB}^A = (Q_m)^{1/z}$, and when $z = 1$, i.e. for uni-univalent
ion exchange, all separation factors and selectivity coefficients are numerically identical. If, however, $z_A > z_B$, then

$$S_{mB}^A = Q_m^{1/z_A}(\bar{m}_A/m_A)^{(z_A - z_B)/z_A} \tag{4C8}$$

and even in ideal systems where Q_m is constant, the separation factor
will vary with concentration. Ion A will be preferred the more, the more
dilute the solution, just because of its higher charge, without the resin
showing any special selectivity for this ion (i.e. even for $Q_m = 1$).

The thermodynamic equilibrium constant for reaction (4C1) is

$$K_m = K_r^{z_A z_B} = \bar{a}_A^{z_B} a_B^{z_A} / a_A^{z_B} \bar{a}_B^{z_A} \tag{4C9}$$

It may be related to the selectivity coefficients by means of the appropriate activity coefficients

$$\log K_m = \log Q_m + \log(\bar{\gamma}_A^{z_B}/\bar{\gamma}_B^{z_A}) - \log(\gamma_A^{z_B}/\gamma_B^{z_A}) \qquad (4C10)$$

$$\log K_r = \log Q_r + \log(\bar{f}_A^{1/z_A}/\bar{f}_B^{1/z_B}) - \log(\gamma_A^{1/z_A}/\gamma_B^{1/z_B}) \qquad (4C11)$$

For uni-univalent exchange, multiplication of the individual ionic activity coefficients in both numerator and denominator with that of the fixed ion R for the second term, and that of the co-ion Y for the third, yields

$$\log K_m = \log Q_m + \log (\bar{\gamma}_{AR}/\bar{\gamma}_{BR}) - 2 \log (\gamma_{\pm AY}/\gamma_{\pm BY}) \qquad (4C12)$$

$$\log K_r = \log Q_r + \log (\bar{f}_{AR}/\bar{f}_{BR}) - 2 \log (\gamma_{\pm AY}/\gamma_{\pm BY}) \qquad (4C13)$$

The mean molar activity coefficients of the third term are those in the mixed electrolyte solution, containing both AY and BY, at molalities m_A and m_B, with $m = m_Y = m_A + m_B$.

b. Ion Exchange Selectivity

The exchange of pairs of ions A and B according to (4Cl), described by Q_B^A, has been measured by a great number of authors as a function of resin crosslinking, composition and capacity, resin functional group, solution ionic strength and composition, temperature, pressure, and some other variables. It is rather difficult to summarize the wealth of information accumulated, and sometimes inconsistent results are obtained among different authors. Then, again, some authors present their results in terms of the rational scale, others in terms of the molal or still another scale, sometimes as selectivity coefficients at given resin compositions or else as integral selectivity constants. Some general rules may, however, be formulated from the comparison of the accumulated data.

The generalization may be made that there is a good inverse correlation between the preference towards an ion, and the swelling of the resin in that particular ionic form. Increased swelling goes along with lowered preference. It is, however, difficult to say which is the primary cause and which the result. For a given pair of ions A and B, of which A is preferred, i.e. $Q_B^A > 1$, the selectivity increases, i.e. Q_B^A increases, for conditions that decrease the swelling, such as increasing crosslinking, decreasing capacity or ionization of the functional group, increasing solution concentration, etc. A direct measure of the swelling is the molality of the resin sites or fixed ions $\bar{m}^0$. Figure 4Cl shows the selectivity of Dowex–50 polystyrene sulphonate cation exchanger for potassium against lithium ions[112,113–116] and of Dowex–2 polystyrene methylene-dimethyl-ethanol-ammonium anion exchanger for bromide against chloride ions[112,113,117], as a function of the average resin molality, $\bar{m}^0$. The relationships of log K_r vs. log $\bar{m}^0$ are seen to be linear for both cases, with a common slope of 0.61, although this uniformity is certainly fortuitous since it does not hold for other pairs of ions. With the

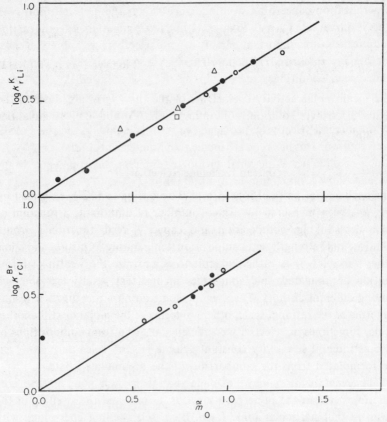

Figure 4C1. The selectivity of ion exchangers as a function of mean molality in the resin phase. Upper portion: Dowex–50 cation exchanger, the K^+–Li^+ exchange. Lower portion: Dowex–2 anion exchanger, the Br^-–Cl^- exchange. For both, log K_r plotted against m^0, with data taken from the references: ● ref. 113, ⊙ and ○ ref. 114, △ ref. 115, and □ ref. 116.

anion exchange system, it should be remarked, the selectivity does not vanish as the resin concentration decreases to low values, contrary to the behaviour of cation exchangers[112,113]. The figure shows, however, that changes in swelling, brought about by differences in crosslinking and capacity, are accompanied by changes in selectivity.

Less clear is the dependence of selectivity on the composition of the resin, i.e. the mole fraction of, say, the preferred ion A, $\bar{x}_A$. Many systems obey the rule[118] that Q_{rB}^A increases as $\bar{x}_A$ decreases ($Q_{rB}^A > K_{rB}^A$ at $\bar{x}_A = 0$ and $Q_{rB}^A < K_{rB}^A$ at $\bar{x}_A = 1$), while some show the opposite ('abnormal')[118] trend. The 'normal' behaviour is, however, contrary to the generalization concerning swelling, since if A is the preferred ion, swelling is larger at $\bar{x}_A = 0$ than at $\bar{x}_A = 1$. Changes in selectivity with composition are important mainly where selectivity is appreciable, i.e. for highly crosslinked, little-swollen resins. Reversals, even, can occur under such conditions, such as in the Na^+–H^+ system[112–114]. Under some conditions selectivity may decrease with increasing crosslinking (and average molality), contrary to the generalization illustrated above by the integral selectivity, provided the resin is mainly in the form of the preferred ion. This has been attributed to inhomogeneity of the resin[112–114], which contains the functional group in different environments. Many systems show a maximum or a minimum in Q_{rB}^A-vs.-$\bar{x}_A$ curves, and the best way of stating the selectivity is by integration, which is equivalent to calculating K_r (see below, section 4.C.c). Another point, which should be considered, is that multivalent ions act as crosslinks in the resin, reducing the swelling, and in many cases of monovalent–divalent exchange there is sharp increase in selectivity as the fraction of the divalent ion in the resin increases towards unity, but this, again, is not universal[119, 120].

Even more complicated is the effect of reduced capacity on the behaviour of Q_r with changing composition. Resins with lowered capacity at given crosslinking may be obtained either by desulphonation (for cation exchangers[13]) or by direct synthesis[14] for both kinds of exchangers. The complicating factor is that the water content of fully swollen resins does not increase in all cases with increasing capacity as expected (i.e. $\bar{n}_W$ is not constant[124]), and inconsistent trends are obtained when plotting selectivities against capacity (at constant crosslinking) or against resin molality or water content (irrespective of crosslinking and capacity). For the Na^+–H^+ exchange, Q_{mH}^{Na} is so near unity that reversals readily occur, and minimal selectivity ($Q_m \sim 1$) is obtained for intermediate values of capacity, the resin preferring Na^+ at high, and H^+ at low capacities. For other pairs ($B = Na^+$, $A = NH_4^+$, Cs^+, Ag^+, $B = Cl^-$, $A = Br^-$) K_{mB}^A usually increases with decreasing capacity, although for $B = NH_4^+$ and $A = Cs^+$ it decreases in some cases, increases in others. Selectivity among pairs of Li^+, Na^+ and K^+ correlates well with equivalent water content[121], as illustrated in Figure 4C1 for a fully sulphonated resin.

The total solution concentration has also a complicated effect on selectivity. One effect is, of course, to decrease the water activity and the swelling, resulting in general in increased selectivity[122]. As long as resin invasion (section 4.B) is unimportant, the effect of ionic strength in the outer solution is slight. At very high solution concentrations special effects are observed, including selectivity reversals, which will be discussed in detail in section 4.D. Although most of the work on selectivity, and particularly all theoretical discussions, are on ion exchange from aqueous solutions, it is appropriate to consider there also some of the effects nonaqueous solvents have on selectivity.

Temperature also, has a complicated effect on ion exchange selectivities. The heat of exchange can be obtained either calorimetrically[74,123–128] or from the temperature coefficient of the selectivity constant. In general the heats are very small, especially near room temperature. The enthalpies obtained are approximately those expected from mixing, dilution (hydration) and concentration (dehydration) of the ions in solution and in the resin due to the ion exchange reaction[123–125]. Wherever specific binding of counter ions to fixed ions occurs, the heat effect may be expected to be larger, as found with the sodium–hydrogen exchange[125], or the caesium–lithium exchange on a polymetacrylate exchanger[128]. Large negative entropy changes point to site-binding of the preferred ion[128]. The standard heats of exchange increase or decrease with temperature, and may change sign[129, 130] and it is very difficult to predict what the temperature effect would be for an unkown system. Table 4C1 shows some representative values of ΔH_B^{oA} at 25 and 100 °C.

The functional group of the resin, i.e. the fixed ion, has an important influence on selectivity, because of specific interactions. Thus, a counter ion such as H^+ forms a bond with a fixed ion such as $-COO^-$ in a carboxylate resin, and, therefore, this resin is very specific for hydrogen ions. The interaction may also be more subtle, and ion-pairs, or oriented dipoles formed by polarization forces, may account for special specificities in special resins such as a phosphonate resin[131,132]. The originally used polystyrene sulphonate cation exchangers and polystyrene methylene-

TABLE 4C1. Heats of ion exchange on polystyrene-type resins (Cal/equivs.).

A–B		X crosslinking	K_{rB}^{A}	ΔH_{B}^{OA}	K_{rB}^{A}	ΔH_{B}^{OA}	Reference
			25°C		100°C		
Li^+	H^+	4	0.77	$+ 244$			a
Li^+	H^+	8	0.79	$+ 528$			a
Li^+	H^+	10	0.82	$+ 387$			b
Li^+	Na^+	16	0.46	$+1700$			h
Li^+	H^+	16	0.69	$- + 516$			a
Na^+	H^+	4	1.24	$- 860$			h
Na^+	H^+	10	1.78	-1255			b
Na^+	H^+	12	1.30	-1110	1.00	$- 410$	c
Na^+	H^+	16	1.57	-1265	1.23	$- 668$	d
Na^+	H^+	16	1.59	-1250			h
Cs^+	H^+	12	7.83	$- 340$	2.98	$- 224$	c
Cs^+	Na^+	16	1.85	-1190			h
Ba^{2+}	H^+	12	5.66	$- 109$	4.83	$+ 210$	c
Ba^{2+}	Na^+	12	4.14	$+ 220$	5.04	$+1020$	c
La^{3+}	H^+	12	5.82	$+ 570$	8.94	$+2040$	c
Cu^{2+}	H^+	16	2.69	$+ 471$	3.28	$+ 622$	d
Cu^{2+}	Zn^{2+}	16	1.22	0	1.11	0	e
Cu^{2+}	Mg^{2+}	8	1.12	$+ 114$	1.13	$- 50$	f
Cu^{2+}	Mg^{2+}	16	1.28	$- 183$	1.17	$- 339$	e
F^-	Br^-	4	0.072	$+3300$			i
F^-	Br^-	10	0.023	$+3580$			i
Cl^-	Br^-	4	0.38	$+1330$			i
Cl^-	Br^-	10	0.31	$+1360$			i
Cl^-	Br^-	10	0.33	$+1530$	0.52	$+1160$	g
I^-	Br^-	4	4.37	-2020			i
I^-	Br^-	16	5.55	-1700			i

a. O. D. Bonner and J. R. Overton, *J. Phys. Chem.*, **65**, 1599 (1961).
b. E. H. Cruickshank and P. Mears, *Trans. Faraday Soc.*, **53**, 1289 (1957).
c. K. A. Kraus and R. J. Raridon, *J. Phys. Chem.*, **63**, 1901 (1959).
d. O. D. Bonner and L. L. Smith, *J. Phys. Chem.*, **61**, 1614 (1957).
e. O. D. Bonner and R. R. Pruett, *J. Phys. Chem.*, **63**, 1420 (1959).
f. O. D. Bonner, G. Dickel and H. Brummer, *Z. Phys. Chem. (Frankfurt)*, **25**, 81 (1960).
g. K. A. Kraus, R. J. Raridon and D. L. Holcomb, *J. Chromatog.*, **3**, 178 (1960).
h. G. E. Boyd, F. Vaslow and S. Lindenbaum, *J. Phys. Chem.*, **68**, 590 (1964).
i. F. Vaslow and G. E. Boyd, *J. Phys. Chem.*, **70**, 2507 (1966).

trimethylammonium anion exchangers are relatively free from such effects in dilute solutions. The crosslinking and swelling properties of the resins have, of course, also the effects of making the resin act as a sieve, permitting only sufficiently small ions to enter the pores in the swollen resin and to undergo ion exchange. Whereas $2X$ resins readily admit ions of 10 Å diameter, a $10X$ resin discriminates against ions larger (in hydrated form) than 5 Å, such ions not being able to saturate the resin to its full capacity[133]. Interaction through London forces between organic ions with the fixed ions or the aromatic rings of the skeleton obviously also depends on the structure of the resin.

The dependence of the selectivity on the properties of the ions themselves is, of course, of the highest interest. There are a few rules-of-thumb which act as useful guides:

a) Selectivity (at least as measured by the distribution coefficient or the separation factor) increases with increasing charge on the ion.

b) Selectivity increases with decreasing (hydrated) radius of the ion.

c) Selectivity increases with increasing polarizability of the ion.

Table 4C2 summarizes available data as K_B^A or Q_B^A at given $\bar{x}_A$ for B $=$ H$^+$ for polystyrene sulphonate cation exchanger, and for B$=$Cl$^-$ for Dowex–1 (trimethyl-methylene-ammonium) and Dowex–2 (ethanol-dimethyl-methylene-ammonium) polystyrene anion exchangers, both of nominal 8–10 X crosslinking.

Some remarks are in place in considering the special features shown by the data in Table 4C2. There is usually fairly good agreement between values given by various authors, within 0.1 units in log K. Bonner[115] gives seemingly too small values for log K_{rH}^M for potassium, rubidium, caesium and uranyl ions, while the values calculated from the distribution factors given by Strelow[135] deviate for lithium and lead ions. The value for zirconium, higher than that for thorium[134], also seems deviating. Some of the values given by Strelow and others are possibly affected by complex formation in the aqueous phase. As regards anions, agreement usually is attained, even between values for the two resins (excluding the special case of the hydroxide ion, which Dowex–2 prefers by a factor of 7 relative to Dowex–1), except for some controversial results, such as those of Gregor, Belle and Marcus[135] for trichloroacetate, thiocyanate and perchlorate, which could not be confirmed by other authors[136,137], in particular as regards the trend of $Q_{Cl}^{ClO_4}$ with x_{Cl}[136], on which Gregor and coworkers[135] based their clustering theory regarding ion-pairs in anion exchangers (cf. ref. 138, p. 175). The data for the halates found by Aveston and coworkers[139] also are in disagreement with other data which seem more consistent.

TABLE 4C2a. Cation exchange selectivities for exchange of M^{m+} for H^+
(approx. 8–10 X Dowex–50).

Cation, M^{m+}	Log Q_r^a	Log Q_m^d	Log K_r^e	Log K_r	Log Q_r^f
Li^+	0.08		-0.10	$-0.06g$	-0.09
H^+	0.00	0.00	0.00	0.00	0.00
Na^+	0.25	0.32	0.20	0.22g	0.25
K^+	0.64	0.65	0.36	0.40g	0.50
NH_4^+			0.31		
Rb^+	0.69	0.74	0.40		
Cs^+	0.79	0.89	0.41	0.48g	0.78
Ag^+	0.76		0.82		
Tl^+	0.85		0.99		
UO_2^{2+}	0.51		0.29	0.49b	
Be^{2+}	0.42	0.33	0.60h		
Mn^{2+}	0.52		0.61h		
Fe^{2+}	0.51				
Co^{2+}	0.52	0.23	0.47		
Ni^{2+}	0.52		0.49		
Cu^{2+}	0.48		0.49		
Zn^{2+}	0.46	0.19	0.44	0.52c	
Cd^{2+}	0.42		0.49	0.50b	
Hg^{2+}	0.47				
Mg^{2+}	0.55		0.42		
Ca^{2+}	0.67		0.62		
Sr^{2+}	0.73		0.71		0.65
Ba^{2+}	0.91	0.77	0.96	0.73a	
Pb^{2+}	0.64		0.90	0.73a	

Cation, M^{m+}	Log Q_r^a	Log Q_m^d	Log K_r^h
Hg^{2+}	0.74		
Cr^{3+}	0.45		0.88
Fe^{3+}	0.50		
Ga^{3+}	0.52		
Al^{3+}	0.57		
Y^{3+}	0.70		
La^{3+}	0.79	0.78	1.02
Ce^{3+}	0.79		1.02
Eu^{3+}		0.76	
V^{4+}	0.26		
Ti^{4+}	0.31		
Zr^{4+}	1.00		
Th^{4+}	0.87		

a. Ref. 134, for hydrochloric acid solutions and $\overline{X}_M = 0.4$. Data are available also for nitric and sulphuric acid solutions: F. W. E. Strelow, R. Rethemeyer and C. J. C. Bothma, *Anal. Chem.*, **37**, 106 (1965); F. W. E. Strelow and C. J. C. Bothma, *Anal. Chem.*, **39**, 595 (1967).

b. H. F. Walton, D. E. Jordan, S. R. Samedy and W. N. McKay, *J. Phys. Chem.*, **65**, 1477 (1961).

c. N. J. Meyer, W. J. Argersinger, Jr. and A. W. Davidson, *J. Am. Chem. Soc.*, **79**, 1024 (1957).

d. Ref. 131, for $\overline{x}_M = 0.0$.

e. Ref. 115.

f. Ref. 114.

g. Refs. 112 and 113.

h. Ref. 120.

TABLE 4C2b. Anion exchange selectivities for exchange of B^{b-} for Cl^-
(approx. 8–10 X resins).

Anion, B^{b-}	Dowex-1 (Amberlite-400)			Dowex-2 (Amberlite-410)			
	Log K_r[a]	Log Q[b]	Log Q	Log Q[b]	Log Q[g]	Log Q[d]	Log Q
OH^-		−1.05		−0.18	−0.28	−0.85	−0.16[c]
F^-	−0.90	−1.05		−0.89	−1.06	−1.20	−0.98[c]
							−1.17[i]
$CH_3CH_2^-$		−0.77	−0.91[h]	−0.74	−0.81	−0.85	−0.77[c]
HCO_2^-		−0.66	−0.77[h]	−0.66			
$C_3H_7CO_2^-$			−0.64[h]				
IO_3^-	0.44				−0.61	−0.60	−0.68[c]
							−0.66[e]
$H_2PO_4^-$		−0.60		−0.47			
HCO_3^-		−0.49		−0.27			
Cl^-	0.00	0.00	0.00	0.00	0.00	0.00	0.00
HSO_3^-		0.11		0.11			
CN^-		0.20		0.11			
NO_2^-		0.08		0.11			0.50[c]
BrO_3^-	0.55			0.00			0.07[e]
ClO_3^-	1.14						0.48[e]
Br^-	0.62	0.45	0.57[f]	0.36	0.40	0.50	0.53[c]
							0.47[i]
NO_3^-		0.58		0.52	0.48	0.50	0.52[c]
HSO_4^-		0.61		0.79			
I^-	1.28	0.93	1.10[f]	0.86	1.25	1.10	1.12[c]
							1.24[i]
CNS^-	1.86				0.66	1.35	1.27[c]
							1.28[j]
$CCl_3CO_2^-$			1.71[h]		0.37		1.26[c]
BF_4^-	1.42						
ClO_4^-	2.18		1.57[f]		1.05	1.55	1.68[j]
							1.51[c]
ReO_4^-	2.26						
MnO_4^-	3.12						

a. Ref. 139.
b. R. M. Wheaton and W. C. Bauman, *Ind. Eng. Chem.*, 45, 228 (1953), for $\bar{x}_B = 0.2$ to 0.7.
c. J. Belle, *Ph.D. Thesis*, Brooklyn Polytechnic Institute, 1954.
d. S. Lindenbaum, G. E. Boyd and G. E. Myers, *J. Phys. Chem.*, 62, 995 (1958), $\bar{x}_B = 0.0$.
e. M. Kikindale, *Compt. Rend.*, 237, 250 (1953).
f. I. Eliezer and Y. Marcus, *J. Inorg. Nucl. Chem.*, 25, 1465 (1963).
g. Ref. 135, $\bar{x}_B = 0.5$.
h. Ref. 137.
i. Refs. 112, 113.
j. Ref. 136.

Figures 4C2a and 4C2b illustrate rules a and b. More or less consistent values of log K_{rH}^M are plotted against r_h, the hydrated radius as given by Monk[140] (cf. Table 1B2) in Figure 4C2a. It is necessary to compare ions of similar form, and rule b seems to hold for the alkali metal, alkaline earth, rare earth and trivalent actinide ions. With the latter, however, inversions have been noted[141], which cannot be readily explained. Unfortunately, hydrated radii are available only for a few

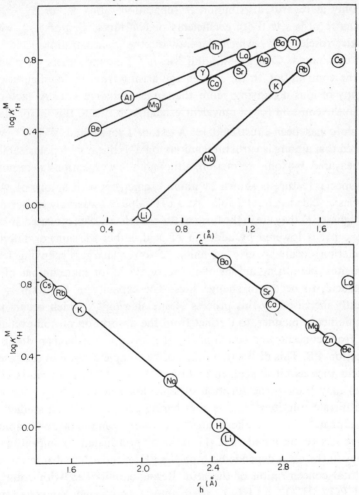

Figure 4C2. The effects of ion size and charge on selectivity, for exchange of metal ions against hydrogen ions on a cation exchanger of Dowex–50 type. Upper portion, log K_r plotted against the crystalline radius r_c, lower portion, log K_r plotted against the hydrated radius r_h.

cations, and the values are by no means definite (section 1.B). Figure 4C2a does show, that apparently, for a given size, the ion with the higher charge is preferred by the resin. The same is seen from a correlation of the selectivity constants with the crystal radii of the ions, Figure 4C2b, and in the sequence of $Q_{r\,La}^{M}$ for complex Co^{III} ions of roughly similar sizes but charges $+3$, $+2$ and $+1^{142}$. Comparative data for anion exchangers are, unfortunately, not available. Rule a above seems to hold not only as regards distribution or separation ratios D_A or $S_B^{A\,138}$, but also as regards selectivity coefficients or constants, $Q_{r\,B}^{A}$ or $K_{r\,B}^{A}$, where the electroselectivity, which depends on the concentration scale, has been eliminated. Cruickshank and Mears[123], among others[143], pointed out the contribution to the free energy arising from the configurational entropy of ions occupying more than one exchange site. A factor of two in the constant for bi-univalent exchange, is due to this factor alone.

Rule c has been illustrated by Aveston, Everest and Wells[139], who showed that among isostructural anions $\log K_r$ is linear with polarizability, as measured by ionic refraction, although a few exceptions were noted.

A special feature is shown by anion exchangers with anions of weak polybasic acids as counter ions. At a pH value considerably higher than the highest pK, they are in the form of the completely deprotonated anions. As the pH is lowered by addition of acid, either the same or different, association occurs to give an anion of lower charge, occupying fewer resin sites, permitting anions of the second acid, or more anions of the same acid, to occupy exchange sites. The capacity of the resin is apparently increased by this process of site sharing[144], which occurs in a stoichiometric manner, as distinct from the absorption of non-exchange acid from concentrated solutions, by the invasion mechanism discussed in section 4.B. This distinction may not be complete, since in the case of an acid such as nitric acid, formation of $H(NO_3)_2^-$ in the resin (section 6.C.e) may lead to the sorption of stoichiometric amounts of the acid from nitrate solutions[145]. The site-sharing process has been studied for the sulphate[98, 146, 147], phosphate[148], citrate[148] and oxalate[149] anions, and it has to be noted that the ratio of protonated to unprotonated anions in the resin depends both on the pH of the external solution and the total concentration of the acid. Resins equilibrated with solutions of $H_2PO_4^-$, HSO_4^- or HCO_3^- need not be completely converted into these forms. Since the pH and the pK of the acid in the resin are not the same as outside, it is difficult to select conditions to convert the resin exactly to the desired protonated form[98]. This point must be remembered when dealing with anion exchange involving anions of polyvalent weak acids.

c. Thermodynamic Theories

A number of approaches have been proposed to interpret the observed selectivity relationships, and to predict ion exchange selectivities from independent data. The problems are to relate K_m or K_r to measured quantities, and to calculate Q_m or Q_r for systems of varying composition and total concentrations from experiments not involving actual exchange.

Let us discuss first the formal thermodynamic approach proposed by Ekedahl, Högfeldt and Sillèn[150] and by Argersinger, Davidson and Bonner[151]. These authors chose as components the hydrated resinates AR and BR, i.e. assigned all the water in the resin to the resinates, without considering any free water. This is tantamount to assuming a linear variation of the water content with the composition, from $\bar{n}_W(A)$ to $\bar{n}_W(B)$ molecules of water per equivalent of resin[152]. The standard states are chosen so that when $\bar{x}_A = 1$, $\bar{a}_A = 1$, $\bar{f}_A = 1$, and when $\bar{x}_B = 1$, $\bar{a}_B = 1$ and $\bar{f}_B = 1$. The selectivity constant is given by equation (4C13), the last term being zero when only very dilute aqueous solutions are considered. Differentiation of (4C13) yields $d \ln Q_r = d \ln \bar{f}_{BR} - d \ln \bar{f}_{AR}$, and the Gibbs–Duhem equation is $\bar{x}_{AR} d \ln \bar{f}_{AR} + \bar{x}_{BR} d \ln \bar{f}_{BR} = 0$. From these and (4C13) one obtains

$$\log K_r = \int_0^1 \log Q_r \, d\bar{x}_{AR} \qquad (4C14)$$

This result is valid also if $z_A \neq z_B$, provided equivalent fractions are used. It should be noted that the thermodynamic rational selectivity constant K_r, calculated according to (4C14), is specifically defined as pertaining to the resinates as components using the standard states defined above.

This treatment has been modified to include the effect of changing water activities, by Davidson and Argersinger[122] and by Högfeldt[153]. A more general treatment was proposed by Gains and Thomas[154]. Holm[155], and Kakihana[93] have later compared these various treatments. Davidson and Argersinger[122] considered the changes in hydration with changes in water activity, so that $\bar{f}_{AR,n_W(A)} = 1$ for $\bar{x}_{AR} = 1$ of a resin in equilibrium with a solution of molality m, instead of pure water. Therefore, (for uni-univalent exchange), as Holm[155] has shown

$$\log K_r = \int_0^1 \log Q_r \, \mathrm{d}\, \bar{x}_{AR, n_W(A)} - 2\log(\gamma_{\pm AY}/\gamma_{\pm BY})$$

$$+ (\bar{n}_W(B) - \bar{n}_W(A)) \left[\log a_W(\bar{x}_A = 1) - \int_{a_W(\bar{x}_A = 0)}^{a_W(\bar{x}_A = 1)} \bar{x}_A \, \mathrm{d}\log a_W \right] \quad (4C15)$$

Davidson and Argersinger themselves use Harned's rule, and the corresponding expression for the water activity (equation 1E20), to obtain an expression for K_r, which improved the constancy of this quantity as m changed from 0.1 to 1.0 m for the H^+–Na^+ exchange, over that achieved with (4C14). Högfeldt[153] does not use hydrated components, but considers the water as a separate component with $\bar{a}_W = a_W$ at equilibrium, and $\bar{f}_{AR} = 1$ for $\bar{x}_{AR} = 1$ of the resin in equilibrium with the m molal solution of water activity a_W. The expression obtained is

$$\log K_r = \int_0^1 \log Q_r \, \mathrm{d}\bar{x}_{AR} - 2\log(\gamma_{\pm AY}/\gamma_{\pm BY}) + \int_{a_W(\bar{x}_{AR} = 0)}^{a_W(\bar{x}_{AR} = 1)} \bar{n}_W \, \mathrm{d}\log a_W$$

$$(4C16)$$

where the restriction of the linearity of $\bar{n}_W$ with $\bar{x}_{AR}$ has been removed. The expression obtained by Gains and Thomas[154] is thermodynamically the most rigorous, but also the most complicated. Again the standard state of water in the exchanger is pure water, but $\bar{f}_{AR} = 1$ for $\bar{x}_{AR} = 1$ of a resin in equilibrium with pure water. Here integrations from unit water activity to that in the resin must be performed for both pure A and pure B resins, in addition to the last integral in equation (4C16).

Another formal treatment that merits consideration is that of Scatchard and Freeman[152], using excess free energy functions[156]. Terminating the expansion after the quadratic term in $\bar{x}_A$ yields an expression

$$\log Q_r = \log K_r + (G_0 - G_1) + (4G_1 - G_0)\bar{x}_A - 4G_1\bar{\gamma}_A^2 \quad (4C17)$$

from which K_r and the coefficients G_0 and G_1 can be obtained by curve-fitting.

For the comparison of the selectivity of an ion exchanger towards various ions it is sufficient to use (4C14), at a given low total molality m. It must also be remembered that $\log K_{rC}^A = \log K_{rB}^A + \log K_{rC}^B$. This relationship has been confirmed experimentally[115, 116, 120, 136] for a number of ternary systems. The formal thermodynamic treatment discussed above has been found useful for describing the ion exchange selectivity, by expressing it as a single constant, K_r, independent of resin composition or solution concentration. The equations cannot, however, be used for predicting the selectivity under given conditions, from independent data.

An approach which attempts such a prediction, starting from the Donnan equilibrium previously discussed (sections 4.B.d and e), has been developed through the efforts of a number of authors[63,70,157]. Basically, equation (4B8) is used, with i being one time ion A and the other the ion B, while $\bar{\mu}_i = \bar{\mu}_i^0 + RT \ln \bar{a}_i + \bar{\pi} \bar{V}_i'$. Standard states are selected so that activity coefficients are unity at infinite dilution, making $\bar{\mu}_i^0 = \mu_i^0$, so that the expression

$$z_B RT \ln \bar{a}_A/a_A + z_B \bar{\pi} \bar{V}_A' = z_A RT \ln \bar{a}_B/a_B + z_A \bar{\pi} \bar{V}_B' \quad (4C18)$$

is obtained. Equation (4C18) may be rewritten so as to give the thermodynamic selectivity constant for reaction (4C1), as defined in (4C9), the absolute values of z_A and z_B being used

$$\log K_m = \log (\bar{a}_A/a_A)^{z_B} (a_B/\bar{a}_B)^{z_A} = (\bar{\pi}/2.303 \, R \, T)(z_A \bar{V}_B' - z_B \bar{V}_A') \quad (4C19)$$

Using (4B16) to substitute for $\bar{\pi}$ gives the alternative expression

$$\log K_m = \frac{z_A V_B' - z_B V_A'}{\bar{V}_W'} \times \log (a_W/\bar{a}_W) \quad (4C20)$$

Using molalities and molal activity coefficients and (4C3) for the molal selectivity coefficient, this quantity becomes

$$\log Q_m = \log (\bar{\gamma}_B^{z_A}/\bar{\gamma}_A^{z_B}) + \log (\gamma_A^{z_B}/\gamma_B^{z_A}) + (\bar{\pi}/2.303 \, R \, T)(z_A \bar{V}_B' - z_B \bar{V}_A') \quad (4C21)$$

If, for simplicity, only the case $z_A = z_B = 1$ is discussed, and mean molal activity coefficients are used, (remembering that $\bar{v} = 1$, section 4.B.b), this equation becomes

$$\log Q_m = \log (\bar{\gamma}_{\pm BR}/\bar{\gamma}_{\pm AR}) + 2 \log (\gamma_{\pm AY}/\gamma_{\pm BY}) + (\bar{\pi}/2.303 \, R \, T)(\bar{V}_{BR}' - \bar{V}_{AR}') \quad (4C22)$$

This equation may be rewritten in terms of hydrated ions, as proposed by Gregor[70], noting that $Q_{mh} \neq Q_m$ when $z_A \neq z_B$, Q_{mh} being calculated with molalities based on free water. Different activity coefficients $\bar{\gamma}_{\pm hi}$ and partial molal volumes $\bar{V}_{hi}$ must then be used for the hydrated ions. Gregor assumed that the first two terms on the right of (4C22), modified for hydration, are effectively zero, which makes the selectivity depend on the pressure–volume term only, and increase with increasing difference between the hydrated volumes of the ions. In cases where changes in solvation become important (at very high crosslinking, in concentrated solutions with considerable resin invasion, or when a large external pressure is applied on the system[158]), Gregor's treatment cannot be used, and (4C22) should be used for the usual, anhydrous, components.

The question of the relative importance of the terms in (4C22) has led to some controversy, but the equation has proved to be very valuable for describing and predicting ion exchange equilibria, as shown by Glueckauf[157], Gregor[70], Boyd and Soldano[63] and others. In evaluating the terms of (4C23) it must be remembered that the quantities $\bar{\gamma}_{\pm}$, $\gamma_{\pm}$ and $\bar{V}'$ pertain to the mixed components AY and BY in the solution and AR and BR in the resin. Approximations which use quantities for the pure components only, denoted by superscript[0]: $\log(\bar{\gamma}_{\pm BY}^{0}\gamma_{\pm AY}^{0}/\gamma_{\pm BY}^{0}\bar{\gamma}_{\pm AY}^{0})$, used by Bonhoeffer[97] or as suggested by Glueckauf[157] $0.4343(\bar{\phi}_{BR}^{0}-\bar{\phi}_{AR}^{0})$ $+ \log(\gamma_{\pm BY}^{0}/\gamma_{\pm AY}^{0})$, instead of the first two terms on the right in (4C20), fail to show the dependence of Q_{m} on resin composition.

Boyd, Myers and Lindenbaum[112,113] proposed a thermodynamic method for evaluating the terms of (4C22). The resin is again considered as a mixture of the resinates AR and BR and of water. Using the crossdifferentiation relationship, they obtained

$$0.018\,(\partial \ln(\bar{\gamma}_{\pm AR}/\bar{\gamma}_{\pm BR})/\partial \ln \bar{a}_{W})_{\bar{x}_{AR}} = -\,(\partial(1/\bar{m})/\partial \bar{x}_{BR})_{\bar{a}_{w}} \qquad (4C23)$$

This gives on integration at constant $\bar{x}_{AR}$

$$\log(\bar{\gamma}_{\pm AR}/\bar{\gamma}_{\pm BR}) = \log(\bar{\gamma}_{\pm AR}^{*}/\bar{\gamma}_{\pm BR}^{*}) - 55.5 \int_{0}^{\log a_{W}} (\partial(1/\bar{m})/\partial \bar{x}_{BR})_{a_{w}}\,d\log \bar{a}_{W} \qquad (4C24)$$

where $\bar{\gamma}_{\pm}^{*}$ is the activity coefficient in a mixed 'strainless' resin. This is a resin of such low crosslinking that $\bar{\pi} = 0$, so that $\log(\bar{\gamma}_{\pm AR}^{*}/\bar{\gamma}_{\pm BR}^{*})$ can be equated to $-\log Q_{m}$ in very dilute solution, neglecting the last two terms in (4C22). The introduction of the coefficients $\bar{\gamma}_{\pm}^{*}$ comes from the necessity to extrapolate in the integration to $\bar{m} = 0$, i.e. infinite dilution in the resin phase, which is an experimental impossibility. Use of a 'strainless' resin of low crosslinking is one way to obtain $\bar{\gamma}^*_{\pm AR}/\bar{\gamma}^*_{\pm BR}$. Another was suggested by Bonner[65], and involves the ratio of the activity coefficients of salts of A and B with the monomer of the resin. Boyd and coworkers used substituted sulphonic acids for the cation exchangers and quaternary benzylammonium salts for the anion exchangers[159], taking into account the variety of environments of the functional groups in the resin by using appropriate mixtures. The data required for evaluating (4C24) are these coefficients $\bar{\gamma}^*_\pm$, a knowledge of the change of the equivalent water content $1000/\bar{m}$ with resin composition $\bar{x}_{BR}$, and the integral of this derivative with respect to resin-water activity. It was found experimentally[112,113] that the term in $\bar{\gamma}^*_\pm$ is relatively unimportant for cation exchangers, while being important for anion exchangers.

An alternative approach, proposed by Glueckauf[157] and developed by Soldano[117], uses an empirical expression for the osmotic coefficient of the mixed resin $\bar{\phi}\,(\bar{m})$. Use of the Gibbs–Duhem relationship permits the evaluation of $\bar{\gamma}^0_\pm$ from the empirical fit of $\bar{\phi}(\bar{m})$. Application of Harned's rule (1E9), in the form

$$\log(\bar{\gamma}_{\pm AR}/\bar{\gamma}_{\pm BR}) = \log(\bar{\gamma}^0_{\pm AR}/\bar{\gamma}^0_{\pm BR}) + \alpha_{AB}\bar{m}_{AR} - \alpha_{BA}\bar{m}_{BR} \quad (4C25)$$

with α_{AB} and α_{BA} themselves calculable from the empirical relationship for $\bar{\phi}^0_{AR}$ and $\bar{\phi}^0_{BR}$, then permits calculation of the activity coefficient ratio from water-uptake measurements only, without the need for selectivity measurements as in Boyd and Myers' method[112,113].

The second term in (4C22) pertains to the aqueous solution, and may be evaluated by use of Harned's rule from independent data, as discussed in section 1.E.

The third term in (4C22) contains the swelling pressure $\bar{\pi}$ and the partial molal volumes of the ions (or resinates) $\bar{V}'$ as unknowns. The swelling pressure may be obtained isopiestically from the uptake of water on the resin studied, compared to a 'strainless' resin, as described in section 4.B.b. Alternatively, the empirical Hooke's law expressions relating $\bar{\pi}$ to equivalent volume of the resin $\bar{V}_s$ may be used (section 4.B.b). A rough approximation to $\bar{V}'_{BR} - \bar{V}'_{AR}$ may be taken as the difference between the partial molal volumes of the ions at infinite dilution. An exact value may be calculated[112,113] from the equivalent volume

$$\bar{V}_s = \bar{x}_W \bar{V}'_W + \bar{x}_{AR} \bar{V}'_{AR} + (1 - \bar{x}_{AR}) \bar{V}_{BR} \qquad (4C26)$$

which, upon differentiation, application of the cross-differentiation identity, and integration, yields at a given $\bar{x}_{AR}$

$$\bar{V}'_{BR} - \bar{V}'_{AR} = (\bar{V}_{dBR} - \bar{V}_{dAR}) - \int_0^{\bar{x}_W} (\partial(\partial \bar{V}_s/\partial \bar{x}_W)_{\bar{x}_{AR}}/\partial \bar{x}_{AR})_{\bar{x}_W} d\bar{x}_W \qquad (4C27)$$

where $\bar{V}_d$ is the dry molar volume of the pure resinate. The required data are the change of the equivalent volume with both water content and composition, holding the other factor constant. Experimentally[112,113] the third term in (4C22) was found to contribute only a small factor to the calculated Q_m, except for exchange of ions varying greatly in size, such as sodium with tetramethylammonium[160].

Fairly good agreement between calculated and experimental Q_m values has been obtained by the treatments of both Boyd and Myers[112,113] and Soldano[117]. However, many measurements of water uptake and equivalent volumes as functions of the composition must be made, so that the above approaches can be used as a description of the ion exchange process, rather than for predicting selectivities.

d. Other Theoretical Treatments

Insight in to the selectivity of ion exchangers can be obtained not only from the phenomenological–thermodynamic approach developed above, but also from statistical and structural considerations. An important concept first developed by Katchalsky and coworkers[161] involves the configurational free energy of the charged polymer, which may be treated in terms of polyelectrolyte theory[162]. This theory has proved to be of value when discussing the behaviour, of weakly crosslinked ion exchangers, particular of the weak-acid type where the dissociation can be successfully correlated with the osmotic properties[64]. Rice and Harris[163] modified this theory in adding to the configurational and interaction-along-the-chain terms of the free energy, a term depending on ion-pairing of the counter ions with the fixed ions

$$G = G_{\text{configuration}}(V) + G_{\text{interaction}}(V, \bar{N}_i) + \sum_{i \neq 0} N_i R T(\ln \bar{K}_i - \ln c_i) \qquad (4C28)$$

The first term depends on the total volume of the system, V, while the second, both on this and on the number $\bar{N}_i$ of moles of sites either free ($i = 0$), or bound in ion pairs with ions i. The last term expresses the contribution of the reactions (ion-pair)$_i$ $\rightleftharpoons$ (fixed ion) + (ion$_i$), the intrinsic equilibrium constant of which is K_i. Ions i not bound with the sites as pairs are free, and are assumed to have the same concentration as in the external solution, c_i. Minimization of G with respect to $\bar{N}_i$ determines the equilibrium composition of the resin. When the solution is very dilute, the number of free counter ions is equal to the number of free sites $\bar{N}_0$. The final result for the selectivity coefficient Q (for uni-univalent exchange, immaterial of concentration scale) is

$$Q_B^A = \frac{K_B + Z}{K_A + Z}, \quad Z = \bar{x}_0(K_B c_A + K_A c_B)/c(1 + \bar{x}_0) \qquad (4C29)$$

Unfortunately the evaluation of $\bar{x}_0 = \bar{N}_0/(\bar{N}_0 + \sum \bar{N}_i)$, i.e. the fraction of free sites, is rather complicated, and there is no direct means for evaluating $\bar{K}_A$ and $\bar{K}_B$. Recourse can be made indirectly by using the values of K_i^* for the dissociation of an ion-pair made up from ion i and the monomeric unit of the ion exchanger, carrying the fixed ion. However, for any value of $\bar{K}_i$, the higher the total amount of ion-pairing, i.e. the smaller $\bar{x}_0$, the more will Q_B^A differ from unity, and the higher the selectivity.

Pauley[164] assumed the simple coulomb law to hold for the electrostatic interaction between counter ion and fixed ion, arriving at the free energy change for the reaction (4C1): $\Delta G^0 = e^2 \bar{\varepsilon}^{-1}(z_B^2/\mathring{a}_{BR} - z_A^2/\mathring{a}_{AR})$, where a is the distance of closest approach between counter and fixed ion. There are, of course, the difficulties of pairing resin sites with multivalent ions, and the value to be used for the dielectric constant in the resin phase, $\bar{\varepsilon}$, is not known. A development of this idea was proposed by Marcus and Maydan[165], who considered the reaction to have three stages

$$\overline{BR}_{z_B} \rightarrow z_B \bar{R} + \bar{B}^{z_B},$$

$$A^{z_A} + (z_A/z_B)\bar{B}^{z_B} \rightarrow \bar{A}^{z_A} + (z_A/z_B)B^{z_B}, \qquad (4C30)$$

and $\qquad \bar{A}^{z_A} + z_A \bar{R} \rightarrow \overline{AR}_{z_A}.$

The first and last stages are accompanied by free energy changes as given by Pauley[164], while the free energy change for the second reaction is calculated with a Born cycle, in which ion A is transferred through the gas phase from a medium (water) with dielectric constant ε to one of dielectric constant $\bar{\varepsilon}$, while ion B is transferred in the opposite direction. Changes in the (hydrated) ion radii during this process are also taken into account. The final equation obtained is

$$\Delta G^0 = \frac{z_A z_B e^2}{2\bar{\varepsilon}} \left[\frac{2z_B}{\bar{r}_R + \bar{r}_B} - \frac{2z_A}{\bar{r}_R + \bar{r}_A} + \left(\frac{z_A}{\bar{r}_A} - \frac{z_B}{\bar{r}_B} \right) - \frac{\bar{\varepsilon}}{\varepsilon} \left(\frac{z_A}{r_{Ah}} - \frac{z_B}{r_{Bh}} \right) \right]$$
(4C31)

Estimates of the dielectric constant in the resin[166, 167] were made from those in aqueous dioxane solution of the same water content (on weight basis, though a volume basis would be more appropriate) as the swollen resin[165]. The values obtained for Dowex–1 anion exchanger ranged from 59 for $2X$ to 33 for $16X$ material. Other estimates of this quantity are roughly similar (e.g. 40 for $5-25$ X polystyrene sulphonate resin[163]), while the measured values found were 38 for the swollen calcium form of the polystyrene sulphonate resin Amberlite IR–120, ca. $8X$[166]. Ionic radii in the resin $\bar{r}$ were considered not to be equal to those in solution for hydrated ions. It was assumed that the volume of the hydration shell is proportional to the water content, and the radii were calculated from

$$\bar{r} = r_c + (W_s/100)^{1/3} (r_h - r_c)$$
(4C32)

where r_h is the hydrated, r_c the crystalline and $\bar{r}$ the resin-phase radii of the ions, and W_s is the percent water content. Application of (4C31) to a number of ions and anion exchangers with different crosslinking showed good agreement (see however section 6.C.f).

Many qualitative discussions of the selectivity of ion exchangers and its causes have been presented. It has been shown that ion-pairing (treatments of Rice and Harris[163], Pauley[164], Marcus and Maydan[165] etc.), or size effects, coupled with their influence on swelling (treatments of Gregor[70], Glueckauf[157] and Boyd[63,112,113] etc.) alone cannot account for all selectivity effects exhibited by various resins (depending mainly on functional group and crosslinking) and pairs of ions. Two important additional factors have been proposed: ion polarization and changes in water structure. The former is chiefly important in anion exchange[135, 139], where ionic size seems to play a relatively minor role. Some correlation has been found[139] between the selectivity coefficient and the polarizability, as measured by ionic refraction, although the shape of the anion is also important. There are, however, some reversals, so that polarizability cannot be the sole factor. Polarizability of the ionic group has also been invoked[168, 169] to explain the reverse order of the selectivity of weakly acid (carboxylic and phosphonic) cation exchangers in neutral or alkaline solutions towards the alkali metal cations, compared to the 'normal' order $Cs > Rb > K > Na > Li$ shown by the strongly acid sulphonate exchangers. Since the polarizability sequence is POO_2^{2-}, $COO^- > H_2O > SO_3^-$, the former anions can displace water from the more highly polarizing alkali metal cations, leading eventually to site binding[74], so that the order of affinity increases with decreasing crystal radius. For even more highly polarizing ions, such as H^+ or alkaline earths, $HPOO^-$ and $PO(OH)O^-$ also are more polarized than water, so that with resins having these functional groups reversals occur too[169].

Another idea concerning selectivity, which has not yet been formulated quantitatively, but which seems to explain a number of inconsistencies in the above theories, is the effect of the ions on the structure of water (section 1.B). Some authors[70, 93, 122] have tried to include the change of the hydration of ions in their models, formulating (4C1) as

$$z_B A(H_2O)_{h_A}^{z_A} + z_A \overline{B(H_2O)_{h_B}^{z_B}} \rightleftharpoons z_B \overline{A(H_2O)_{h_A}^{z_A}} + z_A B(H_2O)_{h_B}^{z_B}$$

$$(4C31)$$

$$+ [z_B(h_A - \bar{h}_A) - z_A(h_B - \bar{h}_B)]H_2O$$

In addition to the change in the water content of the resin by $(h_A - \bar{h}_A)/z_A - (h_B - \bar{h}_B)/z_B$ moles per equivalent, with the accompanying changes in the mechanical energy of swelling, there is also the free energy of the changes in the hydration of the ions as a driving force for reaction (4C33), as pointed out by Diamond[137]. Hydration in the resin is not the same as in the dilute aqueous solution for a number of reasons. The resin phase can be considered as a concentrated solution, the molalities being of the order of 5–10, where hydration is less than in dilute solutions (section 1.B.e). The water in the resin is constrained by the intrusion of large volumes of organic material from taking the most favourable positions around ions, and although primary hydration of strongly solvated ions such as chromium(III) seems not to be affected[59], secondary solvation must suffer greatly from this. Still another feature is the interference with the ordinary hydrogen bonded structure of the water, that the high concentration of ions and organic material produces[137]. One effect of this is the lowering of the dielectric constant, as discussed above, because of the diminished cooperative effect of the water dipoles.

Another effect is the preference[137] for large ions, which are less concerned with water structure around them, compared with small ions which are strongly hydrated, and prefer the structured dilute solution. The more hydrated and hydrogen bonded an ion is, therefore, the less it will prefer the resin over the outside solution (cf. also lower affinity of $Co(NH_3)_4(H_2O)_2^{3+}$ compared to $Co(NH_3)_6^{3+}$ [142]). For anions, the concept of localized hydrolysis[170, 171] should be considered, leading to the conclusion that for similar anions, that derived from the stronger acid should be preferred by the resin, because of its smaller interaction with water. This has been confirmed experimentally in a number of cases[137]. Another formulation of the same ideas, based on the order-disorder concept of Gurney[172] has been presented by Holm[173]. The principle involved is that ions different in their ordering-disordering properties repulse, while those that are similar attract, each other, when forced to such great proximity that their hydration spheres overlap. The functional groups of the resin ($-SO_3^-$ for cation exchangers and $-CH_2NR_3^+$ for anion exchangers) are disordering, because of their size, as shown also by the sequence of the activity coefficients of alkali metal salts of the sulphonate monomers. Hence, large singly charged ions which are disordering will be preferred to small, highly charged, or strongly basic ions, which are ordering, as indeed is found experimentally[173, 174].

D. ION EXCHANGE IN CONCENTRATED SOLUTIONS AND MIXED SOLVENTS

a. Introduction

Ion exchange in concentrated solutions or mixed solvents presents some special problems which are not encountered in dilute aqueous solutions. Activity coefficient ratios in the external solution, which could be neglected at high dilution, become an important factor at higher concentrations. Since measurements are usually carried out as distribution measurements of a trace ion between the resin and a bulk electrolyte, the use of the effective activity function a for the bulk electrolyte (section 1.D) becomes attractive. The similar function for the resin, $\bar{a}$ (section 4.B), also finds its uses, particularly as resin invasion becomes appreciable above ca. 1 M, so much so, that non-exchange electrolyte concentrations may become higher than the concentration of exchange sites when the external solution becomes very concentrated or when nonaqueous solvents are employed. Water activities decrease, and therewith the resin swelling, so that the pressure–volume term becomes unimportant. Finally, dehydration of hydrated ions may set in, differing in magnitude for various ions. All these effects will cause changes in the relative selectivity of the resin, compared to the values in dilute solution. Complex formation is another important factor, which will be dealt with in the next two chapters. This section will deal with cases, where a simple ionic species is stable over the whole concentration range.

b. Cation Exchange

The simplest systems that have been studied involve ions with rare gas configuration, of the alkali metal, alkaline earth, lanthanide and actinide series. The data have been reported almost exclusively in terms of distribution coefficients rather than selectivity coefficients or constants, since trace quantities of the ions studied have been used, in conjunction with a range of concentration (usually 0.1 M to saturation) of the bulk electrolyte. A common co-ion (anion) is implied. Unfortunately, data from different authors, concerning different ions, are rarely available for resins of the same crosslinking, so that comparison is difficult. Figure 4D1 reproduces results[175] obtained in lithium perchlorate and perchloric acid solutions, and Table 4D1 compares some more results[176–181] for perchloric acid (cf. Appendix F). The abscissa in Figure 4D1 is the logarithm of the effective bulk cation activity, log a, which takes account of variations of activity coefficients of the bulk electrolyte in the external solution. The distribution coefficients are nearly the same in both electrolytes. The crossover of the curves for sodium and for caesium (presumably also rubidium) is noteworthy, as is also the markedly decreasing slope of the curve for sodium which shows the expected slope of -1 in dilute solutions. (The same effect is obtained when log $c_{ClO_4^-}$ is plotted as the abcissa). The curve for the heavy alkali metal is S-shaped, and does not conform to the expected slope even in dilute solutions. This may be connected with the strong affinity of this ion to the perchlorate ion in the aqueous solution only (reducing D) until invasion is sufficient to produce the same effect in the resin phase too.

Some results[177–179] for hydrochloric acid and 8X resin are shown in Figure 4D2 (cf. Appendix F). Here again some curves (notably those for sodium, calcium and strontium) show a marked increase in the distribution coefficient as the bulk electrolyte concentration increases. Results

TABLE 4D1. Distribution coefficients ($\log D$) of cations with Dowex–50 X 4.

Cation	HClO$_4$ Molarity			Reference
	1	3	9	
Na$^+$	0.67	0.41	0.12	176
Cs$^+$	0.75	0.27	−0.19	176
Mg^{2+}	1.20	0.60	0.60	177,178
UO$_2^{2+}$	1.40	1.05	3.85	181
Mn^{2+}	1.57	0.88	1.63	181
Ca^{2+}	1.87	1.45	2.48	176,177
Sr^{2+}	2.08	1.83	1.81	176
Hg^{2+}	2.08	1.58	1.30	181
Ba^{2+}	2.42	2.25	1.57	176
Fe^{3+}	1.60	0.30	1.80	180
Ga^{3+}	2.25	1.05	1.45	181
Tm^{3+}	2.40	1.20	2.55	176
Y^{3+}	2.4	1.4	3.0	181
Eu^{3+}	2.55	1.45	3.59	176
Pm^{3+}	2.61	1.57	3.63	176
Ce^{3+}	2.75	1.65	3.68	176
Am^{3+}	2.42	1.13	3.63	176
Sc^{3+}	2.70	1.76	6.2	181
La^{3+}	3.1	1.8	3.8	181
Th^{4+}	4.6	3.0	>7.0	181

for trivalent ions are available only for $4X$ and $12X$ resin[141,176,181,182] and show similar trends: the distribution coefficients for the actinides[182], Am^{3+} to Es^{3+}, show a continued decrease with increasing hydrochloric acid concentration, those for the lanthanides[41,176] turning up above ca. 6 M. Part of the effect is due to the measured quantity D_v (the volume-volume-to distribution coefficient) not being proportional to D (the volume-to-mass distribution coefficient), since as the acid concentration increases the resin shrinks and the proportionality constant ρ_s increases[182,183].

Information is also available concerning other electrolytes[175,176,184–187] with results generally confirming the above pattern, and some results with cation exchangers other than the polystyrene sulphonate used in the work discussed above, namely a synthetic zeolite[188], and a carboxylate resin[175] are also in line. The data for some chloride bulk electrolytes are shown in Figure 4D3 as separation factors S_{Na}^{Cs} (except for caesium chloride solutions where S_{Na}^{Rb} is shown) for resins of different crosslinking.

The results shown in Figures 4D1–3 and Table 4D1, may be summarized by a number of generalizations.

a) Ions that are moderately strongly hydrated (sodium, calcium strontium, scandium, lanthanides, thorium) show an increase in the distribution coefficients at high electrolyte concentrations.

b) Ions that are only slight hydrated (rubidium, caesium, barium, radium) or that are very strongly hydrated (beryllium, magnesium aluminium) show a continued decrease or levelling off at high concentrations.

c) Electrolytes containing a highly solvated ion are most effective in causing inversions or increasing distribution coefficients.

d) For monovalent ions the increase in the distribution coefficient is higher, the higher the crosslinking, while for divalent or trivalent ions it is higher the lower the crosslinking.

These observations can be partly explained by considerations of the hydration of ions, resin invasion, interaction of the counter ions with

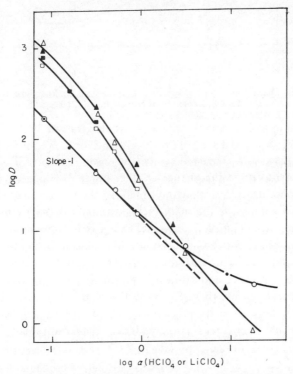

Figure 4D1. Distribution coefficients of cations in perchloric acid and lithium perchlorate solutions on Dowex-50 X 12. Empty symbols: $HClO_4$, filled symbols: $LiClO_4$; △ Cs, □ Rb and ○ Na.

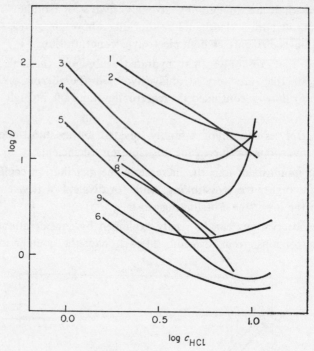

Figure 4D2. Distribution coefficients of cations in hydrochloric acid solutions on
Dowex–50 X 8. Labelling of curves: 1: Ra, 2: Ba, 3: Sr, 4: Ca, 5: Mg,
6: Be, 7: Cs, 8: Rb, 9: Na.

the fixed ions, and the swelling of the resin[175, 181, 187]. For a tight,
highly crosslinked resin, change in the internal water content with
external concentration is relatively small, so the change in hydration in
the external solution is the main factor, the internal solution behaving
as a concentrated solution in any case. Resins of low crosslinking, on the
other hand, are deswollen, and the internal solution becomes more and
more concentrated, along with the external solution. High concentrations
produced by invasion, and a low water content produce a low effective
dielectric constant, and these factors lead to extensive association in the
resin phase. This association becomes stronger the nearer the ions can
approach each other (that is, the smaller the radius of the ion), both for
usually slightly hydrated ions (Rb^+, Ba^{2+}, Ra^{2+}) or for dehydrated
ions (Na^+, Ca^{2+}, Sr^{2+}). The high concentration of co-ions obtained
by invasion leads to association with them in the low dielectric constant
medium in the resin, an effect which also increases the smaller the

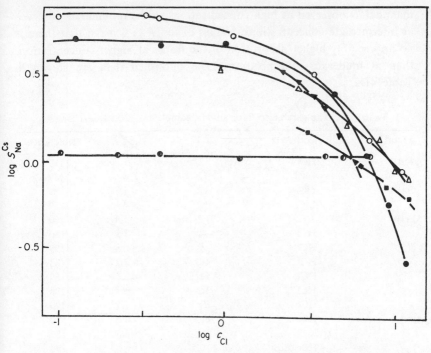

Figure 4D3. The relative distribution coefficients of Cs^+ and Na^+ in concentrated chloride solutions, plotted as $\log S_{Na}^{Cs}$ against $\log c_{Cl}$. Empty symbols: LiCl solutions, half-filled: CsCl solutions, filled: HCl solutions; crosslinking: □ 4X, △ 8X, ○ 12X and ▽ 16X.

effective radius. It is impossible at this stage to give an exact quantitative treatment of dehydration of the ions, although approximate qualitative expressions have been proposed[165].

c. Anion Absorption in Cation Exchangers

An interesting phenomenon observed in concentrated solutions is the sorption of a number of anionic metal complexes on cation exchangers. Along with the invasion of the resin with bulk electrolyte at high concentrations, and the large increases in the distribution of some trace cations noted above (e.g. alkaline and rare earth cations in perchloric acid), a large increase in the distribution coefficient of a number of trace anions, in particular anionic complexes, has been observed[183, 189-193]. It is difficult to compare the various systems studied, since it is impossible to express the increase in the distribution coefficient D by a parameter valid for all the cases. The most suitable parameter is R, the ratio of the maximal D observed at high concentrations to the minimal D observed at intermediate concentrations. Values of log $R \simeq 0$ mean that there is no appreciably higher absorption of the tracer at higher concentrations than at moderate concentrations. The published data are shown in Table 4D2.

TABLE 4D2. The parameter R for anionic complexes on cation exchangers.

Metal	Bulk electrolyte	X	Log R	Reference
Au^{III}	HCl	12	> 1	189
	LiCl	12	> 4	189
Ga^{III}	HCl	12	> 1.7	189
	LiCl	12	> 4	189
Fe^{III}	HCl	2	1.5	190
	HCl	4	3.5	180
	HCl	8a	> 0.8	199
	HCl	10b	> 2.2	183
	HCl	12	> 2.7	189
	HCl	16c	0	190
	HCl	d	> 1.7	190
	HCl	e	0	190
	LiCl	12	> 3.5	189
	HBr, LiBr	0.5	f	193
In^{III}	LiBr	10	1.8	191
	NaBr	10	> 1.2	191
	KBr	10	0	191
	MgBr$_2$	10	1.0	191
	LiI	10	1.5	191
	NaI	10	0.8	191
	KI	10	> 0.9	191
	MgI$_2$	10	0.8	191
Hg^{II}	LiCl	10	> 1.5	191
Tl^{III}	HCl	12	> 1.6	189
	LiCl	12	> 0.5	189
Sb^{V}	HCl	12	> 1.0	189
	LiCl	12	> 1.0	189
Zr^{IV}	HCl	10?	> 1.2	192

For the following systems log $R \lesssim 0$:
Ref. 189 ($X = 12$): Mn^{II}, Co^{II}, Zn^{II}, In^{III}, Bi^{III}, Sn^{IV} with HCl and LiCl.
Ref. 192 ($X = 10$): Cd^{II}, Zn^{II} with LiCl, LiBr, LiI; In^{III} with HCl, LiCl, KCl, MgCl$_2$
a. Also with 50% desulphonated resin, approximately same R.
b. Wofatit KS, presumably similar to Dowex-50 X10.
c. Very slow attainment of equilibrium, R quoted for case where equilibrium is more nearly achieved.
d. Chelating resin, Dowex A-1.
e. Carboxylic resin (polymethacrylate) Amberlite IRC–50.
f. Values for R for resin not given, but should be comparable with those reported for solutions of sodium dinonylnaphthalene sulphonate in toluene (cf. section 2.D and 8.A).

There seems to be no general unifying principle to the observations, except, perhaps, that a positive effect (high R) is more likely, the lower the negative charge on the predominant species (mononegative charge seems not to be a necessary prerequisite: $HgCl_4^{2-}$ and $TlCl_5^{2-}$ or $TlCl_6^{3-}$ seem to be the species absorbed for these elements). There are insufficient data to conclude about the role of crosslinking or the nature of the resin. Since the effect is so similar to the absorption of trace cations at high electrolyte concentrations, the opinion[191] that the best guide to the strong absorption of trace elements on cation exchangers at high concentrations is strong absorption on anion exchangers under similar conditions, is probably incorrect. The extractability of the complexes with appreciable R (Table 4D2) by ether, noted by Kraus, Michaelson and Nelson[189], is, however, significant. Spectral data[193] for the iron(III) bromide system show that it is the ion pairs $Li^+FeBr_4^-$ and $H^+FeBr_4^-$ which are involved in the large distribution ratios observed. This conclusion seems to be generally valid, i.e. also for other ion pairs, of similar nature.

d. Anion Exchange

The study of anion exchange in concentrated solutions is intimately connected with the study of metal complex formation, a subject treated in detail in Chapter 6. Here will be dissussed only the distribution of trace stable anions between a bulk electrolyte solution and an anion exchanger. It is not easy to find many stable anions, i.e. anions that do not dissociate or interact with cations in the solution, but remain as the same species over the widest concentration range. The perrhenate anion seems to be 'well-behaved', being a large anion of a strong acid. (It probably does form ion-pairs with large cations, such as potassium). Its distribution has been studied with a number of electrolytes, with results shown in Figure 4D4. It is instructive to compare the behaviour of other stable anions with that of perrhenate. In addition to the halide ions, also those of tetrahalomercurate, tetrachloroaurate, tetracyanozincate and sulphate, among the anions and anionic complexes studied, may be considered as stable anions, at least over a certain range of concentrations. As seen in Table 4D3, over this range there is an approximately constant ratio between the distribution coefficients of these anions and perrhenate, independent of bulk electrolyte concentration. Beyond this range some anions show their individual peculiarities.

The distribution coefficient of a tracer anion has been related by

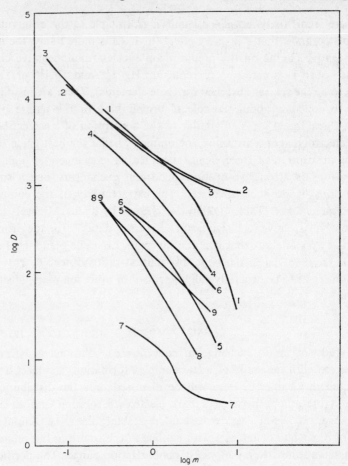

Figure 4D4. Distribution coefficients of tracer perrhenate ions with an anion exchanger. Labelling of curves, for solutions of concentration m from
which the perrhenate is sorbed: 1 HCl, 2 LiCl, 3 NaCl, 4 $(C_2H_5)_4$NCl,
5 HBr, 6 LiBr, 7 LiI, 8 HNO_3, 9 $LiNO_3$.

Marcus and Coryell[103] to the effective activity of the bulk anion in
solution, a, and to the invasion of the resin, using an invasion function
$_rF_a$. Later Marcus and Maydan[165], and Marcus and Eliezer[106], employed
the more general effective activity of the anion in the resin, $\bar{a}$, defined in
equation (4B28). These authors have shown that the invasion function

$$F_B = b^{-1}\log D_B + \log a = b^{-1}\log D_{B(a=1)} + \log\bar{a} - \log\bar{a}_{(a=1)} \quad (4D1)$$

is a function of the bulk electrolyte only, and not the anion B^b. Here

TABLE 4D3. Anion exchange distribution of anions compared to ReO_4^-.

Electrolyte	Anion B^{b-}	Range studied m	Range of constancy m	$\log D_{ReO_4^-}$ $-1/b \log D_B$	Reference
HCl	Br^-	0.1 –13	0.1–4	1.90 ± 0.05	194,201
	Br^-	0.1 –16	0.1–4	1.90 ± 0.05	202
	Br^-	0.1 –10	0.1–4	2.20 ± 0.06	195,198
	$AuCl_4^-$	1 –13	1.6–13	-1.15 ± 0.09	194
LiCl	Br^-	0.1 –20	0.1–13	2.14 ± 0.05	194,201
	Br^-	0.1 –20	0.1–13	2.14 ± 0.05	202
	Br^-	0.1 –20	0.1–13	2.17 ± 0.03	200
	Br^-	0.1 –13	0.1–13	2.23 ± 0.03	195,198
	Br^-	0.3 –13	0.3–13	2.82 ± 0.03	205a
	Br^-	0.01–3	0.1–3	2.19 ± 0.04	207b
	$AuCl_4^-$	0.1 –10	0.1–1.5	-2.90 ± 0.03	194
	SO_4^{2-}	0.1 –0.35	0.1–0.35	3.72 ± 0.02	205a
	$HgCl_4^{2-}$	0.3 –16	1.5–12	0.79 ± 0.10	106
	$Zn(CN)_4^{2-}$	1 –11	1 –11	$0.99 + 0.03$	207b
NaCl	$.Br^-$	0.3 –3	0.3–2	1.33 ± 0.06	195,198
	$HgCl_4^{2-}$	0.3 –5	0.8–5	0.94 ± 0.06	106
HBr	Cl^-	1.5 –5	1.5–5	1.23 ± 0.09	195
	$HgBr_4^{2-}$	0.5 –8	0.5–8	0.20 ± 0.05	204
LiBr	Cl^-	1 –5	1 –3	1.53 ± 0.03	195
	$HgBr_4^{2-}$	0.3 –8	0.8–8	0.27 ± 0.07	106

a. From R_f on SB-2 paper loaded with Amberlite IR–400 ($\sim$ Dowex-1 X 10).
b. Amberlite IR–410 ($\sim$ Dowex-2 X 10).

the first and third terms on the right-hand side pertain to an effective bulk anion activity of unity, and $\log \bar{a} - \log \bar{a}_{(a=1)} = {}_r F_a$. The data shown in Table 4D3 help to validate the supposition that $F_{ReO_4^-}$ (Figure 4D4) can be used as a universal invasion function, at least for those electrolytes where this has been tested.

A final test for the usefulness of $F_{ReO_4^-}$ as an invasion function is a direct comparison with $\bar{a}$, wherever it has been measured directly, since from (4D1) $F_{ReO_4^-} - \log \bar{a} = \log D_{ReO_4^- (a=1)} - \log \bar{a}_{(a=1)} = constant$, independent of a. Such comparisons are shown in Figure 4D5.

In addition to the data on chloride and bromide solutions, collected in Table 4D3[106,194–208], for which $F_{ReO_4^-}$ serves as an invasion function, there are some data on other electrolytes, where comparisons cannot be made, but which can serve as preliminary invasion functions, permitting the interpretation of distribution data on metal complexes (Chapter 6).

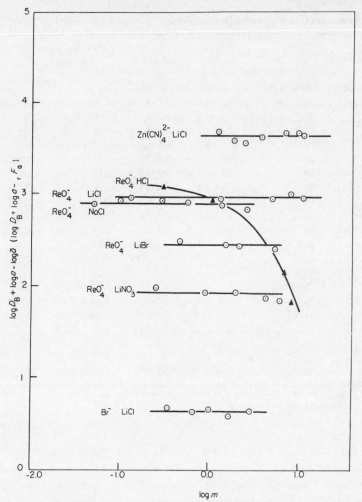

Figure 4D5. Comparison of D_B with $\bar{a}$ for various electrolytes for an anion exchanger. Plotted is $\log D_B + \log a - \log \bar{a}$ or $\log D_B + \log a - {}_rF_a$, as the case may be, against the electrolyte molality, $\log m$, for the trace anion B and the electrolyte designated on each curve.

These are the data on sodium bromide and iodide[106], (using iodide, perrhenate and tetrahalomercurate tracers), ammonium sulphate[194] (using chromate tracer), potassium, caesium and tetraalkylammonium chloride, nitric acid and lithium nitrate[195] (using perrhenate tracer), potassium carbonate[196] (using sulphate tracer), hydrofluoric acid[197] (using tetrafluoroborate tracer), potassium hydroxide and tetramethylammonium chloride (using halide tracers)[199] and potassium thiocyanate[206] and sodium, potassium and ammonium thiocyanates[208] (using iodide tracer, and comparing the results, successfully, with the directly measured invasion). More comparison of different tracer stable anions on these and other electrolytes would be very useful.

Tetrachloroaurate(III) is an anion that shows non-regular behaviour in lithium chloride solutions at high concentrations[194]. Whereas in hydrochloric acid, where non-regular behaviour is also shown, one may

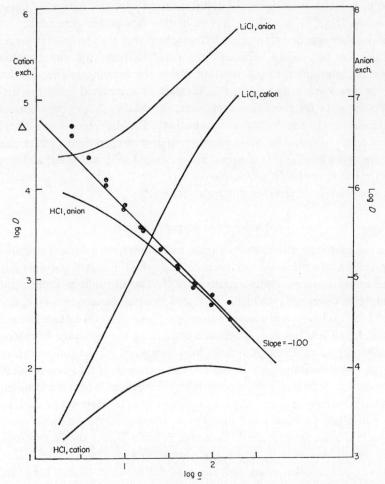

Figure 4D6. Distribution coefficients of tracer $AuCl_4^-$ on a cation exchanger and an anion exchanger from concentrated HCl and LiCl. The points are $\varDelta = \log D_{anion} - \log D_{cation}$ for the two electrolytes, falling near a straight line drawn with a slope of -1.00.

argue that association with hydrogen ions, or some other 'HCl effect'[195] occurs, such anomaly should be absent in lithium chloride solutions. Tetrachloroaurate is, however, the only stable anion for which distribution data were published also for a cation exchanger[189], comparable in structure to the anion exchanger used for the experiments discussed above. The data are shown in Figure 4D6, where also are shown the differences Δ between the curves for the two exchangers for each electrolyte, lithium chloride and hydrochloric acid. It is seen that the differences are nearly the same for both solutions, and that the difference curve $\Delta = \log D_{anion} - \log D_{cation}$ has an inverse first-power dependency on $\log a$. This can be interpreted by assuming that the cation exchanger results for this anion represent the invasion from the concentrated chloride solutions into the solution inside the exchangers (irrespective of the fixed ions), and when this invasion is subtracted from the distribution data for the anion exchanger, the dilute solution behaviour is followed, up to the highest concentrations. The difference curve obeys the correct slope only at sufficiently high concentrations, where the nature of the fixed ions is of minor importance. The invading tetrachloroaurate(III) is probably ion-paired with H^+ or Li^+ ions, as observed[193] for the analogous tetrabromoferrate(III) system.

e. Exchange in Mixed Solvents

The ion exchange distribution of trace metal ions from solutions containing complex-forming anions (particularly acids) in mixed solvents has been studied extensively in recent years[209]. The sorption of such metal complexes on anion exchangers is treated in detail in section 6.D. Combined ion exchange and solvent extraction[210] seems to take place in such cases. Much less has been published concerning the exchange of macro concentrations of simple ions from mixed solvents. As the concentration of the organic component in the solvent increases, the degree of ionic dissociation decreases, until a point is reached beyond which the Donnan potential, acting on ions only, is no longer able to exclude the co-ions even in dilute solutions, and invasion of the electrolyte becomes more important than stoichiometric exchange of ions. Under such conditions it is more convenient to regard the process as a partition of the solute between the swollen resin and the external solution. Since, however, water is preferred in the resin over the nonaqueous component of the mixed solvent (section 4.B.d.), the internal solution in the swollen resin produces higher dissociation of the electrolyte, and therefore ionic interactions, than does the external solution. A competition between the preference of ions for the more solvating, more aqueous, phase in the resin, and the exclusion of co-ions by the Donnan potential is then observed.

The selectivity of ions on ion exchangers in mixed solvents depends on similar factors as in aqueous solutions, except for the special effects introduced by selective solvation and a lower dielectric constant. The generalization that selectivity increases as the swelling decreases (section 4.C.b) seems to hold. Swelling is usually less in organic or mixed solvents than in water (section 4.B.d), and decreases as the organic fraction in the mixture increases. Application of the Born equation to transfer of ions from water to a mixed solvent leads to the equation[211]

$$\log(K_{r(mixed)}/K_{r(water)}) = (e^2 N/RT)(z_A/r_{hA} - z_B/r_{hB})(1/\varepsilon_{mixed} - 1/\varepsilon_W)$$
$$- \sum \Delta A/RT \tag{4D2}$$

provided the organic-solvent content is sufficiently small, so that the resin is predominantly swollen by water, i.e. is the same second medium for both solvents. The last term takes care of specific solvation effects of the two ions A and B, but the first term on the right fails to recognize the differences in the solvated radii r_h in the two media.

Still, for a sulphonate cation exchanger, $\log Q_H^M$ was found to increase linearly with the reciprocal of the dielectric constant of several mixed solvents, as required by (4D2), down to $\varepsilon = 55$. For a given composition, Q increases in the order methanol, ethanol and acetone[212]. The selectivity increases for each solvent with the organic content[212, 216] up to a certain maximum[215]. Many reversals of the selectivity order in aqueous solutions are however observed in mixed solvents. Exchanges involving the hydrogen ion show reversals, as well as the silver–potassium (above 30% ethanol)[219], the potassium–caesium and the sodium–ammonium (in aqueous methanol)[215] exchanges. On a carboxylate exchanger reversals are more pronounced, occurring in acetone and alcohols[217] and ethanolamine[218] for the exchange of alkali metal ions against the ammonium ion[217], and Q_{Na}^{Ba} is found to decrease as the ethanol content of a mixture increases[232]. The selectivity was also found to increase with the equivalent fraction of the preferred ion for the silver–hydrogen exchange in aqueous ethanol or dioxane, i.e. in the 'abnormal' manner (section 4.C.b)[85].

With an anion exchanger (Dowex-1) the selectivity coefficients $Q_{Cl}^{NO_3}$, $Q_{Br}^{NO_3}$ and $Q_{NO_3}^{I}$ decreased in methanol, ethanol and acetone, the effect being more pronounced the lower the crosslinking, i.e. the higher the total swelling[220]. This is contrary to what would be expected for the electrostatic (polarization) effects with changing dielectric constant. There is, unfortunately, little additional information on the selectivity of simple anions for anion exchangers in mixed solvents. A considerable amount of information has, on the other hand, been published on the sorption of metal complexes on anion exchangers (e.g. ref. 209). This subject is discussed in detail in section 6.D.

This section ought to be concluded with the comment that a great deal of fundamental information on the ion exchange properties of resins in mixed and nonaqueous solvents is wanted.

E. REFERENCES

1. B. A. Adams and E. J. Holmes, *J. Soc. Chem. Ind. (London)*, **54**, 1T (1935).
2. G. F. D'Alelio, *U.S. Pat.*, 2,366,007 (1942).
3. W. C. Bauman and R. McKellar (The Dow Chemical Co.),U.S. Pat., 2,614,099 (1952); G. W. Bodamer (Rohm and Haas Co.), U.S. Pat., 2,597,440 (1952).
4. S. A. Rice and M. Nagasawa, *Polyelectrolyte Solutions*, Academic Press, Inc., New York, 1961.
5. H. P. Gregor, K. M. Held and J. Bellin, *Anal. Chem.*, **23**, 620 (1951).
6. W. K. Lowen, R. W. Stoenner, W. J. Argersinger, Jr., A. W. Davidson and D. N. Hume, *J. Am. Chem. Soc.*, **73**, 2666 (1951).
7. E. Högfeldt, *Arkiv Kemi*, **13**, 491 (1959).
8. O. D. Bonner and R. R. Pruett, *Z. Physik. Chem. (Frankfurt)*, **25**, 75 (1960).
9. M. G. Suryaraman, *Ph.D. Thesis*, Univ. Colorado, 1962; M. G. Suryaraman and H. F. Walton, *Science*, **131**, 829 (1960).
10. W. D. Moseley, Jr. and D. H. Freeman, *J. Phys. Chem.*, **67**, 2225 (1963); G. E. Boyd and K. Bunzl, *J. Am. Chem. Soc.*, **89**, 1776 (1967).
11. E. Glueckauf and R. E. Watts, *Nature*, **191**, 904 (1961); *Proc. Roy. Soc. (London)*, **A268**, 339 (1962).
12. R. Kunin, E. Metzner and N. Bostnick, *J. Am. Chem. Soc.*, **84**, 305 (1962).
13. G. E. Boyd, B. A. Soldano and O. D. Bonner, *J. Phys Chem.*, **58**, 456 (1954).
14. S. Lindenbaum, C. F. Jumper and G. E. Boyd, *J. Phys. Chem.*, **63**, 1924 (1959).
15. T. R. E. Kressman and F. L. Tye (Permutit Co. Ltd.), *Brit. Pat.*, 726,918 (1955).
16. G. F. D'Alelio (General Electric Co.), *U.S. Pat.*, 2,340,111 (1944).
17. K. W. Pepper, H. M. Paisley and M. A. Young, *J. Chem. Soc.*, 4097 (1953).

18. K. Berger, *J. Pract. Chem.*, **12**, 146 (1961).
19. A. Heller, Y. Marcus and I. Eliezer, *J. Chem. Soc.*, 1579 (1963).
20. T. R. E. Kressman and J. A. Kitchener, *J. Chem. Soc.*, 1190 (1949); Y. Ohtsuka and S. Umezawa, *J. Chem. Soc. Japan, Ind. Chem. Sect.*, **55**, 230 (1952); *Chem. Abstr.*, **47**, 10766 (1953); B. A. Adams and E. L. Holmes, *Brit. Pat.* 450,309 (1936).
21. Chemiche Farbenfabrik Bundheim, *Brit. Pat*, 727,476 (1955).
22. G. F. D'Alelio (Koppers Co.), *U.S. Pat*, 2,697,079 (1954).
23. J. Kennedy, F. S. Lane and B. K. Robinson, *J. Appl. Chem.*, **8**, 459 (1959).
24. K. W. Pepper and D. K. Hale in *Ion Exchange and its Applications*, Soc. Chem. Ind., London, 1955, p. 13; L. R. Morris and coworkers, *J. Am. Chem. Soc.*, **81**, 377 (1959).
25. A. Schwartz, *Ph.D. Thesis*, Technion, Haifa, 1962.
26. A. Skogseid, *Dissertation*, Oslo, 1948; *U.S. Pat.*, 2,592,350 (1952).
27. J. Kennedy and H. Small, *Brit. Rept.* C/R—1668 (1957); J. Kennedy, *J. Appl. Chem.*, **9**, 26 (1959).
28. J. Kennedy, F. A. Burford and P. G. Sammes, *J. Inorg. Nucl. Chem.*, **14**, 114 (1960).
29. M. Ezrin and H. G. Cassidy, *J. Am. Chem Soc.*, **78**, 2525 (1956); L. Luttiger and H. G. Cassidy, *J. Polymer. Sci.*, **20**, 417 (1956); **22**, 271 (1956).
30. G. Manecke and C. Bahr, *Z. Electrochem.*, **62**, 311 (1958).
31. G. Manecke, *Z. Electrochem.*, **57**, 189 (1953); **58**, 363, 369 (1954).
32. M. Ezrin, H. G. Cassidy and I. H. Updegraff, *J. Am. Chem. Soc.*, **75**, 1610 (1953).
33. B. Sansoni, *Naturwissenschaften*, **41**, 212 (1954).
34. H. P. Gregor, D. Dolar and G. K. Hoeschele, *J. Am. Chem. Soc.*, **77**, 3675 (1955).
35. H. S. Thompson, *J. Roy. Agr. Soc. Engl.*, **11**, 68 (1950); J. T. Way, *J. Roy. Agr. Soc. Engl.*, **11**, 313 (1950).
36. R. Gans, *Jahrb. Preuss. Geol. Landesanstalt (Berlin)*, **26**, 179 (1905); **27**, 63 (1906).
37. R. M. Barrer, *J. Chem. Soc.*, 127, 2158 (1948); 2342 (1950); R. M. Barrer and D. M. McLeod, *Trans. Faraday Soc.*, **51**, 1290 (1955).
38. R. M. Barrer and J. D. Falconer, *Proc. Roy. Soc.*, *(London)*, **A236**, 227 (1956); R. M. Barrer, *Proc. Chem. Soc.*, 99 (1958).
39. S. Van R. Smit, J. J. Jacobs and W. Robb, *J. Inorg. Nucl. Chem.*, **12**, 95,102 (1959).
40. J. Kitil and V. Kourim, *J. Inorg. Nucl. Chem.*, **12**, 367 (1960).
41. C. B. Amphlett, L. A. McDonald and M. J. Redman, *Chem. Ind.(London)*, 1314 (1956); *J. Inorg. Nucl. Chem.*, **6**, 220, (1958).
42. K. A. Kraus and H. O. Phillips, *J. Am. Chem. Soc.*, **78**, 694 (1956).
43. E. Merz, *Z. Elektrochem.*, **63**, 288 (1959).
44. G. Alberti and A. Cante, *J. Chromatog.*, **5**, 244 (1961).
45. G. H. Nancollas, *Brit. Rept.* AERE C/R 2755 (1959).
46. L. Baetsle and J. Pelsmaekers, *J. Inorg. Nucl. Chem.*, **21**, 124 (1961).
47. K. A. Kraus and H. O. Phillips, *J. Am. Chem. Soc.*, **78**, 249 (1956).

48. C. B. Amphlett, L. A. McDonald and M. J. Redman, *Chem. Ind. (London),* 365 (1957); *J. Inorg. Nucl. Chem.,* **6,** 236 (1958).

49. I. Schonfeld, *Israel AEC Rept.* IA-720 (1961).

50. E. J. Duwell and J. W. Sheppard, *J. Phys. Chem.,* **63,** 2044 (1959).

51. Y. Marcus and D. Maydan, *J. Phys. Chem.,* **67,** 983 (1963).

52. R. H. Stokes and R. A. Robinson, *Trans. Faraday Soc.,* **53,** 301 (1957).

53. F. Helfferich, *Ion Exchange,* McGraw Hill Book Co., Inc., New York, 1962.

54. G. E. Boyd, A. W. Adamson and L. S. Myers, Jr., *J. Am. Chem. Soc.,* **69,** 2836 (1947).

55. C. Heitner-Wirguin and V. Urbach, *J. Phys. Chem.,* **69,** 3400 (1965).

56. A. Schwartz, J. A. Marinsky and K. S. Spiegler, *J. Phys. Chem.,* **68,** 918 (1964); C. Heitner-Wirguin and G. Markovitz, *J. Phys. Chem.,* **67,** 2263 (1963).

57. R. Turse and W. Rieman, III, *J. Phys. Chem.,* **65,** 1821 (1961), as corrected by A. Varon and W. Rieman, III. *J. Phys. Chem.,* **68,** 2716 (1964).

58. F. Helfferich, *Discussions Faraday Soc.,* **21,** 83 (1956); F. Helfferich and M. S. Plesset, *J. Chem. Phys.,* **28,** 418 (1956).

59. G. E. Boyd and B. A. Soldano, *J. Am. Chem. Soc.,* **75,** 6091, 6105 (1953).

60. B. A. Soldano and G. E. Boyd, *J. Am. Chem. Soc.,* **75,** 6107 (1953).

61. B. A. Soldano and G. E. Boyd, *J. Am. Chem.,* **75,** 6099 (1953).

62. F. G. Donnan, *Z. Electrochem.,* **17,** 572 (1911); *Z. Physik. Chem. (Leipzig),* **A168,** 369 (1934).

63. G. E. Boyd and B. A. Soldano, *Z. Elektrochem.,* **57,** 162 (1953).

64. A. Chatterjee and J. A. Marinsky, *J. Phys. Chem.,* **67,** 41 (1963); J. A. Marinsky and A. Chatterjee, *J. Phys. Chem.,* **67,** 47 (1963); Z. Alexandrowicz, *J. Polymer Sci.,* **43,** 325 (1960).

65. O. D. Bonner, G. D. Easterling, D. L. West and V. F. Holland, *J. Am. Chem. Soc.,* **77,** 242 (1955); O. D. Bonner, V. F. Holland and L. L. Smith, *J. Phys. Chem.,* **60,** 1102 (1956); O. D. Bonner and O. C. Rogers, *J. Phys. Chem.,* **64,** 1499 (1960); O. D. Bonner and W. C. Rampey *J. Phys. Chem.,* **65,** 1602 (1961); O. D. Bonner and J. R. Overtone, *J. Phys. Chem.,* **67,** 1035 (1963).

66. K. W. Pepper, D. Reichenberg and D. K. Hale, *J. Chem. Soc.,* 3129 (1952); K. W. Pepper and D. Reichenberg, *Z. Elecktrochem.,* **57,** 183 (1953).

67. E. Högfeldt, *Acta Chem. Scand.,* **12,** 182 (1958).

68. H. P. Gregor, D. Nobel and M. H. Gottlieb, *J. Phys. Chem.,* **59,** 10 (1955).

69. D. H. Freeman and G. Scatchard, *J. Phys. Chem.,* **69,** 70 (1965).

70. H. P. Gregor, *J. Am. Chem. Soc.,* **70,** 1293 (1948); **73,** 642 (1951).

71. G. W. Bodamer and R. Kunin, *Ind. Eng. Chem.,* **45,** 2577 (1953).

72. W. R. Heumann and F. D. Rochon, *Anal. Chem.,* **38,** 638 (1966); cf. F. X. Pollio, *Anal. Chem.,* **35,** 2164 (1962).

73. E. Glueckauf and G. P. Kitt, *Proc. Roy. Soc. (London),* **A228,** 322 (1955).

74. V. S. Redhina and J. A. Kitchener, *Trans. Faraday Soc.,* **59,** 515 (1963).

75. H. Ohtaki, K. Gonda and H. Kakihana, *Bull. Chem. Soc. Japan*, **34**, 293 (1961); *Ber. Bunsen Ges. Phys. Chem.*, **67**, 87 (1963).
76. K. W. Kreiselmaier, *Ph.D. Thesis*, M.I.T., 1961.
77. J. E. Gordon, *J. Phys. Chem.*, **66**, 1150 (1962).
78. R. H. Dinius, M. T. Emerson and G. R. Choppin, *J. Phys. Chem.*, **67**, 1178 (1963).
79. G. Zundel, H. Noller and G. M. Schwab, *Z. Elektrochem.*, **66**, 122, 129 (1962).
80. J. Penciner, Y. Marcus and I. Eliezer, unpublished results, 1963; Y. Marcus and E. Eyal, unpublished results, 1967.
81. O. D. Bonner, *J. Chem. Educ.*, **34**, 174 (1957).
82. C. W. Davies and B. D. R. Owen, *J. Chem. Soc.*, 1676, 1681 (1956).
83. E. Sjostrom, L. Nykänen and P. Laitinen, *Acta Chem. Scand.*, **16**, 392 (1962).
84. F. S. Chance, Jr., G. E. Boyd and H. J. Garber, *Ind. Eng. Chem.*, **45**, 1671 (1953).
85. O. D. Bonner and J. C. Moorefield, *J. Phys. Chem.*, **58**, 555 (1954).
86. C. W. Davies and A. Narebska, *J. Chem. Soc.*, 4169 (1964).
87. D. Reichenberg and W. F. Wall, *J. Chem. Soc.*, 3364 (1956).
88. R. A. Stokes and H. F. Walton, *J. Am. Chem. Soc.*, **76**, 3327 (1954).
89. F. Helfferich, *Nature*, **189**, 1001 (1961).
90. M. Ziegler, *Angew. Chem.*, **71**, 283 (1959).
91. J. Aveston and D. A. Everest, *Chem. Ind.* (*London*), 1238 (1957).
92. L. I. Katzin and E. Gebert, *J. Am. Chem. Soc.*, **75**, 801 (1953).
93. H. Kakihana, T. Nomura and H. Ohtaki, *Bull. Tokyo Inst. Techn. Ser. B*, No. 1, 14 (1960).
94. W. C. Bauman and J. Eichhorn, *J. Am. Chem. Soc.*, **69**, 2830 (1947).
95. G. E. Boyd, J. Schubert and A. W. Adamson, *J. Am. Chem. Soc.*, **69**, 2818 (1947).
96. E. Glueckauf, *Proc. Roy. Soc.* (*London*), **A214**, 207 (1952).
97. K. F. Bonhoeffer, L. Miller and U. Schindewolf, *Z. Physik. Chem.* (*Leipzig*), **198**, 279, 281 (1951); G. Manecke and K. F. Bonhoeffer, *Z. Elektrochem.*, **55**, 475 (1951).
98. K. A. Kraus and F. Nelson, *J. Am. Chem. Soc.*, **80**, 4154 (1958).
99. J. Dannon, *J. Phys. Chem.*, **65**, 2039 (1961).
100. D. H. Freeman, *J. Phys. Chem.*, **64**, 1048 (1960).
101. G. Scatchard, personal communications to K. A. Kraus[98] and D. H. Freeman[100].
102. W. A. Plateck and J. A. Marinsky, *J. Am. Chem. Soc.*, **83**, 2583 (1961).
103. Y. Marcus and C. D. Coryell, *Bull. Res. Council Israel*, **8A**, 1 (1959).
104. D. Maydan, *Ph.D. Thesis*, Jerusalem, 1962.
105. Y. Marcus and D. Maydan, *J. Phys. Chem.*, **67**, 979 (1963).
106. Y. Marcus and I. Eliezer, *J. Inorg. Nucl. Chem.*, **25**, 867 (1963).
107. K. A. Kraus and G. E. Moore, *J. Am. Chem. Soc.*, **75**, 1457 (1953).
108. G. Scatchard and N. J. Anderson, *J. Phys. Chem.*, **65**, 1536 (1961).
109. Y. Marcus and M. Givon, *J. Phys. Chem.*, **68**, 2230 (1964).
110. F. L. Tye, *J. Chem. Soc.*, 4784 (1961).

111. E. Glueckauf, *Proc. Roy. Soc.* (*London*), **A268,** 350 (1962).
112. G. E. Myers and G. E. Boyd, *J. Phys. Chem.*, **60,** 521 (1956).
113. G. E. Boyd, S. Lindenbaum and G. E. Myers, *J. Phys. Chem.*, **65,** 577 (1961).
114. D. Reichenberg and D. J. McCauley, *J. Chem. Soc.*, 2741 (1955).
115. O. D. Bonner and L. L. Smith, *J. Phys. Chem.*, **61,** 326 (1957).
116. O. D. Bonner and W. H. Payne, *J. Phys. Chem.*, **58,** 183 (1954).
117. B. A. Soldano and Q. V. Larson, *J. Am. Chem. Soc.*, **77,** 1331 (1955); B. A. Soldano and D. Chestnut, *J. Am. Chem. Soc.*, **77,** 1334 (1955); B. A. Soldano, Q. V. Larson and G. E. Myers, *J. Am. Chem. Soc.*, **77,** 1338 (1955).
118. J. A. Kitchener in *Modern Aspects of Electrochemistry* (Ed. J. O'M. Bockris), No. 2, Butterworths Sci. Publ., London, 1959, p. 119.
119. O. D. Bonner and F. L. Livingstone, *J. Phys. Chem.*, **60,** 530 (1956).
120. O. D. Bonner, C. F. Jumper and O. C. Rogers, *J. Phys. Chem.*, **62,** 250 (1958).
121. W. F. Graydon, private communication from D. Reichenberg, 1961.
122. A. W. Davidson and W. J. Argersinger, Jr., *Ann. N.Y. Acad. Sci.*, **57,** 105 (1953).
123. E. H. Cruickshank and P. Mears, *Trans. Faraday Soc.*, **53,** 1289, 1299 (1957).
124. D. S. Flett and P. Mears, *J. Phys. Chem.*, **70,** 1841 (1966); *Trans. Faraday Soc.*, **62,** 1469 (1966).
125. O. D. Bonner and J. R. Overtone, *J. Phys. Chem.*, **65,** 1599 (1961).
126. F. Vaslow and G. E. Boyd, *J. Phys. Chem.*, **70,** 2295, 2507 (1966).
127. G. E. Boyd, F. Vaslow and S. Lindenbaum, *J. Phys. Chem.*, **68,** 590 (1964).
128. S. Lindenbaum and G. E. Boyd, *J. Phys, Chem.*, **69,** 2374 (1965).
129. K. A. Kraus and R. J. Raridon, *J. Phys. Chem.*, **63,** 1901 (1959).
130. O. D. Bonner, G. Dickel and H. Brummer, *Z. Phys. Chem.* (*Frankfurt*), **25,** 81 (1960).
131. H. Ti. Tien, *J. Phys. Chem.*, **68,** 1021 (1964).
132. K. E. Becker, S. Lindenbaum and G. E. Boyd, *J. Phys. Chem.*, **70,** 3834 (1966).
133. D. K. Hale, D. I. Packham and K. W. Pepper, *J. Chem. Soc.*, 844 (1953).
134. F. W. Strelow, *Anal. Chem.*, **32,** 1185 (1960).
135. H. P. Gregor, J. Belle and R. A. Marcus, *J. Am. Chem. Soc.*, **77,** 2713 (1955).
136. C. D. Coryell, D. H. Freeman and E. Yellin, *Abstr.* 135 *Mtg. Am. Chem. Soc.*, 1959, p. 44R, paper 105; *Progr. Rept. Lab. Nucl. Sci. M.I.T.*, 31 May 1958, p. 21.
137. B. Chu, D. C. Whitney and R. M. Diamond, *J. Inorg. Nucl. Chem.*, **24,** 1405 (1962).
138. F. Helfferich, reference 53, p. 158.
139. J. Aveston, D. A. Everest and R. A. Wells, *J. Chem. Soc.*, 231 (1958).

140. C. B. Monk, *Electrolyte Dissociation*, Academic Press, London, 1961, p. 270.
141. G. R. Choppin and A. Chetham-Strode, *J. Inorg. Nucl. Chem.*, **15**, 377 (1960).
142. B. Baysal, unpublished results, M.I.T., 1958.
143. J. F. Harvey, J. P. Redfern and J. E. Salmon, *Trans. Faraday Soc.*, **62**, 198 (1966).
144. R. E. Macker and D. B. Luten, cf. ref. 53, p. 148–50.
145. Dow Chem. Corp., *Brit. Pat.*, 929,356 (1963); cf. *Chem. Abstr.*, **59**, 8380 e (1963).
146. R. E. Anderson, W. C. Bauman and D. F. Harrington, *Ind. Eng. Chem.*, **47**, 1620 (1955).
147. K. A. Kraus, F. Nelson and J. F. Baxter, *J. Am. Chem. Soc.*, **75**, 2768 (1953).
148. F. Nelson and K. A. Kraus, *J. Am. Chem. Soc.*, **77**, 329 (1955).
149. D. A. Everest, *Research*, **88**, 58,59 (1955).
150. E. Ekedahl, E. Högfeldt and L. G. Sillèn, *Acta Chem. Scand.*, **4**, 556, 828, 1471 (1950).
151. W. J. Argersinger, Jr., A. W. Davidson and O. D. Bonner, *Trans. Kansas Acad. Sci.*, **53**, 404 (1950).
152. D. H. Freeman, *Ph. D. Thesis*, M.I.T., 1957.
153. E. Högfeldt, *Arkiv Kemi*, **5**, 147 (1952).
154. G. L. Gains, Jr. and H. C. Thomas, *J. Chem. Phys.*, **21**, 714 (1953).
155. L. W. Holm, *Arkiv Kemi*, **10**, 151 (1956).
156. E. A. Guggenheim, *Trans. Faraday Soc.*, **33**, 151 (1937).
157. E. Glueckauf, *Proc. Roy. Soc. (London)*, **A214**, 207 (1952).
158. R. A. Horne, R. A. Courant, B. R. Myers and J. H. B. George, *J. Phys. Chem.*, **68**, 2574 (1964).
159. G. E. Boyd, A. Schwartz and S. Lindenbaum, *J. Phys. Chem.*, **70**, 821 (1966); G. E. Boyd, J. W. Chase and F. Vaslow, *J. Phys. Chem.*, **71**, 573 (1967); S. Lindenbaum and G. E. Boyd, *J. Phys. Chem.*, **71**, 581 (1967).
160. A. Schwartz and G. E. Boyd, *J. Phys. Chem.*, **69**, 4268 (1965).
161. A. Katchalsky and S. Lifson, *J. Polymer Sci.*, **11**, 409 (1953); A. Katchalsky, *Progr. Biophys.*, **4**, 1 (1954); I. Michaeli and A. Katchalsky, *J. Polymer Sci.*, **23**, 683 (1957).
162. P. J. Flory, *Principles of Polymer Chemistry*, Cornell Univ. Press. Ithaca, New York 1953.
163. S. A. Rice and F. E. Harris, *Z. Physik. Chem. (Frankfurt)*, **8**, 207 (1956).
164. J. L. Pauley, *J. Am. Chem. Soc.*, **76**, 1422 (1954).
165. Y. Marcus and D. Maydan, *J. Phys. Chem.*, **67**, 983 (1963).
166. S. B. Sachs, *M.Sc. Thesis*, Technion, Haifa, 1961; S. B. Sachs and K. S. Spiegler, *J. Phys. Chem.*, **68**, 1214 (1964).
167. S. Takeshita, *Nippon Kagaku Zasshi.*, **84**, 560 (1963).
168. J. L. Bregman and Y. Murata, *J. Am. Chem. Soc.*, **74**, 1857 (1952); J. L. Bregman, *Ann. N.Y. Acad. Sci.*, **57**, 125 (1953).

169. J. Kennedy, J. Marriot and V. J. Wheeler, *J. Inorg. Nucl. Chem.*, **22**, 269 (1961).
170. R. A. Robinson and H. S. Harned, *Chem. Rev.*, **28**, 419 (1941).
171. R. M. Diamond, *J. Am. Chem. Soc.*, **80**, 4808 (1958).
172. R. W. Gurney, *Ionic Processes in Solution*, McGraw-Hill Book Co., New York, 1953.
173. L. W. Holm, *Arkiv Kemi*, **17**, 461 (1963).
174. R. M. Diamond, *J. Phys. Chem.*, **67**, 2513 (1963).
175. D. C. Whitney and R. M. Diamond, *Inorg. Chem.*, **2**, 1284 (1963).
176. R. H. Dinius, *Ph.D. Thesis*, Florida State Univ., 1960; G. R. Choppin and R. H. Dinius, *Inorg. Chem.*, **1**, 140 (1962).
177. C. K. Mann, *Anal. Chem.*, **32**, 67 (1960).
178. F. Nelson, J. H. Holloway and K. A. Kraus, *J. Chromatog.*, **11**, 258 (1963).
179. R. M. Diamond, *J. Am. Chem. Soc.*, **77**, 2978 (1955).
180. S. Lindenbaum and G. E. Boyd, *USAEC Rept.* ORNL 3320 (1962), p. 77.
181. F. Nelson, T. Murase and K. A. Kraus, *J. Chromatog.*, **13**, 503 (1964).
182. R. M. Diamond, K. Street, Jr. and G. T. Seaborg, *J. Am. Chem. Soc.*, **76**, 1461 (1954).
183. R. Djurfeldt and O. Samuelson, *Acta Chem. Scand.*, **4**, 165 (1950).
184. H. Kakihana, N. Marnichi and K. Yamasaki, *J. Phys. Chem.*, **60**, 36 (1956).
185. H. Ohtaki and K. Yamasaki, *Bull. Chem. Soc. Japan*, **31**, 6, 445 (1958).
186. W. A. Plateck and J. A. Marinsky, *J. Phys. Chem.*, **65**, 2113 (1961).
187. D. C. Whitney and R. M. Diamond, *J. Inorg. Nucl. Chem.*, **27**, 219 (1965).
188. W. A. Plateck, and J. A. Marinsky, *J. Phys. Chem.*, **65**, 2118 (1961).
189. K. A. Kraus, D. C. Michelson and F. Nelson, *J. Am. Chem. Soc.*, **81**, 3204 (1959).
190. H. Titze and O. Samuelson, *Acta Chem. Scand.*, **16**, 678 (1962).
191. H. Irving and G. T. Woods, *J. Chem. Soc.*, 939 (1963).
192. D. Naumann, *Z. Anorg. Allgem. Chem.*, **309**, 37 (1961).
193. G. E. Boyd, S. Lindenbaum and Q. V. Larson, *Inorg. Chem.* **3**, 1437 (1964).
194. K. A. Kraus and F. Nelson in *Structure of Electrolyte Solutions* (Ed. W. J. Hamer), J. Wiley and Sons, Inc., New York 1959, p. 34.
195. B. Chu and R. M. Diamond, *J. Phys. Chem.*, **63**, 2021 (1958).
196. H. H. Sherry and J. A. Marinsky, *Inorg. Chem.*, **2**, 957 (1963).
197. J. P. Faris, *Anal. Chem.*, **32**, 520 (1960).
198. B. Chu, *Ph.D. Thesis*, Cornell University, Ithaca, N.Y., 1959.
199. C. H. Jensen and R. M. Diamond, *J. Phys. Chem.*, **69**, 3440 (1965).
200. K. A. Kraus and F. Nelson, *Spec. Tech. Publ.* No. 195, *Am. Soc. Testing Materials*, 1958, p. 27.
201. K. A. Kraus and F. Nelson, *Proc. First Intern. Conf. Peaceful Uses At. Energy (Geneva)*, **7**, 113 (1956).

202. S. Lindenbaum and G. E. Boyd, *J. Phys. Chem.*, **66**, 1383 (1962).
203. K. A. Kraus and F. Nelson, *J. Am. Chem. Soc.*, **76**, 984 (1954).
204. T. Andersen and A. B. Knutsen, *Acta Chem. Scand.*, **16**, 849 (1962).
205. R. Barbieri, G. H. Guada and G. Rizzardi, *Ric. Sci. Suppl.*, **33**, (II-A), 1033 (1963).
206. R. Barbieri, M. Giustiniani and E. Cervo, *J. Inorg. Nucl. Chem.*, **27**, 1325 (1965).
207. R. Barbieri, *J. Inorg. Nucl. Chem.*, **26**, 1707 (1966).
208. G. Zachariades, W. R. Herrera and C. J. Cuminskey, *J. Inorg. Nucl. Chem.*, **28**, 1707 (1966).
209. J. Korkisch in *Advances in Nuclear Energy, Series IX, Analytical Chemistry*, Pergamon Press, Oxford, Vol. 4 (1966).
210. J. Korkisch, *Separation Sci.*, **1**, 157 (1966).
211. S. Yu. Elovich and L. G. Tonkonog, *Zh. Neorg. Khim.*, **6**, 1795 (1961).
212. T. Sakaki and H. Kakihana, *Kagaku*, **23**, 471 (1953); H. Kakihana and K. Sekiguchi, *J. Pharm. Soc. Japan*, **75**, 111 (1955); Y. Sakaki, *Bull. Chem. Soc. Japan*, **28**, 217 (1955).
213. G. M. Panchenkov, V. I. Gorskov and M. C. Kulanova, *Zh. Fiz. Khim.*, **32**, 361, 616 (1958).
214. R. P. Bhatnagar, R. C. Arora and K. K. Kurian, *J. Indian Chem. Soc.*, **39**, 124 (1963); S. L. Bafna, *J. Sci. Ind. Res. India*, **12B**, 613 (1953).
215. R. G. Fessler and H. A. Strobel, *J. Phys. Chem.*, **67**, 2562 (1963).
216. C. Grigorescu-Sabau, *J. Inorg. Nucl. Chem.*, **24**, 195 (1962).
217. V. T. Athavale, C. V. Krishnan and C. Venkatesvarlu, *Inorg. Chem.*, **3**, 1743 (1964).
218. R. Arnold and S. C. Churms, *J. Chem. Soc.*, 325 (1965).
219. E. A. Materova, Zh. L. Vest and G. P. Grinberg, *Zh. Obshch. Khim.*, **24**, 953 (1954).
220. H. Ogawa and A. Isuji, *Gisei Shikenjo Hokoku*, **76**, 3 (1958).

5

Cation Exchange
of Complexes

Cation exchangers have found much use in the study of metal complexes. As a first approximation, only the hydrated metal ion can be assumed to be sorbed in the resin, while the complexes can be assumed to be excluded.

The fraction of the metal not complexed, α_0, can thus be readily obtained, and hence, by the methods outlined in section 3.B, all the required information concerning complex formation in the aqueous phase. The second approximation considers also the possibility of some of the complexes being sorbed, and the different sorbability of the different cationic species. The differences can be made use of also for estimating the species sorbed in the cation exchange resin. The application of cation exchangers thus falls into two broad categories: stoichiometric methods and distribution methods.

The stoichiometric methods (section 5.A) make use of the equivalent exchange and fixed capacity properties of the resin. Equivalent accounting (Salmon's method) can be used in some simple cases to ascertain the species sorbed in the exchanger. The presence of anionic ligands points directly to the sorption of complexes, and in the absence of ligands, a high equivalent weight points to the sorption of polynuclear hydrolysed species. When the complexes are robust, they can be separated on the exchanger, and identified individually.

Distribution methods are based on the mass-action law (section 5.B), and good use can usually be made of the constant ionic medium principle (section 1.E). The methods of Schubert (useful mainly when chelates are formed) and of Fronaeus (more generally useful) are discussed. In particular, the problem of the sorption of cationic complexes on the exchanger is examined. Some other methods based on distribution measurements, such as the continuous variation and relative absorption curve methods are also reviewed and their limitations are discussed.

The relatively small importance of the cationic complexes in the resin, discussed in section 5.B, is due to their having a lower charge than the hydrated metal ion, provided the ligand is anionic. With uncharged ligands the complexes formed can be strongly sorbed, as discussed in section 5.C. In this section, complex formation in the resin is examined in detail. The ligands bound to a metal ion can be exchanged for others, and the ligand exchange process is here discussed. The main interest, however, centres on complex formation with functional groups of the cation exchanger itself. If, instead of sulphonate groups, the exchanger carries groups capable of functioning as ligands, strong binding and special specificities or selectivities can result. Chelating groups, such as the iminodiacetate one, are particularly effective. The section is concluded with a discussion of resins in which the chelating groups are not charged. These are not ion exchangers, but behave similarly. The mode of behaviour, however, bears resemblance to solvent extraction, in a way discussed more fully in section 6.D.

A. STOICHIOMETRIC METHODS

a. Introduction

The previous chapter discussed in detail the dependence of the exchange of ion A against ion B, or the distribution of ion A in the presence of a bulk electrolyte with ion B of the same charge type, on various parameters. An important subject, which has not yet been discussed, is the effect on the exchange or distribution, of the interaction of the ion A with some other ions or constituents of the solution, to form, in a general way, complexes. This chapter deals with the effect of the formation of complexes on the exchange of metal cations between a cation exchanger and the solution. At moderate concentrations of the metal cation in the solution, it will occupy an appreciable fraction of the exchange sites in the exchanger. Any interaction that decreases the fraction of free metal ions in the solution, or that produces cationic complexes, will be reflected by changes in the number of equivalents of metal and ligand in the resin phase. A study of the stoichiometric composition of the exchanger as a function of the concentrations of metal cation and ligand in the resin, will give information on the species formed in the resin, and some indication on the species and their stability in the solution.

b. Equivalent Accounting

Salmon[1, 2] has proposed and developed a method which consists of an equivalent accounting of the capacity of the resin. The method is applied to the resin loaded completely with the metal to be studied, together with cationic complexes, under conditions of negligible invasion. A known amount of resin in the hydrogen form is equilibrated in a batch process with a solution containing metal ions M^{m+} and ligand ions H_jA^{j-a} at various ratios and pH values. Metal ions and cationic complexes displace the hydrogen ions completely from the resin, but it is assumed that no neutral complexes nor anionic complexes enter the resin. At equilibrium, the total numbers of moles of metal $\bar{n}_M$ and ligand $\bar{n}_A$ per equivalent of resin are measured. The functions $x = \bar{n}_A/\bar{n}_M$ and $y = a + (1 - m\bar{n}_M)/\bar{n}_A$ are then formed. The method is said to be able to determine the species in the resin from the values of x and y, provided that there exist no more than two species.

In the general case there will be the species ij and rs: $M(H_jA)_i^{m+ij-ia}$ and $M(H_sA)_r^{m+rs-ra}$. The mass-balance equations are

$$\bar{n}_M = \bar{n}_{ij} + \bar{n}_{rs} \tag{5A1}$$

$$\bar{n}_A = i\bar{n}_{ij} + r\bar{n}_{rs} \tag{5A2}$$

and the charge balance, accounting for the capacity by adding the equivalents of species in the resin is

$$1 = (m + ij - ia)\bar{n}_{ij} + (m + rs - ra)\bar{n}_{rs} \tag{5A3}$$

However the three equations (5A1), (5A2) and (5A3) are not independent, and elimination of the number of moles of the species per equivalent of resin, $\bar{n}_{ij}$ and $\bar{n}_{rs}$, leaves just one equation

$$m + \frac{ir(s-j)}{i-r}\bar{n}_M - \left(a + \frac{rs-ij}{i-r}\bar{n}_A\right) = 1 \tag{5A4}$$

in the four unknowns i, j, r and s. Without additional information it is, therefore impossible in the general case to know the species absorbed from measurements of $\bar{n}_M$ and $\bar{n}_A$ (or of x and y) only. In particular cases, however, some information can be obtained.

If, for instance, it is found that $x = 0$, it is evident that in the resin $i = r = 0$ and only the free metal ion M^{m+} occupies resin sites.

If it is found that $y = 0$, this requires that $j = s = 0$, i.e. that the ligands in the resin are not protonated (which is the case, of course, when the ligand is the anion of a strong acid). Then the general result of the equivalent accounting is

$$m\bar{n}_M - a\bar{n}_A = 1 \tag{5A5}$$

and the values of i and r cannot be determined from the data.

When it can be assumed that one of the two species in the resin is the free metal cation M^{m+} (i.e. $i = j = 0$), then a result of $0 < y < m$ leads to $s = y$, but r cannot be determined from the data.

When it is certain that the ligands are protonated to the same extent in both species, then $j = s = y$, but, again, the values of i and r cannot be obtained. When, however, it is also certain from independent knowledge that there is only one species in the resin ($i = r$, $j = s$), then the value of $i = r$ is given by x, i.e. simply the ratio of ligand to metal in the resin.

Moreover, if the two species differ only in their degree of protonation ($j \neq s$, $i = r$) again x is equal to the ligand number i, but j and s cannot be determined. As mentioned above, in the general case ($i \neq r \neq 0$ and $j \neq s \neq 0$) nothing can be said about the species present, and this holds even more forcefully if the assumption that there are only two species in the resin cannot be validated *a priori*.

Interesting qualitative conclusions can be drawn, however, if it is found that $y \geqq m^{3,4}$. In this case, neither M^{m+}, nor species $M(H_jA)^{m+ij-ia}$ can account for the results. The assumption of the presence of a poly-nuclear species, such as M_2A^{2m-a}, besides free M^{m+}, does account for the result $y = m$, while for $y > m$ the formation of hydrolysed or mixed polynuclear complexes has been assumed. However, these species cannot account for the capacity of the resin in a unique way. Indeed, a species M_2A^{2m-a} is equivalent on a stoichiometric basis to $M^{m+} + MA^{m-a}$. If $m = a$, as in the cases studied[3,4] ($M = Fe^{3+}$ or Al^{3+}, $A = PO_4^{3-}$), MA would be a neutral species, which could enter the resin, changing $\bar{n}_M$ and $\bar{n}_A$ without occupying resin sites (although in the cases[3,4] of $Fe^{3+}(Al^{3+})$—PO_4^{3-} this is ruled out by the low solubility of $Fe(Al)PO_4$).

The general conclusion from the above analysis is that the equivalent-accounting method using cation exchange is severely limited in its applicability. The need to validate the assumptions of complete loading of metal (or complex), the negligibility of invasion by neutral complex species, ion-pairs or electrolyte, and the presence of only two species in the resin, in addition to the formal difficulties shown above, further restricts the utility of this method.

Still, the 'equivalent-accounting of cation exchange capacity' method has been used in a few cases, where auxiliary evidence encouraged the investigators to make the necessary assumptions. It has been employed by Salmon and his coworkers for studying the formation of phosphate complexes of iron[1,3,5], aluminium[6], chromium[7] and some other elements[8]. Just those species were assumed to be present in the resin, which account for the capacity to within $\pm 3\%$, and it was assumed that these species are also predominant in the external solution, at the pH and ligand-to-metal ratio conditions employed.

In a qualitative manner, the presence of an anionic ligand in an ex-changer is an indication of the formation of a cationic complex. Günther-Schulze[10] has already forty-five years ago interpreted his findings of chloride in a permutite (mineral cation exchanger) in the copper(II) form, as indicating the presence of the species $CuCl^+$ side-by-side with Cu^{2+}.

Lister and McDonald[11] studied the ratio $x = \bar{n}_A/\bar{n}_{Zr}$ for A = chloride, nitrate and perchlorate, with a sulphonic acid cation exchanger. For perchlorate they found $x = 0$, indicative of no complexing, while for the other two anions x increased from about 0.2 at 3 M acid to 0.5 for HCl and 1.0 for HNO_3 at 6 M. However, at 1 M acid, increase of lithium nitrate concentration did not make x increase above 0.05, probably because of hydrolysis. Sorption of electrolytes increasing with dilution is characteristic of multivalent cations and may be indicative of complex formation. The plausibility of this is augmented by noting the increase of sorption with charge, contrary to the expectation from the Donnan potential requirement, as observed on a cation exchange membrane (Nepton CR–61) with the chlorides of potassium, barium, lanthanum and thorium[12].

In a more quantitative way, the charge of a number of cationic species has been determined in loading experiments as the ratio of the capacity to the number of moles of species per unit weight of resin, $1/\bar{m}_M$. Here again, it is assumed that there is only one species predominating in the resin, and that invasion is negligible. Thus, under certain conditions of concentration and pH, the species of Fe^{II} and of Ti^{IV} in sulphuric acid were found [13] to be $FeHSO_4^+$, $Ti(OH)_3^+$, TiO^{2+} and $TiOHSO_4^+$ from loading experiments (the neutral species $FeSO_4$ and $TiOSO_4$ were inferred from elution data). The hydrolytic species of Th^{IV} were found[14] to be $Th(Th(OH)_4)_n^{4+}$ from loading, the value of n not being determinable from the data, as discussed above. Similarly, Be^{II} is absorbed as a polymer with a single charge per beryllium atom, and hydrolysed species are also absorbed from aluminium, copper, lead, titanium(III) and zirconium solutions[15]. The hydrolysed species of Fe^{III} was found from break-through experiments[16], which give the charge per atom absorbed, to be $[Fe(OH)_2^+]_n$.

c. Robust Complexes

In the case of robust complexes, the assumption that there is only one species in each case can be verified by separating the various species on a column of the cation exchanger. The separations are based first of all on the differences of affinity of differently charged ions (section 4.C), and among species with the same charge, such as geometrical isomers, on smaller differences due to different polarizations. King and his co-workers separated and identified in this manner chromium(III) thiocyanate and chloride[17], and cobalt(III) dinitritotetrammine complexes[18].

Connick and coworkers studied in detail the chloride complexes of ruthenium(III)[19, 20], and others studied ruthenium(IV)[21], chromium(III) sulphate[22] and rhodium(III) chloride[23] species. The solvation of chromium(III) by methanol in mixed aqueous–methanol solutions was also studied by this method, and the formation constants of the mixed solvates-hydrates obtained[24].

Connick's method merits closer description. A band containing the ruthenium at the top of a cation exchange column was eluted by a solution of cerium(III) perchlorate[19] for the species Ru^{3+} and $RuCl^{2+}$, and of calcium chloroacetate[20] for $RuCl_2^+$. Analysis of the eluting solution (') and of the eluate (") gave a, the charge per atom ruthenium, formal concentrations being used

$$a = (c'_H - 3c'_{Ce} - c''_H)/c''_{Ru} \qquad (5A6)$$

$$a = [(K_a(2c'_{Ca} + c'_{CH_2ClCO_2H} + c'_H) - c''_H(c''_H + K_a)]/(c''_{Ru} + K_a) \qquad (5A7)$$

where K_a is the ionization quotient for chloroacetic acid. The number of ruthenium atoms per species, or the charge per species, is obtained from experiments involving the mass-action law, described in the next section. These data together, then determine the formulae of the species isolated.

B. MASS-ACTION LAW METHODS

a. Introduction

Cation exchange methods based on the mass-action law provide more information on complex formation in solution than do the methods based on stoichiometry only, discussed above. The selectivity coefficients for the cation exchange reaction, or the distribution coefficients of the metallic ion, follow changes in the ligand concentration through the stability constants of the complexes formed. It is a common practice to study the variation of the distribution of a metal ion at trace concentration, as a function of ligand concentration. This has the advantage that changes in the free-ligand concentrations due to complexation can be neglected.

When only a single complex species $M_pL_q^{m+}$ predominates over a considerable range of ligand concentrations, the charge per species, m, may be obtained in a simple manner[19, 25]. Supposing for simplicity that the 1:1 electrolyte C^+A^- provides the ligand, the distribution coefficient

328   *Ion Exchange and Solvent Extraction of Metal Compexes*

of the tracer metal (D_M, using the molar scale for the solution, the rational for the resin phase, cf. equation (4C6)) will follow the equation

$$\log D_M = \log K_{c\bar{x}} + m \log \bar{x}_C + \log(y_M \bar{f}_C^m / y_C^m \bar{f}_M) - m \log c_C \quad (5B1)$$

where the constant $K_{c\bar{x}}$ includes the thermodynamic constant for the exchange reaction, and a proportionality factor (k, cf. section 4.C.a). If the solutions are sufficiently dilute for the activity coefficient term to remain essentially constant, and since at trace metal concentration $\bar{x}_C \simeq 1$, then

$$\log D_M = const. - m \log c_C \quad (5B2)$$

and the charge per species is obtained from the double logarithmic plot of (5B2). Using a permselective membrane, m is obtained from (5B2) from the distribution of the tracer metal between unequal concentrations of perchloric acid (c_C) on both sides of the membrane[26]. In the more general batch equilibration, if the charge of the cation C is z_C, equation (5B2) is replaced[25] by

$$\log D_M = const. - (m - z_C + 1) \log c_C \quad (5B3)$$

The charge per species has been obtained by this method for a number of complex systems. Welch[27] used it for protactinium, and found a monopositive species, which he thought to be PaO_2^+. The same method was used to identify TiO_2^{2+} [28], $ThCl_2^{2+}$, $Th(NO_3)_2^{2+}$, $Th(HSO_4)^{3+}$ [29], AlF_2^+ and AlF^{2+} [30] (these from paper chromatography with resin-impregnated paper). The sorption of tracer fluoride ion on the aluminium form of a cation exchanger from aluminium chloride solutions has also been studied[31]. The value of $\frac{2}{3}$ for m in a plot of (5B2) shows the fluoride to be present exclusively as AlF^{2+}, in both phases. When the electrolyte concentration is varied over a wide range, changes in activity coefficients and resin invasion can obscure the true behaviour of the system. Fortuitious straight-line portions in the logarithmic plots of (5B2) may result, which do not necessarily correspond to the correct species.

Similarly, if over a certain range a complex ML_n predominates (omitting the charge), and the distribution of tracer M is measured at constant bulk cation concentration c_C, but varying ligand concentration c_L (electroneutrality being provided for an anionic ligand by an 'inert' anion added to make up to the normality of C), then[32]

$$\log D = const. - n \log c_L \quad (5B4)$$

Suitable corrections must be made to c_L for side-reactions (such as with hydrogen ions), and activity coefficients and resin loading with C are assumed constant. In such experiments, L can be also a neutral ligand, provided ML_n is not sorbed on the exchanger.

Some authors have applied cation exchange measurements to obtain the stability constant in the case where only one complex is formed. Mayer and coworkers[32-35] and Bonner and coworkers[36,37] applied the method to study some metal complexes and acid ionization[37]. In the latter case the hydrogen ion takes the role of the metal cation. It is usually assumed that the complex does not exchange with the bulk cation in the resin, but Connick and Mayer[35] made corrections for this effect, equating the selectivity quotient against the hydrogen ion of, for example, $CeCl^{2+}$ to that of Sr^{2+}. Usually however, as the ligand concentration varies, different species will be formed in a stepwise manner, and the general considerations discussed in section 3.B become useful. It should be recalled that for mononuclear complexes the concentration (or activity) of a species ML_n (omitting charges, for simplicity) is $\beta_n(M)(L)^n$, and that the total concentration of metal is given by $c_M = (M) \sum_1^N \beta_n(L)^n$. In order to determine the various β_n it is necessary to find the relationship between the measured distribution coefficient D_M, and the (total) ligand concentration (L) ($= c_L$, in the case of tracer metal concentration, and absence of side-reactions of L).

Of the methods for dealing with this problem that have been proposed, two (those of Schubert and of Fronaeus) have found wide spread acceptance. These will now be discussed in turn. Some other methods, of more limited acceptance and applicability, will be discussed further below.

b. Schubert's Method

The method originally proposed by Schubert[38] also dealt only with a single complex species. However, the method now connected with his name is the modified version[39], which allows for stepwise formation of a series of complexes. The method as is now most often applied has the following features[38]. An ionic medium (section 1.E) is used at constant cation concentration c_C, and variations of the ligand concentration c_L are assumed not to change activity coefficients appreciably. A constant pH (when the ligand is the anion of a weak acid) and trace metal concentrations are used, so that the effective ligand concentration is proportional to c_L. The resin is used completely loaded with C cations. Complex formation is assumed[38] to produce only neutral or anionic species, which are supposed not to be absorbed by the resin. This latter assumption is a distinguishing feature of Schubert's method. Furthermore, if cationic species are produced they too are assumed to be excluded by the resin. This point will be discussed further below.

Keeping in mind the above restrictions, the metal cations M^{m+} may be taken as undergoing the following reactions

$$z_C M^{m+} + m\bar{C}^{z_c} \rightleftharpoons z_C \bar{M}^{m+} + m C^{z_c} \tag{5B5}$$

$$M^{m+} + nL^{l-} \rightleftharpoons ML_n^{m-nl} \quad (n = 1, 2 \cdots N) \tag{5B6}$$

Since activity coefficients are assumed to be constant, and as $\bar{x}_C \simeq 1$, $D_M = k\bar{x}_M/c_M$, equals $D_M^0 = k\bar{x}_M/(M)$ when no ligand is present, which is given by

$$D_M^0 = k(Q_M \bar{x}_C^m/c_C^m)^{1/z_c} \tag{5B7}$$

and will be constant at a given medium concentration. As ligand is added the distribution coefficient changes as (cf. equation 3B28)

$$D_M = D_M^0(M)/c_M = D_M^0/X \tag{5B8}$$

so that the function X is obtained simply as

$$X = D_M^0/D_M \tag{5B9}$$

This can be rewritten in the form in which Schubert's method is most often applied

$$\frac{(D_M^0/D_M) - 1}{c_L} = \beta_1 + \beta_2 c_L + \cdots \beta_N c_L^{N-1} \tag{5B10}$$

which may be solved for the β's by successive extrapolations, as described in section 3.B. Since the method is best applicable to chelating ligands (which usually obey the restrictions that the complexes formed are neutral or anionic), N is a small number, often one or two, and then the association constant is obtained directly, or from a simple linear relationship.

Schubert has shown[38] that it is not necessary to measure D_M^0 directly if $N = 1$, since β_1 can be obtained from equations at two ligand concentrations c_L' and c_L''

$$\beta_1 = \frac{D_M' - D_M''}{c_L D_M' - c_L'' D_M''} \tag{5B11}$$

The value of D_M^0 obtained from the extrapolation of a series of D_M values to $c_L = 0$, should be equal to the value measured directly in the absence of ligand.

Many investigations of metal complex systems utilizing one form or another of Schubert's method, have been published. Schubert and coworkers[39-41] studied many metal complexes of organic acids. Whitaker and Davidson[42] studied the association or iron(III) and sulphate ions. These illustrate the earlier studies reported with organic and inorganic ligands respectively. Later, much work on organic ligands such as oxalate[43], citrate[44, 45], tartrate[44, 46], EDTA[47, 48] and others[49, 50] have been published, the method finding much popularity among Russian[43, 46, 47] and Japanese[44, 49] workers. Work on various inorganic ligands[51] has also been rather prolific. This list of references is necessarily incomplete but may serve as an introduction to the literature employing what is essentially Schubert's method.

In addition to the quantitative use, to determine the stability of the complexes, Schubert[38] has pointed out that the relative stability of the complexes of a metal with various ligands can be obtained simply from the relative decrease in D at a given free-ligand concentration produced by the various ligands. This idea was later proposed again by Russian workers[52].

A number of authors have concerned themselves with the problem of the uptake of cationic complexes by the resin, in the case where they are formed. As mentioned previously, Connick and Mayer[30] corrected their results in an unspecified manner, assuming the affinity of the complex to be similar to that of a simple cation of the same charge. However, an examination of their results shows that inclusion of this correction is not essential for the interpretation of their data in terms of the formation of the first complex they have found. In general it may be expected that with an anionic ligand L^{i-}, the charge of the cationic complexes ML_n^{m-nl} being lower than that of the metal, M^{m+}, they will have lower affinity to the resin than the uncomplexed metal (rule a, section 4.C.b). In the case where $l = m$ even the first complex is neutral and will have a low affinity to the resin, unless conditions for resin invasion are favourable. This has been illustrated with ^{14}C-labelled magnesium oxalate[41], where no oxalate was found in the cation exchanger. With ^3H- or ^{14}C- labelled acetate, even the cationic complexes $MnCH_3CO_2^+$ and $CoCH_3CO_2^+$ were found, from the absence of acetate in the resin[41], to have extremely low affinity for this phase.

Such behaviour is, however, not universal, as the results in section 5.A show, many cationic complexes being absorbed to some extent by the resin.

c. Fronaeus' Method

Because of the inapplicability of Schubert's method in the general case, where cationic complexes can also be absorbed by the resin, Fronaeus[53] devised a method capable of using cation exchange measurements also in this case. Originally[53], the method was proposed in terms of a divalent metal M^{2+} and a monovalent ligand L^-, allowing the cationic complex ML^+ to undergo exchange along with the free metal cation M^{2+}. Later the method was generalized to other valency types[54]. As with Schubert's method, a constant ionic medium, with a constant concentration of cation C^+, and low concentrations of metal (so that $\bar{x}_C \simeq 1$ and $(L) \simeq c_L$) are used. In his early work Fronaeus utilized resin loadings up to 10%, and found that D_M depends on c_M (cf. also section 5.C.b), so that he extrapolated his results to some low and constant resin loading, assuming that at constant $\bar{c}_M = k \sum_{i=0} \bar{x}_{MLi}$ activity coefficients would remain constant. This would presumably hold also for varying c_M, provided loading is lower than 0.5%[55], i.e. with tracer metal concentrations. The general cation exchange reaction is

$$ML_n^{m-nl} + (m-nl)\overline{C}^+ \rightleftharpoons \overline{ML_n^{m-nl}} + (m-nl)C^+$$

(5B12)

$$(n = 0, 1 \cdots m/l)$$

where n takes values to make $m - nl > 0$.

The selectivity coefficient Q_n is constant under the stated conditions. the concentration of a species in the resin is given by

$$k\bar{x}_{ML_n} = kQ_n(\bar{x}_C/c_C)^{m-nl}(ML_n)$$
$$= kQ_n\beta_n(\bar{x}_C/c_C)^{m-nl}(M)c_L^n = D_n(M)c_L^n$$

(5B13)

where $D_n = kQ_n\beta_n(\bar{x}_C/c_C)^{m-nl}$ is a constant. The distribution coefficient is given by

$$D_M = \bar{c}_M/c_M = \sum_{n=0}^{m-1} D_n(M)c_L^n/c_M = \sum_0^{m-1} D_n c_L^n/X$$

(5B14)

In the case that the only value for n allowed in (5B12) is zero, Schubert's method is obtained. If also $n = 1$ is possible, Fronaeus has shown[53] that in principle the β's can be obtained with no knowledge of D_0 and D_1 from the relation

$$\frac{d^2 D_M}{dc_L^2} + \sum_{i=1}^{N} \left(c_L^i \frac{d^2 D_M}{dc_L^2} + 2ic_L^{i-1} \frac{dD_M}{dc_L} + i(i-1)c_L^{i-2} D_M \right) \beta_i = 0 \qquad (5B15)$$

but the (graphical) differentiations of D_M with respect to c_L were not sufficiently accurate to give meaningful values of β_i for $i > 1$. The method Fronaeus proposed[54] to determine the β's from the general equation (5B14) is rather complicated, but has been proved useful. The value of D_0 is first determined either directly, in absence of ligand, or from extrapolation of (5B14) to $c_L = 0$

$$\lim_{c_L \to 0} D_M = \lim_{c_L \to 0} \sum_{n=0} D_n c_L^n / X = D_0 \qquad (5B16)$$

The auxiliary function

$$F = (D_0 D_M^{-1} - 1)/c_L \qquad (5B17)$$

is then calculated, and extrapolated graphically to $c_L = 0$, to obtain F_0

$$F_0 = \lim_{c_L \to 0} F = \lim_{c_L \to 0} \frac{(\beta_1 - D_1/D_0) + (\beta_2 - D_2/D_0)c_L + \cdots}{1 + (D_1/D_0)c_L + (D_2/D_0)c_L^2 + \cdots} = \beta_1 - D_1/D_0 \qquad (5B18)$$

Another auxiliary function G is now calculated

$$G = [(D_0/D_M)(F_0 c_L - 1) + 1]c_L^{-2} \qquad (5B19)$$

and again extrapolated to $c_L = 0$, to obtain G_0

$$G_0 = \lim_{c_L \to 0} G = \beta_1(\beta_1 - D_1/D_0) - (\beta_2 - D_2/D_0) \qquad (5B20)$$

Now the function $(G - G_0)/c_L$ is plotted against $(F - F_0)/c_L$ to give the value of β_1 as slope (provided $D_n = 0$ for $n > 2$)

$$(G - G_0)/c_L = \beta_1(F - F_0)/c_L + D_2 F/D_0 - \sum_{n=3}^{N} \beta_n c_L^{n-3} \qquad (5B21)$$

Once the value of β_1 is known, it is inserted in (5B18) to obtain $D_1 = (\beta_1 - F_0)D_0$, and to obtain $\beta_2 - D_2 D_0 = \beta_1 F_0 - G_0$. Another auxiliary function H is now calculated

$$H = G - \beta_1 F + D_0 = (\beta_1 F_0 - G_0)/D_M c_L = \beta_2 F - \sum_{n=3}^{N} \beta_n c_L^{n-3} \qquad (5B22)$$

and extrapolation of H plotted against F to $F = 0$, gives β_2. In the case where only two sorbable cationic complexes are formed ($D_n = 0$ for $n > 2$) the rest of the β's are then obtained in a straightforward manner from (5B22). Where $D_n > 0$ also for $n > 2$, (5B21) has to be corrected[56] for terms including D_3 and so on, and it cannot be expected that accurate values for the β's be obtained from such lengthy extrapolations.

The method developed by Fronaeus has been used both by him[53, 54, 57] and by others[58-66] to study many metal complex systems: copper(II)[53, 67] nickel[57], cerium(III)[54], gadolinium[62] and americium[62] acetates, cerium(III)[54], thorium[61], zirconium[61], plutonium(IV)[63] and indium[59] sulphates, nickel thiocyanate[57], indium[58, 59, 64], lead[60], americium[62], plutonium(IV)[56], manganese, copper, cobalt and zinc halides[65], zirconium fluoride[66], and others.

It is evident that the method developed by Fronaeus is much more complicated than that proposed by Schubert. In order to assess the worthwhileness of using the more complicated method, it is necessary to have an idea of the quantitative significance of the exchange of the cationic complexes[68-70]. The relative magnitude of the D_n coefficients in (5B14) compared with D_0, measures the effect of these species in the resin. From the definition of these coefficients, $D_n = k Q_n \beta_n (\bar{x}_C/c_C)^{m-nl}$, it is evident that they become smaller the less stable the complexes and the smaller the selectivity coefficients Q_n. These may be expected to be a function of n, through the charge of the species $m - nl$, in the manner discussed in section 4.C (Figure 4C2, Table 4C2). Values calculated from published data are collected in Table 5B1. Usually, as expected, the Q's decrease markedly as complexation proceeds and the charge falls. The data of Morris and Short[65] are an unexplained exception. The Q's must, however, be multiplied by the ratio of the capacity of the resin (meq/g) to the bulk cation concentration (molar) to the $m - nl$ power to give the coefficients D_n (litre/kg M^n) in (5B14). The $m - nl$ power acts in favour of the cationic complexes compared to the free metal. In order to utilize the differences in affinity to their utmost, it is of advantage to use as small bulk cation concentrations, c_C, as is compatible with other requirements (repression of hydrolysis, constancy of the ionic medium), so that $(\bar{x}_C/c_C)^m$ will be much larger than $(\bar{x}_C/c_C)^{m-nl}$.

The general conclusion from the data in the Table is that the exchange of the first cationic complex is usually of significance and must be allowed for in the calculation of the stability constants in solution. The published work, however, leads to the conclusion that neglect of the absorption of

TABLE 5B1. Selectivity coefficients of cationic complexes.

Metal ion	Ligand	Bulk cation	Q_0	Q_1	Q_2	Reference
Co^{2+}	Cl^-	H^+	2.2	10.9	—	65
Ni^{2+}	CNS^-	Na^+	2.3	1.3	—	57
	$CH_3CO_2^-$	Na^+	5.5	5.2	—	57
Cu^{2+}	Cl^-	H^+	2.2	10.7	—	65
	$CH_3CO_2^-$	Na^+	1.2	1.12	—	53
Zn^{2+}	Cl^-	H^+	1.9	9.1	—	65
	Br^-	H^+	1.9	7.6	—	65
In^{3+}	Cl^-	Na^+	3.0	0.0	0	59
		H^+	28	2.5	0.2	58
		Na^+	13	3.5	0	64
	Br^-	H^+	28	5.9	0.8	58
		Na^+	13	3.9	0	64
	I^-	H^+	28	10.4	2.0	58
		Na^+	13	7.4	0	64
	SO_4^{2-}	Na^+	3.0	0	—	59
Ce^{3+}	$CH_3CO_2^-$	Na^+	14.2	4.2	1.4	54
	SO_4^{2-}	Na^+	13.7	5.2	—	54
Eu^{3+}	$(CH_3)_2C(OH)COO^-$	Na^+	27.7	1.9	0	70
	$C_2O_4^{2-}$	Na^+	27.7	<0.2	0	70
	F^-	Na^+	27.7	5.5	1.16	70
	Cl^-	H^+	3.0	0.0	—	68
Gd^{3+}	$CH_3CO_2^-$	Na^+	5.7	2.4	0.9	62
	$CH_2OHCO_2^-$	Na^+	5.7	1.7	0.5	62
Am^{3+}	Cl^-	H^+	50	2.3	0	63
	$CH_3CO_2^-$	Na^+	5.7	1.9	1.0	63
	$CH_2OHCO_2^-$	Na^+	5.7	1.7	0.7	63
Bi^{3+}	F^-	H^+	250	24.5	8.4	69
	Cl^-	$H^+(X4)$	80	54	4.8	69
	Cl^-	$H^+(X8)$	250	57	0	69
	Cl^-	$H^+(X16)$	880	0	0	69
	Br^-	H^+	250	80	0	69
Zr^{4+}	F^-	H^+	51	11.6	0	66
	SO_4^{2-}	H^+	51	0	—	66
Th^{4+}	SO_4^{2-}	H^+, Na^+	100	0	—	61
Np^{4+}	F^-	H^+	205	230	0	72
	SO_4^{2-}	H^+	205	0	—	72
Pu^{4+}	Cl^-	H^+	900	100	10	56
	NO_3^-	H^+	900	100	10	56

the second cationic complex (i.e. with systems $M^{3+} - L^-$, $M^{4+} - L^-$, where it can be formed) does not lead to significant errors in the calculations, within the precision of the distribution measurements attained.

That means that practically, Fronaeus' method can be modified to include D_1 only, and that then it will yield significantly more nearly correct values of the β's than Schubert's method for cases such as $M^{2+}, L^-; M^{3+}, L^-; M^{3+}, L^{2-}$; etc. If independent evaluation of the relative magnitude of D_1 and D_0 is required, a comparison of the cation exchange data (this section) F_R, with the corresponding function F_S obtained from extraction measurements on the same system with a cation exchanging extractant (Chapter 8), where $D_n = 0$ for $n > 0$, gives $(F_S - F_R)/(1 + F_R c_L)$ $= D_1/D_0^{70}$.

The modifications of Fronaeus' method that have been proposed are as follows[71, 72]. As a first approximation, β_1 is set equal to F_0, i.e. D_1/D_0 is set equal to zero. The function J is then calculated

$$J = (F - F_0)D \simeq - D_1 + D\beta_2 c_L + \cdots \qquad (5B23)$$

An extrapolation of J to $c_L = 0$ leads to a first approximation of $D'_1 = - \lim\limits_{c_L \to 0} J$. Since $D_n = 0$ for $n > 1$, the function

$$D'_M = D_M/(1 + D'_1 c_L) \qquad (5B24)$$

will be a better approximation to $1/X$ than is D_M, and this leads via (5B10) to a better approximation, β'_1, to β_1 than F_0. Inserting β'_1 in (5B23) instead of F_0 leads to J', which on extrapolation yields a second approximation D''_1, etc. With the precision attainable for the data, two iterations should suffice. Once a final value D_1^f is arrived at, D_M^f $:= D_M/(1 + D_1^f c_L) = X^{-1}$ is used to obtain all the constants β_1, β_2, etc.

Another modification[72] utilizes the linear function of c_L

$$c_L^2 [F - (F_0 + c_L(dF/dc_L))]^{-1} = K + K(D_1/D_0)c_L \qquad (5B25)$$

Where $K^{-1} = \beta_3 - (D_1/D_0)(\beta_2 - D_1(\beta_1 D_0 - D_1)/D_0^2)$. The ratio of the slope to the intercept yields D_1/D_0, i.e. a value of D_1 that, when used in (5B24), leads immediately to X^{-1}. The applicability of (5B25) (e.g. in the neptunium(IV) sulphate system) and of the iteration between (5B23) and (5B24) (e.g. in the corresponding fluoride system) depends on the nature of the F function[52]. Use can also be made of the fact that at high cross-linking D_1 becomes even less important than at low ones (cf. values for Bi^{3+} in Table 5B1), or of a comparison of D values at several cross-linkings[69].

Differentiation of $\log D_M$ in (5B14), with respect to $\log c_L$ will give some information on the species formed in the two phases. If the derivative function is a straight line of integral slope, then only one species predominates in each phase, and the slope indicates the difference in ligand number between the two phases. If, instead of using a constant medium, the bulk cation concentration is permitted to vary with the ligand concentration (i.e. with univalent ions using C^+L^- as bulk electrolyte), then the coefficients D_n will become functions of c_L. Writing $D_n = D'_n c_C^{n-m} = D'_n c_L^{n-m}$ with D'_n constant, equation (5B14) may be rewritten as[23]

$$D_M = \frac{D'_0 c_L^{-m} + D'_1 c_L^{2-m} + D'_2 c_L^{4-m} + \cdots}{1 + \beta_1 c_L + \beta_2 c_L^2 + \cdots} \tag{5B26}$$

Again, if only one species predominates in each phase, a constant slope will be obtained on differentiation of $\log D_M$ with respect to $\log c_L$ (provided that activity coefficient ratios remain constant as c_L is varied).

d. Other Methods

Some other methods of using the mass-action law to deduce the stability and nature of complex species from cation exchange measurements have been proposed. Tremillon[74] has shown that frontal analysis can give highly sensitive values of the relative distribution coefficients, which can be used in what is essentially Schubert's method to yield stability constants. He applied his method to the chloride complexes of nickel, cobalt and copper. Trofimov and Stepanova[75] proposed to utilize exchangers of different swollen capacities to determine the charge of tracer zirconium ions absorbed from nitric acid, noting that the distribution coefficients vary more with changing resin molality, the higher the charge (cf. ref. 69 and section 6.B). The method of continuous variation has been adapted for use with a cation exchanger[76]. If the ligand and the single complex formed are assumed to be excluded from the resin, and the exchange isotherm is linear $((\overline{M}) = D_0(M))$, then

$$(M) = M(1 - x) - (ML_n) - D_0(M) \tag{5B27}$$

when x volumes of ligand solution of $c_L = M$ are added to $1 - x$ volumes of metal solution of $c_M = M$, where M is the total molarity. Since furthermore, $(L) = Mx - n(ML_n)$ and $(ML_n) = \beta_n(M)(L)^n$, the condition for maximal formation of ML_n, $d(ML_n)/dx = 0$, leads to $n = x$ (extremum)/$1 - x$ (extremum), as usual in Job's method. Here x (extremum) is the value of x at a maximum or minimum of a plot of the difference between the observed exchange, and the exchange, calculated on the assumption that no complexation occured, as a function of x. The method has been used[76] for the Cu^{2+}–EDTA, Ca^{2+}–EDTA and Ca^{2+}–citrate systems, where the non-exchange of the ligand and the single complex formed are reasonable assumptions, yielding results as expected from measurements by other methods, i.e. $n = 1$ for all three systems.

Paramonova[77], has proposed the method of relative absorption curves, which utilizes a cation exchanger, as well as an anion exchanger, to determine the fraction of the total metal concentration in the form of cationic, neutral and anionic species, and in some favourable cases, also the stability of the complexes formed. Calling the amount of a metallic ion absorbed on a unit weight of a cation exchanger, in the absence of ligand g_+^0, and in its presence g_+, the relative absorption is $\gamma_+ = g_+/g_+^0$. Similarly, the amount of anionic complex per unit weight of an anion exchanger, at such excess concentration of ligand that a sole anionic complex ML_N is stable, is g_-^0, while at lower ligand concentration it is g_-, so that the relative sorption is $\gamma_- = g_-/g_-^0$. At tracer loading of metal, the fraction of metal concentration in the form of cationic species will be $\alpha_+ = \gamma_+$, that in the form of anionic species $\alpha_- = \gamma_-$, while that as neutral species $\alpha_0 = 1 - \gamma_+ - \gamma_-$. At higher metal concentrations, provided a constant ionic medium is used and the exchange isotherms are linear

$$\alpha_+ = \gamma_+ \left(\frac{\bar{C}_+ - g_+^0}{\bar{C}_+ - g_+}\right)^{z_+/z_C} \qquad \alpha_- = \gamma_- \left(\frac{\bar{C}_- - g_-^0}{\bar{C}_- - g_-}\right)^{z_-/z_L} \qquad (5B28)$$

where $\bar{C}_+$ and $\bar{C}_-$ are the capacities of the cation- and anion-exchanging resins, z_+ and z_- the charges of the predominant cationic and anionic species, and z_C and z_L those of the bulk cation and ligand (used as the only anion with the anion exchanger), respectively. When no neutral complexes are formed ($\gamma_+ + \gamma_- = 1$), and there is one predominant anionic complex ML_N, the intersection of the plots of γ_+ and of γ_- against the ligand concentration, c_L, will be at $c_L = \beta_N^{-1/N}$. In the range of formation of cationic complexes there will be straight-line portions in the plot of the left-hand side of the equation

$$-\log(\gamma_+/1 - \gamma_+) = \log\beta_n + n\log c_L \qquad (5B29)$$

against $\log c_L$, of slope n. Extrapolation of these portions to $\log c_L = 0$ will give the value of $\log\beta_n$ as the ordinate.

Paramonova's method has been used by her and her coworkers[78] to study the species formed in sulphuric acid solutions of niobium, of rutherium(IV) in perchloric and hydrochloric acids and of zirconium in nitric acid, as well as the acetate complexes of yttrium and uranium, and the uranium(VI) carbonate complexes.

The applicability of Paramonova's method seems to require more assumptions than the others treated in this chapter, especially as to the predominance of a single species in the solution at a range of ligand concentrations, and the presence of a single species in either resin, irrespective of ligand concentration. These assumptions may not always be realistic, which limits the usefulness of the method.

e. Effect of Complex Formation on Separations

The most spectacular use of ion exchangers in inorganic chemistry is, undoubtedly, the mutual separation of the rare earths. The ordinary cation exchanger used, say $8X$ polystyrenesulphonate, does not show high selectivity among the individual rare earths. (A figure of $S_{Er}^{Nd} = 1.8$, i.e. $S_{Z+1}^{Z} = 1.08$, has been reported[79] for an unspecified resin in 0.025 M perchlorate solutions). The stabilities of several complexes of the rare earths, on the other hand, vary appreciably along the series. For a couple of neighbouring rare earths, with atomic numbers Z and $Z + 1$, the separation factor for a displacement reaction

$$\overline{M}_{(Z+1)} + M_{(Z)}X \rightleftharpoons M_{(Z+1)}X + \overline{M}_{(Z)} \tag{5B30}$$

where X is an appropriate chelating ion, may differ appreciably from unity. The equilibrium quotient for reaction (5B30) will be

$$Q_Z^{Z+1} = \frac{\overline{m}_{(Z)}m_{(Z+1,X)}}{\overline{m}_{(Z+1)}m_{(Z,X)}} = \frac{\overline{m}_{(Z)}m_{(Z+1)}}{\overline{m}_{(Z+1)}m_{(Z)}} \cdot \frac{m_{(Z+1,X)}}{m_{(Z+1)}m_{(X)}} \cdot \frac{m_{(Z)}m_{(X)}}{m_{(Z,X)}}$$

$$\tag{5B31}$$

$$= S_{Z+1}^{Z} k_{(Z+1,X)}/k_{(Z,X)} = S_{Z+1}^{Z} K_{(Z,X)}^{(Z+1,X)}$$

The net separation factor Q is thus the product of the resin factor S, which as stated above, is somewhat larger than, but differs only little from, unity, and the complexing factor K, which is appreciably larger than unity (for EDTA it is on the average about 2.1[80]). Both factors work in the same direction: the heavier rare earth, with the smaller crystalline but the larger hydrated radius has a smaller affinity for the resin, but is complexed the better by the chelating agent. In order to obtain good separation on the macro scale, it is best to push a band of the rare earths down a column loaded with copper ions, using the ammonium salt solution of the chelating agent as eluant. The function of the retaining ion is to redeposit the rare earth ion in the resin, and permit removal of the chelating agent in a soluble form [81]. This practice produces a band with a sharp front, which can be pushed along the resin columns over a large distance, i.e. many theoretical plates, producing long portions of the bands of very high purity.

Solubility limits of the rare earth chelates or the chelating agents in the acid form, limit the practical applicability of chelating agents which have optimal Q_Z^{Z+1} values over the whole rare earth series, from lanthanum to lutecium, including yttrium. The system 0.015 M EDTA–copper retaining ion at pH 8.4 is a practical compromise. For certain specific rare earths better systems can be found [82]. HEDTA (N'–(2–hydroxyethyl)-ethylenediamine–N,N,N'–triacetic acid) is useful for the separation of ytterbium and lutecium, while the pair europium–gadolinium is difficult to separate with any reagent.

Similar considerations can be used for the separation of any similar ions, where resin separation factors S_B^A must be amplified by suitable complexing agents, to obtain larger net separation factors, Q_B^A.

C. COMPLEX FORMATION IN CATION EXCHANGERS

a. Complexes with Uncharged Ligands

It is of interest to examine now the affinity of complexes with neutral ligands to the resin, compared to that of the free metal ion. In fact, it must be remembered that the so-called free metal iom is a complex with the solvent (water) as the neutral ligand. So the problem is rather the effect of different neutral ligands on the affinity. A further problem is the influence of the special environment in the resin on the stability of the complexes. The relatively low dielectric constant in the resin affects ion–dipole interactions, and steric hindrance may affect the coordination.

These problems have received only little attention. Walton and coworkers[83–85] studied the interaction of amines with transition metal cations in cation exchangers. The complex between copper and ammonia[83] in a sulphonate cation exchanger was found to be as strong as in aqueous solutions. The formation curve ($\bar{n}$, the average ligand number in the resin, plotted against $-\log c_{NH_3}$) resembles that in a relatively concentrated (5 M) aqueous solution of ammonium nitrate, rather than in a dilute (0.5 M) one. The same holds for the silver–ammonia complex, although here the formation curve resembled most that in 2 M ammonium nitrate. With other amines, the interaction of the organic part with the resin skeleton, and steric hindrance, are complicating factors. The first factor makes silver–n-butylamine and –benzylamine complexes more stable than in aqueous solution, the second makes them relatively less stable at high amine-to-silver ratios, and it also destabilizes the silver-piperidine complex in the resin. With 2-aminoethanol[84], the nickel complex was found to be somewhat more stable, those of silver and copper somewhat less stable, than in aqueous solution.

The uptake of water and ammonia by the resin loaded with metal cations is a good indication of the stability of amine complexes[86]. If it is assumed that no interaction of ammonia occurs with the ammonium form of the resin, ammonia uptake by this resin can be attributed to non-specific invasion or distribution, and be corrected for in cases where specific interactions occur. It was found[86], as expected, that sodium or potassium do not interact with ammonia in the resin, and lithium does only slightly. In this context it is interesting to note that alkali metal and alkaline earth cations exchange with the ammonium ion in a polystyrene sulphonate resin at a satisfactory rate even from liquid ammonia as solvent at $-40°$ to $-74°C$[87]. Selectivity of potassium and the heavier alkali metal ions versus sodium is greatly enhanced, while there is a complete inversion in the order of lithium, sodium and ammonium ions in this medium, compared to water.

Among the transition metal cations, absorbed from aqueous ammonia solutions[86], silver was found to form the diammine, copper the tetrammine while the nickel complex was found to be less stable in the resin than in the aqueous solution, contrary to the behaviour found[84] with 2-ethanolamine, and to that found by other workers[88]. The formation of stronger complexes in the resin has the effect of increasing the selectivity of the resin towards the complex. Thus, whereas a selectivity coefficient of 0.8 was found[88] for exchanging sodium for $Ni(H_2O)_6^{2+}$ with a Russian resin, a value of 1.8 was found with $Ni(H_2O)(NH_3)_5^{2+}$. Ethanolamine, likewise, was found[84] to shift nickel into the resin. With cobalt(III), the 'free' hydrated metal ion Co^{3+} is unstable, but among the amine complexes, $Co(NH_3)_6^{3+}$ was found[89] to have a larger affinity to the resin than $Co(NH_3)_4(H_2O)_2^{3+}$. An explanation for the larger affinity some complexes have, compared to hydrated ions, was given in section 4.C.d, in terms of an interaction with the water. Since water is at a higher activity in the outer solution than in the resin, the interaction is stronger for the water in the hydration shell than for the amine ligands.

Special effects are shown by bidentate amine ligands. The silver–ethylenediamine complex is much more stable in the resin than in solution[83], even more than accounted for by the interaction of the organic part with the resin skeleton. A possible explanation is the enhanced binding of $H_2NCH_2CH_2NH_3^+$ to the silver ions in the resin, with low effective dielectric constant, than in solution. However, in the resin the cation $(H_3NCH_2CH_2NH_3)^{2+}$ predominates over $H_2NCH_2CH_2NH_3^+$. The ions of nickel, copper and zinc are also drawn into the resin, in the presence of ethylenediamine and $\bar{\bar{n}}$ is greater than $\bar{n}$ for these complexes[84]. Another explanation for the enhanced stability of the silver complex is the formation of polymer chains[85]. This is seen by the ligand number in the resin tending towards unity, while that in solution tends towards two, as the ethylenediamine concentration is increased. These linear polymers $(AgNH_2CH_2CH_2NH_2AgNH_2CH_2CH_2NH_2)_x$ were pictured as associated with the polystyrene chains in the resin, rather than bridging the chains, to explain the non-dependence of the results on the cross-linking of the resin. 1,3–Diaminopropane was found to behave with silver in the same manner as ethylenediamine, as also was found for hydrazine with nickel, giving in the resin a three-dimensional network with a ligand number of about three. The enhanced binding of metal ions in the resin by ethylenediamine, compared with aqueous solutions and monoamines in the resin, was also noted by other authors for nickel[88] and cobalt(III)[89].

The importance of the formation in the resin of protonated cations of the amines, rejected by Walton and Suryaraman[85] as an explanation for the enhanced binding of silver, must still not be underestimated. Even ethylenediamine tetraacetic acid (EDTA) in acid solution is capable of being absorbed by a cation exchanger[90], with distribution coefficients up to 400. At high concentrations, hydrogen ions compete with the protonated EDTA and the distribution coefficient decreases. Such protonated species may play an important role in the separation of rare earths with EDTA on cation exchangers (cf. section 5.B.e).

b. Ligand Exchange

The interaction of amines, or neutral ligands in general, with transition metal cations, present as counter ions in a cation exchange resin, permits the phenomenon of ligand exchange[91] to take place. A ligand of high affinity to a certain metal cation in the resin can displace a ligand of lower affinity, and this exchange reaction proceeds on an equivalent basis, just as the ion exchange itself. One may thus define a 'ligand capacity' as

$$\bar{C}_L = \frac{\bar{C}_R(N_M - Z_R)}{z_M d_L} \qquad (5C1)$$

where $\bar{C}_R$ is the resin capacity, in equivalents per unit weight, N_M the coordination number and z_M the charge of the metal cations, Z_R the number of coordination sites occupied by donor atoms provided by resin functional groups, and d_L the 'denticity' of the ligand (i.e. 1 for monodentate, 2 for bidentate ligands, etc.). For example, nickel has $N_M = 6$ and $z_M = 2$, ethylenediamine has $d_L = 2$, while for a sulphonic acid resin $Z_R = 0$, so that $\bar{C}_L/\bar{C}_R = 1.5$. The stronger the complexes formed, the higher the fraction of $\bar{C}_L$ that can be practically utilized. However, it is necessary to provide conditions that prevent the metal ion from being eluted from the resin by non-complexing cations[91, 92]. This can be done by using a resin with a functional group that complexes the metal ion itself, provided that $Z_R < N_M$, i.e. that not all the coordination sites of the metal ion are blocked (see below). For a carboxylate resin $Z_R = z_M$, so that for the above example of nickel with ethylenediamine $\bar{C}_L/\bar{C}_R$ becomes 1.0.

The distribution of the ligand between the resin and solution, as a function of its concentration in the solution and the presence of metal cations, can be calculated, following Helfferich's derivation[92]. The distribution coefficient of the ligand, in the absence of complexing metal, D_L^0, can be measured independently. The concentration of the ligand in the resin will then be given by $(\bar{L}) = D_L^0 c_L$, assuming no interactions of the ligand in the solution. The total concentration of ligand in the resin $\bar{c}_L$ will be the sum

$$\bar{c}_L = (\bar{L}) + (\overline{ML}) + 2(\overline{ML_2}) + \cdots + (N_M - Z_R)/d_L(\overline{ML_{(N_M - Z_R)/d_L}})$$

$$= (\bar{L}) + (\bar{M})\sum \bar{\beta}_i(\bar{L}) \qquad (5C2)$$

Since (section 3.B) $(\overline{M}) = \bar{c}_M / \sum_i \bar{\beta}_i(\bar{L})^i$, the relationship for the distribution coefficient of the ligand in presence of the metal is

$$D_L = \bar{c}_L/c_L = D_L^0 + \bar{c}_M \sum i\,\bar{\beta}_i(D_L^0 c_L)^{i-1} / \sum \bar{\beta}_i(D_L^0 c_L)^i \qquad (5C3)$$

The unknown $\bar{\beta}_i$ may be approximated by the known respective quantities for the aqueous solution, β_i, and $\bar{c}_M$ for the fully loaded resin is

$\bar{C}_R/z_M$ (concentrations in the resin are expressed as moles per unit weight of resin, and the $\bar{\beta}_i$ must be adjusted accordingly; otherwise the molal scale may be used throughout).

For the calculation of ligand-exchange isotherms, the complication of mixed-complex formation should be taken into account[92], although lack of data even for the aqeuous phase makes this impractical. Recourse could be taken to the approximation of statistical distribution of ligands on the coordination sites of the metal ion. Alternatively, in special cases, the constants $\bar{\beta}_{ij}$ for the mixed complexes may be determined from measurements with trace concentrations of one ligand in presence of bulk concentrations of the other, and vice versa, and used in the appropriate modification of (5C3) to calculate the distribution coefficients of the two ligands, and their mutual separation factor, for any concentration ratio. This has been done[92] for three constant total concentrations, 0.1, 1.0 and 10 N (normality calculated on a denticity basis) for ammonia and 1,3-diamino–2–propanol as ligands, nickel as the cation, and Amberlite IRC–50 as the carboxylate resin ($Z_R = 2$) and the calculated isotherms and separation factors agreed satisfactorily with the experimental data. The main feature of the results is the increasing preference of the resin for the diamine with increasing dilution and with decreasing equivalent fraction of the diamine. This preference for a ligand of higher denticity is analogous to that of ions of higher charge in ordinary cation exchange, and is manifested also with resins loaded with copper or zinc, instead of nickel. With silver, as expected, $N_M - Z_R = 1$, and only one ligand donor atom is bound, so that no preference for the diamine should be exhibited, as is indeed found experimentally[92].

c. Hydrolysis of Cations in Cation Exchangers

Metal ions may hydrolyse inside cation exchangers just as in aqueous solutions. If monomeric hydrolysis products result, they will have a lower charge than that of the non-hydrolysed metal cations, and will be exchanged relatively more easily. Polymeric hydrolysis products, however may not be as easily exchangeable. Indeed, Kakihana and Tsubota[93], have found that iron(III) and aluminium cease to be elutable from a cation exchange resin above pH 5 and 6 respectively. These authors assume that this effect is due to hydrolysis (rather than to precipitation of hydrous oxides), that the extent of hydrolysis is based on the pH of the external solution only (and not on that of the resin phase) and that metal cations are not eluted when the concentration of free, unhydrolysed, metal cation in the resin falls below ca. one millimolar. On this basis, the upper pH limit of elution of certain metals has been given by these authors as 1.5 for bismuth, 3.8 for thorium, 4.0 for indium, 4.2 for tin(II), 5.0 for scandium, 5.1 for uranyl, 6.1 for beryllium, and 6.9 for vanadyl(IV) ions.

Zhukov and coworkers[94] have made similar studies, and have found differences in the elution of a number of metal cations by acidified and by neutral 1 M ammonium chloride or nitrate. No special hydrolytic effects were found with manganese, iron(II), cobalt and nickel, while copper and lead are retarded in neutral eluants, compared to acid ones. Beryllium was found to be absorbed in a cation exchanger above pH 2.2 as a singly charged large-sized polymeric cation. No hydrolytic effects were found with tervalent chromium or vanadyl(V), but tervalent bismuth, iron, aluminium and titanium are hydrolysed, the former two forming apparently the ions $Bi_3(OH)_6^{3+}$ and $Fe(Fe(OH)_2)_n^{3+n}$ while aluminium forms above pH 3.85, a polymeric ion with five aluminium atoms per complex. Among the tetravalent elements, zirconium and thorium were found to hydrolyse, the latter forming $Th(Th(OH)_4)_n^{4+n}$ or $Th(Th(OH)_3)_n^{4+n}$. The highly charged polymeric species are more tightly bound to the resin than the free cations, and are eluted with more difficulty by neutral, than by acidic eluents. It is, however, difficult to distinguish between the various possibilities of polynuclear complexes, and the species proposed need not necessarily be those actually important in the resin at high pH values. Spectroscopy on the resin, using for example the Mössbauer effect for iron[95], may be used to determine the species formed.

d. Complex Formation by Resin Functional Groups

Complex formation is not restricted to ligands brought in from the external solution, or those produced from the solvent by hydrolysis, but can occur also with suitable functional groups of the resin itself. This is sometimes a nuisance when the intention is to work with a strongly acidic cation exchanger, in case weakly acidic impurities are present. Early commercial cation exchangers made from sulphonated phenol–formaldehyde polymers (section 4.A) were bifunctional and contained a high proportion of weakly acidic 'impurities', as did also nominally monofunctional resins produced at this time. Fronaeus[96] concluded that the resin with which he developed his method for studying complex formation (section 5.B.c), Amberlite IR–105, contained about 10% of weakly acidic (presumably carboxylic) groups. These have the effect of causing D_0, the distribution coefficient of the free metal, to vary with the total concentration of the metal in the resin, at pH values even as low as 3, because of preferential binding, as found experimentally for the copper–sodium and nickel–sodium exchanges. On the contrary, with Dowex–50 resin no such effects were found for the copper–sodium exchange, from which it was concluded that the maximal content of weakly acidic impurities in this resin is 1.3 %[96].

Carboxylic acid resins (such as Amberlite IRC–50, based on poly-methacrylic acid) are stated in general to show enhanced affinity, not only to protons, but also to transition metal cations, due to specific binding. Complex formation of a typical carboxylic acid, such as acetic acid, with metal cations, is not very strong as a rule, and may have the character of ion-pairing, rather than covalent bonding. Therefore, there is an observable effect even with the lithium ion, which reverses its selectivity relative to the sodium ion with a carboxylic acid exchanger[97, 98]. Alkaline earth metal ions also show enhanced selectivity with carboxylate, compared to sulphonate cation exchangers[98], which also is attributed to specific interaction with resin functional groups. Enhanced binding of polyvalent metal cations by weakly acid cation exchangers is not confined to carboxylate resins. Hydrous zirconium oxide, which has no capacity for alkali metal cations at low pH[99], shows fifty times as high a capacity for copper and aluminium ions as for strontium ions, while caesium is not at all absorbed. Only cations which can compete successfully with hydrogen ions by specific interaction with the functional groups of the exchanger can be taken up by these exchangers from solutions of low pH.

A similar effect is obtained when citrate, EDTA or similar chelating anions are loaded on an anion exchanger and the ligand-loaded exchanger is used as a cation exchanger. Samuelson and coworkers[100] explored the separation possibilities this method provides, and more recently this method was taken up also by other authors[101]. No detailed information on the stability of the complexes formed in the resin has been given. The rate of diffusion of the metal ion in the resin is a clue to its charge (section 4.A.e), hence to the formula of the species[101].

Attempts were made to produce resins with functional groups more specific, and more strongly binding, than those in carboxylate resins (section 4.A). In particular a resin incorporating in its polymer framework a monomer which is a specifically strong ligand for a particular metal, should be a successful specific exchanger for this metal ion. Skogseid's resin (section 4.A.c)[102] is an early example, using the functional group of dipicrylamine for specific binding of potassium ions. Since then many complexing or chelating resins have been proposed, which may be more or less selective for certain metal ions, or specific for groups of metal ions. Many of these resins are based on condensation polymers which involve *m*-phenylenediamine. One of the amine groups is usually bound in the polymeric skeleton, but the other can be incorporated in strongly complexing functional groups. A resin with *m*-phenylenediglycine in its framework[103] was found to show a good capacity towards metal ions, with a maximum at pH 4 to 6. The maximal attainable loading decreases with decreasing metal ion concentration: at 100 mM it is 2.1, 1.6, 1.9 and 1.0 millimole/g respectively for magnesium, cobalt, nickel and zinc, while at 1 m M it is only 0.33, 0.27 and 0.22 millimole/g, respectively for divalent copper, iron and cobalt, all at pH = 5. Hydrogen ions can, however, displace the metal ions effectively, 0.1 M hydrochloric acid eluting all metal ions. A copolymer of *m*-phenylenediamine tetraacetic acid with formaldehyde and resorcinol was found[104] to show specificity for copper and nickel, compared to alkali metal and alkaline earth or even cobalt ions. The latter can be eluted with 0.01 M acid, while nickel or copper require 2 M acid. Reacting a *m*-phenylenediamine–formaldehyde resin with chloroalkyl carboxylic acids produced complexing resins[105] in which heavy metal ions are complexed, and alkaline earth ions form ion-pairs. Titration curves of the resin without and with metal loading ($\bar{m}_M$) give values of the concentration of the ionized functional group, $\bar{m}_{A^-}$, and the bound functional group $\bar{m}_{A(b)}$. From the mass balance of the functional groups.

$$\bar{m}_{A(total)} = \bar{m}_{A(b)} + \bar{m}_{A^-} + \bar{m}_{HA} \tag{5C4}$$

the concentration of undissociated functional group $\bar{m}_{HA}$ is obtained by adding up the total capacity. The association constants of the metal with the ligand forming the functional group (to form MA and MA_2) can be obtained from the standard treatment according to Bjerrum (section 3.B.c), from

$$\bar{\bar{n}} = \frac{\bar{m}_{A(b)}}{\bar{m}_M} = \frac{\beta_1^H(\bar{m}_{HA}/m_{H^+}) + 2\beta_2^H(\bar{m}_{HA}/m_{H^+})^2}{1 + \beta_1^H(\bar{m}_{HA}/m_{H^+}) + \beta_2^H(\bar{m}_{HA}/m_{H^+})^2} \tag{5C5}$$

The difficulty lies in the ignorance of the hydrogen ion concentration in the resin, and its variation as simultaneous functions of external pH, metal loading and degree of ionization, even disregarding changes in activity coefficient resulting from changes in the resin medium because of varying ionization. The constants $\bar{\beta}_n^H$ obtained from the standard treatment are mixed resin–solution constants of limited significance. For the reaction

$$M^{2+} + 2\overline{HA} \rightleftharpoons \overline{MA_2} + 2H^+ \qquad (5C6)$$

Gärtner and Wanjek[105] found for aminoacetic acid resin $-\log \bar{\beta}_2^H = 7.1$, 10.2 and 11.9 for copper, zinc and nickel, respectively.

A large number of resins with different polymer frameworks have been produced. Reaction of hydroxylamine with chloromethylated polystyrene, yields a hydroxamic acid resin, specific for iron(III) compared to divalent metals such as copper or zinc[106, 107]. At 1 mM total concentration a loading to mole fraction 0.2 for iron(III) in the resin can be obtained, but only less than 0.02 for the divalent metals, over a wide mole-fraction range in solution. A resin containing pyridine–2,6–dicarboxylic acid, with or without a hydroxyl group in the 4 position, was found[108] to be highly selective among the alkaline earth ions, permitting clean separations between calcium and strontium. Resins based on ethyleneimine polymers[109], as obtained by copolymerizing β, β'-di(ethyleneimino)-*p*-diethylbenzene with β-(ethyleneimino)ethyl phosphoric acid diethyl ester or ethyleneimino acetic acid ethyl ester and subsequent hydrolysis of the esters, have shown high capacities and good homogeneity. The former type of resin binds the ions in the order Cu > Zn > Ni > Mg, the latter in the order Cu > Ni > Zn > Mg, with copper showing a maximum in the loading curve at pH = 5. Appreciable absorption of transition metal ions occurs even at pH = 1 for the aminoacetate resin. Other resins, such as ones based on anthranilic acid[103, 110], 8–hydroxyquinoline[111], mercaptans[112], diketones[113] or even chlorophyl or haemin-like groups[114], with a polystyrene[112, 113], or other framework, have been proposed, as being specific for metals forming strong complexes with ligands containing nitrogen, oxygen or sulphur donor atoms. Weakly basic anion exchangers, containing primary, secondary or tertiary amine groups, can bind metal ions as amine complexes[115].

Sometimes the resins do not show the specificity expected from the behaviour of the monomers. Thus a resin based on maleic acid hydrazide[111], of which the monomer precipitates selectively monovalent silver, thallium and mercury, as divalent mercury, was found to be specific only for mercury, not absorbing appreciable quantities of silver or thallium. Similarly, whereas the tertiary nitrogen oxide group is specific for a number of metalions, poly(1–hydroxy–4–vinylpyridinium), the protonated resin containing this group, was not found to exhibit any specificity towards metal ions in the accessible pH range[112].

The best known resin with chelating functional groups is the resin produced by chloromethylating crosslinked polystyrene beads, and reacting the intermediate with iminodiacetic acid (section 4.A.c). The commerical version of this resin, proposed already in 1953[118], is Dowex A1. It has gained some popularity, and is certainly the best studied of all specific resins. The dissociation constant of the functional group has been calculated[119,120] from pH and isotope dilution measurements on the sodium–hydrogen exchange. The intrinsic first dissociation constant is $pK_{diss} = 2.92$[119a], 2.77[119b,c] or 2.91[120] (at 25⁰) in fair agreement with $pK_{diss} = 2.36$ for *N*-benzyliminodiacetic acid (at 25⁰)[121] or $pK_{diss} = 2.54$ (at 30⁰) for iminodiacetic acid itself[122]. Protonation of the weakly acid group leads to $pK_2 = 8.57$[120]. The water content of the resin at various mole fractions of sodium exchanged for hydrogen was determined isopiestically at different water activities. The capacity of the resin for sodium ions is 2.6 meq/gram dry resin, corresponding to about one functional group per 2.5 styrene units[119a,123]. More recent samples of commercial Dowex–A1 have one functional group per 1.5 styrene units, i.e. a higher capacity[125]. The exchange of various transition metal cations with this exchanger has been studied by a number of workers. Values for the dissociation constant of various metallic forms of the resin were obtained by Eger and Marinsky[124] as a concentration quotient

$$K_M = \frac{\overline{(M(HR)_2)}\,(H^+)^2}{(\overline{H_2R})^2(M^{2+})} \times Q_{mH}^M / K_{diss} \qquad (5C7)$$

using $K_{diss} = 0.002$ found previously[119b], and molal concentrations. The values found [124] for the association of M^{2+} and HR^- in the resin are $\log K_M = 2.03$ for nickel, 1.03 for cobalt, 0.8 for copper and 0.4 for zinc. The relative affinity of many ions to Dowex–A1 has been measured, especially for bivalent ions[120,126]. The effect of a complexing medium on the affinity of bivalent copper and lead has also been measured[127], finding $\log Q_{Pb}^{Cu} = 0.96$ in 0.2 M acetate medium, but $\log Q_{Pb}^{Cu} = -0.62$ in a 0.2 M glycine medium, both at pH = 4.5, showing the stronger complexing of copper by glycine in the aqueous phase. The use of Dowex A–1 chelating resin in solutions containing other chelating agents has produced some interesting results. It has been shown[125,128] that glycine, glutamic acid and imidodiacetic acid ligands accompany copper ions into the resin, the following reactions taking place simultaneously

$$Cu^{2+} + \overline{RH^-} \rightleftharpoons \overline{CuR} + H^+ \qquad (5C8a)$$

$$CuL + \overline{RH^-} \rightleftharpoons \overline{CuLR^{2-}} + H^+ \qquad (5C8b)$$

Steric factors prevent the copper from binding two functional groups in the resin, but it can take up another ligand from solution, such as iminodiacetic acid, L^{2-}, to give a coordinately saturated complex in the resin, similar to the species CuL_2^{2-} or the mixed ligand species $CuLL'^-$ in solution[125].

A large amount of work has been published by Hering and coworkers[121,129] on methylene N-alkylaminoacetic acid, methylene N, N-iminodiacetic acid (equivalent to $1X$ Dowex–A1), and methylene (N, N-iminodiacetic acid) acetic acid as functional groups in crosslinked polystyrene resins, and appropriate monomeric models. With bivalent ions, such as copper ions, the first resin forms the complex CuR_2, the second CuR and the third CuR^-, capable of taking up sodium ions as a cation exchanger. Copper is absorbed to a capacity almost as high as indicated by these species. With a trivalent ion, such as that of lanthanum, no absorption takes place with the first type of resin, the imidodiacetic acid resin forms a mixture of LaR_2^- and LaR^+, capable of taking up a small amount of sodium ions, while the nitrilotriacetic acid type resin forms a similar mixture of LaR_2^{3-} and LaR, however with practically no capacity for sodium ions, the negative charge on the 1:2 species being neutralized by hydrolysis to give LaR_2H_3. Other ions, such as silver, can be absorbed by cation exchange instead of sodium, and species such as AgR^-Ag^+ with the diacetate resin, or even $AgR^{2-}(Ag^+, H^+)$ with the triacetate resin, can be formed. The diacetate resin was found to show good selectivity among different bivalent ions (cf. ref. 126), but only poor separations were obtained among the trivalent rare earth ions. With the former group, there is a characteristic pH value for each ion, below which it can be eluted. The relative order of these pH values (1.25 for copper, 2.25 for nickel, 2.60 for zinc and 2.70 for cobalt), is also the decreasing order of their affinity towards the resin. These observations with a low crosslinked (?) chelating resin seem to be in conflict with results obtained with the commercial resin[124,126]. The selectivities are, however, a complicated function of the pH, and crossovers occur. More definite information on the stability of the complexes formed in the resins would be helpful.

A difficulty with the chelating exchanger Dowex– A1 and presumably also with other similar resins, is their slow rate of reaching equilibrium. At a symposium held in 1955[130] (before the commercial resin became available) Kitchener questioned the practical usefulness of chelating resins of iminoacetic acid type, exhibiting rates many times slower than highly crosslinked polystyrene sulphonate, indeed too slow to be put to practical use (cf. refs. 55, 56, 57, Chapter 4). Pepper agreed that selectivity was obtained at the expense of rate of exchange, while Hale stressed their properties of selective absorption, which are much more pronounced than those of carboxylic acid resins, and proposed mixed resins, containing ordinary cation exchange sites beside the chelating functional groups, to ensure higher swelling and more rapid equilibration.

e. Uncharged Resins

Ion exchange processes differ from various other chromatographic methods, such as adsorption, in that a stoichiometric reaction takes place, exchange proceeding on an equivalent basis. Some special materials have been prepared, differing from adsorbents in being capable of reacting with metal ions stoichiometrically, without an exchange reaction occuring. This happens when the material contains donor atoms which form strong coordinative bonds with metal ions, under conditions where an electro-neutral component enters the exchanger phase from the solution

$$M^{m+} + mA^- + n\overline{RX} \rightleftharpoons \overline{(RX)_n MA_m} \qquad (5C9)$$

where R is the polymer framework with functional group X. The number n depends on whether A complexes the metal or not, and if so, how many coordination sites it occupies, on the denticity of X, on geometrical factors, solvation possibilities with the solvent, etc. Two types of this uncharged resin have been prepared: in one X is incorporated by copolymerization in the resin framework, in the other, ligand molecules are strongly adsorbed on some suitable inert carrying material.

The phosphorylated neutral resins proposed by Kennedy[131] are an outgrow from the phosphate exchangers (e.g. one based on a polymer of diallyl phosphate). Instead of incorporating an acid phosphate or phosphonate ester into the resin, a neutral ester is used. Polytriallyl phosphate was found to absorb stoichiometric amounts of uranyl and cobalt nitrates and iron(III) and lithium chlorides from organic solvents such as acetone. An organic solvent is necessary since the non-ionic resin does not swell in water, but does so in a suitable solvent. The capacity of the resin in the case of uranyl nitrate corresponded to formation of $UO_2(NO_3)_2X_2$, where X is the phosphate group bound to the resin framework. Another similar resin is diethyl (polystyrenemethylene) phosphate, which is capable of absorbing uranium(VI) selectively from solutions of dibutyl phosphate in benzene in the presence of iron(III), lanthanum, zirconium, niobium, thorium and various fission products. In this case, ligand exchange occurs

$$UO_2((BuO)_2PO_2HO_2P(OBu)_2)_2 + 2(-CH(CH_2-)C_6H_4CH_2OP(OEt)_2O)$$

$$\rightleftharpoons UO_2((BuO)_2PO_2)_2(-CH(CH_2-)C_6H_4CH_2OP(OEt)_2O)_2$$

$$+ ((BuO)_2POOH)_2 \qquad\qquad (5C10)$$

two resin groups displacing a dimeric molecule of dibutyl phosphate from the uranium. The resin is also capable of absorbing stoichiometric amounts of the monomer of dibutyl phosphate at low uranium loadings

$$\tfrac{1}{2}((BuO)_2POOH)_2 + (-CH(CH_2)C_6H_4CH_2OP(OEt)_2O)$$

$$\rightleftharpoons (-CH(CH_2)C_6H_4OP(OEt)_2OHO_2P(OBu)_2) \qquad (5C11)$$

The difficulty with these exchangers, mentioned already, is that they do not swell in water, and, therefore, do not permit absorption to take place from aqueous solutions. To overcome this, materials of the second type have been proposed which have reactive groups incorporated in a surface layer on a finely divided inert support. Diffusion distances in the organic material are then small, and while aqueous solutions cannot displace the organic reagent from the support, absorption of metal compounds takes place efficiently. The reagents may be either neutral, such as trioctylphosphine oxide, or capable of dissociation, such as di(2-ethylhexyl)phosphoric acid, or of associating a proton, such as triisooctylamine. It is very difficult in this case to draw the line between ion exchange (in particular with the ionic compounds) and solvent extraction. Only a few practical examples will be discussed here, a full treatment of these reagents being given in the chapters dealing with solvent extraction.

Ion exchangers, such as Dowex–50 or Dowex–1, were proposed as the carriers for reagents such as dioctyl phosphate or tributyl phosphate[132]. However, the film of water on the resin beads hinders the diffusion or organic-soluble species. Better supports were made from DVB-crosslinked polystyrene beads, surface sulphonated to prevent their clinging together, and swollen with perchloroethylene containing tributyl phosphate[133]. This material showed good specificity, low solvent loss and high rates of adsorption, as evidenced by the small height of the equivalent theoretical plate in a column containing the material. Many other support–reagent combinations have, since then, been proposed, some of them with superior properties, and the technique is now called extraction (reversed-phase) chromatography. Cerrai and Testa[134-136] have developed this method, using as support materials cellulose, either in sheet form[134] or in columns [135], and recently the polymer of chlorotrifluorethylene, Kel–F [136], has been introduced as an inert support as well as polyvinyl chloride–polyvinylalcohol copolymers [137], diatomaceous earth[138] and other materials. This technique has been proved to be extremely useful in separating the lanthanides from one another [124, 129, 130], utilizing the complex-formation properies of dialkyl phosphoric acid, previously studied by the solvent extraction technique (Chapter 8). Separation factors S_Z^{Z+1}, where Z is the atomic number, of 2.2—2.5, and an equivalent height of a theoretical plate of 2 mm were attained. Under the acid elution conditions used, dialkyl phosphoric acid is not dissociated appreciably, and acts as a solvating coordinating reagent rather than as an anionic ligand.

D. REFERENCES

1. J. E. Salmon, *J. Chem. Soc.*, 2316 (1952).
2. J. E. Salmon, *Rev. Pure Appl. Chem.*, **6**, 24 (1956).
3. A. Holroyd and J. E. Salmon, *J. Chem. Soc.*, 959 (1957).
4. J. E. Salmon and J. G. L. Wall, *J. Chem. Soc.*, 1128 (1958).
5. J. E. Salmon, *J. Chem. Soc.*, 2644 (1953).
6. R. F. Jameson and J. E. Salmon, *J. Chem. Soc.*, 4013 (1954).
7. R. F. Jameson and J. E. Salmon, *J. Chem. Soc.*, 360 (1955).
8. A Holroyd and J. E. Salmon, *J. Chem. Soc.*, 269 (1956); J. A. R. Genge and J. E. Salmon, *J. Chem. Soc.*, 1459 (1959).
9. J. E. Salmon and D. Whyman, *J. Chem. Soc. (A)*, 980 (1966).
10. A. Günther-Schulze, *Z. Elektrochem.*, **28**, 89, 287 (1922).
11. B. A. J. Lister and L. A. McDonald, *J. Chem. Soc.*, 4315 (1952).
12. N. W. Rosenberg, J. H. N. George and D. W. Potter, *J. Electrochem., Soc.*, **104**, 111 (1957).
13. J. Beukenkamp and K. D. Herrington, *J. Am. Chem. Soc.*, **83**, 3022, 3025 (1960).
14. A. I. Zhukov, E. I. Kazantsev and V. N. Onosov, *Zh. Neorg. Khim.*, **7**, 915 (1962).

354   *Ion Exchange and Solvent Extraction of Metal Complexes*

15. A. I. Zhukov, G. P. Baranov, V. G. Shakurov and P. V. Plyasonov. *Zh. Neorg. Khim.*, **7**, 1452, 1458 (1962).
16. D. E. Kramm and L. Pockras, *Abstr. 135th. Mtg. Am. Chem. Soc.*, 16R, paper 36 (1959).
17. E. L. King and E. B. Dismukes, *J. Am. Chem. Soc.*, **74**, 1674 (1952); C. Postmus and E. L. King, *J. Phys. Chem.*, **59**, 1208 (1955); J. T. Hougen, K. Shug and E. L. King, *J. Am. Chem. Soc.*, **79**, 519 (1957); E. L. King, K. J. M. Woods and N. S. Gates, *J. Am. Chem. Soc.*, **80**, 5015 (1958).
18. E. L. King and R. R. Walters, *J. Am. Chem. Soc.*, **74**, 4471 (1952).
19. H. H. Cady and R. E. Connick, *J. Am. Chem. Soc.*, **80**, 2646 (1958).
20. R. E. Connick and D. A. Fine, *J. Am. Chem. Soc.*, **82**, 4187 (1960).
21. F. P. Gortsema and J. W. Cobble, *J. Am. Chem. Soc.*, **83**, 4317 (1961).
22. N. Fogel, J. M. J. Tai and J. Yarborough, *J. Am. Chem. Soc.*, **84**, 1145 (1962).
23. W. C. Wolsey, C. A. Reynolds and J. Kleinberg, *Inorg. Chem.*, **2**, 463 (1963).
24. J. C. Jayne and E. L. King, *J. Am. Chem. Soc.*, **86**, 3989 (1964).
25. J. D. H. Strickland, *Nature*, **169**, 620 (1952).
26. R. M. Wallace, *J. Phys. Chem.*, **68**, 2418 (1964); **70** 3922 (1960).
27. G. A. Welch, *Nature*, **172**, 458 (1953).
28. B. I. Nabivanets, *Zh. Neorg. Khim.*, **7**, 412 (1952).
29. K. F. Schulz and M. J. Herak, *Croat. Chim. Acta*, **29**, 49 (1962).
30. G. Alberti, V. Caglioti and M. Lederer, *J. Chromatog.*, **7**, 242 (1962),
31. S. Mlinko and T. Schönfeld, *Radiochimica Acta*, **4**, 6 (1965).
32. E. R. Tomkins and S. W. Mayer, *J. Am. Chem. Soc.*, **69**, 2859 (1947).
33. S. W. Mayer and S. D. Schwartz, *J. Am. Chem. Soc.*, **72**, 5106 (1950).
34. S. W. Mayer and S. D. Schwartz, *J. Am. Chem. Soc.*, **73**, 222 (1951).
35. R. E. Connick and S. W. Mayer, *J. Am. Chem. Soc.*, **73**, 1176 (1951).
36. O. D. Bonner, H. Dolyniuk, C. F. Jordan and G. C. Hanson, *J. Inorg. Nucl. Chem.*, **24**, 689 (1962).
37. O. D. Bonner, R. Jackson and O. C. Rogers, *J. Chem. Educ.*, **39**, 37 (1962).
38. J. Schubert, *J. Phys. Chem.*, **52**, 340 (1948).
39. J. Schubert, E. R. Russel and L. S. Myers, *J. Biol. Chem.*, **185**, 387 (1950).
40. J. Schubert and J. W. Richter, *J. Phys. Chem.*, **52**, 350 (1948); *J. Am. Chem. Soc.*, **70**, 4259 (1948); J. Schubert and A. Lindenbaum, *Nature*, **166**, 913(1950); J. Schubert, *J. Colloid Sci.*, **5**, 376 (1950); *Anal. Chem.* **22**, 1359 (1950);*J. Phys. Chem.*, **56**, 113 (1952); J. Schubert annd A. Lindenbaum, *J. Am. Chem.Soc.*, **74**, 3529 (1952); *Ann. Rev. Phys. Chem.*, **5**, 436 (1954).
41. N. C. Li, W. M. Westfall, A. Lindenbaum, J. M. White and J. Schubert, *J. Am. Chem., Soc.*, **79**, 5864 (1957); J. Schubert, E. L. Lind, W. M. Westfall, R. Pfleger and N. C. Li, *J. Am. Chem. Soc.*, **80**, 4799 (1958).
42. R. A. Whitaker and N. Davidson, *J. Am. Chem. Soc.*, **75**, 3081 (1953).
43. S. Y. Elovich and N. N. Motorina, *Zh. Fiz. Khim.*, **30**, 383 (1956); G. A. Kuyozev and V. V. Fomin, *Zh. Neorg. Khim.*, **1**, 342 (1956);

I. A. Korshunov, A. P. Pochinailo and V. M. Tikhomirova, *Zh. Neorg. Khim.*, **2**, 68(1957); A. I. Moskvin, G. V. Khalturin and A. D. Gelman, *Radiokhimiya*, **1**, 141 (1959); K. B. Zaborenko, A. V. Zavalskaya and V. V. Fomin, *Radiokhimiya* **1**, 387 (1959); I. A. Lebedev, S. V. Pirozhkov and G. N. Yakovlev, *Radiokhimiya*, **2**, 549 (1960); Yu. A. Zolotov, I. N. Marov and A. I. Moskvin,

44. S. Suzuki, *Sci. Rept. Tohoku Univ.*, **5**, 318 (1953); *J. Chem. Soc. Japan, Pure Chem. Sect.*, **74**, 590 (1953).
45. J. A. Schufle and C. D'Agostino, Jr., *J. Phys. Chem.*, **60**, 1623 (1956).
46. D. I. Ryabchikov, A. N. Ermakov, U. K. Belyaeva and I. N. Marov, *Zh. Neorg. Khim.*, **4**, 818 (1959).
47. A. I. Moskvin and P. I. Artyukhin, *Zh. Neorg. Khim.*, **4**, 269 (1959); B. P. Nikolskii, A. M. Trofimov and N. B. Vysokoostrovskaya, *Zh. Neorg. Khim.*, **4**, 389 (1959); A. D. Gelman, P. I. Artyukhin and D. I. Moskvin, *Zh. Neorg. Khim.*, **4**, 599 (1959).
48. J. K. Foreman and T. D. Smith, *J. Chem. Soc.*, 1752, 1758 (1957); L. Baetsle and E. Bengsch, *J. Chromatog.*, **8**, 265 (1962); J. Fuger, *J. Inorg. Nucl. Chem.*, **18**, 263 (1961).
49. H. Tsubota, *Bull. Chem. Soc. Japan.*, **35**, 640 (1962).
50. B. P. Nikolskii, V. B. Kolychev, A. C. Grekhovich and V. J. Paramonov, *Radiokhimiya*, **2**, 330 (1960).
51. L. A. Blatz, *J. Phys. Chem.*, **66**, 160 (1962); J. C. Sullivan, D. Cohn and J. C. Hindman, *J. Phys. Chem.*, **77**, 6203 (1955): M. Ward and G. A. Welch, *J. Inorg. Nucl. Chem.*, **2**, 395 (1956); I. A. Lebedev, S. V. Pirozhkov and G. N. Yakovlev, *Radiokhimiya*, **2**, 549 (1960); D. Banerjea and K. K. Tripathi, *J. Inorg. Nucl. Chem.*, **18**, 199 (1961); H. Gnepf, O. Gubeli and G. Schwarzenbach, *Helv. Chim. Acta.*, **45**, 1171 (1962); D. I. Ryabchikov, A. N. Ermakov, V. K. Belyaeva, I. N. Marov and Jao K'o-min, *Zh. Neorg. Khim.*, **7**, 69 (1962);I. A. Lebedev and G. N. Yakovlev, *Radiokhimiya*, **4**, 304 (1962).
52. M. M. Seryavin and L. I. Tikhonova, *Zh. Neorg. Khim.*, **7**, 1095 (1962).
53. S. Fronaeus, *Acta Chem. Scand.*, **5**, 859 (1951).
54. S. Fronaeus, *Svensk Kem. Tidskr.*, **64**, 317 (1952); **65**, 19 (1953).
55. E. Högfeldt, *Arkiv Kemi.*, **7**, 561 (1954); H. Irving and G. T. Woods, *J. Chem. Soc.*, 939 (1963).
56. I. Grenthe and B. Norèn, *Acta Chem. Scand.*, **14**, 2216 (1960.)
57. S. Fronaeus, *Acta Chem. Scand.*, **6**, 1200 (1952); **7**, 21 (1953).
58. B. G. F. Carleson and H. Irving, *J. Chem. Soc.*, 4390 (1954).
59. N. Sundèn, *Svensk Kem. Tidskr.*, **66**, 173, 345 (1954).
60. C. Karlson, unpublished results, cf. F.J.C. Rossotti and H. S. Rossotti *Determination of Stability Constants*, McGraw Hill Book Co., New York, 1961,p. 248.
61. A. J. Zielen, *J. Am. Chem. Soc.*, **81**, 5022 (1959).
62. A. Sonesson, *Acta Chem. Scand.*, **13**, 1437 (1959): I. Grenthe, *Acta Chem. Scand.*, **16**, 1695, 2300 (1962).
63. I. N. Marov and M. K. Chmutova, *Zh. Neorg. Khim.*, **6**, 2654 (1961).
64. J. A. Schufle and H. M. Eiland, *J. Am. Chem. Soc.*, **76**, 960 (1954).

65. E. L. Short and D. F. C. Morris, *J. Inorg. Nucl. Chem.*, **18**, 192 (1961);
    D. F. C. Morris and E. L. Short, *J. Chem. Soc.*, 5118 (1961); 2672 (1962);
    Electrochim. Acta, **7**, 385 (1962).
66. S. Ahrland, D. Karipides and B. Norèn, *Acta Chem. Scand.*, **17**, 411 (1963)
67. M. Grimaldi, A. Liberti and M. Vicedomini, *J. Chromatog.*, **11**, 101 (1963).
68. H. M. N. H. Irving and P. K. Khopkar, *J. Inorg. Nucl. Chem.*, **26**,
    1561 (1964).
69. H. Loman and E. van Dalen, *J. Inorg. Nucl. Chem.*, **28**, 2037 (1966);
    29, 699 (1967).
70. S. J. Lyle and S. J. Naqvi, *J. Inorg. Nucl. Chem.*, **28**, 2993 (1966).
71. Y. Marcus, 'Ion exchange of metal complexes', in *Advances in Ion
    Exchange* (Ed. J. A. Marinsky), M. Dekker Inc., New York, 1966,
    p. 111; S. Fronaeus, private communication to S. Ahrland, ref. 72.
72. S. Ahrland and L. Brandt, *Acta Chem. Scand.*, **20**, 328 (1966).
73. G. R. Choppin, D. Moy and L. W. Holm, *Radioisotopes in the Physical
    Sciences and Industry*, Intl. At. Energy Agency, Vienna, 1962, p. 283;
    L. W. Holm, G. R. Choppin and D. Moy., *J. Inorg. Nucl. Chem.*, **19**, 251
    (1961).
74. B. Tremillon, *Bull. Soc. Chim. France.*, 1483 (1958).
75. A. M. Trofimov and L. B. Stepanova, *Radiokhimiya*, **1**, 403 (1959).
76. S. Bukata, *Ph. D. Thesis*, Univ. of Buffalo, 1962; S. Bukata and J. A.
    Marinsky, *J. Phys. Chem.*, **68**, 258 (1964).
77. V. I. Paramonova, *Zh. Neorg. Khim.*, **2**, 523 (1957).
78. V. I. Paramonova and S. A. Bartenev, *Zh. Neorg. Khim.*, **3**, 74 (1958);
    V. I. Paramonova, A. N. Mosevich and A. I. Subbotina, *Zh. Neorg.
    Khim.*, **3**, 38 (1958); V. I. Paramonova, *Zh. Neorg. Khim.*, **3**, 212 (1958);
    B. Nikolskii and V. I. Paramonova, *Proc. Second Intern. Conf. Peaceful
    Uses At. Energy*, (Geneva), 1958, **28**, 512; V. I. Paramonova and Ya. F.
    Latichev, *Radiokhimiya*, **1**, 458 (1959); V. I. Paramonova, O. S. Kreichuk
    and R. A. Shishliakov, *Radiokhimiya.*, **1**, 650 (1959); V. I. Paramonova,
    V. B. Kolychev and A. V. Vikhlantsev, *Radiokhimiya*, **3**, 582 (1961):
    V. I. Paramonova and N. M. Nikolaeva, *Radiokimiya.*, **4**, 84 (1962).
79. J. E. Powell and F. H. Spedding, *Chem. Eng. Progr. Symp. Ser.*, **24**,
    101 (1959).
80. C. J. Wheelwright, F. H. Spedding and G. Schwarzenbach, *J. Am.
    Chem. Soc.*, **75**, 4196 (1953).
81. F. H. Spedding, J. E. Powell and E. J. Wheelwright, *J. Am. Chem. Soc.*,
    76, 2557 (1954).
82. J. E. Powell and F. H. Spedding, *Trans. AIME.*, **215**, 457 (1959).
83. R. Stokes and H. F. Walton, *J. Am. Chem. Soc.*, **76**, 3327 (1954).
84. L. Cockerell and H. F. Walton, *J. Phys. Chem.*, **66**, 75 (1962).
85. M. G. Suryaraman and H. F. Walton, *J. Phys. Chem.*, **66**, 78 (1962).
86. C. W. Davies and V. C. Patel, *J. Chem. Soc.*, 880 (1962).
87. A. M. Phipps and D. N. Hume, *Anal. Chem.*, **39**, 1755 (1967).
88. S. Yu. Elovich and L. G. Tonkonog, *Zh. Fiz. Khim.*, **36**, 37 (1962).
89. B. Baysal, unpublished results M.I.T., 1958.
90. T. Schönfeld and C. Friedmann, *Monatsh. Chem.*, **91**, 1192 (1960).

91. F. Helfferich, *Nature*, **189**, 1001 (1961).
92. F. Helfferich, *J. Am. Chem. Soc.*, **84**, 3237, 3242 (1962).
93. H. T. Tsubota and H. Kakihana, *J. Chem. Soc. Japan. Pure Chem. Sect.*, **82**, 1650 (1961).
94. A. I. Zhukov, E. I. Kazantsev and V. M. Onosov, *Zh. Neorg. Khim.*, **7**, 915 (1962); A. I. Zhukov, G. P. Baranov, V. G. Shakurov and P. V. Plyasonov, *Zh. Neorg. Khim.*, **7**, 1452, 1458 (1962); A. I. Zhukov, V. N. Gareev and V. M. Markova, *Zh. Neorg. Khim.*, **7**, 1724 (1962); A. I. Zhukov and V. N. Muzgin. *Zh. Neorg. Khim.*, **7**, 1730 (1962).
95. J. L. Mackey and R. L. Collins, *J. Inorg. Nucl. Chem.*, **29**, 655 (1967).
96. S. Fronaeus, *Acta Chem. Scand.*, **7**, 469 (1953).
97. J. I. Bregman, *Ann. N.Y. Acad. Sci.*, **57**, 125 (1953).
98. K. Hutschnecker and H. Deuel, *Helv. Chim. Acta.*, **39**, 1038 (1956).
99. C. B. Amphlett and J. Kennedy, *Chem. Ind. (London)* 1200 (1958).
100. O. Samuelson, L. Lunden and K. Schramm, *Z. Anal. Chem.*, **140**, 330 (1953); O. Samuelson and K. Schramm, *Z. Elektrochem.*, **57**, 207 (1953);O. Samuelson and E. Sjostrom, *Anal. Chem.*, **26**, 1908 (1954); *Z. Anal. Chem.*, **148**, 195 (1955); O. Samuelson, E. Sjostrom and S. Forsblom, *Z. Anal. Chem.*, **144**, 323 (1955).
101. J. R. Brannan and G. H. Nancollas, *Chem. Ind. (London).*, **45**, 1415 (1959); C. Bamberger and F. Laguna, *J. Inorg. Nucl. Chem.*, **28**, 1067 (1966); E. Brucker and P. Szarvas, *J. Inorg. Nucl. Chem.*, **28**, 2361 (1966).
102. A. Skogseid, *Dissertation*, Oslo, 1948; *U.S. Pat.* 2, 592, 350 (1952).
103. H. P. Gregor, M. Taifer, L. Citarel and E. I. Becker, *Ind. Eng. Chem.*, **44**, 2834 (1952).
104. E. Blasius and G. Olbrich, *Z. Anal. Chem.*, **151**, 81 (1956).
105. K. Gärtner and H. Wanjek, *Z. Physik. Chem. (Leipzig)*, **221**, 391 (1962).
106. J. P. Cornaz, K. Hutschneker and H. Deuel, *Helv. Chim. Acta.*, **40**, 2015 (1957).
107. G. Petrie, D. Locke and C. E. Meloan, *Anal. Chem.*, **37**, 919 (1965).
108. E. Blasius and B. Brozio, *Z. Anal. Chem.*, **192**, 364 (1963).
109. G. Manecke and H. Heller, *Makromol. Chem.*, **55**, 51 (1962).
110. H. von Lillin, *Angew. Chem.*, **66**, 649 (1954); E. Jenckel and H. von Lillin, *Kolloid-Z.*, **146**, 159 (1956).
111. J. R. Parrish, *Chem. Ind. (London).*, 137 (1956).
112. H. P. Gregor, D. Dolar and G. K. Hoeschele, *J. Am. Chem. Soc.*, **77**, 3675 (1955).
113. C. H. McBurney, *U. S. Pat.* 2, 613,200 (1952).
114. W. Lautsch, W. Broser, W. Rothkegel, W. Biederman, U. Doering and H. Zoeschke, *J. Polymer Sci.*, **8**, 191 (1952).
115. K. M. Saladze, Z. G. Demonterik and Z. V. Klimova, *Research in Ion Exchange Chromatography*, USSR Acad. Sci., 1957 (p. 45 of English translation, 1958).
116. E. Blasius and M. Laser, *J. Chromatog.*, **11**, 84 (1963).
117. A. Heller, Y. Marcus and I. Eliezer, *J. Chem. Soc.*, 1579 (1963).
118. D. K. Hale, K. W. Pepper and and S. L. S. Thomas, *Brit. Pat. Appl.*, 16129/53 (1953).

358   *Ion Exchange and Solvent Extraction of Metal Complexes*

119a. J. A. Marinsky and J. Krasner, *Radioisotopes in the Physical Sciences and Industry*, Intl. Atomic Energy Agency, Vienna, 1962, p. 504.

b. J. Krasner, *M. A. Thesis*, Univ. of Buffalo, 1963.

c. J. Krasner and J. A. Marinsky, *J. Phys. Chem.*, **67**, 2559 (1963).

120. D. E. Leyden and A. L. Underwood, *J. Phys. Chem.*, **68**, 2093 (1964).

121. R. Hering, W. Kruger and G. Kuhn, *Z. Chem.*, **2**, 374 (1962).

122. S. Chaberek, Jr. and A. E. Martell, *J. Am. Chem. Soc.*, **74**, 5052 (1952).

123. A. Schwartz, *Ph. D. Thesis*, Technion, Haifa, 1962.

124. C. Eger and J. A. Marinsky, *Proc. 8th. Intern. Conf. Coord, Chem.*, Vienna, 1964, p. 410.

125. H. Loewenschuss and G. Schmuckler, *Talanta*, **11**, 1399 (1964).

126. R. Rosset, *Bull. Soc. Chim. France*, 59 (1966).

127. G. Schmuckler, *Talanta*, **10**, 745 (1963).

128. H. Friedman and G. Schmuckler, *Israel J. Chem.*, **1**, 318 (1963).

129. L. Wolf and R. Hering, *Chem. Tech. (Berlin)*, **10**, 661 (1958); R. Hering, *J. Prakt. Chem.*, **14**, 285 (1961); R. Hering, *J. Inorg. Nucl. Chem.*, **24**, 1399 (1962); R. Hering, *Z. Chem.*, **3**, 30, 69, 108, 153, 571 (1963); **6**, 142, 228 (1966).

130. *Ion Exchange and its Applications*, Soc. Chem. Ind., **1955**, London.

131. J. Kennedy, *J. Appl. Chem.*, **9**, 26 (1959); J. Kennedy, F. A. Burford and P. G. Sammes, *J. Inorg. Nucl. Chem.*, **14**, 114 (1960).

132. H. Small, *J. Inorg. Nucl. Chem.*, **19**, 160 (1961).

133. H. Small, *J. Inorg. Nucl. Chem.*, **18**, 232 (1961).

134. C. Testa, *Anal. Chem.*, **34**, 1556 (1962); E. Cerrai and C. Testa, *J. Chromatog.*, **7**, 112 (1962).

135. E. Cerrai and C. Testa, *J. Chromatog.*, **6**, 443 (1961); E. Cerrai, C. Testa and C. Triulzi, *Energia Nucl. (Milan)*, **9**, 193 (1962).

136. E. Cerrai and C. Testa, *J. Inorg. Nucl. Chem.*, **25**, 1045 (1963).

137. T. B. Pierce and R. S. Hobbs, *J. Chromatog.*, **12**, 74 (1963); T. B. Pierce, P. F. Peck and R. S. Hobbs, *J. Chromatog.*, **12**, 81 (1963).

138. J. W. Winchester, *J. Chromatog.*, **10**, 502 (1963).

6

Anion Exchange

of Complexes

There is a great difference in the mode of applicability of anion exchangers and cation exchangers, since it is mainly the cationic metal ions which are capable of acting as central groups for complexes, while the ligands are anionic. In principle, anion exchangers can be used when the roles are inverted, as for instance when a polybasic acid dissociates hydrogen ions. The main interest, however, lies in the use of the anion exchangers at high ligand concentrations, where anionic complexes are formed, but where special complications also occur.

In analogy with cation exchangers, stoichiometric methods (section 6.A) can be used also with anion exchangers. The equivalent accounting method of Salmon can be used in special circumstances, as for example for the study of polyanions. More information on species formed in the resin can be obtained from resin loading. This is of great significance in studies by distribution methods.

These methods are discussed in section 6.B, in particular as applied to tracer distribution studies. The early method proposed by Fronaeus, and those later developed by Marcus and Coryell and by Kraus and Nelson are discussed in detail. The central problem of the complex species predominating in the resin phase is then reviewed, and several methods to obtain this information are examined.

There are many factors affecting the distribution of metal ions between aqueous solutions and an anion exchanger. They are discussed in section 6.C. They include the formation of additional species by side-reactions, the application of a constant ionic medium, the 'perchlorate effect' connected with it, the secondary cation effect and the effect of acid, the effect of metal loading, the influence of temperature and the effects of resin structure, such as capacity and crosslinking.

The sorption of metal ions on anion exchangers from nonaqueous or mixed solvents has become recently of great practical importance, especially for separations. Unfortunately, little basic knowledge is available to interpret the multitudinous observations that have been made. The discussions in Chapter 2 (in particular sections 2.C and 2.D) and in section 4.D explain why this is so. Still, some of the observations can be correlated, in terms of the effects of various solvents and in terms of the composition of the solution (partly the effect of varying dielectric constants, but even more the effect of preferential swelling of the resin and solvation of the ions in a mixed solvent). The question of the species important in the resin is of special significance here, since in addition to charged anionic complexes also those ion-paired with the bulk cation, and neutral complexes must be considered, much more than for aqueous solutions. The section concludes with two quantitative treatments that have been proposed for rather special cases. A comprehensive theory is unfortunately still far off.

A. STOICHIOMETRIC METHODS

a. Equivalent Accounting

Anion exchangers may be used to study complex formation by the equivalent-accounting method in a manner similar to that used for cation exchangers, described in section 5.A. It is assumed that only a single anionic complex species, MA_n^{m-na}, is sorbed by the resin, in addition to the ligand A^{a-}, both occupying resin sites, and that resin invasion is negligible. A possible variable that is considered is the degree of protonation of the ligand, which need not be the same for the ligand anion absorbed in the resin, H_iA^{i-a}, and for the complex $MH_jA_n^{m+j-na}$. It is, however, assumed that at constant pH in the external solution, i is constant, irrespective of the presence or absence of the complex. Salmon, the originator of the method, realized[1] that the possibility of the presence of a second complex in the resin, $MH_rA_s^{m+r-sa}$, makes an estimate of the composition of the complexes impossible, and even for the case of the formation of but a single complex it is required to know n, in order to determine j from measurements of i, and the number of moles of complex, $\bar{n}_M$, and of ligand $\bar{n}_A$, per equivalent of resin.

The mass-balance equations for the simultaneous absorption of complex (species jn) and ligand (species i) by the resin are

$$\bar{n}_M = \bar{n}_{jn} \tag{6A1}$$

$$\bar{n}_A = \bar{n}_i + n\,\bar{n}_{jn} \tag{6A2}$$

and the charge balance is

$$-1 = (m + j - na)\bar{n}_{jn} + (i - a)\bar{n}_i \tag{6A3}$$

These three equations yield one equation with the two unknowns j and n

$$ni - j = (1 + m\bar{n}_M + (i - a)\bar{n}_A)/\bar{n}_M \tag{6A4}$$

Salmon proposed to assume a value for n, and to try integral values of j for fitting equation (6A4). Still, with this mathematical difficulty, and the restriction of having only one complex in the resin, the method has been used extensively by Salmon and his coworkers, to study anionic complex formation.

A considerable amount of work was done on phosphate complexes, where it was found[2] that $i = 1.53$. At high phosphoric acid concentration the absorption of iron(III)[3] occurs with a ratio $\bar{n}_A/\bar{n}_M = 2.8$, from which it is concluded that $n = 3$, assuming the presence of only a negligible excess of free phosphate. Of course, the combination $Fe(H_jPO_4)_2^{2j-3} + H_jPO_4^{j-3}$ is equivalent to $Fe(H_jPO_4)_3^{3j-6}$ on an equivalent-accounting basis, and indeed the ratio $\bar{n}_A/\bar{n}_M$ between 2.40 and 2.15 found at lower concentrations suggests that $Fe(H_jPO_4)_2^{2j-3}$ may be an important species. Accepting the value $n = 3$, Salmon[2] shows that his results are consistent with $j = 0$ at 1.5 M phosphoric acid (within 1.7%), and with j values below three for higher acid concentrations. The same arbitrariness regarding the value of n was necessary also in the case of the aluminium phosphate complexes, where the authors[4] again prefer $n = 3$ over $n = 2$. Chromium is absorbed at $40°$, but not at $0°$, the species $Cr(PO_4)_2^{3-}$ accounting neatly for the capacity[5]. Of the other metal ions studied[6], tervalent bismuth and indium sorbed on an anion exchanger in phosphate form, as did trace amounts of divalent manganese, but not divalent alkaline earth, transition metal and group IIB cations. Presence of chloride in the solutions either prevents the metal complex from being absorbed in the case of aluminium[7], or complicates the relationships hopelessly, because of possible mixed complex formation, as in the cases of bismuth[6] or iron[8]. With the copper triphosphate system[9, 10], the presence of chloride or nitrate in a highly crosslinked resin prevents the triphosphate anion from being effectively sorbed, but with a highly swollen, less tightly crosslinked resin this anion, as well as its copper complex, can be sorbed. Again, the charge per copper found[10] can be interpreted either as the species $Cu(HP_3O_{10})_2^{6-}$ or as a combination of the species $Cu(HP_3O_{10})^{2-}$ with free $HP_3O_{10}^{4-}$. It was attempted to explain the higher sorption of copper in the presence of chloride, compared to the case where nitrate is present, by postulating the formation of mixed species containing chloride with a higher charge, and presumably higher affinity to the resin[10]. In the copper pyrophosphate system[11] equivalent accounting leads to the species $CuH_j(P_2O_7)_2^{j-6}$, with j varying from 5 at pH = 1 to 1 at pH = 10, the ratio of two pyrophosphate ions per copper ion in the resin complex again being fixed rather arbitrarily at two, from analogy to the known behaviour in solution. In the germanium oxalate system[12] below pH $= 3$, the complex absorbed is $Ge(C_2O_4)_3^{2-}$ with $n = 3$ and $i = j = 0$, as concluded from the ratio $\bar{n}_A/\bar{n}_M \simeq 3$. At higher pH values negative j values are found to be consistent with the data. Since $-2H^+$ ions are equivalent to one O^{2-} ion, such j values correspond to species such as $GeO(C_2O_4)_2^{2-}$ and $GeO_2(C_2O_4)^{2-}$. For the copper oxalate system, Cockerell and Woods[13] found from similar considerations that the species $Cu(C_2O_4)_2^{2-}$ is converted to $Cu(HC_2O_4)_6^{4-}$ as the pH decreases.

Where the ligand is the anion of a strong acid, e.g. a halide ion, $i = j = 0$, and the equation resulting from (6A4) is

$$a\bar{n}_A - m\bar{n}_M = 1 \qquad (6A5)$$

analogous to equation (5A5) obtained for cation exchangers. This means that no information can be obtained concerning the species absorbed (i.e. the value of n), as long as free ligand occupies resin sites along with the complex. This was demonstrated in the cases of tin(IV) and germanium chlorides[14], where both the species $SnCl_5^-$ and $SnCl_6^{2-}$, and $Ge(OH)_3Cl_2^-$ and $Ge(OH)_4Cl^-$, respectively, could account for the capacity of the resin.

b. Polyanions

A variation on the equivalent-accounting method for metal complexes is the study, especially by Everest, Salmon and coworkers, of the formation of polyanions $M_pO_q^{mp-2q}$ in the resin. Mean values of p and q for the complexes absorbed are sought, the usual assumption of no invasion, and negligible amounts of hydroxide loading, being made. The general procedure is to let the polyanion be absorbed on the resin, in, say, chloride form, and measure the number of gram atoms of M and moles of chloride per equivalent of resin, $\bar{n}_M$ and $\bar{n}_{Cl}$ respectively, as functions of pH and total M concentration. The ratio R is determined, defined as the number of atoms of metal per equivalent of complex

$$R = \bar{n}_M/(1 - \bar{n}_{Cl}) \qquad (6A6)$$

and since $R = p/(2q - mp)$, integral values of p and q are sought to account for the value of R found. Again, it must be remembered that one oxide ion is equivalent to -2 hydrogen ions.

In the germanate system[15], for example, at pH $= 4$ $\bar{n}_{Cl} \simeq 1$, and almost no germanium is absorbed, but as the pH increases towards 14, $\bar{n}_M$ increases towards 2 and $\bar{n}_{Cl}$ decreases. Between pH 8.5 and 9.4 it was found that $R = 2.5 \pm 0.2$, indicating the complex $Ge_5O_{11}^{2-}$. At pH $= 11$, R reaches the value 0.5, and the monomeric GeO_3^{2-} is the species sorbed. At higher germanium concentrations, and with low crosslinking of the resin, R may reach a value of 3.5, indicating the formation of a heptagermanate, $Ge_7O_{15}^{2-}$ (or $H_2Ge_7O_{16}^{2-}$). From dilute borate solutions[16] monoborate anions $B(OH)_4^-$ are sorbed above pH $= 7$, as shown by R values slightly lower than one, going down to 0.29 at pH 11.4, indicating the anion BO_3^{3-}, provided coabsorption of hydroxide ions at this pH is ignored. At higher borate concentrations, and above pH $= 10$, again R tends towards unity, indicating sorption of $B(OH)_4^-$, but at lower pH values R values as high as 5 are obtained, signifying absorption of polyborates. A detailed analysis of the results shows them to be consistent with the presence of $B_4O_7^{2-}$ and $B_5O_8^-$ (or $HB_5O_9^{2-}$) in the resin. From solutions of arsenites[17] the absorption of AsO_3^{3-}, $HAsO_3^{2-}$, $H_2AsO_3^-$, $As_2O_4^{2-}$ and $As_3O_5^-$ occurs as the pH is lowered from 13 towards 5, and the arsenic concentration increases. Similarly, R values decreasing[18] from 4 to 0.5 as the pH of the solution is raised from 4 to 11 indicate the sequence of tellurium(VI) species $HTe_4O_{13}^-$, $Te_4O_{13}^{2-}$, $Te_3O_{10}^{2-}$, $H_5TeO_6^-$ and $H_4TeO_6^{2-}$; R values decreasing[19] from 3.33 to 0.66 as the pH increases from 2 to 11 and the swelling of the resin decreases indicate the sequence of vanadium(V) species sorbed as $HV_{10}O_{27}^{3-}$, $HV_6O_{17}^{3-}$, $V_3O_9^{3-}$ and $HV_2O_7^{3-}$; and R values decreasing[20] from 5 to 2 as the pH is raised indicate formation of $H_2Mo_{10}O_{32}^{2-}$ and $Mo_4O_{13}^{2-}$. An important point to notice with the polyanions is the important effect of resin swelling, dictated mainly by the crosslinking, but also by the pH. The resin acts as a molecular sieve, and the lower the swelling, the more difficult it is for large polyanions (with high R values) to enter the resin, and depolymerization may set in. Although in general it is assumed that a good correlation exists between the species important in the resin and in solution[17], this obviously is not so when the resin is too tight to admit the large polyanion existing in solution.

In order to demonstrate that the R values yield the correct stoichiometries of the ions involved, some anions of known charge were tested[21]. The correct R values were indeed found for $C_2O_4^{2-}$, SO_4^{2-}, SeO_4^{2-}, AsO_4^{3-}, CrO_4^{2-} and $H_3IO_6^{2-}$, at the various pH values characteristic for these anions, and also for the corresponding hydrogenated anions at the pertinent lower pH values. On the other hand, there are differences between the pH in the resin and in solution, activity coefficients differ, and the resin shows selectivity towards the different anions. This causes the transition between one R value, characteristic for the hydrogenated anion, to the other, characteristic of a different state of hydrogenation, not to occur at the pH value characteristic for the transition in solution.

c. Loading Experiments

The equivalent-accounting method discussed above was developed mainly to deal with the degree of protonation of the complex sorbed in the resin. With ligands which are the anions of strong acids, the main problem is the instability of the complexes, which necessitates the presence of a considerable excess of ligand in the solution. Under these conditions the resin is only partly loaded by the complex to be studied, and it is difficult, in the general case, to analyse the results in terms of a definite species. Another difficulty, which must be considered, is the invasion of the resin by electrolyte, both the ligand salt or acid and the complex, so that the resin capacity varies. Finally, at high metal loadings the swelling of the resin may be reduced, and sorption therefore considerably slowed down. If complete equilibrium has not been assured, apparent low capacity and high charge on the complex may result.

These points may be illustrated by the work on tin(IV) chloride[14]. From hydrochloric acid solutions more dilute than 1 M, less than 0.5 mole of tin is absorbed per equivalent of resin sites, i.e. a complex $SnCl_6^{2-}$ could account for the data, along with excess chloride. At higher concentrations, more than 0.5 mole of tin is sorbed per equivalent of resin, and it is necessary to postulate some sorption of $SnCl_5^-$, provided resin invasion is ignored, as was done by the authors[14]. Since $SnCl_6^{2-}$ is equivalent to the sum of $SnCl_5^- + Cl^-$, as regards resin site occupation, the results remain inconclusive. Not taking into account the invasion of the resin makes the data inadequate for concluding which tin(IV) species are sorbed.

At tracer concentrations, the affinity of many metal complexes towards the anion exchange resin is much greater than that of the ligand anion. As the metal concentration increases, the affinity usually decreases (see for example the case of zinc chloro-complexes[22]), but may still be appreciably higher than that of the ligand. If a large excess of such a solution, containing enough ligand to stabilize the metal complex and varying metal concentrations, but of sufficiently low concentration that invasion can be ignored, is passed through a column of resin, the resin will ultimately become completely loaded. The loaded resin will be in equilibrium with a given concentration of metal, and the degree of loading (i.e. number of formula weights of complex per equivalent of resin sites) may be plotted against the metal concentration in solution. Where only one species is absorbed in the resin, a straight line parallel to the abscissa is obtained, or, if complete loading has not been obtained at the lower metal concentrations, by not using sufficient solution, a curve is obtained tending towards a limiting value at higher concentrations of metal. At still higher concentrations, invasion of the resin sets in, and the curve turns upwards. Unfortunately, only a few experimental studies have been made according to this scheme.

Loading studies of the chloride complexes of zinc[22, 26], cadmium[23, 26], indium[24] and uranium(VI)[25] gave results shown in Figure 6A1. The curve for indium shows an inflexion point near 1.0 moles per equivalent of resin, indicating a mononegative species. The curves for the other systems show inflexion points near 0.5 moles per equivalent, showing that the complexes in the resin carry a double negative charge. The shape

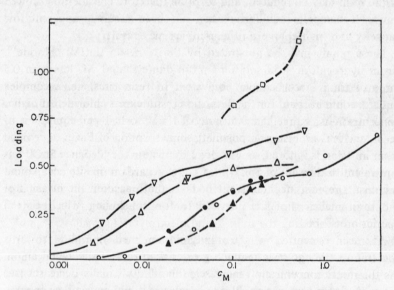

FIGURE 6A1. Anion exchange loading curves for chloride complexes. Plotted is loading, in moles per equivalent of resin, against the total metal concentration in the equilibrium solution. $\square$ In^{3+} in equimolar HCl, $\triangledown$ UO_2^{2+} in 10 M HCl, $\triangle$ Zn^{2+} in up to 2 M HCl, $\circ$ Cd^{2+} with no excess chloride, $\bullet$$Cd^{2+}$ in 0.2 M NaCl, $\blacktriangle$ Zn^{2+} in 0.2 M NaCl.

of the curves, however, leaves much to be desired for making this indication a certainty.

Many other systems were studied by the loading technique, and where complexes are sufficiently stable so that no large excess of ligand is necessary, no detailed loading curves were as a rule obtained. Instead, simply loading the resin up to its capacity gives the charge of the resin complex, as the ratio of the equivalent capacity of the resin to the number of moles of metal sorbed per unit amount. Among the systems for which loading curves or saturation capacity have been measured are the uranium(VI)[27] and plutonium(IV)[28] nitrate systems, the uranium(VI) sulphate[29] and fluoride[30] systems, and the vanadium(IV) thiocyanate[31] system. The predominant species found are $UO_2(NO_3)_4^{2-}$, $Pu(NO_3)_6^{2-}$, $UO_2(SO_4)_3^{4-}$, $UO_2F_3^-$ and $VO(SCN)_4^{2-}$ respectively. It should be stressed that the loading results give only the species likely to predominate in the resin at high metal concentration, but do not prove that the selected species actually does so, in particular at trace concentrations.

Results of loading studies from nonaqueous or mixed media are discussed in section 6.D. The same reservations mentioned above should apply there too.

B. TRACER DISTRIBUTION STUDIES

a. Introduction

Anion exchange resins have been used to separate metal complexes since the late forties, but a method for studying complex formation, using anion exchange distribution data was first proposed by Fronaeus[32]. More sophisticated methods have later been developed by other authors[33, 34], and brought to a state where the theory can account for most observed effects for a huge body of data on many metal–ligand systems. The basis for these methods is the measurement of the distribution coefficient of a trace metal (or central group) M^{m+}, between a solution containing the bulk ligand electrolyte $C_l L_c$, i.e. lC^{c+}, cL^{l-}, and a resin converted fully to the same anionic form, $R_l^+ L^{l-}$, as a function of the aqueous ligand concentration, c_L or m_L. The effects on the distribution coefficient of the trace metal, D_M, of variables such as the bulk cation C^{c+}, the acidity, the resin crosslinking, the temperature, etc., will be dealt with in section 6.C. The present section will describe the methods used to relate D_M to c_L (or m_L), in order to obtain information on the complex species in the aqueous phase.

It was thought at first that the mere fact that a metal absorbs on an anion exchanger proves that it is in the form of an anionic complex[35]. It was later realized that this criterion is insufficient, and that such sorption may be the result of the presence of impurities in the surface of the resin beads[32], or in the bulk of the resin (e.g. for potassium ions sorbed from chloride solutions[36]) which interact with the metal cation. Sorption may also result from resin invasion, at high bulk electrolyte concentration, as is the case, probably, with sorption of barium ions from concentrated lithium nitrate solutions[37]. In these cases, however, the distribution coefficients are usually low, rarely exceeding a value of 10 l/kg, and it may be safely concluded that for cations showing distribution coefficients above 100 l/kg, anionic complexes are formed, at least in the resin phase.

b. Fronaeus' Treatment

Fronaeus has proposed[32] a method for analysing distribution curves, which is applicable in the ideal case, where activity coefficients and the concentration of ligand in the resin remain independent of ligand concentration changes in solution. In this case concentrations can be used in the mass-action equations, with activity coefficients included in the concentration quotients, which are assumed to remain constant. The distribution of the metal can be studied from two approaches: the Donnan distribution law, and the Nernst distribution law (sections 4.B.e and 7.B.a, respectively). From the former follows

$$\overline{(M^{m+})}\,\overline{(L^{l-})}^{m/l} = K(M^{m+})(L^{l-})^{m/l} \qquad (6B1)$$

while from the latter follows, in case m/l is integral,

$$\overline{(ML)}_{m/l} = K'(ML_{m/l}) \qquad (6B2)$$

Since the 'constants' K and K' contain activity coefficients, and since the complex formation 'constants' $\beta_{m/l}$ and $\bar{\beta}_{m/l}$ for the two phases need not be equal, even if the same standard state is used, the 'constants' K and K' are different, contrary to the implicit supposition of Fronaeus. The balances of metal concentrations are $c_M = (M^{m+})\,X$ and $\bar{c}_M = \overline{(M^{m+})}\,\bar{X}$, where X is the complexity function defined in equation (3B28),

$\sum_0^N \beta_n(L^{l-})^n$. The distribution coefficient D_M (in units of volume per mass, with a constant coefficient k, converting from mass of resin to volume of water in the resin implied) is $k\bar{c}_M/c_M$. Since at tracer metal concentrations $(L^{l-}) = c_L$, this yields with the definition of X and (6B1)

$$D = k\,\frac{\bar{c}_M}{c_M} = k\,\frac{(\bar{M}^{m+})\bar{X}}{(M^{m+})X} = kK\,\frac{c_L^{m/l}\bar{X}}{\bar{c}_L^{m/l}X} \tag{6B3}$$

if m/l is integral, the fraction of M as the uncharged complex is $\alpha_{m/l} = (ML_{m/l})/c_M = \beta_{m/l}c_L^{m/l}/X$ and similarly for the resin phase, so that equation (6B2) leads to

$$D = kK'\,\frac{\alpha_{m/l}}{\bar{\alpha}_{m/l}} = kK'(\beta_{m/l}/\bar{\beta}_{m/l})\,\frac{c_L^{m/l}\bar{X}}{\bar{c}_L^{m/l}X} \tag{6B4}$$

from which it is seen that $K' = K\bar{\beta}_{m/l}/\beta_{m/l}$.

An essential assumption in Fronaeus' treatment, besides the constancy of activity coefficients, is the constancy of the ligand concentration in the resin phase, $\overline{(L^{l-})} = \bar{c}_L$ from which follows also the constancy of $\bar{X}$. Collecting all the constant terms in (6B3) or (6B4) into the constant $K'' = kK\bar{X}\bar{c}_L^{-m/l} = kK'\beta_{m/l}\bar{\beta}_{m/l}^{-1}\bar{X}\bar{c}_L^{-m/l}$, the equation

$$D = K''c_L^{m/l}/X = K'''\alpha_{m/l} \tag{6B5}$$

results, with $K''' = K''\beta_{m/l}^{-1}$, i.e. the distribution coefficient is proportional to the fraction of the metal which is in the form of the uncharged complex. Putting (6B5) in logarithmic form, accepting the constancy of K'', and remembering that d $\log X$/d $\log c_L = \tilde{n}$ (cf. equation 3B34), yields the result that there is a maximum in the distribution curve when $\tilde{n} = m/l$

$$\frac{d\log D}{d\log c_L} = (m/l) - \tilde{n} \tag{6B6}$$

and that the slope of the distribution curve yields the value of the ligand number $\tilde{n}$. Since on extrapolation to vanishing c_L values X tends to the limit unity, it is possible to obtain

$$K'' = \lim_{c_L \to 0} D / c_L^{m/l} \tag{6B7}$$

and hence, in principle, using the methods outlined in section 3.B, the complex formation constants β_n can be calculated through $\tilde{n}$ or X from (6B5) or (6B6).

Equation (6B6) does not depend on m/l being integral, hence the maximum in D will be obtained for $\tilde{n} = m/l$ irrespective of the existence of an uncharged complex. Equations (6B5) and (6B6) also show that a plot of log D against log c_L will have an initial rising portion, with initial slope m/l, and after reaching the maximum a declining portion, with final slope $m/l - N$, where N is the coordination number of M with respect to L. A pronounced maximum, according to this treatment, leads to the conclusion that anionic complexes are present in the aqueous phase, at ligand concentrations beyond c_L (max), while the converse conclusion—no anionic complexes existing in solution when there is no maximum—is implied.

Fronaeus also presented[32] the hypothesis that for systems of the same type (e.g. cadmium bromide and iodide), the values of K' and the maximal values of $\alpha_{m/l}$ should be approximately equal, so that the considerable differences observed for the maximal D values, should, according to (6B4), be due to differences in $\bar{\alpha}_{m/l}$, that is in $\bar{\beta}_{m/l}/\bar{X}$. If one assumes that the coordinatively saturated complex ML_N^{m-Nl} predominates in the resin, then $\bar{X} \simeq \bar{\beta}_N \bar{c}_L^N$, and $\bar{\alpha}_{m/l} \simeq \bar{\beta}_{m/l}\bar{\beta}_N^{-1}\bar{c}_L^{m/l-N}$. It is then the ratio of the complexity constants in the resin $\bar{\beta}_N$ and $\bar{\beta}_{m/l}$ that is mainly responsible for the resin selectivity towards the complexes. Since D is directly proportional to $\bar{\beta}_N/\bar{\beta}_{m/l}$, it can be said that the tendency of the metal to add ligands to the neutral complex in the resin phase is the measure of its affinity to the resin.

c. The Marcus–Coryell Treatment

The assumptions involved in Fronaeus' treatment are rather restrictive, and so methods were sought to interpret anion exchange data in a way approaching more nearly the actual conditions. An important objection to the above treatment is the assumption of the constancy of the ligand concentration in the resin, since resin invasion by electrolyte is a readily observable phenomenon. Since a large portion of the available data on anion exchange of metal complexes pertains to concentrated solutions, the treatment should also take into account the considerable deviations of ligand activities from ideality, which is not done in the above treatment. Marcus and Coryell have proposed a treatment of the problem of anion exchange of complexes[33] which explicitly deals with these points, among others.

A rigorous treatment considers the formation of a complex ML_n^{m-nl} from metal cation M^{m+} and ligand L^{l-} in electrolyte (C_lL_c) solutions, described by the thermodynamic stability constant β_n^T, so that

$$\beta_n^T = a_{ML_n}/a_M a_L^n = m_n \gamma_n / m_0 a^n \qquad (6B8)$$

where a with subscript means the activity of the species, m_n is the molal concentration of ML_n^{m-nl}, m_0 is that of M^{m+}, a is defined as (section 1.D.f)

$$a = m_L \gamma_{\pm C_l L_c} \qquad (6B9)$$

and γ_n is defined as

$$\gamma_n = \gamma_{\pm[lML_n,(m-nl)L]}^{1-n+ml} \gamma_{\pm[lM,mL]}^{-(m+l)/l} \gamma_{\pm[lC,cL]}^n \qquad (6B10a)$$

$$\text{for } m \geqq nl$$

$$\gamma_n = \gamma_{\pm[(nl-m)C,cML_n]}^{1+(nl-m)/c} \gamma_{\pm[lM,mL]}^{-(m+l)/l} \gamma_{\pm[lC,cL]}^{[m(l+c)-nl^2]/lc} \qquad (6B10b)$$

$$\text{for } m \leqq nl$$

All the activity coefficients are defined in terms of mean ionic activity coefficients of electrolytes, hence the treatment is thermodynamically rigorous. Simpler expressions are obtained if C_lL_c is a 1:1 electrolyte (i.e. $c = l = 1$)

$$\gamma_n = \gamma_{\pm[ML_n,(m-n)L]}^{1+m-n} \gamma_{\pm[M,mL]}^{-(m+1)} \gamma_{\pm C,L} \qquad (6B11a)$$

$$\text{for } n \leqq m$$

$$\gamma_n = \gamma_{\pm[(n-m)C,ML_n]}^{1-m+n} \gamma_{\pm[M,mL]}^{-(m+1)} \gamma_{\pm C,L}^{2m-n} \qquad (6B11b)$$

$$\text{for } n \geqq m$$

Anionic complexes with index $q > m/l$ can undergo exchange with the ligand in the resin

$$ML_q^{m-ql} + p\overline{L^{l-}} \rightleftharpoons \overline{ML_q^{m-ql}} + pL^{l-} \qquad (6B12)$$

where p is required by electroneutrality to equal $q - m/l$. The thermodynamic constant for this reaction is

$$K_p^T = \bar{a}_{ML_n} a_L^p / a_{ML_n} \bar{a}_L^p = \bar{m}_q \Gamma_p a^p / m_q \bar{a}^p \tag{6B13}$$

where

$$\bar{a} = \bar{m}_L \bar{\gamma}_{\pm C_l L_c} \tag{6B14}$$

$$\Gamma_p = \left(\frac{\bar{\gamma}_\pm}{\gamma_\pm}\right)_{[pl/cC,ML_q]}^{1+pl/c} \left(\frac{\gamma_\pm}{\bar{\gamma}_\pm}\right)_{[C_{l/c}L]}^{[pl/c]} \tag{6B15}$$

or, for a 1:1 electrolyte CL

$$\Gamma_p = \left(\frac{\bar{\gamma}_\pm}{\gamma_\pm}\right)_{[pC,ML_q]}^{1+p} \left(\frac{\gamma_\pm}{\bar{\gamma}_\pm}\right)_{[C,L]}^{p} \tag{6B16}$$

Concentrations of species $\overline{ML_n^{m-nl}}$ in the resin phase and ML_n^{m-nl} in the aqueous phase can be summed, and the ratio of the sums is the molal distribution coefficient for M

$$D_m = \frac{\sum \bar{m}_n}{\sum m_n} = \frac{\bar{m}_0 \sum_n (\bar{\beta}_n^T / \bar{\gamma}_n) \bar{a}^n}{m_0 \sum_n (\beta_n^T / \gamma_n) a^n} \tag{6B17}$$

The ratio $(\bar{m}_0 / m_0)$ can be eliminated from (6B17) using (6B13) for just one complex ML_q^{m-ql}, selected arbitrarily, as a liaison between the two phases. Thermodynamically equivalent expressions are obtained whatever value for q is selected (in principle even values $q < m/l$, although these will require some alterations in the expression). The resulting general equation is

$$D_m = (K_p^T \bar{\beta}_q^T / \beta_q^T) \frac{\bar{\gamma}_a a^{m/l}}{\Gamma_p \gamma_q \bar{a}^{m/l}} \frac{\sum_n (\bar{\beta}_n^T / \bar{\gamma}_n) \bar{a}^n}{\sum_n (\beta_n^T / \gamma_n) a^n} \tag{6B18}$$

Where only a single complex exists in the resin phase (for instance for a system such as gold(III)–chloride, where anionic species other than $AuCl_4^-$ are improbable), only $n = q$ need be considered in the resin in (6B18), which becomes

$$D_m = (K_p^T \bar{\beta}_q^T / \Gamma_p) a^{m/l} \bar{a}^p / \sum_{n=0}^{N} (\beta_n^T \gamma_q / \gamma_n) a^n \tag{6B19}$$

remembering that $p = q - m/l$.

Where a neutral complex $ML_{m/l}$ exists (i.e. m/l is integral), an alternative treatment, which is as rigorous but more elegant, is possible. Complexes ML_n^{m-nl} are related to $ML_{m/l}$ instead of to M^{m+} (cf. section 3.B.c)

$$\text{ML}_{m/l} + i\text{L}^{l-} \rightleftharpoons \text{ML}_{(m/l)+i} \tag{6B20}$$

where i may be both positive or negative. The thermodynamic constant $\beta_i'^{\text{T}}$ (primed symbols being used to distinguish quantities used in this treatment) is defined as

$$\beta_i'^{\text{T}} = a_{\text{ML}_{(m/l)+i}}/a_{\text{ML}_{m/l}} a_{\text{L}}^i = m_i \gamma_i'/a_{\text{ML}_{m/l}} a^i \tag{6B21}$$

where the activity coefficient functions are

$$\gamma_i' = \gamma_{\pm[\text{ML}_{(m/l)+i,i\text{L}]}}^{1-il/c} \gamma_{\pm[l\text{C},c\text{L}]} \qquad \text{for } i \leqq 0 \tag{6B22a}$$

$$\gamma_i' = \gamma_{\pm[il/c\text{C},\text{ML}_{(m/l)+i}]}^{1+il/c} \gamma_{\pm[l\text{C},c\text{L}]}^{il/c} \qquad \text{for } i \geqq 0 \tag{6B22b}$$

or, when $l = c = 1$

$$\gamma_i' = \gamma_{\pm[\text{ML}_{m+i,i\text{L}]}}^{1-i} \gamma_{\pm[\text{C},\text{L}]}^{i} \qquad \text{for } i \leqq 0 \tag{6B22c}$$

$$\gamma_i' = \gamma_{\pm[i\text{C},\text{ML}_{m+i}]}^{1+i} \gamma_{\pm[\text{C},\text{L}]}^{-i} \qquad \text{for } i \geqq 0 \tag{6B22d}$$

At equilibrium with the resin, the simple relationship (cf. section 4.B.e)

$$\bar{a}_{\text{ML}_{m/l}} = a_{\text{ML}_{m/l}} \tag{6B23}$$

holds. Summing in both phases over the various species yields

$$D_m = \frac{\sum_i (\bar{\beta}_i'^{\text{T}}/\bar{\gamma}_i)\bar{a}^i}{\sum_i (\beta_i'^{\text{T}}/\gamma_i)a^i} \tag{6B24}$$

and if only a single complex ML_q^{m-ql} exists in the resin phase, the sum in the numerator is replaced by $(\bar{\beta}_p'^{\text{T}}/\bar{\gamma}_p)\bar{a}^p$, where again $p = q - m/l$.

There are a number of quantities in (6B18) and (6B24) which are not directly accessible experimentally. Thus, although thermodynamically rigorous, these equations are not immediately useful. The somewhat special definitions of the activity coefficient functions γ_n (6B12) or γ_i' (6B22) and Γ_p (6B16), however, have been introduced in order to derive from (6B18) or (6B24) expressions which can be directly related to measured quantities. For this purpose a number of working assumptions are necessary.

Equation (6B11) defines a, the effective ligand activity. It was shown earlier (in section 1.D.f) that this quantity is a realistic estimate of the ligand activity in solution, and that starred complex formation constants defined as

$$\beta_n^* = \beta_n^{\text{T}}/\gamma_n \tag{6B25}$$

or in terms of the uncharged complex, with primed symbols

$$\beta_i'^* = \beta_i'^{\mathrm{T}}/\gamma_i' \tag{6B26}$$

can be considered as constant, independent of a, over the ranges of concentration where the complex ML_n^{m-nl} is of importance (section 3.B.b). It is further assumed that this constancy of β^* holds also in the resin phase, although in general $\bar{\beta}_n^* \neq \beta_n^*$ and $\bar{\beta}_i'^* \neq \beta_i'^*$. This assumption is plausible in view of the high concentration involved in the resin phase. The assumption that β_n^* does vary with concentration, but that $\bar{\beta}_n^*$ does not[38], seems to be less plausible than the one used here, and has not yet been sufficiently tested.

A second assumption involves the use of $\bar{a}$ (6B15) in the resin phase. In this phase there is a solution of a mixed electrolyte: the resinate $l\mathrm{R}^+, \mathrm{L}^{l-}$, and imbibed $l\mathrm{C}^{c+}, c\mathrm{L}^{l-}$. It is assumed that $\bar{\gamma}_{\pm \mathrm{C}_l \mathrm{L}_e}$ expresses the effective activity coefficient of the ligand anion L^{l-}, irrespective of the presence of two different cations in the resin-phase solution. It has been shown that Harned's rule is obeyed by the resin solution (section 4.B.e), and in principle it is possible to calculate a mean value of $\bar{\gamma}_{\pm [(\mathrm{C}^{c+}, \mathrm{R}^+), \mathrm{L}^{l-}]}$ for the mixture, using properly weighted values of $\bar{\gamma}_{\pm \mathrm{C}_l \mathrm{L}_e}$ and $\bar{\gamma}_{\pm \mathrm{R}_l \mathrm{L}}$. This, however, is rather difficult, because data on the activity coefficient of the resinate are usually not easily available. Values of $\bar{\gamma}_{\pm \mathrm{C}_l \mathrm{L}_e}$, on the other hand, can readily be measured from the Donnan law expression

$$\bar{a}_{\mathrm{C}_l \mathrm{L}_e} = a_{\mathrm{C}_l \mathrm{L}_e} \tag{6B27}$$

The effective resin ligand activity is calculated from

$$\log \bar{a} = \log a + \frac{l}{c+l}(\log \bar{m}_\mathrm{L} - \log \bar{m}_\mathrm{C} + \log(l/c)) \tag{6B28}$$

from measurements of $\bar{m}_\mathrm{L}$ and $\bar{m}_\mathrm{C}$ (cf. section 4.B.e). An alternative method for obtaining $\bar{a}$ has been discussed in sections 4.B and 4.D, employing anion exchange distribution results for stable anions.

A third assumption is made, which is explicit for the treatment represented by equations (6B10) to (6B19), but has been absorbed in the treatment represented by equations (6B20–6B24), by extending the first assumption also to the resin phase. It is that the expression $\bar{\gamma}_q/\bar{\Gamma}_p \gamma_q$ (which equals

$$(\gamma_{\pm M,mL}/\bar{\gamma}_{\pm M,mL})^{1+m}(\bar{\gamma}_{\pm C,L}/\gamma_{\pm C,L})^m \quad \text{for } c = l = 1,$$

and a similar expression also, for other types of electrolyte $C_l L_c$), is independent of a (and of $\bar{a}$). To justify this assumption it should be noted that this expression involves only ratios of the activity coefficients of the same species in the two phases. It has been shown for a number of electrolytes that this ratio approaches constancy over a wide range for moderate to high values of α (section 4.B.e). This assumption can be expressed by taking

$$K^* = (K_p^T \beta_q^{T}/\bar{\beta}_q^T)\bar{\gamma}_q/\Gamma_p \gamma_q \tag{6B29}$$

to be constant.

It is further convenient to assume, at least as a first hypothesis, that only one complex species in the resin needs to be considered. At the high ligand activities in the resin this is usually, but not always (see section 6.B.f), the coordinatively saturated complex. One of the problems outstanding in the Marcus–Coryell treatment is finding which complex represents the predominant species in the resin, i.e. what is the value of p. The following treatment will be in terms of a definite species ML_q^{m-ql}, when only one species can exist in the resin, or in terms of a mean, or a predominant, species, if more than one must be considered.

A final assumption is that the proportionality coefficient, which relates concentrations in the resin phase, expressed per unit weight of resin matrix (e.g. per gram of weighed-in air-dry resin), to the resin molality is constant, independent of a. This is tantamount to assuming constancy of the equivalent water content of the resin. For distribution coefficients determined by the column method, the alternative assumption can be made that the volumes of the resin matrix and of the imbibed solutions do not change with changing a. These assumptions are often far from realistic, and moreover, the changes in water content or volume can rather easily be corrected for, so that the assumptions are really not necessary. Still, in order to interpret the many published results, where corrections cannot be calculated, they can be made as a first approximation. If they are made, the measured distribution coefficients for the tracer metal, D, are simply equated in the following with D_m, the proportionality factors being absorbed by the constants.

If the above assumptions are accepted, and using (6B25), (6B26) and (6B29), expression (6B19) can be written as

$$\log D = \log K^* + p\log \bar{a} + m/l \log a - \log \sum_{n=0}^{N} \beta_n^* a^n \qquad (6B30)$$

and expression (6B24) as

$$\log D = \log \beta_p'^* + p\log \bar{a} - \log \sum_{i=-m/l}^{N-m/l} \beta_i'^* a^i \qquad (6B31)$$

treating K^*, β^*, β'^* and p as constants. Evidently K^* can be indentified with $\beta_p'^* \beta_{m/l}^*$. The distribution curve, $\log D$ as a function of $\log a$, can be plotted from the experimental data, and analysed according to equations (6B30) or (6B31). Differentiation yields the slope

$$(\partial \log D / \partial \log a) = p(\partial \log \bar{a} / \partial \log a) + m/l - \bar{n}$$
$$\qquad (6B32)$$
$$= p(\partial \log \bar{a} / \partial \log a) + \bar{\imath}$$

Partial differentiation is used in (6B32) since it is necessary to keep variables other than a (e.g. m_C) constant. The derivatives of the last terms in (6B30) and (6B31) are equal to $\bar{n}$ and $\bar{\imath}$ respectively (cf. equations 3B34 and 3B36), the average charge number $\bar{\imath}$ being equal to $m/l - \bar{n}$, i.e. simply related to the average ligand number $\bar{n}$.

It is clear that the distribution curve will *not* show a maximum for $\bar{\imath} = 0$ or $\bar{n} = m/l$, i.e. for a value of a that produces the maximal concentration of the uncharged complex, as presumed by Fronaeus[32], unless the derivative $(\partial \log \bar{a} / \partial \log a)$ is vanishingly small (p is not zero for the sorption of anionic complexes). Since $\bar{a}$ is an experimentally determinable function of a, the derivative can be calculated and used in (6B32), or, alternatively, a 'corrected' distribution function

$$\log D^0 = \log D - p \log \bar{a} \qquad (6B33)$$

can be used, differentiation of which yields the charge number $\bar{\imath}$ directly. The slope of the 'corrected' distribution curve is, thus, equal to the average charge of the complexes in solution, expressed in units of the ligand charge, l. The treatment proposed by Fronaeus[32] is seen to apply to the 'corrected' distribution curve, rather than directly to the experimental one, the main difference being allowance for resin invasion in the Marcus–Coryell treatment[33].

The 'corrected' distribution function may be written as the ratio of a constant, K^* or $\beta_p'^*$, and a complexity function, $X = \sum \beta_n^* a^n$ or $X' = \sum \beta_i'^* a^i$, and can be solved for the effective complex formation constants β_n^* or $\beta_i'^*$ by the standard methods discussed in section 3.B. The fraction of the metal in the form of the coordinatively saturated complex, ML_N^{m-Nl}, in the aqueous phase, may be calculated from

$$\alpha_N = D^0 a^{Nl-m} / \lim_{a \to \infty} D^0 a^{Nl-m} \tag{6B34}$$

The denominator of the last equation being equal to K^*/β_N^*. The constant K^* may be obtained by extrapolating $D^0/a^{m/l}$ to $a = 0$ (cf. equation 6B7).

d. The Kraus–Nelson Treatment

Kraus and Nelson[34] have proposed a treatment of the problem of analysing distribution curves for anion exchange of metal complexes in a somewhat different manner, although thermodynamically as rigorous as the above discussion. Considering, for the sake of simplicity, that the bulk electrolyte is of 1:1 type, CL, and that only $ML_q^{p-}(p = q - m)$ is sorbed, they write the mass-action expression[38] for the exchange reaction (6B13)

$$\frac{K_p^T}{G_p} = \frac{\overline{(ML_q^{p-})} m_L^p}{(ML_q^{p-}) \bar{m}_L^p} = \frac{\overline{(ML_q^{p-})}}{\sum \bar{m}_n} \times \frac{\sum m_n}{(ML_q^{p-})} \times \frac{m_L^p}{\bar{m}_L^p}$$

$$= D_m \frac{1}{\alpha_q} \times \frac{m_L^p}{\bar{m}_L^p} \tag{6B35}$$

or in the form

$$\alpha_q = D_m \left(\frac{G_p}{K_p^T} \right) \left(\frac{m_L}{\bar{m}_L} \right)^p \tag{6B36}$$

where the only new symbol is G_p, the appropriate product of the ionic activity coefficients of the ions involved in the exchange (6B13). Thus, if $\bar{m}_L$ and D_m are known as functions of the ligand concentration, m_L, and if the charge p of the resin complex is known, and the values of (G_p/K_p^T) as a function of m_L can be estimated, the fraction of the metal in the form of the complex ML_q^{p-} in the aqueous phase, α_q, can be determined. The ion fractions α_q are the primary results of Kraus and Nelson's treatment, and complex formation constants can be calculated from them by the standard methods of section 3.B.

Under conditions of high ligand concentrations, the fraction α_q approaches unity, and the parameter (K_p^T/G_p) may be evaluated by plotting $D_m(m_L/\bar{m}_L)^p$ as a function of m_L. It is observed that in some cases (K_p^T/G_p) does not vary with m_L (e.g. for absorption of trace chromate ion from 0.02–0.3 m ammonium sulphate) or varies linearly with it (e.g. for absorption of trace tetrachlorogallate ion from hydrochloric acid solutions above 8 m). If such behaviour is assumed to hold also for other, similar, systems (e.g. trace disulphatouranate(VI) in 0.02–0.3 m ammonium sulphate, or tetrachlorogallate in hydrochloric acid below 8 m), the ion fractions α_q can be evaluated from the experimental data.

It is seen that this treatment differs from that of Marcus and Coryell in assuming a different combination of activity coefficient to remain constant, or to vary regularly with ligand concentration in a manner obtainable from data on other systems. When $q = N$ (the cases actually discussed by Kraus and Nelson), it is possible to compare the results of the two methods, using (6B34) and (6B36), such a comparison being shown in Figure 6B1. Different assumptions are involved in the two methods, and it is difficult to judge which more nearly represents the 'true' behaviour.

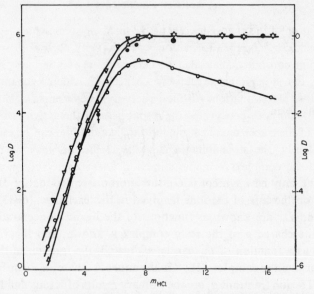

FIGURE 6B1. Comparison of the fraction α_4 for the gallium chloride system obtained by different methods. ⊙ Experimental distribution coefficients, D_{Ga}, ● α_4 from Raman data, both from ref. 34; △ α_4 calculated from equation (6B36) (the Kraus and Nelson method); ▽ α_4 calculated from equation (6B34) (the Marcus and Coryell method).

Differentiation of (6B35) yields after rearrangement

$$\frac{d\log D_m}{d\log m_L} = p\frac{d\log \bar{m}_L}{d\log m_L} - \frac{d\log G_p}{d\log m_L} + \frac{d\log \alpha_q}{d\log m_L} - p \qquad (6B37)$$

The last two terms on the right-hand side of (6B37) may be converted into a term representing the average charge of the complexes in solution $\sum_{n=0}^{N}(m - n)\alpha_n$, and an activity coefficient term

$$\sum_{n=0}^{N} \alpha_n(d\log\gamma_{ML_n}\gamma_L^{m-n}/d\log m_L).$$

If this term and the first two terms on the right (i.e. activity coefficient ratios in the resin phase and variation of the invasion by the electrolyte) are negligible, the slope yields the mean charge, or when the ligand carries a charge l^-, the charge number $\bar{i}$. If these restrictive conditions are assumed to hold[34], it is unnecessary to assume that only one species is absorbed in the resin for calculating the mean charge in solution, as shown already above (section 6.B.c), and by Fronaeus[32].

e. Activity Coefficients

The variation of the properties of a metal–ligand system, which is expressed explicitly by complex formation reactions with their corresponding equilibrium constants, may also be expressed in another, less detailed way, by the stoichiometric activity coefficient. Anion exchange data permit the determination of these activity coefficients, again using certain assumptions concerning the resin phase. If the activity coefficient product of the tracer metal cation and the ligand anion can be estimated independently, the stoichiometric activity coefficients may be used to calculate the complex formation function and stability constants.

Fomin[39] applied these considerations to the anion exchange of tracer cadmium in potassium chloride solutions. The distribution coefficient is expressed as

$$D = K_p^T \beta_{2+p}(\overline{\mathrm{Cl}})^p y_{Cd^{2+}} y_{Cl^-}^{2} / \sum \beta_n \left(\frac{y_{Cd^{2+}} y_{Cl^-}^n}{y_{CdCl_n}^{2-n}}\right)(\mathrm{Cl})^{n-2} \qquad (6B38)$$

where molar concentrations and molar activity coefficients y are used.

Fomin assumed activity coefficients and the ligand concentration in the resin to be independent of the variation of the aqueous potassium chloride concentration. (It is then immaterial what complex undergoes exchange, whether $CdCl_3^-$ as assumed by Fomin, or another). He also, implicitly, assumed the activity coefficient product in the denominator of (6B37) to remain constant, and used literature data[40] to calculate the value of this denominator. Finally, from the ratio of D for two chloride concentrations, the ratio of the products $y_{Cd^{2+}} y_{Cl^-}^2$ for these concentrations can be calculated. Using data for 1 M and 2 M KCl, he obtained the value 0.67 for the ratio of the activity coefficient products, which he considered 'reasonable'.

The stoichiometric activity coefficient of a salt can, of course, be obtained directly from the Donnan law (section 4.B), without consideration of complex formation reactions in either phase. Such an approach was used by Harris[41], to calculate the mean ionic activity coefficient of cadmium chloride, bromide and iodide in the resin, from the distribution of cadmium halide (without excess of ligand) between two phases, and the known coefficients for the aqueous phase. Using the ionic activity coefficients given by Kielland (section 1.C.c), and published complex stability constants, he estimated ionic activity coefficients for the hypothetical uncomplexed cadmium halide in the resin, and found them to be approximately similar to those of barium or cobalt at similar concentrations. The stoichiometric activity coefficients in the resin, on the other hand, were found to be about twenty to forty times smaller than for the other two cations. This may be interpreted as indicating extensive complex formation in the resin phase. The significantly different effect of benzyltrimethylammonium chloride on the activity coefficient of cadmium chloride in aqueous solutions from that shown by hydrochloric acid[42] is interesting in this connexion.

Activity coefficient ratios $\Gamma = \bar{\gamma}_{\pm MCl_2}/\gamma_{\pm MCl_2}$ were obtained (cf. 4B18) by Nelson and Kraus[43] for cobalt (at low molalities) and barium chloride, and they are sufficiently similar to those of completly dissociated alkali metal chlorides, and higher than those for complexed cadmium chloride, to conclude that no appreciable complex formation in the resin phase occurs in these cases. At external cobalt chloride molalities above 1 m, the resin becomes intensely blue coloured, and Γ_{CoCl_2} decreases considerably below the values for Γ_{BaCl_2}, so that complex formation in the resin is clearly indicated.

f. The Resin-phase Complex

Distribution measurements, which constitute most of the information on the sorption of metal complexes on anion exchangers, can at best indicate the average species in the two phases, but not the details of the sorption mechanism. Nevertheless, there have been some discussions in the literature concerning the so-called 'major sorption reaction'. Some conceivable reactions are

$$ML_{\tilde{n}}^{m-\tilde{n}} + (q - \tilde{n})L^- + p\bar{R}^+ \rightleftharpoons \overline{R_pML_q} \qquad (6B39)$$

$$ML_q^{q-m} + p(\bar{R}^+, \bar{L}^+) \rightleftharpoons \overline{R_pML_q} + pL^- \qquad (6B40)$$

$$ML_m \rightleftharpoons \overline{ML_m} \qquad (6B41a)$$

$$\overline{ML_m} + p(\bar{R}^+, \bar{L}^-) \rightleftharpoons \overline{R_pML_q} \qquad (6B41b)$$

where, for the sake of simplicity, a monovalent ligand L^- is considered, and $\overline{ML_q^{q-m}}$ is the major species in the resin, probably ion-paired with resin sites $\bar{R}^+$ or invading cations $\bar{C}^+$. Reaction (6B39) considers the reaction of the generalized species predominant in the aqueous phase, $ML_{\tilde{n}}^{m-\tilde{n}}$, with sufficient additional ligands and resin cations (generalized by $\bar{R}^+$) to give the resin species, if $\tilde{n} < q$, or, if $\tilde{n} > q$, the generalized species is sorbed with concurrent release of $\tilde{n} - q$ ligands from the resin into the solution phase. In reaction (6B40) the attention is focused on the solution species with q ligands, even if it is a minor constituent of the metal complexes in this phase, and considers the anion exchange of this anionic complex with the ligand in the resin. Finally, reaction (6B41) is considered to proceed in two stages: first the neutral complex, again irrespective of its relative importance, distributes between the two phases, unhindered by the Donnan potentials and governed by a Nernst distribution law, and then the neutral species reacts with ligands in the resin to form the major resin species in the second stage. Such a mechanism is, of course, impossible with polyvalent ligands L^{l-}, if m/l is non-integral, unless neutral polynuclear species M_lL_m are stable.

Such a discussion is not very meaningful, since the thermodynamic considerations underlying the distribution measurements are concerned only with the initial and final states of the system, and not with the path the reaction takes. Speculations on this path[26,38,44-47], though interesting, cannot be based on the distribution results. In any case, it cannot be said that the Fronaeus[32] or Marcus–Coryell[33] treatments imply a mechanism according to (6B41), or that the Kraus–Nelson[34] treatment implies a mechanism according to (6B40), the treatments being thermodynamic. Proposed 'reactions' could at most serve as a device to make the calculations clearer[33], without representing the actual reaction mechanism, or affecting the final results. As long as there is no question concerning the actual predominant species in the resin phase, and when relevant kinetic data, such as those showing that the robust Ir^{IV} and Pt^{IV} complexes are indeed sorbed according to equation $(6B40)^{48}$, are not available also for other systems, a discussion of the mechanism of the sorption or elution reactions is meaningless.

An outstanding problem in all the above-mentioned treatments of the anion exchange of metal complexes, is the determination of the metal complex species predominating in the resin. In deriving the 'working equations' (6B19), (6B30), (6B31) and (6B36) from the more general equations (6B18) and (6B24) and the corresponding expression for the Kraus–Nelson treatment, it is assumed that one complex species, with q ligands, predominates in the resin phase. Alternatively, it may be considered that there are more than one species in the resin, of comparable importance, but the mean composition, as measured by a ligand number q, is independent of the ligand concentration. In this case q need not be an integer. The distribution measurements as usually carried out (i.e. using tracer metal concentration, a given resin converted to the ligand form, and varying ligand concentration) do not permit a direct evaluation of q, and it is difficult, in the general case, to obtain a definite value for this important parameter.

In some cases there is no problem, namely when $q = N = (m/l) + 1$, i.e. when there exists only one anionic complex, the coordinatively saturated one, as for example with $GaCl_4^-$ and $Ge(C_2O_4)_3^{2-}$. It is also often argued that the high ligand concentration occuring in the resin will convert the metal cation to the coordinatively saturated complex, irrespective of the existence of lower anionic complexes. Such an argument is probably valid for cases such as HgI_4^{2-}, $Pu(NO_3)_6^{2-}$ and $Fe(C_2O_4)_3^{3-}$. There is, however, often doubt concerning the coordination number, so that it is not always safe to rely on these considerations. An attempt has been made to identify the resin complex for tracer metal experiments from loading results at higher metal concentrations^{24,44,45} (section 6.A.c). This approach is, of course, as inconclusive as the previous ones, because of the uncertainty that the metal complexing will be the same at the two concentration levels.

The difficulty can be illustrated by pointing out cases where the data can be best explained by assuming that the resin species is not the coordinatively saturated one. One case is the silver chloride system^{49}, where the resin species is most probably $AgCl_3^{2-}$, although $AgCl_4^{3-}$ can exist in the aqueous phase. Other examples are that of silver thiosulphate[50], where $AgS_2O_3^-$ seems to be the resin species, compared with $Ag(S_2O_3)_3^{5-}$ in the aqueous phase, and silver cyanide, where an infrared investigation[51] showed the predominant species in the resin to be $Ag(CN)_3^{2-}$, rather than the coordinatively saturated $Ag(CN)_4^{3-}$, predominant in solution. That the effect is not confined to silver complexes may be seen by considering the cases of lanthanum thiosulphate[52], where $La(S_2O_3)_2^-$ is the resin species, while $La(S_2O_3)_3^{3-}$ can be formed in solution, and that of gadolinium glycolate, where the species are $Gd(HOCH_2CO_2)_4^-$ in the resin and $Gd(HOCH_2CO_2)_6^{3-}$ in solution[53]. Steric hindrance may be a factor which inhibits the formation of the highest complex in the resin. Another factor may be the relatively low effective dielectric constant in the resin (section 4.C.d), which will hinder association of an anionic complex with a ligand anion to give a more highly charged complex, at least compared with the reaction in the aqueous phase.

The best method to determine the composition of the predominant metal complex in the resin, is to study its formation as a function of the resin ligand concentration. This, however, must be done at a constant average composition of the aqueous complexes, i.e. without varying the ligand concentration in the aqueous phase, which is possible only when the resin itself is varied. By using resins of different swelling properties (crosslinking) or capacities, it is, in principle, possible to vary $\bar{m}_L$ or $\bar{a}_L$ independently of m_L. A variation of $\bar{m}_L$ can be brought about by varying either: a) the capacity, b) the crosslinking, hence the swelling, c) the degree of proton association in a weakly basic resin by varying the pH, or d) the degree of proton association with the ligand, reducing the free-ligand concentration. Examples of actual experimental applications of these possibilities are not abundant. Trofimov and Stepanova[54] used resins of different swelling capacities to determine the charge of the zirconium nitrate complex absorbed. A more complete description of this method may be found in the work of Marcus and Maydan[55]. These authors studied the distribution of zinc, indium, iridium(III) and iodide tracers between sodium chloride solutions and anion exchange resins with 2, 4, 8, 10 and 16% crosslinking. The slope of a plot of log D against the crosslinking at constant a was found to be a function of p, which permits the distinction at least of $p = 1$ from $p > 1$. For the former case the slope does not exceed 0.02 units in log D per 1% increases in crosslinking (the value being 0.006 for $Ir(H_2O)_2Cl_4^-$, 0.010 for $In(H_2O)_2Cl_4^-$ and 0.018 for I^-), while for the latter a much larger slope was obtained

(0.070 for $ZnCl_4^{2-}$). Further details of this work will be discussed in the next section.

The acid effect (section 6.C.e) provides another means of varying $\bar{m}_L$, where species such as HL_2^- form in the resin, which do not interact to form metal complexes. For the lanthanide nitrates it was shown[56] that $p = 4$ for La^{III} and $p = 2$ for Eu^{III} and Yb^{III} from the decrease of D with acidity at constant m_L.

The effect of a varying capacity on the selectivity of an anion exchanger towards complexes of different charges has been studied[57] in relation to complex cyanides. It was shown that whereas the sorption of monovalent $Au(CN)_2^-$ or $Ag(CN)_2^-$ did not decrease to a great extent, when a weakly basic resin at increasing pH values, or strongly basic resins of decreasing capacities, were used, sorption of multivalent nickel, copper and iron cyanides decreased very markedly. This was explained by noting that a monovalent ion can approach a resin site to the optimal distance irrespective of the spacing of the groups in the resin, while a multivalent ion must seek an optimal position, at mean distances from resin sites which increase as the site density decreases.

In a few cases it is possible to determine the predominant complex in the resin by spectroscopic means, although this naturally necessitates much higher concentrations than the trace concentration considered hitherto, so that, again, the results need not be applicable at the lower concentration level. An infrared absorption study[51] showed that the gold(I) cyanide species sorbed (to an extent of 47 weight %) on an anion exchanger is $Au(CN)_2^-$, the same as in aqueous solutions. Spectrophotometry in the visible region has been applied using both the reflectance and the absorption techniques. The former was used[58] to identify the absorbed cobalt species on a resin in chloride form as $CoCl_4^{2-}$, from the similarity of the spectrum to the absorption spectrum in aqueous solution. Although this identification may be mistaken, since considerable doubt has been cast[59] on the identification of the highest aqueous solution complex with $CoCl_4^{2-}$ (the actual species, although definitely tetrahedral in structure, being possibly $Co(H_2O)Cl_3^-$) new evidence[60] supports the species being $CoCl_4^{2-}$. The data presented[58] for identification of the resin species for the iron(III) chloride system with $FeCl_4^-$ are too few to be conclusive. Some recent results point to the copper(II) species on an anion exchanger being $CuCl_4^{2-}$, again from comparison of spectra of the loaded resin with those of definitely established species in solids and in solution[61]. A considerable amount of work was done on the spectrophotometric identification of the tetra- and hexavalent uranium, neptunium and plutonium chloride and nitrate species absorbed cn anion exchangers[62], from comparison with aqueous and long-chain amine organic solutions and solid compounds. The tetravalent species in chloride-form resin were found to be MCl_6^{2-}, the hexavalent species mainly $MO_2Cl_4^{2-}$ and hydrogen-bonded $H_3O^+(H_2O)_3(MO_2Cl_4^{2-})_2$, when sorbed from alcoholic and aqueous solutions respectively. In nitrate-form resin both tetranitrato and trinitrato species, $MO_2(NO_3)_4^{2-}$ and $MO_2(NO_3)_3^-$ were found, the former predominating.

A final method to be mentioned, of which little use has been made as yet[63], is the application of the dependence of the diffusion rate in the resin on the charge of the diffusing species, for determining the predominant species in the resin. There are not too many data available concerning the rate of diffusion of simple anions, and its dependence on ion charge and size (Table 4A1), but this source of information should be explored.

C. FACTORS AFFECTING THE DISTRIBUTION

a. Introduction

In the previous section, the dependence of the distribution coefficient D of a tracer metal between an anion exchanger and aqueous solutions, on the ligand concentration, has been discussed. A variation in ligand concentration causes changes in the distribution coefficient through a number of mechanisms, expressible as chemical reactions. On the one hand, it affects the equilibria in the aqueous phase (e.g. reaction 6B20), causing complexes with more and more ligands to predominate as the ligand concentration increases. Since D is proportional to the concentration of the species of mean composition $ML_{m/l}$, it will increase to a maximum and then decrease, parallel to the relative concentration of $ML_{m/l}$, unless other factors operate. The main other factor is the variation of the ligand concentration in the resin with its concentration in the solutions. Since D depends on the pth power of the ligand concentration in the resin, through reactions such as (6B30) or (6B31), another source for the dependence of D on the ligand concentration is apparent.

These are more or less obvious effects on D which may be written in terms of chemical reactions with definite stoichiometric coefficients. Beside these it has been noted in the previous section that the non-ideality of the solutions in both phases has important effects, which have been discussed in detail. There are, however, still many other factors which can affect the magnitude of the distribution coefficients. These will be discussed in this section. Some of them, again, may be described in terms of chemical reactions: side-reactions of the metal ion (e.g. hydrolysis) or of the ligand (e.g. acid association) in solution, association of anionic complexes in solution with hydrogen ions to form acids, association of the anionic complexes in the resin with mobile cations or with the functional group, etc. Other effects are more complicated: the ionic medium, the secondary cation effect, the metal loading on the resin, the degree of crosslinking and swelling, the temperature, etc. The ability to predict the distribution coefficients of a given metal–ligand system depends strongly on the understanding and quantitative knowledge of all these factors.

b. Side-reactions

Even if there is just one ligand in the system, there still exists the possibility of competing complexation reactions, namely with the hydroxide anion, through hydrolysis of the metal ion. Some metal cations are strongly hydrolysed even in acid solutions, and a variation of the degree of hydrolysis may accompany changes in the complexes formed, as the ligand concentration varies. This is the case with highly charged cations, such as Mo^{VI}, Sn^{IV}, Pa^{V}, etc., with ligands such as fluoride and chloride. The case of Mo^{VI} in hydrochloric acid[64, 65] may serve as an illustration. The distribution coefficient D of molybdenum, present at low concentrations, between Amberlite IR–400 anion exchange resin in chloride form, and hydrochloric acid solutions, has been measured as a function of the concentration of the latter. D was found to decrease initially from 105 at 0.25 M HCl to a minimum of 11 at 1.0 M HCl, to increase hence to a maximum of 105 at 5 M HCl, and to decrease again slightly at higher concentrations. The equations derived in the previous section, such as (6B32) or (6B37), describing the change of D with ligand concentration, cannot account for the observed behaviour. A minimum in the distribution curve can be obtained, however, if terms that decrease with increasing ligand concentration are included in the numerator of the expression for the distribution coefficient. Such terms may arise from the absorption on the resin of hydrolysed anionic species at low ligand concentration (which in hydrochloric acid solutions mean also low acidities), the relative concentration of which decreases faster than the relative concentration of the absorbed chloride complex increases, as the concentration of hydrochloric acid increases. These hydrolysed species may be the tetrameric $Mo_4O_{13}^{2-}$ or other polymeric species, which interconvert and shift towards chloride complexes as the hydrocholoric acid concentration increases[66, 67], e.g.

$$MoO_4^{2-} \rightleftharpoons Mo_4O_{13}^{2-} \rightleftharpoons HMo_4O_{13}^{-} \rightleftharpoons MoO_2^{2+} \rightleftharpoons MoO_2Cl_2 \rightleftharpoons MoO_2Cl_3^{-}$$

As the acid concentration increases, the concentration of absorbable species, such as MoO_4^{2-} and $Mo_4O_{13}^{2-}$ first decreases[67], until enough $MoO_2Cl_3^{-}$ is formed to increase the resin absorption again.

With ligands, which are the anions of weak acids, it is possible to vary the ligand anion concentration at constant total ligand concentration by varying the pH. The most serious difficulty, however, arises from the possibility of variously associated ligands existing in the resin phase. It has been noted previously (section 4.C.b) that especially with polyvalent weak acids, such as sulphuric or phosphoric, the resin will contain mixtures of ligand species, depending on the pH. The distribution of metal complexes under these conditions is, again, a complicated function of ligand concentration and pH. As an illustration the case of the U^{VI}–phosphate system[68] may be used. This system has been studied at the low pH range, where only H_3PO_4 and $H_2PO_4^{-}$ need be considered. It was, indeed, found that a variety of complex species, containing both forms of phosphate as ligand, exist in the aqueous solution. In the resin, however, the uncharged H_3PO_4 is salted-out by the high concentration of ions, to such an extent that its concentration becomes negligible, and the only uranyl complex that has to be considered in the resin phase is $UO_2(H_2PO_4)_3^{-}$. In other cases it has been found possible to select pH conditions under which one or the other hydrogen-ion-associated forms of the ligand predominates in both phases to the virtual exclusion of the other. This seems to have been the case for the studies of the rare earths in carbonate[69] and bicarbonate[70] solutions. Each form seems to behave as a simple stable anion in the range of its application. A general mathematical treatment of the hydrolysed-metal and associated-ligand cases has been reported recently by Coleman[71].

In solutions containing more than one ligand, complexes of the metal with each ligand, and mixed complexes may be formed, not all of which are absorbed by the resin with high affinity. This is the case in mixed hydrochloric–hydrofluoric acid solutions. The concentration of fluoride ion in both phases is very low, and chloride complexes seem to be preferred to fluoride complexes in the resin[72]. In solution, however, fluoride complexes are formed, sometimes even more readily than chloride complexes, as for zirconium or niobium, and the distribution coefficients, affected by formation of fluoride complexes, may be used to estimate their stability[72]. Distribution coefficients in the absence (D_{Cl}) and in the presence (D_F) of certain fixed concentrations (e.g. 1.0 M) of hydrofluoric acid and at various hydrochloric acid concentrations are compared, and a function

$$F = 1 - D_F/D_{Cl}R \qquad (6C1)$$

is calculated. Even in systems where fluoride complexes are not formed to any appreciable extent, it has been found that D_F and D_{Cl} are not equivalent, probably because of the effect of invading HF on activity coefficients in the resin, but the two curves are parallel, so that a factor $R = 0.63$, independent of hydrochloric acid concentration, has been introduced. It can be shown that the extent of fluoride complex formation may be obtained from data at different fluoride concentrations from the equation

$$\frac{1}{1-F} = 1 + \sum_{i=1}^{N} \beta_{i(F)} \frac{(HF)^i}{(H^+)^i} \Big/ \sum_{i=0}^{N} \beta_{i(Cl)}(Cl)^i \qquad (6C2)$$

c. The Constant Ionic Medium

In general, if two anions are present in the solution together, they will be present also in the resin. If, however, one of the anions does not form complexes with the metal, some information may be gained from the dependence of the distribution coefficients on the concentration of this anion. In this case, the composition of the resin phase varies with that of the external solution, and moreover, conditions of low ligand concentration in the resin may be obtained, so that a stepwise formation of complexes in the resin can, in principle, be studied. If an inert anion is permitted in the system, the advantages of the constant ionic medium method (section 1.E.e) are, naturally, exploited, the ionic strength in the external solution being held constant. Whether a constant medium obtains also in the resin phase depends on the degree of exchange, and on the mode of expressing concentrations. The water content of the resin may change appreciably with resin composition, and with it the swelling, so that both the molality and molarity of ions in the resin vary—even if invasion is negligible—although the total number of equivalents (per equivalent of resin sites) remains constant. Eliezer and Marcus[73] discussed the general problem, and pointed out that only at small substitutions of ligand anions for medium anions in the resin can the medium be considered constant. The choice of perchlorate as the 'inert' anion is helpful in this respect, because of its high relative affinity for the resin (Table 4C2), which assures a small degree of substitution in the resin, even for considerable degrees of substitution in the external solution. This choice, however, requires that the metal complexes have a high affinity for the resin, as measured by high values of K_p^T of equation (6B19) or of $\bar{\beta}_p'^T$ of equation (6B24), as otherwise the distribution coefficients will become too small for accurate measurement.

A fairly constant medium in the resin phase has been achieved in the study of the mercury halide complexes[73] at 0.3, 0.5 and 3.0 M total ionic strength. While the degree of substitution of halide for perchlorate in the external solution extended up to 33, 30 and 8% respectively, the maximal degree of substitution in the resin for the most strongly absorbed ligand, iodide, was only 12%. It can, therefore, be safely assumed that variations in activity coefficients, as well as in the equivalent water content and swelling, may be neglected. The distribution coefficient for mercury tracer (where $(L^-) = c_L$) may, therefore, be expressed as

$$D = \frac{\overline{(HgL_2)} + \overline{(HgL_3^-)} + \overline{(HgL_4^{2-})}}{(HgL_2) + (HgL_3^-) + (HgL_4^{2-})} = K' \frac{1 + \bar{k}_3(\bar{c}_L) + \bar{k}_3\bar{k}_3(\bar{c}_L)^2}{1 + k_3(c_L) + k_3k_3(c_L)^2} \quad (6C3)$$

where K' is the distribution constant of the neutral species HgL_2 (cf. equation 6B2). Knowledge of the variation of $\bar{c}_L$ with c_L, which proceeds according to

$$\bar{c}_L = K_{ClO_4}^L \, \bar{C} \, c_L / [C - c_L(1 - K_{ClO_4}^L)] \quad (6C4)$$

where $K_{ClO_4}^L$ is the selectivity coefficient, $\bar{C}$ the resin capacity and C the medium concentration, permits, in principle, estimation of the five parameters K', $\bar{k}_3$, $\bar{k}_4$, k_3 and k_4 from the variation of D with c_L. In practice it turned out that equation (6C3) is not sufficiently sensitive to the variation of c_L to permit the evaluation of five parameters. The course taken was to use published values of k_3 and k_4 for the $C = 0.5$ M medium, and measured values of K' for the three media to calculate the remaining parameters $\bar{k}_3$ and $\bar{k}_4$, and from these, in turn, to calculate k_3 and k_4 for the remaining media $C = 0.3$ M and $C = 3.0$ M, assuming invariance of the two resin-phase constants with external concentration.

Another approach was taken by Fronaeus, Lundqvist and Sonesson[74], who studied the cadmium bromide system. They used equation (6B4), which is formally valid irrespective of the presence of inert medium anions. Differentiation with respect to $\bar{c}_L$ yields

$$\frac{d\log D}{d\log \bar{c}_{L}} = \bar{\bar{n}} - m/l + (m/l - \bar{n})(d\log c_{L}/d\log \bar{c}_{L}) \qquad (6C5)$$

Knowing $\bar{n}$ from published data for the external solution of 0.75 M medium concentration, and the variation of $\bar{c}_{L}$ with c_{L}, the derivative in (6C5) yields values of $\bar{\bar{n}}$—the average ligand number in the resin—as a function of $\bar{c}_{L}$, the ligand concentration in the resin. Standard methods are then used to obtain the complex formation constants. Alternatively, from knowledge of X as function of c_{L}, equation (6B3) may be rewritten as

$$kK\bar{X} = DX(\bar{c}_{L}/c_{L})^{m/l} \qquad (6C6)$$

from which values of $kK\bar{\beta}_{n}$ may be obtained. If accurate values of D are available for sufficiently low $\bar{c}_{L}$ that $\bar{X}$ approaches unity, kK may be evaluated, since $\bar{\beta}_{0} = 1$. This is, usually, impossible, so that only the ratios $\bar{k}_{n} = \bar{\beta}_{n}/\bar{\beta}_{n-1}$ may be obtained.

In a number of other investigations, the medium in the resin was not kept constant. In a study of the gadolinium glycolate system[75] at external ionic strength of 0.21 M, substitution of glycolate for perchlorate in the resin was permitted to cover the whole range of 0–100%. As a result the ionic strength in the resin varied between 8.5 and 2.5 M, since hydration of the resin in perchlorate and glycolate forms is very different (0.40 and 1.20 g water/g resin respectively). Under these conditions the unlikely result was obtained that $\bar{n}$ decreases from 4 to 3, as $\bar{c}_{L}$ increases. Only very approximate values of the constants $K\bar{\beta}_{n}$ may thus be obtained.

Waki[76] has derived an equation similar to (6C5), using the Kraus and Nelson treatment (section 6.B.d) as a starting point. The resulting equation is

$$\log D = \log(K_{p}^{T}/G_{p}) + (\bar{\bar{n}} - m/l)\log \bar{c}_{L} - (\bar{n} - m/l)\log c_{L} \qquad (6C7)$$

using molar concentrations for the aqueous phase and mole fractions for the resin phase, redefining D, K_{p}^{T} and G_{p} accordingly. Waki assumed K_{p}^{T} and G_{p} to be independent of $\bar{c}_{L}$, and calculated the derivative of $\log D + (\bar{n} - m/l)\log c_{L}$ with respect to $\log \bar{c}_{L}$. For the mercury(II) nitrate system the value of $\bar{n}$ was assumed to be zero, so that $\bar{\bar{n}} = 2 + d$ $(\log D - 2\log c_{L})/d\log \bar{c}_{L}$. Values ranging from 3.2 to 3.8 were obtained, indicating sorption of $Hg(NO_{3})_{3}^{-}$ and $Hg(NO_{3})_{4}^{2-}$ on the resin, as the nitrate concentration increases. Similarly, for the silver nitrate system[77], assuming $\bar{n}$ to be zero for the aqueous phase, $\bar{\bar{n}} = 2$ was obtained for the resin phase. A second series of experiments was made by loading the resin with the silver nitrate complex, instead of using an inert anion. The resin ligand concentration is then decreased since exchange sites are taken up by the complexes (cf. section 6.A)

$$\bar{c}_L = \bar{C} - (\bar{n} - m)l\overline{(MX_{\bar{n}}^{(m - \bar{n})l})} = \bar{C} - (\bar{n} - m)l\bar{c}_M \qquad (6C8)$$

This value of $\bar{c}_L$ is introduced into (6C7), and using an iteration method, a value of n may be obtained. For the silver nitrate system again a value of $n = 2$ was obtained, indicating that the absorbed complex is $Ag(NO_3)_2^-$. For the uranyl nitrate system[78] Waki and coworkers obtained, by similar methods, that $\bar{n}$ increases from 3.1 to 4.0 as the nitrate ion concentration increases, in agreement with Ryan's spectrophotometric results[27]. For the zinc chloride system these authors[78] assumed $\bar{n} = 4$, and found that the solution species have $\tilde{n}$ values varying from zero to four. Results at low chloride concentrations indicate the presence of $ZnCl_3^-$ in the resin, besides $ZnCl_4^{2-}$.

It must be said, in criticism of Waki's method, that some of the assumptions made seem far from realistic. The function G_p can hardly remain constant as perchlorate ions in the resin are substituted to a large and varying extent by nitrate and chloride ions. The assumption of zero $\tilde{n}$ in solution also seems not to be valid for the systems studied. The derivative d log c_L/d log $\bar{c}_L$ is not negligible (it is somewhat larger than unity), so that a variation in $\tilde{n}$ causes appreciable variations in $\bar{n}$. It is apparent that the equations (6C8), (6C5) and (6C3) are too complicated, because of the non-linear dependence of $\bar{c}_L$ on c_L, to be solved for the required parameters in the general case. The only procedure which seems to be valid is to use known complex formation constants for the external phase to calculate X or $\tilde{n}$, and obtain from the distribution measurements the constants for the resin phase. Some of these constants, obtained for a resin mainly in perchlorate form, are compared in Table 6C1 with the corresponding constants for aqueous 3 M sodium perchlorate medium. It is seen that the constants for the resin phase, although larger than those for aqueous solutions, are not very much larger. Therefore, in estimating the species present, known information for the aqueous solutions may serve as a guide.

d. The Perchlorate Effect

Perchlorate ions have been considered frequently as suitable for providing 'inert' medium anions. The high selectivity of the resin towards this ion,

although an advantage from one point of view (section 6.C.c) has also
its drawbacks. The perchlorate ion is known for its strong depressant
action on the distribution coefficients of metal complexes, the so-called
'perchlorate effect', which may be illustrated by the following examples.

TABLE 6C1. Comparison of resin and aqueous complex formation constants.

Complex	Constant	Resin perchlorate	Aqueous 3 M NaClO₄	Reference
$HgCl_3^-$	$\log k_3$	1.55	0.70	79
$HgCl_4^{2-}$	$\log k_4$	3.10	1.30	79
$HgBr_3^-$	$\log k_3$	2.1	1.6	79
$HgBr_4^{2-}$	$\log k_4$	3.5	2.6	79
HgI_3^-	$\log k_3$	3.1	3.0	79
HgI_4^{2-}	$\log k_4$	4.4	4.4	79
$Gd(HOCH_2COO)_2^+$	$\log k_2$	> 2	2.1	75
$Gd(HOCH_2COO)_3$	$\log k_3$	1.0	0.51	75
$Gd(HOCH_2COO)_4^-$	$\log k_4$	1.2	0.70	75
$CdBr_2$	$\log k_2$	1.9	0.75	74
$CdBr_3^-$	$\log k_3$	1.7	0.88	74
$CdBr_4^{2-}$	$\log k_4$	0.5	0.22	74

Whereas in the absence of perchlorate, D of tracer Hg^{II} extrapolates to
200,000 at 0.1 M NaCl[79], it is only 220, 40 and 5 when 0.2, 0.4 and 2.9 M
NaClO₄ are present in addition to the 0.1 M NaCl[73]. This depression is
somewhat less for the bromide complexes and still less for the iodide
complexes[73, 79]. Similarly, at a total concentration of 3 M HCl + HClO₄
D of tracer Zn^{II} does not change in the range 0.1 to 3 mM HClO₄,
but decreases from 1100 at these concentrations to 160 at 0.1 M and to
5 at 1 M HClO₄[45]. For Tl^{III} tracer again similar observations have been
made[80]. The distribution coefficient of U^{VI} from 1.3 M Al(NO₃)₃ is 300
and it is decreased to 70 when 2.1 M HNO₃ is added (the 'acid effect', see below), but only 0.17 M HClO₄ is necessary to obtain the same depression[81].
A strong depressant action of perchlorate has also been noted for the
sorption of U^{VI} from solutions containing chloride, nitrate and sulphate
ions, made up to a constant ionic strength with perchlorate ions[82]. Even
the very highly absorbed platinum-metals chloride complexes which
cannot be displaced from the resin even with high concentrations of
chloride ions[83, 84] can be effectively removed from the resin with 2 M
HClO₄[84].

The 'perchlorate effect' has thus been observed in a great number of systems, both in chloride and in nitrate solutions, and presumably also in solutions of other ligands. Horne[45] attributed the perchlorate effect to the greater selectivity of the resin towards perchlorate ions, compared with the chloride ligand ions (Table 4C2) in the first place, and to the lower water content of perchlorate-form resin (Table 4B1), in the second. Indeed, with increasing perchlorate concentrations in the aqueous phase, the water content of the resin may be expected to decrease, and the uncharged complex to be salted-out from the resin into the solution phase. On the other hand, complex formation in the resin may be expected to be enhanced as the water content in the resin decreases[55], the more so, the more dehydratable the ligand, and this was indeed observed in the sequence of relative complex stability in the resin, with a perchlorate concentration of around 12 M, compared with 3 M $NaClO_4$, increasing from iodide to bromide to chloride, (Table 6C1).

A complete, quantitative, explanation of the 'perchlorate effect' is still lacking, and would be desirable for any work where perchlorate ions are used to regulate the ionic medium.

e. The Secondary Cation Effect

It has been observed already in the early days of the application of anion exchange to the study and separation of metal complexes, that the distribution coefficients depend on the cation C^{c+} of the bulk electrolyte C_lL_c, supplying the ligand. Even at equal concentrations of ligand there can be large differences between values of D for a tracer metal, for two different cations C^{c+}. This difference is particularly pronounced if one of the cations is the hydrogen and the other the lithium ion, and two examples may suffice to illustrate this point. The distribution coefficient for manganese(II) tracer between Dowex–1 chloride and 12 M lithium chloride is 550, while for 12 M hydrochloric acid it is 4[85]. The distribution coefficient of americium(III) between Dowex–1 nitrate and 8 M lithium nitrate is 220[89], whereas it is below unity for 8 M nitric acid[87]. The effect is, however, not confined to the hydrogen and lithium ions, but is shown among many pairs of bulk cations (e.g. for 5.5 m lithium and ammonium nitrates D for Ce^{III} is 88 and 4 respectively[86], for 5 M sodium and caesium chlorides D for Zn^{II} is 2500 and 60 respectively[44]. Attention has first been drawn to this effect at a time when most data were available for hydrochloric acid solutions, and the 'anomalous' behaviour of lithium chloride solutions[85, 88] caused the phenomenon to be called the 'LiCl effect'. Later, when it became evident that the behaviour of lithium chloride solutions is similar to that of the other alkali chloride solutions, but that hydrochloric acid behaves differently, it was called the 'HCl effect'[89]. The generality of the differences between various cations, however, justifies the name of 'secondary cation effect[44], for this phenomenon.

Many authors have considered the possible causes for the secondary cation effect [44, 85, 89-91]. A large part of it is due to differences in the non-ideality in the two phases among the various cations. If the cations are compared, not at equal concentrations (or equal equivalent concentrations for cations of different charge[86]), but at equal effective activities, and if the distribution curves are corrected for invasion by the different electrolytes (section 6.B.c), using the respective $\bar{a}$ values, then the resulting corrected distribution curve, D^0 as a function of a, becomes independent of the bulk cation for a large number of cases. Illustrations of this may be found for the chloride ligand in studies of the distribution of Zn^{II} tracer between an anion exchanger and lithium, sodium, potassium, ammonium and caesium chloride solutions[86], of iodide tracer with sodium and caesium chloride solutions[91], of Hg^{II} tracer with lithium and sodium chloride solutions[79], and for bromide ligand in studies of the distribution of Hg^{II} and Cd^{II} tracers with lithium and sodium bromide solutions[79]. In all these cases the D^0 curves for the different bulk cations coincide. Unfortunately, activity and invasion data are not available for all the systems, where the variation of the distribution of the tracer metal with bulk cation has been studied, in order to test the above statements further, and apparent discrepancies have been noted, e.g. for the cadmium thiocyanate system[92]. In the case of unsymmetrical electrolytes, the application of the concept of effective activity a is complicated, since it is not immediately justified to suppose that the mean activity coefficient is a realistic estimate of the effective activity coefficient of the ligand[50, 86], so that there is no good basis for a rigorous comparison of the distribution of tracer between an anion exchanger and solutions of electrolytes such as lithium, calcium and aluminium chlorides.

The treatment of the secondary cation effect in terms of differences of a and of $\bar{a}$ at equal concentrations of ligand has thus been shown to be successful for the alkali metal halides, but there remains the problem of the acid effect. For instance, in the Zn^{II} and Hg^{II} chloride systems[79, 91], although good agreement between the D^0 curves for lithium and sodium chlorides was obtained, the D^0 curve for hydrochloric acid deviates. Since the activity coefficients for both phases for lithium chloride and hydrochloric acid are rather similar, the deviation of even the D curve, not to mention the D^0 curve, is unexpected[55], unless some special acid effect is operative. As mentioned earlier, this deviation may attain values of two to three orders of magnitude for concentrated lithium chloride and hydrochloric acid[55,89], or lithium nitrate and nitric acid[85,86]. The acid effect is demonstrated most clearly when small amounts of acid are substituted for the lithium salt, at constant total concentration. For the chloride system such data are available for Zn^{II} tracer at 6.0 M total concentration[44], for Fe^{III} at several total concentrations between 2.9 and 11.4 M^{93}, and for Au^{III} tracer at 10 m total concentration[90]. For the nitrate system there are data for La^{III}, Eu^{III} and Yb^{III} tracers at 4.6 m total concentration[93], (Figure 6C1). The Figure shows that the acid effect is much stronger in the nitrate than in the chloride system, much lower concentrations of acid being required to effect the same relative decrease in the distribution coefficient. The same conclusion may also be drawn from the data for perrhenate tracer[89], where at, say, 4 m macro anion L concentration $\Delta = \log D$ (in LiL) $- \log D$ (in HL) is 0.44 for bromide, 0.58 for chloride and 1.02 for nitrate, Δ increasing with concentration. This order of the acid effect may be correlated with the strength of the acids (section 3.C.b).

Several explanations have been advanced for the specific acid effect. The occurence of the effect for so many metal chloride systems led Kraus and coworkers at first[85] to consider the formation of undissociated, non-absorbable chloro complex acids as an unlikely explanation, and to favour as explanation the occurence of differences in the activity coefficients. Later, however, Kraus and Nelson[90] noted that there are no large differences between the activity coefficients of lithium chloride and of hydrochloric acid in the resin phase, so that this cause is too small to account for the large acid effect. Furthermore, they noted that whereas a large acid effect occurs for Zn^{II}, Ga^{III} and Cd^{II} chlorides, it does not occur for Ag^I chloride. It has been pointed out by Marcus and Maydan[91] that the divergence between the D^0 curves for hydrochloric acid and lithium chloride starts at 0.2 M chloride for Cd^{II}, at 2 M for Hg^{II} [79], at 4 M for Br^- ion[89,90], at 5 M for Zn^{II} and at 9 M for Fe^{III}, and it does not occur before 12 M for Ag^I. An explanation based solely on the properties of the lithium chloride and hydrochloric acid resin media obviously cannot account for this diversity of behaviour. The

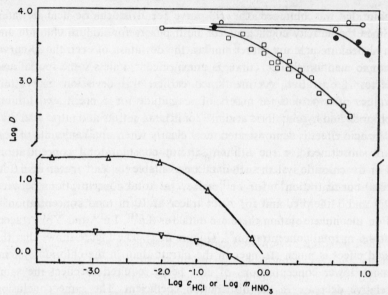

FIGURE 6C1. The acid effect on anion exchange distribution of complexes at a constant ligand concentration. ∇ Yb^{III} in 4.6 m LiNO$_3$, $\triangle$ La in 4.6 m LiNO$_3$, $\square$ Zn in 6.0 M LiCl, $\bigcirc$ Fe^{III} in 8.7 M LiCl, $\bullet$ Fe^{III} in 11.4 M LiCl.

possibility of the formation of undissociated chlorometallic acids, which for some reason are not absorbable in the resin, even not as invading non-electrolyte (section 4.B.d), was therefore advanced by Kraus and Nelson with reasonable confidence[90]. Still later[94], Kraus and Nelson reconsidered the case, in the light of new observations. They reported measurements for the solubility of certain chloroaurates in lithium chloride and hydrochloric acid solutions, and did not find very large differences. They also reported on a Raman study, which showed similar intensity of the characteristic line for Ga^{III} in lithium chloride and hydrochloric acid solutions. These results do not support the conclusion that chloroauric and chlorogallic acids are associated in hydrochloric acid solutions, but neither do they disprove this possibility, since ion-pairing between a hydrogen ion and a tetrachlorogallate ion is not expected to change the symmetry of the bonds of the complex, and affect the Raman line, although it changes the thermodynamic activity of the free tetrachlorogallate ion. Indeed, Ryan[95] has concluded from solubility and spectrophotometric experiments that the undissociated acids HM(NO$_3$)$_6^-$ and H$_2$M(NO$_3$)$_6$, where M is tetravalent Th, U, Np and Pu, are formed in nitric acid solutions, but that these species are not absorbed on an anion exchange resin. Chu and Diamond[89], on the other hand, argue that the chlorometallic acids have been found to be stronger than hydrochloric acid in oxygenated organic solvents and should not be associated in aqueous hydrochloric acid solutions, and certainly hydrobromic acid is stronger than hydrochloric acid in aqueous solutions (section 3.C.b). Still, even for the latter, resin absorption is higher for lithium chloride than for hydrochloric acid, although much less so than for most chloro acids. This argument still does not disprove that selective association, at least to some extent, occurs for the different anions in hydrochloric acid, which is absent in lithium chloride. This could be a partial explanation of the observed results, especially of the diversity of behaviour of the various tracer anions. As pointed out by Horne[44], this is probably the predominant effect in low dielectric constant media, such as in alcoholic solutions (section 6.D.d).

The formation of non-absorbable, undissociated acids in the aqueous phase, however, cannot be a complete explanation for the acid effect. In addition to the difficulty mentioned by Chu and Diamond[89], that some of the proposed species are expected to be rather strong acids in the aqueous phase, there are good indications for the occurence of association in the resin phase in some cases, which provides a good alternative explanation for the acid effect. The low effective dielectric constant in the resin phase (cf. section 4.C.d) should facilitate the association of even as strong an acid as hydrochloric acid, and certainly of acids such as nitric acid[44, 89]. Recently there has accumulated evidence[56] that association may not cease at the stage $H^+ + L^- \rightleftharpoons HL$, which involves invading electrolyte, but continues with anions present as counter ions to form triple ions $HL + L^- \rightleftharpoons HL_2^-$. Of course, it is impossible to distinguish between L^- anions present 'originally', i.e. 'before' association, as counter ions and as invading electrolyte (cf. section 4.C.c). The triple ions may also associate further with the exchange sites, or cations of invading electrolyte C^+, to form ion-pairs $C^+ HL_2^-$. This presupposes that the bonding in HL_2^- is stronger than in an ion-pair, and, indeed, hydrogen bonding has been suggested for this species[96,97]. The stability of the ion-triplet may be seen in the very low values of the ratio of the stoichiometric activity coefficients of the acid in the two phases $\bar{\gamma}_{\pm HL}/\gamma_{\pm HL}$, 0.14 for HCl absorbed from 9.9 m total chloride (LiCl + HCl)[43], and 0.13 for HNO_3 absorbed from 3.9 m total nitrate $(LiNO_3 + HNO_3)$[101], compared with 0.60 and 0.36 for the two acids respectively in absence of the lithium salt, and 0.77 and 0.66 respectively for the lithium salts themselves. This association, whether it stops at the stage of HL, or proceeds to HL_2^-, decreases the concentration of ligand anion L^- available for complex formation with, and binding of, the metal complex, as pointed out by Marcus and Givon[56], and also of the cation H^+ available for associating with and binding of the invading stable anions and anionic metal complexes, a point of view proposed by Chu and Diamond[89]. The latter authors illustrated this view by pointing out that whereas perrhenate and bromide ions, which are anions of stronger acids than hydrochloric acid (as possibly are also the chloro-complex anions), are absorbed less from hydrochloric acid than from lithium chloride solutions, chloride ions, which are anions of an acid weaker than hydrobromic acid, are more absorbed from hydrobromic acid than from lithium bromide solutions. The presence of a high concentration of acid in the exchanger due to invasion will therefore enhance the absorption of tracer ions B^- which are anions of weaker acids, through an exchange reaction $HL + B^- \rightleftharpoons HB + L^-$ in the resin, and repress the absorption of tracer ions which are anions of stronger acids, by decreasing the cation concentration in the resin, compared with what it is in the presence of invading salt only.

The quantitative aspect of the acid effect has been studied on the assumption that non-absorbable species are formed. The parameters obtained from this treatment, e.g. according to the Marcus–Coryell method (section 6.B.c), permit the fitting of the experimental curve for acid solutions, or for acid–salt mixed solutions, adding a term with the acid association constant to the denominator of the expression of the distribution coefficient (6B19 or 6B24). Such a calculation has been made for a number of systems, such as Au^{III}[90], Cd^{II}[98] and Fe^{III}[93] chlorides.

f. Effect of Metal Loading

The discussion in section 6.B and in the paragraphs above pertains mainly to tracer concentration of the metal ion. A considerable amount of information has been accumulated on the anion exchange of metal complexes at macro concentration levels and at appreciable loading of the resin. The loading of the resin is usually expressed as the percent fraction of the equivalent capacity of the resin taken up by the anionic metal complex. Up to 1% loading there is usually very little effect on the distribution coefficients (e.g. ref. 92), but at higher loadings there is often a sharp decrease in D. Sometimes the independence of D of the loading is preserved up to fairly high values, as for example for the Ag^I–chloride system[49]. For 0.125 M HCl log D was found to be 3.80 at 0.5%

and 3.77 at 5.8% loading, and at 4.8 M HCl log D was found to be 1.34 at 0.02% and 1.38 at 2.4% loading. Horne and coworkers[22] found for Zn^{II} log D values not varying between 2 and 7% loading. Above 7% loading, however, the plot of log D against log c_{Zn} is approximately linear, with a slope of -1.25, instead of -1, expected if there is no change in the species saturating the resin. The more negative slope is ascribed to dissociation of $ZnCl_4^{2-}$ to the less strongly sorbed $ZnCl_2$ as the ratio of chloride to zinc in the resin decreases (cf. section 6.B.f). The general shape of the distribution curves was found to be similar for 0.01 M and for tracer zinc chloride[22] and for macro (1 mg metal ion per ca. 0.2 ml resin) and for tracer concentrations of Co^{II}[99] in hydrochloric acid solutions. Jentzsch[100] studied the effect of metal loading on the distribution of zinc, cadmium and copper from hydrochloric acid, and found that the dependence of D on loading is least in the region of the maximum in D, i.e. at highest loading. The effect is much more pronounced at the steep parts of the curve, even though the loading is smaller there. In a study of the sorption of Zn, Cd and Hg^{II} chlorides on an anion exchanger, in the presence and absence of small fixed concentrations of sodium chloride, Tremillon[26] found that whereas zinc and cadmium concentrations in the resin tend towards a fixed, saturated value, those of mercury reached much higher values. This may indicate sorption of neutral $HgCl_2$ in addition to anionic $HgCl_4^{2-}$, which is the main species at tracer concentrations of mercury, when the resin is in chloride form[79]. Also, for the sorption of uranyl sulphate on sulphate-form resin it was necessary[29] to assume that uncharged UO_2SO_4 invades the resin at high concentrations of uranium, in addition to the exchanged $UO_2(SO_4)_3^{4-}$.

In all the above cases, and in some others, where the data have been presented in a form that distribution curves cannot be drawn as a function of loading (as with U^{VI} in 10 M HCl[25] and in 2.0 M $Al(NO_3)_3$[27], it is clear that the sorption isotherm starts out linearly, with a finite slope, i.e. the distribution coefficient at zero (practically, at tracer) metal concentration. At some higher metal concentration, the curve bends and approaches a line of zero slope, as saturation is approached. The distribution coefficient is then inversely proportional to the metal concentration. In some cases, however, the neutral complex species is sufficiently stable to invade the resin as such, and then the bending of the isotherm is slowed down. In other cases, where the complexes are not so stable, the presence of the neutral species in the resin is at the expense of the anionic complexes, and the bending of the isotherm is more rapid. Rather complicated relationships must govern the sorption, since the composition of the resin phase changes drastically as the loading increases, and so do the activity coefficients of the ions and molecules in the resin, as discussed in section 4.C. No rigorous theory capable of describing the changes in sorption with loading has as yet been proposed and tested.

An interesting attempt, however, has been made by James[28] to treat this problem in an approximate, simple manner. Consider a case where a metal complex ML_q^{pl-} (where $p = m/l - q$) predominates in the resin phase. It is taken up by an exchange reaction (cf. 6B13, 6B40)

$$ML_q^{pl-} + p\overline{L^{l-}} \rightleftharpoons \overline{ML_q^{pl-}} + pL^{l-} \qquad (6C9)$$

with an equilibrium quotient

$$K_p = (\overline{ML_q^{pl-}})(L^{l-})^p / (ML_q^{pl-})(\overline{L^{l-}})^p \qquad (6C10)$$

where concentrations in the resin are expressed in the same units as $\bar{C}$, the capacity of the resin, and as used in D, the distribution coefficient of the metal, e.g. moles per litre resin. The concentration of the metal in the resin is the product of its total concentration in solution, c_M, and D. The concentration of the free ligand in the resin is therefore

$$(\overline{L^{l-}}) = \bar{C}/l - (p/l)Dc_M \quad \text{these hold only for } l = 1. \qquad (6C11)$$

Using (6C11) and the expression $\alpha_M = (ML_q^{pl-})/c_M$, the quotient K_p is given as

$$K_p = Dl^p(L^{l-})^p / \alpha_q(\bar{C} - pDc_M)^p \quad \text{these hold only for } l = 1. \qquad (6C12)$$

As c_M approaches zero, (L^{l-}) approaches c_L, (L^{l-}) approaches $\bar{C}/l$, and D approaches D^0, the parameters K_p and α_q may be eliminated from (6C12) to give

$$D = D^0(c_L/(L^{l-}))^p(1 - pDc_M/\bar{C})^p \quad \text{these hold only for } l = 1. \qquad (6C13)$$

assuming K_p and α_q not to vary with c_M. If, in addition, c_M is kept sufficiently low compared with c_L so that (L^{l-}) may be approximated with c_L, (6C13) may be further simplified for $p = 1$ to

$$D = D^0/(1 + D^0 c_M/\bar{C}) \quad \text{these hold only for } l = 1. \qquad (6C14)$$

and for $p = 2$ to

$$D = \frac{\bar{C}}{8D^0 c_{M^l}} (\bar{C} + 4D^0 c_M - \sqrt{\bar{C}(\bar{C} + 4D^0 c_M)}) \text{ these hold only for } l = 1.$$

(6C15)

Equations (6C14) and (6C15) were applied by James[28] to data for the distribution of Th^{IV}, U^{VI}, Pu^{IV} and Au^{III}, sorbed as $Th(NO_3)_6^{3-}$, $UO_2(NO_3)_4^{2-}$, $Pu(NO_3)_4^{2-}$ and $Au(NO_3)_4^{-}$ from $c_L = 7$ M HNO_3, and good agreement was found, considering the assumptions made of constant activity coefficients, α_q, K_p and $\bar{C}$, as c_M varied. For plutonium, the resin loading varied between 1.7 and 99%, and although average deviations of about 25% from the calculated curve were obtained, the deviations did not show any trend. James pointed out that equations (6C14) etc. permitted the prediction of the entire loading curve from a single determination at tracer loading, if p and $\bar{C}$ are known, provided the assumptions made above hold.

The case of mixtures of complex anions distributing at macro concentrations is even more complicated. Results of James[28, 101] indicate that the simplified treatment shown above is far from adequate in this case. He studied the sorption of various elements from 7 M HNO_3 solutions containing thorium, at thorium loadings of 10–50%, and interpreted his results in terms of specific interactions of $Th(NO_3)_6^{2-}$ and the other nitrate complex anions in the resin, apart from the diminution of free nitrate concentration in the resin according to (6C11). These supposed interactions, or any other equivalent non-ideality in the resin phase, caused D of the other elements to be markedly higher than calculated from the simple equations. A more sophisticated treatment of this problem must await the accumulation of much more knowledge and insight into the behaviour of ions in ion exchange resins.

g. Variation of the Distribution with Temperature

Little attention has been given to the variation of the anion exchange distribution of metal complexes with temperature. Most of the experimental data on the distribution has been obtained by the 'batch' method at room temperature, but a certain portion has been obtained from column elutions, some of them at elevated temperatures. A high temperature is of advantage when the diffusion-controlled exchange or sorption reaction (section 4.A) is slow, as with highly crosslinked resins, or with a resin loaded with an anion with high affinity for the resin (which is therefore only slightly swollen), such as perchlorate, or even nitrate. Although in some cases little change in D is observed as the temperature varies between room temperature and 80–90°C (e.g. for La^{III}, Eu^{III} and Am^{III} tracers in lithium nitrate[102], for La^{III} tracer in lithium chloride[103], etc.), for many other cases there is an appreciable dependence of D on temperature (e.g. Am^{III}, Cf^{III} and Lu^{III} tracers in lithium chloride[103] and Zn^{II} tracer in HCl and LiCl[103]). The most extensive study of this problem has been made by Kraus and Raridon[104], who studied the anion exchange absorption of Zn^{II} and Ga^{III} tracers from hydrochloric acid in the temperature range 25 to 150°C (Figure 6C2). Since the distribution coefficient data were not accompanied with resin invasion data, nor with a knowledge of the variation of activity coefficients in the two phases with temperature, only an approximate treatment could be given. Neglecting these factors, the zinc data were interpreted in terms of the average charge $\bar{i}$ (cf. equations (6B32 and 6B37), and the gallium data in terms of the ion fraction α_4 of the complex $GaCl_4^-$ in the aqueous phase (cf. equation 6B36). The concentration of hydrochloric acid at which given values of $\bar{i}$ or of α_4 were obtained, was plotted as a function of the reciprocal of the absolute temperature, yielding a family of curves, the slope of which is equal to the enthalpy change of the complexation and ion exchange reactions respectively. The striking features in the data (Figure 6C2) are the broadening of the distribution curve, the shifting

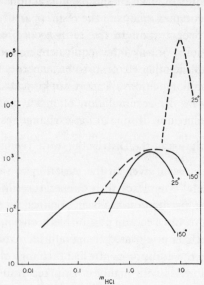

FIGURE 6C2. Temperature effect on sorption of complexes on anion exchangers. —— Zn^{II}, — — — Ga^{III}, in hydrochloric acid at 25° and 150°C.

of the maximum towards lower chloride concentrations, and the general lowering of the distribution curve, as the temperature increases. This is interpreted in terms of a large negative enthalpy change of the ion exchange process, noted also by Horne[22], and a large positive enthalpy of the association process in the aqueous phase leading to the uncharged species, or in other words a large enthalpy change for the process that minimizes the net charge on the species. This result is reasonable in view of the decrease of the dielectric constant of water with increasing temperature, which alone could account for the minimization of the net charge on the species.

h. The Effect of Resin Structure

The distribution coefficients of tracer metal anionic complexes between an anion exchange resin and ligand solutions is strongly dependent on the properties of the resin, just as in anion exchange reactions involving simple anions. In Chapter 4 the dependence of the selectivity coefficient of different ions on the resin structure has been discussed at length. The functional group of the resin may play a role, if it is capable of interacting with the metal ions or complexes[105]. This has been discussed in section 5.C.d. With quaternary ammonium strongly basic resin, such as Dowex-1 or Dowex-2, such interactions are absent, and commercial resins of different manufacturers often behave very similarly[106]. The capacity of these strongly basic resins cannot easily be varied independently of other factors. The effect of the variation of capacity on the relative sorption of complexes of different charges has already been discussed in section 6.B.f. A variable that can be practically varied is the (nominal) crosslinking of the resin. This factor determines the amount of swelling of the resin, and this, in turn, the relative mass or volume fractions of the organic skeleton of the resin, the fixed ions, the imbibed water and the invading electrolytes. The dependence of these factors on the crosslinking has been discussed in section 4.B, and the following discussion, although in terms of the effect of crosslinking, is to be understood as pertaining to the combined effect of a change in all these factors.

Distribution coefficients of tracer ions usually increase quite appreciably with increasing resin crosslinking, just as do selectivities of simple ions. Herber, Tonguc and Irvine[107] studied this problem for Mn^{II}, Co^{II}, Zn^{II} and Cu^{II} tracers in hydrochloric acid. The volume distribution coefficient of the former three increased by a factor of 2.5 to 3.0 as the crosslinking varied from 2 to 12% DVB, while that for copper increased only by 18%. The concentration of chloride in the resin also varied with the changes in crosslinking, but since the different elements were compared at different hydrochloric acid concentrations (12, 5.8, 0.15 and 3.0 M HCl respectively), and no invasion data were obtained, the actual molalities of the resins in the elution experiments are not known. The reluctance of D for copper to increase with crosslinking has not been explained. The distribution coefficient of Pa^V was observed[108] to increase by a factor of 4, as the crosslinking increased from 2 to 10% DVB, while that of Rh^{III} increased only slightly, or even decreased as the crosslinking increased from 4 to 8% DVB[109]. Elution of Am^{III} with 9.9 M LiCl -0.1 m HCl eluant showed[103] an increase by a factor of 2.4 in D at room temperature and by a factor of 4.6 at 87°C as crosslinking was increased from 2 to 10% DVB. Similar results, ranging from only a slight change to a considerable increase of the distribution coefficients with increasing crosslinking, have also been obtained by other authors[110, 111].

A systematic study of this problem has been made by Marcus and Maydan[24, 55, 91, 112]. These authors have studied the distribution of Zn^{II}, In^{III}, Ir^{III} and iodide tracers between sodium chloride solutions and Dowex-1 X 2, 4, 8, 10 and 16, and also Zn^{II} tracer between hydrochloric acid, lithium, potassium, ammonium and caesium chlorides and the same resins, as well as the Donnan invasion characteristics of the electrolytes themselves. Some data available in the work of Chu[111] were found to be in agreement. The distribution coefficients obtained for 1 m NaCl are shown in Table 6C2. In order to appreciate the crosslinking effect it is necessary first to eliminate the obvious dependence of the distribution coefficient on the effective ligand activity in the resin, $\bar{a}$. This variable depends on the effective ligand activity in the solution, a, on the one hand, and on the crosslinking, through the variations of swelling and internal resin molality, on the other. At high crosslinking, $\bar{a}$

TABLE 6C2. Effect of crosslinking on anion exchange distribution from 1 m NaCl.

| | | log D | | | | log K^* | | | |
X	Zn^{II}	In^{III}	Ir^{III}	I^-	Zn^{II}	In^{III}	Ir^{III}	I^-	$\bar{\varepsilon}$
2	1.43	0.75	0.86	1.73	1.85	1.17	0.61	1.44	59
4	1.90	0.78			2.11	1.21			49
8	2.13	0.86	1.06	1.85	2.24	1.26	0.64	1.50	42
10	2.34	0.90	1.07	2.00	2.33	1.29	0.67	1.58	38
16	2.92	0.95		2.25	2.86	1.31		1.78	33

depends also on the reduced capacity, because of incomplete chloro-methylation (section 4.A.b). This can be done by calculating the para-meter K^*, equation (6B30), from the measured distribution coefficients D and electrolyte invasion data, $\bar{a}$, and for those systems where stepwise complex formation occurs, also from known values of β_n^*

$$\log K^* = \log D - p \log \bar{a} - m/l \log a - \log \sum_{n=0}^{N} \beta_n^* a^n \qquad (6C16)$$

These calculations showed[55] the independence of K^* of the sodium chloride concentration, as expected from the Marcus–Coryell treatment (section 6.B.c), and the values obtained are shown in Table 6C2. The dependence of K^* on the crosslinking, which is seen to be slight for In^{III} and Ir^{III} tracers (absorbed on the resin as $In(OH_2)_2Cl_4^{-24}$ and $Ir(OH_2)_2Cl_4^{-55}$) somewhat higher for I^- tracer, and appreciable for Zn^{II} tracer (absorbed on the resin as $ZnCl_4^{2-22}$), has to be explained by considering the resin structure.

It has been found[55, 112] that the variation of inter-site distances neither directly[55], nor through polarization forces[57], can account in a quanti-tative manner for the variation of K^* with crosslinking. The major factor responsible for the crosslinking effect seems to be the lowered effective dielectric constant, $\bar{\varepsilon}$, of the resin, determined by the relative amounts of organic material and water. This property has been considered in section 4.C as leading to ion-pairing in the resin, in a discussion on the sources of the selectivity of the ion exchange resins. The quantity $\bar{\varepsilon}$ has been calculated by using values for dioxane–water mixtures of the same water content as the resin[55], and is shown in Table 6C2. Another important factor, also connected with the water content of the resin, is the dehydra-tion of highly hydrated ions. This is somewhat less important for the anion exchange of complexes, which are often unhydrated, but may play some role for an ion such as iodide, and it certainly does so for the chloride ion, displaced from the resin in the exchange reaction with the trace anions. In the general case the following set of reactions may be taken as expressing the overall exchange of tracer anion P^{p-} (whether simple, like I^-, or complex, like $ZnCl_4^{2-}$) for the bulk anion A^{a-}

$$\overline{AR_a} \rightleftharpoons a\overline{R^+} + \overline{A^{a-}}, \quad K_A \qquad (6C17)$$

$$p\overline{A^{a-}} + aP^{p-} \rightleftharpoons pA^{a-} + a\overline{P^{p-}}, K_E \qquad (6C18)$$

$$\overline{P^{p-}} + p\overline{R^+} \rightleftharpoons \overline{PR_p}, \qquad K_P \qquad (6C19)$$

where K_A, K_E and K_P are the equilibrium constants for the dissociation of the ion-pair of A^{a-} with the resin sites, for the exchange reaction between the two phases, and for the formation of the ion-pair of P^{p-} with the resin sites. The general equation for the free energy change involved in such an exchange has been given previously (4C31). The parameter K^* involves the thermodynamic constant K_p^T for the overall exchange reaction, equation (6B29), hence K^* will be proportional to $K_E K_P^a K_A^{-p}$. For the particular case of the exchange against the chloride anion $a = 1$, and collecting all terms independent of the radii and the dielectric constant into a term K^0, and giving numerical values to the physical constants, yields for 25°C

$$\log K^* = \log K^0$$

$$+\left(\frac{2}{\bar{r}_P + \bar{r}_R} - \frac{2}{\bar{r}_{Cl} + \bar{r}_R} - \left[\frac{\bar{\varepsilon}}{\varepsilon}\left(\frac{p}{r_P} - \frac{1}{r_{Cl}}\right) - \left(\frac{p}{\bar{r}_P} - \frac{1}{\bar{r}_{Cl}}\right)\right]\right)\frac{121p}{\bar{\varepsilon}} \qquad (6C20)$$

The radii of the ions in the resin, $\bar{r}$, can be calculated from equation (4C32), assuming dehydration to proceed in such a manner that the volume of the hydration shell is proportional to the water content of the resin. Equation (6C20) was applied[55] to the chloride complexes of Zn^{II}, In^{III} and Ir^{III} and to the iodide ion. (The term in $\bar{r}_{Cl} + \bar{r}_R$, pertaining to equation (6C17) has been omitted, since it was assumed that chloride ions are not ion-paired in the resin). Good agreement was obtained between K^* values plotted against $\bar{\varepsilon}$, and the slopes calculated from the equation. Quantitative agreement is less good, although the trends are retained, when the term for the ion-pairing of chloride ions is included. A further difficulty with the orginal treatment of Marcus and Maydan is the possibility of ion-pairing occuring not only with the quaternary ammonium groups[113] of the resin sites, but also with imbibed cations C^{c+}, as suggested by Ryan[25,96] and by Diamond[89]. These cations are also progressively dehydrated in the resins as the water content decreases with increasing crosslinking, so that the constant $\bar{r}_R$ in equation (6C20) must be exchanged for a variable parameter, weighted according to the relative importance of pairing with the different cations $\bar{r} = i\bar{r}_R + j\bar{r}_C$. Present knowledge of the reactions in the resin is not sufficient for taking this into account, hence a more rigorous test of equation (6C20) cannot yet be made.

Qualitatively, equation (6C20) may still serve as a basis for predicting the effect of crosslinking, It is evident that the nearer the radius of the tracer ion to that of the displaced ion, both in water and in the partially dehydrated state, the smaller the effect. This was indeed observed by Chu[111] in the series of stable tracer anions Br^-, I^- and ReO_4^-, sorbed from chloride solutions. At 0.1 m chloride, the spread in log D between $X\,2$ and $X\,16$ resin is 0.07, 0.55 and 0.88 respectively. The effect should also increase strongly with the charge p, of the absorbed complex. This, again, is illustrated by the results for divalent $ZnCl_4^{2-}$ and the monovalent I^-, $In(OH_2)_2Cl_4^-$ and $Ir(OH_2)_2Cl_4^-$, where the spread in log K^* between $X\,2$ and $X\,10$ is 0.50, 0.15, 0.10 and 0.06 respectively. As mentioned previously (section 6.B.f), this may serve for distinguishing the charge of the species absorbed on the resin. If this criterion is accepted, the results of Herber, Tonguc and Irvine[107] may be interpreted as indicating the sorption of $MnCl_4^{2-}$ and $CoCl_4^{2-}$ on the resin, in analogy with $ZnCl_4^{2-}$, but of the complex with unit charge, $CuCl_3^-$, for Cu^{II} tracer.

A large part of the crosslinking effect is seen to be due to the differences between the effective dielectric constant in the resin, $\bar{\varepsilon}$, and that in solution. If solvents of lower dielectric constants are used, the effective constants in the resin will vary less as the relative amount of solvent and organic material in the resin varies with swelling, i.e. with crosslinking. Hence, the crosslinking effect is expected to be lower, as indeed observed for ethanol solvent and Zn^{II} and In^{III} tracers[112]. The problem of the role of the solvent in the anion exchange of metal complexes is treated in the next section.

D. DISTRIBUTION FROM ORGANIC SOLVENTS

a. Introduction

The distribution of metal ions between an anion exchange resin and anhydrous or partially nonaqueous solvents has only recently been studied in detail. The property of the organic solvents which most obviously differs from that of water is their lower dielectric constant, which favours ion-pairing in both solution and resin phases. With water-containing solutions, however, a not less important point is the selective solvation of the ions, and the selectivity of the resin phase for water, depending on the composition of the solvent and on its nature. These, as well as some other factors, may be expected to affect the distribution of metal ions.

The sorption of organic solvents on ion exchange resins has been discussed briefly in section 4.B.d. An anion exchange resin may swell even in non-polar solvents, such as benzene, or even kerosene[114], probably because of interaction of the polystyrene skeleton and of the alkylammonium functional group with the solvent, through London forces. The resin may sorb polar solutes from non-polar solvents, as has been demonstrated for nitric acid, which may be removed from solutions of tributyl phosphate in kerosene[115] and of trilaurylamine in benzene[116]. In the latter case the resin also sorbs some trilaurylammonium nitrate. Most of the work done, however, deals with sorption, exchange and distribution of electrolytes or anionic complexes from polar solvents, mainly water-miscible alcohols, ketones, etc.

The distribution of certain metal complexes at trace concentrations has been studied rather extensively. In many cases the distribution coefficients were found to be much higher in mixed solutions, than in purely aqueous ones. Some elements not sorbed at all appreciably from aqueous solutions, even at the highest ligand concentration, show fairly high distribution coefficients from organic solutions. The limited solubility of inorganic salts in such solvents (section 2.C) may not permit the use of as high ligand concentrations in these as with aqueous solutions, but this is usually not an obstacle to the attainment of high distribution coefficients. The choice of the nature and concentration of the organic component gives the experimentalist an important tool for carrying out difficult separations[117]. It would therefore be very useful to have a comprehensive theory, capable of predicting the behaviour of anion exchange systems with various organic solvents. Unfortunately, such a theory has not yet been developed. The lack of knowledge of the behaviour of electrolytes in anhydrous or partially aqueous organic solvents (Chapter 2) is a strong obstacle to the formulation of such a theory. The present fragmentary ideas that have been proposed to account for the observations are discussed below.

The problem of the representation of the results of the distribution studies has not yet been solved satisfactorily. It is necessary to show the variation of the distribution coefficient D as a function of two independent variables of the ternary system water–organic solvent–ligand electrolyte. It is difficult to compare results obtained by different authors because of the various conventions used when reporting the composition of the solution phase. Some authors quote the volume of organic solvent mixed with a volume of aqueous electrolyte of given concentration. This method leads to the most easily reproduced conditions, but without knowledge of the volume change in the system, does not represent the equilibrium composition. The initial water content of the resin and the organic solvent employed must be taken into account. It seems that the best mode of representing composition is in terms of mole or weight fractions, where no ambiguities can occur. A triangular diagram, as usually employed for ternary phase diagrams, with contour lines drawn for certain values of D (say for $\log D = 1, 2, 3 ..$) may then serve for reporting the distribution results, without any ambiguities[118]. Unfortunately, only few of the results in the literature can be recalculated and shown in this form (cf. Figure 6D2).

b. Effect of Various Solvents

The anion exchange distribution of metal ions was most extensively studied in the lower aliphatic alcohols (methanol, MeOH, ethanol, EtOH, normal- and isopropanol, n-PrOH and i-PrOH, etc.), and acetone (Me_2CO). Data for a few systems have also been obtained for a variety of other solvents, such as ethers (diethyl ether and dioxane), ketones, etc. The solvents have rarely been used without admixture of water and only few data are available for completely nonaqueous systems. A discussion of the effect of the solvents, such as the order of change of the distribution coefficients with carbon chain-length of organic solvent, must therefore specify the water content, even when $> 95\%$ solvents (by volume) are used. This is so partly because of the low molecular weight and volume of water, increasing its mole $\%$ manyfold over its volume or weight $\%$, and mainly because of specific solvation effects.

As mentioned in the introduction, the most obvious property that could affect the distribution is the dielectric constant of the solvent or solvent mixture. In completely nonaqueous systems this should indeed be of prime importance. If the ion-pairing with resin functional groups or invading cations is accepted as the major contribution to the affinity of the complexes to the resin, then the lower the dielectric constant of the solvent, the lower will be the effective dielectric constant of the resin, and the higher the distribution coefficients are expected to be. In water, with $\varepsilon = 78$, the effective value in the resin varies between $\bar\varepsilon = 59$ and $\bar\varepsilon = 33$ as the crosslinking varies between $X\,2$ and $X\,16$[55] (cf. section 4.C.d and 6.C.h). In experiments with anhydrous MeOH, with $\varepsilon = 32$, $\bar\varepsilon$ is expected to vary roughly between 24 and 14, while for EtOH, with $\varepsilon = 24$, $\bar\varepsilon$ may vary between 18 and 10, at the same crosslinkings. Since swelling is only slightly less than in water, roughly the same composition of the swollen resins are assumed. These figures show that whereas $\bar{\varepsilon}$ for the organic solvent-swollen resin is considerably lower than for water-swollen resin, the differences between the two solvents are small. Furthermore, the change in $\bar{\varepsilon}$ with crosslinking is smaller than in water. It is therefore expected that D is higher for MeOH and EtOH than for water, only somewhat higher for EtOH than for MeOH (if at all, considering also the enhanced ion-pairing in the solution phase), and that D changes less with crosslinking for these solvents than for water. These expectations are borne out by experiments with In^{III} and Zn^{II} tracers in LiCl solutions[112], Figure 6D1. The singly charged tetrachloroindate anion in aqueous solutions shows a small dependence of D on crosslinking, but in the alcoholic solutions no

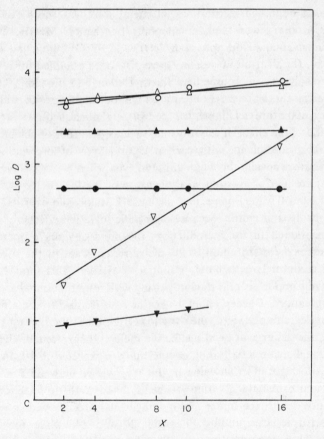

FIGURE 6D1. The crosslinking effect on the distribution of complexes between an anion exchanger and nonaqueous solvents, log D against X. Open symbols: Zn^{II}, filled symbols: In^{III}, $\triangledown$ solvent water, $\bigcirc$ solvent methanol, $\triangle$ solvent ethanol, all 1 M in LiCl with Dowex–1.

dependence at all is found. The chlorozincate anion which is doubly charged in aqueous solutions, shows considerable dependence of D on crosslinking. In alcoholic solutions the dependence is slight. Possibly the low $\bar{\varepsilon}$ causes the species in the resin to be the lower charged $ZnCl_3^-$ or $LiZnCl_4^-$, with an expected smaller slope for the dependence of D on X. On this basis the observed independence of D on X for In^{III} may indicate the species sorbed on the resin to be neutral (solvated) $InCl_3$. These hypotheses should be checked experimentally.

A feature that should be pointed out is that whereas for In^{III} there is a regular progression of increasing D in the order H_2O < MeOH < EtOH (decreasing ε), for Zn^{II} the D's for both alcohols are very similar, although considerably higher than for water. Such irregularities have been observed in many other cases (mostly with mixed aqueous-organic solvents). Korkisch and Janauer[46] have summarized their studies on the distribution of U^{VI} and Th^{IV} between alcohols and an anion exchange resin, finding for 99 % alcohol a steep increase of D with $100/\varepsilon$ (using ε for the pure solvent) for the U^{VI} chloride system, invariance of D with $100/\varepsilon$ for the U^{VI} sulphate system, and a 100–fold decrease of D with $100/\varepsilon$ for the Th^{IV} nitrate system, the alcohols varied from MeOH to pentanol. In all cases, however, sorption from the alcoholic medium was considerably higher than from aqueous solutions. The U^{VI} nitrate, Th^{IV}, V^{IV} and V^{V} chloride systems behave like the U^{VI} chloride, and Th^{IV} sulphate like U^{VI} sulhpate described above, while cerium showed in nitric acid a behaviour like that of Th^{IV}[119, 120]. Conflicting data for Eu^{III} in nitric acid i.e. both an increase[119] and a decrease[120] of D with chain length, have been reported. It is therefore difficult to decide whether the cerium studied has been Ce^{IV}, expected to behave like Th^{VI}, or Ce^{III}, and whether lanthanides show increasing[119] or decreasing[120] D with chain length. Tera and Korkisch[121] showed that for U^{VI} chloride, 99 % ethylene glycol and acetone fit into a more or less smooth curve drawn through the points for the 99 % alcohols in a plot of log D against $100/\varepsilon$. Dioxane, with the very low dielectric constant of $\varepsilon = 2$ does not fit in, showing a value of D like that of an alcohol with $\varepsilon = 20$ approximately. A comparison of normal- with isopropanol and isobutanol for 95 % alcohols showed very similar D values for the two isomers for a large number of metal nitrate systems[122], whereas at lower alcohol concentrations an increase of D with branching was found[119, 120]. Pietrzyk and Fritz[122] found for 90–95 % alcohols the 'normal' order of D increasing with chain length for a large number of metal chloride systems, but Joshino and Kurinura[123] found D increasing in the 'garbled' order PrOH, MeOH, EtOH, Me_2CO for Zn^{II} and Cu^{II}. In mixtures of organic solvents (such as MeOH–EtOH or MeOH–i-PrOH, with 4% water (aqueous HCl) present) log D values were found[122] to be approximately linear with solvent composition.

These largely uncorrelatable results show that although in a large number of cases D indeed increases with the reciporcal of ε, the dielectric constant is by no means the only important factor in determining the effect of the organic solvent. The swelling of the resin in different solvents and aqueous–organic solvent mixtures is another possible important factor, although it may not be directly responsible for the observed changes in D. The correlation of the selectivity of ion exchange resins with their swelling has been discussed in section 4.C.b, and the general conclusion reached was that the smaller the swelling, the larger, usually, the selectivity. Swelling increases normally with the polarity of the solvent, hence with decreasing chain length of the alcohols used for the studies considered above. Swelling is known, however, to vary irregularly with the composition of mixed solvents: for cation exchangers a maximum swelling has been observed for certain alcohol–water mixtures, higher than for each solvent alone (section 4.B.d). A direct correlation of D with swelling in different solvents has, however, not been attempted.

c. Effect of the Composition of the Solution

The effect of the various solvents on the distribution is closely connected with the dependence of the distribution on the composition of the solution phase, since most of the experiments have been done on mixed aqueous–organic solvents. As mentioned in the introduction, the most clear presentation of distribution results is in ternary diagrams, such as in Figure 6D2. Since the solubility of the electrolyte providing the ligand in the solvents is limited, it is not necessary to show the complete triangle, but only the homogeneous part of it. Contours with equal D values are shown for the Zn^{II}–LiCl–EtOH[45, 124], Cu^{II}–HCl–i-PrOH[122] and the U^{VI}–HCl–EtOH[46, 122] systems. The former system shows a maximum in D as the ligand concentration increases at low EtOH concentrations, but at higher EtOH concentrations the maximum is displaced to lower and lower chloride concentrations, so that at very low water concentrations, and in absolute EtOH, only a decrease in D with increasing salt concentration can be found experimentally. For the other two series D

increases with increasing ligand concentration, not reaching a maximum in the experimentally studied region.

For all systems, D generally increases going from the right- (water) to the left- (alcohol) hand side of the diagram, although this behaviour is not universally true, as may be seen in the diagrams. In some cases a small admixture of water to an anhydrous solvent enhances D several fold. This kind of behaviour is often only apparent, when D is plotted against the organic solvent percentage, when the other component is considered to be an aqueous solution of the ligand electrolyte (e.g. 90% EtOH–10% 6 M HCl), so that a slight decrease of the alcohol concentration from 100% is equivalent to an enormous increase not only of the water content but also of the ligand concentration. Maxima are therefore often found in such plots in the region near 100% organic solvent. In some cases, however, the effect is real, when the ligand concentration (in either the molar, molal or mole fraction scale) is kept constant, as seen, for example, in the case of the Zn system in Figure 6D2. Especially,

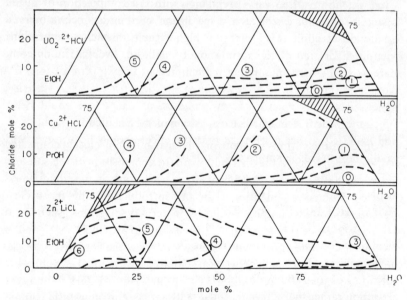

FIGURE 6D2. Anion exchange distribution of complexes in mixed solvents. Ternary diagrams with solvent composition on the base, and the acid or salt at the apex, only the lower, homogeneous region being shown (shaded area: heterogeneous region). Dashed lines are equi-distribution-coefficient lines with log D designated. Numbers on ordinates are mole percents. Upper portion: U^{VI} tracer in HCl–EtOH–H_2O system, middle portion Cu^{II} tracer in HCl–i-PrOH–H_2O system, lower portion Zn^{II} tracer in LiCl–EtOH–H_2O system.

however, with the weakly sorbed metals, such as Ca^{II}[122], Mn^{II}[118], Cr^{III}[122] and Sc^{III}[118] in hydrochloric acid, there is a maximum in D at constant chloride concentration when a few percent water are present. Maxima at mole fractions of 0.2 or 0.4 of water at constant nitric acid molarity were also obtained for the sorption of lanthanide ions in aqueous ethanol or isopropanol[125].

In some cases[46] admixture of small concentrations (up to mole fractions of about 0.3) of organic solvents has little effect on the distribution coefficients. Again, in series of experiments where given volumes of solvent and aqueous electrolyte are mixed, the ligand concentration decreases as the organic solvent content increases, and minima in D have been observed (e.g. the U^{VI}–nitric acid–EtOH system). At constant ligand concentrations this has not been observed. It has been argued, however, that not the nominal (stoichiometric) concentration of ligand should be kept constant as the solvent is varied, but the actual concentration, as measured, e.g. by conductivity[127]. This point should be given attention, although the method proposed[127] need not be the best one. In most cases, at a given ligand concentration, the increase usually observed in D is steepest at high organic solvent concentrations, where dehydration of the resin and of small, i.e. hydrated, ions could start to set in. This may be an indication that dehydration of ions is a more important factor than lowering of the dielectric constant, since it is more selective as regards the two phases, and therefore could better affect the distribution coefficients. Metal complex anions are however non-hydrated to begin with, so that the effect should be stronger with small anions. This has not yet been tested.

For a number of systems (such as Th and U^{VI} in HCl or HNO_3–alcohol, and Pb^{II} in HNO_3–i-PrOH)[126], but not in others (such as Hg^{II} in HCl–EtOH[126] and Zn^{II} in HCl or LiCl–EtOH[45], or La^{III} in HNO_3–i-PrOH[125]) log D was found to be linear with the mole fraction of alcohol at constant ligand concentration over large concentration ranges of the solvent. On grounds of resin selectivity towards water, and selective solvation of ions and molecules, such linear relationships should be the exception rather than the rule, and it is necessary to assume rather special conditions under which linearity of log D with $x_{alcohol}$ could be obtained.

In conclusion, the most generally observed type of behaviour is a large increase of D when organic solvents are substituted for water at a given ligand concentration[46, 122], which is the greater, the less water is present in the solvent mixture, at least down to several percent.

d. The Species in the Resin

It can be expected that the species of metal complexes found in anion exchangers equilibrated with nonaqueous or mixed solvents are different from those sorbed from aqueous solutions. The latter have been discussed in section 6.B.f, where the difficulties of determining the predominant species in the resin have been pointed out. It has also been stressed that distribution or loading data cannot give information on the path the sorption reaction takes. The word 'mechanism' has, however, been (mis)used[46] in discussing the real question of what are the major species found in the resin at equilibrium with an organic or mixed solvent phase. The large variety of types of behaviour of the distribution with varying solvents, solvent composition and ligand concentration noted above, suggests that there are several types of species, each dominating under different conditions. The following types have been proposed[46,122]: a) anionic complexes ML_q^{ql-m}, ion-paired to some extent with resin sites, $\bar{R}^+$, b) neutral complexes $ML_{m/l}$, and c) neutral proton- (or, in general, cation-)associated species $C_{ql-m}ML_q$.

It is natural to consider the anionic complexes ML_q^{m-ql} as the predominant species sorbed on the resin from solutions of relatively high dielectric constant[45], such as mixed solutions of low organic content, and for metal–ligand systems, where there is a pronounced tendency for anionic complex formation. Purely aqueous solutions of stable complexes, such as $PtCl_6^{2-}$ or $Au(CN)_2^-$, are an extreme example. Sorption of such species has been discussed in section 6.B.f. As the dielectric constant decreases, when a series of solvents or mixed solutions are considered, it becomes more and more difficult to add an anionic ligand to an already negatively charged complex. Either cations are then associated along with ligand anions to produce species with a high ligand number but a low charge, or the lowest anionic complex becomes relatively stabilized. The interaction between a neutral complex, especially when polar, and anionic ligands is enhanced by lowering the dielectric constant, hence the lowest anionic complex should be an important species under these conditions. This is presumably what happens in the resin phase for a few systems, even for completely aqueous media (section 6.B.f). For systems where anionic complex formation is not so pronounced in aqueous solutions, dehydration by the use of predominantly non-aqueous solvents is important for promoting the formation of anionic complexes.

From tracer distribution studies it is difficult to prove that a metal is sorbed as an anionic complex although this has been assumed in a number of cases, e.g. $Ag(NO_3)_2^-$ and $Ag(NO_3)_3^{2-}$ in isopropanol[127]. This is so, because it is impossible to measure the extent of the dissociation reaction in the resin phase

$$\overline{ML_q^{m-ql}} \rightleftharpoons \overline{ML_{m/l}} + p\overline{L^{l-}} \qquad (6D1)$$

without varying the activity of the ligand in the resin over a considerable range, which in turn is difficult to do without varying at the same time other properties of the resin. Resin loading experiments (section 6.A.c) on the other hand, may be helpful[26,46]. If there exists a simple stoichiometric relationship between the number of moles of metal sorbed at saturation per unit amount of resin to its equivalent capacity, there are good grounds to believe that the metal is sorbed predominantly as an anionic complex. This is, however, by no means a proof that an anionic complex of this stoichiometry actually predominates. The species $UO_2Cl_4^{2-}$, $UO_2(NO_3)_6^{4-}$ and $UO_2(SO_4)_3^{4-}$, deduced from loading experiments[46], are not in every case those most likely to be predominant. Where spectral data can differentiate between complexes of different composition, spectrophotometric measurements may be applied to the resin phase, to obtain the major species present. The species $UO_2Cl_4^{2-}$ has thus been confirmed[25] in the case of a resin loaded from dilute hydrochloric acid in ethanol. For uranyl nitrate sorbed from acetone solutions, spectral measurements showed that $UO_2(NO_3)_3^-$ along with some higher nitrate are the predominant species[27]. A coordination of the uranyl ion with six ligands as bulky as the nitrate anion, however, seems unlikely. An alternative explanation for the low capacity towards uranyl nitrate shown by the resin (only 25%) may be poor sorption kinetics[27], producing only partial saturation, and a fortuitious stoichiometry.

Metals which have only a moderate tendency to form anionic complexes in aqueous solutions may be expected to form the neutral species in media of not too high dielectric constant. This would most probably be the highest complex formed at low dielectric constants for those metals with very weak complex formation tendencies, such as thorium, nickel and chromium with chloride, or those forming only ion-pairs, such as calcium and lithium, or, in general, alkali and alkaline earth cations, in solutions of very low dielectric constant, such as dioxane with low concentrations of water[126]. Sorption of traces of these metals on an anion exchanger from organic or mixed media containing excess ligand may be thought to involve the neutral species as the predominant species in the resin. This is also expected to be the case when the metal, at small but higher than tracer concentrations, is sorbed on the resin from solutions with no excess ligand, above that neutralizing the charge of the metal[122]. Katzin and Gebert[129] were the first to consider the sorption on the resin of electrolytes from organic (acetone) solutions without the presence of excess ligand. They found, indeed, that the electrolyte as a whole disappears from the solution, sorbing on the resin, and that nitrate-form resin may sorb a metal chloride from solution, and vice versa, without the release of exchanged anions into the solution. These observations, however, do not preclude the occurence of the reverse of reaction (6D1) in the resin, as these authors themselves pointed out. They even went further than considering such a reaction for the divalent transition metals, by proposing the formation of species such as $LiCl_2^-$ or $Li(NO_3)_2^-$ in the resin. This may be unnecessary, since invasion by LiCl and LiNO$_3$ even from aqueous solutions is readily occuring, and the activity coefficients in the resin phase (section 6.C.d) do not suggest the formation of higher ionic aggregates, which would be even less likely in an acetone medium.

Sorption of this type may be considered as the distribution of a neutral species between two liquid phases: the solvent medium inside the resin pores, and the outside solution[124]. In a concentration range where the neutral complex is predominant in solution, a variation of the ligand concentration should not affect the distribution coefficient (apart from salting effects). Furthermore, loading experiments would not indicate any simple stoichiometry, and in particular, sorption of quantitites above the capacity of the resin is indicative of predominant species with zero charge (e.g. in the case of thorium nitrate in alcoholic solutions[46] or of mercury chloride even in aqueous solutions[127]). A further criterion, proposed by Marple[125], is that the variation of $\log D$ with $x_{solvent}$ should be linear (cf. section 6.D.c), if neutral-complex sorption occurs. Although possibly this is the case for a number of systems, no convincing evidence for the generality of this type of behaviour has been shown.

Finally, with metals forming very stable anionic complexes, and of course also for stable anions, such as ReO_4^-, the only way to reduce the charge as the dielectric constant decreases is to associate with cations. Such association has been proposed to occur in the resin phase even for completely aqueous systems in order to explain the secondary cation effect[22,44,89], (section 6.C.e). For nonaqueous or mixed media, such association should be important in both phases, as pointed out by

Horne[22,44]. It is difficult to separate this type of a sorption process from the sorption of anionic complexes, since these also associate with cations—with resin sites, if there are no imbibed secondary cations—at the low dielectric constant in the resin. If, however, there is considerable invasion of the resin with salt or acid, association with mobile cations becomes probable. It should be possible to differentiate between the two types of association by kinetic (diffusion) measurements, but no detailed information has yet been presented.

Except for some spectrophotometric data[25,27], and some cases of loading beyond the capacity[26,46], no convincing evidence has till now been produced to permit a decision to be made on the actual predominating species in the resin, in the presence of an organic medium. Analogy with solvent extraction systems, or with anion exchange sorption from aqueous solutions, has been used to indicate the likely species, but alternative formulations are always not less probable. This is the case also for sorption on resin-impregnated papers[130], where the behaviour of cationic, anionic and uncharged fixed groups may be compared under very similar conditions. Since, however, the concentration of the groups on the paper cannot be conveniently varied, association reactions in the stationary phase, such as $\overline{R}^+ + \overline{ML_q^{m-ql}} \rightleftharpoons \overline{RML_q^{m-ql+1}}$ or $\overline{ML_m} + L^- \rightleftharpoons \overline{ML_{m+1}^-}$ cannot be studied experimentally, although some may be ruled out on general grounds. A great advance to our understanding of the anion exchanger sorption from organic solvents would be reliable knowledge of the species in the resin, and its change as the organic constituent varies in nature and concentration.

e. Quantitative Treatments

The problem presented in the treatment of trace-metal distribution data between an anion exchange resin and nonaqueous or mixed solutions depends mainly on the ion-pairing tendency of the ions in these media.

The discussion of this problem by Fritz and Waki[127] concerns specifically a mixture of the ligand-producing electrolyte (e.g. nitric acid) and an inert electrolyte (e.g. perchloric acid) in certain proportions in a mixed aqueous–organic solvent (e.g. aqueous isopropanol) of varying composition. The proportions of the electrolytes and their concentrations are selected so as to have a constant ionic strength (as given by a selected hydrogen ion concentration, controlled to give a certain constant conductivity in the solution). The association of nitric acid in both phases is taken into account by writing

$$(NO_3^-) = c_{NO_3} / (1 + \beta_H(H^+)) \qquad (6D2)$$

for the solution phase, and similarly, with barred symbols, for the resin phase. The activity of the neutral metal complex is equal in the two phases (cf. equation 6B23), e.g. for the silver nitrate system

$$\overline{(AgNO_3)} = G(AgNO_3) \qquad (6D3)$$

where G is the ratio of the activity coefficients, assumed to be constant, thus independent of the nitric acid concentration at a given solvent composition. The distribution coefficient is given by

$$D = \frac{\sum \overline{(Ag(NO_3)_n^{i-})}}{\sum (Ag(NO_3)_n^{i-})} = G \frac{\sum \bar{\beta}_i^* \bar{c}_{NO_3^i}}{\sum \beta^* c_{NO_3^i}} \qquad (6D4)$$

where $i = 1 - n$, and $\beta_i^* = (Ag(NO_3)_n^{i-})/(AgNO_3)(NO_3^-)^i(1+\beta_H(H^+))^i$ (and similarly for the resin phase, with barred symbols). At constant conductivity it is assumed that the β_i^* are constant in each phase. If, now, the β_i^* are known for the solution phase from independent data, the average charge number is obtained in the usual way,

$$\bar{i} = d \log \sum_{-1}^{N-1} \beta_i^* c_{NO_3}^i / d \log c_{NO_3} \qquad (6D5)$$

The charge number for the resin phase is then obtained from

$$d(\log D + \bar{i} \log c_{NO_3}) / d \log \bar{c}_{NO_3} = \bar{i} \qquad (6D6)$$

since G is considered as a constant.

The results obtained for the silver system led to $\bar{i} = 2$ at low concentrations of isopropanol, and somewhat higher (up to 2.3) at higher concentrations, up to 90%. This is surprising, since the interpretation of $\bar{i} = 2.3$ in terms of a mixture of $\overline{Ag(NO_3)_2^-}$ and $\overline{Ag(NO_3)_3^{2-}}$ in the resin[127] leads to the unusual conclusion that the higher the organic-solvent content, the higher the charge on the species in the resin. The authors[127] did not discuss this difficulty, which is not removed even if species of silver ion-paired with the resin functional groups are considered, which, again, the authors did not. Some of the assumptions made in the derivation may, therefore, be unrealistic.

A different approach has been used by Penciner, Eliezer and Marcus[124] who chose to work with an anhydrous system, in order to avoid the problem of preferential solvation of ions and prefential swelling of the resin in mixed solvents. In ethanol solutions the mean activity of hydrochloric acid is known, hence it can be obtained also for the resin phase from invasion measurements and the Donnan equation (sections 4.B.e and 6.B.c). The species in either phase are considered to be formed from the neutral species (e.g. MCl_2 for zinc or mercury) and the hydrogen chloride ion-pairs. The stability constants for both phases are then defined by the expression (cf. section 6.B.c)

$$(H_i MCl_{2+i}) = \beta_i a_{MCl_2} a^{2i} \gamma_i^{-(1+i)} = \beta_i^* a_{MCl_2} a^{2i} \qquad (6D7)$$

where, as usual, $a = m_{HCl} \gamma_{\pm HCl}$, and $\beta_i^* = \beta_i \gamma_i^{-(1+i)}$ is considered independent of a, assuming the activity coefficient quotient γ_i to be so. The distribution coefficient for the tracer metal ions is then given by the expression

$$D = \sum \bar{\beta}_i^* \bar{a}^{2i} / \sum \beta_i^* a^{2i} \qquad (6D8)$$

The observed independence of D from a for a considerable range where $\bar{a}$ is strongly dependent on a requires that in this range, terms with $i \neq 0$ and $\bar{i} \neq 0$ are negligible, so that in this range MCl_2 predominates inboth phases, and the distribution can be represented as a liquid–liquid partitions (sections 7.B and 7.C)

$$D = \bar{\beta}_0^* / \beta_0^* = \gamma_{MCl_2} / \bar{\gamma}_{MCl_2} = P_{MCl_2} \qquad (6D9)$$

At higher hydrogen chloride concentrations D becomes strongly dependent on a, indeed the slope d log D/d log $a = -4$ is obtained at high concentrations, again for a considerable range. Supposing that a species $\overline{H_p MCl_{2+p}}$ is predominant in the resin phase, (6D8) can be rewritten as (cf. section 6.B.c)

$$\log D = \log K^* + 2 p \log \bar{a} - \log \sum \beta_i^* a^{2i} \qquad (6D10)$$

where $K^* = P_{MCl_2} \bar{\beta}_p \bar{\gamma}_p^{-1}$. Since d log $\bar{a}$/d log $a \simeq 0.8$ in this concentration range leads to $\bar{i} \approx 1.6\, p - $ (d log D/d log a) $\simeq 1.6\, p + 4$. For non-zero values of p this means species with unreasonable coordination numbers (e.g. $H_3 MCl_5$). It was concluded[124] that the most reasonable interpretation of the data is, therefore, with $p = 0$ also at higher ligand concentrations, that is that $\overline{MCl_2}$ predominates in the resin, while in the solution $\bar{i} = 4$ corresponds to $H_2 MCl_4$.

It must be stressed that as long as no definite knowledge of the species in one phase (e.g. the resin phase) is available, the distribution data for trace metal ions cannot be interpreted unequivocally in terms of the species formed in the other (e.g. the solution). This has been pointed out by the authors of both quantitative treatments presented[124, 127]. As discussed in section 6.D.d, independent knowledge of the species in the resin phase is even more difficult to obtain in the case of mixed or nonaqueous solvents than in the case of aqueous solutions, discussed in section 6.B.f.

E. REFERENCES

1. J. E. Salmon, *Rev. Pure Appl. Chem.*, **6**, 24 (1956).
2. R. F. Jameson and J. E. Salmon, *J. Chem. Soc.*, 28 (1954).
3. J. E. Salmon, *J. Chem. Soc.*, 2316 (1952).
4. R. F. Jameson and J. E. Salmon, *J. Chem. Soc.*, 4013 (1954).
5. R. F. Jameson and J. E. Salmon, *J. Chem. Soc.*, 360 (1955).
6. A. Holroyd and J. E. Salmon, *J. Chem. Soc.*, 269 (1956).
7. J. E. Salmon and J. G. L. Wall, *J. Chem. Soc.*, 1128 (1958).
8. J. E. Salmon, *J. Chem. Soc.*, 2644 (1953).
9. C. Heitner-Wirguin, B. E. Mayer and J. E. Salmon, *J. Chem. Soc.*, 460 (1960).
10. C. Heitner-Wirguin, *Bull. Res. Council Israel*, **9A**, 37 (1960).
11. C. Heitner-Wirguin and J. Kendy, *J. Inorg. Nucl. Chem.*, **22**, 253 (1962).
12. D. A. Everest, *J. Chem. Soc.*, 4415 (1955).
13. L. D. Cockerell and P. H. Woods, *J. Am. Chem. Soc.*, **80**, 3856 (1958).
14. D. A. Everest and J. H. Harrison, *J. Chem. Soc.*, 1439, 1820 (1957).
15. D. A. Everest and J. E. Salmon, *J. Chem. Soc.*, 2438 (1954); D. A. Everest and J. C. Harrison, *J. Chem. Soc.*, 4319 (1957); 2178 (1959).
16. D. A. Everest and W. J. Popiel, *J. Chem. Soc.*, 3183 (1956).
17. D. A. Everest and W. J. Popiel, *J. Chem. Soc.*, 2433 (1957).
18. D. A. Everest and W. J. Popiel, *J. Inorg. Nucl. Chem.*, **6**, 153 (1958).
19. R. U. Russell and J. E. Salmon, *J. Chem. Soc.*, 4708 (1958).
20. B. E. Mayer and J. E. Salmon, *J. Chem. Soc.*, 2009 (1962).
21. J. F. Harvey, J. P. Redfern and J. E. Salmon, *J. Chem. Soc.*, 2861 (1963).
22. R. A. Horne, R. H. Holm and M. D. Meyers, *J. Phys. Chem.*, **61**, 1656 (1957).
23. K. A. Kraus and F. Nelson in *The Structure of Electrolyte Solutions* (Ed. W. J. Hamer), J. Wiley and Sons, Inc., New York, 1959, p. 363.
24. D. Maydan and Y. Marcus, *J. Phys. Chem.*, **67**, 987 (1963).
25. J. L. Ryan, *Inorg. Chem.*, **2**, 348 (1963).
26. B. Tremillon, *Bull. Soc. Chim. France*, 275 (1961).
27. J. L. Ryan, *J. Phys. Chem.*, **65**, 1099 (1961).
28. D. B. James, *J. Inorg. Nucl. Chem.*, **25**, 711 (1963).
29. T. V. Arden and G. A. Wood, *J. Chem. Soc.*, 1596 (1956); T. V. Arden and M. Rowley, *J. Chem. Soc.*, 1709 (1957).

30. T. R. Bhat and J. W. Gokhale, *J. Sci. Ind. Res. (India)*, **18B**, 532 (1959).
31. L. D. C. Bok and V. C. O. Schuler, *J. S. African Chem. Inst.*, **13**, 82 (1960).
32. S. Fronaeus, *Svensk Kem. Tidskr.*, **65**, 1 (1953).
33. Y. Marcus and C. D. Coryell, *Bull. Res. Council Israel.*, **8A**, 1 (1959).
34. K. A. Kraus and F. Nelson, ref. 23. p. 340
35. I. Leden, *Svensk Kem. Tidskr.*, **64**, 145 (1952).
36. D. H. Freeman, *J. Phys. Chem.*, **64**, 1048 (1960).
37. Y. Marcus and F. Nelson, *J. Phys. Chem.*, **63**, 77 (1959).
38. L. W. Marple, *J. Inorg. Nucl. Chem.*, **28**, 1319 (1966).
39. V. V. Fomin, L. N. Fedomova, V. V. Sinkovskii and M. A. Andreeva, *Zh. Fiz. Khim.*, **29**, 2042 (1955).
40. C. E. Vanderzee and H. J. Dawson, Jr., *J. Am. Chem. Soc.*, **75**, 5659 (1953).
41. E. L. Harris, *Ph. D. Thesis*, Louisiana State Univ., 1961.
42. E. L. Harris, *J. Phys. Chem.*, **69**, 681 (1965).
43. F. Nelson and K. A. Kraus, *J. Am. Chem. Soc.*, **80**, 4154 (1958).
44. R. A. Horne, *J. Phys. Chem.*, **61**, 1651 (1957).
45. R. A. Horne, R. H. Holm and M. D. Meyers, *J. Phys. Chem.*, **61**, 1661 (1957).
46. J. Korkisch and G. E. Janauer, *Talanta*, **9**, 957 (1962).
47. L. W. Marple, *J. Inorg. Nucl. Chem.*, **27**, 1693 (1965).
48. E. Blasius, W. Preetz and R. Schmitt, *J. Inorg. Nucl. Chem.*, **19**, 115 (1961).
49. Y. Marcus, *Bull. Res. Council Israel.*, **8A**, 17 (1959).
50. Y. Marcus, *Acta Chem. Scand.*, **11**, 619 (1957).
51. L. H. Jones and R. A. Penneman, *J. Chem. Phys.*, **22**, 965 (1954).
52. Y. Marcus, *Israel AEC Rept.*, R/20 (1959).
53. A. Sonesson, *Acta Chem. Scand.*, **15**, 1 (1961).
54. A. M. Trofimov and L. N. Stepanova, *Radiokhimiya*, **1**, 403 (1959).
55. Y. Marcus and D. Maydan, *J. Phys. Chem.*, **67**, 983 (1963).
56. Y. Marcus and M. Givon, *J. Phys. Chem.*, **68**, 2230 (1964).
57. J. Aveston, D. A. Everest and R. A. Wells, *J. Chem. Soc.*, 231 (1958); J. Aveston, D. A. Everest, N. A. Kember and R. A. Wells, *J. Appl. Chem.*, **8**, 77 (1958).
58. E. Rutner, *J. Phys. Chem.*, **65**, 1027 (1961).
59. F. A. Cotton, D. M. L. Goodgame and M. Goodgame, *J. Am. Chem. Soc.*, **83**, 4690 (1961).
60. J. S. Coleman, *J. Inorg. Nucl. Chem.*, **28**, 2371 (1966).
61. C. Heitner-Wirguin and R. Cohen, *J. Phys. Chem.*, **71**, 2556 (1967).
62. J. L. Ryan, *J. Phys. Chem.*, **65**, 1856 (1961).
63. E. Brücher and P. Szarvas, *J. Inorg. Nucl. Chem.*, **28**, 2361 (1966).
64. V. W. Meloche and A. F. Preuss, *Anal. Chem.*, **26**, 1911 (1954).
65. C. Heitner-Wirguin and R. Cohen, *J. Inorg. Nucl. Chem.*, **27**, 1989 (1965).
66. G. W. Latimer, Jr. and N. H. Furman, *J. Inorg. Nucl. Chem.*, **24**, 729 (1962); R. M. Diamond, *J. Phys. Chem.*, **61**, 75 (1957); F. G. Zharovskii, *Zh. Neorg. Khim.*, **2**, 2085, 2096 (1957).

67. C. Heitner-Wirguin and R. Cohen, *J. Inorg. Nucl. Chem.*, **26**, 161 (1964).
68. Y. Marcus, *Proc. 2nd Intern Conf. Peaceful Uses At. Energy, Geneva*, 1958, Vol. **3**, p. 465.
69. H. S. Sherry and J. A. Marinsky, *Inorg. Chem.*, **2**, 957 (1963)
70. H. S. Sherry and J. A. Marinsky, *Inorg. Chem.*, **3**, 330 (1964).
71. J. S. Coleman, *USAEC Rept.*, LA–3819 (1967).
72. F. Nelson, R. M. Rush, and K. A. Kraus, *J. Am. Chem. Soc.*, **82**, 339 (1960).
73. I. Eliezer and Y. Marcus, *J. Inorg. Nucl. Chem.*, **25**, 1465 (1963).
74. S. Fronaeus, I. Lundqvist and A. Sonesson, *Acta Chem. Scand.*, **16**, 1936 (1962).
75. A Sonesson, *Acta Chem. Scand.*, **15**, 1 (1961).
76. H. Waki, *Bull. Chem. Soc. Japan.*, **33**, 1469 (1960); **34**, 829 (1961).
77. H. Waki, *Bull. Chem. Soc. Japan.*, **34**, 1842 (1961).
78. J. Yoshimura, H. Waki and S. Tashiro, *Bull. Chem. Soc. Japan.*, **35**, 412 (1962).
79. Y. Marcus and I. Eliezer, *J. Inorg. Nucl. Chem.*, **25**, 867 (1963).
80. R. A. Horne, *J. Inorg. Nucl. Chem.*, **6**, 338 (1958).
81. J. K. Foreman, I. R. McGowan and T. D. Smith, *J. Chem. Soc.*, 738 (1959).
82. J. A. Alexa, J. Maly, M. Marhol, M. Novak and D. Wagnerova, *Jaderna Energia*, **3**, 200 (1957).
83. K. A. Kraus, F. Nelson and G. W. Smith, *J. Phys. Chem.*, **58**, 11 (1954).
84. S. S. Berman and W. A. E. McBryde, *Can. J. Chem.*, **36**, 835, 845 (1958).
85. K. A. Kraus, F. Nelson, F. B. Clough and R. C. Carlson, *J. Am. Chem. Soc.*, **77**, 1391 (1955).
86. Y. Marcus and I. Abrahamer, *J. Inorg. Nucl. Chem.*, **22**, 141 (1961).
87. R. F. Buchanan and J. P. Faris, *Intern Conf. Uses Radioisotopes in Physical Sciences and Industry*, Copenhagen, 1960, paper RICC/173.
88. Y. Marcus, *Bull. Res. Council. Israel*, **4**, 326 (1954).
89. B. Chu and R. M. Diamond, *J. Phys. Chem.*, **63**, 2021 (1959).
90. K. A. Kraus and F. Nelson, *Proc. 1st Intern Conf. Peaceful Uses At. Energy, U. N. Geneva*, **7**, 113 (1956).
91. Y. Marcus and D. Maydan, *J. Phys. Chem.*, **67**, 979 (1963).
92. G. Zachariades, W. R. Herrera and C. J. Cummiskey, *J. Inorg. Nucl. Chem.*, **28**, 1707 (1966); G. Alexandrides and C. J. Cummiskey, *J. Inorg. Nucl. Chem.*, **28**, 2025 (1966).
93. Y. Marcus, *J. Inorg. Nucl. Chem.*, **12**, 287 (1960).
94. K. A. Kraus and F. Nelson, ref. 23, p. 350.
95. J. L. Ryan, *J. Phys. Chem.*, **64**, 1375 (1960).
96. J. L. Ryan, *Inorg. Chem.*, **3**, 211 (1964).
97. J. A. Salthouse and T. C. Waddington, *J. Chem. Soc. (A)*, 28 (1966).
98. Y. Marcus, *J. Phys. Chem.*, **63**, 1000 (1959).
99. G. E. Moore and K. A. Kraus, *J. Am. Chem. Soc.*, **74**, 843 (1952).
100. D. Jentzsch, *Z. Anal. Chem.*, **148**, 325 (1955).
101. D. B. James, *USAEC Rept.* LA–2836 (1963).

102. Y. Marcus, M. Givon and G. R. Choppin, *J. Inorg. Nucl. Chem.*, **25**, 1457 (1963); Y. Marcus, unpublished results.
103. E. K. Hulet, R. G. Gutmacher and M. S. Coops, *J. Inorg. Nucl. Chem.* **17**, 350 (1961).
104. K. A. Kraus and R. J. Raridon, *J. Am. Chem. Soc.*, **82**, 3271 (1960).
105. J. Kennedy and R. V. Davies, *J. Inorg. Nucl. Chem.*, **12**, 193 (1959).
106. L. R. Bunney, N. E. Ballou, J. Pascual and S. Foti, *Anal. Chem.*, **31**, 324 (1959).
107. R. H. Herber, K. Tonguc and J. W. Irvine, Jr., *J. Am. Chem. Soc.*, **77**, 5840 (1955).
108. C. Keller, *Radiochimica Acta*, **1**, 147 (1963).
109. H. Shimogiva, *J. Chem. Soc. Japan, Pure Chem. Sect.*, **81**, 564 (1960).
110. K. A. Kraus, F. Nelson and G. E. Moore, *J. Am. Chem. Soc.*, **77**, 3972 (1955).
111. B. T. Chu, *Ph. D. Thesis*, Cornell Univ., 1959.
112. D. Maydan, *Ph. D. Thesis*, Hebrew University, Jerusalem, 1962.
113. R. M. Diamond, *J. Phys. Chem.*, **67**, 2513 (1963).
114. G. W. Bodamer and R. Kunin, *Ind. Eng. Chem.*, **45**, 2577 (1953).
115. V. M. Mikhailov, *Zh. Neorg. Khim.*, **6**, 2809 (1962).
116. G. Klotz, G. Scibona and M. Zifferero, *Italian Report*, C.N.E.N., 107 (1961).
117. J. Korkisch in *Progress in Nuclear Energy, Series IX Analytical Chemistry*, Vol. 6, Pergamon Press, London, 1966, p. 1.
118. D. H. Wilkins and G. E. Smith, *Talanta*, **8**, 138 (1961).
119. J. Korkisch, I. Hazan and G. Arrhenius, *Talanta*, **10**, 865 (1963).
120. R. A. Edge, *J. Chromatog.*, **8**, 419 (1962).
121. F. Tera and J. Korkisch, *J. Inorg. Nucl. Chem.*, **10**, 335 (1960).
122. D. J. Pietrzyk, *Ph. D. Thesis*, Iowa State Univ. 1960; J. S. Fritz and D. J. Pietrzyk, *Talanta*, **8**, 143 (1961).
123. J. Joshino and Y. Kurinura, *Bull. Chem. Soc. Japan.*, **30**, 563 (1957).
124. J. Penciner and I. Eliezer, *Israel J. Chem.*, **1**, 259 (1963); J. Penciner, I. Eliezer and Y. Marcus, *J. Phys. Chem.*, **69**, 2955 (1965).
125. L. W. Marple, *J. Inorg. Nucl. Chem.*, **26**, 859 (1964).
126. L. W. Marple, *J. Inorg. Nucl. Chem.*, **26**, 635, 643 (1964).
127. J. S. Fritz and H. Waki, *J. Inorg. Nucl. Chem.*, **26**, 865 (1964).
128. R. R. Ruch, F. Tera and G. H. Morrison, *Anal. Chem.*, **36**, 2311 (1964).
129. L. I. Katzin and E. Gebert, *J. Am. Chem. Soc.*, **75**, 801 (1953).
130. M. Lederer and F. Rallo, *J. Chromatog.*, **7**, 552 (1962); M. Lederer, V. Moscatelli and C. Padiglione, *J. Chromotag.*, **10**, 82 (1963).

7.

Principles of Solvent

Extraction

The wide field of solvent extraction has several unifying principles, which should be considered before a detailed study of specific extraction systems is made. The general properties of organic solvents, both physical and chemical, have been discussed in section 2.A, and the requirements for a solvent for forming a two-phase system with water and aqueous solutions have been discussed in section 2.B. It remains therefore to examine the behaviour of solutes, present as minor components, in the two-phase water–solvent system, in order to understand solvent extraction processes.

It is in order to start section 7.A, dealing with the classification of distribution systems, with definitions of terms pertinent to solvent extraction, before proceeding with a discussion of the nature of the extractable species, of types of extraction processes and the broad classes of solvents. Each of these classifications can serve as a basis for organizing the rest of the Chapters dealing with solvent extraction. The organization used here (Chapters 8–10) has been found to be convenient, but not exclusive. The section is concluded by a discussion of the factors affecting the extraction rates.

Extraction processes are discussed in general terms in section 7.B. It is shown that the phase rule is not a very useful guide to the behaviour of distribution systems, but that Nernst's distribution law is. Unambiguous definitions of the standard states and the concentration scales used are important for a rigorous treatment, and for the interpretation of experimentally obtained distribution coefficients. The simplest distribution systems, which can be expected to conform to the distribution law as a first approximation, are the 'physical distribution' systems, where the same species predominates in both phases. Solubility parameters (section 2.B) can be used for predicting distribution constants in such systems. A detailed illustration of the problems occuring with these systems is provided by a discussion of the distribution behaviour of the mercury halides.

There are two general lines of attack on distribution systems, used for ascertaining the important species found in the two phases, and their relative stabilities.

Section 7.C discusses the quantitative treatment of distribution coefficient data. The distribution coefficient D is related to the important solution variables, such as solute concentration, and the concentrations of ligand and extractant. Several extraction processes are discussed: distribution of simple molecules, of chelates (use of acidic extractants),

Principles of Solvent Extraction

427

of solutes solvated in the organic phase, of acido complexes (where hydrogen ions are solvated), and of long-chain amine and related salts. Information is obtainable also from ligand distribution studies, and by the continuous-variation (Job's) method, which is critically discussed.

Section 7.D, on the other hand, discusses the information obtained by relating the properties of the organic phase with its composition. The general method of examining property-vs.-composition curves, in particular for sharp breaks, is criticized. Other sources of information, such as the extraction stoichiometry, especially with regards to hydration in the organic phase, and the interpretation of the physical and chemical properties of the solutions, are reviewed briefly. Detailed discussion of the methods, as applied to specific illustrating examples, is given in Chapters 8–10.

A. CLASSIFICATION OF DISTRIBUTION SYSTEMS

a. Introduction and Definitions

Solvent extraction is a process in which a substance is transferred from one liquid phase into another liquid phase. The final (i.e. equilibrium) stage of the process is called (liquid–liquid) *partition* or *distribution*, and the substance or species, the distribution of which is studied, may be called the *distribuend*. The two liquid phases are generally in contact, hence they should be reasonably immiscible. Systems are however conceivable, and have actually been studied, where the distribuend is transferred through a third phase (e.g. another liquid[1], the gas phase[2], a membrane, etc.), so that the process may be generalized also to miscible liquids. The vast majority of studies made, and considered in this book, concerns systems where one liquid phase is an aqueous solution, and the second phase is a solvent (hence the term solvent extraction)[3]. The *solvent* is a substance in the liquid state, usually organic, which is practically immiscible with the second liquid (i.e. the aqueous phase) and is capable of dissolving the distribuend to some extent. An *extractant* is a substance with solvent properties (although it may be solid when pure), used in solution in a suitable *diluent*. The extractant reacts with the distribuend by solvation, chelation, ion-pairing, ion exchange, etc., to extract it from the second phase. The diluent is used to dissolve the extractant and improve its physical properties, practically without having extractant properties itself. The extractant and diluent together act as a solvent.

References pp. 492–498

Although some work has been done on the kinetics and mechanism of the solvent extraction process, most workers are interested in the equilibrium state of the process. At equilibrium the concentrations of the distribuends and other components of the system in each phase have values independent of the initial presence of the distribuends in one or the other of the phases, hence the equilibrium state should be more properly termed distribution or partition than extraction. The distribuend is frequently at low concentration relative to the other components of the system, and its partition is studied as a function of such variables as its concentration, the concentrations of the other solutes in the system and the temperature. The variation of total pressure on the system has not played an important role in solvent extraction studies. Such studies are made in order to ascertain the species in which the distribuend occurs in the two phases, and their relative stability. In some cases this information is available from independent measurements for one phase and then it is possible to infer it for the other from distribution measurements only. Otherwise, it is often possible to obtain only the relative composition differences of the average species in the two phases. In any case, a detailed knowledge of the extracted species in the organic phase is desirable.

Distribution systems may be classified according to two concepts a) the nature of the extracted species, and b) the mechanism of the extraction process.

b. Nature of the Extractable Species

There are two main factors which cause the distribution to favour the organic phase: a low affinity of the distribuend to the aqueous phase and a high affinity of it to the organic phase. By affinity is meant the total interaction energy, which is made up of terms such as solvation energy, electrostatic interaction energy among ions, dipole-interaction energy, hole-formation energy, entropy factors, etc. These have been discussed in detail for aqueous solutions in Chapter 1, and for organic solutions in Chapter 2. In a qualitative manner, the factors that cause a low affinity of a species for the aqueous phase are:

a) Zero or low charge. Hydration of ions increases rapidly with increasing charge.

b) Large size. Hydration of ions decreases and the hole formation energy increases with increasing size.

c) Non-polar nature. The lower the polarity of a molecule, the less its interaction with the water dipoles.

d) Absence of electronegative atoms at the surface. Hydrogen bonding to the water will not occur unless there are electronegative atoms at the surface of the species.

e) Low water activity and a highly ordered water structure. Salting-out agents decrease the availability of water for interaction with the distribuend.

These factors usually also cause a high affinity for the organic phase, but the variety of solvents of different properties makes a simple general statement impossible. Thus solvents of high (e.g. nitrobenzene) or moderate (e.g. bis-β,β'-dichlorodiethyl ether) dielectric constants will favour distribuends capable of ionic dissociation over those that are not, and solvents of a highly ordered hydrogen-bonded structure (e.g. alcohols) will favour small distribuends, which require a low hole-formation energy, over large ones.

Distribuends may be classified very broadly into 'molecular' ones and 'ionic' ones. The former category includes:

1a) Simple inorganic molecules, such as noble gases, or other simple species, e.g. I_2.

1b) Partially covalent metal salts, such as $HgCl_2$ or RuO_4.

1c) Weak acids, such as CH_3COOH.

1d) Metal chelates, such as $Bi[(C_6H_5NN)_2CSH]_3$, bismuth dithizonate.

1e) Solvated metal salts in low dielectric constant solvents, such as $UO_2(NO_3)_2 . 2TBP$ in kerosene.

The second category includes:

2a) Ion-pairs consisting of large ions, such as $(C_6H_5)_4As^+ReO_4^-$ or $Cs^+(C_6H_5)_4B^-$.

2b) Ion pairs, the cation of which is an 'onium' cation, such as $(C_2H_5)_2OH^+NO_3^-$ or $[(C_8H_{17})_3NH^+]_2PuCl_6^{2-}$ in benzene.

2c) Metal salts of large organic anions, such as Fe^{3+} tris(dinonyl-naphthalene sulphonate).

2d) Solvated metal salts in high dielectric constant solvents, such as $Cs^+I_3^-$ in nitromethane.

It must be made clear that these categories are not mutually exclusive, and a distribuend may belong to either depending on the circumstances, such as the dielectric constant, the solvating power or the basicity of the solvent, or the ligand concentration in the aqueous phase. Different classifications of distribuends have also been proposed, for example based on the presence of the solvating solvent in the first coordination sphere or outside it[4], or on the type of complexes formed, i.e. solvation, chelation and ion-association complexes[5].

c. Types of Extraction Processes

Extraction processes may be classified in principle according to the mechanism of the extraction reaction. However, in only a few cases is the mechanism known from kinetic measurements, and in general only the equilibrium state is known. The classification according to the extraction reactions is therefore mainly conventional[6]. The following types may be distinguished[3-6].

a) Simple physical distribution. This probably occurs only for inert systems, such as the distribution of argon between nitromethane and water, or of benzene between cyclohexane and water.

b) Distribution involving solvation in either one or both phases. Distribution of mercury(II) bromide between cyclohexane and water, or between toluene and water is an illustration of these processes, where coordinative solvation occurs in one or both phases.

c) Distribution of ion-associates, involving ionic dissociation at least in the aqueous phase. These systems are simplest when the distribuend is mainly in a molecular (or ionic but undissociated) form in the organic solvent and dissociated in the aqueous phase. Metal chelates are often of this form.

d) Distribution involving reaction with excess ligand, usually in the aqueous phase. This was observed for the extraction of mercury(II) halides from aqueous halide solutions, or of iodine from iodide solutions.

e) Distribution involving aggregation, in particular in the organic phase. Distribution of benzoic acid between benzene and highly acidic aqueous solutions involves the dimer in the organic phase and a simple undissociated molecular species in the aqueous phase.

f) Ion exchange reactions, where either a cation is exchanged, usually for a hydrogen ion, as in the extraction of iron(III) with dinonyl-naphthalene sulphonic acid, or an anion is exchanged, as in the extraction of pertechnetate ions with a quaternary ammonium nitrate.

The above examples were selected to illustrate cases where the distribution is governed mainly by one predominant reaction. In most cases more than one reaction occurs, and the equilibria become more complicated. Thus ionic dissociation is usually accompanied by solvation, and either may be accompanied by reaction with excess ligand or by aggregation, etc. Examples for processes where more than one reaction type is of importance are the extraction of uranyl nitrate from aqueous nitric acid into tri-n-butyl phosphate solutions in xylene (types b + c + d), the extraction of uranyl sulphate from aqueous sulphuric acid into tri-isooctylammonium sulphate–bisulphate solutions in kerosene (types c + d + e), and the extraction of uranyl chloride from dilute aqueous solutions into di(2-ethylhexyl)phosphoric acid in hexane (types c + e + f).

It is often convenient to classify extraction systems according to the extractant or solvent types used[7]. These can be classified according to basicity, polarity, dielectric constant, etc. The following list may be useful.

a) Inert solvents. These extract only non-polar species, such as iodine or chelates. They are graded according to polarity (compare non-polar carbon tetrachloride with polar 1,1-dichloroethane), polarizability (compare benzene with cyclohexane), hydrogen–bonding ability (compare chloroform with carbon tetrachloride), etc. They are used also as diluents.

b) Basic solvents. These extract mainly by solvating either hydrogen ions, when they extract acids or anionic metal complexes, as in the extraction of plutonium(IV) from hydrochloric acid by trioctylamine in xylene, or metal ions, as in the extraction of uranyl nitrate with tri-n-butyl phosphate. They are graded according to basicity, e.g. from long-chain tertiary amines R_3N, through phosphine oxides R_3PO, phosphates $(RO)_3PO$, ketones R_2CO, to ethers R_2O.

c) Acidic extractants. They extract usually by the cation exchange reaction, exchanging hydrogen ions for the extracted cations. They are graded according to their acidity from e.g. thenoyltrifluoroacetone through dialkyl phosphates to carboxylic acids.

d) Chelating extractants. They form reasonably strong coordinate bonds with the metal ions and may be more basic (e.g. 8-hydroxyquinoline) or more acidic (e.g. thenoyltrifluoroacetone), and may be neutral or charged, of various denticity, and containing zero, one or several displaceable hydrogen ions.

e) Ionic extractants. These occur either as dissociated ions or as undissociated electrolytes, depending on the dielectric constant of the diluent, and the active extractant species may be either the cation (e.g. tetraphenyl arsonium) or the anion (e.g. tetraphenylboride).

Some extractants belong to more than one class, according to the environment where they are used. Long-chain tertiary amines may be classified under (b) in acid solutions, or in basic solutions, but under (e) when they are neutralized, i.e. used as salts, in neutral solutions. Dialkyl phosphates can be classified under (c), or under (d) when chelate formation is proven.

In this book it was found convenient to classify distribution systems both by processes and by extractants, as follows[7]:

a) Distribution of simple molecules (sections 7.B.c and d).

b) Extraction by compound formation (Chapter 8): chelating agents, acidic phosphorus esters, carboxylic and other acids.

c) Extraction by solvation (Chapter 9): extraction of acids by solvating the hydrogen ion, extraction of salts by solvating the metal cation.

d) Extraction by ion-pair formation (Chapter 10): bulky ionic extractants, long-chain amines.

d. The Rate of Extraction

This book is concerned primarily with the equilibrium state of the extraction systems, but interesting information can also be obtained from kinetic experiments. However, little has been published on the rates of extraction processes. Ordinarily, the rate of extraction is governed by the convection of the distribuend through the bulk of one phase, its diffusion through a thin surface layer of this phase, the crossing of the interface, and diffusion and convection processes in the second phase. These processes depend on the temperature and the viscosities of the two solvents, and on the rate of agitation, or mutual velocity of the two phases, as found, for example for the extraction of nitric acid by TBP[8,9]. This parameter determines the rate of convection and of eddy diffusion, and the thickness of the diffusion films[5,10]. These processes are very rapid (for the nitric acid extraction 12 millisecond contact time suffices)[8], and usually occur well before phase separation occurs after a short vigorous shaking, i.e. within one minute. Careful examination of the process in uranyl nitrate extraction, showed increasing slowness of the process stages in the following order: eddy diffusion, surface barrier crossing and molecular diffusion. With stirred phases eddy diffusion is at first rate controlling[11], but in certain cases a surface barrier rapidly builds up and is then rate controlling[12]. Although this was ascribed to a real effect, not due to surface contamination, no proof of this was offered, and adsorption of the distribuend, or other substances, at the interface cannot be excluded as contributing to the barrier. With unstirred solutions molecular diffusion becomes rate controlling, as it is slowest, except in the initial stage[13]. At this stage, even without mechanical stirring, the heat effect of the extraction reaction causes turbulence and fast eddy diffusion. After this effect has been disposed of by the creation of a steady-state concentration gradient, the rate of mass transfer was indeed found to be diffusion controlled, and in agreement with calculations using self-diffusion coefficient for the two phases determined independently.

In certain cases, desorption of surface-active agents from the interface can be the slow step, especially if a large interface has been created by vigorous shaking. Although physical disengagement of the phases has occurred, the surfactant is still not in its equilibrium state for some time. Differences in the distribution thus occur when contact of the phases has been gentle or violent, and true equilibrium has not been reached in the second case. This effect was first observed for certain long-chain ammonium sulphate extraction systems[14], and has also been found for other extractants. Burger[9] found that surface-active agents usually reduce the rate of transfer of metallic species even if present in small amounts in the system. The effect was attributed to a 'mechanical blocking' of the interface. The controlling resistance to transfer in this 'blocking' is apparently the rate of diffusion of ion-pairs from the interface into the organic phase[15]. In a study of the rate of extraction of hydrochloric and nitric acid into a toluene solution of tridodecylamine, Felton found that the diffusion from the water-side film into the aqueous phase is about three times faster than that from the organic-side film into the bulk of the organic phase. The contribution of the aqueous-side diffusion to the overall rate of transfer seems to be more apparent at high distribution ratios of the metallic species[16, 17]. The transfer of an indicator dye (thymol blue) between water and triethylamine is similarly dependent on the diffusion rates[18]. Aqueous solubility of lower alkylamines frequently causes emulsification in one or both phases. The tendency for emulsion formation depends on the type of diluent. Chloroform and other water-immiscible polar solvents show the smallest tendency. Low extractant and high aqueous electrolyte concentrations and higher temperatures usually prevent emulsion formation, because of lower mutual solubilities.

In most cases the chemical reactions leading to distribution of a distribuend between two phases are rapid, even compared with the rapid eddy-diffusion processes. Sometimes, however, slow chemical reactions may be rate determining, e.g. with some robust complexes, some chelates and some other systems. The distribution method may then be of use to study these relatively slow chemical reactions. A common case is a species which is in rapid distribution equilibrium, but undergoes slow side-reactions

$$A \rightleftharpoons A \qquad \text{Rapid} \qquad (7A1)$$

$$A + B \rightleftharpoons C \qquad \text{Slow} \qquad (7A2)$$

During a short extraction period, the slow equilibrum may be considered frozen, and D_A is a measure of 'free' A in the system. Repeated extractions of the same system[19] provide a progressive picture of the reversible reaction (7A2 which is simple for small D_A, where little A is removed from the aqueous phase, but is somewhat more complicated for high D_A where (7A2) is considerably disturbed by each equilibration (7A1). This is the case with the distribution of the nitrosylruthenium trinitrate complex between TBP and nitric acid[19] where slow interconversion of the nitrate complexes occurs. Slow hydrolytic reactions of protactinium have been blamed for differences in D_{Pa} for solutions in hydrochloric acid aged for different periods, and then extracted with solvents such as diisobutyl ketone[20]. Obviously, extraction for short periods in this system occurs under non-equilibrium conditions, and either a long aging period previous to extraction, or a prolonged extraction, are necessary to attain equilibrium. Prolonged extraction may be harmful in some cases where the solvent can undergo decomposition[21]. An alternative procedure might be a preliminary extraction into the organic solvent of the 'extractable' portion of the solute, and then the distribution may be studied, starting with the solute in the organic phase, and a fresh aqueous solution.

The rate of transfer of a distribuend from one phase into the other has seldom been studied in systems where it is not diffusion controlled. The distribution of mercury(II) bromide between a polyphenyl eutectic and a molten potassium lithium nitrate eutectic is such a case[22]. Distribution between the organic solvent and an aqueous nitrate solution is rapid, and the rate of diffusion is expected to increase from room temperature to 150°C, but the distribution between the molten salt and organic solvent is relatively slow, requiring thirty minutes to approach equilibrium. This is ascribed to the need for the mercury bromide to shed its two solvating aromatic molecules and cross the interface (rapid reactions, since they occur rapidly with an aqueous solvent as the second phase), and then change the configuration and take on two nitrate ions to form a tetrahedral dibromodinitrato complex (a slow step). This mechanism is confirmed by following both D_{Hg} and D_{Br} for a doubly labelled tracer as a function of time. When excess potassium bromide is present in the salt melt the curves for D_{Hg} and D_{Br} no longer agree, it taking longer for D_{Br} to reach its equilibrium value. This is due to a relatively slow exchange of bromide between the mercury complexes and the free ions in the melt. The rate constant of the reaction $\overline{HgBr_2}$. solvate $\rightarrow \overline{HgBr_2(NO_3)_2^{2-}}$ is 0.0027 sec^{-1}, while that of the reverse reaction is 0.025, the equilibrium constant being $D = \bar{m}_{Hg}/m_{Hg} = 9.3$.

The overall rate of metal chelate extraction depends on the rate of dissociation of the chelating acid and the rate of metal chelate formation, rather than on the rate of their transfer, though exceptions are known[23]. Mainly because of slow keto–enol conversions, some reactions involving chelating agents are relatively slow, even with metal ions not forming ordinarily robust complexes. Thus while it is not surprising that chromium(III) must be refluxed with acetylacetone in order to convert it to an extractable species[24], it has also been found that the extraction of zinc(II) with dithizone is slow[25,26]. It was found that the rate of extraction of zinc dithizonate is faster into that solvent, which has the lower distribution coefficient for the reagent itself. This means that a higher equilibrium concentration of dithizone in the aqueous phase promotes faster extraction, i.e. that the reaction between zinc and dithizone in the aqueous phase is rate determining[25]. It was later shown[26] that the rate of extraction is first order with respect to zinc and to dithizone, and inverse first power with respect to hydrogen ions. Although zinc bis-dithizonate is the distribuend, apparently only a slow (at least a rate-determining) addition of the first dithizonate ligand can explain the rate data. The reaction is envisaged as proceeding in the following steps: the hexaaquazinc cation loses a water molecule and a dithizonate ligand is attached through sulphur, another water molecule is lost and the ligand attaches also a nitrogen, two more water molecules are lost and the complex rearranges to a tetrahedral configuration (one of these steps is rate controlling), finally the second ligand attaches stepwise as the first, as two more water molecules are lost. This mechanism still raises some questions, since dithizone extraction of mercury(II) and silver(I) are fast, and that of copper(II) is said to be diffusion controlled[27], although similar substitutions and rearrangements as with zinc would be involved, and zinc does not show specially slow kinetics in many other similar chelation reactions.

Another system where slow kinetics have been observed is that in-volving 2-thenoyltrifluoroacetone (TTA). Although it has been con-cluded that the extraction of plutonium(IV) is diffusion controlled[28], the dependence of the rate on the distribution coefficient shows an important role of the rate of chemical reaction. The forward extraction reaction is fast, but the stripping of the benzene phase, even with immediate reduc-tion in the aqueous phase to plutonium(III), is slow. If chemical reactions control the rate, a high distribution coefficient in favour of the aqueous phase, coupled with a fast forward reaction (extraction from aqueous to organic phase), requires a slow stripping step. The effect of TTA con-centration on extraction rate was noted by Bolomey and Wish[29], who showed slow extraction with 0.01 M TTA in benzene, and fast extraction with 0.1 M solutions. In certain cases the rate of enolization of TTA in the aqueous phase can be the controlling factor[30], in others the reaction with the metal (with Fe^{III}), or dissociation of the complex (with Pu^{IV}). It has been found possible to catalyse the extraction reaction by the use of a synergistic reagent, which itself causes fast, but low distribution, as found in the iron(III)–thiocyanate–TTA system[31].

The rate of extraction with acid organophosphorus esters is usually fast, but a few exceptions have been noted, such as iron(III)[32], alumin-ium[32, 33] and beryllium[34–36]. In the latter case at least, the rate of attainment of equilibrium depends on the structure of the extractant and its concentration, as well as on the acidity of the aqueous phase. The rate decreases when the initial acidity of the aquous phase is higher. A lower rate has been noted with extractants having long and/or branched alkyl chains, the rate being higher with monoalkyl acids. The phenomenon has been attributed to the steric hindrance due to attaching bulky organic molecules around the small beryllium ion[35, 36]. The explanation seems to be at least partially true, since the rate of the metal transfer depends also on the temperature. For example, when extracting beryllium from a sulphate solution by 0.5 M di-(2-ethylhexyl)phosphoric acid, HDEHP, equilibrium has been reached after contacting for 300 min at 25 °C or for 100 min at 45 °C[34].

The rate of uranium(VI) extraction by HDEHP in kerosene is first order with respect to the metal concentration in the aqueous phase[35]. Over a similar range of experimental conditions the rate of iron(III) extraction is also first order with respect to the aqueous iron concentration[32], but the rate of beryllium extraction does not show a simple first-order dependence[35].

In the case of europium extraction by the same ester, the phase transfer was rate determining at reagent concentrations over 0.8 M, but diffusion into the organic phase was critical below this concentration[37]. In the case of iron the rate constant varies inversely with the first power of per-chloric acid concentration and directly with the extractant concentration but to a changing degree: 1.4 above and 0.3 below 0.4 M HDEHP in the n-hexane phase[32]. The rate is inversely proportional to acidity also in the case of beryllium and strontium, though for the relatively fast extraction of uranyl, europium and zinc the rates are nearly independent of acidity[32].

In conclusion, the amount of information that has been published to date on the rate of solvent extraction is not large. It is difficult to make general statements on the factors which affect these rates, and each class of systems seems to require different approaches. The above review could therefore not lead to general correlations, and more work on this problem is justified.

B. THE EXTRACTION PROCESS

a. The Phase Rule

Distribution equilibria between two liquid phases, like all other heterogeneous equilibria, are governed by the phase rule of Gibbs

$$V = 2 + C - P \qquad (7B1)$$

where V is the variance of the system, C the number of components and P the number of phases. The variance of the systems is the number of independent variables: pressure, temperature and concentrations that must be specified in order to describe the state of the system. The number of components is intended in the thermodynamic sense, i.e. the number of chemical substances that can be added independently to the system less the number of independent chemical reactions and the stoichiometric restrictions (e.g. electroneutrality) connecting them.

The gas phase is commonly disregarded in solvent extraction systems, but then the pressure is left to be determined by the equilibrium requirements, i.e. the vapour pressures, and cannot be selected arbitarily. To work at a constant pressure requires the addition of another component, e.g. the atmosphere, when working in open vessels at 1 atm pressure.

Ordinarily both the gas phase and the pressure variable are disregarded, but the variance must be decreased by unity, while the restriction of a constant temperature reduces the variance by one more, yielding

$$V' = C - P \qquad (7B2)$$

Consider a binary system made up of two liquid phases. The variance according to (7B2) is zero, hence at a given temperature the system has a fixed composition: saturated solutions of each of the liquids in the other. At temperatures where complete miscibility occurs there is only one phase, and at a given temperature the system is univariant, the concentrations can be freely selected. If there is a temperature below which complete miscibility no longer occurs, it is called an upper critical consolute temperature, and conversely for a lower critical consolute temperature (cf. section 2.B).

If a solute is added to the binary two-phase system, $C = 3$ and $P = 2$, hence $V' = 1$, and at a given temperature the concentration of the solute in one of the phases may be arbitrarily selected. Its concentration in the other phase, however, is then determined by the equilibrium requirement, hence a definite relationship must exist between the concentrations of the solute in the two phases. Each added solute—provided it is a component in the phase-rule sense defined above — adds another degree of freedom, i.e. its concentration in one phase can be selected at will. If so much of a solute is added to a two-solvent system that a saturated solution in one phase results, and excess undissolved solute exists in equilibrium, there is one more phase and $V' = 0$. The system becomes invariant, and also the second phase must be saturated with the solute. If the solute is a liquid, three liquid phases may exist in equilibrium, but their composition is definite at a given temperature.

With a quaternary system with two liquid phases (solvents) the concentrations of the two solutes can be fixed arbitrarily in one phase. Under certain conditions a third liquid phase appears, and the system with $C = 4$ and $P = 3$ becomes according to (7B2) univariant. This means that the 'third phase' need not have a definite composition, but its composition (as well as those of the other phases) will be given by specifying the concentration of one component only of the system in one phase. Such relationships were discussed e.g. for the water–hydrochloric acid–extractant–diluent systems for tri-n-butyl phosphate[38,39] or a tertiary amine[39] as extractants.

The phase rule is seen to have limited applicability, and to provide only very general guidance as to what variability to expect.

b. The Distribution Law

In a ternary system with a solute distributing between two partially immiscible solvents, it is seen that the phase rule predicts the existence of a definite relationship between the concentrations of the solute in the two phases. This relationship can be derived from thermodynamic considerations[40]. At equilibrium the chemical potentials of the solute in both phases are the same and depend only on the temperature

$$\mu_I = \bar{\mu}_I \tag{7B3}$$

where a bar above a symbol designates one (e.g. the organic) phase and its absence over a corresponding symbol designates the other (e.g. aqueous) phase, and I is the solute. Depending on the selection of the standard state, the relationship between the concentrations of I in the two phases can be derived in different ways. If the standard state of the solute selected is the pure solute, the result is

$$\mu_I = \mu^0_{I(\text{pure})} + RT \ln a'_I = \bar{\mu}^0_{I(\text{pure})} + RT \ln \bar{a}'_I = \bar{\mu}_I \tag{7B4}$$

with $\bar{\mu}^0_I \equiv \mu^0_I$, hence $\bar{a}'_I/a'_I = 1$.

Writing activities in terms of mole fractions and rational activity coefficients this leads to

$$\bar{x}_I/x_I = f'_I/\bar{f}'_I \tag{7B5}$$

The activity coefficient f'_I is related to pure I as standard, and may be determined at one point — a saturated solution of I in a solvent, made up of a saturated solution of one liquid solvent in the other. Since (7B4) is true at saturation, the saturated solution may be selected as a standard state, and μ^0_I identified with $- RT \ln s_I$, hence (7B4) can be rewritten as

$$\mu_I = RT \ln (a''_I / s_I) = RT \ln (\bar{a}''_I / \bar{s}_I) = \bar{\mu}_I$$

$$\bar{a}''_I/a''_I = \bar{s}_I/s_I = S_I \tag{7B6}$$

$$\bar{x}_I/x_I = (f''_I/\bar{f}''_I) S_I$$

The ratio of mole fractions in the two phases is then proportional to the inverse ratio of activity coefficients, defined with saturated solutions as standard states, the proportionality factor being the constant solubility ratio S_I. This selection has been recommended by Milicevic[41]. There are no general and simple means to determine either f_I' or f_I'' over a wide concentration range, hence the usefulness of (7B5) or of (7B6) is limited, and the selection of other standard states is indicated.

If one turns to an infinitely dilute solution as a reference state (cf. section 1.A), one may write (7B4) as

$$\mu_I = \mu^0_{I\,(\text{infinitely dilute})} + RT \ln a_I$$

$$= \bar{\mu}^0_{I\,(\text{infinitely dilute})} + RT \ln \bar{a}_I = \bar{\mu}_I \qquad (7B7)$$

$$\bar{a}_I/a_I = (1/RT)\exp \Delta \mu^0{}_{I(\text{infinitely dilute})} = P_I$$

The quantity $\Delta \mu^0_I$ is the free energy of transfer of one mole of I from one solvent to the other in the infinitely dilute state (i.e. the primary solvent effect). Equation (7B7) may again be rewritten in terms of mole fractions, and the limit, as x_I tends to zero, will be

$$\lim_{x_I \to 0} (\bar{x}_I/x_I) = \lim_{x_I \to 0} (f_I/\bar{f}_I)P_{I(x)} = P_{I(x)} \qquad (7B8)$$

since the limit of the activity coefficient ratio, defined with infinite dilution as the reference state, is unity. From (7B7) it follows that P_I is constant at a constant temperature, and it is called the *partition constant*[40]. For ideal systems where the ratio $f_I/\bar{f}_I$ is constant (or unity) over a practical concentration range, the ratio $\bar{x}_I/x_I$, the *partition ratio*, is also constant as was found empirically by Berthelot and Jungfleisch[42]. Since Henry's law for ideal solutions can be stated also for other concentration scales (cf. section 1.A.a), partition ratios and partition constants may be given in terms of molalities or molarities too. When a *distribution isotherm* is drawn, i.e. the concentration of I in one phase is plotted against that in the other, the initial portion is linear, and the slope is $P_{I(x)}$. However, a distribution isotherm which is linear for one concentration scale over a practical concentration range, cannot be linear for another concentration scale, as pointed out by Milicevic[41]. From the relationship $x = m/(m + 1000/M)$, where M is the molecular weight of the solvent, it follows that only at the limit where $m \ll 1000/M$ is there a constant relationship between $P_{I(x)}$ and $P_{I(m)}$, viz. $P_{I(m)} = P_{I(x)}M/\bar{M}$. It makes, therefore, little sense to compare $P_{I(m)}$ for a given distribuend I for different solvents, but $P_{I(x)}$ may be so compared. Calling the ratio $\bar{x}_I/x_I$ the *distribution coefficient* $D_{I(x)}$, then if $D_{I(x)}$ is constant $D_{I(m)} = D_{I(x)}$ $(\bar{m}_I + 1000/\bar{M})/(m_I + 1000/M)$ cannot be constant. Such a condition is obtained at relatively low concentrations for solvents of high M. A practical demonstration of this effect has been given by Shevchenko and coworkers[43]. They showed that even for ideal solutions, if concentrated, the change in the mole fraction of the solvent with changing solute concentrations prevents the use of distribution coefficients based on molar (or molal) concentrations in mass-action law equations.

Returning to dilute solutions, the 'Nernst distribution law' may be written in terms of molar concentrations

$$\lim_{c_I \to 0} \bar{c}_I/c_I = \lim_{c_I \to 0} D_I = \lim_{c_I \to 0} (y_I/\bar{y}_I)P_I = P_I \tag{7B9}$$

where the distribution coefficient D and the partition constant P without suffix pertain to the molar scale. If the solubility of the solute in both solvents is very low, S_I of equation (7B6), expressed on the molar scale, will approximate to P_I, but in general, saturated solutions are far from ideal, and these quantities are not equal (cf. the case of the mercury halides in benzene and water[44]).

The distribution law (7B9) has been verified in many cases, where the distribution of a thermodynamic component I could be determined directly. A direct verification of (7B7) has also been made[45], in the case where the distribuend is hydrogen chloride and the two solvents are water and tri-n-butyl phosphate (TBP), and a_{HCl} is determined directly by potentiometry in each phase. A water-saturated solution of hydrogen chloride in TBP is presumably used, and the solubility of TBP in aqueous hydrochloric acid is small enough, so that the use of tabulated a_{HCl} for this phase is permissible.

In general, it is possible to measure experimentally the stoichiometric distribution coefficient of distribuend I

$$D_I = \sum_j \bar{c}_{I,j} / \sum_j c_{I,j} \tag{7B10}$$

where the summation covers all the species j in which I occurs and is determined analytically. If the species j are all in equilibrium, there exists a definite relationship between D_I and P_I, where P_I relates to a species which is a component in the thermodynamic sense. The relationship may be written as

$$D_I = P_I \bar{F}_I / F_I \qquad (7B11)$$

where the F's are functions relating the various species j to the component selected. The main problem of solvent extraction as applied to the study of metal complexes is to determine the exact form of the functions F from the experimental data, to obtain P_I, and to relate P_I to structural features of the complexes and solvents.

c. Physical Distribution

Among the numerous distribution systems studied, there is a class of systems involving the least amount of complexity as regards interactions both in the organic and in the aqueous phase. This is the class of systems involving the 'physical' distribution of 'simple' molecules between 'inert' solvents and aqueous solutions. Quotation marks have been used since naturally the distribution processes are simple only as a first approximation. The solvents concerned in this section do not contain atoms with pronounced donor properties. They range from the very inert fluorocarbons through non-polar aliphatic hydrocarbons, non-polar but polarizable chlorocarbons (halocarbons) and aromatic hydrocarbons, to polar and polarizable chlorocarbons and aromatic hydrocarbons, up to polar, polarizable and hydrogen-bonding solvents such as chloroform. Since these solvents lack donor atoms, they do not form coordinative bonds with metal or hydrogen ions, and do not solvate them. Still, they do interact through dispersion forces, through dipole (induced-dipole) interactions, and in the case of aromatic hydrocarbons, through π bonding[49]. Since the free energy released in such interactions is relatively small it will not compensate for the loss of free energy of solvation of most potential distribuends in the aqueous phase (cf. solubility of silver perchlorate in toluene, compared with its non-extractability from aqueous solutions). Therefore, the distribution coefficients of most distribuends will be extremely small, in fact negligible. Of the distribuends which are soluble in water, only those which interact only slightly with the aqueous solvent, i.e. are practically non-solvated, will be extracted. Distribuends which have a high lattice energy in the solid state, i.e. which are ionic and only weakly solvated, will be neither soluble in water, nor in the inert organic solvents, since again there is no sufficient interaction energy to compensate for loss of lattice energy.

It has been attempted to apply the solubility-parameter concept[46] to distribution systems[38, 47–49]. There is little justification to do it in the case of the distribution of a polar distribuend, such as hydrogen chloride[38, 47], or even of a chelating agent such as a diketone[48] (section 2.B), but in the present case, where the extraction of non-polar distribuends by inert solvents is studied, this approach may be fruitful. Since water is not a liquid which can be treated in terms of solubility parameters, the aqueous phase is treated empirically. Actually, only the distribution of a given distribuend between a given aqueous phase and a series of organic solvents can be discussed, as was done by Siekierski[49]. The following formal relationship is obtained, for the distribution coefficient in terms of mole fractions, D_x

$$RT \ln D_x = V_I[(\delta_I - \delta_{aq})^2 - (\delta_I - \delta_S)^2] \qquad (7B12)$$

where V_I is the molar volume of the distribuend I, δ_I its solubility parameter $(\delta = (L/V)^{1/2}$, where L is the heat of vaporization), δ_S is the solubility parameter of the solvent and δ_{aq} an empirical 'solubility parameter' for the aqueous phase. When a series of solvents is compared, one of them can be treated as a reference solvent, having a solubility parameter δ_S^0, and then the empirical δ_{aq} can be eliminated

$$\log D_x/D_x^0 = V_I (2.303RT)^{-1} [(\delta_I - \delta_S^0)^2 - (\delta_I - \delta_S)^2] \quad (7B13)$$

When distribution coefficients are defined on other bases than mole fractions, e.g. volume fractions, molarities or molalities, terms involving the molar volumes of the solvents or their molecular weights must be added (section 2.B). To distribuends which are not liquids or do not volatilize, a value of δ_I may be assigned from distribution data and equation (7B13).

These considerations leave only two categories of distribuends, which may show considerable solubility in organic solvents, and appreciable distribution coefficients. To one category belong organic substances, including metal chelates with organic ligands, which will not be treated here (see Chapter 8). The other includes a handful of elements and compounds, that show measurable distribution coefficients: certain gases, mercury, halogens and interhalogen compounds, mercury halides,

halides of group IVB and VB metals, ruthenium and osmium tetroxides, and a few others. Since it is impossible in this monograph to review all the systems that fall into this class, the behaviour of 'simple distribution' systems will be illustrated by one example, the mercury halides.

d. Mercury Halides, an Illustration

The distribution of mercury halides between inert organic solvents and aqueous solutions was studied long ago. A comprehensive review of early results has been published by Marcus[44], and the data are given in Table 7B1, together with a few recent data. Experiments made with excess halide present gave data which could be analysed in terms of equations (7C19) and (7C20), i.e. in terms of the formation of HgA_3^- and HgA_4^{2-} from HgA_2. The constants obtained[50-54] agree well with those determined by independent methods. A similar method was also used[55] to study the halide complexes formed with mercury halides in a molten potassium–lithium nitrate eutectic at 150°C. Distribution constants measured directly were compared with those calculated from the solubility ratio, equations (7B6-9)[44]. Good agreement was obtained for mercury bromide and iodide (and within a large experimental error also for the thiocyanate), but a discrepancy is noted for the chloride. This discrepancy is due to the appreciable dimerization of mercury chloride in both phases[56], with an equilibrium constant $K_2 = \bar{K}_2 = 0.3$. Because of the low concentration of mercuric chloride in a saturated benzene solution, only dimerization in the aqueous phase needs to be taken into account. Using an equation analogous to (7C14), good agreement between P_0 measured at tracer concentrations and $D_{\text{corr}} = S_{\text{corr}}$, the solubility ratio corrected for dimerization, is obtained[55]. The salting-out of mercury halides by sodium perchlorate from aqueous solution into benzene has been measured[44]. A Setchenov constant (section 1.F.a) $k_s = 0.14$ was found for all three halides, within experimental error. This constant agrees with calculated values, using the McDevit and Long[57] or a modified[58] Debye method. The ligand distribution method (section 7.C.g), employing the distribution of iodine between CCl_4 or CS_2 and aqueous iodide solutions, has been used to study mercury iodide complex formation, and gave results similar to those obtained from mercury iodide distribution, using the same solvents[59]. The distribution of mercury halides was also used to study the disproportionation of the monoligand complex HgA^+ to give Hg^{2+} and HgA_2. The equation obtained is

Principles of Solvent Extraction

445

TABLE 7B1. Distribution constants P of mercury halides between solvents and water at 25°.

Distributed	Solvent	Log P	Method	Reference
HgCl$_2$	Benzene	—1.08	Distr., extrap.[a]	52, 56
		—1.12	Solub.	51
		—1.057	Distr.	44
		—1.142	Solub.	44
		—1.084	Solub., corr.[b]	44
	Toluene	—1.02	Distr., extrap.[a]	52, 56
		—1.00	Distr.	53
		—1.10	Solub.	c
		—1.04	Distr.	d
		—1.03	Distr., extrap.[a]	e
	Cyclohexane	—3.22	Solub., corr.[b]	63
	p-Xylene	—.093	Solub., corr.[b]	63
	Polyphenyl/(K, Li)NO$_3$	0.38	Distr.[f]	55
HgClBr	Benzene	—0.53	Distr.[g]	61
	Polyphenyl/(K, Li)NO$_3$	0.59	Distr.[f]	62
HgBr$_2$	Benzene	0.05	Distr.	51
		0.04	Distr.	44
		0.049	Solub.	44
	Toluene	0.01	Distr.	50
	Cyclohexane	—1.79	Solub.	63
	p-Xylene	0.18	Solub.	63
	Polyphenyl/(K, Li)NO$_3$	0.96	Distr.[f]	55
HgBrI	Benzene	0.17	Distr.[g]	61
	Polyphenyl/(K, Li)NO$_3$	2.6	Distr.[f]	62
HgClI	Benzene	0.68	Distr.[g]	61
	Polyphenyl/(K, Li)NO$_3$	2.0	Distr.[f]	62
HgI$_2$	Benzene	1.58	Solub.	61
		1.68	Solub.	44
		1.66	Distr.	44
		1.63	Distr.	54
	Polyphenyl/(K, Li)NO$_3$	1.58	Distr.[f]	55
	Toluene	1.70	Distr.[g]	h
	Cyclohexane	0.51	Solub.	63
	p-Xyelene	1.86	Solub.	63
Hg(SCN)$_2$	Benzene	—2.5	Solub.	44
		—2.3	Distr.[g]	44

a. Extrapolated to zero HgA$_2$.
b. Corrected for dimerization see text.
c. O. W. Brown, *J. Phys. Chem.*, 2, 51 (1898).
d. K. Drucker, *Z. Elektrochem.*, 18, 246 (1912).
e. A. Hantzsch and A. Vagt, *Z. Physik. Chem.* (*Leipzig*) 38, 735 (1901).
f. At 150°C.
g. Corrected for salting-out.
h. L. G. Sillèn and G. Biedermann, *Svensk. Kem. Tidskr.*, 61, 63 (1949).

$$K_{0-2} = \frac{D[1 - \tilde{n})(1 + D)P + D(1 + P)]}{[\tilde{n}(1 + D)P - 2D(1 + P)]^2} \qquad (7B14)$$

where K_{0-2} is the disproportionation constant, and it simplifies to $K = D^2(1 + P)/(P - D(P + 2))^2$ for the special case $\tilde{n} = 1$. The results obtained[53] agree with potentiometric values. The disproportionation of HgA_2 to HgA^+ and HgA_3^-, on the other hand, need not to be taken into account, since the expression for the distribution coefficient when $\tilde{n} = 2.00$ is

$$D = D_0/1 + 2\sqrt{K_{1-3}} \approx 0.9982 D_0 \qquad (7B15)$$

so that within experimental error the measured distribution coefficient equals the true one[53]. At extremely low concentrations, however, the dissociation of HgA_2 to HgA^+ and A^- becomes appreciable, and k_2 for the inverse, association, reaction can be calculated from

$$k_2 = PD(1 + D)/(P - D)^2 c_{Hg}$$

which is very sensitive to the small differences $(P - D)$ observed, hence requires extremely low c_{Hg} values (total mercury in both phases) to give acceptable values[53].

The extraction of mercury halides from solutions containing ligands other than halide ions has rarely been used to study complex formation. This is because of the possibility of mixed complex formation. Thus extraction of mercury chloride with benzene from oxalate solutions[60] shows that $HgCl_2C_2O_4^{2-}$ is formed in the aqueous phase. Mixed complex formation can very well be studied by the extraction method, which has been applied to the mixed neutral and charged complexes, both in aqueous solutions[53,61] and in a molten nitrate eutectic at $150°$ C^{62}. Stability constants could be calculated for all the complexes formed. In the organic (benzene for aqueous[61] and a biphenyl-terphenyl eutectic for the molten salt[62] studies) the species HgA_2, $HgAB$ and HgB_2 are present, while in the other phase, when excess halide (over $\tilde{n} = 2.00$) is present, trihalide and tetrahalide binary and ternary (mixed) complexes are also formed. For the aqueous systems and the chloride-bromide molten salt system, the constant for the reaction $HgA_2 + HgB_2 \rightleftharpoons 2\,HgAB$ can be obtained for $R = c_{B\,(total)}/c_{A\,(total)} = 1$ from

$$K = \frac{[(P_A - D)(1 + P_B) - (D - P_B)(1 + P_A)]^2}{(1 + P_A)(1 + P_B)(D - P_{AB})^2} \qquad (7B17)$$

where P_{AB} is obtained from the regions where either HgA_2 can be neglected compared with HgB_2 and $HgAB$ ($R \gg 1$) or vice versa

$$P_A = \frac{2(1 + P_B)D + (P_B - D)(R - 1)}{2(1 + P_B) + (P_B - D)(R - 1)} \quad \text{for } R \gg 1 \qquad (7B18a)$$

$$= \frac{2(1 + P_A)RD + (D - P_A)(1 - R)}{2(1 + P_A)R + (D - P_A)(1 - R)} \quad \text{for } R \ll 1 \qquad (7B18b)$$

For the case of mixed mercury halides involving iodide in the molten salt study (at 150°C), the curve of D against R exhibits a maximum, and (7B17b) cannot be used, but instead

$$K = \frac{4R_{max}(P_B - P_A)^2}{(R_{max} - 1)^2(P_{AB} - P_A)(P_{AB} - P_B)} \qquad (7B19a)$$

$$P_{AB} = \frac{R_{max}P_A(D_{max} - P_B) - P_B(D_{max} - P_A)}{R_{max}(D_{max} - P_B) - (D_{max} - P_A)} \qquad (7B19b)$$

must be used, where R_{max} and D_{max} are the R and D values pertaining to the maximum in the curve[62]. The equations that have to be solved for calculating the constants for the formation of mixed tri- and tetrahalide complexes are still more complicated.

The solubility of the mercury halides in a variety of inert solvents has been measured, and it is concluded that in aromatic (and olefinic) hydrocarbons π bonding between the mercury halides and the solvent occurs[63]. The mercury halides are probably bent in the organic solutions[64]. This explains the higher stability noted for the mixed dihalide complexes in water, compared to aromatic solvents[65]. The interaction also causes the solubility, and hence the distribution constants, to be about two orders of magnitude higher in the aromatic hydrocarbons than in aliphatic hydrocarbons, CCl_4, CS_2, etc.

Zinc halides were also found to be extractable into aromatic hydrocarbons from strongly salted aqueous nitrate solutions, as well as from molten nitrates. The behaviour is generally similar to that of the mercury halides[66].

C. QUANTITATIVE TREATMENT OF DISTRIBUTION EQUILIBRIA

a. Introduction

Distribution equilibria can be studied in a great variety of ways. Thus extraction isotherms of the distribuend I, plots of $\bar{c}_I$ against c_I, may be drawn and interpreted, or a property P of the organic phase may be studied as a function of the change in concentration of the distribuend, the extractant, or other ligands, holding the other concentrations constant, or the change in P may be studied as a function of the stoichiometric ratios of extractant to distribuend, etc. A very common approach is the measurement of distribution coefficients D_I of the distribuend as functions of the above mentioned concentrations. The distribution coefficient is defined as the ratio of the analytical (i.e. total or stoichiometric) concentration of I in one phase (the organic phase) to that in the other (the aqueous phase).

$$D_I = \bar{c}_I/c_I \qquad (7C1)$$

Sometimes it is the percentage extraction E_I that is measured, i.e. the percent of the total quantity of I, that is present in the organic phase. This is obviously dependent on the ratio of the amounts of the two phases, which is usually expressed as the volume ratio $\bar{V}/V$, thus

$$E = \frac{100\,D}{D + 1/(\bar{V}/V)} \text{ or } D = \frac{E}{(100 - E)(\bar{V}\,V)} \qquad (7C2)$$

These equations are reduced to $E = 100\,D/(1 + D)$ and $D = E/(100-E)$ for the most common case of phases of equal volume, i.e. $\bar{V} = V$. For this particular case the point for 50% extraction coincides with $D = 1$. As extraction approaches completeness, or E approaches 100%, D increases very rapidly (for $\bar{V} = V$, when $E = 90$, 98.04 and 99.9% $D = 9$, 50 and 1000 respectively).

In this section the information obtainable from an analysis of the variation of the distribution coefficients as functions of different concentration variables will be derived in a general mathematical way. Specific applications of the equations will be discussed in subsequent Chapters. In general, in each phase there exist a number of species containing I, which are related by equilibria, so that the concentrations may be interrelated by certain functions F_I (cf. section 7.B.b). The major problem of solvent extraction as applied to the study of metal complexes, is to ascertain the form of the functions F_I and F_I in (7B11) separately, and the way they vary with concentration. In principle, it should be possible to set up a completely comprehensive and rigorous equation, covering all cases, and then to make simplifying approximations, and apply restrictions. In practice, it will be necessary to treat each class of systems separately, in order to have more meaningful equations, in particular, equations whose terms can be obtained independently.

In every case, it will be necessary to distinguish between concentration effects expressible by mass-action law equations, and effects related to non-specific interactions, non-idealities, etc.[7]. The former can be given a relatively simple mathematical form, the latter are much more complicated, although in some cases it will be justified to ignore them.

It is useful to recapitulate here some of the concepts discussed in section 3.B. A stepwise complex formation system between a central group M and ligands L (with charges omitted for the sake of generality) producing mononuclear complexes ML_n, is characterized by overall stability constants β_n and stepwise stability constants k_n, defined by

$$(ML_n) = \beta_n(M)(L)^n = k_1 k_2 \dots k_n (M)(L)^n \qquad (7C3)$$

in terms of concentrations. The complexity function X is the ratio of the total concentration of M to the concentration of free M, (M), and is given by

$$X = \sum_0^N (ML_n)/(M) = \sum_0^N \beta_n(L)^n \qquad (7C4)$$

where N is the maximal value that the index n assumes, the coordination number. The degree of formation of the nth complex, α_n is the fraction of total M appearing as the species ML_n, and is given by

$$\alpha_n = (ML_n)/\sum_0^N (ML_n) = \beta_n(L)^n/\sum_0^N \beta_n(L)^n = \beta_n(L)^n/X \qquad (7C5)$$

It should be noted that $\alpha_0 = 1/X$ and that $\sum_0^N \alpha_n = 1$. The ligand number $\bar{n}$ is the average number of ligands L bound per M, and is given by

$$\bar{n} = \sum_1^N n\beta_n(L)^n / \sum_0^N \beta_n(L)^n = d\log X/d\log(L) \qquad (7C6)$$

Complex formation may be related to the uncharged complex MA_m for a central ion M^{m+} and ligands A^-, e.g. by the reaction

$$MA_m \rightleftharpoons MA_{m-i}^i + i\,A^- \tag{7C7}$$

which is a dissociation reaction for positive i and an addition reaction for negative i values. The corresponding equilibrium constants are primed, $\beta_i' = \beta_n/\beta_m$ for $n = m - i$. An average charge number $\bar{i}$ is defined by

$$\bar{i} = d\log \alpha_m/d\log(A) = m - \tilde{n} \tag{7C8}$$

These basic formal equations and definitions together with the partition equation are sufficient for describing most solvent extraction situations. It should be remembered however, that β_n or β_i' are not thermodynamic equilibrium constants, but products of the appropriate true constant β_n^T or β_i^T with activity coefficient quotients $G_n = y_M y_L^n/y_{ML_n}$ or $G_i' = y_{MA_m}/y_{MA_{m-i}} y_A^i$ respectively. Thus (7C4) and (7C6) are correct provided that G_n and G_i do not vary with the ligand concentration.

Complicating side-reactions, such as metal ion hydrolysis or reactions with masking agents or competing ligands can simply be taken into account, by modifying X in (7C4) to include terms in the ligands involved

$$X = 1 + \beta_{1(L)}(L) + \beta_{2(L)}(L)^2 + \cdots \beta_{N(L)}(L)^{N(L)} + \beta_{1(OH)}(OH) + \cdots$$

$$+ \cdots \beta_{N(OH)}(OH)^{N(OH)} + \beta_{1(A)}(A) + \cdots \beta_{N(A)}(A)^{N(A)} + \cdots \tag{7C9}$$

$$= 1 + X_L + X_{(OH,A,\ldots)}$$

where $N(L) + N(OH) + N(A) + \ldots$ sum up to the coordination number N if all the ligands are monodentate. Corresponding modifications may also be used in α_n and $\tilde{n}$, which may be defined separately for each ligand: $\tilde{n}_L = (\partial \log X/\partial \log (L))_{(OH,A\ldots)} = d\log(1 + X_L)/d(L)$.

These equations hold for any solvent, including of course aqueous solutions. In many organic solvents, however, many of the reactions leading to these equations are of little importance. Because of the low dielectric constants of these solvents, the concentrations of ions are very low, and ionic species are negligible compared with neutral ion-pairs or molecular species. Other reactions, such as aggregation, are of importance in these systems, which will now be discussed in turn.

b. Distribution of Simple Molecules

'Simple molecules', which will be symbolized by I, will be defined as distribuends which exist in both phases to a large extent as the same molecular species, I. The distribution equation (7B7) then applies

$$\bar{a}_I = P a_I \tag{7C10}$$

If the distribuend occurs only in the form I in both phases, then as derived before

$$D = (\bar{I})/(I) = \bar{c}_I/c_I = P(y_I/\bar{y}_I) \tag{7C11}$$

The distribution constant P may be obtained from

$$P = \lim_{c_I \to 0} D = \lim_{c_I \to 0} d\bar{c}_I/dc_I \tag{7C12}$$

A common reaction that I may undergo is aggregation in the organic phase to I_q, with an equilibrium constant $\bar{K}_q$ applying to the reaction

$$q\bar{I} \rightleftharpoons \bar{I}_q \tag{7C13}$$

If the solutions are ideal, then $\bar{c}_I = \sum_1^Q q(\bar{I}_q) = (\bar{I}) \sum_1^Q q\bar{K}_q(\bar{I})^{q-1}$, and if I remains monomeric in the aqueous phase

$$D = P \sum_1^Q q \bar{K}_q(\bar{I})^{q-1} = \sum_1^Q q \bar{K}_q P (\bar{I})^{q-1} \tag{7C14}$$

For non-ideal solutions, $\bar{K}_q$ is the product $\bar{K}_q^T y_I^q/y_{I_q}$, where $\bar{K}_q^T$ is the thermodynamic equilibrium constant. If D is plotted against $(I) = c_I$, the intercept will be $\lim D$ as $c_I \to 0 = P$, as before, (since $\bar{K}_1 = 1$), but as c_I increases D increases (Figure 7C1), the initial slope being finite if dimers exist

$$\lim_{c_I \to 0} dD/dc_I = 2\bar{K}_2 P^2 \tag{7C15}$$

The degree of aggregation $\bar{q}$ may be defined as[67]

$$\bar{q} = \sum_1^Q q(\bar{I}_q) / \sum_1^Q (\bar{I}_q) = \sum_1^Q q\bar{K}_q P^q c_I^q / \sum_1^Q \bar{K}_q P^q c_I^q \tag{7C16}$$

and may be obtained, graphically or otherwise, from

$$\tilde{q} = c_I^2(D - P) \Big/ \int_0^{c_I} c_I(D - P)dc_I \qquad (7C17)$$

If only a monomer and a dimer exist in the organic phase, the dimerization constant $\bar{K}_2$ may be obtained from (7C15). If only a Q-mer exists, the

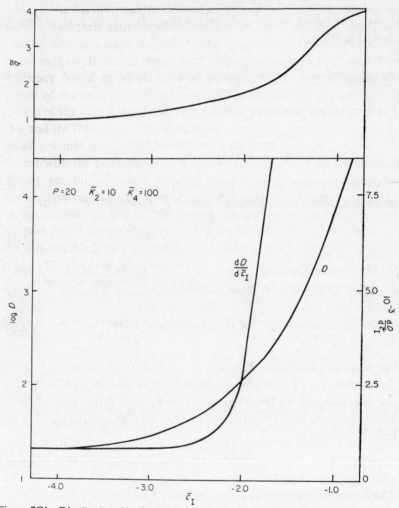

Figure 7C1. Distribution in the case where the distribuend is aggregated in the organic phase. Upper part: mean aggregation number $\tilde{q}$, lower part: log D and d $D/d\bar{c}_1$, all against total distribuend concentration in the organic phase. The parameters for this hypothetical system: $P = 20$, $\bar{K}_2 = 10$ and $\bar{K}_4 = 100$.

initial slope of D against c_I will be zero, since $\bar{K}_2 = \bar{K}_{q \neq Q} = 0$. A plot of $\log D$ against $\log c_I$ will have the slope $Q - 1$, since

$$\log D = \log(QK_Q P^Q) + (Q - 1) \log c_I \qquad (7C18)$$

and the value of the constant $K_Q P^Q$ may be obtained from the intercept at $\log c_I = 0$. Graphical solutions for other cases have been proposed[67] utilizing the variation both of D and of $\int_0^{c_I} c_I (D - P) \, dc_I$ with c_I.

Another common reaction occurs between I and ligands L in the aqueous phase. The ligands may be uncharged or anionic, hydrogen ions, if I is basic, hydroxide ions if acidic (this case may be treated in terms of splitting off of hydrogen ions using negative ligand numbers), etc. The general reaction in the aqueous phase is then

$$I + nL \rightleftharpoons IL_n \qquad (7C19)$$

governed by an equilibrium constant K_n. The distribution coefficient is given by

$$D = (\bar{I}) / \sum_0^N (IL_n) = P / \sum_0^N K_n(L)^n = P/X = P\alpha_0 \qquad (7C20)$$

where X is the function defined before (equation 7C2) and α_0 is the fraction of I in the free form in the aqueous phase. The constants K_n may be obtained from $X = P/D$ as described in section 3.B. The partition constant is obtained as before, as $P = \lim D$ as $c_L \to 0$. In the simple case of association with one hydrogen ion, $I + H^+ \rightleftharpoons IH^+$ replaces (7C19) and

$$D = P/(1 + K_{b1}(H^+)) \qquad (7C21)$$

A plot of $\log D$ against pH will show an initial increase with a limiting slope of unity at high acidities, and a gradual levelling off to a constant value $D = P$ (Figure 7C2a). In the case of simple acidic dissociation $I + H_2O \rightleftharpoons I' + H_3O^+$ or $IOH^- + H^+$, the equation for the distribution coefficient becomes

$$D = P/(1 + K_{a1}/(H^+)) \qquad (7C22)$$

In this case the curve starts out as $D = P$ at high acidities and the distribution coefficient decreases with a final slope of -1 as pH increases (Figure 7C2b). For certain distribuends, both reactions may occur, and the equations can be easily modified accordingly (Figure 7C2c).

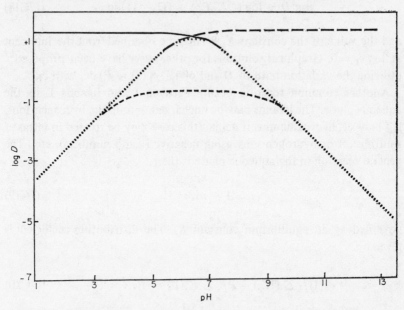

Figure 7C2. Distribution in the case of acid–base reactions in the aqueous phase. For curve —— ····· $K_{a1} = 10^{-7}$, for curve ···· ——— $K_{b1} = 10^6$, for curve ·········· $K_{a1} = 10^{-7}$ and $K_{b1} = 10^6$ (for all three $P = 20$), for curve ··· ——— ··· $K_{a1} = 10^{-9}$, $K_{b1} = 10^4$ and $P = 0.20$.

c. Distribution involving Chelating, and Acidic Extractants

Quantitative treatments of the extraction of metal chelates have been presented by numerous authors[5,68–72]. The most common case is the extraction of a metal M^{m+} with a chelating agent HX from an aqueous solution of a (relatively non-complexed) salt of the metal into an 'inert' organic solvent, such as carbon tetrachloride or benzene, or even chloroform. As will be shown later (next Chapter), the most useful chelating agents for extraction purposes have a denticity d and a charge x, so that $m/x = N/d$. The ratio $N/m = 2$ is very common, thus both a tetra-coordinating bivalent metal, and a hexacoordinating tervalent metal may be extracted by a bidentate mononegative chelate anion X^-, forming the complex MX_m which is electrically neutral and coordinatively saturated. The extractant HX is also usually much more soluble in the organic diluent than in the aqueous phase, so that $P_{HX} \gg 1$. Extraction can be formally written to occur in the following steps

$$\overline{HX} \rightleftharpoons HX \qquad 1/P_{HX} \qquad (7C23a)$$

$$HX \rightleftharpoons H^+ + X^- \qquad K_{aX} \qquad (7C23b)$$

$$M^{m+} + mX^- \rightleftharpoons MX_m \qquad \beta_m \qquad (7C23c)$$

$$MX_m \rightleftharpoons \overline{MX_m} \qquad P_M \qquad (7C23d)$$

The quantities on the right are the equilibrium constants of the various steps. It is possible to view the extraction as a complete process, involving only the major species in each phase: ionic M^{m+} and H^+ in the aqueous phase and non-ionized $\overline{HX}$ and $\overline{MX_m}$ in the organic phase:

$$M^{m+} + m\overline{HX} \rightleftharpoons \overline{MX_m} + mH^+ \quad K_X \qquad (7C24)$$

The equilibrium constant of this reaction is connected with those of the different steps as

$$K_X = P_{HX}^{-m} K_{aX}^m \beta_m P_M \qquad (7C25)$$

Since $\overline{MX_m}$ and M^{m+} are the only species involving M considered hitherto, the ratio of their concentration is the distribution coefficient D

$$D = K_X G_X (\overline{HX})^m (H^+)^{-m} \qquad (7C26)$$

where $G_X = y_{M^{m+}} y_{H^+}^{-m} \bar{y}_{HX}^m \bar{y}_{MX_m}^{-1}$ is the pertinent activity coefficient quotient. Under conditions of constant activity coefficients (e.g. a constant ionic medium in the aqueous phase, a low concentration of extractant in the organic phase and at tracer metal concentrations) G_X is constant and $K_X G_X$ may be substituted by another constant K', while $(\overline{HX})$ may be substituted by $\bar{c}_{HX}$. The distribution coefficient may be expressed logarithmically as

$$\log D = \log K' + m \log (\bar{c}_{HX}/(H^+)) = \log K' + m \log \bar{c}_{HX} + m \text{ pH} \qquad (7C27)$$

Therefore, when it is found experimentally that D is a function of the ratio $\bar{c}_{HX}/(H^+)$ only, but not of $\bar{c}_{HX}$ or the pH individually at a constant value of this ratio, then the extracted species can be taken to be MX_m. Furthermore, $\log D$ is a linear function of the pH. In many cases, however, the curve is found to bend over and tend to a plateau with zero slope, or even show a maximum with subsequent decrease as $\bar{c}_{HX}/(H^+)$ is increased. This is due to formation of complexes MX_n^{m-n} in the aqueous phase, up to $MX_{N/d}^{m-N/d}$, the coordinatively saturated one, which is the same as MX_m when $N/d = m$ (resulting in a plateau in the curve) or an anionic complex (when a maximum is obtained)[69]. Equation (7C27) must then be replaced by

$$D = K_X(\overline{HX})^m(H^+)^{-m} / \sum_0^{N/d} \beta_n(K_{ax}/P_{HX})^n(\overline{HX})^n(H^+)^{-n} = K_X'\alpha_m \quad (7C28)$$

The slope $\tilde{n} - m$ of the curve $\log D$ against $\log(\bar{c}_{HX}/(H^+))$ will range from $+ m$ through zero for $\tilde{n} = m$ to $-(N/d - m)$.

The results of distribution experiments are however often expressed as curves of E against the pH. At unit phase volume ratio, a combination of (7C2) and (7C27) yields

$$E = \frac{100}{1 + (H^+)^m(K'\bar{c}_{HX})^{-m}} \quad (7C29)$$

Using a normalized variable $pH' = -\log(H^+) + \log(K'\bar{c}_{HX})$, it is seen that E is a unique function of this variable for each value of the charge m: $E = 100/(1 + 10^{-mpH'})$. A family of curves may, therefore, be drawn using one template for each value of m. They may be compared with the experimental values for a simple evaluation of K' (Figure 7C3). The pH value for which $E = 50\%$ at a given extractant concentration is called pH_{50}

$$pH_{50} = -(1/m)\log K'\bar{c}_{HX} \quad (7C30)$$

Calling the difference $pH - pH_{50} = \Delta pH$ for a given extractant concentration and metal, it is easily seen from (7C29) and (7C30) that $m\Delta pH = \log(100 - E)/E$. This can be transformed[73] into

$$E = 50(1 - \tanh\frac{2.303}{2}m\Delta pH) \quad (7C31)$$

The family of curves (Figure 7C3) may be drawn from this relationship

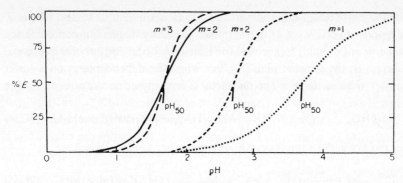

Figure 7C3. Percent extraction against normalized pH, for chelates of metals of different charges m. The values of the parameter $K'\bar{c}_{HX}$ are 0.02 for the curves ———— and – – –, 0.002 for the curve – – – – – –, 0.0002 for the curve ....

If in the extraction of two metals of equal charge m with a given extractant $\left| pH_{50}^{(1)} - pH_{50}^{(2)} \right| \geqq 3$, the metals are readily separated at equal phase volumes with maximal 1% mutual contamination, provided that the pH is selected in the middle of the pH_{50} interval. Selecting a pH value 1.5 units higher than pH_{50} for a given metal will ensure practically quantitative extraction of this, and all the more extractable metals. Stripping with an aqueous solution of pH 1.5 units less than pH_{50}, ensures that only the wanted metal, with those approximately equally extractable, will return to the aqueous phase, while the much more extractable metals will remain in the organic phase.

Complicating factors which are often encountered can be readily taken into account. If for instance $dm < N$, the complex MX_m is coordinatively unsatured, and it often adds on HX molecules, to give $MX_m(HX)_n$. The overall extraction reaction

$$M^{m+} + (m + n)\overline{HX} \rightleftharpoons \overline{MX_m(HX)_n} + mH^+ \qquad (7C32)$$

has an equilibrium constant $K_{X,HX}$, and the distribution coefficient will be

$$D = K_{X,HX}G_{X,HX}(HX)^{m+n}(H^+)^{-m} \qquad (7C33)$$

The formulation $MX_m(HX)_n$ for the organic species deduced from an experimentally found dependence according to (7C33) is only a representation of the stoichiometry of the complex. Often it is found that $m = n$, and the species may then be $M(XHX)_m$, with an internally hydrogen-bonded dimeric extractant (section 8.C). Cases where D depends inversely on a higher power of the hydrogen ion concentration than m are difficult to account for (this may occur for polymerized metal species in the organic phase[69]), but when the dependence is on a lower power than m, then either the metal is hydrolysed in the aqueous phase

$$M(OH)_p^{m-p} + (m + n)\overline{HX} \rightleftharpoons \overline{MX_m(HX)_n} + (m-p)H^+ + nH_2O \quad (7C34)$$

or the extracted complex contains some other anionic ligand e.g. A^-

$$M^{m+} + (m + n - p)\overline{HX} + pA^- \rightleftharpoons \overline{MA_pX_{m-p}(HX)_n} + (m - p)H^+ \quad (7C35)$$

A decision between these possibilities can made by analysing the extractant dependence, or noting the coextraction of the anion (section 8. C). Other complicating factors, such as polymerization of the metal ions or complexing by non-extracted ligands in the aqueous phase, etc., can be readily taken into account. This latter case occurs in the many investigations by means of an extractant HX, of complexing between the metal M^{m+} and ligands A^{a-}, in which stable mixed complexes MA_pX_m do not form in either phase. In the organic phase MX_m or $MX_m(HX)_n$ will predominate, while in the aqueous phase only species MA_p^{m-pa} are important. The distribution coefficient will then be given by

$$D = K'(\overline{HX})^{m+n}(H^+)^{-m} / \sum \beta_p(A^{a-})^p = K'(\overline{HX})^{m+n}(H^+)^{-m}/X_A \quad (7C36)$$

The complex stability constants for the species MA_p^{m-pa} in the aqueous phase are then obtained by the usual graphical or computer[74] methods discussed in section 3.B.

Different complications occur when $(\overline{HX})$, the concentration of available extractant in the organic phase, cannot be taken to be equal to $\bar{c}_{HX}$, the total concentration (which usually is also equal to the intitial concentration). This can occur for a number of reasons: the distribution of HX between the two phases (equation 7C23a, b) can result in an appreciable fraction in the aqueous phase, the extractant may polymerize to $(\overline{HX})_q$ in the organic phase, at high metal ion concentrations an appreciable fraction of the extractant may be bound to the metal, etc. In such cases, it is necessary to take into account the mass balance of the extractant[70], and calculate the concentration of free monomeric extractant $(\overline{HX})$ entering in equations (7C26), (7C33), etc. Assuming equal phase volumes, the total amount of extractant present initially in the organic phase, $\bar{c}_{HX}$, is apportioned between the various species as follows

$$\bar{c}_{HX} = (\overline{HX}) + (HX) + (X^-) + q(\overline{H_qX_q}) + (m+n)(\overline{MX_m(HX)_n}) + \ldots \quad (7C37)$$

Concentrations in the aqueous phase must be divided by the phase volume ratio $\bar{V}/V$ when it differs from unity at equilibrium. Utilizing (7C23a), (7C23b), (7C13), (7C33), etc., to relate the concentrations of the various species to $(\overline{HX})$, leads finally to

$$(\overline{HX}) = \bar{c}_{HX}\bigg/\bigg(1 + P_{HX}^{-1}(1 + K_{aX}(H^+)^{-1}) + qK_q(\overline{HX})^{q-1}$$
$$+ (m+n)K_{X,HX}c_M(H^+)^{-m}(\overline{HX})^{m+n-1} + \cdots\bigg) \quad (7C38)$$

The terms in the large brackets involving $(\overline{HX})$ and c_M (the equilibrium metal ion concentration in the aqueous phase) can be conveniently calculated by successive approximations. The constants P_{HX}, K_{aX} and K_q can be determined independently, in experiments that do not involve the metal[75].

The various equilibrium constants for the extraction and complexation reactions (K_X, $K_{X,HX}$, β_n, etc.) may be obtained from the data $D(\bar{c}_{HX}, (H^+), (A^-) \ldots)$ by the methods discussed in section 3.B. Sometimes it is impossible to obtain distribution data over a sufficiently wide concentration range, usually because D becomes smaller than 10^{-3} or larger than 10^3. Only within this range can D be determined accurately, even if the phase volume ratio is chosen optimally[76]. If a metal ion may be assumed to be extracted as the same species with different diluents, only the values of P_{HX} and P_M will be affected, i.e. the value of K' in (7C36), but not the shapes of the curves log D against log $(\bar{c}_{HX}/(H^+))$, which will be shifted parallely. Either a diluent can then be selected which permits the evaluation of D over the complete range, or D may be reconstructed from fragments of curves for a number of diluents which fall in the experimentally accessible range[76] (Figure 7C4).

In extractions involving chelating agents or other acidic extractants, the concentrations of the reacting constituents in the solutions (metal ions, extractant, ligands, hydrogen ions, etc.) are usually sufficiently small, for variations in the activity coefficient quotients G appearing in equations (7C26) and (7C36), and ignored in other equations, to be neglected. These quotients may be incorporated in the overall extrac-

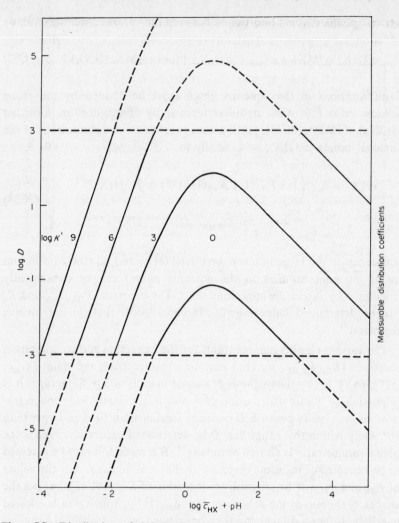

Figure 7C4. Distribution of chelates with very large or very small distribution constants, $\log K'$ for a hypothetical system of metal of charge $m = 3$, forming chelates in the aqueous phase with $\beta_1 = 10$, $\beta_2 = 20$, $\beta_3 = 4$ and $\beta_4 = 1$ (equation 7C28).

tion constants K' (e.g. equations 7C27, 7C28 or 7C36). A constant ionic medium can often be applied in the aqueous phase to ensure approximate constancy of the activity coefficients in this phase. Activity coefficients of the extractant, or ratios of activity coefficients of the extractant and the complex can sometimes be determined (cf. Chapter 8).

d. Distribution involving Solvation in the Organic Phase

Most species soluble in water are solvated by it to some extent. Not so in organic solvents: many distribution systems need no solvation in the organic phase in order to exhibit high distribution coefficients (e.g. chelates extracted into inert solvents). Still, a large number of species are stabilized in the organic phase by solvation. This solvation involves dative, coordinate bonds, which can be distinguished from covalent bonds on the one hand, and ionic interactions on the other. Ordinarily, solvation in the organic phase is a specific reaction, involving definite stoichiometric relations, for example a metal ion M^{m+} can make up its first coordination sphere with a coordination number N with m monovalent anionic ligands A^-, and with $N - m$ solvent molecules S, each with one donating atom (usually oxygen). In some cases solvation can occur by different means, involving for instance, the π-electron system of aromatic molecules, or strong dipoles of polar solvents.

Molecular distribuends I, considered earlier in this section, may undergo solvation in the organic phase

$$I + p\bar{S} \rightleftharpoons \overline{I\,S_p} \tag{7C39}$$

The distribution coefficient will depend on the concentration of the extractant S

$$D = PK_S\bar{G}_S f\,(c_I, c_L, \ldots)\,\bar{c}_S^p\,\bar{y}_S^p \tag{7C40}$$

where K_S is the thermodynamic equilibrium constant of (7C39), P the partition constant of unsolvated I, $\bar{G}_S$ the activity coefficient ratio $\bar{y}_{IS_p}/\bar{y}_I$, $f\,(c_I, c_L, \ldots)$ a function of the distribuend, ligand, etc., concentrations, in cases where I undergoes the appropriate reactions (7C13), (7C19), etc. and $\bar{c}_S$ and $\bar{y}_S$ are the concentration and activity coefficient of the extractant respectively. The total concentration $\bar{c}_S$ is used instead of the more appropriate free concentration $(\bar{S})$, on the understanding that S is present in large excess, and only a negligible portion is consumed by solvating I. On partial differentiation

$$(\partial \log D/\partial \log \bar{c}_S)_{c_L\cdots} = p + p\,(\partial \log \bar{y}_S/\partial \log \bar{c}_S) \tag{7C41}$$

provided the ratio $\bar{G}_S$ can be treated as constant and no polymerization

462   *Ion Exchange and Solvent Extraction of Metal Complexes*

of I occurs, and that c_L, etc., can be kept constant, instead of the rigorously required (L), etc. For ideal solutions of the extractant, the second term on the right vanishes, and the *solvation number p* is obtained directly from (7C41). Distribution experiments are, however, often conducted with appreciable extractant concentrations and non-idealities must be taken into account, including deviations caused by the concentration scale used (cf. section 7.B.b). For some solvent systems log $\bar{y}_S$ is proportional to $\bar{c}_S$ over a practical concentration range[77], and then $(\partial \log D / \partial \log \bar{c}_S) = p (1 + k_S \bar{c}_S)$, k_S being the proportionality factor. Solvation numbers should, therefore, be determined at the smallest $\bar{c}_S$ practicable. If I participates in a polymerization reaction in the organic phase, (7C13), in addition to being solvated, a change in $\bar{c}_S$ will change the concentration of free I, $(\bar{I})$, hence will disturb the aggregation equilibrium, and it is impossible to perform the partial differentiation (7C41) at a constant value of the function $f(c_I, c_L, ...)$ in (7C40) without more complicated terms.

Sometimes chelates or compounds MX_m extracted by a solution of an extractant HX in a diluent or solvent S can be further solvated. This occurs when $md < N$, and then MX_m is coordinatively unsaturated and M is directly solvated by S, or even in some cases when M is coordinatively saturated, S can bond to an active site in the ligand X, without a direct bond to M. The reaction

$$\overline{MX_m} + p\overline{S} \rightleftharpoons \overline{MX_mS_p} \tag{7C42}$$

leads to an expression for the distribution coefficient analogous to (7C40)

$$D = K_X K_S G_X \bar{G}_S (\overline{HX})^m (H^+)^{-m} (\overline{S})^p \bar{y}_S^p \tag{7C43a}$$

or under conditions of constant G_X and $\bar{G}_S$ and $\bar{y}_S$, and when $(\overline{HX})$ and $(\overline{S})$ can be equated with $\bar{c}_{HX}$ and $\bar{c}_S$ respectively

$$\log D = \log K' + m \log \bar{c}_{HX} + m \, \mathrm{pH} + p \log \bar{c}_S \tag{7C43b}$$

Partial differentiation again gives the solvation number

$$(\partial \log D / \partial \log \bar{c}_S)_{\bar{c}_{HX}, \mathrm{pH}} = p. \tag{7C43c}$$

Of greater importance is the large class of distribution systems where the distribuend is an associated metal salt MA_m, solvated by a solvent or extractant S. Extraction of salts involving divalent anions is very rare, hence the equations will be derived in terms of a monovalent anion A^-, representing the common cases of halide, thiocyanate, nitrate, etc. The metal M will be considered to undergo stepwise complex formation, forming a series of species MA_n^{m-n} up to MA_N^{m-N} in the aqueous phase, including MA_m, the uncharged species. This species is capable of distributing between the two phases, and will be solvated in the organic phase, forming MA_mS_p. Depending on the nature of the extractant or solvent S and the diluent if present, $MA_m\,S_p$ may dissociate to form ion-pairs or ions MS_p^{m+}, mA^- (solvated). Two cases will now be considered: in one, only species involving M, but not the electrolyte CA providing A^-, can be extracted into the organic phase, in the other, both metal species and CA (e.g. HA or LiA) can be extracted.

In the first case, the pertinent steps in the distribution equilibria are

$$M^{m+} + mA^- \rightleftharpoons MA_m \qquad \beta_m \qquad (7C44a)$$

$$MA_m \rightleftharpoons \overline{MA_m} \qquad P \qquad (7C44b)$$

$$\overline{MA_m} + p\overline{S} \rightleftharpoons \overline{MA_mS_p} \qquad K_S \qquad (7C44c)$$

which may be summarized in an overall extraction equilibrium

$$M^{m+} + mA^- + p\overline{S} \rightleftharpoons \overline{MA_mS_p} \qquad K_M \qquad (7C45)$$

disregarding the possibility of ionic dissociation in the organic phase. The overall thermodynamic equilibrium constant of (7C45) is $K_M = \beta_m P K_S$. Since $\overline{MA_mS_p}$ is the only species in the organic phase containing the metal, the distribution coefficient is

$$D = K_M G_M(A^-)^m(\overline{S})^p/X = PK_S G_M(\overline{S})^p\alpha_m = K'_M G'_M(\overline{S})^p/X' \quad (7C46)$$

where $\quad X = \sum_0^N \beta_n G_n(A^-)^n$, while $X' = \sum_{-m}^{N-m} \beta'_i G'_i(A^-)^i$ applies to

the formulation relating the aqueous complexes to the uncharged species MA_m e.g. (7C7) (cf. section 3.B.c). The activity coefficient quotients $G_M = y_A^m \bar{y}_S^p \bar{y}_{MA_mS_p}^{-1}$, $G_n = y_A^n y_{MA_n}^{-1}$, $G'_M = \bar{y}_S^p \bar{y}_{MA_mS_p}^{-1}$ and $G'_i = y_A^i y_{MA_{m+}}^{-1}$, may be considered constant ass a first approximation. If effective activities of the ligands are used (sections 1.D.f and 3.B.b), equation (7C46) becomes

$$\log D = \log K_M^* + p\log \bar{c}_S - \log \sum \beta_i'^* a^i \qquad (7C47)$$

Substituting c_A for (A^-) and $\bar{c}_S$ for $(\overline{S})$ where this is permissible, yields on partial differerentiation the average ligand number $\tilde{n}$ and charge $\bar{\imath}$ of the species in the aqueous phase, and the (average) solvation number $\tilde{p}$ in the organic phase

$$(\partial \log D/\partial \log a)_{\bar{c}_S} = m - \tilde{n} = \bar{\imath} \qquad (7C48)$$

$$(\partial \log D/\partial \log \bar{c}_S)_{c_A} = \tilde{p} \qquad (7C49)$$

An average solvation number is given by (7C49) if solvation in reaction (7C44c) occurs stepwise. If a unique solvate MA_mS_p is formed, a constant and integral value for p is obtained from (7C49), provided the approximations leading to it are valid.

If excess ligand is not added, and only the salt MA_m is used in the aqueous phase, the distribution equation can sometimes be simplified. Steps (7C44b) and (7C44c) then represent the whole reaction, and the distribution coefficient is given by

$$D = P\,K_S\,\bar{y}_{MA_mS_p}^{-1}\,\bar{c}_S^{\,p}\,\bar{y}_S^{\,p}\,c_M^{\,m}\,y_{\pm MA_m}^{\,m+1} \qquad (7C50)$$

At very low concentrations, where $\bar{y}_{MA_mS_p}$ and $y_{\pm MA_m}$ equal unity, the simple m law (e.g. square law for uranyl nitrate) holds, as well as a p law regarding the solvation number[7.7, 78].

In solvents of sufficiently high dielectric constant, partial dissociation of the extracted salt may occur. This affects the distribution coefficient, multiplying D by a factor

$$(1 + K(c_S\bar{y}_S)^{-(p+t)/2}(c_M y_{\pm MA_m})^{-(m+1)/2}\,\bar{y}_{MA_mS_p}\bar{y}_\pm^{-1}),$$

where $K = \sqrt{K_d/K_S P}$, K_d is the dissociation constant of $[\overline{MA_mS_p}]$ into $[\overline{MA_{m-1}S_r^{m-1}}] + [\overline{AS_s^-}] + t\overline{S}$ (with $r + s + t = p$ and with t possibly negative), and $\bar{y}_\pm$ is the mean ionic activity coefficient of these ions in the organic phase. As the total metal concentration increases, c_M increases, and D increases, according to equation (7C50), but not as fast as it would if no dissociation occurred.

At high metal concentrations an appreciable fraction of the extracted S becomes bound to the metal and in every case where $\bar{c}_S$ occurs above, it must be replaced by $(\overline{S}) = \bar{c}_S - pDc_M$, which may be calculated by successive approximations. If a number of solvates are in equilibrium, it can be expected that $\tilde{p}$ decreases as $(\overline{S})$ decreases, and the problem becomes very complicated.

Besides being bound by the metal ions, the extractant can be bound by coextracted electrolyte C^+A^-, added to provide the anions A^- for complexing the metal. This electrolyte is often the acid H^+A^-, which is well extracted, and since it is often present at high concentrations, a large fraction of the extractant is thus bound. As the acid concentration increases, the term $(A^-)^m$ in the numerator of (7C46) increases, but also terms with $(A^-)^n$ in the denominator increase, and $(\overline{S})^p$ in the numerator decreases, so that usually a maximum in the curve of D against c_{CA} occurs (Figure 7C5). The hydrogen ions extracted sometimes carry with them water of hydration, so that their solvation number for S, h, need not equal the maximum coordination number 4, and both ion-pairs $[HS_h(H_2O)_{4-h}]^+A^-$ (where the anion may also be hydrated), or solvated molecules (often hydrogen bonded) AHS may occur. Disregarding the structure of the solvated acid or salt C^+A^-, the stoichiometric solvation number being t, the concentration of free extractant is $\bar{c}_S - t\bar{c}_{CA}$. The distribution equation for C^+A^- being

$$C^+ + A^- + t\overline{S} \rightleftharpoons \overline{CAS_t} \qquad K_C \qquad (7C51)$$

with an equilibrium constant K_C, the concentration of free extractant is

$$(\overline{S}) = \bar{c}_S/(1 + tK_C a_{\pm CA}^2 (\overline{S})^{t-1} \bar{y}_S^t \bar{y}_{CAS_t}^{-1}) \qquad (7C52)$$

which for $t = 1$ (e.g. for S = tri-n-butyl phosphate (TBP) and CA = nitric acid) can be approximated by $(\overline{S}) = \bar{c}_S/(1 + K_C' a_{\pm CA}^2)$ at concentrations where the activity coefficient in the organic phase can be considered to remain constant.

If coextraction of CA occurs, the partial differentiations (7C48) and (7C49) will not yield $\bar{i}$ and $\bar{p}$, since a and $(\overline{S})$ cannot be varied independently. However, the differentiation $(\partial \log D/\partial \log(\overline{S}))_a$ can still be performed, with $(\overline{S})$ calculated from (7C52) (noting that $c_{\pm CA}^2 = a^2$). In order to find the ligand dependence, a corrected distribution function D^0, defined as

$$
\begin{aligned}
\log D^0 &= \log D - p\log(\overline{S})/\bar{c}_S \\
&= \log D + p\log(1 + tK_C a^2(\overline{S})^{t-1} \bar{y}_S^t \bar{y}_{CAS_t}^{-1})
\end{aligned}
\qquad (7C53)
$$

can be differentiated with respect to a to obtain $\bar{i}$. For $t = 1$ the distribution coefficient may be calculated from (7C46) and (7C52) as

$$\log D = \log K'_M + p \log \bar{c}_S + \log(\bar{y}^p_S / \bar{y}_{MA_mS_p}) - \log \sum_{-m}^{N-m} \beta^*_i a^i \qquad (7C54a)$$

at low a, where the fraction of S bound to CA may be neglected, and as

$$\log D = \log(K'_M / K_C) + p \log \bar{c}_S + \log(\bar{y}^p_{CAS} / \bar{y}_{MA_mS_p}) - \log \sum_{-m}^{N-m} \beta^*_i a^{i+2p} \qquad (7C54b)$$

at such high a, that most of the S is bound to CA, i.e. the second term in parenthesis in equation (7C52) is much larger than unity[77, 79].

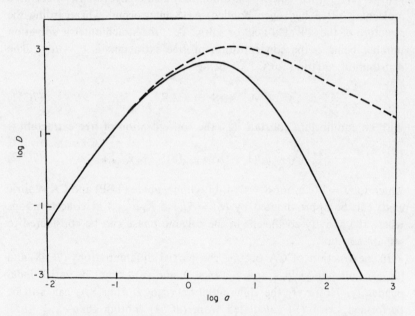

Figure 7C5. Distribution curves for cases where the extractant is bound to coextracted ligand acid. Hypothetical case with $m = 3$, $p = 3$, $\bar{c}_S = 2$ M, $K'_M = 10^3$, $\beta_1 = 10$, $\beta_2 = 20$, $\beta_3 = 4$, $\beta_4 = 1$. Curve - - -: equation (7C46); curve ——— corrected for equation (7C51), with $t = 1$ and $K_c = 0.1$.

In these derivations it is assumed that although CA is coextracted with MA_m into the organic phase, both being solvated there, there is no interaction between them to yield solvated ion-pairs $C^+_n \, MA^{n-}_{m+n}$ in the organic phase. Such an interaction probably does not occur in the extraction of uranyl nitrate from nitric acid with TBP, but it seems to occur in the extraction of uranyl chloride from hydrochloric acid with the same extractant. Such cases belong to the class of acido-complex distribution to be discussed below.

e. Distribution of Acido Complexes

Many metal complexes are extracted as anions of an 'acido complex', i.e. as a (partly or practically) undissociated ion-pair. Solvated (or solvated and hydrated) hydrogen ions constitute the cations of the these ion-associates, and anionic metal complexes constitute the anions. The quantitative treatment of these systems depends strongly on the dielectric constant range of the solvent. In solvents of very low dielectric constants the ion-associates aggregate to ion-triplets, -quadruplets, etc., i.e. to species containing more than one metal atom. In solvents of an intermediate dielectric constant, simple monomeric ion-pairs are the predominating metal species in the organic phase. In solvents of relatively high dielectric constant, ionization to solvated hydrogen ions and complex metal anions occurs. In this case the presence of a non-complexing but extracting acid introduces considerable complications, absent with solvents of lower dielectric constant. In all cases, the metal may also be coextracted partly as a solvated neutral complex, and this introduces further complications.

Consider again the metal M^{m+} forming in the aqueous phase mononuclear complexes MA_n^{m-n} with a ligand A^- provided by an acid HA. The acid is extracted into the organic phase as $\overline{H^+A^-}$ with a distribution constant P_{HA}, while the metal is extracted as the ion-pair $\overline{H^+MA_{m+1}^-}$, which may be aggregated to $\overline{(HMA_{m+1})_q}$. All species, and in particular their cations, are solvated to some extent, which is not indicated in the following.

In a solvent of low dielectric constant the following reactions are important (see section 2.D on ion aggregation in organic solvents).

$$H^+ + MA_{m+1}^- \rightleftharpoons \overline{HMA_{m+1}} \qquad K_M \qquad (7C55a)$$

$$q\overline{HMA_{m+1}} \rightleftharpoons \overline{(HMA_{m+1})_q} \qquad K_q \qquad (7C55b)$$

The total metal concentration in the organic phase is given by $\bar{c}_M = \overline{(HMA_{m+1})} + q\overline{(HMA_{m+1})_q}$, and that in the aqueous phase by $c_M = \Sigma(MA_n^{m-n})$. Using the equilibrium constants of (7C55a) and

(7C55b), appropriate activity coefficient quotients G, and the approximation $(H^+) = (A^-) = c_{HA}$, the expression for the distribution coefficient is obtained

$$D = K_M G_M \beta_{m+1} c_{HA}^{m+2} [1 + q K_q G_q (K_M G_M \beta_{m+1} c_{HA}^{m+2} X^{-1})^{q-1} c_M^{q-1}] X^{-1}$$

(7C56)

where X has its usual meaning. Assuming that activity coefficients may be considered constant, and absorbing them in the equilibrium constants, yields on differentiation, the slope

$$(\partial \log D/\partial \log c_M)_{c_{HA}} = (q-1) K c_M^{q-1}/(1 + K c_M^{q-1})$$

(7C57)

where K is a constant. If only monomers are important $q = 1$ and the derivative equals zero, while for higher values of q the slope varies from zero at low c_M to $q - 1$ at high c_M. Equations (7C56) and (7C57) are correct provided a negligible fraction only of c_{HA} and the solvent are consumed by being bound to the metal species. Otherwise, free (A^-) and (H^+) must be used in the equations, and $(\bar{S})$ must be introduced explicitly. If the aggregation reaction (7C55b) does not produce a unique ion-multiplet but a series of aggregated species, a varying average $\tilde{q}$ should be used in (7C56), or a sum of terms, and since $\tilde{q}$ varies with c_M, the derivative (7C57) is no longer correct and must be modified.

At tracer level c_M and at intermediate dielectric constants of the solvent, the term involving the aggregates becomes negligible. Partial differentiation of (7C56) with respect to $\log c_{HA}$ (assuming constant activity coefficients) yields

$$(\partial \log D/\partial \log c_{HA})_{c_M \to 0} = m + 2 - \tilde{n}$$

(7C58)

It is usually better to analyse the variation of D with the effective ligand activity $a = c_{HA} y_{\pm HA}$, which should replace c_{HA} in (7C58) at high ligand concentrations, when only HA is present at macro concentrations.

If a generalized metal species $H_p MA_{m+p}$ is extracted, and if $HA + HB$ are used at constant c_H^+ (where B^- is a non-complexing anion) or $HA + CA$ are used at constant c_A^-, it is possible to evaluate the partial differentials with regard to hydrogen ions and ligands separately. Neglecting aggregation effects and variations in activity coefficients (which may be small in the aqueous phase if constant media are used) leads to

$$(\partial \log D/\partial \log c_{H+})_{c_{A-}} = p$$

(7C59a)

$$(\partial \log D/\partial \log c_{A-})_{c_{H+}} = m + p - \tilde{n}$$

(7C59b)

This result may be generalized to include various species in the organic phase which are uncharged and have an average composition $H_{\bar{p}}^{-}MA_{m+\bar{p}}$, and the species in the aqueous phase which may be charged and have an average composition $H_{\tilde{p}}MA_{\tilde{n}}$. This leads to a distribution coefficient[80]

$$\log D = \log K + (\bar{p} - \tilde{p})\log c_{H^+} + (m + \bar{p} - \tilde{n})\log c_{A^-} \quad (7C60)$$

which yields the difference in the average protonation or ligand numbers in the two phases, on partial differentiation. Aggregation of the metal species in the aqueous phase is rare in acid solutions from which appreciable extraction occurs, while aggregation in the organic phase has been discussed above. Equation (7C60) may be used in solvents of intermediate dielectric constants where $(\partial \log D/\partial \log c_M)_{c_{A^-}, c_{H^+}} = 0$.

Solvents of a relatively high dielectric constant promote ionic dissociation, while aggregation to ion-multiplets is not important. Reaction (7C55b) may therefore be replaced by

$$\overline{HMA_{m+1}} \rightleftharpoons \overline{H^+} + \overline{MA_{m+1}^-} \qquad K_{DM} \qquad (7C61)$$

The extracted acid HA may also dissociate in the organic phase, and the two reactions

$$H^+ + A^- \rightleftharpoons \overline{HA} \qquad K_A \qquad (7C62a)$$

$$\overline{HA} \rightleftharpoons \overline{H^+} + \overline{A^-} \qquad K_{DA} \qquad (7C62b)$$

must also be considered. The important aspect of this case is that hydrogen ions in the organic phase $\overline{H^+}$ are provided by both the metal complex and the ligand anion extracted. Depending on the relative magnitudes of K_{DM} and K_{DA}, and on the relative concentrations of ligand and metal in the organic phase different types of behaviour are observed[81, 82]. The concentration of hydrogen ions in the organic phase is given by

$$
\begin{aligned}
(\overline{H^+}) &= (\overline{MA_{m+1}^-}) + (\overline{A^-}) \\
&= K_{DM}G_M'(\overline{HMA_{m+1}})(\overline{H^+})^{-1} + K_{DH}K_AG_A(H^+)(A^-)(\overline{H^+})^{-1} \\
&= [K_{DM}G_M''\beta_{m+1}(A^-)^m(M^{m+}) + K_{DH}K_AG_A]^{1/2}[(H^+)(A^-)]^{1/2} \\
&= (K_{DM}G_M^*\alpha_m c_M + K_{DA}G_A')^{1/2} c_{HA} \qquad (7C63)
\end{aligned}
$$

where G_M^* and G_A' are constants, provided activity coefficients are constant and provided $(H^+) = (A^-) = c_{HA}$. The distribution coefficient of the metal is given by

$$D = K_M G_M \beta_{m+1} c_{HA}^{m+2}[1 + K_{DM}G'_M(\overline{H^+})]X^{-1} \qquad (7C64)$$

and this expression is in the form[85] (Figure 7C6)

$$D = P + Q(R + Sc_M)^{-1/2} \qquad (7C65)$$

where P, Q, R and S are independent of c_M, but depend on c_{HA}:

$$P = K_M G_M \beta_{m+1} c_{HA}^{m+2}/X, \quad Q = K_M K_{DM}\beta_{m+1}G_M G'_M c_{HA}^{m+1}/X, \quad R = K_{DA}G'_A$$

and $S = K_{DM}G_M^* \alpha_A$. Partial differentiation shows that when $Sc_M \gg R$ and for $Q \gg P$, i.e. at high metal concentrations for a metal-complex acid which is dissociated more than the ligand acid, the slope $(\partial \log D/\partial \log c_M)_{c_{HA}} = -1/2$ is obtained. Otherwise, i.e. at very low metal concentrations, or for a ligand acid which is well extracted and highly dissociated in the organic phase, the slope approaches zero. The dependence of D on c_{HA} at constant c_M is also complicated, since the term in square brackets depends on c_{HA}. For vanishingly small metal concentrations, all the hydrogen ions are contributed by the extracted acid, $(\overline{H^+}) = c_{HA}(K_{DA}K_A G'_A)^{1/2}$, and if this is introduced into (7C64), the resulting equation is of the form

$$D = T c_{HA}^{m+2}(1 + Uc_{HA}^{-1})X^{-1} \qquad (7C66)$$

where T and U do not depend on c_{HA}, provided activity coefficients remain constant (a rather unrealistic assumption, when only HA is the bulk electrolyte present). The slope of a plot of $\log D$ against $\log c_{HA}$ (or $\log a$) will have a slope $m + 2 - \tilde{n} - (U/(U + c_{HA}))$ where $U = K_D\,G'_M \times (K_{DA}K_A G'_A)^{-1/2}$, varying from $m + 1 - \tilde{n}$ at very low c_{HA} to $m + 2 - \tilde{n}$ at very high c_{HA}, the latter value of the slope being that obtained (equation 7C58) if no ionic dissociation occurred.

The presence in the system of another acid HB, which dissociates in the aqueous phase to give a non-complexing anion B^-, but which is capable of being extracted and of dissociating in the organic phase according to equations similar to (7C62), introduces further complications[81,82]. Hydrogen ions in the organic phase may then be chiefly contributed by HB, and the parameter R in (7C65) must now be written as $R = (K_{DA}G'_A + K_{DB}G'_B a_{HB}c_{HA}^{-1})$, while U in (7C66) must be written as $U = K_{DM}G'_M R^{-1/2}$. Experiments are often made at constant $c_{HA} + c_{HB} = c_{H\,total}$ in order to keep the medium constant, and it must be remembered when applying (7C65) and (7C66) with the modified values of R and of U

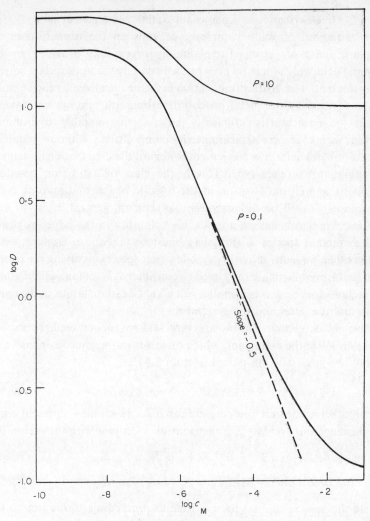

Figure 7C6. Effect of metal concentration on distribution of dissociating acido complex. Hypothetical case, according to equation (7C65), with $Q = 200$, $R = 100$, $S = 10^9$ and two different values of P, marked on curves.

that $\partial c_{HA} = - \partial c_{HB}$. A detailed analysis of this case and of the case where $c_{HA} + c_{CA} = c_{A\,total}$ is kept constant, has been presented[82]. It should be emphasized that the common-ion effect discussed by Saldick[81] and by Diamond[82] are important only for solvents of high dielectric constant where ionic dissociation of the solutes is important. Otherwise the general equations derived by Irving, Rossotti and Williams[80] apply.

f. Distribution of Long-chain Amine and Related Salts

The extraction of acido complexes depends on the formation in the organic phase of cations (involving hydrogen ions provided by the mineral acid which must be present), solvated by the appropriate solvent or extractant. The 'hydronium' cation is then neutralized by the complex anion formed by the metal on coordination with appropriate ligands. When the extractant is sufficiently basic, so that a stable 'hydronium' cation forms at the stoichiometric composition, without requiring excess mineral acid, it is convenient to regard the distribution systems as belonging to another type. This is the class of extraction reactions involving a preformed cation in the organic phase, and anionic metal complexes. It will be discussed as two separate general cases. In one, the extractant salt has an appreciable solubility in the aqueous phase, and extractant species in this phase must be taken into account, while in the other, aqueous solubility is so low that species involving the extractant in the organic phase only need be considered. With long-chain amine extractants the former case applies below ca. 20 carbon atoms in the molecules and the latter above this number.

Bivalent extractant cations are very seldom encountered, hence R^+ will symbolize the extractant, which dissolves in the aqueous phase as a salt R^+A^-, partially ion-paired (section 3.A) as

$$R^+ + A^- \rightleftharpoons R^+A^- \qquad K_A \qquad (7C67)$$

When other anions are present, such as B^{b-} (which may represent e.g. a perrhenate anion) or MA_{m+p}^{p-}, they may also ion-pair with the cation R^+

$$bR^+ + B^{b-} \rightleftharpoons R_b^+ B^{b-} \qquad K_B \qquad (7C68a)$$

$$pR^+ + MA_{m+p}^{p-} \rightleftharpoons R_p^+ MA_{m+p}^{p-} \qquad K_M \qquad (7C68b)$$

where the association has been written as proceeding in one step to the neutral species. Each of the neutral species may distribute into the organic phase, which ordinarily is an 'inert' diluent, such as chloroform, toluene or dodecane, with its characteristic partition constant, P_A, P_B or P_M respectively. When B^{b-}, M^{m+}, etc. are present at tracer concentrations, and only R^+A^- and possibly C^+A^- (where C^+ is a non-complexed cation, say Na^+) are present at bulk concentrations, the concentration of R^+ in the aqueous phase is given by the following expression, c_R being the total concentration introduced initially in the aqueous phase

$$(R^+) = c_R/[1 + K_A(A^-)y_{\pm RA}^2(y_{RA}^{-1} + (\bar{V}/V)P_A^- y_{RA}^{-1})] \qquad (7C69)$$

For the particular cases where $(\bar{V}/V)P_A \gg \bar{y}_{RA}/y_{RA}$ and activity coefficients are constant in both phases $(R^+) = c_R/(1 + K'_A(A^-))$, while if no excess A^- is added, and all the anion A^- derives from the salt R^+A^-, then $(R^+) = [(1 + 4K'_A c_R)^{1/2} - 1]/2K'_A$. The quantity (R^+) is used in the mass-action law expressions for the distribution of B^{b-} and M. The distribution coefficient of B^{b-} is

$$D_B = P_B K_B (R^+)^b y_{\pm R_b B}^{b+1} \bar{y}_{R_b B}^{-1}/(1 + K_B(R^+)^b y_{\pm R_b B}^{b+1} y_{R_b B}^{-1}) \qquad (7C70)$$

If the second term in the denominator is much larger than unity, i.e. when $R_b B$ is largely associated, (7C70) reduces simply to $D_B = P_B y_{R_b B} \bar{y}_{R_b B}^{-1}$ (cf. 7C11), and is independent of the concentration of extractant, which is not consumed by tracer concentrations of B^{b-}. When the ion-pair involving B^{b-} is largely dissociated, D_B will approximately equal the numerator of (7C70), and if activity coefficients remain constant, and the concentration of A^- is held constant, $(\partial \log D_B/\partial \log c_R)_{(A^-)} = b$ and the charge of B^{b-} may be determined from the slope.

The distribution coefficient of the metal M is

$$D_M = \frac{P_M K_M \beta_{MA_{m+p}}(A^-)^{m+p}(R^+)^p y_{\pm RA}^{2p} \bar{y}_{R_p MA_{m+p}}^{-1}}{K_M \beta_{MA_{m+p}}(A^-)^{m+p}(R^+)^p y_{\pm RA}^{2p} y_{R_p MA_{m+p}}^{-1} + \sum \beta_n (A^-)^n y_A^{n-m} y_{MA_n}^{-1}} \qquad (7C71)$$

The first term in the denominator pertains to the association reaction (7C68b), while the sum in the second term pertains to all other complexes MA_n^{m-n} that the metal forms with the ligand A^-. If activity coefficients remain constant, and K_M is very small (i.e. (7C68b) proceeds only slightly), the expression $D_M = K'_M(R^+)^p(A^-)^{m+p}/X_A$ holds. Since however (R^+) varies at a given c_R with (A^-), according to (7C69), no simple partial differentiation with respect to (A^-) can be made to yield the ligand number.

When the basic extractant salt does not dissolve appreciably in the aqueous phase, equations (7C67) and (7C68) should be replaced by an equation involving the ion-pair R^+A^- in the organic phase only. The reaction may be written in a number of alternative forms: as an ion exchange reaction

$$B^{b-} + b\overline{R^+A^-} \rightleftharpoons \overline{R_b^+B^{b-}} + bA^- \qquad (7C72a)$$

$$MA_{m+p}^{p-} + p\overline{R^+A^-} \rightleftharpoons \overline{R_p^+ MA_{m+p}^{p-}} + pA^- \qquad (7C72b)$$

as an extraction involving ion-association in the organic phase

$$M^{m+} + mA^- + p\overline{R^+A^-} \rightleftharpoons \overline{R_p^+MA_{m+p}^{p-}} \tag{7C73}$$

or as an extraction of an uncharged species

$$MA_m + p\overline{R^+A^-} \rightleftharpoons \overline{R_pMA_{m+p}^{p-}} \tag{7C74}$$

Thermodynamically the formulations (7C72b), (7C73) and (7C74) are equivalent, and only a detailed analysis of rate measurements can give the actually important reaction steps. If only the equilibrium state is considered as done here, the expression chosen will be dictated by convenience, and (7C74) will be used henceforth.

The formal similarity of the reactions (7C72) and (7C74) to those occuring with resin anion exchangers, e.g. (6B39), (6B40) and (6B41b), is striking. Completely similar formal equations can be written for the distribution coefficient of tracer M in this system[83]. Using the quantity $X' = \sum \beta_i' G_i'(A^-)^i = \sum \beta_i'^* a^i = c_M/(MA_m)$ for the aqueous phase (section 7.C.a), and an equilibrium constant K_M for reaction (7C74), yields the distribution coefficient

$$D = K_M(\overline{R^+A^-})^p \bar{y}_{RA}^p y_{MA_m} \bar{y}_{R_pMA_{m+p}}^{-1}/X' \tag{7C75}$$

if only the species $\overline{R_p^+MA_{m+p}^{p-}}$ is important in the organic phase, i.e. if p has a unique value. Otherwise, either an average $\tilde{p}$ is used in (7C75), or else the term $(\overline{R^+A^-})^p \bar{y}_{RA}^p \bar{y}_{R_pMA_{m+p}}^{-1}$ must be replaced by the appropriate sum of terms. Collecting the activity coefficient terms in (7C75) in one term G_M, and assuming that $K' = K_M G_M$ does not vary with the concentrations of the extractant and the ligand, permits partial differentiation of the resulting equation $\log D = \log K' - p \log \bar{c}_R - \log X'$ to give

$$(\partial \log D/\partial \log \bar{c}_R)_a = p \tag{7C76a}$$

$$(\partial \log D/\partial \log a)_{\bar{c}_R} = \tilde{i} \tag{7C76b}$$

A common complicating factor is the formation of acidic species of the ligand in the organic phase e.g. $R^+HA_2^-$. Even if excess H^+A^- extracted into a solution of R^+A^- in the diluent does not form the species $R^+HA_2^-$, as long as it is present at a lower concentration than R^+A^-, the excess acid can be treated thermodynamically as if it were extracted by the reaction

$$H^+ + A^- + \overline{R^+A^-} \rightleftharpoons \overline{R^+HA_2^-} \tag{7C77}$$

with an equilibrium constant K_{A2}. Assuming only 'free' ligands R^+A^-, but not acid species $R^+HA_2^-$, to participate in metal extraction, it is necessary to calculate the concentration of the proper value of (R^+A^-) to insert in (7C75)[84, 85].

$$\overline{(R^+A^-)} = \bar{c}_R/(1 + K_{A2}(H^+)(A^-)y_{\pm HA}^2 \bar{y}_{RA} \bar{y}_{RHA_2}^{-1}) \qquad (7C78)$$

More complicated corrections must be applied if more acid is extracted, and obviously, if the metal concentration is sufficiently high so that an appreciable amount of R^+A^- is consumed in reaction (7C74), an appropriate correction for this must be made. At a constant concentration of ligand (A^-), and for small substitution of acid for salt, $y_{\pm HA}$ may be considered constant as may the ratio $\bar{y}_{RA}/\bar{y}_{RHA_2}$, so that (7C78) may be simplified to $\overline{(R^+A^-)} = \bar{c}_R/(1 + K'(H^+))$. Insertion of this in (7C75) and assuming constancy of the other activity coefficients there too, yields

$$\lim_{(H^+) \to 0} (\partial \log D/\partial \log(H^+))_{(A^-), \bar{c}_R} = 0 \qquad (7C79a)$$

$$\lim_{(A^-) \gg (H^+) \gg 1/K'} (\partial \log D/\partial \log(H^+))_{(A^-), \bar{c}_R} = -p \qquad (7C79b)$$

A very common and important complication occurring with distribution systems involving long-chain amines is the aggregation of both the extractant and the metal complex in the organic phase. Our very imperfect knowledge of these aggregated systems (Chapter 10) makes a quantitative description of the distribution equilibria impossible. Qualitatively it may be stated that if the extractant is largely associated to a q-mer, then $\overline{(R^+A^-)}$, the concentration of the monomer, is proportional to $\bar{c}_R^{1/q}$, and if q is a large number, $\overline{(R^+A^-)}$ varies very little even for large variations of $\bar{c}_R$. Such behaviour has been observed[86]. If only the monomers are capable of reacting with the metal complex, but not the aggregated species, then D for the metal should be almost invariant with $\bar{c}_R$, as observed in certain cases. On the other hand, if the aggregates are open structured so that the ion-pair end of the long-chain amine is mobile and relatively independent of its neighbour, the amine phase resembles an anion exchange resin with variable capicity, and equation (7C75) with the approximation $\overline{(R^+A^-)} = \bar{c}_R$ remains valid. Again, such behaviour, i.e. a dependence of D on the pth power of $\bar{c}_R$, as if no aggregation occurred, has been observed in a number of cases, where aggregation of the amine salt R^+A^- has been proven. Aggregation of the metal complexes is even more difficult to handle quantitatively, but is less likely than aggregation of the simple salts, since aggregation occurs most readily when one ion of the ion-pair is large and the other small, and in any case can be neglected at trace metal concentrations. Mixed aggregates, however, remain important even in this case.

General equations for extraction with long-chain amines have been derived by van Ipenburg[87], taking into account consumption of extractant by the metal complex, possible polyvalency of the ligand, effect of water molecules involved in the extraction, and also as special cases, association of the metal complex or of the extractant to aggregates. These equations however, are rather cumbersome, and for the cases discussed above they are equivalent to the present formulation.

g. Ligand Distribution

The above discussion dealt mainly with distribution of the central cation M, which forms complexes. Some distribution studies of complex formation involve instead the distribution of the ligand L, or of a species not involving M that is in equilibrium with the ligand. This has the advantage of being applicable also in cases where the complexes formed by M are non-extractable, or where there is no convenient analytical method (i.e. a radioisotope), for measuring the distribution of M. Of course, if M and L form but one complex, ML, or a series of polynuclear complexes (with respect to M), M_nL, with charges omitted for the sake of generality, the roles of central group and ligand are reversed, and the equations discussed above apply. Thus the extraction of silver with isobutene from aqueous nitrate solution[88] belongs to this category. The distribution of isobutene was measured, and the species found are $C_4H_8AgNO_3$ in the CCl_4 extracts, and $C_4H_8Ag^+$, $C_4H_8Ag_2^{2+}$ and $(C_4H_8)_2Ag^+$ in the aqueous phase.

The distribution of a ligand to study metal complex formation was first applied by Dawson[59], who studied the distribution of iodine between CCl_4 or CS_2 and aqueous solutions containing iodide and mercury iodide. The ligand participates in the following reactions

$$nA^- + M^{m+} \rightleftharpoons MA_n^{m-n} \qquad \beta_n \qquad \text{(7C80a)}$$

$$A^- + A_2 \rightleftharpoons A_3^- \qquad K_3 \qquad \text{(7C80b)}$$

$$A_2 \rightleftharpoons \overline{A}_2 \qquad P_A \qquad \text{(7C80c)}$$

The concentration of free ligand is given by

$$(A^-) = (P_A - D_A)/K_3 D_A \qquad \text{(7C81)}$$

where D_A is the distribution coefficient of total iodine. At sufficiently small ligand concentrations, where $(A^-) \ll c_{A^-}$, $\tilde{n}$ can be calculated from $\tilde{n} = (c_A - (A^-))/c_M$, and the constants β_n from the relationship between $\tilde{n}$ and (A^-).

A different way of utilizing ligand distribution measurements involves radiometric measurement of total-ligand distribution between the two phases, again used for the mercury iodide system[61]. The species in the organic phase is MA_m, those in the aqueous phase are free A^- and complexes MA_n^{m-n}, so that the distribution coefficient is

$$D_A = m(\overline{MA_m})/\left[(A^-) + \sum_{1}^{N} n\beta_n(M)(A)^n \right] = mP_m/\left[\beta_m^{-1} c_M^{-1} \sum_{1-m}^{N-m} \beta_i'(A^-)^{i+1} \right.$$

$$\left. + \sum_{1-m}^{N-m} (i+m)\beta_i'(A^-)^i \right] \tag{7C82}$$

where P_m is the partition constant of the uncharged complex MA_m. For very low metal concentrations, c_M, the second term in the denominator of (7C82) becomes negligible compared with the first, while c_A may be used instead of (A^-). At constant c_M the limiting slope $d \log D_A/d \log c_A$ becomes $m - (N + 1)$. At very low values of the ligand concentration the first term in the denominator becomes negligible compared with the second, and if, as for the mercury iodide system studied[61], species with $i \leqq m$ may be disregarded, D_A is equal at the limit to P_m. For intermediate ligand concentrations, the approximation $(A^-) = c_A - \tilde{n}' c_M$ may be used, where $\tilde{n}'$ is an estimate of the true ligand number. Using the value of P_m determined at the lowest c_A, and the estimates of (A^-), and if only values of $m = 2 < n = 2 + i < N = 4$ need be considered, (7C82) can be transformed into

$$\frac{2(P_m - D_A)}{D_A(A^-)} = A(1 + B(A^-) + C(A^-)^2) \tag{7C83}$$

where $A = (3\beta_1' + 1/c_M)$, $B = (\beta_1' + 4\beta_2' c_M)/(1 + 3\beta_1' c_M)$ and $C = \beta_2'/(1 + 3\beta_3')$. The three-parameter equation can be solved by graphical or other means (Chapter 3). The constants β_1' and β_2' may be used to calculate improved values of $\tilde{n}$ and of (A^-), if required to calculate better values of the constants.

The more difficult problem of formation of mixed bromo-iodo complexes was solved in a similar manner[61], while the distribution of bromine ligand was used in studies of the kinetics of extraction of mercury bromide from a molten nitrate medium[22].

h. The Continuous-variation Method

The method of continuous variations, or Job's method[89], has been applied also to two-liquid phase systems. This was first proposed by Babko and Pilipenko[90] and applied to the extraction of metal dithizonates, and was later applied to the extraction of other chelates by Irving and Pierce[91]. The equations are similar also for extraction of solvates, ion-associates, etc. The criticism raised against the indiscriminate use of the continuous-variation method in a single phase applies, of course, also to two-phase systems[92,93]. In particular, when stepwise complex formation occurs, or strong non-idealities can be expected as the composition of the system varies, the method is inapplicable. Therefore, apart from simple cases which usually can be studied better by other means, use of the method should be confined to chelates, where it can often be safely assumed that no interfering factors are encountered, and there is only one predominant complex extracted.

When a chelating reagent HX reacts with a metal ion to form a chelate MX_m which is extractable, the concentration of this complex in the organic phase can serve as the property which is treated by the continuous-variation method. The total concentrations of metal ions c_M and of chelating agent c_{HX} in the equal-volume two-phase system are varied in such a manner that the sum $c_M + c_{HX} = c$ remains constant with a fraction $x = c_M/c$ of metal, and $(1 - x)$ of chelating agent. The partition constants of MX_n and of HX are P_M and P_X respectively, the association constant of the ligand acid is $K_X = (HX)/(X^-)(H^+)$, and of the metal complex is $K_M = (MX_n)/(M)(X^-)^n$. Considering the extraction at equal phase volumes and at a given hydrogen ion concentration, the material balance equations are

$$xc = c_M = (M) + (MX_n)(1 + P_M) \tag{7C84a}$$

$$(1 - x)c = c_X = (X^-) + (HX)(1 + P_X)) + n(MX_n)(1 + P_M) \tag{7C84b}$$

$$= (X^-)(1 + K_X(H^+)(1 + P_X)) + n(MX_n)(1 + P_M)$$

and the concentration of the complex in the organic phase is (Figure 7C7)

$$(\overline{MX_n}) = P_M(MX_n)$$

$$= P_M K_M (1 + K_X(H^+)(1 + P_X))^{-1}[xc - (MX_n)(1 + P_M)] \times$$

$$[(1 - x)c - n(MX_n)(1 + P_M)]^n \tag{7C85}$$

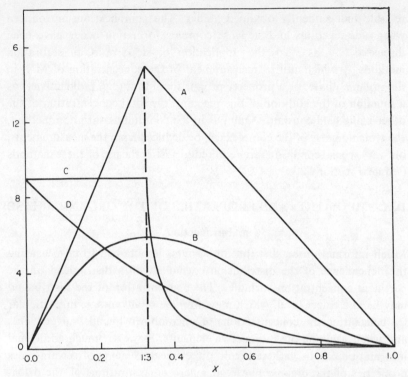

Figure 7C7. The method of continuous variation applied to distribution systems. For curves A and C: $K = \infty$, for curves B and D: $K = 1.4$. In curve A is plotted 500 $(\overline{MX_n})$, in curve B, 5000 $(\overline{MX_n})$, in curve C is plotted D, and in curve D is plotted 100 D. The parameters for the hypothetical system are $c = 0.1$ M, $P_X = 3$, $K_X(H^+) = 0.1$, $P_M = 9$ and $n = 2$.

Differentiation of $(\overline{MX_n})$ with respect to x and setting the derivative equal to zero, for obtaining an extremum in the continuous-variation plot, leads to

$$0 = c[(1-x)c - n(1 + P_M)(MX_n)] + n(-c)[xc - (1 + P_M)(MX_n)] \quad (7C86)$$

from which the usual relation

$$n = (1 - x)/x \quad (7C87)$$

is readily obtained. The maximum in the concentration of MX_n in the organic phase does not coincide with a maximum in the distribution coefficient (Figure 7C7). The method has been applied to a number of systems[90, 91, 94, 95], with satisfactory results, i.e. values of n agreeing with reliable independently obtained values. The complications introduced when side-reactions and stepwise complex formation occur have been discussed[91,96] as also the application of slope ratio, or saturation methods[97], which utilize measurement of the concentration of MX_n in the organic phase, or a property proportional to it (e.g. radioactivity) as a function of the addition of one reactant, the total concentration of the other being held constant. Again, as in the continuous-variation method, the stoichiometry of the complex can be deduced from the measurements, but very special conditions are required to justify the use of these methods (see next section).

D. COMPOSITION AND PROPERTIES OF THE ORGANIC PHASE

a. Introduction

Much information on distribution systems is obtained from measuring the dependence of the distribution coefficient of a distribuend on the pertinent concentration variables. The concentration of the distribuend may be at a tracer level, which may often be of advantage, in particular for calculating the concentrations of uncombined ligand, solvent, etc., and when studying the radioactive elements. However, much additional information can be gathered from direct measurements of certain gross properties of the organic phase, at macro concentrations of the distri-buend. The organic phase usually contains fewer different species than the aqueous phase, so that the properties measured are more directly dependent on the species present. The composition, and sometimes the structure of the species, may therefore be deduced from data.

b. Property-vs.-Composition Curves

The method of physicochemical analysis (of binary mixtures) has been ascribed to Kurnakov[98], and it involves plotting the values of a gross property of a phase (e.g. specific volume, conductivity, light absorbance, etc.) against the composition of the phase, in terms of the mole fraction of one of its components. Other variations of the method are possible, e.g. plotting the values of the property against the ratio of concentrations of two reactants, the mutual interaction of which is being studied. There are two approaches to obtaining information from these property-vs.-composition curves. One is to concentrate on abrupt breaks in the curves, and to attribute them to the complete formation of species of definite stoichiometry given by the composition or reactant ratio where the breaks occur. Another approach is to analyse the magnitude of the property as it varies with composition, and attribute physical meanings to such quantities as the slope of the curve.

The first approach, attempting to analyse breaks in the property-vs.-composition curve, used mainly for extraction with undiluted extractants, may be illustrated by the work of Kertes and coworkers[99-103] and others[104-109]. The systems studied involve tributyl phosphate and dibutylphosphoric acid and the mineral acids—hydrochloric, hydrobromic, perchloric and nitric, as well as uranyl chloride, as distribuends between aqueous solutions and the organic phase. The properties of the organic phase studied include density, volume changes, viscosity, conductivity and the coextraction of water. Two examples of the data obtained are shown in Figure 7D1. In the papers quoted[99-109], it was taken as self-evident that the breaks must correspond to the formation of definite species. Indeed, many methods, both analytical (e.g. conductometry) and for studying complex formation (e.g. the molar-ratio[110] or the continuous-variation[89] methods) are based on the interpretation of breaks in similar curves.

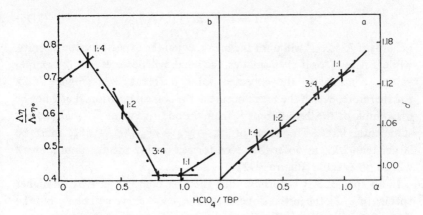

Figure 7D1. Property–vs.–composition curves for the extraction of perchloric acid by TBP. Ordinates: ϱ(right-hand part, a) and $\Lambda\eta/\Lambda_0\eta_0$ (left-hand part, b); abcissa: ratio $\alpha = \bar{c}_{HClO_4}/\bar{c}_{TBP}$. Data (experimental points, straight-line segments and breaks) from ref. 101.

A recent analysis of this problem throws light on the applicability of this method[111]. Consider the simple case of having two reactants A and B, which form a single complex AB_n. The total concentration of A is held constant at c_A, while increments of B are added, to yield total concentrations $c_B = \alpha\, c_A$. A property P of the mixture is measured, which is linear with the concentrations of the species present

$$P = P_A(A) + P_B(B) + P_n(AB_n) \qquad (7D1)$$

whese P_I is the partial molar property of species I. The material balance equations are

$$c_A = (A) + (AB_n) = (A) + K(A)(B)^n = (A)(1 + Kx^n) \qquad (7D2)$$

where K is the equilibrium constant for the formation of the complex, and x is written for the concentration of free B, (B), and

$$c_B = (B) + n(AB_n) = x + n\,K(A)x^n = x + nKc_Ax^n(1 + Kx^n)^{-1} \qquad (7D3)$$

The value of the property of the solution per unit concentration of A is then obtained as a function of x as

$$P/c_A = P_B\alpha + (P_A + P_nKx^n - nP_BKx^n)(1 + Kx^n)^{-1} \qquad (7D4)$$

For the particular case $n = 1$, it can easily be shown that at the break

$$\alpha = 1 + (Kc_A)^{-1} \qquad (7D5)$$

i.e. only if $K \gg c_A^{-1}$ will α for the break, equal the expected value of unity, while if $K \simeq c_A^{-1}$ or if even smaller, the break will occur at $\alpha = 2$ or larger values, contrary to the expected value α (break) $= n$ (Figure 7D2). Furthermore, even if the break occurs at the correct position, it will not be discernible, if the deviation of P_n from the additive value $P_A + nP_B$ is not sufficiently large. The value of $[P_n - (P_A + nP_B)]/(P_A + nP_B)$ must be larger than 20% in order that even for reasonably precise data a break will be observable (Figure 7D2).

If no species AB is formed, but the first complex is AB_2 or higher species ($n \geqq 2$), the initial slope of the P/c_A-vs.-α curve will be P_B, but the slope changes very rapidly. The asymptote or α values tending to zero has the same slope as that for very high α values, and the curve must be S-shaped. A really straight section is approached only at the inflexion, but if produced, it does not meet the asymptote for high α values at the expected point $\alpha = n$. For the more usual case, where complexes form consecutively, the species AB_2 may be looked upon as formed from (AB) + B, so that for sufficiently high K values and favourable $P_2 - (P_1 + P_B)$, the second break will occur at the correct point[111].

The above analysis shows that for some favourable cases, provided the

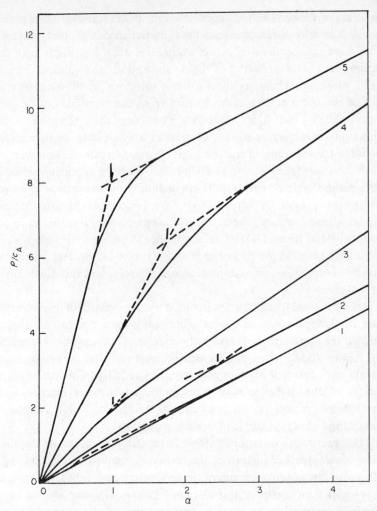

Figure 7D2. Calculated property-vs.-composition curves for several hypothetical cases. For all curves $c_A = 0.01$, $n = 1$, $P_A = 0$ and $P_B = 1$. For curve 1: $K = 10^4$, $P_n = 1.1$; curve 2: $K = 10^2$, $P_n = 2$; curve 3: $K = 10^4$, $P_n = 2$; curve 4: $K = 10^2$, $P_n = 8$; curves 5: $K = 10^4$, $P_n = 8$. Breaks occur at 'correct' point only for curves 3 and 5. (By permission from *Israel J. Chem.*, **5**, 147 (1967)).

property P is linear with the concentrations of the species, the molar ratio property-vs.-composition curve does yield the stoichiometries of the species formed from the breaks in the curve[111,112]. The restrictions that the system behaves ideally, and that the mass-action law is obeyed in terms of concentrations apply here too. It has however been pointed out[93] that only a few methods have partial molar property values P_I which are independent of concentration, i.e. that P is linear with concentration. Measurements of light absorption and optical rotation yield properties characteristic in dilute solutions of the reactants and not of the solvent. The data obtained from measurements of magnetic susceptibility and light refraction (but not the refractive index), although characteristic of the solution as a whole, can be apportioned between the solvent and the reactants. In many methods, however, P is characteristic of the solution as a whole but *cannot* be apportioned linearly between the various reactants. These include[93] measurements of various colligative properties (cryoscopy, etc.), viscosity, fluidity, density, electrical conductance, surface tension, compressibility, dielectric constant, etc. For these, (7D1) does not hold, so that even for high K values the curves in principle do not approach straight asymptotes, so that the breaks observed in curves do not necessarily represent the stoichiometry of real complexes.

An additional defect in the method is the hazard of bias creeping into the interpretation of curves, where the slopes change only slightly, and if the data are not extremely precise. This may be exemplified by Figure 7D1a, where breaks can be read into the figure, based on breaks in a curve of another property, Figure 7D1b. Thus although the weight of the evidence may support the conclusions reached about the species formed in the interactions studied[99-109] (cf. Chapter 9), the method always requires independent support.

The continuous-variation method is similar in concept to the mole-ratio method, where instead of the concentration of one reactant (c_A in the above discussion), the sum of the concentrations of the two reactants, $c_A + c_B$, is kept constant in the organic phase. The application of the continuous-variation method to measurements of the concentration of solute in the organic phase has been discussed above (section 7.C.h). The method has also been applied to the measurement of physical properties of the organic phase in a number of studies. The properties measured include light absorption in the ultraviolet[113], visible[114], infrared[115], electrical conductance[94], dielectric constant[106] and vapour pressure[115] of the organic phase. The method has been extended to ternary systems, such as the interaction of uranyl ions, trioctylphosphine oxide and dioctylphosphoric acid[121]. The merits and drawbacks of the continuous variation method, and its limits of applicability have been much discussed[93], and are similar in principle to those discussed above concerning the mole-ratio method.

On a firmer basis stand methods where the information regarding the species formed is obtained from the magnitude of the changes in the property itself, rather than from breaks in the property-vs.-composition curve. The most widely studied property is the swelling of the organic phase, i.e. its change of volume as solute is extracted into it, as a function of the concentration of the extracted solute. Indeed, if the concentration is expressed on the molal scale, i.e. per a given number of moles of solvent, the slope of the swelling-vs.-compocition (or mole-ratio distribuend to solvent) curve is the apparent molal volume. Sometimes a straight line is obtained in such a plot, and this shows that the solute has a constant partial molal volume. In the case of extraction of nitric acid by tributyl phosphate[104] an initial straight portion of slope 23.7 ml/mole HNO_3 was obtained, which changed after a molar ratio of 1:1 HNO_3:TBP was reached, to a different slope, 58.0 ml/mole. The valid deduction drawn from this is that two species are formed in the organic phase, with different molal volumes, the first of which being completely formed at a mole ratio of 1:1, and is probably $TBP.HNO_3$, since its apparent molal volume equals the difference between that of hydrated TBP 271 ml/mole, and of $TBP.HNO_3$, 315 ml/mole[105,109]. In other studies[116,117] the inverse of the slope of the swelling vs. moles acid extracted curve, which is straight over an appreciable concentration range, was taken as the molar concentration of the acid extracted in the water coextracted with it. For example[117], a slope of 118 ml/mole perchloric acid extracted into dibutylcellosolve is interpreted as the extraction of $(118 \text{ ml/mole})^{-1}$ = 8.45 M $HClO_4$ = 56.5% w/w $HClO_4$. Such a solution has a mole ratio of H_2O:$HClO_4$ = 4.3, which is nearly the number of water molecules coextracted with the hydrogen ions of the strong acid.

c. Extraction Stoichiometry

In the previous section it was shown how physical properties of the organic phase in an extraction system relate to the species formed, while the composition of this phase varies. Important information is obtained from the variation of the composition itself, and from the maximal amounts of solutes that may be extracted into a given quantity of solvent. In many cases saturation of the solvent depends on simple solubility limitations, depending on non-specific interactions. In other cases, where a definite species is formed, saturation will occur at the stoichiometry of this species. The extraction isotherm shows the concentration of the solute in the organic phase increasing with its concentration in the aqueous phase, until saturation is reached, while beyond this point it remains constant.

Much use of this method has been made in studies of the extraction of acids and metal complexes into basic solvating solvents, or with basic extractants, where definite complex species are formed, e.g. the $U^{VI}-Cl^--TBP$ system[118], or the $U^{VI}-NO_3^--TOA$ system[119]. Care must however, be used with this method, in particular with aggregated extractants (such as long-chain ammonium salts), since in some cases large loading numbers, that is extractant-to-distribuend ratios, result in the saturated solution, without indicating the true composition of a complex species. In the extraction of uranium(VI) from sulphuric acid with trioctylammonium sulphate solutions[120], uranium can be loaded into the organic phase up to a ratio of TOA : U of 5, while plutonium(IV) can be extracted with this reagent from nitric acid[119] up to a ratio TOA : Pu of 4. In both cases there is good evidence that the species formed are not $(R_3NH)_5UO_2(HSO_4)_7$ and $(R_3NH)_4Pu(NO_3)_8$, as one is tempted to deduce from these ratios.

Three factors were found to affect the loading numbers[121, 122]: a) the coordination number and charge of the metal and the charge of the ligand, b) the steric accessibility of the anion in an amine salt, and of the basic donor atom in a basic solvent[122], and c) the aggregation of the extractant and the metal complex. Polar diluents or diluents capable of hydrogen bonding may interact strongly with the extractant, making it less available for the distribuend, thus decreasing the solubility of a distribuend in the organic phase. More important, however, is the aggregation of the extractant, which, because of the possible coordination of just one member of the aggregate with the metal ion, leaves an excess of unused extractant, and leads to high loading numbers[85]. Exceptionally low loading numbers (e.g. a loading number of 0.55 for the light organic phase separating from a 0.05 M solution of TOA nitrate in dodecane on the dissolution of uranyl nitrate[121]) may result from the formation of polynuclear species, sharing anions and having therefore a lower negative charge, requiring fewer long-chain ammonium cations, than the mononuclear species.

Even in unsaturated solutions, the stoichiometry of the species extracted can often be deduced from analytical measurements on the organic phase at macro concentrations.

A second organic phase, when formed, need not have a constant composition, as shown by phase-rule considerations (section 7.B.a). However, quite often the composition remains approximately constant as concentrations in the aqueous phase vary, it approaches the definite stoichiometry $TBP.HCl.H_2O^{38}$ and $[H(TBP)_2]^+[UO_2(NO_3)_3]^{-123}$ in systems involving these components and a kerosene diluent). Such a 'third phase' may be considered to be an uncrystallized insoluble compound.

Much attention has been given to the coextraction of water with other distribuends, in particular acids, and the molar ratio H_2O : acid in the organic phase has been determined. In some cases acids were found to displace water hydrating the solvent, in others water is coextracted with the acid so that the concentration of water in the solvent increases.

The stoichiometry of metal complex hydrates extracted at macro concentrations has also been determined directly[124, 125]. The results of these studies are discussed fully in Chapter 9.

d. Interpretation of Chemical and Physical Properties

The above discussion concerned the stoichiometry of the species in the organic phase, which may in favourable cases be deduced from breaks in property-vs.-composition curves, and from gross-analytical measurements. More information about these species, and a possibility of their identification, is available from the interpretation of the magnitude of certain physical and chemical properties of the organic solutions. These include certain colligative properties (freezing point depression), irreversible thermodynamic properties (viscosity or electrical conductance), and spectroscopic properties (absorption spectra).

An important factor in distribution studies is whether the extractant is monomeric, oligomeric or aggregated, and whether it associates with the diluent. Such information can often be obtained by measuring certain colligative properties of the solution, which yield the vapour pressure of the solvent (diluent), hence the osmolality of the solute (extractant). If solutions are assumed ideal apart from specific chemical interactions, the ratio of the formality to the osmolality yields the average degree of aggregation, and this, times the formal molecular weight of the extractant, gives the average molecular weight. The vapour pressure of the solvent can be measured directly, but it is often done more conveniently isopiestically[117, 126, 127] or by differential vapour-pressure lowering (osmometry)[128, 129], using the cooling effect of drops of solvent and solution evaporating on a pair of thermistors. Indirectly, the activity of the solvent or the osmolality of the solute can be determined ebullioscopically and cryoscopically. The former method has been used only rarely[130, 131], but the latter is rather convenient, especially with benzene as solvent[131, 132]. The solvates of various acids and metal complexes, with ethers as extractants, were obtained in benzene diluents, as well as the association of amine salts. The diluent may not be completely inert, and a maximum in the association-vs.-concentration curve is interpreted[131] as pointing to solvation of the long-chain ammonium nitrate with benzene. Another interpretation is a decrease in the aggregation which is due to the increasingly polar nature of the solution, as the concentration of the polar solute is increased[129].

A method that has often been used to yield qualitiative or semiquantitative information on the degree of aggregation of an extractant or distribuend in the organic phase is the measurement of the viscosity of these solutions. In a typical study[127], to a solution of 0.1 M bis-2-ethylhexylphosphoric acid in n-hexane were added increments of a uranium(VI) salt so that the ratio $\bar{c}_U/\bar{c}_{HX}$ increased up to the limiting value of 0.5, corresponding to saturation. The specific viscosity $\eta_{sp} = (\eta_{soln} - \eta_{solvent})/\eta_{solvent}$ changed little up to a ratio of $\bar{c}_U/\bar{c}_{HX}$ of about 0.45, but beyond this increased by two orders of magnitude. The curve of specific viscosity against average aggregation number $\tilde{n} = 2(\bar{c}_U/\bar{c}_{HX})/(1 - 2(\bar{c}_U/\bar{c}_{HX}))$, was found to obey the following relationship above $\tilde{n} = 10$

$$\eta_{sp} = b(\tilde{n})^a \tag{7D6}$$

which is similar to equations usually applied to polymeric solutions[133]. On the other hand, viscosity measurements on dodecylamine sulphate in benzene[134] show it to be aggregated, in agreement with isopiestic information, but as uranium(VI) is loaded into the organic phase, aggregation decreases until at $\bar{c}_U/\bar{c}_{(R_2NH_2)_2SO_4} \sim 0.15$, monomeric species result. Indeed, the viscosity results, correlated in a manner similar to (7D6), but with lim of $(\eta_{sp}/\bar{c}_{(R_2NH_2)_2SO_4})$ as $\bar{c} \to 0$ against the average molecular weight, showed the presence of monomeric uranium sulphate complex amine salts along with micellar amine sulphate.

Viscosity measurements give direct information on the relative volumes of the components. If for a mixture the viscosity agrees with the Kendall equation[148]

$$\log \eta = x_1 \log \eta_1 + (1 - x_1) \log \eta_2 \qquad (7D7)$$

the molar volumes of the components are approximately equal. This is probably true for dry and hydrated TBP, since the experimental curve shows the ideal behaviour expected[105].

Besides indicating aggregation, viscosity measurements are important in the interpretation of conductivity data in terms of ionic dissociation (section 2.D). Walden's rule in one form or another[135] gives the degree of dissociation

$$\alpha \sim \Lambda \eta / \Lambda_0 \eta_0 \qquad (7D8)$$

where Λ is the molar conductivity and η the viscosity, and subscript zero denotes infinite dilution. In aqueous solutions $\Lambda_0 \eta_0 \sim 0.060$ but in organic solutions a different empirical constant may be more appropriate. For solutions in undiluted TBP, the value $\Lambda_0 \eta_0 = 0.030$ was obtained for perchloric acid, which was assumed to be completely dissociated[136], but was recently corrected to give the value 0.06 as in aqueous solutions, showing that perchloric acid is almost, but not quite, completely dissociated[137]. For this acid α calculated from (7D8) varies little with concentration, while for hydrochloric, sulphuric and nitric acids α varies with concentration up to 0.01 M according to Ostwald's dilution law, with a dissociation constant around 10^{-8} [137].

The use of conductance data to estimate ionization in the organic phase is common, and only a few examples will be quoted for the purpose of illustration. The degree of ionization of uranyl nitrate in various ethers, esters and ketones was determined[124]. Using Walden's rule value $\Lambda_0 \eta_0 = 0.060$, very slight dissociation was found in many solvents, whereas in others, notably isobutanol, appreciable dissociation was found. In some solvents an appreciable solvent correction must be applied. Thus wet dibutyl phosphate has a specific conductance of 6×10^{-4} mhos[110], while up to 0.1 M nitric acid extracted into it does not raise the conductance appreciably. The conductivity of organic solvents and solutions of extractants in diluents varies widely. There is a good correlation of the conductivity of solutions of potentially strong electrolytes with the dielectric constants of the solvents. Thus solutions of long-chain amine salts in diluents such as chloroform or benzene are very poor conductors of electricity, but solutions in nitrobenzene conduct quite well[139]. Similarly, solutions of mineral acids extracted into diluted tributyl phosphate conduct rather poorly, but show much higher conductances in the more polar undiluted TBP[140] (section 9.C). The more polar a solvent becomes by dissolving electrolytes, the higher the conductance: as excess acid is extracted into solutions of amine salts in nonpolar diluents, the conductivity increases[141] As the concentration of an electrolyte in the organic phase increases, a minimum in the molar conductance is often found, because of ionic aggregation (section 2.D.e).

Measurement of electromigration of ions in organic solvents has been rare, because of the low conductivities of most solvents. Important information concerning formation of ionic species can, however, be obtained from such experiments. Electromigration was applied[142] to extracts of iron(III) and gold(III) from hydrochloric acid into benzene, chlorobenzene, *o*-dichlorobenzene and nitrobenzene, and the expected anodic migration of the metals was observed. In some cases, however, ion-triplet formation led to cationic migration of species such as $H_2AuCl_4^+$.

Much use has been made of spectral measurements on organic extracts to give information on the species extracted. On the one hand, data for the visible spectrum (mainly) are of use for identifying the metal complex species extracted. On the other hand, infrared spectra give information on the bonds formed between solvating solvents and the metal salts, on hydrogen bonding in extractant oligomers, or in interaction products of extractant and diluent, etc.

Provided the spectra in the visible region are sensitive to the structure of the complexes, it is usually sufficient to compare the absorption spectrum of the organic extract with that of a solid whose structure is definitely known, in order to identify the extracted species. A good example for such an identification is the case of iron(III) chloride extracted from aqueous hydrochloric acid into solvents such as ethers. Spectrophotometry[131, 143-145] showed that the extracted species has an absorption spectrum similar to that shown by iron(III) in concentrated hydrochloric acid, in fact identical with that of solid $KFeCl_4$ and of anhydrous solutions of $KFeCl_4$ in diethyl ether and in ethylene bromide[146]. There is therefore no doubt that the ether extracts contain the iron in the anion as $FeCl_4^-$. Amine extracts also contain the iron exclusively as the $FeCl_4^-$ species[147], in spite of amine concentration dependence results[148], which would lead to $FeCl_5^{2-}$ (but see section 10.C).

Infrared spectra give information on the bonds formed between the extractant and the diluent, between the extractant and distribuends and between the extractant molecules themselves. Shifts or splits in the bands of certain groups, such as the $P = O$ vibration in TBP or the N—H vibration in a tertiary amine, or the appearance of otherwise forbidden bands, are indicative of more or less strong bonding of a new atom to these groups, changing thereby the symmetry. Interesting in this connexion is the possibility of identifying in polyfunctional extractants, the specific donor atom or group to which a metal ion is attached. Thus in TBP, once the $P = O$ oxygen is bonded to one acceptor atom (e.g. hydrogen in HNO_3), excess of distribuend may be bonded to the P—O—C bonds. There is no definite evidence that this actually occurs. The excess distribuend may be simply dissolved in the mixture, or it may bond to other distribuend molecules. Present knowledge of these spectra, in particular for the rather opaque and complicated solutions treated, is seen to be insufficient for a definite assignment[128,149–151] (section 9.C).

The use of infrared spectra will be illustrated by one more example: the extraction of iron(III) chloride into ketones. The spectrum shows the same shift in the $C = O$ band when $HFeCl_4$ is extracted, as in HCl extracts[152], but no shift in solutions of anhydrous $KFeCl_4$. A similar shift is also observed[153] in extracts of $HClO_4$. The conclusion is that the hydrogen ions (possibly hydrated) are solvated by the extractant, and ion-paired with non-solvated chloroferrate ions.

The application of Raman spectra should be quite similar to that of infrared spectra, and indeed they have been used to identify certain metal complex species in organic solvents. Thus in ether extracts from hydrochloric acid solutions of iron(III), gallium(III) and indium(III) only tetrahedral $FeCl_4^-$, $GaCl_4^-$ and $InCl_4^-$ could be identified[154], irrespective of the existence of this ion at appreciable concentration in the aqueous phase at equilibrium. Similarly, pyramidal $SnCl_3^-$ and $SnBr_3^-$ are the species in ether extracts of tin(II)[155]. In TBP solutions of zinc(II) halides, however, both linear ZnA_2 (hydrated and solvated in the plane around this axis), and tetrahedral ZnA_4^{2-} can be obtained, depending on the presence of excess lithium halide[156]. Mercury(II) halides extracted from concentrated hydrohalic acid into TBP shows only tetrahedral HgA_4^{2-}, but solutions of $LiHgA_3$ show characteristic Raman lines for this configuration[157]. The Raman spectroscopic method is far from being fully exploited for solvent extraction studies, and more definite information concerning extracted species may be expected in the future.

Another method which has not as yet been much used for solvent extraction studies, but which shows much promise, is nuclear magnetic resonance. The few studies reporting the use of n.m.r.[151, 158, 159] show that much information concerning the bonding of acids or metal complexes extracted into organic solvents can be obtained from observations of chemical shifts, line widths and other features of the spectrum. Other, non-spectroscopical, optical methods, such as light scattering[140] give valuable additional information concerning the degree of aggregation of extractants and distribuends in the organic phase.

In every case where the distribution of species at macro concentrations can be studied, and for all new and not sufficiently well-known extractants, physical measurements on the organic phase can yield much important information, and should always be carried out.

E. REFERENCES

1. E. Glueckauf, H. A. C. McKay and A. R. Mathieson, *J. Chem. Soc. Supp.*, 2, 299 (1949).

2. L. Brewer, T. R. Simonson and L. K. J. Tong, *J. Phys. Chem.*, **65**, 420 (1961); D. B. Scaife and H. J. V. Tyrrell, *J. Chem. Soc.*, 386 (1958).

3. For extraction from molten salts, see Y. Marcus, in *Solvent Extraction Chemistry* (Eds. D. Dyrssen J. O. Liljenzin and J. Rydberg), North Holland Publ. Co., Amsterdam, 1967, p. 555.

4. R. M. Diamond and D. G. Tuck in *Progress in Inorganic Chemistry*, Vol. 2, Interscience Publ. Co., New York, 1960, p. 109.

5. G. H. Morrison and H. Freiser, *Solvent Extraction in Analytical Chemistry*, J. Wiley & Sons, New York, 1957.

6. Yu. G. Frolov and A. V. Ochkin, *Zh. Neorg. Khim.*, **7**, 766 (1962); V. V. Fomin, *Zh. Neorg. Khim.*, **7**, 769 (1962); F. S. Martin and R. J. W. Holt, *Quart. Rev. (London)*, **13**, 327 (1959).

7. Y. Marcus, *Chem. Rev.*, **63**, 139 (1963).

8. R. P. Wischow, *Dissertation Abstr.*, **19**, 1937 (1959); K. K. Lunes and R. P. Wischow, *USAEC Rept.*, TID-11414 (1958).

9. L. L. Burger, *USAEC. Rept.*, HW-62087 (1959).

10. R. E. Treybal, *Liquid Extraction*, McGraw Hill Book Co., New York, 1951.

11. J. B. Lewis and H. R. C. Pratt, *Nature*, **171**, 1155 (1953); R. Murdoch and H. R. C. Pratt, *Trans. Inst. Chem. Engrs. (London)*, **31**, 307 (1953); J. B. Lewis, *Chem. Eng. Sci.*, **3**, 248, 260 (1954).

12. J. B. Lewis, *Nature*, **178**, 274 (1956).
13. H. T. Hahn, *J. Am. Chem. Soc.*, **79**, 4625 (1957).
14. K. A. Allen and W. J. McDowell, *J. Phys. Chem.*, **64**, 877 (1964).
15. L. D. Felton, *Ph. D. Thesis*, Mass. Inst. Techn., Cambridge, 1959.
16. W. Knoch, *J. Chem. Eng. Data*, **9**, 60 (1964).
17. J. B. West, *USAEC Rept.*, TID-11844 (1961).
18. Y. Marcus and I. Even-Sapir, unpublished results, Jerusalem, 1966.
19. G. Rudstam, *Acta Chem. Scand.*, **13**, 1481 (1959).
20. A. Goble, J. Golden and A. G. Maddock, *Can. J. Chem.*, **34**, 284 (1956); A. Goble and A. G. Maddock, *J. Inorg. Nucl. Chem.*, **7**, 94 (1958).
21. A. S. Kertes and M. Halperin, *J. Inorg. Nucl. Chem.*, **19**, 359 (1961); **20**, 117 (1961).
22. M. Zangen, *J. Phys. Chem.*, **69**, 1835 (1965).
23. G. K. Schweitzer and J. R. Rimstidt, Jr., *Anal. Chim. Acta*, **27**, 389 (1962).
24. J. P. McKaveney and H. Freiser, *Anal. Chem.*, **30**, 1965 (1958); H. E. Hellwege and G. K. Schweitzer, *Anal. Chim. Acta*, **29**, 46 (1963).
25. H. Irving, G. Andrew and E. J. Risdan, *J. Chem. Soc.*, 541 (1949); H. Irving and R. J. P. Williams, *J. Chem. Soc.*, 1845 (1949); H. Irving, C. F. Bell and R. J. P. Williams, *J. Chem. Soc.*, 356 (1952).
26. C. B. Honaker and H. Freiser, *J. Phys. Chem.*, **66**, 127 (1962).
27. R. W. Geiger, *Ph.D. Thesis*, Univ. of Minnesota, 1951; R. W. Geiger and E. B. Sandell, *Anal. Chim. Acta*, **8**, 197 (1953).
28. B. Rubin and T. E. Hicks, *USAEC Rept.*, UCRL-126 (1955), based partly on data of J. R. Thomas and H. W. Crandall, *USAEC Rept.*, CN-3733 (1946).
29. R. A. Bolomey and L. Wish, *J. Am. Chem. Soc.*, **72**, 4486 (1950).
30. R. W. Taft and E. H. Cook, *J. Am. Chem. Soc.*, **81**, 46 (1959).
31. H. L. Finston and Y. Inoue, *J. Inorg. Nucl. Chem.*, **29**, 199, 2431 (1967).
32. *Chem. Tech. Div. Ann. Progr. Rept. USAEC Rept.*, ORNL-3627, (1964), p. 204; ORNL-3830, (1965) p. 217; ORNL-3945, (1966) p. 185.
33. C. A. Blake, Jr., K. B. Brown, C. F. Coleman, D. E. Horner and J. M. Schmitt, *USAEC Rept.*, ORNL-1903 (1955).
34. R. O. Dannenberg, D. W. Bridges and J. B. Rosenbaum, *Bur. Mines, U. S. Government Rept.*, RI-5941 (1962).
35. R. A. Wells, D. A. Everest and A. A. North, *Nucl. Sci. Eng.*, **17**, 259 (1963).
36. C. J. Hardy, B. F. Greenfield and D. Scargill, *J. Chem. Soc.*, 174 (1961).
37. J. Bosholm and W. Pippel, *Z. Physik. Chem.* (*Leipzig*), **227**, 217 (1964).

38. E. Foa, N. Rosintal and Y. Marcus, *J. Inorg. Nucl. Chem.*, **23**, 109 (1961).

39. A. S. Kertes and Y. Elhanan Habousha, *J. Inorg. Nucl. Chem.*, **25**, 1531 (1963).

40. W. Nernst, *Z. Physik. Chem.* (*Leipzig*), **8**, 110 (1891).

41. B. Milicevic, *Helv. Chim. Acta*, **46**, 1466 (1963).

42. M. Berthelot and J. Jungfleisch, *Ann. Chim. Phys.*, **26**, (4), 396 (1872).

43. V. B. Shevchenko, A. S. Solovkin, I. V. Shilin, L. M. Kirilov, A. V. Radionov and V. V. Balandina, *Radiokhimiya*, **1**, 257 (1959).

44. Y. Marcus, *Acta Chem. Scand.*, **11**, 329 (1957).

45. H. A. C. McKay, and J. H. Miles, *Nature*, **199**, 65 (1963).

46. J. H. Hildebrand and R. L. Scott, *The Solubility of Non-electrolytes*, A.C.S. Monograph, 3rd ed., Reinhold Publ. Co., New York, 1950.

47. A. S. Kertes, *J. Inorg. Nucl. Chem.*, **26**, 1794 (1964).

48. T. Wakahayashi, S. Oki, T. Omori and N. Suzuki, *J. Inorg. Nucl. Chem.*, **26**, 2255 (1964); T. Omori, T. Wakahayashi, S. Oki and N. Suzuki, *J. Inorg. Nucl. Chem.*, **26**, 2265 (1964).

49. S. Siekierski and R. Olszer, *J. Inorg. Nucl. Chem.*, **25**, 1351 (1963); R. Olszer and S. Siekierski, *J. Inorg. Nucl. Chem.*, **28**, 1991 (1966).

50. H. Morse, *Z. Physik. Chem.* (*Leipzig*), **41**, 709 (1902).

51. M. S. Sherrill, *Z. Physik. Chem.* (*Leipzig*), **43**, 705 (1903); M. S. Sherrill and R. Abbegg, *Z. Electrochem.*, **9**, 549 (1905).

52. G. A. Linhart, *J. Am. Chem. Soc.*, **38**, 1272 (1916).

53. Y. Marcus, *Acta Chem. Scand.*, **11**, 610 (1957).

54. H. C. Moser and A. F. Voigt, *J. Inorg. Nucl. Chem.*, **4**, 354 (1957).

55. M. Zangen and Y. Marcus, *Israel J. Chem.*, **2**, 49 (1964).

56. G. A. Linhart, *J. Am. Chem. Soc.*, **37**, 258 (1915).

57. W. F. McDevit and F. A. Long, *Chem. Rev.*, **51**, 119 (1952).

58. M. Givon, Y. Marcus and M. Shiloh, *J. Phys. Chem.*, **67**, 2495 (1963).

59. H. M. Dawson, *J. Chem. Soc.*, **95**, 870 (1909).

60. G. H. Cartledge and S. L. Goldheim, *J. Am. Chem. Soc.*, **55**, 3585 (1933).

61. Y. Marcus, *Acta Chem. Scand.*, **11**, 811 (1957).

62. M. Zangen, *Israel J. Chem.*, **2**, 91 (1964); M. Zangen and Y. Marcus, *Israel J. Chem.*, **2**, 155 (1964).

63. I. Eliezer, *J. Chem. Phys.*, **42**, 3625 (1965).

64. I. Eliezer, *J. Chem. Phys.*, **41**, 3276. (1965)

65. Y. Marcus, I. Eliezer and M. Zangen, *Israel AEC Rept.*, IA–929 (1964); *Proc. Tihany Conf. Coord. Chem.*, Akademiai Kiado, Budapest, 1965, p. 409.

66. M. Zangen, unpublished results, 1966.

67. F. J. C. Rossotti and H. Rossotti, *J. Phys. Chem.*, **65**, 926 (1961).

68. I. M. Kolthoff and E. B. Sandell, *J. Am. Chem. Soc.*, **63**, 1906 (1941).

69. J. Rydberg, *Arkiv Kemi*, **8**, 101 (1955).

70. M. Oosting, *Anal. Chim. Acta*, **21**, 301, 397, 505 (1959).

71. G. K. Schweitzer, *Anal. Chim. Acta*, **30**, 68 (1964).

72. J. Stary, *Chem. Listy*, **53**, 556 (1959); *The Solvent Extraction of Metal Chelates*, Pergamon Press, London, 1964.

73. H. Irving and R. J. P. Williams, *J. Chem. Soc.*, 1841 (1949).

74. J. Rydberg and J. C. Sullivan, *Acta. Chem. Scand.*, **13**, 2057 (1959).

75. D. Dyrssen and D. H. Liem, *Acta Chem. Scand.*, **14**, 1091 (1960); D. Dyrssen, S. Ekberg and D. H. Liem, *Acta Chem. Scand.*, **18**, 235 (1964).

76. J. Rydberg, *Arkiv. Kemi*, **5**, 517 (1953).

77. E. Hesford and H. A. C. McKay, *Trans. Faraday Soc.*, **54**, 573 (1958).

78. E. Glueckauf, H. A. C. McKay and A. R. Mathieson, *Trans. Faraday Soc.*, **47**, 437 (1951).

79. Y. Marcus, *J. Phys. Chem.*, **65**, 1647 (1961); *Israel A.E.C. Rept.*, IA–582 (1960).

80. H. Irving, F. J. C. Rossotti and R. J. P. Williams, *J. Chem. Soc.*, 1906 (1955).

81. J. Saldick, *J. Phys. Chem.*, **60**, 500 (1956).

82. R. M. Diamond, *J. Phys. Chem.*, **61**, 69 (1957).

83. D. Maydan and Y. Marcus, *J. Phys. Chem.*, **67**, 983 (1963).

84. G. Duyckaerts, J. Fuger and W. Müller, *Euratom Rept.*, EUR 426. f, (1963).

85. Y. Marcus, *J. Inorg. Nucl. Chem.*, **28**, 209 (1966).

86. G. Markovits and A. S. Kertes, in *Solvent Extraction Chemistry* (Eds. D. Dyrssen, J. O. Liljenzin and J. Rydberg), North Holland Publ. Co., Amsterdam, 1967, p. 390.

87. K. van Ipenburg, *Rec. Trav. Chim.*, **80**, 269 (1961).

88. W. F. Eberz, H. J. Welge, D. M. Yost and H. J. Lucas, *J. Am. Chem. Soc.*, **59**, 45 (1937); S. Winstein and H. J. Lucas, *J. Am. Chem. Soc.*, **60**, 836 (1938).

89. I. Ostromislenskii, *Chem. Ber.*, **44**, 268, 1189 (1911); R. B. Denison, *Trans. Faraday Soc.*, **8**, 20, 35 (1912); P. Job, *Ann. Chim. Phys.*, **9**, 113 (1928).

90. A. K. Babko and A. T. Pilipenko, *Zh. Analit. Khim.*, **2**, 33 (1947); A. T. Pilipenko, *Zh. Analit. Khim.*, **5**, 14 (1950); **8**, 286 (1953).

91. H. Irving and T.B. Pierce, *J. Chem. Soc.*, 2565 (1959).

92. F. Woldbye, *Acta Chem. Scand.*, **9**, 299 (1955).

93. M. M. Jones and K. K. Innes, *J. Phys. Chem.*, **62**, 1005 (1958).

94. H. Specker and E. Jackwirth, *Z. Anal. Chem.*, **167**, 416 (1959).

95. A. H. A. Heyn and Y. D. Soman, *J. Inorg. Nucl. Chem.*, **26**, 287 (1964).

96. N. P. Komar, *Zh. Neorg. Khim.*, **2**, 1015 (1957).

97. T. B. Pierce and P. F. Peck, *Brit. Rept.*, AERE, R4187 (1962).

98. N. S. Kurnakov, *Trudy po Khimii Kompleksnykh Soadinenii* (Eds. I. I. Chernyaev and O. E. Zvyagintsev), Izd. Akad. Nauk SSSR, Moscow, 1963; cf. G. G. Urazov, *Uspekhi Khim*, **21**, 1019 (1952), for a review of Kurnakov's work.

99. A. S. Kertes, *J. Inorg. Nucl. Chem.*, **14**, 104 (1960).

100. A. S. Kertes, and V. Kertes, *Can. J. Chem.*, **38**, 612 (1960).

101. A. S. Kertes and V. Kertes, *J. Appl. Chem.*, **10**, 287 (1960).

102. A. S. Kertes and M. Halpern, *J. Inorg. Nucl. Chem.*, **16**, 308 (1961).

103. A. S. Kertes, A. Beck, and J. Habousha, *J. Inorg. Nucl. Chem.*, **21**, 108 (1961).

104. D. G. Tuck, *J. Chem. Soc.*, 2783 (1958).

105. D. G. Tuck, *Trans. Faraday Soc.*, **57**, 1297 (1961).

106. Z. A. Sheka and E. E. Kriss, *Zh. Neorg. Khim.*, **4**, 2505 (1959).

107. M. Halpern, T. Kim and N. C. Li, *J. Inorg. Nucl. Chem.*, **42**, 1251 (1962).

108. D. F. C. Morris and E. L. Short, *J. Inorg. Nucl. Chem.*, **25**, 291 (1963).

109. J. M. Fletcher and C. J. Hardy, *Nucl. Sci. Eng.*, **16**, 421 (1963); H. A. C. McKay, *Belg. Chem. Ind.*, **12**, 1278 (1964).

110. H. Yoe and A. L. Jones, *Ind. Eng. Chem. Anal. Ed.*, **16**, 111 (1944).

111. Y. Marcus, *Israel J. Chem.*, **5**, 143 (1967).

112. H. L. Schläfer, *Komplexbildung in Lösung*, Springer Verlag, Berlin, 1961, p. 253.

113. T. J. Collopy and J. F. Blum, *J. Phys. Chem.*, **64**, 1324 (1960).

114. H. Ihle, H. Michael and A. Murrenhoff, *J. Inorg. Nucl. Chem.*, **25**, 734 (1963).

115. H. T. Baker and C. F. Baes, Jr., *J. Inorg. Nucl. Chem.*, **24**, 1277 (1962).

116. D. G. Tuck, *J. Chem. Soc.*, 3202 (1957).

117. D. G. Tuck and R. M. Diamond, *Proc. Chem. Soc.*, 236 (1958); *J. Phys. Chem.*, **65**, 193 (1961).

118. V. B. Shevchenko, I. G. Slepchenko, V. S. Schmidt and E. A. Nenarokomov, *Zh. Neorg. Khim.*, **5**, 1095 (1960).

119. W. E. Keder, J. C. Sheppard and A. S. Wilson, *J. Inorg. Nucl. Chem.*, **2**, 327 (1960).

120. K. A. Allen, *J. Am. Chem. Soc.*, **80**, 4133 (1958).

121. J. M. P. J. Verstegen, *J. Inorg. Nucl. Chem.*, **26**, 1589 (1964).

122. T. H. Siddall, III, *J. Inorg. Nucl. Chem.*, **13**, 151 (1960).

123. A. S. Solovkin, N. S. Povitskii and K. P. Kunichkina, *Zh. Neorg. Khim.*, **5**, 2115 (1960).

124. H. A. C. McKay and A. R. Mathieson, *Trans. Faraday Soc.*, **47**, 428 (1951).

125. V. M. Vdovenko and J. A. Smirnova, *Radiokhimiya*, **1**, 36, 521 (1959).

126. C. F. Baes, Jr., *J. Phys. Chem.*, **66**, 1629 (1962).

127. C. F. Baes, Jr., R. A. Zingaro and C. F. Coleman, *J. Phys. Chem.*, **62**, 129 (1958).

128. G. Scibona, S. Basol, P. R. Danesi and F. Orlandini, *J. Inorg. Nucl. Chem.*, **28**, 1441 (1966).

129. W. Müller and R. M. Diamond, *J. Phys. Chem.*, **70**, 3469 (1966).

130. C. E. Higgins and W. H. Baldwin, *J. Inorg. Nucl. Chem.*, **24**, 415 (1962).

131. V. V. Fomin and V. T. Potapova, *Zh. Neorg. Khim.*, **8**, 990 (1963).

132. V. V. Fomin, P. A. Zagorets and A. F. Morgunov, *Zh. Neorg. Khim.*, **4**, 639 (1959); V. V. Fomin, P. A. Zagorets, A. F. Morgunov and I. I. Tertishnik, *Zh. Neorg. Khim.*, **4**, 2276 (1959); V. V. Fomin and A. P. Morgunov, *Zh. Neorg. Khim.*, **5**, 1385 (1960); V. V. Fomin and R. N. Maslova, *Zh. Neorg. Khim.*, **6**, 483 (1961).

133. P. J. Flory, *Principles of Polymer Chemistry*, Cornell University Press, Ithaca, New York, 1953, p. 310.

134. K. A. Allen, *J. Phys. Chem.*, **62**, 1119 (1958).

135. P. Walden, *Electrochemie Nichtwässeriger Lösungen*, Barth, Leipzig, 1924.

136. E. Hesford and H. A. C. McKay, *J. Inorg. Nucl. Chem.*, **13**, 156 (1960).

137. P. Biddle, A. Coe, H. A. C. McKay, J. H. Miles and M. J. Waterman, *J. Inorg. Nucl. Chem.*, **29**, 2615 (1967).

138. H. Kendall, *Meddel. Vetenskapsakad. Nobelinst.*, **2**, 25 (1913).

139. J. M. P. J. Verstegen, *Trans. Faraday Soc.*, **58**, 1878 (1962).

140. K. Alcock, S. S. Grimley, T. V. Healy, J. Kennedy and H. A. C. McKay, *Trans. Faraday Soc.*, **52**, 39 (1956).

141. G. Scibona, B. Scuppa and M. Zifferero, *Italian Rept.*, CNEN, No. 28 (1963); A. S. Kertes and I. Platzner, *J. Inorg. Nucl. Chem.*, **24**, 1417 (1962).

142. A. G. Maddock, W. Smulek and A. J. Tench, *Trans. Faraday Soc.*, **58**, 973 (1957).

143. S. Kato and R. Ishii, *Sci. Papers Inst. Phys. Chem. Res., Tokyo*, **36**, 82 (1939).

144. D. E. Metzler and R. J. Myers, *J. Am. Chem. Soc.*, **72**, 3776 (1950).

145. N. H. Nachtrieb and J. G. Conway, *J. Am. Chem. Soc.* **70**, 3547 (1948).

146. H. L. Friedman, *J. Am. Chem. Soc.*, **74**, 5 (1952).

147. M. L. Good and S. E. Bryan, *J. Am. Chem. Soc.*, **82**, 5636 (1960).

148. J. M. White, P. Kelly and N. C. Li, *J. Inorg. Nucl. Chem.*, **16**, 337 (1961).

149. J. Bullock, S. Choi, J. Goodrick, D. Tuck and E. Woodhouse, *J. Phys. Chem.*, **68**, 2687 (1964).

150. J. M. P. J. Verstegen, *J. Inorg. Nucl. Chem.*, **26**, 2311 (1964).

151. W. E. Keder and A. S. Wilson, *Nucl. Sci. Eng.*, **17**, 287 (1963); W. E. Keder and L. L. Burger, *J. Phys. Chem.*, **70**, 3025 (1965).

152. I. V. Seryakova, Yu. A. Zolotov, A. V. Karyakin, L. A. Gribov and M. E. Zubrilina, *Zh. Neorg. Khim.*, **7**, 2013 (1962).

153. I. V. Seryakova, Yu. A. Zolotov, A. V. Karyakin and L. A. Gribov, *Zh. Neorg. Khim.*, **8**, 474 (1963).

154. L. A. Woodward and M. J. Taylor, *J. Chem. Soc.*, 4473 (1960).

155. L. A. Woodward and M. J. Taylor, *J. Chem. Soc.*, 406 (1962).

156. D. F. C. Morris, E. L. Short and D. N. Waters, *J. Inorg. Nucl. Chem.*, **25**, 975 (1963).

157. E. L. Short, D. N. Waters and D. F. C. Morris, *J. Inorg. Nucl. Chem.*, **26**, 902 (1964).

158. W. E. Shuler, *USAEC Rept.*, DP-513 (1960).

159. W. Knoch, *J. Inorg. Nucl. Chem.*, **27**, 2075 (1965).

8.
Extraction by Compound Formation

The type of extractants dealt with in this Chapter are frequently called
liquid cation exchangers. Indeed, these reagents operate, at least
formally, by interchange of hydrogen ions of the acidic organic reagent
for the cation in the aqueous phase. Despite the similarity of the extraction reaction, there are important differences between the various
acidic organic reagente in their interaction with metals. Taking these
differences into account, cation exchange extraction systems will
be treated under three different headings: extraction by chelating
agents (8.A), extraction by acidic organophosphorus esters (8.C), and

(c) extraction by carboxylic and sulphonic acids (8.D). The main differences are between extraction processes involving chelates on the one hand, and phosphorus and carboxylic acids on the other.

There are obviously a number of differences between the behaviour of inner complexes and that of salts of organic acids. The acids, HX, forming the inner complex MX_m are usually weak and more organic-than water-soluble. The chelates MX_m are stable, monomeric compounds which can frequently be isolated in the solid state. They are water insoluble but freely soluble in, or extracted into, various organic solvents. Their composition is usually unaffected by the nature of the diluent used. In view of their stability, the composition is usually unaffected by the ligand concentration in the organic phase, even when it is varied sufficiently to change the metal distribution ratio over a wide range. The overall heterogeneous extraction reaction is well described by simple mass-action law equations derived in section 7.C, since the behaviour of these systems closely approaches ideality.

On the other hand, extraction equilibria in systems involving carboxylic and phosphorus acids are far less amenable to description by distribution law equations. The extraction reaction itself is usually well understood, but it is the non-ideal behaviour of the solutes in the organic phase which complicates the description of the extraction process by simple equations. Weak complexes are involved in such systems, and a high ligand-to-meta concentration ratio is required for their formation. The extracted species are frequently solvated to a varying extent, and only exceptionally can these complexes be isolated. Their composition varies with the ligand concentration in the organic phase and the nature of the diluent used. The acids, from which the anions of the extracted salts are derived, are usually stronger than those forming the inner complexes. They are frequently also somewhat water soluble. Since the composition of the aqueous phase affects the distribution of the free-ligand acid between the two phases, the systems usually deviate from an ideal behaviour.

The formation of uncharged chelates only, will be considered in this chapter. Chelating agents, such as 1,10-phenanthroline and its derivatives, which usually form positively charged metal chelates, or the negatively charged chelates formed between a metal ion and a chelating acid with multiple negative charge, such as citric or tartaric acid, will not be reviewed here. The prerequisite for the extraction of such charged chelates is their coupling with an (preferentially bulky) anion or cation respectively, into an electrically neutral ion-pair. In both these cases the nature of the extractable species is much different from the nature of those involving hydrogen atom replacement and coordination to form a neutral chelate. Such complexes will be discussed in Chapter 10.

A. EXTRACTION BY CHELATING AGENTS

a. Introduction

Many organic reagents known as metal-precipitating agents in aqueous solutions form chelates which are frequently extractable into an organic solvent. In chelating extractive systems, the organic chelating agents act as weak acids and contain a donor group so as to form with the metal a bidentate chelate. The metal-bearing organic phase complex must contain a number of chelating molecules equal to the actual charge on the metal ion, in order to yield an electrically neutral species. Now, it seems to be a rule, that whenever the coordination number of the metal equals double its ionic charge, the chelate formed satisfies the coordination requirements of the metal and the species will be extractable to an appreciable extent. Such is usually the case with polyvalent cations, whereas some bivalent metal ions having a coordination number higher than four may retain one or more water molecules in their first coordination sphere. Hydrated metal chelates are poorly extractable into non-polar solvents which do not have the tendency to replace the water of hydration. Thus it is not invariably true that metal chelates which are insoluble in aqueous solutions are extractable into water-immiscible non-polar organic solvents.

The extraction of metal chelates, as discussed in detail in the previous chapter, can be represented by the heterogeneous phase reaction

$$M^{m+} + m\,\overline{HX} \rightleftarrows \overline{MX_m} + m\,H^+ \qquad (8A1)$$

with an equilibrium constant

$$K_{11} = [\overline{MX_m}][H^+]^m\,[M^{m+}]^{-1}\,[\overline{HX}]^{-m} \qquad (8A2)$$

In terms of the mass-action law treatment of the extraction process, metal extraction by chelating acids is by far the simplest of the possible heterogeneous extraction equilibria, and is essentially the only type of inorganic extraction system that yields easily to quantitative treatment. There are several reasons for the simplicity of such systems. Since metal-chelate extraction is usually achieved from aqueous solutions of low electrolyte concentration, there are no complicating thermodynamic factors affecting the activity coefficient of the solutes in the aqueous phase. Under such conditions metal complexing in the aqueous phase can readily be taken into account if the equilibrium constants of the aqueous metal-bearing species are known, which is usually the case. As to the organic phase, the lack of molecular association and dissociation of the chelating acids makes the solutes behave ideally, and the distribution of the metal follows the equilibria based on the mass-action law, free from complicating side-reactions occurring in the organic medium. It reflects the apparent ideal behaviour of chelate extraction systems that the distribution ratio of metals is close, and frequently identical, to the ratio of solubilities in the organic and aqueous phases.

In some cases, however, a mass-action law treatment of the distribution data becomes difficult because of the relatively high loss of the reagent to the aqueous solution. Such is the case, for example, when cupferron or oxine are used as the chelating agents for extraction of metals from aqueous solutions of unfavourable pH. Their relatively low distribution ratios under such conditions, enables aqueous complexing of the metal with the reagents. A simple set of equations usually does not take into account either the amount of the reagent dissolved in the aqueous phase, or the amount of metal complexed in that phase by the chelating agent.

The magnitude of the equilibrium constant K_{11} is a measure of the stability of the metal chelate MX_m, and also a measure of its formation at a given aqueous pH. For a given series of closely related chelating extractants there is a relationship between the dissociation constant K_{HX} of the acid and the stability constants of the complexes formed with any given metal, as predicted by equation (7C25).

A number of typical chelating extractant groups are listed in Table 8A1, but many more have been shown to be fairly good extracting agents under specified experimental conditions[1-5].

TABLE 8A1. Some versatile chelating extractants.

β-Diketones

$$-\underset{\underset{OH}{|}}{C}=CH-\underset{\underset{O}{\|}}{C}- \qquad -\underset{\underset{O}{\|}}{C}-CH_2-\underset{\underset{O}{\|}}{C}-$$

enol   keto

Dithizone

$$HS-\underset{\diagdown}{\overset{\diagup N=N-}{C}} \qquad S=\underset{\diagdown}{\overset{\diagup N=N-}{C}}$$
$$\qquad N-NH \qquad\qquad NH-NH-$$

enol   keto

8-Hydroxyquinoline

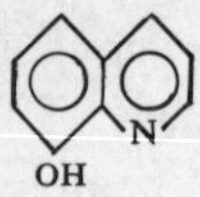

Monoximes

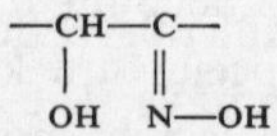

$$-\underset{\underset{OH}{|}}{CH}-\underset{\underset{N-OH}{\|}}{C}-$$

Dioximes

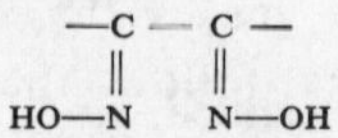

$$-\underset{\underset{HO-N}{\|}}{C}-\underset{\underset{N-OH}{\|}}{C}-$$

Nitrosophenols

$$-\underset{\underset{HO}{|}}{C}-\underset{\underset{N-OH}{\|}}{C}-$$

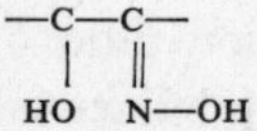

Nitrosarylhydroxylamines

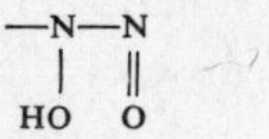

$$-\underset{\underset{HO}{|}}{N}-\underset{\underset{O}{\|}}{N}$$

Hydroxamic acids

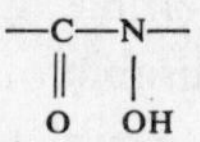

$$-\underset{\underset{O}{\|}}{C}-\underset{\underset{OH}{|}}{N}-$$

Dithiocarbamates

$$\diagdown N - C \diagup^S_{\diagdown SH}$$

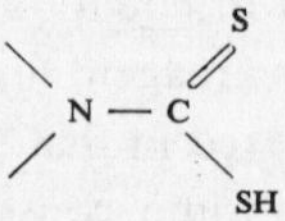

Though solvent extraction was primarily designed as a method of separation, recent trends and interest in chelating solvent extraction systems are in both branches of analytical chemistry: for separation and concentration of a metal and its quantitative determination. This is an obvious trend, since in many cases the extracts are directly suitable for quantitative determination of the extracted metal using spectral methods. Metal extraction by dithizone, for example, is often more important as a method of quantitative determination than separation. Dithizonates are highly coloured compounds, thus a spectrophotometric determination of the metal in the extract is a relatively simple analytical method.

A brief account is given in the following paragraphs of the characteristics of chelating extraction systems generally, and of the more important chelating agents which have found application, almost exclusively, in the field of analytical chemistry. Because of the many review articles and monographs now available[1-16], some of them very recent, on the chemistry of solvent extraction involving chelating agents and their application to analytical problems, no attempt has been made in this section to cover the topic comprehensively. It is, furthermore, beyond the scope of this monograph to give detailed information about conditions of metal extraction, such as the optimum aqueous-phase acidity, the nature of the reagent and its concentration, the use of masking agents for extractive separation of metals etc. For that purpose the existing monographs are well suited, and detailed information may readily be obtained from the sources indicated there.

b. Extraction Characteristics of Chelate Systems

(i) *Aqueous-phase parameters* Although almost all the chelating agents which form extractable metal complexes react with a large number of metals, the reaction may be rendered selective, or even specific for certain metals, by a proper adjustment of the initial aqueous solution. Since all the reagents reviewed in this section are weak or very weak acids, the formation of extractable complexes will much depend on the pH of the aqueous solution. Thus the simplest way is to affect equilibrium (8A1) by adjustment of the aqueous pH. This procedure is useful for extractive separation of metals whenever the equilibrium constants are sufficiently different for the metals in question (section 7.C). Since the extractability of the metal is directly proportional to the mth power of the hydrogen ion concentration

$$D = K_{11}\overline{[HX]}^m[H^+]^{-m} \qquad (8A3)$$

(neglecting activity coefficients) at any given organic-phase composition, the distribution ratio can be evaluated from the known equilibrium constant K_{11}. The effect of pH will obviously be different with a different charge on the extractable metal ion, m. The distribution ratio will increase by a factor of 10 for a monovalent cation, by a factor of 10^2 for a divalent, 10^3 for tervalent and 10^4 for a tetravalent cation when the pH of the aqueous solution is altered by one unit (section 7.C).

Selectivity can frequently be achieved by factors other than the pH control of the aqueous medium. They involve the use of masking agents which readily form water-soluble metal complexes of a higher stability than those formed by the extracting chelating agents. Numerous analytical procedures have been worked out for extractive separations using masking agents such as the various complexons, tartaric and salicylic acids and cyanide.

(ii) *Organic-phase parameters* Chelating agents which form uncharged and extractable metal chelates have at least two functional groups entering into reaction with the metal ion. One of those groups of the chelating molecule is an OH (or SH) group not necessarily on a carbon atom (Table 8A1). The other is a basic functional group capable of coordinating by its donor properties to the metal. The hydrogen of the OH is replaced by an equivalent of metal, and a stable ring, usually five- or six-membered, is formed. The stability of the chelate will depend of course, on the basic nature of the coordination site and the acidity of the OH group, in addition to the acidic nature, coordination number and coordination ability of the metal. The latter factor is usually common to all chelating and complex-forming agents and will not be discussed here.

Structural factors affect both the acidity of the OH group and the basic strength of the donor atom. Additionally, the relative position of the two active sites on the chelating molecule will affect the extractive effectivness of the agent by determining the size of the chelate ring formed.

As to the effect of acidity of the chelating agent on its extractive usefulness, two different aspects have to be considered. A more basic chelating agent will form a more stable metal chelate. On the other hand, and this is the second aspect, a more acidic chelating agent with a lower pK value, will be more useful for the extraction of metals from acidic aqueous solutions. Thus, for example, acetylacetone ($pK = 8.93$) will not extract thorium quantitatively from an aqueous solution below pH 5, whereas thenoyltrifluoroacetone ($pK = 6.31$) will extract the metal completely at a pH as low as 1. A low pH may often be needed in order to avoid hydrolysis of metal ions (Table 3C2).

The donor properties of the basic nitrogen, sulphur or oxygen atoms depend on their electronegativity. increasing in that order. More stable metal chelates are formed by participation of atoms of lower electronegativity.

The extent of extractability is much affected by solubility characteristics of chelates in organic solvents. Substitution of a methyl group on the chelating agent molecule by a benzene ring, for example, will result in an increased extractability of its metal chelate in a non-polar solvent. Or, more generally, rendering the chelating molecule more hydrophobic will contribute to the transfer of the metal chelate. Chelating agents possessing a free acidic or basic group may exhibit only a limited solubility in organic solvents of low or zero polarity. For the metal chelates formed by such chelating extractants, polar alcohols or ketones serve as diluents.

Certain uncharged metal chelates, especially those of bivalent metals having a coordination number higher than four, are non-extractable into non-polar organic solvents. The phenomenon is due to the retention of water molecules coordinatively bound to the metal in order to satisfy its coordination requirements. An increased extraction of such metal chelates may be achieved by use of oxygen-containing diluents, polar in nature, such as ketones, esters and alcohols, especially the latter[17-19]. Their contribution to an increased extraction is due to the ability of these solvents to replace the water molecules. Such a cooperative effect of a chelating agent and a solvent with solvating properties, enhances the transfer of the metal in a synergistic way. These extraction systems will be treated separately in Chapter 11.

c. Chelating Agents and Metal Chelates

(i) *β-Diketones* Undoubtedly the most versatile and most widely used chelating agents in solvent extraction are the β-diketones. In their enol form they have a hydrogen replaceable by a metal equivalent, and a ketonic oxygen for coordination. The acidity of that hydrogen is affected by the nature and structure of the alkyl or aryl groups. The simplest members of the group, the symmetrical acetylacetone and dibenzoyl-methane, are the least acidic. Introduction of trifluoromethyl groups increases the acidity of the enol form in water by several orders of magnitude, as can be seen from the values compiled in Table 8A2. A higher acidity of the enol form makes metal extraction possible from low pH solutions. The distribution constants of several β-diketones between water and chloroform and benzene are included in Table 8A2. The highest values are those for dibenzoylmethane, due to the two benzoyl rings in the molecule; otherwise the distribution constants increase with the hydrophobic character of the alkyl or aryl radicals in the reagent.

β-Diketones form unionized chelates according to equation (8A1) with about sixty metals. The equilibrium constants (8A2) are compiled in Table 8A3 for the three most investigated β-diketones[5, 20, 21]. The values refer to benzene, the most widely used solvent for β-diketone systems. Provided the aqueous phase is free from other complexing anions, the equilibrium constants are affected by the nature of the diluent only. A comparison of the constants is given in Table 8A4, along with the aqueous pH value necessary for a 50% extraction of the metal.

TABLE 8A2. Dissociation constant K_{HX} of β-diketones in water, at 25°C, and their distribution constants between water and chloroform or benzene, at 20–25°C.

β-Diketone	Log K_{HX}	P_{CHCl_3}	$P_{C_6H_6}$
Acetylacetone	− 8.93	25	6
Benzoylacetone	− 8.74	1760	1380
Dibenzoylmethane	− 9.20	250,000	224,000
Trifluoroacetylacetone	− 6.30		1.5
Benzoyltrifluoroacetone	− 6.30		
Furoyltrifluoroacetone			7.4
Thenoyltrifluoroacetone	− 6.30		4.9
Selenoylacetone	− 8.55	833	1000
Selenoyltrifluoroacetone	− 6.32	83	
Benzoylselenoylmethane	− 7.86	7950	4580
Furoylselenoylmethane	− 2.51	1260	480
Thenoylselenoylmethane	− 2.53	1480	563
Diselenoylmethane	− 2.50	1480	1410

Thenoyltrifluoroacetone, TTA, is perphaps the most versatile β-diketone for metal extraction. It can be obtained in pure state, has a high acidity in its enol form, and is therefore, useful at low aqueous pH. In aqueous solution it forms a keto hydrate

which has a low distribution ratio into chloroform and benzene. The enolization of the keto hydrate of the aqueous TTA is nearly quantitative and rapid at high pH.

An unusually large number of diketonates has been prepared using various methods[22-29]. The majority of these crystalline substances contain one or more water molecules. Some of them exhibit *cis–trans* isomerism[28], and the existence of optical isomers is also known[26,27].

Table 8A3. Equilibrium constants, log K_{11}, of metal diketonates in benzene at room temperature.

Metal diketonate	Acetylacetone	Benzoylacetone	Thenoyltri-fluoroacetone
AgX		− 7.8	
TlX			− 4.65
BeX$_2$	− 2.79		− 3.20
MgX$_2$		− 16.65	
CaX$_2$		− 18.28	− 12.00
SrX$_2$		− 20.00	− 14.13
BaX$_2$			− 14.64
ZnX$_2$		− 10.79	
CdX$_2$		− 14.92	
CuX$_2$	− 3.93	− 4.17	− 1.32
CoX$_2$		− 11.11	− 6.70
NiX$_2$		− 12.12	
PbX$_2$	− 10.15	− 9.61	− 5.24
MnX$_2$		− 14.63	
PdX$_2$		− 1.2	
AlX$_3$	− 6.48	− 7.6	− 5.23
FeX$_3$	− 1.39	− 0.5	+ 3.30
GaX$_3$	− 5.51	− 6.30	
InX$_3$	− 7.20	− 9.30	− 4.34
BiX$_3$			− 3.21
ScX$_3$	− 5.83	− 5.99	− 0.77
YX$_3$		− 16.95	− 7.39
LaX$_3$		− 20.46	− 10.51
CeX$_3$			− 9.43
PrX$_3$			− 8.86
NdX$_3$			− 8.57
PmX$_3$			− 8.06
SmX$_3$			− 7.68
EuX$_3$		− 18.9	− 7.66
GdX$_3$			− 7.57
TbX$_3$			− 7.51
DyX$_3$			− 7.03
HoX$_3$			− 7.25
TmX$_3$			− 6.96
YbX$_3$			− 6.72
LuX$_3$			− 6.77
PuX$_3$			− 4.44
AmX$_3$			− 7.46
CmX$_3$			− 8.60a
BkX$_3$			− 7.52a
CfX$_3$			− 7.80a
EsX$_3$			− 7.90a
FmX$_3$			− 7.70a
ThX$_4$	− 12.16	− 7.68	+ 0.90
ZrX$_4$ (?)			+ 8.11
HfX$_4$ (?)			+ 6.30
U^{4+}X$_4$			+ 4.18
N$_p^{4+}$X$_4$			+ 5.58
P$_u^{4+}$X$_4$			+ 6.34
PaX$_4$ (?)		− 4.68	− 1.37
UO$_2$X$_2$.HX		− 4.68	− 1.37
PuO$_2$X$_2$			− 1.83

a) In toluene rather than benzene.

TABLE 8A4. Effect of diluent on the equilibrium constant of some diketonates.

	Undiluted		Benzene		Chloroform		Carbon tetrachloride	
	pH_{50}	$\log K_{11}$	pH_{50}	$\log K_{11}$	pH_{50}	$\log K_{11}$	pH_{50}	$\log K_{11}$
Be–acetylacetone	0.67	-3.3	2.45	-2.79				
In–acetylacetone	1.7	-8.1	3.95	-7.2	4.55	-9.09	4.15	-7.2
Cd–benzoylacetone			8.48	-14.92	8.93	-15.83	8.48	-14.90
In–benzoylacetone			4.14	-9.30	4.60	-10.65	4.13	-9.24
Y–benzoylacetone			6.86	-16.95	7.31	-18.30	6.89	-17.04
La–benzoylacetone			7.96	-20.46	8.41	-21.81	7.95	-20.34
UVI–benzoylacetone			3.82	-4.68	3.72	-5.06	4.03	-4.44

(ii) *Dithizone and its derivatives* Dithizone, diphenylthiocarbazone, H_2Dz, and its dihalogen derivatives dissolve in most organic solvents, to a lesser degree in non-polar hydrocarbons, as seen from the solubility data compiled in Table 8A5. In the most practical solvents, $CHCl_3$ and CCl_4, dithizone gives a green-coloured solution, whereas in polar solvents, such as nitrobenzene, the reagent is yellow. In water and aqueous acid solutions it is only sparingly soluble, but in alkali solutions its solubility is higher than in chloroform. At pH > 7, the alkali salt is completely dissociated, the anion HDz^- colouring the aqueous solution yellow. The yellow tinge of dithizone solutions is frequently due to the presence of diphenylcarbodiazone, an oxidation product of the easily oxidizable reagent.

For the dissociation constant $K_{H_2Dz} = [HDz^-][H^+][H_2Dz]^{-1}$ in water, values ranging between 1.5×10^{-5} and 3.2×10^{-5} have been reported[10], and the second ionization constant $K_{HDz} = [Dz^{2-}][H^+]$

TABLE 8A5. Solubility of dithizone and its derivatives at room temperature.

Reagent	Solvent	Solubility, (g H_2Dz/l)
Dithizone	Water	5×10^{-5}
	Carbon tetrachloride	0.64
	Chloroform	17.8
	Benzene	1.64
	n-Heptane	0.04
	Diethyl ether	0.40
p,p-Dimethyldithizone	Carbon tetrachloride	1.79
p,p-Dichlorodithizone	Carbon tetrachloride	0.42
p,p-Dibromodithizone	Carbon tetrachloride	0.62
p,p-Diiododithizone	Carbon tetrachloride	0.76

$[HDz^-]^{-1}$ has been estimated to have a value ranging between 10^{-5} and 10^{-12}; K_{H_2Dz} values in carbon tetrachloride, chloroform and benzene are 2×10^{-9}, 2.4×10^{-11} and 1×10^{-11} respectively.

The extractability of dithizone depends very much on the pH of the aqueous phase, in sense of equations (7C23a, b), as shown in Figure 8A1, reproduced from Irving and coworkers[30]. The distribution constant of H_2Dz at pH ~ 1, has the value of 1.1×10^4 in carbon tetrachloride and 2×10^5 in chloroform.

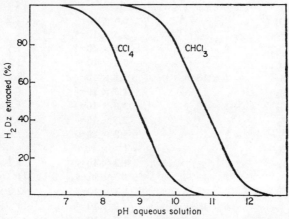

Figure 8A1. Extraction of dithizone into carbon tetrachloride and chloroform as a function of the hydrogen ion concentration in the aqueous phase.

Dithizone and its derivatives form intensely coloured chelates with about twenty metals in the middle of the periodic system. All of them are extractable into $CHCl_3$ and CCl_4. The reagents form two types of compounds: primary dithizonates, bidentate chelates, formed with the reagent in its keto form by the replacement of the hydrogen in the amino group and coordination to the double-bonded nitrogen are well known and many of them have been isolated in solid crystalline form. Some of them are listed in Table 8A6, along with their solubilities

TABLE 8A6. Solubility of metal dithizonates, at room temperature.

Metal dithizonate	Solubility (mole/l)		
	H_2O	CCl_4	$CHCl_3$
AgHDz		$>2 \times 10^{-3}$	4×10^{-2}
TlHDz			$>1 \times 10^{-5}$
Hg(HDz)$_2$		6.6×10^{-5}	2.8×10^{-4}
Pd(HDz)$_2$		4.5×10^{-4}	
Cu(HDz)$_2$	8×10^{-9}	$>1.2 \times 10^{-3}$	$>1.5 \times 10^{-3}$
In(HDz)$_3$		7.8×10^{-4}	1.1×10^{-3}
Zn(HDz)$_2$	$\sim 1 \times 10^{-4}$	$>2.8 \times 10^{-3}$	$>1.6 \times 10^{-3}$
Cd(HDz)$_2$		1.3×10^{-5}	1.3×10^{-4}
Co(HDz)$_2$	$\sim 1 \times 10^{-4}$	1.6×10^{-4}	1.4×10^{-3}
Pb(HDz)$_2$		2.5×10^{-5}	4.3×10^{-4}
Ni(HDz)$_2$	$\sim 1 \times 10^{-4}$	1.2×10^{-3}	1.6×10^{-3}
Au(HDz)$_3$		$\sim 1 \times 10^{-5}$	
Bi(HDz)$_3$		1.3×10^{-5}	1.2×10^{-5}
Ag$_2$Dz		$>1 \times 10^{-6}$	$>1 \times 10^{-6}$
HgDz		1.3×10^{-3}	2.8×10^{-3}
CuDz	6×10^{-8}	2.1×10^{-3}	3×10^{-3}
CoDz		$>2.5 \times 10^{-5}$	

in water, chloroform and carbon tetrachloride. They are stable in acid and neutral media, and as a rule, more extractable than the secondary dithizonates. The latter, formed with the reagent in its enol form by the additional replacement of the hydrogen of the sulphohydryl group,

contains twice as much metal as the primary complex per dithizone molecule. Some of them are included in Table 8A6. The less numerous secondary dithizonates are stable in alkaline solutions and extracted as such under conditions of high organic-phase loading. By changing the extraction conditions, either by acidifying the aqueous solution or providing an excess of the reagent, secondary dithizonates can be transformed into the primary metal chelates.

The equilibrium constants for metal dithizonates, calculated by equation (8A2) represent fairly well the order of extractability of metal dithizonates. Some values of K_{11}, compiled from the data tabulated by Iwantscheff[10], are given in Table 8A7. Chelates having the highest values of K_{11}, those of group B metals, i.e. silver, mercury, copper bismuth, palladium and gold are readily extractable from acid solutions whereas those of zinc, cobalt and cadmium will require a neutral medium, but pH > 7 is needed for a quantitative transfer of lead and nickel[5, 6, 13].

TABLE 8A7. Equilibrium constants of dithizonates, at room temperature.

Metal dithizonate	CCl$_4$	CHCl$_3$
	log K_{11}	
AgHDz	7.60	6.00
TlHDz	−3.34	
Hg(HDz)$_2$	26.81	
Cu(HDz)$_2$	10.48	6.51
Zn(HDz)$_2$	2.18	0.64
Co(HDz)$_2$	1.56	−1.52
Ni(HDz)$_2$	−1.19	−2.92
Cd(HDz)$_2$	1.18	0.08
Pb(HDz)$_2$	−0.05	−1.20
Bi(HDz)$_3$	9.75	5.36
In(HDz)$_3$	4.84	
	log K_{12}	
Ag$_2$Dz	5.70	
HgDz	~ −3.00	
CuDz	2.30	

Secondary dithizonates are assumed to be formed according to the equilibrium

$$M^{m+} + m/2\,\overline{H_2Dz} \rightleftarrows \overline{MDz_{m/2}} + m\,H^+ \qquad (8A4)$$

Some equilibrium constants, K_{12}, for this reaction are included in Table 8A7. Whenever a metal forms both types of complexes, there should exist an equilibrium between the primary and secondary dithizonates, which can be represented by an organic-phase reaction

$$\overline{M(HDz)_m} \rightleftarrows \overline{MDz_{m/2}} + m/2\,\overline{H_2Dz} \tag{8A5}$$

with the equilibrium constant

$$K_{22} = [\overline{MDz_{m/2}}]\,[\overline{H_2Dz}]^{m/2}[\overline{M(HDz)_m}]^{-1} = K_{12}K_{11}^{-1} \tag{8A6}$$

The equilibrium between primary and secondary dithizonates will depend on the aqueous-phase conditions; transformation of the primary into secondary dithizonates is not favoured unless the organic phase is in equilibrium with a strongly basic aqueous solution, in order to facilitate the second ionization of H_2Dz.

(iii) 8-*Hydroxyquinoline and its derivatives* Oxine or oxyquinoline, a white crystalline powder, is a widely used precipitating agent in analytical chemistry[31]. Its solubility in water depends sharply on the pH of the aqueous solution[32], as illustrated in Figure 8A2. In the pH range 5.5-8.5 the reagent is difficultly soluble, and the steep increase in the solubility both in acid and alkali range is due to the amphoteric character of the reagent. The solubility of oxine is also low in ether and aliphatic hydrocarbons, but it is readily soluble in benzene, chloroform and alcohols, as seen from the data in Table 8A8.

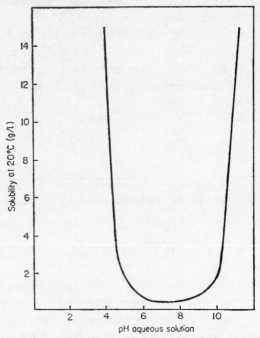

Figure 8A2. Solubility of oxine in water as a function of pH, 20°C.

TABLE 8A8. pK_{OH} and pK_{NH} of oxine and its derivatives in water and their solubilities in various solvents, at 25°C.

Reagent	Solvent	Solubility (g/l)	pK_{OH}	pK_{NH}
8-Hydroxyquinoline	Water	0.516	9.66	5.00
	Chloroform	381.3		
	Benzene[a]	172.6		
2-Methyl-8-hydroxyquinoline	Chloroform	80.41	10.30	5.55
5-Methyl-8-hydroxyquinoline	Chloroform		9.93	5.29
5,7-Dichloro-8-hydroxyquinoline	Chloroform	12.56	7.4	2.9
5,7-Dibromo-8-hydroxyquinoline	Chloroform	9.15	7.3	2.6
5,7-Diiodo-8-hydroxyquinoline	Chloroform	4.33	8.0	2.7

a) At 18°C.

Due to its amphoteric character, in acid solution the hydroxyquinolinium ion is formed $HOx + H^+ \rightleftarrows H_2Ox^+$, and in basic solution HOx undergoes dissociation according to $HOx \rightleftarrows Ox^- + H^+$. The values for the equilibrium constants $K_{OH} = [Ox^-][H^+][HOx]^{-1}$ and $K_{NH} = [HOx][H^+][H_2Ox^+]^{-1}$ are given in Table 8A8 for oxine and some of its methyl and dihalogen derivatives.

The distribution ratio of oxine is also affected by the amphoteric character of the reagent (cf. equations 7C23, a, b). Figure 8A3 illustrates the pH effect on the partition of oxine between chloroform and water. The extraction of the reagent is quantitative only in the pH limits between

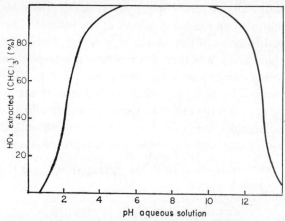

Figure 8A3. Extraction of oxine into chloroform as a function of pH.

6 and 9; more rigorous calculation[31, 33] indeed shows that around pH 7 the aqueous-phase species of the reagent is the molecular undissociated HOx, which is the only extractable form of it. The distribution constant, P_{HX} of oxine and its derivatives has recently been determined in a number of organic solvents[34]. The data are reproduced in Table 8A9.

TABLE 8A9. Distribution constant of oxine and its derivatives between water and chloroform.

Reagent	P_{HX}
8-Hydroxyquinoline	433
2-Methyl-8-hydroxyquinoline	1670
4-Methyl-8-hydroxyquinoline	1860
5-Methyl-8-hydroxyquinoline	1910
5-Acetyl-8-hydroxyquinoline	630
5,7-Dichloro-8-hydroxyquinoline	7240
5,7-Dibromo-8-hydroxyquinoline	14,100
5,7-Diiodo-8-hydroxyquinoline	14,100
5-Chloro-7-iodo-8-hydroxyquinoline	7600

The reactivity of oxine with a large number of metals is due to the formation of stable five-membered chelates through the replacement of the hydrogen from the acidic phenolic group and coordination to the nitrogen. The reaction is not selective and such chelates are formed with about fifty different metals. The reagent may be rendered somewhat more specific by introduction of either a methyl group in various positions to the nitrogen or of two halogens in the *ortho* and *para* positions to the phenolic group. The increased selectivity is due to steric hindrance, to chelation and/or to an increased acidity of the reagent.

Some metal oxinates of the normal type MOx_m, formed according to the equilibrium (8A1), have been shown to be non-extractable into chloroform and other water-immiscible non-polar solvents. They may be rendered extractable, however, under somewhat different experimental conditions, e.g. by a large excess of the reagent in the organic phase. In this case, the organic-phase complex has the composition $MOx_m(HOx)_n$. The formation of such chelate-solvates can be represented by the reaction

$$M^{m+} + (m + n) \overline{HOx} \rightleftarrows \overline{MOx_m(HOx)_n} + mH^+ \qquad (8A7)$$

with an equilibrium constant

$$K_{111} = [\overline{MOx_m(HOx)_n}] \ [H^+]^m [M^{m+}]^{-1} [\overline{HOx}]^{-(m+n)} \qquad (8A8)$$

The solvation of the MOx_m chelates with at most two HOx molecules, may be due to the basic character of the pyridine group, the basic nitrogen competing under certain conditions favourably with water for the coordination sites on the metal. There is apparently a steric restriction to this replacement, since in the alkaline earth series the stoichiometry of the organic-phase complexes is: $MgOx_2$, $CaOx_2.HOx$, $SrOx_2.(HOx)_2$ and $BaOx_2.(HOx)_2$.

The equilibrium constants K_{11} and K_{111} as determined by Stary[5,35] are given in Table 8A10.

TABLE 8A10. Equilibrium constants of oxinates in chloroform, at room temperature.

Oxinate	$\log K_{11}$	$\log K_{111}$
$BeOx_2$	$- 9.62$	
$MgOx_2$	-15.13	
$MnOx_2$	$- 9.32$	
$NiOx_2$	$- 2.18$	
$CuOx_2$	$+ 1.77$	
$PbOx_2$	$- 8.04$	
$PdOx_2$	$+15.00$	
$AlOx_3$	$- 5.22$	
$FeOx_3$	$+ 4.11$	
$GaOx_3$	$+ 3.72$	
$InOx_3$	$+ 0.89$	
$TlOx_3$	$+ 5.0$	
$BiOx_3$	$- 1.2$	
$LaOx_3$	-16.37	
$ThOx_4$	$- 7.18$	
$TiOOx_2$	$+ 0.9$	
$ZrOOx_2$	$+ 2.71$	
VO_2Ox	$+ 1.67$	
MoO_2Ox_2	$+ 9.88$	
$AgOx.HOx$		$- 4.51$
$CaOx_2.HOx$		-17.89
$ScOx_3.HOx$		$- 6.64$
$UO_2Ox_2.HOx$		$- 1.60$
$SrOx_2.(HOx)_2$		-19.7
$BaOx_2.(HOx)_2$		-20.9
$CoOx_2.(HOx)_2$		$- 2.16$
$ZnOx_2.(HOx)_2$		$- 2.41$
$CdOx_2.(HOx)_2$		$- 5.29$

The distribution constant of the oxinates is as much dependent on the nature of the solvent as is that of the reagent alone. Mottola and Freiser[34] have reported data for the copper chelate to show this effect. Their numerical values for $P = [\overline{\text{CuOx}_2}] [\text{CuOx}_2]^{-1}$ are given in Table 8A11. The distribution constant for the simple chelate species MOx_m is apparently only slightly temperature dependent but the equilibrium constant of the solvated species such as the $\text{CdOx}_2.(\text{HOx})_2$ is very much affected by temperature: $\log K_{111}$ has a value of -4.4 at $0°\text{C}$ and -7.6 at $50°\text{C}$[37].

Table 8A11. Distribution constant of oxine and copper oxinate in various solvents, at 25°C.

Solvent	P_{HOx}	P_{CuOx_2}
Toluene	162	181
o–Dichlorobenzene	303	580
Chloroform	433	3020
Carbon tetrachloride	116	112
Isobutyl methyl ketone	135	92
1–Butanol	45	142
3–Methyl–1–butanol	62	274
1–Octanol	92	466
Isoamyl acetate	175	95

(iv) *Oximes* Monooximes, such as α-benzoinoxime and salicylal-doxime, and the *vic*-dioximes, containing the rather amphoteric N—OH group are more selective extractants than the β-diketones or oxines, The best known is perhaps dimethylglyoxime, which is only sparingly soluble in acidic aqueous solutions. The solubility in organic solvents is also low, in non-polar solvents even lower than in water, as seen from the data given in Table 8A12. The solubility of α-benzildioxime is even lower,

TABLE 8A12. Solubility of dimethylglyoxime in water and organic solvents, at 25°C.

Solvent	Solubility (g/l)
Water	0.6264
Chloroform	0.0522
Carbon tetrachloride	0.0069
Butyl alcohol	6.496
Isoamyl alcohol	5.220

though that of α-furyldioxime is generally higher in both water and organic solvents. The dissociation constant of dimethylglyoxime in water is $pK_{HX} = 10.6$.

The *vic*-dioximes are sensitive reagents toward nickel, palladium, rhenium, molybdenum, tungsten, niobium, cobalt and copper.

(v) *Nitrosophenols and nitrosonaphthols* This group of reagents contains a C—OH group coupled with the N=O group and is a powerful chelating extractant, rather selective for certain metals. Mention should be made of *o*-nitrosophenol, *p*-nitrosocresol, *o*-nitrosoresorcinol, monomethyl ether, 1-nitroso-2-naphthol and 2-nitroso-1-naphthol. The hydroxyl group is acidic, the acid dissociation constants pK_{HX} in water of the last two are 7.63 and 7.24 respectively. The tautometric form containing —C=O and —C=N—OH may exist under certain conditions, but the extractable complex is unlikely to be derived from that form.

The usually highly coloured chelates of cobalt, iron, uranium(VI), palladium and thorium extracted into a variety of solvents[5] have a rather uncertain stoichiometry[13].

(vi) *Nitrosoarylhydroxylamines* The combination of an oxime group with a nitroso group on an aryl radical yields a useful chelating acid, e.g. *N*-nitrosophenylhydroxylamine or *N*-nitrosonaphthylhydroxylamine. Their ammonium salts, cupferron and neocupferron, are used as extractants. The reagents may exist in two forms,

$$\begin{array}{ccc} \text{—N—N} & & \text{—}^+\text{N=N} \\ |\quad\|| & \text{and} & |\quad\,| \\ \text{HO}\ \ \text{O} & & \text{O}^-\ \ \text{OH} \end{array}$$

nitrosohydroxy   amine oxide

the acidic character being mainly due to the amine oxide tautomer, and is the predominating species in aqueous solution at pH > 4. Above that hydrogen ion concentration the fraction of the nitrosohydroxy form increases. The ionization constant of cupferron at 25°C is $pK_{HX} = 4.16$. It is considerably water soluble (~ 0.4 M) at room temperature, and is non-extractable into chloroform and other organic solvents above pH ~ 8, as can be seen in Figure 8A4. The distribution constant of the unionized nitrosohydroxy form into chloroform is ~ 118 from $HClO_4$ solutions and ~ 132 from HCl, and may be even higher from H_2SO_4 aqueous solutions[37]. Above pH 3 the constant is practically independent of the nature and composition of the aqueous solution. It has a value of

142 into chloroform, 284 into ethyl acetate, 217 into butyl acetate, 65 into diethyl ether and 2200 into carbon tetrachloride. Neocupferron is expected to have higher distribution constants.

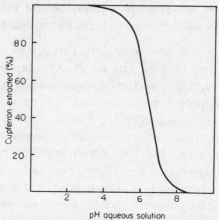

Figure 8A4. Distribution of cupferron between chloroform and water as a function of pH.

Metal extraction into chloroform and other solvents has been found to be quantitative for many metals under defined experimental conditions[5]. Separation of metals is frequently possible by a control of the aqueous pH and the reagent concentration. From acidic media the extraction of antimony and bismuth, copper and iron, molybdenum, thorium, tin, titanium, vanadium and zirconium can be made quantitative. Aluminium, zinc, cadmium, manganese, mercury and nickel are extractable from slightly acidic or neutral solutions. The equilibrium constants for extraction of some MX_m complexes into chloroform are listed in Table 8A13.

TABLE 8A13. Equilibrium constants for extraction of cupferrates into chloroform, at room temperature.

Metal cupferrate	Log K_{11}	Metal cupferrate	Log K_{11}
BeX_2	-1.54	TlX_3	$\sim +3.0$
CoX_2	-3.56	BiX_3	$+5.07$
PbX_2	-1.53	ScX_3	$+3.32$
CuX_2	$+2.66$	LaX_3	-6.22
HgX_2	$+0.91$	CeX_4	$+4.6^a$
AlX_3	-3.50	ThX_4	$+4.4$
FeX_3	$+9.85$	HfX_4	$> +8.0$
GaX_3	$+4.92$	UX_4	$+8.0^b$
InX_3	$+2.42$	PuX_4	$+7.0$

a) Butyl acetate.
b) Diethyl ether.

(vii) *Hydroxamic acids* Hydroxamic acids may have either one or two alkyl or aryl radicals in the molecules, depending on whether there is a hydrogen or an organic radical on the nitrogen. With hydrogen on it, the acidity of N—OH is usually weakened and the reagent becomes more aqueous soluble and much less organic soluble. This is the reason that the metal chelates of benzhydroxamic, salicylhydroxamic, aminobenz-hydroxamic and quinaldinohydroxamic acids are only sparingly extractable into chloroform or benzene, and more polar solvents such as alcohols and esters have to be used to improve the organic solubility of the complexes. On the other hand, when two alkyl or aryl radicals are on the molecules, as in *N*-benzoyl-*N*-thenoyl- or *N*-cinnamoyl-*N*-phenylhydroxylamines, their metal chelates are freely extractable into chloroform.

N-Benzoyl-*N*-phenylhydroxylamine is perhaps the most employed reagent of this group. It has a dissociation constant of $pK_{HX} = 8.15$ in water at 25°C.

(viii) *Dithiocarbamates* These extracting agents are frequently an alternative to dithizones. In certain cases dithiocarbamates may possess advantages over dithizones, arising from their higher aqueous solubility and a higher rate of chelate formation. Also metal dithiocarbamates have often a markedly higher solubility in the organic diluent, which enables the use of small volumes of solvent. This is advantageous for isolation and simultaneous concentration of the metal.

Sodium diethyldithiocarbamate is readily soluble in water (~ 0.2 M), and has an acid dissociation constant of $pK_{HX} = 3.35$. Its partition between an aqueous solution and chloroform or carbon tetrachloride is similar to that of cupferron shown in Figure 8A4; at low pH values the reagent will be completely in the organic phase, whereas at pH > 8 it remains entirely in the aqueous solution. The distribution constant of its undissociated molecule is 342 in CCl_4 and 2360 in $CHCl_3$.

The reagent decomposes rapidly in acid solutions, the rate of decomposition decreasing with increasing pH.

B. EXTRACTION BY ACIDIC ORGANOPHOSPHORUS COMPOUNDS

a. Introduction

Acidic organophosphorus compounds extract metals by a cation exchange reaction between one or two acidic hydrogens of the extractant and the extractable metal ion. In spite of the simple exchange reaction, basically identical with that valid for cation exchange resins (Chapter 5), the experimental data fit only occasionally into the mass-action equation of a simple exchange equilibrium. The main reason for a deviation from the ideal behaviour lies in the particular state of these extractants in solution. Unlike the acidic extractants, reviewed in the preceding section, which dissociate ionically in the aqueous phase and do not associate molecularly in the organic phase, organophosphorus acids have a very pronounced tendency toward association into dimers and larger aggregates. A no less complicating phenomenon is, furthermore, the solvation of the metal-extractant complex by one or more additional extractant molecules.

It should be emphasized that both the extent of polymerization of the extractant and the degree of solvation of the metal species are usually rather sensitive to various experimental parameters. These include the nature of the diluent (enhanced aggregation in aliphatic hydrocarbons), the concentration of the extractant and the metal loading of the organic phase (saturation results in lowering the solvation of the complex and even deaggregation of the extractant). These, basically unpredictable side-reactions can only seldom be satisfactorily expressed in the form of a mass-action equation and mathematically unequivocally treated.

Organophosphorus acids are often used to study the physical and complex chemistry of cations in aqueous solutions. Although the extraction by these acids is frequently difficult to control for reasons mentioned above, they have the advantage of a rapid extractions and usually high extraction capacity. Numerous analytical separations of metals are based on the use of these reagents and even separation of closely related elements was successful in some instances. The main interest in the development of these reagents was, however, closely connected with the object of finding organophosphorus compounds with considerable promise for practicable solvent extraction applications in the field of processing of spent reactor fuel. At an early stage of tributyl phosphate use for fuel reprocesing, it was recognized that the acid hydrolysis products of this reagent, mono- and dibutylphosphoric acids, were themselves effective metal extractants. This observation prompted a systematic survey of phosphoric acid derivatives as solvent extraction reagents.

b. The Extractants

The nomenclature of organophosphorus compounds is rather confused and apparently there is no commonly accepted terminology for the organophosphorus acids and their salts[38-40]. Without attempting to favour any of the terminologies in use, and without intending to add to the already existing confusion, we adopted for use in this monograph the best established names currently in use in the solvent extraction literature. Incidentally, these names are mostly in accordance with the 1952 Committee's recommendations[40]. Members, alkyl or aryl, or mixed alkyl–aryl of all the classes of compounds listed in Table 8B1 have been examined as potential solvent extractants.

A variety of these classes have been shown to be practical. Their preparation usually follows conventional organic synthetic techniques[38,39,41]. The synthesis of a number of new mono- and dialkylphosphoric, phosphonic and phosphinic acids has been described in the recent solvent extraction literature, and their physical properties[42-44] (density, viscosity, refractive index, melting or boiling point) reported. Both mono- and dialkylphosphorus esters have been prepared by dealkylation of the corresponding trialkyl derivatives[45-52].

TABLE 8B1. Acidic organophosphorus compounds.

Dialkylphosphoric acid	RO–P(=O)(OH)–OR	$(RO)_2PO(OH)$
Alkyl alkylphosphonic acid	RO–P(=O)(OH)–R	$(RO)RPO(OH)$
Dialkylphosphinic acid	R–P(=O)(OH)–R	$(R)_2PO(OH)$
Monoalkylphosphoric acid	RO–P(=O)(OH)–HO	$ROPO(OH)_2$
Monoalkylphosphonic acid	R–P(=O)(OH)–HO	$RPO(OH)_2$

1,2-Dialkylpyrophosphoric acid

$$O \leftarrow \underset{\underset{OH}{|}}{\overset{\overset{OR}{|}}{P}} - O - \underset{\underset{OH}{|}}{\overset{\overset{OR}{|}}{P}} \rightarrow O$$

$[ROPO(OH)]_2O$

Trialkylmethylenediphosphonic acid

$$O \leftarrow \underset{\underset{OR}{|}}{\overset{\overset{OR}{|}}{P}} - CH_2 - \underset{\underset{OH}{|}}{\overset{\overset{OR}{|}}{P}} \rightarrow O$$

$(RO)_2P(O)CH_2P(O)(OR)(OH)$

Dialkylmethylenediphosphonic acid

$$O \leftarrow \underset{\underset{OH}{|}}{\overset{\overset{OR}{|}}{P}} - CH_2 - \underset{\underset{OH}{|}}{\overset{\overset{OR}{|}}{P}} \rightarrow O$$

$[ROPO(OH)]_2CH_2$

Dialkylphosphorothioic acid

$$\overset{RO}{\underset{RO}{>}} P \overset{\nearrow O}{\underset{\searrow SH}{}}$$

$(RO)_2PO(SH)$

Dialkylphosphorodithioic acid

$$\overset{RO}{\underset{RO}{>}} P \overset{\nearrow S}{\underset{\searrow SH}{}}$$

$(RO)_2PS(SH)$

Dialkyl pyrophosphates, ranging from diethyl to didodecyl, have been synthesized from the corresponding alcohols and phosphoric acid anhydride[53-62]. The product obtained is a mixture of alcoholysis derivatives of phosphorus pentoxide rather than pure pyrophosphoric ester[55]. The yield in pure pyrophosphoric ester depends strongly on the experimental conditions, including the P_2O_5: alcohol ratio, the nature of the organic solvent used as the reactant medium[54,55], the extent to which water has been excluded and the temperature[58].

Only little is known on the preparation of the last four extractant types listed in the Table. Diphosphonates were synthesized by Grdenic[63,64] and Kennedy[65]. The preparation of the sulphur derivatives is apparently better known and some of them are even commercially available[66-69].

The phosphoric, phosphonic and phosphinic acids, both the mono- and diesters, are usually viscous yellowish liquids. Although they have a high boiling point, monobasic derivatives can be distilled under reduced pressure without excessive decomposition. The weakest point of the molecule with respect to thermal decomposition is the oxygen–aliphatic carbon bond.

The pyrophosphoric acids and the sulphur derivatives of the ortho-phosphoric acids are relatively unstable compounds deteriorating on standing. Pyrophosphates are easily hydrolysed when in contact with the aqueous phase, due to the hydrolysable pyrophosphate–oxygen bond. The methylene diphosphonates are, thus, expected to be more stable. The sulphur-containing esters are reported to be more stable than the dithiocarbamates, to which they are closely related by their chemical behaviour[68].

The organic synthetic procedures for the preparation of acidic organo-phosphorus compounds do not yield usually the desired reactant only. In addition to the simultaneous formation of both mono- and dibasic acids, the products are frequently contaminated with impurities such as the unreacted alcohols, neutral trialkyl phosphates and some other polyphosphorus or pyrophosphorus compounds[55]. These impurities, obviously, affect greatly the extractive properties of the reactants. Much attention has, thus, been given to the separation of the acids and their purification. Mono- and dialkyl phosphates have been separated by various methods. A fractional crystallization of their barium salts has been proved successful but time consuming[70,71]. Various ion exchange procedures using Dowex–50[72–74], Dowex–1[75,76] and Dowex–2[77] ion exchangers yielded, under well-defined experimental conditions, reagents of 99% purity or better. Dialkyl esters were shown to be preferentially absorbed on some metal oxides, such as Al_2O_3, ZrO_2 and SiO_2, through equilibration from organic solutions[78]. The method of purification most employed is based on the hydrolysis of polymeric materials and their subsequent extractive separation, which, in turn, takes advantage of the different water solubility of the acids. The general procedure of Stewart and Crandall[79] has been repeatedly modified to meet specific require-ments[42, 45, 80–83].

Mono- and dibasc alkyl acids can be identified by simple paper-chromatographic methods[55,81,84] and quantitatively determined using a cation exchange resin[85], potentiometrically[86] or colorimetrically with cacotheline[87].

The solubility of water in undiluted di-n-butylphosphoric acid at 25°C is 99.4 g/l[88–90], and 102 g/l in di-n-butylphosphonic acid[43]. The water content of the organic phase in equilibrium with water is much less if the extractant is dissolved in a water-immiscible diluent[88]. The water content of an organophosphorus acid organic phase often increases with increasing conversion of the acid to its salt by the extraction of metal ions. For example, the transfer of water into a di-(2-ethylhexyl)phosphoric acid–benzene phase increases linearly with increasing conversion of the acid to its sodium salt[91].

Low molecular weight acidic esters, those containing only methyl or ethyl groups, are miscible with water. Dibasic esters, even with long alkyl chain are still readily water soluble unlike the corresponding monobasic derivatives. The solubility of di-n-butylphosphoric acid in water is 0.082 mole/l at 25°C[89,90,92], and the presence of a branched alkyl chain in the extractant will keep its water solubility even lower[93]. The water solubility decreases, furthermore, when the extractant is diluted with an inert diluent[90], and apparently the extent of the extractant solubility depends on the nature of the inert solvent[86].

The solubility of the esters is lower in aqueous acid solutions than in pure water. In the case of dialkyl esters the aqueous solubility usually falls, up to a certain limit, as the ionic strength increases. This is illustrated by dibutylphosphoric acid which is soluble in 0.1 M nitric acid to the extent of about one third of that in pure water[90,92]. Solubility in sulphuric[95], phosphoric[95], hydrochloric and perchloric[94] acids is even lower. Branching in the alkyl chain of the dialkyl extractant again lowers its loss to the aqueous phase[49,61]. This decreased solubility in aqueous acid solutions can readily be understood as being due to a repressed dissociation of the extractant acid in the presence of hydrogen ions. Their increased solubility in the aqueous solution of lithium nitrate[96], instead of nitric acid, and their practically complete solubility in sodium hydroxide or carbonate[96], are in line with the above explanation. The distribution ratio of dibutylphosphoric acid between kerosene and aqueous alkali or carbonate solutions does not exceed 5×10^{-4}[96]. On the other hand, the solubility of monoalkyl esters, already high in pure water, does not seem to be much affected by the presence of either acids or alkali salts in the aqueous solution[86,96].

The replacement of a hydrogen by a non-acidic alkyl chain in the orthophosphoric acid molecule increases the acid strength of the derivative. The acidity of these esters (and also that of acidic phosphonic and phosphinic esters[98,99]) within a given class of reagents, as shown in Table 8B2, decreases slightly with increasing number of carbon atoms in the alkyl chain[71]. Their apparent pK values in ethanol–water mixtures decreases roughly with increased branching of the alkyl chain near the phosphorus atom[93,99].

TABLE 8B2. Acid dissociation constants of mono- and dialkyl esters
of orthophosphoric acid at 25°C.

Acid	pK_1	pK_2	Reference
In water			
Orthophosphoric	2.16	7.16	
Monomethyl	1.54	6.31	71
Monoethyl	1.60	6.62	71
Mono-n-propyl	1.88	6.67	71
Mono-n-butyl	1.89	6.84	71
Dimethyl	1.29		71
Diethyl	1.39		71
Diethyl	0.73		97
Di-n-propyl	1.59		71
Di-n-butyl	1.72		71
In 75% ethanol in water			
Orthophosphoric	4.0	8.4	98
Orthophosphoric	4.17		100
Mono–2–ethylhexyl	3.1	8.4	98
Mono-n-octyl	3.4	8.2	98
Mono–2–octyl	3.8	9.0	98
Monodiisobutylmethyl	4.1	9.3	98
Mono–3,5,5–trimethylhexyl	3.7	8.6	98
Mono–2,6,8–trimethylnonyl	4.2	9.6	98
Mono–4–ethyl–1–isobutyloctyl	4.2	9.6	98
Mono–3,9–diethyltridecyl	4.1	9.4	98
Monocyclohexyl	4.80		100
Monophenyl	3.96		100
Monophenoxy	3.13		100
Di–(2–ethylhexyl)	3.2		98
Di-n-octyl	2.9		98
Di–2–octyl	3.6		98
Di(diisobutylmethyl)	4.4		98
Di(di–3,5,5–trimethylhexyl)	3.0		98
Dicyclohexyl	5.92		100
Cyclohexyl-cyclohexoxy	4.73		100
Dicyclohexoxy	3.81		100
Cyclohexyl-phenyl	5.02		100
Cyclohexyl-phenoxy	3.60		100
Cyclohexoxy-phenyl	3.83		100
Cyclohexoxy-phenoxy	2.64		100
Diphenyl	4.10		100
Phenyl-phenoxy	2.85		100
Diphenoxy	2.28		100

c. Interactions of the Acidic Organophosphorus Compounds.

In the last few years a considerable amount of work has contributed to a fair understanding of acidic organophosphorus extractants, their nature and behaviour in solvents. Various physicochemical methods including cryoscopy, isopiestic measurements, spectral studies in the ultraviolet and infrared ranges, Raman and n.m.r. spectroscopy and distribution between two immiscible phases have provided a solid body of evidence that monobasic compounds are dimeric, while dibasic are predominantly polymeric in organic less solvents of low polarity. In accordance with the behaviour of hydroxylic compounds in a nonionizing medium, it is the hydrogen-bond association which gives rise to the formation of molecular aggregates.

Extensive freezing point studies on solutions of both alkyl and aryl esters[44, 51, 101—104], fortified by data obtained by an isothermal distillation method[102] and isopiestic measurements[93, 101, 105, 106] have revealed that the monobasic phosphoric, phosphonic and phosphinic esters are dimeric in the non-polar hexane, cyclohexane, benzene and carbon tetrachloride. They are monomeric in methanol and acetic acid but both species are present in acetone and chloroform at a solute concentration range of 0.01–0.06 M. The presence of water in the organic phase increases the cryoscopically determined molecular weight of monobasic acids, apparently participating in the hydrogen bonding.

Dibasic acids are highly polymeric in aliphatic hydrocarbons, the degree of aggregation apparently increasing with the solute concentration. In solvents such as benzene and carbon tetrachloride with equally low polarity and dielectric constant but a certain solvating power, the dibasic acids are apparently only hexameric, while they are dimeric in acetone, hexone and chloroform, and monomeric in methanol.

The extraordinarily strong intermolecular hydrogen bond in the acidic organophosphorus ester aggregates, hence their stability, has been repeatedly confirmed by i.r. spectral measurements under a variety of experimental conditions[44, 103, 107—119]. Hydrogen-bond association gives rise to a displacement toward lower frequencies of the ground-state stretching vibration in the i.r. spectral region of the free OH bond and of the phosphoryl oxygen. The free undisturbed $P \rightarrow O$ frequency in neutral trialkyl phosphates is about 1275 cm^{-1}. This band shows usually a slight displacement in case of participation of the phosphoryl oxygen in the hydrogen bond. This can be illustrated with the butyl derivatives: tributyl phosphate 1275 cm^{-1}, dibutylphosphoric acid 1230 cm^{-1}, monobutylphosphoric acid 1140 cm^{-1} and in orthophosphoric acid the band appears at 1120 cm^{-1}. The results of several monobasic alkyl- and arylphosphoric acids suggest that this band decreases—becomes more P$^+$ → O$^-$ in character—with increasing electronegativity of the alkyl group. For the acidic phosphonates, however, the opposite effect is observed with increasing electronegativity—an increase in the P = O character is apparent[103]. The acid strength increases with increasing electronegativity of the alkyl group for both types of extractants. The free OH frequency located at about 3500 cm^{-1} could not be observed in any of the compounds studied. Instead all show an absorption characteristic to bonded OH in the 2300–2700 cm^{-1} region. This indication for a rather strong intermolecular hydrogen bonding persists even at high dilution (5 × 10^{-5} M in carbon tetrachloride) and elevated temperatures (135°C). As could be expected, this band disappears from the spectrum of the salts of the corresponding acids.

On the other hand, hydrogen bonding in the sulphur analogues of the acidic organophosphorus compounds is considerably weaker. No cryoscopic evidence could be found for the dimerization of dibutyl phosphorodithioic acid[66, 120], while i.r. data indicate[116] that the dimer of dibutylphosphorothioic acid can be broken by dilution.

Raman spectra tend to support the above findings[109]. However, proton magnetic resonance studies[121] indicate a different state of solute in a carbon tetrachloride solution. The variation of the chemical shift as a function of the monobasic phosphate, phosphonate and phosphinate concentration has been explained in terms of a monomer ⇄ dimer ⇄ aggregate equilibrium. At mole fractions less than 0.005 the breaking of hydrogen bonds, while at high concentrations further growth of the dimer, has been postulated. A similar dependency exists with temperature. Comparison of several organophosphorus acids at identical experimental conditions reveals that the tendency toward aggregation increases with the acid strength of the extractant. A value of 5.5–5.8 kcal/mole per hydrogen bond has been estimated, which is higher than in the dimers of acetic acid, alcohols or phenols.

Actually, the possibility for an aggregation of acidic organophosphorus compounds beyond a dimer has not been excluded. Some cryoscopic[51, 101] and isopiestic measurements[101, 102, 106] and distribution data[105, 122, 123] indicate average molecular weights greater than the dimer in several acidic ester–diluent systems. On the other hand, no discontinuities or inflections were observed in the dielectric-constant values of a n-octane solution of di-(2-ethylhexyl)phosphoric acid measured over a range of concentrations[124]. The majority of these studies reveal a deviation from an ideal dimer behaviour not larger than 20%, so the actual amount of higher polymer must be very small. Consequently, a non-specific non-ideal behaviour of dimers rather than further association may be responsible for such a slight deviation.

There is a fairly general agreement[99, 103, 123] as to the structure of the unhydrated orthophosphorus compound dimers illustrated in 1 of Figure 8B1. While the water solubility in undiluted dibutylphosphoric[88–90] and dibutylphosphonic[43] acids amounts to an acid-to-water molar ratio of roughly unity, there is no direct evidence concerning the extent of water–ester interaction and the degree of hydration of such dimers. Structure 2 has been proposed for a dimer dihydrate[90]. For a hypothetical trimer[105] of monobasic extractants the ring structure 3, and for the conversion of the dimer into higher aggregates in form of an extended-chain[123] structure 4 have been proposed.

For the polymeric dibasic $(RO)PO(OH)_2$ and the analogues $(R)PO(OH)_2$ in non-polar solvents several structures have been discussed. One possibility, based on i.r. spectral considerations[113], assumes that each ester molecule is linked to its four neighbours (structure 5). Another suggestion[125] assumes that the polymers of dibasic esters are built up from 8-membered ring dimeric units which are apparently the basic units in the polar hexone[106]. The polymer of dimeric units might be either linear, as shown in structure 6, with two terminal OH groups, or, say, three dimeric units might be joined, end-to-end to form a cyclic hexamer. If such a polymer contains water molecules in its dimeric units, the structure illustrated in 7 is proposed[125]. Structures 5–7 have to be regarded as purely speculative. Nevertheless, the experimental fact of high water solubility of dibasic esters speaks in favour of the linear polymeric aggregate[104].

d. Distribution of Acidic Organophosphorus Compounds

The state of the organophosphorus extractant in an organic solvent affects its distribution between that phase and an aqueous electrolyte solution. The distribution, in turn, has an important influence on the overall distribution ratio of a cation between the two given phases. This early-recognized effect has been studied in recent years to a considerable extent and attempts have been made to interpret it in terms of mass-action law equations[48, 79, 88, 89, 92, 93, 96, 97, 122, 123, 126–131].

Figure 8B1. Proposed structures for organophosphoric acid polymers.

A number of equilibria have to be assumed in any two-phase dialkyl-phosphoric acid–diluent–aqueous acid system. At low acid concentration in the aqueous phase (where experience shows that HA.HX—HX is the extractant and HA the aqueous mineral acid—would be the only acid-bearing species in the organic phase) these equilibria could schematically be presented as in Figure 8B2, as similarly suggested by Hardy[132]. While there exists a reasonable amount of information on the values of K_2 and K_d for many dialkyl esters under a variety of conditions, little is known about the other equilibrium constants.

Figure 8B2. Schematic diagram of equilibria in an organophosphoric acid (HX)–diluent–aqueous acid (HA) system.

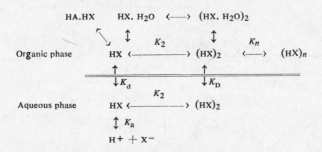

By fitting experimentally determined distribution ratios of dibutylphosphoric acid (obtained under such aqueous conditions that the fraction of HX bonded as the HA.HX complex is negligible in the organic phase) into a set of mass-action equations and assuming an ideal behaviour of the solute species in the organic phase, the dimerization constants K_2 have been evaluated in a number of water-immiscible organic solvents. Additionally, the graphical curve-fitting method enabled the authors to calculate the distribution constant K_d of the monomer. These values, listed in Table 8B3 reveal that K_2 decreases as K_d increases if the organic diluent is the only variable[129]. The highest K_2 values are characteristic of non-solvating inert diluents of the aliphatic hydrocarbon type. If log K_2 values are plotted against log K_d, as was done by Baes[122] and Dyrssen[123], the points fall close to a straight line of slope 2, so that the product $K_2K_d^2$ is fairly constant ($\sim 5 \times 10^3$) for a number of 'inert' aliphatic and aromatic solvents. The constancy of this product implies, by definition of the equilibrium constants involved, that the distribution constant of the dimeric species K_D should be independent of the nature of the organic diluent.

The correlation $K_2K_d^2 = const.$ is apparently true whenever the equilibria assumed in Figure 8B2 represent the actually occurring interactions, and the assumption of ideal behaviour in the organic phase is fair. In a number of systems, however, either of the assumptions, or both, are not true. An additional interaction, not shown in the scheme, $\overline{HX} + \overline{S} \rightleftharpoons \overline{HXS}$ with an equilibrium constant K_S may occur in the organic phase between the extractant acid, HX, and the organic diluent, S. Indeed, in a number of cases a mass-action law analysis of the distribution data suggests that a complex (usually 1:1) is formed with solvents such as chloroform or isobutyl methyl ketone, presumably through hydrogen bonding. Such interaction certainly reduces the concentration of the

TABLE 8B3. Dimerization and distribution constants for di-n-butylphosphoric acid in various two-phase systems.

Solvent	Aqueous phase	$\mathrm{Log}\,K_2$	$\mathrm{Log}\,K_d$	Reference
Kerosene	1 M HNO_3	5.78	−1.96	92
n-Hexane	0.1 M HNO_3	6.87	−2.34	123
Carbon tetrachloride	0.1 M HNO_3	6.48	−1.44	123
Carbon tetrachloride	1 M HNO_3	5.33	−0.92	92
Carbon tetrachloride	0.1 M $HClO_4$	4.85	−0.70	127
Benzene	1 M HNO_3	4.80	−0.42	92
Toluene	1 M HNO_3	5.09	−0.70	88
Chloroform	0.1 M $HClO_4$	4.20	0.34	92
Chloroform	0.1 M $HClO_4$	4.48	0.34	89
Chloroform	1 M $HClO_4$	4.36	0.28	92
Chloroform	0.1 M HNO_3	4.61	0.24	123
Nitrobenzene	1 M HNO_3	3.55	−0.13	92
Isopropyl ether	0.1 M HNO_3	2.29	0.52	123
Dibutyl ether	0.1 M $HClO_4$	3.18	−0.15	126
Isobutyl methyl ketone	0.1 M $HClO_4$	1.19	1.36	89
Isobutyl methyl carbinol	0.1 M HNO_3	< −1	2.21	123
Tri-n-butyl phosphate	0.1 M HNO_3		1.97	123
n-Octane(a)	0.1 M $HClO_4$	4.48		93
Water		1.11		89

(a) Di-(2-ethylhexyl)phosphoric acid.

dimer and may reduce even that of the uncomplexed monomer if $K_S > K_2$, as is the case of isobutyl methyl ketone and carbinol[123]. In all these cases the stabilization of the monomer, expressed through an increase in

K_d, was shown to be increasingly effective in polar solvents with donor or acceptor properties for hydrogen bonding.

Such a combination of an acidic extractant with a neutral compound, irrespective of its extractive power, is known to produce, frequently, enhanced extraction of cations. The phenomenon—termed synergism—plays an important role in many practical solvent extraction processes. It has received also wide attention from its fundamental physicochemical aspects. We will review and discuss the phenomenon separately in Chapter 11.

C. METAL EXTRACTION BY ACIDIC ORGANOPHOSPHORUS COMPOUNDS

a. Introduction

Although acidic organophosphorus reagents extract metals, essentially by a cation exchange reaction between the replacable hydrogen of the extractant and the aqueous metal ion, the individual extraction reactions, unlike the case of cation exchange resins, are only occasionally stoichiometrically equivalent. In the majority of extraction processes the dimeric dialkyl ester molecules entering into the exchange reaction are only monoionized, i.e. one of the hydrogens remains unreplacable. The total number of extractant molecules involved in the organic-phase metal complex depends—in a not always predictable way—on the charge and coordination number of the metal ion. While this is usually true under very low organic-phase loading conditions, under different conditions it is the charge on the metal ion which determines the number of extractant molecules in the organic phase complex. Less frequently the aqueous-phase anion may also affect the number of extractant molecules. This is apparent whenever the aqueous-phase anion competes favourably with the extractant's anion for the available cation. In such cases the extracted organic species is a mixed complex, in so far the metal ion is complexed by both the aqueous and the extractant anion. An extreme case may also occur, at high electrolyte concentration and high acidity: the charge of the metal is satisfied with the aqueous anion, meaning that no cation exchange reaction takes place. The extraction occurs in these cases via a solvation mechanism characteristic of non-ionic phosphorylated extractants, dealt with in section 9.D.

As the experimental information on the reaction by which acidic organophosphorus compounds extract metals has been accumulating, a systematization of the available material became possible and a generalization of the extraction processes justified. Each type of reaction results in a different organic-phase metal complex. The composition of these complexes, and the reactions leading to their formation will be discussed separately for mono- and dibasic alkyl esters in the following two sections.

In order to evaluate organophosphorus compounds as solvent extraction reagents, a number of survey studies has been carried out in recent years[49,127,132-135]. These investigations revealed a wide variety of possibilities for separation of cations, which, in turn, has made these reagents versatile and especially useful in the field of nuclear-fuel reprocessing.

Generally speaking, extractability of metals increases with increasing charge on the ion, while within a group of equally charged ions the distribution ratio varies inversely with respect to the ion radius. This is the general picture which emerges from the survey studies carried out with dialkyl monobasic extractants—di-(2-ethylhexyl)- and di-n-butylphosphoric acids—from various aqueous media. The monovalent ions thus, show the lowest extractability. The affinity of alkali metal ions for di-(2-ethylhexyl)phosphoric acid, DEHPA, in benzene increases with decreasing ionic radius, the reverse of that found between these ions and solid sulphonic acid type ion exchangers (Chapter 4). The distribution of monovalent ions seems to be largely controlled by the ease of ionic dissociation, since both silver and thallium(I) have distribution ratios at least one order of magnitude higher than the alkali ions. The alkaline earth ions yield under comparable experimental conditions values of D two orders of magnitude higher, their relative extractability decreasing in the order Ca > Mg > Sr > Ba. Divalent transition metal ions have usually somewhat higher D's, while divalent oxy-ions UO_2^{2+}, NpO_2^{2+} and PuO_2^{2+} have a distribution ratio about 10^4 greater than the alkaline earths. The tervalent lanthanide and actinide ions show a distribution ratio a few orders of magnitude higher ($> 10^3$) than the divalents, while the tetravalent zirconium, hafnium and thorium have a distribution ratio of at least 10^5 under identical conditions. Ce^{4+} and Bk^{4+}, however, have D values higher than the tervalent ions by a factor of as high as 10^6 when extracted from nitric acid medium by DEHPA in heptane. The extractability of both lanthanides and actinides increases with increasing atomic number though apparently not uniformly.

For example, the separation factor for the light adjacent lanthanides was found to be 1.95 as compared to 2.6 for the heavy ones[127, 142, 151]. Similar differences were obtained when rare earths were eluted from chromatographic columns in which di-(2-ethylhexyl)phosphoric acid has been retained on a solid support[152–154].

When high distribution ratios are measured, the extraction of metals remains unaffected by changing the nature of the macro electrolyte in the aqueous phase, though the extractability of transition metals and some actinides varies with the aqueous anion, sometimes to a considerable extent. As a rule, extraction of metals is better from such aqueous medium where the inorganic ligand does not compete, or competes less, against the extractant, for the metal ion. The situation is well illustrated by the distribution ratio of californium from 2 N perchloric, hydrochloric or nitric acid into 2-ethylhexylphenylphosphonic acid in diethylbenzene where the D values of 9.5, 1.1 and 0.9 were measured, respectively. Evidently, nitrate complexing is the most noticeable. Or, $D_{Fe^{3+}}$ at a given ionic strength and acidity, is about 10^4 from perchlorate, 2×10^3 from chloride but only 135 from sulphate solution.

b. Isolated Metal Complexes of Organophosphorus Compounds

A great variety of metal complexes with dialkyl esters of phosphoric acid (HX) have been prepared and isolated[57, 63, 85, 155–170]. In the majority of cases they are the salts of di-n-butylphosphoric acid containing the stoichiometric amount of HDBP required by the valence of the cation. They are prepared from the components, the metal carbonate or hydroxide and the acid, in alcoholic solutions. Frequently the method of hydrolysis of a metal-loaded tri-n-butyl phosphate phase results in a higher yield.

Apparently very stable, mixed nitrato–dialkyl ester complexes of zirconium have been prepared with di-n-butyl-[171], and of thorium, zirconium, hafnium and cerium(IV) with di-(2-ethylhexyl)phosphoric acid[168]. The first compound was analysed as $Zr(NO_3)_2 [(C_4H_9O)_2POO]_2$ and was thermally stable up to 165°C. The decomposition of the organic ligand occurs first, followed by the loss of nitrate, resulting finally in zirconium pyrophosphate.

Infrared studies carried out on great many[156, 160–162, 168] of these solid complexes reveal a number of interesting features. The shift of the asymmetric and symmetric frequencies of POO^- indicates a predominantly ionic link in the alkali and alkaline earth salts. A linear correlation has been found to exist between the shift in these frequencies and the ratio of charge/ionic radius. Lower shifts were observed as the ratio increases, which in turn coincides with an increase in the electronegativity of the metal. The correlation holds, however, only for the simplest cases, where the bond is ionic in character. The spectra of the majority of the salts investigated point to the covalent character of this link. Where a high degree of covalency is introduced in the metal-to-phosphate bond the frequencies are shifted toward higher values. The interpretation of the i.r. spectra becomes less dependable in the case of complexes involving metals with high charge. There is, namely, a strong tendency of uranium, zirconium, hafnium and thorium complexes to form polymeric species, apparently through the metal being bound to the oxygen of the P—O—C group.

Cryoscopic measurements on UO_2X_2 and CuX_2 (X = di-n-butyl phosphate anion) indicate that these complexes may indeed be highly associated into aggregates of rather complicated structure and a molecular weight higher than 10,000[85].

The fragmentary information available[57,158,159,161,163–165,167,169–171] on the solubility of metal–dialkyl ester complexes in water and aqueous electrolyte solutions points to a solubility of about 10^{-4} M or lower. Solubility of rare earth dialkyl phosphates[158, 161] in water decreases with increasing chain length of the alkyl radical: 1.734, 0.0161, 0.0033 and 0.00039 mole/l being the solubilities of LaX_3 at 25°, where X represents the dialkyl phosphate anion with 1, 3, 4 and 5 carbon atoms per chain respectively. The monoalkyl compounds exhibit, usually, a higher solubility. The sodium, and other alkali metal salts of both acids are considerably more water soluble while practically insoluble in aliphatic hydrocarbons. The solubility of other metal complexes in non-polar solvents is similarly low: UO_2X_2 is soluble at 25° to the extent of 1×10^{-5} M in anhydrous diethyl ether and n-hexane; saturated solution of YbX_3 in carbon tetrachloride is about 10^{-7} M at room temperature. The uranyl compound (more soluble) in benzene forms a viscous solution. The high viscosity was taken as evidence for a polymeric structure of the solute in which the uranium is octacoordinated[159].

The solubility of metal–dibutyl phosphate complexes is enormously increased in organic solvents upon addition of free dibutylphosphoric acid[159, 163–165]. In the case of uranyl dibutyl phosphate in benzene, the addition of HDBP is accompanied by a drastic reduction of the viscosity. The phenomenon is attributed to depolymerization of the solute and the formation of monomeric complexes of a composition similar to those formed in the organic phase during extraction. In a more quantitative investigation[163, 164], the solubility of YbX_3 in carbon tetrachloride was followed as a function of HDBP concentration. The solubilizing effect, expressed through the reaction

$$YbX_3 + m(HX)_2 \rightleftharpoons YbX_3.(HX)_{2m} \qquad (8C1)$$

shows on a log–log plot of the solubility vs. the dimer $(HX)_2$ concentration, a slope of 1.5, indicating the complex $Yb(HX_2)_3$, with a formation constant of 0.13. Further evidence for the correctness of the above solubilization reaction has been found in the effect the diluent has on it. Kriss[165], measuring the solubility of NdX_3 in a series of solvents at equal HDBP concentration, has found that the solubility decreased in the order: carbon tetrachloride > n-hexane > isooctane > benzene > diethyl ether > dichloroethane > n-octanol > n-butanol > n-propanol > pyridine > ethanol. There is roughly a two orders of magnitude difference in the solubility of NdX_3 in carbon tetrachloride and ethanol. The above sequence of solvents is similar to that established for the extent of interaction taking place between them and HDBP, as shown in Table 8B3. Thus, the stronger the interaction of the diluent with HDBP, the more successfully will the diluent compete with $Nd(DBP)_3$ for the available HDBP molecules.

c. Organic-phase Metal Complexes with Monobasic Dialkyl Phosphates

(i) *Complexes at low organic loading* The extraction of metals by dimeric dialkyl esters is usually fairly well represented by the reaction (cf. equation 7C32)

$$M^{m+} + m\overline{(HX)_2} \rightleftharpoons \overline{M(X.HX)_m} + mH^+ \qquad (8C2)$$

i.e. the extraction is proportional to the m-th power of the dimer concentration and inversely proportional to the m-th power of proton concentration, when m represents the actual positive charge on the metal ion. Assuming that $M(X.HX)_m$ is the only metal-bearing species in the organic phase and that M^{m+} is the only one in the aqueous phase, the stoichiometric equilibrium constant of the above extraction equilibrium

$$K_m = \overline{[M(X.HX)_m]} \, [H^+]^m [M^{m+}]^{-1} \, [\overline{(HX)_2}]^{-m} \qquad (8C3)$$

has been successfully evaluated for a large number of cations. Evidence has been reported to show that beryllium[132, 172, 173], iron[179], the lanthanons[80, 127, 149, 151, 164, 174–179], and the actinides[68, 144, 159, 180] extract from aqueous solutions by di-n-butyl- or di-(2-ethylhexyl) phosphoric acid in accordance with equation (8C2). Thus the results of many of the systems investigated substantiated the above assumptions on the metal-bearing investigated substantiated the above assumptions on the metal-bearing species in the equilibrium phases. However, in systems where ions were extracted from solutions in which complex formation takes place, the assumption that M^{m+} is the only metal-containing aqueous species is no longer warranted. In these cases the experimentally determined distribution data do not fit into equation (8C3), and a set of equations which takes into account aqueous complexing of the metal ion has to be used. Such set a of equations, (7C34–7C36), has been derived rigorously in section 7.C. Indeed, in several cases[173, 183, 186, 191] constants for metal complexing, occuring in the aqueous phase, have been introduced to account correctly for the term $[M^{m+}]$ in equation (8C3). More often, however, there is only qualitative evidence for this aqueous complexing effect[49, 133, 147, 193, 194].

It should be noted here, that in extreme cases aqueous complexing of metal ions may govern the extraction to such an extent that the resulting organic-phase species is a mixed complex. The mechanism of extraction of such metal complexes is considered to be qualitatively different and will be discussed separately in this section.

The stoichiometric equilibrium constant given in equation (8C3) has been evaluated for a number of lanthanide, actinide and other metal complexes. The available information is compiled in Table 8C1. Some reported K values have not been included because there was doubt about the correctness of assumptions made in their evaluation[164, 174, 175, 183, 184]. Baes[147] reported K values for a large number of cations, evaluating them frequently from fragmentary experimental data. In a number of cases the values were additionally corrected and 'standardized' for di-(2-ethylhexyl)phosphoric acid as extractant and toluene as diluent, when the original data refered to another extractant or diluent, or both. Baes' tabulation of the K values, although numerically not necesarily correct is very informative. The increasing trend in the stability of complexes in going from mono- to di- to tervalent ions is well illustrated.

The monomeric organic-phase complex, $M(X.HX)_m$, resulting from the interaction of m single-ionized HX-dimers and the metal ion has been postulated to have an 8-membered ring structure

TABLE 8C1. Values of the equilibrium constant K_m for various metal ions with monobasic dialkylphosphoric acids (HX).

Ion	HX	Diluent	Concentration range (M) Aqueous		Organic	$LogK_m$	Ref.
Be^{2+}	Dibutyl	Toluene	0.25	HNO_3	0.05–0.5	1.55	173
Be^{2+}	Di–(2–ethylhexyl)	Toluene	0.25	HNO_3	0.05–0.5	1.28	173
Be^{2+}	Diisooctyl	Toluene	0.25	HNO_3	0.05–0.5	1.45	173
Be^{2+}	Di-n-octyl	Toluene	0.25	HNO_3	0.05–0.5	1.71	173
Be^{2+}	Diisodecyl	Toluene	0.25	HNO_3	0.05–0.5	1.48	173
Be^{2+}	Di-n-decyl	Toluene	0.25	HNO_3	0.05–0.5	1.73	173
Be^{2+}	Di–(1–isobutyl–3,5-dimethylhexyl)	Toluene	0.25	HNO_3	0.05–0.5	1.78	173
Fe^{3+}	Di–(2–ethylhexyl)	n-Octane	0.5–2	$HClO_4$	0.01–0.3	6.54	179
Y^{3+}	Di-n-butyl	Chloroform	0.1	HNO_3	0.003–0.1	3.23	176
Y^{3+}	Diisoamyl	Benzene	0.2–2	HNO_3	0.5–1	2.49	178
Y^{3+}	Di-n-butyl	Ether	0.1	$HClO_4$	0.001–0.5	5.08	149
La^{3+}	Di-n-butyl	Ether	0.1	$HClO_4$	0.001–0.5	1.30	149
La^{3+}	Di-n-butyl	CCl_4	0.1	$HClO_4$	0.001–0.5	1.70	149
Ce^{3+}	Di-n-butyl	Ether	0.1	$HClO_4$	0.001–0.5	1.90	149
Ce^{3+}	Di-n-butyl	CCl_4	0.1	$HClO_4$	0.001–0.5	2 00	149
Pr^{3+}	Di-n-butyl	Ether	0 1	$HClO_4$	0.001–0.5	2.00	149
Pm^{3+}	Di-n-butyl	Ether	0.1	$HClO_4$	0.001–0.5	2.40	149
Sm^{3+}	Di-n-butyl	Ether	0.1	$HClO_4$	0.001–0.5	2.70	149
Eu^{3+}	Di-n-butyl	Ether	0.1	$HClO_4$	0.001–0.5	3.08	149
Eu^{3+}	Di-n-butyl	CCl_4	0.1	$HClO_4$	0.001–0.5	3.60	149
Gd^{3+}	Di-n-butyl	Ether	0.1	$HClO_4$	0.001–0.5	3.30	149
Gd^{3+}	Di-n-butyl	CCl_4	0.1	$HClO_4$	0.001–0.5	3.81	149
Tb^{3+}	Di-n-butyl	Ether	0.1	$HClO_4$	0.001–0.5	4.08	149
Ho^{3+}	Di-n-butyl	Ether	0.1	$HClO_4$	0.001–0.5	4.90	149
Ho^{3+}	Di-n-butyl	CCl_4	0.1	$HClO_4$	0.001–0.5	5.40	149
Tm^{3+}	Di-n-butyl	Ether	0.1	$HClO_4$	0.001–0.5	5.90	149
Lu^{3+}	Di-n-butyl	Ether	0.1	$HClO_4$	0.001–0.5	6.81	149
Lu^{3+}	Di-n-butyl	CCl_4	0.1	$HClO_4$	0.001–0.5	7.40	149
Pu^{3+}	Di-n-butyl	Ether	0.1	$HClO_4$	0.001–0.5	2.60	149
Pu^{3+}	Di-n-butyl	CCl_4	0.1	$HClO_4$	0.001–0.5	2.90	149
Am^{3+}	Di-n-butyl	Ether	0.1	$HClO_4$	0.001–0.5	1.89	149
Am^{3+}	Di-n-butyl	CCl_4	0.1	$HClO_4$	0.001–0.5	2.30	149
Cm^{3+}	Di-n-butyl	Ether	0.1	$HClO_4$	0.001–0.5	1.89	149
Cf^{3+}	Di-n-butyl	Ether	0.1	$HClO_4$	0.001–0.5	3.50	149
Es^{3+}	Di-n-butyl	Ether	0.1	$HClO_4$	0.001–0.5	3.30	149
UO_2^{2+}	Di–(2–ethylhexyl)	n-Hexane	0.4–2	$HClO_4$	0.05–1.0	4.60	185
UO_2^{2+}	Di–(2–ethylhexyl)	Kerosene				4.53	189
UO_2^{2+}	Di-n-butyl	Toluene	1	HNO_3	0.001–0.1	4.56	189
UO_2^{2+}	Di-n-butyl	Chloroform	1	$HClO_4$	0.001–0.5	3.58	185
UO_2^{2+}	Di-n-butyl	Hexanone	1	$HClO_4$	0.001–0.5	4.50	185
UO_2^{2+}	Di–(β–naphthyl)	Chloroform	1	$HClO_4$	0.0001–0.001	5.24	188
UO_2^{2+}	Dibenzyl	Chloroform	1	$HClO_4$	0.0001–0.001	4.88	188
UO_2^{2+}	Di–(p-tolyl)	Chloroform	1	$HClO_4$	0.0001–0.001	4.60	188
UO_2^{2+}	Diphenyl	Chloroform	1	$HClO_4$	0.0001–0.001	4.87	188

$$\left[\begin{array}{c} \text{(OR)}_2 \\ | \\ \swarrow P \diagdown \\ O \qquad O \\ \vdots \qquad | \\ M^{m+} \qquad\qquad H \\ \diagdown O \qquad \vdots \\ \qquad O \\ \diagup \\ P \nearrow \\ | \\ \text{(OR)}_2 \end{array} \right]_m$$

This chelate structure accounts for the stoichiometry deduced from the observed extraction-ratio dependencies expressed in equation (8C2). It satisfies also both the charge and coordination number requirements of the metal ion as long as the latter is $2m$. This is, however, not always the case.

Indeed, data are now available[148, 195–198] to show that the extractant dependency for certain metals may be greater than the m-th power. Distribution of tracer amounts of alkali and alkaline earth metals between an aqueous chloride, nitrate or perchlorate solution of constant ionic strength and xylene or benzene solution of various dialkyl- or alkylarylphosphoric acid and phosphonic acids indicates an extractant dependence markedly different from the usual stoichiometry of equation (8C2) found in most extracted metal species. D_{Na} shows a second-power dependence on $[(\overline{HX})_2)]$ and a negative first-power on $[H^+]$[197]; D for calcium[195], strontium[195, 197], barium and europium(II)[195, 196] varies with the third power of $[(\overline{HX})_2]$ and inversely with the second power of $[H^+]$, when di-(2-ethylhexyl)phosphoric acid was used as extractant. While the hydrogen ion dependency is not affected by using (1,1,3,3-tetramethylbutyl)phenylphosphoric acid rather than HDEHP, the extractant dependency is reduced to a 2.5 power for D_{Ca} and D_{Sr} but not for D_{Ba}[195].

Peppard and coworkers[141, 195, 196] suggest that the extraction involving a third-power extractant-dependence takes place according to the reaction

$$M^{2+} + 3\overline{(HX)_2} \rightleftarrows \overline{M(HX_2)_2(HX)_2} + 2H^+ \qquad (8C4)$$

while in the case of a 2.5-**power** dependence observed for calcium and strontium in some of **the system**s investigated, the organic complex formed should be formulated as $M(HX)_2(HX)$. The lower-power dependence has been explained in terms of steric factors.

McDowell and Coleman[197] interpret their identical experimental data in terms of equations

$$Na^+ + 2\overline{(HX)_2} \rightleftarrows \overline{NaX.3HX} + H^+ \qquad (8C5)$$

and

$$Sr^{2+} + 3\overline{(HX)_2} \rightleftarrows \overline{SrX_2.4HX} + 2H^+ \qquad (8C6)$$

or in a generalized form

$$M^{m+} + (n/2)\,\overline{(HX)_2} \rightleftarrows \overline{MX_m.(n-m)HX} + mH^+ \qquad (8C7)$$

pointing out that in the formulation suggested 'no special significance is given to hydrogen bonding between ligands within the complex', which may be implicitly assumed in Peppard's formulation. On the other hand, the model of McDowell and Coleman advert to the importance of satisfying the coordination requirements of the metal in addition to satisfying its charge requirements.

In other cases, where the experimental data do not satisfy the requirements of equation (8C3) the reason still lies with the assumptions made on the organic-phase metal-bearing species. Apart from the non-specific non-idealities of the organic-phase solutes $(HX)_2$ and $M(X.HX)_m$, it has been demonstrated that the main reason for the departure from ideality is due to formation of additional or other metal-bearing organic-phase complexes. Experience has repeatedly shown that equilibrium (8C2) governs the transfer of the metal ion only when a) its concentration is very low in comparison to that of the extractant in the organic phase, b) the metal is extracted from low ionic strength aqueous phases, and c) the predominating species of the acidic ester in the diluent is dimeric.

Practically, these conditions are met only in systems where the metal is present as tracer, and the concentration of the supporting mineral acid in the aqueous phase is kept considerably below 1 M. Under changed experimental conditions, where the ratio $[\overline{M(X.HX)_m}]/[\overline{(HX)_2}]$ in the system is higher than 10^{-3} and/or the composition of the aqueous electrolyte solution is such as to affect the concentration of $(HX)_2$ in the equilibrium organic phase, the extraction of metal ions is at variance with equilibria (8C2) and (8C3).

The nature of the diluent may drastically affect the dimerization of dialkylphosphoric acids in the sense discussed in the previous section. Whenever the monomer is the predominating species in the organic phase, and this is the case when polar diluents are employed, metal extraction follows the reaction

$$M^{m+} + m\,\overline{HX} \rightleftarrows \overline{MX_m} + m\,H^+ \tag{8C8}$$

As a rule, the extractability of MX_m is markedly lower than that of $M(X.HX)_m$[177, 189, 192]. For example, D_{Eu} is five orders of magnitude lower when the metal is extracted by dibutylphosphoric acid into hexanol than when extracted into hexane, under otherwise identical experimental conditions[177]. As one would expect, the monomer $\rightleftarrows$ dimer equilibrium of the extractant may bring about the coextraction of both types of metal complexes, MX_m and $M(X.HX)_m$. The phenomenon will be discussed in more detail later on (solvent effect).

(ii) *Complexes at saturation* It has been mentioned that at low extractant concentration in the organic phase, and/or an increased metal transfer into it, the species $M(X.HX)_m$, formed according to reaction (8C2) will, no longer be the predominant metal-bearing species in that phase. Approaching saturation of the organic phase other monomeric and polymeric species will coexist in equilibrium

There is not sufficient evidence as to the nature of the organic-phase reaction as it becomes increasingly loaded with metal. In a variety of systems, however, the composition of the saturated organic phase has been determined. The combining ratio of ester-to-metal approaches 2:1 when beryllium[172], strontium[197], copper, cobalt and nickel[193] and uranium(VI)[185] are extracted from mineral acid solutions. The ratio is three for iron[179] and the lanthanides[166, 174], and four in the case of thorium[168, 200]. Under certain experimental conditions a gel or a precipitate is formed in the systems when approaching these limiting ratios. The empirical composition of the solids isolated corresponds to that of simple metal salts discussed earlier in this section.

The process of saturation of the organic phase occurs probably via a mechanism of replacement of the protons in the single-ionized dimers in $M(X.HX)_m$ by a metal atom. The ejection of a solid phase from such systems when approaching saturation suggests that the MX_m compound is polymeric. Indeed, isopiestic measurements supplied evidence that the $UO_2(DEHP)_2$ compound exists in a hexane solution as a chain polymer[186]. A similar structure has been attributed to the beryllium compound of the same acid in kerosene[172]. Isopiestic measurements have shown[148, 197–199] that an increasing polymerization occurs in a mixture of di-(2-ethylhexyl)phosphoric acid and its sodium salt in benzene as the fraction of the salt increases. NaX has an average polymerization number of 13 at 0.1 M, but about 50 at 0.25 M concentration in benzene at 20°C. It may be instructive to note the unusually high water content of 5 H_2O: 1 NaX in the organic phase; the amount of water in the benzene phase is proportional to its NaX concentration. It would be of interest to have hydration data on other MX_m complexes in organic solvents, in as much as the $M(X.HX)_m$ complexes are apparently unhydrated in the organic phase[147].

(iii) *Mixed complexes* The influence of aqueous anion complexing of the extractable metal, resulting usually in lowering the *D* value, has been illustrated earlier in this section. It has been mentioned also that in cases where the aqueous complexing leads to the formation of strong cationic complexes the extraction of the metal cannot be adequately described by equation (8C2). If, now, such a cationic complex is the predominant metal-bearing species in the aqueous phase, the metal may be extracted in the form of a mixed complex, containing both the aqueous anion and the dialkyl phosphate anion in terms of equations (7C34) and (7C35) discussed in section 7.C. While a few exceptions are known[172, 179, 201], the existence of mixed organic complexes are characteristic to tetra- and pentavalent cations[42,103,132,141,181,182,184,191,202—209] readily forming MA^{3+}, MA_2^{2+}, MA_3^+ or similar complexes (A = Cl^-, NO_3^-, ClO_4^- or OH^-) in aqueous acid solution.

The extraction of thorium has been investigated in considerable detail under a variety of conditions, to allow an illustration of the possible ways mixed complexes are formed in the ester phase[103,141,168,182,191,202—205]. Apparently, the stoichiometry of the organic-phase complex, the number of inorganic and ester ligands participating in it, will depend on all the possible parameters of such multicomponent extraction systems:

a) The in aqueous-phase parameters will define the central ion in terms discussed in section 3.C. Though the inorganic ligands are simple monodentate anions A^- a series of cationic complex species ThA_n^{4-n}, and even mixed hydroxo – ligand complexes, can be formed, and consequently, should be taken into account.

b) When the aqueous metal concentration exceeds about 0.1 M the results suggest that polymerization of the organic-phase species may occur[204, 205]. For a rigorous treatment of distribution data obtained at macro thorium concentrations, activity coefficients should be well controlled.

c) The structure of the dialkyl ester will primarily affect its acidity but has also a slight effect from purely steric considerations.

d) The diluent, its polarity and donor capacity, will affect the extent of dimerization of the extractant molecules and their association with the diluent molecules. The K_a, K_d and K_2 values, defined in section 8.B, for an ester–diluent pair, in equilibrium with the appropriate aqueous solution, should also be involved in a rigorous interpretation of the distribution data.

None of the published reports considers these complexities fully. The closest is perhaps that of Dyrssen and Liem[205], where the experiments were carried out at constant aqueous medium, but the aqueous complex chemistry of thorium has not been taken into account. As a result, a wide variety of mixed complexes has been claimed to exist in the organic phase.

Qualitatively speaking, the concentration of the inorganic ligand will affect the number of A^- anions in the organic-phase thorium complex. The more pronounced nitrate complexing of thorium is apparent from a larger number of nitrate ligands attached to the metal, than either chloride or perchlorate at equal ionic strength of the aqueous solution.

When polar diluents are employed, it is predominantly the monomeric extractant, while in kerosene and hexane, it is apparently the monoionized dimer, which neutralizes the remaining charge on the ThA_n^{4-n} complex. At saturation, the thorium-bearing species is the solvated ThA_4 from chloride or nitrate solutions, but it is ThX_4 when the extraction from sulphate solutions is carried out.

The available information on the distribution of zirconium[103, 157, 168, 171, 182, 206], hafnium[103, 207] and protactinium[42, 182, 208–209] from nitric acid, is similar to that of thorium in so far as several nitrato species have been identified in the extract. The situation is probably more complex due to the hydrolysed species of the elements in the aqueous solution.

(iv) *Solvated complexes* It is also possible for some metal nitrates and various mineral acids to be extracted via a solvation mechanism characteristic of non-ionic neutral phosphorylated extractants. Beryllium[132, 173] and uranyl[132, 159, 189] nitrate have been shown to extract from aqueous solutions of > 3 M HNO_3 by dialkylphosphoric acid, through solvation by a dimer, as $Be(NO_3)_2(HX)_2$ and $UO_2(NO_3)_2(HX)_2$. This in spite of the fact that at lower acidities the organic complex of both metals has been identified as $M(X.HX)_2$. Rhenium(VII)[210, 211] and technetium(VII)[211–213] are well extracted from nitric acid solutions by di-n-butylphosphoric acid in the form of their solvated oxy acids. Phosphoric acid[214] is transferred into the HDBP phase forming a solvate complex with a 1:1 combining ratio which is apparently dimerized. Surprisingly enough, hydrochloric acid seems not to form such solvates. Its distribution ratio does not exceed 0.005 in an aqueous acid range between 1 and 8 M. This is the case at least with (2-ethylhexyl)phenylphosphonic acid as the extractant[144].

The relatively high extractability of nitric acid, being of particular importance, has been investigated into considerable detail[88,90,215,216]. The composition of the organic-phase solvate varies with the amount of the acid extracted. At low acid loadings the hydrated dimer of di-n-butylphosphoric acid solvates the acid, and the complex $HNO_3.(HX.H_2O)_2$ is the predominant species. The extent of hydration, however, of both the free dimer and the nitric acid solvate is markedly dependent upon the nature of the organic diluent and the concentration of HDBP. In equilibrium with up to about 13.5 M acid the organic-phase species is essentially $HNO_3.HX.H_2O$, again the extent of hydration depending on the diluent and the extractant concentration. At still higher acid concentrations, $(HNO_3)_2.HX.H_2O$ becomes the important species in the organic phase. Infrared spectral data on di-(2-ethylhexyl)phosphoric acid–nitric acid extracts indicate that the nitric acid transferred into such organic phase is predominantly of a molecular nature, with bondings to the phosphoryl oxygen[217].

d. Organic-phase Metal Complexes with Dibasic Monoalkyl Phosphates

While extraction by dialkyl phosphorus compounds is reasonably well understood with respect to the reaction involved, not much progress towards an adequate understanding of the basic chemistry of extraction by dibasic monoalkyl phosphorus compounds has been noted. In view of the nature of this type of extractant and its state in solution, discussed in section 8.B, extraction studies have proven especially difficult.

Qualitatively speaking, water-insoluble monoalkyl derivatives of phosphoric acid can be reasonably powerful extractants under appropriate experimental conditions. Beryllium[173] from chloride, nitrate and sulphate solutions, zinc, cadmium, scandium, iron(III), indium(III) and tin(II) and (IV) from hydrochloric acid and sulphuric acid[218–220] have been extracted, and a method for the extractive separation of indium from various contaminating elements described[220]. Various rare earths from chloride[226] and nitrate[125,225] solutions and the actinides in their various oxidation states[42,125,206,207,209,222–224,228,229] have been surveyed for their extractability by different monoalkylphosphoric acids (Appendix F).

The distribution ratio of beryllium[173], the lanthanides[125,226,229] and several ter-, tetra- and hexavalent actinides[125,140,141,222—224,228,229] when extracted from dilute acid solutions ($< 3M$) by monoalkylphosphoric acid dissolved in aliphatic or aromatic hydrocarbon diluents, varies with the first power of the extractant concentration irrespective of the charge on the cation, m, but with the negative mth-power of hydrogen ion concentration. Thus the generalized equation

$$M^{m+} + \overline{(H_2X)_x} \rightleftarrows \overline{M(X_m.H_{2x-m}X_{x-m})} + mH^+ \qquad (8C9)$$

describes the extraction of metals by the polymerized species of monoalkylphosphoric acid. The postulated empirical formula $M(X_m.H_{2x-m}X_{x-m})$ implies that the polymeric species of the extractant contains an unknown and undetermined number of monomers, and, more important, that the number of monoionized dibasic acids in the polymeric entity is equal to the charge, m, of the extractable metal cation. Evidence that the latter assumption is at least formally true can be found in extraction experiments where the extractant has been dissolved in polar diluents, such as hexanone and the alcohols, rather than in hydrocarbons. It will be recalled that in alcoholic solvents the extractant is predominantly monomeric, while it is dimeric in hexone. While $D_{Eu^{3+}}$ and $D_{Am^{3+}}$ varies with the first power of $ROPO(OH)_2$ or $RPO(OH)_2$ concentration if toluene is the diluent, the dependence becomes third power using n-decanol or hexone as the organic solvent[106,125]. Similarly, Warren and Suttle[225], using amyl alcohol solvent for a number of mono-(n-alkyl)phosphoric acids ranging from propyl to octyl, found that D_{Sc} varies with the third power of the extractant concentration*. In all these cases $D_{Me^{3+}}$ varied with the negative third power of hydrogen ion concentration. The extraction of Eu, Sc and Am by monomeric extractant can thus be represented by the equation

$$M^{3+} + 3\overline{H_2X} \rightleftarrows \overline{M(HX)_3} + 3H^+ \qquad (8C10)$$

---

* In similar systems D_{La} and D_Y vary with the fourth power of the extractant concentration. The results of lanthanum and yttrium, however, were obtained using a range of extractant concentration higher by about two orders of magnitude. In view of the effect the extractant concentration has on the state of aggregation and dissociation, the latter results should be not directly comparable with those of scandium.

suggesting again an interaction of the metal cation with monoionized dibasic alkyl phosphates.

Under experimental conditions of high organic-phase loadings the available information on the organic-phase complex is contradictory. Monododecyl- or heptadecylphosphoric acid dissolved in kerosene can be loaded up to an ester-to-thorium ratio of four, a limiting ratio identical to that obtained with the monobasic di-(2-ethylhexyl)phosphoric acid[200]. Similarly, the extractant: uranium(VI) ratio at saturation appears to be two[230]. The results imply that the participating dibasic acid is only monoionized. On the other hand, dodecylphosphoric acid in kerosene can be saturated after a long equilibration time with uranyl sulphate up to a ratio of 1.65 acid per uranium, and with ferric iron from nitrate, chloride or sulphate media, up to a ratio corresponding to the compound Fe_2X_3, where X is the dinegative dodecyl phosphate ion[231,232]. Spectrophotometrically and conductometrically a 1 : 1 complex has been identified between UO_2^{2+} and monoisoamylphosphoric acid in alcoholic solutions[166].

Mention has to be made of acidic diphosphorly reagents of the type OH(RO)OP—Y—PO(OR)OH, where Y may be oxygen or methylene, which act, at least formally, as dibasic extractants. It was legitimately assumed that such bidentate extractants will be capable of forming chelated complexes, which in turn should be readily extractable. In many events, indeed, high distribution ratios have been obtained[57,58,60,64,233—235]. While apparently some of the diphosphoryl reagents form bidentate chelates[57,58,64,234] in the organic phase with a stoichiometric composition of $M(DP)_{m/2}(H_2DP)_x$, where DP stands for the dinegative diphosphoryl ion, m is the charge on the metal ion and x may have a value of 0, 1 or 2, the enhanced distribution ratios obtained are not necessarily due to a chelated structure of the extracted complex. It has been shown[55,236] namely, that the reagents prepared by the conventional technique are not pure pyrophosphates but a complex mixture containing often as many as ten components. Since pure dioctylpyrophosphoric acid is less effective by two orders of magnitude, as far as uranium(VI) extraction is concerned, than the mixture obtained by synthesis, a synergistic effect may be responsible for the high D values measured (cf. Chapter 11).

e. Extraction Characteristics of the Organic Phase

(i) *Solvent effect* In section 8.B attention has been drawn to the effect of polarity and the dielectric constant properties of the carrier organic solvent on the state of mono- and dialkyl phosphorus extractants in solution. It has been shown that the extent of polymerization or dimerization, respectively, is related to the extent of extractant–diluent interaction. For the dialkyl acids the interaction has been expressed quantitatively through the values of K_2 and K_d, the dimerization and monomer distribution constants of the acid ester. There is now ample evidence [42,125,147, 162,165,173, 177,189,192,229, 236,236–241] that the extent of this interaction affects in turn the extraction of metallic species. The effect is certainly complex, depending to an unknown extent on the dielectric constant, dipole moment and polarizability of both the diluent and the complex, and their hydrogen-bond donor or acceptor abilities.

As mentioned earlier, the effect of diluent may be so drastic as to influence the composition of the organic-phase metal complex. In diluents where the extractant is predominantly monomeric, the composition of the complex is invariably MX_m rather than $(M(X.HX)_m$[177,189,192,236] which is the complex composition in non-polar solvents. Since the composition of the species for a given metal depends on K_2 of the dialkylphosphoric acid in the diluent, one is tempted to assume that in diluents where K_2 has a relatively small value (methyl isobutyl ketone) HX and $(HX)_2$ coexist in equilibrium, and both species may extract M^{m+} independently. The overall $\log D$ dependence upon the extractant concentration may have in this case a non-integral value ranging numerically between $m/2$ and m. Kolarik[240] has found from slope analysis that when europium and terbium are extracted by dioctylphosphoric acid the composition of the complex is $M(HX_2)_3$ (or the way we prefer to write, $M(X.HX)_3$) in aromatic hydrocarbons, $MX(HX_2)_2$ in aliphatics and $MX_2(HX_2)$ in methyl isobutyl ketone. A similar variation in the distribution-ratio dependence occurs when various esters are compared in a given solvent. Peppard and coworkers[50,51,101], recently reported that europium (and other lanthanides and actinides(III) are extracted into benzene solution by several alkyl phosphorus acids, $R_2PO(OH)$ or $RHPO(OH)$ and di-(hexoxy-ethyl)phosphoric acid $(n-C_6H_{13}OCH_2CH_2O)_2PO(OH)$*, as $MX(HX_2)_2$ rather than $M(X.HX)_3$, as it does into di-(2-ethylhexyl)phos-

---

* The 2.5-power dependence has been explained by the possible availability of the ethereal oxygen for coordination. It should be noted, however, that participation of ethoxy oxygens under similar experimental conditions with a trialkyl phosphate has not been detected (M. Halpern, T. Kim and N. C. Li, *J. Inorg. Nucl. Chem.*, **24**, 1251 (1962) and M. Halpern, T. Kim, A. S. Kertes and N. C. Li, *Can. J. Chem.* **42**, 878 (1964)).

phoric acid. This stoichiometry has been deduced from a 2.5-power dependence of D upon the ester concentration, and a minus 3-power dependence upon the hydrogen ion concentration. Surprisingly, however, the stoichiometry of the uranium(VI) complex is identical to that found with other monobasic alkylphosphoric acids, $UO_2(X.HX)_2$. In no case has the K_2 value been invoked in determining the concentration of the dimeric ester species in benzene.

Hardly any information on the diluent effect on the extractive capacity of monoalkylphosphoric acids is available. It seems that the effect is considerably smaller[125,230]. Under certain conditions even an opposite trend has been noted[230] for uranium(VI) extraction by mono-(2,6,8-trimethyl-4-nonyl)phosphoric acid. This has been explained as being due to partial depolymerization of the extractant as the dielectric constant increases, thus more extractant molecules becoming available for complex formation.

(ii) *Reagent structure effect* The extractive power of various acidic phosphorus extractants has frequently been compared, in an attempt to correlate it with the structure of the esters. In spite of much effort[42,49,50,147,173,188,190,225,230,237,242] no simple correlations have been formulated so far. For example, straight-chain dialkylphosphoric acid will extract aluminium from sulphate solution, whereas a branched-chain compound will extract it only poorly, under otherwise identical conditions. Or, (2-ethylhexyl)ethylhexyl phosphonate is a more effective extractant for uranium(VI) giving a D value twice as high as di-(2-ethylhexyl) phosphate. On the other hand, the former is less effective for the rare earths and actinides in similar systems, the ratio being 1.9 for uranium, and 0.01 for thorium and being the tervalent lanthanides and actinides[190]. A combination of such factors as the extractant acidity and steric hindrance may lead to a difference in the D values as high as 10^3. Such was the case in extraction of calcium from chloride solutions by di-(2-ethylhexyl)phosphoric acid, and the more successful octylphenyl-phosphoric acid[195].

Metal-extraction power generally decreases with increasing branching of the alkyl chain. It has been pointed out in section 8.B that the apparent acid strength of the extractant decreases roughly in the same order, D increases apparently also with the length of the normal alkyl chains in the dialkyl compounds, but that influence is much less pronounced when monoalkyl derivatives of phosphoric acid are compared[207,225].

A good correlation between D and K_a, K_d and K_2 of the esters has been shown to exist in extraction of uranium(VI) by a number of diaryl-phosphoric acids[188]. Having established that the stoichiometry of the organic-phase species is identical with all the esters (see Table 8C1) investigated, the formation constants K_m (listed in Table 8C1) of $UO_2(X.HX)_2$ have been correlated through the equation

$$\log K_m = 12.00 - 2 \log K_2 + 0.5 \log K_d - 0.5 \log K_a \quad (8C11)$$

to the acidity (K_a), dimerization (K_2) and distribution (K_d) of the ester. Log K_m in turn, is related to the distribution ratio through

$$\log D = \log K_m - 2 \log [H^+] + 2 \log [\overline{(HX)_2}] \quad (8C12)$$

under the assumption that equation (8C2) is valid.

D. EXTRACTION BY CARBOXYLATES AND SULPHONATES

a. Introduction

While chelate extraction systems have been widely used in the analytical laboratory, and extraction processes for fuel reprocessing have been developed using acidic phosphorus extractants, extraction systems using acidic carboxylic and sulphonic compounds are not much in demand. There are several reasons accounting for their lack of popularity. The relatively high solubility of these acidic extractants and their salts in aqueous electrolyte solutions represents a disadvantage in plant processes. Their extractive power, being restricted usually to slightly acidic aqueous media, is something of a limitation for both their analytical and process applicability. Finally, the tendency of their salts for aggregation and micelle formation in both aqueous and organic media complicates the physicochemical interpretation of the extraction equilibrium data and their mathematical treatment.

Nevertheless, solutions of carboxylic and sulphonic acids in non-polar organic liquids being analogues of crosslinked resnous cation exchangers frequently offer several advantages over them. They have been shown to be useful, under certain experimental conditions, in studies of equilibria involved in formation of various aqueous metal complexes[245-253]. They have the advantage, furthermore, of requiring often only a pH adjustment of the aqueous phase to achieve a neat separation of metals having similar chemical properties[254–259].

Various carboxylic, alkyl or aryl-alkyl sulphonic and sulphuric acids are known for their ability to form salts with a number of metals, including the alkali and alkaline earths. These salts are frequently only slightly soluble in aqueous solutions but exhibit an increased solubility in some organic liquids. The extractability of metals by these acids, commonly called liquid cation exchanger, parallels in general, their basicity. Metals forming insoluble hydroxides at comparatively low pH values are separable from those giving insoluble hydroxides at higher pH's.

The theoretical treatment of metal extraction by the extractants is once again complicated by their state in solution. Polymerization of these acids, although usually not going beyond the dimer, and the extensive aggregation of their metallic salts, plays a significant role in the extraction of metals[253] (section 2.D).

b. The Extractant Acids and their Salts. Their State in Solution

Fragmentary data show the usefulness of a great variety of carboxylic acids and their derivatives as extractants. The bulk of the available information refers to normal or branched-chain aliphatic monocarboxylic acids with seven or more carbon atoms in the chain. Under well-defined experimental conditions the aliphatic and aromatic hydroxy carboxylic and dicarboxylic acids can also possess extractive properties towards certain metals. Mention has to be made, further, of various benzoic acid derivatives, dinitrobenzoic, cinnamic and phenylacetic acids, and amino and perfluoro derivatives of n-carboxylic acids.

Aliphatic and aromatic sulphonic acids are represented by the formula RSO_2OH, in which R may be a straight- or branched-chain or aromatic saturated radical. A wide variety of commercial products is available. They are usually crystalline solids, rather hygroscopic and water soluble. Thus for solvent extraction more useful reagents are the alkylated benzene and naphthalene sulphonates. Examples of this class are the dodecylbenzene and dinonylnaphthalene sulphonic acids. In the majority of cases these reagents contain two alkyl groups, although some surface-active properties of such and similar compounds would indicate that those with three alkyl groups may be superior extractants. Akin to the alkylbenzene or naphthalene sulphonates are the alkylated petroleum sulphonates, the petroleum fraction being chiefly aromatic.

Aliphatic sulphuric acids with the general formula $ROSO_3H$, where R is an alkyl chain containing usually at least 12 carbon atoms, have also been shown to be potentially useful extractants. They are more soluble in water than the sulphonates and less stable chemically when in contact with aqueous acid solutions, undergoing hydrolysis to aliphatic alcohols.

A great variety of metal carboxylates, metal soaps, has been prepared and isolated in a crystalline form[255,259-266]. Magnetic susceptibility measurements on a number of copper n-alkanoates revealed that these solid salts contain binuclear molecules[263]. Various alkali and alkaline earth salts of dinonylnaphthalene sulphonic acid have been crystallized and some of their properties investigated[267,268]. It has to be noted that the metallic salts of carboxylic and sulphonic acids are markedly surface active in solution, and it is in this capacity mainly that they have been prepared and studied.

Solubility of normal fatty acids in water decreases markedly with increasing chain length. They show also a decreased dissociation in water with increasing molecular weight. Their derivatives are soluble in water to a varying extent, but there is no simple correlation between their solubility and acid strength. Lower homologues of sulphonic and sulphuric acids are water soluble and possess a strong acid reaction. Higher homologues are less water soluble, but not necessarily less dissociated.

Monocarboxylic acids, liquids at room temperature (those up to nine carbon atoms in the chain), are miscible with the majority of organic solvents, polar and non-polar. Table 8D1 shows the solubility of some higher n-carboxylic acids in various water-immiscible organic solvents at $20°$ and $30°C$[269,270]. The distribution of normal fatty acids, up to hexanoic, between water and some ketones, ethers and alcohols[272] and of a mixture of fatty acids with seven, eight and nine carbon atoms between water and kerosene[255,297] has been measured at 25°C. The distribution ratio in all solvents increases with increasing number of carbon atoms in the molecule. An early investigation on the distribution of a large number of mono- and dicarboxylic acids and their derivatives[273,274] between water and diethyl ether, chloroform and xylene shows that stronger acids, those with a dissociation constant of $K \geq 10^{-4}$ have lower D values than the weaker acids. Furthermore, the distribution ratio of these acids is apparently higher in organic solvents capable of donor–acceptor interaction. Namely D decreases in the order ether > chloroform > xylene. Some of these early data are shown in Figure 8D1.

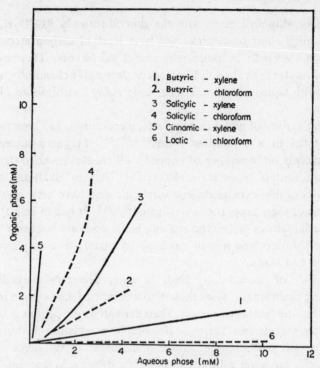

Figure 8D1. Distribution of carboxylic acids between water and xylene or chloroform, at 25°C.

Carboxylic, naphthenic and sulphonic acids are associated in organic solvents. This is indeed what should be expected from consideration of the medium effect upon the extent and strength of intermolecular hydrogen bonding in such compounds. Cryoscopic and ebullioscopic determinations of molecular weight of fatty acids in solvents practical for solvent extraction indicate that the extent of self-association decreases with increasing length of the alkyl chain in the molecule. At least this is the case in benzene and cyclohexane[269,270], where association decreases up to myristic and lauric acids, respectively. Association constants for alkanoic acids are apparently greater in anhydrous than in wet solvents, implying a competing hydrogen bonding between the dissolved water and the acid[275].

Aggregation of sulphonic acids has been reported in ethyl acetate[276], heptane[253] and benzene[268]. While in dry benzene, dinonylnaphthalene-sulphonic acid is dimeric, in the presence of water the existence of some larger aggregates has been claimed[268].

TABEL 8D1. Solubility of some n-carboxylic acids in various water-immiscible solvents (g acid/100 g solvent).

Solvent		Octa-noic	Deca-noic	Lauric	Myristic	Palmitic	Stearic
n-Hexane	20°	∞	290	47.7	11.9	3.1	0.5
	30°	∞	5150	193	41.8	14.5	4.3
Cyclohexane	20°	∞	342	68	21.5	6.5	2.4
	30°	∞	7600	215	72	27.4	10.5
Carbon tetrachloride	20°	∞	210	53	17.6	5.8	2.4
	30°	∞	4650	160	55	21.5	10.7
Chloroform	20°	∞	326	83	32.5	15.1	6.0
	30°	∞	6550	207	78	36.4	17.5
1,2–Dichloroethane	20°	∞	260	36.5	5.0	0.6	
	30°	∞	4060	170	35.5	6.0	1.0
Benzene	20°	∞	398	93.6	29.2	7.3	2.5
	30°	∞	8230	260	87.4	34.8	12.4
Toluene	20°	∞	323	97	30.4	8.7	2.0
	30°	∞	4100	251	82.1	30.0	10.6
o-Xylene	20°	∞	316	92	26.1	7.9	1.7
	30°	∞	4050	238	75.4	25.2	9.1
Chlorobenzene	20°	∞	305	87	23.6	7.8	2.2
	30°	∞	4500	239	72.4	25.8	10.8
Nitrobenzene	20°	∞	131	8.8	3.0	0.1	
	30°	∞	3550	66.3	7.1	1.2	< 0.1
n-Butanol	20°	∞	280	83	28.7	10.5	1.6
	30°	∞	4650	217	71	30	9.0
Water	20°	0.068	0.015	0.0055	0.0020	0.00072	0.00029
	30°	0.079	0.018	0.0063	0.0024	0.00083	0.00034

Most of the heavy metal alkanoates have only limited solubility in water, but they are appreciably soluble in organic solvents. The aqueous solubility of copper soaps, for example, depends also on the length of the alkyl radical: while the salts of acids with more than five carbon atoms in the chain are insoluble in water, the lower homologues are insoluble in benzene[277]. The solubility of metal carboxylates in water is also closely related to the nature of the metal, increasing with the basicity of the metal. The order of solubilities of metal caprylates is $FeR_3 < CuR_2 < NiR_2 < CoR_2 < MnR_2 < NaR$ on a mole-fraction basis[258,259].

Aqueous solutions of salts of higher fatty acids at low concentrations exhibit properties characteristic of ordinary 1:1 electrolytes. Above a certain concentration, specific for each solute, the colligative properties of solutions deviate appreciably from those expected for normal electrolytes. We are concerned here with micelle formation, one of the distinctive features of colloidal electrolytes, discussed in detail in section 2.D.

The chemistry in aqueous solutions of anionic surface-active agents[278–284] now under consideration is not of immediate importance to our main topic, except, perhaps, for the ability of their aqueous solutions to solubilize substances which have only slight water solubility. This solubilizing action may be of interest to us in so far that aqueous soap solutions may, and often do to a considerable extent, dissolve hydrocarbons and their derivatives (diluents in solvent extraction systems) to such an extent that the equilibrium is markedly affected.

The solubility of metal alkanoates in slightly polar organic liquids drops off with increasing chain length of the hydrophobic alkyl radicals[260,262,277,280,285,286]. For example, saturated ethylene glycol solution of sodium acetate at 25° is 3.54 M, while that of sodium stearate is only 0.016 M^{280}. Similar is the trend in the solubilities of calcium, magnesium, barium, zinc and lead carboxylates in chloroform and propylene glycol[285]. The salts are usually less soluble in non-polar hydrocarbons and ethers, and the solubility differences are small. For example, the solubilities of the lithium salts of lauric, myristic, palmitic and stearic acids in ether at 15.8°C are 0.011, 0.013, 0.007 and 0.011 in grams per 100 g of solvent, respectively[287].

The amount of information available on the *cmc* and the size of the aggregates in nonaqueous systems is unfortunately much smaller than that for aqueous systems. The length of the alkyl chain in metal alkanoates does not affect the *cmc* so much in organic solvents, but it is markedly higher in slightly polar liquids than in aromatic or aliphatic hydrocarbons and their non-polar derivatives. On the other hand, it seems that the number of molecules per micelle increases with decreasing number of carbon atoms in the alkyl chain. The zinc salts of fatty acids with 8, 10, 12, 14 and 18 carbon atoms per chain have an average aggregation number in toluene at 111°C of 6.3, 5.4, 4.8, 4.2 and 3.2 respectively[262,278]. The same is true for copper alkanoates in benzene[277] and several other metal soaps in toluene and xylene[260–262]. A marked reduction in the micellar size is observed in alcohols and other polar organic liquids[260,262]. It has been fouud, furthermore, that the number of monomers in sodium or lithium carboxylate soap aggregates is considerably higher in anhydrous organic solvents[288]. The average aggregation number depends also, apparently, on the nature of the cation, ferric laurate forming a smaller aggregate than either zinc or copper salts[278].

The critical range for micelle formation of metal dinoylnaphthalene sulphonates in benzene is lower than 10^{-6} M. The size of the micelle in benzene does not depend much on the cation, the average aggregation number of caesium, magnesium, calcium, zinc and aluminium varying between 9–14. The micellar weight of sodium dinonylnaphthalene sulphonate is about 2400 in dry heptane at $37°C^{252}$. Unlike the carboxylate soaps, the number of sulphonate molecules in the micelle is rather independent of concentration and moisture[267,268].

c. Metal Extraction Characteristics

(i) *Aqueous phase* The extractability of metals by carboxylic and sulphonic acids depends to a considerable extent on the composition of the aqueous phase. In section A of this Chapter we have dealt with the effect of pH on metal extraction. There is no qualitative difference, as far as this variable is concerned, between the chelating-type extractants treated previously, and the acidic extractants discussed in this section. It remains only to show experimental evidence for the validity of the equation (identical to 7C24)

$$M^{m+} + m\overline{HX} \rightleftarrows \overline{MX_m} + mH^+ \qquad K_X \qquad (8D1)$$

when HX is a monobasic carboxylic or sulphonic acid.

It has been shown that fatty acids, or their solutions in water-immiscible organic solvents, extract metals preferentially from aqueous solutions containing alkali[247,255,258,259,289–294]. There is almost no extraction in absence of it. This, and the fact that an alternative way to achieve high D values is the use of alkali salt of the fatty acid instead of the free acid itself, are good evidence for the qualitative validity of equation (8D1).

For a large number of metal ions, equation (8D1) has been shown to be valid for a wide range of hydrogen ion concentration[255,259]. The experimentally determined distribution ratio, under conditions of constant activity coefficient, fitted into the equation

$$\log D = \log K_X + m \log [HX] + m\text{pH} \qquad (8D2)$$

derived previously in section 7.C. In all these cases m was equal to the nominal charge of the metal, proving that MX_m and M^{m+} are the predominant species in the organic and aqueous phases, respectively. Under identical experimental conditions the pH-vs.-log D curves for a large number of equally charged cations gave a family of parallel lines[255], implying that the separation factor between two equally charged metals is independent of total reagent concentration and the pH of the equilibrium aqueous phase.

The sequence and the spacing between the curves for various cations depends on the basicity and, obviously enough, on the charge of the metal. D at a given acidity increases with increasing charge and decreases as the basic character of the metal increases. On the semilogarithmic plot (Figures 8D2 and 8D3), reproduced from the paper of Gindin and coworkers[259], a typical family of curves for metal carboxylates are presented. Thus, in spite of the fact that carboxylic acids react with a large number of metals, extraction can be made selective for certain of them by adjusting the pH of the aqueous phase. Moreover, the effect of electronegativity on the order of metal extraction is so striking that a cation exchange of the type

$$\overline{M_I X_m} + M_{II}^{m+} \rightleftarrows \overline{M_{II} X_m} + M_I^{m+} \tag{8D3}$$

can be made practically complete. The less electronegative sodium was shown to be quantitatively exchanged for any more electronegative metal, or cobalt can be exchanged for copper, copper for iron or iron for bismuth[255-258]. The constant of the exchange equilibrium (8D3) should be equal to the ratio of the respective K_X's to the appropriate powers. This has been proved to be true for many exchange systems studied[259].

As mentioned in the introduction, the usefulness of carboxylic extractants for aqueous complexing studies is greatly extended by the fact that such systems frequently behave ideally and the mass-action law expression (8D2) describes the extraction reaction fairly well. In some cases, however, the outcome is likely to be much less ideal than this equation would predict. This is actually the case whenever the equilibrium aqueous pH is such as to cause hydrolysis of the metal. For example, above a pH of 5.8 and 4.6 the log D–pH plots for cobalt and copper extraction have a slope of one rather than two[255]. Or, on the pH range between 3 and 5 non-integral slopes with the values 0.4–1.1, 2.5, 1.5 and 2.1 were observed for zirconium, niobium, uranyl and ferric iron, respectively[258]. A more elaborate set of equations which takes into account the varying degrees of metal hydrolysis has to be used for quantitative treatment of the experimental data.

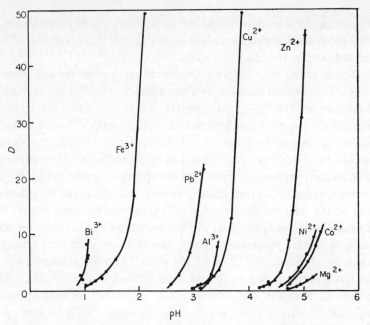

Figure 8D2. Distribution of metals between water and undiluted fatty acid ($C_7 - C_9$) as a function of aqueous pH.

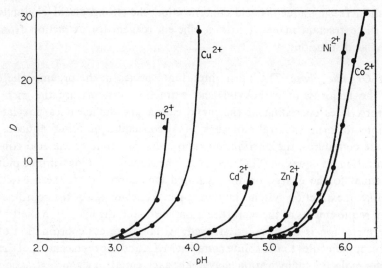

Figure 8D3. Distribution of metals between water and 2.8 M kerosene solution of fatty acid ($C_7 - C_9$) as function of aqueous pH.

References pp. 563–574

The effect of the aqueous-phase parameters on the extraction of metals using alkylsulphuric or alkyl–arylsulphonic acids is similar to that discussed above for the carboxylic extractants[250–253,276,295–300]. The distribution ratios are − 2 power dependent on the equilibrium aqueous hydrogen ion concentration for divalent metals[251,252,256], and uranium(II) and neptunium(VI)[298]; − 3 for iron(III), indium[251,253] and the tervalent lanthanides[250], and − 4 for thorium and neptunium (IV)[298] when dodecylbenzene or dinonylnaphthalene sulphonic acids are the extractants. However, the extraction of alkali metals by dinonylnapthalene sulphonic acid into isopropylbenzene is low and not appreciably affected by the pH between 1 and 13[300]. From dilute aqueous acid solutions the observed slopes of the distribution ratio for zirconium and hafnium are -2[298], and similarly, slopes lower than would be required by the oxidation state of the metals were observed when thorium and zirconium were extracted by dodecylsulphuric acid[296]. A survey study[297] on the extractability of some 60 elements by this extractant from 0.01—2 N aqueous HCl solutions shows that the distribution ratios of metals forming chloride complexes is considerably lower than equation (8D2) would predict (Appendix F).

The major difference between the sulphonic (and sulphuric) and carboxylic acid extractants is the capacity of the former to extract cations from solutions of pH < 1. While, as a rule, the distribution ratios decrease with increasing aqueous acid concentration, the strongly acid character of the extractant makes it still an efficient reagent for extraction from 2 M acid solutions.

(ii) *Organic phase* The most important species in the organic phase in the majority of carboxylic acid extraction systems are the metal carboxylates consisting of the metal cation and sufficient carboxylate anions to form a neutral molecule[246]. At constant equilibrium aqueous-phase conditions, the distribution ratio is the function of the acid concentration, as equation (8D2) requires. The slope m of a logarithmic plot is equal to the charge on the hydrated metal ion. In all these cases, unlike in extraction systems involving metal chelates (8.A), the coordination requirements of the metal are unsatisfied, and the hydration shell of the metal ion is usually undestroyed. Although direct coordination of solvent molecules to the solute complex MX_m, and the replacement of the water molecules of hydration, may not be as essential for a good extraction to take place, as was the case with the more acidic hydrogen organophosphorus extractants, evidence has been frequently presented for the importance of the solvation power in achieving high distribution ratios. The perfluorooctanoate of scandium and strontium, for example, are not extractable into the non-polar carbon tetarchloride but are readily extractable into hexone or ether, i.e. when, solvent molecules are apparently attached to the MX_m species[301]. Or beryllium, aluminium and iron butyrate are well extracted into chloroform but poorly into benzene or carbon tetrachloride[293,296]. On the other hand, zirconium and hafnium p-bromomandelates, ZrX_4 and HfX_4, were found to be unsolvated in highly solvating organic liquids such as isopentyl and n-butyl alcohols[248]. This is conceivable, however, considering the structure of the ligand with its sites available to satisfy the coordination requirements of metals. Thus, although there is practically no direct information as to the uptake of water by the organic-phase complexes, the fact that metal alkanoates are apparently more extractable into polar solvents than into hydrocarbons should suggest that the metal does transport water into the organic phase.

Another type of organic-phase complex formation may take place under certain experimental conditions, depending mainly on the structure of the carboxylic ligand and the nature and properties of the metal. In such cases intermolecular interaction leads to the formation of monomeric or dimeric[302] complexes of the type $MX_m(HX)_n$, where the sum $m + n$ is not necessarily equal to the coordination number of the metal. Unfortunately, there is not enough experimental information available to allow even a rough generalization of the factors affecting the formation of such complexes[302]. The strength of metal hydration, the extent of intermolecular hydrogen bonding of the extractant molecules and steric considerations are factors dominating to varying extents. The uranium(VI) and thorium complexes of salicylic acid extracted into isobutyl methyl ketone or chloroform have the composition $UO_2X_2.HX$ and $ThX_4.HX$, while the corresponding complexes with methoxybenzoic acid correspond to the formula UO_2X_2 and ThX_4[245]. There is a good reason to believe that the plutonium(IV)–salicylic acid complex in amyl acetate has a similar composition[303]. When iron(III) is extracted by a carboxylic acid mixture containing 7–9 carbon atoms, the organic-phase complexes identified were FeX_3 and $FeX_3.HX$. A solid compound isolated from the extract had a composition corresponding to the second formula[290]. The solubility data of various copper alkanoates in chloroform–conjugate acid mixture points strongly to the fact that the solubility is due to a complex formation, possibly of the type discussed here[264]. Cobalt and nickel extracted by naphthenic acid, a commercial alicyclic monocarboxylic acid, into benzene, form organic-phase complexes of a formula $M_2X_4.4HX$[304]. The copper[304] and cobalt[305] salts of this acid in benzene[304], and those of propionic acid in chloroform[306] have a similar stoichiometric composition. A log D vs. log [HX] plot of the extraction data of cacsium by the carboxylic acid mixture into benzene or isooctane had a slope of eight. This made the authors[289] suggest the formula CsX.7HX. Such a high solvation number deduced from slope analysis may not be very convincing.

Finally, in some of the carboxylic acid extraction systems there may be a considerable degree of complex formation in the aqueous phase also. Such is the case whenever a log D–log [HX] plot has a slope lower than m. The resulting species in the organic phase may be a mixed-ligand complex. In the carboxylic acid systems the most important are the metal–hydroxo complexes when extraction is carried out under conditions of metal hydrolysis in the aqueous phase. Since the degrees of hydrolysis are specific properties of metals differing from one to another, the composition of the organic-phase complex will largely depend on this[254,255,258].

A mass-action law treatment of extraction equilibria leading to the formation of the last two types of organic-phase complexes discussed above, i.e. whenever the slope value of the log D vs. log [HX] is higher or lower than m, has been derived for the simplest cases in section 7.C.

The complexes formed by alkylsulphuric acid are apparently similar to those formed by carboxylic acids[296].

The composition of the organic-phase metal complexes in alkylarylsulphonic acid systems is, however, markedly different. The main reason is the aggregated state of the extractant in the organic solution. The extraction of various lanthanides[250], and cobalt, zinc, manganese, iron(III) and indium[251–253,256] from dilute perchloric acid solution by dinonylnaphthalene sulphonic acid into heptane, and zinc, cerium(III) and caesium into benzene[295], and the alkalies into diisopropylbenzene[300] has been investigated under varying experimental conditions. Regardless of the charge of the metal, the log D – log [HX] plots were of a unit slope in all cases. The exchange reaction, represented as

$$M^{m+} + \overline{(HX)_n} \rightleftarrows \overline{M(H_{n-m}X_n)} + mH^+ \qquad (8D4)$$

is similar to that assumed to be operative in extraction systems involving the polymeric dibasic monoalkylphosphoric acids discussed in the previous section. Surprisingly, the group of Japanese workers[276,297-299] investigating the extraction of metals by dodecylbenzene sulphonic acid under comparable experimental conditions found a close to second-power dependence of D rather than the first-power dependence with the similar naphthalene sulphonic acid. There is no explanation in view, since the difference in the degree of the extractant aggregation is unlikely to account for this discrepancy.

E. REFERENCES

1. T. S. West, *Metallurgia*, **53**, 91, 132, 185, 234, 292 (1955), **54**, 47, 103 (1956).

2. G. H. Morrison and H. Freiser, *Solvent Extraction in Analytical Chemistry*, J. Wiley & Sons, New York, 1957.

3. G. H. Morrison and H. Freiser, *Anal. Chem.*, **30**, 632 (1958); **32**, 37R (1960); **34**, 64R (1962); **36**, 93R (1964); **38**, 131R (1966).

4. H. Freiser, *Chemist-Analyst*, **50**, 62 (1961).

5. J. Stary, *The Solvent Extraction of Metal Chelates*, Pergamon Press, London, 1964.

6. L. I. Katzin in *The Chemistry of Non-Aqueous Solvents* (Ed. J. J. Lagowski), Academic Press, New York, 1966.

7. H. L. Finston, 1963 ACS Anal. Symp. on *Metal Chelates in Chemical Analysis*, Tucson, Arizona, June 1963.

8. G. H. Morrison and H. Freiser in *Comprehensive Analytical Chemistry* (Eds. C. L. Wilson and D. W. Wilson), Vol. 1, Elsevier Publ. Co., Amsterdam, 1959.

9. A. K. De, *Separation of Heavy Metals*, Pergamon Press, London, 1961.

10. G. Iwantscheff, *Das Dithizon*, Verlag Chemie, Weinheim, 1958.

11. R. Belcher, C. L. Wilson and T. S. West in *New Methods in Analytical Chemistry*, 2nd ed., Chapman and Hall, London, 1964.

12. *Symposium on Solvent Extraction in Analysis of Metals*, ASTM Spec. Techn. Publ. No. 238, 1958.

13. E. D. Sandell, *Colorimetric Determination of Traces of Metals*, Interscience Publ. Inc., New York, 1959.

14. T. S. West, *Anal. Chim. Acta*, **25**, 401 (1961).

15. H. Irving and R. J. P. Williams in *Treatise on Analytical Chemistry* (Eds. I. M. Kolthoff and P. J. Elving), Part 1, Vol. 3. Interscience Publ. Inc., New York, 1959, p. 1309.

16. H. Freiser and G. H. Morrison in *Annual Review of Nuclear Science*, Vol. 9, 1959.

17. I. P. Alimarin and Y. A. Zolotov, *Talanta*, **9**, 891 (1962).

18. Y. A. Zolotov and I. P. Alimarin, *Radiokhimiya*, **4**, 272 (1962).

19. G. K. Schweitzer, *Anal. Chim. Acta*, **30**, 68 (1964).

20. A. M. Poskanzer and B. M. Foreman, Jr., *J. Inorg. Nucl. Chem.*, **16**, 323 (1961).

21. A. K. De and S. M. Khopkar, *J. Sci. Ind. Res. (India)*, **21A**, 131 (1962).

22. R. H. Holm and F. A. Cotton, *J. Am. Chem. Soc.*, **80**, 5658 (1958).

23. G. W. Pope, J. F. Steinbach and W. F. Wagner, *J. Inorg. Nucl. Chem.*, **20**, 304 (1961).

24. K. Nakamoto, Y. Morimoto and A. E. Martell, *J. Am. Chem. Soc.*, **83**, 4533 (1961).

25. J. R. Ferraro and T. V. Healy, *J. Inorg. Nucl. Chem.*, **24**, 1463 (1962).

26. R. E. Sievers, R. W. Moshier and M. L. Morris, *Inorg. Chem.*, **1**, 966 (1962).

27. R. E. Sievers, B. W. Ponder, M. L. Morris and R. W. Moshier, *Inorg. Chem.*, **2**, 693 (1963).

28. R. C. Fay and T. S. Piper, *J. Am. Chem. Soc.*, **85**, 500 (1963).

29. T. J. Pinnavaid and R. C. Fay, *Inorg. Chem.*, **5**, 233 (1966).

30. H. Irving, S. J. H. Cooke, S. C. Woodger and R. J. P. Williams, *J. Chem. Soc.*, 1847 (1949).

31. R. G. W. Hillingshead, *Oxine and its Derivatives*, Vol. 1, Butterworths Sci. Publ., London, 1954.

32. H. Irving, J. A. D. Ewart and J. T. Wilson, *J. Chem. Soc.*, 2672 (1949).

33. T. Moeller and F. L. Pundsack, *J. Am. Chem. Soc.*, **75**, 2258 (1953).

34. H. Mottola and H. Freiser, *XXth IUPAC Conference*, Moscow, 1965, Abstract No. E 82.

35. J. Stary, *Anal. Chim. Acta*, **28**, 132 (1963).

36. H. E. Hellwege and G. K. Schweitzer, *Anal. Chim. Acta*, **28**, 236 (1963).

37. D. M. Kemp, *Anal. Chim. Acta*, **27**, 480 (1962).

38. G. M. Kosolapoff, *Organophosphorus Compounds*, J. Wiley & Sons, Inc., New York, 1950.

39. J. R. Van Wazer, *Phosphorus and its Compounds*, Vol. I., Interscience Publ. Inc., New York, 1958.

40. W. C. Fernelius, *Chemical Nomenclature, Advances in Chemistry Series*, No. 8, Amer. Chem. Soc., Washington, 1953.

41. C. R. Dutton and C. R. Noller, *Organic Syntheses*, Collective Volume II, J. Wiley & Sons, Inc. New York, 1943.

42. V. B. Shevchenko, V. A. Mikhailov and Y. P. Zavalyskii, *Zh. Neorg. Khim.*, **3**, 1955 (1958).

43. L. L. Burger and R. M. Wagner, *J. Chem. Eng. Data*, **3**, 310 (1958).

44. D. F. Peppard, J. R. Ferraro and G. W. Mason, *J. Inorg. Nucl. Chem.*, 4, 371 (1957); 7, 231 (1958); 10, 275 (1959); 12, 60 (1959); 16, 246 (1961).

45. W. H. Baldwin and C. E. Higgins, *Anal. Chem.*, 30, 446 (1958).

46. T. D. Smith, *J. Inorg. Nucl. Chem.*, 15, 95 (1960).

47. E. K. Dukes, *USAEC Rept.*, DP–250 (1957).

48. F. Krašovec and J. Jan, *Croat. Chem. Acta*, 35, 183 (1963).

49. Z. Kolarik and H. Pankova, *J. Inorg. Nucl. Chem.*, 28, 2325 (1966).

50. D. F. Peppard, G. W. Mason and S. Lewey, *J. Inorg. Nucl. Chem.*, 27, 2065 (1965).

51. D. F. Peppard, G. W. Mason and C. Andrejasich, *J. Inorg. Nucl. Chem.*, 28, 2347 (1966).

52. J. F. Ferraro, D. F. Peppard and G. W. Mason, *J. Inorg. Nucl. Chem.*, 27, 2055 (1965).

53. R. S. Long, D. A. Ellis and R. H. Bailes, *Proc. U.N. Intern. Conf. Peaceful Uses At. Energy*, 1st, Geneva, 8, 77 (1955).

54. R. S. Long, D. A. Ellis, R. M. Ragen, G. R. Moore, J. E. Magner, R. S. Olson and J. Joldersma, *USAEC Rept.*, DOW–84 (1952); DOW–83 (1952); DOW–160 (1957).

55. M. Zangen, *J. Inorg. Nucl. Chem.*, 16, 165 (1960); *Israel AEC Rept.* IA–594 (1961); M. Zangen, E. D. Bergman and Y. Marcus, *Separation Science*, 3, 1 (1968).

56. W. B. Smith and J. Dewry, *Analyst*, 86, 178 (1961).

57. D. Grdenic and B. Korpar, *J. Inorg. Nucl. Chem.*, 12, 149 (1959).

58. F. Habashi, *J. Inorg. Nucl. Chem.*, 13, 125 (1960).

59. B. F. Green, O. W. Allen and D. E. Tynan, *Ind. Eng. Chem.*, 49, 628 (1957).

60. V. M. Vdovenko, A. A. Lipovskii and S. A. Nikitina, *Radiokhimiya*, 3, 396 (1961).

61. B. N. Laskorin, D. I. Skorovarov, E. A. Philipov and A. I. Shilin, *Radiokhimiya*, 5, 424 (1963).

62. S. A. Potapova, *Radiokhimiya*, 3, 422 (1961).

63. H. Gorican and D. Grdenic, *Proc. Chem. Soc.*, 288 (1960).

64. D. Grdenic and V. Jagodic, *J. Inorg. Nucl. Chem.*, 26, 167 (1964).

65. J. Kennedy, *Chem. Ind. (London)*, 950 (1958).

66. W. H. Baldwin, *USAEC Rept.*, ORNL–3320 (1962).

67. T. H. Handley and J. A. Dean, *Anal. Chem.*, 34, 1312 (1962).

68. T. H. Handley, *Nucl. Sci. Eng.*, 16, 440 (1963).

69. H. Bode and W. Arnswald, *Z. Anal. Chem.*, 185, 99 (1962).

70. R. H. A. Plimmer and W. J. N. Bunch, *J. Chem. Soc.*, 292 (1929).

71. W. D. Kumler and J. J. Eiler, *J. Am. Chem. Soc.*, 65, 2355 (1943).

72. C. Cesarano and C. Lepscky, *J. Inorg. Nucl. Chem.*, 14, 276 (1960).

73. E. Cerrai and F. Gadda, *Nature*, **183**, 1528 (1959).

74. R. B. Lew, H. Gard and F. Jacob, *Talanta*, **10**, 911 (1963).

75. D. A. Ellis, R. M. Ragen, J. E. Magner, R. R. Grinstead. R. S. Long, J. Joldersma and R. H. Bailes, *USAEC Rept.*, DOW-85 (1952).

76. A. Varon, F. Jacob, K. C. Park, J. Ciric and Wm. Rieman III, *Talanta*, **9**, 573 (1962).

77. C. E. Higgins and W. Ii. Baldwin, *J. Org. Chem.*, **21**, 1156 (1956).

78. I. J. Gal and D. M. Petkovic, *J. Inorg. Nucl. Chem.*, **25**, 129 (1963).

79. D. C. Stewart and H. W. Crandall, *J. Am. Chem. Soc.*, **73**, 1377 (1951).

80. D. F. Peppard, G. W. Mason, J. L. Maier and W. J. Driscoll, *J. Inorg. Nucl. Chem.*, **4**, 334 (1957).

81. C. J. Hardy and D. Scargill, *J. Inorg. Nucl. Chem.*, **10**, 323 (1959).

82. T. V. Healy, *Radiochim. Acta*, **2**, 52 (1963).

83. J. M. Schmitt and C. A. Blake, *USAEC Rept.*, ORNL-3548 (1964).

84. E. Cerrai, C. Cesarano and F. Gadda, *Energia Nucl. (Milan)*, **4**, 405 (1957).

85. W. H. Baldwin and C. E. Higgins, *J. Inorg. Nucl. Chem.*, **17**, 334 (1961).

86. C. A. Blake, D. J. Crouse, C. F. Coleman and K. B. Brown, *USAEC Rept.*, ORNL-2172 (1957).

87. T. D. Smith, *Anal. Chim. Acta*, **22**, 249 (1960).

88. B. F. Greenfield and C. J. Hardy, *Brit. Rept. AERE*-R 3686 (1961).

89. D. Dyrssen, *Acta Chem. Scand.*, **11**, 1771 (1957).

90. A. S. Kertes, A. Beck and Y. Habousha, *J. Inorg. Nucl. Chem.*, **21**, 108 (1961).

91. A. L. Myers, W. J. McDowell and C. F. Coleman, *J. Inorg. Nucl. Chem.*, **26**, 2005 (1964); *USAEC Rept.*, ORNL-TM-181 (1962).

92. C. J. Hardy and D. Scargill, *J. Inorg. Nucl. Chem.*, **11**, 128 (1959).

93. V. S. Uljanov and R. A. Sviridova, *Radiokhimiya*, **5**, 419 (1963).

94. N. I. Ampelogova, *Radiokhimiya*, **5**, 562 (1963).

95. W. H. Baldwin, *USAEC Rept.*, CF-52-11-57 (1957).

96. V. B. Shevchenko and V. S. Smelov, *At. Energ. (USSR)*, **5**, 542 (1958); **6**, 140 (1959).

97. D. Dyrssen, S. Ekberg and D. H. Liem, *Acta Chem Scand.*, **18**, 135 (1964).

98. C. A. Blake, K. B. Brown, C. F. Coleman, D. E. Horner and J. M. Schmitt, *USAEC Rept.*, ORNL-1903 (1955).

99. C. A. Blake, C. F. Baes, K. B. Brown, C. F. Coleman and J. C. White, *Proc. U.N. Intern. Conf. Peaceful Uses At. Energy, 2nd, Geneva*, **28**, 289 (1958).

100. D. F. Peppard, G. W. Mason and C. M. Andrejasich, *J. Inorg. Nucl. Chem.*, **27**, 697 (1965).

101. D. F. Peppard, G. W. Mason and G. Griffin, *J. Inorg. Nucl. Chem.*, **27**, 1683 (1965).

102. J. R. Ferraro, G. W. Mason and D. F. Peppard, *J. Inorg. Nucl. Chem.*, **22**, 285 (1961).

103. J. R. Ferraro and D. F. Peppard, *Nucl. Sci. Eng.*, **16**, 389 (1963).

104. G. M. Kosolapoff and J. S. Powell, *J. Chem. Soc.*, 3535 (1950).

105. C. F. Baes, Jr., *USAEC Rept.*, ORNL–2737 (1959); *J. Phys. Chem.*, **66**, 1629 (1962); C. F. Baes, Jr. and H. T. Baker, *J. Phys. Chem.*, **64**, 89 (1960).

106. G. S. Rao, G. W. Mason and D. F. Peppard, *J. Inorg. Nucl. Chem.*, **28**, 887 (1966).

107. L. W. Daasch and D. C. Smith, *Anal. Chem.*, **23**, 853 (1951).

108. C. I. Meyrick and H. W. Thompson, *J. Chem. Soc.*, 225 (1950).

109. J. W. Maarsen, M. C. Smit and J. Matze, *Rec. Trav. Chim.*, **76**, 713 (1957).

110. L. J. Bellamy and L. Beecher, *J. Chem. Soc.*, 1701 (1952); 728 (1953).

111. J. R. Ferraro, *J. Inorg. Nucl. Chem.*, **24**, 475 (1962).

112. J. R. Ferraro and C. M. Andrejasich, *J. Inorg. Nucl. Chem.*, **26**, 377 (1964).

113. L. Winarnd and Ph. Dreze, *Bull. Soc. Chim. Belges.*, **71**, 410 (1962).

114. G. Aksnes, *Acta Chem. Scand.*, **14**, 1475 (1960).

115. J. T. Braunholtz, G. E. Hall, F. G. Mann and N. Sheppard, *J. Chem. Soc.*, 868 (1959).

116. L. C. Thomas, R. A. Chittenden and H. E. R. Hartley, *Nature*, **192**, 1283 (1961).

117. J. Gaunt and G. M. Meaburn, *Brit. Rept.*, AERE–C/R–785 (1951).

118. C. A. Horton and J. C. White, *Talanta*, **7**, 215 (1961).

119. J. Kennedy and A. M. Deane, *Brit. Rept.*, AERE–R–3410 (1961).

120. R. H. Zucal, J. A. Dean and T. H. Handley, *Anal. Chem.*, **35**, 988 (1963).

121. J. R. Ferraro and D. F. Peppard, *J. Phys. Chem.*, **67**, 2639 (1963).

122. C. F. Baes, Jr., *J. Inorg. Nucl. Chem.*, **24**, 707 (1962).

123. D. Dyrssen and H. D. Liem, *Acta Chem. Scand.*, **14**, 1091 (1960); H. D. Liem in *Solvent Extraction Chemistry* (Eds. D. Dyrssen, J. O. Liljenzin and J. Rydberg), North-Holland, Amsterdam, 1967 p. 264.

124. *Ann. Rept.*, Chemical Technology Division, Oak Ridge National Lab., *USAEC Rept.*, ORNL–3627 (1964).

125. G. W. Mason, S. McCarty and D. F. Peppard, *J. Inorg. Nucl. Chem.*, **24**, 967 (1962).

126. Ph. Dreze, *Bull. Soc. Chim. Belges*, **68**, 674 (1959).

127. G. Duyckaerts, Ph. Dreze and A. Simon, *J. Inorg. Nucl. Chem.*, **13**, 332 (1960).

128. D. Dyrssen and S. Ekberg, *J. Inorg. Nucl. Chem.*, **13,** 1909 (1959).

129. P. Courtemanche and J. C. Merlin, *Compt. Rend.*, **260,** 3053 (1965).

130. S. A. Potapova and V. V. Fomin, *Radiokhimiya*, **7,** 14 (1965).

131. V. B. Shevchenko and V. S. Smelov, *Ekstraktziya*, Vol. 2, Gosatomizdat, Moscow. 1962, p. 58.

132. C. J. Hardy, *Nucl. Sci. Eng.*, **16,** 401 (1963).

133. K. Kimura, *Nippon Genshiryoku Gakaishi.*, **2,** 585 (1960).

134. K. Kimura, *Bull. Chem. Soc. Japan.*, **33,** 1038 (1960).

135. J. J. McCown and R. P. Larson, *Anal. Chem.*, **32,** 597 (1960); **33,** 1003 (1961).

136. T. Goto and M. Smutz, *J. Inorg. Nucl. Chem.*, **27,** 1369 (1965).

137. C. F. Coleman, C. A. Blake, Jr. and K. B. Brown, *Talanta*, **9,** 297 (1962).

138. F. E. Butler, *Anal. Chem.*, **35,** 2069 (1963).

139. T. Ishimori and E. Nakamara, *Japanese Rept.*, JAERI 1047 (1963).

140. G. W. Mason and D. F. Peppard, *Nucl. Sci. Eng.*, **17,** 247 (1963).

141. D. F. Peppard and G. W. Mason, *Nucl. Sci. Eng.*, **16,** 382 (1963).

142. D. F. Peppard, S. W. Moline and G. W. Mason, *J. Inorg. Nucl. Chem.*, **4,** 344 (1957).

143. D. F. Peppard, G. W. Mason and S. W. Moline, *J. Inorg. Nucl. Chem.*, **5,** 141 (1957).

144. R. D. Baybarz, *USAEC Rept.*, ORNL–3273 (1962).

145. T. V. Healy, *Radiochim. Acta*, **2,** 52 (1963).

146. W. J. McDowell and G. N. Case, *USAEC Rept.*, ORNL–TM–181 (1962).

147. C. F. Baes, Jr., *J. Inorg. Nucl. Chem.*, **24,** 707 (1962).

148. *Ann. Rept.*, Chemical Technology Division, Oak Ridge National Lab., *USAEC Rept.*, ORNL–3627, (1964) p. 195.

149. G. Duyckaerts and Ph. Dreze, *Bull. Soc. Chim. Belges.*, **11,** 306 (1962).

150. D. E. Horner, D. J. Crouse, K. B. Brown and B. Weaver, *Nucl. Sci. Eng.*, **17,** 234 (1963).

151. T. B. Pierce and P. F. Peck, *Analyst*, **88,** 217 (1963).

152. T. B. Pierce and P. F. Peck, *Nature*, **194,** 84 (1962); **195,** 597 (1962).

153. E. Cerrai and C. Testa, *J. Chromatography*, **8,** 232 (1962).

154. E. Cerrai, C. Testa and C. Triulzi, *Energia Nucl. (Milan)*, **9,** 193, 377 (1962).

155. R. W. Catrall, *Australian J. Chem.*, **14,** 163 (1961).

156. T. D. Smith, *J. Inorg. Nucl. Chem.*, **9,** 150 (1959).

157. A. S. Kertes and M. Halpern, *J. Inorg. Nucl. Chem.*, **19,** 359 (1961).

158. Z. A. Sheka and E. I. Sinyavskaya, *Zh. Neorg. Khim.*, **9,** 1974 (1964).

159. T. V. Healy and J. Kennedy, *J. Inorg. Nucl. Chem.*, **10**, 128 (1959).

160. J. Kennedy and A. M. Deane, *J. Inorg. Nucl. Chem.*, **19**, 142 (1961); **20**, 295 (1961).

161. J. Marsh, *J. Chem. Soc.*, 554 (1939).

162. J. R. Ferraro, *J. Inorg. Nucl. Chem.*, **24**, 475 (1962).

163. E. E. Kriss and Z. A. Sheka, *Dokl. Akad. Nauk SSSR*, **138**, 846 (1961).

164. Z. A. Sheka and E. E. Kriss, *Zh. Neorg. Khim.*, **7**, 658 (1962).

165. E. E. Kriss, *Zh. Neorg. Khim.*, **8**, 1505, 1513 (1963).

166. B. Jezowska-Trzebiatowska and coworkers, *Proc. U.N. Intern. Conf. Peaceful Uses At. Energy*, 2nd, Geneva., **28**, 253 (1958).

167. W. Davis, Jr., *USAEC Rept.*, ORNL–3084 (1961).

168. D. F. Peppard and J. R. Ferraro, *J. Inorg. Nucl. Chem.*, **10**, 275 (1959).

169. P. G. M. Brown, J. M. Fletcher, C. J. Hardy, J. Kennedy, D. Scargill, A. G. Wain and J. L. Woodhead, *Proc. U.N. Intern. Conf. Peaceful Uses At. Energy*, 2nd, Geneva, **17**, 118 (1958).

170. H. T. Hahn and E. M. Vander Wall, *USAEC Rept.*, IDO–14560 (1961).

171. E. M. Vander Wall, *USAEC Rept.*, IDO–14567 (1961).

172. R. A. Wells, D. A. Everest and A. A. North, *Nucl. Sci. Eng.*, **17**, 259 (1963).

173. C. J. Hardy, B. F. Greenfield and D. Scargill, *J. Chem. Soc.*, 174 (1961).

174. E. E. Kriss and Z. A. Sheka, *Zh. Neorg. Khim.*, **5**, 2819 (1960); **6**, 1930 (1961).

175. V. B. Shevchenko and V. S. Smelov, *Zh. Neorg. Khim.*, **6**, 732 (1961).

176. D. Dyrssen, *Acta Chem. Scand.*, **11**, 1277 (1957).

177. D. Dyrssen and H. D. Liem, *Acta Chem. Scand.*, **14**, 1100 (1960).

178. E. N. Patrusheva, N. E. Brezhneva and G. V. Korpusov, *Radiokhimiya*, **2**, 541 (1960).

179. C. F. Baes, Jr. and H. T. Baker, *J. Phys. Chem.*, **64**, 89 (1960).

180. D. F. Peppard, G. W. Mason and I. Hucher, *J. Inorg. Nucl. Chem.*, **18**, 245 (1961).

181. B. Weaver and D. E. Horner, *J. Chem. Eng. Data*, **5**, 260 (1960).

182. E. Nakamura, *Bull. Chem. Soc. Japan.*, **34**, 402 (1961).

183. Z. A. Sheka and E. I. Siniavskaya, *Radiokhimiya*, **5**, 485 (1963).

184. N. I. Ampelogova, *Radiokhimiya*, **5**, 562 (1963).

185. C. F. Baes, Jr., R. A. Zingaro and C. F. Coleman, *J. Phys. Chem.*, **62**, 129 (1958).

186. B. J. Thamer, *USAEC Rept.*, LA–1996 (1956).

187. D. Dyrssen and F. Krašovec, *Acta Chem. Scand.*, **13**, 561 (1959).

188. F. Krašovec and C. Klofutar, *J. Inorg. Nucl. Chem.*, **27**, 2431 (1965).

189. C. J. Hardy, *J. Inorg. Nucl. Chem.*, **21**, 348 (1961).

190. D. F. Peppard, G. W. Mason, I. Hucher and F. A. J. A. Brandao, *J. Inorg. Nucl. Chem.*, **24**, 1387 (1962).

191. D. F. Peppard, G. W. Mason and S. McCarty, *J. Inorg. Nucl. Chem.*, **13**, 138 (1960).

192. T. G. Lenz and M. Smutz, *J. Inorg. Nucl. Chem.*, **28**, 1119 (1966).

193. D. C. Madigan, *Australian J. Chem.*, **13**, 58 (1960).

194. K. B. Brown, D. J. Crouse, D. E. Horner and W. B. Howerton, *USAEC Rept.* ORNL–TM–449 (1963).

195. D. F. Peppard, G. W. Mason, S. McCarty and F. D. Johnson, *J. Inorg. Nucl. Chem.*, **24**, 321 (1962).

196. D. F. Peppard, E. P. Horwitz and G. W. Mason, *J. Inorg. Nucl. Chem.*, **24**, 429 (1962).

197. W. J. McDowell and C. F. Coleman, *J. Inorg. Nucl. Chem.*, **25**, 234 (1963); **27**, 1117 (1965); **28**, 1083 (1966).

198. *Ann. Rept.*, Chemical Technology Division, Oak Ridge Natl. Lab., *USAEC Rept.* ORNL–3452, (1963) p. 182; ORNL–3945, (1966) p. 183.

199. A. L. Myers, W. J. McDowell and C. F. Coleman, *J. Inorg. Nucl. Chem.*, **26**, 2005 (1964).

200. D. C. Madigan, *J. Appl. Chem.*, **9**, 252 (1959).

201. E. V. Ukraintsev and N. F. Kashcheer, *Radiokhimiya*, **4**, 279 (1962).

202. D. F. Peppard, M. N. Namboodiri and G. W. Mason, *J. Inorg. Nucl. Chem.*, **24**, 979 (1962).

203. P. H. Tedesco, V. B. de Rumi and J. A. Gonzalez Quintana, *J. Inorg. Nucl. Chem.*, **28**, 3027 (1966).

204. T. Sato, *J. Inorg. Nucl. Chem.*, **27**, 1395 (1965).

205. D. Dyrssen and D. H. Liem, *Acta Chem. Scand.*, **18**, 224 (1964).

206. C. J. Hardy and D. Scargill, *J. Inorg. Nucl. Chem.*, **17**, 337 (1961).

207. O. Navratil in *Solvent Extraction Chemistry* (Eds. D. Dyrssen, J. O. Liljenzin and J. Rydberg), North-Holland, Amsterdam, 1967, p. 256.

208. C. J. Hardy, D. Scargill and J. M. Fletcher, *J. Inorg. Nucl. Chem.*, **7**, 257 (1958).

209. V. A. Mikhailov, V. B. Shevchenko and V. A. Kolganov, *Zh. Neorg. Khim.*, **3**, 1959 (1958).

210. A. S. Kertes and A. Beck, *J. Chem. Soc.*, 5046 (1961).

211. A. Beck, *Ph.D. Thesis*, The Hebrew University of Jerusalem, 1962.

212. A. S. Kertes and A. Beck, *Proc. 7th. Intern. Conf. Coordination Chem.*, *Stockholm*, 352 (1962).

213. G. E. Boyd and Q. V. Larson, *J. Phys. Chem.*, **64**, 988 (1960).

214. B. J. Thamer, *J. Phys. Chem.*, **64**, 694 (1960).

215. Y. Habousha, *M.Sc. Thesis*, The Hebrew University of Jerusalem, 1963.

216. B. F. Greenfield and C. J. Hardy, *J. Inorg. Nucl. Chem.*, **21**, 359 (1961).

217. D. F. Peppard and J. R. Ferraro, *J. Inorg. Nucl. Chem.*, **15**, 365 (1960).

218. A. P. Samodelov, *Zh. Neorg. Khim.*, **10**, 1723 (1965).

219. I. S. Levin and V. A. Mikhailov, *Dokl. Akad. Nauk SSSR*, **138**, 1392 (1961).

220. I. S. Levin and T. V. Zabolotskii, *Dokl. Akad. Nauk SSSR*, **139**, 158 (1961).

221. E. M. Scadden and N. E. Ballou, *Anal. Chem.*, **25**, 1602 (1953).

222. D. F. Peppard, G. W. Mason, W. J. Driscoll and R. J. Sironen, *J. Inorg. Nucl. Chem.*, **7**, 276 (1958).

223 D. F. Peppard, G. W. Mason and R. J. Sironen, *J. Inorg. Nucl. Chem.*, **10**, 117 (1959).

224. D. F. Peppard, G. W. Mason, W. J. Driscoll and S. McCarty, *J. Inorg. Nucl. Chem.*, **12**, 141 (1959).

225. C. G. Warren and J. F. Suttle, *J. Inorg. Nucl. Chem.*, **12**, 336 (1960).

226. *Ann. Rept.*, Chemical Technology Division, Oak Ridge Natl. Lab., *USAEC Rept.*, ORNL–2993, (1960) p. 140.

227. C. A. Blake, Jr., C. F. Baes, Jr. and K. B. Brown, *Ind. Eng. Chem.*, **50**, 1763 (1958).

228. T. S. Urbanski, *Nukleonika*, **5**, 831 (1950).

229. R. D. Baybarz and R. E. Leuze, *Nucl. Sci. Eng.*, **11**, 90 (1961).

230. C. A. Blake, Jr., C.F. Baes, Jr., K. B. Brown, C. F. Coleman and J. C. White, *Proc. U.N. Intern. Conf. Peaceful Uses At. Energy*, 2nd, Geneva, **28**, 289 (1958).

231. T. S. Urbanski, *Nukleonika*, **5**, 341 (1960); **6**, 299 (1961).

232. T. S. Urbanski and S. Minc, *Nukleonika*, **6**, 765 (1961).

233. B. N. Laskorin, V. S. Ulyianov and R. A. Sviridova in *Ekstraktziya*, Vol. 1, Gosatomizdat, Moscow, 1962, p. 171.

234. M. Zangen, *Bull. Res. Council Israel*, **7A**, 153 (1958).

235. C. S. Cronan, *Chem. Eng.*, **66**, 108 (1959).

236. G. W. Mason, S. Lewey and D. F. Peppard, *J. Inorg. Nucl. Chem.*, **26**, 2271 (1964).

237. C. A. Blake, D. J. Crouse, C. F. Coleman, K. B. Brown and A. D. Kelmers, *USAEC Rept.*, ORNL–1172 (1957).

238. R. E. McHenry and J. C. Posey, *Ind. Eng. Chem.*, **53**, 647 (1961).

239. Ann. Prog. Report, *USAEC Rept.*, ORNL–3452, (1963) p. 111.

240. Z. Kolarik in *Solvent Extraction Chemistry* (Eds. D. Dyrssen, J. O. Liljenzin and J. Rydberg, North-Holland, Amsterdam 1967, p. 250.

241. T. Sato, *J. Inorg. Nucl. Chem.*, **27**, 1853 (1965).

242. M. Taube, *J. Inorg. Nucl. Chem.*, **12**, 174 (1959).

243. J. Kennedy, *Chem. Ind. (London)*, 950 (1958).

244. C. A. Blake, Jr., C. F. Baes, Jr. and K. B. Brown, *Ind. Eng. Chem.*, **50**, 1763 (1958).

245. B. Hoek-Bernstrom, *Acta Chem. Scand.*, **10**, 163, 174 (1956).

246. M. J. Jaycock and A. D. Jone in *Solvent Extraction Chemistry* (Eds. D. Dyrssen, J. O. Liljenzin and J. Rydberg), North-Holland, Amsterdam, 1967, p. 160.

247. P. G. Manning and C. B. Monk, *Trans. Faraday Soc.*, **57**, 1996 (1961).

248. I. P. Alimarin and S. Han-hsi, *Zh. Neorg. Khim.*, **6**, 2062 (1961).

249. J. Stary and V. Balek, *Collection Czech. Chem. Commun.*, **27**, 809 (1962).

250. G. R. Choppin and P. J. Unrein, *J. Inorg. Nucl. Chem.*, **25**, 387 (1963).

251. J. M. White, P. Tang and N. C. Li, *J. Inorg. Nucl. Chem.*, **14**, 255 (1960).

252. S. M. Wang and N.C. Li, *J. Inorg. Nucl. Chem.*, **27**, 2093 (1965).

253. J. M. White, P. Kelly and N. C. Li, *J. Inorg. Nucl. Chem.*, **16**, 337 (1961).

254. I. A. Tserkovnitskaya and A. K. Charykov, *Radiokhimiya*, **2**, 222 (1960).

255. L. M. Gindin, P. I. Bobikov, E. F. Kouba and A. V. Bugaeva, *Zh. Neorg. Khim.*, **5**, 1868, 2366 (1960).

256. D. F. C. Morris and M. W. Jones, *J. Inorg. Nucl. Chem.*, **27**, 2454 (1965).

257. L. M. Gindin, A. A. Vasilyeva and I. M. Ivanov, *Zh. Neorg. Khim.*, **10**, 497 (1965).

258. N. M. Adamski, S. M. Karpacheva and A. M. Rozen in *Ekstraktziya*, Vol. 2, Gosatomizdat, Moscow, 1962, p. 80.

259. L. M. Gindin, P. I. Bobikov, G. M. Patinkov and A. M. Rozen in *Ekstraktziya*, Vol. 2, Gosatomizdat, Moscow, 1962, p. 87.

260. E. P. Martin and R. C. Pink, *J. Chem. Soc.*, 1750 (1948).

261. V. D. Tughan and R. C. Pink, *J. Chem. Soc.*, 1804 (1951).

262. S. M. Nelson and R. C. Pink, *J. Chem. Soc.*, 1744 (1952).

263. R. L. Martin and H. Waterman, *J. Chem. Soc.*, 2545 (1957)

264. D. P. Graddon, *Nature*, **186**, 715 (1960).

265. C. M. French and E. R. Monks, *J. Chem. Soc.*, 466 (1961).

266. W. V. Malik and R. Haque, *Z. Anal. Chem.*, **189**, 179 (1962).

267. S. Kaufman and C. R. Singleterry, *J. Colloid Sci.*, **10**, 139 (1955).

268. S. Kaufman and C. R. Singleterry, *J. Colloid Sci.*, **12**, 465 (1957).

269. A. W. Ralston, *Fatty Acids and their Derivatives*, J. Wiley & Sons, New York, 1948, p. 376.

270. W. S. Singleton in *Fatty Acids* (Ed. K. S. Markeley), Interscience Publ. Inc., New York, 1960, Part I, p. 609.

271. N. M. Adamski, S. M. Karpacheva and S. I. Sorokin, *Radiokhimiya*, 3, 284 (1961).

272. R. C. Archibald, *J. Am. Chem. Soc.*, 54, 3178 (1932).

273. H. W. Smith, *J. Phys. Chem.*, 21, 204 (1921).

274. H. W. Smith, *J. Phys. Chem.*, 21, 616 (1921).

275. S. D. Christian and M. W. C. Dharmawarchana, *J. Phys. Chem.*, 66, 1187 (1962).

276. T. Ishimori, E. Nakamura and H. Murakami, *Nippon Genshiryoku Gakkaishi*, 3, 193 (1961).

277. R. L. Martin and A. Whitley, *J. Chem. Soc.*, 1394 (1958).

278. K. Shinoda, T. Nakagawa, B. Tamamushi and T. Isemura, *Colloidal Surfactants*, Academic Press, New York, 1963.

279. J. L. Moilliet and B. Collie, *Surface Activity*, Spon Ltd., London, 1951.

280. M. E. L. McBain and E. Hutchinson, *Solubilization*, Academic Press, New York, 1955.

281. A. M. Schwartz, J. W. Perry and J. Berch, *Surface-active Agents and Detergents*, Interscience Publ. Inc., New York, 1958.

282. G. S. Hartley, *Quart. Rev.*, (London) 2, 152 (1948).

283. K. S. Markley, *Fatty Acids*, Vol. 2, Interscience Publ. Inc., New York, 1960, Vol. 2, p. 715.

284. Ref. 269, p. 887.

285. S. R. Palit and J. W. McBain, *J. Am. Oil Chem. Soc.*, 24, 190 (1947).

286. S. R. Palit and J. W. McBain, *Ind. Eng. Chem.*, 38, 741 (1946).

287. Ref. 269, p. 284.

288. L. Arkin and C. R. Singleterry, *J. Colloid Sci.*, 4, 537 (1949).

289. S. M. Karpacheva, N. M. Adamskii and V. V. Borisov, *Radiokhimiya*, 3, 272 (1961).

290. S. M. Karpacheva, N. M. Adamskii and V. V. Borisov, *Radiokhimiya*, 3, 291 (1961).

291. *Ann. Progr. Rept.*, *Chemical Technology Division*, Oak Ridge Natl. Lab., *USAEC Rept.*, ORNL–3627, (1964) p. 213.

292. P. W. West, T. C. Lyons and J. K. Carlton, *Anal. Chim. Acta*, 6, 400 (1952).

293. A. K. Sundaram and S. Banerjee, *Anal. Chim. Acta.*, 8, 526 (1953).

294. S. Banerjee, A. K. Sundaram and H. D. Sharma, *Anal. Chim. Acta*, 10, 256 (1956).

295. G. E. Boyd and S. Lindenbaum, *USAEC Rept.*, ORNL–2782, (1959) p. 44.

296. S. Miyamoto, *Mem. Fac. Sci. Kyushu Univ. Ser. C.*, 3, 93 (1960).

297. T. Ishimori, E. Nakamura and H. Murakami, *Nippon Genshiryoku Gakkaishi*, 3, 590 (1961).

298. E. Nakamura, *Nippon Genshiryoku Gakkaishi*, **3**, 684 (1961).
299. T. Ishimori and E. Nakamura, *Nippon Genshiryoku Kenkyusho*, JAERI 1047 (1963).
300. *Ann. Progr. Rept. Chemical Technology Division*, Oak Ridge Natl. Lab., *USAEC Rept.*, ORNL–3452, (1963) p. 175.
301. D. Dyrssen, *J. Inorg. Nucl. Chem.*, **8**, 291 (1958).
302. M. Tanaka and T. Niinomi, *J. Inorg. Nucl. Chem.*, **27**, 431 (1965); M. Tanaka, N. Nakasuka and S. Goto in *Solvent Extraction Chemistry* (Eds. D. Dyrssen, J. O. Liljenzin and J. Rydberg), North-Holland, Amsterdam 1967 p. 154.
303. B. G. Harvey, H. G. Heal, A. G. Maddock and E. L. Rowle, *J. Chem. Soc.*, 1010 (1947).
304. A. W. Fletcher and D. S. Flett, *J. Appl. Chem.*, **14**, 250 (1964).
305. L. V. Shikheeva, *Zh. Neorg. Khim.*, **10**, 1486 (1965).
306. D. P. Graddon, *J. Inorg. Nucl. Chem.*, **11**, 337 (1959).

9.

Extraction by Solvation

According to the classification of extraction systnms adopted in this monograph, we consider all oxygen-bearing organic solvents to extract electrically neutral inorganic species by virtue of solvation. This definition may appear to some extent arbitrary insofar as under certain conditions an acid extraction system involving an oxygenated extractant may form a cation with the proton, and thus extract the acid by virtue of ion association. Nevertheless, we wish to make a clear distinction between systems discussed in this and in the following chapter, where the more basic, mostly nitrogen-containing extractants, by virtue of their basicity, readily form ion-pairs.

There are strong attractive forces between solvent and solute in the organic phase (section 2.C). Inorganic species are solvated by coordination of oxygen-bearing solvent molecules to the central atom, often through a water-molecule bridge. It is thus debatable whether ion-association species involving oxygen-bearing solvents have any physically real distinction from the more common interaction involving the solvation of the inorganic molecule or the central atom. Oxygen-containing extractants generally do not meet the criterion of being able to form a cation with the central atom, essential for an oxonium mechanism.

The electronegativity of the extractant, affects the extractability of inorganic species. Both inorganic and organic acids, for example, are easily extractable by strongly basic amines, whereas they are less readily extracted by organophosphorus compounds, and only moderately extractable by the even less basic compounds such as ethers and ketones. These differences between the extractive efficiency of the various solvent groups, are what should be expected as a result of the competition between water, solvent and the anion for the available proton, or more generally, for the available coordinating positions around the metal ion. It is obvious that the strongly basic amine will compete favourably with water and the anion in extracting the bare proton. In fact, it is the anhydrous acid which is extracted into an amine-bearing organic phase. On the other hand, organophosphorus and weakly basic carbonyl compounds extract acids along with a varying number of water molecules, especially, in the latter case, where more than one aqueous–solvate complex may exist in equilibrium.

Despite the similar extraction reaction, taking place between the inorganic species and oxygenated organic solvents, the extraction systems considered in this chapter are certainly not identical. Several features emphasize the difference between ethers, ketones and other oxygenated solvents (class I), and the neutral organophosphorus esters (class II). It is the strongly polar character of the latter which is responsible for the differences.

One of the most striking differences between the two classes of extractants lies in the specific role ascribed to water. In organophosphorus ester-containing systems, water is frequently eliminated from the organic-phase metal complex, whereas in ethers and ketones, water is a necessary part of the complex, usually acting as a bond between the solvent and salt molecules. For example, uranyl nitrate is transferred into dibutyl-carbitol along with four water molecules, whereas the salt alone will go into tributyl phosphate. A comparison of the thermodynamic data for the extraction of the salt by these two solvents shows their very different extracting power (as indicated by the thermodynamic partition coefficient P).

TABLE 9A1. Data for the extraction of uranyl nitrate into tri-n-butyl phosphate (TBP) and dibutylcarbitol (DBC)[1].

	TBP	DBC
P	2230	1.53
ΔG	-4580	-2530
ΔH	-3615	-7700
ΔS	$+3.24$	-17.3

The heat content change is bigger for DBC, the weak-donor solvent, most probably due to the transfer of the hydrated salt. Moreover, the uranyl nitrate complex isolated from TBP has the composition $UO_2(NO_3)_2 \cdot 2TBP$, whereas that from a DBC solution in carbon tetrachloride the complex has the formula $UO_2(NO_3)_2 \cdot DBC \cdot 2H_2O$. The water in the stable crystalline solid is held very tenaciously and could not be removed without destroying the complex.

Another dominant role is played by the extent of solvation of the extracted species, which is again, different for the two classes of solvents. Just because of the first hand importance of this parameter, numerous efforts have been made to evaluate the degree of solvation. While these efforts were generally successful when phosphorus compounds were used as extractants, in systems involving ethers, ketones and alcohols the task proved to be formidable. In a few cases the solvation number has been estimated, but more frequently a dilution of the class I extractant by an inert diluent does not lead to an unequivocal assignment of the solvation number. In reality, in such systems a number of mixed hydrate-solvates coexist in equilibrium, as there is no complete dehydration of the electrolyte during its transfer from the aqueous into the organic phase.

Owing to a high extractive capacity of the phosphorus extractants, their solutions may be rendered almost ideal when brought to high dilutions by an 'inert' solvent. This enables such systems to be handled mathematically, in contrast to systems where class I extractants are employed. In the latter systems, the contribution to the overall extraction by the extractive capacity of the inert diluent may, in certain cases, become important, in view of the low distribution ratios obtained by ethers, ketones and alcohols alone. Thus in the majority of cases where these are involved there is hardly any theoretical assistance to predict the behaviour of any particular system.

Finally, one would expect the extractive capacity of these two classes of extractants, and the composition of the organic-phase complexes, to show a different dependence on the electrostatic energy of hydration of the metal or the metal salt. One may attempt to define this relationship in very general terms. The affinity of the solvent is usually saturated by the metal ion, and vice versa, the electropositivity of the metal is saturated by the solvent whenever strong extractant (class II) and readily extractable electrolytes are combined. Such complexes are usually unhydrated and their solvation number is low, as for example in the system tributyl phosphate and hexavalent actinide nitrates. For the same extractant but a less strongly electropositive metal, the complex in the organic phase is hydrated, frequently with two or three water molecules. This may be explained by the persisting attraction of the metal for water. Similarly, hydrated organic-phase complexes are obtained when a less eletronegative extractant extracts a stongly electropositive metal. Here again, the metal salt is primarily hydrated and the solvent becomes attached to the complex either on the coordination sites left over by the water, or perferentially, in a second solvation shell, occurring through hydrogen bonding to the water molecules in the primary shell. In the latter case the number of organic molecules attached loosely to the electrolyte is indeed, very large, as for example, in extraction of plutonium (IV) nitrate by ketones.

A. CARBON-BONDED OXYGEN-DONOR EXTRACTANTS

a. Introduction

The extractants to be discussed here are all electron donors. The relative chemical inertness and relative water immiscibility of alkyl ethers and polyethers, ketones and some esters, placed these organic compounds among the first solvents to be used and studied in solvent extraction of inorganic species. Water-immiscible higher alcohols, though included in the group of solvents dealt with in this section, represent a special group of solvents compared to the ethers, ketones and esters. Their particular extractive behaviour is caused by the amphoteric nature of the hydroxyl group, most similar to that of water, showing both donor and acceptor properties. This characteristic of alcohols affects their solvating capacity, which in turn is the major parameter governing the solubility of inorganic salts in, and their extractability into, them. Other solvents of this class, lacking hydrogen, are only available for coordination through their basic oxygen. It should be noted, however, that the order of solvent efficiency found experimentally is not necessarily correlated with the power of solvents to form hydrogen bonds.

One of the outstanding features of the extraction systems discussed in this and the following section is an appreciable change in the individual phase volumes when water or aqueous electrolyte solutions are equilibrated with the polar oxygen-containing extractants. This behaviour is almost general, whenever solvents capable of hydrogen bonding are involved (section 2.B). Frequently it is the aqueous phase which increases at the expense of that of the solvent phase. The mutual solubility is affected by all the constituents of the extraction system. In many cases it is appreciably higher when aqueous acid solutions are equilibrated, and frequently there is a certain aqueous acid concentration where the phases become completely miscible.

Though the main extraction reaction in the systems under consideration is the solvation of electrically neutral species, a possible dissociation of the solute in the organic phase must be taken into account since some of these solvents are sufficiently polar to bring about ionic reactions. An alternative explanation for the observed electrical conductivity in the solvent phase in the case of acid extraction is the oxonium hypothesis, postulating the direct addition of the proton of a strong acid to the basic oxygen of the solvent.

Mainly, it is these two reasons—the unusually high mutual miscibility, and the organic-phase dissociation—which make the majority of the systems considered here show a thermodynamically non-ideal behaviour. Only in exceptional cases, elaborate sets of equations, taking into account organic-phase reactions of both dissociation and association, in addition to the aqueous-phase considerations, were successful enough to allow a mathematical treatment of the partition data. For this reason we preferred to discuss the extraction systems, in this and the following section, according to their chemical behaviour and character, rather than by strict thermodynamic criteria.

b. The Extractants

Organic solvents discussed in this chapter fall into the groups of alcohols, ethers, ketones and esters, each group comprising usually several homologous series. Instances are known where some solvents containing more than one functional group are successful extractants. However, the polyethers were practically the only polyfunctional solvents finding a wide application. Others, such as the glycol ethers and the disubstituted amides, which have been shown to be in some cases extraordinarily powerful solvents[2], are not much in demand.

Each of these classes of solvents is characterized by its functional group, the characteristic properties being exhibited most strongly by the lower homologues. For the purpose considered here, the most important property of the solvents is their basicity, which generally decreases as the molecular weight increases, or as the proportion of the functional group to the hydrocarbon radical of the molecule decreases. The effect of this factor upon the extractive characteristics of solvents will be discussed in the following sections, together with other structural and physical considerations.

Certain characteristics of a solvent, such as stability, viscosity, volatility, water immiscibility, dielectric constant, dipole moment, inflammability, toxicity and cost, should all be known so that a solvent suitable for a particular purpose can be chosen. Several of these characteristics are frequently interrelated. Thus, for instance, the molecular weight of a solvent governs to a considerable extent its water miscibility and toxicity. Or, in a homologous series, the increase in the molecular weight is accompanied by an increase in solvent viscosity and density.

Some of the physical properties of the most popular solvents used in solvent extraction are compiled in Appendix C. It was attempted to avoid the vexing subject of nomenclature by using names commonly found in solvent extraction literature. In questionable cases, or when commercial designations are used, skeletal structural formula have been given also. The values quoted are taken mainly from standard reference sources[3-9] or pamphlets supplied by solvent manufactures. The latter frequently refer to products of commercial purity, thus some of them are of value only as rough guides.

Some of the solvents, especially the ethers, decompose on standing. Peroxidation of polyethers in light and air under normal laboratory conditions is appreciable[10]. The peroxides are usually removed by standard methods[3-9,11], and the purified solvent stored in solid CO_2 until required[12], or addition of $10^{-5}M$ hydroquinone or catechol keeps the ether peroxide-free for several months[13]. In this regard Irvine and coworkers[14,15] note that a simple passage of once-distilled ether through a column of activated alumina results in a product with a significantly higher extraction power than triple-distilled ether, and three times greater distribution ratios than the commercial solvent as supplied by the manufacturer. So purified ether deteriorates more slowly and can be restored to its original state by passing through the column before use. The authors[14,15] believe that the increased stability of their product is due to the removal, by alumina, of impurities which catalyse its decomposition.

c. Solubility of the Extractants in Water and Aqueous Electrolyte Solutions

For any mass-action treatment of experimental partition data (section 7.B) it is important to be certain that the equilibrium aqueous phase contains no solvent in either free or complexed form. With the type of extractants discussed in this section this is, however, only exceptionally the case. It becomes important in this respect to take into account the losses of organic-phase species to the aqueous phase.

The extent of solvent solubility in water is shown in Appendix D for a number of alcohols, ethers, ketones and esters most frequently used. Whenever possible data from standard reference books[16-20] were selected without attempting to compare critically older data with new. The values must be regarded as merely indicative, good for a broad comparison only. The reason is that frequently values quoted by various authors show remarkably wide discrepancies. These discrepancies are often due to an inevitable element of uncertainty, on account of the different sources, qualities and purities of solvents used in determinations as well as the different experimental techniques, or differences in the conditions of measurements. For illustration, several literature data for the solubility of diethyl ether in water (and water in the ether) are given in Table 9A2. Similar discrepancies were noted for n-hexyl alcohol[16,19,26], diisopropyl ether[16,21,27] and many other solvents.

An examination of the tabulated data (Appendix D) reveals that the higher molecular weight solvents, due to their increased size, hydrocarbon-

TABLE 9A2. Mutual solubilities of diethyl ether and water (weight %).

Ether in water	Water in ether	Temperature (°C)	Reference
6.896	1.264	20	16
6.02	1.30	25	21
6.50	3.01	20	22
7.30		20—22	23
5.28	3.95	22	24
6.95	1.20	20	25

like nature and lower basicity, are less soluble in, and dissolve less water. Irrespective of the class, the solubility of a solvent in water is influenced by its ability to form hydrogen bonds. Acting against water solubility of solvents, in addition to size, nature and basicity, is the attraction between the molecules of the solvent itself. A comparison of solubilities (Table 9A3) of normal aliphatic compounds with six carbon atoms in the molecule will illustrate this effect. Normal hexanol, being able to form double hydrogen bonds with water by virtue of its functional group, would be expected to exhibit the highest solubility in water. This is, however, not the case. Association of its molecules in the pure liquid state interferes apparently with the expected order of solubilities. As to the mutual attraction due to hydrogen bonding between water and the solvent, Hildebrand and Scott[29] comparing the solubility of some solvents in water, interpreted the differences as due to the different fugacities they introduced into water. The authors[29] noted that 'it is the hydrogen-bonding character rather than the dipole moment which determines the ability of the substance to dissolve in water'. These examples show that an interpretation of water solubilities in organic solvents and their prediction is by no means simple (section 2.B).

TABLE 9A3. Water solubility of some normal aliphatic compounds with six carbon atoms in the molecule.

Solvent	B.p. (°C)	Temperature (°C)	Vapour pressure[28] (mm Hg)	Solubility (weight %) solvent in water	Solubility (weight %) water in solvent	Dipole moment (debyes)
n-Butyl methyl ketone	127	25	5	3.5	3.7	2.70
n-Butyl acetate	127	25	8.2	1.0	1.37	1.85
n-Hexanol	158	20	1	0.58	7.2	1.64
Di-n-propyl ether	90	20	70	0.25	0.68	1.10

Burger[30] investigated the physical properties of a large number of linear dialkyl diethers and their mutual solubility with water. The structure variables were the terminal groups, the oxygen–oxygen spacing and the carbon-to-oxygen ratio. The data show a decrease in solubility with increase in carbon-to-oxygen ratio. A further point of interest is the decreased solubility with increased molecular weight but the same carbon-to-oxygen ratio (diethyl ether, diethoxybutane and dibutylcarbitol).

Turning now to the solubility of these non-electrolytes in aqueous electrolyte solutions, it is desirable to discuss the solubility in acid and salt solutions separately.

The effect acids have on the mutual solubility of water and solvent can best be evaluated from ternary diagrams of water–acid–solvent. Such diagrams have been constructed for many ternary systems pertinent to solvent extraction, and some of them will be mentioned in the next section. For most practical purposes, however, a direct determination of solvent solubility in aqueous electrolyte solutions provides the information needed for a correct evaluation of the partition data.

As the aqueous acid concentration increases, the solubility of most solvents in water increases slowly at first, then rises sharply. In many acid–solvent systems the incseased solubility is accompanied by very marked changes in phase volumes. Illustrative of this are the studies on the solubility of diethyl ether in hydrochloric[31,32] and hydrobromic[33] acid, sulphuric, perchloric and phosphoric acids[31,34]; of isopropyl ether in hydrochloric[13,27,32,35,36] and hydrobromic[36] acid; of isobutyl- and isopentyl alcohol in aqueous solutions of hydrogen halides[37,38] except HF; of isopropyl ketone in HCl and HBr[36]; and n-butanol[38], methyl hexyl ketone[32], cyclohexanone[39], amyl formate and acetate[32], ethyl, propyl and butyl benzoate[39] in aqueous hydrochloric acid solutions. Some of the data have been plotted in Figure 9A1. As can be seen from the curves of isopropyl ether, the reproducibility is rather poor. This is probably due to the different techniques employed rather than to the slight temperature differences at which the measurements were made. Some of the techniques, generally those measuring volumes, are unlikely to be very reliable.

When diluted by an inert diluent the solubility of the solvent in the aqueous phase decreases. This has been shown to be the case with cyclohexanone in hydrochloric acid[39], but it is believed to be a more general phenomenon for oxygen-bearing solvents in different aqueous acid solutions.

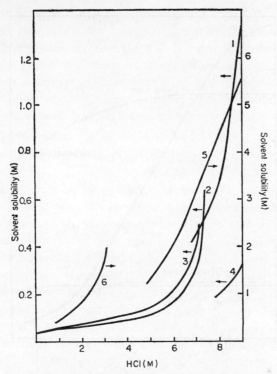

Figure 9A1. Solubility of solvents in aqueous HCl solutions at 22 ± 2°C.
1) Isopropyl ether, ref. 32;　　2) Isopropyl ether, ref. 35;
3) Isopropyl ether, ref. 27;　　4) Isoamyl acetate, ref. 32;
5) Diethyl ether, ref. 32;　　　6) Cyclohexanone, ref. 39.

An opposite solvent solubility effect is observed when the aqueous electrolyte is a salt rather than an acid. Various dialkyl diethers are salted-out from aqueous solutions of ammonium nitrate[40], and the solubility of cyclohexanone in hydrochloric acid decreases when the acid is replaced partially by lithium chloride[39]. A good illustration of the effect was provided by McKay and coworkers[23,41] who measured the solubility of various ethers and ketones in aqueous solutions of uranyl nitrate. Some of their data are plotted in Figure 9A2. The data were obtained by vapour-pressure measurements and were accurate enough to enable the authors to calculate the effect of solvent on the activity of the salt and water in the aqueous solution.

An interpretation of the phenomena described above is certainly not simple. A change in phase volumes due to a change of solvent solubility in the aqueous acid solution results obviously in a modification of the

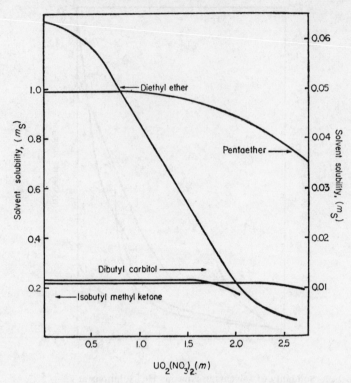

Figure 9A2. Solubility of solvents in aqueous solutions of uranyl nitrate at $21 \pm 1\,°C$.

media, both the aqueous and the organic, since acid is simultaneously extracted into the solvent phase as well. An increase of the mutual solubility indicates a progressively increased similarity in the characteristics of the phases, which often become eventually completely miscible.

Attempting a qualitative interpretation, it seems safe to assume that the sharp increase in the solubility of the solvent at a certain aqueous acid concentration should be related to the formation of acid–solvent adducts in *both* phases. In the case of hydrochloric acid solution and equal volumes of the phases, for example, the sharp increase in solubility is noted at about 7 M initial concentration. At this acidity the aqueous HCl solution becomes water deficient, when calculated on the basis of the tetrahydrated proton. It becomes then tempting to assume that solvent molecules are transferred to the aqueous phase in order to provide solvating molecules for the hydronium ion. At the very same initial acid concentration the extractability of the acid increases sharply (this will be shown in the following sections), again presumably due to the incomplete hydration of the proton in the aqueous solution. Consequently, the fraction of mixed hydrato-solvated acid species increases in both phases, exhibiting a high mutual solubility.

The greatly different solubility behaviour of the solvents in neutral aqueous salt solutions is perhaps related to a lower degree and energy of hydration of metal salts, as compared to that of the proton. If this is true, the solubility behaviour of solvents in aqueous solutions of strongly hydrated salts, such as those of beryllium and aluminium, should be similar to that in acids. There is, however, no evidence available to validate this assumption, though the salting-out efficiency of these 'dehydrating' electrolytes tends to support the above view. We shall return to this topic when dealing with the extraction of metals.

Despite the lack of theoretical instruments to treat the phenomenon, and evaluate its consequences in studies of complex formation by distribution measurements, the mutual solubility of the phases must be taken into account. The practical and simple way appears to be to precondition the phases. This technique obviously eliminates the changes in phase volumes provided the concentration of the extractable metal species is kept low.

d. Solubility of Water in the Extractant

The frequently appreciable solubility of the solvent in water is not necessarily accompanied by a significant phase volume change upon equilibration. The reason is the solubility of water in the organic phase, usually of a comparable magnitude to that of the solvent solubility. For many alcohols, however, the organic phase increases in volume, indicating a higher solubility of water in the solvent than vice versa. The solubility of water in alcohols is sometimes higher by as much as an order of magnitude than that in ethers, ketones and esters of comparable aliphatic chain length. For example, at 20°C a saturated solution of isopentyl alcohol is 4.34 molar in water, whereas isoamyl acetate is 0.49 M, and diisopropyl ether is only 0.25 M[42]. Compare also data in Table 9A3.

Solubility measurements with various ethers and polyethers indicate that the uptake of water is a function of the carbon-to-oxygen ratio, increasing with decreasing ratio[43-45]. Glueckauf's data are plotted in Figure 9A3. As originally plotted by the author, log $\bar{m}_W$ is a linear function of the oxygen-to-carbon ratio. It should be noted that the latter relationship was found to hold considerably well for the affinity of the extractants towards other electrolytes extracted from aqueous solutions[44].

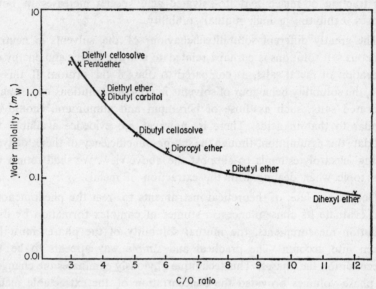

Figure 9A3. Extraction of water into ethers as a function of the carbon-to-oxygen ratio.

Water solubility is, furthermore, a linear function of the activity of water for a number of different solvents. This is generally true for those whose solubility for water lies below a molar water-to-solvent ratio of five. Such is the case with dibutyl[45], β,β'-dichlorethyl[46] and diisoamyl ether[46], methyl and isoamyl butyrate[46] and dibutylcarbitol[43,46,47]. The experimental data for another polyether, the dibutyl tetraethylene glycol, pentaether, conform to the linear dependence for values of a_W up to about 0.7. Pentaether, however, has a high affinity for water, reaching at $a_W = 1$, a mole ratio of water-to-solvent of 0.67^{43}. Some of the data are shown in Figure 9A4.

These results show that the solubility of water in solvents throughout the range of linearity conforms to Henry's law

$$\bar{a}_W = k\bar{m}_W/(1 + 0.001 M_S \bar{m}_W) \qquad (9A1)$$

where M_S is the molecular weight of the solvent, and Henry's law constant has a value of 15.9 for diisoamyl ether, 13.1 for dibutyl ether, 3.7 for β,β'-dichlorethyl ether and 1.6 for dibutylcarbitol. For pentaether, up to a point of $\bar{m}_W = 1$, k has a value of 0.95.

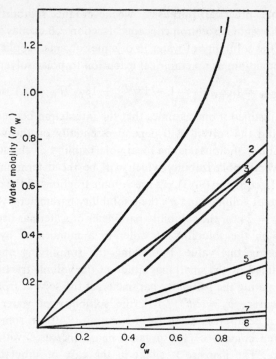

Figure 9A4. Solubility of water in organic solvents at 25°C as a function of water activity.
1) Pentaether, ref. 47;   2) Methylbutyrate, ref. 46;
3) Dibutyl carbitol, ref. 47;   4) Dibutyl carbitol, ref. 46;
5) Isoamylbutyrate, ref. 46;   6) β, β'-Dichlorethyl ether, ref. 46;
7) Dibutyl ether, ref. 45;   8) Diisopentyl ether, ref. 46.

Additionally, as McKay and coworkers[47] have pointed out, in all these cases the solvent activity should obey Raoult's law

$$\bar{a}_S = 1/(1 + 0.001\, M_S \bar{m}_w) \qquad (9A2)$$

Such apparently ideal behaviour of solvent–water binary systems that obey Henry's law is supported by some spectral[48] and vapour-pressure measurements[1]. I.r. spectra of water-saturated ethyl and butyl acetate are characterized by three sharp bands in the 3500–3600 cm^{-1} spectral region, indicating loosely bound water, presumably through weak hydrogen bonding. Vapour-pressure measurements at 25°C on a series of water-saturated dibutyl carbitol–carbon tetrachloride mixtures, after allowing for the water vapour pressure itself, gave the same vapour pressure as the corresponding dry mixture.

For some practical purposes, whenever the solubility of water is small, the regular solution concept[29] (section 2.B.c) may be useful in evaluating the solubility of water in organic solvents. Application of the solubility equation, as an empirical extension to polar solvents,

$$\ln \bar{a}_{W} = \ln \varphi_{W} + \varphi_{S}(1 - 18/V_{S}) + 18\varphi_{S}^{2}(\delta_{W} - \delta_{S})^{2}/RT \qquad (9A3)$$

may seem justified if one assumes that the interaction between the water molecule and the solvent will depend essentially on the water dipole. Thus, in dilute solutions it is the local polarizability of the solvent rather than its overall polarizability which will be the determining factor of solubility. In equation (9A3), φ's are volume fractions of the components, V's their molal volumes and δ's their solubility parameters. An empirical value of $\delta_{W} = 24$ for the solubility parameter of water has been evaluated from data on the solubility of water in a number of hydrocarbons. Now, a) using this value, b) making the simplifying approximation $\varphi_{S} \cong 1$, justified by the small magnitude of the volume fraction of water, and c) estimating the solubility parameter of the solvent from its energy of vaporization, $\delta_{S} = (\Delta E^{V}/V_{S})^{1/2}$, the solubility of water in volume fraction can be approximated from equation (9A3), in some cases.

The water content of the organic phase increases with increasing temperature. The increase is slight in the case of dibutycarbitol and pentaether[47], but is considerable (from 0.27 wt% at 0°C to 0.73 wt% at 55°C) when β,β'-dichlorethyl ether is equilibrated with water[49].

The solubility of water in solvents when extracted from aqueous electrolyte solutions along with acids or metal salts will be discussed in the following sections separately.

e. Extraction of Acids

In most practical solvent extraction studies, when tracer or even microgram amounts of metal salt are used, the overall behaviour of the system will not differ significantly from that of the simpler basic system containing the aqueous acid (and perhaps a non-extractable electrolyte) and the organic solvent only. Thus, the behaviour of such systems can be regarded in most cases as a common feature for the extraction of any metal, at least as far as the physical properties of the phases and the nature of the acid–solvent interaction are concerned.

As to the physical properties, there is, first of all, the effect due to the mutual miscibility of the phases. Extraction of acids is usually accompanied by substantial changes in the phase volumes. One may even state that a readiness for miscibility of an organic solvent with an aqueous acid solution is apparently a condition for the extractability of the acid[42]. Its extent, thus, is determined by the solvent and the nature of the aqueous phase. For example, the extractability of HCl (and also HBr and HI) is much higher into diethyl and diisopropyl ether than into β,β'-dichlorethyl and di-n-butyl ether, the latter two solvents exhibiting little or no solubility in the aqueous phase. Or, as a rule, alcohols extract the acids much more efficiently than esters, ketones and ethers do, usually in that order. This is true for any mineral acid[50-52].

Mutual miscibility increases with increasing acid concentration, but at equinormal acidities, in a given solvent, the extent of miscibility will be different for different acids. For example, the volume changes upon equilibration increase in the order HI > HBr > HCl for many ketones and ethers. A complete miscibility is reached at lower aqueous acid concentration in the same order. Interestingly, however, as will be discussed later, these changes are not necessarily due to an increased normality of either acid or water in the solvent phase.

This brings us to the second factor affecting the changes in physical properties of the phases. That is the extractability of water. Great attention has been paid to the connection between acid extractability and water distribution. Indeed, the extraction of acids provides a typical example of the effect of hydration upon extractability in general, and that of the proton in particular. This aspect of acid extraction has been thoroughly investigated[12,36,53] and discussed[54] by Diamond and Tuck. Briefly, the extractability of acids will depend upon the competition between molecules of water and of solvent, and the anion for the proton. With strong acids, the strong hydration energy of the proton causes the acid transferred into the organic phase to be almost always hydrated, and they take more water along into the other phase than do weak acids. The composition of the acidic species suggested in diisopropyl ketone (DIPK) and dibutyl cellosolve (DBC), are DBC. H_2O. HNO_3, DBC. $H_2O.CCl_3COOH$, DBC. $(H_2O)_4$. $HClO_4$ and DIPK .$(H_2O)_4.HClO_4$. They illustrate the effect of acid strength upon the hydration of the proton.

The low hydration of relatively weak acids such as nitric and trichloracetic, is due to the high fraction of the undissociated acid in the aqueous phase, which is the one preferentially extracted. On the other hand, strong acids extract essentially either a) by the addition of one or more oxygen-containing organic solvents to the trihydrated hydronium cation

$H_3O^+(H_2O)_3$ or b) as in the case of more basic solvents, by the direct addition of the solvent to H_3O^+ replacing thus one or more water molecules. In spite of these differences, both weak and strong acids extract essentially according to the same reaction of solvation, in terms formulated in the introduction to this Chapter, and not through the formation of ion-association complexes.

Since extraction by ion-associate formation requires the direct addition of the oxygen-containing solvent to the bare proton of the acid, we feel that the strongest argument in favour of extraction by solvation and not by the ion-association is the role water plays in the structure of the organic-phase complex. If there were no water extracted—as in the case of ion-association systems to be discussed in the following Chapter—one could assume that the ion-association type complex $R_2OH^+X^-$ (with an ether, for example) would exist in the organic phase. However, in the presence of water infrared spectra of many acid-solvent systems indicate that the stretching vibrations of water are those due to strongly bound water. Since these vibrations are absent from the spectra of water-saturated solutions in absence of acid, we believe that the effect is due to the water being strongly bound to the proton as the oxonium ion H_3O^+. The additional water hydrates this ion. Now, it is unlikely that the solvent molecules will be bound in a manner different from that of the water molecule attached in excess to the oxonium ion. We thus feel that the extracted water is associated with the acid, and we deal in these extraction systems with the solvation of the hydrated acid rather than hydration of the solvated acid.

In order to interpret the distribution of an acid between water and solvent, complete information can be obtained by the isotherms of a ternary diagram, since any change of one component in a ternary system will alter the concentration of the other two. Such ternary diagrams, however, are available for only a few systems. Among them are n-butyl, isobutyl and isoamyl alcohol with aqueous HCl, HBr and HI[37], cyclohexanone[37], diethyl[42] and diisopropyl ether[27] with hydrochloric acid, the system methyl isobutyl ketone–hydrofluoric acid–water[64], and di-n-butyl ether–nitric acid–water[74]. For a complete construction of such diagrams, involving the determination of positions of the tie lines and their correlation, and the determination of the critical mixing point by construction from the conjugate lines, the reader is referred to the more specific literature[57]. The quantitative aspects of the phase rule as applied to such systems and a thermodynamic approach to the equilibria involved, have been recently discussed on the hydrochloric acid–water–diethyl ether system. These considerations will not be repeated here and the reader is referred to the original report[58].

(i) *Nitric acid* Most of the practical solvents have been screened for their extractive capacity towards nitric acid. The more recent sources of information are given in Table 9A4.

In Figure 9A5 the curves drawn represent the distribution of nitric acid between some of the solvents and water. The data on these systems from other sources agree with the curves drawn, though slight deviations are a rule rather than exception. Alcohols have a higher extractive power than any other type of solvent. In any given homologous series, the extent of acid extracted decreases with increasing molecular weight of the solvent. Figure 9A6 illustrates this effect for normal symmetrical dialkyl ethers and a series of normal alkyl methyl ketones[61]. The aqueous solution in all these acids was kept constant at 5 M HNO_3.

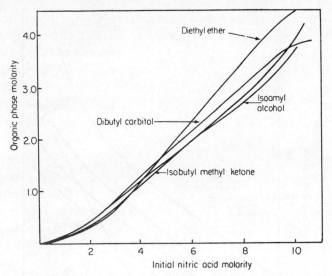

Figure 9A5. Extraction of nitric acid by various solvents[52,64,72,82].

It is characteristic of nitric acid–organic solvent systems that the volume of the equilibrium organic phase increases with the aqueous acidity, regardless of the nature of the oxygenated solvent employed. Figure 9A7 demonstrates the phenomenon. Alcohols show the highest tendency for mutual solubility during acid extraction, as the shape of the isoamyl alcohol curve implies. One carbon atom lower alcohols, n-butyl

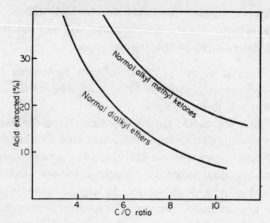

Figure 9A6. Extraction of nitric acid as a function of carbon-to-oxygen ratio in the solvent.

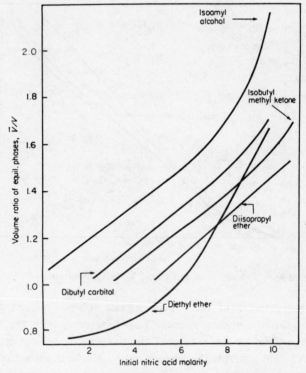

Figure 9A7. Volume changes on equilibration of nitric acid with organic solvents[64,66,82].

TABLE 9A4. Extraction of nitric acid by carbon-bonded
oxygen type extractants.

Solvent	References
Ethers	
Diethyl	51, 52, 59–64
Ethyl n-propyl	65
Di-n-propyl	61, 65
Diisopropyl	61, 66
Di-n-butyl	43, 56, 61, 67–71
Di-n-pentyl	61
Di-n-hexyl	43
β, β'-Dichlorethyl	49, 61
Dimethoxypentane	40
Dibutyl cellosolve	10, 36, 53
Dibutylcarbitol	10, 12, 36, 40, 43, 64, 67, 72
Pentaether	43, 67
Ketones	
Diethyl	61
Diisopropyl	36, 61
Methyl n-propyl	61, 63, 73
Methyl isobutyl	61, 64, 73–78
Methyl *tert*-butyl	61, 73
Methyl-n-pentyl	61, 73
Methyl-n-hexyl	61, 73
Methyl-n-nonyl	61
Cyclohexanone	61, 74
Methylcyclohexanone	73, 79
Alcohols	
n-Butyl	52
Isobutyl	80
Isopentyl	51, 52, 81, 82
n-Hexyl	70
n-Octyl	70
Benzyl	52, 81
Esters	
Isobutyl acetate	52
Isopentyl acetate	52

and isobutyl alcohol are completely miscible already with a 5 M nitric acid aqueous solution[52,80]. For nitric acid systems the behaviour of diethyl ether at low acidities is quite an exception. Generally speaking, the acidity at which sharp changes in volume of the phases occur depends on the structure and the molecular weight of the solvent. Since this acid concentration frequently coincides with the acidity at which the solvent exhibits its maximum extractive efficiency for any metallic species, for reasons yet to be discussed, it is of a practical importance to have that information in choosing the optimum extraction conditions.

The transfer of nitric acid to the organic solvent is increased in the presence of non-extracted nitrates because of both common-ion and salting-out effects. The extractability increases with the concentration of the nitrate, as shown in Figure 9A8, and the effect is more pronounced at higher acid concentrations as seen in Figure 9A9. However, in order to have any significant salting-out effect, the concentration of the salt must be rather high. In the methyl isobutyl ketone system the acid distribution is only slightly affected when the concentration of

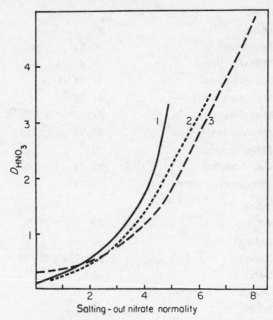

Salting-out nitrate normality

Figure 9A8. Effect of salting-out nitrates on the distribution of nitric acid in various systems.
1) $Fe(NO_3)_3$ at constant 1.5 N HNO_3 into diethel ether[83].
2) $LiNO_3$ at constant 1.0 N HNO_3 into methyl isobutyl ketone[76].
3) $Ca(NO_3)_2$ at constant 2.0 N HNO_3 into isoamyl alcohol[51].

$LiNO_3^{63,76}$,$Ca(NO_3)_2^{78}$or $Al(NO_3)_3^{40}$ are kept below 2 N, when compared to its distribution in absence of the salts. The salting-out action is different for different nitrates, increasing, for example in a diethyl ether system, from ammonium to zinc to cadmium to aluminium nitrate[52]. The order of efficiency is identical to that for the extraction of metals, and is apparently determined by the same factors. This, again, will be discussed later in more detail.

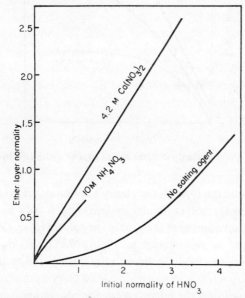

Figure 9A9. Extraction of nitric acid from water and aqueous electrolyte solutions into diethyl ether[62].

The water content in some normal ethers when in equilibrium with aqueous nitric acid solutions is shown in Figure 9A10. In many other solvents investigated, dihexyl ether[43], β,β'-dichlorethyl ether[49], dibutyl-carbitol[12,43,64,72], dibutyl cellosolve[53], pentaether[43], diisopropyl ketone[36] and methyl isobutyl ketone[64], the presence of nitric acid increases the water uptake of the organic phase. Apparently, the effect of acid is greater for solvents exhibiting a higher affinity towards water when in equilibrium with pure water.

Several authors[43,49,67,71] found that the water uptake is proportional to the activity of water

$$\bar{m}_w = (k\bar{m}_S + h\bar{m}_A)\bar{a}_w \qquad (9A4)$$

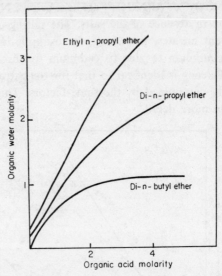

Figure. 9A10. Water molarity in some ethers against nitric acid molarity[65,71].

where k, from equation (9A1) is the Henry's law constant for the solubility of water in a binary solvent–water system, and h may be regarded as the mean hydration number of the acid in the solvent layer. Glueckauf and Davies[43] plotted, as reproduced in Figure 9A11, $\bar{m}_W/\bar{a}_W$ against $\bar{m}_A$ for various ethers to show the linear relationship, indicating that were $\bar{a}_W$ constant, every nitric acid molecule would carry a constant, h, amount of water molecules into the ether, except for pentaether.

Along the same line of reasoning, and applying simple thermodynamics, several authors have derived a mean hydration number for nitric acid in various solvents from equally good fits as those shown in Figure 9A11. For di-n-butyl ether, for example, Glueckauf's[43] value for h is 0.83, while McKay's[67] value is 0.88, and Vdovenko[56,69] found that up to $\bar{m}_A = 1.5$, h has a value of 0.6, but 0.15 at higher acid loadings. Or, for dibutyl-carbitol several different values for the mean hydration number have been found experimentally: $h = 1.0^{12,64}$, $h = 1.75^{72}$ and $h = 1.84^{43}$. Other values quoted for similar systems were: diisopropyl ketone[36] $h = 1$; methyl isobutyl ketone[64] $h \cong 1$; dihexyl ether[43] $h = 0.5$; pentaether $h = 8^{43}$ and $h = 6.4^{67}$, and for β,β'-dichlorethyl ether[49] the value for h decreased from 1.7 to 1.0, when $\bar{m}_A$ increased from 0.03 to 0.9.

The variations in the derived values of h in a given solvent, and the wide range through which the values vary in going from one solvent to

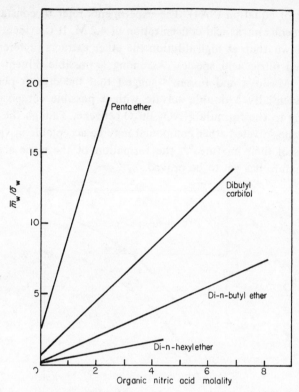

Figure 9A11. Water uptake in ethers as function of water activity and nitric acid content[43]. (By permission from U.K. Atomic Energy Authority).

another, throw the assumption of true hydrate formation into considerable doubt. This, and two more experimental facts, that of the temperature effect on h and the effect of solvent concentration in the organic phase upon it, tend to indicate that the acid does not form any specific hydrate but the organic solution is rather a mixture of species with varying amounts of solvent and water.

At least in the case of di-n-butyl ether[43], the hydration number derived by equation (9A4) was reported to be temperature dependent. The values found were 0.95, 0.83 and 0.66 for 0°, 25° and 50°C respectively, indicating an exothermic interaction between the solutes in the ether.

Even more striking is the effect upon h when the solvent is diluted by a non-extracing diluent. In Figure 9A12, $\bar{m}_{Wcorr}/\bar{m}_A$ is plotted for di-n-butyl and di-n-propyl ether mixtures with benzene[65,71], where $\bar{m}_{Wcorr}$ is the water molality of the ether phase when corrected for the first term on the right side of equation (9A4). The experiments refer to constant equilibrium aqueous nitric acid concentration of 4.2 M. It is evident from the curves shown that at high dilution the ether extracts preferentially an unhydrated nitric acid species. Assuming a possible diluent–ether interaction, Maslova and Fomin[65] suggest that the organic-phase nitric acid is essentially a double solvate with a possible composition corresponding to the formula $HNO_3.Bu_2O.benzene$. Though the existence of a benzene-solvated ether compound may be acceptable in view of the properties of their mixtures[84], the formation of the nitric acid adduct in this mixture has yet to be proved.

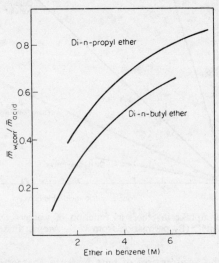

Figure 9A12. The influence of benzene diluent on the water content of nitric acid extracts in ethers.

Additional data on these[69] and other systems, including dibutylcarbitol–carbon tetrachloride[43,64], ethyl n-propyl ether–benzene[65], diethyl ether–benzene[64] and methyl isobutyl ketone–benzene[64] support the general conclusion that when the solvent is diluted with an inert diluent, h is dependent also on the mole fraction of the extracting solvent component.

It is doubtful whether the interaction in such two-phase four-component systems can be treated thermodynamically without being open to serious objections. The basic assumption which has been made is that in the extraction of acid and water, the concentration of the unbound water (to the acid) will be dependent on the water activity in the aqueous solution and the concentration of the solvent in the diluent. However, this is unlikely to be a fair assumption, since the unbound water must be proportional to the activity of the uncomplexed solvent rather than its total concentration. Thus, the term $\bar{a}_S = \bar{m}_S \bar{\gamma}_S$ has to be corrected for the proportion to which the concentration of the free solvent falls as it is progressively converted to the acid solvate. Furthermore, the amount of unbound water extracted by the uncomplexed ether will be affected by the changes in the dielectric properties of the mixture brought about by an increased fraction of the polar acid–solvate adduct in it. To demonstrate these changes, which actually apply to ternary systems (without diluent) as well, would be a formidable task.

To obtain a clearer picture about the nature of the acid–solvent interaction, and perhaps for its quantitative understanding, dilution experiments were carried out for a number of systems. Without taking into account activity coefficients of the organic species, but taking the precaution of keeping their mole fraction as low as practical for measurements, the value of n in the solvation equilibrium.

$$\text{HNO}_3 + n\bar{S} \rightleftharpoons \overline{\text{HNO}_3 \cdot S_n} \qquad\qquad K_{11} \qquad (9A5)$$

has been found to be unity when mono- and polyethers were used as extractants. The reported values for the equilibrium quotient K_{11} of reaction (9A5) are complied in Table 9A5. They cannot be considered accurate in view of the assumptions made about the activities involved[65,70]. Furthermore, in the majority of the cases the quotients had a constant value in a limited range of solute concentrations only.

TABLE 9A5. Equilibrium quotients K_{11} (equation 9A5) and K_{111} (equation 9A6) for the formation of nitric acid solvates at room temperature.

Extractant	Diluent	K_{11}	K_{111}	Reference
Diethyl ether	CCl$_4$	0.061		64
Ethyl n-propyl ether	Benzene		0.0078	65
Di-n-propyl ether	Benzene		0.0040	65
Di-n-butyl ether	Benzene	0.00088	0.0022	71
	CCl$_4$		0.0020	43
Dibutylcarbitol	CCl$_4$		0.064	43
	CCl$_4$	0.40		64
Methyl isobutyl ketone	CCl$_4$	0.040		64

There is independent experimental evidence available to show that there is a rather strong hydrogen bond existing between the acid and the solvent. Glueckauf and Davies[43] calculated the heat of solution of 100% nitric acid in dry di-n-butyl ether from direct calorimetric measurements, and found that the value of 7.25 kcal/mole of HNO_3 at infinite dilution, corresponds to the heat of solution of the acid in water. This solvation energy is definitely high. Or, the partial vapour pressure of nitric acid measured in diethyl ether[60] as a function of acid concentration, shows that up to a 1:1 molar ratio of acid to ether, the partial vapour pressure of acid remains practically zero, while that of the ether decreases to zero. Thus, in equimolar proportion there is a vapour pressure close to zero at room temperature, believed to be due to a specific compound formation. Fragmentary information of the infrared absorption spectrum of a dibutylcarbitol extract shows that the absorption peak at 1124 cm^{-1} due to the free ethereal oxygen linkage is shifted to lower values in the acid adduct[12]. The hydroxyl group frequencies in HNO_3 are also absent in such solutions. Cryoscopic measurements on di-n-butyl ether–nitric acid in benzene were interpreted in terms of monosolvate formation[71]. Finally, though not necessarily evidence for compound formation, the electrical conductivity measurements on extracts in ethers indicate only slight dissociation of the solvate, at least at low acid loadings[43]. Similar conclusions were drawn from distribution data in a dibutylcarbitol system.[12] The extent of dissociation is, however, more marked in ketones and it is appreciable in alcohols. Conductance of the acid in methyl isobutyl ketone is compared with that of other acids in Figure 9A13[85]. There is a two orders of magnitude

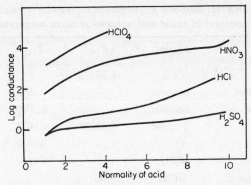

Figure 9A13. Conductance of methyl isobutyl ketone after extraction of various acid solutions[85]. (By permission from Elsevier Publ. Comp.)

difference, in the conductivity of HNO_3 compared to that of HCl. In hexyl alcohol nitric acid is considerably ionized, even when diluted with benzene[70].

With the exception of polyethers and a few other extraction systems the total amount of bound water entering the organic phase remains less than the quantity of acid. The simplest explanation, made by several authors[36,43,65,71], is that both anhydrous acid-solvates and the monohydrate acid-solvates are present in the organic phase in different proportions, depending on the nature of the solvent and the acid concentration. In some cases, making a simplifying assumption as to the activity coefficients, the equilibrium quotient, K_{111}, for the reaction

$$HNO_3 + H_2O + \overline{S} \rightleftharpoons \overline{HNO_3.H_2O.S} \qquad K_{111} \qquad (9A6)$$

has been estimated. These results are included in Table 9A5. Tuck[36] in the diisopropyl ketone system estimated the ratio between the hydrated and unhydrated species to be around 20, in good agreement with previous estimates in dibutyl cellosolve and dibutylcarbitol[72]. As to the hydrated acid solvate, the author suggested a double hydrogen–bonded formula

In view of the fact that in a number of nitric acid systems the amount of acid extracted is beyond that corresponding to an acid-to-solvent ratio of unity, one or more new species may be present in the organic phase. While the existence of specific complexes with more than one nitric acid per solvent may seem plausible with the polyethers, their formation with the simpler oxygenated solvents is questionable. Glueckauf and Davies[43] calculated the apparent molar heat content of di-n-butyl ether in an extract at high acid loadings, from heat of dilution and heat of mixing experiments, and found that the excess acid is essentially in the form of associated nitric acid in strong interaction with the simple, probably hydrated, acid–ether adduct. Glueckauf[44] suggested that the new species consisting of three HNO_3 molecules per molecule of di-n-butyl ether has a quadrupolar arrangement of the type

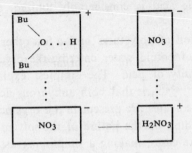

for which a heat of formation of 4.8 kcal per mole HNO_3 has been calculated. In the case of dibutylcarbitol, a polyether with three ethereal oxygens, the authors[43] found that species with second and third HNO_3 molecules exist when the HNO_3/DBC ratio exceeds 0.5. This conclusion was indicated essentially by heat of dilution experiments, and allowed the authors to calculate the formation constants of the various species in equilibrium, containing one, two and three nitric acid molecules per solvent. There is no experimental indication that dibutylcarbitol is able to extract more acid than that corresponding to the trinitrate. Highly loaded extracts are rather unstable.

For the sake of a rigorous thermodynamic treatment of partition data, several authors have attempted to evaluate the activity coefficient of nitric acid in various solvents including diethyl ether[64], dibutyl ether[43,70], dibutylcarbitol[12,64], methyl isobutyl ketone[64] and hexyl alcohol[70]. The calculations were based on a formal mass-action law treatment of the experimental data after several assumptions had been made. Among them are: a) that the solvent dissolved in the aqueous phase has no influence on the mean activity coefficient of aqueous nitric acid; b) that the acid transferred into the organic phase is essentially undissociated; and c) that only one type of acid complex exists in the organic phase. Apparently, these assumptions are not fully justified, and the activity coefficients reported vary over a wide range.

The stability of the majority of organic solvents towards nitric acid solutions at higher concentrations is unsatisfactory. Isopentyl alcohol[82], diisopropyl ether[66], dibutylcarbitol[10,12,40], methyl isobutyl ketone[75,68,87] and methyl cyclohexanone[79] have been reported to be especially unstable at high acid loadings. The reaction between the ethers, even freshly distilled, and the acid, is violent, with liberation of nitrogen oxides. The chief oxidation degradation products of dibutylcarbitol are glyoxal and butanol, but butyric and acetic acid, and especially oxalic acid are always present in variable quantities. With isopropyl ether, the time required for this reaction depends on the acid concentration, being instantaneous when in contact with 14 M aqueous acid solution. The ketones undergo similar reaction when in contact with acid solutions higher than about 5 M. The time necessary for the reaction to be visible depends, again, on the acid concentration and the amounts of nitrogen oxide vapours above the aqueous acid solution. Freshly distilled nitric acid solutions yield more stable extracts.

The instability of the organic phases should be regarded as an important factor, since the various decomposition products frequently affect the extractive properties of the solvent. The alcohols formed from ketones and ethers may markedly change the initial selectivity of the solvent. Or, traces of oxalic acid, for example, formed in degradation of the highly vulnerable ethylene glycol linkage in the polyethers will react with the metal ion, complex them or precipitate them in the aqueous phase.

(ii) *Halogen acids* Aqueous halogen acid–oxygenated solvent systems, with the possible exception of some alcohols, are characterized by a low degree of acid extractability. More recent experiments on systems involving hydrochloric acid extraction (Table 9A6) show that while alcohols are fairly good extractants, all other classes of solvents extract the acid to an appreciable extent only from highly concentrated aqueous solutions. Hydrobromic and hydriodic acids are somewhat more easily extracted than hydrochloric acid.

The extractability of hydrogen fluoride from aqueous solutions up to 20 M, increases in the order di-n-butyl ether < pentyl acetate < pentyl alcohol < methyl isobutyl ketone < diethyl ether[50,55]. The extractability of this acid is at least one order of magnitude higher than that of other hydrogen halides.

Into ethers, ketones and esters the order of extractability for other acids is HCl < HBr < HI, the differences often being considerable. The differences in their extractabilities into alcohols are less marked. Generally speaking, the partition of the acids increases with their aqueous concentration. With HCl and HBr the distribution ratios into ethers and ketones are only occasionally higher than $10^{-3} - 10^{-2}$ up to 6–7 M initial concentration; the organic acid being measurable to any reasonable precision only above an initial concentration of about 3 M. There is frequently a maximum in the distribution ratio of the acids with their increasing initial concentration. Such is the case, for example, in diethyl and diisopropyl ether systems, but there is no maximum with di-n-butyl and β,β'-dichlorethyl ether. The phenomenon was explained[54] as being due to an increased solubility in the aqueous phase of the former ethers above a certain initial acid concentration, whereas dibutyl and β,β'-dichlorethyl ether show little or no aqueous solubility.

In contrast to the nature of the volume changes in nitric acid extraction systems, equilibration of hydrochloric, hydrobromic and hydriodic, but not hydrofluoric acid, with the majority of ethers and ketones results in an increase of the aqueous phase at the expense of that of the organic phase. Fig. 9A14 gives the volume changes $(\bar{V}/V)$ which occur on equilibrating diethyl ether at 20°C with initially equal volumes of the four halo acids[50,58]. Mutual miscibility increases with increasing acidity, and at equimolar acidities it is greatest for hydriodic acid. Complete miscibility is reached at lower acid concentration in the same order. Figure 9A15 shows some

TABLE 9A6. References for the extraction of halogen acids by oxygen-bearing solvents.

Solvent	Acids			
	HF	HCl	HBr	HI
Ethers				
Diethyl	50, 55	42, 48, 51, 54, 88, 96	33, 42, 54, 58, 98	54, 58, 99
Diisopropyl		13, 27, 42, 54, 88-90, 96	42, 54, 58	54, 58
Di-*n*-butyl	50, 55	54, 91, 96	54	54
β,β'-Dichlorethyl		49, 54, 58, 90	49, 54, 58	54, 58, 90
Dibutyl cellosolve		36		
Ketones				
Diisopropyl		36, 92	36	
Methyl isobutyl	50, 55			
Cyclohexanone		37		
Alcohols				
Isobutyl		37	37	37
Isopentyl	50, 55	37, 42, 51, 81, 93, 94	37, 42, 51	37
n-Hexyl		94		
n-Heptyl		94		
n-Octyl		81, 94, 95		
Benzyl		81		
Esters				
Ethyl acetate		42, 48		
Butyl acetate		48, 97		
Isopentyl acetate	50, 55	42, 51, 96	51	

typical results for the volume changes occurring with various solvents in HCl systems. The tendency for mutual solubility is greater with the lower members of a homologous series. The $\bar{V}/V$ ratios are 2.92, 1.70 and 1.12 for diethyl, di-n-propyl and di-n-butyl ethers, respectively, when equilibrated with 20 M HF solution. As for the other acids, the sequence in Figure 9A14 prevails for any given solvent. While there is only a slight volume change with the chlorinated ether in the HCl system, a somewhat greater mutual solubility may be observed with HI[90]. As mentioned, when there is no volume change, there is no maximum in the distribution ratio of acids with their increasing aqueous concentration. As will be shown later, in such cases there is no maximum in the distribution ratios of metal halides extracted from aqueous hydrohalide solutions, either.

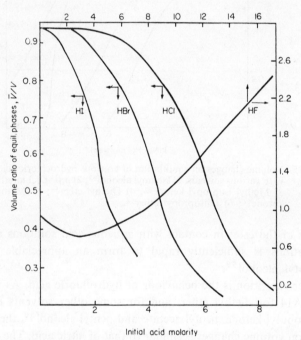

Initial acid molarity

Figure 9A14. Volume changes on equilibrium of diethyl ether with aqueous hydrogen halide solutions.

An exception to the above order of volume changes is exhibited by alcohols, and, to some extent by the esters. As shown in Figure 9A15, isopentyl alcohol increases in volume at the expense of the equilibrium aqueous phase. The behaviour of isopentyl acetate is similar, where

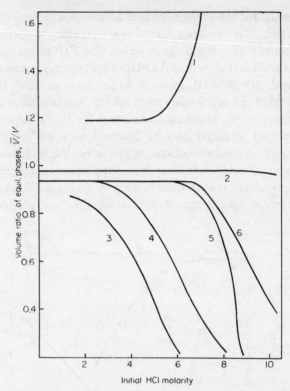

Figure 9A15.   Volume changes on equilibrium of aqueous hydrochloric acid solutions with various solvents. 1) Isoamyl alcohol[93]; 2) β,β'-Dichlorethyl ether[90], 3) Methyl isopropyl ketone[42]; 4) Diethyl ether[58]; 5) Methyl isobutyl ketone[42]; 6) Diisopropyl ether[58].

acidolysis of the ester in contact with aqueous acid solutions at higher concentrations is sufficiently rapid to form an appreciable quantity of isopentyl alcohol[58].

Another exception is the behaviour of hydrofluoric acid. As shown in Figure 9A14 for diethyl ether, and for some other solvents including methyl isobutyl ketone, pentyl acetate and pentyl alcohol[50], the effect of this acid on volume changes is similar to that of nitric acid. The direction of volume changes is apparently connected with the extent of extractability of the acid, since both HNO_3 and HF are more extractable than other hydrogen halides by several orders of magnitude. Though the former are much weaker acids in water, it is not necessarily the acid strength which governs acid extractability and, thus, the direction of volume changes. This will be evident from a comparison of the extractive behaviour of perchloric acid (similar to HNO_3 and HF) and phosphoric acid (similar to HCl and HBr). The reasons for the volume change, its direction and extent, should be taken into consideration when evaluating the various factors affecting the relative extractability of acids into oxygenated solvents.

As would be expected, the extraction of acids is greater when the aqueous solution contains an essentially unextractable salt with anion in common, and is higher with higher aqueous electrolyte concentration[33,89,93,100]. Equally expected, the amount of acid extracted is lower when the solvent is diluted by a non-extracting hydrocarbon. For example, when dibutyl ether is diluted, the HCl content is almost directly proportional to the ether concentration[91]. Due to changes in the activity coefficient of the ether in the diluent, there is a slight positive deviation (after correction for the HCl extracted by the benzene alone) from proportionality on a molarity scale when benzene is the diluent, and a negative one when carbon tetrachloride is employed.

Turning now to the transfer of water into the solvent phase during the extraction of acids, there is again a clear distinction to be made between the behaviour of ethers and ketones, and that of alcohols (and esters to some extent). The direction of volume changes is, of course, related to the extent of water solubility in the organic phase. As the concentration of HCl and HBr in the aqueous phase increases, the water uptake of the ethereal phase decreases[13,33,42,49,83,91,96]. The opposite is true when alcohols are used as solvents[37,42]. The solubility of water in esters first decreases with increasing acidity but eventually rises again, as the hydrolysis of the esters proceeds[96]. For illustration, Irving and Rossotti's data[42] for a number of HCl systems are plotted in Figure 9A16.

Bearing in mind the generally low extractability of acids, the trend of water solubility in ethers does not imply by any means that the organic acid species is unhydrated. The solubility of water in β,β'-dichlorethyl ether, for example, is, by about an order of magnitude, higher (on a molar basis) than that of either HCl or HBr, when the solvent is in equilibrium with an 8 M, or lower, aqueous solution. Though, under these conditions the total organic water content is about three times lower than that when in contact with pure water, it is still higher when allowance is made for a decreased water solubility, due to a decreased water activity, as required by equation (9A4). Qualitatively, similar are the observations in the diisopropyl ether–HCl system[13], though here the

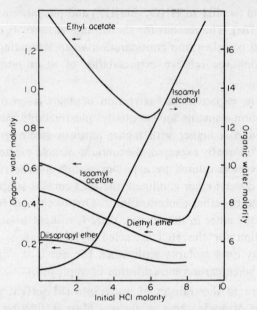

Figure 9A16. Organic water molarity as a function of hydrochloric acid molarity in the initial aqueous solution.

effect is complicated by the solubility of the ether in the acid phase. Irving and Rossotti[42] explained the connexion between water extractability and acid distribution in terms of mixed hydration–solvation of the acid in the aqueous phase. Taking the hydration number of these acids in water as 8.0, 8.6 and 10.6 for HCl, HBr and HI respectively, as a basis for calculation, at an aqueous acid concentration of 6.0, 5.5 and 4.4 M respectively, all the water will be bound to the acid. At higher concentrations then, these 'halogen acids could not be fully hydrated and the solvation sheaths [around the oxonium cation] would be completed by organic molecules withdrawn from the organic phase'. This is visible from the ever diminishing organic phase as demonstrated in Figures 9A14 and 9A15. Thus, for the range of acid concentration where the above mechanism applies, it is apparent that the hydrated–solvated acid complex is preferentially water soluble. This is the explanation for the low extractability of both acid and water. For illustration, in the system of 6 M initial HBr, with an initially equal volume of diethyl ether, the aqueous phase is increased by about 60%, though only 0.17% of HBr are transfered into the organic phase[98]. As the aqueous acid concentration increases even more, the equilibrium aqueous concentration reaches a certain maximum value which is lower than the initial acidity, and is the maximum acidity attainable. Obviously, this is primarily due to the marked dilution by the ether, and much less to the transfer of acid into the organic phase, as illustrated above. An increased initial acidity beyond this maximum value will cause the complete miscibility of the phases[42].

Now, we suggest that when this process is prevailing in the system, the acid complex becomes increasingly richer in solvating than in hydrating molecules, thus more 'ether like', and its partition will now be increasing in favour of the ether phase. This process is likely to contribute to the changes in the dielectric properties of the, so far, almost pure ether, to enhance its complete miscibility with the aqueous solution. In fact, the water content of diisopropyl ether, for example, reaches a minimum (not shown in Figure 9A16) then increases with increasing HCl in the ether[27]. An alternative explanation for that stage of the process, offered by Irving and Rossotti[42], considers that the decrease in the dielectric constant of the aqueous phase, due to the increased ether transfer, will also decrease the dissociation of the acid. The undissociated acid will be more extractable into the anhydrous ether.

The above considerations imply that hydrochloric acid, and also HBr and HI, do not exist as any specific hydrate in ethereal phases. There is rather a mixture of various hydrate-solvates which can be represented by the general formula $HCl.(H_2O)_n.S_m$. The values of n and m will vary throughout the range of acidities used, and will certainly vary from one solvent to another.

For the reasons outlined previously, as far as water coextraction is concerned, the behaviour of HF–ether (–ketone) systems is expected to be similar to those of nitric acid.

The water solubility isotherm and the nature of volume changes in alcohol systems cannot be explained along the lines described above. The solubility of alcohols in the aqueous phase is built up very slowly as compared to the increase in the solubility of water in the alcohols. At high halogen acid concentration, the solubility of isobutyl and isopentyl alcohols in the aqueous phase increases very rapidly[37], and a complete miscibility of the phases is achieved usually at much lower overall acidities than in the case of ethers and ketones. Though no quantitative data are available, it seems reasonable to assume[42], that the acids extract into alcohols in a fully hydrated form (a trihydrated oxonium ion[93]?).

Electrical conductivity on HCl extracts of isopentyl alcohol indicate an almost complete dissociation of the acid[93]. Distribution data, fitted into a set of mass-action law equations, confirm these findings.[93] In ketones and ethers the dissociation of HCl and HBr must be considerably lower. This implies the conductivity curve shown in Figure 9A13; the conductance of HCl in methyl isobutyl ketone is lower than that of nitric acid at equinormal concentrations by two powers of ten. While the acids extracted from dilute aqueous solutions into diethyl and diisopropyl ether were believed to be substantially ionized[88], in β,β'-dichlorethyl ether, despite its relatively high dielectric constant of 19.5, both HCl and HBr behave as weak acids. By conductivity measurements and distribution data, Irvine and coworkers[90,101] found a pK value of about 7.0 for HCl and 5.1 for HBr. For comparison, pK_{HCl} in nitrobenzene (dielec. const. 34.8) is 7.4.

Infrared spectra[48,97] of anhydrous hydrogen chloride saturated diethyl ether and ethyl and butyl acetate, compared with solutions where water has been added, indicate that some type of ionization, but not necessarily dissociation, might be present in wet solutions. By stepwise addition of water, the stretching vibration around 3200 cm^{-1}, characteristic of strongly hydrogen-bonded water, increases relative to the total water (free water at $\sim$ 3500 cm^{-1}). In view of the possible acidolytic degradation, caution should be exercized in interpretation of the spectra in the acetates.

(iii) *Other mineral acids* Here we shall briefly review the extraction of, mainly, perchloric, sulphuric and phosphoric acids. While the latter two acids have much in common as far as their extractive behaviour is concerned, perchloric acid differs from them in many respects. Unfortunately, there is not much information available on these systems to help to explain these differences.

Perchloric acid, generally speaking, has an extractive behaviour similar in many aspects to that of hydrofluoric and nitric acid. The main features would be: a) its relatively high extractability into ethers[49,51,53], ketones[36], alcohols[51,81,102] and esters[51]; b) volume changes upon equilibrium occurring at the expense of the aqueous phase; c) the acid is accompanied into the organic phase by considerable amounts of water; d) the acid is dissociated in ethers, ketones and, certainly, alcohols, behaving even in ether as a reasonably strong acid.

On the other hand, the overall extractive behaviour of sulphuric and phosphoric acids is closer to that of hydrochloric acid: a) low extractability into ethers[36,51,52,103-106] and ketones [36,104,106] from any concentration, and into esters[51,52,106] from not too concentrated ($<$7-8 N) aqueous solutions. High extractability into alcohols[51,52 81,104,106] and into esters from initially highly acidic aqueous solutions. This is due, presumably, as in HCl systems, to an increased hydrolysis of the esters; b) the volume changes occur in favour of the aqueous phase with ethers and ketones, and the reverse is the case when alcohols are employed as extractants; c) apparently, very little water crosses along with the acid into ethers and ketones.

Figure 9A17 shows the extraction of $HClO_4$ into ethers and a ketone. No volume changes occur with the chlorinated ether in the concentration range plotted, whereas for the other two solvents shown in the figure, the organic-phase volume increases. At low acid loadings in β,β'-dichlorethyl ether, the water content increases almost linearly with the acid concentration in the ether, though at first its concentration decreases, and starts to increase only as acid starts to be extracted, at about 3 M initial $HClO_4$. In dibutyl cellosolve the water content, higher than in the previous ether, is also greater at low organic loadings, the $H_2O/HClO_4$ ratio decreasing from 4.4 to 3.6 when the organic acid concentration increases from ~ 0.1 M to ~ 2.7 M [53].

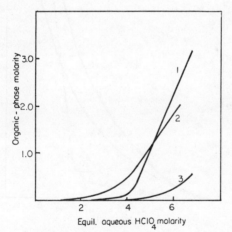

Figure 9A17. Extraction of perchloric acid into various solvents.
1) Dibutyl cellosolve[36];
2) Diisopropyl ketone[36];
3) β,β'-Dichlorethyl ether[49].

The relatively high water content in the organic phase when $HClO_4$ is extracted, contributes to the dissociation of the acid. In isobutyl methyl ketone, perchloric acid shows conductivity of an order of magnitude higher than nitric acid, and three powers of ten greater than HCl^{85} (Figure 9A13). The dissociation constant in β,β'-dichlorethyl ether, as determined conductometrically, has a value of about three orders of magnitude higher than that for hydrochloric acid in the same solvent[14].

The extraction curves of sulphuric and phosphoric acids into diethyl ether and isopentyl alcohol are shown in Figure 9A18, and the volume changes accompanying their extraction in Figure 9A19. The extractability of the acids is higher, by at least two orders of magnitude, into alcohols than into ethers. For a number of ethers, esters and alcohols their extractibility increases with the dielectric constant of the solvent[52]. The mole ratio of water to acid is close to unity in the phosphoric acid–diethyl ether system[105].

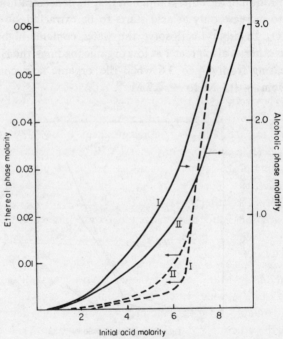

Figure 9A18. Distribution of sulphuric acid (I) and phosphoric acid (II) between water and diethyl ether (dashed lines) and isoamyl alcohol[52].

Infrared spectra of ethyl and butyl acetate extracts of sulphuric acid exhibit a band due to strongly bound water, absent from the spectra of a saturated solution of water alone in these solvents. This may be interpreted as being due to water molecules forming part of the extracted acid species[48].

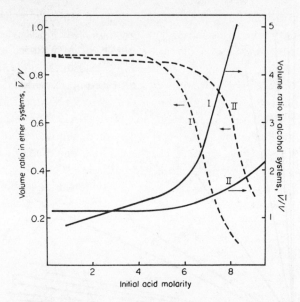

Figure 9A19. Volume changes on equilibrium in extraction of sulphuric acid (I) and phosphoric acid (II) by diethyl ether (dashed lines) and isoamyl alcohol[52].

The good extractability of thiocyanic acid into many oxygenated solvents has been known for many years[107]. From an aqueous solution of NH$_4$CNS containing a mineral acid, the, probably unhydrated HCNS is the organic-phase species. Its extractability decreases in the order alcohol > ketone > ether[108].

The distribution ratio of hydrogen peroxide into ethyl ether[109] and many other solvents[110] increases with its aqueous concentration, but is generally low.

(iv) *Organic acids* Many organic acids have been screened for their extractability into a wide variety of solvents, but especially rich information is available on diethyl ether systems. Three detailed survey studies with this ether[111-113] indicate that the distribution ratio is strongly dependent upon the acid strength and tendency of the acid towards dimerization. From an 0.1 normal aqueous solution, the distribution ratio decreases from 300 to 10 in the following order: benzenesulphonic, tartaric, citric, glycine, malic, sulphosalicylic, diglycolic, glycolic, oxalic, lactic, malonic and maleic[113]. Halogen derivatives of carboxylic acids, being fairly strong acids, show generally low distribution ratios. Distribution data for some of the acids are plotted in Figure 9A20[111].

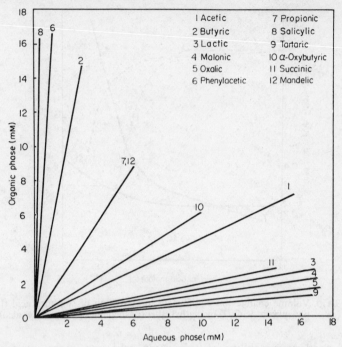

1 Acetic	7 Propionic
2 Butyric	8 Salicylic
3 Lactic	9 Tartaric
4 Malonic	10 α-Oxybutyric
5 Oxalic	11 Succinic
6 Phenylacetic	12 Mandelic

Figure 9A20. Distribution of organic acids between water and diethyl ether.

For any given homologous series the distribution ratio increases with increasing molecular weight of the acid. This effect, shown for diethyl ether in Figure 9A20, is equally valid for ketones and alcohols[21,114–116] and is still marked with hexanoic acid[115]. An interesting feature of the majority of the organic acid extraction systems is that the distribution ratio is only slightly affected by the initial aqueous concentration, unlike in mineral acid extraction systems.

The effect of acid strength and dimerization on the distribution ratio has been demonstrated on the series of acetic acid and its three chloro derivatives[117]. A comparison of the ratios in a number of organic solvents reveals that the monomer–dimer equilibrium is affected by solute–solvent interaction due to hydrogen bonding. Distribution constants have been calculated on the assumption that the experimental distribution ratios are governed by the extent of acid dimerization. While for some of the solvents, ethers for example[53], the assumption may be justified, in alcohols the possibility of acid dissociation in the organic phase may affect the partition of the stronger acids.

In a series of papers Geankoplis and coworkers[104,114,116,118] reported the influence of mineral acids upon the extractability of carboxylic acids into methyl isobutyl ketone. As one would expect, the extractability of the organic acid is increased in the presence of a strong acid, primarily due to a common-ion effect. To achieve a certain salting-out effect, less H_2SO_4 is needed for propionic acid than for acetic than for formic acid, the differences being expressed numerically through the Setchenow constant (section 1.F.a).

B. METAL EXTRACTION WITH CARBON-BONDED OXYGEN-DONOR EXTRACTANTS

a. Introduction

Oxygenated solvents, such as ethers, ketones and alcohols, dealt with in this section, are primarily extractants for metal halides and halo-metallic complexes, and have been investigated and used for some years. Irving and Williams[119] included in their review a comprehensive survey of the earlier literature on the extraction of inorganic substances by these solvents, as a part of the history of the development of inorganic solvent extraction chemistry. Diethyl and diisopropyl ether, for example, have been the accepted extractants for ferric chloride in analytical procedures already in the last century, and have remained to this day the most popular ones for its extractive separation. On the other hand, application of these solvents for extraction of metal nitrates is essentially a recent practice, initiated by the needs for an efficient separation technique in the technology of fuel reprocessing. Metal sulphates and perchlorates are generally non-extractable into these solvents, with the exception of some transition metal perchlorates which are extractable into alcohols[120].

One of the features of metal-extraction systems is the high degree of hydration of the solute in the organic phase, a phenomenon in common with the previously reviewed acid-extraction systems involving these solvents. $FeCl_3$, for example, needs about five molecules of water to become ether soluble, otherwise it separates out in the form of a second organic phase. Because of the low basicity of the solvents in comparison to that of class (II) extractants such as tributyl phosphate, expressed through the weak donor properties of their oxygen, water may remain directly bound to the metal ion in its transfer from the aqueous into the organic phase.

The solvent molecules are apparently hydrogen bonded to the water molecules of the primary hydration shell. The extent of solvation varies widely. Solvation numbers as high as 9 and 12, reported for plutonium(IV) in ketones[121] make the existence of specific solvates of any reasonable stability questionable. On the other hand, Specker and coworkers[122] suggested that efficient extractants, such as cyclohexanone (and the class(II)phosphorous esters) have less inorganic ligand molecules in the organic-phase complex, than do the less basic extractants such as ethers and the hydrogen-bonded alcohols, but that the sum of inorganic ligands and organic solvate molecules is constant. The species $Fe(SCN)_3$. 3S and $FeCl_3.3S$ have been claimed[122] to exist in the cyclohexanone (and TBP) extract, but $HFe(SCN)_4.2S$ in ether, under otherwise identical experimental conditions. Nevertheless, we believe that the real existence of such well-defined and specific complexes has yet to be confirmed. Unless this is done for this and many more systems, one is hardly justified in speaking in terms of specific solutes involving the class of extractants under consideration.

The great number of organic-phase complexes, exceeding that existing in simple acid extracts, makes the interpretation of the distribution data of metal salts, in terms of thermodynamic equations, rather complicated. In the majority of cases, these systems deviate from an ideal behaviour predicted by mass-action law equations derived in Chapter 7.

Finally, a number of very useful separation processes of analytical importance have been developed using the type of systems discussed here. Analytical aspects of solvent extraction are beyond the scope of this monograph, the reader is thus referred to the review articles appearing in the *Annual Reviews of Analytical Chemistry*[123].

b. Solution of Electrolytes in Oxygenated Solvents

The knowledge of the solubility of salts in solvents and especially the behaviour of such solutions, has direct bearing on similar extraction systems. Frequently, a good solubility of a hydrated salt in a water-immiscible solvent may serve as a basis for the prediction of extractability of the salt by this solvent. It is not only from the formal point of the extent of solubility, that dissolution of salts in organic solvents bears on the process of extraction, but more important is the formation of soluble solvates. Indeed, the composition of the solute species often resembles the one extracted into an organic solvent, and this is of a primary importance to the topic under consideration.

In this respect water plays an important role. In the case of uranyl nitrate, for example, a fair correlation of solubilities with extractabilities can be obtained only when the solubility of the hydrated salt has been compared, in view of the fact that the extractable uranyl nitrate complex is hydrated in the organic phase[124]. It is not surprising that water, at least part of it, is retained efficiently by the salt, since it is the strongest electron donor present in extraction systems involving ethers, ketones and frequently also alcohols. This, however, is not necessarily true when the solvent is a class (II) extractant, an organophosphorus compound. As will be shown (section D of this Chapter), uranyl nitrate is essentially unhydrated in tributyl phosphate systems. Thus the ratio of water to uranyl nitrate can serve as a measure of the relative competitive strengths of the various oxygenated extractants, though the order may be masked due to steric factors[54,100].

(i) *Solubility related to extraction* The solubility of electrolytes in active organic solvents has already been discussed in section 2.C. It has been shown that inorganic salts soluble in water-immiscible oxygen-bearing organic solvents are usually weak electrolytes, or they involve a complex ion. However, for the topic under consideration the coordination ability of the solvent becomes the important general factor governing the solubility of electrolytes.

The Born equation (2C5) and the various modifications of it[125,126] fail to describe the principal equilibria where account must be taken of specific chemical interactions as factors governing the solubility of electrolytes. There is, as yet, no adequate theory to express the forces that determine solubility when the solute–solvent interaction, expressed in the form of the solvation energy, is of a greater importance than the effect of dielectric constant upon the electrolytic dissociation. Or in other words, in terms of the solubility equations, the concept of dielectric continuum of solvent breaks down when complex formation becomes the important factor. It is in these terms that the solubility in oxygenated water-immiscible solvents has to be considered. In this respect, empirical relationships between solubilities and other properties of the salts and/or solutions have frequently been suggested, as will be shown in the forthcoming paragraphs, but a comprehensive theoretical treatment has yet to be provided.

Turning now to the effect of the solvent on the solubility of an ionic salt, a certain regularity becomes apparent. In an extensive series of investigations, Katzin and coworkers[127,128] and also Vdovenko[129,130] have studied the various factors affecting the solubility of several electrolytes, among them uranyl nitrate. They concluded[129,131] that the solubility of a salt is affected by the basicity of the solvent, its electron-donating ability, and the possibility of the metal ion maintaining its coordination number.

Alcohols exhibit a markedly higher dissolving power than either ketones or ethers. For $LiClO_4$, even water is a poorer solvent than many of the alcohols, the solubility of the salt in water at $25°C$ being only 0.092 mole fraction, as compared with a remarkable constancy of solubility found in many alcohols of about 0.35 mole fraction.

These unusually high dissolving power of alcohols made some authors believe that salts are only soluble in solvents having a hydroxyl group, thereby explaining the negligible solubility of, for example, $CoCl_2$ and $NiCl_2$[100] in ethers and ketones. This is, however, unlikely to be the only, though perhaps the main, factor affecting the solubilities of ionic salts in alcohols.

The solubilizing power of oxygenated solvents is related to steric factors also[541,32]. This is especially the case when the solubility of salts of small cations is considered, the difficulty arising from arranging the solvent molecules around the ion. The solubility of a salt in any particular homologous series of solvents is determined by the length of the hydrocarbon skeleton rather than the strength of the metal–oxygen bond, which differs comparatively little, though changing in the same direction. Solubilities, as a rule, decrease with increasing carbon-to-oxygen ratio, as shown for uranyl nitrate in Figure 9B1[133]. The steric effect becomes even more apparent when solubilities in solvents of essentially the same basicities are compared[134].

Numerous measurements of the heat of solution of ionic salts in a series of organic solvents permitted evaluation of the energy of interaction between the salt, or its ions, and the oxygenated solvent[54,131,135–140]. The strength of this interaction, in turn, affects the solubilizing power of the solvent. The heat values in the case of uranyl, thorium and cobaltous nitrate and cobaltous chloride have shown that ethers and alcohols are generally stronger solvating molecules than the ketones and esters, and compete distinctly more strongly with water of hydration initially bound to the salt.

The sequence of basic strengths of oxygenated solvents as estimated by the infrared data of Gordy and Stanford[134], shows that esters are weaker bases than the ketones, and they are weaker than ethers and

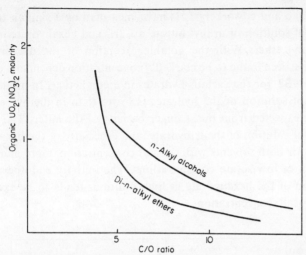

Figure 9B1.   Solubility of $UO_2(NO_3)_2.6H_2O$ in organic solvents as a function of the carbon-to-oxygen ratio.

alcohols. The heats of solution of the salts investigated in the various solvents are essentially in good agreement with the above order of donor strength, the only noticeable exception being isobutyl alcohol, the only simple aliphatic alcohol studied. The alcohol appears to be a weaker solvent than it should. The discrepancy has been explained by the tendency of the alcohol to hexasolvate the metal atom, involving the necessary break of the cation–anion bond to form, for example, the $UO_2Alc_6^{2+}$ complex. This process, against the electrostatic attraction between UO_2^{2+} and the nitrate ion, apparently absorbs energy which might otherwise appear as heat. In all other solvents investigated, including among the simple solvents, diethyl ether, methyl ethyl ketone, methyl isobutyl ketone and some esters, the heats of solution of the salts followed the order of their base strengths. Moreover, the heats of solution of the dihydrate of uranyl and cobaltous nitrate in any given solvent, differ by a nearly constant value of about 6 kcal/mole. The explanation for this observation 'was based on the correspondence in type of the solute in both sets of solutions (i.e. the entities $UO_2(NO_3)_2.2H_2O.2S$ and $Co(NO_3)_2.2H_2O.2S$), and an apparent equality of energy released in binding the two solvent groups'[137]. In view of the effect of dilution upon the heat of solution, such comparison should not be pushed too far. Nevertheless, with certain reasonable assumptions, it has been possible to estimate the energies of binding of solvent molecules by the metal atom[136].

Vdovenko and coworkers[39,140] have measured by a similar technique the heat of solution of uranyl nitrate di-, tri- and hexahydrate in diethyl and dibutyl ether. With the notable exception of the dihydrate, ΔH values have been found to be markedly concentration dependent, as shown in Figure 9B2, for the various hydrates in diethyl ether. In dibutyl ether, the heat of solution of the hydrates is lower than in diethyl ether, as would be expected from the stronger basicity of the latter. The exothermic heat of solution of the trihydrate, and especially of the hexahydrate, increases for both solvents with the salt concentration, more markedly in the region of low solute concentrations. The initially endothermic heat of solution of the hexahydrate at high dilutions tends to be exothermic at higher salt concentrations.

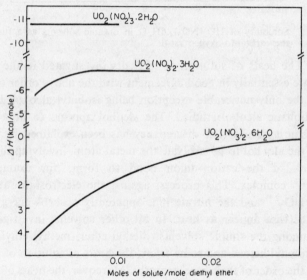

Figure 9B2.   Heat of solution of uranyl nitrate hydrates in diethyl ether at 25°C.

By comparing the shape of the curves for the various hydrates in Figure 9B2, it becomes apparent that there is a competition between the donor groups, water and ether, for the coordination positions around the metal atom in uranyl nitrate. In the process of a partial substitution of the initial water molecules by molecules of ether, the differences in their binding energies affect the observed heat of solution values. Ternary-phase studies on the system uranyl nitrate–water–diethyl ether[141] reveal that ethers are generally unable to displace the last two water molecules of hydration attached to the salt, either in solution or in the isolated solid ethereates. The constant value of ΔH for the dihydrate indicates thus, that a) the solution of this salt yields largely one species $UO_2(NO_3)_2.2H_2O.2Et_2O$, and b) the large exothermic numerical value should be ascribed to the solvent association process of putting two solvent molecules in the vacant coordination positions. On the other hand, the process of dissolution of the two higher hydrates in the ethers (and probably other basic solvents, too) entails loss of a fraction of their water, and addition of another number of solvation molecules. The shape of the corresponding curves in Figure 9B2 may be explained in terms of this process occurring to a varying extent upon dilution. The extent of dissociation of the hydrate into, let us say, a dihydrate and free water, will be more complete at lower overall solute concentrations. Apparently, judging again from the shape of the curve, this process is endothermic, and the measured heat is that corresponding to the difference between the binding energies of the solvent groups and water. Obviously, it will also depend upon the total number of hydrating and solvating molecules attached at equilibrium.

(ii) *State of the solute* It has been mentioned that the anomalous heat of solution values of salts in alcohols have been attributed to the nature of the metal–solvent interaction and its extent. This observation, coupled with the fact that in various salt–water–alcohol ternary systems[141–143] the existence of solid anhydrous alcoholates has been shown, led Diamond and Tuck[54] to suggest that 'it seems that all salts which have strong primary solvation in aqueous solutions will be solvated by alcohols, and by a similar interaction'. Though this is likely to be true with lower, generally water-miscible, alcohols, a complete dehydration of the cation, which Diamond and Tuck's generalization implies, may not be as general with higher or more complex alcohols, as already emphasised by Katzin and coworkers[135]. It is unlikely that a solvated equivalent of cobaltous hexahydrate ion, for example, such as $Co(Alc)_6^{2+}$ may be formed, from purely steric considerations. With n-hexyl alcohol as solvent, Templeton[144,145] found hydrated species as the alcohol-phase complex in ternary systems involving nickel or aluminium nitrate. It remains true, nevertheless, that the coordination of the alcohol to the metal is stronger than that of any other oxygenated solvent, including the ethers.

For this discussion of the state of solute, it is, perhaps, most pertinent to investigate the influence of water carried along with the electrolyte in the process of dissolution, because of its immediate importance in the solvent extraction of these electrolytes. This is, however, difficult to do for the lack of explicit experimental data. Some essentially qualitative correlation may still throw light on the problem.

In addition to what has been said on the heat of solution of salts hydrated to a varying extent, it is evident from solubility data that electrolytes retain some water in the organic phase. Though exceptions are known, for example, anhydrous $MgBr_2$ and $Mg(ClO_4)_2$ in diethyl ether[146,147], salts which normally crystallize as hydrates, become only sparingly soluble in ethers, ketones and esters when rendered anhydrous. Alcohols, for the reasons outlined above, may behave differently. On the other hand, when highly hydrated salts are dissolved, for example the uranyl nitrate hexahydrate in diethyl or dibutyl ether, water is ejected from the organic solution when the concentration of the solute exceeds 0.026 and 0.004 mole per mole of solvent, respectively[139-141]. These observations may be taken as rather straightforward evidence for the participation of water in the dissolved complex. The extent to which it actually participates in the complex will vary from one salt to another, though not necessarily from one solvent to another. Uranyl nitrate forms in several ethers and ketones[148-150] a trihydrate-monosolvate, due to the tendency of the metal to prererve a coordination number of eight. Neptunyl and plutonyl nitrates behave similarly. In dibutyl-carbitol, however, the soluble complex has been reported to have the composition $UO_2(NO_3)_2.4H_2O.2DBC^1$. It should be noted here, that the solid ethereates, isolated from such solutions, have not necessarily the same composition as the complexes predominating in solution (see following heading).

A detailed infrared investigation of diethyl ether and methyl ethyl ketone solutions of uranyl and thorium nitrate[151] reveals that the metal and the nitrate ions are in direct contact, regardless of the original degree of hydration. Furthermore, two molecules of water remain also coordinated to the uranyl ion, again regardless of its original hydration. That observation is in good support of the findings of Katzin[135] and Vdovenko[140] on the heat of solution of these salts in these and similar solvents. The position of the excess of water molecules remains obscure, though it is quite clear that they are bound considerably less strongly. High hydration is perhaps the reason that uranyl nitrate shows an appreciable conductivity in diethyl ether when dissolved in the form of the hexahydrate, whereas the lower hydrates remain essentially undissociated under comparable experimental conditions[140,152].

On the other hand, the infrared spectrum of a solution of anhydrous uranyl nitrate indicates that the solvent molecules are very strongly bound to the salt[48]. Similarly, spectral and physical observations on an anhydrous solution of gallium chloride in isopropyl ether have been interpreted in terms of a strong interaction leading to the formation of the tetrahedral $GaCl_3.i-Pr_2O$ complex[153]. On addition of water, the ether is replaced by water to give $GaCl_3.H_2O$. A comparison of the spectra of an anhydrous $FeCl_3$ solution in butyl acetate, with that of its hexahydrate, also supports the above conclusions[48]. In the spectrum of the anhydrous salt, the carbonyl stretching band of the solvent (1740 cm^{-1}) has undergone an 80 cm^{-1} displacement to lower values, suggesting that the solvent molecule is attached to the iron as ligand. The spectrum of $FeCl_3.6H_2O$ shows a shift of only 40 cm^{-1}, where the water molecule is that primarily occupying the coordination site of the metal. In this case, the solvent molecules are not directly solvating the metal, but are assumed to be in the primary hydration shell.

Direct and differential vapour-pressure measurements indicate that salts dissolved in ethers may be associated to various extents, depending on concentration. $UO_2(NO_3)_2.2H_2O$ in diethyl ether has been found to be dimeric when the solute concentration exceeds 2.5 molal[154]. NH_4GaCl_4 is apparently a trimer, in the same solvent, at a concentration as low as 0.15 molal[155], but $LiClO_4$ may form clusters of varying size[126].

(iii) *Formation of solid adducts* A number of mono- and polyethers (and some alcohols and ketones) readily form ethereates with many inorganic compounds. In the more recent literature the methods of preparation of ethereates of uranyl nitrate[1,2,30,139,148,149,156], cobalt[148] and copper nitrate[157], chlorides of cobalt[158], tin[159], beryllium[160,161] and several more complex chlorides of the $HMX_4.mR_2O$ type have been described[162-164]. The older literature, summarized in Gmelin, contains data on many other metal halides and perchlorates for elements from each group in the periodic system.

When the uranyl nitrate hydrate is dissolved in diethyl ether and cooled to about zero, a pale-yellow solid separates out, which analysed as $UO_2(NO_3)_2.2H_2O.2R_2O$. From solutions which have been saturated at lower temperatures, about $-30°C$, the compound crystallizes as a tetraethereate rather than the diethereate[139,140]. The first compound has a melting point of $45 \pm 2°C$, and the second melts at $2.2 \pm 0.8°C$.

The latter loses easily two ether molecules, having a partial vapour pressure of 102 mm at 0°C, as compared to 16 mm of the diethereate. With methoxy ethoxy pentane, a dialkyl diether, the anhydrous salt forms a yellow solid $UO_2(NO_3)_2.2$ ether. The adduct is not soluble in excess ether, but dissolves readily upon addition of water[30]. From a dibutylcarbitol extract, the isolated crystalline solid has the composition $UO_2(NO_3)_2.DBC.2H_2O$[1]. In both cases, apparently, two ethereal oxygens of the same molecule participate in coordination.

A comparison of the chemical properties of a number of metal halide ethereates indicates the adducts formed vary widely in stability. Some of these compounds, $MgCl_2.R_2O$ for example, decompose upon heating, losing ether with the formation of free magnesium chloride[165]. The stability of the ethereates increases from chloride to iodide, $MgI_2.R_2O$ melts without decomposition. On the other hand, $AlCl_3.R_2O$ exhibits an extraordinarily high thermal stability, and the compound can be distilled *in vacuo* without decomposition. Its eventual decomposition at high temperatures is not its deetherization but the rupture of the Al—Cl bonds[166,167]. A similarly high stability of the metal halide ethereates has been demonstrated for several other salts, including beryllium[160,161], gallium[168], titanium[169], tin[170], vanadium[171], niobium and tantalum[172], chromium[173], molybdenum[174] and iron[170].

Judging from the displacement of the C–O–C frequencies in the infrared spectra of these ethereates, the most stable are the adducts formed by aluminium, boron, zirconium, titanium and beryllium. Unlike the low-stability ethereates, these stable metal halide ethereates show no significant differences in stability in going from chlorides to iodides. Thus, for example, the heats of the reaction

$$AlX_{3(gas)} + (C_2H_5)_2O_{(gas)} \rightleftharpoons AlX_3.(C_2H_5)_2O_{(solid)} \qquad (9B1)$$

for the three aluminium halides are almost identical: $AlCl_3$ 35.7 kcal/mole, $AlBr_3$ 37.6 and AlI_3 37.8 kcal/mole[165]. Metal fluorides form adducts only exceptionally[175].

From the fragmentary information available, it is apparent that the complex-forming tendency decreases in a homologous series with increasing molecular weight of the ether and the donor power increases from aliphatic to alicyclic ethers[165].

Several metal halides form more than one stable ethereate. Depending upon the experimental conditions, $BeCl_2$ and $BeBr_2$ form either di- or triethereates. While triethereates are stable only at low temperatures, below $-2°C$, the diethereates are well-formed hexagonal prisms with sharp melting points of $43°C$ and $53°C$, respectively. For these compounds the formula

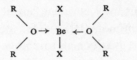

has been suggested[160,161]. Despite their symmetrical arrangement the complexes are highly polar, having a dipole moment of about 7 debyes[176]. Similar behaviour has been noted with the, previously mentioned, two possible $MgBr_2$ adducts, the mono- and diethereates[165,177]. Here again, the lower ethereate is the more stable, the diethereate changing at $28°C$, without melting, into the monoethereate, which is stable up to fairly high temperatures.

In some cases a replacement of the halide ion by the solvent molecules is also possible. Dissolving $SnCl_4$ in C_1 to C_5 alcohols, the solid products formed are of the type $SnCl_4.ROH$, $SnCl_3.(OR).ROH$ or $SnCl_3(OR)$[159]. Similar mixed complexes may be formed when cobalt chloride is dissolved in anhydrous solvents[158].

Complex chloro and bromo acids of the type $HMeX_4$ of aluminium, gallium, indium, iron and thallium(III) form diethereates under certain experimental conditions. Anhydrous $FeCl_3$, for example, saturates dry isopropyl ether to the extent of 0.367 M. Passing dry hydrogen chloride gas through the system, a solid separates out. It is light yellow in colour, crystalline, extremely hygroscopic, and forms a dark-green syrupy liquid immediately upon exposure to atmospheric moisture. Analysis gave the composition $HFeCl_4.2(i-Pr_2O)$[164]. The complex is only sparingly soluble in ether, and to solubilize it, water must be added to an extent corresponding to 5 moles of H_2O per mole of complex. The u.v. spectrum of such a solution is identical to that observed for an ether layer obtained in extracting ferric iron from aqueous hydrochloric acid solutions[89,164,178]. The behaviour of other complex acids is similar. They are all very hygroscopic and not very soluble in ether or hydrocarbons, but have considerable solubility in chloroform or nitrobenzene. They have low melting points[163]:

$HAlCl_4.2Et_2O$	$\sim 46°C$	$HAlBr_4.2Et_2O$	$\sim -30°C$
$HGaCl_4.2Et_2O$	$\sim 25°C$	$HGaBr_4.2Et_2O$	$\sim -70°C$
$HInCl_4.2Et_2O$	$\sim -35°C$	$HInBr_4.2Et_2O$	$\sim -80°C$
$HTlCl_4.2Et_2O$	$\sim -80°C$		

References pp. 716–736

Though the behaviour of alkali or ammonium salts of these acids in ethers is similar to that of the corresponding acids, no well defined ethereate of any of these compounds has been isolated[155,179].

Non-ionic, predominantly covalent, metal halides, such as mercuric halides, tin (II) and antimony(III) iodides, germanium and silicon tetrachlorides, despite their, usually high solubility in ethers and other oxygenated solvents, have no tendency to yield adducts which can be isolated in crystalline form[180].

c. Metal Nitrates

Two aspects of metal nitrate extraction have received considerable attention, the role of sslting-out agent and that of water in the composition in of the organic-phase complex.

Generally, it has been accepted that the water content of the organic phase consists either of water hydrating the electrolyte or of free water, due to its solubility in the solvent or both. In nitrate systems both types of water are present, regardless of the type and basicity of the solvent used. Under conditions of extraction, the nitrate ion is basic enough to compete favourably with either water or solvent for the coordination sites around the metal atom. This seems to be true even in the case of alcohol extractants, despite what has been said in the previous section in discussing the composition of the complex in an alcoholic solution when in equilibrium with a hydrated solid salt. The different behaviour can easily be understood if one bears in mind that a successful extractive transfer of a metal nitrate can be achieved only from aqueous solutions containing a high concentration of salting-out nitrates. This in turn, will favour a metal-nitrate association in the aqueous solution, from a purely mass-action law consideration. A good piece of evidence for that aqueous-phase reaction is that metal nitrates are extracted, at least partially, under such conditions, in the form of their anionic complexes.

As in the case of nitric acid adducts in the organic phase, it is again reasonable to speak in terms of solvation of hydrated species rather than hydration of solvated species. Experimental evidence points strongly to the fact that water is possibly acting as a bond between the solvent and the salt. This, however, is not necessarily the case in alcoholic solvents. Indeed, it seems that the amount of bound water in alcoholic systems is lower than that in other solvents[152], and this despite the higher total content of water in alcoholic systems[145,181,182]. Alcohol molecules, apparently, may replace water of hydration at the coordination sites, left over by the associated nitrate ions, more efficiently than ethers or ketones.

Several ternary systems of metal nitrate–water–organic solvent have been investigated at room temperature. Among them are aluminium nitrate[144] and nickel nitrate[145] with n-hexanol, uranyl nitrate with diethyl ether[141], dibutyl ether[45], dihexyl ether[141], isobutyl alcohol[141] and methyl isobutyl ketone[141], and the quinary system $UO_2(NO_3)_2–NH_4NO_3–HNO_3–H_2O$–diethyl ether, investigated by Kurnakov and his co-workers[183]. Though the information which can be derived from such phase diagrams is not of immediate use to predict the behaviour of similar systems in a solvent extraction process, it may often be relevant to it. The main dissimilarity arises from the fact that metal nitrates are only poorly extractable from their aqueous solutions in the absence of salting agent.

The presence of nitrate ions in the aqueous solution is necessary to achieve any extraction of metal nitrates, though the extractable salt may serve often as its own salting-out agent[124,138,152,184]. Reasonably high distribution ratios have been obtained under such conditions when diethyl ether, dibutylcarbitol, methyl ethyl ketone, methyl isobutyl ketone and cyclohexanone were used as solvents. For a practical, good extraction the aqueous phase needs to be 2–3 molar in an unextractable nitrate, or 3–4 molar in a partially or well-extractable nitrate[62,83,185–190].

The effect of nitrate concentration is illustrated in the charts in Appendix F, where the distribution ratios of over 60 elements into diethyl ether, diethyl cellosolve, pentaether and methyl isobutyl ketone are shown[191].

Despite the numerous attempts, based on different approaches, to correlate the salting-out effect with the nature and properties of the salting agent, the problem remains essentially unresolved. There are several reasons for the lack of success, for a number of thermodynamic factors are unknown and probably will not be known for some time.

While exceptions have been noted, as for example in the case of cobalt and nickel extraction by alcohols[14,5181,182], in the majority of metal nitrate extractions the distribution ratios are affected by the nature of the salting nitrate. It has been found, that at a given mole fraction of the salting nitrate, the extractability of uranyl nitrate increases with an increase in the charge and a decrease in the radius of the cation[83,188,189]. This can be illustrated by the curves shown in Figure 9B3. Theories of thermal motion of water molecules in aqueous solutions of electrolytes have also been invoked in an attempt to correlate the salting effect of essentially non-extractable metal nitrates[190]. A thermodynamic approach has beendeveloped by the Harwell group[184,187]. A large number of experiments, involving a dozen or more metal nitrates, mostly in the dibutylcarbitol system, but also compared with other solvents, revealed an apparent correlation between the activity coefficient of the metal nitrate in a mixed, $MNO_3 + HNO_3$ nitrate solution, and its salting efficiency. Up to an ionic strength of about 5 molal, the larger the activity coefficient of the metal nitrate the greater is its salting-out power*. However, an equally detailed investigation[62] on such systems, into diethyl ether, indicates that the effectiveness of salts of the same valence type is not necessarily in the same order as their activity coefficient. For example, in the 1–2 molar range, copper nitrate gives higher $D_{UO_2^+}$ values than does magnesium nitrate, though the activity coefficient of the latter is appreciably greater at the same aqueous concentration.

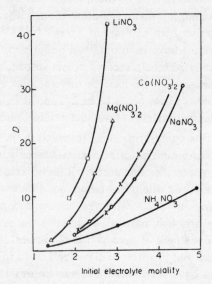

Figure 9B3.   Effect of salting-out nitrates on the distribution of uranyl nitrate into cyclohexanone.

A different, more empirical and more complicated relation seems to exist between the water-binding capacity of the non-extractable metal nitrate and its salting efficiency[62,185,190,192]. Under comparable experi-

---

* Additionally, a set of mass-action law equations allowed the authors to evaluate the mean molal activity coefficient of traces of uranium nitrate in these solutions.

mental conditions, nitrates of metals which crystallize with water are better salting agents, as a comparison of the data[83] in Table 9B1 tends to indicate. It has been suggested[185] that the main effect is probably due to a preferential extraction of uranium nitrate as a dihydrate, as compared to its possible higher hydrates. A salting-out agent with high water-binding capacity will cause the uranyl ion to be less hydrated, and in addition to the common-ion effect at higher concentrations, will render the aqueous solution more deficient in free-water.

This correlation, though not necessarily of a general validity, brings us to the problem of hydration of metal nitrates in the solvent phase which has received an unusually wide coverage[1,30,45-47,62,67,124,127,138,141,152, 184,192-200].

In the case of uranyl nitrate extraction a simple plot of the total organic water content against organic-salt concentration usually shows a minimum in dilute salt region, regardless of the solvent used[127,141,193]. The minimum is due, most probably, to a lowering of the water activity in the aqueous phase as salt is dissolved in it. This, in turn, will lower the partition of free, non-hydrate, water before a sufficient quantity of water

TABLE 9B1. Extraction of uranyl nitrate from 3N aqueous electrolyte solutions into diethyl ether.

Electrolyte	Water of crystallization	$UO_2(NO_3)_2$ extracted (%)
NH_4NO_3	None	30
$NaNO_3$	Hygroscopic	38
$Ca(NO_3)_2$	$4H_2O$	54
$Ce(NO_3)_3$	$6H_2O$	57
$Mg(NO_3)_2$	$6H_2O$	62
$Fe(NO_3)_3$	$9H_2O$	69

associated with the salt is extracted. In view of the fact that the bulk of the total water is indeed often unbound, the slope of such plots only exceptionally represents the mean hydration number of the salt. At an early stage, the Harwell group[152,184] developed a simple approach to estimate the degree of solvation of uranyl nitrate in a number of solvents. For many solvents the water solubility could be represented by a linear function

$$\bar{m}_W = \bar{m}_W^\circ + h\bar{m}_{UO_2} \qquad (9B2)$$

where $\bar{m}_W$ and $\bar{m}_{UO_2}$ are the water and uranyl nitrate molalities, $\bar{m}_W^o$ the molality of water in the solvent in absence of the salt, and h has been identified as the average number of water molecules hydrating the salt. Except at high organic loadings, h has been found to be satisfactorily constant in many ethers, ketones and esters, but not in alcohols. Though h often had a value of 4, it varied from 2 up to 6–7, when replacing ethers by ketones, or diethyl ether by dihexyl ether. The lack of constancy in alcoholic solvents has been explained by the assumption that alcohol molecules are bound directly to the metal ion rather than through a hydrating water molecule.

A reinvestigation of these systems by isopiestic vapour-pressure measurements[47,67] revealed that a series of hydrates, starting with the dihydrate, may exist in a given solvent, depending on the experimental conditions. The water content will change as a function of the water activity. At low a_W the uranyl nitrate will be a dihydrate, which may build up at high a_W to a hexahydrate in dibutylcarbitol, or even higher in other polyethers. A hydration number of four is then a mean value, rather than representing a specific hydrate. It is, thus, more correct to speak in terms of several hydrates of comparable stability in equilibrium with one another.

By a similar approach, taking into account the variation of $\bar{m}_W^o$ as a function of a_W, Vdovenko and coworkers[45,46] evaluated a mean hydration number for uranyl nitrate in several ethers and esters. The lowest values, 2.2–2.5, were found in dibutyl and diisopentyl ether, whereas in diethyl and diisopropyl ether and in dibutylcarbitol, the mean hydration number was around 4. In several alkyl benzoates, butyrates and phthalates the values varied between 2.3 and 5.2. With both types of solvents, the mean hydration number decreased in a homologous series with increasing carbon-to-oxygen ratio. In solvents allowing a high degree of salt hydration, only slightly lower h values have been calculated on the assumption that $\bar{m}_W^o$ remained constant and unaffected by the water activity in the aqueous phase. This is rather surprising in view of the high variation of water solubility in solvents under changing a_W conditions.

When the oxygenated extractant is diluted with benzene, carbon tetrachloride or chloroform, the hydration number evaluated by equation (9B2) decreases with increasing dilution by the hydrocarbon[124,193]. As shown in Figure 9B4 the mean hydration number is a linear function of the volume percent of the extractant. This relationship, however, has no quantitative meaning. It is more significant that infrared spectra of ether extracts in CCl_4, even at high dilution, indicate that the hydrogen bond between water and the solvent molecules is not destroyed[194]. Thus the decrease in water content, must occur at the expense of the unbound water, or water molecules loosely bound at a farther solvation layer. Namely, spectra of diethyl ether solutions of uranyl nitrate dihydrate, when compared to those containing water in excess of the dihydrate, reveal that the positions of the first two of the subsequent water molecules in the coordination sphere of uraniumare not equivalent. There is a similar situation in various dialkyl diethers, where the O–H stretching band characteristic of water-saturated solvent is shifted from 3520 cm^{-1} to a lower 3170 cm^{-1} in the presence of uranium[30].

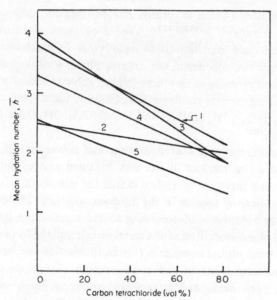

Figure 9B4.   Effect of carbon tetrachloride on the mean hydration number of $UO_2(NO_3)_2$ in various solvents[193].
1) Diethyl ether; 2) Dibutyl ether; 3) Methyl butyrate; 4) Ethyl butyrate; 5) Isoamyl butyrate.

Despite the inaccuracy in the method of calculating the mean hydration number of uranyl nitrate, the data may still be informative insofar as they provide one aspect of the variation of metal-salt distribution ratio for the various systems investigated: the salt carries water into the organic phase, the lowest hydrate to be considered is probably the dihydrate. In certain polyethers the salt may be hydrated by as much as seven water molecules, but in any solvent there is a mixture of hydrates. The non-integral values for h, evaluated in many systems, should be taken as evidence for the coexistence of several hydrates.

In connexion with what has been said on the effect of 'dehydrating salting-out agents' upon hydration of uranyl nitrate, the question arises whether different hydrates have different extractabilities into a given solvent. A careful examination of the uranium extraction ratio into various solvents and its extent of hydration in them, suggests that higher hydrates of uranyl nitrate are less extractable than the dihydrate. To prove this in a quantitiative way does not seem feasible, since knowledge of the stoichiometry of every single mixed hydrate-solvate species would be required for a mass-action law treatment.

The extent of solvation of uranyl nitrate hydrates differs from one solvent to another[1,2,41,45,64,201]. Vapour-pressure measurements in various ethers and partition data into ether–hydrocarbon mixtures indicate at least two solvates in the organic phase, a di- and a trisolvate. In ketones, and probaby some ethers, higher solvates are also formed[41]. Analysis of ether extracts confirms these findings, identifying at least two species, $UO_2(NO_3)_2.2H_2O.2R_2O$ and $UO_2(NO_3)_2. 3H_2O.3R_2O$, existing in equilibrium[201].

In conclusion, the extent of hydration and solvation will affect the extraction of metal nitrates in the way outlined above, but the main factor governing their extractability is still the equilibrium between the various metal-bearing species in the aqueous solution. The presence of nitrate ions in solution will, according to the mass-action law requirements, affect the dissociation of the metal nitrate and the formation of the extractable metal nitrate complexes (for an illustration see section 12.C).

Spectral measurements in the visible region of various extracts of actinides in both hexa- and tetravalent oxidation state, suggest that both neutral and anionic species are present, depending upon the aqueous-phase nitrate concentration. The species extracted from dilute nitrate solutions are $MO_2(NO_3)_2$ and $M(NO_3)_4$ respectively, whereas from concentrated aqueous solutions the $MO_2(NO_2)_3^-$ and $M(NO_3)_6^{2-}$ ions are predominant in the extract in their protonated form[30,64,68,121,186,196, 202-204].

d. Metal Halides

The main characteristics of extraction from aqueous halide solutions by oxygen-containing solvents is the extraction of protonated metal complex anions, prefentially single-charged, An equally important feature of the systems is the high extractability of essentially covalent metal halides (cf. section 7.B).

There is, however, a striking difference in the extractive behaviour of these halides, and that of a complex halo acid. While the latter are extractable into oxygenated solvents only, the covalent halides can be transferred to a comparable extent, into non-polar hydrocarbons also[90,205–209].

With some restriction as to the aqueous complex chemistry of metals in halide solutions, high extractabilities have been observed of monovalent complex anions of the bivalent and trivalent transition metals. On the other hand, despite the pronounced complexing tendency of bivalent elements, it appears that the high charge on the usual aqueous-phase complex MX^{2-} prevents their extractability even into relatively high dielectric constant solvents such as butanol[210] and octanol[100,211,212], and that, despite their usually high solubilities in these solvents[100,135,142]. An example[154] of the effect of aqueous complex chemistry upon metal extractability, is the fact that only a few metals are extracted from fluoride solutions, or more precisely, only a few metals form extractable fluoride complexes. With the small fluoride as ligand, most metals form polynegative anions such as FeF_6^{3-}, as compared to $FeCl_4^-$ or $FeBr_4^-$. Additionally, even if mononegative fluoride complexes are formed, they are small in size, and thus, generally less extractable.

Based on this and similar characteristics of halide extraction systems, a large variety of separation processes have been developed in the past, and there is certainly the possibility of many more practical applications. As a fair guide of these the survey studies on the extractability of many metals into ethers and ketones from fluoride[50,213], chloride[205,214–218], bromide[98,205], and iodide[205] aqueous solutions may be useful. Charts in Appendix F show the distribution ratios of about sixty elements from hydrochloric acid solution into diisopropyl ketone and methyl isobutyl ketone[191].

It is perhaps interesting to note that in some cases the distribution ratios of poorly extractable metal chlorides and bromides may be affected by the presence of well-extractable salts. The phenomenon, observed in ether systems[219–221], is characteristic for systems with a combination of elements in which the macro component has a higher distribution ratio than the micro component. Though this coextraction is selective in character, for it is governed by the nature of the ion, the effect is usually greater the greater the difference in the individual distribution ratios.

Several sets of mass action law equations have been developed to explain the extractive behaviour of halometal complexes under various conditions[14,42,54,58,222–225]. These equations, summarized in Chapter 7, based on the stepwise formation of complex ions of the metal in the aqueous solution, take into account the effect of ionic and/or molecular association of the solute in the organic phase, upon the overall distribution ratio. Considerations have also been given to the effect of the extractable supporting hydrogen halide in the system and to the influence of the metal ion concentration in the aqueous phase. Under highly idealized experimental conditions, and in the most favourable cases where the usual assumptions on the organic-phase activity coefficients were realistic, the experimental data could be fitted into the derived equations.

This mathematical treatment fails, and this is a substantial failure in the overwhelming majority of cases, where deviations from ideality are too large in the aqueous or in the organic phase, or in both. In the practical solvent extraction systems under consideration, we are not dealing with ideally immiscible liquids, and they do not keep their individual properties throughout the changes in the concentrations of the various solutes. Dielectric constant changes in the organic phase due to the solute and water dissolved, and in the qaueous phase due to the solvent dissolved, affect the equilibrium to an unpredictable extent. The composition of the organic-phase complex frequently changes due to the above effects, and the evaluation of its solvation, though attempted[225] usually cannot be taken care of mathematically.

The extraction of iron(III) chloride by diethyl ether is one of the classic examples of extraction of inorganic species by organic solvents. The complex chemistry of ferric chloride as studied by the solvent extraction technique is fully discussed in section 12.C. Here, the solvent extraction characteristics of iron(III) and other tetrahalometallates(III) will be dealt with.

The extractive behaviour of several tervalent metal halides closely parallels that of iron[42,54,222,223]. The shape of the extraction curves, shown in Figure 9B5 for the HBr–diethyl ether system[226], may be illustrative of this effect. The curves for iron, indium and gallium are strikingly similar, those of thallium and gold showing only a quantitative difference caused by their aqueous-phase complex chemistry.

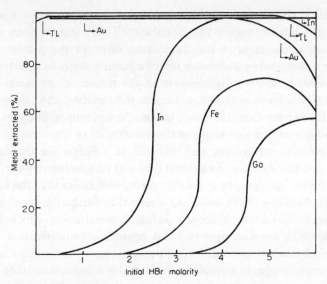

Figure 9B5.   Extraction of metal(III) bromides into diethyl ether.

In the extraction of these metals from aqueous hydrogen halide solutions there is an initial steep rise with increasing acid concentration. The halogen ion concentration, at the point where the extractability rises steeply, depends on the complexity constants of the metals. In the case of thallium and gold this increase is very abrupt and appears at much lower ligand concentration. The high complexing ability of thallic and auric ions will cause a large fraction of the metals to be in an extractable tetrahalometallic acid form already at 0.1 M concentration of the supporting acid. In diethyl ether, the extraction curves show a maximum for all the metals discussed here. The appearance of maxima, however, is by no means a general phenomenon in extraction systems, as illustrated in Figure 9B6 for ferric chloride in ethers, esters and a ketone[54,227]. The phenomenon is connected with the changes brought about by equilibration of some solvents with aqueous acid solutions. It has been shown previously, that both acid and water are normally extracted into the organic solvent, it may thus be expected that these solutes will also exist in the extract when a metal halide is distributed between the phases, apart from that associated with the metal. Generally speaking, the amount of free acid and water in any solvent will depend on the supporting acid concentration in the aqueous layer and, in the case of macro concentration of the metal halide in the system, on its distribution also. Now, a comparison of extraction curves for a number of metals from any one of the haloacids[42,227-230] into a given solvent exhibiting a maximum in the distribution ratio (or the percentage extraction), reveals that a decrease in extractability starts at a certain acid concentration, which is independent of the nature of the metal and is specific to a solvent–acid pair, as long as the metal extacted is in tracer or small concentrations. In other words, the aqueous halide concentration will govern the *extent* of metal extractability in the sense discussed in the previous paragraph, and this will be different for the different metals, but the *decrease* in extractability will be governed by the nature of the solvent and the type of the supporting acid rather than the aqueous complex chemistry of the metal. As a rule, the distribution ratio starts to decrease at a certain hydrogen halide concentration which always coincides with gross changes in phase volumes on equilibration. It will be recalled, that the volume changes occur in these systems at the expense of the organic phase, a large amount of the solvent dissolving in the aqueous phase. This type of increased miscibility of the phases will dilute the aqueous phase drastically and affect the complex-formation equilibrium between the metal and the ligand. Though this may not be the only reason for the decreased metal extractability, as it is not alone responsible for a similar extractive behaviour of simple mineral acids, it is apparently the key factor.

Discussing similar phenomena in mineral acid extraction systems, it has been noted that not all solvents become soluble in aqueous hydrogen halide solutions at high concentrations of the latter. With an increasing carbon-to-oxygen ratio in ethers, ketones and esters, the solvent becomes less water soluble, and the concentration of the supporting acid, where a decrease in metal extractability starts, is shifted to higher acid concentrations in the initial solution[227,231,232]. With still less soluble solvents, such as di-n-propyl ketone, β,β'-dichlorethyl ether and isopentyl acetate shown in Figure 9B6, the volume of the aqueous phase remains unchanged upon equilibration, and there are no maxima in the distribution curves of metal halides. Again, as in the case of extraction of hydrogen halides, there are no maxima when alcohols are used as solvents. The reason being that alcohols are not water soluble, and if volume changes occur, as with the lower alcohols, they are at the expense of the aqueous rather than the organic phase.

In several instances, attempts have been made to correlate the physical properties of solvents with their extractive power toward tervalent metal

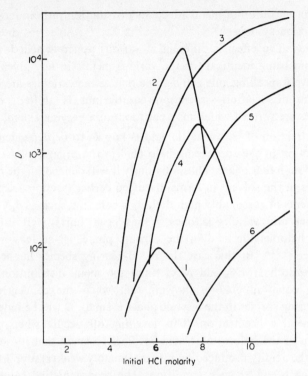

Figure 9B6.   Extraction of iron(III) chloride into various solvents.
1) Diethyl ether; 2) Methyl isobutyl ketone; 3) Di-n-propyl ketone;
4) n-Butyl acetate; 5) Isoamyl acetate; 6) β,β'-Dichlorethyl ether.

halides. The most explicit investigation with this aim is perhaps that of Irving and Rossotti[42], using the extraction of indium chloride, bromide and iodide (in trace concentrations) as the model. Any suggested order of solvent efficiency can only be purely empirical, for a quantitative treatment requires knowledge of thermodynamic quantities pertaining to the solvent and the solutes in such multicomponent systems.

It is frequently difficult to establish even an empirical order. As ʔn illustration of the difficulties involved, is the fact that ferric iron is extracted more completely by diethyl ether than by diisopropyl ether, from a 4 M hydrocloric acid solution, whereas the order is reversed at 9 M[233]. This change in the sequence, common to many comparable systems, can be well understood in the light of what has been said in the previous paragraphs concerning the appearance of maxima in the extraction curves.

The concentration of metal affects its distribution ratio in most practical extraction systems[13,14,33,88,91,178,225,227,234,235] in terms discussed in section 7.C. The effect is different at various aqueous acid levels, and is far from being identical for the various metals in any given solvent. Qualitatively speaking, one can assume that as the variables are changed, either the dissociation–association equilibrium is shifted, or other complexes are formed which may partition to a varying extent, or both.

The extraction of halometallic acid is known to be dependent on the concentration of the supporting halo acid in the organic phase, and its dissociation in that phase. Thus, obviously it will depend on the dielectric properties of the solvent in terms discussed earlier (section 7.C). Briefly, with solvents of reasonably high dielectric constant, such as β,β'-dichloroethyl ether and various ketones, it has been fairly well established that the halometallic acid in the organic phase, is, at least partially, dissociated[54 225]. In this case the metal-bearing species in the organic phase are both HMX_4 and MX_4^- and the metal distribution ratio is given by equation (7C64), showing that when the concentration of metal-bearing species in the organic phase is small, D will be independent of the metal concentration. This levelling off occurs when the concentration of the halometallic acid becomes so small that its ionization is set by the acidity produced by the (presumably weaker) supporting acid HX, which is, under such conditions, the predominating solute in the organic phase. On the other hand, when the metal concentration is substantially increased, D will become dependent on its concentration and will decrease with increasing metal concentration in high dielectric constant solvents[101,224,225,236]. The particular metal concentration where the point of inflexion occurs in a given solvent will depend on the strength of the halometallic acid. When the extraction conditions are such as to depress substantially, or prevent, the dissociation of the acid in the organic phase (at high metal or high HX concentrations, or both), D becomes, again, independent of the metal concentration, but at a lower value.

Where the distribution data indicate appreciable dissociation in the extract, conductometric measurements[225] and migration experiments[224] have substantiated the dissociation of HMX_4 species. For example, the apparent dissociation constants for $HAuCl_4$ in benzene, chlorobenzene, o-dichlorobenzene and nitrobenzene have the values $4.5 \times 10^{-6}, 4 \times 10^{-7}, 4 \times 10^{-4}$ and 2.5×10^{-1} respectively[224]. The constants for $HFeCl_4$ follow the same trend.

In the case of low dielectric constant solvents, the distribution ratio, given by equation (7C56), shows that D will increase with increasing metal concentration due to molecular association of the extracted species, and/or changes in the activity coefficients assumed in deriving the equation. Numerous investigations of such behaviour of metal extraction systems have been made[13,33,178,225,235,236]. Conductivity measurements made on ether extracts of ferric chloride and bromide transferred under a variety of conditions, have been interpreted in terms of association of neutral ion-pairs[33,237]. The shape of the conductivity curves, similar to that of NH_4GaCl_4[155], with a minimum, is characteristic of associated polar compounds in media of low dielectric constant[238]. An association of triple ions (section 2.D) such as $[H_3O^+.4H_2O.$ $FeCl_4^-.H_3O^+4H_2O]^+$ and $[FeCl_4^-.H_3O^+.4H_2O.FeCl_4^-]^-$ into an undissociated entity has been suggested to be a reasonable form of the polymers in the organic phase of a low dielectricum. Similar conclusions are drawn from isopiestic measurements on an isopropyl ether layer containing ferric chloride in the 10^{-3} M concentration range[228]. The polymerization number reaches a value of about three at 5×10^{-2} M $FeCl_3$.

The polymeric state of the solute under consideration has frequently been questioned[88,89,91,235], and the shape of the distribution curve as a function of total metal concentration has been ascribed to the low activity coefficient of the salt in the ether phase rather than to polymerization. There is not much doubt that changes of activity functions occur, but such a thermodynamic argument confirms, rather than invalidates the hypothesis of polymerization.

Finally, on this point, direct experimental evidence has been provided to show that it is indeed the dielectric constant of the medium which is primarily responsible for the way D varies with the concentration of the metal in the system. Fomin and Morgunov[225] investigating the extraction of ferric iron from HCl solutions into dibutyl ether and into its mixtures with dichloroethane (dielec. const. 10.3) found that at a high dilution of the ether, D decreases with increasing iron concentration, just as in β,β'-dichlorethyl ether and ketones of high dielectric constant. Inversely, when methyl hexyl ketone is diluted by benzene and the dielectric constant is reduced significantly, D becomes independent of the metal concentration again, as in many low dielectric constant solvents[54,91].

There is a wealth of information available to show that the tetrahalometallic acids extracted are highly hydrated[13,33,48,89,96,153,164,227,236,239–241].

Water content varies with the acid and metal concentration, indicating that there is no single value of hydration of the extracted species in any given solvent. In diethyl ether, which has a high water content, and the $FeBr_3$–HBr system, the water-to-proton ratio decreases with increasing initial acid concentration from 8.45 to 6, when HBr increases from 2.7 to 5.4 M[33]. The anhydrous $HFeCl_4.2(i-Pr_2O)$ complex becomes ether soluble only after 5 moles of water per mole of complex have been added to the ethereal phase[164]. Indeed, that number of water molecules seems to be just about the minimum number of hydration of $HFeCl_4$ in this and other ethers as well[162,227,237,240,242]. In the water-tolerant diethyl ether, analysis of the extract indicates the presence of decahydrates of both iron[89] and indium[236] chloro complexes.

Infrared[48,243,244] and n.m.r.[239] spectra of ferric (and thallic) chloride extracts in a variety of oxygenated solvents, including ethers, ketones and esters, provide good evidence to show that the $FeCl_4^-$ anion is neither hydrated nor solvated in the organic phase. The spectra of $FeCl_4^-$ are identical with that of an anhydrous solution of $KFeCl_4$ in ether. Furthermore, i.r. spectra of the extracts show stretching frequencies characteristic of strongly bound water (~ 3200 cm^{-1}) in contrast to loosely bound water when the solvent is saturated by water alone.

When the supporting electrolyte is a metal halide rather than hydrogen halide, the number of water molecules decreases sharply[33] up to the point where the organic phase splits into two. From mixed acid–alkali chloride solutions, even when the solution is predominantly the latter, the extracting species is chiefly the protonated species[39,222]. However, when the aqueous-phase conditions are such as not to allow the possibility of $HFeCl_4$ extraction, (proton-deficient) complexes such as $LiFeCl_4$ or $Ca(FeCl_4)_2$ have been identified in the organic phase[39,222,243,245]. The extractability of these and similar complexes enhances the extraction of otherwise non-extractable metals such as calcium, and is the probable reason for the coextraction phenomenon mentioned earlier in this section[219-221]. As to the extent of their hydration, the i.r. spectra indicate that the iron complex extracted from aqueous solutions of LiCl, $CaCl_2$, $MgCl_2$ and $AlCl_3$ shows an increased tendency for hydration in that order. This is again good evidence that the water adds on directly to the cation, the total water content and the fraction of bound water depend on the hydration capacity of the cation.

From the available information it is hardly possible to draw any conclusion as to the extent of solvation of the hydrated halometallic acid in the organic solution. In line with what has been said previously on solvation of simple inorganic acids, it is unlikely that there would be a single value for the number of solvent molecules associated with the extracting species for the various conditions of extraction. Nevertheless, the metal distribution ratio is sometimes a linear function of the concentration (molarity, or mole fraction) of the oxygenated solvent when diluted with a non-extracting hydrocarbon, such as benzene, cyclohexane or carbon tetrachloride[224,242,246]. However, when brought to high dilutions the distribution ratio deviates from linearity since the diluents themselves extract the complex under certain conditions[224,242,247]. The curves in Fig. 9B7 are illustrative of this effect[224]. The extractive contribution of diluent becomes unimportant at high mole fractions of the active oxygen, and the plot is linear in that region. The dilution at which deviation becomes apparent will depend on the extractive efficiency of the oxygenated extractant, higher dilution being needed for ethers and ketones than for alcohols, for deviation to start. By slope analysis, a solvation number of about 3 has been deduced for $HFeCl_4$ in dibutyl ether–benzene or carbon tetrachloride system[91,240]. In the diethyl ether–benzene system

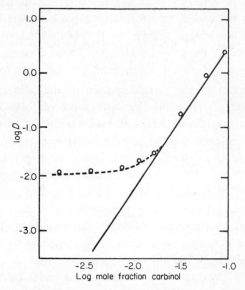

Figure 9B7. Extraction of gold(III) from 6 M HCl into diisopropyl carbinol in benzene. Dotted line experimental data; full line corrected data using the distribution ratio in pure benzene[224]. (By permission from the Faraday Society).

an even higher solvation number has been claimed[242]. The number is decreasing with increasing carbon-to-oxygen ratio in the extractant molecule, and approaches two when diisopentyl ether in used.

The shifts in the infrared spectra of the $C=O$ group frequencies at ~ 1700 cm^{-1} have been taken as evidence that the solvent molecules are bound to water and not directly to the proton[48,243,244]. This conclusion is based on the observation that there is no change in the shift or intensity caused by replacing the proton with lithium or another cation.

At aqueous acidities higher than about 8 M HCl, the analysis of the extract reveals that the HCl:FeCl$_3$ ratio is frequently higher than unity[89,149,248] when low dielectric constant ethers are used as extractants. A similar observation was made in the isopropyl ether–GaCl$_3$–HCl system[249]. The absorption spectrum of the ethereal layer does not change, suggesting that the iron remains in the tetrachloroferrate form and that the excess acid is but loosely associated with the complex, if at all. Under certain experimental conditions the ethereal layer might split into two[250], but the absorption spectra of both layers are identical, though different from that of aqueous or ethereal FeCl$_3$ and from that of ferric iron in aqueous hydrochloric acid. Diamond and Tuck[54] suggested that the excess acid forms with the halometallic acid a mixed ion-association complex such as $H^+FeCl_4^- H^+Cl^-$ instead of the $H^+FeCl_4^- H^+FeCl_4^-$ quadrupoles assumed to be formed under similar conditions in low dielectric constant media at high iron loadings.

The comparison of the visible spectra of the pale greenish-yellow ethereal extracts of ferric chloride with those of solid KFeCl$_4$ or its anhydrous ethereal solution, and with a solution of FeCl$_3$ in concentrated hydrochloric acid, provided the first direct evidence that the FeCl$_4^-$ ion, with a tetrahedral sp^3 configuration is the principal iron-containing species[88,89,149,164,227,228,251]. Various other spectral measurements in the ultraviolet[164,243], and infrared region[243], Raman spectra[153,252,253], and electron magnetic resonance measurements[239] have confirmed the earliest findings, and provided evidence that this form of the extracted complex is not restricted to ferric iron, but appears to hold generally for the tervalent halides in any oxygenated solvent.

The reported spectra of tetrachloroferric acid extracts obtained under a variety of conditions and its molar extinction coefficients in the visible region are generally in fair agreement. Beer's law is obeyed, usually over quite a wide range of iron concentration[88,228]. Raman spectra of diethyl ether and methyl isobutyl ketone extracts of indium chloride and bromide and of diisopropyl ether extract of gallium chloride, are those of the regular tetrahedral species. They are identical irrespective of the aqueous-phase conditions, and despite their relatively low concentration in aqueous solutions when compared to the concentration of higher chloro complexes. These observations are equally valid for a ferric chloride–diethyl ether system.

Except at very high acidities, the only stoichiometry computed by direct analysis of the extract is that corresponding to HMX_4. The lower MX_3 or higher MX_5^{2-} or MX_6^{3-} species in the organic phase, despite their known presence in many cases in the aqueous solutions, have not been identified[98,99,226,244,249,254]. Identical results were obtained by slope analysis (the variation of D_M as a function of ligand concentration) in a variety of solvents over a wide range of electrolyte concentration[42,255]. At low Cl^- concentration in a $TlCl_3$–isopropyl ether system[255,256], the possibility of $TlCl_3$ existing in the ether phase could not be excluded. Similar conclusions were drawn again from distribution data, when the metal concentration was varied in iron[224,257] and gold[224] extraction systems. $FeCl_3$ and H_3AuCl_6 may be extractable to some extent. Katzin[153] has assigned some of the Raman peaks observed in a diisopropyl ether extract of gallium chloride to a tetrahedral species $GaCl_3X$, where X is probably water, though it might be ether. By the continuous-variation method Specker and coworkers[205,245] have claimed to have identified, in addition to a series of various metal MX_4^- complexes with chloride, bromide and iodide, the neutral complex FeX_3.

e. Organic-phase Characteristics

(i) *Structure of the solvent* It is of both theoretical and practical interest to know the way in which the extractive power changes as the result of the class and structure of the solvent molecule. As to the class of the solvent, it has been shown in the previous paragraphs that the order of solvent efficiency remains a specific characteristic of any particular system. While, generally speaking, higher distribution ratios have been observed in ethers than in ketones for the hexavalent actinide nitrates, the trend is opposite when tervalent metal halides are extracted from aqueous solutions. However, to achieve a change in the efficiency of a solvent it is not necessary to affect the nature of the complexing group. It is frequently enough in a particular class of solvent to alter the structure in order to influence metal extraction capacity.

The variation in the extractive properties of solvents brought about by structural changes in the molecule are caused essentially by three factors: the concentration of the active functional group, the polarity of the molecule and the steric hindrance of the molecule. These factors will be demonstrated in the following paragraphs.

The most frequently observed phenomenon is the lowering of metal extraction (and also acid extraction as demonstrated in the previous section), within an homologous series, on increasing the hydrocarbon character of the organic solvent. Thus, lengthening the alkyl chain in alcohols[258,259], ketones[121,124,231,260] and ethers[30,124,184] by introducing further methylene groups will cause a decrease in the extractive power of the molecule. This dependence is usually expressed through the carbon-to-oxygen ratio in the molecule, though such correlation is not necessarily always indicative in view of possible density differences in a homologous series. A more important correlation should be based on the variation of the metal distribution ratio with the molarity of the solvent. This actually has been done by Boyd and Larson[259] in comparing the extractability of technetium(VII) from sulphuric acid solutions into n-hexanol and n-hexanone, and by Specker and Hovermann[261] for the model system CdI_2, $FeCl_3$ and $Zn(SCN)_2$ in heptanone and octanone. Log D is a linear function of the molar concentration of the hydroxyl group, as shown in Figure 9B8[259], for technetium in hexanol diluted with n-hexane and compared with those measured in other straight-chain alcohols.

This correlation cannot be extended beyond a particular and strictly homologous series. For example, in polyethers the carbon-to-oxygen ratio can be used as a guide to the extraction power, but only when the spacing of the oxygen atoms is kept unchanged in the series. Comparison of D_{UO_2} values by members of the homologous series like $R'O(CH_2)_2OR$ will reflect the regularity, but a comparison between series with varying number of CH_2 groups between the oxygens will show little connexion with the carbon-to-oxygen ratio[240].

The dipole moments of ethers and ketones are affected by the symmetry of their molecules. As a rule, asymmetric ethers[232,259] and ketones[259,262] have a greater efficiency than the symmetric ones. Methyl alkyl ketones have been shown to be more powerful solvents for $HTeO_4$ and $Co(SCN)_2$ than the symmetrical dialkyl ketones of equal carbon-to-oxygen ratio, under otherwise identical conditions. Similarly, the asymmetric ethers extract gallium and indium chloride with a higher distribution ratio

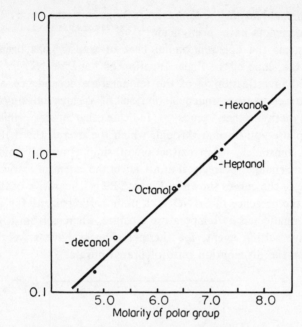

Figure 9B8.   Extraction of Tc^{VII} by undiluted n-alcohols (open circles), and by n-hexanol in n-hexane (filled circles)[259]. (By permission from the American Chemical Society).

than the symmetric ones. It may be assumed that the greater dipole moment of the asymmetric extractants leads to an increased electron density at the donor oxygen, and hence the increased tendency to coordinate[232].

The distribution ratio of metals decreases with increasing branching of the alkyl group, the effect being especially marked when the branching is near the active oxygen. The phenomenon has been observed with ethers, ketones and alcohols for a variety of metals and for various aqueous media[30,121,124,259,261,263]. For example, a comparison of a number of cyclohexanone derivatives for their extractive power towards several metals revealed that a substitution in the vicinity of the carbonyl group decreased the distribution ratio, but a substitution on the fourth position showed only a small effect. The effect is larger with longer-chain derivatives. This and other similar observations point strongly to a possible steric hindrance, caused by crowding near the oxygen, interfering with the extractive efficiency of the solvent.

(ii) *Temperature* In all the extraction systems discussed in this section, the distribution ratio of the metals decreases with increasing temperature, at least in the temperature range between 0° and 60°C where the measurements have been made[62,79,89,227,229,255,259,260,264,265]. For many systems the apparent molal heat of transfer has been evaluated from the slope of the linear function of $\log D$ vs. $1/T$[255,259,260,265]. A detailed investigation[265] of the temperature dependence of uranyl nitrate extraction from aqueous solutions of various salting-out nitrates into diethyl ether, tends to indicate that it is primarily the aqueous-phase equilibrium through which the overall thermal effect is governed. Inasmuch, as in extraction of simple mineral acids alone, temperature changes, scarcely, if at all, affect the extent of extraction[37,94]. The shape of the curves shown in Figure 9B9 is illustrative of this point. D_{UO_2} in the presence of $Sr(NO_3)_2$, a poor salting agent for uranium, is almost unaffected by temperature changes, whereas from $Mg(NO_3)_2$, an effective salting agent, the thermal changes have a very marked effect upon the distribution ratio of uranyl nitrate.

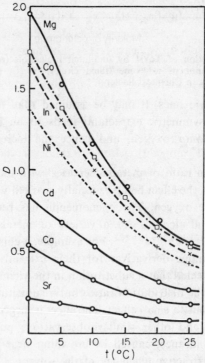

Figure 9B9.   Extraction of uranyl nitrate from various aqueous nitrate solutions of designated metals, of equal normality, into diethyl ethyl, as a function of temperature[265].

(iii) *Third phase* In many systems where third-phase formation has been observed, it seems to be associated with a high concentration of the metal salt transferred into the organic phase[13,33,54,96,162,164,178,235 249,250,255]. However, this is not a rule. It is known that the third phase may be formed when the metal is present in micro concentrations, or not present at all[36,42,255], thus the formation of a second organic phase cannot be explained by the presence of metal alone.

There is good reason to believe that the formation of two organic phases is due to changes occurring in the composition of the organic-phase complexes when parameters of an extraction system are varied[266]. In the system under consideration it is the extent of hydration of the solvent-phase solute which causes the organic phase to split. The systems are, thus, affected by the water activity, which in turn depends on the concentration and nature of the electrolytes in the aqueous phase (cf. section 2.C). At very high acid or salting-out electrolyte concentrations in the aqueous solution, the acid acts as a dehydrating agent upon the water-saturated ether containing the highly hydrated metal complex. Removal of this bound water from the ethereal phase will render the phase unstable, and will cause it to split into a heavy layer containing the bulk of the metal, and water, and a light one containing the ether with perhaps some traces of the adduct. As a matter of fact, addition of inorganic salts with dehydrating properties, to an ethereal solution of the metal complex, will deprive the complex of its water molecules and two phases will be formed. Spectra and density determinations of the phases reveal an essentially identical composition irrespective of the initial conditions in which they were formed. This is at least the case in the $TlCl_3$[96] and $FeCl_3$–diisopropyl ether systems[13,164,178,235,240], but is believed to have a more general validity. Similar considerations about the hydrated acid species in the organic phase may explain why systems split even in the absence of metals, usually when the aqueous phase contains an electrolyte of dehydrating properties[33].

The mere fact that splitting has been observed in ethers suggests that the nature and properties of the organic solvent play an important role in the process. In media of relatively high dielectric constant, such as the ketones and alcohols, the phenomenon does not occur. Furthermore, under identical experimental conditions, diisopropyl ether will form two phases, but not β,β'-dichlorethyl ether, which has a markedly higher dielectric constant, 4 and 21 respectively[54,162]. There is, apparently, a greater tendency for high dielectric constant solvent molecules to replace the water of hydration in the first coordination shell of the metal ion when water is removed from the organic phase.

C. PHOSPHORUS-BONDED OXYGEN-DONOR EXTRACTANTS

a. Introduction

Introduction of neutral organophosphorus extractants marks a notable advance in the refinement of the theories on the nature of the extraction processes and on the effect of the experimental variables upon it. The vast amount of experimental data, including work done with tools such as nuclear magnetic resonance and infrared spectroscopy and other physicochemical methods, accumulated in the last fifteen years contributed enormously to widening the scope of solvent extraction processes. From the technology side, this is certainly reflected in a complete replacement of ethers and ketones in raw material processing and nuclear fuel reprocessing. Indeed, the primary stimulus for developing neutral organophosphorus compounds arose from the need for nuclear fuel reprocessing, initiated by the pioneering work done at the Oak Ridge National Laboratory.

Tri-n-butyl phosphate is, of course, by far the most familiar extractant of this type, and has received by far the widest coverage. An annotated bibliography[267] on the use of TBP covering the period up to 1960 contains one thousand references from periodicals, reports and patents. Other neutral organophsophorus compounds are similar in their extractive behaviour, and the type of complexes they form. The great majority of alternative reagents, however, offer a greatly increased range of extraction power. Trioctylphosphine oxide, for example, extracts uranium(VI) with distribution ratios higher by five orders of magnitude than does TBP. As a rule, the extraction power increases markedly as the number of direct C–P linkages increases in the series phosphate–phosphonate–phosphinate–phosphine oxide. Although alkyl- and alkoxyphosphines have been mentioned as potential extractants[268], they seem to be impractical in view of their low stability.

b. Preparation, Purification and Properties of the Extractants

The neutral organophosphorus compounds of interest to solvent extraction of acids and salts are the esters of the simple oxyacids of phosphorus. The four main groups of extractants are the following alkyl or alkoxy derivatives of orthophosphoric acid: the phosphates, with the general formula $(RO)_3PO$; the phosphonates $(RO)_2(R)PO$; phosphinates $(RO)(R)_2PO$ and the phosphine oxides $(R)_3PO$, where R is an alkyl or aryl radical. A large number of the neutral monofunctional organophosphorus compounds, the symmetrical and unsymmetrical phosphates, phosphonates and phosphinates and symmetrical phosphine oxides, has been synthesized, most of them by conventional procedures[268-278]. The methods usually combine features of several techniques described in the literature, and are only exceptionally typical for a specific compound. Preparation of ^{32}P-labelled tri-n-butyl phosphate[279], and of di-(2,4-diethyl-n-octyl) methylphosphonate[275] may be two of them.

Some polyphosphorus compounds, such as the pyrophosphates, diphosphonates and diphosphine oxides, have obtained considerable attention, being powerful chelating extractants for a number of metals. All of them are bidentate extractants that can form five- or six-membered ring chelates. Various tetraalkyl alkylene diphosphonates, diphosphinates and diphosphine oxides of the general formula $(R)_2P(O)R'P(O)(R)_2$, where R is an alkyl or alkoxy group with three or more carbon atoms, and R' ranges from methylene to propylene, have been synthesized[280] and their extractive capacity surveyed[268-271,281,282]. Also prepared were[282] dialkyl N,N-dialkylcarbamyl phosphonates $(RO)_2P(O)C(O)N(R)_2$ and dialkyl N,N-dialkylcarbamylmethylenephosphonates $(RO)_2P(O)CH_2$-$C(O)N(R)_2$ and tetraphenyldiphosphine oxide[283].

Triphenyl phosphite[284] $(RO)_3P$, and other derivatives of phosphine[268], and sulphur analogues of neutral organophosphorus compounds[285], the trialkyl phosphorothioates, $(RO)_3PS$, and trialkylphosphine sulphides $(R)_3PS$, have been prepared and screened for their applicability in solvent extraction. The phosphine derivatives are subject to a readily occurring oxidation to phosphine oxide.

In all these compounds some impurities are usually present as by-products of the original synthesis, others may be formed by chemical- and radiation-induced deterioration. The profound effect of impurities, acid phosphates, for example, on the extractive behaviour of the neutral esters has been early recognized[286-290]. Purification of the neutral esters usually involves multistage alkaline and acid washes[272,273,291,292], extraction based on differential distribution of the product and the impurities between carbon tetrachloride and water[294] or vacuum distillation[278].

A variety of analytical methods for inspection of the products, mainly tri-n-butylphosphate, has been developed[295-300].

Proton magnetic resonance[268,272,273,301] and infrared[268,278,279,281,283,302-307] spectroscopy have been used to characterize a large number of neutral organophosphorus compounds. The bonds to alkyl or alkoxy substituents are apparently primarily sigma in character and the spectra can be interpreted without assuming π bonds to these groups. The P−O stretching frequency suggests increasing single-bond character in the order of increasing number of direct carbon-to-phosphorus bonds in the molecule. Thus, the electronegativity of the substituents around the phosphorus is the most notable factor affecting the extractive capacity of the molecule. The more electronegative alkoxy group decreases the availability of electrons on the phosphoryl oxygen, and its hydrogen-bonding ability, the molecule becoming thereby a weaker extractant. Similarly, each methylene group inserted between two P → O groups in tetraalkyl diphosphates and diphosphinates steps down the extractive capacity of the molecule. The extent to which a change of structure brings about variations of the extractive power toward metal salts will be discussed later on in more detail.

Tri-n-butyl phosphate has a dielectric constant of 8.05 ± 0.05[292,293], and that of tri-n-butyl thiophosphate is 6.82 ± 0.03[308] at 25°C. From dielectric constant measurements, mostly in carbon tetrachloride solution, the dipole moment of various esters shown in Table 9C1 have been reported. Various other physical properties of tributyl phosphate[292,311] and some other organophosphorus compounds have also been reported[277,278,293,312,313].

TABLE 9C1. Dipole moments of some neutral organophosphorus compounds at 25°C.

Compound	Dipole Moment (debyes)	References
Trimethyl phosphate	3.02	309
Triethyl phosphate	3.07, 3.08	309, 308
Tri-n-propyl phosphate	3.09	309
Triisopropyl phosphate	2.85	309
Tri-n-butyl phosphate	3.05, 3.07, 3.5, 3.10	309, 308, 310, 311
Triphenyl phosphate	2.89	309
Tetrahexyl methylene-diphosphonate	3.6	281
Tri-n-propyl phosphite	1.99	309
Tri-n-butyl phosphite	1.92	309
Triphenyl phosphite	1.59	309
Triethyl thiophosphate	2.82	308
Tri-n-butyl thiophosphate	2.84	308

c. Solubility of the Extractants

Low water solubility of neutral organophosphorus compounds is one of their most important characteristics. Compounds containing a toral of less than 10–12 carbon atoms are not practical for solvent extraction application, because of high solubility losses to the aqueous phase. The solubilities of a large number of pure undiluted compounds in water at 25°C have been determined[277,278,281,313-316] and are listed in Appendix D. Solubility in water increases in the predictable order: phosphates < phosphonates < phosphinates < phosphine oxides, in line with the increasing polar properties of the phosphoryl group in the molecule. Again, as expected, solubility decreases with increasing chain length: solubility in water of dialkyl phenyl phosphonates, for example, ranges from 0.11 M for the diethyl to about 5×10^{-6} M for the dioctyl derivative. Less understandable however, is the increased solubility with chain branching.

The effect of electrolyte in the aqueous solution upon the solubility of the esters is generally such as to bring about an overall decrease[277,292, 314-320]. The bulk of the available information refers to the salting-out characteristics of tri-n-butyl phosphate. Figure 9C1 shows this effect in mineral acid. The British data[317,318] for nitric and hydrochloric acids refer to initial aqueous concentrations, while the Oak Ridge[315] and Hanford[277] data are those for equilibrium acid molarities. They are not strictly comparable in view of the acid extracted by the undiluted TBP.

In aqueous solutions of neutral salts, the solubility of TBP decreases as a rule. Higgins and coworkers[315] studied the effect of various electrolytes in the concentration range up to about 3 moles/litre, and found a decreasing solubility of TBP in the following order of electrolytes: $LiI > NaI > AgNO_3 > KI > LiBr > LiCl > CsCl > KCl > NaCl > NaOH$. Kennedy and Grimley's[317] data, extended to more concentrated aqueous solutions of NaCl, $NaNO_3$ and $UO_2(NO_3)_2$, suggest also a continuous decrease in TBP solubility with increasing NaCl and $NaNO_3$ content in the aqueous phase, but a slight rise is discernable at low concentrations of uranyl nitrate.

The effect of mineral acids is more complicated. Hydrochloric, nitric and phosphoric acids decrease the aqueous solubility of TBP at low concentrations. At high acid concentrations, however, the solubility begins to rise and eventually exceeds the solubility in water alone. The electrolyte concentration at which the TBP solubility has a minimum depends apparently on the electrolyte. It is low for uranyl nitrate

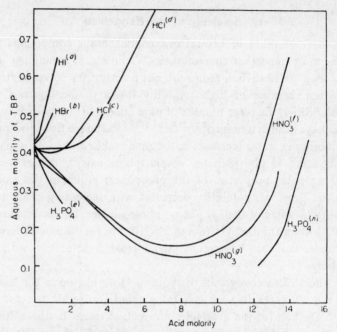

Figure 9C1.   Solubility of TBP in aqeous acid solutions. References: (a) (b) (c)[315], (d)[317], (e)[314], (f)[277], (g)[318], (h)[314].

(0.02–0.03 M) and hydrochloric acid (~ 1 M) and high for nitric (7–8 M) and phosphoric acid.

Undiluted diisopentyl methylphosphonate[320] shows a similar solubility behaviour in nitric acid, as does TBP. A minimum in the solubility curve appears at about 2.5 M acid; at 16 M acid its solubility is 54 g/l as compared to 1.9 g/l in pure water. Tri-n-butylphosphine oxide shows a stronger specific interaction with aqueous electrolyte solutions, than either diisopentyl methylphosphonate or tributyl phosphate, as it does with water. Its solubility increases in hydrochloric acid[316,319] but decreases in nitric acid[316].

An interpretation of these results is certainly not simple. Attempts have been made[315,317,318] to evaluate the activity coefficients of TBP applying the empirical Setchenov equation (section 1.F) and to express the salting-out or salting-in effect of the electrolyte through a salting coefficient. While the approach is formally justified, it fails with some of the neutral salts. Whenever the Setchenov equation holds reasonably well, as for example at low concentrations of sodium nitrate or chloride, and perhaps some other alkali halides, one may assume the distribution of one single non-reacting species, that of TBP alone. No such assumption can be made, however, in systems where the aqueous phase contains mineral acids, uranyl nitrate or even lithium chloride[321]. There is ample evidence for compound formation of TBP with the aqueous electrolytes used. Understandably, in such cases, and they are the majority of practical solvent extraction systems, it is difficult to treat the effect of the electrolyte on the solubility of TBP independently from that of its composite species. It seems, thus, unreasonable to speak in terms of a simple salting effect. It is unlikely that TBP, or any other neutral organophosphorus ester, is salted-out from the aqueous phase, for example at low nitric acid concentrations, but salted-in at higher acidities. At 8 M HNO_3, where the curves show increasing aqueous solubility, the equilibrium organic phase does not contain free TBP molecules any more. Instead, a higher aqueous solubility of the various hydrated and unhydrated $TBP-HNO_3$ complexes is more likely to be a correct qualitative interpretation of the solubility curves shown in Figure 9C1.

Only a small amount of data is available relative to the temperature effect on the solubility of esters in water and aqueous electrolyte solutions[315–317,322]. Figure 9C2 shows the effect of temperature on the solubility of tributyl phosphate and tributylphosphine oxide. The Oak Ridge data[315] for TBP are somewhat higher, due probably to a different experimental technique employed. In the light of the considerable negative temperature coefficient of solubility of these esters in water and aqueous electrolye solution, it is interesting to note the phenomenon associated with the usual procedure of centrifugation of the equilibrated solution[323]. Centrifugation may affect the solution equilibrium by raising the solution temperature by several degrees, due mainly to the motor heat.

Undiluted tri-n-butylphosphine oxide is completely miscible with water below 13°C. For TBP no complete mutual miscibility temperature has been observed. Several of the trialkyl phosphates, on the other hand, show a minimum aqueous solubility around 70°C.

The solubility of TBP in water from a TBP–diluent mixture, or more precisely, the distribution of the ester between water and various water-immiscible organic solvents has also been investigated[268,292,317,318,324–326]. Some of the graphically presented results[324] are reproduced here in Figure 9C3 as a function of the volume fraction of the dry TBP–diluent solution prior to contacting with water. For some other diluents, not

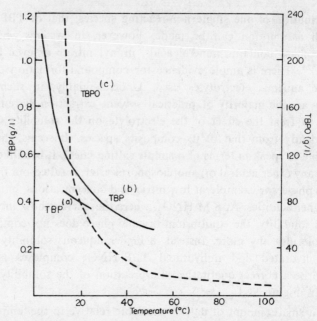

Figure 9C2.   Solubility of tri-n-butyl phosphate (TBP) and tri-n-butylphosphine oxide (TBPO) in water at various temperatures. References: (a)[318], (b)[322], (c)[315].

shown in the figure, decaline, n-heptane and toluene, solubility data have been tabulated[292,317,324]. The solubility of di-n-butyl butylphosphonate in water from a mixture with dodecane is similar to that of TBP under comparable conditions[327].

The solubility behaviour of TBP from a diluent mixture in aqueous electrolyte solutions is known mainly[292] for the TBP–kerosene-nitric acid system[317,318,326]. The results are those expected from Figures 9C1 and 9C3.

Finally, there is not much to be said about the solubility of neutral organophosphorus esters in organic solvents. With the exception of phosphine oxides, the esters are high boiling-point liquids[277] mutually miscible with water-immiscible organic solvents practical for solvent extraction application. Trialkyl- or triarylphosphine oxides, low melting point waxy crystals, have usually only limited organic solubility. Tri-n-octylphosphine oxide, for example, dissolves at 25° in cyclohexane to give a 0.92 molar solution, the solubility increasing with temperature[312,327].

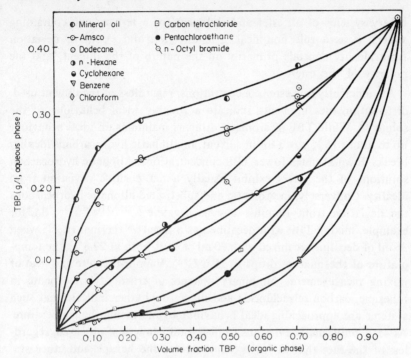

Figure 9C3.   Solubility of TBP in water from diluent solutions vs. volume fraction TBP in organic phase[324]. (By permission from the USAEC)

The partial molal volume of TBP has been calculated from densities of its solutions in a number of organic solvents, and was found to be reasonably constant and identical to the molal volume of TBP which is 273.7[292,324,327,328]. Variations were between 272.6 in hexane and 275.3 in chloroform, which is surprising in view of the non-ideal behaviour of such binary mixtures.

d. Organic Solutions of Extractants

By various experimental tools, such as dielectric constant[329], vapour pressure[330-333] and heat of mixing[333-337] measurements, infrared spectra[338,339,341], solubility and distribution[268,317,318,324,325] and apparent molal volume[324,328] determinations, several attempts have been made to evaluate the nature of the interaction between neutral organophosphorus esters and various organic solvents used as diluents in solvent extraction systems.

This information points strongly to the fact that the behaviour of binary systems of an ester and a diluent (or a ternary one containing water), is as a rule non-ideal. The direction and extent of deviation from ideality depends primarily on the nature of the diluent, and the polarity of the ester.

At sufficiently low ester concentrations, regardless of the diluent used, distribution measurements indicate a roughly ideal behaviour of the solute[317]. While TBP in aromatic diluents maintains an ideal behaviour up to about 15% w/v TBP in solvent, in aliphatic hydrocarbons ideality breaks down at much lower TBP concentrations. Aliphatic hydrocarbon solutions of the esters exhibit usually a net positive deviation from ideality. Conversely, in aromatic and substituted aliphatic hydrocarbons the deviation from Raoult's law is negative[277,317,318,332,333]. So, for example, mixing TBP with decaline has a negative thermal effect: when 35 ml of decaline are mixed with 70 ml of TBP, both at 22.8°C, the temperature of the mixture drops to 20.6°C[334]. Vapour-pressure and heat of mixing measurements on binary systems of triisobutyl phosphate in n-hexane, carbon tetrachloride and diisopropyl ether indicate that such systems are approaching ideal behaviour by increasing the temperature. At 45°C only small deviations from Raoult's law were observed, regardless of the fact that at lower temperatures the hexane and ether systems exhibited a positive deviation, while for carbon tetrachloride the deviaton was negative[333]. Most discussed is the negative deviation of systems where the diluent is a polar organic solvent[325,334]. Heat of mixing[334,337] and infrared spectral[338,340] data indicate that a TBP–chloroform mixture deviates strongly from the properties of ideal solution. The formation of a 1:1 complex between various organophosphorus compounds and hexone, chloroform and ether solvents has been demonstrated[339 340]. Simlarly, viscosity measurements on both wet and dry solutions of TBP in pentachloroethane at varying molar ratios suggest the existence of a 1:1 compound between the components[324].

Infrared, heat of mixing and vapour-pressure data indicate that a change in the polarity of the ester brings about variations in the extent of interaction between the esters and the diluent[331,332,337,341]. Hydrogen bonding of esters to chloroform, for example, increases by replacement of alkoxy by alkyl radicals in the molecule, as can be seen from the data in Table 9C3. In other diluents the activity coefficient increases with increasing number of direct P–C bonds in the ester at equal mole fractions of the solutes. The difference is more pronounced in hexane than in carbon tetrachloride.

Activity coefficients of tributyl phosphate[317,318,324,325,328,330-332,343] triisobutyl phosphate[333], dibutyl butylphosphonate[331], butyl dibutylphosphinate[331], tributylphosphine oxide[331], and dibutyl phenyl phosphate[332] in a number of solvents have been either determined by direct vapour-pressure and heat of mixing measurements or evaluated from distribution data of the esters between water and the diluent. Highest values of the activity coefficients have been obtained in aliphatic hydrocarbons. They decrease with decreasing molal volume of the diluent, which is roughly that of decreasing chain length. The lowest values of activity coefficients are characteristic for the relatively polar chloroform, bromoform and pentachloroethane, while intermediate activities have been measured in aromatic and substituted aliphatic hydrocarbons. No simple correlation was found to exist between activity coefficient of TBP and various physical properties of the organic solvent. However, Siekierski[325] has found a linear relationship between the activity of a 5% solution of TBP in twelve solvents and their solubility parameter. Although the trend of increasing activity with decreasing solubility parameter of the solvent, as observed, is perhaps correct, the linear relationship found is probably fortuitous, since volume percentage is not a suitable concentration unit for comparison. Generally speaking, while there is a reasonable qualitative agreement of the results reported for identical systems by various authors, the numerical agreement is generally poor[268,318,324,325,330,332].

TABLE 9C3. Association constant and energy of the hydrogen bond between tributyl phosphorus compounds and deuterochloroform[340].

Compound	K_{ass}	Energy of hydrogen bond (kcal)
Tributyl phosphate	1.0 ± 0.1	0.9
Dibutyl butylphosphonate	1.6 ± 0.2	1.1
Butyl dibutylphospinate	2.4 ± 0.3	1.4
Tributylphosphine oxide	4.8 ± 0.3	1.8

Finally, association of TBP molecule by the formation of a system of dipole-dipole bonds has also been suggested[317,334,339,344,346]. From i.r. data, the self-association constant $K_{(TBP)_2}$ has been estimated to be 2.9 ± 0.1, but is affected by the nature of the diluent[339].

It is easily understood that such extractant–diluent interactions as outlined above will frequently affect the extractive capacity of a neutral organophosphorus ester towards metal salts. This influence will be discussed in the forthcoming paragraphs, and more specifically in section 9.D.

e. Solubility of Water in the Extractants and their Interaction

The two-phase system of water and a neutral organophosphorus extractant has been investigated[268,277,279,281,282,292,304,313,318,322,324,342,347–372] in considerable detail and still there is no firm conclusion about the extent of the organic-phase interaction leading to the formulation of the complexes existing. The role that water of hydration has on the extractability of metallic species, a fact recognized at an early stage in TBP extraction systems, prompted more fundamental investigations on the solubility of water in TBP and other phosphorus extractants under a variety of conditions. The earliest view that a water-saturated TBP is essentially the hydrate $TBP.H_2O$ seems today to be a mere over-simplification, even at room temperature where the analytically determined stoichiometry tempts this view. More recent information on the water solubility, i.r. and n.m.r. spectra of water-saturated extractants and other properties of such systems indicate that the water content is not due solely to the formation of a single addition compound.

The available data on the solubility of water in a number of phosphates, phosphonates, phosphinates, phosphine oxides and diphosphonates at 25°C are compiled in Table 9C2, together with the water-to-ester mole ratios at saturation.

The water content of undiluted TBP at 25°C deserves some comments. It was that content of approximately 3.6 M at saturation that caused the early belief that the $TBP.H_2O$ is a stable complex, quantitatively formed upon equilibration of the ester with water or dilute aqueous electrolyte solutions (undiluted TBP is $\sim$3.6M, and has a molar volume of 273.7 ml). Very careful recent works from Harwell[348] and Oak Ridge[342] on the water concentration of undiluted TBP at 25°C indicate that the difference between the actually determined $TBP.H_2O$ mole ratio and unity is statistically significant, thus, the mole ratio of unity at room temperature is fortuitous rather than indicative of a strong complex.

The solubility of water in TBP is definitely temperature dependent. The existing information, however, is not much in agreement, as shown in Figure 9C4. Water is soluble in tributylphosphine oxide[322] at 13°C

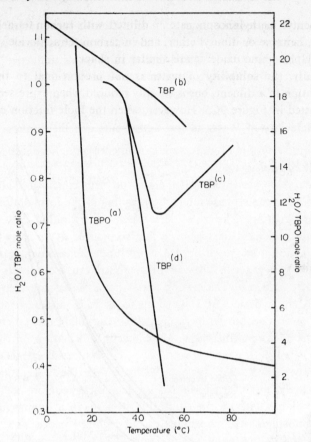

Figure 9C4.   Solubility of water in esters as a function of temperature. References:
(a)[322], (b)[348], (c)[359], (d)[357].

to the extent of complete miscibility, and decreases with increasing
temperature, the sharpest changes occuring below 25°C. Diisopentyl
methylphosphonate[313,372] dissolves water to the extent of an ester:
water ratio of 0.55 at room temperature, which changes only little with
temperature, having the value 0.51 at 50°C.

   As could be expected, an important variable in the extraction of water
by esters is the presence of a diluent in the system. Solubilities of water
in binary mixtures composed of TBP and various diluents have been
measured by a number of different methods and under a variety of
equilibration conditions[292,318,342,347,348,351,352,355,360,363]. Some of the
data are shown in Figure 9C5. Plots representing the solubility of water

in diisopentyl methylphosphonate[313] diluted with carbon tetrachloride, kerosene, benzene or dibutyl ether, and in carbon tetrachloride solution of tributylphosphine oxide[322] are similar in shape.

Generally, the solubility of water is not proportional to the ester concentration in a diluent, but is sharply reduced when expressed in the units plotted in Figure 9C5. However, when the mole fraction of water

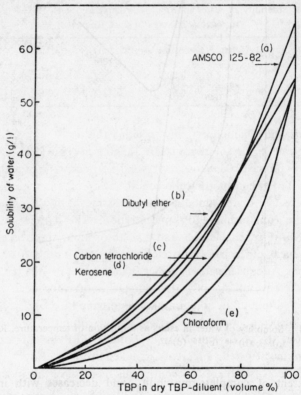

Figure 9C5.   Variation of solubility of water in TBP-diluent solutions with volume % of TBP in dry TBP-diluent mixture at 25°C. References: (a)[342], (b) (c)[355], (d)[318], (e)[355].

in the saturated organic phase is plotted against the mole fraction of TBP in the initial, dry TBP–diluent mixture, the solubility is very nearly a linear function. Davis[342] has plotted his data in this way, for the TBP–Amsco 152–82, a mixture of aliphatic hydrocarbons, system and others, for the similar TBP–odourless kerosene[360] and TBP–hexane[363] systems. It is difficult to draw a firm conclusion relative to the effect of the diluent on the solubility of water in the organic phase, mainly owing to uncertainties about the correct choice of concentration scale. The effect is apparently similar regardless of the structure of the ester, as indicated by the data compiled in Table 9C4 for 20% by volume solutions of tributyl phosphate[355] and diisopropyl methylphosphonate[313] (DiPMP).

TABLE 9C4. Solubility of water in 20% by volume solutions of TBP and DiPMP in various solvents at 25°C (mole/l).

	Dibutyl ether	Kerosene	Benzene	CCl_4	$CHCl_3$
TBP	0.28	0.20	0.16	0.10	0.11
DiPMP	0.63	0.52	0.42	0.33	

Whitney and Diamond[352], investigating the transfer of water into a carbon tetrachloride solution of TBP ranging in concentration from 0.1 to 60% by volume (0.0366 to 2.2 M) found a straight line of unit slope up to 0.1 M TBP when plotting the concentration of water in the organic phase, corrected for the amount of water extracted by the diluent alone, against the equilibrium TBP concentration. The data were obtained by the Karl-Fischer method and i.r. absorption spectroscopy using the 2.72 μ peak of the free OH stretch and the 2.90 μ peak of the bonded-OH stretch. At concentrations higher than 0.1 M TBP the data indicate a greater relative uptake of water. Similar, apparently ideal, behaviour was observed in a dilute solution of TBP in benzene[354].

The influence of water activity, a_W, on the solubility of water in undiluted TBP[347,348,357] and 20% by volume TBP in kerosene[348] has also been investigated. Figure 9C6, taken from McKay[347], indicates that as a_W increases there is a continuous increase in the water content of the TBP phase with no sign of saturation at the molar composition of 1:1.

Infrared spectra of water in esters indicate a moderately strong hydrogen bonding of water to the phosphoryl oxygen in TBP[304, 318,348-350,352,360,362] and a number of phenyl phosphonates[278] and diphosphonates[372]. Of particular interest is the P → O stretching frequency. In undiluted, dry TBP it appears at 1270–1280 cm^{-1}, and in phenyl phosphonates around 1250 cm^{-1}. There is a 10–12 cm^{-1} shift to lower frequencies in the case of TBP and lower phenyl phosphonates when saturated with water, but the shifts are barely perceptible for the higher phosphonates. There is no i.r. evidence, however, for

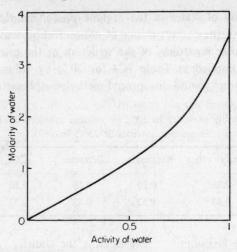

Figure 9C6. The activity of water dissolved in TBP at 25°C [347]. (By permission from Ind. Chim. Belge)

water being bound to the butoxy oxygens, since the absorption bands due to P–O–C at 1057, 1026 and 990 cm^{-1} seem not to be affected upon saturation of TBP with water[322,368]. The changes in the absorption bands of water in the 3500 cm^{-1} spectral region also support the existence of hydrogen bonding in water–TBP mixtures[348,350,352]. Water at low concentrations in pure TBP shows peaks at $\sim$3480 cm^{-1} and $\sim$3550 cm^{-1} assigned to symmetric and antisymmetric stretching vibrations of the OH group. When TBP is diluted by carbon tetrachloride the two bands are more separated. At high water concentrations (> 1 M), however, in both pure and CCl$_4$-diluted TBP the two bands become broader and the higher frequency band cannot be resolved. The mean stretching frequency of the OH groups is lower by about 180 cm^{-1} than that in water vapour, but higher by about 100 cm^{-1} from that of liquid water.

Nuclear magnetic resonance offers supplementary information on an ester–water interaction[268,357,364,366]. The single resonance shown in the n.m.r. spectrum is reproduced in Figure 9C7. Murray and Axtmann[364] and Burger[268] found that the chemical shift of the water proton in TBP solutions varies linearly with the water content throughout the concentration range between 0.44 and 6.3 weight % water, while Bullock and Tuck's[357] measurements show a linearity only up to about 2.6 weight% (40% of saturation with water). The latter authors were apparently unaware of the earlier n.m.r. measurements, thus no explanation for the apparent disagreement has been offered. When TBP saturated with water is diluted with carbon tetrachloride the shift is not any more a linear function of the water content [268], as shown in Figure 9C8. Extrapolation of the water signals places the infinite dilution shift for water in TBP displaced from the pure water signal to higher fields by about 2 p.p.m. These observations are characteristic of hydrogen-bonded solutions but do not support the existence of simple hydrated TBP species.

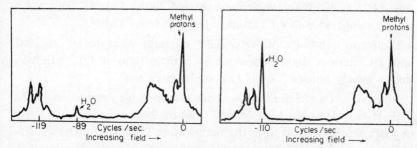

Figure 9C7. Nuclear magnetic resonance spectrum of tributyl phosphate containing 1.7% water (left), and of tributyl phosphate saturated with water (right)[364]. (By permission from the Amer. Chem. Soc.).

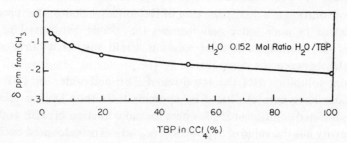

Figure 9C8. Chemical shift of solute protons in TBP solutions of water[268]. (By permission from Amer. Nucl. Soc.)

Axtmann[366], using n.m.r. data[364], evaluated the strength of the TBP–H_2O bond and found a value of about 5 kcal/mole, which is rather close to that of a single water-to-water bond. This value has to be compared with that of 4 kcal/mole, estimated[366] from equilibrium data[342].

Incidently, methods were developed for determination of water in the organic phase by infrared[278,352] and n.m.r.[382] spectroscopy.

Viscosity and density measurements on the liquid phases of the TBP–H_2O system have also been carried out in an attempt to identify the interaction between the components[292,348,356,357,365,370]. The density curve at 25°C for the system undiluted TBP–H_2O in the full range of water concentration does not reveal any singular point which could support compound formation[356]. The increase in TBP viscosity with an increase in water content, though small, indicates an interaction in the system, but again does not give evidence for the existence of a definite compound. The variation of viscosity with temperature up to 55°C allowed the estimation of an apparent energy of activation for viscous flow at 25°C of 9.5 kcal/mole for hydrated 100% TBP[348]. This value can be compared with 4.1 kcal/mole for anhydrous TBP[365].

Dielectric constants of water–TBP mixtures measured[370] at 25°C and 2°C show a slight interaction at a mole ratio of 1:1, indicating that a weakly bonded TBP.H_2O complex may exist.

Another piece of information about the probable interaction in the TBP–H_2O systems has been derived from the heat of solution of water in TBP. ΔH_W evaluated from the temperature coefficient of water solubility in undiluted TBP has the value of -290 cal/mole. It is strongly dependent on the water activity, becoming endothermic, $+280$ cal/mole in the region of $a_W = 0.55$–0.59[348]. Olander and coworkers[363] reported the value of -1050 cal/mole. Reasoning, as Axtmann[366] did, that the heat of solution is actually the sum of the enthalpy changes for breaking the bonds in pure water and forming the bonds between TBP and water, the exothermic heat of solution would suggest a fairly stable complex between the components.

From solubility data the activities of TBP and water in TBP–H_2O solutions have been evaluated by applying the Gibbs–Duhem relation. In a saturated solution at 25°C, where the mole fraction of TBP is 0.488, its activity has the value of $a_{TBP} = 0.498$ ($a_W = 1$). The calculated activities a_W and a_{TBP} are roughly linear functions of their mole fractions in the mixture.

It is appropriate to discuss now the state of water and the species present in an ester–water mixture. Since the bulk of the data refers to tributyl phosphate, the following discussion must be limited to it. However, with certain restrictions as to the extent of interaction, it is believed that its nature is similar to that of all the neutral organophosphorus esters. The affinity of esters towards water will depend on the basicity of the esters, their polarity, structure, and especially the degree of steric hindrance around the polar group.

The belief that in a saturated TBP solution at room temperature, the only species present is the simple molecular adduct TBP.H_2O is unlikely to be true. The arguments against the exclusive existence of the 1:1 compound include the dependence of this ratio on temperature, water activity and presence of diluent in the system. While all these factors affect the TBP: water ratio, as has been discussed above, they do not contradict the existence of the monohydrate complex. They merely make its *quantitative* formation at room temperature questionable.

Infrared and n.m.r. spectral data supply direct physical evidence for the formation of a hydrogen bond between the phosphoryl oxygen and the water protons. The energy of the bond, 4–5 kcal/mole, is rather low. The only direct evidence that this hydrogen bonding leads to the formation of a 1:1 compound originates from distribution data. At low TBP concentrations in some diluents, the solubility of water corresponds to a 1:1 ratio. For the formation constant of $TBP.H_2O$ various values have been reported: 0.15 in carbon tetrachloride[352], ~0.1 in kerosene[353], but as high as 14.4 in benzene[354]. The most probable structure of this complex should be

$$\rangle P \rightarrow O \cdots \cdots H - O - H$$

where one OH group is bound and the other free.

While this is apparently the water-bearing complex predominating when the concentration of the solutes, TBP and water, are kept sufficiently low to approach the requirements of an ideal solution, the situation is considerably different when either of the components, or both, increase in concentration.

When the TBP: H_2O ratio is high in the solution, the intensity of the spectral band corresponding to free OH is markedly lowered, or even disappears. Each water molecule may be doubly hydrogen bonded to two TBP molecules

$$\rangle P \rightarrow O \cdots \cdots H - O - H \cdots \cdots O \leftarrow P \langle$$

as proposed by Bullock and Tuck[357] and agreea upon by Hardy and coworkers[348].

When, on the other hand, the experimental conditions are such that the amount of water in the TBP phase is higher than, ca. 0.1 M, spectral data[348,352] indicate that the water molecules, on the average, are becoming more highly hydrogen bonded. Whitney and Diamond[352] suggest that some water molecules may dissolve without being bound to the phosphoryl oxygen but to the water already hydrating the TBP molecule instead. Whether such a 'dihydrate' can be formed, may depend upon the competition between a) an unhydrated phosphoryl oxygen, (free TBP) and a free OH group for the available water molecule, or b) a water and a diluent molecule to hydrate or solvate, respectively the free TBP in the solution.

It is tempting to attribute, partially at least, the high (above a 1:1 ratio) water uptake by a number of esters listed in Table 9C2 so a similar di-[313,359] or higher-hydrate formation through hydrogen bonding to the first water molecule initially bound to the phosphoryl oxygen. It is, however, questionable if such hydrates can be recognized as true chemical species. Inasmuch as n.m.r. spectra show only one single signal from the water protons in TBP under a variety of conditions, this indicates that the various hydrates, if formed at all, must be in rapid equilibrium.

This rapid equilibrium applies to any complex formed in a TBP–H_2O mixture, several of them having been suggested by Bullock and Tuck[357].

f. Extraction of Acids

The most interesting feature in acid extraction by neutral organophosphorus compounds is perhaps their capacity to extract mineral acids well over an acid-to-ester ratio of unity. This excess over a 1:1 formula becomes proportionally larger as the concentration of the ester increases in the diluent. One has now the choice of interpreting the experimental data as dissolved, hydrated or unhydrated, acid in the ester–acid complex, or of postulating additional solvates of mineral acids containing two or, as in case of nitric acid, for example, even three or four acid molecules per molecule of extractant. In our opinion, the weight of evidence is in favour of dissolution, rather than higher solvate formation.

In this connexion the question of alkoxy-oxygen participation in extraction is becoming important. The question has frequently been touched upon and it seems that alkoxy oxygens in phosphates, phosphonates and phosphinates *do not* provide a coordination site in the extractant molecule, and they *do not* have acid-binding properties.

The extractability of acids by an ester depends on factors such as the acid strength, size and hydration of the anion. The order of extractability depends apparently on the acid concentration in the aqueous phase.

Trioctylphosphine oxide[312], for example, extracts acids in the order $HNO_3 > HClO_4 > HCl > H_3PO_4 > H_2SO_4$ up to 2 M acid in the aqueous phase, while at 6 M level the order is $HCl > HNO_3 > H_2SO_4 > H_3PO_4$ ($HClO_4$ forms a third phase above 3 M). With tributyl phosphate[373] the order of extractability is: oxalic $\sim$ acetic $> HClO_4 > HNO_3 > H_3PO_4 > HCl > H_2SO_4$, which is roughly the order of hydration energy of the anion. Such comparison is unlikely to be factual. Distribution ratios of acids should only be compared when acids have the same solvation and hydration state in the organic phase.

Extraction of strong mineral acids by TBP provides evidence on the hydration of the proton in aqueous solutions. Strong acids are extracted usually with the hydronium ion H_3O^+ retaining a primary hydration shell of three water molecules, similar to the structure prevailing in water. The ester molecules, with their phosphoryl oxygen, coordinate to this shell of water and are usually not directly coordinated to the proton. The situation is different, of course, in anhydrous systems. HCl, HBr, $HClO_4$ and H_2SO_4 are present as molecular adducts to the ester through its phosphoryl-oxygen site. On addition of water to an anhydrous solution ion-pairs may be formed[374].

The descriptive concept of Diamond and coworkers[53,352,375-378] provides a plausible explanation of the experimentally observed phenomena related to the coextraction of water along with acids. The approach implies a competition between the basic extractant, water and the anion for the available proton. In the case of strong acids, the anion is too weak a base to enter successfully into competition, and only the competition between H_2O and TBP needs to be considered. The moderately basic nitrate ion, however, will enter into the competition with water and the extractant for association with the proton. The absolute water content is thus peculiar to the acid itself. When TBP is the extractant, the amount of water coextracted with each acid appears to decrease in the order $H_3PO_4 > HI > HBr > HCl > HClO_4 > H_2SO_4 \gg HNO_3$. Nitric acid forces out water from TBP at low acidities, while the others have a characteristic acid concentration when water content in the organic phase is at maximum.

(i) *Nitric acid* The distribution of nitric acid between its aqueous solution and a neutral ester has been studied more extensively and, certainly more fundamentally than any other acid with any given class of extractants. While tributyl phosphate has received most attention[281, 292,298,310,318,330,342,351,360,362,371,373,374,378-398,407-420], many other phosphates[399,402,407], phosphonates[274,320,327,399–402,407], phosphine oxides[312,328,381,403–405] and diphosphonates[282,328,394,405] have been investigated also. A comparison of nitric acid extraction by various phosphates and phosphonates shows that the distribution is only slightly affected by the class of extractant or by the differences in the alkyl chains within a class. Figure 9C9, taken from Siddall[399], shows these effects. The extractants were all diluted to 1.09 M with n-dodecane and the data were obtained at 30°C. At this, relatively high extractant level, the distribution ratios at low nitric acid loading are about twice as high with the phosphonates as with the phosphates. At higher acid concentrations the differences diminish. At low extractant level, as seen from the data in Table 9C5[407], the distribution ratio of nitric acid is practically insensitive to changes in both, the class of extractant and the alkyl substituent.

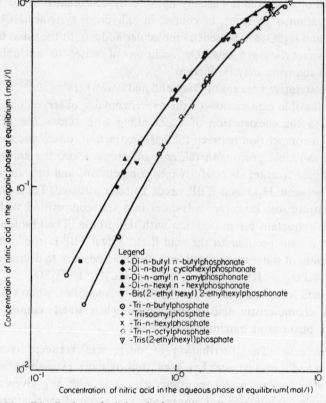

Legend
● -Di-n-butyl n-butylphosphonate
◆ -Di-n-butyl cyclohexylphosphonate
■ -Di-n-amyl n-amylphosphonate
▲ -Di-n-hexyl n-hexylphosphonate
▼ -Bis(2-ethyl hexyl) 2-ethylhexylphosphonate

○ -Tri-n-butylphosphate
+ -Triisoamylphosphate
× -Tri-n-hexylphosphate
◇ -Tri-n-octylphosphate
▽ -Tris(2-ethylhexyl)phosphate

Concentration of nitric acid in the aqueous phase at equilibrium(mol/l)

Figure 9C9.   Distribution of nitric acid for dialkyl alkyl phosphonates and trialkyl phosphates[399]. (By permission from the Amer. Chem. Soc.).

TABLE 9C5. D_{HNO_3} by 0.1 M phosphate and phosphonate esters at room temperature[407].

Ester	Diluent	Initial aq. HNO_3 2.01 M	10.05 M
Tributyl phosphate	Amsco	0.21	0.12
Tributyl phosphate	Xylene	0.21	0.12
Tris(2-ethylhexyl) phosphate	Amsco	0.21	0.12
Dibutyl butylphosphonate	Amsco	0.28	0.12
Dibutyl phenylphosphonate	Xylene	0.22	0.12
Di-*sec*-butyl phenylphosphonate	Xylene	0.25	0.13

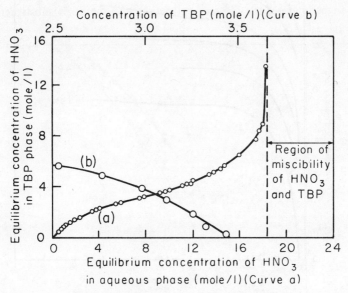

Figure 9C10.   a. Equilibrium distribution of nitric acid between TBP and water at 25°C; b. Concentration of TBP and nitric acid in the organic phase[351]. (By permission from USAEC).

The extraction of nitric acid by TBP is a complex phenomenon in the entire concentration range of both solutes. The interpretations of distribution data by various authors are unaminous in their belief that TBP (and other esters, also) tends to saturate at a one-to-one ratio with nitric acid. Under the majority of experimental conditions used, however, more acid is forced into the organic phase when in equilibrium with sufficiently concentrated aqueous acid, until finally a 96% nitric acid, becomes completely miscible with undiluted TBP (and other esters also). A typical distribution curve of nitric acid between water and undiluted TBP at 25°C, and the resulting equilibrium organic-phase concentrations of TBP and the acid are shown in Figure 9C10[351]. Partition of the acid between water and TBP diluted with kerosene is shown in Figure 9C11[318,360].

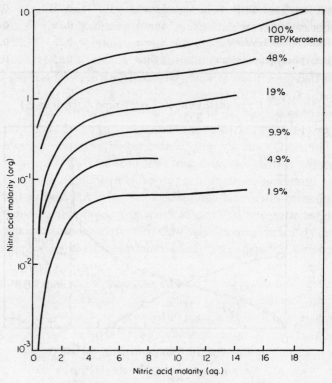

Figure 9C11. Partition of HNO_3 between water and TBP in kerosene[360]. (By permission from the UKAEA).

As mentioned, the form in which the excess acid exists in the organic phase, is in dispute. Depending primarily upon the experimental conditions but also on the method by which the distribution data were obtained, a wide array of different hydrated and unhydrated ester-acid species have been claimed to exist. Others have indicated that the equilibrium curves account for a dissolution of the excess acid rather than complex formation.

As for the equimolar complex, it is apparent from a large number of independent investigations, carried out under a variety of experimental conditions, that the formation constant of the $TBP.HNO_3$ complex $(H^+ + NO_3^- + \overline{TBP} \rightleftharpoons \overline{TBP \cdot HNO_3})$ has a value between 0.15 and $0.20^{292,298,318,351,360,371,384-388,390,391,395,398,408,418}$. These apparent equilibrium constants have been evaluated without taking into account the activity coefficients of the organic-phase species. Their validity is thus, limited to low TBP concentration ranges. High dilution of TBP introduces minor variations of the constant due to a diluent effect^{320,330,393,395,416}. Vapour-pressure measurements of TBP over two–phase water–nitric acid–undiluted TBP system^{371} at 25°C have indeed shown that the activity of water-saturated TBP decreases from 1 to 0.1 as the ratio $HNO_3:TBP$ increases from 0 to 1. Nitric acid concentration in the aqueous phase up to 4–6 M has seemingly little effect on the non-ideality of the organic solution. This is perhaps due to such a variation of the $TBP.HNO_3$ activity with concentration as to compensate for the decreasing activity of unreacted TBP or $TBP:H_2O^{318}$.

The formation constant of a (probably hydrated) diisopentyl methyl-phosphonate–nitric acid equimolar complex^{320} has a value of 16.2 with undiluted ester, and 10.9 and 22.3 in a 30% solution in carbon tetrachloride and benzene, respectively. With trioctylphosphine oxide the reported values are 9.1^{328} and 10.2^{405}, and with tetrahexyl methylenediphosphonate 15.2^{423}.

Collopy and coworkers$^{382-384}$ have chosen to evaluate the equilibrium constant for the reaction taking place between tributyl and trioctyl phosphate and the associated nitric acid in the *organic phase*. They found that their K values are linearly dependent on the reciprocal of the ester concentration in the diluent, according to the empirical relationships

$$K_{TBP.HNO_3} = 98.53\,[TBP]^{-1} + 1.07,$$

and

$$K_{TOP.HNO_3} = 91.00\,[TOP]^{-1} + 44.10$$

Several authors explained the shape of their equilibrium curve and analytical data in the region where more acid is transferred than needed for the equimolar complex, in terms of additional complexes such as hydrated or unhydrated $TBP.2HNO_3^{298,342,351,379,381,386,387,390,409,411,413}$

$TBP.3HNO_3$[351,385,413] and $TBP.4HNO_3$[351,413]. The values calculated for the formation constant of $TBP.2HNO_3$ varied between 0.0004 and 0.004, depending again, on the experimental conditions chosen and the way of their computation[298,386,387,390,409]. The equilibrium constant for the formation of $TBP.3HNO_3$ has been estimated[385] to be 0.04.

For diisopentyl methylphosphonate the existence of similar complexes, with two and three nitric acid molecules, has been claimed[320]. The formation constant of the lower complex has the value of 1.1 in undiluted DiPMP, and 0.4 and 1.4 in 30% solution of carbon tetrachloride and benzene, respectively.

Some authors[382–384,404] chose to consider all excess acid in the ester phase above the equimolar complex to be the result of mere solubility distribution of associated, molecular, nitric acid. The distribution constant of the associated nitric acid between its aqueous solution and TBP has been shown to be linearly dependent on the ester concentration according to the relationship[382]

$$K_{d(HNO_3)ass} = 0.437 \, [TBP] - 0.0017$$

For trioctyl phosphate the relationship is

$$K_{d(HNO_3)ass} = 0.0331 \, [TOP] + 0.003$$

The salting out of nitric acid from an aqueous electrolyte solution into TBP is of considerable theoretical interest and technological importance. In most cases it has been ascertained that the salting-out electrolyte itself is extractable to a varying extent into the organic phase[360,395,410,414]. In nearly all cases, however, the extraction of the acid is enhanced, usually ba y simple common-ion effect.

A detailed study[360] on the salting-out effect of various nitrates shows that the tervalent aluminium has only a slightly improved effect over that of calcium or lithium in the higher acidity region, while they all have a very similar effect at lower acidities. As the acidity increases the salting-out electrolyte has a lower effect, until, at 7 M HNO_3 the distribution ratio, even in the presence of a high aluminium concentration, is approaching that in pure nitric acid itself. A similar study has been made of the salting-out effects of calcium nitrate from 0 to 4 M[360,410] and lithium nitrate from 0 to 6 M[360].

The salting-in effect of uranyl nitrate upon nitric acid is a common feature for salts having a greater affinity for TBP than HNO_3. Such are the nitrates of the actinides and some other tetravalent nitrates, particularly ceric nitrate. In their presence D_{HNO_3} is never as large as in their absence. This much more complex equilibrium will be discussed later.

A special case of double-electrolyte systems is that where the aqueous phase is a mixture of two mineral acids, independently well extractable. The general picture of preferential extractability holds again, the degree of extraction of the more extractable acid is enhanced by the addition of the less extractable one. The phenomenon has been demonstrated for several systems: mixture of HNO_3 and HCl by TBP[400,419] and diisopentyl methylphosphonate[400]; HNO_3 and $HClO_4$ by TBP[364,497]; and mixture of HNO_3 and H_2SO_4 by trioctylphosphine oxide[312]. In all these cases it is the activity of nitric acid which is increased in the presence of another acid.

Numerous studies have dealt with the water content of the ester phase resulting from the extraction of nitric acid. The bulk of the available information refers, again, to tributyl phosphate[36,292,318,342,351,356,360,371,374,378,379,381,392,418,420-422], but dibutyl butyl-[327] and diisopentyl methylphosphonate[320], trioctylphosphine oxide and various diphosphates[281] have also been investigated.

Typical data on the water content in the equilibrium organic phase are shown in Figure 9C12 for undiluted TBP. To explain this and similar water-content results the existence of a number of various species has been postulated. Since the water content may decrease, remain constant or increase with transfer of acid depending on the TBP-to-diluent ratio and the nature of the diluent, the presence of water does not appear to affect the reaction of nitric acid with either TBP or $TBP.H_2O$. Indeed, a mathematical treatment[342] of the majority of the literature data is consistent with the interpretation that HNO_3 reacts with both TBP and $TBP.H_2O$ to form $TBP.HNO_3$ and $TBP.H_2O.HNO_3$. Thus at lower acidities the extraction process is equivalent for both species. At medium acidities, where the water content falls steeply, obviously hydrated species give place to unhydrated species. Nitric acid may replace water from $TBP.H_2O$ to increase the fraction of $TBP.HNO_3$. At higher acidities, in the region of $HNO_3:TBP > 1$, the water content increases, maintaining an acid: water ratio of roughly 3:1, the same ratio as that at the minimum water-content point.

Direct calorimetric measurements[379] and distribution data over the

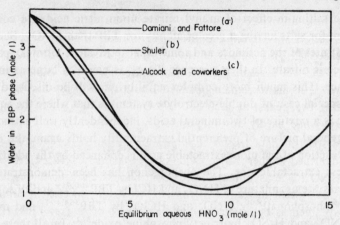

Figure 9C12. Extraction of water into TBP from nitric acid solutions. (a) Ref. 409, (b) Ref. 351, (c) Ref. 318.

temperature range of 0°—50°C permitted the calculation of the thermodynamic quantities, ΔH, ΔF and ΔS for the extraction of nitric acid by a large number of neutral organophosphorus compounds[360,402,403]. The entropy changes for extraction of the acid into phosphates and phosphonates varied from −5 to −7 e.u., and had a value of −8.8 e.u. for trioctylphosphine oxide. ΔH varied between −0.1 to −0.7 kcal/mole for the phosphates and phosphonates, and $\Delta H = -3.9$ kcal/mole for trioctylphosphine oxide. The values, −0.6 kcal/mol[402] and −3.6 kcal/mole[360], reported for tributyl phosphate are very much at variance.

Infrared spectra have shown that the P → O stretching frequency in free TBP undergoes a moderately strong shifting to lower values upon equilibration with acid[268,304,310,349,351,362,392,420,423]. The shift of 50–80 cm⁻¹ is still more characteristic of hydrogen-bonded compounds than P → O ⋯ H⁺ ions, thus, one may assume that it is the molecular acid which is extracted, the bonding being chiefly P → O ⋯ HNO_3 with some contribution from ion-pairs. In the case of the more basic trioctylphosphine oxide, however, the shift of 84 cm⁻¹ was believed[328] to be sufficiently large to indicate a considerable fraction of ion-pair structure ($P^+ \rightarrow OH$)(NO_3^-). The shift towards lower frequencies increases with the concentration of nitric acid both in wet[304,362] and anhydrous[351] TBP systems.

As to the participation of the butoxy oxygen in TBP, the triplet band assigned to the P—O—C group is only slightly affected when TBP is equilibrated with aqueous nitric acid[304,362,392,423]. In an anhydrous system these absorption bands at 980 to 1080 cm^{-1} are observed to shift to longer wavelengths as the concentration of nitric acid in TBP is increased[351]. This shift is of the same magnitude as the phosphoryl shift. In the wet system, the most intense band of the triplet, that at 1030 cm^{-1}, seems to remain unaffected both in intensity and position. The other two peaks, at 1050 and 990 cm^{-1}, show a slight change in intensity, while hardly any in position.

From spectral data in the near-infrared region Shuler[351] claimed to have identified higher nitric acid species, namely TBP.3HNO$_3$ and TBP.4HNO$_3$. The asymmetry of his continuous-variation curves for an anhydrous system measured in the 1023–1820 mμ spectral region, has been explained as being due to the coexistence of several TBP.nHNO$_3$ species, where n has a value of 1 to 4. In contrast to this interpretation of infrared spectral data in terms of specific complexes existing in an anhydrous TBP–HNO$_3$ system, no direct infrared spectral evidence shows that past the 1:1 complex the additional HNO$_3$ forms complexes in a wet TBP solution. Smutz and coworkers[349,420] have interpreted their i.r. data in terms of simple dissolution of the associated acid in the TBP.HNO$_3$ complex.

Nuclear magnetic resonance spectra showed only one peak for anhydrous nitric acid in TBP[268,351]. There is still a rapid exchange of the protons producing a single sharp line in the spectrum when the organic phase is prepared by equilibration with aqueous nitric acid[268]. The shift is not proportional to the acid content when a dry TBP–HNO$_3$ solution is diluted with carbon tetrachloride or TBP, as can be seen from Figure 9C13[268]. In wet systems the position of the line depends not only on the amount of acid extracted but also on the amount of water transferred along with the acid. Figure 9C14, also taken from Burger's[268] work, shows this shift for undiluted TBP solutions. The maximum in the chemical shift appears when the stoichiometric composition of the organic phase is about 3.1:3.1:1.0 mole ratio TBP:HNO$_3$:H$_2$O. The water content is at a minimum at this point[381]. In the lower acid region, where the shift varies nearly linearly with concentration of the acid, an equilibrium between TBP.H$_2$O and TBP.HNO$_3$ species may be considered.

The dry TBP–HNO$_3$ system has been investigated spectrophotometrically also in the ultraviolet region[382,383]. The method of continuous variation yielded absorption curves with a maximum at a 1:1 ratio. The shape of the curves, identical in carbon tetrachloride, chloroform

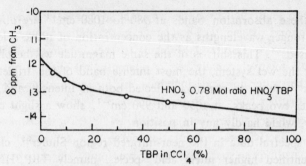

Figure 9C13. Chemical shift of solute protons in TBP solutions of nitric acid[268].
(By permission from the Amer. Nucl. Soc.)

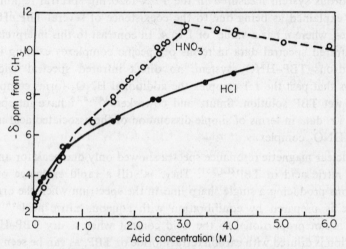

Figure 9C14. Concentration of water in TBP equilibrated with aqueous nitric acid[268]
(By permission from the Amer. Nucl. Soc.)

and an isooctane–ether mixture, was unsymmetrical on the excess-acid
side. It was suggested that the excess acid over that of a 1:1 ratio is due
to the solubility of associated HNO_3 in the $TBP.HNO_3$ complex. On
the other hand, the existence of several complexes with varying ester:
acid: water ratios has been identified in a diisopentyl methylphosphonate–nitric acid system using ultraviolet spectrometry[320].

Dissolving anhydrous nitric acid in undiluted TBP does not affect its
viscosity[347,379]. When water is added to the binary TBP–HNO_3 mixture[347] or when acid is extracted into undiluted TBP there is a substantial
change in viscosity[347,356,365,379,381,418]. The quite sharp change in viscosity
is interpreted in terms of a quantitative formation of a 1:1 compound[347].
The unchanged viscosity in the dry system, on the other hand, is appar-

ently due to small changes in the size of the species as the acid is bound to the phosphoryl oxygen[347]. The change in the molar volume, from dry TBP (273 ml) or TBP.H$_2$O (291 ml) to the unhydrated TBP.HNO$_3$ (315 ml), is relatively small[292,365]. Apparently, the effect is not additive since neither water nor acid alone have much effect on the viscosity, but the two together achieve the marked effect.

Conductivity of the extracts in undiluted TBP shows a break in the specific conductivity-vs-concentration curve at a 1:1 ratio[347,381]. The change is again quite sharp and this discontinuity suggests the quantitative formation of a strong complex.

Fletcher and Hardy[381], on the basis of viscosity and conductivity data in an undiluted TBP system, propose that the first complex of considerable stability which is formed as nitric acid is transferred into a hydrated TBP is one with a HNO$_3$: TBP ratio of ~0.5 or lower. Such a di-TBP complex, which may be an ion-pair, has been found to exist in other TBP–acid systems.

The ionization estimated by Walden's rule shows that the apparent ionization constant of nitric acid in undiluted TBP is of the order of 10^{-4} [318,360,381,418]. The ionization of the acid is even lower when the organic phase contains kerosene as diluent[318].

The dielectric constant of a series of TBP–HNO$_3$ mixtures in carbon tetrachloride[385] indicates an equimolar complex, and an additional compound, possibly TBP.3HNO$_3$, but other acid species cannot be excluded either.

Summarizing the above experimental evidence the following discussion can be offered. Several authors[328,374,381,392,415] have advocated that a small fraction of the total acid is in the form of ionic species in an undiluted TBP–nitric acid system. In the presence of non-polar diluents the postulated ions are associated into ion-pairs, indistinguishable from unionized species. An initial increase in the conductivity of the organic phase upon transfer of the acid is the supporting evidence for the existence of ionic species, postulated to be (TBP.H$_2$O.H)$_3$O$^+$ and (TBP.H)$_2$(H$_2$O.H)O$^+$. Distribution curves, such as those in Figures 9C10 and 9C11, show an initial steep increase of the acid content in the organic phase at low aqueous acid concentrations. This would indicate that the acid is not neccessarily extracted in its associated molecular state only, as some authors believe to be the case[351,382–384], supporting the existence of ions at very low organic loadings.

An association of these cations with the nitrate ion should result in complexes containing three or two ester molecules per nitric acid. While such higher solvates are very characteristic of systems with hydrohalic and other strong mineral acids, their existence in nitric acid systems in the form of well-defined complex species is yet to be proved[310,328,373]. This lack of direct evidence is not surprising. The equimolar complex between TBP and HNO_3 is stronger by an order of magnitude or more than the corresponding solvates of either hydrohalic or perchloric acids. The tendency to form the unionized $TBP.HNO_3$ dominates the equilibrium in the TBP–nitric acid system. This tendency is apparently markedly less pronounced in the HCl or $HClO_4$ systems, allowing, thus, higher solvates of these acids of stability comparable to that of the corresponding equimolar complex, to appear.

If direct evidence has been presented in favour of alkoxy-oxygen participation in acid bonding, it comes from infrared spectral measurements. None of the many interpretations of the spectra involved are so firm on the point as that of Shuler[351]. The high extractability of acid is not even circumstantial evidence, since phosphine oxides, in which there is no such alkoxy oxygen, readily extract a second nitric acid molecule per ester[312,328,403-405]. This simple argument is believed to outweigh by far any evidence brought up in favour of alkoxy-oxygen bonds.

However, the lack of additional coordination sites in the ester molecule does not necessarily imply that higher nitric acid complexes may not be formed in the system. A considerable number of investigators believe, indeed, that such complexes do exist under certain experimental conditions. The majority of the information from infrared and ultraviolet spectra and physical measurements failed to show unequivocally that such species are present in solution. There is still only a single n.m.r. signal from the acid proton indicating a rapid equilibrium. It seems to us that the concept of simple distribution of excess molecular nitric acid is more likely to be true.

(ii) *Hydrochloric acid* Systems involving hydrochloric acid as extraction-promoting agent have only limited technological application. Understandably, publications dealing with the distribution of hydrochloric acid are both less numerous and more limited in scope. Tributyl phosphate–HCl systems have received most attention[373,379,415,418,419,424-429], but some other neutral organophosphorus compounds[361,405,430,431] have also been investigated for their extractive capacity towards hydrochloric acid.

A typical distribution curve of HCl between water and undiluted

TBP is shown in Figure 9C15[427]. The curve shows that extractability of HCl is markedly lower than that of HNO_3. Undiluted TBP is not miscible with a concentrated aqueous hydrochloric acid solution, the acid content of the organic phase when in equilibrium with it amounts to an HCl:TBP ratio of only 1.3. Under comparable experimental conditions the extractability of the acid into tributoxyethyl phosphate[361], diisopentyl methylphosphonate[430], tributylphosphine oxide[431] and tetrahexyl alkylenediphosphonate[405] is similar.

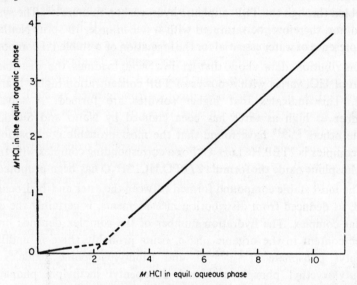

M HCl in equil. aqueous phase

Figure 9C15. HCl molarity in the equilibrium organic phase versus its molarity in the equilibrium aqueous phase, for extraction by undiluted TBP[427]. (By permission from Pergamon Press).

On the other hand, anhydrous gaseous HCl is soluble in tributyl phosphate[347] and tributylphosphine oxide[431] to give an acid: ester ratio higher than that attainable by equilibration. Whatever the compound formed by introducing dry HCl into a carbon tetrachloride solution of the phosphine oxide in excess to a unit ratio, it is apparently rather unstable. Nitrogen gas can push out the excess HCl. Addition of water up to a ratio of 1 TBPO: 1 HCl:1 H_2O causes the organic phase to split. The heavier, carbon tetrachloride layer, contains the solute in a stoichiometry corresponding to $2TBPO.HCl.3H_2O$, while the upper phase is predominantly the compound $TBPO.HCl.H_2O$. Adding one mole of TBPO the system again becomes homogeneous, and then it can accept water up to a ratio of 2 TBPO : 1 HCl : 3 H_2O. Similarly, at high acid loadings TBP–aliphatic hydrocarbon mixtures split into two under a variety of experimental conditions[425,428]. The second HCl molecule escapes by bubbling nitrogen gas through the layer containing the bulk of TBP. The remaining solution analyses as 1 TBP:1 HCl:1 H_2O with only small amounts of the diluent. It is apparent, thus, that the splitting is caused in this case by the incompatability of the diluent with a solution of hydrogen chloride in the hydrated TBP.HCl complex. However, two organic phases are also formed in the absence of water, when HCl gas is bubbled through the TBP–aliphatic hydrocarbon mixture[428]. The system need not, therefore, be saturated with water in order to split. Neither is the presence of water essential for the formation of a stable 1:1 compound.

Distribution data show that at low acid loadings the distribution ratio of HCl varies with a power of TBP concentration higher than the first. This indicates that higher solvates are formed. A solvation number as high as three has been claimed by Naito and Susuki[373], while others[424,427] have found that the most probable composition of the complex is 2TBP.HCl.6H_2O. For a corresponding complex with tributylphosphine oxide the formula 2TBPO.HCl.3H_2O has been proposed[431].

The most stable compound formed between the ester and hydrochloric acid, as deduced from distribution measurements, is certainly the equimolar complex. The hydration number of the complex depends on the ester content in the organic phase, being probably three in undiluted tributyl phosphate[422,424,425,427] and tributylphosphine oxide[431]. In tributyloxyethyl phosphate[361] and diisopentyl methylphosphonate[430] systems, both dissolving considerably more water than TBP, the apparent, anlaytically determined, hydration number of the 1:1 complex is higher than three. The formation constant of TBP.HCl has been calculated to be about 10^{-3} [418].

When the initial aqueous acid solution is above 10 M the equilibrium undiluted TBP phase contains an excess of HCl. It has been suggested[424,425,429] that this excess is bound in the form of a complex 2HCl.TBP. In view of the observations described above, the compound is unlikely to be stable, and it is perhaps safer to consider the excess as dissolved acid[415,427].

The extraction of water into a given quantity of undiluted TBP along with the acid is shown in Figure 9C16, plotted against the extracted acid, both in mmoles[427]. The first point on the plot is that corresponding to 0.12M initial aqueous acid concentration, and the last to 12.15 M. The data reported by others for TBP systems[53,415,422,424,425] are similar. The slope of the first part of the curve is equal to four, indicating that the proton is hydrated by four water molecules on passing from dilute HCl solutions into the ester. The sharp break in the plot appears at a TBP:HCl ratio of two in the extract (see section 7.D).

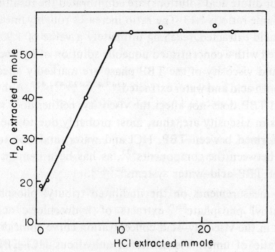

Figure 9C16. Coextraction of water and HCl into undiluted TBP. Mmoles of H_2O vs. mmoles of HCl, both in the equilibrium organic phases[427]. (By permission from Pergamon Press).

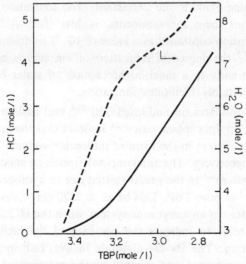

Figure 9C17. Concentration of TBP, HCl and H_2O in the organic phase, tor the system TBP-H_2O-HCl.

References pp. 716-736

A plot of acid and water content in an undiluted TBP phase is shown in Figure 9C17 as a function of the equilibrium ester concentration[415].

Obviously, such an extensive transfer of water into the organic phase causes large volume changes. When equal volumes of dry undiluted TBP and water (or dilute acid solutions) are equilibrated the resulting organic: aqueous volume ratio is 1.13. The ratio increases roughly linearly with the amount of acid extracted, reaching ultimately a value of 1.90 when TBP is equilibrated with a concentrated aqueous solution of hydrochloric acid.

Density and viscosity of the TBP phase are markedly affected by the amount of both acid and water extracted[347,425,427,432,433]. Anhydrous HCl in undiluted TBP does not affect the viscosity, neither does water alone. The changes in viscosity are, thus, most probably due to the size of the complexes formed beween TBP, HCl and water, and not to the mere interaction between the components[432], as has been demonstrated with a number of TBP–acid–water systems[347,433].

Viscosity measurements on the undiluted tributyl phosphate[427] and tributyloxyethyl phosphate[361] extracts of hydrochloric acid show a sharp break in the viscosity–acid concentration curve coinciding with an ester–acid ratio of unity. Viscosities of anhydrous HCl–TBP mixtures, however, are not very much greater than that of pure TBP[379]. Data on the equivalent conductance of the organic solutions show, in the case of TBP, a break at a TBP:HCl ratio of two, but a ratio of one when tributylphosphine oxide is the extractant. The ionization of the acid, as deduced from these measurements, is low[379,418,424,425,427], the estimated ionization constant has a value of 10^{-4}. In diisopentyl methylphosphonate[430] the degree of ionization of the acid is higher than in TBP, due apparently to a much higher uptake of water by the organic phase which certainly facilitates ionization.

Infrared spectral data on undiluted TBP[374], and carbon tetrachloride solutions of tributylphosphine oxide[431] indicate that the anhydrous HCl is present in the system in the form of molecular adducts TBP.HCl and TBPO.HCl, respectively. The fundamental stretching mode of the HCl molecule at 2880 cm^{-1} in the gas, is shifted due to hydrogen bonding, by about 600 cm^{-1} in pure TBP. This band, at 2270 cm^{-1}, disappears upon addition of water. In an anhydrous system, when the HCl:TBPO ratio is two, i.r. data seem to indicate the presence of the bichloride anion, HCl_2^-, in solution. The P—O—C bands in wet TBP in the 950–1100 cm^{-1} spectral region are not affected by the presence of hydrochloric acid[374].

Summarizing, one can safely conclude that the first molecules of hydrochloric acid pass into the ester phase along with the undisturbed primary hydration shell of the proton. Though the spectral evidence for or against the presence of the hydronium ion is inconclusive[374], it is reasonable to assume that at low acid content in undiluted ester the hydronium ion may be partially solvated with ester molecules to yield mixed hydrated-solvated H_3O^+. Such a complex containing two TBP molecules per proton seems to be of considerable stability. To account for the high water content of the organic phase under these conditions it is necessary to postulate the hydration of the chloride ion and a possible secondary hydration shell around the hydrated-solvated hydronium ion[415].

The equimolar complex in an anhydrous system is most probably a hydrogen-bonded molecular adduct, while for the corresponding complex formed in the organic phase in the process of extraction, an ion-pair structure $\underset{\nearrow}{\searrow} P \rightarrow OH^+(H_2O)_n Cl^-$ has been suggested by a number of authors and for various esters investigated[347,415,428,430,431].

In anhydrous systems, infrared spectra indicate that the excess acid is bound in the form of a bichloride ion in the case of tributylphosphine oxide[431]. Its formation under conditions of extraction cannot be excluded either, especially in view of the fact that there is good evidence that it can be formed when long-chain amine extractants are equilibrated with aqueous HCl of high concentration (Chapter 10). The decreased conductivity[427] and ionization[418] of the TBP phase at high acid loadings may be explained either by strong ion association of the ion-pair $\underset{\nearrow}{\searrow} P \rightarrow OH^+(H_2O)_n HCl_2^-$ as compared to that of $\underset{\nearrow}{\searrow} P \rightarrow OH^+(H_2O)_n Cl^-$ or by the molecular state of the dissolved excess hydrogen chloride.

(iii) *Perchloric acid* A considerable amount of work has been published on extraction of $HClO_4$ by tributyl phosphate[348,373,379,388,397,415,418, 434–439] and some other esters[281,405,440]. The characteristic features of this system are a) the high solvation number of the acid, b) its unusually high dissociation in the organic phase and c) its decreasing distribution ratio into concentrated ester solutions, with increasing aqueous acid concentrations. None of these phenomena have been encountered in systems with other simple mineral acids.

In undiluted TBP the distribution ratio increases at first, reaching a maximum of $D_{HClO_4} \sim 2.5$ at about 0.05 M aqueous acid concentration. The subsequent decrease of D_{HClO_4} is a trend opposite to that observed with either nitric or hydrochloric acids. As shown in Figure 9C18, taken

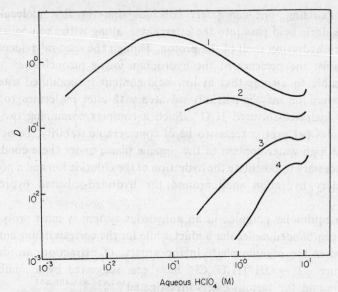

Figure 9C18. Partition coefficient for $HClO_4$ as function of the concentration in the equilibrium aqueous phase 1—100% TBP; 2—50% TBP; 3—20% TBP; 4—10% TBP, all in benzene.

from Siekierski and Gwozdz[436], this unusual shape of the distribution curve is still maintained in a 50% TBP–benzene mixture.

At low acid concentrations in pure or concentrated TBP (when $c_{TBP} \gg c_{HClO_4}$), distribution data suggest the presence of highly solvated acid molecules. A solvation number of four has been proposed[437,438], and a value of 0.16 has been calculated as the formation constant of $HClO_4 \cdot 4TBP$. Others[373,436] have found that the TBP-dependence slope is around three. At low TBP concentrations, a slope analysis of the distribution data indicates that the limiting solvation number is three[352]. As the acid content increases in the organic phase, the stoichiometric ratio of the components approaches unity. But prior to reaching this ratio, distribution data[388,418,437,438] show that the compound $HClO_4 \cdot 2TBP$ is a possible intermediate of considerable stability. The influence of the ester concentration on the solvation number, which is very significant at low acidities, becomes increasingly unimportant at higher acidities, where the tendency to saturate the organic phase by the formation of an equimolar complex governs the equilibrium[352,418,436]. The value for the formation constant of $HClO_4 \cdot TBP$ depends apparently on the experimental conditions, and varies with the ionic strength of the aqueous phase. At $I = 3$ it is 0.067, at $I = 2$ it is 0.085, but at lower I's the calculated values are not constant[434,435].

For the extraction from concentrated perchloric acid solutions into pure TBP, the acid:TBP ratio actually exceeds unity. This excess, however, is smaller than with other mineral acids.

The apparent hydration number of the extracted acid–TBP species varies greatly with the acid concentration[415,418,436,437]. Figure 9C19 represents the plot of acid and water molarity as a function of the equilibrium TBP concentration in an undiluted system[415]. At low TBP concentrations the shape of the plot is similar[352,436]. The first acid molecules are extracted into undiluted TBP along with four water molecules per acid molecule[352 436,437], while in dilute TBP a slope of three has been observed[352]. From more concentrated aqueous perchloric acid solutions, the average number of water molecules attached to the acid approaches two. Finally, as the TBP phase approaches saturation, the water content remains practically unchanged and independent of an increase in the acid concentration in the equilibrium organic phase[436,437].

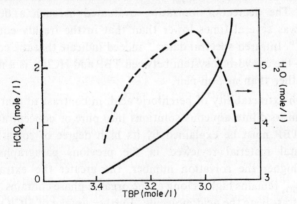

Figure 9C19. Concentration of TBP, HClO₄ and H₂O in the organic phase, for the system TBP–H₂O–HClO₄.

The system TBP–HClO₄ has also been investigated by various physicochemical methods. Freezing point measurements using benzene as the cryoscopic medium, showed that at low organic-phase loadings the acid is tetrasolvated. With an increase in the acid concentration, the total number of moles in the organic phase corresponds to the formation of the compound 2HClO₄.4TBP. Ultimately, a further reduction of the freezing point indicates the existence of a compound with a TBP:HClO₄ ratio of less than two[438]. Curves representing the swelling of the organic phase, its conductivity and viscosity as a function of acid (and water) extracted into undiluted TBP, show sharp intersections, which were interpreted as indicating the following molar composition of the extracted species: $HClO_4.4TBP$, $HClO_4.2TBP$ and $HClO_4.TBP$[437].

The estimated ionization, by Walden's rule, shows that perchloric acid, unlike other mineral acids, is dissociated to an appreciable extent [379,415,418,436,437]. A Kohlrausch plot of the conductivity data[418] suggests that the acid may be virtually completely dissociated below 10^{-2} M in undiluted TBP. In dilute TBP, ionization is markedly decreasing, becoming negligible at 10% TBP in benzene[436]. It is interesting to note in this connexion that the high degree of ionization of $HClO_4$ at low acid loadings in undiluted TBP, is primarily due to the high water content of the organic phase. At acid concentrations below ~ 1 M, the molarity of water is twice as high as that of undiluted TBP. Now, removing water from the phase, either by slow evaporation from an open vessel or keeping the solution over anhydrous $CaCl_2$, reduced its conductivity and viscosity. The degree of ionization evaluated in such a dehydrated solution was several times lower than that in the freshly equilibrated solution[436]. Infrared spectral data[374] indeed indicate that the compound formed in the anhydrous system between TBP and $HClO_4$ is a molecular adduct rather than an ion-pair.

The high extractability of perchloric acid, in contrast to that of other acids, from its dilute aqueous solutions into pure or concentrated solutions of TBP must be explained by its high degree of solvation. The experimental material reviewed in the previous paragraphs, shows that the higher the solvation number, the greater the extractability. Thus D_{HClO_4} remains high as long as the organic phase contains sufficient TBP to tetrasolvate the acid molecule. Further uptake of $HClO_4$ by TBP will obviously result in formation of progressively lower solvates, until, finally all the ester is bound in the form of the monosolvate[437]. High solvation numbers cannot be reached at low concentrations of the extracting agent, tributyl phosphate and trioctylphosphine oxide, as shown by the distribution data of Diamond and coworkers[352,440].

Free ions must exist to explain the appreciable conductivity of a dilute solution in pure TBP. Such free ions can be represented as $[H(TBP)_n(H_2O)_x]^+$. where n may have any value between 4 and 1, and x a value of 8 or less[437]. Several authors have suggested[352,415,436] that in pure or concentrated TBP solutions the hydronium ion is possibly surrounded by a complete primary hydration shell, which acts as a bridge between the hydronium ion and the TBP molecules in the secondary solvation shell

```
 ⎡  R3PO · · · · H            H           ⎤+
 ⎢                \O      O/              ⎥
 ⎢           H  /    ·. .·   \H · · · · OPR3⎥
 ⎢              H   H                     ⎥
 ⎢              O                         ⎥
 ⎢              H                         ⎥
 ⎢              ⋮                         ⎥
 ⎢              O                         ⎥
 ⎢            /   \                       ⎥
 ⎣  R3PO · · · H      H                   ⎦
```

Such secondary solvation shells may accomodate more, unless sterically hindered, or less TBP molecules, thus the model may be common to all dilute solutions of mineral acids, provided the hydronium ion retains its hydration characteristic in its aqueous solution.

The lack of detectable ionization in dilute ester solutions is due to an increased ion-pair association, as an effect of the non-ionizing medium of the diluent. As the TBP concentration decreases the uptake of water by the organic phase decreases also, and the primary hydration shell of the hydronium ion must be disturbed and water replaced by TBP molecules[352,415,436]. From purely steric considerations it is questionable, however, whether all three water molecules surrounding the hydronium ion can be replaced by the ester molecules, as suggested by Whitney and Diamond[352]. It is more likely that under such experimental conditions lower solvates become more stable. Once again, it is an experimental fact that in dilute ester solutions high solvation numbers are not attained.

(iv) *Other acids* The experimental information on the other acids, mineral and organic, comes mainly from comparative studies, and only exceptionally have they been investigated on their own merit.

The extraction of nitrous acid resembles that of nitric acid in many respects[268,441]. It is extracted with a higher distribution ratio than HNO_3 into TBP, due probably to its weaker dissociation in aqueous solutions. At an equivalent concentration of unionized molecules, however, D_{HNO_2} is higher by a factor of two. The organic phase has the ultraviolet light-absorption characteristics of molecular nitrous acid. Distribution and infrared spectral data indicate a stable 1:1 complex, where the hydrogen bonding to the phosphoryl oxygen modifies the bonding in nitrous acid towards an ionic structure. Only a fraction of the complex is monohydrated, like that of nitric acid, and it is unionized in benzene or isooctane.

Hydrofluoric acid is well extracted by undiluted TBP, better than HCl or even $HClO_4$[418]. This is most probably a result of its low dissociation in the aqueous phase. Conductivity measurements indicate an equally low ionization in the organic phase.

As would be expected, the extractive behaviour of hydrobromic acid is almost identical to that of HCl[376,421,422,442]. Its extractability into TBP is higher than that of HCl from dilute aqueous solutions, but lower in the high acid region. At high aqueous HBr concentrations decomposition has been noted[376]. With an exact analogy to perchloric acid, Whitney and Diamond[376] interpreted their distribution data in terms of a trisolvate formation. This is a surprisingly high solvation number for halo acids (compare HCl), especially when dilute TBP solutions are used. Their experimental data do not exclude, however, the existence of lower solvates, the most probable being the equimolar complex. On the other hand, based mainly on physicochemical measurements, conductivity and viscosity, of the organic phase in the undiluted TBP system, Kertes and Kertes[442] suggested the presence of $2TBP.HBr.6H_2O$ and $TBP.HBr.3H_2O$, in complete analogy with the HCl system.

Hydriodic acid is extracted more effectively than either HBr or HCl into undiluted TBP, at least from dilute or moderately concentrated aqueous solutions[421].

The extraction of sulphuric acid into TBP follows the general patterns observed with other strong mineral acids[276,347,353,373,415,418,439,443,444]. Its extractability into diisopentyl methylphosphonate[445], butyl dibutylphosphinate[276], tributyl-[276] and trioctylphosphine oxide[405], and tetrahexyl methylenediphosphonate[405] is similar to the more extensively studied TBP systems. The equilibrium acid and water concentrations in undiluted TBP are shown in Figure 9C20[415].

Distribution data indicate that at low organic acidities a sulphuric acid disolvate is formed with TBP[347,379,443-445]. For the corresponding species involving the diisopentyl methylphosphonate the compositions suggested were the mono- and dihydrated $H_2SO_4.2DiPMP$. Increasing acid transfer leads to saturation of the organic phase in the sense of the formation of the equimolar complex[363,415,419]. Brauer and Högfeldt[353] proposed the formula $TBP.H_2SO_4.8H_2O$ for the hydrated compound, and have calculated the value of $\sim10^{-2}$ as its formation constant. As

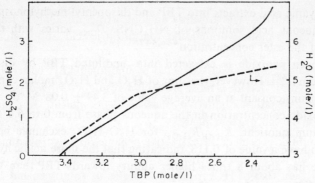

Figure 9C20. Concentration of TBP, H_2SO_4 and H_2O in the organic phase, for the system $TBP-H_2O-H_2SO_4$.

deduced from the infrared spectra[362,347], in an anhydrous system the equimolar complex is a molecular adduct which forms an ion-pair upon addition of water. The estimated ionization of the acid is, nevertheless, low[379], even in pure TBP, similar to that of HCl and HNO_3[418].

Anhydrous orthophosphoric acid is completely miscible with un-diluted TBP, or removes all the TBP from a TBP–diluent mixture[314]. The extractability of the acid is high[276,373], comparable to that of nitric acid. Undiluted TBP in contact with 9 M aqueous H_3PO_4 extracts acid to reach a 1:1 ratio; at 13.6 M, the acid: ester ratio is 2, and continues to increase rapidly to a ratio of ~4 when in equilibrium with an initial aqueous solution of 16 M H_3PO_4. At very low concentrations of TBP in cyclohexane, $D_{H_3PO_4}$ varies with a power of the ester concentration, higher than 3.3 but becomes gradually first power in the range between 3% and 100%[314]. Phosphoric acid carries water into undiluted TBP to an extent higher than other mineral acids. The maximum water content of roughly 2 H_2O:1 TBP is reached at a $TBP:H_3PO_4$ molar ratio of 3, corresponding to a stoichiometric composition of 3TBP. $H_3PO_4.6H_2O$. Further acid transfer releases water from the organic phase, but not below a H_2O:TBP ratio of unity, the ratio remaining constant regardless of the acid content[314].

In the presence of H_3PO_4 the P $\rightarrow$ O vibration in TBP is degenerated to a wide band, from ca. 1100 cm^{-1} to 1220 cm^{-1}. The shift of about 110 cm^{-1} should be characteristic of a rather strong hydrogen bond[362].

From ebullioscopic molecular weight determinations, Higgins and Baldwin[314] concluded that the acid is a trimer in TBP solutions, but no specific compound is formed between the anhydrous acid and the undiluted ester.

Thiocyanic acid extracts into TBP and diisopentyl methylphosphonate from aqueous acid solutions of NH_4CNS. D_{CNS} varies with the first power of the ester concentration[446,447].

Hydrogen peroxide is extracted into undiluted TBP by displacing water from it[423]. However, the sum of H_2O and H_2O_2 molar concentrations remains consant at an average value of 3.49 ± 0.08 M by varying the peroxide concentration in the aqueous phase from 0 to 6.5 M. The equilibrium quotient, $K_{H_2O}/K_{H_2O_2}$, for $H_2O-H_2O_2$ exchange has been found to have a value of 0.118, indicating that the more acidic hydrogen peroxide has about 8 times higher affinity towards TBP than the less acidic water.

Neutral organophosphorus compounds are excellent extractants for practically all metal oxy acids such as perrhenic, pertechnetic, chromic, vanadic, and complex metal acids of the tetrachloroferric acid type. The extraction of these acids to any appreciable extent requires, as a rule, the presence of another strong simple mineral acid. Their extraction chemistry will be discussed in the next section.

Finally, some data are available on the extractive power of the esters toward the weak organic acids, which are generally well extractable[36,276,373,448-452] into TBP. It is an especially effective extractant for the aliphatic acids, their extractability increasing with the length of the carbon chain. Monobasic acids are more extractable than the dibasic ones with an equal number of carbon atoms. Hydroxyl groups strongly depress extraction, but chloro or phenyl groups increase it. A double bond in the molecule has a depressant effect.

Both mono- and dicarboxylic acids are extracted by TBP with a higher distribution ratio than any of the mineral acids investigated under comparable experimental conditions[36,373]. Their distribution ratio decreases with increasing aqueous acid concentration, and increases with temperature. The composition of the extracted species for several organic acids has been deduced by slope analysis. Acetic acid is extracted as a monosolvate of TBP, which is only poorly hydrated, if at all[36]; tartaric malic, and probably also lactic acid form with TBP disolvates, which seem to be monohydrates[451]. Their respective formation constants were 0.039, 0.109 and 0.26. The citric acid complex has the composition $3TBP.Citr.2H_2O$ and a formation constant 0.038[452]. Trihydroxyglutaric acid forms an unhydrated disolvate, and has a formation constant of 0.016[450]. Infrared spectra[453] indicate strong hydrogen bonds between phosphine oxides and acetic acid derivatives.

(v) *The effect of diluent and temperature* The effect of diluent on the extractive properties of neutral organophosphorus compounds towards acids is less marked than one would expect. This is primarily true at low acid loadings. A plausible explanation for these observations is probably the fact that the acid bonding of the phosphoryl oxygen is of a considerable strength. The interaction between the ester and the diluent, discussed earlier, is apparently too weak to interfere successfully with the high affinity of the esters towards acids.

The most drastic effect diluents have on extraction systems, including those under consideration, is certainly the phenomenon of splitting of the organic phase, resulting in third-phase formation. The phenomenon will be discussed in the forthcoming section, when dealing with the extraction of metal salts. In such systems the phenomenon is more frequent and more experimental material is available. Here, a brief mention will be made of acid-extraction systems where splitting occurs when aliphatic solvents are used as diluents.

The second organic phase usually appears first as a cloudiness in the organic phase. As the experimental variables enhancing the appearance of the phase intensify, the initial cloudiness separates out in the form of an ejected phase, usually between the lower aqueous and the upper organic phase containing the bulk of the low-density diluent. Its appearance depends practically on all the variables of an acid-extraction system. There is first, the type of the acid and its concentration in the aqueous phase, then the concentration of the ester and the nature of the diluent, and, finally, the temperature[314,317,413,418,425,428,435,454].

The critical acid concentration at which a given organic phase splits is generally lower in HCl, $HClO_4$ and H_2SO_4 systems than in HNO_3. For example, while a 20% TBP–kerosene mixture splits when in contact with 6 M aqueous hydrochloric acid at room temperature, an aqueous concentration of as high as 16 M HNO_3 is needed to observe the effect, under otherwise identical conditions[317]. Also a 0.5 M aqueous perchloric acid solution will cause similar TBP–kerosene mixtures to split[435].

A comparison of five normal aliphatic hydrocarbons, ranging from hexane to tridecane, as diluents for a 20% TBP solution, as to their behaviour when equilibrated with aqueous HCl solutions, shows that the greater the molecular weight of the hydrocarbon, the lower the aqueous HCl concentration which causes a third phase to appear. Or, at a given HCl level, the concentration of TBP at which splitting occurs, decreases as the molecular weight of the hydrocarbon increases.

Finally for a given diluent, the critical TBP concentration falls as the acid concentration is raised, and reaches a minimum value at 11–12 M HCl[454].

Decreasing the temperature causes the second organic phase to appear at lower critical concentrations of the solutes, both acid and ester. Additionally, the lower and upper critical solution temperatures depend on the diluent. In a TBP–HCl system, for example[428], the upper critical solution temperatures (above which no third phase is formed at a defined composition of the system) were −6°C for hexane, 0°C for petroleum ether, +5° for isooctane, 10° for kerosene and 44° for heavy kerosene. The critical temperature increases apparently with increasing chain length of the aliphatic hydrocarbon.

The fragmentary information available on the composition of the ejected phases does not allow general conclusions to be drawn[266]. In hydrochloric[428,431] and perchloric acid[435] systems it is apparently an acid monosolvate which constitutes the bulk of the third phase, while a ratio of 1 TBP: 2 H_3PO_4 was found in the third phase, formed over a wide range of initial acid and TBP concentration[314]. TBP.4HNO$_3$ has been suggested to be the composition of the third phase in a nitric acid system[413].

Very little systematic information is available on the effect of temperature in the systems under consideration. From an aqueous solution of up to 8 M nitric acid, its extraction by tributyl phosphate[327,390,455] and other neutral esters[274,327,328] is not very temperature dependent. From more concentrated solutions a negative temperature coefficient of extraction is characteristic of the TBP–HNO$_3$ systems[97,389]. Phosphoric acid behaves differently. At lower aqueous acid concentrations $D_{H_3PO_4}$ decreases by a factor of two when the temperature is increased from 2° to 55°C. The distribution ratio is unaffected when the acid is extracted from concentrated aqueous solution, in the same temperature range[314]. The distribution ratio of some organic acids into undiluted TBP decreases with increasing temperature[449].

D.   METAL EXTRACTION WITH PHOSPHORUS-BONDED OXYGEN-DONOR EXTRACTANTS

a.   Introduction

The high extractive power of neutral organophosphorus esters has been demonstrated for a large number of metal salts, the nitrates and chlorides having received most attention. Since the extractability by these reagents involves the solvation of electrically neutral metal salts, formed by their depressed ionization in the aqueous solution, the reagents function satisfactorily only in the presence of a high concentration of salting-out electrolytes in the aqueous phase. The extent of extraction will, thus, depend on the degree of formation of the extractable species and the solvation number of the metal salt.

The amount of water coextracted with the metal salt affects its extractability. While acids usually retain the water of hydration of the proton, as demonstrated in the previous section, in the majority of cases where metal salts are extracted, there is a strong competition between the water and extractant molecules for the coordination sites of the metal ion. The extractive capacity will, thus, depend largely on its capacity to replace water in forming the unhydrated adduct with the metal salts. This capacity, of course, depends on the type and structure of the neutral organophosphorus extractant. Nearly all metal extractions are increased as the number of direct C–P linkages increases in the series phosphate, phosphonate, phosphinate, phosphine oxide. Under a separate heading this section attention will be focused on these structural aspects.

Among the many variables affecting extractability of metal salts, that of the diluent certainly deserves also a careful consideration. The nature of the diluent affects the activity of the solute in the organic phase, and the extent of its interaction with the extractant ester. Both effects are common, though not necessarily identical, to all neutral organophosphorus compounds.

On the other hand, it is well beyond the scope of this monograph to deal with the immense fields of radiochemical separations, analytical chemistry, the general field of ore processing by hydrometallurgical methods, atomic fuel reprocessing and reactor technology, uranium and other metals production and refining technology, fused-salt chemistry and others, where intensive exploitation of neutral organophosphorus extractants suggests many potential uses.

b. Solubility of Metal Salts in Esters, Metal Salt–Ester Complexes and their Behaviour in Solution

Several studies have been carried out on the solubilities of metal salts in pure[311,360,456–462] and dilute[460,462] tributyl phosphate, diisopentyl methyl-phosphonate[463] and tributyl thiophosphate[459]. The solubilities of some salts in undiluted TBP at 25°C (refs. [458,459] in the temperature range 25°–27°C) are compiled in Table 9D1. The effect of diluent on the solubilities of $Cr(NO_3)_3$ and $Cu(NO_3)_2.3H_2O$ in TBP–kerosene and TBP–benzene mixtures, respectively, is plotted in Figure 9D1. In diiso-pentyl methylphosphonate the solubilities of caesium and barium nitrates are greater by about an order of magnitude than in TBP, whereas that of lithium nitrate is lower by one order of magnitude. The solubility of uranyl nitrate is about the same in these two esters. In tributyl thio-phosphate the solubilities of copper, thorium and uranyl nitrate are markedly lower than in TBP.

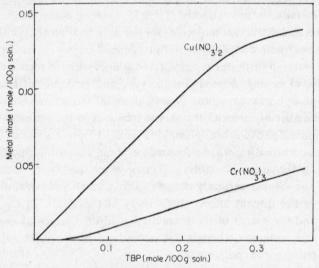

Figure 9D1.   Solubilities of $Cr(NO_3)_3$ in TBP–kerosene and $Cu(NO_3)_2.3H_2O$ in TBP–benzene mixtures, at 25°C.

Few data are available on the effect of temperature on the solubilities of salts. In the range from 0° to 50°C there is practically no change in the solubility of $UO_2(NO_3)_2.6H_2O$ in either tributyl phosphate[311] or diisopentyl methylphosphonate[463]. Neither is the solubility of thorium nitrate in TBP affected in the wide range from 25° to 130°C[311]. An increasing solubility has been noted, however, for lithium, calcium and cobalt nit-rates in TBP in the 25°–50°C temperature range[311], and for $UO_2(NO_3)_2$ above 50°C[461].

The solubility of metal salts in any one group decreases with increasing atomic weight of the metal. Alkali perchlorates are more soluble than the corresponding nitrates, and the trend is increasing solubility on going from chloride to iodide. The solubility of alkali sulphates is very low. Apparently, the solubilities of hydrated ionic metal salts are appreciably higher than those of anhydrous salts. In the case of hydrated salts, the system in equilibrium has three phases, as water separates out. The amount of water ejected from the organic phase, or not entering the organic phase, is particular to the salt. Water-content measurements on the organic phase indicate that, for example, lithium, calcium and copper nitrates are hydrated in undiluted TBP to an extent of about two moles of water per mole of salt, whereas thorium and uranyl nitrates are essentially unhydrated.

In the system $UO_2(NO_3)_2.6H_2O$–diisopentyl methylphosphonate, the amount of water appearing as the third phase is exactly that corresponding to the six water molecules of hydration.

The heat of solution of the salts at 25°C in undiluted TBP depends on the extent of their hydration[135–137], as shown in Table 9D2.

TABLE 9D1. Solubility of metal salts in undiluted tri-n-butyl phosphate at 25°C.

Salt	Concentration of salt (mole/l)	(g/100 g of soln.[a])	Concentration of water (mole/l)	(g/100 g of soln.)	Reference
$LiNO_3 . H_2O$	1.32	8.7	3.14	5.65	311, 456
$LiNO_3 . 3H_2O$		11.9			458
$LiCl . H_2O$	1.90	8.0			311, 360
$LiBr . 2H_2O$	2.16	17.5			311, 360
$LiI . 3H_2O$	1.99	22.1			311, 360
$LiClO_4 . 3H_2O$	2.52	23.3			311, 360
Li_2SO_4		< 0.04			360
$LiOH$		< 0.03			360
$NaNO_3$	0.054	0.46			311, 360, 456
$NaCl$	0.0042	0.025			311, 360
$NaBr . 2H_2O$	0.42				311, 360
$NaI . 2H_2O$	1.30	19.2			311, 360
$NaClO_4 . H_2O$	1.04				311, 360
$Na_2SO_4 . ?H_2O$		< 0.003			360
KNO_3	0.0025	0.026			311, 360, 456
KCl	0.0015	0.011			311, 360
KBr	0.0043	0.051			311, 360
KI	0.0041	0.069			311, 360
$KClO_4$	0.0048				311
K_2SO_4		< 0.003			360
$CsNO_3$	0.0019	0.038			311
$AgNO_3$		2.59			458

TABLE 9D1 (Cont.)

Salt	Concentration of salt		Concentration of water		Reference
	(mole/1)	(g/100 g of soln.[a])	(mole/1)	(g/100 g of soln.)	
$Mg(NO_3)_2 . 6H_2O$		11.5			458
$Ca(NO_3)_2.4H_2O$	0.99	15.0	1.84	3.03	311, 457
$CaCl_2 . 6H_2O$	1.60	17.3			311
$CaBr_2 . 6H_2O$	1.40	13.9			311
$Sr(NO_3)_2$	0.16	0.34			311, 456
		0.81			458
$Sr(NO_3)_2.4H_2O$	0.194	4.07	3.19	5.7	311, 456, 457
$SrCl_2 . 6H_2O$	< 0.001	< 0.016			311
$SrBr_2 . 6H_2O$	0.44				311
$Ba(NO_3)_2$	0.00075	0.020			311, 356, 457
$BaCl_2 . 2H_2O$	< 0.001	≪0.016			311
$BaBr_2 . 2H_2O$	0.058				311
$Co(NO_3)_2.6H_2O$	0.83	13.9	3.20	5.25	311, 456
		13.5			458
$CoCl_2.6H_2O$	0.32	4.15			311
$Ni(NO_3)_2.6H_2O$		10.5			458
$Cu(NO_3)_2.3H_2O$		2.08			460
$Cu(NO_3)_2.6H_2O$		21.4			458, 459
$Cu(NO_3)_2. ?H_2O$	1.14	19.8	2.28	3.75	311, 456
$CuCl_2.2H_2O$	0.63	8.10			311
$Zn(NO_3)_2.6H_2O$		20.6			458
$Cd(NO_3)_2.4H_2O$		21.5			458
$Hg(NO_3)_2$		18.5			458
$Pb(NO_3)_2$	0.078	2.55			311, 457
		0.39			458
$Al(NO_3)_3.9H_2O$	0.54	10.9	4.80	8.16	311
		9.41			458
$AlCl_3.6H_2O$	0.24	3.23			311
$Cr(NO_3)_3. ?H_2O$		11.4			462
$Fe(NO_3)_3.9H_2O$		34.4			458
$Bi(NO_3)_3.5H_2O$		28.6			458
$La(NO_3)_3.6H_2O$		28.4			458
$Ce(NO_3)_3.6H_2O$		28.0		0.19	311
$Ce(NH_4)_2(NO_3)_6$		45.8		0.0	311
$Th(NO_3)_4.4H_2O$		42.5			458, 459
$Th(NO_3)_4. ?5H_2O$		20.1			311
$UO_2(NO_3)_2.6H_2O$	1.60	42.7	0.33	0.40	311, 456
		43.5			458, 459
UO_2Cl_2	1.50	39.6			311

a of anhydrous salt

In some cases the solubility of metal salts in esters indicates the formation of molecular adducts of defined stoichiometry between the salt and the ester. Saturated solutions of lithium salts in TBP all have the approximate composition LiA.2TBP. The complex of uranyl and thorium nitrate has an ester: salt ratio of two, in tributyl phosphate, diisopentyl methylphosphonate and trioctylphosphine oxide, but not in tributyl thiophosphate. This limiting value of two remains unchanged with temperature increase or dilution of the ester with carbon tetrachloride[463]. Diphospates accordingly, form monosolvates with these salts[406, 465].

Infrared vibrational spectra of a large number of metal salts in ester solutions have indeed shown that complexes are formed[362,466-473]. A decrease in the position of the P → O stretching frequencies takes place due to a bond being formed between the metal atom and the phosphoryl oxygen. Saturated solutions of thorium, cerium(IV) and uranyl nitrates in undiluted TBP show the band at ~ 1180 cm^{-1} as compared to the undisturbed position at 1280 cm^{-1}. The large shift indicates that the interaction with these metals is stronger than with the hydrogen ion[362,469]. For the tervalent lanthanide nitrates the shift is somewhat smaller[469], but when the ligand is triphenylphosphine oxide rather than TBP, the shifts are again large[473]. A comparison of these spectra with those in aqueous solutions suggests that the nitrate groups in the TBP solvates are not purely ionic but are held coordinatively by the cation[468,469]. The P—O—C bond vibration at ca. 1030 cm^{-1} shows a slight change in TBP solutions of thorium and uranyl nitrate. Thus, the possibility of participation of the butoxy oxygens has been mentioned[362]. The spectral behaviour of some metal chloride solutions in TBP is similar[467].

TABLE 9D2. Heats of solution in TBP at 25°C (kcal/mole).

Salt	H_2O	$2H_2O$	$3H_2O$	$4H_2O$	$6H_2O$
$UO_2(NO_3)_2$		−10.46			0.8
$Co(NO_3)_2$		−4.74	0.99	2.15	5.18
$CoCl_2$	−4.16	−2.09			9.39

When solid $Cs_2U(NO_3)_6$ or $Cs_2Pu(NO_3)_6$ are dissolved in TBP containing lithium nitrate, or when acid-free uranyl nitrate solution is stirred with a lithium nitrate-saturated TBP, the spectra reveal the presence of anionic $M(NO_3)_6^{2-}$ and $UO_2(NO_3)_3^-$ complexes[474]. However, when $Cs_2U(NO_3)_6$ is contacted with undiluted TBP in the absence of lithium nitrate, only $U(NO_3)_4$ is dissolved, leaving $CsNO_3$ behind[347]. Rare earth nitrates behave similarly. Only $M(NO_3)_3$ is dissolved out of $Rb_2M(NO_3)_5$, though in the presence of $LiNO_3$ new species are formed, as found spectrophotometrically[475]. The spectrum of uranium(IV) undergoes changes as nitric acid is dissolved in the organic phase. This suggests that several species may exist in a saturated TBP solution. $HUO_2(NO_3)_3.2TBP$ has been proposed as one possibility in an acid-deficient TBP.

A wide variety of metal salt–ester complexes have been prepared and isolated in the form of pure crystalline solids[302,368,403,456, 463,466,470–473,476,484]. They are usually prepared by mixing alcoholic solutions of the components, and recrystallized from alcohols or some solvent mixtures. The melting points and the dipole moments of some of the greenish-yellow uranyl nitrate adducts with a general formula $UO_2(NO_3)_2.2S$ are listed in Table 9D3. Water is not present in these

TABLE 9D3.   Melting points and dipole moments of some $UO_2(NO_3)_2.2S$ complexes.

Ester (S)	Melting point (°C)	Dipole moment (debyes)	Reference
Tri-n-butyl phosphate	-9.75 ± 0.1		483
	-6		456
		3.1 ± 0.02	481
		3.2	310
Triisobutyl phosphate	72		476
Diethyl phenylphosphonate	82		368
Di-n-propyl phenylphosphonate	51		368
Diisopropyl phenylphosphonate	117		368
Di-n-butyl phenylphosphonate	42–43		368, 480
Diisobutyl phenylphosphonate	93		368
Di-sec-butyl phenylphosphonate	102		368
Di-tert-butyl phenylphosphonate	> 260		368
Diisopentyl methylphosphonate		3.5 ± 0.02	463
Butyl dibutylphosphinate	56–57		480
Tributylphosphine oxide	51–53		480
Trioctylphosphine oxide		7.8	403
Triphenylphosphine oxide	289–293		480
Bis(di-2-ethylbutylphosphinyl)methane (monosolvate)	275		

compounds. Di-*tert*-butyl phenylphosphonate, triphenylphosphine oxide and several diphosphine oxide–uranyl nitrate complexes[484] have unusually high thermal stability, suggesting a very high lattice energy. In the triethyl phosphate adduct, x-ray examination showed that the mean $P \rightarrow O$ distance is 1.52 Å[482]. The similarity of the dipole moments of $UO_2(NO_3)_2$. 2TBP and pure TBP is rather surprising. The structure of the uranyl nitrate adduct is apparently such as to compensate partially for the dipoles of the two ester molecules. The solubility of this adduct in normal hydrocarbons decreases with increasing chain length and increases with temperature[476].

No crystalline uranyl nitrate compounds could be obtained with alkyl phosphates having six or more carbon atoms per chain[368], except with the diphosphates[484].

Solid adducts of thorium, zirconium and cerium(IV) nitrate with triisobutyl phosphate[484], triphenylphosphine oxide[472] and diphosphonates[484] have been prepared, also. $Th(NO_3)_4$.2TiBP is white and melts at 102°C, $Ce(NO_3)_4$.2TiBP is an orange crystalline compound which melts at 85°C without decomposition[476]. The nitrates of cobalt, nickel, copper and zinc and the lanthanides[473] form crystalline adducts with two triphenylphosphine oxide molecules. Infrared spectra of their solutions in nitrobenzene indicate that in the monomeric compound the nitrate ions enter into the coordination shell so that the structure of the complexes should be assigned as $(Me^{2+}(TPPO)_2(NO_3)_2)$[466].

Trimethyl- and triphenylphosphine oxide gave crystalline disolvates with $CoCl_2$, $CoBr_2$, CoI_2, CdI_2 and ZnI_2. The perchlorates of these metals, and also those of copper, nickel, manganese and iron(III) are, however, tetrasolvates, the metals satisfying their coordination requirements with the extractants[302,466].

c. Metal Nitrates

Though the bulk of the information on the extraction of metal nitrates refers to the actinides, lanthanides and fission-product elements, nitrates of other metals have also been included in various survey studies. The extractability of aluminium, iron, cobalt and nickel, for example, from aqueous nitric acid solution in the range from 1 to 15 M, into tri-n-butyl phosphate[410,485,486], trioctylphosphine oxide[312,487-489], tetrabutyl methylene- and ethylenediphosphonate[486,489] has been demonstrated. However, their extractability by these and similar organophosphorus compounds is lower and occurs only with high salting. In Appendix F, charts are reproduced from Japanese publications[486] to show the extractability of about sixty nitrates into undiluted TBP and a 5% TOPO–toluene solution.

In the extraction of tetra- and hexavalent actinide nitrates it is the disolvate which is the most stable organic-phase complex, whereas actinide(III) and lanthanide nitrates are extracted as trisolvates under a variety of experimental conditions and involving all classes of neutral esters. An understandable exception is the solvation by diphosphonates and pyrophosphates[327,405,464,490,491]. In the case of uranium(VI) for example, D varies with the second power of the extractant concentration when it is in a large excess, though when the excess is not large (loading experiments) the diphosphates act apparently as bidentate ligands, forming the monosolvate.

An ideal second-(or third-) power dependence of D on $\bar{c}_S$, according to equation (7C49), is however limited to relatively low ester and aqueous acid concentrations. Though the limit of the ester concentration, $\bar{c}_S$, depends both on the nature of the diluent and the type of the ester, it is usually below 10% by volume. At higher $\bar{c}_S$ the deviation from a quasi-ideal behaviour becomes too large. As to the aqueous-phase acid concentration, at low hydrogen ion concentration the plots of equation (7C49) are frequently linear throughout a reasonably wide range of $\bar{c}_S$. At high acid concentrations, but not necessarily at high salting nitrate concentration, c_A, the deviation from linearity starts at markedly lower $\bar{c}_S$. This can be understood in terms of competitive coextraction of the acid, according to equilibrium (7C51) in addition to the non-specific non-ideal behaviour of the solutes. Acid, water and the metal compete for the same coordination position of the phosphoryl oxygen. Though no quantitative correlation of this competition is at hand, infrared studies of a large number of nearly saturated solutions of metal nitrates[349,350,469] suggest that the $P \rightarrow O$ shift is larger when metals are bound to the oxygen, and is larger the higher the ionic potential (charge/ionic radius) of the cation.

In several metal nitrate systems a higher solvate is favoured at tracer level of the metal, or more precisely, when $c_M \ll \bar{c}_S$, whereas at high metal loadings the more stable disolvate is the predominating species. For example, $Th(NO_3)_4$ forms a trisolvate and a disolvate, regardless of the class of the extractant[327,399,491–497]. Siddall[399] has suggested that an equilibrium

$$\overline{2Th(NO_3)_4.3S} + Th(NO_3)_4 \rightleftharpoons \overline{3Th(NO_3)_4.2S} \qquad (9D1)$$

may be responsible for the above experimental findings. Even the trisolvate need not be the limiting solvation number of $Th(NO_3)_4$. Under certain experimental conditions a tetrasolvate has been claimed to be the primary form in which the salt is extracted[266,498].

The state of the metal nitrate in the aqueous solution may govern primarily its extractability. The rather complicated chemistry of some actinides(IV), zirconium, hafnium, protactinium and ruthenium in aqueous nitric acid is affected by hydrolysis, polymerization and the formation of polynuclear hydrated species of the metal. It is quite possible that all the various metal-bearing species formed in the acid-dependent interconversion reactions are actually extractable, but to a varying extent, with a different overall rate (including the aqueous phase process) and possibly even with a different solvation number[196,387,426, 499–501].

Though in the majority of cases it is the neutral metal nitrate which is the axtractable species, in some cases at low HNO_3 but high $LiNO_3$ concentrations as the supporting electrolyte, the organic phase contains species such as the $Li_2U(NO_3)_6$ complex[499]. The extractability of this complex is however slight, and apparently in HNO_3 systems similar $H_2M(NO_3)_6$ species with uranium(IV) or plutonium(IV) do not exist in the extract[381,502,503]. Glueckauf[44], on the other hand, interpreted the extraction curve of cerium(III) nitrate from nitric acid solutions in terms of the protonated anionic complex, $HCe(NO_3)_4$.

Spectral measurements on the extract and its water content suggest that actinide and lanthanide nitrates are transferred into the ester phase essentially without water[278,313,332,355,463,503–505]. The water content of the organic phase varies over the range of loading, inversely with the salt content, which at saturation replaces all the water from the organic phase, as illustrated for the uranyl nitrate–TBP–kerosene system[355] in Figure 9D2. On the other hand, if the salt carries water along through retaining the hydration of the metal, as in the case of the slightly extractable lithium, magnesium or nickel nitrates, the water concentration in the organic phase is greater than when that phase is saturated with water alone[350]. Indeed, low extractability of some transition metal nitrates, even into undiluted extractants, is apparently due to the high hydration energy of the metal[460,462,501]. It is due to the water content, at least partially, that the degree of ionization[43,44] of $LiNO_3$ in TBP is higher by an order of magnitude than that of $UO_2(NO_3)_2$ at equal solute concentration, as shown in Figure 9D3. Uranyl nitrate disolvate is essentially unionized[463,483,493].

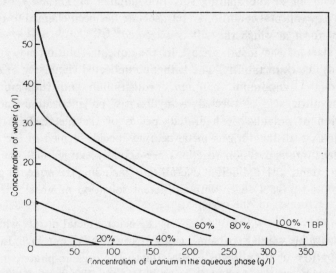

Figure 9D2.   Displacement of water from undiluted TBP and TBP–kerosene mixtures.

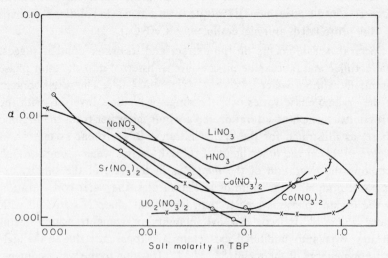

Figure 9D3.   Estimated degree of ionization of nitrates in TBP, $a = \lambda\eta/60$, from conductivity measurements[43]. (By permission from the UKAEA).

d. Metal Chlorides

Several neutral organophosphorus extractants have been screened for their extractive capacity towards metal chlorides. Such survey studies include tributyl phosphate[425,483,506], tributylphosphine oxide[486,507] and trioctylphosphine oxide[312,486-488]. Charts presenting distribution data of many elements are given in Appendix F for undiluted TBP and 5% TOPO in toluene[486].

A comparison of the corresponding charts in Appendix F for the D values in TBP and TOPO for extraction from HCl and HNO_3 solutions, reveals the effect of the anion on the extractability of the metal. Without discussing these differences in detail, one feature is immediately apparent: the high extractability of protonated anionic chloro complexes, similarly to the previously discussed ether and ketone extractants (section 9.B). With a possible exception of uranium(IV)[499], there was no specific evidence to show that anionic nitrate complexes may be transferred into, or formed in, the organic phase.

Distribution and loading experiments show that tetra- and hexavalent actinides are extracted in the form of their neutral anhydrous disolvate, irrespective of the diluent[508] and the type of the ester[509]. A notable exception[510] is the loading of tri-n-butoxyethyl phosphate, which has an ethereal oxygen in addition to each butoxy oxygen. Upon saturation, the three ethereal-oxygen atoms become available for participation in complexation. However, when lithium chloride is the supporting electrolyte rather than the acid, some lithium is also extracted and a fraction of uranium is in the form of UCl_6^{2-} and $UO_2Cl_3^-$ respectively. Equally, the latter is the predominating uranium-bearing species when anhydrous UO_2Cl_2 is dissolved in dry TBP and HCl gas is bubbled through the solution. Adding water will decrease the concentration of the complex anion, which disappears when the organic phase becomes water saturated[512].

The above specific effect of lithium chloride on the extractability (or the formation of) anionic complexes is apparently restricted to the actinides. Bi- and tervalent transition metals are readily extractable as their protonated anionic complexes, though lithium may replace the hydrogen in the complex $HMCl_4$, and $HMCl_3$ or H_2MCl_4[513,514]. The similarity of the distribution curves of iron(III) into TBP from HCl and LiCl suggests that $HFeCl_4 \cdot 2TBP$ and $LiFeCl_4 \cdot 2TBP$ are of a comparable stability[507,515].

In the case of divalent metals, whatever the metal-bearing species is, MCl_2, $HMCl_3$ or H_2MCl_4, the solvation number is two, whereas for the tervalent metals the neutral chloride is a trisolvate, $MCl_3.3S$, rather than the disolvate, as for the anionic $HMCl_4.2S$ species. The solvation number is apparently regulated to a considerable extent by the coordination requirements of the metal, suggesting a direct metal-to-oxygen bond in these and similar solvates. Indeed, infrared spectra of the organic phase show that the strength of the metal-to-oxygen bond is unaltered when chloride is the ligand rather than nitrate, the shifts in the $P \rightarrow O$ frequencies being identical[349,516-518]. As to the $MCl_2.2S$ and $HMCl_3.2S$ species, they may be hydrated to satisfy coordination requirements[425,519].

e. Other Metal Salts

The extraction of metals from perchlorate media gives an example of the lack of general validity of the distribution equation (7C47), discussed in section 7.C. The complexities encountered in perchloric acid extraction by neutral esters (section 9.C) are clearly reflected in extraction of metal perchlorates. Figure 9D4 gives an example of the effect of the nature and composition of the aqueous solution upon the metal distribution ratio[520]. The reason for such a behaviour of perchlorate systems is the high extractability of the supporting electolyte itself, CA in terms of the distribution equation (7C51). Consequently, a partial differentiation of equation (7C49) leads to different values of the mean solvation number $\bar{p}$ for $UO_2(ClO_4)_2$, when $HClO_4$ or $LiClO_4$ or $NaClO_4$ were used as the salting electrolyte CA. Again, different $\bar{p}$ values were evaluated when c_{CA} was varied[405,434,518]. The high extractability of the supporting perchlorate (acid or salt) makes the value of $\bar{p}$ for $UO_2(ClO_4)_2$ increase with decreasing c_{CA}, and it reaches a value as high as 6.6 at 0.1 M $HClO_4$.

Two more features are noteworthy in perchlorate systems: water extractability and high ionization of the solutes. $UO_2(ClO_4)_2$ for example[521,522], carries water into a TBP phase instead of replacing it, as does UO_2Cl_2 or $UO_2(NO_3)_2$, and the complex has the composition $UO_2(ClO_4)_2.2H_2O.2TBP$. It is mainly due to the high water content of uranium(VI), and other metal extracts, that the salts are almost completely ionized in a TBP extract[476,522].

Extraction of metals from sulphate and phosphate solutions is generally poor, as can be seen from the chart given in Appendix F for extraction of various metal sulphates into undiluted TBP[191]. The extractability of metals from halide solutions other than chloride, follows closely that o

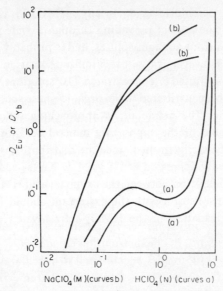

Figure 9D4. Variation of D with perchlorate concentration for extraction of europium (upper curve of pair) and ytterbium (lower curve of pair) with undiluted TBP. Curves (a) for $HClO_4$, curves (b) for $0.1 M \ HClO_4-$ variable $NaClO_4$.

the latter[504,523,524]. A large number of metal thiocyanates can be quantitatively transferred from their acidified aqueous solutions into various neutral esters[446,523,525].

The extraction of metals that form oxyanions in their highest oxidation state, resembles that of simple mineral acids. Chromium(VI)[523,526,527] rhenium(VII) and technetium(VII)[259,352,440,528–531] and other metal ions[526] have been shown to be extracted from different acid media in the form of the potonated oxyanion. Their distribution depends primarily on the type of the supporting acid and its concentration in the aqueous solution. While at low concentrations all the simple acids enhance the transfer of the oxymetal acids, a competition for the available extractant becomes increasingly marked at high initial acidities. The extent of competition depends on the strength of the mineral acid-ester adduct, and at a given concentration of acids it decreases in the order $HClO_4 > HNO_3 > H_2SO_4 > HCl$.

The extraction of pertechnetic and perrhenic acids is no less complex than that of perchloric acid discussed in the previous section (9.C). Their solvation number depends, in addition to the aqueous-phase parameters, also on the concentration of the extractants, as in the case of the similar perchloric acid. Briefly, $HMO_4.3S$ is apparently the composition of the adduct prevailing through a wide range of concentrations of the various aqueous solutes, and is probably of a considerable stability[352,528]. However, it is not the limiting solvation number, since on extraction into undiluted or concentrated TBP solutions, a solvation number of four has been attributed to the organic-phase species[529,530]. On the other hand, when the extractant is approaching saturation, by either the metal oxy acid or the supporting mineral acid, or by both, lower solvates are necessarily formed since the distribution ratio of the oxy-metals is markedly depressed[352,440,529-531]. Kertes and Beck[529] suggested that the competition between the oxymetal acid HMO_4 and the mineral acid HA for solvation with the extractant should be regarded as a step by step degradation of the HMO_4–tetrasolvate, represented as

$$\overline{HMO_4.4TBP} + HA \rightleftharpoons \overline{HMO_4.3TBP} + \overline{HA.TBP}$$

$$\overline{HMO_4.3TBP} + HA \rightleftharpoons \overline{HMO_4.2TBP} + \overline{HA.TBP}$$

$$\tag{9D2}$$

$$\overline{HMO_4.2TBP} + HA \rightleftharpoons \overline{HMO_4.TBP} + \overline{HA.TBP}$$

$$\overline{HMO_4.TBP} + HA \rightleftharpoons H^+ + MO_4^- + \overline{HA.TBP}$$

This formation of progressively lower solvates is consistent with the behaviour of the $HClO_4$–TBP system. In the systems under consideration, instead of increasing the perchloric acid concentration, the concentration of nitric acid has been increased. The effect of nitric acid, for example, is even more drastic, since it has a higher affinity for TBP through the formation of the stable monosolvate. The different solvation numbers of chromic acid could be explained by a similar set of reactions[527].

It is a quite frequent phenomenon that metal ion distribution ratios are increased or decreased by the complexing action of two anionic ligands in the aqueous phase. The extraction of metals from mixed-ligand aqueous solutions arose from practical needs to overcome the lack of extractability of metals encountered in some systems. Such are, for example, the poor extractability of thorium from sulphate media, or of zirconium from aqueous solutions of hydrofluoric acid.

Thorium and uranium are not extractable, or very poorly so, by neutral esters from sulphuric acid solutions. Thus, addition of sulphate ions to a nitrate system drastically reduces the distribution ratio of these metals through their aqueous complexing effect[512,532-535]. Indeed, various mixed complexes of the type $Th(NO_3)_{4-2i}(SO_4)_i$ have been found in the aqueous phase. On the other hand, when $TBP.HNO_3$ is used to extract uranium(VI) from chloride, sulphate, phosphate or fluoride media rather than pure TBP, several times more uranium is transferred into the organic phase[534].

Zirconium and hafnium are unextractable from pure hydrofluoric acid media, due presumably, to the presence of the hexafluoro anion. Addition of nitrate[400,536] or perchlorate[537] results in a markedly enhanced extractability. The extracted species were found to contain two anions in varying ratios. Some of the mixed fluoro-nitrato complexes are extractable, the most extractable being, apparently, that with one fluoride ion in the complex. If the metals are extracted from $HCl-HNO_3$ mixtures rather than $HF-HNO_3$, at a maximum distriibution ratio, $i = 2$ in the complex $M(NO_3)_{4-i}Cl_i. 2TBP$[538]. The best fit of experimental data in an $HF-HClO_4$ system has been given by the model assuming a mono- and a disolvate adduct (with trioctylphosphine oxide) of the mixed complex $HfF_2(ClO_4)_2$[537]. Presumably, the complex in an $HF-H_2SO_4$ system has a similar composition[539].

f. Factors Affecting Exractability of Metal Salts

(i) *Cation of the salting agent* As has been pointed out in the previous sections, it is the most common phenomenon that as the concentration of the supporting acid in aqueous solutions increases the metal distribution ratio increases, passes through a maximum and then decreases. Such a shape of the distribution curves is explained by an increased competition for the extractant molecule by the transferred acid. This undesirable condition can be overcome, usually, by partially replacing the acid by its alkali salt, which greatly reduces the excessive transfer of the supporting electrolyte.

Several attempts have been made[360,497,540-543] to classify the various salting agents according to their capacity to drive the metal salt into the organic phase. The cation of the salting agent has a major effect on the extractability of metals. The metal distribution ratio increases as the radius of the cation decreases in the series of alkali and alkaline earth metal salting agents. There is, however, no relationship between the two groups. The effectiveness increases also with increasing charge on the cation. For example, the disrtibution ratios of europium from unacidified nitrate solutions at a 5.6 N level into TBP, range from 270 to 2 in the order: Al > Mg > Zn > Li > Cu > Na, Fe, Ca > NH_4. With slight changes in this order, D_{Eu} falls from 3000 to 40 when dipentyl pentylphosphonate is the extractant rather than TBP[543].

(ii) *Temperature* Temperature is perhaps the most complex factor affecting the equilibrium in a solvent extraction process, and is certainly the least understood. The many fragmentary data on its effect upon the extractability are mainly process oriented, with the sole aim of increasing separation factors in fuel reprocessing technology[290,327,389,401,446,541, 544–546].

Temperature affects the complexing equilibrium of the metal in the aqueous phase, which, in turn, depends on the composition of the aqueous solution.

Fletcher and Hardy[381], relying on their results in a zirconium nitrate-tributyl phosphate system, proposed the generalization that the temperature coefficient of extraction is positive with those metals where the distribution ratio does not decrease above 7 M HNO_3. Otherwise, with metals which have D values decreasing beyond that aqueous concentration, it is usually negative. Just one step further, one may assume that the temperature coefficient of extraction for strong metal salt adducts is positive, whereas for weak solvate complexes, it is negative. It will not be easy to provide evidence for this extended generalization of Fletcher and Hardy in view of the variety of adducts which may be, and frequently are, formed in one given system as a function of the solute concentrations in both phases.

(iii) *Structure of the extractant* Although the extractability of a metal salt with neutral organophosphorus esters will primarily depend on the composition of the aqueous phase and the chemical equilibrium to form the extractable species, the importance of the structure of the esters on their extractive power towards metals has been long recognized. The observation that compounds containing direct carbon-to-phosphorus bonds gave higher distribution ratios, and that the distribution ratio increases with the number of such C–P bonds, has been shown to be a very general phenomenon in the series $(RO)_n R_{3-n} PO$ with decreasing value of n, and valid for all extractable salts.

Numerous attempts to express the relationship between extractive capacity of the esters and their structure in general terms, except the one mentioned above, were unsuccessful so far. Some qualitiative observations may be valid for a larger number of compounds than others, but it is clear that other extraction variables often mask slight structural effects. For example, when extraction takes place under such aqueous conditions as to result in high organic-phase loadings, there is hardly any noticeable structural effect at all. The only exceptions known are the tetraalkyl alkylenediphosphonates and pyrophosphates[44], and esters with ethereal oxygens in the alkoxy chain[510] which possess coordination sites in addition to the one phosphoryl oxygen.

Under different experimental conditions, however, the structural effect may be drastic. There is a difference of eight orders of magnitude in the distribution ratio of tracer uranium(VI) from 1 M nitric acid solution into the esters shown in Figure 9D5[44]. The qualitiative observations on structural effects may be summarized as follows: a) The distribution ratio either increases or is unaffected as the alkyl or alkoxy groups are made larger. The effect is more pronounced with lower chain-length esters due to their marked solubility in aqueous solution; b) Extraction generally improves with branched alkyl or alkoxy groups, the effect being most marked with branching near the alkoxy oxygen. This is connected with the basicity of the ester, which increases with branching. Thus, if such an alteration of the substituent results in decreased extraction, as in the case of thorium nitrate[268,399], the negative effect must be steric; c) Attaching an unsaturated group to esters decreases the metal distribution ratio, and a still more pronounced reduction is observed with phenyl groups as substituents. A phenyl group linked through oxygen to the phosphorus appears to reduce solvent strength by a larger factor than when direct carbon-to-phosphorus bonds exist; d) More electronegative groups than phenyl, chlorine for example, nearly destroy the solvent strength. Substitution of sulphur for oxygen in the neutral esters reduces considerably the complex-binding properties of the derivatives with respect to metal ions. $(RO)_3PS$ and R_3PS compounds have been shown to be selectively reactive towards metal ions that form insoluble sulphides[285].

Burger[480] has attempted to correlate the extractive power of the ester and the electronegativity of the phosphoryl oxygen as measured by the stretching frequency of the phosphoryl bond for a large number of various neutral phosphates. Generally speaking, the determined $P \rightarrow O$ bond stretching frequencies and D's could be correlated as expected. The relationship is valid, however, only to distinguish between the four classes of neutral esters but within a class of compounds the scatter is too great for meaningful comparison. The presence of chlorine atoms

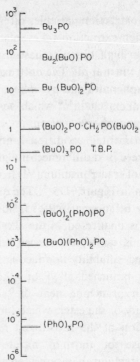

Figure 9D5.   Distribution ratio of tracer $UO_2(NO_3)_2$ between solvent and 1M MNO_3.

seems to have a greater effect in depressing metal distribution ratio than the $P \rightarrow O$ shift would predict.

(iv) *Diluent* The differences in the extractive properties of an ester in various diluents are usually ascribed to a varying extent of non-ideality of the solutes, the free extractant and its metal salt adduct, in the diluent. From a formal point of view this is certainly true. The thermodynamic approach, however, does not reveal the causes leading to deviations from the ideal partition law. Experience has shown, that the distribution ratio of uranyl nitrate, for example, varies with the diluent in some cases by as much as 50 times, but that the direction and size of the deviation from ideality is different in each case when changing any other parameter in such multicomponent systems. A quantitative determination of these effects is at present hardly possible, because of the lack of data on the physicochemical properties of even the less complicated systems, such as a binary system composed of the extractant and diluent, not to mention the lack of activity data in the more pertinent ternary systems of immediate interest to us, such as H_2O–extsactant–diluent.

For this reason, the available data have been presented by the less elegant phenomenological approach, which is the only possible one in the present state of our knowledge.

The effect of the diluent[334,446,481,544,546,548] on the variation of the metal distribution ratio may be summarized as follows: At low metal loadings, the metal distribution ratio is generally higher in aromatic and aliphatic hydrocarbons than in ethers and ketones.

Especially low values are obtained in the polar alcohols and chloroform. With diluents of zero or low polarities, D's are higher than those calculated with allowance for dilution of TBP alone. Polar diluents cause the opposite effect. It still depends frequently on the aqueous-phase conditions whether the unsubstituted aromatic or aliphatic hydrocarbons will show the higher distribution ratio.

In most cases, a comparison of metal distribution ratios measured in various diluents shows a lack of correlation with such physical properties of the diluents as their dielectric constant, dipole moment and polarizability. Taube's concept[549] is based on the electrostatic interaction between the complex and the dipole of the solvent molecules, and on the work needed to form a hole in the organic medium. Since non-polar hydrocarbons are more structureless, the extraction into these liquids is higher than into the more ordered polar ones. This rough differentiation correlates well with the dielectric constant of polar and non-polar diluents but does not reflect the observed differences within a given class of diluents. The molecular polarization was successfully correlated with the distribution ratio of uranyl nitrate when the polar alcohols were used as diluents for TBP[547]. The correlation was less clear with less or non-polar liquids[329,334,481]. With the polar chloroform and alcohols, for example, hydrogen-bond interaction certainly affects the extractive capacity of the ester, in the way discussed in section 9.C. That this hydrogen bonding is of a comparable strength to that of the metal–oxygen linkage is evident from the drastically decreased metal distribution ratios observed in such systems[334,547]. The extent of this competition obviously depends, on the strength of the metal–ester bond, and is thus particular to the metal-bearing complex.

To account for the observed differences in the metal distribution ratios when non-polar hydrocarbon diluents are compared, Kertes[266] has suggested that the solvation ability of the diluent may be responsible for the more positive deviation from the ideal partition law in aliphatic hydrocarbons than in aromatics.

Another serious difficulty is the choice of the proper concentration unit of the esters to be correlated. From various types of plots, Shevchenko and coworkers[329] concluded that the best comparison of the influence of diluents on metal distribution ratios is obtained when expressing concentrations of the extracting agent in mole fraction or mole percents. While these units may give a better fit than others in dilute ester solutions, activities should be used rather than concentrations.

In spite of these differences in the extractive power of the ester occurring as an overall effect of changing the medium, the stoichiometric composition of the adduct remains usually unchanged. This is not surprising in view of the strong bonds formed between the metal and the phosphoryl oxygen. In a few cases, however, when metals with low affinity towards the ester are extracted, the diluent–extractant interaction may interfere with the solvation of the metal by the ester. This is illustrated in the extraction of copper nitrate by TBP[460].

The formation of a third phase when some salts are extracted in macro quantities into aliphatic hydrocarbon solutions of esters is connected with the influence of the diluent on the heterogeneous equilibrium. The phenomenon has been noted when uranium[399,463,550] and thorium[266,401,493,551] are extracted, but is common to the majority of extraction systems under similar conditions. Formally, the splitting of the organic phase into two may be explained as due to the limited solubility of a metal salt–ester adduct in the non-polar hydrocarbon, since one of the two layers usually contains the almost pure solvate, and the other is predominantly the diluent with only traces of ester or adduct. This simple mechanism is supported by the fact that the miscibility gap in the system is temperature dependent, the formation of the third phase being a reversible reaction.

Kertes[266] has formulated the limited solubility of the adduct in aliphatic hydrocarbon solvents, in contrast to its unlimited solubilities in polar solvents, and its high solubilities in equally non-polar aromatic hydrocarbons, in terms of solvation of the adduct by diluent molecules in the latter case, and the absence of such dipole-dipole interaction when aliphatic diluents are employed.

(v) *Stability* An important factor which makes neutral phosphorus esters one of the most widely used extractants in fuel reprocessing is their high chemical stability. The only possible changes lead to their hydrolytic degradation.

It is beyond the scope of this monograph to review and discuss the various aspects stability of the extractants has in process chemistry[289-291,552-560]. Neither is it our intention to do that with the vast field of radiation chemistry of these compounds[289,368,555,559-563]. Instead the reader is referred to the references mentioned here.

The acid-induced decomposition of tributyl phosphate leads to the formation of dibutyl and monobutyl phosphates, butyl nitrate (chloride or perchlorate, depending on the acid producing acidolysis), or the phosphoric acid and butanol in roughly this order of yield, butyl ether, tetrabutyl pyrophosphate and various other, mostly polymeric, compounds in much lower yields. The fundamental reaction in the hydrolysis is the clevage of the O—C linkage, the primary products being, thus acid esters and butyl nitrate. The rate of decomposition and the yields of these products, identified by gas[557] or paper chromatography[563], zirconium extraction[289,290] and other methods[300,564], depends largely on the nature and amount of the dissolved inorganic acid and salt. For example, TBP is degraded 1000 times faster by the TBP–Zr reaction than by the TBP–HNO$_3$[557] one. The reaction is faster with HBr, HCl or LiCl than with HNO$_3$[321,362,565]. The differences in rate of decomposition by the various acids have been attributed to the water content of the organic phase, the higher the water content the higher the rate[565].

Normal paraffinic hydrocarbon diluents have no effect on the rate of decomposition, while carbon tetrachloride greatly increases it, and benzene reduces it, at least when acidolysis occurs in the presence of nitric acid[554].

Small changes in structure of the phosphorus compound may alter its chemical stability[313,361,368,555,558,566]. Phosphonates are more resistant towards acidolytic degradation than TBP. Evidently the P—C linkage is stronger than that in P—O—C, in agreement with the reaction of hydrolysis outlined above.

There is some information available on the stability of TBP towards oxidizing agents. Extracts of chromium(VI) in TBP are reduced to chromium(III)[567]. Uranium(VI) is photochemically reduced to uranium(IV) in the presence of HCl, HClO$_4$ and HNO$_3$[568-570]. The ester is degraded in all these cases to its usual degradation products.

Finally, in recent years reference has been made to metal complexing by degradation products arising from the contact of diluents, mainly of the paraffinic type, with inorganic acids in an extraction system[571–574]. Since degradation of diluents requires a prolonged contact with the acids, the problem is more of a technological concern[291].

E. REFERENCES

1. T. V. Healy and P. E. Brown, *Brit. Rept., AERE* C/R 1738 (1956).

2. T. H. Siddall, III, *J. Phys. Chem.*, **64**, 1963 (1960).

3. J. Timmermans, *Physico-Chemical Constants of Pure Organic Compounds*, Elsevier Publ. Co., Amsterdam, 1950.

4. T. H. Durrans, *Solvents*, 7th ed., Van Nostrand, New York, 1957.

5. L. Scheffan and M. B. Jacobs, *The Handbook of Solvents*, Van Nostrand, New York, 1953.

6. R. R. Dreisbach, *Physical Properties of Chemical Compounds*, Vols. I, II, III., Advances in Chemistry Series, Nos. 15, 22 and 29, A.C.S., Washington, 1961.

7. A. A. Maryott and E. R. Smith, *Tables of Dielectric Constants of Pure Liquids*, Natl. Bur. Std., Circular No. 514, Washington, D. C., 1951.

8. O. Fuchs and K. L. Wolf, *Dielektrische Polarisation*, Akad. Verlag., Leipzig, 1935, Appendix I.

9. A. Weissberger, E. S. Proskauer, J. A. Riddick and E. E. Toops, Jr., *Organic Solvents*, Interscience Publ., New York, 1955.

10. E. S. Lane, A. Pilbeam and J. M. Fletcher, *Brit. Rept., AERE* R–4440 (1965).

11. A. Weissberger, *Separation and Purification*, Vol. II, Part I, Interscience Publ., New York, 1956.

12. D. G. Tuck, *J. Chem. Soc.*, 3202 (1957).

13. R. J. Myers, D. E. Metzler and E. H. Swift, *J. Am. Chem. Soc.*, **72**, 3767 (1950).

14. A. M. Poskanzer, R. J. Dietz, E. Rudzitis, J. W. Irvine Jr. and C.D· Coryell, *Radioisotopes in Scientific Research*, Vol II, Pergamon Press, London, 1958, p. 518.

15. R. J. Dietz, Jr., *Ph.D. Thesis*, Department of Chemistry, Massachusetts Institute of Technology, Cambridge, 1958.

16. C. Marsden, *Solvents Manual*, Cleaver-Hume Press, London, 1954.

17. *International Critical Tables*, Vol. III, p. 386.

18. H. Stephen and T. Stephen, *Solubilities of Inorganic and Organic Compounds*, Vol. I. Part I, Pergamon Press, London, 1963.

19. A. Seidel, *Solubility of Organic Compounds*, Vol. II, 3rd ed., Van Nostrand, New York, 1941.

20. W. F. Linke, *Solubilities*, Vol. I, Van Nostrand, New York, 1958, p. 1136.

21. C. H. Werkman, *Anal. Chem.*, **20**, 1094 (1948).

22. M. Rylek and J. Vondrak, *Collection Czech. Chem. Commun.*, **25**, 2497 (1960).

23. A. W. Gardner and H. A. C. McKay, *Trans. Faraday Soc.*, **48**, 1099 (1952).

24. W. Herz, *Chem. Ber.*, **31**, 2671 (1898).

25. D. C. Jones, *J. Chem. Soc.*, 799 (1929).

26. C. C. Templeton, *J. Am. Chem. Soc.*, **71**, 2187 (1949).

27. D. E. Campbell, A. H. Laurene and H. M. Clark, *J. Am. Chem. Soc.*, **74**, 6193 (1952).

28. T. E. Jordan, *Vapor Pressure of Organic Compounds*, Interscience Publ., New York, 1954.

29. J. H. Hildebrand and R. L. Scott, *The Solubility of Non-electrolytes*, 3rd ed., Reinhold Publ. Co. New York, 1950, p. 264.

30. L. L. Burger, *J. Chem. Eng. Data*, **9**, 112 (1964).

31. C. Marie and G. Lejeune, *Monatsh. Chem.*, **53**, 69 (1929).

32. K. V. Troitskii, *Zh. Neorg. Khim.*, **3**, 1457 (1958).

33. G. S. Golden and H. M. Clark. *J. Phys. Chem.*, **65**, 1932 (1961).

34. G. Lejeune, *Compt. Rend.*, **208**, 1225 (1939).

35. D. G. Tuck, *Anal. Chim. Acta*, **20**, 159 (1959).

36. D. G. Tuck, *J. Chem. Soc.*, 2736 (1963).

37. W. T. Reburn and W. N. Shearer, *J. Am. Chem. Soc.*, **55**, 1774 (1933).

38. F. G. Zharovskii, *Zh. Neorg. Khim.*, **2**, 623 (1957).

39. V. I. Kuznetsov and I. V. Seriakova in *Ekstraktziya*, Vol. I, Gosatomizdat, Moscow, 1960, p. 104.

40. J. M. Googin, W. L. Harper, L. R. Phillips and F. W. Postman, *USAEC Rept.*, Y–DA–202; *Solvent Extraction Chemistry Symposium*, Gatlinburg, 1962.

41. H. A. C. McKay, *Trans. Faraday Soc.*, **48**, 1103 (1952).

42. H. Irving and F. J. C. Rossotti, *J. Chem. Soc.*, 1946 (1955).

43. E. Glueckauf and B. Davies, *Brit. Rept.*, *AERE* C/R 2029 (1956).

44. E. Glueckauf, *Ind. Chim. Belge*, **23**, 1215 (1958).

45. V. M. Vdovenko and I. G. Suglobova, *Zh. Neorg. Khim.*, **3**, 1403 (1958).

46. V. M. Vdovenko and E. A. Smirnova, *Radiokhimiya*, **1**, 36 (1959).

47. A. W. Gardner, H. A. C. McKay and D. T. Warren, *Trans. Faraday Soc.*, **48**, 997 (1952).

48. I. V. Seryakova, Y. A. Zolotov, A. V. Karyakin and L. A. Gribov, *Zh. Neorg. Khim.*, **8**, 474 (1963).

49. R. A. Horne, *J. Chem. Eng. Data.*, **7**, 1 (1962).

50. R. Bock and M. Herrmann, *Z. Anorg. Allgem. Chem.*, **284**, 288 (1956).

51. A. A. Grinberg and G. S. Lozhkina, *Zh. Neorg. Khim.*, **5**, 738 (1960).

52. F. G. Zharovskii and V. F. Melnik, *Zh. Neorg. Khim.*, **6**, 1466 (1961).

53. D. G. Tuck and R. M. Diamond, *J. Phys. Chem.*, **65**, 193 (1961).

54. R. M. Diamond and D. G. Tuck in *Progress in Inorganic Chemistry* (Ed. F. A. Cotton), Vol. 2, Interscience Publ., New York 1960, p. 109.

55. M. Cockbill and J. Magid, *Chem. Ind. (London)*, 2009 (1962).

56. V. M. Vdovenko, D. N. Suglobov and A. I. Skoblo, *Zh. Neorg. Khim.*, **4**, 2376 (1959).

57. L. Alders, *Liquid—Liquid Extraction*, Elsevier Publ. Co., Amsterdam 1959; A. W. Francis in *Solubilities* (Eds. A. Seidell and W. F. Linke), Suppl. 3rd. ed., Van Nostrand, New York, 1952.

58. H. Irving, F. J. C. Rossotti and R. J. P. Williams, *J. Chem. Soc.*, 1906 (1955).

59. A. Norström and L. G. Sillèn, *Svensk. Kem. Tidskr.*, **60**, 227 (1948).

60. J. Desmarouk, R. Dalmon and R. Vandoni, *Compt. Rend.*, **214**, 352 (1942).

61. A. W. Wylie, *J. Chem. Soc.*, 1474 (1951).

62. N. H. Furman, R. J. Mundy and G. H. Morrison, *USAEC. Rept.*, AECD–2938 (1955).

63. R. Bock and E. Bock, *Z. Anorg. Allgem. Chem.*, **263**, 146 (1950).

64. J. Kooi, *JENER Publ. Rept.*, No. 11 (1956).

65. R. N. Maslova and V. V. Fomin, *Zh. Neorg. Khim.*, **6**, 738 (1961).

66. A. S. Kertes and A. Beck, *J. Chromatog.*, **1**, 496 (1958).

67. H. A. C. McKay, *J. Inorg. Nucl. Chem.*, **4**, 375 (1957).

68. V. M. Vdovenko, A. A. Lipovskii and M. G. Kuzina, *Zh. Neorg. Khim.*, **2**, 975 (1957).

69. V. M. Vdovenko and A. S. Krivokhatskii, *Radiokhimiya*, **1**, 454 (1959).

70. V. V. Fomin, R. N. Maslova and L. L. Zaitseva, *Zh. Neorg. Khim.*, **5**, 1383 (1960).

71. V. V. Fomin and R. N. Maslova, *Zh. Neorg. Khim.*, **6**, 481 (1961).

72. V. M. Vdovenko and N. A. Alekseeva, *Radiokhimiya*, **1**, 450 (1959).

73. V. Vesely, H. Beranova and J. Maly, *Collection Czech. Chem. Commun.*, **25**, 2622 (1960).

74. V. V. Fomin, A. F. Morgunov and I. V. Korobov, *Zh. Neorg. Khim.*, **5**, 1846 (1960).

75. A. S. Kertes, *J. Chromatog.*, **1**, 62 (1958).

76. A. S. Kertes and A. H. I. Ben-Bassat, *J. Chromatog.*, **1**, 489 (1958).

77. J. Kooi, *Rec. Trav. Chim.*, **74**, 137 (1955).

78. J. Rydberg and B. Bernström, *Acta Chem. Scand.*, **11**, 86 (1957).

79. F. Havlicek and V. Sraier, *Collection Czech. Chem. Commun.*, **28,** 2251 (1963).

80. G. Almassy, *Acta Chim. Acad. Sci. Hung.*, **25,** 391 (1960).

81. S. Tribalat, *Ann. Chim.*, **8,** 642 (1953).

82. A. S. Kertes and A. Beck, *Bull. Res. Council Israel*, **7A,** 138 (1958).

83. T. R. Scott, *Analyst*, **74,** 486 (1949).

84. A. Schulze, *Z. Phys. Chem. (Leipzig)*, **97,** 388 (1921).

85. S. A. Katz, W. M. McNabb and J. F. Hazel, *Anal. Chim. Acta*, **27,** 405 (1962).

86. L. E. Glendenin, K. F. Flynn, R. F. Buchanan and E. P. Steinberg, *Anal. Chem.*, **27,** 59 (1955).

87. F. L. Culler, *Proc. U. N. Intern. Conf. Peaceful Uses At. Energy, 1st Geneva*, **9,** 464 (1956).

88. D. E. Chalkley and R. J. P. Williams, *J. Chem. Soc.*, 1920 (1955).

89. N. H. Nachtrieb and J. G. Conway, *J. Am. Chem. Soc.*, **70,** 3547 (1948).

90. G. O. Brink, P. Kafalas, R. A. Sharp, E. L. Weiss and J. W. Irvine, Jr., *J. Am. Chem. Soc.*, **79,** 1303 (1957).

91. V. V. Fomin, P. A. Zagorets, A. F. Morgunov and I. I. Tertishnik, *Zh. Neorg. Khim.*, **4,** 2276 (1959).

92. J. Golden and A. G. Maddock, *J. Inorg. Nucl. Chem.*, **2,** 46 (1956).

93. L. M. Gindin, I. F. Kopp, A. M. Rozen and E. F. Kouba, *Zh. Neorg. Khim.*, **5,** 139 (1960).

94. E. D. Crittenden, Jr. and A. N. Hixson, *Ind. Eng. Chem.*, **46,** 265 (1954).

95. L. Garwin and A. N. Hixson, *Ind. Eng. Chem.*, **41,** 2303 (1949).

96. A. I. Sukhanovskaya, I. P. Alimarin and Y. A. Zolotov, *Zh. Neorg. Khim.*, **10,** 707 (1965).

97. Y. A. Zolotov, I. V. Seryakova, A. V. Karyakin, L. A. Gribov and M. E. Zubrilina, *Zh. Neorg. Khim.*, **8,** 481 (1963).

98. R. Bock, H. Kusche and E. Bock, *Z. Anal. Chem.*, **138,** 167 (1953).

99. J. Vondrak and M. Rylek, *Collection Czech. Chem. Commun.*, **26,** 307 (1961).

100. L. Garwin and A. N. Hixson, *Ind. Eng. Chem.*, **41,** 2298 (1949).

101. R. J. Dietz, Jr., J. Mendez and J. W. Irvine, Jr., *Radioisotopes in the Physical Science and Industry*, IAEA, Vienna, 1962, p. 415.

102. O. E. Zvyagintsev and O. I. Zakharov-Nartsissov, *Zh. Neorg. Khim.*, **5,** 124 (1960).

103. M. Bachelet, E. Cheylan and J. Le Bris, *J. Chim. Phys.*, **44,** 302 (1947).

104. K. E. Whitehead and C. J. Geankoplis, *Ind. Eng. Chem.*, **47,** 2114 (1955).

105. B. Helferich and U. Baumann, *Chem. Ber.*, **85,** 461 (1952).

106. Z. S. Golynko, D. I. Skorovarov, V. F. Smienov and N. V. Skovortsov *Zh. Prikl. Khim.*, **38,** 271 (1965).

107. R. Bock, *Z. Anal. Chem.*, **133**, 110 (1951).

108. A. S. Kertes and A. Beck, *J. Chromatog.*, **2**, 362 (1959).

109. N. de Kolossovsky, *Bull. Soc. Chim. Belges*, **28**, 257 (1919).

110. J. H. Walton and H. A. Lewis, *J. Am. Chem. Soc.*, **38**, 633 (1916).

111. H. W. Smith, *J. Phys. Chem.*, **25**, 616 (1921).

112. H. A. C. McKay, K. Alcock and D. Scargill, *Brit. Rept.*, *AERE C/R* 2221 (1956).

113. O. C. Dermer and V. H. Dermer, *J. Am. Chem. Soc.*, **65**, 1653 (1943).

114. H. J. Vogt and C. J. Geankoplis, *Ind. Eng. Chem.*, **45**, 2119 (1953).

115. R. C. Archibald, *J. Am. Chem. Soc.*, **54**, 3178 (1932).

116. R. D. Harris and C. J. Geankoplis, *J. Chem. Eng. Data*, **7**, 218 (1962).

117. C. P. Brown and A. R. Mathieson, *J. Phys. Chem.*, **58**, 1057 (1954).

118. P. J. McAteer, R. W. Cox and C. J. Geankoplis, *A. I. Ch. E. J.*, **7**, 456 (1961).

119. H. Irving and R. J. P. Williams in *Treatise on Analytical Chemistry* (Eds. I. M. Kolthoff and P. J. Elving); Vol. III Interscience Publ. New York, 1959, Part I, p. 1309.

120. T. E. Moore, R. J. Laran and P. C. Yates, *J. Phys. Chem.*, **59**, 90 (1955).

121. L. Kuča, *Collection Czech. Chem. Commun.*, **27**, 2372 (1962).

122. H. Specker, E. Jackwerth and G. Hovermann, *Z. Anal. Chem.*, **177**, 10 (1960).

123. G. H. Morrison and H. Freiser, *Anal. Chem.*, **34**, 64R (1962); **36**, 93R (1964); **38**, 131R (1966); **40**, 522R (1968).

124. R. K. Warner, *Australian J. Appl. Sci.*, **3**, 156 (1952).

125. R. G. Bates in *Solubility of Inorganic and Organic Compounds* (Eds. A. Seidel and W. F. Linke), Suppl. to 3rd. ed. Van Nostrand, New York, 1951.

126. D. L. Leussig in *Treatise on Analytical Chemistry* (Eds. I. M. Kolthoff and P. J. Elving) Vol. I, Interscience, New York, 1959, Part I., p. 675.

127. L. I. Katzin, *J. Inorg. Nucl. Chem.*, **4**, 187 (1957).

128. J. R. Ferraro, L. I. Katzin and G. Gibson, *J. Inorg. Nucl. Chem.*, **2**, 118 (1956).

129. V. M. Vdovenko, I. G. Suglobova, I. Y. Wang and D. N. Suglobov, *Radiokhimiya*, **6**, 532 (1964).

130. V. M. Vdovenko, I. G. Suglobova and D. N. Suglobov, *Radiokhimiya*, **6**, 539 (1964).

131. L. I. Katzin in *Advances in Transition Metal Chemistry* (Ed. R. L. Carlin), Vol. 3, M. Dekker, New York, 1967.

132. C. C. Templeton and N. F. Hall, *J. Phys. Colloid Chem.*, **51**, 1441 (1947).

133. L. Yaffe, *Can. J. Res.*, **B27**, 638 (1949).

134. W. Gordy and S. C. Stanford, *J. Chem. Phys.*, **8**, 170 (1940).

135. L. I. Katzin, D. M. Simon and J. R. Ferraro, *J. Am. Chem. Soc.*, **74**, 1191 (1952).

136. L. I. Katzin and J. R. Ferraro, *J. Am. Chem. Soc.*, **74**, 6040 (1952).

137. L. I. Katzin and J. R. Ferraro, *J. Am. Chem. Soc.*, **75**, 3821 (1953).

138. L. I. Katzin, J. R. Ferraro, W. W. Wendlandt and R. L. McBeth, *J. Am. Chem. Soc.*, **78**, 5139 (1956).

139. V. M. Vdovenko, M. P. Kovalskaya and T. V. Kovaleva, *Zh. Neorg. Khim.*, **2**, 1677 (1957).

140. V. M. Vdovenko and I. G. Suglovoba, *Zh. Neorg. Khim.*, **3**, 1573 (1958).

141. L. I. Katzin and J. C. Sullivan, *J. Phys. Colloid Chem.*, **55**, 346 (1951).

142. L. I. Katzin and J. R. Ferraro, *J. Am. Chem. Soc.*, **72**, 5451 (1950).

143. L. I. Katzin and J. R. Ferraro, *J. Am. Chem. Soc.*, **75**, 3825 (1953).

144. C. C. Templeton, *J. Phys. Colloid Chem.*, **54**, 1255 (1950).

145. C. C. Templeton and L. K. Daly, *J. Phys. Chem.*, **56**, 215 (1952).

146. H. H. Rowley and W. R. Reed, *Proc. Oklahoma Acad. Sci.*, **31**, 129 (1950); *Chem. Abstr.*, **46**, 5946 (1952).

147. U. Berglund and L. G. Sillèn, *Acta Chem. Scand.*, **2**, 116 (1948).

148. A. R. Mathieson, *J. Chem. Soc.*, S 294 (1956).

149. H. M. Irving, *Quart. Rev.*, **5**, 200 (1951).

150. L. I. Katzin and J. C. Sullivan, *USAEC Rept.*, AECD–2537 (1948).

151. Y. I. Riskin, V. P. Shvedov and A. A. Soloveva, *Zh. Neorg. Khim.*, **4**, 2268 (1959).

152. H. A. C. McKay and A. R. Mathieson, *Trans. Faraday Soc.*, **47**, 428 (1951).

153. K. Schug and L. I. Katzin, *J. Phys. Chem.*, **66**, 907 (1962).

154. V. M. Vdovenko, D. N. Suglobov and L. G. Mashirov, *Radiokhimiya*, **3**, 173 (1961).

155. H. L. Friedman and H. Taube, *J. Am. Chem. Soc.*, **72**, 3362 (1950).

156. M. Bachelet and E. Cheylan, *J. Chim. Phys.*, **44**, 248 (1947).

157. C. C. Addison, B. J. Hathaway, N. Logan and A. Walker, *J. Chem. Soc.*, 4308 (1960).

158. S. Buffagni and T. M. Dunn, *J. Chem. Soc.*, 5105 (1961).

159. D. C. Bradley, E. V. Caldwell and W. Wardlaw, *J. Chem. Soc.*, 3039 (1957).

160. N. Y. Turova, A. V. Novoselova and K. N. Semeneko, *Zh. Neorg. Khim.*, **5**, 117 (1960).

161. N. Y. Turova, A. V. Novoselova and K. N. Semeneko, *Zh. Neorg. Khim.*, **5**, 941 (1960).

162. J. Houben and W. Fischer, *J. Prakt. Chem.*, **123**, 99 (1929).

163. E. Wiberg, M. Schmidt and A. S. Galinos, *Z. Angew. Chem.*, **65**, 443 (1954).

164. A. H. Laurene, D. E. Campbell, S. E. Wiberley and H. M. Clark, *J. Phys. Chem.*, **60**, 901 (1956).

165. N. Y. Turova, K. N. Semeneko and A. V. Novoselova, *Zh. Neorg. Khim.*, **8**, 882 (1963).

166. W. Menzel and M. Fröhlich, *Chem. Ber.*, **75B**, 1055 (1942).

167. P. Hagenmüller and J. Rouxel, *Compt. Rend.*, **247**, 1623 (1958).

168. R. E. van Dyke and H. E. Crawford, *J. Am. Chem. Soc.*, **72**, 2829 (1950).

169. H. P. Hamilton, R. McBeth, W. Bekelrede and H. H. Sisler, *J. Am. Chem. Soc.*, **75**, 2881 (1953).

170. F. Klages, H. Menresch and W. Steppich, *Ann. Chem.*, **592**, 81 (1955).

171. D. Cozzi and S. Cecconi, *Ric. Sci. Suppl.*, **23**, 609 (1953).

172. A. Cowley, F. Fairbrother and N. Scott, *J. Chem. Soc.*, 3133 (1958).

173. W. Herwig and H. H. Zeiss, *J. Org. Chem.*, **23**, 1404 (1958).

174. W. Wardlaw and H. W. Webb, *J. Chem. Soc.*, 2103 (1930).

175. D. E. McLaughlin, M. Tamres and S. Searles, *J. Am. Chem. Soc.*, **82**, 5621 (1960).

176. W. Nespital, *Z. Physik. Chem.*, **B16**, 153 (1932).

177. W. V. Evans and H. H. Rowley, *J. Am. Chem. Soc.*, **52**, 3523 (1930).

178. R. J. Myers and D. E. Metzler, *J. Am. Chem. Soc.*, **72**, 3772 (1950).

179. H. L. Friedman and H. Taube, *J. Am. Chem. Soc.*, **72**, 2236 (1950).

180. H. H. Sisler, H. H. Batey, B. Pfahler and R. Mattair, *J. Am. Chem. Soc.*, **70**, 3821 (1948).

181. W. J. McManamey, *J. Appl. Chem.*, **11**, 44 (1961).

182. W. J. McManamey, *J. Phys. Chem.*, **65**, 1053 (1961).

183. A. G. Kurnakov and A. V. Nikolaev, *Zh. Neorg. Khim.*, **3**, 1028 (1958); A. V. Nikolaev and A. G. Kurnakov, *Zh. Neorg. Khim.*, **3**, 1037 (1958).

184. E. Glueckauf, H. A. C. McKay and A. R. Mathieson, *Trans. Faraday Soc.*, **47**, 437 (1951).

185. V. M. Vdovenko and E. A. Smirnova, *Radiokhimiya*, **2**, 291 (1960).

186. L. Kaplan, R. A. Hildebrandt and M. Ader, *J. Inorg. Nucl. Chem.*, **2**, 153 (1956).

187. I. L. Jenkins and H. A. C. McKay, *Trans. Faraday Soc.*, **50**, 107 (1954).

188. S. Minc and A. Ugnevskaya, *Radiokhimiya*, **3**, 137 (1961).

189. V. M. Vdovenko, T. V. Kovaleva and V. G. Potapov, *Radiokhimiya*, **4**, 34 (1962).

190. O. Y. Samoilov and V. I. Tikhomirov, *Radiokhimiya*, **2**, 183 (1960).

191. T. Ishimori, E. Akatsu, K. Tsukuechi, T. Kobune, Y. Usuba, K. Kimura, G. Onawa and H. Uchiyama, *JAERI Rept.*, 1106 (1966).

192. N. M. Adamskii, *Radiokhimiya*, 2, 653 (1960).

193. V. M. Vdovenko and E. A. Smirnova, *Radiokhimiya*, 1, 521 (1959).

194. V. M. Vdovenko, D. N. Suglobov and E. A. Smirnova, *Radiokhimiya*, 2, 296 (1960).

195. D. G. Tuck, *J. Inorg. Nucl. Chem.*, 6, 252 (1958).

196. H. A. C. McKay and R. J. W. Streeton, *J. Inorg. Nucl. Chem.*, 27, 879 (1965).

197. R. Bock and E. Bock, *Naturwissenschaften*, 36, 344 (1949).

198. B. F. Rothschild, C. C. Templeton and N. F. Hall, *J. Phys. Colloid Chem.*, 52, 1006 (1948).

199. K. Seidl and M. Beranek, *Collection Czech. Chem. Commun.*, 24, 298 (1959).

200. A. V. Nikolaev, A. G. Kurnakova and I. I. Yakovlev, *Zh. Neorg. Khim.*, 5, 1832 (1960).

201. V. V. Fomin in *Ekstraktziya*, Vol. 1, Gosatomizdat, Moscow, 1960, p. 104.

202. T. V. Healy and A. W. Gardner, *J. Inorg. Nucl. Chem.*, 1, 245 (1958).

203. V. M. Vdovenko, A. A. Lipovskii and M. G. Kuzina, *Zh. Neorg. Khim.*, 4, 2502 (1959).

204. S. I. Sinyakova and N. S. Klassova, *Zh. Neorg. Khim.*, 4, 2000 (1959).

205. H. Specker, M. Cremer and E. Jackwerth, *Angew. Chem.*, 71, 492 (1959).

206. K. Študlar, *Collection Czech. Chem. Commun.*, 31, 1999 (1966).

207. D. I. Riyabchikov and M. M. Provalova, *Zh. Neorg. Khim.*, 3, 1694 (1958); in *Ekstratziya*, Vol. II, Gosatomizdat, Moscow, 1960, p. 165.

208. F. E. Edwards and A. F. Voigt, *Anal. Chem.*, 21, 1204 (1949).

209. B. Z. Ioffa, K. P. Mitrofanov, M. V. Plotnikova and S. Kopach, *Radiokhimiya*, 6, 419 (1964).

210. W. J. McManamey, *J. Appl. Chem.*, 13, 207 (1963).

211. L. I. Katzin and E. Gebert, *J. Am. Chem. Soc.*, 72, 5464 (1950).

212. W. D. Beaver, L. E. Trevorow, W. E. Estill, P. C. Yates and T. E. Moore, *J. Am. Chem. Soc.*, 75, 4556 (1953).

213. S. Kitahara, *Kagaku Kenkyusho Hokoku*, 25, 165 (1950).

214. F. L. Moore and S. A. Reynolds, *Anal. Chem.*, 29, 1596 (1957).

215. A. Classen and L. Bastings, *Z. Anal. Chem.*, 160, 403 (1958).

216. W. Doll and H. Specker, *Z. Anal. Chem.*, 161, 354 (1958).

217. C. R. Boswell and R. R. Brooks, *Anal. Chim. Acta.*, 33, 117 (1965).

218. K. A. Orlandini, M. O. Wahlgren and J. Barclay, *Anal. Chem.*, 37, 1148 (1965).

219. A. G. Karabash, L. I. Moseev and V. A. Kuznetsov, *Zh. Neorg. Khim.*, **5**, 1358 (1960).

220. L. I. Moseev and A. G. Karabash, *Zh. Neorg. Khim.*, **6**, 1944 (1961).

221. A. B. Sokolov, L. I. Moseev and A. G. Karabash, *Zh. Neorg. Khim.*, **6**, 994 (1961).

222. R. M. Diamond, *J. Phys. Chem.*, **61**, 69, 1522 (1957); **63**, 659 (1959).

223. J. Saldick, *J. Phys. Chem.*, **60**, 500 (1956).

224. A. G. Maddock, W. Smulek and A. J. Trench, *Trans. Faraday Soc.* **58**, 923 (1962).

225. A. F. Morgunov and V. V. Fomin, *Zh. Neorg. Khim.*, **8**, 508 (1963).

226. I. Wada and R. Ishii, *Sci. Paper Inst. Phys. Chem. Res. Tokyo*, **24**, 135 (1934).

227. E. Bankmann and H. Specker, *Z. Anal. Chem.*, **162**, 18 (1958).

228. D. E. Metzler and R. J. Myers, *J. Am. Chem. Soc.*, **72**, 3776 (1950).

229. I. Nelidov and R. M. Diamond, *J. Phys. Chem.*, **59**, 710 (1955).

230. H. M. Irving and F. J. C. Rossotti, *Analyst*, **77**, 801 (1952).

231. R. W. Dodson, C. J. Forney and E. H. Swift, *J. Am. Chem. Soc.*, **58**, 2573 (1936).

232. R. R. Brooks and P. J. Lloyd, *Nature*, **189**, 375 (1961).

233. V. I. Kuznetsov, *Zh. Obshch. Khim.*, **17**, 175 (1947).

234. R. M. Walters and R. W. Dodson in *Solvent Extraction Chemistry* (Eds. D. Dyrssen, J. O. Liljenzin and J. Rydberg) North-Holland, Amsterdam, 1967, p. 71.

235. N. H. Nachtrieb and R. E. Fryxell, *J. Am. Chem. Soc.*, **74**, 897 (1952).

236. H. Irving and F. J. C. Rossotti, *J. Chem. Soc.*, 1938 (1955).

237. D. E. Campbell, H. M. Clark and W. H. Bauer, *J. Phys. Chem.*, **62**, 506 (1958).

238. C. A. Kraus and R. M. Fuoss, *J. Am. Chem. Soc.*, **55**, 2387 (1933).

239. G. R. Hertel and H. M. Clark, *J. Phys. Chem.*, **65**, 1930 (1961).

240. V. V. Fomin and A. F. Morgunov, *Zh. Neorg. Khim.*, **5**, 1385 (1960).

241. R. L. Erickson and R. L. McDonald, *J. Am. Chem. Soc.*, **88**, 2099 (1966).

242. D. A. Meyers and R. L. McDonald, *J. Am. Chem. Soc.*, **89**, 486 (1967).

243. Y. A. Zolotov, I. V. Seryakova, A. V. Karyakin, L. A. Girbov and M. E. Zubrilina, *Zh. Neorg. Khim.*, **7**, 1197, 7, 2013 (1962); **8**, 481 (1963).

244. I. P. Alimarin, Y. A. Zolotov, A. V. Karyakin, A. V. Petrov and A. I. Sukhanovskaya, *Zh. Neorg. Khim.*, **10**, 524 (1965).

245. H. Specker and E. Jackwerth, *Z. Anal. Chem.*, **167**, 416 (1959).

246. H. Irving and F. J. C. Rossotti, *J. Chem. Soc.*, 1475 (1956).

247. A. T. Casey and A. G. Maddock, *Trans. Faraday Soc.*, **58**, 918 (1962).

248. B. V. Nekrasov and V. V. Ovsyankina, *Zh. Obshch. Khim.*, **11**, 573 (1941).

249. N. H. Nachtrieb and R. E. Fryxell, *J. Am. Chem. Soc.*, **71**, 4035 (1949).

250. A. G. Maddock and Z. B. Maksimovic, in *Solven Extraction Chemistry* (Eds. D. Dyrssen, J. D. Liljenzin and J. Rydberg), North-Holland, Amsterdam, 1967, p. 79.

251. H. L. Friedman, *J. Am. Chem. Soc.*, **74**, 5 (1952).

252. L. A. Woodward and P. T. Bill, *J. Chem. Soc.*, 1699 (1955).

253. L. A. Woodward and M. J. Taylor, *J. Chem. Soc.*, 4473 (1960).

254. Y. A. Zolotov, I. P. Alimarin and A. I. Sukhanovskaya, *Zh. Analit. Khim.*, **20**, 165 (1965).

255. D. L. Horroks and A. F. Voigt, *J. Am. Chem. Soc.*, **79**, 1440 (1957).

256. G. Nord (Waind) and J. Ulstrup, *Acta Chem. Scand.*, **18**, 307 (1964).

257. E. Rabinovich, *J. Am. Chem. Soc.*, **64**, 334 (1942).

258. L. M. Gindin, I. F. Kopp, A. M. Rozen, P. I. Bobikov, E. F. Kouba and N. A. Ter-oganesov, *Zh. Neorg. Khim.*, **5**, 149 (1960).

259. G. E. Boyd and Q. V. Larson, *J. Phys. Chem.*, **64**, 988 (1960).

260. D. I. Ryabchikov, Yu. B. Gerlit, A. V. Karyakin, V. A. Zarinskii and M. E. Zubrilina, *Dokl. Akad. Nauk SSSR*, **144**, 585 (1962).

261. H. Specker and G. Hovermann, *Z. Anorg. Allgem. Chem.*, **316**, 247 (1962).

262. C. H. Brubaker, Jr. and C. E. Johnson, *J. Inorg. Nucl. Chem.*, **9**, 189 (1959).

263. H. Specker and E. Jackwerth, *Naturwissenschaften*, **46**, 262 (1959).

264. R. A. Sharp and G. Wilkinson, *J. Am. Chem. Soc.*, **77**, 6519 (1955).

265. R. T. Golovatenko and O. Y. Samoilov, *Radiokhimiya*, **4**, 25 (1962).

266. A. S. Kertes in *Solvent Extraction Chemistry of Metals* (Eds. H.A.C. McKay, T. V. Healy, I. L. Jenkins and A. Naylor), Macmillan, London, 1966 p. 377.

267. K. S. Wallach, *Israel AEC Rept.*, LS/70 (1963).

268. L. L. Burger, *Nucl. Sci. Eng.*, **16**, 428 (1963).

269. J. R. Van Waser, *Phosphorus and its Compounds*, Interscience Publ., New York, 1958.

270. G. M. Kosolapoff, *Organophosphorus Compounds*, J. Wiley and Sons, New York, 1958.

271. P. C. Crofts, *Quart. Rev. (London)*, **12**, 341 (1948).

272. T. H. Siddall, III and C. A. Prohaska, *J. Am. Chem. Soc.*, **84**, 2502 (1962).

273. T. H. Siddall, III and C. A. Prohaska, *J. Am. Chem. Soc.*, **84**, 3467 (1962).

274. T. H. Siddall, III, *USAEC Rept.*, DP-548 (1961).

275. B. N. Laskorin, D. I. Skorovarov, E. A. Philipov and A. I. Shilin, *Radiokhimiya*, **5**, 424 (1963).

276. C. E. Higgins, W. H. Baldwin and J. M. Ruth, *USAEC Rept.*, ORNL–1338 (1952).

277. L. L. Burger and R. M. Wagner, *Ind. Eng. Data Series*, **3**, 310 (1958).

278. H. Bostian and M. Smutz, *J. Inorg. Nucl. Chem.*, **26**, 825 (1964); H. Bostian, *Dissertation Abstr.*, **20**, 2723 (1960).

279. C. E. Higgins and W. H. Baldwin, *J. Org. Chem.*, **21**, 1156 (1956).

280. J. W. O'Laughlin in *Progress in Nuclear Energy, Series IX* (Eds. D. C. Stewart and H. A. Elion), Vol. 6, Pergamon Press, London, 1966, Chap. 2.

281. J. W. O'Laughlin and C. V. Banks, in *Solvent Extraction Chemistry* (Eds. D. Dyrssen, J. O. Liljenzin and J. Rydberg), North-Holland, Amsterdam, 1967, p. 270.

282. T. H. Siddall, III, *J. Inorg. Nucl. Chem.*, **25**, 883 (1963); in *Solvent Extraction Chemistry* (Eds. D. Dyrssen, J. O. Liljenzin and J. Rydberg), North-Holland Amsterdam, 1967, p. 501; T. H. Siddall, III and C. A. Prohaska, *Inorg. Chem.*, **4**, 783 (1965).

283. G. P. Nikitina and M. F. Pushlenkov, *Radiokhimiya*, **5**, 456 (1963).

284. T. H. Handley and J. A. Dean, *Anal. Chem.*, **33**, 1087 (1961).

285. T. H. Handley, *Nucl. Sci. Eng.*, **16**, 440 (1963).

286. A. J. Fudge and J. L. Woodhead, *Analyst.*, **81**, 417 (1956).

287. A. J. Fudge and J. L. Woodhead, *Chem. Ind. (London)*, 1122 (1957).

288. A. G. Goble and A. G. Maddock, *J. Inorg. Nucl. Chem.*, **7**, 94 (1958).

289. T. H. Siddall, III and R. M. Wallace, *USAEC Rept.*, DP–286 (1958).

290. N. M. Adamskii, S. M. Karpacheva, I. N. Melnikov and A. M. Rozen, *Radiokhimiya*, **2**, 400 (1960).

291. C. A. Blake, A. T. Gresky and J. M. Schmitt, *USAEC Rept.*, ORNL CF–60–5–37 (1960).

292. P. Leroy, *French Rept.*, CEA-R 3207 (1967).

293. W. J. McDowell, *USAEC Rept.*, ORNL-TM–1893 (1967).

294. J. Kennedy and S. S. Grimley, *Brit. Rept. AERE*, CE/R–968 (1957).

295. Y. A. Pentin, E. G. Teterin, N. N. Shesterikov, *Zh. Analit. Khim.*, **17**, 239 (1962).

296. U. A. Sant and H. Sankar Das, *Anal. Chim. Acta*, **19**, 202 (1958).

297. Anonymous, *Brit. Rept.*, PG 116 (W) (1960).

298. R. J. Allen and M. A. De Sesa, *Nucleonics*, **15**, 88 (1957).

299. W. E. Shuler, *USAEC Rept.*, DP–449 (1959).

300. W. H. Baldwin and C. E. Higgins, *Anal. Chem.*, **30**, 446 (1958).

301. W. E. Shuler and R. C. Axtmann, *USAEC Rept.*, DP–474 (1960).

302. F. A. Cotton, R. D. Barnes and E. Bannister, *J. Chem. Soc.*, 2199 (1960).
303. G. P. Nikitina and M. F. Pushlenkov, *Radiokhimiya*, **5**, 436 (1963).
304. K. Nukada, K. Naito and U. Maeda, *Bull. Chem. Soc. Japan*, **33**, 894 (1960).
305. L. C. Thomas and R. A. Chittenden, *Chem. Ind. (London)*, 1913 (1961).
306. L. C. Thomas and R. A. Chittenden, *Talanta*, **9**, 86 (1962).
307. C. A. Horton and J. C. White, *Talanta*, **7**, 215 (1961).
308. G. K. Estok and W. W. Wendlandt, *J. Am. Chem. Soc.*, **77**, 4767 (1955).
309. A. E. Arbuzov and P. I. Rakov, *Izv. Akad. Nauk. SSSR, Otd. Khim. Nauk*, 237 (1950).
310. K. Oshima, *Nippon Genshiryoku Gakkaishi*, **4**, 8 (1962).
311. H. A. McKay and T. V. Healy, *Progr. Nucl. Energy, Process Chemistry*, Series III, Vol. 2, Pergamon Press, London, 1959. p. 546.
312. J. C. White and W. J. Ross, *Nucl. Sci. Ser.*, NAS–NS 3102 (1961).
313. A. S. Solovkin, *Zh. Neorg. Khim.*, **5**, 1107 (1960).
314. C. E. Higgins and W. H. Baldwin, *J. Inorg. Nucl. Chem.*, **24**, 112 (1962).
315. C. E. Higgins, W. H. Baldwin and B. A. Soldano, *J. Phys. Chem.*, **63**, 113 (1959).
316. C. E. Higgins, W. H. Baldwin and R. W. Higgins, *USAEC Rept.*, ORNL–2782 (1959).
317. J. Kennedy and S. S. Grimley, *Brit. Rept. AERE*, CE/R–1283 (1953).
318. K. Alcock, S. S. Grimley, T. V. Healy and J. Kennedy, *Trans. Faraday Soc.*, **52**, 39 (1956).
319. T. Ishimori and T. Fujino, *Nippon Genshiryoku Gakkaishi*, **3**, 276 (1961).
320. A. S. Solovkin, *Zh. Neorg. Khim.*, **5**, 1345 (1960).
321. D. F. C. Morris and E. L. Short, *J. Inorg. Nucl. Chem.*, **25**, 291 (1963).
322. C. E. Higgins and W. H. Baldwin, *Anal. Chem.*, **32**, 233 (1960).
323. C. E. Higgins and W. H. Baldwin, *Anal. Chem.*, **32**, 236 (1960).
324. W. F. Johnson and R. L. Dillon, *USAEC Rept.*, HW–29086 (1953).
325. S. Siekierski, *J. Inorg. Nucl. Chem.*, **24**, 205 (1962).
326. Y. Marcus, *J. Phys. Chem.*, **65**, 1647 (1961).
327. R. A. Zingaro and J. C. White, *J. Inorg. Nucl. Chem.*, **12**, 315 (1960).
328. W. Davis, Jr., J. Mrochek and C. J. Hardy, *J. Inorg. Nucl. Chem.*, **28**, 2001 (1966); W. Davis, Jr. and J. Mrochek in *Solvent Extraction Chemistry* (Eds. D. Dyrssen ,J. O. Liljenzin and J. Rydberg), North-Holland, Amsterdam, 1967, p. 283.
329. V. B. Shevchenko, A. S. Solovkin, I. V. Shilin, L. M. Kirilov, A. V. Rodionov and V. V. Balandina, *Radiokhimiya*, **2**, 281 (1960).
330. M. F. Pushlenkov and O. N. Shuvalov, *Radiokhimiya*, **5**, 536 (1963).
331. M. F. Pushlenkov and O. N. Shuvalov, *Radiokhimiya*, **5**, 651 (1963).

332. T. V. Healy, J. Kennedy and G. M. Waind, *J. Inorg. Nucl. Chem.*, **10**, 137 (1959).

333. K. Shwabe and K. Wiesener, *Z. Elektrochem.*, **68**, 39 (1962).

334. A. A. Nemodruk and L. P. Glukhova, *Zh. Neorg. Khim.*, **8**, 2618 (1963).

335. S. Siekierski, *Nukleonika*, **9**, 601 (1964).

336. V. M. Vdovenko, T. V. Kovaleva and M. A. Ryazanov, *Radiokhimiya* **7**, 133 (1965).

337. G. M. Kosolapoff and J. F. McCullough, *J. Am. Chem. Soc.*, **73**, 5392 (1951).

338. M. F. Pushlenkov, E. V. Komarov and O. N. Shuvalov, *Radiokhimiya*, **2**, 537 (1960).

339. D. Dyrssen and Dj. M. Petkovic, *J. Inorg. Nucl. Chem.*, **27**, 1381 (1965); Dj. M. Petkovic, in *Solvent Extraction Chemistry* (Eds. D. Dyrssen, J. O. Liljenzin and J. Rydberg) North-Holland, Amsterdam, 1967, p. 305; *J. Inorg. Nucl. Chem.*, **30**, 603 (1968).

340. M. F. Pushlenkov and E. V. Komarov, *Radiokhimiya*, **6**, 426 (1964).

341. A. L. Geddes, *J. Phys. Chem.*, **58**, 1062 (1959).

342. W. Davis, Jr., *Nucl. Sci. Eng.*, **14**, 159, 169, 174 (1962).

343. A. Apelblat and A. Hornik, in *Solvent Extraction Chemistry* (Eds. D. Dyrssen, J. O. Liljenzin and J. Rydberg) North-Holland, Amsterdam, 1967, p. 296.

344. V. V. Fomin and T. I. Rudenko, *Radiokhimiya*, **7**, 33 (1965).

345. C. D. Miller, C. R. Miller and W. Rodgers, Jr., *J. Am. Chem. Soc.*, **80**, 1562 (1958).

346. J. Michalczyk, *Nukleonika*, **8**, 237 (1963).

347. H. A. C. McKay, *Ind. Chim. Belge*, **12**, 1278 (1964).

348. C. J. Hardy, D. Fairhurst, H. A. C. McKay and A. M. Willson, *Trans. Faraday Soc.*, **60**, 1626 (1964).

349. E. W. Nadig and M. Smutz, *USAEC Rept.*, IS-595 (1963).

350. L. I. Katzin, *J. Inorg. Nucl. Chem.*, **20**, 300 (1961).

351. W. E. Shuler, *USAEC Rept.*, DP-513 (1960).

352. D. C. Whitney and R. M. Diamond, *J. Phys. Chem.*, **67**, 209 (1963).

353. E. Brauer and E. Hogfeldt, *J. Inorg. Nucl. Chem.*, **23**, 115 (1961).

354. V. M. Vdovenko, L. M. Balov and A. A. Chaikhovskii, *Radiokhimiya*, **1**, 439 (1959).

355. A. M. Rozen in *Ekstraktsiya*, Vol. 1, Gosatomisdat, Moscow, 1962, p. 6.

356. A. V. Nikolaev, K. E. Mironov and E. V. Karaseva, *Dokl. Akad. Nauk SSSR*, **147**, 380 (1962).

357. E. Bullock and D. G. Tuck, *Trans. Faraday Soc.*, **59**, 1293 (1963).

358. D. R. Olander and M. Benedict, *Nucl. Sci. Eng.*, **14**, 287 (1962).

359. T. H. Siddall III, *J. Phys. Chem.*, **64**, 1340 (1960).

360. T. V. Healy and P. E. Brown, *Brit. Rept.*, *AERE* C/R 1970 (1956).

361. M. Halpern, T. Kim and N. C. Li, *J. Inorg. Nucl. Chem.*, **24**, 1251 (1962).

362. A. L. Mills and W. R. Logan, *J. Inorg. Nucl. Chem.*, **26**, 2191 (1964).

363. D. R. Olander, L. Donadien and M. Benedict, *A. I. Ch. E. J.*, **7**, 152 (1961).

364. B. B. Murray and R. C. Axtmann, *Anal. Chem.*, **31**, 450 (1959).

365. D. G. Tuck, *Trans. Faraday Soc.*, **57**, 1297 (1961).

366. R. C. Axtmann, *Nucl. Sci. Eng.*, **16**, 241 (1963).

367. D. R. Olander, *Nucl. Sci. Eng.*, **16**, 243 (1963).

368. L. L. Burger and E. C. Martin, 139*th Natl. Mtg. Am. Chem. Soc.*, 1961.

369. J. J. Richard, K. E. Buike, J. W. O'Laughlin and C. V. Banks, *J. Am. Chem. Soc.*, **83**, 1722 (1961).

370. J. W. Roddy and J. Mrochek, *J. Inorg. Nucl. Chem.*, **28**, 3019 (1966).

371. W. Davis, Jr. and H. J. de Bruin, *J. Inorg. Nucl. Chem.*, **26**, 1069 (1964); W. Davis, Jr., *USAEC Rept.*, ORNL–3452, (1963) p. 217.

372. J. E. Mrochek, J. J. Richard and C. V. Banks, *J. Inorg. Nucl. Chem.*, **27**, 625 (1965).

373. K. Naito and T. Suzuki, *J. Phys. Chem.*, **66**, 983 (1962).

374. J. H. Milles, *J. Inorg. Nucl. Chem.*, **27**, 711 (1965).

375. D. G. Tuck and R. M. Diamond, *J. Phys. Chem.*, **64**, 886 (1960).

376. D. C. Whitney and R. M. Diamond, *J. Phys. Chem.*, **67**, 2583 (1963).

377. M. I. Tocher, D. C. Whitney and R. M. Diamond, *J. Phys. Chem.*, **68**, 368 (1964).

378. D. C. Whitney and R. M. Diamond, *USAEC Rept.*, UCRL–10914 (1963).

379. P. Biddle, A. Coe, H. A. C. McKay, J. H. Miles and M. J. Waterman, *J. Inorg. Nucl. Chem.*, **29**, 2615 (1967).

380. T. H. Siddall, III, *USAEC Rept.*, DP–181 (1956).

381. J. M. Fletcher and C. J. Hardy, *Nucl. Sci. Eng.*, **16**, 421 (1963).

382. R. K. Klopfenstein, W. P. Tolos and T. J. Collopy, *USAEC Rept.*, NLCO–825 (1961).

383. T. J. Collopy and J. F. Blum, *J. Phys. Chem.*, **64**, 1324 (1960).

384. T. J. Collopy and J. H. Cavendish, *J. Phys. Chem.*, **64**, 1328 (1960).

385. Z. A. Sheka and E. E. Kriss, *Zh. Neorg. Khim.*, **4**, 2505 (1959).

386. V. V. Fomin and E. P. Maiorova, *Zh. Neorg. Khim.*, **3**, 540 (1958).

387. G. F. Egorov, V. V. Fomin, Yu. G. Frolov and G. A. Yagodin, *Zh. Neorg. Khim.*, **5**, 1044 (1960).

388. V. V. Fomin, R. E. Kartushova and E. P. Maiorova, *Zh. Neorg. Khim.*, **5**, 1337 (1960).

389. V. B. Shevchenko and I. A. Fedorov, *Radiokhimiya*, **2**, 6 (1960).

390. N. M. Adamskii, S. M. Karpacheva, I. N. Melnikov and A. M. Rozen, *Radiokhimiya*, **2**, 13 (1960).

391. A. P. Smirnov-Averin, G. S. Kovalenko and N. N. Krot, *Zh. Neorg. Khim.*, **8**, 2400 (1963).

392. D. F. Peppard and J. R. Ferraro, *J. Inorg. Nucl. Chem.*, **15**, 365 (1960).

393. M. F. Pushlenkov and O. N. Shuvalov, *Radiokhimiya*, **5**, 543 (1963).

394. V. I. Zemlyanukhin, G. P. Savoskina and M. F. Pushlenkov, *Zh. Neorg. Khim.*, **6**, 694 (1964).

395. V. B. Shevchenko, N. S. Povitzkii, A. S. Solovkin, I. V. Shilin, K. P. Lunichkina and Z. N. Tzvetkova, *Zh. Neorg. Khim.*, **3**, 2109 (1958).

396. A. M. Rozen and L. P. Khorkhorina, *Zh. Neorg. Khim.*, **2**, 1956 (1957); **3**, 549 (1958).

397. S. S. Korovin, E. N. Lebedeva, K. Dedich, A. M. Reznik and A. M. Rozen, *Zh. Neorg. Khim.*, **10**, 518 (1965).

398. V. B. Shevchenko, I. G. Slepchenko, V. S. Schmidt and E. A. Nenarokomov, *At. Energ. (USSR)*, **7**, 236 (1959).

399. T. H. Siddall, III, *Ind. Eng. Chem.*, **51**, 41 (1959); *J. Inorg. Nucl. Chem.*, **13**, 151 (1960).

400. G. A. Yagodin, G. E. Kaplan, O. A. Mostovaya, S. D. Moiseev and L. P. Dmitrieva, *Zh. Neorg. Khim.*, **8**, 1973 (1963).

401. T. H. Siddall, III. *Proc. U.N. Intern. Conf. Peaceful Uses At. Energy*, *2nd*, **17**, p. 339 (1958); USAEC Rept., DP-548 (1961).

402. T. H. Siddall, III, *J. Am. Chem. Soc.*, **81**, 4176 (1959).

403. A. H. A. Heyn and Y. D. Soman, *J. Inorg. Nucl. Chem.*, **26**, 287 (1964).

404. B. Martin, D. W. Ockenden and J. K. Foreman, *J. Inorg. Nucl. Chem.*, **21**, 96 (1961).

405. J. E. Mrochek, J. W. O'Laughlin, H. Sakurai and C. V. Banks, *J. Inorg. Nucl. Chem.*, **25**, 955 (1963).

406. J. E. Mrochek and C. V. Banks, *J. Inorg. Nucl. Chem.*, **27**, 589 (1965).

407. D. E. Horner and C. F. Coleman, *USAEC Rept.*, ORNL-3051 (1961).

408. P. E. Burns and C. Hanson, *J. Appl. Chem.*, **14**, 117 (1964).

409. L. Damiani and V. Fattore, *Energia Nucl. (Milan)*, **6**, 793 (1959).

410. B. Bernstrom and J. Rydberg, *Acta Chem. Scand.*, **11**, 1173 (1957).

411. S. M. Karpacheva, L. P. Khorkhorina and A. M. Rozen, *Zh. Neorg. Khim.*, **2**, 1441 (1957).

412. T. J. Collopy, *USAEC Rept.*, NLCO-749 (1958).

413. W. Korpak and C. Deptula, *Nukleonika*, **5**, 63 (1960).

414. A. S. Solovkin, N. S. Povitskii and J. V. Shilin. *Zh. Neorg. Khim.*, **4**, 1454 (1959).

415. C. J. Hardy, *Brit. Rept.*, AERE R-3124 (1959).

416. Z. Dizdar and J. Rajnvajn, *Bull. Inst. Nucl. Sci. 'Boris Kidrich'*, **11**, 181 (1961).

417. D. R. Olander and M. Benedict, *Nucl. Sci. Eng.*, **15**, 354 (1963).

418. E. Hesford and H. A. C. McKay, *J. Inorg. Nucl. Chem.*, **13**, 156 (1960).

419. A. M. Rozen, A. M. Reznik, S. S. Korovin and Z. A. Metonidze, *Zh. Neorg. Khim.*, **8**, 1003 (1963).

420. P. J. Kinney and M. Smutz, *USAEC Rept.*, IS–728 (1963).

421. W. H. Baldwin, C. E. Higgins and B. A. Soldano, *J. Phys. Chem.*, **63**, 118 (1959).

422. D. G. Tuck and R. M. Diamond, *Proc. Chem. Soc.*, 236 (1958).

423. D. G. Tuck, *J. Chem. Soc.*, 2783 (1958).

424. I. A. Apraksin, S. S. Korovin, A. M. Reznik and A. M. Rozen, *Zh. Neorg. Khim.*, **8**, 137 (1963).

425. H. Irving and D. N. Edgington, *J. Inorg. Nucl. Chem.*, **10**, 306 (1959).

426. D. F. Peppard, G. W. Mason and J. L. Maier, *J. Inorg. Nucl. Chem.*, **3**, 215 (1956).

427. A. S. Kertes, *J. Inorg. Nucl. Chem.*, **14**, 104 (1960).

428. E. Foa, N. Rosintal and Y. Marcus, *J. Inorg. Nucl. Chem.*, **23**, 103 (1961).

429. I. J. Gal and A. Ruvarac, *Bull. Inst. Nucl. Sci. 'Boris Kidrich'*, **8**, 67 (1958).

430. E. A. Nenarokomov, A. S. Solovkin and V. S. Shmidt, *Zh. Neorg. Khim.*, **6**, 509 (1961).

431. G. P. Nikitina and M. F. Pushlenkov, *Radiokhimiya*, **5**, 445 (1963).

432. D. G. Tuck, *J. Chem. Soc.*, 3905 (1963).

433. A. S. Kertes and P. J. Lloyd, *J. Chem. Soc.*, 3477 (1965).

434. V. B. Shevchenko, I. V. Shilin and A. S. Solovkin, *Zh. Neorg. Khim.*, **3**, 225 (1958).

435. N. S. Povitskii, A. S. Solovkin and I. V. Shilin, *Zh. Neorg. Khim.*, **3**, 222 (1958).

436. S. Siekierski and R. Gwozdz, *Nukleonika*, **5**, 205 (1960).

437. A. S. Kertes and V. Kertes, *J. Appl. Chem.*, **10**, 287 (1960).

438. V. V. Fomin and E. P. Maiorova, *Zh. Neorg. Khim.*, **5**, 1100 (1960).

439. V. M. Vdovenko, A. A. Lipovski and S. A. Nikitina, *Zh. Neorg. Khim.* **5**, 1337 (1960).

440. T. J. Conocchioli, M. I. Tocher and R. M. Diamond, *USAEC Rept.*, UCRL–10913 (1963); *J. Phys. Chem..* **69**, 1106 (1965).

441. J. M. Fletcher, D. Scargill and J. L. Woodhead, *J. Chem. Soc.*, 1705 (1961).

442. A. S. Kertes and V. Kertes, *Can. J. Chem.*, **38**, 612 (1960).

443. V. B. Shevchenko and Yu. F. Zhdanov, *Radiokhimiya*, **3**, 7 (1961).

732   *Ion Exchange and Solvent Extraction of Metal Complexes*

444. R. Mitamura, S. Nishimura and Y. Kondo, *Mem. Fac. Eng. Kyoto Univ.*, **28**, 198 (1966).
445. A. S. Solovkin, *Zh. Neorg. Khim.*, **5**, 1857 (1960).
446. H. Yoshida, *J. Inorg. Nucl. Chem.*, **24**, 1257 (1962).
447. O. A. Sinegribova and G. A. Yagodin, *Zh. Neorg. Khim.*, **10**, 1250 (1965).
448. H. A. Pagel and F. W. McLafferty, *Anal. Chem.*, **20**, 272 (1948).
449. H. A. Pagel and K. D. Schwab, *Anal. Chem.*, **22**, 1207 (1950).
450. V. B. Shevchenko, E. V. Renard and A. S. Solovkin, *Zh. Neorg. Khim.*, **5**, 2350 (1960).
451. V. B. Shevchenko and E. V. Renard, *Zh. Neorg. Khim.*, **8**, 516 (1963).
452. Z. N. Tsvetkova and N. S. Povitskii, *Zh. Neorg. Khim.*, **5**, 2827 (1960).
453. D. Hadzi and N. Kobilarov. *J. Chem. Soc. (A).* 439 (1966).
454. E. M. Indikov, A. S. Solovkin, E. G. Teterin and N. N. Shesterikov, *Zh. Neorg. Khim.*, **8**, 2187 (1963).
455. B. Weaver, F. A. Kappelmann and A. C. Topp, *J. Am. Chem. Soc.*, **75**, 3943 (1953).
456. B. O. Field and C. J. Hardy, *Quart. Rev. (London)*, **16**, 361 (1964).
457. T. V. Healy and B. L. Davies, *Brit. Rept.*, *AERE* C/R 1971 (1956).
458. W. W. Wendlandt and J. M. Bryant, *J. Phys. Chem.*, **60**, 1145 (1956).
459. W. W. Wendlandt and J. M. Bryant, *Science*, **123**, 1121 (1956).
460. V. B. Shevchenko, I. V. Shilin and Yu. F. Zhdanov, *Zh. Neorg. Khim.*, **5**, 1366 (1960).
461. A. V. Nikolaev, Y. A. Dyadin and I. I. Yakovlev, *Dokl. Akad. Nauk SSSR*, **158**, 1130 (1964).
462. V. B. Shevchenko, I. V. Shilin and Yu. F. Zhadanov, *Zh. Neorg. Khim.*, **5**, 2832 (1960).
463. A. S. Solovkin, M. I. Konarev and D. P. Adaev, *Zh. Neorg. Khim.*, **5**, 1861 (1960).
464. J. E. Mrochek, J. W. O'Laughlin and C. V. Banks, *J. Inorg. Nucl. Chem.*, **27**, 603 (1965).
465. J R. Parker and C. V. Banks, *J. Inorg. Nucl. Chem.*, **27**, 631 (1965).
466. F. A. Cotton and E. Bannister, *J. Chem. Soc.*, 1873, 1878, 2276 (1960).
467. O. A. Osipov, V. I. Gaivoronski and A. A. Shvets, *Zh. Neorg. Khim.*, **8**, 2190 (1963).
468. L. I. Katzin, *J. Inorg. Nucl. Chem.*, **24**, 245 (1962).
469. J. R. Ferraro, *J. Inorg. Nucl. Chem.*, **10**, 319 (1959).
470. K. W. Bagnall, D. Brown and J. G. H. du Precz, *J. Chem. Soc.*, 5217 (1965).
471. D. Brown, J. F. Easey and J. G. H. du Precz, *J. Chem. Soc. (A)*, 258 (1966).

472. A. K. Majumdar and R. G. Bhattachryya, *J. Inorg. Nucl. Chem.*, **29**, 2359 (1967).

473. D. R. Cousins and F. A. Hart, *J. Inorg. Nucl. Chem.*, **29**, 1745 (1967).

474. J. L. Woodhead, *J. Inorg. Nucl. Chem.*, **26**, 1472 (1964).

475. I. Abrahamer and Y. Marcus, *Inorg. Chem.*, **6**, 1203 (1967).

476. S. Niese, *Kernenergie*, **3**, 554 (1960).

477. D. G. Tuck, *Coordin. Chem. Rev.*, **1**, 286 (1966).

478. A. J. Carty and D. G. Tuck, *J. Chem. Soc. (A)*, 1081 (1966).

479. A. V. Nikolaev and S. M. Shubina, *Zh. Neorg. Khim.*, **6**, 799 (1961).

480. L. L. Burger, *J. Phys. Chem.*, **62**, 590 (1958).

481. V. B. Shevchenko, A. V. Radionov, A. S. Solovkin, I. V. Shilin, L. M. Kirilov and U. V. Balandina, *Radiokhimiya*, **1**, 257 (1959).

482. J. E. Fleming and H. Lynton, *Chem. Ind. (London)*, 1415 (1960).

483. J. J. v. Aartsen and A. E. Korvezee, *Trans. Faraday Soc.*, **60**, 510 (1964).

484. J. R. Parker and C. V. Banks, *J. Inorg. Nucl. Chem.*, **27**, 583 (1965).

485. N. R. Geary, *Brit. Rept.*, 8142 (1959).

486. T. Ishimori and E. Nakamura, *Japan. Rept.*, JAERI–1047 (1963).

487. E. Cerrai and C. Testa, *J. Chromatog.*, **7**, 112 (1962).

488. T. Ishimori, K. Kimura, T. Fujino and H. Murakami, *Nippon Genshiryoku Gakkaishi*, **4**, 117 (1962).

489. T. Ishimori, K. Kimura, E. Nakamura, J. Akatsu and T. Kobune, *Nippon. Genshiryoku Gakkaishi*, **5**, 633 (1963).

490. T. V. Healy and J. Kennedy, *J. Inorg. Nucl. Chem.*, **10**, 128 (1959).

491. H. Saisho, *Bull. Chem. Soc. Japan.*, **34**, 1254 (1961).

492. K. Alcock, G. F. Best, E. Hesford and H. A. C. McKay, *J. Inorg. Nucl. Chem.*, **6**, 328 (1958).

493. T. V. Healy and H. A. C. McKay, *Trans. Faraday Soc.*, **52**, 633 (1956).

494. E. Hesford, H. A. C. McKay and D. Scargill, *J. Inorg. Nucl. Chem.*, **4**, 321 (1957).

495. D. C. Madigan and R. W. Cattrall, *J. Inorg. Nucl. Chem.*, **21**, 334 (1961).

496. V. V. Fomin and E. P. Maiorova, *Zh. Neorg. Khim.*, **1**, 1703 (1956).

497. V. P. Ionov and V. I. Tichomirov, *Radiokhimiya*, **5**, 559 (1963).

498. J. E. Savolainen, *USAEC Rept.*, CF–52–2–113 (1952).

499. J. L. Woodhead and H. A. C. McKay, *J. Inorg. Nucl. Chem.*, **27**, 2244 (1965).

500. J. M. Fletcher, C. E. Lyon and A. G. Wain, *J. Inorg. Nucl. Chem.*, **27**, 1841 (1965).

501. V. D. Nikolskii and V. S. Shmidt, *Zh. Neorg. Khim.*, **3**, 2467 (1958).

502. H. A. C. McKay and R. J. W. Streeton, *J. Inorg. Nucl. Chem.*, **27**, 879 (1965).

503. E. Hesford and H. A. C. McKay, *Trans. Faraday Soc.*, **54**, 573 (1958).

504. E. V. Komarov and M. F. Pushlenkov, *Radiokhimiya*, **3**, 567, 575 (1961).

505. G. F. Best, E. Hesford and H. A. C. McKay, *J. Inorg. Nucl. Chem.*, **12**, 136 (1959).

506. T. Ishimori, K. Watanabe and E. Nakamura, *Bull. Chem. Soc. Japan*, **33**, 636 (1960).

507. T. Ishimori, K. Watanabe and T. Fujino, *Nippon Genshiryoku Gakkaishi* **3**, 19 (1961).

508. V. B. Shevchenko, I. G. Shepchenko, V. S. Smidt and E. A. Nenarokomov, *Zh. Neorg. Khim.*, **5**, 1095 (1960).

509. V. B. Shevchenko, V. S. Shmidt and E. A. Nenarokomov, *Radiokhimiya* **3**, 129 (1961).

510. M. Halpern, T. Kim, A. S. Kertes and N. C. Li, *Can. J. Chem.*, **42**, 878 (1964).

511. V. M. Vdovenko and A. A. Lipovskii, in *Solvent Extraction Chemistry* (Eds. D. Dyrssen J. O. Liljenzin and J. Rydberg), North-Holland, Amsterdam, 1967, p. 309.

512. V. M. Vdovenko, A. A. Lipovskii and S. A. Nikitina, *Zh. Neorg. Khim.*, **5**, 935 (1960).

513. H. Specker, E. Jackwerth and H. G. Kloppenburg, *Z. Anal. Chem.*, **183**, 81 (1964).

514. D. F. C. Morris, E. L. Short and D. N. Slater, *Electrochim. Acta*, **8**, 289 (1963).

515. V. I. Levin, I. V. Meshcherova and U. K. Samrov, *Radiokhimiya*, **3**, 417 (1961).

516. N. Saito and A. Yamasaki, *Bull, Chem. Soc. Japan*, **36**, 1055 (1963).

517. M. F. Pushlenkov, G. P. Nikitina and N. M. Simitsyn, *Radiokhimiya*, **2**, 215 (1960).

518. K. Naito and T. Suzuki, *J. Phys. Chem.*, **66**, 989 (1962).

519. D. F. C. Morris and C. F. Bell, *J. Inorg. Nucl. Chem.*, **10**, 337 (1959).

520. H. Yoshida, *J. Inorg. Nucl. Chem.*, **26**, 619 (1964).

521. V. B. Shevchenko, A. S. Solovkin and I. V. Shilin, *Zh. Neorg. Khim.*, **3**, 1965 (1958).

522. E. Hesford and H. A. C. McKay, *J. Inorg. Nucl. Chem.*, **13**, 165 (1960).

523. H. Specker, *Z. Anal. Chem.*, **197**, 109 (1963).

524. D. F. C. Morris, E. L. Short and D. N. Slater, *J. Inorg. Nucl. Chem.*, **26**, 627 (1964).

525. E. W. Berg and E. Y. Lan, *Anal. Chim. Acta*, **27**, 248 (1962).

526. K. H. Arend and H. Specker, *Z. Anorg. Allgem. Chem.*, **333**, 18 (1964).

527. D. G. Tuck and R. M. Walters, *J. Chem. Soc.*, 1111 (1963).

528. R. Colton, *UKAEA Rept.*, AERE-R 3823 (1961).

529. A. S. Kertes and A. Beck *J. Chem. Soc.*, 1921 (1961).

530. A. S. Kertes and A. Beck, *Proc. 7th. Intern. Conf. Coord. Chem.*, Stockholm, 1962 p. 352; A. Beck, *Ph. D. Thesis*, Hebrew University, Jerusalem, 1961.

531. A. V. Petrov, A. V. Karyakin and K. V. Marunova, *Zh. Neorg. Khim.*, **10**, 986 (1965).

532. E. P. Maiorova and V. V. Fomin, *Zh. Neorg. Khim.*, **3**, 1937 (1958).

533. R. Simard, A. J. Gilmore, U. M. McNamara, H. W. Parsons and H. W. Smith, *Can. J. Chem. Eng.*, **39**, 229 (1961).

534. T. J. Collopy, J. H. Cavendish, W. S. Miller and D. A. Stock, *USAEC Rept.*, NLCO-750 (1958).

535. T. J. Collopy and D. A. Stock, *USAEC Rept.*, NLCO-801 (1960).

536. S. S. Korovin, A. M. Reznik and I. A. Apraksin, *Zh. Neorg. Khim.*, **7**, 1483 (1962).

537. L. P. Varga and D. N. Hume, *Inorg. Chem.*, **3**, 77 (1964).

538. A. M. Reznik, A. M. Rozen, S. S. Korovin and I. A. Apraksin, *Radiokhimiya*, **5**, 49 (1963).

539. W. J. Maeck, G. L. Booman, M. C. Elliott and J. E. Rein, *Anal. Chem.*, **32**, 922 (1960).

540. A. S. Solovkin, *Zh. Neorg. Khim.*, **5**, 2119 (1960).

541. R. T. Golovatenko, *Zh. Neorg. Khim.*, **8**, 2395 (1963).

542. T. Sato, *J. Inorg. Nucl. Chem.*, **16**, 155 (1960).

543. K. B. Brown, *USAEC Rept.*, CF-59-9-85 (1959).

544. A. M. Rozen and E. I. Moiseenko in *Ekstraktsiya*, Vol. 2, Gosatomizdat, Moscow, 1962 p. 235.

545. S. Siekierski, *J. Inorg. Nucl. Chem.*, **12**, 129 (1959).

546. M. Tarnero, *French Rept.*, CEA-R 3206 (1967).

547. V. B. Shevchenko, A. S. Solovkin, L. M. Kirilov and A. I. Ivantsev, *Radiokhimiya*, **3**, 503 (1961).

548. Z. I. Dizdar, J. K. Rajnvajn and O. S. Gal, *Bull. Inst. Nucl. Sci. 'Boris Kidrich'*, **3**, 368 (1956).

549. M. Taube, *J. Inorg. Nucl. Chem.*, **12**, 174 (1959).

550. A. S. Solovkin, N. S. Povitskii and K. P. Lunichkina, *Zh. Neorg. Khim.*, **5**, 2115 (1960).

551. A. L. Mills and W. R. Logan, in *Solvent Extraction Cpemistry* (Eds. D. Dyrssen, J. D. Liljenzin and J. Rydberg), North-Holland, Amsterdam, 1967, p. 322.

552. L. L. Burger, *Prog. Nucl. Energy, Ser. III, Process Chem.*, Vol. 2, Pergamon Press, London, 1958, p. 307.

553. L. Salomon and E. Lopez Menchero, *Eurochemic Rept.*, ETR 203 (1967).

554. C. A. Blake, Jr. and J. M. Schmitt in *Solvent Extraction Chemistry of Metals*, (Eds. H. A. C. McKay, T. V. Healy, I. L. Jenkins and A. Naylor), Macmillan, London, 1966, p. 161; C. A. Black Jr., *USAEC Rept.*, ORNL–4212 (1968.

555. R. M. Wagner and L. H. Towle, *USAEC Rept.*, AECU–4300 (1959).

556. C. E. Higgins and W. H. Baldwin, *USAEC Rept.*, ORNL–2983 (1960), p. 42,

557. A. J. Moffat and R. D. Thompson, *J. Inorg. Nucl. Chem.*, **16**, 365 (1961).

558. E. Cherbuliez, *Chimia*, **15**, 327 (1961).

559. A. H. Samuel and W. E. Wilson, *USAEC Rept.*, SRIA–62 (1962).

560. T. Rigg and W. Wild, *Prog. Nucl. Energy, Ser. III, Process Chem.*, Vol. 2, Pergamon Press, London, 1958, p. 320.

561. J. G. Burr, *Radiation Res.* **8**, 214 (1958).

562. T. F. Williams and R. W. Wilkinson, *Nature*, **179**, 540 (1957).

563. V. P. Shvedov and S. P. Rosyanov, *Zh. Fiz. Khim.*, **35**, 569 (1961).

564. R. C. Propst and J. C. Simms, *USAEC Rept.*, DP–772 (1962).

565. A. S. Kertes and M. Halpern, *J. Inorg. Nucl. Chem.*, **20**, 117 (1961).

566. R. M. Wagner, E. M. Kinderman and L. H. Towle, *Ind. Eng. Chem.*, **51**, 45 (1959).

567. D. G. Tuck and R. M. Walters, *J. Chem. Soc.*, 4712 (1962).

568. S. Minc and T. Bryl, *Nukleonika*, **5**, 33 (1960).

569. C. F. Baes, Jr., R. A. Zingaro and C. F. Coleman, *J. Phys. Chem.*, **62**, 129 (1958).

570. A. S. Kertes and M. Halpern, *J. Inorg. Nucl. Chem.*, **19**, 359 (1961).

571. F. Baroncelli and G. Grossi in *Solvent Extraction Chemistry of Metals* (Eds. H. A. C. McKay, T. V. Healy, I. L. Jenkins and A. Naylor), Macmillan, London, 1966, p. 197.

572. J. Kennedy, *Brit. Rept.*, AERE M–1064 (1962).

573. D. A. Landsman and E. S. Lane, *J. Appl. Chem.*, **12**, 24 (1962).

574. E. S. Lane, B. O. Field and J. L. Williams, *J. Appl. Chem.*, **12**, 391 (1962).

10.
Extraction by Ion-pair
Formation

737

The systems discussed in this chapter involve extractable species formed by virtue of an interaction between a neutral or anionic metallic species in the aqueous phase and the salt of an organic base or its cation in either the organic or the aqueous phase. Extraction by such an interaction is considered, as the heading of this chapter states, to be an extraction of ion-associates, the cationic portion of the ion-pair being usually an alkylammonium, -arsonium or -phosphonium cation.

Solvent extraction by high molecular weight organic bases has become increasingly popular in recent years in studying metal complexes. A plausible reason could be found in the belief that the chemistry of extraction by these organic compounds is comparable to that underlying the absorption of metal complexes on anion exchange resins (Chapter 6). An additional reason for the increased interest in the type of extractants discussed in this chapter, especially the tertiary alkylamines, is the excellent performance they show in process chemistry for recovering fertile and fissionable metals from irradiated fuel elements.

The main interest of metal extraction by the reagents reviewed in the forthcoming sections lies in their selectivity towards anionic metal complexes, reversibly formed in an aqueous solution, rather than towards simple anions. Thus the extractability of the anionic species depends upon the aqueous-phase conditions more than on the differences in the specific affinities of anions for the bulky alkyl cation. As far as this factor is concerned, the extractions by all these reagents are similar in general aspects, and closely parallel to the aqueous conditions affecting the sorption of anionic complexes on anion exchange resins. Since metals are extracted from aqueous solutions preferentially when they exist as anions (dichromate, vanadate, perrhenate, pertechnetate) or from anionic complexes (tetrachloroferrate), the parameters controlling the extraction of metal ions are still those affecting the formation of such extractable species in the aqueous phase. These parameters include the aqueous ligand concentration, concentration of other complexants that may compete for the metal ion, presence of other anions that compete for the bulky cation, the hydrogen ion concentration, the concentration of the metal ion itself and such physical parameters as temperature. A simple aqueous system is expected to conform to equations derived from the mass-action law, and if no complication arises from the state of solute in the organic phase, the overall heterogeneous extraction equilibrium conforms also to simple mass-action equations, discussed in detail in section 7.C. In the majority of the actual extraction systems, however, simple mass-action equations fail to describe the underlying reaction of extraction. This is due mainly to the fact that the state of the the solute in the organic phase satisfies only exceptionally the requirements of ideal behaviour. Lack of proper understanding of the non-ideal nature of the solute in the organic solvent, and even more so for its quantitative presentation, renders even the most elaborate set of mass-action equations of little use for a more general application.

The problem of selectivity of the extraction process by the reagents discussed in this section remains essentially unsolved. The responsibility for the specificity of the process has been placed on factors such as the degree of electrostatic interaction in ion-pairing, degree of hydration of the ions, their polarizability and others. Empirically, the trend of selectivity of the extraction process was successfully correlated with the degree of hydration of the anion, which is in turn a function of its size. Large, single-charged ones, such as the perchlorate, apparently without a primary hydration shell in aqueous solutions, extract readily with the equally unhydrated large organic cations. Smaller simple anions show decreased extractability.

In addition to the parameters of the aqueous phase, ion-association extraction systems are especially sensitive to organic-phase parameters. These variables include the nature, structure and size of the organic base, its concentration and type of the organic solvent used as diluent. Each of these factors affects the overall extraction of a metallic species to an almost unpredictable extent. This phenomenon of the ion-association extraction systems is undoubtedly due to a state of solute of the extractants in an organic solvent.

References pp. 800–814

A. HIGH MOLECULAR WEIGHT AMINE EXTRACTANTS

a. Introduction

The extraction by high molecular weight amines is considered to take place by the formation of an ammonium salt, by the reaction

$$\overline{R_3N} + H^+ + A^- \rightleftharpoons \overline{R_3NH^+A^-} \tag{10A1}$$

$\overline{R_3N}$ is a water-insoluble, high molecular weight tertiary alkylamine, and the amine salt $\overline{R_3NH^+A^-}$ is a polar ion-pair with a high ion-association constant characteristic of ion-pairs composed of bulky unhydrated ions in a non-ionizing medium (section 2.D). The amine salt may undergo, under proper experimental conditions, an anion exchange reaction

$$\overline{R_3NH^+A^-} + B^- \rightleftharpoons \overline{R_3NH^+B^-} + A^- \tag{10A2}$$

bearing apparent similarity to the process on solid anion exchange resins (section 4.C).

The exchange reaction, readily occuring in acid solutions, is feasible under certain conditions also in neutral or even basic solutions, provided the complex formed is strong enough to prevent the hydrolysis of the ammonium cation. Caution must, however, be exercised in applying the known correlations to ion-associations of the type considered here, since at higher concentrations of the solutes in the organic medium a further molecular association, to give dipole aggregates, often takes place. This has been shown to cause complications in applying mass-action law equations to describe quantitatively the process of extraction.

b. Amines

Numerous amine compounds of all three classes have been screened and evaluated as extractants for inorganic species[1-5]. Experience has shown that for practical solvent extraction applications, an amine has to fulfill certain basic requirements, such as compatibility with a practical diluent, sufficient extraction power, rapid phase separation, low aqueous solubility and sufficient chemical stability. Although exceptions are known[6], usually only saturated nitrogen compounds fulfill these requirements of a useful extractant.

Higher homologues of saturated primary aliphatic amines are low-melting solids[7,8]. Secondary amines usually possess higher melting and boiling points, dinonylamine being already solid at room temperature[9]. Symmetrical secondary amines have higher freezing points than the corresponding tertiary one. The unsymmetrical tertiary amines are either liquids or low-melting solids[8,10-12]. The majority of the amines decompose upon heating, only the lower members can be purified by distillation without decomposition.

Lower molecular weight primary amines are markedly water soluble. The higher, although water insoluble, exhibit a solubilization phenomenon, their solution in a water-immiscible organic solvent will dissolve substantial quantities of water at room temperature[13,14] (section 2.D). The phenomenon, actually not restricted to the primary amines, has been successfully exploited for demineralization of saline water by solvent extraction[15,16]. Highly branched aliphatic primary amines, however, with desirable physical properties, such as higher solubility in hydrocarbons and nearly complete immiscibility with water, have been used with considerable success in extraction of metals. The presence of a second fatty chain in the amine molecule markedly reduces the solubility of the secondary amines in polar solvents, and increases the solubility in nonpolar ones as compared to the primaries. The tertiary amines, even those with straight fatty chains, are virtually water insoluble, the solubility being below 5 p.p.m. for amines with more than eight carbon atoms per chain.

Primary and secondary amines form a series of hydrates; the degree of hydration and the stability of the hydrates increases with the solubility of the free amines in water[17]. Stable crystalline hydrates of the lower homologues can be isolated[18].

The solubility of primary long-chain amines in non-polar solvents increases with increasing chain length. No correlation could be observed, however, between the solubility and polarity of the solvent. On the other hand, the generally more organic-soluble secondary amines show a marked correlation, being only sparingly soluble in highly polar solvents. Tertiary amines are completely miscible with non-polar solvents at room temperature and sparingly soluble in alcohols and other polar solvents, their solubility in polar solvents increasing with decreasing chain length. Tri-n-octylamine forms a 1:1 hydrogen-bond complex with n-octanol[19]. All three classes of amine form eutectics with non-polar solvents such as benzene, carbon tetrachloride and cyclohexane[20].

The solubilities of some normal primary, symmetrical secondary and tertiary amines are given in Table 10A1.

TABLE 10A1. Solubility of normal aliphatic amines in organic solvents at 20°C (g/100g).

Amine	Benzene	Cyclo-hexane	Carbon tetrachlo-ride	Chloro-form	Hexane	Reference
$C_{10}H_{21}NH_2$	∞	∞	∞	∞	∞	b
$C_{12}H_{25}NH_2$	277	230	148	315	196	b
$C_{14}H_{29}NH_2$	83	68	56	110	50.8	b
$C_{16}H_{33}NH_2$	30.7	26.6	21.2	56	18.3	b
$C_{18}H_{37}NH_2$	14.8	13.2	7.7	31.9	5.4	b
$(C_8H_{17})_2NH$	710	650	355	435	390	c
$(C_{12}H_{25})_2NH$	14.6	12.4	15.0	37.4	5.6	c
$(C_{14}H_{29})_2NH$	4.3	3.1	6.3	15.7		c
$(C_{18}H_{37})_2NH$	0.2	<0.1	1.2	2.8		c
$(C_{18}H_{37})_3N$*	4.2	18.6	13.0	13.5	5.4	d

* Lower tertiary amines are completely miscible with these solvents at 20°C.
a. C. W. Hoerr and H. J. Harwood, *J. Org. Chem.*, 16, 779 (1951).
b. A. W. Ralston, C. W. Hoerr, W. O. Pool and H. J. Harwood, *J. Org. Chem.*, 9, 102 (1944).
c. C. W, Hoerr, H. J. Harwood and A. W. Ralston, *J. Org. Chem.*, 9, 201 (1944).
d. A. W. Ralston, C. W. Hoerr and P. L. DuBrow, *J. Org. Chem.*, 9, 258 (1944).

It has been frequently observed that several amines, but especially the tertiary ones, readily react with carbon tetrachloride and chloroform. The reaction product is apparently hydrogen chloride[21].

The solubility parameter[22] of a number of normal primary, secondary and tertiary amines has been evaluated[23] either from their estimated heats of vaporization[8] or their solubilities in non-polar organic solvents, or both (section 2.B). Primary hexyl- and heptylamine have a solubility parameter of 8.7, higher primaries up to twenty carbon atoms in the normal alkyl chain, have $\delta_2 = 8.6$. Secondary amines with $C_6 - C_8$ per chain have a $\delta_2 = 8.1$, $C_{10}-C_{12}$ 8.3, $C_{13}-C_{16}$ 8.4, $C_{18}-C_{20}$ 8.3. Normal tertiary amines with $C_6 - C_{12}$ per chain have $\delta_2 = 7.6$, $C_{14}-C_{16}$ 7.5 and $C_{18}-C_{20}$ 7.4.

Hyrdogen bonding with like molecules is known to occur through the NH group of primary and secondary amines, the self-association of the primary amines being the higher. Spectral data reveal that the extent of NH hydrogen bonding, like that involving OH, is a function both of the solvent and of the concentrativn, in addition to temperature and the length of the alkyl chain[24].

Proton association constants of high molecular weight amine bases in aqueous solution show that secondary amines are stronger bases than the primary and tertiary amines, in accordance with the behaviour of the lower homologues[25]. While the basicity of primary and secondary amines changes only little with increasing chain length in the molecule, that of the tertary amines increases with the number of carbon atoms[26,27]. The K_B values of primary and secondary amines, for which the alkyl groups have more than two carbon atoms, are close to those of ethylamine, 5.6×10^{-4}, and diethylamine 1.3×10^{-3} at $25°C$[28,29]. The higher basicity of secondary and primary alkylamines is now believed to be due to the effect which solvation of the ammonium ion has in determining base strength[25,30]. It has been suggested[30] that the primary amine is hydrated by three, the secondary by two and the tertiary amine by one water molecule. The observed trend of basicity has also been explained by a different degree of steric hindrance[31]. An empirical relationship has been proposed[32] to correlare the relative basicities of amines with their ability to form hydrogen bonds with the solvent molecules. Although the general applicability of such a correlation has been questioned[33], it seems useful at least for closely related compounds.

The basicity constant of many aliphatic amines has been evaluated[25,34–36] using the values of Taft's induction constants[30] of substituents attached to the nitrogen atom. For many arylamines a reasorable correlation has been found[26,38,39] between basic strength and the Hammett function[40].

In polar organic solvents, the order of amine basicity is that in aqueous solutions, presumably due to a similar solvation effect[41,42].

In aprotic organic solvents, however, the basicity usually decreases from tertiary to secondary to primary[26,43,44], though the basic strength of (lower) amines may be considerably solvent dependent[45]. It may be true even in non-ionizing media that the specific solvation of the alkylammonium cation, or its proton, is an important factor affecting base strength. For example, addition of alcohol, or any solvent capable of hydrogen bonding, to a non-polar solvent will affect the basicity of primary and secondary amines, but has practically no effect on that of the tertiary amines[45].

The basicity of some long-chain aliphatic amines in various organic solvents has been determined spectrophotometrically using the acid 2,4-dinitrophenol[46]. Some of the data are compiled in Tables 10A2 and 10A3. The information available on the dielectrics propertie of long-chain

TABLE 10A2. K_B values of some n-alkyl primary amines in benzene and n-alkyl tertiary amines in chloroform, 25°C[46].

	C_5	C_6	C_7	C_8	C_9	C_{10}	C_{11}
Primary, $K_B \times 10^3$	5.4	4.5	5.7	5.5	4.5	2.6	9.0

	C_2	C_4	C_8	C_9	C_{10}
Tertiary, $K_B \times 10^5$	5.3	5.7	5.0	3.9	4.0

TABLE 10A3. pK_B values of n-octylamines in various organic solvents, 25°C[46].

Solvent	$C_8H_{17}NH_2$	$(C_8H_{17})_2NH$	$(C_8H_{17})_3N$
Chloroform	2.6	4.2	4.1
Benzene	2.3	3.1	3.0
Methyl ethyl ketone	4.2	5.5	4.9
Methyl isobutyl ketone	3.9	5.2	4.4
Isopropyl ether	3.9	4.3	3.2
Ethyl acetate	4.6	5.1	3.8

amines is very limited. The dipole moments decrease from primary to secondary to tertiary amines. In the case of primary and secondary amines the dipole moment increases with the chain length of the alkyl group, while in the case of the tertiary amines the opposite seem to be the case[26]. The lower primary amines, upto five carbon atoms, have a dipole moment of 1.2–1.3 debye, the corresponding secondaries between 0.9–1.0 and that of the tertiaries range between 0.8–0.9 debye[47]. Di-n-decylamine has a dipole moment of 2.05 and tri-n-octylamine only 0.28 debye[48]. Apparently, the position of the polar group does not affect markedly the dipole moment of the molecule, which is also the case for various alkyl halides and alkanols[49]. The dielectric constant of tri-n-dodecylamine is 2.20[50].

Under the usual experimental conditions the chemical stability of higher alkylamines towards radiation-induced decomposition and oxidation has been found satisfactory[51,52]. However, it has recently been found that the radiation damage to tricaprylamine under certain conditions (40 Mrads) may cause up to 4% amine to be damaged. The degradation products are chiefly mono- and dialkylamines, carboxylic acids and nitroparaffins (when nitric acid is present)[53].

The increased application of amines in technology has resulted in a renewal of interest in their analytical chemistry. Several new methods based on various principles have been recently suggested for identification and quantitative determination of the different amine classes. Apart from the fact that the commercially available products are usually mixtures of various isomers, they are frequently contaminated with undesired amine classes and variable amounts of non-amine constituents. Reference is made here to the biannually appearing reviews on analytical aspects of solvent extraction where the analytical procedures can be located[54].

c. Amine Salts

The most characteristic reaction of the amine bases is that with acids to form salts. Since long-chain aliphatic amines exhibit a poor water solubility, frequently different reaction conditions are necessary than for those of their lower homologues.

Primary amine hydrochlorides are best prepared by dissolving the amine base in benzene or ether and bubbling hydrogen chloride through the solution[55-57]. Several recrystallization cycles from organic solvent mixtures leave an anhydrous and non-hygroscopic white crystalline substance[58-60]. Secondary amine chlorides, due to their lower solubility in polar solvents, can be prepared by neutralizing an alcoholic solution of the amine with aqueous hydrochloric acid[61,62], and subsequent recrystallization from alcohol or acetone. A higher yield can be obtained using non-polar solvents rather than alcohols as the solvent for the amines[60]. Hydrochlorides of triheptyl- and trioctylamine were prepared by passing hydrogen chloride gas into a petroleum ether solution of the amine cooled in dry ice and acetone. Higher tertiary amine derivatives may be obtained by equilibrating their petroleum ether solution with aqueous hydrochloric acid. Recrystallization from petroleum ether leaves anhydrous salts. Tertiary amine hydrobromides can be prepared essentially by the same procedure[63,64,66].

White anhydrous crystalline secondary and tertiary amine nitrates can be prepared by bringing a $\sim 10\%$ excess of the concentrated aqueous acid into contact with an acetone solution of the amines. Several recrystallization cycles are needed to obtain the pure compound[65,66]. In a similar way, solid trilaurylamine perchlorate and bisulphate were isolated from an acetone solution. The precipitate formed upon cooling was filtered off rapidly on a Buchner funnel. The white crystalline salts, recrystallized three times from acetone, are non-hygroscopic and stable[65,66]. Under slightly different conditions a mixture of amine sulphate and amine bisulphate is formed[67,68]. When an amine solution in aliphatic hydrocarbon is equilibrated with dilute aqueous sulphuric acid, the normal sulphate is formed at the interface[68].

Neutralization of trialkylamines with HF solution leads directly to the formation of the amine bifluoride[67,69,70]. White and non-hygroscopic tridecylamine hydrofluoride has been prepared from the bifluoride when kept in a dry atmosphere for a prolonged period[69], or by careful crystallization from very dilute aqueous solutions[70].

The melting points of some alkylamine salts are given in Table 10A4[60,63,65–67].

TABLE 10A4. Melting point of some alkylamine salts (°C).

Ammonium salt	Number of C atoms per chain					
	C_6	C_7	C_8	C_{10}	C_{12}	$C(C_8)_2^*$
n-Primary chlorides	216–218	201–204	195–197	188–191	178–180	
n-Secondary chlorides	262–264	249–251	236–238	198–201	199–202	
n-Secondary nitrates	~185		180–182	178–180	164–166	
n-Tertiary chlorides		73–74	68–69	78–79	84–85	149–150
n-Tertiary bromides		75–76	72–73	82–83	86–87	150–151
n-Tertiary iodide					52	118–119
n-Tertiary nitrate					51–52	63
n-Tertiary perchlorate					58–59	
n-Tertiary bisulphate					64–65	

* Methyldioctylamine.

The salts of the long-chain primary amines are the best known of the amine salts. Various fatty acid salts of primary amines have been prepared and their physical properties investigated[71,72]. These salts are as a rule appreciably water soluble and so are the salts of several mineral acids. In aqueous solution such salts function as cationic surface-active agents (section 2.D). Fragmentary data indicate that the solubility of mineral acid salts of long-chain symmetrical secondary and tertiary amines in water and aqueous electrolyte solutions is very low, generally not exceeding 10^{-4}M at ordinary temperature.

The solubility of primary amine salts in organic solvents decreases with decreasing polarity of the solvent, the salts being only very sparingly soluble in aliphatic hydrocarbons[73]. The solubility of secondary, but especially that of tertiary amine hydrochlorides in non-polar solvents is higher on a molar basis, than that of the primary ones, as can be deduced from the values compiled in Table 10A5. The solubility of tertiary amine chlorides in aromatic hydrocarbons and carbon tetrachloride is very high[60,74]. The solubility of trilaurylamine hydrochloride in some dry and some wet organic solvents is shown in Table 10A6[74].

TABLE 10A5. Solubility of normal alkylamine hydrochlorides in various solvents[60] at 25°C.

Ammonium salt	Solvent	Solubility (g/100 g solvent)
Octyl	Benzene	0.55
Decyl	Benzene	0.13
Dodecyl	Benzene	0.02
	Chloroform[a,c]	4.05
	n-Butanol[a,c]	7.98
Dihexyl	Benzene	0.34
	Carbon tetrachloride	0.87
	n-Heptane	<0.001
Diheptyl	Benzene	0.39
	Carbon tetrachloride	1.82
	n-Heptane	<0.002
Dioctyl	Benzene	0.77
	Carbon tetrachloride	2.45
	n-Heptane	<0.002
Didecyl	Benzene	0.09
	Carbon tetrachloride	0.76
	n-Heptane	<0.002
Didodecyl	Benzene	0.08
	Carbon tetrachloride	0.01
	n-Heptane	<0.002
Methyl-dodecyl	Benzene[b,d]	0.16
Triheptyl	n-Heptane	0.3
Trioctyl	n-Heptane	0.23
Tridecyl	n-Heptane	0.24
Tridodecyl	n-Heptane	0.16
Dimethyl-dodecyl	Benzene[b,e]	0.30

a. C. W. Hoerr and H. J. Harwood, *J. Am. Chem. Soc.*, **74**, 4290 (1952).
b. F. K. Brown and H. J. Harwood, *J. Am. Chem. Soc.*, **72**, 3257 (1950).
c. 28°C.
d. 28.4°C.
e. 27°C.

TABLE 10A6. Solubility of trilaurylamine hydrochloride in some dry and wet* organic solvents (mole/l)[74].

| | at 25°C | | at 35°C | |
	Dry	Wet	Dry	Wet
Benzene	0.6	0.7	0.9	1.1
Toluene	0.4	0.6	0.8	1.0
Diethylbenzene	0.04	0.09	0.2	0.8
Diisopropylbenzene	0.02	0.05	0.1	0.5
Cyclohexane	0.02	0.08		
Chloroform	1.0	1.2		
Carbon tetrachloride	0.5	0.7	0.9	1.0

* Saturated solution prepared by dissolving TLA.HCl in solvents in the presence of water.

Tertiary amine nitrates and sulphates show a similar solubility behaviour. Sulphates are generally more soluble than the corresponding bisulphates, and the secondary amine bisulphate is less soluble, at least in benzene, than the salt of the tertrary amine[68].

Some alkylamine salts are more soluble in non-polar solvents when alcohol is added, presumably due to solvation[19,75]. Cryoscopic data on ternary systems of benzene–alcohol–trioctylamine nitrate, $TOA.HNO_3$, indicate specific solvates by hydrogen bonding. With methanol and ethanol the compound formed was proposed to be a trisolvate $TOA.HNO_3.3ROH$, while with the higher alcohols, ranging from butanol to decanol, a disolvate[75] or a monosolvate[76] was suggested. The difference in the solvation number was assumed to be due to steric hindrance.

Hydrogen bonding of the type $N^+ - H \cdots X^-$ between the ions of the ammonium salts has been observed for a large number of various alkyl[66,77–91], aryl[77–81,83] and heterocylic[77–81,88] amines. The presence of hydrogen bond, both in solids and their organic solutions, has been detected by i.r. and n.m.r. spectra.

Formation of intramolecular hydrogen bonds in amine salts is pronounced mainly when the amine is a weak base or the acid is relatively weak, and the complete transfer of the proton from the acid to the base is considerably reduced. Such is the case in tribenzylamine–picric acid[92], pyridine–acetic acid[93] and aniline–acetic acid[94] pairs. The number of possible hydrogen bonds is the greatest with the primary amines and decreases with the number of hydrogen atoms attached to the nitrogen[78,95].

There is some doubt as to the value which should be taken for the infrared normal N—H stretching frequency. Herzberg[96], by analogy with ammonia, indicates that the value is near 3400 cm^{-1}. For solid $NH_4{}^+$ salts Waddington[97,98] found the 3100–3200 cm^{-1} range. Different intermediate values have also been suggested[77,81,99] but a survey of a number of various ammonium salts[100] indicates that the values 3250 and 3350 cm^{-1} should be assigned in the free ammonium ion to the symmetric and degenerate stretching frequencies, respectively. The position of the unperturbed N—H frequency in hydrogen-bonded compounds is expected to be shifted to lower values, the greatest shift indicating the strongest hydrogen bonding. Generally speaking, the hydrogen-bond strength increases on going from primary to secondary to tertiary amine salts. There are large shifts, of the order of 1000 cm^{-1} for the tertiaries, indicating strong hydrogen bonding[66]. The extent of the shift increases apparently with increasing chain length in the aliphatic mixtures, indicating a stronger hydrogen-bond formation with the longer-chain amine salts at least up to a certain number of carbon atoms per chain[91]. Furthermore, there is a difference in the hydrogen-bond strength depending on the anion of the amine salt. The bond strength decreases in the halide series in going from fluoride to iodide, as could be expected. With trilaurylamine salts a linear relationship has been observed between the radii of the anions and the strength of the bond, increasing with decreasing radius[66].

Infrared spectra of these amine salts in organic solvents indicate that the characteristic bands appear roughly at the same position as in the solids, though somewhat broadened. Even in high dielectric constant solvents, such as nitromethane and nitrobenzene, stable ion-pairs are formed when the structure involved permits strong hydrogen bonding between the cation and anion. The stability will depend on the number of effective hydrogen bonds[101]. In non-polar solvents, the bands are shifted to lower frequencies, indicating an increased strength of the hydrogen bond, at least in the case of tertiary amine halides and nitrates[90,91,102–106].

The data for the N—H and N—D stretching vibrations for a number of octylamine salts in carbon tetrachloride[104] indicate that large anions such as ClO_4^-, IO_3^- and I_3^- exhibit weak hydrogen bonding. Surprisingly, the i.r. shift in the spectrum of the trioctylammonium fluoride, which might be expected to form the strongest hydrogen bonds, is too small. The phenomenon is apparently due to the fact that under the experimental conditions of equilibrating an amine-base solution in carbon tetrachloride with aqueous hydrogen fluoride solution, the organic-phase compound is the amine bifluoride rather than the normal salt.

It is perhaps interesting to note in this connexion, that the strength of hydrogen bonding between protonated ammonium and metal (or the bihalide) complexes is weaker than for the salts of the simple acids[66,104,107-109]

In aqueous solutions, hydration of the alkylammonium cation interferes with the formation of intramolecular hydrogen bonds; hydrogen atoms which would have formed bonds with the anion will form bonds with the water molecule.

N.m.r. measurements on amine extracts of various acids indicate a rapid exchange of protons between extracted water and the acid in excess to the amine. The chemical shifts observed are due to strong hydrogen bonding between the alkylammonium cation and the acid radical. In carbon tetrachloride solution the observed chemical shifts decrease in the order $Cl^- > Br^- > NO_3^- > I^-$ and exhibit an approximately linear relation with the observed i.r. shifts[104,110].

The solubility parameters of long-chain normal aliphatic primary, secondary and tertiary amine hydrochlorides, evaluated from their solubilities in hydrocarbons, show that they decrease markedly on going from the primary to the tertiary amines. The high δ_2 values for the primaries (13.5–14) are typical of solutes which possess low solubility in organic solvents. The secondary amines have a δ_2 value around 12.5 and the tertiary amines have one around 10.5 — practically irrespective of the chain length[60,111] (see section 2.B).

The limited data available on the dielectric properties of high molecular weight ammonium salts indicate that they are highly polar. McDowell and Allan[48] found that the dipole moment of amine sulphates and bisulphates is in the range of 4.6–4.8 and 3.6–5.4 debyes respectively. Tertiary amine salts have a higher moment than either the primary or secondary ones, in qualitative accord with the tendency of these compounds towards aggregation. Picrate, acetate and halide salts of tributylamine have a higher dipole moment[112,113] than that of trioctylamine sulphate. The dipole moment apparently increases with both the size of the anion and length of the alkyl chain[112,113]. The possible correlation between the polarity of the amine salt, and the structure and basicity of the amine has frequently been discussed[26,27,114]. There is a serious limitation to the polarity–basicity relationship due to the fact that basicity depends strongly on the medium. Structural factors, though affecting basicity also, have usually a tremendous influence on the polarity of the salt.

d. Quaternary Ammonium Compounds

High molecular weight, preferentially water-insoluble, quaternary ammonium salts are useful in the separation of simple or complex anions, just as strong-base anion exchange resins of the quaternary ammonium type. Although acidic media are generally preferred, several publications have demonstrated their applicability at higher pH's also. This can readily be understood in view of the strong electrolyte character of these compounds, which retain their ionic character in either acidic or basic ionizing media.

Quaternary ammonium salts, fall into two rather sharply distinguishable groups[115] according to their solubility, surface active properties and tendency towards molecular association. If the number of carbon atoms in the molecule is 24 or higher, the compound is soluble in organic solvents and gives, as a rule, a high extraction ratio when distributed between an aqueous electrolyte solution and a water-immiscible solvent. The second group, those with five or less carbon atoms per chain, are readily water soluble. Measurements of the equilibrium amine concentration in the phases show[116] that these lower amines are excessively distributed to the aqueous phase, as shown in Table 10A7, which accounrs at least partially, for their low extraction power. Thus, practically all ion-association extractions involving quaternary ammonium compounds

TABLE 10A7. Distribution ratio, log D, of various quaternary ammonium halides between chloroform and various hydrochloric acid solutions. Organic phase contained initially 0.1 M quaternary ammonium halide.

HCl(M)	Hyamine* 1622	Arquad** 2HT–75	$(C_7H_{15})_4NI$	$(C_4H_9)_4NBr$	$C_{18}H_{37}(CH_3)_3NCl$
0.1	3.6	4.4	4.6	−4.2	−4.5
1	4.3	4.6	4.6	−3.3	−3.5
6	4.2	5.0	4.0	−4.1	−4.4
12.2	3.9	5.0	4.5	−3.2	−3.4

* Benzyl diisobutylphenoxyethoxy ethyldimethylammonium chloride.
** Dimethyl dioctadecylammonium chloride.

have used those having at least one, but preferably more, long-chain alkyl radicals in the molecule.

Lower, symmetrical quaternary ammonium salts, those with five or less carbon atoms per chain, have been screened for their applicability in solvent extraction processes[116,117]. Although no straightforward comparison is possible, due to somewhat different experimental conditions, these survey studies of a number of metals from various acid and sodium hydroxide solutions seem to indicate that the degree of extraction increases with increasing molecular weight of the extractant, under otherwise comparable conditions of extraction.

Quaternary ammonium compounds, in contrast to the various organic reagents capable of forming ion-association complexes yet to be discussed in the following sections, have shown potential utility in nuclear fuel processing and fission-product recovery and other separation processes, in addition to their widespread applicability for analytical purposes. New substituted quaternary ammonium salts are developed for a specific purpose in radiochemical processing. Useful extractants have been synthesized as a result of the continuous programme of evaluation, at Oak Ridge National Laboratory and reported in the form of Annual Reports of the Chemical Technology Division. Extraction can be controlled by the appropriate choice of the reagent structure. Thus a symmetrical cation shows usually a higher extractive capacity than an unsymmetrical one with an equal number of carbon atoms. Substitution of an aromatic group for an aliphatic frequently enhances the extractive power. The most frequently used quaternary ammonium compounds are known by their commercial designation and their exact composition has not always been precisely determined.

High molecular weight quaternary ammonium compounds are only sparingly water soluble, their solubility decreasing in presence of electrolytes in aqueous solutions. It is only reasonable to assume that they behave in aqueous solutions as strong electrolytes.

At low concentrations short-chain alkyl quaternary ammonium salts are present in aqueous solutions chiefly as monomers. The properties of these solutions have been briefly discussed in section 1.B. Above a certain concentration, termed the critical micelle concentration, the *cmc* (see section 2.D and the following sections), most of the solute added to a solution goes into the formation of large aggregates — micelles. Generally speaking, this concentration depends on the nature of the solute molecule, length of the hydrocarbon chains, and temperature; for a given colloidal electrolyte the critical micelle concentration depends on the presence of other electrolytes in the solution, and their concentration.

It has been repeatedly demonstrated that the addition of an electrolyte with common ion to the solution of quaternary ammonium compound causes a decrease in *cmc* and an increase in the apparent molecular weight. The effect is shown for some chlorides and bromides in Table 10A8. The micellar molecular weights listed in the Table cannot be regarded as true values, since they depend markedly on the way they have been determined. The aggregation number of dodecyltrimethylammonium suphate, $[C_{12}H_{25}(CH_3)_3N]_2SO_4$, for example, was found to be 25 by conductivity[118] but as high as 65 by light-scattering[119] measurements. The effect of added electrolyte upon the *cmc* cannot be described in terms of the principle of ionic strength or predicted by the mass-action law; the lower the *cmc*, the greater the lowering of *cmc* by equal amounts of an electrolyte in solution[120]. The prominent role of temperature on the tendency of quaternary ammonium salts toward aggregation is not well understood. Small temperature changes produce marked changes in the system. In the case of Hyamine 1622[121], for example, at a given temperature there is a critical electrolyte concentration above which the system separates into two layers, one practically free from the amine.

The *cmc* is slightly affected by the anion associated with the quaternary ammonium cation. Some typical data are compiled in Table 10A9. Usually, salts of polyvalent anions have a lower *cmc*, the sulphate being an apparent exception[122]. A comparison of the *cmc* values of octadecyltrimethylammonium with those of octadecylpyridinium chlorides and nitrates makes the effect of the organic groups apparent; the extent of ion-pairing being generally higher with the higher molecular weight members of a homologous series[123].

The information available on aqueous solutions of watrr-soluble, low molecular weight quaternary ammonium compounds indicates a number of instances of anomalous thermodynamic behaviour (see section 1.B). The apparent molal heat capacity[124] and the osmotic and activity coefficients calculated from isopiestic measurements[125-128] and by the e.m.f. method[129] and calorimetric data[132] are the evidence for abnormal behaviour, although the experimentally determined activity coefficients of tetraethylammonium iodide are only slightly less than those calculated by the Debye–Hückel limiting equation[127,131].

TABLE 10A8. Micellar molecular weights of some quaternary ammonium salts in aqueous solutions of electrolytes.

Quaternary ammonium salt	Medium, t°C	cmc(M)	Aggregation number	Micelle (M.Wt.)	Reference
$(C_{12}H_{25})(CH_3)_3NCl$					
	Water, 23°	0.021	36	9.900	a
	0.02 M NaCl	0.015	37	10.200	
	0.04 M NaCl	0.011	40	11.000	
	0.1 M NaCl	0.007	60	16.400	
$(C_{14}H_{29})(CH_3)_3NCl$					
	Water, 23°	0.004	34	10.100	a
	0.1 M NaCl	0.0010	85	25.500	
Hyamine 1622					
	Water		13.3	5.500	b
	0.01 M NaCl		75.4	31.200	
	0.02 M NaCl		78	32.400	
	0.03 M NaCl		83	34.400	
	0.04 M NaCl		88	36.500	
	0.05 M NaCl		100	41.500	
$(C_{10}H_{21})(CH_3)_3NBr$					
	Water		32	10.200	c
	Water	0.064			d
	0.05 M NaCl	0.057			
	2.19 M NaCl	0.044			
	5.65 M NaCl	0.032			
	0.1 M BaCl₂	0.045			
	2.42 M BaCl₂	0.033			
	0.013 M KBr		34	10.700	c
$(C_{12}H_{25})(CH_3)_3NBr$					
	Water	0.00014	44	15.500	c
	0.013 M KBr		50	17.400	c
$(C_{14}H_{29})(CH_3)_3NBr$					
	Water		67	25.300	c
	0.013 M KBr		84	32.100	
$(C_{16}H_{33})(CH_3)_3NBr$					
	0.013 M KBr		150	61.700	c

a. L. M. Kushner, W. D. Hubbard and R. A. Parker, J. Res. Natl. Bur. Std., 59, 113 (1957).
b. J. Cohen and T. Vassiliades, J. Phys. Chem., 65, 1774, 1781 (1961).
c. P. Debye, J. Phys. Chem., 53, 1 (1949).
d. M. L. Corrin and W. D. Harkins, J .Am. Chem. Soc., 69, 683 (1947).

TABLE 10A9. Critical micelle concentration of quaternary ammonium salts in water at 25°C (from conductometric measurements).

Salt		$cmc \times 10^4$ (m)	Reference
$C_{18}H_{37}(CH_3)_3N$	— chloride	3.4	a
	— chloride	3.4	b
	— nitrate	2.3	a
	— bromate	3.1	a
	— formate	4.4	a
$[C_{18}H_{37}(CH_3)_3N]_2$	— oxalate	< 1	a
$[C_{12}H_{25}(CH_3)_3N]_2$	— sulphate	1.6	c
$C_{12}H_{25}(CH_3)_3N$	— bromide	1.4	c
$C_{18}H_{37}C_5H_5N$	— chloride	2.4	a
	— nitrate	1.3	a

a. P. F. Grieger and C. A. Kraus, *J. Am. Chem. Soc.*, **70**, 3803 (1948).
b. H. Kimizuka and I. Satake, *Bull. Chem. Soc. Japan*, **35**, 251 (1962).
c. H. V. Tartar and A. L. M. Lelong, *J. Phys. Chem.*, **59**, 1185 (1955).

In non-ionizing solvents, quaternary ammonium compounds are in the form of highly polar ion-pairs. Both the dipole moment and the ion-pairs contact distance, a, depend on the symmetry of the ions and the medium[112,133-138]. For large, symmetrical and highly polarizable ions considerable variations might be expected. Some values are shown in Table 10A10.

The information available on nonaqueous solutions tends to indicate that a monomeric state of solute in non-polar media is rather exceptional or, more precisely, the solute concentration at which monomers are the predominant species is extremely low in such solvents. Low molecular weight quaternary ammonium salts show a tendency for molecular association, though only exceptionally beyond the dimer (quadrupole)[133].

The extent of aggregation in non-ionizing organic solvents increases with the molecular weight of the solute. This was shown to be the case at least in the homologous series from tetramethyl- to tetrabutylammonium compounds, but the trend is probably even more marked with the high molecular weight water-insoluble compounds. For the time being however, no experimental evidence has been presented to substanttate the prediction and we shall assume that their behaviour is similar to that of high molecular weight tertiary amine salts discussed in the next section. As a rule, these salts are readily soluble in non-polar aromatic hydrocarbons and carbon tetrachloride, and polar solvents such as nitrobenzene, chloroform or ethylene dichloride, but only sparingly soluble in aliphatic hydrocarbons[139].

TABLE 10A10. Dipole moment and contact distance of quaternary ammonium ion-pairs in organic solvents at 25°C.

Ion-pair	Solvent	Dipole moment ($\mu \times 10^{18}$)	a (Å)	Reference
(Isoamyl)$_4$N –picrate	Benzene	18.3	3.82	112
–thiocyanate	Benzene	15.4	3.23	112
(n-Butyl)$_4$N –bromide	Benzene	11.6	2.43	112
–bromide	Benzene	13.9	2.90	134
–bromide	Benzene	12.2	4.5	135
–bromide	Benzene–methanol (0.021 mole fraction)	14.0	2.92	134
–bromide	Benzene–methanol (0.073)	14.4	3.00	134
–bromide	Benzene–methanol (0.138)	15.8	3.29	134
–iodide	Benzene	12.7	4.6	135
–perchlorate	Benzene	14.1	2.96	112
–perchlorate	Benzene	17.2	5.1	135
–acetate	Benzene	11.2	2.35	112
–chloroacetate	Benzene	14.8	5.2	135
–benzoate	Benzene	12.1	5.1	135
–picrate	Benzene	17.8	3.73	112
–picrate	Benzene	20.8	4.34	134
–picrate	Benzene	15.3	5.4	135
–picrate	Toluene	17.8	3.71	134
–picrate	Dioxane	19.9	4.15	134

Returning now to the state of tetrabutyl- (and lower alkyl-) ammonium salts in organic solvents, it is apparent that the extent of dimerization depends on the dielectric properties of the solvent (see section 2.D). Addition of a polar solvent, such as methanol, to a solution of, for example, tetrabutylammonium bromide in benzene or carbon tetrachloride causes the aggregates to break down to ion-pairs, which are more polar than quadrupoles or higher aggregates[134]. Spectral and polarization measurements of tetrabutylammonium iodide in various mixtures of non-polar and polar solvents have been interpreted in terms of solute–polar-solvent interaction[140-142]. It has been postulated that ion-pairs in the presence of polar solvents are essentially solvent-shared or solvent-separated ion-pairs, as distinguished from short distance or contact ion-pairs in non-polar solvents.

Thus, addition of methanol to a carbon tetrachloride solution of $(C_4H_9)_4NI$ converts the contact ion-pair solute to a solvent-shared ion-pair solute, which is the state of solute in the polar chloroform, for example. As seen from the numerical values in Table 10A4, the methanol-shared ion-pairs have a higher dipole moment, due to mutual polarization of methanol and the solute which, in turn, tends to increase the charge separation in both. Stern[134] found that there is a critical molar ratio, moles methanol/moles ammonium salt, i.e. the number of moles of polar component, necessary for a complete breakdown of the higher aggregates into individual ion-pairs at a particular concentration. Further addition of methanol has no effect on the dipole moment of the solute, although the bulk dielectric constant of the solution is obviously increasing. Conductance measurements on tetrabutylammonium picrate in various isodielectric mixtures have shown[143] that the polar constituent of binary mixtures influences the association into ion-pairs, 'both by controlling the macroscopic dielectric constant and by specific interaction with ions' (cf. section 2.D). This specific interaction may be formulated as being of a type which leads to solvent-shared ion-pairs[140,142], as discussed above. A different explanation has been suggested by Bodenseh and Ramsey[144] and Hyne[145], based on a model which assumes preferential accumulation of polar solvent molecules around the highly polar solute molecules. Again, a different one is based on interaction of the ion with a definite number of solvent molecules by ion–dipole attraction, as proposed by D'Aprano and Fuoss[143,146].

Bauge and Smith[135] evaluated the dipole moments of some dimers of tetrabutylammonium salts in nitrobenzene. Their dielectric constant measurements of such binary systems indicate that the quadrupoles of bromide, iodide and perchlorate have a zero dipole moment, whereas those of chloroacetate, picrate and benzoate have dipole moments of about 3, 4 and 4 debyes respectively. They suggest that these unusually high dipole moments of dimers are due to their flexible structure, leading to high atom polarizations.

e. Aggregation of Amine Salts

The proper understanding of the reaction governing the extraction of metal complexes by long-chain amines is hampered by failure to appreciate the importance of the behaviour of these surfactants in solution. Although several attempts have been made in recent years to approach the problem and treat it by what may be called principles of the chemistry of surface-active agents, there is, nevertheless, a tendency to underestimate its significance. In section 2.D a general discussion of the aggregation of colloidal electrolytes, to which class the high molecular weight ammonium salts belong, has been presented.

The principles and theories of the chemistry of surface-active agents implicitly assume that the formation of micelles at the critical micelle concentration takes place by aggregation of monomers, without intermediate, lower, aggregates being formed. This has been the way the abrupt changes with concentration of many physical properties of solutions of surface-active agents have been interpreted. Consequently, a closer study of the state of solute in the concentration range between the monomeric ion-pairs and the micelles has been usually disregarded.

Investigations of the various aspects of possible aggregation of long-chain aliphatic amine salts in water-immiscible non-polar solvents, in connexion with solvent extraction equilibria, revealed that, in such systems at least, the monomer–micelle equilibrium involves intermediate stages. Information is now available to show that amine salts form dimers, trimers and higher multimers prior to micellization. Furthermore, the stepwise formation of these low polymers in dry and wet amine salt–organic solvent systems satisfies the requirements of the mass-action law.

In view of the existence of such dimers, trimers, tetramers, etc., the question will arise what should be the size of a polymeric unit to be regarded as a micelle. The question can hardly be answered before more thermodynamic data is available on such systems (section 2.D.g). Högfeldt's[147–151] interpretation of distribution data seems to indicate that in addition to low polymers, aggregates containing as many as $\bar{q}=$ 20–40 monomers can be formed at high enough amine salt concentrations ($\sim$ 0.3M). Such big units, in a mass-action law equilibrium with the lower polymers, should perhaps be regarded as micellar units. Allen's[152] early light-scattering data on the aggregation of some alkylamine sulphates and bisulphates, among them those of di-n-decylamine in benzene, recently confirmed by McDowell and Coleman[153] using interfacial tension measurements, also points strongly to the possibility of micellar aggregates with as much as 40 monomers being formed. These findings, however, are contradicted by the cryoscopic data of Fomin and Potapova[154]. The average molecular weights of extracts of several trioctylamine salts in benzene, in the concentration range $\sim$ 0.1 – 1.7 M, correspond to a mean aggregation number not exceeding 8. There is a maximum in $\tilde{q}$ around ~ 0.5 M salt concentration[154], and $\tilde{q}$ decreases to two at a concentration close to saturation, due perhaps to an increased polarity of the solution as it becomes increasingly loaded with the salt[172].

For a mass-action law treatment of distribution data the most important quantitative information needed is exactly in the range between the monomer and the micelle in the equilibrium

$$R_3NH^+ + X^- \rightleftharpoons R_3NA^+X^- \rightleftharpoons (R_3NH^+ \, X^-)_2$$

$$\text{(10A3)}$$

$$\rightleftharpoons (R_3NH^+ \, X^-)_n \rightleftharpoons (R_3NH^+X^-)_{\text{micelle}}$$

While the dissociation of ion-pairs in organic diluents can easily be disregarded (see section 10.B), the electrostatic molecular association into dimers and other multimers frequently greatly affects metal extraction.

The available information leads to several generalizations concerning the aggregation of alkylamine salts. Within these generalizations however, there are considerable differences in the behaviour of the salts as will be shown in the forthcoming paragraphs.

a) The nature of the organic diluent and its solvation power are the most important parameters affecting both the extent (number of aggregated units) and degree (size of aggregated units) of aggregation.

Not until the recent solvent extraction studies has it been realized that aromatic and aliphatic hydrocarbons, with equally low dielectric constant and zero dipole moment, have markedly different effects upon the aggregation of amine salts. The striking difference in the solvent behaviour of aromatic (and certain derivatives of aliphatic) and aliphatic hydrocarbons can be explained by the solvation power of the aromatics[155,156]. This increased solvation power is due to the π electrons in the case of benzene, or to $Cl = C^+$ bonds in the case of carbon tetrachloride[157]. The solvation is exhibited by a solute–solvent interaction leading to a shielding of the high dipole moment of the ion-pair, stabilizing it and increasing the compatibility of the polar monomer with the solvent. This effect obviously interferes with the aggregation of the amine salt.

The way solvation affects the aggregation of amine salts is even more apparent in polar solvents of high dielectric constant. It has been shown that alcohols, for example, in addition to their decidedly high solvation power, will form specific solvates with the amine ion-pairs[19,75,76,158]. No evidence has been found for an aggregation in such solvents. There is even the possibility, drawing on similar systems investigated previously[140,142], that the alcohol molecules, or polar molecules generally, are oriented towards the polar amine ion-pairs, so that one, or at the most two, alcohol molecules squeeze in between the ions in contact and form a solvent-shared or solvent-separated ion-pair.

A similar interaction is probably the reason that no aggregates of amine salts are formed in hydrocarbons as long as free amine base is present in the system. This seems to be a general phenomenon, at least in nitrate[50,154,159] and sulphate systems[160]. The phenomenon may well be explained by a specific solvation of the amine salt by the amine base. Such an organic phase can be regarded as a solution of the amine salt in a binary solvent mixture of amine base and the diluent. The selective solvation is a dipole–dipole interaction, where the amine base molecule, with its low but permanent dipole, has a greater tendency than the diluent to accumulate in the vicinity of the large dipole of the amine salt. Such a shielding of the polarity of the amine ion-pair will stabilize it and prevent its aggregation[161,162].

In aliphatic hydrocarbons, on the other hand, the complete lack of solvating power will facilitate formation of aggregates and their growth.

b) Both the extent and degree depends, as could be expected, on the concentration of the solute in the non-polar solvent. However, the extent of solubility of the salt is not necessarily related to its tendency towards aggregation. For example, the solubility of tertiary amine salt is generally higher in aromatic hydrocarbons than in the aliphatics by roughly two order of magnitude. Nevertheless, the salts are undoubtedly more aggregated, both in extent and degree, in the aliphatic hydrocarbons for the reasons outlined in a). Similarly, the general trend is an increasing solubility from primary to secondary to tertiary to quaternary amine salts in water-immiscible organic solvents. The phenomenon is due, at least partially, to an increasing number of hydrophobic CH_2 groups. Still the secondary amine salts are more aggregated in a given solvent, again both in extent and degree, than those of the tertiary amines[152].

c) It should be noted that some observations point to the fact that the extent of aggregation, though not necessartly the degree, of a given class of amine salts will depend on the chain length in a homologous series. Increasing chain length favours the extent of aggregation, but only up to a certain number of carbon atoms in the chain, and afterwards the opposite seems to be true[50,154].

This structural effect upon aggregation is closely related to the influence of size and symmetry upon the dipole moment of the salts[163]. Symmetrical, quaternary alkylammonium cations and small anion ion-pairs have a very pronounced tendency toward polymerization. Similar salts, but with a tertiary or secondary ammonium cation, with a lower dipole moment, will exhibit also a lower aggregation number[68,152,163,164].

d) In a given solvent and for a given alkylammonium radical, the aggregation of bisulphate and perchlorate is more pronounced than that of chloride and nitrate[154]. The extent of dimerization of trilaurylamine salts in benzene increases in the order $Cl^- < NO_3^- < Br^- < ClO_4^-$ $< HSO_4^-$, and the degree of their aggregation increases roughly in the same order, also[65].

e) As could be predicted, at higher temperatures the degree, and certainly the extent, of aggregation is lowered[165,166]. Tetraheptylammonium nitrate is still predominantly dimeric in a mixture of polyphenyls at $\sim 140°C$[167].

Before turning to a review of the information available on specific amine salt–diluent systems, mention should be made of the experimental techniques employed. For the detailed treatment of the experimental data the sources referred to should be consulted (cf. section 2.D).

Work has been done[65,165] to determine osmometrically the degree and extent of aggregation of several salts in the temperature range between 25°–65°C in a number of practical diluents. The technique seems to be reliable in the optimal concentration range between $\sim 0.02–0.1$ M.

Light-scattering measurements[152,168] on a number of amine sulphate and bisulphate extracts in benzene gave reproducible data in a solute concentration range between 0.02–0.2 M, though for the exceedingly aggregated salts of di-n-decylamine a lower concentration range could be investigated. More recently, a few experiments carried out for trilaurylamine chloride, bromide, nitrate and bisulphate in dry benzene at 25°C indicate that, 0.01 M is the absolute lower limit to obtain any significant data from light-scattering measurements[169].

Cryoscopic (and ebullioscopic) measurements on a variety of amine salt extracts have been carried out mainly in benzene[154,159]. It seems that the method is of no value below 0.1 M solute concentration. The main objection to that technique, however, is that it provides the information at and around the freezing point of the cryoscopic medium. There is no justification to assume that the information on aggregation will be valid at other temperatures (see section 2.D).

The two-phase e.m.f. titration technique of the Stockholm school[147–151] is probably the only reliable one based on distribution data. It allows one to investigate, with high precision, a wide range of solute concentration (0.002–0.4 M).

The physical methods could be used for both dry, strictly binary systems of the amine salt and the diluent, and for wet systems, where the organic phase is saturated with water. While the dry systems are usually obtained by dissolving the solid amine salt in anhydrous organic solvents, the wet systems are essentially the extracts obtained during an actual extraction of the acid from an aqueous solution into the amine base–organic diluent phase. Obviously, all the variations of the distribution technique provide the information on aggregation in the wet extracts.

It is interesting to note in this connexion that a comparison of aggregation data obtained in dry and wet systems, under otherwise identical conditions[74,170], suggest that the coextracted water does not play a decisive role in aggregation of amine salts, though Gourisse[50] claims that in the case of trilaurylamine nitrate the presence of water lowers the extent of aggregation. The association constants for various amine salts compiled in Table 10A11 indicate the minor role played by water.

Quantitative data available on the extent and degree of aggregation of various amine salts are given in Table 10A11, in the form of the equilibrium constant, β_n, of the association reaction

$$q\,\overline{RHX} \rightleftharpoons \overline{(RHX)}_q \qquad \beta_q \qquad (10A4)$$

Whenever possible equilibrium constants K_q of reactions

$$q\overline{R} + qH^+ + qX^- \rightleftharpoons \overline{(RHX)}_q \qquad K_q \qquad (10A5)$$

have been expressed in the form of β_q through the relationship

$$\beta_q = K_q/K_{11}^q \qquad (10A6)$$

TABLE 10A11. Association constants of trilaurylamine salts in various diluents at 25C°.

Amine salt	Solvent	Log association constant				Method	Reference
		β_2	β_3	β_4	β_n		
Nitrate	Benzene		2.48			Osmom.	65
	Carbontetra chloride	1.34	3.43	5.64		Osmom.	65
	Benzene	1.82	2.60			Osmom.	65,165
	Benzene (wet)	2.09				Distr.	178
	o-Xylene (wet)	1.59	2.48		$\beta_{10}=12.8$	E.m.f.	147
	m-Xylene (wet)	1.82	3.40		$\beta_{12}=27.8$	E.m.f.	147
	n-Octane (wet)			9.31	$\beta_{18}=39.01$	E.m.f.	149
Chloride	Benzene	1.72				Osmom.	165,180
	Benzene (wet)	1.3				Distr.	172
	o-Xylene (wet)	1.17	2.89		$\beta_{48}=70.61$	E.m.f.	148
Bromide	Benzene	1.34	2.53	(2.76)		Osmom.	65
	Cyclohexane	2.91		7.95		Osmom.	65
	o-Xylene (wet)	1.49	3.13		$\beta_{30}=46.10$	E.m.f.	151
	Benzene (wet)	1.6				Distr.	172
Iodide	Benzene (wet)	1.9				Distr.	172
Thiocyanate	Benzene (wet)	1.9				Distr.	172
Perchlorate	Benzene	2.38			$\beta_8=14.00$	Osmom.	65
	Cyclohexane	3.25			$\beta_8=20.17$	Osmom.	65
	Benzene (wet)	2.26				Distr.	156
	Dodecane (wet)	6.56			$\beta_8=35.40$	E.m.f.	179
	Benzene	2.27			$\beta_8=14.07$	Osmom.	180
Bisulphate	Benzene	3.46			$\beta_6=12.28$	Osmom.	65

where K_{11} is the equilibrium constant for the formation of the monomeric ion-pair (ion-pair association constant).

Many more systems have been investigated but only qualitative results have been reported. In benzene and xylene cryoscopic measurements[154,159] indicate that the average aggregation number, $\tilde{q}$, of trialkylamine nitrates (trihexyl to tridodecyl) does not exceed three, the trimer predominating in the concentration range between 0.3–0.8 M. Further increase in the solute concentration lowers $\tilde{q}$; in a ~ 1.5 M solution the salts are dimeric.

Trioctylamine hydrochloride in benzene is at most dimeric[154]; but diisononyl hydrochloride in chloroform is only dimeric[171], no evidence of a monomer could be found in the concentration range between 0.006–0.3 M. Osmometric measurements in wet chloroform, benzene, carbon tetrachloride and cyclohexane have been semi-quantitatively evaluated to show that the average aggregation number of trilaurylamine halides increases in the above order of solvents[172]. No molecular association has been found for these salts in chloroform, while $\tilde{q}$ has a value of 2.85, 3.56, 4.50 and 3.85 in cyclohexane for TLA.HCl, TLA.HBr, TLA.HI and TLA.HSCN respectively at 0.1 M concentration.

Light-scattering measurements on the benzene extract of trioctylamine sulphate and bisulphate indicate that the sulphate is monomeric
up to 0.5 M concentration, whereas the bisulphate is dimeric in the
concentration range between 0.02—0.2 M, but becomes tetrameric at
~ 0.5 M[152,160,173]. Branched-chain tertiary amine sulphates are aggregated[152]. The highest degree of aggregation ($\tilde{q} = 38$–40) has been found
for di-n-decyl- and di-n-dodecylamine sulphate in benzene[152,153] in the
concentration range between 0.0003 $-$ 0.002 M. Distribution data of
trihexyl- and trioctylamine sulphate suggest high aggregation in kerosene[174], though in benzene their monomeric state has been confirmed[175].
Cryoscopic measurements on similar extracts in benzene, in the concentration range 0.12 $-$ 0.25 M show the presence of dimers and perhaps
even higher polymers[164].

Trialkylamine bisulphates are, as a rule, more aggregated[152,168,176]
than the corresponding sulphates. Probably because of this difference in
their tendency towards aggregation, the sulphate–bisulphate exchange
reaction is extremely sensitive to the medium. So, for example[19], in
alcohols, and presumably also in other polar solvents[177], the solvation
of trialkylamine sulphate interferes with the extraction of more acid,
to form the amine bisulphate.

In Figures 10A1 and 10A2 the concentrations of monomers and

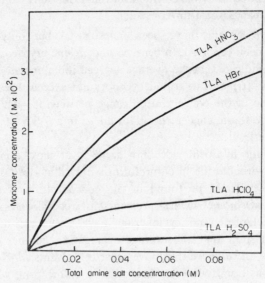

Figure 10A1. Variation of the monomer concentration with the total amine salt concentration in benzene at 25°C[65]. (By permission from North-Holland Publ. Comp.)

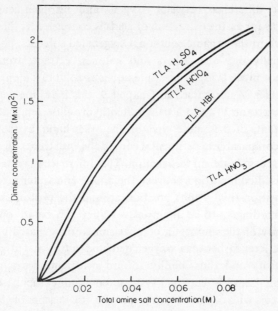

Figure 10A2. Variation of the dimer concentration with the total amine salt concen
tration in benzene at 25°C[65]. (By permission from North-Holland Publ. Comp.)

dimers, respectively, of various triaurylamine salts in benzene as a
function of the total salt concentration, is given. The curves have been
computed from the dimerization constants[65] given in Table 10A11.

B. EXTRACTION OF ACIDS BY AMINES AND THEIR SALTS

a. Introduction

The distribution of both organic and mineral acids between water and a
water-immiscible organic solution of amine favours the organic phase.
The extractability of organic acids from an aqueous into an organic
phase is not complete even in the presence of a stoichiometric excess of
amine base in the organic phase. Under similar experimental conditions,
however, the extraction of aqueous mineral acids is practically quantitative, leaving the equilibrium aqueous phase acid deficient. Moreover,
the extraction of mineral acids is usually not limited to formation
of simple amine salts in the organic phase; when the initial acid concentration in the aqueous phase is high the acid-to-amine ratio in the
organic phase frequently exceeds unity.

A striking difference between oxygen- and nitrogen-bearing basic extractants, as far as the extraction of acids is concerned, is the behaviour of the proton in the organic solution. Oxygenated solvents, phosphorous esters especially but also ethers and ketones, extract strong mineral acids (but not nitric acid) from dilute aqueous solutions along with four water molecules. As discussed in Chapter 9, this is in line with the tendency of the proton to form a trihydrated hydronium ion. Nitrogenated basic extractants of the amine type, on the other hand, extract the same acids under comparable experimental conditions with traces of water only, as will be discussed later in this section. The difference may be explained by an incseased competition between the amine and water molecules for the first solvation (hydration) position around the proton and may be reduced, accordingly, to a quantitative difference based only on the higher basicity of the amines in comparison to oxygenated extractants. The large difference between oxygenated and nitrogenated extractants in their basicity and thus affinity toward acids, implies a qualitative differentiation between between systems containing the bases. In extraction systems with oxygen-bearing solvents, it is reasonable to consider the extent to which an extracted acid dissociates in the organic phase, refering to the dissociation of the acid into a (usually hydrated) proton and the anion (see Chapter 9). On the other hand, the acid extracted into the amine-containing organic phase is no longer regarded as an acid. It is rather the extent of ion-pairs association between the alkylammonium cation and the anion, which is the measure of the stability of the organic-phase complex.

b. Extraction of Nitric Acid

The extraction of nitric acid by all four classes of amines has been studied under a wide variety of experimental conditions[50,67,103,106,110,150,154,159,181-206]. All data unanimously agree that the acid transferred from the aqueous into the organic phase results in the quantitative formation of the amine nitrate as long as free amine base is present in the system. Values for the association constant of the amine nitrates

$$K_{11} = [\overline{R.HNO_3}]\,[\overline{R}]^{-1}[H^+]^{-1}[NO_3]^{-1} \qquad (10B1)$$

where R stands for a primary, secondary or tertiary amine, are shown in Table 10B1.

TABLE 10B1. Equilibrium constants K_{11} for amine nitrates.

Amine	Diluent	Aqueous acid range (M)	Organic amine range (M)	$K_{11} \times 10^{-5}$ M^{-2}	Reference
Tri-n-octyl	Carbon tetrachloride	0.13–0.20	0.225	10	187
Tri-n-octyl	Nitrobenzene (40°)	0.01–2.0	0.02–0.1	500	106
Tri-n-octyl	Chloroform (40°)	0.01–2.0	0.02–0.1	62	106
Tri-n-octyl	Benzene (40°)	0.01–2.0	0.02–0.1	1.1	106
Tri-n-octyl	Benzene (30°)	0.01–2.0	0.02–0.1	2.7	106
Tri-n-octyl	Benzene (25°)	0.01–2.0	0.02–0.1	3.8	106
Tri-n-octyl	Xylene (25°)	0.004–0.04	0.05	1.4	194
Triisooctyl	Xylene (25°)	0.11–0.43	0.425	1.4	194
Triisooctyl	Toluene	0.002–0.05	0.05–0.2	41	203
Tri-n-nonyl	Xylene (25°)	0.01–0.20	0.21	1.5	194
Tri-n-dodecyl	m-Xylene (25°)		0.002–0.1	0.17	192
Tri-n-dodecyl	o-Xylene (25°)		0.002–0.3	0.35	192
Tri-n-dodecyl	Xylene (30°)		0.05–0.3	12	50
Tri-n-dodecyl	Chlorobenzene (30°)		0.05–0.3	27	50
Tri-n-dodecyl	n-Octane (25°)		0.002–0.3	0.002	192
Tri-n-dodecyl	n-Dodecane (30°)		0.05–0.3	2.5	50
Amberlite LA–1	Benzene	>0.05		5.4	190
Amberlite LA–1	Carbon tetrachloride	>0.05		3.8	190

As to the uptake of further nitric acid in excess of that necsseary for the stoichiometric neutralization of the amine base, an empirical relationship, apparently first suggested by Bertocci[181], has repeatedly been confirmed[50,184,185,190,191,201]. The excess nitric acid has been found to be proportional to the concentration of both the acid and of the amine. The empirical proportionality factor Q_{12}, in

$$[\overline{HNO_3}]_{excess} = Q_{12}[\overline{R}] \, [HNO_3] \tag{10B2}$$

has a value of $0.16-0.18$ for the tertiary, $0.10-0.14$ for the secondary and $0.06-0.11$ for the primary amines. The experimental parameter can be evaluated graphically[185], but if the diluent itself extracts the acid, as benzene does at high acidities, an appropriate correction has to be introduced. This simple relationship is followed for a wide range of acid loadings and amine concentrations.

Since the empirical relationship cannot be interpreted in terms of an extraction reaction, attempts have been made to fit the experimental data into simple mass-action equations. Assuming the formation of an adduct according to the reaction

$$\overline{RH^+NO_3^-} + H^+ + NO_3^- \rightleftharpoons \overline{RH^+NO_3 . HNO_3} \qquad (10B3)$$

Shevchenko[188] and Mezhov[189] have evaluated its formation constant K_{12}. In an aqueous acid range from 1 to 5 M, and an amine concentration of about 0.5 M, the values $K_{12} = 0.9$ for trioctylamine in carbon tetrachloride and 0.13 in xylene, and 0.033 for dioctylamine in xylene have been calculated. Other attempts[187,190,193] to fit the experimental data into the above equation and evaluate K_{12} were unsuccessful, the values were not satisfactorily constant. Good results were obtained, however, when allowance was made for the equilibrium concentration of undissociated acid in the aqueous phase. K_{12} of the adduct formation reaction

$$\overline{RH^+NO_3^-} + HNO_3 \rightleftharpoons \overline{RH^+NO_3^- . HNO_3} \qquad (10B4)$$

gave a value of $K_{12} = 7.2$ for triheptylamine in benzene[187], and 2.24 and 1.85 for Amberlite LA–1 in benzene and carbon tetrachloride respectively[190]. Knoch[194] found that the K_{12} is only slightly affected by the total amine concentration: for triisooctylamine in xylene $K_{12} = 0.23$ at a 0.05 M amine level, and 0.26 at 0.43 M. For trinonylamine in xylene the value is 0.18.

Without having evaluated the K_{12} values, Mason and coworkers[169,198] have shown that the excess nitric acid in the organic phase is a function of the activity of the undissociated nitric acid in the equilibrium aqueous phase, both in the presence and absence of salting-out nitrates. Nitrate salting causes a substantial increase in the amount of excess nitric acid in the organic phase[50,193].

In view of an almost certain non-ideality of the solutes in the organic phase[192,199], the partial success of fitting experimental data into simple mass-action equations cannot be regarded as evidence that the second molecule of nitric acid (per amine) does not simply dissolve in the organic phase but combines chemically with the amine nitrate in solution. The excess nitric acid in such systems can easily be washed back by water[67,183] or even pumped off by bubbling nitrogen gas through the solution[190], indicating the relative instability of the adduct.

When the acid-to-amine ratio exceeds unity, i.r. absorption spectra of the extract reveal a weakening of the hydrogen bond in the amine nitrate[67,103,194,201]. In some cases peaks corresponding to the presence of free, unbound nitric acid[50,106,185,197,202] have also been observed. The spectrum is very different from that of the simple amine nitrate[67,103], believed to be due to hydrogen bonding of HNO_3 to NO_3^- to give in chloroform solution the reasonably stable methyloctylamine binitrate

$$R^1R_2^2NH^+ \qquad \left[\begin{matrix} O \\ O \end{matrix} \Big> N-O \cdots H-O-N \Big< \begin{matrix} O \\ O \end{matrix} \right]^-$$

N.m.r. spectral data tend to confirm these findings[194].

Neither an amine nor an amine nitrate solution in low dielectric constant solvents show measurable conductivity, but the presence of excess nitric acid cause such solutions to be fairly conducting[50,154,190,201]. Both density and viscosity of the extract change with the amine-to-acid ratio[50,106,187,190,198,201].

At very high aqueous acid concentrations the acid-to-amine ratio in the organic phase exceeds two. This additional excess of nitric acid has been variously attributed to simple distribution of molecular, perhaps hydrated, HNO_3[50,190,194] without formation of any specific amine–acid complex richer in nitric acid than $R_3N(HNO_3)_2$, and to the formation of a series of specific amine–acid adducts $R_3NHNO_3(HNO_3)_n$[183,204].

From the data available it is difficult to draw even qualitative conclusions as to the role which coextraction of water plays in the transfer of nitric acid into the organic phase. It has been shown[50,196,199] that the amount of water extracted along with nitric acid depends, for a given amine content, on the amount of acid transferred. In a plot of water content vs. acid content the initial slope varies between 0.2 and 0.33, more water being coextracted at higher amine nitrate concentrations[196,205]. An excess of HNO_3 (over 1:1) in trilaurylamine replaces some water, not necessarily one for one[50,199], while more water is coextracted as the third molecule (per amine) of acid passes into benzene or xylene solutions of tertiary amines. Gourisse[50] suggests that the increased water content is due to dissolution of the hydrated dimer, $(HNO_3)_2.H_2O$. Similar results were reported for triheptylamine in benzene[187], and trioctylamine in xylene[186] and carbon tetrachloride[110]. Högfeldt's data[192] for tridodecylamine in dodecane and octane indicate that the water content depends upon the concentration of the amine, roughly 1/3 mole water per mole of amine, and is practically independent of the acid content in the region investigated up to an acid-to-amine ratio of

2.5. The amount of water extracted per amine decreases with increasing chain length of the normal tertiary amines in dodecane and xylene[192]. The amount of water in the triheptylamine–carbon tetrachloride system seems to be acid independent[187] also. Högfeldt's observations[205] are similar for the trilaurylamine–dodecane (or octane)–octanol systems. The amine and the alcohol extract, both the acid and the water independently of each other. In view of the alcohol–amine interaction, discussed earlier in this chapter, such a conclusion seems to be an oversimplification of the interactions occuring in these multicomponent systems.

The effect of diluents on the extraction of nitric acid follows the general patterns outlined in the preceeding section. Trioctylamine nitrate may be dissociated to some extent in nitrobenzene[206], and association into higher aggregates is inhibited in chloroform. For the effect of solvent on the extent and degree of aggregation of alkylamine nitrates see section 10.A.e.

However, the degree of aggregation is affected by the amount of nitric acid in the organic phase. No aggregation of the amine nitrate is apparent in benzene or xylene as long as free amine base is present in the system, though no association of the salt with the corresponding amine could be detected even in aliphatic hydrocarbons[50]. On the other hand, as the second nitric acid molecule per amine is transferred into the organic solution, the amine dinitrate, presumably formed, is more aggregated than the simple amine nitrate. The degree of aggregation of the hypothetical dinitrate increases with its concentration in benzene and passes through a maximum of $\bar{q} = 6$ at ~ 0.5 M, and then falls. As the third acid molecule per amine enters the organic phase, the apparent molecular weight of the organic solute drops, corresponding now to a mixture of monomers and dimers[154].

The activity coefficient of trilaurylamine nitrate in various diluents has been calculated from osmometric data[111].

Third-phase formation (see following section) has been observed when nitric acid is extracted into tertiary amines dissolved in dodecane or cyclohexane[158,207].

c. Extraction of Halo Acids

With certain restrictions as to the diluent, primary, secondary and tertiary amines extract the halo acids quantitatively (provided they are in a stoichiometric excess to the acid) leaving the equilibrium aqueous phase with a pH 3–5, depending on the amine concentration. The affinity of tertiary amines for the halo acids, which is in the order of their extractability, decreases from HI > HBr > HCl > HF. This order seems to be independent of the nature of the diluent.

Of the halo acids hydrochloric acid has received the most attention[63,67,102,104,105,107,110,148,154,155,161,171,172,186,187,195,196,203,208–215, 218]

The information available on the extractability of hydrobromic and hydriodic acids[67,107,151,172,187,214,215] indicates that their distribution curves are comparable to that of hydrochloric acid. The extractive behaviour of hydrofluoric acid, however, has been found to be quite different from that of other halo acids[69,107,186,216,217] as the shapes of the distribution curves in Figure 10B1 show[186]. The reported association

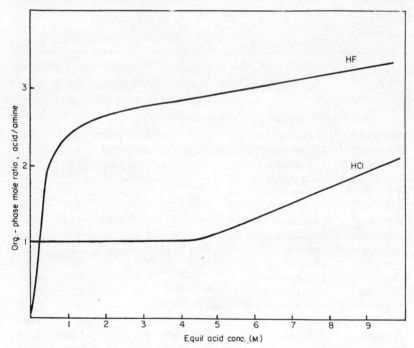

Figure 10B1. Distribution of HCl and HF between water and trioctylamine in xylene[186].

constants K_{11} for the amine hydrochlorides are compiled in Table 10B2 and for some other amine hydrohalides in Table 10B3. Although the K_{11} value generally depends on both the amine and acid concentration, in addition to the effect of diluent, most of the values listed in the Tables have been averaged to simplify tabulation. Though the temperature of the experiments for their determination are not always explicitly given, it is assumed to have been about 25°C.

Extraction of excess acid (to that necessary for the formation of the simple amine hydrochloride salt) varies considerably with the different halo acids (Figure 10B1). Excess hydrochloric acid extraction is slight from aqueous acid concentrations below 2–3 M. Excess-acid extraction is immediate with hydrofluoric acid, but there is no excess acid extracted with hydriodic acid up to ~ 7 M in the initial aqueous solution[67]. The extractive behaviour of hydrobromic acid is intermediate between that of HCl and HI.

TABLE 10B2. Equilibrium constants K_{11} for amine hydrochlorides.

Amine	Diluent	Aqueous acid range (M)	Organic amine range (M)	$K_{11} \times 10^{-5}$ M^{-2}	Reference
Di-n-hexyl	Carbon tetrachloride		0.001–0.02	7.9	211
Di-n-heptyl	Carbon tetrachloride		0.001–0.02	8.7	211
Di-n-octyl	Carbon tetrachloride		0.001–0.02	15.8	211
Di-n-decyl	Carbon tetrachloride		0.001–0.02	17.2	211
Di-n-dodecyl	Carbon tetrachloride		0.001–0.02	9.1	211
Tri-n-octyl	Carbon tetrachloride	0.13–0.2	0.225	0.35	187
Tri-n-octyl	Carbon tetrachloride		0.02–0.08	0.10	209
Tri-n-octyl	Benzene		0.002–0.02	0.13	208
Tri-n-octyl	Benzene	0.08–0.21	0.227	0.55	187
Tri-n-octyl	Toluene	~ 0.05	0.1	0.07	110
Tri-n-octyl	Cyclohexane	~ 0.05	0.1	0.0011	110
Tri-n-octyl	Nitrobenzene			56	218
Tri-n-octyl	Nitrobenzene	~ 0.05	0.1	400	110
Tri-n-octyl	2-Nitropropane	~ 0.05	0.1	280	110
Triisooctyl	Toluene	0.008–0.04	0.2	0.81	203
Tri-n-dodecyl	Benzene (23°)		0.015	0.45	214
Tri-n-dodecyl	Benzene (25°)		0.0085	0.13	215
Tri-n-dodecyl	Benzene (37°)		0.0085	0.089	215
Tri-n-dodecyl	Benzene (64°)		0.0085	0.066	215
Tri-n-dodecyl	o-Xylene (25°)		0.016–0.31	0.056	148
Tri-n-dodecyl	Toluene	0.004–0.21	0.01–0.4	0.16	105
Primene JMT	Benzene		0.015	74	214
Amberlite LA–2	Benzene		0.015	5.7	214

TABLE 10B3. Equilibrium constants K_{11} for amine hydrobromides and hydroiodides.

Amine salt	Diluent	Aqueous acid range (M)	Organic amine range	$K_{11} \times 10^{-5}$ M^{-2}	Reference
Tri-n-octyl bromide	Carbon tetrachloride	0.10–0.17	0.225	7.3	187
Tri-n-octyl bromide	Benzene	0.08–0.2	0.227	5.4	187
Tri-n-octyl bromide	Toluene	~0.05	0.1	1.2	110
Tri-n-octyl bromide	Cyclohexane	~0.05	0.1	0.013	110
Tri-n-octyl bromide	Nitrobenzene	~0.05	0.1	5000	110
Tri-n-octyl bromide	2–Nitropropane	~0.05	0.1	800	110
Tri-n-dodecyl bromide	Benzene (23°)		0.015	1.1	214
Tri-n-dodecyl bromide	Benzene (25°)		0.0085	0.6	215
Tri-n-dodecyl bromide	Benzene (37°)		0.0085	0.31	215
Tri-n-dodecyl bromide	Benzene (64°)		0.0085	0.067	215
Tri-n-dodecyl bromide	Xylene			0.16	151
Primene JMT bromide	Benzene		0.015	200	214
Amberlite LA–2 bromide	Benzene		0.015	16	214
Tri-n-octyl iodide	Toluene	~0.05	0.1	90	110
Tri-n-octyl iodide	Cyclohexane	~0.05	0.1	0.9	110
Tri-n-octyl iodide	Nitrobenzene	~0.05	0.1	220.000	110
Tri-n-octyl iodide	2–Nitropropane	~0.05	0.1	80.000	110
Tri-n-dodecyl iodide	Benzene (23°)		0.015	13	214
Tri-n-dodecyl iodide	Benzene (25°)		0.0085	28	215
Tri-n-dodecyl iodide	Benzene (37°)		0.0085	13	215
Primene JMT iodide	Benzene		0.015	470	214
Amberlite LA–2 iodide	Benzene		0.015	84	214

The extractability of hydrochloric acid by quaternary ammonium halides is markedly lower than that of nitric acid. As shown in Figure 10B2, the extraction of hydrochloric acid increases linearly with its concentration in the aqueous phase[219,220].

The excess acid extracted by all four classes of amines seems to be interpreted by the majority of the authors in terms of hydrogen dihalide formation.

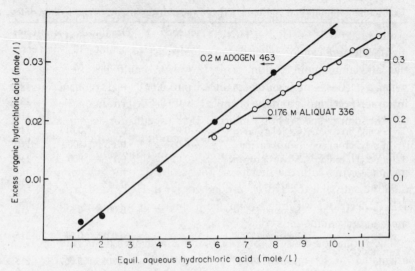

Figure 10B2. Extraction of hydrochloric acid by quaternary ammonium chlorides in toluene[220].

Fifty years ago, there were reports[221] of a number of hydrogen dihalides of lower tetraalkylamines having been prepared. They were obtained by passing hydrogen halide gas over the amine salt, and were identified by chemical analysis. Some of these findings have been recently confirmed. Similar salts of hydrogen dihalides, R_4NHCl_2[222–224], R_4NHBr_2[223–226] and R_4NHI_2[223,224,227] (where R may be any alkyl chain from methyl to butyl, but not an aryl radical) have been isolated. X-ray powder photography, i.r. and n.m.r. spectra led Waddington to conclude that in HCl_2^- the symmetrical hydrogen bond is almost identical to that of the unequivocal and well-documented hydrogen difluoride[228]. Similar conclusions were reached by Chang and Westrum[229] based on decomposition pressure measurements, the crystal entropy being characteristic of a linear and symmetrical bond. From the pressure–composition isotherms obtained by treating dry tetraalkylammonium halide with anhydrous hydrogen halide gas, McDaniel and Vallée[224] calculated the hydrogen-bond energies of 14.2, 12.8 and 12.4 kcal/mole for $ClHCl^-$, $BrHBr^-$ and IHI^- species respectively.

Tetramethyl- and tetraethylammonium dichlorides were shown to form when the corresponding normal chlorides are dissolved in nitrobenzene containing hydrogen chloride[230]. The solid dichlorides could be solated from the solution. The equilibrium constant in nitrobenzene at 25°C for the formation of the dichloride ion has been calculated to be around 500 ($Cl^- + HCl \rightleftharpoons HCl_2^-$). It has been postulated that the dichloride ion cannot be free but must exist in a tight ion-pair with the alkylammonium cation. This, however, could not be confirmed when a tertiary ammonium is the cation. The N—H frequency of the infrared spectrum[102] and the chemical shift for the proton on the nitrogen for the n.m.r. spectrum[104] indicate that the dichloride anion is only weakly hydrogen bonded to the alkylammonium cation. In addition, conductivity measurements[105] on the extract indicate that less ion-pairing occurs with the dichloride than with the chloride anion. N.m.r. data indicate that the hydrogen-bond energy in HCl_2^-, and probably also HBr_2^-, decreases with increasing chain length, at least in the quaternary ammonium compounds[223].

The formation constant of trioctylamine hydrogen difluoride has been estimated to have a value of 1.8×10^3 in toluene at 25°C[110].

Once formed, under the conditions of solvent extraction, the tertiary amine dichloride seems to be surprisingly stable in such solvents as benzene or carbon tetrachloride. Bubbling of air through an amine solution that had an acid-to-amine ratio greater than two, as a result of equilibration with gaseous HCl or concentrated HCl solution, removes from the solution only the acid that is in excess to that necessary for the formation of the dichloride[102]. It should be noted, however, that only exceptionally the acid concentration in the organic phase exceeds that corresponding to $R_3NH \cdot HCl_2$. This excess may be due to the simple dissolution of molecular HCl in the diluent. The transfer of the acid is markedly affected by the presence of alkali chlorides[231]. The tendency to extract hydrobromic and hydroiodic acid, in excess of that needed to form the simple salt, is so much lower than in the case of a parallel HCl system that the formation of dibromide (and obviously diiodide) ion under experimental conditions of solvent extraction has been questioned[63,102].

The extractive behaviour of hydrofluoric acid is quite different from that of other halo acids, the main differences being a) a less complete extraction of the acid when the amine is in excess, and b) an increased solubility of excess acid in the amine hydrofluoride. The phenomena can be explained by the incomplete dissociation, and tendency for dimerization of the acid, respectively. In other words, the ability of the fluoride ion to add on HF with the formation of HF_2^- exceeds the proton-accepting power of the nitrogen atom. In the extraction of HF the acid-to-amine mole ratio may be greater than four, and the formation of a series of compounds $R_3NH_nF_{n+1}$ may be anticipated[96], though in these compounds the strength of hydrogen bonds may be slightly less than in HF_2^{-}[232]. Indeed, infrared spectra of hydrofluoric acid extracts[233] tend to indicate that simple R_3NHF is under no conditions the predominating species; the shift in the N—H frequencies in these extracts is too small[104].

Extraction of HCl by amine base seems to be accompanied by the extraction of much more water than occurs in the extraction of any other monobasic strong mineral acid[74,110,172,186,187,196]. Tertiary amine hydrochlorides[74,105,148,172,196] are monohydrates in toluene and benzene, and probably other aromatic hydrocarbon diluents. Less water is coextracted into chloroform[74,172] and much less into aliphatic hydrocarbons[105]. Duyckaerts and Müller[105] have demonstrated that the amount of water extracted into a given amine–diluent system depends on the activity of water in the aqueous phase. The water content decreases with increasing acid content of the organic phase only if the water activity decreases in the aqueous solution, but is dependent on the acid transferred if the aqueous activity of water is kept reasonably constant.

Under comparable experimental conditions less water is coextracted along with hydrobromic and hydriodic acids, and much less along with HF[172,186,216].

Activity coefficients of trilaurylamine hydrochloride and hydrobromide in benzene have been evaluated using vapour pressure lowering measurements[111]. While other attempts to evaluate the activity coefficient reported, may not have a general validity, either because of the questionable experimental procedure used[234], or because of the limitations of the assumptions made[60], they may nevertheless be useful in *comparing* the effect of diluent on the activity of the amine salt. Studies on distribution of trace halide ions[65,235] indicate that the activity coefficient ratios of similar amine salts are reasonably constant. This was the case for amine chloride–bromide and perchlorate–perrhenate pairs, indicating a similar deviation of the salt-pairs from an ideal behaviour.

The molecular aggregation of amine halides has been discussed earlier in this chapter, and evidence has been presented to show that the presence of water does not significantly affect the aggregation of at least, trilaurylamine hydrochloride in benzene and toluene[74].

Fomin and Potapova[154] found that the amine–dichloride ion-pairs show a higher degree of molecular aggregation than the simple trioctylamine hydrochloride at a given amine concentration. Duyckaerts and

Müller[105], however, have found no evidence for molecular aggregation of $R_3NH^+HCl_2^-$ ion-pairs (using cryoscopy also) under comparable experimental conditions.

d. Extraction of Sulphuric Acid

The distribution of sulphuric acid between water and organic solutions of amines has received considerable attention with a special emphasis on the amine sulphate–bisulphate equilibrium in the organic phase[67,68,76,110,152,153,160,164,168,173–177,186,195,196,236–239].

Generally, an interpretation of distribution curves for sulphuric acid systems in terms of the mass-action law is apparently more difficult than in the case of monobasic acids. The contradicting interpretations sugges-ted reveal the complexity of such systems due to the equilibria in amine sulphate and amine bisulphate formation, polymerization of sulphate- and bisulphate-bearing species, the equilibria in the sulphate–bisulphate exchange in the aggregates, and in the of partition equilibria of the various sulphate–bisulphate aggregates. In sulphuric acid systems the nature of both the amine and the diluent seems to play a more decisive role than in other amine–acid systems, because of a much more pronoun-ced tendency toward molecular aggregation of the amine sulphates and especially bisulphates. A comparison of the equilibrium-constant values, in Table 10B4 for the formation of amine sulphates, illustrates the effect.

Regardless of the state of the amine sulphate and bisulphate in the organic solutions, which has been discussed previously, the amine sulphate, which is formed initially by the transfer of sulphuric acid, becomes gradually replaced by the amine bisulphate. Apparently, amine bisulphate is formed even before the amine becomes quantitatively converted to the sulphate, $(R_3NH)_2SO_4$. The sulphate–bisulphate equilibrium depends much on the medium[177,239]. For example, a higher dielectric constant of the solvent will favour extraction of the acid but only as long as the normal sulphate is formed. An opposite trend is apparent when the bisulphate, R_3NHHSO_4, is formed predominantly. When octanol is added to a benzene solution of trioctylamine, the amine sulphate formed was found to be $(R_3NH)_2SO_4 . 2ROH$. This solvate seems to interfere with the subsequent formation of bisulphate, as it inteferes also with the extraction of metals[239].

Infrared spectra[67,196] of the extracts obtained under various conditions reveal marked differences between those containing sulphate and bi-sulphate.

TABLE 10B4. Equilibrium constant K_{21}^a for amine sulphates and K_{11}^b for amine bisulphates.

Amine	Diluent	Aqueous acid range (M)	Organic amine range (M)	$K_{21} \times 10^{-7}$ M^{-4}	K_{11} M^{-2}	Reference
Didecyl	Benzene			40.000		237
Didecyl	Carbon tetrachloride			3000		237
Methyl dioctyl	Carbon tetrachloride			0.3		237
Methyl dioctyl	Benzene		0.005–0.49	20		237
Methyl dioctyl	Benzene	0.2–0.8	0.12		0.26	237
Methyl didecyl	Benzene			40.000		237
Methyl didecyl	Carbon tetrachloride			3000		237
Tri-n-hexyl	Benzene			0.5		174
Tri-n-octyl	Benzene	0.001–0.1	0.05–0.25	19		68
Tri-n-octyl	Benzene		0.005–0.49	30		237
Tri-n-octyl	Benzene		<0.02	19		174
Tri-n-octyl	Carbon tetrachloride			0.5		237
Tri-n-octyl	Benzene	0.001–0.1	0.05–0.25		1500	68
Tri-n-octyl	Benzene	0.05–0.1	0.05–0.1		30	237
Tri-n-octyl	Benzene		0.09		15	237
Tri-n-decyl	Carbon tetrachloride				1	237
Tri-n-decyl	Benzene				0.02	237
Tri-n-decyl	Kerosene				0.02	237

a $K_{21} = [\overline{R_2H_2SO_4}][\overline{R}]^{-2}[H^+]^{-2}[SO_4^{2-}]^{-1}$

b $K_{11} = [\overline{RH_2SO_4}][\overline{R}]^{-1}[H^+]^{-1}[HSO_4^-]^{-1}$

The amount of water coextracted with H_2SO_4 seems to be larger than with any monobasic mineral acid[196,239]. A benzene phase of trioctylamine sulphate contains between two and four water molecules per sulphate molecule, depending on the salt concentration. The formation of bisulphate, due to an increased transfer of the acid, is accompanied by ejection of water from the benzene phase.

e. Extraction of other Inorganic Acids

The efficient extraction of a number of acids has been described in the survey studies of Moore[217] and Smith and Page[240].

The extraction of perchloric acid has been studied under a variety of conditions[67,110,154,156,179,187,195,196]. In low dielectric constant solvents, as would be expected, the perchlorate ion has a high affinity towards the alkylammonium cation, and the equilibrium constant K_{11} in carbon tetrachloride has a value of 2.1×10^{8} [110]. On the other hand, i.r. spectra indicate that the tertiary amine perchlorate is dissociated to some extent in chloroform[67] and nitrobenzene[156]. In a solvent of low dielectric constant the salt is highly aggregated, as discussed earlier. The amount of coextracted water is small[110,196].

The extractability of nitrous acid by tertiary amines has been found to be dependent upon the nitric acid content of the aqueous solution[67,181,241,242]. Exact distribution data have not been obtained, since the acid is rather unstable under the usual experimental conditions.

Phosphoric acid is much better extracted by tertiary alkyl- than arylamines, but even by the first with a markedly lower distribution ratio than other mineral acids[217]. Infrared spectra of trioctylamine extracts in benzene indicate that the amine phosphate, initially formed, is gradually replaced by hydrogen phosphate and even dihydrogen phosphate, as the acid content of the organic phase increases[196]. The alkylamine phosphate is highly hydrated.

Cyanic acid is readily extractable by tertiary amines[67]. The salt formed is easily hydrolysed. The extractability of thiocyanic acid, and the behaviour of trilaurylamine thiocyanate are similar to that of hydriodic acid and its salt[172].

f. Extraction of Organic Acids

The extractability of organic acids by amines is much higher than by any other type of extractants[243,244]. This is evidently due to the fact that the more basic amines compete favourably with the organic anion for the proton. Judging only from the limited amount of data available[217,240] it seems that the higher basicity of the organic anion affects adversely the extractability: as would be expected, the extractability increases with acid strength. This is, furthermore, in line with a higher extractability of mineral acids as compared to that of fatty acids and their derivatives. The conclusion, apparent from the above trend, is that the main factor affecting extractability of acids is the basicity of the anion, since all other factors, such as the size of the anion, the dipole moment of the extracted ion pair and the extent of proton hydration should favour the extractactability of organic acids. Thus, for example, the weak amino acids are practically unextractable. The only possible hydrogen bond which may be formed between the nitrogen atom and the acid, without any proton transfer, is apparently too weak a bond to bring about a transfer of the weak acid from the aqueous into the organic phase.

Extraction of oxalic acid by tertiary amines into chloroform[67,245], of lactic acid by various amines into polar solvents[244], of acetic acid[246] and trichloroacetic acid[247] into heptane and o-xylene has been investigated in more detail. Based on distribution data, the existence of a 1:1 amine: oxalic acid complex was obtained at low (< 0.05 M) acidities. At high acidities, the more stable 2:1 complex is apparently the predominating organic-phase species. Depending upon the overall solute concentration, a white precipitate, analysing with a 1:1 ratio, may be formed at the interface. In the case of lactic acid, trioctylamine in chloroform extracts a larger amount of the acid, due apparently to the extractability of the acid by the chloroform itself.

Several complexes of the type $R_3N(HAc)_n$, where $n = 1-4$, are formed in the extraction of acetic acid by trilaurylamine dissolved in heptane or o-xylene[246]. As two-phase e.m.f. itration data indicate, no aggregation is apparent in these systems. The organic-phase complexes in the trichloracetic acid–triaurylamine–o-xylene system are $R_3N.HA$ and $R_3N(HA)_2$ at lower aqueous acidities, but at higher acid concentrations $(1-4$ M) the formation of complexes such as $R_3N(HA)_3.H_2O$ and $R_3N(HA)_7.(H_2O)_7$ were found to fit the experimental data. The equilibrium constants for the first two complexes were log $K_{11} = 6.93$ and log $K_{12} = 9.44$.

C. EXTRACTION OF METALS BY AMINES AND THEIR SALTS

a. Introduction

In less than ten years, metal extraction by long-chain aliphatic amines has grown into one of the most promising tools in separation chemistry. The study of such solvent extraction systems, and their immediate technological application, is still increasing extensively. In the field of nuclear fuel reprocessing several potential processes have been reviewed by Coleman[52], and many recent developments in amine extraction systems have been summarized in papers presented at the Brussels Symposium in 1963[248], the 3rd Geneva Conference in 1964 and the International Solvent Extraction Conference in Harwell in 1965[249].

In addition to the new technological procedures and analytical applications[250] of amine extraction systems, a considerable number of equilibrium studies on metal extraction is available to provide information on the nature of the extracted species and the extraction reaction involved. These studies mainly, are reviewed in the forthcoming paragraphs, while the use of amine extraction systems for studying metal complex formation in aqueous solutions will be illustrated in Chapter 12.

b. Extraction Characteristics

This section concerns the choice of the physical and chemical parameters that should be considered in attempting to use amine extraction systems sucessfully.

(i) *Diluent.* In the case of extraction systems discussed in the pre-ceeding two chapters, a change of diluent has been shown to have a certain influence upon the numerical values of the distribution ratio of an inorganic species. In ion-association extraction systems enormous changes in D are produced by changes in the nature and structure of the organic diluent. The prime importance in the selection of the organic diluent in such systems was early emphasized and numerous efforts have been made to establish an empirical order of solvent efficiency[251–258] and to explain the effect of diluent on the extractive capacity of various amines[116,117,156,200,251–263]. The great difficulty in attempting to rational-ize this distinctive feature of amine extraction systems, and the lack of correlation between the distribution ratio of a metal and such basic physical characteristics of a solvent as its dipole moment and dielectric constant, led Coleman and coworkers[1] to conclude that 'the amine–diluent combination rather than the amine alone should be considered as the effective extractant'.

Based on a large amount of experimental information Taube[260] con-sidered three main parameters affecting the distribution ratio of a metallic species: a) degree of aggregation of the extracted species; b) its dipole moment and c) the dielectric constant of the solution. Ion-associa-tion extraction systems involve as a rule large (usually coordinatively unhydrated) ions with mutual dipole interaction (Chapter 2). Unless aggregated, such species will favour solvents with a high dielectric constant[251,262]. In an opposite case, when aggregated, the low dipole moment clusters will be better extracted into low dielectric constant solvents.

The concept is certainly helpful as a first orientation but is too gross an over-simplification to be useful for any quantitative correlation[200,261]. It assumes a medium of uniform and unchanged dielectric constant and a constant dipole moment for the solute, irrespective of the various experimental parameters. This is, of course, not valid for an actual solvent extraction system. Under certain experimental conditions a solute may fall clearly into the class of unaggregated ion-pa⁺⁻s, while an increased concentration of the solute, thus enhanced aggregation, will place the same solute into the class of those with low dipole moment. The way in which such a progressive modification of the solute's state occurs is far from being known. Thus where there are departures from the ideal behaviour of the solute (expressed through the failure of the mass-action equation (7C15) when applied to the experimentally determined distribution ratios) no mathematical treatment would be possible. In such complicated systems changes occurring in the state of the organic solute should first be taken care of. For this, however, little theoretical assistance is yet available (section 10.A).

It has been pointed out earlier in this chapter that the extent, and probably also the degree, of aggregation of an amine salt at a given concentration is considerably greater in hydrocarbons than in solvents with relatively high dielectric constant. Furthermore, aggregation is more pronounced in aliphatic hydrocarbons (considered to be ideally inert) than in aromatics of equally low dielectric constant and zero dipole moment, but with a considerable solvating power[60,156]. The phenomenon is believed to be due to a stabilization of single non-aggregated ion-pairs by the more ordered, high dielectric constant solvents, (and to some extent by the aromatics) and lack of such a stabilization in the completely structureless aliphatics. Thus, the differences in D occuring with various solvents, under otherwise identical experimental conditions, are the result of a purely organic-phase reaction involving a monomer $\rightleftharpoons$ polymer equilibrium, specific for any given solute–solvent pair, rather than a result of a different compensation for the loss in solvation energy of the ionic species in their transfer from the aqueous solution in terms postulated by Diamond and Tuck[262].

The above concept of the diluent effect implies that not much variation of D for a given species should be expected in various solvents as long as the concentration of the total amine-bearing species in each of the solvents is kept constant and *below* the concentration where aggregation starts. Data are yet lacking to illustrate this point, though a careful examination of the available information tends to support it. Experiments to substantiate this concept must await independent information on binary systems of amine salts in various diluents (see section 10.A). For many systems, especially those involving quaternary ammonium salts, this concentration may be as low as $10^{-7} - 10^{-6}$ M. Additionally, experience has shown that departure from a monomeric state of solute in actual solvent extraction systems is also affected by the aqueous-phase conditions[259,264]. This effect results usually in a decrease of that concentration with increased ionic strength of the aqueous phase, but the effect may be different in different solvents.

(ii) *Amine structure* One of the many variables that makes amine extraction versatile is the type of the amine extractant itself. It is apparently a general phenomenon that the extractive power of alkylamines increases from primary to secondary to tertiary to quaternary amines. A notable exception to this rule is perhaps the extraction by bulky amines, where steric factors play a predominating role.

Structural changes in the amine molecule, however, are frequently sufficiently effective to reverse that order of amine classes. Steric factors generally affect the basicity of amines, which in turn affects the stability and polarity of their salts. For example, when an aromatic substituent becomes directly attached to nitrogen, the amine is a rather weak extractant. The effect of the alkyl-chain length on the extractive power has been studied for several model systems[177,265,266]. Under the usual experimental conditions it appears that for tertiary amines, the optimum number of carbon atoms is eight per chain. Further increase in the chain-length frequently causes a decrease in the extractive capacity, believed to be due to steric interferences. The same effect becomes apparent when amines with branched chains near the nitrogen are used.

(iii) *Third-phase formation* One of the distinctive features of the organic phase is the phenomenon of the formation of a second organic phase, and the effect of added 'modifiers' for its elimination.

A serious limitation in the use of alkylamines is the frequently occurring limited compatibility of their common inorganic salts with a number of organic diluents[1,52,155,158,161,162,186,206,207,237,267-272]. Addition of a long-chain alcohol to the aliphatic or aromatic hydrocarbon increases, however, the compatibility and prevents third-phase formation, but often reduces the extractive capacity of the amine extractants[158,162,184,192,241, 254,273-278]

A review of the fragmentary data available leads to several generalizations: a) The formation of the third phase, a second organic one, is more common in systems where the diluent is an aliphatic hydrocarbon. Aromatic, and some derivatives of aliphatic hydrocarbons, show the phenomenon usually at high organic-phase loading; b) Although a marked diluent effect may be obtained, the splitting of the organic phase is less common when straight-chain alkylamines are used; c) The formation of a second organic phase is most characteristic of amine sulphate systems. The compatibility increases in the order sulphate $<$ bisulphate $<$ chloride $<$ nitrate; d) The formation of a third phase is markedly temperature dependent, there being a particular temperature for a given system below which the previously homogeneous organic phare turns reversibly heterogeneous. Within these generalities, however, there are large differences in the behaviour of specific extraction systems, indicating that the phenomenon of an amine salt–diluent system splitting into two partially miscible phases is by no means simple, and apparently involves the simultaneous operation of a number of factors. Hence, it is not surprising that the literature on third-phase formation is confined to qualitative statements concerning specific systems and conditions for elimination of the second phase[162,279].

Though an adequate treatment of the problem will not be possible until more extensive work has been done in the field, it has been suggested[155,161,162] that the splitting of the organic phase can be rationalized in terms of specific interactions taking place between the organic-phase species. The results of a careful examination of the two organic phases in several acid-extraction systems, among them triisononylamine–hexane–hydrochloric acid and methyl dioctylamine–cyclohexane–nitric acid, and metal-extraction systems such as triisononylamine hydrochloride–toluene–uranyl chloride–hydrochloric acid–water, indicate that the miscibility gap prevails as long as the organic phases contain at least two different ion-pairs. In the simpler acid extraction systems the two possible different ion-pairs were the $R_3NH^+Cl^-$ (or $R_3NH^+NO_3^-$) and $R_3NH^+HCl_2^-$ (or $R_3NH^+H(NO_3)_2^-$), whereas in the above metal extraction system the ion-pairs present were the normal amine hydrochloride $R_3NH^+Cl^-$ and the metal-bearing amine species $(R_3NH^+)_2UO_2Cl_4^{2-}$. Now, results reveal that two organic phases coexist in equilibrium when the organic phase is a mixture of the different ion-pairs, but one single phase prevails when the extraction conditions are such that only one type of ion-pair predominates in the organic phase. In the above metal extraction system, for example, the accumulation of $(R_3NH^+)_2UO_2Cl_4^{2-}$ due to the extraction of metal will cause the splitting of the previously homogeneous organic phase containing $R_3NH^+Cl^-$. As the concentration of that simple ion-pair increasingly diminishes owing to additional transfer of the metal, the volume of the ejected phase first increases and then the gap decreases. At a point where enough UO_2Cl_2 has been transferred to tie up all the amine in the form of the amine tetrachlorouranyl ion-pairs, the solution again becomes homogeneous.

This rather complex behaviour of the miscibility gap could probably be explained in terms similar to the dipole–dielectric constant interaction concept suggested by Friedman[280] for binary systems of electrolytes in nonaqueous solvents that split to form two liquid phases. The high dipole moments of the two different ion-pairs may interact and rearrange into more polar or less polar forms, depending upon whether they line up in a parallel or antiparallel fashion. Whether or not this interaction occurs and, should it occur, the extent to which it does so, depends on the dipole moment of the amine salt and the polarizability of the hydrocarbon diluent.

This interaction and rearrangement is believed to be closely related to the tendency of amine salts towards aggregation and eventual micellization. It has been mentioned in Chapter 2 that amine salt type colloidal electrolytes form micelles at high enough concentrations. Polar solvents (and to some extent aromatic hydrocarbons) in addition to their retarding effect upon the *cmc* of the solute may prevent (or retard in the case of aromatic diluents) the formation of large micelles and their eventual ejection from homogeneous solution. Aliphatic hydrocarbons, on the other hand, are ideally inert solvents with no solvating power, and will facilitate the growth of the aggregates and their ejection.

The above picture appears to have a justification in the solubilizing effect long-chain alcohols have on such three-phase systems. The phenomenon of solubilization of the third phase, or more precisely, the elimination of its formation by addition of small volumes of long-chain aliphatic alcohols to an amine salt–aliphatic hydrocarbon system is known and has received attention from the practical point of view. Although the low volume fraction of alcohols employed (usually 0.03–0.05) cannot change appreciably the overall dielectric properties and the non-polar nature of the hydrocarbon–alcohol mixture, the interaction of alcohol molecules with the amine salt ion-pairs may still be plausible. It has been shown[144,145,281] that the extent of association of quaternary amine ion-pairs in a binary solvent mixture of organic solvents with different dielectric constants is lower than to be expected from the Denison-Ramsey relationship (section 2.D). This has been explained by the assumption that molecules of the more polar components are preferentially accumulated in the neighbourhood of the polar solute, thus the portion of the mixture in vicinity of the solute has different dielectric properties than the bulk of the solvent mixture. Similarly, one may assume that the alcohol molecules, due to their high solvation power and dipole–dipole interaction, are oriented towards the highly polar amine ion-pairs, thereby stabilizing them and retarding their subsequent aggregation. This assumption that the alcohol molecule is readily taken up by micelles, thus playing an important role in prevention of extensive micellization, seems to be justified by the findings[241,273] that the alcohol has to be considered as a part of the reagent, and the amine–alcohol ratio must be kept within certain limits in order to achieve the solubilizing effect of the alcohol.

The above findings, on the other hand, may reveal certain similarities to a cosolvency effect. Although salts of long-chain amines are solids, and can be isolated as such, the third-phase formation is a phenomenon in purely liquid systems and the appearance of a solid is an exception. Thus, in terms of the systems under consideration, the alcohols play the function of a solubilizing cosolvent for two partially miscible phases, an amine salt-rich and a diluent-rich. Winsor's concept[13] of cosolvency in terms of intermolecular forces promoting miscibility postulates that for the case under consideration a succesful solubilization can be achieved by the coeffect of the lipophilic (non-polar) and hydrophilic (polar) affinities of the alcohols. The miscibility-promoting interactions are therefore, those between the hydroxyl group of the alcoholic modifier and the amine salt–rich phase, and the paraffin-chain radical of the alcohol and the non-polar solvent–rich phase. The amount of alcohol required to cause mixing is fixed hy the amount of the amine salt present in the system, provided the aliphatic chain has a great enough lipophilic affinity. Indeed, the maximum effect of cosolvency has been observed with branched-chain (greater) lipophilic affinity) and low molecular weight (higher dielectric constant) alcohols. The above concept was successfully applied in the correlation of data on the solubility action of some quaternary ammonium salts on water in several water-immiscible solvents[282,283].

Finally, our concept of the phenomenon of solubilization of the third phase, by addition of polar modifiers to amine extraction systems is in line with the discussion offered previously (section 10.A) concerning the general problem of the influence of the diluent on the extent and degree of aggregation of amine salts. This mechanism, disagrees with what has been previously suggested by others. Talat-Erben[284] assumed a change of the overall dielectric properties of the solvent mixture upon the addition of the alcohol, whereas Verstegen[158] explained the modified effect by hydrogen bonding between the alcohol and the amine salt.

c. Metal Nitrates

The order of extractability of tetravalent and hexavalent actinide nitrates, and the tervalent lanthanides is determined by the tendency of the elements to form anionic nitrato complexes. In the case of the actinides, both (IV) and (VI), uranium is less extractable than plutonium[285-288], while the lanthanides exhibit an odd–even effect, explained by the alternating affinity of the elements towards the ligand[289]. The spectra of extracts suggest indeed that the organic-phase species are predominantly those of $MO_2(NO_3)_3^-$ [287,290,291] or $M(NO_3)_6^{2-}$ [287,290,292-294] regardless of the aqueous-phase conditions[290,292]. It is impossible to discriminate, from their spectra, between nitrate groups originating from the amine nitrate and from the actinide[293] or lanthanide[289,295] nitrate.

The distribution ratio of metal nitrates increases by replacing nitric acid with metal nitrate as the macro electrolyte in the aqueous phase[263,290-299], though above $\sim 7M$ overall nitrate concentration the D values become usually independent of the salting-out cation. The phenomenon is associated with the formation of acidic species, described in the previous section (10.B), according to the formal equilibrium (7C77). Due to this, and another complicating factor originating from the effect of diluent on the extent of solute aggregation, equations (7C76a, b) only occasionally describe the extraction equilibrium satisfactorily. This and similar mass-action law equations give widely varying and inconsistent results as to the stoichiometric composition of the solute in the organic phase[52,286,291,296,298,300-305]. A notable example is perhaps that given by Lloyd and Mason[306], where the distribution data cf uranyl nitrate between acidic nitrate solutions and tridodecylamine in toluene have been quantitatively correlated by extraction equilibria involving monomeric and dimeric amine nitrate species, discussed in more detail in section 12.C.

d. Metal Chlorides

Numerous survey studies[56,117,238,259,307-311] on the extractability of metal chlorides from hydrochloric acid or a mixture of acid and alkali chloride solutions indicate a similar order of metal extractability irrespective of the amine class. However, quaternary amines are usually better extractants of metal chlorides than are tertiary amines, and these are better than secondary and primary amines. The shape of the extraction curves is usually very similar[238,259,311] as shown for several metals in Figure 10C1. Charts on the extraction of some sixty elements from HCl solutions into amines of the four classes are given in Appendix F. Polyvalent metals, readily forming anionic chloro complexes, exhibit the highest distribution ratios, followed by metals forming oxy anions and the bivalent transition metals. The lanthanides are only slightly extractable even at high amine concentrations.

Most frequently, the nature of the supporting electrolyte in the aqueous phase affects metal extractability, especially at high ionic strengths[210,211,219,238,259,299,307,311-313]. Generally, higher D values are obtained from alkali chloride solutions than from hydrochloric acid. A comparison of the curves in Figure 10C2 illustrates that effect[238,259,311]. The HCl effect of the appearance of a maximum in the extraction isotherms, is usually attributed, at least partially, to the

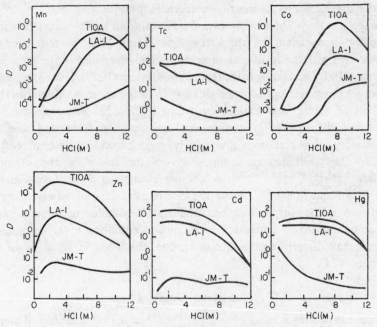

Figure 10C1. Comparison of extractability by various amine classes[311].

JM-T   0.28 M Primene JM-T in xylene
LA-1   0.20 M Amberlite LA-1 in xylene
TIOA   0.11 M Triiso-octylamine in xylene
(By permission from the USAEC)

interaction of the coextracted acid with the amine hydrochloride to form the HCl_2^- anion in terms of the equilibrium (7C77), on the assumption that the latter is taken up preferentially by the ammonium cation. Recent data[74] indicate, however, that, for example, the transfer of macro amounts of $FeCl_4^-$ into the amine phase decreases the amount of dichloride, and the amine is complexed to the tetrachloroferrate. Another possibility would be the lack of dissociation of the protonated anionic chlorometalate in concentrated HCl solutions, in contrast to an

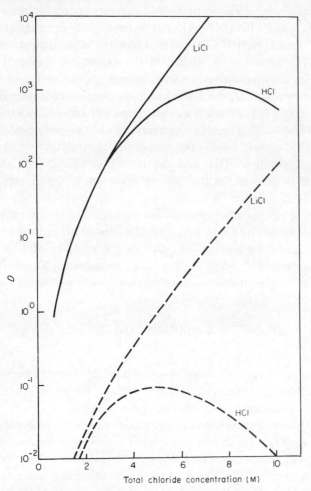

Figure 10C2. Extraction of iron (III) and vanadium (IV) (dashed lines) from hydrochloric acid or lithium chloride (0.2 M HCl) solutions into 0.1 M
Aliquate 336 in diethylbenzene (3 % tridecanol).

alkali salt of the chlorometal acid. For example, distribution data and analysis of a trioctylamine extract in xylene[234], showing the presence of one Li^+ ion per one Am^{3+} ion in the organic phase, suggest that Am^{3+} is extracted in the form of the complex $[(R_3NH^+)_2Li^+AmCl_6^{3-}]$, since low D's are obtained when the aqueous phase does not contain LiCl, though LiCl alone is non-extractable[314,315]. On the other hand, under similar aqueous-phase conditions, where D_{Am} has equally a second-power dependence upon the amine content (equation 7C76a), and is equally inversely proportional to approximately the second power of $[H^+]$, the ion-pair $[(R_3NH^+)_2AmCl_5^{2-}]$ or the quadrupole $[R_3NH^+AmCl_4^-R_3NH^+Cl^-]$ were suggested the organic-phase species[74]. The situation is similar in the extraction of uranyl chloride though its distribution ratio value depends on the nature of the supporting electrolyte[299], it is always second-power dependent on the amine concentration. However, the extracts from salt and acid solutions have different spectra[108]. The visible spectrum of a tertiary amine extract is that of $UO_2Cl_4^{2-}$ only if LiCl is the supporting electrolyte[108,301]. Hydrogen bonding between R_3NH^+ and the tetrachlorouranate(VI) anion was suggested to account for the different spectrum of the extract from an acid chloride solution[108].

This and similar discrepancies are perhaps due to the effect of the aqueous solution, its nature and concentration, on the process of aggregation of the amine-bearing ion-pairs in the organic solution. In this connexion Marcus[314] suggests that even an aggregated amine salt may still be only monoionized, and will exchange only one anion

$$(R_3NH^+A^-)_q \rightleftharpoons [(R_3NH^+)_q(A^-)_{q-1}]^+ + A^- \qquad (10C1)$$

Indeed, it has frequently been observed that the distribution ratio of a complex MA_{m+p}^{p-} varies with the pth power of the amine concentration, whereas a q-merization would require a p/q power.

Under given aqueous- and organic-phase conditions the order of metal extractability is what should be expected from the tendency of metals to form anionic chloro complexes. When the actinides are in their tetravalent oxidation state[210,301,316,317] the order of increasing extractability is Th < Pa < U < Np < Pu. In all these systems, D has a nearly second-power dependence on the amine concentration (equation 7C76a), and the spectra of uranium, neptunium and plutonium extracts are identical to those known to be of the MCl_2^{6-} anion[108,301]. Extractability of the tervalent transition metals is also governed by their tendency to form stable tetrachlorometallate ions in aqueous chloride solutions. The observed order being gold > gallium > iron > indium[238,259,307,311]. Indeed, the spectra of the metal species in tertiary amine extracts are invariably those of MCl_4^-. Distribution and loading data, however, are inconsistent with this spectral information, in addition to not being self-consistent[74,218,234,238,247,259,311,313,318–325]. The distribution ratios for iron and gallium show most frequently a second-power dependence with a tertiary amine, while a first-power dependence with the quaternary amines is found. Or, while thallium shows a first-power dependence, in the case of indium, the amine dependence tends towards limiting values of one and three, but with a second-power dependence over a wide range of amine concentration. These apparent discrepancies are believed to be due to the formation of mixed associates of the quadrupole type, $[R_3NH^+MCl_4^-R_3NH^+Cl^-]$, and possibly also to the existence of higher aggregates $[R_3NH^+Cl^-R_3NH^+MCl_4^-R_3NH^+Cl^-]$.

In quite a few cases the existence of an extractable chlorometalate anion in the aqueous solution is not necessarily a prerequisite for the extractability of the metal. Distribution data of zinc, and also of cobalt, under a variety of experimental conditions[313,326,327], loading experiments and spectral information, suggest the presence of four-coordinated complex ions in the extract, though in the aqueous phase tetrahedrally coordinated species could not be detected, $ZnCl^+$ being the predominating aqueous metal-bearing species[325,328]. Similarly, spectral evidence[314,329] has been presented to show that several tervalent actinides are in the form of cationic species MCl^{2+} even in concentrated aqueous lithium chloride solution, whereas anionic complexes prevail in the extract[74,314].

e. Metal Sulphates

A number of survey studies on the extractability of metals from sulphate media[1,2,310] indicate that the effect of hydrogen ion concentration in the aqueous solution of sulphates is much more important than in extractions from either chloride or nitrate media. This can be understood in view of the effect of pH on the sulphate–bisulphate equilibrium in an aqueous solution and if one knows that the bisulphate competes strongly with metal sulphate complexes for the available alkylammonium cation. This latter generalization can be deduced from the fact that the distribution ratio of metals, as a rule, decreases with increasing bisulphate loading of the amine phase. Charts in Appendix F illustrate these effects[310].

While primary and secondary amines are generally poor extractants from aqueous chloride or nitrate solutions, as compared to the tertiary amines, from sulphate solutions their extractive capacity is of a comparable magnitude or frequently even higher[266,330,331].

Irrespective of the amine class, a maximum in the distribution ratio of metal sulphates is observed at acidities lower than 0.1M which is usually followed by a very abrupt decrease as the bisulphate/sulphate ratio increases. If the acid is replaced by alkali or ammonium sulphate in the aqueous phase, the initially higher D's decrease much less with increased electrolyte concentration[332-337].

Distribution data suggest that there are several uranium(VI) sulphate species in the amine phase depending again on the acidity of the aqueous solution[110,153,168,177,239,265,337-339]. Earlier spectral, loading and distribution measurements tend to indicate that the composition of the organic-phase complex in equilibrium with dilute aqueous sulphuric acid solution is close to $[(R_3NH)_4UO_2(SO_4)_3]$. This is apparently true at least for the tertiary amines[168], in spite of the fact that the fraction of uranium complexed in the trisulphate form in an aqueous solution of below 1 M is negligible[168,339]. Kinetic studies on the individual rate constants of extraction of various aqueous uranium(VI)-bearing species under comparable experimental conditions indicate, however, that the UO^{2+} ion is about six times as effective as either UO_2SO_4 or $UO_2(SO_4)_2^{2-}$ in transferring uranium from an aqueous phase[236]. On the other hand, more recent interfacial tension measurements have been interpreted in terms of uranium(VI) extraction (by di-n-dodecylamine sulphate) via an exchange

$$UO_2(SO_4)_2^{2-} + 3\overline{(R_2NH_2)_2SO_4} \rightleftharpoons \overline{(R_2NH_2)_6UO_2(SO_4)_4} + SO_4^{2-}$$

(10C2)

and a neutral salt-transfer mechanism

$$UO_2SO_4 + 3\overline{(R_2NH_2)_2SO_4} \rightleftharpoons \overline{(R_2NH_2)_6UO_2(SO_4)_4},$$ (10C3)

the uranium(VI)-bearing species being monomeric[153].

With increasing aqueous sulphate activity, the distribution ratio of uranium decreases. The spectrum of the extract differs from that obtained by equilibration with dilute sulphate solutions[110]. The lack of agreement between distribution and loading data, obtained at high aqueous electrolyte concentrations, are presumably due to an extensive aggregation of the amine bisulphate.

f. Other Metal Salts

While the extraction of unprotonated monovalent oxyanions of the VII B group is readily achieved under a variety of conditions (pH. 1–13)[251,253, 274,350–355], that of polyvalent oxanions of the VI B group is insignificant from either HCl or H_2SO_4 solutions[117,311,343,345]. The extraction of MoO_4^{2-} from dilute nitric acid, and of chromium(VI) from any acid media (through the formation of $HCrO_4^-$ or $HCr_2O_7^-$ [346]) being notable exceptions.

Though the extractability of TcO_4^- and ReO_4^- is feasible from any mineral acid solution, the severity of competition by the macro-electrolyte anion is higher by nitrate than by either chloride or sulphate. Even from nitric acid media, the nitrate can quantitatively be exchanged by perrhenate in the organic phase[341]. Spectra of the extracts and distribution data suggest that the metal-bearing species in the organic phase has an amine-to-metal ratio of unity and is monomeric, though the formation of mixed quadrupoles, formed between the amine nitrate and amine pertechnetate or perrhenate ion-pairs $[R_3NH^+NO_3^-R_3NH^+TcO_4^-]$, has also been suggested[342].

D. EXTRACTION BY QUATERNARY PHOSPHONIUM, ARSONIUM AND SIMILAR COMPOUNDS

a. Introduction

Certain quaternary phosphonium and arsonium compounds have found wide use in extraction of simple acid radicals, oxyanions of metals, anionic inorganic coordination complexes and anionic coordination complexes involving organic ligands. Although various organic reagents have been investigated as potential extractants, only a few, those which are conjugate salts of strong acids and bases, came close to satisfying the requirements of an extractant, and even these seem to be practical for analytical purposes mainly. These strong electrolytes, usually water soluble and not free from emulsion-forming tendencies, are not particularly well suited to solutions of high concentration.

For this particular type of extractants, used in the past in analytical chemistry for the precipitation and preparation of sparingly soluble metal complexes, water solubility is not necessarily objectionable, since the majority of the ion-pairs formed in the course of the extraction process with these bulky cations, are preferentially distributed into a water-immiscible organic solvent. The metal complexes formed when the reagent is added to a neutral or weakly acidic aqueous solution have usually a well-defined stoichiometric composition, and as such they are transferred into a slightly polar organic phase such as chloroform, *o*-dichlorobenzene, methyl isobutyl ketone and others.

The extraction process is usually represented as the equivalent exchange of the complex ion for free-ligand ion, and the species existing in the organic phase are characterized as ion-pairs. There exist indeed many results which point to the fact that in all these non-ionizing solvents ion-pair formation is anticipated as the interaction between the bulky cations and the acid radical, simple or complex. Although neither the reaction to produce neutral species through ion-association, nor the degree of ion-association has been so far considered explicitly, the results on the relative extractability of these ion-pairs are consistent with the correlation of ion-association with the dielectric constant of the medium discussed earlier (section 2.D).

Additionally, these ion-association systems do not suffer from complexities occuring with systems involving the similar quaternary ammonium compounds. Since these compounds are soluble to a practical extent preferentially in polar solvents, there is no molecular association either of the extractant or of the extracted metal-bearing species. This is actually the main reason for their being reviewed separately.

b. Extraction by Salts of Polyphenyl-onium Bases

The process of extraction can be rationalized in simple terms of competition between the anion, usually chloride, originally associated with a bulky organic cation, such as tetraphenyl-arsonium or-phosphonium, and the anionic metal complex (MCl_4^-) or metal oxyanion (MO_4^-) for the available Ph_4As^+ or Ph_4P^+ to form closely associated ion-pairs. As discussed in detail earlier in section 7.C, this competition is quantitatively expressed through the equilibrium constants of two competing reactions (7C68a) and (7C68b), or in terms of the above example, and neglecting activity coefficients, through the equilibrium quotients

$$Ph_4As^+ + Cl^- \rightleftharpoons \overline{Ph_4As^+Cl^-} \quad K_{Cl} \tag{10D1}$$

$$Ph_4As^+ + MCl_4^- \rightleftharpoons \overline{Ph_4As^+MCl_4^-} \quad K_{MCl_4} \tag{10D2}$$

where the K's are the combined measure of stability of the ion-pairs and their distribution between the two phases. Obviously, if the quaternary organic salt is water soluble, which is usually the case, K_{Cl} will have a small value, and if MCl_4^- is large enough to enter an ion-pair with high enough degree of association, K_{MCl_4} will have a high value. In this case the metal distribution is given by equation (7C71).

Apart from the effect of the anion on the extractability of an ion-pair, which is similar for all of the ion-association systems, the size of the organic cation bears markedly on the extractive property of the ion-pair. The effect is qualitatively similar to that of quaternary ammonium salts, discussed in the previous section, but with two differences: an increased size of the extractant does not limit markedly its solubility in water, and does not affect its state of solute in either of the phases[347]. $(CH_3)_4As^+$ and $(CH_3)_3PhAs^+$ cations, for example, failed to form chloroform-extractable ion-pairs with the $Co(NCS)_4^{2-}$ complex, or o-dichlorobenzene-extractable ion-pairs with the SbI_4^- complex, whereas $(CH_3)_2Ph_2As^+$ and the higher phenyl derivatives, $CH_3Ph_3As^+$ and Ph_4As^+ show comparable extractability[348,349]. Surprisingly enough, even slight structural modifications, such as the substitution of the normal-alkyl chain in the triphenyl-n-propyl-phosphonium compound for the isomeric triphenylisopropyl-phosphonium may bring about marked differences in extractability of ion-pairs into certain solvents[350]. The former shows greater tendency to form extractable ion-pairs. By increasing the alkyl chain from methyl to heptyl there is apparently a steady increase in the extractability, but it remains constant, or even decreases slightly with further increase in the number of carbon atoms in the chain[350].

On the other hand, a generally close extractive property of similar large organic cations, at least toward simple anions and oxyanions, has been demonstrated by Bock and coworkers[351-354].

Distribution ratios for some anions with Ph_4As^+, Ph_4P^+, Ph_3S^+ and Ph_3Sn^+ cations are compiled in Table 10D1[351-354].

The distribution ratio of simple or complex anions is affected by the nature of the diluent used. Distribution ratios higher by several orders of magnitude are obtained in polar solvents, as shown in Table 10D2[355].

TABLE 10D1. Distribution ratio, D, of anions by large organic cations from acidic and basic aqueous solutions into chloroform.

	Ph$_4$As$^+$		Ph$_4$P$^+$		Ph$_3$S$^+$		Ph$_3$Sn$^+$	
pH*	~1.5	~12	~1.5	~12	~1.5	~12	~1.5	~4
F$^-$	<0.01	<0.01	<0.005	<0.005	<0.01	<0.01		
Cl$^-$	0.22	0.18	0.18	0.20	0.24	0.25	26	3.5
Br$^-$	4.6	4.7	3.3	3.0	0.64	0.53	41	2.7
I$^-$	>300	>300	54	54	17.1	14.4	420	10.5
CNS$^-$	34.7	32.3	>380	>380	~1.05	~0.9	1100	2.7
ClO$_3^-$	>150	>150	>100	>100				
BrO$_3^-$	0.9	0.8	0.48	0.46	0.40	0.39		
IO$_3^-$	<0.004	<0.004	<0.005	<0.005	0.003	0.003		
ClO$_4^-$	>200	>200	>200	>200	>100	>100		
MnO$_4^-$	>300	>300	>300	>300	>200	>200		
ReO$_4^-$	>200	>200	>600	>600	>100	>100	0.01	
IO$_4^-$		0.02		0.02				
NO$_3^-$	20.3	42.5	4.8	5.6	0.65	0.51	<0.002	
NO$_2^-$		0.22		0.11		0.13	34	6
SO$_4^{2-}$	<0.001	<0.001	<0.001	<0.001	>0.002	<0.002	<0.001	
SO$_3^{2-}$		0.015		<0.002				
S$_2$O$_3^{2-}$		0.017		<0.003				
PO$_4^{3-}$	~0.01	~0.01	0.031	0.025	0.11	0.08	11.1	0.12
P$_2$O$_7^{4-}$		0.003					3.1	0.01
CrO$_4^{2-}$	71	0.025	24.7	0.005	31.2	0.024		
MoO$_4^{2-}$		<0.005	<0.002	<0.002		<0.001		
WO$_4^{2-}$		<0.005		<0.003		<0.001		
VO$_4^{3-}$		<0.005		<0.001	<0.002	<0.002	0.66	
AsO$_3^{3-}$	0.006	0.047	<0.002	<0.002	<0.003	<0.003	1.75	2.5
AsO$_4^{5-}$	<0.01	<0.01	0.014	0.013	0.007	0.007	3.7	0.02
SeO$_3^{2-}$	<0.01	0.09	<0.002	<0.002	<0.003	0.3	18.7	2.9
TeO$_3^{2-}$	<0.01	0.09	<0.002	<0.002		0.015	0.013	

* Values of the appropriate pH of the aqueous phase at equilibrium.

As expected, no marked influence of the aqueous pH in the range of pH 1.3–12.5 on the extraction has been observed, with the obvious exception of weakly basic anions, such as the fluoride[356], and where the hydrogen ion concentration affects the equilibrium of the species in the

TABLE 10D2. Distribution ratios of chromium (VI) when extracted by triphenyl-tetrazolium into various organic solvents[355].

Solvent	Log $D_{Cr^{VI}}$
Cyclohexane	−3.19
Tetrachloroethylene	−3.07
Carbon tetrachloride	−2.69
Benzene	−2.66
Trichloroethylene	−2.62
Butyl acetate	−2.04
Chloroform	−0.48
n-Hexanol	+0.55
Nitrobenzene	+0.51

aqueous solution (chromate–dichromate)[355]. Large monovalent anions such as ReO_4^-, MnO_4^-, ClO_4^- and TcO_4^- [357] are practically completely stripped from the aqueous phase if the hydrogen ion concentration is kept above 0.1 M. Under similar conditions polyvalent oxyanions, due presumably to their greater hydration energy, remain preferentially in the aqueous phase.

Tetraphenyl arsonium and -phosphonium chlorides are, generally speaking, better extractants than the other two in Table 10D1, and better than several others screened for the purpose, such as triphenylselenonium or -telluronium, diphenyl-iodinium or tetraphenyl-stibonium[358,359]. Moreover, both triphenyl-sulphonium and triphenyltin are unstable in the presence of oxidizing agents, and form water- and benzene-insoluble compounds with a number of anions[351,354,360]. Triphenyllead, -tin and -antimony, and triphenylarsenic dihydroxide are capable of extracting also weak carboxylic acids[351,361].

A chart presenting the extractability of a large number of elements by Ph_4AsCl into chloroform as a function of hydrochloric acid concentration is given in Appendix F[362]. The distribution data obtained from concentrated HCl solutions are probably not reliable, as the reagent may decompose when in contact with concentrated acid[347].

Generally speaking, mononegative chlorometallic complexes are highly extractable; increasing HCl concentration does not affect their complete extractability. This, however, is not necessarily true for the equally mononegative oxyanions, where at high supporting-acid concentrations the depressed dissociation of HMO_4 reduces the extractability of the metal-bearing anion. The distribution ratio of dinegative chloro complexes is lower, though under somewhat different conditions the extractability of anions such as $HgCl_4^{2-}$, $SnCl_6^{2-}$, $OsCl_8^{2-}$, $RuCl_6^{3-}$ and other platinum group metal chloro complexes can be made quantitative[363-365]. Mono- and polyvalent thiocyanate complexes are more extractable than the corresponding chloro complexes[419-421].

c. Extraction by Other Types of Bulky Cations and Anions

There are several additional ways in which organic compounds can participate in formation of extractable ion-pairs. The majority of organic chelating agents, described in Chapter 8, form uncharged, thus extractable, complexes. Some complexants, however, possess reactive groups which enable them to form charged species. Such complex anions or cations containing a metal atom, may, by the proper use of a counter ion, form ion-associates. These ion-pairs are in many instances highly coloured and show a low solubility in water. In fact, most of these ion-pair extraction systems found application in photometric determination of the metal, incorporated into the charged complex anion or cation.

From the large variety of specific atomic groupings capable of forming chelates with metal ions, only those having the ferroin

$$= N - C - C - N =$$
$$\parallel \quad \parallel$$

and the cuproin

$$R - C = N - C - C - N = C - R$$
$$\mid \qquad \parallel \quad \parallel \qquad \mid$$

groupings seem to be the charged chelate-forming ligands. 1,10-Phenanthroline, 2,2'-dipyridyl, 2,2'2''-terpyridyl and their derivatives combine with metal ions not through the replacement of hydrogen atoms but through secondary valence forces, thus forming charged chelates. The coloured complex ions, such as $[Fe(2,2'-dip)_3]^{2+}$ and $[Fe(1,10-phen)_3]^{2+}$ or $[Cu(2,2'-dip)_2]^+$ and $[Cu(1,10-phen)_2]^+$ are well-known examples of such charged chelates. The coordination capability of these heterocyclic nitrogen compounds has been shown towards a number of bivalent metal ions[258,369,370], though various derivatives of these ligands exhibit quite a different behaviour towards metal ions. The differences have been attributed both to the coordination number of the metals and to the steric effect of the ligands[369,371]. Both tris- and bis-type molecular complexes are extraordinarily stable in both acid and alkalies.

The extent of extraction of these complexes from aqueous solutions depends on the anion associated with the complex cation to form the extractable ion-pair. Association with the apparently too small nitrate, chloride, sulphate or acetate anions will not cause extraction[369,372,373], but a nearly 100% extraction into chloroform or nitrobenzene was demonstrated from aqueous solution containing perchlorate, thiocyanate or iodide anions[258,373]. The range of pH for the complete extraction is between 2 and 10, which is the pH range of stability of the reagents. Altering the molecule of the ligand but without destroying its inherent capacity to react with the metal in forming a charged chelate, may improve some of the extractive characteristics of the compound formed[374-379]. The effect of solvent may also be of importance for widening the conditions favourable for separation of metals by the extractive technique[370,373,376]. However, many of the common solvents used in solvent extraction do not extract this type of ion-pairs[258].

An interesting example of extractable ion pairs is that involving the tetraphenyl borate as anion and the heavy alkali metals, rubidium, caesium, francium and also radium. The formation of ion-pairs is favoured by the increasing radius of the cation. Fr and Ra are extracted into nitrobenzene (at pH 9[380]) and also caesium and rubidium with as high distribution ratio as $> 10^3$ and $> 10^2$ respectively[381]. High D's are obtained for caesium tetraphenyl borate extraction into hexone, with little variation between pH 3-12[382].

A number of triphenylmethane dyes have been found to have a widespread application as extracting agents in analytical chemistry. The enormous variety of synthetic dyes makes it possible to choose a proper derivative with a favourable effect on extractability. Kuznetsov and Bolyshakova[383] have demonstrated, for example, that the length of the alkyl chain introduced in esterification of Rhodamine B, tetraethylrhodamine, affects the solubility and the extractability of the ion-pair.

Mention has to be made also of more simple heterocyclic compounds such as pyridinium and its derivatives lutidinium and anilinium cations. They form under certain conditions extractable ion-associates with oxyanions[251], metal–thiocyanate complex anions[384] and negatively charged metal complexes with organic ligands such as salicylic[385] and acetic acid[386]. It has to be noted, however, that such cyclic nitrogen bases as pyridine may likewise function as a neutral portion of some halo- and thiocyanato–metal complexes. These large mixed complexes are only sparingly water soluble and some of them such as $[Cu(py)_2X]$, $[Cu(py)_3X]$ where $X = Cl^-$, Br^- or I^-, and $[Cu (py)_2(SCN)_2]$ are extractable even into the non-polar carbon tetrachloride, benzene and hexane, but with a higher distribution ratio into chloroform[387,388]. The requirements for the transfer of the metals into an organic phase are similar to those discussed in Chapter 8 for the extraction of uncharged chelates.

Among the five-membered ring systems with two hetero atoms, the pyrazole derivatives of the antipyrine series have shown potentialities as extractants[252,389,390]. More complicated heterocyclic compounds of two- or three-ring condensed systems, quinoline[385,391] and acridine[252] and phenyl derivatives of triazole and tetrazole[252] have also been among those screened for extraction of inorganic anions.

E. REFERENCES

1. C. F. Coleman, K. B. Brown, J. G. Moore and D. J. Crouse, *Ind. Eng. Chem.*, **50**, 1756 (1958).

2. C. F. Coleman, C. A. Blake, Jr. and K. B. Brown, *Talanta*, **9**, 297 (1962).

3. K. B. Brown, D. J. Crouse and W. D. Arnold, *USAEC Rept.*, ORNL-TM-265 (1962).

4. H. G. Petrov, O. A. Nietzel and J. C. Apidianakis, *USAEC Rept.*, WIN–61 (1957).

5. A. C. Rice and C. A. Stone, *US Bur. Mines Rept. Invest.*, RI–5923 (1962).

6. J. A. Brothers, R. G. Hart and W. G. Mathers, *J. Inorg. Nucl. Chem.*, **7**, 85 (1958).

7. A. W. Ralston, C. W. Hoerr, W. O. Pool and H. J. Harwood, *J. Org. Chem.*, **9**, 102 (1944).

8. R. R. Dreisbach, *Physical Properties of Chemical Compounds*, Vol. III, ACS, Advances in Chemistry Series, No. 29, Washington, 1961.

9. C. W. Hoerr, H. J. Harwood and A. W. Ralston, *J. Org. Chem.*, **11**, 199 (1946).

10. O. Westphal and D. Jerchel, *Chem. Ber.*, **73B**, 1002 (1940).

11. O. Westphal, *Chem. Ber.*, **74B**, 1365 (1941).

12. J. v. Braun and R. Klar, *Chem. Ber.*, **73B**, 1417 (1940).

13. P. A. Winsor, *Solvent Properties of Amphiphilic Compounds*, Butterworths Sci. Publ., London, 1954.

14. M. E. L. McBain and E. Hutchinson, *Solubilization*, Academic Press, New York, 1955.

15. R. R. Davison, A. F. Isbell, W. H. Smith, Jr. and D. W. Hood, *Ann. Rept., Office Saline Water*, U. S. Department of Interior, 1958 and 1961.

16. R. R. Davison, W. H. Smith, Jr. and D. W. Hood, *J. Chem. Eng. Data*, **5**, 420 (1960); *ACS, Advances in Chemistry Series*, No. 27, Washington, 1960; *J. Chem. Eng. Data*, **11**, 304 (1966).

17. M. D. Gregory, S. D. Christian and H. E. Affsprung, *J. Phys. Chem.*, **71**, 2283 (1964) and references therein.

18. A. W. Ralston, C. W. Hoerr and E. J. Hoffman, *J. Am. Chem. Soc.*, **64**, 1516 (1942).

19. V. S. Shmidt and V. N. Shesterikov, *Zh. Fiz. Khim.*, **39**, 440 (1965).

20. A. W. Ralston, *Fatty Acids and their Derivatives*, J. Wiley & Sons, New York 1948, p. 658.

21. R. Foster, *Chem. Ind.* (*London*), 1354 (1960) and references therein.

22. J. H. Hildebrand and R. L. Scott, *Regular Solutions*, Prentice-Hall, Englewood Cliffs, N. J., 1963.

23. A. S. Kertes, *J. Inorg. Nucl. Chem.*, **26**, 1764 (1964).

24. H. Wolff and D. Staschewski, *Z. Electrochem.*, **65**, 840 (1961).

25. H. K. Hall, *J. Am. Chem. Soc.*, **79**, 5441 (1957).

26. V. S. Shmidt and E. A. Mezhov, *Usp. Khim.*, **34**, 1388 (1965).

27. J. Clark and D. D. Perrin, *Quart. Rev.* (*London*), **18**, 295 (1964).

28. D. D. Perrin, *Dissociation Constants of Organic Bases in Aqueous Solutions*, Supl. Pure and Appl. Chem., 1965.

29. C. W. Hoerr, M. R. McCorkle and A. W. Ralston, *J. Am. Chem. Soc.*, **65**, 328 (1943).

30. A. F. Trotman-Dickenson, *J. Chem. Soc.*, 1293 (1949); G. Briegleb, *Z. Elektrochem.*, **53**, 350 (1949); R. G. Pearson, *J. Am. Chem. Soc.*, **70**, 204 (1948); H. A. Staab, *Einführung in die theoretische Organische Chemie*, Verlag Chemie, Weinheim, 1959, p. 630; W. S. Muney and J. T. Coetzee, *J. Phys. Chem.*, **66**, 89 (1962).

31. H. C. Brown and M. D. Taylor, *J. Am. Chem. Soc.*, **69**, 1332 (1947).

32. W. Gordy and S. C. Stanford, *J. Chem. Phys.*, **9**, 204 (1941).

33. M. Tamres, S. Searles, E. M. Leighly and D. W. Mohram, *J. Am. Chem. Soc.*, **76**, 3983 (1954).

34. H. K. Hall, Jr., *J. Phys. Chem.*, **60**, 63 (1956).

35. G. N. Chremos and H. K. Zimmermann, *Chimia*, **18**, 265 (1964).

36. R. Stewart and J. O'Donnel, *Can. J. Chem.*, **42**, 1694 (1964).

37. R. W. Taft, Jr., *Steric Effects in Organic Chemistry* (Ed. M. S. Newman), J. Wiley & Sons, New York, 1956, Chap. 13.

38. N. A. Palm, *Usp. Khim.*, **30**, 1069 (1961).

39. J. Roberts and H. H. Jaffe, *Tetrahedron*, **19**, 455 (1963).

40. L. P. Hammett, *Physical Organic Chemistry*, McGraw-Hill Book Co., New York, 1940, Chap. 7.

41. N. A. Izmailov and T. V. Mozharova, *Zh. Fiz. Khim.*, **34**, 1709 (1960).

42. J. W. Bayles and A. F. Taylor, *J. Chem. Soc.*, 417, (1961).

43. R. R. Grinstead in *Solvent Extraction Chemistry* (Eds. D. Dyrssen, J. O. Liljenzin and J. Rydberg), North-Holland, Amsterdam, 1967, p. 426.

44. R. P. Bell and J. W. Bayles, *J. Chem. Soc.*, 1518 (1952); J. W. Bayles and A. Chetwyn, *J. Chem. Soc.*, 2328 (1958); R. P. Bell and J. E. Crooks, *J. Chem. Soc.*, 3513 (1962).

45. R. G. Pearson and D. C. Vogelsong, *J. Am. Chem. Soc.*, **80**, 1038 (1958).

46. A. Rieure, M. Pumeau and B. Tremillon, *Bull. Soc. Chim. France*, 1053 (1964).

47. W. Ostwald, *Kolloid-Z.*, **45**, 56 (1928); O. Fuchs and K. L. Wolf, *Dielektrische Polarisation*, Akad. Verlag, Leipzig, 1935.

48. W. J. McDowell and K. A. Allen, *J. Phys. Chem.*, **63**, 747 (1959).

49. J. Errera and M. L. Sherrill in *The Dipole Moment and Chemical Structure* (Ed. P. Debye), Blackie and Son, London, 1931, p. 39.

50. D. Gourisse, *French Rept.*, CEA–R–3005 (1966); in *Solvent Extraction Chemistry* (Eds. D. Dyrssen, J. O. Liljenzin and J. Rydberg), North-Holland, Amsterdam, 1967, p. 373.

51. J. M. Atwood and W. A. Snyder, *USAEC Rept.*, HW–62000 (1959); T. Ichikawa and S. Urono, *Bull. Chem. Soc. Japan*, **33**, 569 (1960); T. Ishihara, T. Tsujino and Y. Komaki, *Nippon Genshiryoku Gakkaishi*, **4**, 307 (1962); G. Smith and G. A. Swan, *J. Chem. Soc.*, 886 (1962) and references therein; A. Chesne, G. Koehly and A. Bathellier, *Nucl. Sci. Eng.*, **17**, 557 (1963); F. Baroncelli, G. Calleri, A. Moccia, G. Scibona and M. Zifferero, *Nucl. Sci. Eng.*, **17**, 298 (1963).

52. C. F. Coleman, *At. Energy Rev.*, **2**, 3 (1964).

53. F. Baroncelli and G. Grossi, private communication (1966).

54. H. Freiser, *Anal. Chem.*, **34**, 64R (1962); **36**, 93R (1964); **38**, 131R (1966).

55. A.W. Ralston and D. N. Eggenberger, *J. Am. Chem. Soc.*, **70**, 436 (1948).

56. J. T. Hazel and H. O. Strange, *J. Colloid Sci.*, **12**, 529 (1958).

57. M. Shirai and B. Tamamushi, *Bull. Chem. Soc. Japan*, **29**, 733 (1956).

58. L. M. Kushner, W. D. Bubberd and R. A. Parker, *J. Res. Natl. Bur. Stds.*, **59**, 113 (1951).

59. E. Hutchinson and C. Winslow, *Z. Physik. Chem. (Frankfurt)*, **11**, 165 (1957).

60. A. S. Kertes, *J. Inorg. Nucl. Chem.*, **27**, 209 (1965).

61. H. King and T. S. Work, *J. Chem. Soc.*, 1307 (1940).

62. V. C. Petrillo, *J. Am. Chem. Soc.*, **73**, 2381 (1951).

63. C. D. Strehlow, *M.Sc. Thesis.*, Mass. Inst. of Techn., Cambridge, 1964.

64. G. Scibona, R. A. Nathan, A. S. Kertes and J. W. Irvine, Jr., *J. Phys. Chem.*, **70**, 735 (1966).

65. G. Markovits and A. S. Kertes in *Solvent Extraction Chemistry* (Eds. D. Dyrssen, J. O. Liljenzin and J. Rydberg), North-Holland, Amsterdam, 1967, p. 390

66. A. S. Kertes, G. Gutman, O. Levy and G. Markovits, *Israel J. Chem.*, **6**, (1968).

67. J. I. Bullock, S. S. Choi, D. A. Goodrick, D. G. Tuck and E. J. Woodhouse, *J. Phys. Chem.*, **68**, 2687 (1964).

68. K. A. Allen, *J. Phys. Chem.*, **60**, 239, 943 (1956).

69. A. A. Lipovskii and S. A. Nikitina, *Zh. Neorg. Khim.*, **10**, 176 (1965).

70. J. Soriano, unpublished results, Jerusalem, 1966.

71. W. O. Pool, H. J. Harwood and A.W. Ralston, *J. Am. Chem. Soc.*, **67**, 775 (1945).

72. C. W. Hoerr and A. W. Ralston, *J. Am. Chem. Soc.*, **64**, 2824 (1942).

73. A. S. Kertes, unpublished results, 1962.

74. W. Müller, G. Duyckaerts and J. Fuger, *Solvent Extraction of Metals* (Eds. H. A. C. McKay, T. V. Healy, I. L. Jenkins and A. Naylor), Macmillan, London, 1966, p. 233.

75. V. N. Shesterikov and V. S. Shmidt, *Zh. Fiz. Kim.*, **39**, 3007 (1965).

76. Yu. G. Frolov and V. V. Sergievskii, *Zh. Neorg. Khim.*, **10**, 697 (1965).

77. B. Chenon and C. Sandorfy, *Can. J. Chem.*, **36**, 1181 (1958).

78. C. Brissette and C. Sandorfy, *Can. J. Chem.*, **38**, 34 (1960).

79. A. Cabana and C. Sandorfy, *Spectrochim. Acta.* **18**, 843 (1962).

80. D. J. Stone, J. C. Craig and H. W. Thompson, *J. Chem. Soc.*, **52**, (1958).

81. R. A. Heacock and L. Marion, *Can. J. Chem.*, **34**, 1782 (1956).

82. R. D. Waldron, *J. Chem. Phys.*, **21**, 734 (1953).

83. K. Nakanishi, T. Goto and M. Ohashi, *Bull. Chem. Soc. Japan*, **30**, 403 (1957).

84. J. Bellanato, *Spectrochim. Acta*, **16**, 1344 (1960).

85. J. Bellanato and J. R. Matutano, *Am. Soc. Espan. Fis. Quim.*, **52B**, 469 (1956).

86. J. Despas, J. Khaladji and R. Vergoz, *Bull. Soc. Chim. France*, 1105 (1953).

87. R. S. Silas, *Ph. D. Thesis*, University of Florida, Tallahassee, 1955.

88. R. C. Lord and R. E. Merrifield, *J. Chem. Phys.*, **21**, 166 (1953); R. E. Merrifield, *Ph. D. Thesis*, Mass. Inst. Techn., Cambridge, 1953.

89. R. S. McDowell, *Ph. D. Thesis*, Mass. Inst. Techn., Cambridge, 1960.

90. C. G. Swain and Y. Okamoto, *Lab. Nucl. Sci. Progr. Rep.*, Mass. Inst. Techn., November 1960.

91. C. D. Strehlow, A. S. Kertes and J. W. Irvine, *Lab. Nucl. Sci. Progr. Rep.*, Mass. Inst. Techn., May 1964.

92. K. J. Pedersen, *J. Am. Chem. Soc.*, **56**, 2615 (1934); C. R. Witschouke and C. A. Kraus, *J. Am. Chem. Soc.*, **69**, 2472 (1947).

93. V. M. Kazakova and L. S. Feldshtein, *Zh. Fiz. Khim.*, **35**, 488 (1961).

94. P. P. Shorygin and A. K. Khalidov, *Zh. Fiz. Khim.*, **25**, 1475 (1951).

95. M. M. Davis and M. Paabo, *J. Am. Chem. Soc.*, **82**, 5081 (1960).

96. G. Herzberg, *Infrared and Raman Spectra*, Van Nostrand, New York, 1945, p. 295.

97. T. C. Waddington, *J. Chem. Soc.*, 4340 (1958).

98. R. H. Nuttall, D. W. A. Sharp and T. C. Waddington, *J. Chem. Soc.*, 4965 (1960).

99. K. Nakamoto, M. Margoshes and R. E. Rundle, *J. Am. Chem. Soc.*, **77**, 6480 (1955).

100. J. P. Mathieu and H. Poulet, *Spectrochim. Acta*, **16**, 696 (1960).

101. M. M. Davis, *J. Res. Natl. Bur Std.*, **31**, 221 (1947); M. M. Davis and M. B. Hetzer, *J. Res. Natl. Bur. Std.*, **48**, 381 (1952); G. M. Barrow and E. A. Yerger, *J. Am. Chem. Soc.*, **76**, 5211 (1954); E. H. Yerger and G. M. Barrow, *J. Am. Chem. Soc.*, **77**, 6206 (1955); H. van Looy and L. P. Hammet, *J. Am. Chem. Soc.*, **81**, 3872 (1959).

102. V. M. Vdovenko, A. A. Lipovskii and S. A. Nikitina, *Radiokhimiya*, **6**, 56 (1964).

103. V. M. Vdovenko, T. A. Demiyanova, M. G. Kuzina and A. A. Lipovskii, *Radiokhimiya*, **6**, 49 (1964).

104. W. E. Keder, A. S. Wilson and L. L. Burger, *USAEC Rept.*, HW–SA–2959 (1963).

105. G. Duyckaerts and W. Müller, *Euratom Rept.*, EUR–426f (1963).

106. J. M. P. J. Verstegen, *Trans. Faraday Soc.*, **58**, 1878 (1962).

107. A. S. Wilson and N. A. Wogman, *J. Phys. Chem.*, **66**, 1552 (1962).

108. J. L. Ryan, *Inorg. Chem.*, **2**, 348 (1963); **3**, 211 (1964).

109. J. M. P. J. Verstegen, *J. Inorg. Nucl. Chem.*, **26**, 25 (1964).

110. W. E. Keder and A. S. Wilson, *Nucl. Sci. Eng.*, **17**, 287 (1963).

111. A. S. Kertes and G. Markovits in *Thermodynamics of Nuclear Materials*, IAEA, Vienna, 1968, p. 227.

112. J. A. Geddes and C. A. Kraus, *Trans. Faraday Soc.*, **32**, 585 (1936).

113. C. A. Kraus, *J. Phys. Chem.*, **60**, 129 (1956).

114. N. Pilpel, *Chem. Rev.*, **63**, 221 (1963).

115. M. J. Rosen and H. A. Goldsmith, *Systematic Analysis of Surface-Active Agents*, Interscience Publ., New York, 1960.

116. A. M. Wilson, L. Churchill, K. Kiluk and P. Hovsepian, *Anal. Chem.*, **34**, 203 (1962).

117. W. J. Maeck, G. L. Booman, M. E. Kussy and J. E. Rein, *Anal. Chem.* **33**, 1775 (1961); **34**, 212 (1962).

118. H. Kimizuka and I. Satake, *Bull. Chem. Soc. Japan*, **35**, 251 (1962).

119. H. V. Tartar and A. L. M. Lelong, *J. Phys. Chem.*, **59**, 1185 (1955).

120. M. L. Corrin and W. D. Harkins, *J. Am. Chem. Soc.*, **69**, 683 (1947).

121. I. Cohen and T. Vassiliades, *J. Phys. Chem.*, **65**, 1774 (1961).

122. J. F. Voeks and H. V. Tartar, *J. Phys. Chem.*, **59**, 1190 (1955).

123. C. A. Kraus, *Ann. N. Y. Acad. Sci.*, **51**, 789 (1949).

124. H. S. Frank and W. Y. Wen, *Discussions Faraday Soc.*, **24**, 133 (1957).

125. R. H. Stokes, *Trans. Faraday Soc.*, **59**, 761 (1963) and references therein.

126. H. S. Frank, *J. Phys. Chem.*, **67**, 1554 (1963).

127. S. Lindenbaum and G. E. Boyd, *J. Phys. Chem.*, **68**, 911 (1964).

128. G. E. Boyd, A. Schwartz and S. Lindenbaum, *J. Phys. Chem.*, **70**, 821 (1966).

129. M. A. V. Devanathan and M. J. Fernando, *Trans. Faraday Soc.*, **58**, 784 (1962).

130. V. E. Bower and R. E. Robinson, *Trans. Faraday Soc.*, **59**, 1717 (1963).

131. R. M. Diamond, *J. Phys. Chem.*, **67**, 2513 (1963).

132. S. Lindenbaum, *J. Phys. Chem.*, **70**, 814 (1966).

133. F. M. Batson and C. A. Kraus, *J. Am. Chem. Soc.*, **56**, 2017 (1934).

134. E. A. Richardson and R. H. Stern, *J. Am. Chem. Soc.*, **82**, 1296 (1960).

135. K. Bange and J. W. Smith, *J. Chem. Soc.*, 4244 (1964).

136. R. L. Buckson and S. G. Smith, *J. Phys. Chem.*, **68**, 1875 (1964).

137. H. Sadeck and R. M. Fuoss, *J. Am. Chem. Soc.*, **81**, 4507 (1959).

138. E. Hirsch and R. M. Fuoss, *J. Am. Chem. Soc.*, **82**, 1018 (1960).

139. S. Basol, A. S. Kertes and J. W. Irvine, Jr., *USAEC Rept.*, NYO–10066 (1964).

140. T. R. Griffiths and M. C. R. Symons, *Mol. Phys.*, **3**, 90 (1960).

141. D. Nicholls and M. Szwarc, *J. Phys. Chem.*, **71**, 2828 (1968).

142. F. S. Larkin, *Trans. Faraday Soc.*, **59**, 403 (1963).

143. A. D'Aprano and R. M. Fuoss, *J. Phys. Chem.*, **67**, 1704 (1963).

144. A. K. Bodenseh and J. B. Ramsey, *J. Phys. Chem.*, **67**, 140 (1963).

145. J. B. Hyne, *J. Am. Chem. Soc.*, **85**, 304 (1963).

146. A. D'Aprano and R. M. Fuoss, *J. Phys. Chem.*, **67**, 1871 (1963).

147. E. Högfeldt and F. Fredlund, *Trans. Roy. Inst. Techn.*, Stockholm, Nos. 226 and 227 (1964).

148. E. Högfeldt and M. de Jesus Tavares, *Trans Roy. Inst. Techn.*, Stockholm, No. 228 (1964).

149. E. Högfeldt, F. Fredlund and K. Rasmussen, *Trans. Roy. Inst. Techn.*, Stockholm, No. 229 (1964).

150. E. Högfeldt and F. Fredlund, *Acta Chem. Scand.*, **18**, 543 (1964).

151. M. A. Lodhi and E. Högfeldt in *Solvent Extraction Chemistry* (Eds. D. Dyrssen, J. O. Liljenzin and J. Rydberg), North-Holland, Amsterdam, 1967, p. 421; M. A. Lodhi, *Arkiv Kemi*, **27**, 309 (1967).

152. K. A. Allen, *J. Phys. Chem.*, **62**, 1119 (1958).

153. W. J. McDowell and C. F. Coleman in *Solvent Extraction Chemistry* (Eds. D. Dyrssen, J. O. Liljenzin and J. Rydberg), North-Holland, Amsterdam, 1967, p. 540.

154. V. V. Fomin and V. T. Potapova, *Zh. Neorg. Khim.*, **8**, 990 (1963).

155. A. S. Kertes and Y. E. Habousha, *J. Inorg. Nucl. Chem.*, **25**, 1531 (1963).

156. J. J. Bucher and R. M. Diamond, *J. Phys. Chem.*, **69**, 1565 (1965).

157. M. M. Davis, *J. Am. Chem. Soc.*, **84**, 3625 (1962).

158. J. M. P. J. Verstegen, *J. Inorg. Nucl. Chem.*, **27**, 201 (1965).

159. V. M. Vdovenko, B. Y. Galkin and A. A. Chaikorskii, *Radiokhimiya*, **3**, 448 (1961).

160. C. F. Coleman and J. W. Roddy in *Solvent Extraction Chemistry* (Eds. D. Dyrssen, J. O. Liljenzin and J. Rydberg), North-Holland, Amsterdam, 1967, p. 362.

161. A. S. Kertes and Y. Habousha, *Proc. U.N. Intern. Conf. Peaceful Uses At. Energy Geneva, 3rd.*, **10**, 392 (1964).

162. A. S. Kertes in *Solvent Extraction Chemistry of Metals* (Eds. H. A. C. McKay, T. V. Healy, I. L. Jenkins and A. Naylor), Macmillan, London, 1966, p. 377.

163. D. T. Copenhafer and C. A. Kraus, *J. Am. Chem. Soc.*, **73**, 4557 (1951).

164. V. V. Fomin, P. A. Zagorets and A. F. Morgunov, *Zh. Neorg. Khim.*, **4**, 700 (1959).

165. G. Scibona, S. Basol, P. R. Danesi and F. Orlandini, *J. Inorg. Nucl. Chem.*, **28**, 1441 (1966).

166. A. S. Kertes and G. Markovits, unpublished data, 1966.

167. I. J. Gal, private communication, 1966.

168. W. J. McDowell and C. F. Baes, Jr., *J. Phys. Chem.*, **62**, 777 (1958).

169. S. Y. Tyree, Jr. and J. H. Patterson, private communication, 1966.

170. P. R. Danesi and F. Orlandini, private communication, 1966.

171. B. Warnqvist, *Acta Chem. Scand.*, **21**, 1353 (1967).

172. W. Müller and R. M. Diamond, *J. Phys. Chem.*, **70**, 3469 (1966).

173. *Chem. Techn. Div.*, *Ann. Progr. Rept.*, ORNL-3945, (1966) p. 186.

174. J. M. P. J. Verstegen and J. A. A. Ketelaar, *Trans. Faraday Soc.*, **57**, 1527 (1961).

175. J. M. P. J. Verstegen and J. A. A. Ketelaar, *J. Phys. Chem.*, **66**, 216 (1962).

176. A. S. Wilson in *Solvent Extraction Chemistry* (Eds. D. Dyrssen, J. O. Liljenzin and J. Rydberg), North-Holland, Amsterdam, 1967, p. 369.

177. O. E. Zvyagintsev, Yu. G. Frolov, Ch. Chin-pang and A. V. Valkov, *Zh. Neorg. Khim.*, **10**, 981 (1965).

178. P. J. Lloyd and E. A. Mason, *J. Phys. Chem.*, **68**, 3120 (1964).

179. P. R. Danesi and E. Högfeldt, *Italian Rept.*, CNEN RT/CHI(66) 38 (1966).

180. F. Orlandini, P. R. Danesi, S. Basoł and G. Scibona in *Solvent Extraction Chemistry*, (Eds. D. Dyrssen, J. O. Liljenzin and J. Rydberg) North-Holland, Amsterdam, 1967. p. 408.

181. U. Bertocci, *Brit. Rept.*, *AERE*, R 2933 (1959).

182. A. S. Wilson, *Proc. U. N. Intern. Conf. Peaceful Uses At. Energy, Geneva, 2nd*, **17**, 348 (1958).

183. D. J. Carswell and J. J. Lawrence, *J. Inorg. Nucl. Chem.*, **11**, 69 (1959).

184. V.C.A. Vaughen and E. A. Mason, *USAEC Rept.*, TID–12665 (1960).

185. F. Baroncelli, G. Scibona and M. Zifferero, *J. Inorg. Nucl. Chem.*, **24**, 405 (1962).

186. U. Bertocci and G. Rolandi, *J. Inorg. Nucl. Chem.*, **23**, 323 (1961).

187. O. I. Zakharov-Nartsissov and A. V. Ochkin, *Zh. Neorg. Khim.*, **6**, 1936 (1961); **7**, 665 (1962).

188. V. B. Shevchenko, V. S. Shmidt, E. A. Nenarokomov and K. A. Petrov, *Zh. Neorg. Khim.*, **5**, 1852 (1960).

189. E. A. Mezhov, A. A. Pushkov and V. S. Shmidt, *Zh. Neorg. Khim.*, **7**, 932 (1962).

190. A. S. Kertes and I. T. Platzner, *J. Inorg. Nucl. Chem.*, **24**, 1417 (1962).

191. D. E. Horner and C. F. Coleman, *USAEC Rept.*, ORNL–3051 (1961).

192. E. Högfeldt, B. Bolander and F. Fredlund, *Trans. Roy. Inst. Techn., Stockholm*, Nos. 224–227 (1964), E. Högfeldt, F. Fredlund and K. Rasmusson, *Trans. Roy. Inst. Techn., Stockholm*, No. 229 (1964).

193. R. E. Skavdahl and E. A. Mason, *Trans. Am. Nucl. Soc.*, **5**, 463 (1962), *USAEC Rept.*, MITNE–20 (1962).

194. W. Knoch, *J. Inorg. Nucl. Chem.*, **27**, 2075 (1965).

195. O. E. Zvyagintsev, Yu. G. Frolov, A. A. Pushkov and B. Dushek, *Zh. Neorg. Khim.*, **10**, 512 (1965).

196. T. Sato, *J. Appl. Chem.*, **15**, 10 (1965).

197. J. M. P. J. Verstegen, *J. Inorg. Nucl. Chem.*, **26**, 2311 (1964).

198. T. H. Timmins and E. A. Mason, *USAEC Rept.*, MITNE–30 (1963).

199. V. M. Vdovenko, M. P. Kovalskaya and E. A. Smirnova, *Radiokhimiya*, **3**, 403 (1961).

200. V. M. Vdovenko, A. A. Lipovskii and M. G. Kuzina, *Radiokhimiya*, **3**, 555 (1961).

201. J. M. P. J. Verstegen, *J. Inorg. Nucl. Chem.*, **26**, 1085 (1964).

202. V. M. Vdovenko, A. A. Lipovskii and S. A. Nikitina, *Radiokhimiya*, **6**, 44 (1964).

203. J. C. Peak, *M. Sc. Thesis*, Mass. Inst. Techn., Cambridge, (1959).

204. M. Talat-Erben, *Eurochemic Rept.*, ETR–83 (1960).

205. E. Högfeldt and B. Bolander, *Acta. Chem. Scand.*, **18**, 548 (1964).

206. J. M. P. J. Verstegen, *J. Inorg. Nucl. Chem.*, **26**, 1589 (1964).

207. S. S. Choi and D. G. Tuck, *J. Phys. Chem.*, **68**, 2712 (1964).

208. L. Newman and P. Klotz, *J. Phys. Chem.*, **65**, 796 (1961).

209. J. Bizot and B. Tremillon, *Bull. Soc. Chim. France*, 122 (1959).

210. V. B. Shevchenko, V. S. Shmidt and E. A. Mezhov, *Zh. Neorg. Khim.*, **5**, 1911 (1960).

211. A. S. Kertes, K. Kimura, C. R. Schlaijer and J. W. Irvine, *Jr.*, *USAEC Rept.*, NYO–10063 (1963).

212. C. M. Davidson and R. F. Jameson, *Trans. Faraday Soc.*, **60**, 2845 (1964).

213. S. Lindenbaum and G. E. Boyd, *USAEC Rept.*, ORNL–3320 (1962).

214. G. Scibona, F. Orlandini and P. R. Danesi, *J. Inorg. Nucl. Chem.*, **28**, 1701 (1966).

215. P. R. Danesi and F. Orlandini, *CNEN Internal Rept.*, No. 232 (1965).

216. V. M. Vdovenko, M. P. Kovalskaya and E. A. Smirnova, *Radiokhimiya*, **4**, 610 (1962).

217. F. L. Moore, *Anal. Chem.*, **29**, 1660 (1957).

218. A. D. Nelson, J. L. Fasching and R. L. McDonald, *J. Inorg. Nucl. Chem.*, **27**, 439 (1965).

219. M. L. Good, S. E. Bryan, F. F. Holland and G. J. Maus, *J. Inorg. Nucl. Chem.*, **25**, 1167 (1963).

220. C. D. Strehlow, J. W. Irvine, Jr. and A. S. Kertes, unpublished data, 1964.

221. F. Kaufler and E. Kunz, *Chem. Ber.*, **42**, 385, 2482 (1909); F. Ephraim, *Chem. Ber.*, **47**, 1828 (1914).

222. T. C. Waddington, *J. Chem. Soc.*, 343, 1708 (1958).

223. J. D. Cotton, J. A. Salthouse and T. C. Waddington, *Proc. 8th Intern. Conf. Coord. Chem.*, Vienna, 1964.

224. D. H. McDaniel and R. E. Vallée, *Inorg. Chem.*, **2**, 996 (1963).

225. D. G. Tuck and E. J. Woodhouse, *Proc. Chem. Soc.*, 53 (1963).

226. T. C. Waddington and J. A. White, *J. Chem. Soc.*, 502, 2701 (1963).

227. K. M. Harmon and P. A. Gebauer, *Inorg. Chem.*, **2**, 1319 (1963).

228. L. Pauling, *The Nature of the Chemical Bond*, 3rd ed., Cornell University Press, Ithaca, 1960, p. 460.

229. S. Chang and E. F. Westrum, *J. Chem. Phys.*, **36**, 2571 (1962).

230. H. F. Herbrandson, R. T. Dickerson, Jr. and J. Weinstein, *J. Am. Chem. Soc.*, **76**, 4046 (1954).

231. *Oak Ridge National Laboratory, Chem, Techn. Div., Ann. Progr. Rept.,* USAEC Rept., ORNL–3452 (1963).
232. I. D. Forrester, M. E. Senko, A. Zalkin and D. H. Templeton, *Acta Cryst.*, **16**, 58 (1963).
233. C. D. Strehlow, A. S. Kertes and J. W. Irvine, Jr., *LNS Progr. Rept.*, Mass. Inst. Techn. Cambridge, May 1964; J. Soriano, Soreq Nuclear Research Centre, private communication (1967).
234. W. Smulek and S. Siekierski, *J. Inorg. Nucl. Chem.*, **24**, 1651 (1962).
235. S. Lindenbaum and G. E. Boyd, *J. Phys. Chem.*, **66**, 1383 (1962).
236. K. A. Allen, *J. Phys. Chem.*, **64**, 667 (1960).
237. C. Boirie, *French Rept.*, CEA–1262 (1960); *Bull. Soc. Chim. France*, 1088 (1958).
238. K. B. Brown, D. J. Crouse and W. B. Howertone, *USAEC Rept.*, ORNL–TM–265 (1962).
239. T. Sato, *J. Inorg. Nucl. Chem.*, **24**, 1267 (1962); **25**, 441 (1963).
240. E. L. Smith and J. E. Page, *J. Soc. Chem. Ind.*, **67**, 48 (1948).
241. F. Baroncelli, G. Scibona and M. Zifferero, *J. Inorg. Nucl. Chem.*, **25**, 205 (1963).
242. A. Bathellier in *Solvent Extraction of Metals* (Eds. H. A. C. McKay, T. V. Healy, I. L. Jenkins and A. Naylor), Macmillan, London, 1966, p. 295.
243. F. L. Moore, *USAEC Rept.*, NAS–NS–3101 (1960).
244. W. P. Ratchford, E. H. Harris, Jr., C. H. Fisher and C. O. Willits, *Ind. Eng. Chem.*, **43**, 778 (1951).
245. V. S. Smelov and A. V. Strahova, *Radiokhimiya*, **5**, 509 (1963).
246. E. Högfeldt and F. Fredlund, unpublished data, 1966.
247. L. Kuča and E. Högfeldt, *Acta Chem. Scand.*, **21**, 1017 (1967).
248. Eurochemic Symp., *Aqueous Reprocessing Chemistry for Irradiated Fuels*, Brussels, 1963.
249. *Solvent Extraction Chemistry of Metals* (Eds. H. A. C. McKay, T. V. Healy, I. L. Jenkins, and A. Naylor), Macmillan, London, 1966.
250. H. Freiser, *Anal. Chem.*, **38**, 121R (1966).
251. G. E. Boyd and Q. V. Larson, *J. Phys. Chem.*, **64**, 988 (1960).
252. O. Navratil and J. Toul, *Collection Czech. Chem. Commun.*, **28**, 1848 (1963).
253. A. A. Pozdnyakov, N. N. Basargin and Y. B. Gerlit, *Dokl. Akad. Nauk SSSR*, **144**, 861 (1962).
254. E. Cerrai and C. Testa, *Energia Nucl. Milan*, **6**, 707, 768 (1959).
255. H. A. Mahlman, G. W. Leddicotte and F. L. Moore, *Anal. Chem.*, **26**, 1939 (1954).
256. F. L. Moore, *Anal. Chem.*, **30**, 908 (1958); **33**, 748 (1961).

257. K. Beyerman, Z. Anal. Chem., 183, 91 (1961).

258. F. Vydra and R. Pribil, Talanta, 3, 72 (1959).

259. K. B. Brown, D. J. Crouse and F. G. Seeley, USAEC Rept., ORNL–TM–181 (1962).

260. M. Taube, Nukleonika, 5, 531 (1960); J. Inorg. Nucl. Chem., 12, 134 (1960).

261. H. Marchart and F. Hecht, Mikrochim. Acta, 1152 (1962).

262. R. M. Diamond and D. G. Tuck, Progress in Inorganic Chemistry, (Ed. F. A. Cotton), Vol. 2, Interscience Publ., New York 1960, p. 109.

263. E. P. Horwitz, C. A. A. Bloomquist, L. J. Sauro and D. S. Henderson, J. Inorg. Nucl. Chem., 28, 2313 (1966).

264. A. S. Kertes, J. W. Irvine, Jr. and S. Basol, unpublished data, Mass. Inst. Techn., Cambridge, 1964.

265. B. Tremillon, Bull. Soc. Chim. France, 1057 (1964).

266. D. J. Crouse, K. B. Brown and F. G. Seeley, Solvent Extraction Chemistry of Metals (Eds. H. A. C. McKay, T. V. Healy, I. L. Jenkins and A. Naylor), Macmillan, London, 1966, p. 327.

267. I. L. Jenkins and A. G. Wain, Brit. Rept., AERE–M 537 (1959).

268. M. L. Good and S. E. Bryan, J. Inorg. Nucl. Chem., 20, 140 (1961).

269. M. de Trentinian and A. Chesne, French Rept., CEA–1426 (1960).

270. A. S. Wilson, USAEC Rept., HW–68207 (1961).

271. V. M. Vdovenko, T. W. Kovaleva and M. A. Ryazanov, Radiokhimiya, 4, 609 (1962).

272. W. Kraak and R. Bac, Solvent Extraction Chemistry of Metals, (Eds. H. A. C. McKay, T. V. Healy, I. L. Jenkins and A. Naylor), Macmillan, London, 1966, p. 267.

273. K. B. Brown, A. Faure, B. Weaver and D. R. Collins, USAEC Rept., ORNL–TM–107 (1962).

274. C. F. Coleman, F. A. Kappelman and B. Weaver, Nucl. Sci. Eng., 8, 507 (1960).

275. F. Baroncelli, G. Scibona and M. Zifferero, Radiochimica Acta, 1, 75 (1963).

276. A. S. Kertes and M. Haendel, unpublished results, 1968.

277. C. Deptula and W. Korpak, Nukleonika, 5, 845 (1960).

278. B. N. Laskorin and V. A. Kuznetsov in Ekstraktsiya, Vol. 2, Gosatomizdat, Moscow, 1962, p. 209.

279. W. Müller, Actinides Review, 1, 71 (1967).

280. H. L. Friedman, J. Phys. Chem., 66, 1595 (1962).

281. E. Hirsch and R. M. Fuoss, J. Am. Chem. Soc., 82, 1013, 1021 (1960).

282. S. R. Palit, V. A. Moghe and B. Biswas, Trans. Faraday Soc., 55, 463 (1959).

283. A. S. C. Lawrence and R. Stenson in *Proceedings of the Second Intl. Congress of Surface Activity*, Academic Press, N. Y., 1957, p. 388.

284. M. Talat-Erben, *Eurochemic Rept.*, ETR–129 (1961).

285. B. Weaver and D. E. Horner, *J. Chem. Eng. Data*, **5**, 200 (1960).

286. P. R. Danesi, F. Orlandini and G. Scibona, *J. Inorg. Nucl. Chem.*, **27**, 449 (1965).

287. W. E. Keder, J. L. Ryan and A. S. Wilson, *J. Inorg. Nucl. Chem.*, **20**, 131 (1961).

288. G. Koch in *Solvent Extraction Chemistry of Metals* (Eds. H. A. C. McKay, T. V. Healy, I. L. Jenkins and A. Naylor), Macmillan, London, 1966, p. 247.

289. I. Abrahamer, Y. Marcus and I. Eliezer, *Israel AEC Rept.*, IA–1014 (1964).

290. F. Baroncelli, G. Scibona and M. Zifferero, *J. Inorg. Nucl. Chem.*, **24**, 541, 547 (1962).

291. V. M. Vdovenko, M. G. Kuzina and A. A. Lipovskii, *Radiokhimiya*, **6**, 121 (1964).

292. A. S. Wilson and W. E. Keder, *J. Inorg. Nucl. Chem.*, **18**, 259 (1961).

293. R. Bac, *J. Inorg. Nucl. Chem.*, **28**, 2335 (1966).

294. G. Koch and E. Schwind, *J. Inorg. Nucl. Chem.*, **28**, 571 (1966).

295. I. Abrahamer and Y. Marcus, *Israel AEC Rept.*, IA–809 (1963).

296. Y. Marcus, M. Givon and C. R. Choppin, *J. Inorg. Nucl. Chem.*, **25**, 1457 (1963).

297. S. F. Marsh, W. J. March, G. L. Booman and J. E. Rein, *Anal. Chem.*, **34**, 1406 (1963).

298. W. E. Keder, J. C. Sheppard and A. S. Wilson, *J. Inorg. Nucl. Chem.*, **12**, 327 (1960).

299. V. I. Tikhomirov, A. A. Kuznetsov and E. D. Batorovskaya, *Radiokhimiya*, **6**, 173 (1964).

300. T. Sato, *J. Appl. Chem.*, **15**, 92 (1965).

301. W. E. Keder, *J. Inorg. Nucl. Chem.*, **24**, 561 (1962).

302. G. Calleri and M. Talat-Erben, *Eurochemic Rept.*, ERT–90 (1960).

303. W. Knoch, *Z. Naturforsch.*, **16a**, 525 (1961).

304. J. van Ooyen, in *Solvent Extraction Chemistry* (Eds. D. Dyrssen, J. O. Liljenzin and J. Rydberg), North-Holland, Amsterdam, 1967, p. 485.

305. C. J. Hardy, D. Scargill and J. M. Fletcher, *J. Inorg. Nucl. Chem.*, **1**, 257 (1958).

306. P. J. Lloyd and E. A. Mason, *J. Phys. Chem.*, **68**, 3120. (1964).

307. *Oak Ridge National Laboratory, Chem. Techn. Div., Ann. Progr. Rept.*, ORNL–3314 (1962), p. 104; ORNL–3452 (1963), p. 170; ORNL–3627 (1964), p. 179; ORNL–3830 (1965), p. 193; ORNL–3945 (1966), p. 176.

308. R. D. Baybarz, B. S. Weaver and H. B. Kinser, *Nucl. Sci. Eng.*, **17**, 457 (1963).

309. G. B. Fasolo, R. Maloano and A. Massaglia, *Anal. Chim. Acta.*, **29**, 509 (1963).

310. T. Ishimori, *Data of Inorganic Solvent Extraction*, JAERI Rept., 1047 (1963); 1062 (1964); 1106 (1966) and references therein.

311. K. B. Brown, D. J. Crouse, F. G. Seeley and K. G. Caulton, *USAEC Rept.*, ORNL-TM-449 (1963).

312. M. Shiloh and Y. Marcus, *Israel AEC Rept.*, IA-781 (1962).

313. M. L. Good and S. C. Srivastava, *J. Inorg. Nucl. Chem.*, **27**, 2429 (1965).

314. Y. Marcus, *J. Inorg. Nucl. Chem.*, **28**, 209 (1966).

315. A. A. Mazurova and L. M. Gindin, *Zh. Neorg. Khim.*, **10**, 489 (1965).

316. R. Muxart and H. Arapaki-Strapelias, *Bull. Soc. Chim. France*, 888 (1963).

317. J. L. Ryan, *J. Phys. Chem.*, **65**, 1856 (1961).

318. D. Maydan and Y. Marcus, *J. Phys. Chem.*, **67**, 987 (1963).

319. K. S. Venkateswarlu and P. Charan Das, *J. Inorg. Nucl. Chem.*, **25**, 730 (1963).

320. J. M. White, P. Kelly and N. C. Li, *J. Inorg. Nucl. Chem.*, **16**, 337 (1961).

321. M. L. Good and F. F. Holland, *J. Inorg. Nucl. Chem.*, **26**, 321 (1964).

322. D. G. Tuck and E. J. Woodhouse, *J. Chem. Soc.*, 6017 (1964).

323. C. V. Kopp and R. C. McDonald in *Solvent Extraction Chemistry* (Eds. D. Dyrssen, J. O. Liljenzin and J. Rydberg), North-Holland, Amsterdam, 1967, p. 447.

324. A. S. Kertes, P. J. Lloyd, K. Kimura, J. F. Byrum and J. W. Irvine, Jr., *USAEC Rept.*, TID-20451 (1964).

325. S. Lindenbaum and G. E. Boyd, *J. Phys. Chem.*, **67**, 1238 (1963).

326. V. Schindewolf, *Z. Elektrochem.*, **62**, 335 (1958).

327. D. Dyrssen and M. de Jesus Tavares in *Solvent Extraction Chemistry* (Eds. D. Dyrssen, J. O. Liljenzin and J. Rydberg), North-Holland Amsterdam, 1967, p. 465.

328. P. J. Lloyd in *Solvent Extraction Chemistry* (Eds. D. Dyrssen, J. O. Liljenzin and J. Rydberg), North-Holland, Amsterdam, 1967, p. 458.

329. M. Shiloh and Y. Marcus, *Israel J. Chem.*, **3**, 123 (1965); *J. Inorg. Nucl. Chem.*, **28**, 2725 (1966).

330. D. J. Crouse and K. B. Brown, *Ind. Eng. Chem.*, **51**, 1461 (1959).

331. F. W. Bruenger, B. J. Stover and D. R. Atherton, *Anal. Chem.*, **35**, 1671 (1963).

332. M. L. Good, S. E. Bryan and F. Inge, *Inorg. Chem.*, **2**, 963 (1963).

333. J. I. Bullock and D. G. Tuck, *J. Inorg. Nucl. Chem.*, **28,** 1103 (1966).
334. W. J. McDowell and K. A. Allen, *J. Phys. Chem.*, **65,** 1358 (1961).
335. D. Brown, T. Sato, A. J. Smith and R. G. Wilkins, *J. Inorg. Nucl. Chem.*, **23,** 91 (1961).
336. V. M. Vdovenko, M. P. Kovalskaya and E. V. Shirvinskii, *Radiokhimiya*, **3,** 3 (1961).
337. V. B. Shevchenko and Y. F. Zhdanov, *Radiokhimiya*, **3,** 676 (1961).
338. V. I. Tikhomirov, A. A. Kuznetsov and E. D. Batorovskaya, *Radiokhimiya*, **6,** 187 (1964).
339. K. A. Allen, *J. Am. Chem. Soc.*, **80,** 4133 (1958).
340. G. B. S. Salaria, C. L. Rulfs and P. J. Elving, *Anal. Chem.*, **35,** 983 (1963).
341. A. S. Kertes and A. Beck, *J. Chem. Soc.*, 1926 (1961); A. Beck, *Ph.D. Thesis*, Hebrew University, Jerusalem, 1961.
342. A. Beck, D. Dyrssen and S. Ekberg, *Acta Chem. Scand.*, **18,** 1695 (1964).
343. A. A. Zaitsev, I. A. Lebedev, S. V. Pirozhkov and G. N. Yakovlev, *Zh. Neorg.. Khim.*, **8,** 2184 (1963).
344. A. S. Kertes and A. Beck, *Proc. 7th Intern. Conf. Coord. Chem.*, *Stockholm*, 1962, p. 352.
345. R. H. Bailes, *USAEC Rept.*, DOW–159 (1957).
346. W. J. Maeck, M. E. Kussy and J. E. Rein, *Anal. Chem.*, **34,** 1602 (1962).
347. S. Tribalat, *Anal. Chim. Acta*, **4,** 228 (1950).
348. A. J. Cameron and N. A. Gibson, *Anal. Chim. Acta*, **25,** 24 (1961).
349. B. Figgis and N. A. Gibson, *Anal. Chim. Acta.*, **7,** 313 (1952).
350. P. Senise in *Analytical Chemistry 1962, Proc. Intern. Symposium* (Ed. P. W. West), Elsevier Publ. Co., Amsterdam, 1963, p. 171.
351. R. Bock, H. T. Niederauer and K. Behrends, *Z. Anal. Chem.*, **190,** 33 (1962).
352. R. Bock and G. M. Beilstein, *Z. Anal. Chem.*, **192,** 44 (1963).
353. R. Bock and J. Jainz, *Z. Anal. Chem.*, **198,** 21 (1963).
554. R. Bock and C. Hummel, *Z. Anal. Chem.*, **198,** 176 (1963).
355. J. Hala, O. Navratil and V. Nechuta, *J. Inorg. Nucl. Chem.*, **28,** 553 (1966).
356. L. H. Bowen and R. T. Rood, *J. Inorg. Nucl. Chem.*, **28,** 1985 (1966).
357. S. Tribalat and J. Beydon, *Anal. Chim. Acta.*, **6,** 96 (1952); **8,** 22 (1953).
358. H. E. Affsprung, N. A. Barnes and H. A. Potratz, *Anal. Chem.*, **23,** 1680 (1951).
359. K. D. Moffet, J. R. Simmler and H. A. Potratz, *Anal. Chem.*, **28,** 1356 (1956).
360. N. Allen and N. H. Furman, *J. Am. Chem. Soc.*, **54,** 4625 (1932).
361. G. K. Schweitzer and S. W. McCarty, *J. Inorg. Nucl. Chem.*, **27,** 191 (1965).

814   *Ion Exchange and Solvent Extraction of Metal Complexes*

362. K. Veno and C. Chang, *Nippon Genshiryoku Gakkaishi*, **4**, 457 (1962).
363. H. H. Willard and L. R. Perkins, *Anal. Chem.*, **25**, 1634 (1953).
364. W. Ceilmann and R. Neeb, *Z. Anal. Chem.*, **156**, 420 (1957).
365. J. W. Murphy and H. E. Affsprung, *Anal. Chem.*, **33**, 1658 (1961).
366. P. Senise and L. R. M. Pitombo, *Anal. Chim. Acta*, **26**, 89 (1962).
367. R. Neeb, *Z. Anal. Chem.*, **182**, 10 (1961).
368. H. E. Affsprung and J. W. Murphy, *Anal. Chim. Acta*, **30**, 501 (1964); **32**, 381 (1965).
369. F. Feigl, *Chemistry of Specific, Selective and Sensitive Reactions*, Academic Press, New York, 1949.
370. D. W. Margerum and C. V. Banks, *Anal. Chem.*, **26**, 200 (1954).
371. H. Irving and D. L. Pettit in *Analytical Chemistry 1962, Proc. Intern. Symp.*, Elsevier Publ. Co., Amsterdam, 1963, p. 122.
372. L. Ducret and L. Pateau, *Anal. Chim. Acta*, **20**, 568 (1959).
373. R. R. Miller and W. W. Brandt, *Anal. Chem.*, **26**, 1968 (1954).
374. G. F. Smith and W. H. McCurdy, *Anal. Chem.*, **24**, 371 (1952).
375. P. Collins and H. Diehl, *Anal. Chem.*, **31**, 1692 (1959).
376. A. R. Gahler, *Anal. Chem.*, **26**, 577 (1954).
377. J. L. Walter and H. Freiser, *Anal. Chem.*, **26**, 217 (1954).
378. J. Hoste, *Anal. Chim. Acta*, **4**, 23 (1950).
379. E. M. Penner and W. R. Imman, *Talanta*, **10**, 407 (1963).
380. R. Muxart, M. Levi and G. Bouissieres, *Compt. Rend.*, **239**, 1000 (1959).
381. R. C. Fix and J. W. Irvine, Jr., *LNS Progr. Rept.*, Mass. Inst. Techn., Cambridge, Aug. 1955 and May 1956.
382. K. B. Brown *USAEC Rept.*, CF–60–5–114 (1960); CF–61–3–141 (1961).
383. V. I. Kuznetsov and L. I. Bolyshakova, *Zh. Analit. Khim.*, **15**, 523 (1960).
384. G. H. Ayres and S. S. Baird, *Anal. Chim. Acta*, **23**, 446 (1960).
385. A. K. Babko and A. I. Volkova, *Zh. Analit. Khim.*, **15**, 587 (1960).
386. V. M. Vdovenko and L. N. Lazarev, *Zh. Neorg. Khim.*, **3**, 155 (1958).
387. T. Moeller and R. E. Zogg, *Anal. Chem.*, **22**, 612 (1950).
388. F. Jancik and J. Korbl, *Talanta*, **1**, 55 (1958).
389. V. P. Zhivopistsev and M. N. Chelnokova, *Zh. Analit. Khim.*, **18**, 717 (1963).
390. A. I. Busev and L. M. Skrebkova, *Zh. Analit. Khim.*, **17**, 56 (1962).
391. V. I. Spitsyn, A. F. Kuzina, N. N. Zamoshnikow and T. S. Tagily, *Dokl. Akad. Nauk SSSR*, **144**, 1066 (1962).

11.

Synergistic Extraticon

It has been observed that certain combinations of two extractants, S_1 and S_2, give, under well-defined experimental conditions, an enhanced extraction for a number of metals than could be expected from the distribution ratios of the components separately, and

$$D_{exp} = D_{S_1} + D_{S_2} + \Delta D \qquad (11\text{-}1)$$

where ΔD is the magnitude of the synergistic enhancement of extraction.

The phenomenon of a definite enhancement has been apparently first mentioned in the open literature by Cunningham and coworkers[1]. The Harwell investigators observed that a mixture of thenoyltrifluoroacetone with tributyl phosphate in benzene would extract rare earth nitrates from nitric acid solutions to a markedly greater extent than either of the reagents alone. Shortly after, a similar phenomenon has been noted by Schmitt in Oak Ridge[2], during a survey study in which various acidic and neutral phosphorus alkyl esters have been screened for their extractive capacity towards uranyl nitrate. The term synergistic extraction was introduced by Coleman from ORNL.

The large enhancement of extraction has subsequently been widely investigated both from fundamental, physicochemical, and applied, technological point of view. It has become clear that the phenomenon of synergism is far more general than was thought after the first reports from Oak Ridge[3,4]. It has been observed in a wide variety of systems including chelating agents, carboxylic acids or acidic phosphorus esters in combination with ethers, ketones, amines, alcohols, phenols, carboxylic esters, but especially neutral alkyl phosphates, phosphonates and phosphine oxides. Such combinations produce enhanced extraction of many ions, under a variety of experimental conditions.

The diversity of systems where synergism has been observed indicates that the mechanism of the phenomenon cannot be identical in all cases. It still remains generally true, however, that synergism is essentially an organic-phase reaction, due to at least one of two fundamental factors: a) the extractive power of the extractants changes when in the presence of one another, or in thermodynamic terms, the activity of the extractants is affected, and b) the composition of the metal-bearing species in the organic phase is not the same as in the case of a one extracant system.

Obviously, within these generalities, synergistic systems differ in many ways. In line with the classification of extraction systems adopted in this monograph, there are four different types of synergistic combinations. Two of them involve one acidic and one neutral extractant, while the third and fourth groups involve the combination of two neutral and two chelating extracting agents respectively. The first group to be discussed, that of an acidic chelating agent and a neutral solvating extractant, is the simplest and best understood. The synergism caused by such a combination of reactants is rather general, in the sense that it applies to a sizeable number of metals in different oxidation states. These synergistic systems usually behave ideally under a variety of conditions, and the experimental data can relatively easily be interpreted in terms of simple mass-action equations. In many respects these systems resemble those of metal extraction by chelating agents alone.

The second group of synergistic systems, involving mainly alkylphosphoric acids, but also carboxylic and sulphonic acids, rather than chelating acids as the acidic component, exhibit, as a rule, markedly lower synergistic effects. The number of metals which can be synergistically extracted in such systems is apparently more limited also. The main difference in the synergistic behaviour of these two types of combinations arises from the composition of the synergistic adduct formed, and its stability. While the neutral ligand may be hydrogen bonded to the acid extractant in the latter systems, a direct coordination of the neutral ligand to the cation probably occurs in the former. However, since in the case where hydrogen bonding is involved, the structure of the suggested synergistic adduct leaves the metal frequently coordinatively unsaturated, the possibility, here too, of the ligand being coordinated to the metal cannot be excluded either. It remains true, nevertheless, that the stability of synergistic adducts formed between a chelated metal and the neutral ligand is generally much higher. Many such adducts have been isolated in solid crystalline form, whereas none has been isolated from synergistic systems of organophosphorus acid.

An additional difference in the extent of their synergistic enhancement is due to the interaction between the synergistic components in the organophosphorus acid–neutral ligand systems. Such specific interaction interferes with the extractive capacity of both components. This behaviour is definitely in contrast to that of the former systems. If there is a similar interaction between a chelating agent and a neutral ligand at all, it is very weak.

The synergistic enhancement produced by the combination of extractants belonging to the third and fourth groups is low. Little experimental material is available on two-neutral extractant systems and they are not well understood. In case of two chelating acids participating in the complex, group four, the reaction is essentially that of mixed ligand–complex formation, which may or may not lead to an enhanced metal extraction.

References pp. 854–858

A. CHELATING AGENT–NEUTRAL

LIGAND SYSTEMS

a. Introduction

As experimental data accumulate, it becomes increasingly evident that the phenomenon of synergism involving a chelating acid and a neutral extractant lacks specificity, in that it is apparently not a distinctive feature of any particular metal or group of metals. This is true to almost the same extent as metals form extractable complexes with the vast number of available bidentate chelating agents (section 8.A).

The discovery of the phenomenon of a definite enhancement by the combination of thenoyltrifluoroacetone, HTTA, and tributyl phosphate was first described by Cunningham and coworkers[1] for praesodynium and neodynium extraction from nitric acid solutions. Subsequently, it has been demonstrated that this synergistic system is by no means restricted to the tervalent rare earths, but it is a common property of divalent metals and the actinides in their different oxidation states. The degree of enhancement, however, will vary from one metal to another.

Though HTTA has been most widely used as the chelating acid in synergistic systems, a variety of other β-diketones have been found to be fairly good alternatives. Among them, mention should be made of the similar tri- and hexafluoroacetylacetones, acetylacetone and benzoylacetone, but some nitrogen chelating agents such as cupferron, 8-hydroxyquinoline and dimethylglyoxime have also shown the effect. For many metal chelates formed with these agents, it is known that they readily react with a neutral ligand to form coordinatively saturated adducts.

It has been also demonstrated that the neutral component of the synergistic system has not to be an organophosphorus ester. It may well be an alcohol, ketone, amine or amide, or more generally, any of the neutral extractants considered in Chapter 9. The extent of synergistic enhancement, however, will depend on the nature and properties of the component. Figure 11A1 shows typical synergistic curves obtained for metals in various thenoyltrifluoroacetone–neutral phosphorus ester systems[5-7]. A synergistic enhancement by a factor of 10^5 is very common in these and similar systems.

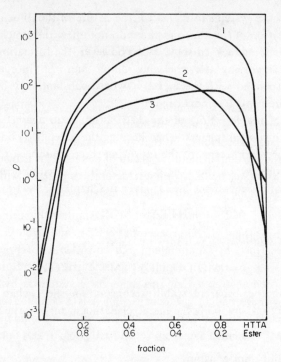

fraction

Figure 11A1. Synergic effect in theonyltrifluoroacetone–neutral organophosphorus ester systems.
(1) Promethenum–trioctylphosphine oxide–benzene[5] (aqueous phase 0.01 N HCl)
(2) Zinc–tributoxyethyl phosphate–carbon tetrachloride[6] (aqueous phase pH 5)
(3) Uranium-(VI)–tributylphosphate–cyclohexane[7] (aqueous phase 0.01 N HCl)

b. Distribution Studies

The composition of the synergistic adduct formed during the extractive transfer of the metal from an aqueous solution has been determined by the conventional technique of changing the extraction variables and applying equations (7C48) and (7C49). The metal is usually present in tracer amounts, thus reducing the synergistic parameters to three: the concentration of the hydrogen ion in the aqueous solution, and that of the chelating acid and the neutral ligand in the organic phase. When experimental conditions are chosen so as to avoid interference from possible side-reactions, such as metal hydrolysis in the aqueous phase, or formation of polynuclear species in either of the phases, the systems are generally simple and conform reasonably well to mass-action law requirements of the simple distribution law (equation 7C46). In the majority of the systems investigated, the nature of the diluent used and the composition of the aqueous phase do not affect the stoichiometry of the synergistic adduct. In some cases, however, the formation of mixed complexes, involving the chelating anion and the inorganic anion initially associated with the metal, has been substantiated by distribution measurements (see next heading).

Results of log–log plots of the distribution ratio against one of the three synergistic parameters, when keeping the other two constant, have been interpreted in terms of the following three equilibria, representing the stoichiometry of the synergistic adduct. The equilibrium for the formation of the metal chelate, as given in Chapter 7, is

$$M^{m+} + m\overline{\text{HTTA}} \rightleftharpoons \overline{M(\text{TTA})_m} + mH^+ \qquad (11A2)$$

and has the equilibrium constant

$$K_{12} = \overline{[M(\text{TTA})_m]}\,[H^+]^m[M^{m+}]^{-1}\overline{[\text{HTTA}]}^{-m} \qquad (11A3)$$

representing the apparent stability constant of the chelate. The synergistic reaction, involving the metal, chelating agent and the ligand

$$M^{m+} + m\overline{\text{HTTA}} + x\overline{S} \rightleftharpoons \overline{M(\text{TTA})_m \cdot S_x} + mH^+ \qquad (11A4)$$

has the equilibrium constant

$$K_{121} = \overline{[M(\text{TTA})_m \cdot S_x]}\,[H^+]^m[M^{m+}]^{-1}\overline{[\text{HTTA}]}^{-m}[\overline{S}]^{-x} \qquad (11A5)$$

The ratio $K_{121}/K_{12} = K_{111}$ represents the equilibrium constant of the organic-phase synergistic reaction

$$\overline{M(\text{TTA})_m} + x\overline{S} \rightleftharpoons \overline{M(\text{TTA})_m S_x}. \qquad (11A6)$$

In the following tables the values of $\log K_{12}$, $\log K_{111}$ and $\log K_{121}$ for a number of metal–thenoyltrifluoroacetone adducts have been compiled. Table 11A4 lists these values for some other chelate adducts.

From a careful examination of the tabulated values, conclusions can be drawn on the factors affecting the synergic equilibrium and the stability of the adducts formed. It becomes evident that a) the basicity and the structure of the neutral ligand play a decisive role, as does b) the nature of the diluent, c) the coordination ability of the metal, and finally, d) the complexing power of the chelating agent. Let us now examine these factors separately.

a) With organophosphorus compounds as the neutral ligand, the order of synergistic enhancement is that of increasing base strength, i.e.

phosphate < phosphonate < phosphinate < phosphine oxide[5, 6, 10, 33, 34, 35] (Table 11A5). This is the order of decreasing ester linkage and increasing carbon-to-phosphorus linkage as well as that of extractive efficiency (section 9.D).Triphenyl phosphate, for example, which is, alone, a poor extractant for metal salts, functions also as a poor synergist adduct. Tributoxyethyl phosphate is somewhat stronger than tributyl phosphate, the enthalpies of hydrogen-bond formation being -1.55 and -1.21 kcal mole^{-1} respectively[6].

The difference in the stability of the adducts becomes more pronounced when the neutral ligand is a nitrogen-bearing base[17, 18, 32, 33]. With considerable restrictions as to steric factors, the stability increases with increased donor properties of the molecule.

TABLE 11A1. Equilibrium constants of thenoyltrifluoroacetone adducts of some actinides.

Adduct	Neutral ligand	Diluent	Log K_{12}	Log K_{111}	Log K_{121}	Reference
Pu(TTA)$_3$. S$_2$	TBP	Cyclohexane	-7.22	12.52	5.30	24
Am(TTA)$_3$. S	TBP	CCl$_4$	-8.88	5.06	-3.82	30
Am(TTA)$_3$. S$_2$	TBP	Cyclohexane	-6.62	10.03	3.41	24
	TBP	Cyclohexane		9.8	2.34	26
	TBP	Cyclohexane	-6.80	9.23	2.43	27
	TBP	Benzene	-6.80	5.84	-0.96	5
	TBP	Benzene	-7.46	6.50	-0.96	26
	TBP	Hexane	-6.80	7.83	1.03	27
	TBP	CCl$_4$	-8.88	8.89	0.01	30
	TBP	CCl$_4$	-6.80	7.23	0.43	27
	TBP	Chloroform	-6.80	3.52	-3.28	27
	TPP	Benzene	-6.80	2.16	-4.64	5
	DBBP	Benzene	-6.80	7.30	0.50	5
	TOPO	Cyclohexane	-6.80	11.94	5.14	27
	TOPO	Cyclohexane		12.7	5.20	26
	TOPO	Benzene	-7.46	9.97	2.51	26
	TOPO	Benzene	-6.80	9.30	2.50	5
	TPPO	Benzene	-6.80	7.65	0.85	5
Cm(TTA)$_3$. S$_2$	TBP	Cyclohexane	-6.60	9.10	2.50	27
	TBP	Cyclohexane		9.70	2.60	26
	TBP	Benzene	-7.10	6.40	-0.70	26
	TBP	Benzene	-6.60	5.90	-0.70	5
	TOPO	Cyclohexane	-6.60	12.04	5.44	27
	TOPO	Cyclohexane		12.60	5.53	26
	TOPO	Benzene	-7.10	9.93	2.83	26
	TOPO	Benzene	-6.60	9.43	2.83	5

TABLE 11A1. (Cont.)

Adduct	Neutral ligand	Diluent	Log K_{12}	Log K_{111}	Log K_{121}	Reference
Th(TTA)$_4$. S	TBP	Cyclohexane	1.67	6.28	7.95	26
	TBP	Cyclohexane		6.90	7.85	26
	TBP	Benzene	1.00	4.70	5.70	26
	TBP	Chloroform		3.30		30
	TBP	Benzene	1.67	6.18	7.85	5
	TBP	CCl$_4$		5.18		30
	TPP	Cyclohexane	1.67	3.91	5.58	27
	TPP	Benzene	1.67	3.87	5.54	5
	TOPO	Cyclohexane	1.67	8.91	10.58	27
	TOPO	Cyclohexane		9.50	10.45	26
	TOPO	Benzene	1.00	6.70	7.70	26
	TOPO	Benzene	1.67	8.78	10.45	5
	BAA	Cyclohexane	1.67	5.23	6.90	27
	EHA	Cyclohexane	1.67	3.03	4.70	27
HNp(TTA)$_2$.. S	TBP	Cyclohexane	1.00	4.30	5.30	37
UO$_2$(TTA)$_2$. S	TBP	Cyclohexane	-2.82	6.52	3.70	27
	TBP	Cyclohexane		5.80	3.56	26
	TBP	Cyclohexane	-2.82	6.77	3.88	37
UO$_2$(TTA)$_2$. S	TBP	Benzene	-2.26	4.74	2.48	26
	TBP	Benzene	-2.80	5.28	2.48	5
	TPP	Cyclohexane	-2.82	3.43	0.61	27
	TPP	Benzene	-2.80	2.48	-0.32	5
	TOPO	Cyclohexane	-2.82	6.62	3.80	27
	TPPO	Benzene	-2.80	6.34	3.54	5
	BAA	Cyclohexane	-2.82	3.25	0.43	27
UO$_2$(TTA)$_2$. S$_3$	TBPO	Cyclohexane	-2.82	17.22	14.40	7
	TOPO	Cyclohexane	-2.82	15.72	12.90	27
	TOPO	Benzene	-2.82	11.43	8.61	5
PuO$_2$(TTA)$_2$. S	TBP	Cyclohexane	-1.54	4.67	3.13	37

BAA—*N*-n-Butyl acetanilide
EHA—Ethylhexyl alcohol
TPP—Triphenyl phosphate
TPPO—Triphenylphosphine oxide
TOPO—Trioctylphosphine oxide
DBBP—Di-n-butyl n–butylphosphonate
TBPO—Tributylphosphine oxide

TABLE 11A2. Equilibrium constants of thenoyltrifluoroacetonate adducts of some lanthanides.

Adduct	Neutral ligand	Diluent	Log K_{12}	Log K_{111}	Log K_{121}	Reference
La(TTA)$_3$. S	TBP	CCl$_4$	-10.95	4.83	-6.12	30
	MIBK	CCl$_4$	-10.95	2.0	-8.95	30
La(TTA)$_3$. S$_2$	TBP	CCl$_4$	-10.95	9.33	-1.62	30
	MIBK	CCl$_4$	-10.95	2.9	-8.05	30
Ce(TTA)$_3$. S$_2$	DBBP	Kerosene	-9.43[a]	12.36[b]	2.93	35
	D2EH2EHP	Kerosene	-9.43[a]	11.60[b]	2.17	35

TABLE 11A2. (Cont.)

Adduct	Neutral ligand	Diluent	Log K_{12}	Log K_{111}	Log K_{121}	Reference
Pm(TTA)$_3$. S$_2$	TBP	Cyclohexane	−7.00	9.30	2.30	27
	TBP	Cyclohexane		10.00	2.26	26
	TBP	Benzene	−7.00	5.96	−1.04	5
	TBP	Benzene	−7.77	6.70	−1.04	26
	TPP	Benzene	−7.00	2.30	−4.70	5
	TPP	Cyclohexane	−7.00	5.1	−1.92	27
	DBBP	Benzene	−7.00	7.43	0.43	5
	TOPO	Cyclohexane	−7.00	12.00	5.00	27
	TOPO	Benzene	−7.00	9.43	2.43	5
	TOPO	Cyclohexane		12.9	5.13	26
	TOPO	Benzene	−7.77	10.2	2.43	26
	TPPO	Benzene	−7.00	7.78	0.78	5
	EHA	Cyclohexane	−7.00	4.15	−2.85	27
Eu(TTA)$_3$. S	TBP	CCl$_4$	−8.57	5.15	−3.42	30
	TBP	Chloroform		3.63		30
	TOPO	CCl$_4$	−8.57	7.49	−1.08	30
Eu(TTA)$_3$. S	TOPO	Chloroform		5.40		30
	MIBK	CCl$_4$	−8.57	1.71	−6.86	30
	MIBK	Chloroform		1.16		30
	QUIN	CCl$_4$	−8.57	3.48	−4.09	30
	QUIN	Chloroform		3.29		30
	DBSO	CCl$_4$	−8.57	5.09	−3.48	30
Eu(TTA)$_3$. S$_2$	TBP	Cyclohexane	−7.66[a]	9.44	1.78	24
	TBP	CCl$_4$	−8.57	8.89	0.32	30
	TBP	Chloroform		5.40		30
	DBBP	Kerosene	−7.66[a]	11.62[b]	3.96	35
	D2EH2EHP	Kerosene	−7.66[a]	10.89[b]	3.23	35
	TOPO	CCl$_4$	−8.57	12.26	3.69	30
	TOPO	Chloroform		7.60		30
	MIBK	CCl$_4$	−8.57	2.34	−6.23	30
	MIBK	Chloroform		1.52		30
	QUIN	CCl$_4$	−8.57	5.16	−3.41	30
	DBSO	CCl$_4$	−8.57	8.58	0.01	30
Tb(TTA)$_3$. S$_2$	DBBP	Kerosene	−7.51[a]	11.55[b]	4.04	35
	D2EH2EHP	Kerosene	−7.51[a]	10.83[b]	3.32	35
Tm(TTA)$_3$. S$_2$	TBP	Cyclohexane	−5.6	8.2	2.62	27
	TBP	Cyclohexane		9.9	2.96	26
	TBP	Benzene	−6.96	6.62	−0.34	26
	TBP	Benzene	−5.6	5.26	−0.34	5
	DBBP	Kerosene	−6.96[a]	11.00[b]	4.04	35
	TOPO	Cyclohexane	−5.6	11.32	5.72	27
Tm(TTA)$_3$. S	TBP	Cyclohexane		5.2	−0.42	27
	TOPO	Cyclohexane		6.84	1.24	27
Lu(TTA)$_3$. S$_2$	TBP	CCl$_4$	−7.34	6.67	−0.67	30
	DBBP	Kerosene	−6.77[a]	10.20[b]	3.43	35
	D2EHEHP	Kerosene	−6.77[a]	9.81[b]	3.04	35
Lu(TTA)$_3$. S	TBP	CCl$_4$	−7.34	5.69	−1.65	30
Y(TTA)$_3$. S$_2$	DBBP	Kerosene	−7.39[a]	11.17[b]	3.78	35
Sc(TTA)$_3$. S$_2$	D2EH2EHP	Kerosene	−0.77[a]	7.38[b]	6.61	35

a Data from ref. 36.
b Calculated from K_{12} reported here, rather than in ref. 35.
D2EH2EHP — Di–(2–ethylhexyl) 2–ethylhexylphosphonate
QUIN — Quinoline
DBSO — Dibutyl sulphoxide
MIBK — Methyl isobutyl ketone
For other abbreviations see Table 11A1.

TABLE 11A3. Equilibrium constants of thenoyltrifluoroacetone
adducts of some divalent metals.

Adduct	Neutral ligand	Diluent	Log K_{12}	Log K_{111}	Log K_{121}	Reference
Ca(TTA)$_2$. S	TBP	CCl$_4$	−13.40	4.11	−8.29	30
	TOPO	CCl$_4$	−13.40	5.64	−6.76	30
	MIBK	CCl$_4$	−13.40	1.83	−11.57	30
Ca(TTA)$_2$. S$_2$	TBP	CCl$_4$	−13.40	8.22	−5.18	30
	TBP	Benzene	−12.00	6.7	−5.27	5,26
	TBP	Cyclohexane	−12.00	8.8	−3.27	26,27
	TOPO	CCl$_4$	−13.40	10.68	−2.72	30
	TOPO	Benzene	−12.00	8.7	−3.27	5,26
	TOPO	Cyclohexane	−12.00	10.1	−1.99	26,27
	MIBK	CCl$_4$	−13.40	2.66	−10.74	30
Sr(TTA)$_2$. S	TBP	CCl$_4$	−15.30	3.76	−11.54	30
	TOPO	CCl$_4$	−15.30	5.39	−9.91	30
	MIBK	CCl$_4$	−15.30	1.80	−13.50	30
Sr(TTA)$_2$. S$_2$	TBP	CCl$_4$	−15.30	7.52	−7.78	30
	TOPO	CCl$_4$	−15.30	9.78	−5.52	30
	MIBK	CCl$_4$	−15.30	2.60	−12.70	30
Cu(TTA)$_2$. S	TBP	CCl$_4$		2.23		25
	TBP	Benzene		1.56		25
	TBP	Cyclohexane	−2.84	3.32	0.48	28
	TBP	CCl$_4$	−1.08	2.27	1.19	29,32
	TBP	Chloroform	−1.25	0.65	−0.60	29,32
	TBP	Benzene		1.66		13
	TPP	Benzene		0.47		13
	TOPO	Benzene		3.23		13
	TPPO	Benzene		2.34		13
	TBPO	Cyclohexane	−2.84	3.98	1.14	28
	MIBK	CCl$_4$	−1.08	0.47	−0.61	29,32
Zn(TTA)$_2$. S	TBP	Cyclohexane	−8.33	4.77	−3.56	28
	TBP	CCl$_4$	−8.64	4.38	−4.26	6
	TBP	CCl$_4$	−8.04	4.34	−3.70	29,32
	TBP	Chloroform	−8.13	2.69	−5.44	29,32
	TBEP	CCl$_4$	−8.64	4.15	−4.49	6
	TBEP	Chloroform		2.50		6
	TOPO	Cyclohexane			−1.0	20

TABLE 11A3. (Cont.)

Adduct	Neutral ligand	Diluent	Log K_{12}	Log K_{111}	Log K_{121}	Reference
Zn(TTA)$_2$. S	MIBK	CCl$_4$	−8.04	1.22	−6.82	29,32
	MIBK	Chloroform	−8.13	1.00	−7.13	29,32
Cd(TTA)$_2$. S	TOPO	CCl$_4$		1.02		11
Co(TTA)$_2$. S	TBP	Cyclohexane	−8.37	4.59	−3.78	28
Co(TTA)$_2$. S$_2$	TBP	Cyclohexane	−8.37	7.26	−1.11	28

TBEP — Tri–n–butoxyethyl phosphate.
For other abbreviations see previous Tables.

TABLE 11A4. Equilibrium constants of some metal–chelate adducts.

Adduct	Neutral ligand	Diluent	Log K_{12}	Log K_{111}	Log K_{121}	Reference
Cu(Acac)$_2$. S	Pyridine	Benzene		0.94		38
	3-Acetylpyridine	Benzene		1.01		38
	4–Benzoylpyridine	Benzene		1.26		38
	3–Picoline	Benzene		0.78		38
	4–Picoline	Benzene		1.04		38
	Quinoline	Benzene		0.47		16
	Isoquinoline	Benzene		0.72		16
Zn(IPT)$_2$. S	Tributyl phosphate	CCl$_4$	−7.00	3.21	−3.79	29,32
	Tributyl phosphate	Chloroform	−7.01	2.40	−4.61	29,32
Cu(DMG)$_2$. S	Pyridine	Benzene		3.50		32
	Dodecylamine	Benzene		3.6		32
	Dodecylamine	Chloroform		3.36		29,32
	Butylamine	Benzene		3.8		32
	Dibutylamine	Benzene		3.5		32
	Dibutylamine	Chloroform		2.09		32
	Triethylamine	Benzene		<0.5		32
	Aniline	Chloroform		1.92		32
	Dimethylaniline	Chloroform		0.51		32
	Quinoline	Chloroform		2.04		29,32
Th(IPT)$_4$. S	Tributyl phosphate	Chloroform		1.2		30
	Tributyl phosphate	CCl$_4$		1.7		30
Th(IPT)$_4$. S$_2$	Tributyl phosphate	Chloroform		2.4		30
	Tributyl phosphate	CCl$_4$		2.8		30

Acac — Acetylacetone.
IPT — β-Isopropyltropolon,
DMG — Dimethylglyoxime.

References pp. 854–858

Similar conclusions are reached from heat of solution measurements of copper acetylacetonate in a solution of pyridine bases in dry benzene[38]. Under such experimental conditions, the heat of reaction is due to the addition of the base and it is a direct measure of the energy of the adduct coordination bond formed. ΔH varies between -1.5 and -11.5 kcal/mole, with a trend of higher heat of reaction for the stronger bases, and levels off around -1.5 kcal/mole as the basicity decreases. It was found that the bond energy decreases in the order 4-picoline > 3-picoline > pyridine > 3-acetylpyridine > 4-benzoylpyridine. This order nearly follows that of their basicity, though the overall energy of the bond formed is essentially low in all these cases.

These relatively small effects upon basicity are probably the reason that the synergistic effect is frequently overpowered by steric effects[16,32,35,38] For many metal–TTA chelates investigated, K_{111} values are higher with dibutyl butylphosphonate than with di-(2-ethylhexyl) 2-ethylhexyl-phosphonate, indicating that branching of the alkyl groups of the neutral adduct molecule offers steric obstruction[35]. Also, the distribution ratio of thorium is depressed when tributyl phosphate is replaced by a more bulky neutral phosphorus ester such as the tris(4-methyl-2-pentyl) phosphate[39]. The order of K_{111} values for the extraction of copper acetylace-tonate[18,32] by tributyl phosphate and pyridine supports the view that steric hindrance may play an important role in the extractability of synergistic adducts.

b) Many fragmentary data point strongly to the fact that the extent of synergistic enhancement depends also on the diluent employed, and is higher the lower its polarity. Under comparable experimental conditions, the measured distribution ratio for many metal chelates is higher in benzene or carbon tetrachloride than in chloroform, probably due to an interaction of the latter with the neutral ligand in the system[6,18,30,40,41]. Healy's[27,42] extensive data for promethium and americium(III) thenoyl-trifluoroacetonates and TBP show that the diluent may have an effect on synergism greater than the initial synergic effect. For Pm extraction, a synergic enhancement of 1 to 500 has been observed in benzene, whereas in cyclohexane the effect is 1 to 10^6, a difference by a factor of 2000 only for changing the diluent. The distribution ratio decreased in the order cyclohexane > hexane > carbon tetrachloride > benzene > chloroform. In chloroform, a much higher concentration of TBP is needed to achieve any synergistic effect. There is a similar effect of diluent on the synergistic enhancement in the extraction of thorium, calcium and other rare earths, but not of uranium(VI). The lack of diluent effect in the latter case, Healy believes to be due to the weakly polar character of the symmetrical $UO_2(TTA)_2$ complex.

It should be pointed out that despite these marked differences in synergistic extractability, the composition of the adduct remains unchanged.

c) Though one would not expect large differences in K_{111} values for closely related metal ions where the stoichiometric composition of the adducts is also similar, there is still a small but finite decrease in the synergistic enhancement with decreasing ionic radius of the metal[30,35]. Manning's data [35] for a series of lanthanides clearly demonstrate this effect. Though the stability of the metal–TTA chelates is usually greater the smaller the ionic radius[29,36], the K_{111} is apparently dominated by the stability of the adduct formed; changes in K_{12} are overcompensated by the K_{121} values. This is evident from the fact that K_{111} increases with increasing ionic radius in many systems[24,26,30,35,37,43] (see the rare earths series, Table 11A3.). At least partially, the phenomenon can be attributed to a lower energy needed to accomodate the neutral ligand with increased ionic radius of the central metal atom.

d) As to the complexing power of the chelating agent, many comparative studies[9,13,31] indicate that a stronger chelating agent, which forms a stable metal complex, has a lower tendency to facilitate the binding of the neutral ligand to the metal, than a weaker chelating agent. The results of Li and coworkers[13] shown in Table 11A5, demonstrate this effect. The acidity of the copper–β–diketone chelates increases in the order: acetylacetone, (AA) < trifluoroacetylacetone, TFA < thenoyltrifluoroacetone, TTA < hexafluoroacetylacetone, HFA, whereas decreasing electron density around the central Cu^{2+} ion will enhance its ability to take on a further monodentate ligand in that order. The above results also show that with a given copper chelate the stability of the adduct formed increases with increasing basicity of the neutral organophosphorus compound. The sequence is similar for these copper chelates and their adducts with pyridine and several of its derivatives[33], and of europium chelates with TBP[31].

Though the bulk of the information available on the composition of the synergistic adducts refers to systems containing the metal in tracer amounts, there is evidence that the systems behave essentially in the same way when macro concentrations of the metal are extracted[41,43]. Spectral data support the fact, and confirm that the composition of the adduct

TABLE 11A5. Log K_{111} of copper–β-diketonate chelates with neutral
organophosphorus esters in benzene, $25°C$[13].

	TOPO	TPPO	TBP	TPP
Cu(HFA)$_2$. S		4.08	3.68	2.11
Cu(TTA)$_2$. S	3.23	2.34	1.66	0.48
Cu(TFA)$_2$. S	2.97	2.25	1.55	0.48
Cu(AA)$_2$. S	1.65			

remains unchanged at macro metal levels[7,8]. In the visible region, the
spectra of solid $UO_2(TTA)_2$ and $UO_2(TTA)_2$.S dissolved in benzene,
differ only slightly, and the spectrum of the uranium(VI) extract is
identical to them.

c. Extraction of Mixed Adducts

Cunningham and coworkers[1], but mainly Irving and Edgington[24,43]
have demonstrated that ter- and tetravalent actinides and europium form,
under certain conditions, various mixed complexes, with the participation
of the inorganic anion in the aqueous solution. It seems that when
adducts are formed with strongly donor molecules such as a phosphine
oxide, the inorganic anion tends to replace the chelating molecule of TTA
to a greater extent than the less basic TBP. For example, while TBP with
$Am(TTA)_3$ or $Eu(TTA)_3$ forms synergistic adducts of the $M(TTA)_3.S_2$
type, under similar experimental conditions, TBPO will form the synergistic adduct $M(TTA)_2NO_3.S_2$ when extracted from nitric acid media[24].
When the latter complex is the predominating species, distribution data
indicate that the formation of $M(TTA)_3.S_2$ does not take place to any
considerable extent. Additionally, the number of TTA molecules replaced
in the complex by nitrate ions — to preserve neutrality — increases also
as the stability of the metal–TTA chelate decreases. However, such a
difference in TBP and TBPO may partly be due to steric factors as well.
Namely, at high neutral ligand concentrations in the system, metal-
chelate adducts are formed with two neutral ligand molecules rather than
one. With the tetravalent actinides apparently, the addition of two TBP
molecules does not require the removal of the chelating molecule from
the organic-phase adduct, whereas that of TBPO does.

For the formation of such mixed nitrato–chelate adducts, the equilibrium reactions can be represented in a manner analogous to that
given for the simpler chelate adducts (cf. equations 7C34–7C36). The
stability constant of the chelate K_{12} is defined as in equation (11A3).
The reaction involving the synergistic adduct

$$M^{n+} + (m-n)\overline{HTTA} + nNO_3^- + x\overline{S} \rightleftharpoons \qquad (11A7)$$

$$\overline{M(TTA)_{m-n}(NO_3)_nS_x} + (m-n)H^+$$

has an equilibrium constant

$$K_{121n} = \overline{[M(TTA)_{m-n}(NO_3)_n \cdot S_x]}[H^+]^{(m-n)} \times \qquad (11A8)$$

$$[M^{m+}]^{-1}\overline{[HTTA]}^{(m-n)}[NO_3^-]^{-n}\overline{[S]}^{-x}$$

As before, $K_{121n}.K_{12}^{-1}$ gives the equilibrium constant for the synergistic reaction

$$\overline{M(TTA)_m} + nH^+ + nNO_3^- + x\overline{S} \rightleftharpoons \qquad (11A9)$$

$$\overline{M(TTA)_{m-n}(NO_3^-)_n \cdot S_x} + n\overline{HTTA}$$

with the constant

$$K_{111n} = \overline{[M(TTA)_{m-n}(NO_3^-)_n \cdot S_x]}\; \overline{[HTTA]}^m \times \qquad (11A10)$$

$$\overline{[M(TTA)_m]}^{-1}[H^+]^{-n}[NO_3^-]^{-n}\overline{[S]}^{-x} = K_{121n}K_{12}^{-1}$$

The stoichiometry of some of these mixed nitrato–TTA adducts found by Irving and Edgington, along with their formation constants are listed in Table 11A6. Similar complexes have been reported for thorium[43], but there is an indication that the formation of mixed complexes is not

TABLE 11A6. Equilibrium constants of nitrato–thenoyltrifluoroacetone adducts in cyclohexane.

Adduct	Neutral ligand	Log K_{12}	Log K_{111n}	Log K_{121n}	Reference
Eu(TTA)$_2$NO$_3$. S$_2$	TBPO	−8.8	14.8	6.0	25
Am(TTA)$_2$NO$_3$. S$_2$	TBPO	−6.6	14.0	7.4	25
Np(TTA)$_3$NO$_3$. S	TBP	5.2	2.6	7.8	24
	TBPO	5.2	3.8	9.0	24
Np(TTA)$_2$(NO$_3$)$_2$. S$_2$	TBPO		2.5	11.5	24
Pu(TTA)$_3$NO$_3$. S	TBP	6.4	2.3	8.7	24
	TBPO	6.4	3.5	9.9	24
Pu(TTA)$_2$(NO$_3$)$_2$. S$_2$	TBPO		2.0	11.9	24

restricted to the participation of the nitrate ion[44,45]. Szabo and co-workers[41] have reported that uranium(VI) is synergistically extracted from $(H,Na)ClO_4$ aqueous solutions by acetylacetone and a neutral phosphorus ester, and that the composition of the adduct is close to $UO_2.Acac.ClO_4.S$.

d. Structure of the Adducts

As has been shown in Chapter 8, metal β-diketonates and many other metal chelates are usually insoluble in non-polar organic solvents. The unhydrated thenoyltrifluoroacetonates of copper, zinc and uranium(VI), for example, are coordinatively unsaturated, thus insoluble in solvents incapable of electron donation. Addition of neutral esters—or any electron-donor molecule—to the non-polar solvent, will enhance the solubility of these electrically neutral chelates[8,9]. The amount of the basic donor component needed in a given chelate–non-polar solvent system to achieve a complete dissolution will depend on the basic strength of the donor. From such three-component systems, in many instances, solid adducts have been isolated and their structure and properties investigated. Though the stoichiometry of crystalline substances isolated from organic extracts is not necessarily identical to that in solution, it has been shown that in the systems under consideration this was the case.

The general method of preparation, similar to that for solid metal chelates[9-16], consists of equilibration of an aqueous metal salt solution with an organic solution containing the chelating acid and the neutral ligand. The extract, frequently in a low-boiling point aliphatic hydrocarbon, is partially evaporated, filtered, washed and recrystallized from the same solvent. The composition and melting point of some metal-thenoyltrifluoroacetonates adducts with organophosphorus esters are given in Table 11A1. The melting point of the adduct is usually lower by 50–80°C than that of the corresponding metal chelate. Many others have an even lower melting point and they are oils at room temperature, as for example $Pb(TTA)_2$ and $Ni(TTA)_2$ with TOPO, the $Sc(TTA)_3$ adduct with TBP, TOPO or TPPO, and $Nd(TTA)_3$ with TOPO[15]. A similar procedure was unsuccessful in preparing the adduct of aluminium, copper and zinc with the less basic TBP. In these cases it is the $M(TTA)_2$ which precipitates[9,11]. The solids listed in the Table are quite stable, not losing the adduct molecule at the melting point. All of them are anhydrous and non-hygroscopic[5,8,15]. Similar compounds have been

TABLE 11A7. Composition and melting point of metal-thenoyltrifluoroacetone adducts with neutral organophosphorus compounds.

Compound	Melting point (°C)	Reference
Cu(TTA)$_2$. TOPO	73	9
Zn(TTA)$_2$. TOPO	67	9
Cd(TTA)$_2$. TOPO	92	11
Cd(TTA)$_2$. TOPO	159	11
Nd(TTA)$_3$. 2TPPO	285	8
Th(TTA)$_4$. TBP	178	8
Th(TTA)$_4$. TOPO	110	8
Th(TTA)$_4$. TPPO	305	8
UO$_2$(TTA)$_2$. TBP	112	8, 9
UO$_2$(TTA)$_2$. TPPO	238	8
UO$_2$(TTA)$_2$. T2EHPO	60	8
UO$_2$(TTA)$_2$. DBBP	68	8
UO$_2$(TTA)$_2$. TOPO	57	8, 9
UO$_2$(TTA)$_2$. 3TOPO	37	8

T2EHPO—Tri-(2-ethylhexyl)phosphine oxide

prepared with n-butylacetanilide[8] and pyridine[17] as the adducts, or by crystallization of copper acetylacetonate from solutions containing pyridine or quinoline[9,10,16,18,19].

The reaction between metal diketonates and organic bases in anhydrous non-polar organic solvents may involve the formation of several adducts in solution[16,18-22]. Though in the majority of cases the 1:1 compound is preferentially formed, in very large excess of the donor component, 1:2 species have frequently been identified, especially when the neutral ligand is a strong base. Vapour-pressure osmometric measurements in dry benzene or carbon tetrachloride indicate that the species are monomeric[9,15], though some adducts containing more than one ligand may be associated in non-coordinating solvents[23].

It has been discussed in Chapter 8 that HTTA and many other β-diketones may exist in keto, enol and ketohydrate form. Obviously the medium affects the keto–enol equilibrium, the enol form prevailing in a non-ionizing medium where the internally hydrogen-bonded molecule is apparently less polar than the keto form[46]. A normal enol structure of an organic solution will still prevail in the presence of a donor such as tributyl phosphate, provided the concentration of the latter in a hexane medium is kept below 0.1 M^{47}. In the presence of higher TBP concentrations, however, the spectrum changes due to hydrogen bonding of TBP to the enolic form of the chelating acid

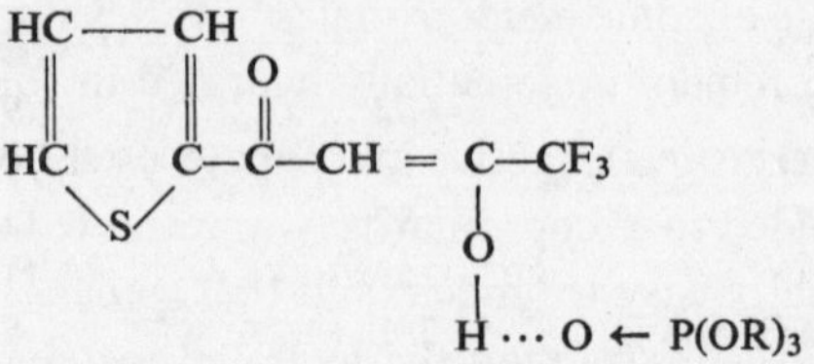

N.m.r. studies[48] on the extent of interaction between various β-diketones and neutral organophosphorus esters in dry carbon tetrachloride, suggest that the hydrogen-bond strength will depend on the acid strength of the chelating acid and the basicity of the phosphoryl oxygen. The formation constants K_{11} of the reaction

$$\overline{HX} + \overline{S} \rightleftharpoons \overline{HX.S} \qquad (11A11)$$

evaluated from spectral data are compiled in Table 11A8 for several chelating acid–ester pairs, along with ΔH values estimated from measurements in the temperature range from -2 to $37°C$. Temperature has not much effect on K_{11}, in line with the monomeric state of the β-diketones in the solvent. Similar conclusions were drawn on HTTA interaction with quinoline and dodecylamine[49] or trioctylamine[50] ($\log K_{11} = 3.15$).

TABLE 11A8. Formation constants, $\log K_{11}$ and ΔH kcal/mole of the 1:1 hydrogen-bonded complexes between β-diketones and neutral esters in carbon tetrachloride at 21°C[48].

	TBP		DEEP	
	$\log K_{11}$	ΔH	$\log K_{11}$	ΔH
HHFA	0.26	3.4	0.44	3.7
HTTA	-0.72	1.3	-0.57	1.3
HTFA	-0.65	0.9	-0.52	0.7

DEEP—Diethyl ethylphosphonate
HHFA—Hexafluoroacetylacetone
HTFA—Trifluoroacetylacetone

Many metal β-diketonates, whose spectra show the normal enol structure of the chelating acid, readily take up one or two molecules of water to form coordinatively saturated complexes. For that matter, especially in the case of divalent metals, the neutral ligand may be an alcohol, organic amine or any other donor molecule. Since HTTA, or β-diketones generally, are known to be dimerized only exceptionally[30,44], and an undissociated HTTA molecule would not solvate the metal–TTA chelate (as in the case of some metal oxinates[51], for example) the chelates under consideration remain preferentially hydrated in solution.

Now, how a synergistically active phosphate ester, or pyridine and similar bases, are added to a coordinatively unsaturated metal is readily visualized. It is equally easy to understand that under the experimental conditions of synergistic extraction the hydrated metal chelate becomes continuously less hydrophilic by substitution of the ligand for the water as the concentration of the neutral ligand in the system increases. This mechanism implies that the organic species $M(TTA)_m.(H_2O)_y$ is replaced gradually by $M(TTA)_m.S_x$. Infrared spectra provide evidence that the chelate rings remain intact during that reaction[8,10,16], though the water–neutral ligand exchange is not necessarily molecule for molecule. The water content of a hexane solution is very low in the presence of metal, though in the same mixture of HTTA and TBP, but without the metal, the hexane phase contains water[15,27]. Copper(II) is most probably hexacoordinated when its β-diketonate is a dihydrate. With tributyl phosphate, methyl isobutyl ketone, pyridine and quinoline, distribution and spectral data unequivocally indicate a monoligand synergistic adduct, to which a pentacoordination of the metal has been ascribed[13,16,18,29,49,52,53]. While cobalt(II) and zinc behave similarly in the process of synergistic extraction, and the two water molecules in the $Co(TTA)_2.2H_2O$ are readily replaced by heterocyclic bases to form the extractable $Co(TTA)_2.S_2$[6,17,28,32,33], a nickel chelate, with its preferentially tetracoordinated saturation, show no tendency to coordinate water, quinoline or dodecylamine[49].

The structure of the synergistic adduct and the mechanism of its formation becomes less understandable when the complex-forming metal is coordinatively saturated by the attached chelating anions alone. Such is the case, for example, with the usually hexacoordinated tervalent rare earths and the octacoordinated tetravalent actinides. Two possibilities have been suggested to explain the phenomenon.

Healy[15,27], basing his statement on infrared data, advocates that in the synergistic adduct of a metal which becomes coordinatively saturated by the chelating molecules alone, one or more chelate rings may open so as to form an adduct which contains some TTA molecules as monodentate and some as bidentate ligands. For example, in the adduct $UO_2(TTA)_2.TOPO$ and $UO_2(TTA)_2.3TOPO$, where uranium has the coordination number of 6 and 8 respectively, one TTA is monodentate and the other a bidentate. Similarly, thorium with its coordination number of eight, has three bidentate TTA groups and one monodentate, or the hexacoordinated rare earth metal adduct contains two monodentate and one bidentate chelating molecules in the complex $RE(TTA)_3.2TBP$. Also, Li[6] suggested that in the $Zn(TTA)_2.TBP$ adduct one of the chelating agents is monodentate. The possibility of a β-diketone figuring as a monodentate has been seriously questioned by Graddon[18].

On the other hand, Irving and Edgington[24,33,43] argued that the coordination number of tervalent lanthanides and actinides may rise to eight in order to accomodate the neutral ligand molecules, and still preserve the bidentate structure of the thenoyltrifluoroacetonate. The possibility of the existence of a hydrated $Eu(TTA)_3.2H_2O$ has not been excluded ($Eu(TTA)_3.HTTA$ has also been claimed[30]) thus the mechanism for synergistic adduct formation may be either the replacement of water or direct addition of TBP (or TBPO) to the non-hydrated chelate. They consider that for HTTA to act as a monodentate ligand is 'a process which would involve a considerable loss of chelation energy'[24]. Their further argument is the fact that in some synergistic systems involving the lanthanides and tri- and tetravalent actinides, but especially the latter, one or more chelating molecules are displaced from a $M(TTA)_3$ or $M(TTA)_4$ chelate by the unidentate nitrate ion, thus enabling the neutral ester molecule to be coordinatively bound to the metal

$$M(TTA)_4 \qquad\qquad M(TTA)_3NO_3 . OPR_3$$

Newman[26], on the basis of the similar K_{111} values for several tervalent metals (evaluated from Healy's[5] experiments) suggested a third possibility; the TBP molecules are bound to TTA and not directly to the metal, otherwise there would be a large difference in K_{111} values, as there is one in the corresponding K_{12} constants.

e. Destruction of Synergism

From the shape of the curves in Figure 11A1 it is evident that there is an optimal HTTA: ester ratio for a maximum synergistic enhancement. Under different experimental conditions, when keeping $[\overline{\text{HTTA}}]$ and the aqueous $[H^+]$ constant, and increasing $[\overline{S}]$, as shown in Figure 11A2, the metal distribution ratio will decrease after a short range of a nearly constant value. Such a decrease in synergism has been noted in many systems.

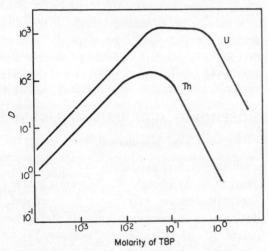

Figure 11A2. Synergistic and relative antagonistic effect of TBP[47]
U—uranium(VI)–0.015 M HTTA–cyclohexane (aqueous phase 0.01 N HCl)
Th—thorium–0.01 M HTTA–cyclohexane (aqueous phase 0.1 N HCl)

Like synergism, its destruction is not confined to phosphorus esters either; any neutral oxygenated extractant exhibiting synergism is likely to produce its destruction when in excess. The role of the excess ligand in causing the effect is apparently interchangeable: synergism built up by one ligand can be broken down by another ligand. Furthermore, as illustrated in Figure 11A2, the metal distribution ratios have a positive synergistic effect of $+1$ power dependency on the neutral ligand concentrations, and a -2 power antagonistic dependency. The slopes appear to be independent of the nature of the neutral ligand, and have the same value for a given metal. The synergistic and antagonistic solvent powers are $+1$ and -2 respectively for thorium, but are $+2$ and -4 for the tervalent actinides and lanthanides, again regardless of the nature of the neutral ligand (phosphorus ester, alcohol, amide, ketone or ether).

It is remarkable that throughout both the synergistic and antagonistic stages, the $M(TTA)_m$ chelate remains an entity in the organic phase. Healy and coworkers[27,47] believe 'that water in the organic phase (the amounts transferred increase with an increasing fraction of the neutral ligand in the organic mixture) has an appreciable effect on the breakdown of the anhydrous synergistic complex $M(TTA)_m.S_x$'. Indeed, Li and coworkers[20,54] have interpreted their n.m.r. data in terms of ketohydrate formation from the enolic β-diketones. The latter interpretation of the antagonistic effect, however, implies the destruction of the metal β-diketonate, which has not been found by Healy. It seems to be safer to assume that at high neutral ester[47] or alcohol[20] concentration the antagonistic effect is due to competition between water or alcohol and the metal diketonate for the neutral ligand molecules in the system.

B. ALKYLPHOSPHORUS ACID NEUTRAL LIGAND SYSTEMS

a. Introduction

The synergistic enhancement of extraction was initially thought to be a distinctive property of dialkylphosphoric acids when used in combination with neutral organophosphorus extractants[3,4]. Based on fragmentary experimental information it seemed, at an early stage, that the phenomenon was limited to the extraction of uranium(VI) from aqueous solutions. Subsequent work revealed, however, that the phenomenon can be realized in a variety of extractant combinations, and be effective for a number of metals.

Though in the bulk of the synergistic studies a dialkylphosphoric acid has been used as the salt-forming acidic component, information is available to show that it can be effectively replaced by various carboxylic acids[55,56] or dialkylnaphthalene sulphonic acid[3,57,58] in its role of neutralizing the charge on the metal. Neither is the neutral ligand restricted to neutral organophosphorus esters, though again, they have been most widely used. Under certain experimental conditions alkylamines[4,40,59], dodecanol[56] and butyl-(methylbenzyl-) phenol[56,57] have been shown to enhance the extractability of metal–acidic extractant complexes. As to the metals, it is certainly the synergistic extraction of uranium(VI) which has received the widest coverage, but the phenomenon has been demonstrated also in the extraction of other actinides[4,40,60,61], the lanthanides[61,62] the heavy alkali metals[56-58], calcium[61] and strontium[2,56,57].

Before making a generalization, it should be stressed that the ability of both the acidic and the neutral phosphates to contribute to synergism can vary considerably with the properties and structure of the reagents. The extraction of uranium(VI) has been used as a model to compare the synergistic enhancement of various dialkylphosphoric acids and dialkyl hydrogen phosphonates[4,61,63,64]. Their synergistic contribution followed the same trend as when these reagents were used alone: the extraction ability increases with acid strength and with decreased branching of the alkyl chain. Zangen[63] suggests that this is perhaps due to a lower tendency towards dimerization with increasing acid strength. The results of several comprehesive survey studies[4,64,65] for the evaluation of a large number of neutral phosphorus esters in their effectiveness for synergistic combination, indicate an increasing effect with increasing basicity. The distribution ratio increases as the number of carbon-to-phosphorus bonds increases, phosphate < phosphonate < phosphinate < phosphine oxide, as illustrated in Figure 11B1, reproduced from the Oak Ridge report[4]. Here again, normal-alkyl chain compounds give higher D's than branched. It is a general phenomenon that the metal distribution ratio increases with increasing concentration of the neutral ligand to a maximum and then decreases. That maximum is reached with lower concentrations of those neutral esters which give the strongest synergistic effect, as can be seen from the curves in Figure 11B1. Finally, the degree of synergistic enhancement depends on the diluent employed[4,56,57,64]. Highest effects are obtained in aliphatic hydrocarbons, the gross effect being lower in benzene and carbon tetrachloride, but markedly lower in chloroform. The concentration of the neutral synergist required to reach the maximum in D decreased as more favourable diluents were used.

b. Interaction Between Acidic and Neutral Extractants

The experimental observations described so far leave little doubt that in these systems too, the synergistic enhancement is due to an organic-phase interaction. In addition to the equilibria between the metal, the acid (HX) and the neutral extractant (S) to yield a synergistic metal adduct, there is an interaction between HX and S to yield one or more associated species. The nature of this interaction between the two extractants, and especially its extent, is frequently the governing factor of the synergistic enhancement in metal extraction. It is thus pertinent to discuss this inter-extractant association prior to considering the possible synergistic reactions.

It is believed that the association between the acidic and the neutral

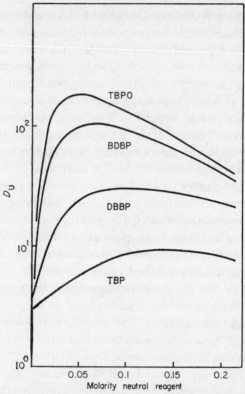

Figure 11B1. Effect of tributylphosphate (TBP), dibutyl butylphosphonate (DBBP), butyl dibutylphosphinate (BDBP) and tributylphosphine oxide (TBPO) on synergistic extraction of uranium(VI) when added to 0.1 M di-(2–ethyl–hexyl)phosphoric acid in kerosene. Aqueous phase: 1.5 M H_2SO_4 and 0.004 M uranium[4]. (By permission from the USAEC).

components is relatively strong, since it may take place through the monomerization of the dimeric alkylphosphoric acid, according to the reaction

$$\overline{(HX)_2} + 2\overline{S} \rightleftharpoons 2\overline{HX.S} \tag{11B1}$$

or

$$\overline{HX} + \overline{S} \rightleftharpoons \overline{HX.S} \tag{11B2}$$

with the corresponding equilibrium constants

$$K_{22} = \overline{[HX.S]}^2 \overline{[(HX)_2]}^{-1} \overline{[S]}^{-2} \text{ and } K_{11} = \overline{[HX.S]}\,\overline{[HX]}^{-1}\overline{[S]}^{-1}$$

In either case, the reaction involves the rupture of two hydrogen bonds in the de-dimerization of the dimeric acidic phosphorus ester, and the formation of two new ones between two pairs of HX and S molecules. Since the stability of the dimer depends on the diluent employed, as has been shown in section 8.B, that of the mixed adduct between two given components will also depend on the diluent. This has been shown to be true in a variety of systems, involving various acidic and neutral phosphorus reagents[53,63,66-73], and also for systems where the neutral component is an amine[59] or a phenol[56,74]. Baes[2] has compared the stability of the HX.S adduct between di-(2-ethylhexyl)phosphoric acid, HDEHP, and tributyl phosphate in benzene, carbon tetrachloride, hexane and kerosene, and found it increasing in that order of solvents in a ratio 1:1.2:2.5:4. The extent of formation of mixed associates will depend, furthermore, on the basicity of S, a higher stability constant being expected with more basic neutral ligands. This is illustrated again by Baes' data, where the stability of HDEHP adducts in kerosene with TBP, DBBP, BDBP (n-butyl di-n-butylphosphinate) and TBPO has been found to relate as 1:1.5:4.5:8.3.

Table 11B1 shows the reported association constants for various extractant pairs. The K_{11} values have been evaluated from experimental data from sources indicated, whereas the self-association constant K_2 of the reaction

$$2\overline{HX} \rightleftharpoons \overline{(HX)_2} \qquad (11B3)$$

TABLE 11B1. Association constants of HX.S adducts formed between acidic and neutral extractants.

Acidic component	Neutral component	Diluent	Log K_2	Log K_{11}	Log K_{22}	Reference
HDBP	TBP	Kerosene	5.78	2.83	−0.12	67
HDBP	TBP	Hexane	6.66	2.94	−0.78	69
HDBP	TBP	Chloroform	4.61	1.60	−1.41	75
HDBP	TBP	CCl$_4$	6.37	2.65	−1.07	69
HDBP	TBP	TBP	−0.12	2.84	5.80	68
HDBP	TOPO	Hexane	6.66	4.88	3.10	69
HDBP	TOPO	CCl$_4$	6.37	4.36	2.35	69
HDBP	MIBK	MIBK	1.19	1.86	2.53	68
HDBP	DIPE	DIPE	2.29	1.06	−0.17	68
HDBP	Chloroform	Chloroform	4.61	0.53	−3.55	68
HDBP	Nitrobenzene	Nitrobenzene	3.55	0.25	−3.05	68
HDBP	EPMP	Benzene			−0.92	66
HDEHP	TOPO	n-Octane	4.47	4.28	4.09	70

HDBP — Di–n–butylphosphoric acid
EPMP — Diethylpolystyrene methylenephosphonate resin
DEHPD — Di–(2–ethylhexyl) phosphoric acid
DIPE—Diisopropyl ether
For other abbreviations see previous Tables in this Chapter.

has been taken from Table 8B3, and K_{22} if not determined in source, has been calculated from the relationship $K_{22} = K_{11}^2 K_2^{-1}$. Baes' constants, mentioned above, are not included in the Table, since they are not based on direct experimental data[2].

It has been suggested also that the association between the non-identical molecules of HX and S may not necessarily involve the monomerization of the dimeric alkylphosphoric acid[2,4,40,63,69,70]. It may well be possible that the adduct formation proceeds, at least partially, according to the reaction

$$\overline{(HX)_2} + \overline{S} \rightleftharpoons \overline{(HX)_2.S} \qquad (11B4)$$

$$K_{21} = [\overline{(HX)_2S}] \,[\overline{(HX)_2}]^{-1}[\overline{S}]^{-1} \qquad (11B5)$$

The interaction is assumed to take place again through hydrogen bonding between a P—OH and the phosphoryl oxygen of the neutral ligand, but this time the adduct is formed upon the rupture of one hydrogen bond only in the dimeric $(HX)_2$

$$(RO)_2(O)P—OH \cdots OP(OR)_2—OH \cdots O \leftarrow P(R)_3$$

Using the continuous-variation method for a mixture of HDEHP and TOPO in n-octane, Baker and Baes[70] found the experimental data to be consistent with the formation of both types of associates, the HX.S and $(HX)_2.S$. From the intensity of the phosphoryl bands in the i.r. spectra of such solutions, the authors evaluated the equilibrium quotient of reaction (11B4) to be 100 ± 25. The stability of the 2:1 adducts is apparently affected by the same factors as that of the 1:1 associates, as can be seen from the $\log K_{21}$ values compiled in Table 11B2. The stability increases in the order of increasing base strength of S, and increases with the same order of diluents.

Liem and Dyrssen[69] found more recently that the best fit of their distribution data on labelled HDBP between hexane and 0.1 M H_2SO_4 aqueous solution in the presence of TBP, as evaluated by the Letagrop Vrid computer programme, could be obtained when in addition to equilibria (11B1) and (11B4) the reaction

$$\overline{HX} + 2\overline{S} \rightleftharpoons \overline{HX.S_2} \qquad (11B6)$$

is taken into account. The equilibrium constant evaluated had a value of

TABLE 11B2. Association constants of $(HX)_2 \cdot S$ adducts formed between acidic and neutral extractants.

Acidic component	Neutral component	Diluent	Log K_{21}	Reference
HEH2EHP	D2EH2EHP	Cyclohexane	1.30	63
HEH2EHP	DBBP	Cyclohexane	1.30	63
HEH2EHP	TOPO	Cyclohexane	2.14	63
HDEHP	D2EH2EHP	Cyclohexane	1.30	63
HDEHP	DBBP	Cyclohexane	1.30	63
HDEHP	TOPO	Cyclohexane	2.30	63
HDMBPP	D2EH2EHP	Cyclohexane	2.60	63
HDMBPP	DBBP	Cyclohexane	2.74	63
HDEHP	TBPO	Kerosene	1.52	2
HDEHP	BDBP	Kerosene	1.25	2
HDEHP	DBBP	Kerosene	0.78	2
HDEHP	TBP	Kerosene	0.60	2
HDEHP	TBP	Hexane	0.40	2
HDEHP	TBP	CCl$_4$	0.04	2
HDEHP	TBP	Benzene	0.00	2
HDBP	TBP	Hexane	1.27	69
HDBP	TBP	CCl$_4$	0.22	69
HDBP	TOPO	Hexane	2.60	69

H2EH2EHP — 2-Ethylhexyl 2-ethylhexylphosphoric acid
HDMBPPA — Di-(1,1,3,3-tetramethylbutyl) phenylphosphoric acid
BDBP — n-Butyl di-n-butylphosphinate

$\log K_{12} = 3.42$ at 25°C. It should be noted howevee, that when hexane was replaced by carbon tetrachloride, or TBP by TOPO, reaction (11B6) did not occur.

The above picture of the interaction between the two extractants in a synergistic system, derived mainly from distribution data, received independent support by infrared and isopiestic measurements[70], and, to some extent, by dielectric constant determinations[72]. The latter technique revealed an interaction between HDEHP and TOPO, but failed to show a deviation from additivity when the less basic TBP was used as the neutral component rather than TOPO.

c. Mechanism of Synergism

The mechanism of synergistic enhancement of metal extraction has for some time been in dispute. There have been essentially two versions offered: one proposing that the neutral ligand S is added to the species formed between the metal and the acid component, and the other, postulating a substitution of one or more acidic extractant molecules HX in the metal–HX species by an equal number of neutral ligand molecules. From the experimental material available, it seems to us that an addition product is responsible for the enhanced metal transfer, in a way similar to that discussed previously in chelating acid–neutral ligand systems.

It has been shown in section 8.C that dialkylphosphoric acids, which are dimeric in practical diluents, extract metals essentially in two different ways. Under very low organic-phase loading conditions the extraction is given by the reaction

$$M^{m+} + m\overline{(HX)_2} \rightleftharpoons \overline{M(X.HX)_m} + mH^+ \qquad (11B7)$$

with an equilibrium constant

$$K_{12} = \overline{[M(X.HX)_m]}\,[H^+]^m[M^{m+}]^{-1}\overline{[(HX)_2]}^{-m} \qquad (11B8)$$

and the distribution ratio is proportional to the mth power of the dimer concentration and inversely proportional to the same power of hydrogen ion concentration. The dimeric dialkyl ester molecule entering into a formally cation exchange reaction, is only monoionized, i.e. a hydrogen in one of the OH groups remains unreplaceable.

On the other hand, under different loading conditions, the species $M(X.HX)_m$ will not be the predominant metal-bearing species in the organic phase, and the charge on the metal becomes apparently satisfied by monomeric extractant molecules. In many systems, due to the de-dimerization of the dialkylphosphoric acid and the formation of simple metal salts, the composition of the complex is MX_m. Further, in polar diluents which minimize or even prevent the dimerization of HX acids, the composition of the organic-phase metal species is again MX_m rather than $M(X.HX)_m$.

Now, when a neutral extractant is added to a system containing tracer amounts of metal, the distribution ratio increases due to the higher extractability of a synergistic adduct formed by the organic-phase reaction

$$\overline{M(X.HX)_m} + x\overline{S} \rightleftharpoons \overline{M(X.HX)_m.S_x} \qquad (11B9)$$

with the corresponding equilibrium constant

$$K_{111} = \overline{[M(X.HX)_m.S_x]}\,\overline{[M(X.HX)_m]}^{-1}\overline{[S]}^{-x} \qquad (11B10)$$

The overall heterogeneous-system reaction

$$M^{m+} + m\overline{(HX)_2} + x\overline{S} \rightleftharpoons \overline{M(X.HX)_m.S_x} + mH^+ \qquad (11B11)$$

has an equilibrium constant

$$K_{121} = \overline{[M(X.HX)_n.S_x]}[H^+][M^{m+}]^{-1}\overline{[(HX)_2]}^{-m}\overline{[S]}^{-x} \qquad (11B12)$$

The earliest Oak Ridge data[3,4] on the synergistic extraction of uranium(VI) from sulphuric acid solutions have been explained by the addition mechanism represented in equation (11B11). In these experiments D_U varied with an approximate second-power dependence on $\overline{[(HX)_2]}$ when $\overline{[S]}$ was kept constant, in a manner identical to systems in the absence of S. This has been shown to be true for many synergistic additives and at any of the synergist concentration (up to a maximum D). The distribution ratio had a first-power dependence on $\overline{[S]}$ at a particular fixed HX level. The synergistic adduct, thus, was suggested to have the composition $UO_2(X.HX)_2.S$. Zangen[63], too, has found in a large number of systems that D_U, for extracion into cyclohexane from dilute HCl solutions, has in the synergistic region* the same dependence on $\overline{[(HX)_2]}$ and on $\overline{[S]}$. Baes[2], in what is essentially a review article, reported a variety of new synergistic data on tracer uranium extraction in complete accordance with the formation of adducts $UO_2(X.HX)_2.S$. Shevchenko and coworkers[64,76] found also an enhanced extraction of uranium(VI) from nitric acid media by a variety of synergistic combinations. Distribution data could be fitted into a (11B9)-type equation, but assuming a monomeric HX, they were practically independent of the metal concentration in the system. A comparison of the absorption spectra[77] of a kerosene solution of UO_2^{2+}–HDEHP in the visible region revealed only small differences when TOPO had been added. The results of the continuous-variation method, as followed

---

*Zangen, however, suggested the ion-paired structure $[HX.HS^+]\,[UO_2X_3^-]$ for the synergistic adduct. Though a similar ion-paired entity $H^+UO_2X_3^-$ has been previously proposed by Kennedy and Deane[66] to account for some of their infrared observations, such structure would contradict the widely held belief that interaction between the metal and the dimerized acidic extractant is through monoionization of the latter. See section 8.C.

by spectral measurements, are consistent with a U:TOPO ratio of unity in the synergistic adduct.

The reported equilibrium constants K_{111} or K_{121} for uranium are compiled in Table 11B3, along with the available K_{12} values taken from Table 8C1. It is immediately apparent that the stability of the synergistic adduct depends on the nature and properties of the organic-phase components: more stable adducts are formed by more basic S, more acidic HX and in aliphatic hydrocarbons.

TABLE 11B3. Equilibrium constants in the synergistic extraction of uranium(VI).

Acidic component	Neutral component	Diluent	Log K_{12}	Log K_{111}	Log K_{121}	Reference
HDEHP	TBPO	Kerosene	4.53	4.28	8.81	2
HDEHP	BDBP	Kerosene	4.53	3.78	8.31	2
HDEHP	DBBP	Kerosene	4.53	2.78	7.31	2
HDEHP	TBP	Kerosene	4.53	2.18	6.17	2
HDEHP	TBP	Hexane	4.60	1.85	6.45	2
HDEHP	TBP	CCl$_4$		1.60		2
HDEHP	TBP	Benzene		1.20		2
H2EH2EHP	D2EH2EHP	Cyclohexane			6.64	63
H2EH2EHP	DBBP	Cyclohexane			6.78	63
H2EH2EHP	TOPO	Cyclohexane			8.78	63
HD2EHP	D2EH2EHP	Cyclohexane			7.40	63
HD2EHP	DBBP	Cyclohexane			7.60	63
HD2EHP	TOPO	Cyclohexane			9.56	63
HDMBPP	D2EH2EHP	Cyclohexane			9.74	63
HDMBPP	DBBP	Cyclohexane			10.04	63
HD2EHP	TOPO	Kerosene	4.53	3.85	8.38	77
HDBP	TBP	Benzene	4.69	4.95	9.64	76
HDAP	TBP	Benzene	4.34	4.14	8.48	76
HDAP	DAMP	Benzene		4.01		64
HDAP	TBPO	Benzene		6.15		64
HAMP	TBP	Benzene	3.15	3.20	6.35	64
HAMP	DAMP	Benzene	3.15	3.70	6.85	64
HAMP	TBPO	Benzene	3.15	6.08	9.23	64

HDAP — Diisopentylphosphoric acid
DAMP — Diisopentyl methylphosphonate
HAMP — Isopentyl methylphosphonic acid

Zangen[61,80] has further found that the lighter lanthanides, Cm and Am, but not the heavy lanthanides (Z > 64), are extracted synergistically by a variety of acidic phosphorus–neutral phosphorus ligand combinations into carbon tetrachloride, benzene or cyclohexane. The metal distribution ratio for tervalent metal ions has a third-power dependence on $\overline{[(HX)_2]}$, and a first-power dependence on $\overline{[S]}$, corresponding to a $M(X.HX)_3.S$ stoichiometry of the synergistic adduct. The different behaviour of the light and heavy lanthanides has been explained by the possible octacoordination of the former, but not of the latter. Several K_{121} values have been reported for Am^{3+} and Am^{6+} in carbon tetrachloride[80].

A further example of an assumed addition mechanism has been provided in the synergistic extraction of some alkali metals by a combination of acid alkyl phosphates or carboxylic acid and a phenol derivative[55,78,79].

As to the structure of the synergistic uranium adduct, its formation probably involves an increase of the metal coordination number. If unhydrated, the complex $UO_2(X.HX)_2$ contains the metal in its hexacoordination state, with the single-ionized dimeric $(X.HX)^-$ acting as a bidentate ligand. Subject to steric restrictions, the neutral ligand may become attached to the metal through the phosphoryl oxygen, making the metal heptacoordinated. Blake[4] and Baes[2], though allowing for such a structure, are closer to believing that the ligand becomes hydrogen bonded to one of the ionized OH groups in the HX dimers. Their main argument is the generally lower stability-constant values of the uranium–acidic phosphorus–neutral phosphorus adducts, in comparison to that of uranium–chelating acid–neutral phosphorus adducts, where hydrogen bonding of S to TTA, for example, is markedly weaker. The more recent experimental observations of Zangen[61,80] on the synergistic enhancement of the lighter lanthanides and americium support the neutral ligand-to-metal coordination structure.

A different view of the synergistic mechanism has been formulated by Kennedy and coworkers[66,81] in postulating a substitution reaction for the synergistic extraction of uranium(VI), represented as

$$\overline{UO_2X_2.2HX} + 2\overline{S} \rightleftharpoons \overline{UO_2X_2.2S} + \overline{(HX)_2} \qquad (11B13)$$

This reaction was tested by adsorption of the metal on a non-ionic phosphorylated resin[81,82] from a benzene solution of $UO_2(DBP)_2.2HDBP$. The experiments show 85–90% of the metal being adsorbed and a coresponding amount of HDBP liberated. Part of that acid reacts with the resin, presumably as HX.R and not as $(HX)_2.R$. The resin-phase complex was assumed to have the composition $UO_2(DBP)_2.xR$, where

x was believed to be two. It was argued that the above substitution mechanism should be energetically favoured, since it avoids the monomerization of the relatively stable dimers of HX. It was assumed namely, that the two HX molecules function as solvating ligands, rendering the metal atom coordinatively saturated. Subsequent work[66], including i.r. spectral data, could not support this hypothesis of octacoordination of the uranium atom in its dibutyl phosphate complex. Neither could the composition $UO_2X_2.2S$ be unequivocally confirmed, though no evidence was found for species such as $UO_2(X.HX)_2.S$ (S being now trioctylphosphine oxide), which should have been formed by the addition reaction. The spectra were explained as showing three DBP ligands in an 'equivalent association' with the metal, and interpreted in terms of another substitution reaction

$$UO_2^{2+} + 2\overline{(HX)_2} + \overline{S} \rightleftharpoons \overline{UO_2X_2.HX.S} + \tfrac{1}{2}\overline{(HX)_2} + 2H^+ \qquad (11B14)$$

Dyrssen and Kuča[83] found also a monosubstituted adduct between HDBP and TBP or TOPO, formed by a reaction similar to that shown in (11B14). With the stronger synergist TOPO, an additional HX, the second one, may be substituted under certain experimental conditions. In carbon tetrachloride, for the equilibrium constant of the reaction

$$\overline{UO_2(DBP)_2(HDBP)_2} + \overline{TBP} \rightleftharpoons \overline{UO_2(DBP)_2.HDBP.TBP} + \overline{HDBP}$$

$$(11B15)$$

a value of 1×10^{-3} was obtained. However, Baes[2] in reexamining the experimental data of Dyrssen and Kuča[83], by taking into account a) the presence of a radioactive contaminant (believed by the authors to be present), and b) the formation of mixed adducts of the type $(HX)_2.S$ instead of HX.S, considered by the authors, found that the data may well be consistent with a mass-action equation representing the extraction of uranium by the addition reaction (11B9).

In more recent reports from the Swedish school[69] distribution data of uranium(VI) were interpreted in terms of both addition and substitution reactions. Whether addition or substitution predominates depends on the extraction conditions. High concentration and basicity of the neutral component will promote substitution, whereas less basic esters at low concentrations seems to favour addition.

In conclusion, we believe that the experimental evidence so far available on the nature of the synergistic enhancement is convincing enough to decide in favour of the 'addition mechanism'. Such an extraction reaction will be favoured energetically whenever the organic-phase complex between the metal and the acidic extractant involves either the mono-ionized dimeric anion $X.HX^-$ (low metal loading), or the anion X^- of the monomeric acid phosphate (high metal loading, or high dielectric constant diluent). As a final argument, the synergistic extraction of strontium by a mixture of HDEHP and TBP is perhaps most conclusive. It has been pointed out in Chapter 8, that some divalent metals, and among them strontium, are extracted by dialkylphosphoric acids via the reaction

$$M^{2+} + 3\overline{(HX)_2} \rightleftharpoons \overline{M(X.HX)_2(HX)_2} + 2H^+ \qquad (11B16)$$

The organic-phase metal complex contains, in addition to the two mono-ionized $X.HX^-$ dimers, an undisssociated dimer molecule to solvate the $Sr(X.HX)_2$ complex. More recent dielectric constant measurements[56,57], suggest that TBP replaces $(HX)_2$ to form the synergistic adduct $Sr(X.HX)_2.2TBP$. At high TBP concentrations in the organic phase, where the de-dimerization of $(HX)_2$ becomes considerable, the synergistic complex may have the composition $SrX_2.4TBP$. The synergistic adducts of sodium may have similar compositions[56].

d. Destruction of Synergism

As mentioned previously, when the neutral ligand is added to a system at a constant concentration of the acidic component, a destruction of synergism follows, after a maximum has been reached in the metal distribution ratio. The effect of an excess of S is understood in terms of the competing reaction between the components, leading to the formation of mixed adducts HX.S and $(HX)_2.S$ previously discussed. Zangen's[61,63] experimental data in this region point to the formation of $(HX)_2.S$, whereas Baes[2] assumes both mixed adducts to be formed. It is not felt that the existing experimental information can conclusively settle the question of the antagonistic mechanism. Inasmuch as under conditions of increasing concentration of the neutral ligand, species other than $M(X.HX)_m.S$ seem to be extractable to an ever increasing extent. Qualitatively speaking, the competition between S and the metal ion for the available $(HX)_2$ molecules may first result in de-dimerization of $(HX)_2$, at least partially. It is known, that as the free $(HX)_2$-to-metal molar ratio decreases, the predominating organic-phase metal complex will be the less extractable MX_m rather than $M(X.HX)_m$. An even higher concentration of S will, according to mass-action law requirements, favour the transfer of the metal in the form of a solvated salt such as $MA_m.S_y$, where A is the inorganic ligand initially associated with the metal in the aqueous solution, as would be the case in extraction systems containing neutral ligand alone. This has been shown to be the case in the extraction of europium[61], yttrium [84] and uranium[85] from nitric acid solution at low overall concentration of the acidic component.

The behaviour of monoalkylphosphoric acids $(OH)_2(RO)PO$, as the acidic component is typically antagonistic. Still, in the course of the first studies when the phenomenon was discovered, attention was directed to dibasic phosphoric acid esters in a hope of an even higher synergistic enhancement, than that observed with the monobasic alkylphosphoric esters. It turned out, however, that in such combinations a decrease in distribution ratio of uranium takes place immediately after addition of the neutral component. This is observed in a variety of systems and with many metals[4,40,86-88]. It has been explained as being due to an interaction between the components resulting in the formation of an association product which removes, effectively, both extractants from participating in the transfer of the metal from the aqueous into the organic phase.

It has been noted in section 8.C that the extractive capacity of monoalkylphosphoric acid is strongly dependent upon the diluent employed. The metal distribution ratio is lower by several orders of magnitude in polar alcohols or chloroform than in non-polar hydrocarbons, under otherwise identical conditions. As a rule, lower extraction is achieved in diluents capable of hydrogen bonding, either by their donor or acceptor properties. Furthermore, using the non-polar diluent, the extractant dependency of D is first power for any polyvalent metal regardless of its actual charge, whereas in polar diluents the same dependence is usually numerically equal to the effective charge on the extractable cation. This has been explained in terms of the monomer $\rightleftharpoons$ polymer equilibrium of the monoalkylphosphoric acid in the diluent. The experimental data indicate a hexamer of the acid in the non-polar benzene, a dodecamer in cyclohexane, but a dimer in some ketones and a monomer in alcohols. Large polymers, of the size formed in hydrocarbons, are apparently capable of losing several protons by ionization, usually a number equal to the charge on the cation.

In the light of these observations it becomes understandable that the metal distribution ratio decreases when tributyl phosphate, or any other neutral synergistic ligand, is added to a benzene or cyclohexane solution of a monoalkylphosphoric acid. Using i.r. spectral data, isopiestic and cryoscopic measurements on organic mixtures of mono-(2-ethylhexyl)-phosphoric acid, H_2MEHP, and either TBP or n-decanol in toluene or cyclohexane, the Argonne workers found that when TBP is the ligand, the acidic ester becomes dimeric to form $(H_2MEHP)_2$. TBP by hydrogen bonding. The formation of such a complex causes antagonism by an effective blocking of the active OH groups of the ester. The strongest effect is due to n-decanol, where there is a very strong interaction between the components to inactivate all reaction sites of the ester, presumably forming $H_2MEHP.(ROH)_2$ species. Though such a picture of specific interactions is perhaps somewhat over-simplified, it still remains reasonable to assume that some of the potential sites for cation extraction are no longer available to the cation when donor molecules are brought into contact with a monoalkylphosphoric acid. This implies that the association between H_2MEHP and TBP is stronger than the self association of the former. That would be in line with the discussion in section 8.C: non-associated monoalkylphosphoric acid is a poor extractant.

C. MIXED COMPLEX FORMATION AND OTHER EFFECTS

a. Systems with Two Neutral Ligands

Under this heading the enhanced metal extraction by a mixture of two extractants, both neutral in nature and usually exhibiting only donor properties, will be considered. Various data show that acid and metal extraction by such mixtures is frequently, higher than that calculated on the assumption of additive extractive power for the pure solvents. Little information is available on such systems and the mechanism of the synergistic enhancement is but little understood.

Data, though mostly descriptive in nature and fragmentary in scope, show that the synergistic combination of two oxygenated neutral extractants of the type considered throughout Chapter 9, is effective for a number of elements. Nitric acid[89-92], hydrochloric acid[93] and several carboxylic acids [94] exhibit an enhanced extractability into mixed solvents. Several rare earths[42,95,96] and actinides, among them uranium(VI)[42,90,91], plutonium(VI)[40], thorium[42] and protactinium[97,98], technetium[99], ferric iron[91,93,97], antimony[93] and zirconium, niobium and ruthenium[99], have been extracted synergistically by combinations of various ethers, ketones, alcohols, carboxylic and phosphorus esters. A typical curve is shown in Figure 11C1. The overall synergistic effect in these extractant combinations is lower than in either of the synergistic systems previously discussed, and the synergistic factor is only exceptionally higher than ten, varying mostly between 2 and 5. Often there is no synergistic enhancement at all.

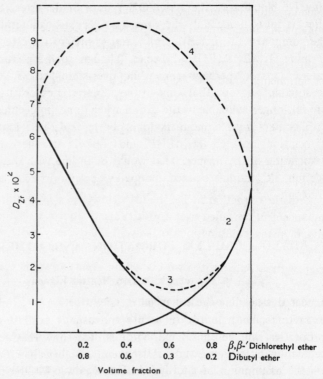

Figure 11C1. Distribution ratio of zirconium[96] from nitrate medium by dibutyl ether–benzene (1), β,β-dichlorethyl ether–benzene (2) and mixture of the two ethers (4). Curve (3) represents the sum of curves (1) and (2).

In certain solvent combinations, mainly when one of the components is inert, in a sense that it does not extract the metal under given experimental conditions, there is a linear relationship between the logarithms of concentration of the active extractant and the metal distribution ratio. Conditions for the linearity of this relationship are the absence of specific interaction between the two components (extractant and diluent) and thermodynamically ideal behaviour of the components in the mixture[97,100]. If these conditions are fulfilled, Irving and Rossotti[101] have shown that the expression

$$\log D_{S_1 S_2} = \log D_{S_1} - x_{S_2} (\log D_{S_1} - \log D_{S_2}) \qquad (11C1)$$

where x is the mole fraction of component and $D_{S_1 S_2}$ the distribution ratio into the mixed solvent, validly represents the experimental data. Such an additivity rule may be valid also in certain cases when both components act independently as extractants, a complete additivity having been shown[101] in the extraction of indium bromide by a mixture of methyl isobutyl ketone and methyl isopropyl ketone. This is, however, an exception and Maddock[97] believes that the linearity is perhaps due to lack of participation of the solvents in the extracting complex of $HInBr_4$. This is unlikely to be the case, in view of the strong combined hydration-solvation of acidic species in ketones and similar oxygen-bearing extractants (see section 9.B).

Maddock and coworkers[89,98], in what is perhaps the most detailed report so far made on the topic, have shown that in no case could the synergistic extraction of Pa^V be fitted into the relationship (11C1). The enhancement prevailed over a wide range of aqueous HCl concentration, and in many combinations of solvents. Especially high was the effect when one of the components, but only one, had a high dielectric constant, even without having its own extraction ability towards the metal, such as nitrobenzene. The mole ratio $S_1 : S_2$, where the synergistic maximum appeared, depends on the particular combination and varies from one to four, frequently in integral numbers.

Data on the synergistic extraction of $FeCl_3$ revealed a very similar behaviour in these systems[97]. Polarity and the dielectric constant of one of the components have a similar effect on the synergistic extraction of organic acids[94]. A high synergism could be observed in a combination of a hydrogen-bond acceptor and a donor molecule. Pairs of ketones, or ethers or alcohols caused usually a small enhancement, but combinations of either ketones or ethers with an alcohol resulted in a several times higher percentage of extraction. An identical effect was shown by the pair heptanone + heptanol in the extraction of technetium[99].

It is easiest to provide an explanation for synergism when the mixture is composed of an ether or ketone and an alcohol. There are probably two effects working parallel to each other. The first is an increased extractive capacity of the alcohol upon addition of the ketone, due to the rupture of its hydrogen-bonded structure, and thus an increased thermodynamic activity. The second factor being due to the dual action of the more polar alcohol, in enabling hydrogen bonding in addition to its solvation properties.

Marcus and Eliezer[102] have discussed the general problem of the stabilization of mixed species. As to the case under consideration, the Russian school assumes formation of mixed solvates to explain the synergistic extraction by two neutral components. An enhanced synergism may occur whenever the soivation number of the inorganic species is higher than one, and when the chemical properties and the size of the components are not very different[91,103]. Fomin[92,104] suggested an empirical relationship

$$k_{12} = 2 \sqrt{k_1 k_2} \tag{11C2}$$

where the k's are the formation constants of the individual and mixed solvates. Rozen[105] suggested that a better form should be

$$k_{12} = \sqrt{k' k_1 k_2} \tag{11C3}$$

where k' would be the disproportionation constant of the exchange reaction

$$\overline{MA_m.2S_1} + \overline{MA_m.2S_2} \rightleftharpoons \overline{2MA_m.S_1.S_2} \tag{11C4}$$

For the synergistic extraction of nitric acid by a mixture of cyclohexanone and methyl isobutyl ketone, Fomin[92] estimated k_{12} to be 4.8×10^{-3}, as compared to the individual formation constants of the species $HNO_3.2S$ of 2.3×10^{-3} and 1.4×10^{-3} for the pure components respectively. Similar formation constants for nitric acid and uranyl nitrate solvates have been evaluated in other solvent pairs[90]. The composition of such mixed solvates cannot be determined with certainty (see section 9.B), though the position of the maximum in D values may be indicative of the mole ratio of the components in the synergistic adduct. Its composition varies from metal to metal, and certainly depends on the components in the mixture[93,96]. In a combination of TBPO and TOPO, the tervalent rare earth nitrates are extracted in the form $M(NO_3)_3.(TBPO)_2.TOPO$, the stoichiometry having been determined by slope analysis[95]. Though the formation of mixed solvates is acceptable, it is not easy to visualize why the formation of mixed solvates should enhance synergistically the extraction of inorganic species, unless there is an entropy effect similar to that discussed under the next heading. As a matter of fact, no explanation has been offered by the authors. Though allowing for the possibility of some type of cooperative solvation, Maddock[97,98] in a more critical experimental approach, could not provide a clear answer to the problem, whether or not mixed complexes are responsible for an enhanced metal extraction. It should be noted, however, that their choice of the model system, diisopropylcarbinol + diisopropyl ketone may not have been very fortunate, since the two solvents do not solvate the species containing protactinium, iron and gold, the three metallic species investigated, to the same extent: the organic-phase solvate contains 5 molecules of the ketone per metal, whereas only 3 of the carbinol.

An alternative reason suggested[94,97,98] may come from changes of solvation energies of the extracted inorganic species brought about by changes in the physical properties of the mixture. Such a situation is likely to arise whenever the two components have widely different dielectric properties. One may visualize that the less polar but donor molecule will solvate the salt and that the more polar solvent will not, but the distribution of such solvates will be favoured into an organic mixture of higher dielectric constant. This explanation could be applied to mixtures of diisopropylcarbinol and nitrobenzene, for example, where the latter has a poor solvating ability towards the proton of the protactinium-bearing species.

The third alternative considered by Maddock is a deviation from ideal behaviour affecting the degree of formation of the extractable solvates. This situation will arise when there is a positive deviation from Raoult's law. The increased activity of the complexing solvent upon addition of a second solvent may cause a maximum in the metal distribution ratio for some solvent combinations (section 2.B).

b. Systems with Two Acidic Ligands

An enhanced extraction of several metals, chelating with two β-diketones, has been observed in a variety of experimental combinations[30,32,106,107]. In the majority of combinations investigated by Newman and Klotz[106], on the model systems of uranium(VI) and copper(II) with acetylacetone and five of its derivatives, there was no marked synergistic effect. This in spite of the fact that the formation of mixed complexes, according to the reaction

$$\overline{MX_2} + \overline{MY_2} \rightleftharpoons \overline{2MXY} \qquad (11C5)$$

has been ascertained. Similarly, the mixed-ligand complexes of indium and europium with thenoyltrifluoroacetone, X and β-isopropyltropolone, Y, MX_2Y and MXY_2 do not exhibit synergistic extraction[30]. The reason may be that the mixed species are of a comparable stability in both phases.

Various solid mixed diketonates have been isolated in the solid state and their i.r. and n.m.r. spectra investigated. Zirconium, for example, forms three different complexes in benzene solution with thenoyltrifluoro-acetone and trifluoroacetylacetone[108]. Several rare earth metals form with acetylacetone, X, and propionylacetone, Y, the complex $LnXY_2$, where X can also be benzoyl- or dibenzoylacetone[109].

In Newman's experiments[106] only the combination of acetylacetone and thenoyltrifluoroacetone indicates a somewhat more favourable formation of the mixed species with the metals investigated. In line with a hypothesis of Marcus and coworkers[34,102], the authors suggest that only the combination of one weak and one strong β-diketone, but not that of two weak or two strong ones, will bring about the stability of the mixed-ligand complex. The greater the difference between the parent binary complexes the stronger the mixed complex. Additionally, or alternatively, Sekine and Dyrssen[107] postulate that 'two ligands provide a more flexible electron system, which may increase the average metal–oxygen bond strength in the complex'.

Finally, a short note on the observation of synergistic extraction by the cooperative complex formation of two metals Healy[79] reports that, for example, caesium is synergistically extracted in the presence of calcium, uranium or other alkali metals by thenoyltrifluoroacetone. The formation of double complexes such as $NaTTA.CsTTA$ or $NaTTA.CsNO_3$ or $Ca(TTA)_2.CsNO_3$ may be responsible for an enhancement in the caesium extraction of several orders of magnitude.

D. REFERENCES

1. J. G. Cunningham, D. Scargill and H. H. Willis, *Brit. Rept.*, AERE, C/M 215 (1954).
2. C. F. Baes, Jr., *Nucl. Sci. Eng.*, **16**, 405 (1963).

3. C. A. Blake, C. F. Coleman, K. B. Brown, C. F. Baes and J. C. White, *Proc. 2nd U. N. Intern. Conf. Peaceful Uses At. Energy, Geneva* **28**, 289 (1959).

4. C. A. Blake, D. E. Horner and J. M. Schmitt, *USAEC Rept.*, ORNL-2259 (1959).

5. T. V. Healy, *J. Inorg. Nucl. Chem.*, **19**, 314 (1961).

6. R. L. Scruggs, T. Kim and N. C. Li, *J. Phys. Chem.*, **67**, 2194 (1963).

7. H. Irving and D. N. Edgington, *J. Inorg. Nucl. Chem.*, **15**, 158 (1960).

8. T. V. Healy and J. R. Ferraro, *J. Inorg. Nucl. Chem.*, **24**, 1449 (1962).

9. W. R. Walker and N. C. Li, *J. Inorg. Nucl. Chem.*, **27**, 411, 2255 (1965).

10. N. C. Li, S. M. Wang and W. R. Walker, *J. Inorg. Nucl. Chem.*, **27**, 2263 (1965).

11. G. N. Rao and N. C. Li, *J. Inorg. Nucl. Chem.*, **28**, 2931 (1966).

12. W. R. Walker and M. S. Farrell, *J. Inorg. Nucl. Chem.*, **28**, 1483 (1966).

13. C. H. Ke and N. C. Li, *J. Inorg. Nucl. Chem.*, **28**, 2255 (1966).

14. W. C. Fernelius, Ed., *Organic Syntheses*, Vol. 2, McGraw Hill Book Co., New York, 1946, p. 15.

15. J. R. Ferraro and T. V. Healy, *J. Inorg. Nucl. Chem.*, **24**, 1463 (1962).

16. H. M. N. H. Irving and N. S. Al-Niaimi, *J. Inorg. Nucl. Chem.*, **27**, 1671, 2231 (1965).

17. H. Irving, *Proc. Symp. Coord. Chem. Tihany*, Akademiai Kiado, Budapest, 1965, p. 219.

18. D. P. Graddon and E. C. Walton, *J. Inorg. Nucl. Chem.*, **21**, 49 (1961).

19. J. P. Fackler, Jr., *Inorg. Chem.*, **2**, 266 (1963).

20. S. M. Wang, W. R. Walker and N. C. Li, *J. Inorg. Nucl. Chem.*, **28**, 875 (1966).

21. L. Sacconi, G. Lombardo and P. Paoletti, *J. Inorg. Nucl. Chem.*, **8**, 217 (1958).

22. L. Sacconi and G. Lombardo, *J. Am. Chem. Soc.*, **82**, 6266 (1960).

23. F. A. Cotton and R. H. Soderberg, *Inorg. Chem.*, **3**, 1 (1964).

24. H. Irving and D. N. Edgington, *J. Inorg. Nucl. Chem.*, **20**, 321 (1961); **21**, 169 (1961).

25. I. J. Gal and R. M. Nikolić, *J. Inorg. Nucl. Chem.*, **28**, 563 (1966).

26. L. Newman, *J. Inorg. Nucl. Chem.*, **25**, 304 (1963) .

27. T. V. Healy, *Nucl. Sci. Eng.*, **16**, 413 (1963).

28. H. M. N. H. Irving and D. N. Edgington, *J. Inorg. Nucl. Chem.*, **27**, 1359 (1965).

29. T. Sekine and D. Dyrssen, *J. Inorg. Nucl. Chem.*, **26**, 1727 (1964).

30. T. Sekine and D. Dyrssen, *J. Inorg. Nucl. Chem.*, **29**, 1457, 1475, 1481. 1489 (1967).

31. T. Sekine and M. Ono, *Bull. Chem. Soc. Japan*, **38**, 2087 (1965).

32. D. Dyrssen, *Proc. Symp. Coord. Chem. Tihany*, Akademiai Kiado, Budapest, 1965, p. 231.

33. H. Irving, in *Solvent Extraction Chemistry* (Eds. D. Dyrssen, J. O. Liljenzin and J. Rydberg) North-Holland, Amsterdam, 1967, p. 91 and references therein.

34. Y. Marcus, *Chem. Rev.*, **63**, 139 (1963).

35. P. G. Manning, *Can. J. Chem.*, **41**, 658 (1963).

36. A. M. Poskanzer and B. M. Foreman, Jr., *J. Inorg. Nucl. Chem.*, **16**, 323 (1961).

37. H. Irving and D. N. Edgington, *J. Inorg. Nucl. Chem.*, **20**, 314 (1961).

38. W. R. May and M. M. Jones, *J. Inorg. Nucl. Chem.*, **25**, 507 (1963).

39. T. H. Siddall, III, *J. Inorg. Nucl. Chem.*, **13**, 151 (1960).

40. M. Taube, *Radiokhimiya*, **4**, 260 (1962).

41. L. Bakos, E. Szabo, L. Andras and N. Yyes, *Proc. Symp. Coord. Chem., Tihany*, Akademiai Kiado, Budapest, 1965, p. 241.

42. T. V. Healy, *J. Inorg. Nucl. Chem.*, **19**, 328 (1961).

43. H. Irving and D. N. Edgington, *Chem. Ind. (London)*, 77 (1961).

44. R. Guillaumont, *Bull. Soc. Chim. France*, 132 (1965).

45. J. Hala, in *Solvent Extraction Chemistry* (Eds. D. Dyrssen, J. O. Liljenzin and J. Rydberg), North-Holland, Amsterdam, 1967, p. 135.

46. J. L. Burdett and M. T. Rogers, *J. Am. Chem. Soc.*, **86**, 2105 (1964).

47. T. V. Healy, D. F. Peppard and G. W. Mason, *J. Inorg. Nucl. Chem.*, **24**, 1429 (1962).

48. G. Pukanic, N. C. Li, W. S. Brey, Jr. and G. B. Savitsky, *J. Phys. Chem.*, **70**, 2899 (1966).

49. D. Dyrssen and M. Hennicks, *Acta. Chem. Scand.*, **15**, 47 (1961).

50. L. Newman and P. Klotz, *J. Phys. Chem.*, **65**, 769 (1961).

51. D. Dyrssen, *J. Inorg. Nucl. Chem.*, **8**, 291 (1958).

52. D. P. Graddon, *Nature*, **183**, 1610 (1959).

53. H. M. N. H. Irving and N. S. Al-Niaimi, *J. Inorg. Nucl. Chem.*, **27**, 717 (1965).

54. S. M. Wang, D. Y. Park and N. C. Li, in *Solvent Extraction Chemistry* (Eds. D. Dyrssen, J. O. Liljenzin and J. Rydberg) North Holland, Amsterdam, 1967, p. 111.

55. V. A. Gordieyeff, *Anal. Chem.*, **22**, 1166 (1950).

56. *Ann. Progr. Rept., Chem. Technol. Div., USAEC Rept.*, ORNL–3830, (1965) p. 211–213.

57. *Ann. Progr. Rept. Chem. Technol. Div., USAEC Rept.*, ORNL–3452, (1963) p. 175–182.

58. S. M. Wang and N. C. Li, *J. Inorg. Nucl. Chem.*, **28**, 1091 (1966).

59. C. Deptula and S. Minc, *Nukleonika*, **6**, 197 (1961).

60. C. J. Hardy, D. Scàrgill and J. M. Fletcher, *J. Inorg. Nucl. Chem.*, **7**, 257 (1958).

61. M. Zangen, *J. Inorg. Nucl. Chem.*, **25**, 1051 (1963).

62. E. V. Ukraintsev and N. F. Kashcheev, *Radiokhimiya*, **4**, 279 (1962).

63. M. Zangen, *J. Inorg. Nucl. Chem.*, **25**, 581 (1963).

64. V. B. Shevchenko, V. S. Smelov and A. V. Strakhova, *Zh. Neorg. Khim.*, **7**, 1736 (1962).

65. B. N. Laskorin, V. S. Uliyanov and R. A. Sviridova in *Ekstraktziya*, Vol. 1, Gosatomizdat, Moscow, 1960, p. 171.

66. J. Kennedy, F. A. Burford and P. G. Sammes, *J. Inorg. Nucl. Chem.*, **14**, 114 (1960); J. Kennedy and A. M. Deane, *J. Inorg. Nucl. Chem.*, **19**, 142 (1961).

67. C. H. Hardy and D. Scargill, *J. Inorg. Nucl. Chem.*, **11**, 128 (1959).

68. D. Dyrssen and D. H. Lien, *Acta Chem. Scand.*, **14**, 1091 (1960); **18**, 224 (1964).

69. D. H. Liem and D. Dyrssen, *Acta Chem. Scand.*, **20**, 272 (1966); D. H. Liem, in *Solvent Extraction Chemistry* (Eds. D. Dyrssen, J. O. Liljenzin and J. Rydberg) North-Holland, Amsterdam, 1967, p. 264.

70. H. T. Baker and C. F. Baes, Jr., *J. Inorg. Nucl. Chem.*, **24**, 1277 (1962).

71. A. P. Ilozhev, I. V. Poddubskaya and A. M. Rozen, *Radiokhimiya*, **2**, 411 (1960).

72. *Ann. Progr. Rept. Chem. Technol. Div., USAEC Rept.*, ORNL–3627, (1964) p. 198.

73. L. Winand and Ph. Dreze, *Bull. Soc. Chim. Belges*, **71**, 410 (1962).

74. G. Aksnes and T. Gramstad, *Acta Chem. Scand.*, **14**, 1485 (1960).

75. D. Dyrssen, *Acta Chem. Scand.*, **11**, 1771 (1957).

76. V. B. Shevchenko, V. S. Smelov and A. V. Strakhova in *Ekstraktziya*, Vol. 2. Gosatomizdat, Moscow, 1960, p. 179.

77. H. Ihle, H. Michael and A. Murrenhoff, *J. Inorg. Nucl. Chem.*, **25**, 734 (1963).

78. W. E. Keder, E. C. Martin and L. A. Bray in *Solvent Extraction Chemistry of Metals* (Eds. H. A. C. McKay, T. V. Healy, I. L. Jenkins and A. Naylor) Macmillan, London, 1966, p. 343.

79. T. V. Healy, in *Solvent Extraction Chemistry* (Eds. D. Dyrssen, J. O. Liljenzin and J. Rydberg) North-Holland, Amsterdam, 1967, p. 119.

80. M. Zangen, *J. Inorg. Nucl. Chem.*, **28**, 1693 (1966).

81. J. Kennedy, *J. Appl. Chem.*, **9**, 26 (1959); *British Rept.*, AERE CM 369 (1958).

82. J. Kennedy, E. S. Lane and B. K. Robinson, *J. Appl. Chem.*, **8**, 459 (1958).

83. D. Dyrssen and L. Kuča, *Acta Chem. Scand.*, **14**, 1945 (1960).

84. D. Dyrssen and S. Ekberg, *Acta Chem. Scand.*, **13**, 1909 (1959).

85. H. T. Hahn and E. M. Vander Wall, *J. Inorg. Nucl. Chem.*, **26**, 191 (1964).

86. D. F. Peppard, G. W. Mason and R. J. Sironen, *J. Inorg. Nucl. Chem.*, **10**, 117 (1959).

87. J. R. Ferraro and D. F. Peppard, *J. Phys. Chem.*, **65**, 539 (1961).

88. G. W. Mason, S. McCarty and D. F. Peppard, *J. Inorg. Nucl. Chem.*, **24**, 967 (1962).

89. Z. Dizdar and J. Rajnvajn, *Bull. Inst. Nucl. Sci. 'Boris Kidrich*, **11**, 181 (1961).

90. V. M. Vdovenko and A. S. Krivokhatskii, *Zh. Neorg. Khim.*, **5**, 494 (1960).

91. V. V. Fomin and A. F. Morgunov, *Zh. Neorg. Khim.*, **5**, 233 (1960).

92. V. V. Fomin, A. F. Morgunov and I. V. Korobov, *Zh. Neorg. Khim.* **5**, 1846 (1960).

93. V. M. Vdovenko, A. S. Krivokhatskii and A. V. Chizov, *Zh. Neorg. Khim.*, **5**, 2363 (1960).

94. D. E. Pearson and M. Levine, *J. Org. Chem.*, **17**, 1356 (1952).

95. C. T. Rhee, *Chem. Abstr.*, **61**, 7766 (1964).

96. V. M. Vdovenko, A. S. Krivokhatskii and Y. K. Gusev, *Radiokhimiya*, **2**, 531 (1960).

97. A. G. Goble and A. G. Maddock, *Trans. Faraday Soc.*, **55**, 591 (1959).

98. A. T. Casey and A. G. Maddock, *Trans. Faraday Soc.*, **58**, 918 (1962).

99. G. E. Boyd and Q. V. Larson, *J. Phys. Chem.*, **64**, 988 (1960).

100. T. V. Healy and H. A. C. McKay, *Rec. Trav. Chim.*, **75**, 730 (1956).

101. H. Irving and F. J. C. Rossotti, *J. Chem. Soc.*, 2475 (1956).

102. Y. Marcus and I. Eliezer, *J. Phys. Chem.*, **66**, 1661 (1962); Y. Marcus, I. Eliezer and M. Zangen, *Proc. Symp. Coord. Chem. Tihany*, 1964, Akademiai Kiado, Budepest, 1965, p. 409.

103. S. Siekerski and M. Taube, *Nukleonika*, **6**, 489 (1961).

104. V. V. Fomin, *Chemistry of Extraction Processess*, Gosatomizdat, Moscow, 1960. Translated by A. S. Kertes, Israel Program for Sci. Trans., Jerusalem, 1962.

105. A. M. Rozen in *Ekstraktziya*, Vol. 1, Goatamoizdat, Moscow, 1960 p. 59.

106. L. Newman and P. Klotz, in *Solvent Extraction Chemistry* (Eds. D. Dyrssen, J. O. Liljenzin and J. Rydberg), North-Holland, Amsterdam, 1967, p. 128.

107. T. Sekine and D. Dyrssen, *J. Inorg. Nucl. Chem.*, **26**, 2013 (1964).

108. T. J. Pinnavaid and R. C. Fay, *Inorg. Chem.*, **5**, 233 (1966).

109. N. K. Dutt and S. Upadhyaya, *J. Inorg. Nucl. Chem.*, **28**, 2719 (1966) and references therein.

12.

Applications of Distribution Methods

It is the theme of this monograph that specific results on the ion exchange and solvent extraction behaviour can be correlated by applying the general concepts of solution chemistry described in Chapters 1–3. Thus the methods discussed in detail in Chapters 4–6 and 7–10, when applied to a given metal-complex system in aqueous solution, should give consistent results, whatever the ion exchanger or extractant used. There are of course many reported examples of inconsistencies, but the general picture emerging is that when valid methods are used, and all special effects are taken into account, well-consistent information is indeed obtained.

In section 12.A a summary of the complex systems treated in this book is given. The aqueous species formed and the organic-phase species (both in exchangers and solvents) in equilibrium with them, are reviewed.

Distribution methods are then compared in section 12.B, on the basis of the most general thermodynamic equations. The relative advantages of the various distribution methods are discussed, in terms of the free choice of system parameters available for each: functional group, solvents, concentration ranges, temperature and pressure. The possible variability of the organic phase for a given system is an important factor, which, together with those above, guides in the selection of a distribution method for solving a particular problem involving metal complexes.

The application of the distribution methods discussed in this monograph to the study of metal complexes in aqueous solutions, and also in organic solvents, is illustrated in section 12.C for a few chosen cases. Information from all methods is correlated for each system. The systems treated are the hydrolysis of zirconium(IV), the fluoride complexes of tantalum(V), the chloride complexes of iron(III), the bromide complexes of indium(III) and the nitrate complexes of uranium(VI). Full use is made in these correlations of the equations developed in Chapters 5, 6 and 7, and the general chemical behaviour discussed in Chapters 8, 9 and 10.

A. COMPLEX SPECIES STUDIED

a. Aqueous-phase Species

Distribution methods are usually applied to aqueous solutions containing the central ion to be studied, ligand molecules or ions, the necessary counter ions, other electrolytes added for special purposes, and, of course, the solvent water. Under certain circumstances the list is different, but it always contains the essential ingredients—a solvent and a solute. It is the purpose of distribution studies to elucidate certain interactions involving the solute. Returning to the usual case described above, the distribution studies apply to interactions of the central ion with the ligands, and certain other components of the solution.

The first problem consists of the definition of the central ion, which we shall write conventionally as M^{m+}. We shall deal here with central cations, although anions or neutral entities have also been subject to study as central 'ions' in distribution investigations. Distribution methods, in which equilibrium concentration relationships are studied, belong to those thermodynamic methods which cannot differentiate between species involving interaction with components of the solution which remain constant in the experiments. Thus if a study involves a constant ionic medium C^+B^-, then variations of central-ion or ligand concentrations over even many orders of magnitude, provided these are small compared to the concentrations of solvent and of C^+B^-, cannot tell us anything concerning participation of the solvent, and of the ions C^+ and B^- in the reacting species. The generalized central-ion species can be written as $M(H_2O)_iC_jB_k^{m+j-k}$. Certain methods do, however, differentiate between ions of different charge, so that the simple central ions will be only those with $j = k$, which still have a net charge of $m+$. If there are several bulk ions $C_{(1)}^+$, $C_{(2)}^+$, $C_{(3)}^+$ ... (where $C_{(n)}^+$ could also be e.g. $\frac{1}{2} Mg^{2+}$) the concentrations of which are kept constant, all of them should be considered in the generalized formula. Furthermore, the concentration of the free or uncomplexed central ion must be considered as the sum of the various possible species. It is conventional to write this concentration simply as (M^{m+}), which is defined as

$$(M^{m+}) = \sum_{i,\,j,\,k,\,\cdots} (M(H_2O)_i^{m+}C_{(n)j}^+B_{(p)k}^- \cdots) \qquad (12A1)$$

The parentheses denote concentration in the scale used: molar, molal, mole fraction, etc.

Even apart from the uncertainty present because of the ignorance of the indices $i, j, k, \ldots$ in (12A1), it is not always known what the central ion to be studied actually is, particularly when present at trace concentrations. Its oxidation state may not be definitely known (ignorance of m), or it may be polymerized strongly, so that it may be present as the q-mer over the whole accessible concentration range (M_q^{qm+}, with q unknown), or it may be strongly bound to another ion (such as an oxygen atom reducing the effective charge), again over the whole accessible concentration range. Examples of the last two cases are Hg^I and U^{VI}, present as Hg_2^{2+} and UO_2^{2+} in aqueous solutions. The first case has occured with the new, artificial elements. In such cases, distribution methods can be of use to determine the net charge on the central ion, and its polymerization number.

As discussed in section 3.C, the hydrated central ion can split off one or more hydrogen ions in reversible reactions

$$M(H_2O)_i^{m+} \rightleftharpoons [M(H_2O)_{i-1}OH]^{m-1} + H^+ \qquad (12A2)$$

and so on.

At sufficiently high hydrogen ion concentrations (low pH values) the equilibrium lies so completely to the left, that species other than $M(H_2O)_i^{m+}$(equivalent to M^{m+}) can be neglected. At lower acidities, (M^{m+}) is decreased because of appreciable hydrolysis. Distribution methods are capable of determining limiting acidity, and the nature and stability of the hydrolysed species. If it is not intended to study the hydrolysis reaction itself, but rather interactions with some ligands, a preliminary study of the conditions required for preventing hydrolysis could be useful. Some extractant, for instance, might be ruled out for use, because its effective use requires a too high pH region. In some cases the presence of hydrolysed species to a small or a large extent, cannot be avoided, and they must be taken into consideration in a study of the interactions of the central ion with ligands. Sometimes even mixed hydroxo–ligand complexes can be formed, and should be taken into account.

When neutral ligands L are added to a central cation M^{m+} they replace water molecules from the first coordination shell. The resulting complexes, written conventionally as ML_n^{m+} still have the charge $m+$. Their properties are appreciably different from those of the simple hydrated cation M^{m+} only if L has special properties, such as bulkiness or hydrophobic nature. Thus although the bonds between M^{m+} and the donor atom of L may be quite different from those between M^{m+} and the oxygen of the hydration water, as indicated by changes e.g. in the absorption spectra, other properties, affecting the free energy of the M^{m+} component, may change little. Thus in some cases distribution methods are not suitable for studying such complexes, formed between M^{m+} and small neutral monodentate ligands L.

The situation is different if the ligand is charged. For many metal cations, the coordination number N is twice the charge $m+$. Addition of bidentate mononegative ligands X^- will cause very strong binding to form a complex with zero net charge

$$M(H_2O)_N^{m+} + m X^- \rightleftharpoons MX_m + N H_2O \qquad (12A3)$$

These chelates have no residual affinity for water, are in fact insoluble in water, and can very easily be extracted into organic solvents. Their formation is thus a natural subject for study with distribution methods. When, however, there is no exact balance between the coordination number N and the charge $m+$ of the central ion M^{m+} and the denticity d and charge x of the ligand X^{x-}, any number n of ligands per central ion will leave the complex species with extra charge and/or coordination sites or donor atoms. In such cases there may remain an affinity to the solvent water, through a molecule of residual hydration water, or the possibility of adding more ligands, in their neutralized form H_xX. Distribution methods usually can unravel the complicated equilibria ensuing in such cases.

A still different situation is encountered when the ligand is a simple monodentate anion A^{a-}. If, as usual, $m > a$, there is the possibility of forming cationic complex species MA_n^{m-na}, as long as $n < m/a$. These species have positive charges different from that of the central cation, and therefore, they have different thermodynamic properties. Although it is generally impossible by means of distribution methods to differentiate between an outer-sphere complex $[M(H_2O)_N^{m+}A^{a-}]$ and an inner-sphere complex $[M(H_2O)_{N-1}A]^{m-a}$, the formation of a species with lower net charge is easily measured. Those exchangers and extractants which have anions as their functional group (e.g. sulphonic acid type cation exchangers or dialkylphosphoric acid extractants) react with the central ion M^{m+}, but also, to a different and often much lower degree, with the cationic complexes MA_n^{m-na} ($n < m/a$). The distribution methods can thus be used to study complex formation leading to these species.

In the special case where $m = va$, v being an integer, a neutral, uncharged, complex MA_v will form at sufficiently high ligand-concentration. Such a complex may not be particularly stable, but it is possible that it could be extracted into certain organic solvents, including the low dielectric constant environment inside anion-exchanging resins. It will most probably undergo further interactions in either phase: it can be solvated with solvent molecules, or it can react with ligands when present in the organic phase, and it will associate with ligands or dissociate in the aqueous phase, or it may form polynuclear species. Schematically, the following reactions can and often do occur

$$MA_v + iA^{a-} \rightleftharpoons MA_{v+i}^{ia} \qquad (i > 0 \text{ or } i < 0) \quad (12A4a)$$

$$MA_v + p\overline{A^a} \rightleftharpoons \overline{MA_{v+p}^{pa-}} \qquad (12A4b)$$

$$MA_v + n\overline{S} \rightleftharpoons \overline{MA_vS_n} \qquad (12A4c)$$

$$qMA_v \rightleftharpoons M_qA_{qv} \qquad (12A4d)$$

etc.

In many distribution reactions the mechanism is known to involve the neutral species MA_v. In other cases the mechanism is unknown, but the distribution equilibrium can be described in terms of the neutral complex, in a way equivalent thermodynamically to the real reaction, whatever it may be. It is, therefore, natural that distribution methods be applied to the study of the neutral complex MA_v, and of the species formed from it by dissociation (e.g. MA_{v-1}^{a+}) or association (e.g. MA_{v+1}^{a-}).

If the concentration of the ligand added is sufficiently high, or in the rare cases when $a > m$, anionic complexes will be formed in the solutions, MA_n^{m-na} ($n > m/a$). Those exchangers and extractants which have cations as their functional groups (e.g. quaternary ammonium type anion exchangers or trialkylammonium salt extractants) react with these anionic species, to a different degree depending on the net charge, and usually much more strongly than with non-anionic species. The distribution methods can, therefore, be applied with advantage to the study of the anionic species.

This survey by no means exhausts the applications of distribution methods to the study of the species formed in aqueous solutions and their relative stabilities. Central ions can be other than metallic cations M^{m+} (they can be anions, such as PO_4^{3-} associating with the ligands H^+, or such as I^- associating with the ligands I_2). Ligands can be intermediate hydrolysis products, links such as (nominally) UO_3 associating with a central core, such as UO_2^{2+}, or metallic cations, such as Ag^+ associating with a central anion, such as CNS^-, etc. The 'aqueous phase' should also be meant as a general term, covering such dissociating solvents as molten salts, mixed aqueous–organic solvents, or high dielectric constant organic solvents. Distribution methods can be, and in fact have been, applied to the study of interactions to form new species in such systems.

b. Organic-phase Species

The 'aqueous' phases discussed above have qualitative attributes which contrast with those of the immiscible 'organic' phase. The latter is distinguished by its low polarity, low dielectric constant, more covalent nature, lesser amount of structure, etc. Again it need not be organic, it could be the borate-rich phase in a B_2O_3–Na_2O–$NaCl$ system, contrasted with the more ionic salt-rich phase.

The attributes of dielectric constant, polarity and basicity often vary in a parallel manner in a series of organic solvents. It is difficult to generalize about such diverse solvents as nitrobenzene, tributyl phosphate, benzene, hydrated polystyrene methylene–trimethylammonium chloride (crosslinked with divinylbenzene), or the above-mentioned borate-rich phase. From the general qualitative attributes listed above, follows the tendency of the species in the organic phase to have a minimal net charge (often no charge) and a minimal polarity, but they can often be bulky.

Although in many cases the 'organic' phase is used in the distribution method only as a tool, in order to study interactions in the 'aqueous' phase which are of main interest, the species formed in the organic solutions and their interactions merit investigation too. Some complex species may indeed be formed only in organic solutions, because of limitations on available and required concentrations of the ligand in the aqueous phase, which have to compete with the water hydrating the central ion. It is therefore, instructive to survey the species that are usually encountered in these organic phases.

Ion exchange resins are in a sense the most water-like of the organic phases. The resins with a low degree of crosslinking are certainly highly swollen with imbibed water, and there are extensive regions inside the resin where water-like conditions prevail, and the structure-breaking, low dielectric constant organic skeleton has little effect. With the ordinary 8–10% crosslinked resins this is not so nearly true, and the organic portion has a higher effect. An effective dielectric constant of 40–60 may be attributed to the hydrated resin. The species in such an environment will thus be ionic although ion-pairing should be much more prevalent than in aqueous solutions. The environment in ion exchangers is conducive to solvent-enforced ion-pairing and to dehydration of ions because of space limitations and swelling pressures.

The functional groups in ion exchangers have a better chance to compete with solvent molecules and ligand ions for association with the central ion than would have similar groups in the external solution. Thus the prevalent species would be partly (or totally) dehydrated ions (complex ions) ion-paired to some extent with the functional groups, or with invading counter ions.

When the swelling liquid is not water (whether coming from a non-aqueous or a mixed solvent), but an organic solvent of low dielectric constant, ion-pairs will predominate over other species, except neutral complexes. In some cases the latter may be the prevalent species in an anion exchange resin. The neutral complexes may, but sometimes do not, react with the ligand, ion-paired with the functional group, to form an anionic complex, again ion-paired with the functional group in the resin.

Anion exchange resins swollen with low dielectric constant solvents, resemble solutions of long-chain ammonium salts in inert diluents. There again an anionic complex exists, ion-paired with the organic cation. The extractant in this case is usually aggregated; it is still unknown whether to micelles or to smaller monodisperse aggregates. It is likely that from such an aggregate only one anion dissociates, e.g. for the tertiary amine R_3N

$$\overline{(R_3NH^+A^-)_q} \rightleftharpoons \overline{[(R_3NH^+A^-)_{q-1}R_3NH]^+} + \cdots \overline{A^-} \qquad (12A5)$$

this anion then being capable of reacting with the metal complex, resulting finally in an ion-paired complex $[(R_3NH^+A^-)_{q-1}R_3NH]_{N-v}^+.MA_N^{v-N}$. Recent evidence is that q in (12A5) is larger than unity and smaller than ten, and has a similar value in the metal-complex species, but definite information is still lacking.

Extractants, which are formally similar to the above class, may behave differently, especially if the diluent is very polar, or the charge on the extractant is sterically hidden. In such cases aggregation yields to direct interaction with the diluent, as with tetraphenyl-arsonium salts in chloroform. Since the latter is present in large excess, the species is again written conventionally as the monomeric organic cation.

Solvents less basic than the amines do not form exclusively 1:1 adducts with hydrogen ions. In most cases water molecules are carried into the organic phase to solvate this ion, and in some cases a few solvent molecules complete the coordination of the hydrogen ion, often to four, or coordinate to the water in a second coordination shell. A metal central ion may be capable under the circumstances of extraction of forming an anionic complex, which is then extracted with the hydrogen ions to ensure no net charge transfer. It must compete, however, with the ligand anions, and it usually does so successfully because of its larger bulk and lower hydration. The resulting species in the organic phase is $[HS_n(H_2O)_{4-n}]^+_{(N-v)a}MA_N^{(v-N)a}$ where, for successful extraction $(N-v)a$ must almost exclusively be unity, i.e. $a = 1$ and $v = m$, since otherwise $MA_N^{(v-N)a}$ interacts so strongly with the water in the aqueous phase, that it cannot be extracted.

A solvent which is sufficiently basic to solvate the hydrogen ion may also solvate the metal ion sufficiently to cause its extraction. It is still unclear what properties are required for the solvent, and what experimental conditions are required, for a metal to be extracted as a species MA_vS_{N-v} rather than the solvated acido complex above. However, such species, or ion-paired $MS_{N-v}^{m+}A_v^-$, can exist in the organic phase, sometimes along with the ligated-metal acids. Again, the thermodynamic distribution methods are incapable of differentiating between the 'isomers' MA_vS_{N-v} and $MS_{N-v}^{m+}A_v^-$, having the same number of ligands and solvent molecules, particularly when the solvent is present in large excess. When the solvent is diluted, some of the coordination sites on the metal may be occupied by water of hydration, so that the exact species, appart from its stoichio-metric composition, is still uncertain.

The interaction of the solvent S with the neutral metal complex MA_v may be strong, so that S enters the first coordination sphere, and even displaces the ligand A^{a-}, or it may be less strong, or even so weak that the mere existence of any interaction may be doubtful. In such cases it appears that MA_v 'simply' dissolves in the inert solvent, although such 'simple' dissolution normally involves some interaction. Still, since the solvent usually is present at a large excess, it is conventional to speak of the species MA_v in the organic phase, disregarding the interactions.

When the anion is an organic chelating agent X^- instead of the monodentate A^{a-} then the species MX_m usually interacts very little with the solvent, and is therefore soluble even in 'inert' solvents. As mentioned above in the discussion of the aqueous-phase species, this is most likely when $N/m = d/x$ $(= 2$ in the majority of cases). Unless $x > m$ (as with ethylenediamine tetraacetate and di- or trivalent metal ions) it is usually possible to modify the chelates formed so as to compensate for unneutralized charge or coordination number, when the above equality does not hold. Thus, if xN/d, which is the product of the number of ligands with their charge, is less than m, then anions A^{a-} may be coextracted to yield neutral species, e.g. for $m = 4$, $x = 1$, $d = 2$ and $N = 6$ and $a = 1$, a species MA_2X_2 could be extractable, whereas MX_3^+ and MA_4X^- (all coordinatively saturated) could not. On the other hand, if $xN/d > m$, extraction can occur if one or more of the ligands is monodentate instead of multidentate. In such cases a ligand may contribute to the occupation of coordination sites, but not to charge neutralization. If, for example, in the last case $m = 2$ instead of 4, a species $MX_2.2HX$ would be stable in the organic phase, if HX is considered to occupy one coordination site only, or when HX is still bidentate, $MX_2.HX$ would be the organic-phase species.

B. INTERCOMPARISON OF DISTRIBUTION METHODS

a. Thermodynamic Equations

When reduced to the barest essentials, a distribution method compares the chemical potential of a neutral component, the distribuend I, in the two phases. Since at equilibrium the chemical potential must be equal

$$\mu_I \equiv \mu_I^0 - RT \ln(I) + RT \ln \gamma_I = \bar{\mu}_I \equiv \bar{\mu}_I^0 + RT \ln(\bar{I}) + RT \ln \bar{\gamma}_I \qquad (12B1)$$

a distribution coefficient

$$D_I = (\bar{I})/(I) \qquad (12B2)$$

different from unity, must be due to an inequality of μ_I^0 and $\bar{\mu}_I^0$ and/or of γ_I and $\bar{\gamma}_I$. For this general consideration, the distribuend I (as long as it is a thermodynamic component) and the concentration scale can be chosen freely for optimal convenience in describing the experimental system. The standard chemical potential of the distribuend may be chosen to be the same in the two phases, $\mu_I^0 = \bar{\mu}_I^0$, from which follows an assignment of $K_I = 1$, K_I being the equilibrium constant for the transfer reaction

$$I \rightleftharpoons \bar{I} \qquad (12B3)$$

Such a choice, however, automatically defines the activity coefficients γ_I and $\bar{\gamma}_I$, or at least their ratio. In other cases it is more convenient to choose values for the activity coefficients (e.g. $\gamma_I \to 1$ as $(I) \to 0$, or

$\gamma_I \rightarrow 1$ as $x_I \rightarrow 1$, on the mole-fraction scale), which automatically results in definitions for the standard chemical potentials which are not necessarily equal for both phases, and from which a value for $K_I^0 = \exp{(\mu_I^0 - \bar{\mu}_I^0)}/RT$, different from unity, results.

The various distribution systems vary in the choice of those parameters listed above which can be chosen freely: the nature of the component I, and whether $K_I = 1$ (equal standard chemical potentials) or the activity coefficients in each phase are individually defined. Most distribution methods can be rewritten in terms of the molal concentration scale, so that the concentration scale chosen is normally not a point of difference. For the sake of simplicity, the molal scale will be used in the following.

There are many points of similarity between distribution methods involving cation exchange with resins, extraction with acidic extractants such as acidic phosphate esters or extraction of chelates. Common to all is the choice of the distribuend. Instead of choosing the neutral component I, the exchange reaction

$$\text{M}^{m+} + m\overline{\text{C}^+} \rightleftharpoons \overline{\text{M}^{m+}} + m\text{C}^+ \tag{12B4}$$

is used, which is thermodynamically equivalent to (12B3), since there is no net transfer of charge. The ions M^{m+} and C^+ need, of course, not be free, they can in each phase be associated with any other ions, e.g. with the chelating ligand. The main point, however, is that (12B4) can be considered the essential phase-transfer step for the distribuend of interest, M^{m+}. The co-ion C^+, of course, need not be monovalent, although in the majority of cases it is the hydrogen or one of the alkali metal ions. Formally, the ratio M/C^m can replace the quantity I in all the relationships discussed above, where the symbols M, C and I stand for the respective concentrations and activity coefficients.

With this relationship common for all the methods mentioned here, the differences occur in the conventions for standard chemical potentials and activity coefficients. For cation exchange with resins in aqueous solutions it is usual to define the same standard state for both phases

$$\mu_{ex}^0 = \bar{\mu}_{ex}^0 \tag{12B5}$$

not necessarily for the individual ions, but for the ratio M/C^m. It follows that K_{ex}^0, the thermodynamic equilibrium constant for (12B4) equals unity. Writing I as a general symbol for concentrations and activity coefficients of species I and similarly M, C, etc. for the respective symbols, substituting now M/C^m for I everywhere in (12B1) and applying (12B5), yields in this case the relationship

$$D = \bar{m}_M/m_M = (\bar{m}_C/m_C)^m (\bar{\gamma}_M/\gamma_M)^{-1} (\bar{\gamma}_C/\gamma_C)^m \qquad (12B6)$$

This equation is equivalent to (5B1), when appropriate concentration scales are substituted. This basic expression holds also for extraction systems; however here the chelating ligand or extractant anion is usually explicitly included by enlarging, the ratio $\overline{M/C^m}$ by $\overline{X^m}$ for the organic phase to yield $\overline{MX^m}/\overline{(CX)^m}$. Furthermore, it is conventional to define the standard states separately for the two solvents, water and the organic diluent, so that $K^0_{ex} \neq 1$. The activity coefficients are then defined accordingly, and will be primed in the following. The expression for the distribution coefficient, resulting by substituting $\overline{M/C^m}$ for I and $\overline{MX^m}/\overline{(CX)^m}$ for I in (12B1) and using the primed activity coefficients is

$$D = \bar{m}_{MX_m}/m_M = K^0_{ex}(\bar{m}_{CX}/m_C)^m (\gamma'_M/\gamma'^m_C)(\bar{\gamma}'^m_{CX}/\bar{\gamma}'_{MX_m}) \qquad (12B7)$$

since $\bar{m}_{MX_m} = \bar{m}_M$, $\bar{m}_{CX} = \bar{m}_C = \bar{m}_X$, $\bar{\gamma}_{MX_m} = \bar{\gamma}_M \bar{\gamma}_X^m$ and $\bar{\gamma}_{CX} = \bar{\gamma}_C \bar{\gamma}_X$. Equation (12B7) is equivalent to equation (7C26), where the activity coefficient term has been collected in a symbol G_X. Although (12B7) and (12B6) look somewhat different, they pertain to the same basic process (12B4), to which the basic thermodynamic relationship (12B1) is applied, illustrating the inherent similarity of cation exchange and chelate or acidic-extractant extractions.

When the metal ions are present at tracer concentrations, it is often possible to keep the quantity $\bar{\gamma}_I$ constant. Thus at a constant temperature it is possible to rewrite equation (12B1) as

$$\ln a_I = \ln(I) + \ln \gamma_I = const. + \ln(\bar{I}) \qquad (12B8)$$

The organic phase is thus an 'electrode' for the distribuend in the aqueous phase, the concentration $(\bar{I})$, measured in the organic phase, playing the role of the electrode potential for measuring the activity of I in the aqueous phase, a_I. A comparison of a_I with the stoichiometric concentration m_I may show deviations, which can be ascribed to the stoichiometric activity coefficient γ_I, or, if there are reasons to believe that γ_I remains essentially constant, to chemical interactions which G undergoes. In this sense the distribution method can be used as a probe, akin to the metal-indicating electrode in potentiometric, polarographic and similar methods, used to study complex formation.

Distribution methods involving anion exchange with resins, extraction with basic extractants, such as long-chain ammonium salts, or extraction of ligand-metallic acids by solvating solvents, also show many similarities. The characteristic reaction for these systems can be written with the cation solvated in the organic phase and as involving the exchange of anions, species in the organic phase being considered as ion-paired

$$\text{MA}^{p-}_{m+p} + p\overline{\text{CS}^+_n\text{A}^-} \rightleftharpoons \overline{(\text{CS}^+_n)_p\text{MA}^{p-}_{m+p}} + p\text{A}^- \qquad (12\text{B}9)$$

An alternative way of looking at this distribution reaction is that involving the extraction of the neutral complex

$$\text{MA}_m + p\overline{\text{CS}^+_n\text{A}^-} \rightleftharpoons \overline{(\text{CS}^+_n)_p\text{MA}^{p-}_{m+p}} \qquad (12\text{B}10)$$

Equation (12B10) is related to equation (12B9) by the expression $a_{\text{MA}^{p-}_{m+p}} a_{\text{A}}^{-p} = a_{\text{MA}_m} \beta'_p$ (section 3.B.b). For the cases under discussion it is seldom useful to define the standard chemical potentials as in (12B5), so that $K^0_{\text{ex}} \neq 1$. Defining $K^{0\prime}_{\text{ex}} = K^0_{\text{ex}} \beta'_p$, the distribution coefficient can be expressed for both (12B9) and (12B10) as

$$D = \bar{m}_{\text{M}}/m_{\text{M(total)}}$$

$$= K^{0\prime}_{\text{ex}} \, \alpha_m \bar{m}_{\text{CS}_n\text{A}}{}^p \, \gamma_{\text{MA}_m} \, \bar{\gamma}_{\text{M}}^{-1} \, \bar{\gamma}_{\text{CS}_n\text{A}}^{\,p} \qquad (12\text{B}11)$$

The quantity α_m, the fraction of total M in the aqueous phase present as the species MA_m, arises from the definition of D in terms of the total concentrations. The corresponding quantity for the organic phase, $\bar{\alpha}_{m+p}$, is assumed to be unity. Equation (12B11) is derived from the fundamental (12B1) and (12B2) in the same manner as (12B6), replacing I by MA_m and $\bar{I}$ by $\bar{M}/\bar{C}^p$, $\bar{C}$ in this case being the concentration or activity coefficient of $\overline{\text{CS}_n\text{A}}$. Equation (12B11) is now compared with the equations derived in previous Chapters for the individual distribution systems.

Fronaeus' treatment for anion exchangers (section 6.B.b) combines the activity coefficients and the organic-phase ligand concentration terms into one parameter, which includes also the exchange constant $K^{0\prime}_{\text{ex}}$, and is considered to be constant: K''', so that (12B11) reduces to $D = K''' \alpha_m$, identical with equation (6B3). Writing $\alpha_q = \alpha_{m+p} = \alpha_m \beta'_p m_{\text{A}}^p$, activity coefficients being included in G_p in the Kraus and Nelson treatment, permits a conversion of (12B11) into the equivalent expression (12B12), which is directly related to the exchange reaction (12B9)

$$D = K_{ex}^0 \alpha_{m+p} \, m_A^{-p} \, \bar{m}_{CS_nA}^p \, \gamma_{MA_m} \bar{\gamma}_M^{-1} \, \bar{\gamma}_{CS_nA}^p \tag{12B12}$$

With the substitution $\bar{m}_A = \bar{m}_{CS_nA}$, (12B12) is equivalent to $\alpha_{m+p} = D(G_p/K_{ex}^0)(\bar{m}_A/m_A)^p$, which is identical with (6B36). It is somewhat more difficult to relate (12B11) or (12B12) with the expressions derived for the Marcus and Coryell treatment, since activity coefficients have been grouped together in a different manner. Disregarding activity coefficients in equation (6B19), the term $a^{m/l}/\sum_0^N (\beta_n^T \gamma_q/\gamma_n) a^n$ is equivalent to α_m, a^{-p} to $\bar{m}_{CS_nA}^p$ and $(K_p^T \beta_q^T/\Gamma_p)$ to K_{ex}^0, thus equation (6B19) is completely equivalent to (12B11).

As regards long-chain ammonium salt extraction, equation (12B12) corresponds directly to equation (7C75), noting that $1/X'$ in that equation equals α_m, and the other terms are the same in the two equations, except, of course, that the long-chain cation is probably not solvated.

The organic reagent $\overline{CS_n^+A^-}$ may have a very low solubility in the aqueous phase, as in the two systems discussed above: resin anion exchangers and long-chain ammonium salt extractants. It may, however, have an appreciable solubility in the aqueous phase, in fact, in an important class of extractions C^+A^- originates in the aqueous phase, and only in the organic phase is it (the cation usually) solvated. In the extraction of ligand-metal acids, equation (12B10) may serve for the extraction reaction. Since α_m is proportional to $m_A^{m-\bar{n}}$, and $\bar{m}_{CS_nA}$ is proportional to $m_{C^+} m_{A^-}$, and assuming no protonation of the aqueous metal species and constant activity coefficients, equation (12B11) becomes equivalent to (7C60), which is correct under these idealized conditions. Sometimes a metal complex extracted by direct solvation of the metal by the solvent, $\overline{MA_m S_{N-m}}$, reacts with extracted ligand acid $\overline{CS_n^+A^-}$ to give a higher complex, $\overline{(CS_n)_p^+ MA_{m+p}^{p-}}$. Such a species would also form when M is extracted from an aqueous phase containing C^+A^- at such a high concentration that the activity of $\overline{CS_n^+A^-}$ becomes sufficiently great. This case can also be described by equation (12B10) and (12B11), and is observed with such extractants as TBP, and such ligands as H^+Cl^-.

It is often advantageous, when working with distribution systems requiring high ligand concentrations, as those discussed above, to use effective ligand activities a (in the aqueous phase) and $\bar{a}$ in the organic phase, and effective equilibrium constants, K^* (section 3.B.b). Equation (12B11) can then be rewritten in a logarithmic form as

$$\log D = \log K_{ex}^* + p \log \bar{a} - \log \sum \beta_i'^* a^i \tag{12B13}$$

For anion exchangers $\bar{a}$ is interpreted as equalling $\bar{m}_A \bar{\gamma}_{\pm C^+ A^-}$ (section 6.B.c), for long-chain ammonium extractants as equalling $\bar{m}_{R^+ A^-} \bar{\gamma}_{R^+ A^-}$, while for extractants such as TBP, $\bar{a} = \bar{m}_{C^+ A^-} = Ka^2$. Using experimental values for $\bar{a}$ as defined here, the same aqueous-phase complex formation function $\sum \beta_i'^* a^i$ should result for a given system for all three distribution methods, provided the correct value of p is selected, and K_{ex}^* is treated as an adjustable parameter.

If the 'organic' phase were invariable, e.g. a crystalline solid, then $K_{ex}^{0'}(\bar{m}_{CS_n A} \bar{\gamma}_{CS_n A})^p \bar{\gamma}_M^{-1} = K_{ex}^* \bar{a}^p = K_{sol}$, a constant, and equations (12B11) and (12B13) pertain directly to solubility equilibria

$$s = K_{sol} \alpha_m \gamma_{MA_m} = K_{sol}^* / \log \sum \beta_i'^* a^i \tag{12B14}$$

for the solubility s of the solid salt MA_m in a solution containing ligand A^-. The solubility method for studying complexes between M^{m+} and A^- (or in general, for solid $M_a A_m$, with ligand A^{a-}) has the great advantage of simplicity in equation (12B14), but only seldom is the solubility in the convenient range for relating s and a. Complications, such as solid-solution formation (the solid not being pure MA_m) or too high solubility, affecting the ligand activity, often occur. Sometimes a salt $C_p MA_{m+p}$ has suitable solubility, and then an expression based on (12B12) and analogous to (12B14) results, involving α_{m+p} and $m_{C^+}^p$.

If the 'solid' is in fact a solution of the neutral component MA_m in an organic solvent S, reacting to form species $MA_m S_{N-m}$, where N may be, but need not be, the coordination number, then provided that S is not affected by the ligand (as for extraction of a metal ion from salt solutions, where the salt, e.g. sodium nitrate, is not extracted) the same equation (12B14) can apply also for the distribution. For a reaction

$$MA_m + (N - m)\bar{S} \rightleftharpoons MA_m S_{N-m} \tag{12B15}$$

the expressions

$$D = K_{ex}^0 \alpha_m \bar{m}_S^{N-m} \bar{\gamma}_S^{N-m} \gamma_{MA_m} \bar{\gamma}_M^{-1} \tag{12B16}$$

$$\log D = \log K_{ex}^* + (N - m)\log \bar{m}_S - \log \sum \beta_i'^* a^i \tag{12B17}$$

the latter the same as (7C41), are readily seen to apply. Obviously, (12B16) can be derived from the fundamental equations (12B1) and (12B2), with $I = MA_m$ and $\bar{I} = \overline{MA_m S_{N-m}}$, recalling that whereas D_I applies to thermodynamic components, D applies to total concentrations.

It is thus seen that the thermodynamic equations describing the distribution in the various systems are all derivable from the simple equations pertaining to neutral components, hence are in a sense equivalent. These equations are independent of the 'real' mechanism of the extraction or exchange reaction, which is often unknown. Whatever the mechanism, the distribution can always be decribed in terms of a convenient chemical reaction, and provided that the major species present are not neglected, the equation obtained for D will not only be formally correct, but also meaningful and fruitful in terms of the information obtainable through it on the system studied.

b. Choice of System Parameters

An important aspect in the comparison of various distribution methods for studying complex formation in aqueous solutions, is the freedom the experimenter has to select suitable parameters for the systems. A few examples may be used to illustrate this point.

The sterically available active carbonyl oxygen of acetone may make it a desirable extractant, but acetone cannot be used as an organic solvent for extraction from dilute aqueous solutions, because of its miscibility with them. It can, in fact, be used in highly salted aqueous solutions, such as nearly saturated aqueous lithium chloride, with which it is not miscible. Diethyl ether, on the other hand, though immiscible with water, becomes completely miscible with concentrated hydrochloric acid, so that it can be used for extractions only from somewhat diluted acid. The salt or acid concentration is thus an important parameter, determining the solubility or miscibility of the solvent with the aqueous phase (sections 2.C and 9.A). Such considerations are absent when using resin ion exchangers, which show negligible solubilities in aqueous solutions.

The concentration of exchange sites in ion exchangers can ordinarily be changed over a narrow range only, perhaps by an order of magnitude, when the swelling, degree of invasion, etc. are changed (section 4.B). Extractants, on the other hand, can usually be diluted with suitable diluents down from the pure extractant to its infinitely dilute solution. Thus concentrations can be chosen freely in the one case, while there is a practical lower limit, and a limited overall range, in the other.

The following are some of the important system parameters, which can be chosen with more or less freedom.

(i) *Functional groups*   The most important variable that can be chosen for a given distribution problem is, of course, the functional group of the extractant or ion exchanger. Thus, once it has been decided to apply to a problem a cation exchange method or a method involving a solvating extractant, choice can be made among for example, sulphonic acid (i.e. strong) or carboxylic acid (i.e. weak) cation exchange resins, or between for example, TBP (strong) or ether (weak) as extractants.

Although there are a variety of functional groups that have been incorporated as exchange sites in ion exchange resins, most of the significant work concerning complex formation has been done with just two types: sulphonate-type cation exchangers and methylenetrimethyl-ammonium-type anion exchangers. This is so because the interaction between the metal ions or complexes and the functional groups should be relatively weak, and of a purely electrostatic ion-association nature, in order not to introduce unnecessary complications. Thus although in principle it is possible to use weakly basic or acidic groups, or chelating groups, in complex formation studies, in practice it has proven to be more convenient to adhere to the strongly basic or acidic groups.

With liquid extractants, however, this is not the case, and for each type of extraction (e.g. metal ion solvation, acido complex extraction, chelation, etc.) many different reagents have been used. It is thus possible to select exctractants so that the distribution coefficients are in a convenient range, or so that there would be less interference from competing reactions, such as the extraction of excess acid, or so that there would be more or less ionic dissociation, as required for a particular problem. In this respect liquid extractants are indeed more versatile than ion exchangers, and full use of this aspect should be made.

(ii) *Solvents or diluents* Another factor which can often be selected freely is the solvent (diluent) that is to be used in a given distribution system. For liquid–liquid extraction systems, where the metal ion, perhaps a ligand anion, and the extractant are prescribed, the diluent can often be selected freely, so as to give the most favourable results. For resin ion exchangers the organic medium is usually not varied: most useful resins are based on a hydrocarbon skeleton, mostly divinylbenzene–crosslinked polystyrene. However, the aqueous phase can be modified by using a miscible organic solvent, without the danger of solubilizing the 'organic' phase (i.e. the resin) in the 'aqueous' (i.e. mixed) phase. There is a danger of extracting the organic component into the resin, but the opposite trend is usually encountered, the aqueous component being preferred. Such mixed aqueous–organic solutions cannot, as a rule, be used in a liquid–liquid distribution system, because of the too great mutual solubility of the two phases that would result.

Selection of a solvent or diluent determines several of the properties of the system, such as the macro dielectric constant, the polar nature, the hydrogen-bonding tendency, etc. These, in turn, determine factors such as degree of ionic dissociation, aggregation, solvation or the rate of attainment of equilibrium (often the rate of diffusion). In ion exchangers, which are in equilibrium with aqueous solutions, the effective dielectric constant is rather high. Depending on the crosslinking, it may have values ranging from 60 (at 2% crosslinking) to 30 (at 16% crosslinking) (sections 4.C and 6.C). This can be modified further by using nonaqueous solvents, or mixed solvents. Some solvents used for extraction have a relatively high dielectric constant, such as nitrobenzene (36), others have intermediate values, such as bis(β-chlorethyl) ether (21) or hexone (13), while some have low dielectric constants, such as hydrocarbons (2–3), chloroform (4.8) or diethyl ether (4.3). Some of these latter liquids can be used only as diluents for more active extractants, or for extracting certain chelates. In general there is considerable latitude possible in selecting the diluent or solvent.

(iii) *Concentration* As mentioned in an example above, the concentration of an extractant usually can be varied over a very wide range. This statement must be amplified, however, since variation of the stoichiometric concentration does not always cause the expected changes in the distribution. A diluent may not be completely inert, thus placing a lower limit on the effective concentration of extractant that can be used (sections 7.C and 9.B). Aggregation of the extractant may lead to a constant thermodynamic activity (i.e. a separate phase) although the concentration is varied (sections 9.B and 10.B.c). A solvent may be a relatively poor extractant, so that its efficiency decreases very rapidly on dilution, and only a narrow range of concentrations may be useful for extraction. It is difficult to measure distribution coefficients outside the range 10^{-3} to 10^3 to any reasonable reliability (section 7.C), and this fact also limits the range of extractant concentrations that can be used effectively. Still, in general it is true that it is possible to vary the concentration of an extractant over a wide range, while the concentrations of exchange sites in resins cannot be varied without causing other significant parameters (such as crosslinking and effective dielectrtic constant) to vary also.

(iv) *Temperature* For extractions or ion exchange from dilute aqueous solutions, the freezing point of water is an obvious lower limit, while for non-pressurized systems, the boiling point is the upper limit. Volatile solvents will have lower high-temperature applicability limits, and bubble formation will limit ion exchange column operation also to somewhat lower temperatures than 100°C. Application of pressure increases the upper temperature limit, ion exchange at 150°C having been reported (section 6.C). The use of mixed or nonaqueous phases also extends the temperature range, thus extraction from a molten nitrate solution at 150–200°C or ion exchange in methanol–water solutions at −40°C become possible. Temperatures are then limited to thermal stability of the extractants or exchangers (about 200–250°C, except for inorganic ion exchangers) on the one hand, and freezing of extractants and slowing down of diffusion in exchangers at low temperatures, on the other.

(v) *Pressure* Few studies have been published concerning the effect of pressure on distribution systems. Since the distribution process is usually accompanied by a volume change of the distribuend, appreciable pressure effects are to be expected. Thus high hydrostatic pressures affect the strontium–hydrogen ion exchange with a polystyrene–sulphonate cation exchanger, although it affects little the potassium–hydrogen ion exchange (section 4.C). Practically nothing is known concerning pressure effects in liquid distribution systems. It is expected that the effects would be even larger than with ion exchangers, since the volume changes connected with the transfer from an aqueous to an organic phase are expected to be rather large.

Having conveniently measurable distribution coefficients, and independence from interfering side-reaction are the prime considerations in selecting the system parameters for a given problem. Thought should also be given to rate considerations. Too slow diffusion in a tightly crosslinked ion exchange resin, slow keto–enol tautomerism of an extractant, or slow metal-chelate reaction kinetics may preclude the use of distribution systems which involve these steps. Sometimes it is possible to select other parameters, in a way which would tend to ameliorate these conditions, such as a higher temperature, a different solvent, etc.

(vi) *Variability of the organic phase* Another important consideration is the desirability of having an organic phase of constant composition and hence of constant thermodynamic properties, when studying complex formation of metal ions (at tracer concentrations) with ligands in the aqueous phase. If such ideal conditions can be attained, the distribution coefficients will reflect only changes occuring in the aqueous phase as the ligand concentration is varied. Otherwise, a significant, and sometimes even predominant, portion of the total dependence of the distribution coefficients on the ligand concentration will be due to changes in the organic phase. This complicates the situation since such changes are often considerably less well understood than corresponding changes in the aqueous phase, and much less supporting information concerning them is available.

When complex formation is sufficiently strong, and only low ligand concentrations are required, the constant ionic medium method can be used to keep activity coefficients constant in the aqueous phase. Cation exchange involving resin or liquid cation exchangers, or chelate extraction can be used in this case, with a reasonable certainty that the organic phase does not change appreciably in character when the ligand concentration is varied. Since the thermodynamic equations are the same for these types of distribution studies (section 12.B.a), considerations of experimental convenience should be predominant.

When high concentrations of ligands are required, it becomes less easy to keep the organic phase constant. Solutions of long-chain ammonium salts in hydrocarbon diluents are perhaps least affected by high concentrations of electrolytes in the aqueous phase, and may provide in some situations nearly constant conditions even for a wide variation of ligand concentration in the high concentration range. Resin anion exchangers and some other liquid extractants may also exhibit nearly ideal behaviour in exceptional cases. In general, however, it is necessary to correct for the variation of the organic phase with variations in the ligand concentrations, as described in detail in Chapters 6–10.

If the interactions occuring in the organic phase are of interest, then it is necessary to vary the properties of the organic phase in a suitable way. This is usually more easy to do with liquid extractants than with ion exchangers, although even with the former the interpretation of the results is not straightforward. Thus at one time, it was believed that the variation of the distribution coefficient of a trace metal ion at a constant aqueous-phase composition, with the concentration of a long-chain ammonium salt extractant, is a direct indication of the species present in an anion exchange resin equilibrated with the same aqueous solution. It has later been shown that not only is this inference incorrect, it is doubtful whether this function can give direct information concerning the species in the long-chain ammonium salt solutions themselves (section 10.C). Caution must be exercised when inferring from one distribution system to another, inspite of external similarities in behaviour.

C. ILLUSTRATIONS OF THE APPLICATION
OF DISTRIBUTION METHODS

a. Introduction

Although there have been a few instances of the use of distribution methods to study the formation of metal complexes in solution in the earlier decades of this century, widespread use has started only about twenty years ago. In recent years, however, the literature has proliferated in publications on this subject, and it would be an enormous task to examine the pertinent publications comprehensively and critically. It is possible here to illustrate for a few chosen cases only, the applicability of distribution methods to the problem of determining the composition and stability of metal complexes in solution.

Most of the important publications are summarized and evaluated in the following sections for these chosen cases, in a way that permits a comparison to be made between the results obtained from different distribution methods. Reference is made to the appropriate equations and discussions in Chapters 5–10 for the various methods used. In no case has the last word been said, and many problems have remained open, awaiting study by better methods with more suitable distribution systems. However, a general view of the 'state of the art' can be obtained from these illustrations.

b. Metal Ion Hydrolysis: Zirconium(IV)

Zirconium(IV) is strongly hydrolysed in aqueous solutions, and both monomeric and polymeric hydrolysis products have been proposed to exist. The methods that have found most use in studying these species have been cation exchange and chelate extraction.

In 1949 Connick and McVey[1] noted that until then there had been no studies of the species formed by zirconium(IV) in solution. They used extraction with thenoyltrifluoroacetone (TTA, symbolized as HX in the following) and zirconium labelled with the tracer ^{95}Zr to study the hydrolysis. Their solution, however, contained an impurity, which caused extraction to be lower at the tracer level (where an appreciable fraction of the zirconium reacted with the impurity) than at the $3 \times 10^{-4}M$ level. At 2 M perchloric acid, they concluded, the predominant zirconium species are Zr^{4+} and $ZrOH^{3+}$ at low zirconium concentrations, and

References pp. 906–912

hydrolytic polymers at high ones. Since that time a number of other studies, utilizing the same reagents and perchlorate solutions of various acidities and ionic strengths, have been made[2-4], and the original data have been recalculated[5]. In the later work, the effect of the impurity has been decreased, and more confidence can be placed in the results[2]. The use of Fomin and Maiorova's method[6] for recalculating the data of Connick and McVey[1] by Solovkin[5] does not seem warranted: the data are too imprecise to permit a calculation of four successive constants for the equilibria $Zr(OH)_{i-1}^{5-i} + H_2O \rightleftharpoons Zr(OH)_i^{4-i} + H^+$ for $i = 1,2,3$ and 4, in particular since the values show only a minimal spread of the successive values: $k_4/k_1 = 9$. At very low zirconium concentrations, where monomeric species predominate, the equations (12C1) and (12C2)

$$Zr^{4+} + 4\overline{HX} \rightleftharpoons \overline{MX_4} + 4H^+ \qquad (12C1)$$

$$D = K_X(\overline{HX})^4(H^+)^{-4} \qquad (12C2)$$

(where K_X is the equilibrium constant for (12C1) and activity coefficients are assumed constant and are included in K_X), which are applications of the general equations (7C24) and (7C26), have been applied to the data. They showed that for 2 or 4 M perchloric acid ZrX_4 is the organic-phase species and Zr^{4+} is the aqueous-phase species[2,4]. At higher concentrations polymerization starts. Since here it is assumed that polymerization occurs only in the aqueous, but not in the organic phase, equations (7C13) to (7C17) nay be applied, but with D^{-1} and P^{-1} instead of D and P. The resulting expression is

$$d\log D/d\log c_{Zr} = -1 + 1/\sum q\alpha_q \qquad (12C3)$$

for the formation of a series of q-mers. In fact, at low c_{Zr} values the slope is zero, so that $\alpha_1 = 1$. The average polymerization number $\tilde{q}$ can be deduced from the slope, and it has been concluded that trimers and tetramers predominate, while dimers are not important even at 1 M or 2 M perchloric acid, above $c_{Zr} \sim 10^{-3}$M[4]. One aspect of the problem, which has not been taken into account, is the formation of TTA complex species in the aqueous phase, which strongly affects the pH dependence of the distribtion data (equation 7C28). It has been demonstrated[3] that even with dilute TTA solutions, where extraction of ZrX_4 is negligible, the aqueous species ZrX^{3+} is formed. The extraction of mixed hydroxide–TTA complexes has not been considered in these studies either, but this may not be an important complication, since in ZrX_4 the four bidentate

TTA groups take care of all the eight coordination sites on the zirconium ion.

Other reagents, besides TTA, have been used to study the hydrolysis of zirconium(IV). Thus benzoylacetone, which reacts similarly to TTA, also being a β-diketone, has been used[7]. At a relatively high acidity (1 M perchloric acid) and low zirconium concentrations (7.4×10^{-7} to 7.4×10^{-6}M) D has been found to depend on $\bar{c}_{HX}/c_{H+}$ only (equation 12C2), and not on c_{H+} and c_{Zr}, confirming the species ZrX_4 in the organic phase and Zr^{4+} in the aqueous phase. The distribution data at lower acidities were analysed according to Fomin's method[6], and again four mononuclear hydrolysis constants were calculated from rather imprecise data, obtaining at an ionic strength of 1 M, values about three times higher than those obtained by Solovkin[5] at an ionic strength of 2 M.

Another group of studies used cation exchange methods[8-12]. The availability of pores of different average diameters in resins of different crosslinking has been used to demonstrate the presence of species having diameters of the same magnitude as the pore size (4–6 Å) in the acidity range 0.3–3.0 M, by measuring the rate of diffusion through the resin particles[8]. These species have been proposed to have the formula $O(Zr(OH)_2)_2^{2+}$, from considerations of specific charge (loading data) and size (diffusion data). At lower acidities larger polymeric species (e.g. $Zr_3(OH)_{10}^{2+}$ or $Zr_4(OH)_{14}^{2+}$) can account for the extremely slow diffusion, and at acidities above 3 M small ionic species account for the rapid diffusion, which, however, is appreciably lower than that of the Th^{4+} ion[8]. Since the rate of diffusion decreases the tighter the ion is held, the facts do not preclude the existence of Zr^{4+}, which is held much more tightly than Th^{4+}, contrary to the authors' conclusions[8]. From strongly acid solutions loading can be as high as 0.96 millimoles zirconium per gram of air-dried resin, having a capacity of 3.87 milliequivalents per gram[9]. Thus full loading to ZrR_4 can be achieved. Perchlorate ions were found not to accompany the zirconium into the resin[8]. The main conclusion from Larsen and Wang's work[10] is that at 1 or 2 M acid and tracer zirconium(IV) concentrations the unhydrolysed species Zr^{4+} predominates in the aqueous phase, while in the resin the metal is hydrolysed. Loading to ZrR_4, achieved at high zirconium concentrations[9], does not prove that at low concentrations, and with low acid invasion into the resin (estimated at about 0.1 M), the zirconium cannot be hydrolysed in the resin. Indeed, the larger ion Th^{4+} has been found capable of undergoing hydrolysis to strongly held polymeric species $Th(Th(OH)_4)_n^{4+}$ in the resin at relatively low pH values[13,14], so that it should not be surprising to find the zirconium hydrolysed in the resin. The nature of the hydrolysed species in the resin could not be deduced from the data, and either monomers, with an average charge of 2.4 (i.e. $Zr(OH)^{3+}$ and $Zr(OH)_2^{2+}$) or polymers can account for the data. The fact that in concentrated acid solutions (4 M perchloric acid) only monomeric species occur has been confirmed also by another study, which also proposes the formation of polymeric species in solution at lower acidities (0.14 M $HClO_4$)[14].

Paramonova[11] has applied her method of relative absorption curves (section 5.B.d) to estimate the degree of hydrolysis of zirconium(IV) in mixed sodium perchlorate–perchloric acid solutions of constant total concentration of 2 M. Above $c_{H^+} = 1.2$ M, γ_+ is constant and $\gamma_- = 0$, signifying that no hydrolysis occurs. Between $c_{H^+} = 1.2$ and $c_{H^+} = 0.5$ M, γ_+ decreases, while γ_- is still very small. This behaviour is attributed to hydrolysis in the aqueous phase. Below $c_{H^+} = 0.5$ M, γ_- becomes appreciable, signifying the formation of anionic species. Equilibration has taken a long time with the tightly held ions, zirconium(IV) and perchlorate, obstructing easy diffusion in both types of resin. Too much reliance should not be put on the results reported in this paper, however, because of the serious shortcomings of the method, as discussed previously (section 5.B.d).

A comparison of the results obtained by the two main methods, TTA extraction and cation exchange, shows good agreement. The main conclusion is that above about 1 M acid and below $c_{Zr} = \sim 10^{-3}$ M unhydrolysed Zr^{4+} is the predominant zirconium(IV) species. At lower acidities and/or higher c_{Zr}, hydrolysed monomeric and especially polymeric (trimer, tetramer) species are formed. Each method suffers from its own complications, such as aqueous TTA complex formation or hydrolysis in the resin phase. Additional work, with better control of conditions (e.g. avoidance of impurities), and advanced computational methods may be worthwhile.

c. Fluoride Complexes: Tantalum(V).

Tantalum(V) is similar to zirconium(IV) in being highly hydrolysed in solution, but in solutions containing fluoride ions, tantalum(V) is stable as mononuclear complex species.

Only qualitative information concerning complex formation in the tantalum(V)–fluoride system has been obtained by the use of cation exchangers. Lederer[15] noted that tantalum is not held on a Dowex–50–impregnated paper, when eluted with 0.45 M HF. This has been ascribed to strong complex formation. Keller[16] obtained the distribution curve for tracer tantalum(V) from batch experiments with Dowex–50 X 8 and 0.01–20 M HF. In the range 0.01–0.1 M HF he found a small increase in D with increasing hydrofluoric acid concentrations, possibly connected with the formation of cationic fluoride complexes, which can be sorbed on the resin, e.g. TaF_3^{2+} or TaF_4^+, from non-sorbable hydrolysed species, possibly polymeric. Beyond 0.1 M HF there is a decrease in D, possibly due to the formation of neutral or anionic complexes, such as TaF_5 or TaF_6^-. Keller[16], however, ascribes the relatively high sorption ($D \sim 40$ at 0.1 M HF and $D \sim 4$ at 20 M HF) to sorption of anionic species on the cation exchange resin, as is the case for a number of anionic chloride complexes (cf. section 4.D.c).

The extraction of tantalum(V) by solvating solvents has yielded more quantitative information. Early workers have studied the extraction from both pure hydrofluoric acid[17,18], and from mixtures containing an additional mineral acid[17,19–21]. For pure hydrofluoric acid solutions and extraction into methyl isobutyl ketone a maximum $D \sim 2$ has been found near 7 M HF[17], while for extraction with diethyl ether (at 0.1 M Ta) D remains constant near 3.8 above ca. 10 M HF[18]. Extraction is enhanced when another acid is added: in the presence of 2 M HCl D reaches 2 for 0.085 M HF with diisopropylcarbinol[20], for about 0.2 M HF with methyl isobutyl ketone[17] and for 0.6 M HF with diisopropyl ketone[19], compared with ca. 6 M HF required when no additional acid is present. In the presence of 6.5 M H_2SO_4 tracer tantalum(V) is extracted so efficiently by diisobutyl ketone in the presence of fluoride, that the extraction can be used for a quantitative estimation of the latter ion[21]. Indeed, the stoichiometry of extraction, when tantalum is present in excess of one seventh of the fluoride concentration, shows[21] that the extracted species is $(H \text{ solvated}^+)_2 TaF_7^{2-}$. Presumably the same species is extracted also in the other systems.

Recently, Varga and coworkers[22,23] published a detailed study of the fluoride complexes of tantalum(V) in 3 M $HClO_4$ using extraction methods, employing three reagents: methyl isobutyl ketone (hexone), *N*-benzoylphenylhydroxylamine (BPHA) and tri-n-octylphosphine oxide (TOPO). For BPHA in chloroform, a second-power reagent dependence of D was found, as well as a mole ratio of F:Ta in the organic phase of 3.8 ± 0.1. These data were used with an equation similar to (7C35), to establish the extracted species as $TaF_4X.HX$ (which could be hepta-coordinated if X^- is bidenate and HX monodetate). The distribution coefficient is given by

$$D = K_X \overline{(BPHA)}^2 (H^+)^{-1} (F^-)^4 / \sum_4^7 \beta_n (F^-)^n \qquad (12C4)$$

similar to the general equation (7C36), but with $(F^-)^4$ in the numerator. While BPHA probably acts as a chelating agent, TOPO acts by solvation, but for reasons not definitely clarified, the species extracted has been stated to be $TaF_4ClO_4.p$TOPO. Values of p between one and two were obtained from reagent concentration dependence in cyclohexane and chloroform solutions. Finally, hexone is considered to solvate the hydrogen ions, coextracted with TaF_6^-, a species deduced from measured mole ratios in the organic phase of F:Ta $= 6.1 \pm 0.2$. The equations used to evaluate the results for extraction with TOPO and hexone were

$$D = K_X (TOPO)^p (ClO_4^-)(F^-)^4 / \sum_3^5 \beta_n (F^-)^n \qquad (12C5)$$

with $p = 2$, and

$$D = K_X (F^-)^6 (H^+) / \sum_4^6 \beta_n (F^-)^n \qquad (12C6)$$

which are essentially the same as the general equations (7C46) and (7C56), without the term in square brackets. The known hydrogen ion concentration (3M) and the calculated fluoride ion concentration were used, instead of c_{HA}. The data and equations were used with a computer programme, which did not yield a consistent set of stability constants for all three systems. The final constants corresponding to the formation of species TaF_{3+j}^{2-j} from TaF_3^{2+} with $j = 1, 2, 3$ and 4, have been selected rather arbitrarily, and the results indicate that in 3 M $HClO_4$, TaF_3^{2+}, TaF_4^+ and TaF_7^{2-} are of major, while TaF_5 and TaF_6^- are of minor importance. The special stability of TaF_4^+, which adds ClO_4^- to be extracted by TOPO, or $(X^- + HX)$ to be extracted by BPHA, seems very remarkable. Not less remarkable is the claim that whereas for 0.01 M TOPO in chloroform the extracted species is $TaF_4ClO_4.p$TOPO, for 0.01–0.05 M TOPO in cyclohexane there are two species formed, as the fluoride concentration increases: $TaF_3(ClO_4)_2.p$TOPO and $HTaF_6.p$TOPO. It seems that too much confidence has been put into the computer results, with too few control experiments, and too little direct experimental evidence, to warrant the remarkable conclusions reached[22]. That the stability constants cannot be very reliable is also apparent from their gross disagreement with results obtained by other methods, and discussed below.

Some more information concerning tantalum(V) fluoride complexes has been obtained from anion exchange studies. Kraus and Moore[24] studied hydrofluoric–hydrochloric acid mixtures and tentatively identified the species formed as $TaX_5F_3^-$ and $HTaX_5F_3^{2-}$ (X = F^-, Cl^- or OH^-) from the acid and fluoride ion concentration dependencies of the distribution coefficients. Buslaev and Nikolaev[25] gave the charge of the species sorbed on the anion exchanger AN–2F as 2.17 ± 0.06, signifying the species $TaOF_5^{2-}$, without any details of the loading data on which this figure is based. The species could also be TaF_7^{2-} or some other suitable complex. Survey-type studies give information on the anion exchange tracer sorption of tantalum(V) from hydrofluoric acid solutions[26,27], and from mixtures of hydrofluoric and hydrochloric or nitric acids[28,29]. For 0–20 M HF, $\log D$ decreases from approximately 3 (for Dowex–1 $X\,10^{26}$) or 4 (Russian resin AV–17 $X\,14^{27}$) to approximately 2 (both resins). More detailed data have been presented by Keller[16], who used Dowex–1 $X\,8$, and found $\log D$ to increases from 3.8 at 0.01 M HF to 4.3 at 0.2 M HF and hence to decrease to 2.0 at 12 M HF. The decrease in D has been ascribed to a decrease in the concentration of fluoride ions, as the concentration of hydrofluoric acid increases. Lacking invasion data for the resin–HF system, and a knowledge of the species sorbed on the resin (see above), no detailed analysis of the results, in terms of the treatments presented in section 6.B, can be made.

A detailed study of the anion exchange behaviour has been made by Varga and Freund[23], who employed a perchlorate-form resin and a 1 M perchloric acid medium, and up to 0.017 M fluoride ion concentrations (corresponding to ca. 2 M HF), with tracer concentrations of tantalum(V). The authors used the equation

$$D = K\bar{c}(F^-)c_{ClO_4^-} \Big/ \sum_{-5}^{N-5} \beta_i'(F^-)^i = K'\bar{c}_F \Big/ \sum_{-5}^{N-5} \beta_i'(F^-)^i \quad (12C7)$$

which is essentially similar to a combination of (6C3) and (6C4). They assumed that the species sorbed is TaF_6^- (without experimental evidence), so that $\bar{n} = 6$. Since $m/l = 5$ (the charge of Ta^{5+}) and $\bar{c}$ is proportional to c (for very high selectivity of perchlorate over fluoride, equation 6C4), they arrived at $\bar{n} = 6 - d\log D/d\log(F^-)$, which is the same as equation (6C7). The data were fitted to a relationship of D and (F^-), which could be differentiated to give $\bar{n} = 9.22 - 0.97 \log (F)$, i.e. a relationship from which it was concluded that the species TaF_4^+, TaF_5 and TaF_6^- exist in the range of fluoride concentrations studied by anion exchange. If, however, it were assumed that the species in the resin is TaF_7^{2-} (see above, Buslaev's results), then the same data would yield $\bar{n}$ values higher by one unit, and the species in solution would then be TaF_5, TaF_6^- and TaF_7^{2-}. The data do not permit a decision to be made, but in either case, the middle species, TaF_5 or TaF_6^- is favoured over a wide fluoride concentration range, as the calculated stepwise formation constants indicate. This is in direct contradiction to the extraction results discussed above, and is unlikely to be explained by the difference in perchloric acid concentration (1 M vs. 3 M).

d. Chloride Complexes: Iron(III)

A comprehensive review of the studies of chloride complexes in solution by ion exchange and solvent extraction methods has been published recently[30]. Since, however, complex formation between iron(III) and chloride ious has received such a great deal of attention, and many different methods of investigation, including ion exchange and solvent extraction methods, have been applied to this problem, it will be used here as an illustration.

Iron(III) is sorbed on cation exchangers from dilute chloride solutions, but complex formation decreases the distribution coefficients, compared to perchlorate solutions. Thus for Dowex–50 X 4 and 1 M acid, $D \sim 28$ for 1 M HCl compared with $D \sim 70$ for 1 M $HClO_4$. The distribution coefficient reaches very low values ($D < 1$ for the above resin) at 3–5 M HCl, but increases again at higher hydrochloric acid and lithium chloride concentrations[31-33]. This behaviour has been discussed earlier (section 4.D.c). No quantitative conclusions concerning the species formed have been reached.

The liquid cation exchanger, di-n-nonylnaphthalene sulphonic acid, has been used to study iron(III) chloride complex formation, in constant $H(Cl,ClO_4)$ ionic media[34]. The equation that has been used to account for the data

$$(D^0/D) = 1 + \beta_1 c_{Cl^-} + \dots \qquad (12C8)$$

is the same as used for resin cation exchange (5B10) according to Schubert's method (section 5.B.b), and for extraction with chelating agents which do not form aqueous-phase chelates (7C36). The values of β_1 obtained at 0.2, 0.3 and 1.0 M ionic media agree well with values obtained by other methods.

A large body of information exists concerning the extraction of iron(III) chloride complexes with organic solvating solvents, especially those containing oxygen donor atoms. These systems have been reviewed previously[35,36], the analysis of Diamond and Tuck[36] being very thorough. The properties and nature of the species formed in the organic phase are discussed in detail in section 9.B.

Solutions of anhydrous iron(III) chloride in anhydrous solvents, such as ethers, ketones, etc., have been shown spectrophotometrically to contain only iron trichloride molecules, solvated with one solvent molecule, $FeCl_3.S$[37,38]. When anhydrous hydrogen chloride is added, the bisolvated tetrachloroferric(III) acid $HFeCl_4.2S$ is formed, having a low solubility in the solvent[38]. Only two species are present in the solution, as the appearance of isosbestic points shows, and equilibrium constants for the reaction $FeCl_3 + HCl \rightleftharpoons HFeCl_4$ have been calculated[37]. When water is added gradually to a suspension of the solvated tetrachloroferric(III) acid in the solvent, the solid dissolves, and complete dissolution occurs when 3–5 moles of water per mole of iron(III) are added[38,39]. The spectrum of this solution is identical with that of iron(III) extracted by the same solvent from hydrochloric acid solutions, and these spectra have been studied by several authors[40–43]. The spectra are similar to those of solid tetrachloroferrates(III) and to solutions of the potassium and ammonium salts in ethers[44]. An infrared study[43] showed the C = O bond shifts for the solvent in the iron(III) extracts to be the same as for pure hydrochloric acid extracts with no other bands appearing, while there is no shift in anhydrous $KFeCl_4$ solutions, for a large number of solvents. A Raman spectroscopic study showed the iron in ether extracts to be bound tetrahedrally to four chlorine atoms[45]. The conclusion from these observations is that the iron(III) species in the extracts is $H(xH_2O,yS)^+FeCl_4^-$, and not $FeCl_3$ or Fe_2Cl_6[46], H_2FeCl_5 or H_3FeCl_6[47] and $H(xH_2O)^+FeCl_4S_2^-$[35], as believed earlier. Only a solvent of low basicity but high polarity, such as nitrobenzene, is an exception to exclusive proton solvation, as it solvates the ion-pair as a whole[48]. The values of x and y must be specified for every system, as well as the degree of dissociation and aggregation of the extracted species.

Hydrogen chloride is extracted into the solvents from concentrated hydrochloric acid solutions, and more is coextracted with iron(III) chloride from very concentrated hydrochloric acid, than is extracted from the acid alone[40,46]. The number of moles of water coextracted is 4–5 per mole of iron(III) under various extraction conditions[39,40,49], but this does not mean that $x = 4$ or 5, since some of the water is associated with the excess hydrogen chloride. The number of solvent molecules, y, has been determined from dilution studies (equation 7C41) and from cryoscopy, and under apparently similar conditions the values $y = 2$ and $y = 3$ have been found[38,50–52]. It is reasonable to suppose that $x + y = 4$, if the hydrogen ion is pictured as $[O(H...OH_2)_{x-1}(H...OR_2)_y]^+$ but the total number that has been obtained seems to exceed four, and the detailed binding has not yet been made clear[43].

The dissociation of the ion-pair $H(xH_2O, yS)^+FeCl_4^-$ and its aggregation depend on the dielectric constant of the solvent when saturated with water and hydrogen chloride[36,42,53–60]. In solvents of low dielectric constant, association to ion-triplets and higher aggregates occurs[53–55], and an increase of c_{Fe} causes an increase in D, equation (7C57). Isopiestic measurements have confirmed the aggregation[42,55], and association with both iron(III) and excess hydrogen chloride has been found. As pointed out above, more hydrogen chloride is extracted into solvents of low dielectric constant, such as ethers, in the presence of iron(III) than in its absence[41,47]. When the solvent is diluted with another one of higher dielectric constant, the conditions for aggregation disappear, and D ceases to increase with increasing c_{Fe}.[60] If the dielectric constant of the solvent is above about 10, some ionic dissociation occurs, as shown by the conductivity of the organic phase and by the anodic migration of the iron(III)[56,58,59]. The observed dependence of D on c_{Fe} conforms to equation (7C65), and to the behaviour depicted in Figure 7C6: at very low and very high c_{Fe} values D is independent of c_{Fe}, while at intermediate c_{Fe} values D varies with $c_{Fe}^{-1/2}$[35,56]. Dilution with a diluent of low dielectric constant causes D to become independent of c_{Fe}[60]. Solvents with donor atoms other than oxygen, such as dibutyl sulphide[61] and adiponitrile[62] behave analogously to the oxygenated solvents, the donor atoms coordinating with the extracted hydrogen ions, whereas solvents with no donor atoms, such as benzene, carbon tetrachloride or dichloroethylene, show only negligible extraction of iron(III)[63].

The solvent TBP has been considered an exception to the generalization that it is hydrogen ion and not the iron(III), which is solvated[51,52,64].

At low hydrochloric acid concentrations a third-power dependence of D on $\bar{c}_{\text{TBP}}$ is observed and at high c_{HCl} a second-power dependence. This has been taken to indicate a completion of the coordination number of iron(III) to six, in the species $FeCl_3.3TBP$ and $H^+[FeCl_4.2TBP]^-$ respectively. It has, however, been shown that in every case the iron(III) species is tetracoordinated, non-solvated $FeCl_4^-$, and that TBP, like the other solvents, associates with the hydrogen ion[43,65]. The change in solvation number, observed also with other solvents, is due to competition of coextracted hydrogen chloride. The published data for extraction of iron(III) at low concentrations from hydrochloric acid solutions into untiluted $TBP^{52,65-68}$ do not agree at all well.

The extraction reaction can be written as

$$FeCl_{3-i}^i + H^+ + (i+1)Cl^- + p\bar{S} \rightleftharpoons \overline{HS_p^+FeCl_4^-} \qquad (12C9)$$

where S stands for the solvent TBP. The distribution coefficient should then be given by the expression

$$\log D = \log K_{\text{ex}} + 2\log a + p\log(\bar{S}) - \log \sum_{-3}^{N} \beta_i'^* a^i \qquad (12C10)$$

according to equations (7C47) and (7C56), where K_{ex} is the equilibrium constant for reaction (12C9) and activity coefficients in the organic phase are assumed constant. Since, however, the hydrogen chloride extracted independently of the tracer iron(III) binds TBP, as $\overline{HS_t^+Cl^-}$, $(\bar{S})$ will be less than $\bar{c}_S$, and will be given by (7C52). For the special case where $t = p = 1$, it can be shown that the concentration of bound HCl (equalling that of bound TBP: $\bar{c}_{\text{HCl}} = K_C(\bar{S})a^2$, where K_C is the equilibrium constant for the extraction of hydrogen chloride) can be substituted in (12C10) to give

$$\log D = \log K + \log \bar{c}_{\text{HCl}} - \log \sum_{-3}^{N} \beta_i'^* a^i \qquad (12C11)$$

This equation expresses the data of Bankman[65] and of Weidemann[67] fairly well, using published data of $\bar{c}_{\text{HCl}}{}^{66,69}$ and of the constants $\beta_i'^{*70}$, the latter obtained by the anion exchange method discussed below (Figure 12C1). The data are, however, too scattered to permit an independent evaluation of the constants, or a check on the assumptions that $t = p$, and that activity coefficient ratios in the organic phase are constant.

The extraction of iron(III) with long-chain ammonium chlorides has recently been studied by several authors[34,71-86]. The spectra of the organic phases show that the iron(III) species is $FeCl_4^{-}$[74,79,81], whether the slope $(\partial \log D / \partial \log \bar{c}_{R_3NHCl})_{c_{HCl}, c_{Fe}}$ is one (at high c_{Fe}[75,78]) or two (at tracer iron(III) concentrations[34,75]). Saturation experiments[74,81,84], as well as the slope at high iron(III) concentrations, indicate (equation 7C76a) the formation of $R_3NH^+FeCl_4^-$ as the major species[81,84]. At lower iron(III) concentrations more free substituted ammonium chloride exists, which can associate with the iron species to give $R_3NH^+Cl^-.R_3NH^+FeCl_4^-$, either as dipole–dipole head-to-tail association of $R_3NH^+Cl^-$ with $R_3NH^+FeCl_4^-$, or as having two R_3NH^+ cations hydrogen bonded to one chloride anion. The alternative formulation, $(R_3NH^+)_2FeCl_5^{2-}$, although capable of accounting for the slope[34], is at variance with the spectrum[81].

In dilute solutions, the dependence of D on c_{HCl} led to the conclusion that up to 0.8 M HCl only Fe^{3+} and $FeCl^{2+}$ coexist in the aqueous phase, since it was found[34] to conform to

$$D = K c_{HCl}^3 \; \bar{c}_{R_3NHCl}^2 \; /(1 + k_1 c_{HCl}) \qquad (12C12)$$

This expression is equivalent to (7C75) under the simplifying assumptions that $\overline{(R_3NHCl)} = \bar{c}_{R_3NHCl}, K_M \bar{y}_{R_3NHCl}^2 \bar{y}_{(R_3NH)_2FeCl_5}^{-1} y_{FeCl_3} = K$ is constant, and activity coefficients in the aqueous phase are constant. For more concentrated hydrochloric acid solutions, these assumptions are no longer realistic. The curves $\log D$ vs. $\log a$ have a slope decreasing from ca. 3 near 0.1 M HCl to zero (a maximum in the curve) near 8 M HCl, and further to a negative value at higher hydrochloric acid concentrations[85]. This behaviour is to be expected from equation (7C75), when the aqueous-phase species Fe^{3+}, $FeCl^{2+}$, $FeCl_2^+$, $FeCl_3$, $FeCl_4^-$ and $HFeCl_4$ are considered to be formed consecutively as the hydrochloric acid concentration is increased. However, if the values of $\beta_i'^*$ that have been used successfully to account for TBP extraction and for anion exchange distribution are employed with (7C75), the D curve for a given $\bar{c}_{R_3NHCl}$ cannot be reproduced. Moreover, the log D-vs.-log a curves for different $\bar{c}_{R_3NHCl}$ are not parallel, as they should be according to equation (7C75), so that the distribution data cannot be used to give a definite set of stability data. The dependence of D on $\bar{c}_{R_3NHCl}$ is also not regular, since it varies from approximately second power to approximately first power as the long-chain ammonium chloride concentration is increased, contrary to expectation. This, of course, can be connected with the aggregation of the amine salt, and association of $FeCl_4^-$ with monoionized aggregates (see equation 12A5)[86], but this should then be rather independent of c_{HCl}, but it is not. The only way in which the data conform to theoretical expectations is the dependence, at constant chloride molarity (10 M (H, Li)Cl), of D on c_H^+: in a region where 1:1 complex formation is excepted, $R_3NH^+FeCl_4^-$, D is directly proportional to $\overline{(R_3NHCl)}$, as calculated from equation (7C78).

$$\overline{(R_3NHCl)} \approx \bar{c}_{R_3NHCl}/(1 + K'c_H^+ + y_{\pm HCl}^2) \qquad (12C13)$$

Strong absorption of iron(III) on anion exchangers from concentrated hydrochloric acid has been demonstrated a while ago[87], and several authors have since studied the distribution of iron(III) and its dependence on c_{Fe}, c_{Cl}^-, c_H^+ and the crosslinking of the resin[70,88,89] giving data with fairly good agreement. The distribution coefficients were found to be independent of c_{Fe}, up to a loading as high as 10%, and independent of acidity at constant chloride concentration in (H,Li)Cl solutions up to 9 M[70]. The reflectance spectrum of the loaded resin[90] shows the presence of $FeCl_4^-$ ions only. The distribution data[70] fitted well the expression (6B31) (Figure 12C1)

$$\log D = \log K_{Fe}^* + p\log \bar{a} - \log\left(\sum_{-3}^{1} \beta_i'^*a^i + K_H^*c_H^+ c_{Cl}^- y_{\pm HCl}^2\right) \qquad (12C14)$$

with $p = 1$, as indicated by the spectral evidence, and the term $K_H^*c_H^+ c_{Cl}^- y_{\pm HCl}^2$ arising from the formation of non-sorbable $HFeCl_4$. This species becomes important only above 9 M chloride, with $\log K_H^* = -4.6$ being the equilibrium constant for $FeCl_3 + H^+ + Cl^- \rightleftharpoons HFeCl_4$. The constant β'^*_{-3} could not be evaluated, since the data do not extend to sufficiently low chloride concentrations, and $\log \beta'^*_{-2} \leqq 2.1$ could be estimated only approximately for the same reason, but the higher constants, $\log \beta'^*_{-1} = 1.40$ and $\log \beta'^*_{-1} = =1.92$ (for $FeCl_3 \rightleftharpoons FeCl_2^+ + Cl^-$ and $FeCl_3 + Cl^- \rightleftharpoons FeCl_4^-$ respectively) could be obtained[70]. The constant K_H^* is valid not only for HCl solutions but also for mixed (H,Li)Cl solutions, and not only for anion exchange distribution but also for light-absorption results. The only data for a lower crosslinking than the usual X–10 published[91] are for an X–1 resin, which shows a 19–fold lower D. No comparison can be made with other data for known species, since usually the lower limit for crosslinking dependence studies is X–2 (section 6.C.h). A 19–fold change in D is large for a mononegative ion, but lacking invasion and capacity data for the resin in question, no explanation of this apparent difficulty is possible.

e. Bromide Complexes: Indium(III)

Indium(III) forms in aqueous solutions fairly strong bromide complexes, which have been studied by a variety of methods. The cation exchange method has first been applied by Schufle and Eiland[92] and subsequently by Sundèn[93] and by Carleson and Irving[94]. The first authors[92] worked at pH = 3.8, where indium(III) is extensively hydrolysed. The formation of species such as $InOH^{2+}$, $InBrOH^+$ and polynuclear hydrolysis products,

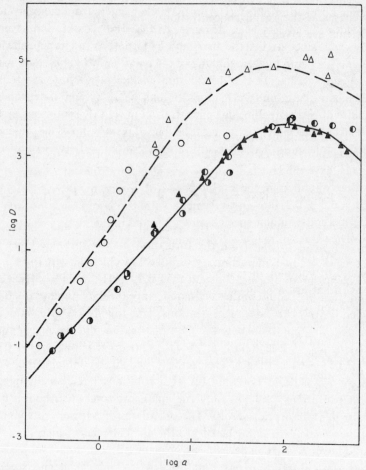

log *a*

Figure 12Cl. The distribution coefficients of tracer iron(III) between TBP, or an anion exchanger (Dowex-1) and aqueous HCl. TBP: ○ ref. 67, △ ref. 65, ———— calculated from equation (12C11); anion exchange: ◑ ref. 87 ◐ ref. 88, ▲ ref. 70 (or 89), ———— calculated from equation (12C14).

some of which are sorbed on the cation exchange resin, above any other considerations concerning the calculations, precludes a meaningful analysis of the distribution data, as has been pointed out[94]. Carleson and Irving[94] used Fronaeus' method (section 5.B.c) to calculate the constants from data obtained at constant acidity (0.691 M) and low constant loading of the resin (0.07%), with up to 0.43 M bromide substituted for perchlorate in the aqueous phase. It was assumed that activity coefficients, as well as hydrogen ion concentrations, remain constant in both phases as the ligand concentration is increased. The distribution coefficients are given by the expression

$$D = D_0(1 + D_1'(\mathrm{Br}^-) + D_2'(\mathrm{Br}^-)^2)/X \qquad (12\mathrm{C}15)$$

which is the same as (5B14), with $D_n' = \beta_n Q_n^{-1}(\mathrm{H}^+)^n(\overline{\mathrm{H}^+})^{-n} = D_n D_0^{-1}$, D_0 being the distribution coefficient in the absence of ligand, and Q_n the selectivity constant for the nth complex against hydrogen ions. Application of equations (5B17) to (5B21) or the equivalent original equations of Fronaeus, gave the constants β_1, β_2 and also the values of Q_1 and Q_2. The values obtained for the β's ($\log \beta_1 = 2.01 \pm 0.01, \log \beta_2 = 3.1 \pm 0.1$, $\log \beta_3 = 3.3 \pm 0.2$) are in agreement with those obtained by other reliable methods; anionic species were not found in the range of measurements. The values of the selectivity constants are reasonable (cf. section 5.B.c. and Table 5B1).

Because of complex formation, the distribution coefficients decrease as the concentration of bromide increases. However, beyond a certain concentration (ca. 2 M for lithium, sodium and magnesium bromides) D increases again somewhat[95], although not to as high values as in the cases of iron(III) and gallium(III) chlorides, for example. Such an increase correlates with sorbability on anion exchangers from concentrated solutions and with extractability by ether. Both conditions hold in the case of the indium(III)–bromide system, as discussed below. The reasons for this behaviour are still unclear, and the phenomenon has been discussed in more detail in section 4.D.

Chelate extraction is akin in many ways to cation exchange, as mentioned above for the case of zirconium(IV) hydrolysis. The reagent α-hydroxynaphthoic acid, dissolved in diisopropyl ether, has been used in this manner[43], to study the indium(III) bromide complexes. For a constant reagent concentration, the term $K'(\overline{\mathrm{HX}})^{m+n} = K$ in equation (7C36) is constant, hence the simple equation

$$X_A = K(H^+)^{-3}D^{-1} \tag{12C16}$$

used by Sundèn, is obtained, provided mixed complexes involving the chelating ligand and bromide ions can be neglected, as proposed by the author. The stability constants obtained: $\log \beta_1 = 1.92 \pm 0.2$ and $\log \beta_2 = 2.60 \pm 0.05$, were confirmed partly by cation exchange[93], which yielded the value $\log \beta_1 = 1.90 \pm 0.05$. These values obtained for a 1 M $NaClO_4$ medium at pH = 2.7–3.0, are somewhat lower than those obtained at higher acidity[94] (0.691 M $HClO_4$), and this again may be due to partial hydrolysis. In the chloride system, a change from 1 M $NaClO_4$ to 1 M $HClO_4$ medium caused an apparent increase in $\beta_1 - (D_1/D_0)$ of almost two-fold[93].

The extraction of indium(III) from hydrobromic acid solutions into solvating solvents, mainly oxygenated solvents, has received a great deal of attention since early days[96,97]. It has been established by analysis[97,99] that the species in the organic phase has the ratio H:In:Br equalling 1:1:4. A Raman spectroscopic study[99] has shown that in diethyl ether and in hexone solutions the species is indeed tetrahedral $InBr_4^-$. Indium(III) is extracted well even from relatively dilute hydrobromic acid solutions, below 1 M, and for a given acidity and solvents, shows higher D for bromide than for chloride, although there is an opposite trend in the stability constants in the aqueous phase.

The most important feature that has been studied with the indium(III)–bromide system is the dependence of D on the metal concentration, c_{In}. The distribution has been studied from carrier-free tracer up to macro concentrations (ca. 0.7 M) of indium(III)[96–105]. It has been found that whereas in solvents of low dielectric constant (such as ether) D increases with c_{In}, in solvents of higher dielectric constant D decreases with c_{In}, but shows plateaus at both high and low values of c_{In}, at constant c_{HBr}. This behaviour has been fully discussed in sections 7.C and 9.B. The explanation offered by Irving and Rossotti[101] for the low dielectric constant solvents, of association in the organic phase causing an increase in D is reasonable. However their explanation for the observed behaviour with high dielectric constants solvents[98], of dimerization in the aqueous phase at high c_{In} to $In_2Br_6^{2-}$ or $In_2Br_7^{3-}$, is not acceptable. This dimerization could not be proven by other methods whereas an alternative explanation has been shown to hold: ionic dissociation in the organic phase, as affected by the common-ion effect (section 7.C.e). The equations used by Poskanzer, Dietz, Irvine and coworkers[103,104] to explain the results

$$D = \dot{K}_{DIn} G_{In}^{-1} \left(K_{DBr}(\overline{HBr}) + K_{DIn} c_{In} \right)^{-1/2} \tag{12C17}$$

are similar to (7C65), but neglect any non-dissociated $HInBr_4$ in the organic phase, and do not express explicitly the dependence of $(\overline{HBr})$ on c_{HBr}. At constant c_{HBr} and not too large c_{In}, equations (12C17) and (7C65) give the same dependence of D on c_{In}, shown in Figure 7C6. It is possible to determine the dissociation constant for HBr in the organic phase from the distribution data at such low c_M that D is independent of it, and for bis(β-chlorethyl) ether log $K_{DBr} = -5.1$[104]. This value is confirmed by parallel but independent measurements with gallium(III).

The dependence of D on c_{HBr}[100,104,106] and on c_H^+ and c_{Br}^- individually[100], has been studied. Dietz[104,106] used the dependence of D on $a\,(= c_{HBr} y_{\pm HBr})$ for 0.25 to 12 M HBr, and Poskanzer's equations[107], which reduce essentially to equation (7C58), to calculate the average ligand number $\tilde{n}$ in the aqueous phase

$$\tilde{n} = 5 - (\partial \log D / \partial \log a)_{c_M \to 0} \tag{12C18}$$

Similar values of $\tilde{n}$ were obtained from data for extraction with nitrobenzene and with bis(β-chlorethyl) ether, confirming the validity of the treatment. Using values of β_1 and β_2 from the literature, values of β_3^* and β_4^* could be calculated from the $\tilde{n}$ data. The values obtained (log $k_3^* = -1.22$ and log $k_4^* = -1.92$) are rather low, compared with the constants for the lower complexes. Irving and Rossotti[101] made similar calculations, but used c_{HBr} instead of a, which does not make an appreciable difference at the range of concentrations used (0.01–1.0 M HBr). Most of their data were obtained at a constant ionic strength made up with sodium perchlorate or potassium nitrate (media from which indium(III) is not extracted by hexone), and some of the data were obtained at constant c_{H+} (and varying c_{Br-}) or constant c_{Br-} (and varying c_{H+}). The equations they use are the same as equation (7C59). The slope = $(\partial \log D/\partial \log c_H^+)c_{Br-} = p$, for the extraction of $H_p^+ InBr_{3+p}^{p-}$ from solutions containing only $InBr_n^{3-n}$ species ($n = 0, 1...N$), was indeed $+ 1$, expected for extraction of $H^+InBr_4^-$ at acidities above ca. 0.1 M, but approached zero at low acidities, signifying extraction of species such as $InBr_3$ and/or $Na^+InBr_4^-$ (or $K^+InBr_4^-$). The slopes $(\partial \log D/\partial \log c_{Br-})c_{H+}$ $= 3 + p - \tilde{n}$ were 1.6 and 2.4 at 0.0016 M H^+ (where $p \simeq 0$) and at 0.25 M H^+ (where $p = 1$) respectively, yielding essentially the same values of $\bar{n}$ (1.4 and 1.6 respectively) for the range of bromide ions used (0.1–1.0 M). The relatively low $\bar{n}$ value agrees with the values computed with the constants given by Dietz above, but not with the higher $\bar{n}$ values (1.6 to 2.6) obtained from the constants suggested by Irving and Rossotti (log $k_3 \sim 0.2$, log $k_4 \sim -1.3$) nor with the values of $\bar{n}$ obtained from the slope (∂ log D/∂ log c_{HBr}) = 5 − $\bar{n}$: about 2.3. Thus results obtained in ionic media which contain foreign ions (Na^+, K^+, NO_3^- or ClO_4^-) are not in agreement with those obtained in their absence (pure hydrobromic acid solutions). This discrepancy has not been satisfactorily resolved by Irving and Rossotti[100], even taking into account coextraction of $InBr_3$ and/or $Na^+InBr_4^-$ (or $K^+InBr_4^-$).

The effect of the solvent used on the magnitude of the extraction constant K_M (equation 7C55a) has been studied by Irving and Rossotti[101,108], among others[48,103–107]. In the absence of complicating factors, the distribution curves log $D = f$ (log c_{HBr}) should be parallel, shifted by differences Δ log K_M for various solvents. Since, however, the hydrobromic acid is also extracted, and the species $\overline{H^+InBr_4^-}$ and $\overline{H^+Br^-}$ dissociate to different degrees in various solvents, the curves become non-parallel. The limiting slopes expected from (12C18), +5 at $c_{HBr} \rightarrow 0$ and + 1 or 0 for $c_{HBr} \rightarrow \infty$, depending on whether $InBr_4^-$ or $H^+InBr_4^-$ are the ultimate aqueous-phase species, are found only in some cases. In other cases negative slopes are found at high acidities, which are due to increasing miscibility of the phases, and to a breakdown of the simplifying assumptions on which (12C18) is based. For very similar solvents, such as methyl isobutyl ketone (hexone) (A) and methyl isopropyl ketone (B), the simple relationship

$$\log D = \log D_A - x_B(\log D_A - \log D_B) \qquad (12C19)$$

holds. With more dissimilar diluents for hexone, such as benzene or even cyclohexanone, the linearity persists only up to $x_B = 0.3$.

Little information has been published concerning the extraction of indium(III) from hydrobromic acid solutions with long-chain ammonium bromides. A 10% solution of Amberite LA–1 (a secondary amine) in xylene extracts indium(III) well, $D > 50$ above 2 M HBr. As with the solvating solvents, D for hydrobromic acid is considerably larger than for hydrochloric acid solutions with the amine extractant[109].

The same relative order of bromide and chloride media is also observed with resin anion exchangers applied to indium(III)[110,111]. As with many other systems, sorption is higher from lithium bromide than from hydrobromic acid solutions[110], and addition of alcohol enhances the distribution coefficients[111,112]. A detailed investigation of the sorption of indium(III) from hydrobromic acid[110] and lithium bromide[113] solutions on Dowex-1 X 10 has been made by Andersen and Knutsen. They used the Marcus-Coryell method (section 6.B.c) to analyse their data for the lithium bromide solutions. They determined the invasion function, or effective ligand activity in the resin, $\bar{a}$, by direct measurement (equation 6B28), and found it to approximate to that for lithium chloride. The equation they used to calculate stability constants (12C20) is identical with (6B31) with the value $p = 3$ for the species $InBr_6^{3-}$ predominating in the resin, and $N = 6$ the coordination number.

$$\log D_{In} = \log K_{In}^* + p \log \bar{a} - \log \sum_{-2}^{N-3} \beta_i'^* a^{\,i} \qquad (12C20)$$

The slopes $(\partial \log D_0 / \partial \log a) = (\partial (\log D - 3 \log \bar{a}) / \partial \log a)$, equations (6B32 and 33), i.e. the average charge $\bar{\imath}$ of the species, were found to range from $\sim +1.6$ to -3.00 in the range 0.1–9.3 M LiBr. Stability constants $\beta_{-2}'^*$ to $\beta_{+3}'^*$ could be computed from the data according to equation (6B31). The hexacoordinated species $InBr_6^{3-}$ in the resin is, however, not very probable, in view of the small tendency of indium(III) to add bromide ligands beyond the third. The data can be expressed in as good a manner by the expression (12C20) with $p = 1$ and $N = 4$, signifying a species $InBr_4(H_2O)_2^-$ in the resin, analogous to the similar species in the chloride system[114], and the same species as the ultimate species in the aqueous phase. The values used by Dietz[104,106] for the constants $\beta_i'^*$ permit a reasonable fit of the data, and the slightly modified values, $\log k_2^* = 1.4$, $\log k_3^* = -1.3$ and $\log k_4^* = -2.2$ fit the data well. That the same data can be fitted adequately by two sets of constants, based on different values of p, is explicable through the rough proportionality of $\bar{a}$ and a at high concentrations. In this range $\log \bar{a} \sim 0.91 \log a$, hence the data can be fitted with the same value of $N = p - 3$, whatever the individual values of N and p. Only an independent determination of N or of p can resolve the problem.

f. Nitrate Complexes: Dioxo Uranium(VI) (Uranyl)

A vast amount of work concerning the ion exchange, and especially the solvent extraction of uranyl nitrate has been published, and a large portion of it deals explicitly with the complex species formed, mainly in the organic phase. It is possible here to review only some of the publications, and emphasis is given to those studies which yield information, through the application of the ion exchange and solvent extraction investigation methods, on the complex species formed by the uranyl cation in aqueous nitrate solutions.

The results of the cation exchange study of Banerjea and Tripathi[115] cannot be relied on for the following reasons. The constants obtained, $k_1 = 0.04$, $k_2 = 1.0$, $k_3 = 8.0$ have the very anomalous trend for complexes between a cation and an anion, of $k_1 < k_2 < k_3$. A rather narrow range of nitrate concentration (about a factor of two) and a very considerable substitution of the 1 M sodium perchlorate medium (from ca. 0.55 to ca. 0.93 M) prevent the distinction of nitrate complexing from general non-ideality, and non-constancy of activity coefficients (section 1.E), in view of the small total effect measured (only ca. 36% change in D at the highest nitrate concentration). Finally, the high pH used, about 3, is in a range where considerable hydrolysis to soluble products occurs (Table 3C3). A reexamination of the experimental distribution results[115] shows that the data can be reproduced quite well with the single parameter $\beta_1 = 0.3 \pm 0.1$, without requiring three constants for three species. In a study made in the absence of a constant ionic medium, Strelow[116] found slopes d log D/d log c_{HNO_3} of ca. -1.4, -1.7 and -2.2 for uranyl, calcium(II) and mercury(II) cations respectively, in the range 0.1–2.0 M HNO_3. A slope of -2 is expected when no complexation occurs, and when ideal behaviour is assumed, while a steeper slope is expected when complex formation occurs (equation 5B23, with $D'_i = 0$ for $i > 0$ and $\beta_i > 0$). Compared with mercury(II), which forms relatively strong complexes, and with calcium(II), which forms weak ion-associates, uranyl seems to be complexed very weakly indeed. Since, however, the effects of non-ideality have not been considered, the observed slope does not exclude some association.

Further information has been obtained from the extraction of uranyl cations with TTA, 0.50 M in benzene, from solutions 0.05 M in acid, and having a constant medium of 2.00 M (sodium perchlorate and nitrate, with up to 98% substitution)[117]. Here again, as the authors themselves point out, it is difficult to separate specific complexation from general non-ideality (non-constancy of activity coefficients), the maximal change in D being ca. 50%, even for up to 7 M $NaNO_3$ solutions. The constant obtained for 25°C is $\beta_1 = 0.24$, in good agreement with the rough estimate given above from cation exchange data, and with results obtained by other methods not involving distribution measurements.

A great many distribution studies involving the extraction of uranyl nitrate with solvating solvents have been published. Only a fraction of them, however, deal specifically with the complex species present in the aqueous phase. Some of them deal with the composition of the extracted species, its solvation by water and the solvent, and its interaction with excess nitrate in the organic phase.

It is doubtful whether publications which consider uranyl nitrate complex formation in the aqueous phase, as affecting the extraction with solvating solvents, are independent sources for the stability constants of these complexes. It can only be said that the distribution data are consistent with the hypothesis that such species are formed, with stability constants of the approximate magnitude $\beta_1 = 0.2$ and $\beta_2 = 0.05$, the latter of importance only in work involving concentrations above ca. 2–3 M HNO_3 or another nitrate. Jenkins and McKay[118] have considered the effect of the formation of $UO_2NO_3^+$ on the ionic strength of uranyl nitrate solutions at high concentrations, or of solutions containing another nitrate, hence on the activity coefficients, and finally on the distribution coefficients. A qualitative argument by Bernström and Rydberg[119] considering the effect of uranyl ion complexing, attempts to explain deviations observed in the extraction of uranyl nitrate from nitric acid with TBP. More explicitly, Hesford and McKay[120], and subsequently Marcus[121], define the function

$$F = (1 + k_1 Y_1 (NO_3^-))^{-1} \qquad (12C21)$$

with $k_1 = 0.2$ being the stability constant of $UO_2NO_3^+$, and Y_1 a ratio of activity coefficients, which Marcus[121] approximated by $(y_{\pm UO_2(NO_3)_2}^t)^3 \times y_{\pm HNO_3}^{-2}$, $y_{\pm X}^t$ being the activity coefficient of trace X in the nitrate solution, and assuming for $X = UO_2NO_3^+ NO_3^-$, $y_{\pm X}^t = y_{\pm HNO_3}$, both being 1:1 electrolytes. Values of $y_{\pm X}^t$ for uranyl nitrate in nitric acid and other nitrates have been published[118], and values of the nitrate ion concentration (NO_3^-) are known for nitric acid solutions (section 3.C.b), hence the function F can be computed. The above authors have shown that if the total uranyl ion concentration is multiplied by F, and introduced in the appropriate expressions in place of the free uranyl ion concentration, a good agreement with distribution data for nitric acid up to 7 M is obtained. Hardy[122] showed that up to 4 M HNO_3 the extraction of uranyl nitrate by TBP conforms to that expected if a similar function

$$F = (1 + \beta_1(NO_3^-) + \beta_2(NO_3^-)^2)^{-1} \qquad (12C22)$$

is used, with the above-mentioned stability constants β_1 and β_2, and without activity coefficient corrections. He showed that such corrections (as in equation 12C21 for β_1) would account, at least partly, for the deviations in the range 5–10 M HNO_3.

Uranyl nitrate is soluble in numerous organic solvents as such, or as a hydrated species, and in both cases it is solvated by the solvent. The information of interest is how many molecules of solvent and water solvate the uranyl nitrate, whether and how these numbers change with conditions, and what other interactions the uranyl nitrate undergoes in the organic phase. A detailed study of solutions in such solvents as alcohols, ketones and ethers has been published by McKay and co-workers[118,124–128]. It was found that anhydrous uranyl nitrate does not dissociate appreciably to ions in these solvents[123,124], but the hexahydrate is more dissociated. Many solvents were found to be 'linear' in the sense that the water content increased linearly with the amount of uranyl nitrate extracted, with the slope being usually four[124,126], signifying species $UO_2^{+2}(NO_3)_2^-(H_2O)_4S_p$, with a solvation number p, as yet undetermined. Alcohols, however, are 'non-linear', showing generally slopes lower than four[124]. The 'linear' solvents show a conformity of the distribution coefficients with (7C50), and at sufficiently low concentrations of uranyl nitrate and the solvent in a suitable diluent[120] with the simple square law[125]

$$D = P K_S c_M^2 \bar{c}_S^p \qquad (12C23)$$

where PK_S is a constant. The solvation number p is then obtained from the slope $(\partial \log D / \partial \log \bar{c}_S)_{c_M}$ (7C41, 7C49). At high c_M and undiluted solvents, the concentration of free solvent, that is that not bound by uranium, must be used. Diethyl ether shows an anomalous behaviour, exhibiting conformation to (12C23) at such high concentrations $c_M \sim 1$ M, that $\bar{y}_M$ must be much greater than unity, and $\bar{y}_S$ much smaller. This is explained by the fact that a_W and y_M for the aqueous phase, which appear in (7C50), are strongly modified by the large solubility of the ether in the aqueous phase[127] (section 9.A). The quantities $\bar{y}_M$ and $\bar{y}_S$ can be estimated from the distribution data[125,128], with the help of auxiliary data such as solubilities of the solvents in the aqueous phase. A great part of the stoichiometric deviations can be ascribed to solvation, with a constant value of p, being two for many solvents, but nearer three for others. Another part is a remaining non-ideality of the organic phase, containing hydrated and solvated uranyl nitrate and excess water and solvent[128]. Changes in activities in the aqueous phase because of dissolved solvent are negligible, except for diethyl ether, as mentioned above[128]. The effect of a non-extractable nitrate is to furnish nitrate ions, and to modify activity coefficients in the aqueous phase, so that equation (7C46) is followed[118], with $X^{-1} = F^{-1}$ as defined in (12C21), to take account of complexation.

The solvent TBP has received special attention. Again, uranyl nitrate is unionized in this solvent[129], and is solvated by a definite number of solvent molecules: $p = 2$. This result was obtained from the TBP concentration dependence[120,121], according to (7C41). Since, however, a part of the TBP is bound to nitric acid, when the uranyl nitrate is extracted from this medium, the differentiation ought to be done with respect to free-ligand concentration. A corrected distribution coefficient D_0 can be calculated (equation 7C53, where for $HNO_3.TBP$ $t = 1$), or else the variation of $(\bar{S})$ can be taken explicitly into account

$$(\partial \log D/\partial \log \bar{c}_S)_{(H^+),(NO_3^-)} =$$

$$p[1 - (K_H a^2_{\pm HNO_3}\bar{c}_S)(1 + K_H a^2_{HNO_3}\bar{y}_S\bar{y}^{-1}_{HNO_3S})^{-1}(\partial(\bar{y}_S\bar{y}^{-1}_{HNO_3S})/\partial\bar{c}_S)_{(H^+),(NO_3^-)}$$

$$+ d\log\bar{y}_S/d\log\bar{c}_S] - d\log\bar{y}_{UO_2(NO_3)_2} \cdot _{pS}/d\log\bar{c}_S \qquad (12C24)$$

At low concentrations of nitric acid, where the term involving $a^2_{\pm HNO_3}$ is negligible, D is too small to measure unless $\bar{c}_S$ is appreciable. But then d $\log \bar{y}_S$/d $\log \bar{c}_S$ is no longer small compared with unity, and can be approximated by $\lambda_S\bar{c}_S$, while the last term becomes $\lambda_{UO_2(NO_3)_2pS}\bar{c}_S$ so that (12C24) becomes

$$(\partial \log D/\partial \log \bar{c}_S) = p + (p\lambda_S - \lambda_{UO_2(NO_3)_2pS})\bar{c}_S \qquad (12C25)$$

At high nitric acid concentrations it has been shown[120] that a similar expression holds, but with λ_{HNO_3S} replacing λ_S, while at intermediate acidities, where a part of the TBP is free, and the rest is bound to nitric acid, the complicated equation (12C24) must be used. Because of the many unknown quantities occuring in (12C24) and (12C25), the value $p = 2$, which has been obtained from experiments not involving co-extraction of nitric acid, has been assumed to hold also for extractions from this acid. With this assumption, it has been shown[121] that the expression

$$D = K\bar{c}_S^2\,\bar{y}_S(\text{NO}_3^-)^2(y^t_{\pm\text{UO}_2(\text{NO}_3)_2})^3/F(1 + K_H a^2_{\pm\text{HNO}_3}\bar{y}_S\,\bar{y}_{\text{HNO}_3\text{S}}^{-1})^2 \quad (12\text{C}26)$$

holds for the extraction of tracer uranium(VI) with the value $K_H\bar{y}_S\bar{y}_{\text{HNO}_3\text{S}} \simeq 0.16\ \text{M}^{-1}$ obtained from published work[120], $y^t_\pm$ and F having the same values as used before (p.899), and $\bar{y}_S$ obtained from distribution measurements of TBP itself between the two phases, assuming the aqueous solutions to be ideal with respect to the very low concentrations of TBP.

The solvation number of uranyl nitrate in TBP has been obtained not only from the concentration dependence of D for diluted TBP, but also from the dependence of the extraction at macro concentrations of uranyl nitrate and undiluted TBP on the exhaustion of free TBP. The same value, $p = 2$, has been obtained[130].

It has been examined as to whether the uranyl nitrate extracted into TBP reacts with other reagents in this solvent. The visible spectrum does not reveal any reaction with excess nitric acid to form $\text{H}^+\text{UO}_2(\text{NO}_3)_3^-$ ion-pairs[129]. However, addition of lithium nitrate does cause the spectrum of the $\text{UO}_2(\text{NO}_3)_3^-$ ion to appear[131]. This effect is connected with the availability of nitrate ions, which increases the less strongly the added nitrate binds TBP[132]. The strength of this interaction can be estimated from the shift to higher energy of the nitrate peak in the u.v. in the TBP solvent[132]. The addition of a polar substance, such as an amide, to the solution of uranyl nitrate in TBP also causes changes: the degree of ionic dissociation increases, and the spectrum shows that some replacement of nitrate and TBP by amide takes place[133]. Acidic extractants are able to replace all of the nitrate or the TBP, and mixed complexes and solvates are formed, e.g. $\text{UO}_2(\text{NO}_3)((\text{C}_4\text{H}_9\text{O})_2\text{PO}_2)$ (TBP)[134]. Essentially similar species and interactions have been observed for a great number of other solvents: acid phosphorylated esters, neutral phosphorylated esters and other solvating solvents, under conditions where direct solvation of the uranyl ion is of prime importance (section 11.B).

A considerable number of publications deal with the extraction of uranium(VI) from nitrate solutions with long-chain ammonium nitrate solutions. It is generally accepted now that the slopes ($\partial \log D/\partial \log \bar{c}_R)_{(\text{H}^+),\,(\text{NO}_3)}$, where $\bar{c}_R$ is the concentration of the extractant in the organic phase, vary between one and two, and may be non-integral,

♥

depending on the extractant and the diluent[135-144]. The nature of the uranium(VI) species in the organic phase cannot thus be inferred from the slopes. A recent systematic study[143] has shown that the stoichiometry of the extracted species $(R_iH_{4-i}NNO_3)_nUO_2(NO_3)_2$, as obtained from the slopes[143] and from loading data in nitrobenzene diluent[141,143], is $n = 1$ for $i = 4$ (i.e. quaternary ammonium extractants), $n = 2$ for $i = 3$ and 2, and has indefinite values $n > 2$ for $i = 1$, and for any i except 4 in other solvents. This stoichiometry, however, does not define the nature of the species formed from uranyl nitrate upon extraction into the organic phase. This information has been obtained from examination of the absorption spectra of the organic extracts. The compound tetraethylammonium trinitratodioxouranate(VI) has been prepared, and its spectrum in nitromethane solution measured[136]. This spectrum, which has been definitely assigned to the $UO_2(NO_3)_3^-$ ion, is very similar to the spectra of the extracts obtained with extractants with $i = 2, 3$ or 4 [136,137,139,143]. The species should, therefore, be formulated $(R_2NH_2)^+[UO_2(NO_3)_3]^-$. $(R_2NH_2^+NO_3^-)$, $(R_3NH)^+[UO_2(NO_3)_3]^-$. $.(R_3NH^+NO_3^-)$ and $(R_4N)^+[UO_2(NO_3)_3]^-$. The same spectrum is also obtained on the dissolution of solid uranyl nitrate hexahydrate in a solution of trilaurylammonium nitrate in xylene. On the other hand, for a primary amine extractant, i.e. $i = 1$, the spectrum is different, and corresponds to the species $UO_2(NO_3)_2.n(RH_3N^+NO_3^-)$.

Under a wide range of conditions the uranium(VI) on extraction into the organic phase brings with it just two nitrate ions. Specifically, no evidence for extraction of $HUO_2(NO_3)_3$, if at all formed in the aqueous phase, could be found, either by direct analysis[142], or by the observation of the OH stretching frequency in the infrared spectrum of an extract from nitric acid solutions[144]. In evident contradiction to this is the observation that two moles nitric acid per mole uranium are salted into the organic phase by high concentrations of sodium nitrate, while none is salted-out in the absence of uranium[144].

The tertiary ammonium nitrate extractants have been most widely studied. Accepting the formulation $(R_3NH)^+[UO_2(NO_3)_3]^-$. $(n - 1)$ $(R_3NH^+NO_3^-)$ for the extracted species, the value of n and the bonding of the uranium-containing ion-pair with the excess extractant ion-pairs have still to be determined. In polar solvents, which themselves interact with the extractant (Chapter 10) and monomerize it, the value of n is two, as for example for nitrobenzene[143]. In toluene, on the other hand, the reactions

$$\text{UO}_2^{2+} + 2\text{NO}_3^- + 2\overline{\text{R}_3\text{NH}^+\text{NO}_3^-} \rightleftharpoons \overline{(\text{R}_3\text{NHNO}_3)_2\text{UO}_2(\text{NO}_3)_2}   \quad (12\text{C}27\text{a})$$

and

$$\text{UO}_2^{2+} + 2\text{NO}_3^- + 2\overline{(\text{R}_3\text{NH}^+\text{NO}_3^-)_2} \rightleftharpoons \overline{[(\text{R}_3\text{NHNO}_3)_2]_2\text{UO}_2(\text{NO}_3)_2} \\ (12\text{C}27\text{b})$$

have been postulated in order to account for the distribution data[142]. The species written on the right-hand side specify only the stoichiometry, not the structure, so that they do not contradict the findings summarized above. The reaction (12C27a) is in line with $n = 2$, in the above formulations, and with a slope of two in the distribution curve, if the extractant is monomeric. The formation of extractant dimers, and their participation in the extraction is taken into account by equation (12C27b). It has not yet been settled whether long-chain ammonium salt aggregates extract large anionic metal complexes without disaggregating (section 10.C). The evidence for (12C27b) comes partly from loading information, the four extractant (monomer) molecules bound per uranium from the total extractant concentration.

In conclusion, the species $(\text{R}_3\text{NH})^+[\text{UO}_2(\text{NO}_3)_3]^-$ has been proved spectrophotometrically, and has been found under certain circumstances to interact with excess tertiary ammonium nitrate. The exact nature of this interaction, the number of molecules involved, and their dependence on the nature of the extractant, the diluent, the temperature and the bulk aqueous nitrate, have not yet been clarified.

The dependence of the distribution coefficients of uranium(VI) on the nitrate concentration has been studied for several bulk cations. With 0.1 M HNO_3 and varying sodium nitrate concentrations D increases with the nitrate concentration[144], while for nitric acid solutions a maximum at 6–7 M HNO_3 is observed[135,137,139,144]. Scibona and coworkers[139] have estimated the relative concentrations of uranium(VI) complex species in the aqueous phase, and concluded that at high nitric acid concentrations the formation of the non-extractable acido complex $\text{HUO}_2(\text{NO}_3)_3$ causes the decline in D. They based their calculations on the debatable results of Banerjea and Tripathi[115], but concluded correctly that nitrate complexing is not very important, so that for calcium nitrate solutions a slope of two for $(\partial \log D / \partial \log c_{\text{NO}_3^-})_{(\text{R}_3\text{NHNO}_3),(\text{H}^+)}$, is found as expected according to equation (7C76b). The assumption of the formation of a non-extractable acido complex, which has not been confirmed independently, is, however, unnecessary, since competition by extraction of excess acid (to form the binitrate, cf. section 10.B) accounts satisfactorily for the observed results. No direct evaluation of distribution data in terms of complexes in the aqueous phase has been published.

The sorption of uranium(VI) on anion exchangers from nitrate solutions has been reported some years ago by Kraus and Nelson[145], and since then by a few other authors[146–150]. The values of D for nitric acid solutions are low, reaching ca. 20 at high concentrations[145–147,149], but D reaches very high values for non-acid nitrates[147,148], up to 8000 in concentrated aluminium nitrate. On an equivalent basis, that is at equal nitrate ion concentration, e.g. 4 N, the order of cations causing increasing D is NH_4^+; Ca^{2+}; Na^+; Fe^{3+}; Cu^{2+}; Zn^{2+}; Ni^{2+}; Al^{3+}; Li^+. This is approximately the order of increasing hydration ability of the cations (section 1.B), and this property has been suggested[147] to be chiefly responsible for the order of cations. At a given nitrate ion concentration, the substitution of ammonium ions for aluminium ions decreases D considerably, but the substitution of hydrogen ions decreases D even more. The effect of the addition of perchloric acid is also considerable and as with addition of nitric acid may be due to the formation of non-exchanging binitrate ions in the resin (section 6.C).

The nature of the uranium(VI) species in the anion exchanger has been studied spectrophotometrically by Ryan[150]. The spectrum of the partly loaded resin, corresponds to the species $UO_2(NO_3)_4^{2-}$ (as shown by its similarity to that of $[(C_2H_5)_4N]_2^+[UO_2(NO_3)_4]^{2-}$) modified to some extent by the presence of the species $UO_2(NO_3)_3^-$. The relative amounts of the two species depend little on the crosslinking, and not at all on the bulk cation in solution, but they do depend on the capacity of the resin and the degree of loading. The higher the loading, the more of the mononegative species is formed, at the expense of the dinegative species, as is expected. The presence of water in the resin makes no difference, the same spectra were obtained for resins loaded from acetone or nitromethane solutions of uranyl nitrate hydrates, or of tetraethylammonium tetranitratodioxouranate, or for dried resins, previously loaded from acetone. The loading curve (section 6.A) shows some levelling off at 0.5 moles uranium per equivalent of resin (corresponding to $UO_2(NO_3)_4^{2-}$), but continues to increase towards loading of $UO_2(NO_3)_3^-$.

The anion exchange results for aqueous solutions are not sufficient to permit analysis in terms of the complex species formed in the aqueous phase.

906   *Ion Exchange and Solvent Extraction of Metal Complexes*

Because of the good solubility of uranyl nitrate in organic solvents, including water-miscible solvents, it has been of interest to study the distribution of uranium(VI) between such solutions and anion exchange resins[151-156]. For solutions of uranyl nitrate, without excess nitric acid, sorption is higher for water-soluble solvents (e.g. ethanol or propanol) than for immiscible solvents (e.g. hexone or TBP)[151]. The better the extraction of uranium into such solvents, in general, the less it is sorbed on the resin. Alcohols enhance the distribution of uranium(VI) from nitric acid[152-156], as does anhydrous acetic acid[155]. It is difficult to load the resin completely from mixed solvents: from 0.1–0.2 M HNO$_3$ in 96–98% propanol, uranium loads only up to one quarter of a mole per equivalent of resin. This has been interpreted in terms of the formation of the species UO$_2$(NO$_3$)$_6^{4-}$ in the resin[153,154]. There is no justification for this unusually high ligand number and charge (low charges characterize species sorbed from organic solvents, section 6.D). The spectral data for acetone solutions do not support such a species. A slow rate of sorption in a relatively unswollen resin, at the high organic solvent content of the system, can account for the low sorption observed. The increase of D with the nitric acid concentration, at relatively low concentration of alcohol in the solution phase, has been cited as evidence for an exchange mechanism being responsible for the sorption of the uranium, while at high alcohol concentrations in the solutions, sorption of the neutral complex takes place[154]. The evidence presented for these mechanisms of sorption is not very definite, although the conclusions are plausible.

D. REFERENCES

1. R. E. Connick and W. H. McVey, *J. Am. Chem. Soc.*, **71**, 3182 (1949).
2. R. E. Connick and W. H. Reas, *J. Am. Chem. Soc.*, **73**, 1171 (1951).
3. A J. Zielen and R. E. Connick, *J. Am. Chem. Soc.*, **78**, 5785 (1956).
4. A. J. Zielen, *Ph.D. Thesis*, Univ. California, Berkeley, 1953.
5. O. S. Solovkin, *Zh. Neorg. Khim.*, **2**, 611 (1951).
6. V. V. Fomin and E. P. Maiorova, *Zh. Neorg. Khim.*, **1**, 1703 (1956).
7. V. M. Peshkova, N. V. Melchakova and S. G. Zhemchuzhin, *Zh. Neorg. Khim.*, **6**, 1233 (1961).
8. B. A. J. Lister and L. A. McDonald, *J. Chem. Soc.*, 4315 (1952).
9. J. T. Benedict, W. C. Schumb and C. D. Coryell, *J. Am. Chem. Soc.*, **76**, 2036 (1954).

10. E. M. Larsen and P. Wang, *J. Am. Chem. Soc.*, **76**, 6223 (1954).

11. V. I. Paramonova and A. N. Serveev, *Zh. Neorg. Khim.*, **3**, 215 (1958).

12. A. K. Babko and B. I. Nabivanets, *Dopovidi Akad. Nauk Ukr. RSR*, 646 (1960).

13. A. I. Zhukov, E. I. Kazantsev and V. N. Onosov, *Zh. Neorg. Khim.*, **7**, 915 (1962).

14. A. I. Zhukov, V. N. Onosov, V. Ya. Kudjakov and B. M. Serzeev, *Zh. Neorg. Khim.*, **8**, 871 (1963).

15. M. Lederer, *J. Chromatog.*, **2**, 209 (1959).

16. C. Keller, *Radiochim. Acta*, **1**, 147 (1963).

17. J. R. Wering, K. B. Higbie, J. T. Grace, B. F. Speece and H. L. Gilbert, *Ind. Eng. Chem.*, **46**, 644 (1954).

18. R. Bock and M. Herrmann, *Z. Anorg. Allgem. Chem.*, **284**, 288 (1956).

19. P. C. Stevenson and H. G. Hicks, *Anal. Chem.*, **25**, 1517 (1953).

20. A. T. Casey and A. G. Maddock, *J. Inorg. Nucl. Chem.*, **10**, 289 (1959).

21. F. L. Moore, *Anal. Chem.*, **35**, 1032 (1963).

22. L. P. Varga, W. D. Walkley, L. S. Nicolson, M. L. Madden and J. Patterson, *Anal. Chem.*, **37**, 1003 (1965).

23. L. P. Varga and H. Freund, *J. Phys. Chem.*, **66**, 21 (1962).

24. K. A. Kraus and G. E. Moore, *J. Am. Chem. Soc.*, **73**, 13, 2900 (1951).

25. Yu. A. Buslaev and N. S. Nikolaev, *Zh. Neorg. Khim.*, **4**, 465 (1959).

26. J. P. Faris, *Anal. Chem.*, **32**, 520 (1960).

27. M. K. Nikitin, *Dokl. Akad. Nauk SSSR*, **148**, 595 (1963).

28. E. J. Dixon and J. B. Headridge, *Analyst*, **89**, 185 (1964).

29. E. A. Huff, *Anal. Chem.*, **36**, 1921 (1964).

30. Y. Marcus, *Coord. Chem. Rev.*, **2**, 195, 257 (1967).

31. K. A. Kraus, D. C. Michelson and F. Nelson, *J. Am. Chem. Soc.*, **81**, 3204 (1959).

32. H. Titze and O. Samuelson, *Acta Chem. Scand.*, **16**, 678 (1962).

33. F. Nelson, T. Murase and K. A. Kraus, *J. Chromatog.*, **13**, 503 (1964).

34. J. M. White, P. Kelly and N. C. Li, *J. Inorg. Nucl. Chem.*, **16**, 337 (1961).

35. G. H. Morrison and H. Freiser, *Solvent Extraction in Analytical Chemistry*, J. Wiley & Sons, New York, 1957, p. 63.

36. R. M. Diamond and D. G. Tuck, *Progr. Inorg. Chem.*, **2**, 109 (1960).

37. P. A. McCusker, and S. M. S. Kenward, *J. Am. Chem. Soc.*, **81**, 2976 (1959).

38. V. V. Fomin and A. F. Morgunov, *Zh. Neorg. Khim.*, **5**, 1385 (1960).

39. A. H. Laurene, D. E. Campbell, S. E. Wiberley and H. M. Clark, *J. Phys. Chem.*, **60**, 901 (1956).

40. S. Kato and K. Ishi, *Sci. Papers Inst. Phys. Chem. Res. (Tokyo)*, **36**, 82 (1939).

41. N. H. Nachtrieb and J. G. Conway, *J. Am. Chem. Soc.*, **70**, 3547 (1948).
42. D. E. Metzler and R. J. Myers, *J. Am. Chem. Soc.*, **72**, 3776 (1950).
43. Yu. A. Zolotov, S. V. Seryakova, I. J. Antipova-Karataeva, Yu. I. Kutsenko and A. V. Karyakin, *Zh. Neorg. Khim.*, **7**, 1197 (1962); S. V. Seryakova, Yu. A. Zolotov, A. V. Karyakin, L. A. Gribov and M. E. Zubrilina, *Zh. Neorg. Khim.*, **7**, 2013 (1962).
44. H. L. Friedman, *J. Am. Chem. Soc.*, **74**, 5 (1952).
45. L. A. Woodward and M. J. Taylor, *J. Chem. Soc.*, 4473 (1960).
46. E. Rabinowitch and W. C. Stockmayer, *J. Am. Chem. Soc.*, **64**, 335 (1942).
47. B. V. Nekrasov and V. V. Ovsyankina, *Zh. Obshch. Khim.*, **11**, 573 (1941).
48. R. L. Erickson and R. L. McDonald, *J. Am. Chem. Soc.*, **88**, 2099 (1966).
49. J. Axelrod and E. H. Swift, *J. Am. Chem. Soc.*, **62**, 33 (1940).
50. V. V. Fomin, P. A. Zagorets, A. F. Morgunov and I. I. Tertishnik, *Zh. Neorg. Khim.*, **4**, 2276 (1959).
51. H. Specker and M. Cremer, *Z. Anal. Chem.*, **167**, 110 (1959).
52. S. K. Majumdar and A. K. De, *Talanta*, **7**, 1 (1960).
53. R. W. Dodson, G. J. Forney and E. H. Swift, *J. Am. Chem. Soc.*, **58**, 2573 (1936).
54. N. H. Nachtrieb and R. E. Fryxell, *J. Am. Chem. Soc.*, **70**, 3552 (1948); **74**, 897 (1952).
55. R. C. Myers, D. E. Metzler and E. H. Swift, *J. Am. Chem. Soc.*, **72**, 3767, 3772 (1950).
56. R. H. Herber, W. E. Bennett, D. R. Benz, L. C. Bogar, R. C. Dietz, Jr., G. S. Golden and J. W. Irvine, Jr., *126th Mtg. Am. Chem. Soc.*, p. 33R (1954).
57. D. E. Chalkley and R. J. P. Williams, *J. Chem. Soc.*, 1920 (1955).
58. D. E. Campbell, H. M. Clark and W. H. Baur, *J. Phys. Chem.*, **62**, 506 (1958).
59. A. G. Maddock, W. Smuleck and A. J. Tench, *Trans. Faraday Soc.*, **58**, 923 (1962).
60. A. F. Morgunov and V. V. Fomin, *Zh. Neorg. Khim.*, **8**, 508 (1963).
61. V. M. Vdovenko and A. S. Krivokhatskii, *Zh. Neorg. Khim.*, **5**, 745 (1960).
62. G. W. Latimer and N. H. Furman, *J. Inorg. Nucl. Chem.*, **24**, 729 (1962).
63. M. Mori and R. Tsuchiya, *Nippon Kagaku Zasshi*, **77**, 1525 (1956).
64. H. Specker, M. Cremer and W. Jackwerth, *Angew. Chem.*, **71**, 492 (1959).
65. E. Bankman and H. Specker, *Z. Anal. Chem.*, **162**, 18 (1959).
66. H. Irving and D. N. Edgington, *J. Inorg. Nucl. Chem.*, **10**, 306 (1959).
67. G. Weidemann, *Can. J. Chem.*, **38**, 459 (1960)
68. A. Musil, G. Haas and G. Weidemann, *Mikrochim. Acta*, 883 (1962).
69. A. S. Kertes and M. Halpern, *J. Inorg. Nucl. Chem.*, **16**, 308 (1961).
70. Y. Marcus, *J. Inorg. Nucl. Chem.*, **12**, 287 (1960).

71. C. V. Kopp and R. L. McDonald in *Solvent Extraction Chemistry* (Eds. D. Dyrssen, J. O. Liljenzin and J. Rydberg), North-Holland, Amsterdam, 1967, p. 447.

72. H. A. Mahlman, G. W. Leddicote and F. L. Moore, *Anal. Chem.*, **26**, 1939 (1954).

73. G. Nakagawa, *J. Chem. Soc. Japan, Pure Chem. Sect.*, **63**, 444 (1960).

74. M. L. Good and S. E. Bryan, *J. Am. Chem. Soc.*, **82**, 5636 (1960).

75. K. Van Ipenburg, *Ph.D. Thesis*, Amsterdam, 1961.

76. W. Smulek and S. Siekierski, *J. Inorg. Nucl. Chem.*, **24**, 1651 (1962).

77. A. v. Baeckmann and O. Glemser, *Z. Anal. Chem.*, **187**, 429 (1962).

78. R. Kunin and A. G. Winger, *Chem. Ing. Tech.*, **34**, 461 (1962).

79. S. Lindenbaum and G. E. Boyd, *J. Phys. Chem.*, **67**, 1238 (1963).

80. T. Omori and M. Suzuki, *Bull. Chem. Soc. Japan*, **36**, 850 (1963).

81. G. Duyckaerts, J. Fuger and W. Müller, *Euratom Rept.*, EUR–426.f (1963).

82. M. Okazaki, T. Horiguchi and Y. Watanabe, *Nippon Kagaku Zasshi*, **84**, 917 (1963).

83. B. E. McClellan and V. B. Benson, *Anal. Chem.*, **36**, 1985 (1964).

84. A. D. Nelson, J. L. Fasching and R. L. McDonald, *J. Inorg. Nucl. Chem.*, **27**, 439 (1965).

85. W. Müller, J. Fuger and G. Duyckaerts, *Euratom Rept.*, EUR–2169e (1964); *Proc. VIII Intern. Conf. Coord. Chem., Vienna*, 406 (1964).

86. L. Kuca, E. Högfeldt and L. G. Sillèn in *Solvent Extraction Chemistry* (Eds. D. Dyrssen, J. O. Liljenzin and J. Rydberg), North-Holland, Amsterdam, 1967, p. 454.

87. G. E. Moore and K. A. Kraus, *J. Am. Chem. Soc.*, **72**, 5792 (1950).

88. D. Jentzsch and I. Frotscher, *Z. Anal. Chem.*, **144**, 17 (1955).

89. F. Nelson, R. M. Rush and K. A. Kraus, *J. Am. Chem. Soc.*, **82**, 339 (1960).

90. E. Rutner, *J. Phys. Chem.*, **65**, 1027 (1961).

91. K. A. Kraus, F. Nelson and G. E. Moore, *J. Am. Chem. Soc.*, **77**, 3972 (1955).

92. J. A. Schufle and H. M. Eiland, *J. Am. Chem. Soc.*, **76**, 960 (1954).

93. N. Sundèn, *Svensk. Kem. Tidskr.*, **66**, 345 (1954).

94. B. G. F. Carleson and H. Irving, *J. Chem. Soc.*, 4390 (1954).

95. H. Irving and G. T. Woods, *J. Chem. Soc.*, 939 (1963).

96. I. Wada and R. Ishi, *Sci. Papers, Inst. Phys. Chem. Res. Tokyo*, **24**, 136 (1934); **34**, 787 (1938).

97. R. Bock, R. Kusche and E. Bock, *Z. Anal. Chem.*, **138**, 167 (1953).

98. H. Irving and F. J. C. Rossotti, *J. Chem. Soc.*, 1938 (1955).

99. L. A. Woodward and P. T. Bill, *J. Chem. Soc.*, 1699 (1955).

100. H. Irving and F. J. C. Rossotti, *J. Chem. Soc.*, 1927 (1955).

101. H. Irving and F. J. C. Rossotti, *J. Chem. Soc.*, 1948 (1955).

102. C. E. Hudgens and J. Nelson, *Anal. Chem.*, **24**, 1472 (1952).

103. A. M. Poskanzer, R. J. Dietz, Jr., E. Rudzitis, J. W. Irvine, Jr. and C. D. Coryell, *Radioisotopes in Scientific Research, Proc. 1st. UNESCO Conf. Paris*, 1957, Vol. II, Pergamon Press, London, 1958, p. 518.

104. R. J. Dietz, Jr., J. Mendez and J. W. Irvine, Jr., *Conf. Uses Radio-Isotopes in Phys. Sci. and Ind.*, Copenhagen, 1960, paper RICC/180.

105. E. B. Owens, *Anal. Chem.*, **32**, 1366 (1960).

106. R. J. Dietz, Jr., *Ph.D. Thesis*, Mass. Inst. Techn., Cambridge, 1958.

107. A. M. Poskanzer, *Ph.D. Thesis*, Mass. Inst. Techn., Cambridge, 1957.

108. H. Irving and F. J. C. Rossotti, *J. Chem. Soc.*, 2475 (1956).

109. T. Suzuki and T. Sotobayashi, *Bunseki Kagaku*, **12**, 910 (1963).

110. T. Andersen and A. B. Knutsen, *Acta Chem. Scand.*, **16**, 849 (1962).

111. E. P. Tsintsevich, I. P. Alimarin and L. I. Nikolaeva, *Vestn. Mosk. Univ. Ser. Mat. Mekhan. Astron. Fiz. i. Khim.*, **14**, 189 (1959).

112. J. Korkisch and I. Hazan, *Anal. Chem.*, **37**, 707 (1965).

113. T. Andersen and A. B. Knutsen, *Acta Chem. Scand.*, **16**, 875 (1962).

114. D. Maydan and Y. Marcus, *J. Phys. Chem.*, **67**, 987 (1963).

115. D. Banerjea and K. K. Tripathi, *J. Inorg. Nucl. Chem.*, **18**, 199 (1961).

116. F. W. E. Strelow, R. Rethemeyer and C. J. C. Bothma, *Anal. Chem.*, **37**, 106 (1965).

117. R. A. Day, Jr. and R. M. Powers, *J. Am. Chem. Soc.*, **76**, 3895 (1954).

118. I. L. Jenkins and H. A. C. McKay, *Trans. Faraday Soc.*, **50**, 107 (1954).

119. B. Bernström and J. Rydberg, *Acta Chem. Scand.*, **11**, 1173 (1957).

120. E. Hesford and H. A. C. McKay, *Trans. Faraday Soc.*, **54**, 513 (1958).

121. Y. Marcus, *Israel AEC Rept.* IA–582 (1960); *J. Phys. Chem.*, **65**, 1647 (1961).

122. C. J. Hardy, *J. Inorg. Nucl. Chem.*, **21**, 348 (1961).

123. B. Jezowska-Trzebiatowska and M. Chimelowska, *J. Inorg. Nucl. Chem.*, **20**, 106 (1961); B. Jezowska-Trzebiatowska and S. Ernst, *J. Inorg. Nucl. Chem.*, **28**, 1435 (1966).

124. H. A. C. McKay and A. R. Mathieson, *Trans. Faraday Soc.*, **47**, 428 (1951).

125. E. Glueckauf, H. A. C. McKay and A. R. Mathieson, *Trans. Faraday Soc.*, **47**, 437 (1951).

126. A. W. Gardner, H. A. C. McKay and D. T. Warren, *Trans. Faraday Soc.*, **48**, 997 (1952).

127. A. W. Gardner and H. A. C. McKay, *Trans. Faraday Soc.*, **48**, 1099 (1952).

128 H. A. C. McKay, *Trans. Faraday Soc.*, **48**, 1103 (1952).

129. T. V. Healy and H. A. C. McKay, *Trans. Faraday Soc.*, **52**, 633 (1956).

130. T. V. Healy, J. Kennedy and G. M. Waind, *J. Inorg. Nucl. Chem.*, **10**, 137 (1959).

131. J. L. Woodhead, *J. Inorg. Nucl. Chem.*, **26**, 1472 (1964).

132. J. L. Woodhead, *J. Inorg. Nucl. Chem.*, **27**, 1111 (1965).

133. S. Minc and L. Werblau, *Roczniki Chem.*, **32**, 1419 (1958).

134. H. T. Hahn and G. M. Vander Wall, *J. Inorg. Nucl. Chem.*, **26**, 191 (1964).

135. W. E. Keder, J. C. Sheppard and A. S. Wilson, *J. Inorg. Nucl. Chem.*, **12**, 327 (1960).

136. W. E. Keder, J. L. Ryan and A. S. Wilson, *J. Inorg. Nucl. Chem.*, **20**, 131 (1961).

137. V. M. Vdovenko, M. P. Kovalskaya and E. A. Smirnova, *Radiokhimiya*, **3**, 403 (1961); V. M. Vdovenko, A. A. Lipovskii and M. G. Kuzina, *Radiokhimiya*, **3**, 555 (1961).

138. V. M. Vdovenko, T. V. Kovaleva and M. A. Ryazanov, *Radiokhimiya*, **4**, 609 (1962).

139. F. Baroncelli, G. Scibona and M. Zifferero, *J. Inorg. Nucl. Chem.*, **24**, 547 (1962).

140. F. Baroncelli, G. Scibona and M. Zifferero, *Radiochim. Acta*, **1**, 75 (1963).

141. J. M. P. J. Verstegen, *Norwegian Rept.*, KR–68 (1964); *J. Inorg. Nucl. Chem.*, **26**, 1589 (1964).

142. P. J. Lloyd and E. M. Mason, *J. Phys. Chem.*, **68**, 3120 (1964).

143. P. R. Danesi, F. Orlandini and G. Scibona, *J. Inorg. Nucl. Chem.*, **27**, 449 (1965).

144. T. Sato, *J. Inorg. Nucl. Chem.*, **26**, 1295 (1964).

145. K. A. Kraus and F. Nelson, *Proc. U. N. Intern. Conf. Peaceful Uses At. Energy, 1st Geneva, 1955*, p. 113.

146. D. J. Carswell, *J. Inorg. Nucl. Chem.*, **3**, 384 (1957).

147. J. K. Foreman, I. R. McGowan and T. D. Smith, *J. Chem. Soc.*, 738 (1959).

148. D. A. Vita, C. F. Trivisonno and C. W. Phipps, *USAEC Rept.*, GAT–283 (1959).

149. R. F. Buchanan and J. P. Faris, *Proc. Conf. Uses Radioisotopes Phys. Sci. Ind., Copenhagen, 1960*, paper RICC/173 (1961); J. P. Faris and R. F. Buchanan, *USAEC Rept.*, ANL–6811 (1964).

150. J. L. Ryan, *J. Phys. Chem.*, **65**, 1099 (1961).

151. V. M. Vdovenko, A. A. Lipovskii and M. G. Kusina, *Radiokhimiya*, **3**, 365 (1961).

152. J. Korkisch and F. Tera, *Z. Anal. Chem.*, **186**, 290 (1962).
153. J. Korkisch and F. Tera, *J. Chromatog.*, **7**, 567 (1962).
154. J. Korkisch and G. E. Janauer, *Talanta*, **9**, 957 (1962).
155. J. Korkisch and G. Arrhenius, *Anal. Chem.*, **36**, 850 (1964).
156. L. W. Marple, *J. Inorg. Nucl. Chem.*, **26**, 635 (1964).

13.
Appendices

A. LIST OF SYMBOLS

(Equation where first used given in parenthesis after some of the symbols. Symbols for species or quantities pertaining to the resin or organic phase have a bar above them).

Universal constants

e — electronic charge $= 4.802 \times 10^{-10}$ e.s.u. $= 1.602 \times 10^{-19}$ abs. coulomb

F — Faraday $= 96493$ abs. coulomb/equiv.

k — Boltzmann's constant $= 1.380 \times 10^{-16}$ erg/deg molecule

N — Avogadro's number $= 6.024 \times 10^{23}$ mole^{-1}

R — gas constant $= 8.314 \times 10^7$ erg/deg mole $= 1.987$ cal/deg mole

Chemical symbols

A^- — mononegative anion

B^{b-} — anion, of charge $b-$

C^{c+} — non-complexing cation, of charge $c+$

D — detergent

G — electrolyte, generalized

I — distribuend, generalized

J — electrolyte, in electrolyte mixtures

L^{l-} — ligand, of charge $l-$

M^{m+} — complex-forming cation, of charge $m+$

N — non-electrolyte

R^+, R^- — exchange site of anion, cation exchanger

S — solvating solvent

W — water

X^{x-} — chelating ligand, of charge $x-$

Z^z — generalized ion, of algebraic charge z

In the following, any one of these is symbolized as a subscript X, to describe properties pertaining to the chemical species X. In the text, subscripts i, j, ⋯ are used to designate general species.

Molar quantities

C_p — heat capacity at constant pressure

E — energy

G — free energy

ΔG — free energy change

H — enthalpy

ΔH	— enthalpy change
P	— polarization, generalized property
S	— entropy
ΔS	— entropy change
V	— volume
δV_{w}	— electrostriction of water
Y	— generalized thermodynamic function

Quantities and functions

A	— cross-sectional area
A	— constant in Debye–Hückel equation (1C21)
A_{M}	— Madelung constant (1D3)
a	— distance of closest approach of ions, diameter of ions ($\overset{\circ}{a}$, when expressed in Å, equation 1C7).
a	— effective activity (of ligand, equation 1D38)
a_{X}	— activity of X
B	— constant in Debye–Hückel equation (1C23)
$B(T)$	— second virial coefficient
b	— linear coefficient in activity coefficient expression (1D25)
b	— parameter in ion-association theory (2D16)
b	— four-times excess free energy for equimolar binary mixture (2B9)
b_{H}	— parameter related to enthalpy of mixing (1D21)
$\bar{C}$	— capacity of resin (section 4.A.e)
C_{ii}	— cohesive energy density (2B23)
c_{X}	— molarity of X
cmc	— critical micelle concentration (section 2.D.e)
D_{X}	— distribution coefficient of X
D	— diffusion coefficient (4A5), self-diffusion coefficient (2D21)
d	— diameter
d	— denticity of chelating ligand (5C1)
E	— potential energy (2A2)
E	— electromotive force of cell, electrode potential
E_{j}	— liquid-junction potential (1E27)
E_{X}	— percentage extraction of X (7C2)
F	— auxiliary function in cation exchange (F_{R}) and extraction with acidic extractants (F_{S}) (5B17)
$F(t)$	— fractional attainment of equilibrium at time t (4A6)
$_{\mathrm{r}}F_{\mathrm{a}}$	— invasion correction function (section 4.D.d)

F_n	— activity coefficient function (1D39)
$f_{\bar{X}}$	— rational activity coefficient of X (1A16a)
G	— free energy of system
G_X	— free energy of X
G	— activity coefficient function (6B35)
g	— coefficient of dipole term for associated molecules (2A17)
$g(r)$	— radial distribution function (2A1)
H	— enthalpy
H_i, H_R	— acidity functions (1D29)
h	— hydration number
I	— ionic strength (1C19)
I_X	— contribution of X to total ionic strength
i	— charge number of complex (3B23)
$\bar{i}$	— average charge number (3B36)
J	— parameter in Debye salting theory (1F17)
J	— coefficient of linear term in conductance equation (2D3)
K	— equilibrium constant
K_X	— acid dissociation constant ($X = I$ for indicator H_iI, $X = R$ for indicator HR, $X = W$ ionic product of water, $X = a1$, etc., first proton of acid, etc., $X = b$ for base, etc.)
K_{ass}, K_{diss}	— association, dissociation constant (of ion-pair)
K_s	— solubility product
K_n^T	— thermodynamic stepwise complex formation constant (1D39)
K_r, K_m	— thermodynamic selectivity constant of an ion exchange resin (rational and molal scales)
K_B^A	— selectivity constant of A over B
k	— proportionality constant relating quantity per unit weight of resin to its mole fraction there
k_n	— stepwise complex formation constant
k_s	— salting coefficient (1F2, 1F5)
M_X	— molecular weight (gram-formula weight) of X
m_X	— molality of X (1A12)
N_X	— number of moles of X (1A2)
N	— coordination number of metal ion
N	— most probable micelle size (2D40)
n	— number of ligands in metal complex
$\bar{n}$	— average ligand number (3B33)
n_X	— number of molecules of X per unit volume (1 cm^3)

n_D, n_W	—	refractive index at sodium D line (of water)
$\bar{n}_X$	—	number of moles of X per equivalent of resin (5A1)
P	—	applied pressure
P	—	generalized property of a system (7D1)
P_X	—	partition constant of X (7B7)
p	—	vapour pressure of solution (1A4)
p_X^0	—	vapour pressure of pure liquid X (1A4)
pH	—	quantity related to hydrogen ion activity (1C31)
pH_{50}	—	pH at 50% extraction
pwH	—	corrected pH value (1C29)
$Q(b)$	—	integral function of b (Table 3A1, equation 3A6)
Q_n	—	stepwise complex formation quotient (same as k_n) (3B8)
Q_B^A	—	selectivity quotient of ion exchanger for A over B
q	—	Bjerrum critical distance (3A3)
q	—	aggregation number (7C13)
$\tilde{q}$	—	average aggregation number (7C16)
R	—	hydrodynamic radius of ions
R	—	number of atoms of metal per equivalent of metal (6A6)
R	—	ratio of maximal value of D to minimal value in cation exchange
r	—	radial distance (1C1)
r	—	radius
r_{ii}	—	coordinate of minimum potential energy (2A3)
r_+, r_-	—	radius of cation, anion
r_c, r_{ci}	—	crystalline ionic radius
r_h	—	radius of hydrate ion
r_n	—	radius of solution volume available per ion (1F23)
r_v	—	ratio of molar volumes of components in mixture (e.g. V_G/V_W, equation 1D12)
S	—	entropy
S_X	—	entropy of X
S	—	Onsager slope, coefficient of square-root term in conductance equation (2D3)
S_K, S_V	—	slope of Masson's compressibility, volume equation (1B28, 1B27)
S_X	—	ratio of solubilities of X in two phases (7B6)
S_B^A	—	separation factor of A from B (4C7)
s_X	—	solubility of X (2C5)
T	—	absolute temperature

T_c	—	critical consolute temperature (2B14)
t	—	time
t_+, t_-	—	transport number of cation, anion
$U(r)$	—	potential energy function in liquids (2A2)
U_{ii}^*	—	minimum potential energy (2A3)
u	—	interaction-energy parameter (3A2)
u	—	number of $-CH_2-$ linkages in aliphatic chain (2D46)
u_+, u_-	—	mobility of cation, anion, at unit field
V	—	volume
ΔV_W	—	total electrostriction
v_X	—	molecular volume of X, V_X/N
v_+, v_-	—	velocity of cation, anion
W	—	water content of resin, weight percent
W_r, W_X	—	interaction energy, total and of X, with ion (1B33, 1F27)
w_X	—	weight fraction of X
X	—	complex formation sum (3B28)
X	—	crosslinking of resin
x_X	—	mole fraction of X (1A9)
Y	—	generalized thermodynamic function (G, H, S, V, etc.)
Y_n	—	activity coefficient quotient (3B8)
y_X	—	molar activity coefficient of X (1A16c)
Z	—	atomic number
Z	—	number of nearest neighbours
z, z_+, z_-	—	algebraic charge of ion, cation, anion, in electron units
α	—	thermal expansion coefficient
α	—	polarizability (2A15)
α	—	fraction of eletrolyte dissociated to free ions (2D3)
α	—	ratio of the total concentrations of two reactants (7D5)
α_n	—	fraction of metal in the form of the nth complex
α_X	—	Harned's rule coefficient (1E9)
β	—	compressibility coefficient
β_n	—	overall complex stability constant of nth complex (3B12)
β_X	—	Harned's rule coefficient of the square term (1E10)
β_i'	—	complex formation constant referred to the neutral species (3B24)
Γ	—	perturbation parameter (2B43)
Γ	—	ratio of molal activity coefficients in two phases $\bar{\gamma}/\gamma$ (section 4.B.e)
Γ	—	product of activity coefficient ratios (6B15)

γ_X	— molal activity coefficient of X
γ	— normalized minimum potential (2B39)
γ	— packing factor (2A8)
γ_+, γ_-	— relative sorption of cations, anions
γ_n, γ_i	— activity coefficient functions (6B10, 6B22)
Δ, δ	— thickness of layer around ion (1B14)
δ	— thickness of diffusion film around resin particle (4A5⁾
δ_X	— molar dielectric decrement, by 1 M of X (1F10)
δ_X	— solubility parameter of X (2B24)
ε	— dielectric constant
ε_X	— molar absorptivity of X
η	— viscosity
θ	— normalized potential in mixture (2B39)
κ	— reciprocal of the diameter of the ionic atmosphere (1C6)
Λ	— equivalent conductivity (2D3)
λ	— lattice length, wavelength
λ_i	— ionic conductivity (1B24)
λ_X	— functions of ionic polarizabilities (1F22)
μ	— dipole moment
μ_X	— dipole moment of X (1B16)
μ_X	— chemical potential of X (1A3)
μ_X^0	— chemical potential of X in standard state (1A3)
ν, ν_+, ν_-	— number of ions, cations, anions, into which electrolyte dissociates (1A1)
$\bar{\pi}$	— swelling pressure of resin (4B2)
ρ	— density, charge density
ρ^0	— density of pure solvent
ρ	— normalized radius (2B39)
σ	— surface tension
σ	— chemical shift in n.m.r. (1B36)
$\sigma(\kappa a)$	— a function defined in (1C15)
$\tau(\kappa a)$	— a function defined in (1C12)
$\emptyset_K$	— apparent molar compressibility
$\emptyset_V$	— apparent molar volume
ϕ	— molal osmotic coefficient (1A21)
φ_0	— ionic activity function (1D42)
φ_X	— volume fraction of X (1D11)
ψ	— electrostatic potential (1C1)

ω_i — contribution of dipole moment to solubility parameter (2B49)

Subscripts, superscripts and other symbols

abs	— absolute
ad	— air-dried
aq	— aqueous
b	— boiling
c	— critical, crystalline
con	— conventional
e,eff	— effective
el	— electrical
f	— free, freezing, of formation
g	— gaseous
h	— hydrated
hyd	— of hydration
i	— of an ion
id	— ideal
ion	— of ionization
m	— molal
neut	— of neutral species
p	— primary (of solvation)
r	— reduced, rational
R	— of reference
ss	— of standard state
std	— of standard
stat	— statistical
s	— swollen, secondary (of solvation)
vac	— in vacuum
0	— at infinite dilution
E	— excess
f	— free
M	— of mixing
t	— trace
T	— thermodynamic
V	— of vaporization
'	— partial molal
*	— effective constant
0	— standard, pure solvent, corrected
—	— pertaining to resin or organic-solvent phase
~	— average

B. ACTIVITIES IN AQUEOUS ELECTROLYTE SOLUTIONS

This appendix contains a compilation of the activities of electrolytes on a molar concentration basis. Most compilations in the literature list only m and γ^{1-3}, while most work on ion exchange and solvent extraction has been done in the c scale. The compilations of Gazith[4] are very valuable for the conversion from the molal to the molar scale, and calculation of the relevant mean ionic activities: $\log a = \log m + \log \gamma_{\pm} + \log \rho_{o}$ for 1 : 1 electrolytes. For the three 1 : 2 electrolytes listed, $\log a$ pertains specifically to the anion, hence $a = 2 \rho_{o} m \gamma_{\pm}$ has been calculated.

Water activities, calculated by Gazith[4] from osmotic or activity coefficient data: $a_{W} = \exp(-\nu m \phi / 55.51)$, are also given, for the electrolyte molarities listed.

The data for water activities have been obtained from ref. 1, except those for nitric acid, for which more recent data were used[5]. Activity coefficients were also taken from ref. 1, except those for nitric acid[5] and sodium thiocyanate[6], and have been completed with data from refs. 2, 3 and 7. Density data for conversion from the molal to the molar scale have been taken from refs. 7 and 8.

REFERENCES

1. R. A. Robinson and R. H. Stokes, *Electrolyte Solutions*, 2nd ed., Butterworths Sci. Publ., London, 1959.

2. H. S. Harned and B. B. Owen, *Physical Chemistry of Electrolyte Solutions*, 3rd ed., Reinhold Publ. Co., New York, 1958.

3. B. E. Conway, *Electrochemical Data*, Elsevier Publ. Co., Amsterdam, 1952.

4. M. Gazith, *Israel AEC Rept.*, IA–1004 (1964); IA–1009 (1965).

5. W. Davis and J. J. DeBruin, *J. Inorg. Nucl. Chem.*, **26**, 1069 (1964).

6. M. L. Miller and C. L. Sheridan, *J. Phys. Chem.*, **60**, 184 (1956).

7. *International Critical Tables*, Vol. III, McGraw Hill Book Co., New York, 1928.

8. *Landoldt-Börnstein Tabellen*, Vol. II, Egb. IIb (1931); Egb. IIIc (1936).

	HF					HCl		
m	γ	Log a	a_W	c	m	γ	Log a	a_W
0.1003	0.077	−2.114	0.9981	0.1	0.1005	0.796	−1.096	0.9966
0.2008	0.059	−1.928	0.9959	0.2	0.201	0.767	−0.812	0.9932
0.302	0.044	−1.852	0.9950	0.3	0.303	0.756	−0.640	0.9897
0.504	0.031	−1.809	0.9928	0.5	0.506	0.757	−0.416	0.9824
0.707	0.027	−1.719	0.9900	0.7	0.712	0.773	−0.259	0.9746
1.015	0.024	−1.620	0.9814	1.0	1.022	0.812	−0.081	0.9625
				1.5	1.549	0.906	0.147	0.9394
2.062				2.0	2.085	1.033	0.333	0.9136
				2.5	2.637	1.193	0.497	0.8846
3.126				3.0	3.198	1.395	0.649	0.8524
4.226				4.0	4.358	1.962	0.933	0.7808
5.36				5.0	5.57	2.83	1.198	0.700
				6.0	6.84	4.30	1.469	0.614
7.72				7.0	8.18	6.23	1.707	0.524
				8.0	9.59	9.42	1.956	0.438
				9.0	11.08	13.8	2.185	0.360
11.55				10.0	12.67	20.3	2.368	0.288
				11.0	14.35	29.7	2.630	0.229
				12.0	16.15	43.9	2.850	0.170
20.56				16.0				

	HBr					HI		
m	γ	Log a	a_W	c	m	γ	Log a	a_W
0.1005	0.805	−1.093	0.9965	0.1	0.1007	0.818	−1.084	0.9966
0.202	0.782	−0.802	0.9930	0.2	0.202	0.807	−0.788	0.9930
0.303	0.777	−0.628	0.9897	0.3	0.304	0.811	−0.608	0.9892
0.508	0.790	−0.397	0.9820	0.5	0.511	0.841	−0.367	0.9814
0.715	0.817	−0.234	0.9739	0.7	0.721	0.888	−0.194	0.9729
1.029	0.877	−0.045	0.9608	1.0	1.041	0.975	0.007	0.9588
1.564	1.016	0.201	0.9360	1.5	1.593	1.170	0.271	0.9317
2.114	1.218	0.410	0.9066	2.0	2.166	1.450	0.497	0.9131
2.679	1.486	0.600	0.8731	2.5	2.760	1.830	0.703	0.863
3.259	1.92	0.796	0.8331	3.0	3.374	2.44	0.915	0.818
4.468	3.40	1.181	0.735	4.0	4.685	3.90	1.262	0.726
5.75	5.60	1.508	0.646	5.0	6.12	6.88	1.625	0.615
7.10	8.50	1.780	0.554	6.0	7.71	13.9	2.030	0.492
8.55	13.4	2.052	0.461	7.0	9.46	41.0	2.589	0.325
10.09	24.0	2.383	0.360	8.0				
11.75	44.0	2.713	0.281	9.0				
13.54	80.0	3.035	0.196	10.0				
15.49				11.0				
17.61				12.0				
22.54				14.0				
28.68				16.0				

	HNO$_3$					HClO$_4$		
m	*γ*	Log *a*	*a*$_\text{W}$	*c*	*m*	*γ*	Log *a*	*a*$_\text{W}$
0.1006	0.790	−1.100	0.9966	0.1	0.1008	0.803	−1.091	0.9966
0.202	0.754	−0.817	0.9933	0.2	0.203	0.778	−0.802	0.9932
0.304	0.735	−0.651	0.9898	0.3	0.305	0.768	−0.630	0.9895
0.509	0.720	−0.436	0.9828	0.5	0.513	0.770	−0.403	0.9821
0.717	0.717	−0.289	0.9759	0.7	0.725	0.787	−0.244	0.9741
1.034	0.725	−0.125	0.9642	1.0	1.051	0.832	−0.059	0.9611
1.575	0.759	0.078	0.9494	1.5	1.613	0.950	0.186	0.9354
2.136	0.809	0.237	0.9207	2.0	2.206	1.125	0.395	0.9052
2.714	0.874	0.376	0.8956	2.5	2.824	1.370	0.588	0.8695
3.311	0.957	0.500	0.8682	3.0	3.476	1.720	0.776	0.828
4.573	1.157	0.723	0.808	4.0	4.887	3.00	1.166	0.724
5.93	1.378	0.912	0.744	5.0	6.44	5.90	1.579	0.597
7.41	1.607	1.076	0.681	6.0	8.16	13.00	2.026	0.456
9.01	1.846	1.221	0.603	7.0	10.09	31.9	2.508	0.313
10.81	2.08	1.352	0.555	8.0	12.26	87.0	3.027	0.191
12.79	2.30	1.468	0.494	9.0	14.72	287	3.584	0.101
15.02	2.52	1.578	0.436	10.0	17.56	976	4.234	0.046
17.58	2.72	1.679	0.381	11.0	20.93			
20.53	2.93	1.779	0.325	12.0	25.03			
28.17	3.32	1.970	0.221	14.0				

	H$_2$SO$_4$				NaOH		
m	*γ*	*a*$_\text{W}$	*c*	*m*	*γ*	Log *a*	*a*$_\text{W}$
0.1006	0.265	0.9963	0.1	0.1002	0.764	−1.116	0.9967
0.202	0.209	0.9928	0.2	0.2004	0.725	−0.838	0.9934
0.304	0.183	0.9890	0.3	0.3005	0.706	−0.674	0.9900
0.511	0.155	0.9815	0.5	0.5005	0.688	−0.462	0.9833
0.720	0.141	0.9735	0.7	0.7006	0.680	−0.322	0.9764
1.041	0.131	0.9600	1.0	1.001	0.677	−0.169	0.9661
1.593	0.127	0.9338	1.5	1.501	0.687	0.013	0.9482
2.169	0.129	0.9028	2.0	2.003	0.707	0.151	0.9295
2.775	0.137	0.8660	2.5	2.508	0.741	0.269	0.9090
3.406	0.152	0.8233	3.0	3.015	0.783	0.373	0.8878
4.775	0.199	0.7207	4.0	4.042	0.908	0.564	0.8396
6.31	0.274	0.6023	5.0	5.09	1.092	0.745	0.7842
8.04	0.389	0.4790	6.0	6.16	1.346	0.918	0.7234
10.00	0.559	0.3612	7.0	7.26	1.705	1.093	0.6573
12.27	0.802	0.2580	8.0	8.40	2.23	1.273	0.5864
14.89	1.139	0.1729	9.0	9.58	2.53	1.384	0.5129
18.01	1.610	0.1076	10.0	10.80	3.84	1.618	0.4413
21.78	2.259	0.0610	11.0	12.06	5.26	1.802	0.3732
			12.0	13.40	6.70	1.953	0.3107
			14.0	16.27	12.06	2.292	0.2109
			16.0	19.46	18.23	2.550	0.1443

	NH$_4$NO$_3$					LiCl		
m	γ	Log a	a_W	c	m	γ	Log a	a_W
0.1008	0.740	−1.128	0.9967	0.1	0.1005	0.789	−1.101	0.9966
0.203	0.676	−0.863	0.9936	0.2	0.201	0.757	−0.818	0.9933
0.305	0.634	−0.714	0.9906	0.3	0.303	0.744	−0.647	0.9897
0.514	0.579	−0.527	0.9843	0.5	0.506	0.739	−0.427	0.9826
0.727	0.541	−0.405	0.9780	0.7	0.711	0.749	−0.274	0.9751
1.054	0.498	−0.279	0.9695	1.0	1.022	0.776	−0.102	0.9632
1.623	0.446	−0.141	0.9547	1.5	1.548	0.845	0.116	0.9412
2.226	0.404	−0.046	0.9398	2.0	2.087	0.938	0.292	0.9168
2.861	0.374	0.029	0.9245	2.5	2.636	1.058	0.445	0.8894
3.532	0.347	0.088	0.9117	3.0	3.198	1.217	0.590	0.8588
5.01	0.302	0.179	0.8829	4.0	4.361	1.675	0.864	0.7888
6.69	0.266	0.250	0.8534	5.0	5.58	2.40	1.127	0.7078
8.62	0.237	0.310	0.8220	6.0	6.86	3.56	1.388	0.6190
10.81	0.213	0.362	0.7890	7.0	8.20	5.43	1.648	0.526
13.51	0.190	0.410	0.7517	8.0	9.62	8.48	1.911	0.435
16.67	0.170	0.452	0.7126	9.0	11.11	12.93	2.158	0.351
				10.0	12.69	19.4	2.391	0.279
				11.0	14.35	28.2	2.607	0.219
				12.0	16.09	38.5	2.792	0.173
				14.0	19.92	62.0	3.092	0.1110

	LiBr					NaCl		
m	γ	Log a	a_W	c	m	γ	Log a	a_W
0.1005	0.796	−1.098	0.9966	0.1	0.1005	0.778	−1.107	0.9966
0.202	0.766	−0.811	0.9931	0.2	0.201	0.735	−0.831	0.9934
0.303	0.756	−0.639	0.9897	0.3	0.303	0.710	−0.667	0.9900
0.508	0.753	−0.416	0.9824	0.5	0.506	0.681	−0.464	0.9833
0.715	0.768	−0.260	0.9746	0.7	0.711	0.666	−0.325	0.9765
1.029	0.807	−0.082	0.9622	1.0	1.022	0.657	−0.173	0.9661
1.564	0.909	0.153	0.9386	1.5	1.549	0.656	0.006	0.9479
2.116	1.046	0.345	0.9114	2.0	2.087	0.668	0.145	0.9284
2.682	1.221	0.515	0.8810	2.5	2.639	0.695	0.263	0.9074
3.265	1.464	0.680	0.8375	3.0	3.203	0.727	0.367	0.8850
4.484	2.267	1.007	0.762	4.0	4.376	0.815	0.552	0.8350
5.78	3.62	1.320	0.667	5.0	5.611	0.941	0.722	0.7783
7.15	6.15	1.643	0.562	6.0				
8.62	11.13	1.981	0.451	7.0				
10.18	21.8	2.346	0.340	8.0				
11.85	43.8	2.715	0.242	9.0				
13.63	90.8	3.093	0.163	10.0				
15.52	171.5	3.425	0.1091	11.0				
17.57	299	3.720	0.0728	12.0				

	KCl					CsCl		
m	γ	Log a	a_W	c	m	γ	Log a	a_W
0.1006	0.769	−1.111	0.9967	0.1	0.1007	0.756	−1.118	0.9967
0.202	0.717	−0.839	0.9934	0.2	0.202	0.694	−0.854	0.9935
0.303	0.687	−0.682	0.9902	0.3	0.305	0.655	−0.700	0.9903
0.509	0.648	−0.482	0.9836	0.5	0.512	0.604	−0.510	0.9842
0.716	0.625	−0.350	0.9771	0.7	0.723	0.572	−0.378	0.9778
1.033	0.602	−0.206	0.9672	1.0	1.046	0.540	−0.249	0.9681
1.574	0.581	−0.039	0.9500	1.5	1.605	0.509	−0.088	0.9516
2.134	0.572	0.086	0.9320	2.0	2.193	0.492	0.032	0.9366
2.714	0.569	0.188	0.9131	2.5	2.808	0.481	0.130	0.9123
3.314	0.571	0.276	0.8933	3.0	3.453	0.476	0.215	0.8939
4.584	0.586	0.429	0.8503	4.0	4.858	0.475	0.363	0.8512
				5.0	6.43	0.483	0.492	0.7973
				6.0	8.20	0.498	0.612	0.7460
				7.0	10.23	0.509	0.716	0.6882

	LiNO$_3$					NaSCN		
m	γ	Log a	a_W	c	m	γ	Log a	a_W
0.1006	0.788	−1.100	0.9966	0.1	0.1005	0.787	−1.101	0.9966
0.202	0.752	−0.818	0.9932	0.2	0.202	0.749	−0.820	0.9933
0.304	0.736	−0.650	0.9898	0.3	0.304	0.730	−0.654	0.9898
0.509	0.726	−0.433	0.9826	0.5	0.511	0.714	−0.438	0.9828
0.717	0.730	−0.282	0.9752	0.7	0.721	0.710	−0.291	0.9756
1.034	0.743	−0.115	0.9636	1.0	1.043	0.713	−0.129	0.9642
1.575	0.803	0.102	0.9415	1.5	1.598	0.730	0.067	0.9439
2.135	0.869	0.268	0.9178	2.0	2.182	0.757	0.218	0.9217
2.714	0.947	0.410	0.8921	2.5	2.792	0.790	0.344	0.8976
3.312	1.040	0.537	0.8643	3.0	3.430	0.847	0.463	0.8714
4.568	1.263	0.761	0.8026	4.0	4.813	0.995	0.680	0.8094
5.93	1.528	0.957	0.7381	5.0	6.37	1.21	0.887	0.7353
7.39	1.86	1.138	0.6686	6.0	8.11	1.51	1.088	0.652
8.96	2.24	1.302	0.643	7.0	10.05	1.91	1.283	0.563
10.66	2.68	1.456	0.529	8.0	12.18	2.32	1.451	0.492
12.46	3.15	1.594	0.461	9.0	14.53	2.72	1.596	0.411
14.29	3.64	1.716	0.404	10.0	17.10	3.08	1.721	0.350
16.17	4.11	1.822	0.353	11.0				
12.99	4.55	1.913	0.310	12.0				

	CaCl$_2$				Ca(NO$_3$)$_2$			
m	γ	Log a	a_W	c	m	γ	Log a	a_W
0.1005	0.518	−0.982	0.9954	0.1	0.1007	0.487	−1.009	0.9955
0.201	0.472	−0.722	0.9907	0.2	0.203	0.428	−0.760	0.9911
0.303	0.455	−0.560	0.9857	0.3	0.304	0.396	−0.618	0.9866
0.507	0.448	−0.343	0.9751	0.5	0.513	0.364	−0.428	0.9774
0.714	0.461	−0.183	0.9633	0.7	0.725	0.348	−0.297	0.9677
1.028	0.505	0.015	0.9430	1.0	1.053	0.338	−0.148	0.9521
1.564	0.634	0.297	0.9016	1.5	1.624	0.339	0.042	0.9227
2.120	0.850	0.557	0.8492	2.0	2.236	0.354	0.200	0.8886
2.695	1.220	0.818	0.7727	2.5	2.885	0.377	0.337	0.8503
3.293	1.900	1.098	0.7131	3.0	3.593	0.413	0.472	0.8056
4.569	4.42	1.618	0.5517	4.0	5.202	0.526	0.738	0.6971
5.98	11.00	2.120	0.3936	5.0				
7.55	22.49	2.531	0.2790	6.0				
9.31	36.9	2.837	0.2030	7.0				

		UO$_2$(NO$_3$)$_2$		
c	m	γ	Log a	a_W
0.1	0.102	0.542	−0.956	0.9952
0.2	0.206	0.512	−0.676	0.9900
0.3	0.312	0.511	−0.494	0.9844
0.5	0.528	0.540	−0.244	0.9716
0.7	0.750	0.594	−0.050	0.9568
1.0	1.099	0.720	0.199	0.9300
1.5	1.728	1.033	0.553	0.8740
2.0	2.448	1.564	0.884	0.8029
2.5	3.327	2.24	1.173	0.7246
3.0	4.415	2.81	1.395	0.6645

C. PHYSICAL PROPERTIES OF SOLVENTS

a. Physical Properties by Water-miscible Solvents

Solvent	M.wt.	Boiling point (°C)	Density (g/ml)	Viscosity (Cp)	Dielectric constant	Dipole moment (debyes)
Methanol	32.0	64.5	0.787	0.55	32.6	1.67
Ethanol	46.1	78.3	0.785	1.08	24.3	1.70
n-Propanol	60.1	97.2	0.804	2.3	22.2	1.66
Isopropanol	60.1	82.4	0.786	2.08	13.8	1.65
Acetone	58.1	56.2	0.785	0.30	20.7	2.8
Dioxane	88.1	101.3	1.036	1.2	2.24	0
Tetrahydrofuran	72.1		0.883	0.45	6.4	1.71
Dimethylsulphoxide	78.1	189	1.096	1.96	46.6	3.96
Sulpholane	120.1	283	1.262	9.87	44	
Ethylene carbonate[a]	88.1	238	1.321		89.6	4.87
Propylene carbonate	102.1	141.7	1.207		64.6	4.94
Acetic acid	60.0	118.1	1.044	1.04	6.2	1.73
Formamide	45.0	193	1.129	3.30	111.3	3.68
N-Methylformamide	59.1		0.999	1.65	185.5	
Dimethylformamide	73.1	153	0.945	0.80	26.6	3.86
N-Methylacetamide[a]	73.1	206	0.942	3.02	165.5	
Dimethylacetamide	87.0	165	0.937	0.92	37.8	3.79
Acetonitrile	41.0	81.6	0.777	0.33	36.2	3.4
Adiponitrile	108.0		0.959	5.99	32.5	3.76
Pyridine	79.1	115.6	0.973	0.83	12.3	2.23

All values valid for 25°C, except those marked [a], valid for 40°C.

b. Physical Properties of Water-immiscible Solvents

Solvent	M.wt.	Boiling point (°C)	Density (g/ml)	Viscosity (Cp)	Dielectric constant	Dipole moment (debyes)
Hydrocarbons						
n-Pentane	72.1	36.15	0.626	0.227	1.845	0.0
n-Hexane	86.2	68.7	0.659	0.294	1.910	0.0
n-Heptane	100.2	98.4	0.684	0.411	1.924	0.0
n-Octane	114.2	125.6	0.703	0.542	1.948	0.0
n-Decane	142.3	173.3	0.730	0.852	1.991	0.0
n-Dodecane	170.3	216.3	0.748	1.353	2.014	0.0
Cyclopentane	70.1	49.5	0.751	0.39	1.96	0.0
Cyclohexane	84.2	80.75	0.778	0.80	2.10	0.0
Methylcyclohexane	98.2	100.9	0.769	0.64	2.02	0.0
Benzene	78.1	80.1	0.881	0.60	2.28	0.0
Toluene	92.3	110.6	0.871	0.55	2.38	0.4
o-Xylene	106.2	144.4	0.876	0.95	2.57	0.5
m-Xylene	106.2	139.1	0.860	0.79	2.37	0.4
p-Xylene	106.2	138.4	0.857	0.80	2.27	0.0
Ethylbenzene	106.2	136.2	0.867	0.58	2.41	0.35
n-Propylbenzene	120.2	159.2	0.862	0.75	2.23	
Halogenated hydrocarbons						
n-Fluorohexane	104.1	91.5	0.799			
n-Fluorooctane	132.2	142.3	0.811			
n-Chlorohexane	120.6	134.5	0.878			
n-Chloroheptane	134.6	159.1	0.876		5.48	1.85
n-Chlorooctane	148.7	182.0	0.873		5.05	
Dichloromethane	84.9	40.7	1.336	0.43	9.14	1.57
1,1-Dichloroethane	98.9	57.3	1.176		10.0	1.99
1,1-Dichloropropane	112.9	88.1	1.132			2.06
1,1-Dichlorobutane	127.0	113.8	1.086			
1,2-Dichloroethane	98.9	83.5	1.253	0.73	10.4	1.40
1,2-Dichloropropane	112.9	96.4	1.156	0.86	8.93	1.85
Trichloromethane	119.4	61.3	1.475	0.54	4.90	1.15
1,1,1-Trichloroethane	133.4	74.1	1.339		7.52	1.57
1,1,2-Trichloroethane	133.4	113.8	1.439	1.2		1.55
Tetrachloromethane	153.8	76.5	1.594	0.88	2.24	0.0
1,1,1,2-Tetrachloroethane	167.8	130.5	1.541			1.2
1,1,2,2-Tetrachloroethane	167.8	146.2	1.595	1.7	8.20	1.9
cis-1,2-Dichloroethylene	96.9	60.3	1.274	0.48	9.20	1.8
trans-1,2-Dichloroethylene	96.9	48.3	1.249	0.41	2.14	0.0
Chlorobenzene	112.6	131.7	1.105	0.79	5.62	1.56
o-Dichlorobenzene	147.0	180.5	1.306		9.93	2.26
m-Dichlorobenzene	147.0	173.1	1.288		5.04	1.48
p-Dichlorobenzene	147.0	174.2	1.247		2.46	0.0
Other derivatives of hydrocarbons						
Nitromethane	61.0	101.1	1.131	0.59	35.8	3.40
Nitrobenzene	123.1	210.7	1.198	1.63	34.8	3.95
Propionitrile	55.1	97.2	0.781	0.39	27.8	3.40
Butyronitrile	69.1	117.5	0.788	0.51	23.3	3.5
Benzonitrile	103.1	191.1	0.999	1.05	25.2	3.92
Carbon disulfide	76.1	46.2	1.293	0.37	2.63	

c. Physical Properties of Oxygenated Extractants

Solvent	M.wt.	Boiling point (°C)	Density (g/ml)	Viscosity (Cp)	Dielectric const.	Dipole moment (debyes)
Ethers						
Diethyl ether	74.1	34.6	0.715	0.22	4.33	1.15
Di-n-propyl ether	102.2	90.4	0.736	0.42	3.39	1.18
Diisopropyl ether	102.2	68.4	0.726	0.38	4.04	1.20
Di-n-butyl ether	130.2	142.4	0.769	1.62	3.08	1.22
Diisopentyl ether	158.3	178.0	0.780		2.84	1.23
Di-n-hexyl ether	186.3	226.2	0.794	1.68		
Methyl butyl ether	100.2				4.20	
Methyl hexyl ether	128.2				3.90	
Ethyl hexyl ether	142.2	137.1	0.833		4.05	
β,β'-Dichlorethyl ether	143.0	178.0	1.222	2.06	21.2	2.58
1,5-Dimethoxypentane (C–O–C–C–C–C–C–O–C)	132.1	155	0.889	0.86		1.79
5-Ethoxy-1-methoxypentane (C–O–C–C–C–C–C–O–C–C)	146.1	173	0.856	1.10		1.83
1,5-Diethoxypentane (C–C–O–C–C–C–C–C–O–C–C)	160.1	187	0.886	1.45		1.69
Diethyl cellosolve (diethyl ethylene glycol) (C–C–O–C–C–O–C–C)	118.2	121.4	0.853	0.63		
Dibutyl cellosolve (dibutyl ethylene glycol) (C–C–C–C–O–C–C–O–C–C–C–C)	174.3	203.3	0.837	1.34		
Dibutylcarbitol (dibutyl diethylene glycol (C–C–C–C–O–C–C–O–C–C–O–C–C–C–C)	218.3	254.6	0.885	2.39		1.83
Ketones						
Diisopropyl ketone	114.2	125.5	0.808	0.75		2.73
Diisobutyl ketone	142.2	168.1	0.809	0.93		2.74
Methyl ethyl ketone	72.1	79.6	0.805	0.42	15.45	2.74
Methyl n-propyl ketone	86.1	103.3	0.809	0.47	15.10	2.71
Methyl isopropyl ketone	86.1					2.76
Methyl n-butyl ketone	100.1	127.5	0.820	0.58	12.20	2.70
Methyl isobutyl ketone (hexone)	100.1	115.9	0.804	0.55	13.11	2.79
Methyl n-pentyl ketone	114.2	150.2	0.817	0.65		2.59
Methyl n-hexyl ketone	128.2				10.7	2.70
Ethyl n-butyl ketone	114.2	147.8	0.820			2.78
Cyclohexanone	98.1	156.7	0.948	2.20	18.2	2.80
Methylcyclohexanone	112.2	169.0	0.914	1.78	12.4	

Solvent	M.wt.	Boiling point (°C)	Density g/ml	Viscosity (Cp)	Dielectric constant	Dipole moment (debyes)
Esters						
Ethyl acetate	88.1	77.1	0.902	0.45	6.4	1.81
Ethyl n-butyrate	116.2	121.0	0.879	0.67	5.1	1.74
Ethyl benzoate	150.2	213.0	1.047	2.2	6.0	1.99
n-Propyl acetate	102.1	101.6	0.888	0.55	8.1	1.78
Isopropyl acetate	102.1	88.8	0.880	0.53		1.85
n-Butyl acetate	116.2	126.5	0.884	0.69	5.1	1.85
Isobutyl acetate	116.2	117.2	0.872	0.71	5.3	1.86
n-Butyl propionate	130.2	146.5	0.882			1.78
Isopentyl formate	116.2	123.5	0.886	0.83	7.7	1.90
Isopentyl acetate	130.2	142.1	0.866	0.87	4.6	1.82
Isopentyl propionate	144.3	160.3	0.872	0.94	4.2	
Benzyl acetate	150.2	215.0	1.062	1.42	5.1	1.80
Alcohols						
n-Butyl alcohol	74.1	117.7	0.81	2.46	16.1	1.68
Isobutyl alcohol	74.1	108.0	0.81	3.90	17.9	1.75
sec-Butyl alcohol	74.1	99.5	0.81	4.21	15.8	1.55
n-Pentyl alcohol	88.2	137.9	0.82	3.31	5.1	1.63
Isopentyl alcohol	88.2	130.0	0.82	4.01	14.7	1.66
n-Hexyl alcohol	102.2	157.8	0.82	5.44	13.3	1.64
n-Heptyl alcohol	116.2	176.0	0.82		12.1	1.70
sec-Heptyl alcohol	116.2	160.4	0.82	5.06	9.2	1.70
n-Octyl alcohol	130.2	195.2	0.82	8.31	10.3	1.68
sec-Octyl alcohol	130.2	178.5	0.82		8.2	
Isooctyl alcohol	130.2	183.5	0.83	7.07		1.66
Isononyl alcohol	144.3	194.0	0.83			1.60
n-Decyl alcohol	158.3	232.9	0.83		8.1	1.63
Cyclohexanol	100.2	160.6	0.94	4.6	15.0	1.92
2-Methylcyclohexanol	114.2	173.0	0.92	28.1	13.3	1.95
Benzyl alcohol	108.1	205.0	1.045	5.6	13.1	1.66

d. Physical Properties of Alkylamine Extractants.

Aminoalkanes	M.wt.	Boiling point, (°C)	Freezing point, (°C)	Density 20° (g/ml)
n-Hexylamine	101.2	132.7		0.762
n-Heptylamine	115.2	156.9		0.775
n-Octylamine	129.2	179.6		0.783
n-Nonylamine	143.3	202.2		0.788
n-Decylamine	157.3	220.5		0.794
n-Dodecylamine	185.3	259.2	28.3	0.801
Di-n-hexylamine	185.3	239.8		0.789
Di-n-heptylamine	213.4	272.0		0.797
Di-n-octylamine	241.4	302		0.804
Di-n-nonylamine	269.5	334	25	0.809
Di-n-decylamine	297.5	359	34	0.813
Di-n-dodecylamine	353.6	403	51	0.819
Methyl-n-octylamine	143.3	186		0.776
Tri-n-octylamine	353.6	357		0.812
Tri-n-decylamine	437.8	406		0.819
Tri-n-dodecylamine	521.9	448	15.7	0.825
Tri-n-tetradecylamine	606.1	484	33	0.829
Dimethyl-n-octylamine	157.3	194		0.766
Dimethyl-n-decylamine	185.3	235		0.778
Dimethyl-n-dodecylamine	213.4	271		0.788

D. MUTUAL SOLUBILITIES OF SOLVENTS AND WATER (at 25°C)

a. Hydrocarbons and Oxygenated Derivatives

	Aqueous phase		Organic phase	
	wt% S	$100x_S$	wt% H_2O	$100x_W$
Hydrocarbons and derivatives				
n-Hexane	0.001	0.0002		
n-Heptane	0.0003	0.00005		
n-Octane	0.0001	0.00002		
Cyclopentane	0.016	0.004		
Cyclohexane	0.006	0.0013		
Methylcyclohexane	0.0014	0.0003		
Benzene	0.178	0.042	0.072	0.31
Toluene	0.050	0.01	0.042	0.22
o-Xylene	0.018	0.003		
Ethylbenzene	0.006	0.001		
Chloroform	0.705	0.107	0.072	0.48
Carbon tetrachloride	0.077	0.009	0.009	0.076
Chlorobenzene	0.18	0.029	0.11	0.69
Nitrobenzene	0.19	0.028	0.25	1.71
Ethers				
Diethyl ether	6.04	1.54	1.26	4.97
Di-n-propyl ether	0.49	0.09	0.45	2.34
Methyl n-propyl ether	3.05	0.76		
Diisopropyl ether	1.2	0.21	0.63	3.48
Methyl isopropyl ether	6.5	1.66	0.19	1.3
Di-n-butyl ether	0.03	0.005		
Di-n-hexyl ether	0.01	0.001	0.12	1.2
β,β'-Dichlorethyl ether	1.12	0.16	0.28	2.1
1,5-Dimethoxypentane	7.2	1.05		
1,5-Diethoxypentane	1.7	0.27		
Diethyl cellosolve	1.0	0.16	3.4	10.1
Dibutyl cellosolve	0.2	0.02	0.6	5.5
Dibutylcarbitol	0.3	0.03	1.4	15.0

a. Oxygenated Derivatives (cont.)

	Aqueous phase		Organic phase	
	wt % S	$100x_S$	wt % H_2O	$100x_W$
Ketones				
Diethyl ketone	4.81	1.07	1.62	7.1
Diisobutyl ketone	0.06	0.008	0.45	3.3
Methyl ethyl ketone	22.6	6.8	9.9	28.9
Methyl n-propyl ketone	6.0	1.32	3.6	15.2
Methyl n-butyl ketone	3.5	0.65	3.7	17.5
Methyl isobutyl ketone	1.7	0.31	1.9	9.8
Methyl n-pentyl ketone	0.43	0.07	1.50	8.8
Ethyl n-butyl ketone	0.43	0.07	0.78	4.8
Cyclohexanone	5.0	0.94	8.7	34.1
Methylcyclohexanone	3.0	0.5		
Esters				
Ethyl formate	8.2	2.12		
Ethyl acetate	7.48	1.63	3.20	13.9
Ethyl n-butyrate	0.68	0.11	0.75	4.6
n-Propyl acetate	2.6	0.44	1.9	9.9
Isopropyl acetate	2.9	0.54	1.9	9.9
n-Butyl acetate	1.0	0.17	1.37	8.2
Isobutyl acetate	0.67	0.10	1.65	9.8
sec-Butyl acetate	0.74	0.12	2.1	12.2
n-Butyl propionate	0.15	0.02	0.58	4.0
Isopentyl formate	0.3	0.05		
Isopentyl acetate	0.2	0.03	1.0	6.8
Isopentyl propionate	0.3	0.04		
Alcohols				
n-Butyl alcohol	7.31	1.88	20.4	48.8
Isobutyl alcohol	8.0	2.07	16.5	45.0
sec-Butyl alcohol	20.3	5.85	38.0	71.5
n-pentyl alcohol	2.19	0.45	7.5	28.3
Isopentyl alcohol	2.5	0.52	9.0	32.7
n-Hexyl alcohol	0.56	0.10	7.2	42.1
sec-Heptyl alcohol	0.35	0.06	5.1	25.9
n-Octyl alcohol	0.03	0.05		
sec-Octyl alcohol	0.05	0.008	0.1	0.7
Isooctyl alcohol	0.1	0.016	2.55	15.9
Isononyl alcohol	0.1	0.014	2.9	19.4
n-Decyl alcohol	0.02	0.002	3.0	21.4
Cyclohexanol	6.0	1.15	11.8	42.6
2-Methylcyclohexanol	1.1	0.19	3.0	16.3

The values pertain to 25°C.

b. Neutral Organophosphorus Esters

	Ester in water (g/l)	Water in ester (g/l)	Mole ratio
Phosphates			
Diethyl pentyl	7.5 (1)	180 (1)	2.2
Diethyl decyl	0.1 (1)	130 (1)	2.1
Di-n-butyl methyl	7.1 (1)	145 (1)	1.8
Di-n-butyl ethyl	3.4 (1)	120 (1)	1.7
Di-n-butyl n-hexyl	0.1 (1)	49 (1)	0.8
Di-n-butyl n-octyl	0.1 (1)	44 (1)	0.8
Di-n-butyl n-decyl	0.1 (1)	42 (1)	0.8
n-Butyl diphenyl	0.2 (1)	16 (1)	0.3
Di-n-butyl ethoxybutyl	0.7 (1)	111 (1)	1.9
Tri-n-butoxyethyl	1.1 (1)	73 (1)	1.3
Tri-n-butyl	0.39 (1)	64 (1)	
	0.414 (2)	54 (3)	
	0.422 (4)	64.372 (5)	1.045
		64.4 (6)	1.05
		64.44 (7)	1.103 (8)
(commercial)			1.084 (8)
Triisobutyl			1.142 (8)
Tri-*sec*-butyl			0.784 (8)
Tri-β-chloroethyl	0.2 (1)	78 (1)	1.2
Phosphonates			
Dimethyl n-octyl		910 (9)a	12.8
Diethyl n-hexyl	0.6 (1)	275 (1)	3.4
Diethyl n-octyl	0.2 (1)	172 (1)	2.4
		235 (9)a	3.9
Diethyl phenyl	23.65 (10)	200 (10)	2.7
Diethyl benzyl		292 (9)a	4.0
Di-n-propyl phenyl	3.63 (10)		
Diisopropyl phenyl	6.56 (10)		
Di-n-butyl methyl		270 (1)	3.1
Di-n-butyl ethyl		175 (1)	2.2
Di-n-butyl n-butyl	0.5 (1)	103 (1)	1.4
		104 (9)a	1.4
Di-n-butyl n-hexyl	0.2 (1)	75 (1)	1.2
Di-n-butyl isooctyl	0.2 (1)	54 (1)	0.9
Di-n-butyl n-decyl	0.2 (1)	55 (1)	1.0
Di-n-butyl n-hexadecyl	0.2 (1)	50 (1)	1.1
Di-n-butyl phenyl	0.43 (10)	64 (1)	1.0
Diisobutyl phenyl	0.46 (10)		
Di-*sec*-butyl phenyl	0.84 (10)		

b. Neutral Organophosphorus Esters (cont.)

	Ester in water (g/l)	Water in ester (g/l)	Mole ratio
Di-*tert*-butyl phenyl	1.90 (10)		
Diisopentyl methyl	1.90 (11)	121 (11)	1.8
Diisopentyl phenyl	0.069 (10)		
Di-n-hexyl phenyl	0.013 (10)		
Di-n-octyl phenyl	0.002 (10)	20 (10)	0.5
		22 (1)	0.5
Diisooctyl phenyl	0.004 (10)	25 (1)	0.6
Di-β-chloroethyl phenyl	2.8 (10)		
Di-2-ethylhexyl phenyl	0.002 (10)		
Phosphinates			
Ethyl di-n-butyl	13 (1)	416 (1)	4.8
Ethyl di-n-hexyl	0.1 (1)	150 (1)	2.2
n-Butyl di-n-butyl	4.5	160 (1)	2.3
n-Octyl diethyl		1260 (9)a	18.4
Phosphine oxide			
Tri-n-butyl	40 (1)	330 (1)	4.0
	55 (12)	530 (13)	7.3
Diphosphonates			
Tetra-n-butyl methylene		132 (14)	3.3
Tetra-n-butyl ethylene		150 (14)	3.9
Tetra-n-butyl propylene		191 (14)	5.1
Tetra-n-butyl butylene		180 (14)	5.0
Tetra-n-hexyl methylene	0.015 (15)		

All values pertain to 25°C, except those marked a, valid at 30°C.

1. L. L. Burger and R. M. Wagner, *Ind. Eng. Data Series*, 3, 310 (1958).
2. C. E. Higgins and W. H. Baldwin, *J. Inorg. Nucl. Chem.*, 24, 112 (1962).
3. A. M. Rozen in *Ekstraktsiya*, Vol. 1, Gosatomizdat, Moscow, 1962, p. 6.
4. C. E. Higgins, W. H. Baldwin and B. A. Soldano, *J. Phys. Chem.*, 63, 113 (1959).
5. W. Davis, Jr., *Nucl. Sci. Eng.*, 14, 159 (1962).
6. C. J. Hardy, D. Fairhurst, H. A. C. McKay and A. M. Willson, *Trans. Faraday Soc.*, 60, 1626 (1964).
7. J. W. Roddy and J. Mrochek, *J. Inorg. Nucl. Chem.*, 28, 3019 (1966).
8. C. E. Higgins and W. H. Baldwin, *J. Org. Chem.*, 21, 1156 (1956).
9. T. H. Siddall, III., *J. Phys. Chem.*, 64, 1340 (1960).
10. L. L. Burger and E. C. Martin, *Proc. 139th Natl. Mtg. ACS*, 1961.
11. A. S. Solovkin, *Zh. Neorg. Khim.*, 5, 1107 (1960).
12. C. E. Higgins, W. H. Baldwin and R. W. Higgins, *USAEC Rept.*, ORNL–2782 (1959).
13. C. E. Higgins and W. H. Baldwin, *Anal. Chem.*, 32, 233 (1960).
14. T. H. Siddall, III and C. A. Prohaska, *Inorg. Chem.*, 4, 783 (1965).
15. J. J. Richard, K. E. Buike, J. W. O'Laughlin and C. V. Banks, *J. Am. Chem. Soc.*, 83, 1722 (1961).

E. PROPERTIES OF COMMERICAL ION EXCHANGERS

Resin designation and type	Crosslinking (%DVB)	Capacity eq./l bed	meq./g dry	Matrix and ionic group
Amberlite IR 4B (A)	Obsolete, weakly basic anion exchanger			
IR 45 (A)		2	5	Polystyrene, weak amine
IR 100 (C)	Obsolete, cation exchanger			
IR 112 (C)		1.9–2.2		Polystyrene, $-SO_3^-$
IR 120 (C)	8	1.9	4.3–5.0	Polystyrene, $-SO_3^-$
IR 122 (C)	10	2.1	4.3–5.0	Polystyrene, $-SO_3^-$
IR 124 (C)	12	2.1	4.3–5.0	Polystyrene, $-SO_3^-$
IRA 400 (A)	8	1.2	2.6	Polystyrene, $-CH_2N(CH_3)_3^+$
IRA 401 (A) Porous		1.0	3.0	Polystyrene, $-CH_2N(CH_3)_3^+$
IRA 402 (A)		1.3		
IRA 410 (A)	6	1.2	3.0	Polystyrene, $-CH_2N(CH_3)_2C_2H_4OH^+$
IRA 411 (A) Porous		0.7	3.0	Polystyrene, $-CH_2N(CH_3)_2C_2H_4OH^+$
IRC 50 (C)		3.5	9.5	Polyacrylate, $-COOH$
XE 100 (C)	4	1.2	4.5	Polystyrene, $-SO_3^-$
Amberlyst 15 (C) Macro-reticular		1.2	4.9	Polystyrene, $-SO_3^-$
Anionit AV 17 (A)			3.4	Polystyrene, $-CH_2NR_3^+$
EDE 10 (A)			7.8	Phenol condensation, weak base
Bio-Rex 40 (C)	Equivalent to Duolite C3, analytical grade			
62 (C)	Equivalent to Duolite C62, analytical grade			
63 (C)	Equivalent to Duolite C63, analytical grade			
70 (C)	Equivalent to Duolite CS 101, analytical grade			
Chelex 100 (C)	Equivalent to Dowex A 1, analytical grade			
De Acidite FF (A)	7 to 9	1.6	4.0	Polystyrene, $-CH_2N(CH_3)_3^+$
G (A)		1.6	4.0	Polystyrene, $-CH_2N(C_2H_5)_2$
H (A)		1.5	3.8	Polystyrene, mixed strong and weak base
M (A)		2.2	5.5	Polystyrene, polyamine
Dowex 1 (A)	1 to 16		3.5	Polystyrene, $-CH_2N(CH_3)_3^+$
2 (A)	2 to 16		3	Polystyrene, $-CH_2N(CH_3)_2C_2H_4OH^+$
3 (A)	3		6	Polystyrene, weak base
50 (C)	1 to 16		4.9–5.2	Polystyrene, $-SO_3^-$
A–1 (C)		0.33	1–1.2	Polystyrene, $-CH_2N(CH_2COOH)_2$
21 K (A)	4, Porous	1.2	4.5	Polystyrene, $-CH_2N(CH_3)_3^+$

e. Properties of Ion Exchangers (cont.)

Resin designation and type		Crosslinking (%DVB)	Capacity eq./l bed	meq./g dry	Matrix and ionic group
Duolite	A 14 (A)		2.5	8	Polystyrene, weak base
	A 40 (A)		1.1	3.7	Polystyrene, $-CH_2N(CH_3)_2C_2H_4OH^+$
	A 42 (A)		0.7	2.3	Polystyrene, $-CH_2N(CH_3)_3{}^+$
	A101 (A)		1.3	4.0	Polystyrene, $-CH_2N(CH_3)_2C_2H_4OH^+$
	C 3 (C)		1.2	2.9	Phenolic consdensation, $-CH_2SO_3{}^-$
	C10 (C)	Porous	0.6	2.9	Phenolic condensation, $-CH_2SO_3{}^-$
	C20 (C)	8	2.2	5.1	Polystyrene, $-SO_3{}^-$
	C25 (C)	Porous	1.7	5.1	Polystyrene, $-SO_3{}^-$
	C62 (C)	Porous	2.6	6.0	Polystyrene, $-HPOOH$
	C63 (C)	6	3.2	6.6	Polystyrene, $-PO(OH)_2$
	CS 100 (C)		0.8	1.9	Phenolic condensation, $-COOH$
	ES 105 (A)		1.2		Polystyrene, $-SR_2{}^+$
Ionac	A300 (A)	Equivalent to Permutit A			
	A315 (A)	Equivalent to Permutit W			
	A540 (A)	Equivalent to Permutit S1			
	A550 (A)	Equivalent to Permutit S2			
	C240 (C)	Equivalent to Permutit Q			
	C265 (C)	Equivalent to Permutit H			
	C270 (C)	Equivalent to Permutit H70			
Kationit	KU1 (C)			4.0	Phenol condensation, $-SO_3{}^-$
	KU2 (C)		1.2	4.6	Polystyrene, $-SO_3{}^-$
	SBS (C)		0.8	2.9	Polystyrene, $-SO_3{}^-$
Lewatit	CNO (C)		2.5	4.0	Phenolic condensation, $-COOH$
	KSN (C)		1.6	4.0	Phenolic condensation, $-SO_3{}^-$
	M500 (A)		1.6	4.0	Polystyrene, $-CH_2N(CH_3)_3{}^+$
	M600 (A)		1.6	4.7	Polystyrene, $-CH_2N(CH_3)_3{}^+$
	S 100 (C)	8	2.5	4.8	Polystyrene, $-SO_3{}^-$
Nalcite	HCR (C)	Equivalent to Dowex 50 X 8			
	HGR (C)	Equivalent to Dowex 50 X 10			
	SAR (A)	Equivalent to Dowex 2			
	SBR (A)	Equivalent to Dowex 1			
	WBR (A)	Equivalent to Dowex 3			

e. Properties of Ion Exchangers (cont.)

Resin designation and type	Crosslinking (% DVB)	Capacity eq./l bed	Capacity meq./g dry	Matrix and ionic group
Permutit	A (A)		2	8	Condensation, strong and weak base
	C (C)		4	10	Polyacrylate, $-COOH$
	ESB (A)		1.2	3.2	Polystyrene, $-CH_2N(CH_3)_3^+$
	ES (A)		1.2	3.2	Polystyrene, $-CH_2N(CH_3)_2C_2H_4OH^+$
Permutit	H (C)		1.9	5.0	Phenolic condensation, $-COOH$
	HC (C)			4.0	Phenolic condensation, $-COOH$
	H70 (C)		3.6	7.9	Polyacrylate, $-COOH$
	Q (C)	8	2.0	4.8	Polystyrene, $-SO_3^-$
	RS (C)			5.5	Polystyrene, $-SO_3^-$
	S1 (A)		0.9	3.1	Polystyrene, $-CH_2N(CH_3)_3^+$
	S2 (A)		1.2	3.3	Polystyrene, $-CH_2N(CH_3)_2C_2H_4OH^+$
	W (A)		2.0	5.7	Polystyrene, weak base
Wofatit	CN (C)			2.0	Phenolic, $-COOH$
	CP (C)			10	Polyacrylate, $-COOH$
	F (C)			2.9	Phenolic, $-SO_3^-$
	KPS (C)			4.5	Polystyrene, $-SO_3^-$
	SBW (A)			3.5	Polystyrene, $-CH_2N(CH_3)_3^+$
Zeokarb	215 (C)		0.9	2.6	Phenolic, $-SO_3^-$
	216 (C)		1.1	2.5	Phenolic, $-COOH$
	225 (C)	1 to 20	2.1	4.8	Polystyrene, $-SO_3^-$
	226 (C)	2.5; 4.5	3.5	10	Polyacrylate, $-COOH$

Manufacturers and Distributors

Amberlite and Amberlyst:	Rohm and Haas Co., U.S.A., marketed by Mallinckrodt Co.
Bio-Rex and Chelex:	Biorad Laboratories Inc., U.S.A.
De Acidite and Zeokarb:	Permutit Co. Ltd., England.
Dowex:	Dow Chemical Co., U.S.A., marketed also by Biorad, J.T. Baker, Fluka.
Duolite:	Chemical Process Co., U.S.A.
Ionac:	Ionac Co., U.S.A.
Kationit and Anionit:	produced in U.S.S.R
Lewatit:	Farbenfabriken Bayer, Germany (West).
Nalcite:	Nalco Chem. Co. (National Aluminate Corp.), U.S.A
Permutit:	Permutit Co., U.S.A. and Permutit A.G., Berlin Germany (West).
Wofatit:	VEB Farbenfabrik, Wolfen, Germany (East).

F. DISTRIBUTION CHARTS

This appendix contains a survey of the distribution ratios of most of the elements at tracer concentrations between aqueous acid or salt solutions and various ion exchangers and extractant solutions.

The ion exchange charts were compiled from data given by Kraus and by Faris and their coworkers[1-4], while those for the extractability of metals were taken from similar charts compiled by the group at JAERI[5-9].

REFERENCES

1. K. A. Kraus and F. Nelson, *Proc. U.N. Intern. Conf. Peaceful Uses At. Energy, 1st, Geneva*, **7**, 113 (1956).

2. F. Nelson, T. Murase and K. A. Kraus, *J. Chromatog.*, **13**, 503 (1964).

3. J. P. Faris, *Anal. Chem.*, **32**, 520 (1960).

4. R. F. Buchanan and J. P. Faris, *Intern. Conf. Uses Radioisotopes in Phys. Sci. and Ind.*, Copenhagen, 1960, paper RICC/173.

5. K. Kimura, *Bull. Chem. Soc. Japan*, **33**, 1038 (1960).

6. T. Ishimori et al., *JAERI Rept.*, 1106 (1966).

7. T. Ishimori and E. Nakamura, *JAERI Rept.*, 1047 (1963).

8. T. Ishimori et al., *JAERI Rept.*, 1062 (1964).

9. K. Ueno and C. Chang, *Nippon Genshiryoku Gakkaishi*, **4**, 457 (1962).

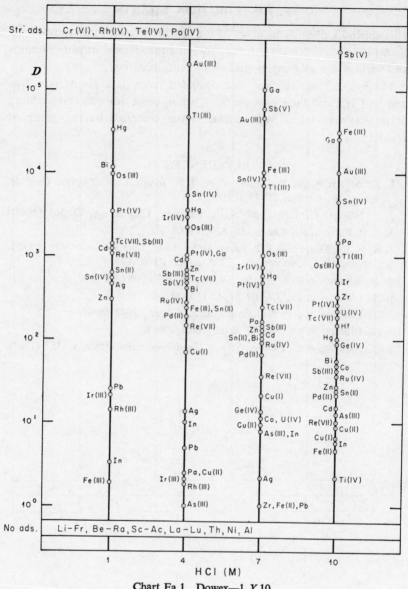

Str. ads. Cr(VI), Rh(IV), Te(IV), Po(IV)

No ads. Li–Fr, Be–Ra, Sc–Ac, La–Lu, Th, Ni, Al

HCl (M)

Chart Fa 1   Dowex—1 X 10

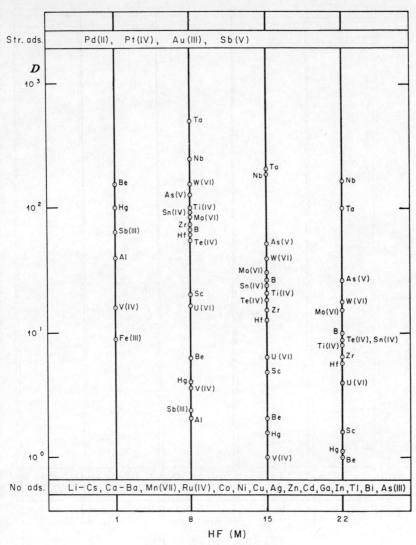

Chart Fa 2   Dowex—1 X 10

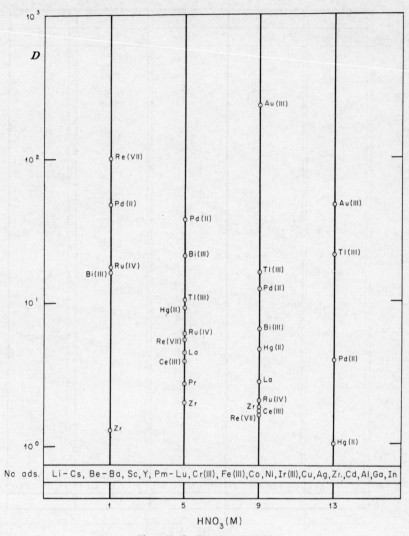

Chart Fa 3　　Dowex—1 X 10

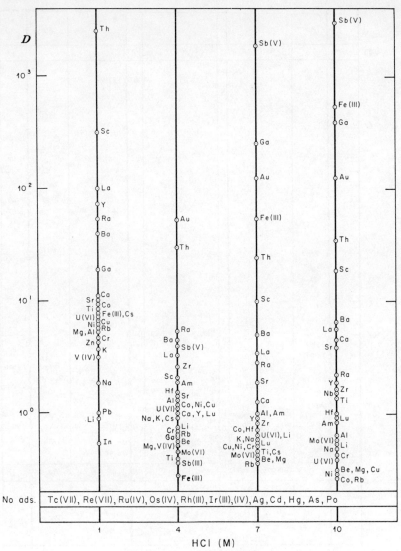

No ads. | Tc(VII), Re(VII), Ru(IV), Os(IV), Rh(III), Ir(III),(IV), Ag, Cd, Hg, As, Po

HCl (M)

Chart Fa 4   Dowex—50 *X* 4

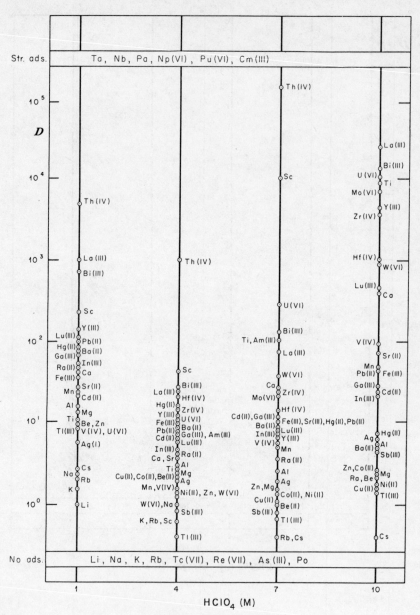

Chart Fa 5   Dowex—50 *X* 4

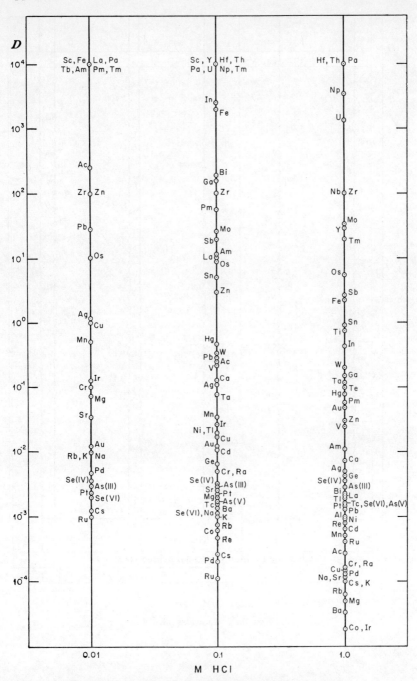

Chart Fb 1   50% Bis-(2-ethylhexyl)phosphoric acid in toluene

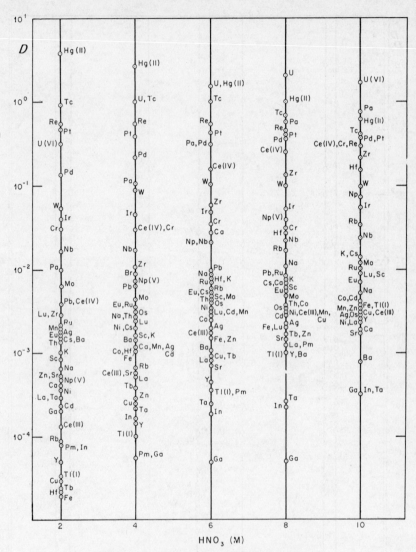

Chart Fc 1   Methyl isobutyl ketone (hexone)

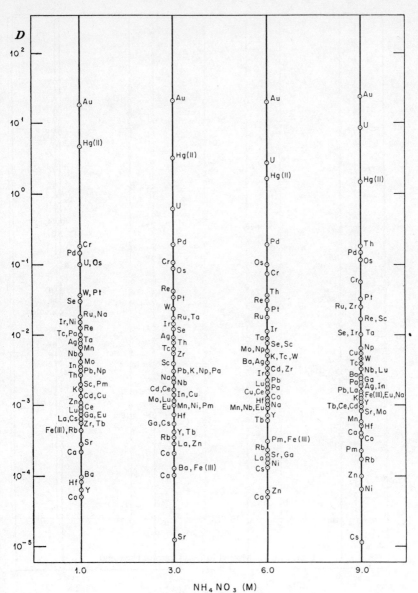

NH₄NO₃ (M)

Chart Fc 2   Diethyl ether

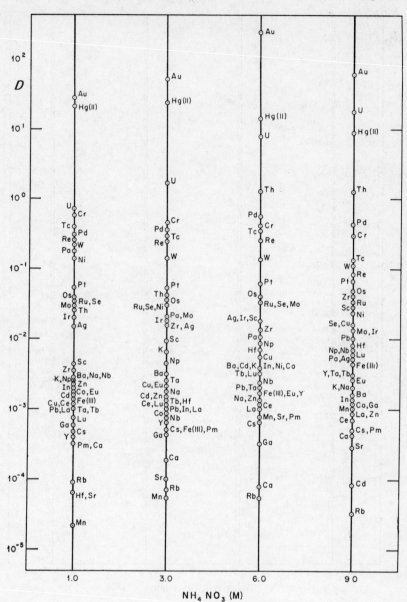

Chart Fc 3   Diethyl cellosolve

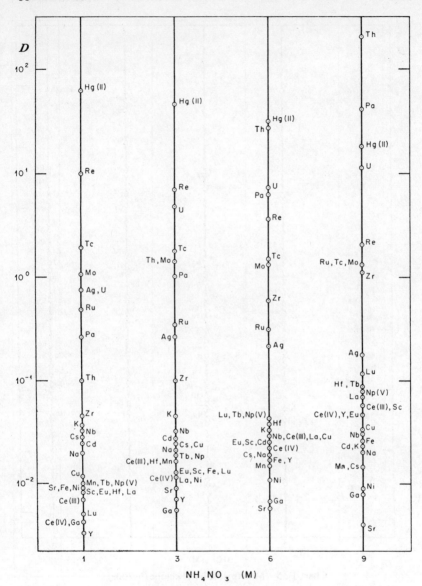

Chart Fc 4   Pentaether

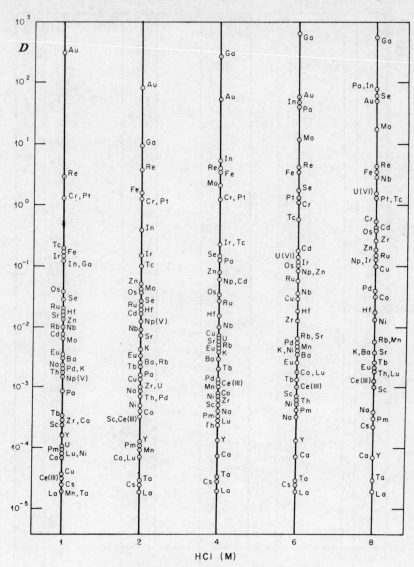

HCl (M)

Chart Fc 5   Mmethyl isobutyl ketone (hexone)

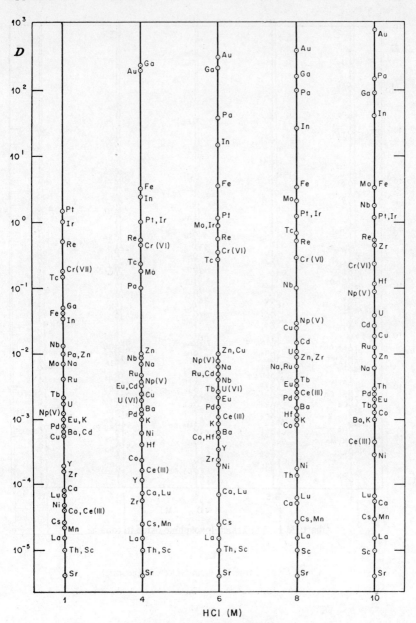

Chart Fc 6   Diisopropyl ketone

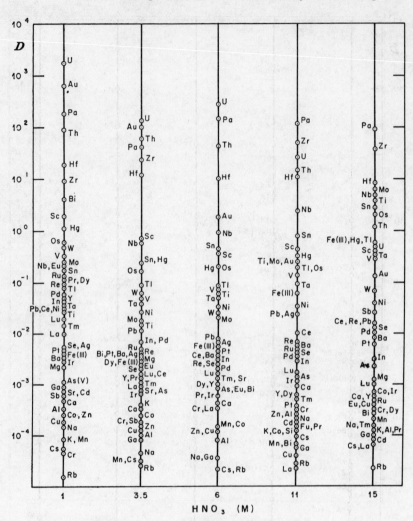

Chart Fd 1   5% Trioctylphosphine oxide in toluene

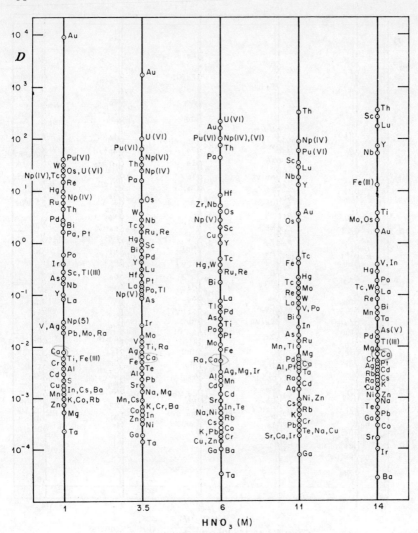

Chart Fd 2   Undiluted tributyl phosphate

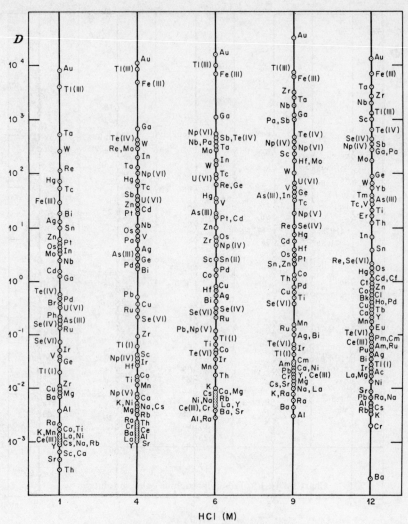

HCl (M)

Chart Fd 3   Undiluted tributyl phosphate

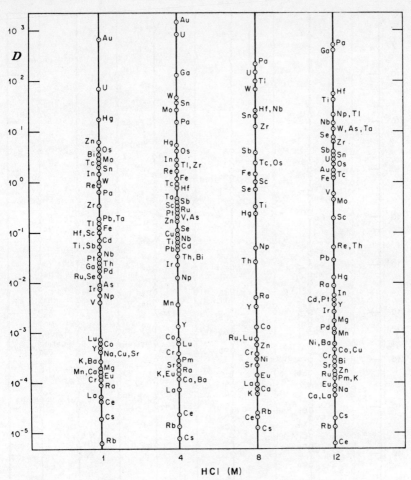

Chart Fd 4   1% Tributylphosphine oxide in toluene

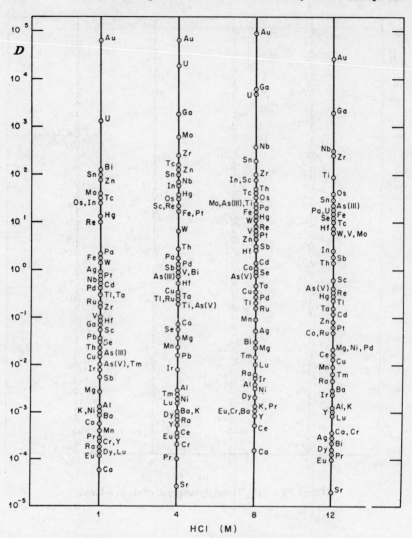

HCl (M)

Chart Fd 5   5% Trioctylphosphine oxide in toluene

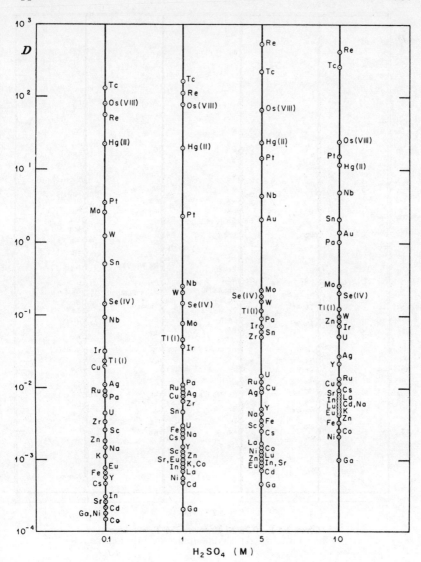

Chart Fd 6   Undiluted tributyl phosphate

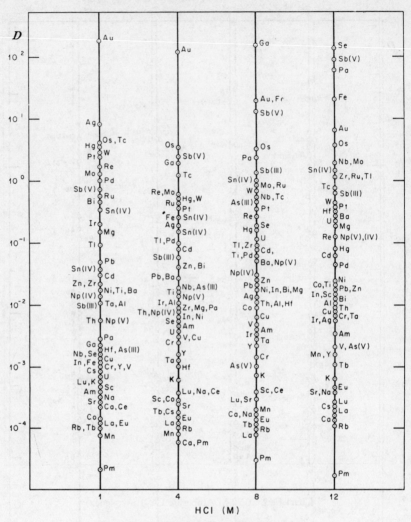

Chart Fe 1   0.28 M Primene JM-T in xylene

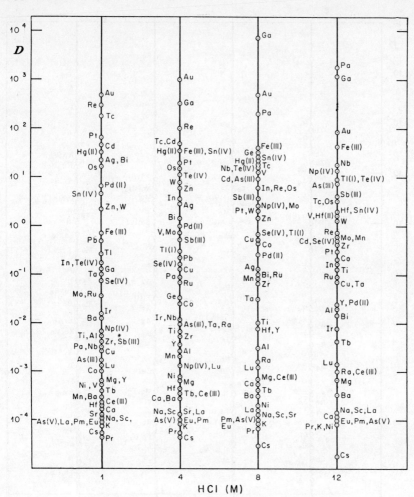

Chart Fe 2   0.20 M Amberlite LA–1 in xylene

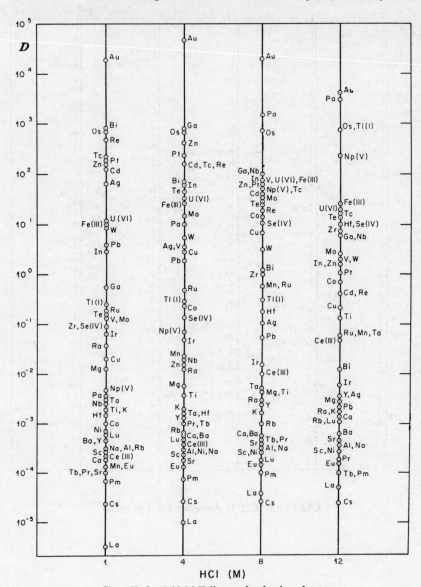

Chart Fe 3   0.11 M Triisooctylamine in xylene

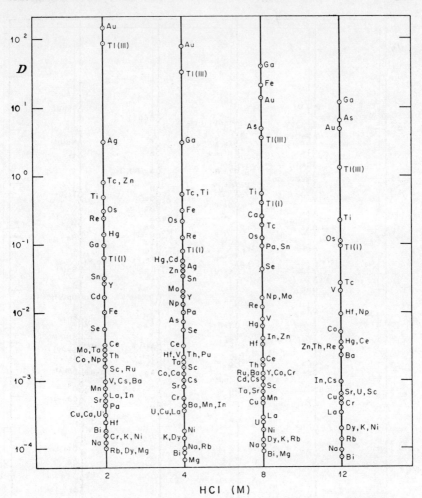

Chart Fe 4   0.1 M Benzyldimethylphenylammonium chloride in chloroform

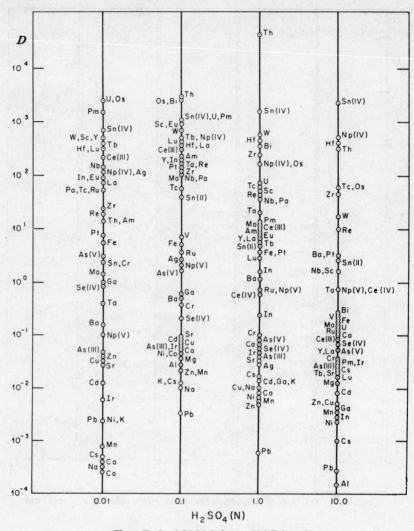

Chart Fe 5   0.28 M Primene JM-T in xylene

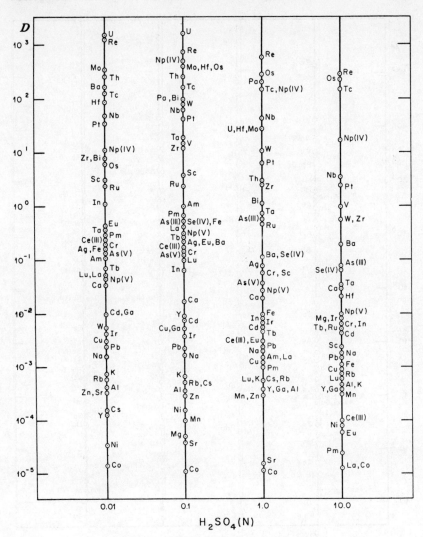

Chart Fe 6   0.20 M Amberlite LA–1 in xylene

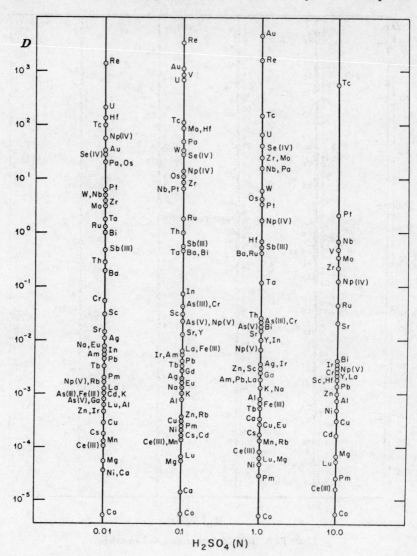

Chart Fe 7   0.11 M Triisooctylamine in xylene

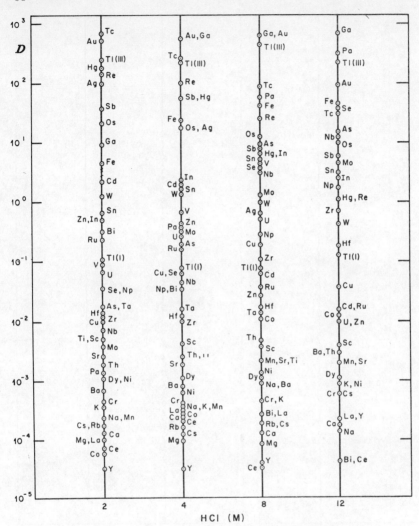

Chart Fe 8   0.05 M Tetraphenyl arsonium chloride in chloroform

Author Index

This author index is designed to enable the reader to locate an author's name and work with the aid of the reference numbers appearing in the text. The page numbers are printed in normal type in ascending numerical order, followed by the reference numbers in brackets. The numbers in *italics* refer to the pages on which the references are actually listed.

If reference is made to the work of the same author in different chapters, the above arrangement is repeated separately for each chapter.

978   *Author Index*

Eliezer, I.—*contd.*
447 (63–65), *494*, 787 (289) *811*,
852, 854 (102), *858*
Elliott, M. C., 709 (539), *735*
Ellis, D. A., 524 (53, 54), 525 (75),
565, *566*
Elovich, S. Yu., 311 (211), *319*, 331
(43), 341, 342 (88), *354*, *356*
Elving, P. J., (340), *813*
Emerson, M. T., 168, 173 (233), *185*,
265 (78), *315*
Ephraim, F., 774 (221), *808*
Erickson, R. L., 641 (241), *724*, 887,
896 (48), *908*
Ermakov, A. N., 331 (46, 51), *355*
Ernst, J., 900 (123), *910*
Errera, J., 123 (48), *179*, 744 (49), *802*
Estill, W. E., 635 (212), *723*
Estok, G. K., 652, 653 (308), *727*
Eucken, A., 82 (219), *92*
Evans, D. F., 20 (43), *86*, 150 (153,
156), 155 (168), 161 (153), *183*
Evans, M. G., 14, 17 (17), *85*
Evans, M. W., 21, 29 (46), *86*
Evans, W., 60 (162), *90*
Evans, W. V., 627 (177), *722*
Even-Sapir, I., 433 (18), *493*
Everest, D. A., 267 (91), 284, 286
(139), 288 (139, 149), 296 (139),
315–317, 363 (12, 14), 364 (15–
18), 365 (14), 384, 405 (57), *421*,
422, 436 (35), *493*, 539, 543, 544
(172), *569*
Ewart, J. A. D., 514 (32), *564*
Eyal, E., 266 (80), *315*
Eyring, H., 7, 8 (3), *85*
Ezrin, M., 250 (29, 32), *313*

Fackler, J. P., Jr., 831 (19), *855*
Fairbrother, F., 626 (172), *722*
Fairhurst, D., 660, 661, 663–667, 685
(348), *728*, (6), *935*
Falconer, J. D., 250 (38), *313*
Falkenhagen, H., 150, 151 (158), *183*
Faris, J. P., 307, 308 (197), *318*, 392,
393 (87), *423*, 885 (26), 905 (149),
907, *911*, (3, 4), *939*
Farrell, M. S., 826, 830 (12), *855*
Fasching, J. L., 771, 772, 791 (218),
808, 889, 890 (84), *909*

Fasolo, G. B., 787 (309), *812*
Fattore, V., 669, 673, 674, 676 (409),
730
Faure, A., 783, 786 (273), *810*
Fay, R. C., 508 (28, 29), *564*, 854
(108), *858*
Feakins, D., 144, 145 (124), *182*
Fedomova, L. N., 379 (39), *422*
Fedorov, I. A., 669, 694, 710 (389),
730
Feigl, F., 798, 799 (369), *814*
Feldshtein, L. S., 748 (93), *804*
Felton, L. D., 433 (15), *493*
Fernando, M. J., 753 (129), *805*
Fernelius, W. C., 523 (40), *564*
Ferraro, J. R., 135 (89–94), 137 (96),
145 (90–94), 147 (91), *181*, 508
(25), 523 (44, 52), 528 (44, 102,
103, 111, 112), 529 (102, 103,
121), 530 (103), 536 (162, 168),
543 (168), 544, 545 (103, 168),
546 (217), 549 (162), *564*, *565*,
567, *569*, *571*, 619 (128), 620
(135–138), 621 (136, 137), 623
(135, 142, 143), 624 (135), 629, 631
(138), 635 (135, 142), 669, 675,
676 (392), 697 (135–137), 699,
702 (469), *720*, *721*, *730*, *732*, 828
(8), 830, 831, 833 (8, 15), 848 (87),
855, *858*
Fessler, R. G., 311 (215), *319*
Fiat, D., 31 (80), *87*
Field, B. O., 695, 697, 698, 700 (456),
716 (574), *732*, *736*
Figgis, B., 795 (349), *813*
Fine, D. A., 327 (20), *354*
Finston, H. L., 436 (31), *493*, 504 (7),
563
Fischer, I., 124 (55), *180*
Fischer, W., 625, 642, 649 (162),
721
Fisher, C. H., 779, 780 (244), *809*
Fix, R. C., 799 (381), *814*
Fleming, J. E., 701 (482), *733*
Fletcher, A. N., 226 (102), *238*
Fletcher, A. W., 562 (304), *574*
Fletcher, J. M., 481, 484, 485 (109),
496, 536, 537 (169), 544, 545
(208), *569*, *570*, 582, 595, 604
(10), 669, 670, 673, 675, 677–679

Seeley, F. G., 781 (259), 783 (259, 266)
787 (259, 266, 311), 788 (259,
311), 791 (259, 311), 792 (266),
793 (311), *810, 812*
Seidel, A., 582 (19), *717*
Seidell, A., 135, 181 (83), *131*
Seidl, K., 631 (199), *723*
Sekiguchi, K., 311 (212), *319*
Sekine, T., 821–823 (30), 824, 825
(29, 30), 826 (30), 827 (29–31),
833 (29, 30), 834 (30), 853, 854
(30, 107), *855, 858*
Semeneko, K. N., 625 (160, 161), 626,
627 (160, 161, 165), *721, 722*
Semmes, R. T., 221 (72), *236*
Senise, P., 793, 795 (350), 798 (366),
813, 814
Senko, M. E., 776 (232), *809*
Sergievskii, V. V., 748, 760, 777 (76),
803
Serveev, A. N., 881, 882 (11), *907*
Seryakova, I. V., 491 (152, 153), *498,*
584, 585 (39), 589 (48), 606, 612
(48, 97), 614, 625, 641 (48), 642
(39, 48, 243), 644 (48, 243), 694
(97), *717, 719, 724*
Seryakova, S. V., 887–889, 893 (43), *908*
Seryavin, M. M., 331, 336 (52), *355*
Serzeev, B. M., 882 (14), *907*
Shakurov, A. I., 345 (94), *357*
Shakurov, V. G., 326 (15), *354*
Shane, N., 175, 178 (245), *186*
Sharma, H. D., 557 (294), *573*
Sharp, D. W. A., 749 (98), *804*
Sharp, R. A., 606–608, 612, 635 (90),
648 (264), *719, 725*
Shearer, W. N., 584, 592, 606, 609,
611, 648 (37), *717*
Shedlovsky, T., 153 (160), *183*
Sheka, Z. A., *233*, 481, 484 (106), *496,*
536, 537 (158, 163, 164), 538 (163,
164), 539 (164, 174, 183), 543
(174), *568, 569,* 669, 673, 674, 679
(385), *729*
Shepchenko, I. G., 705 (508), *734*
Sheppard, J. C., 486 (119), *496,* 787
(298), *811,* 903, 904 (135), *911*
Sheppard, J. W., 251 (50), *314*
Sheppard, N., 528 (115), *567*
Sheridan, C. L., (6), *921*

Sheridan, R. C., 157, 158 (179), *184*
Sherrill, M. L., 744 (49), *802*
Sherrill, M. S., 444, 445 (51), *494*
Sherry, H. H., 307, 308 (196), *318*
Sherry, H. S., 387 (69, 70), *423*
Shesterikov, N. N., 651, 653 (295),
693, 694 (454), *726, 732*
Shesterikov, V. N., 741 (19), 748, 760
(19, 75), 764 (19), *801, 803*
Shevchenko, V. B., 441 (43), 486 (118),
494, 496, 523, 525 (42), 526 (96),
530 (96, 131), 539 (175), 544, 545,
547 (42, 209), 549, 550 (42), *564,*
566, 568–570, 657 (329), 669 (389,
395, 398), 673 (395, 398), 674
(395), 685, 687 (434), 690 (443),
692 (450, 451), 694 (389), 695
(460, 462), 698 (460, 462), 700
(481), 703 (460, 462), 705 (508,
509), 706 (434, 521), 710 (389),
713 (329, 481, 547), 714 (329,
460), *727, 730–735,* 766, 768
(188), 771, 788, 790 (210), 792
(337), *807, 808, 813,* 837 (64),
843, 844 (64, 76), *857*
Shikheeva, L. V., 562 (305), *574*
Shilin, A. I., 524, 526 (61), *565,* 651
(275), *726*
Shilin, I. V., 441 (43), *494,* 657 (329),
669 (395, 414), 673 (395), 674
(395, 414), 685, 687 (434, 435),
693, 694 (435), 695, 698 (460,
462), 700 (481), 703 (460, 462),
706 (434, 521), 713 (329, 481),
714 (329, 460), *727, 730–734*
Shiloh, M., 80, 82, 83 (212), *91,* 444
(58), *494,* 788 (312), 791 (329),
812
Shimogiva, H., 404 (109), *424*
Shinoda, K., 167, 169, 170 (222), *185,*
556, 557 (278), *573*
Shirai, M., 745 (57), *802*
Shirvinskii, E. V., 792 (336), *813*
Shishliakov, R. A., 338 (78), *356*
Shmidt, V. S., 680–682, 684, 685 (430),
703 (501), 705 (509), *731, 733,*
734, 741 (19), 743, 744 (26), 748
(19, 75), 750 (26), 760 (19, 75),
764 (19), 766, 768 (188, 189), 771,
788, 790 (210), *801, 803, 807, 808*

Subject Index

Absorption spectrum, charge transfer, 174, 198, 199
  infrared, in ion exchangers, 384
  infrared, of solvents, 104, 123, 491
  visible, of species in ion exchangers, 305, 384, 416, 905
  visible, of species in solvents, 490
Acetone, properties of, 101, 128, 927
  solubility of salts in, 135–138
Acetonitrile, properties of, 927
  solubility of salts in, 135, 137
Acetylacetone (AA), properties of, 506, 507
  dissociation constant of, 508
  distribution constant of, 508
  metal complexes of, 509
  synergistic extraction with, 818, 825, 853
Acido-complexes, extraction of, 467–471, 871, 887–889, 894–896
Acids, ionic dissociation of, 60, 222–228, 396, 397
Acid Effect, in anion exchange of complexes, 310, 384, 394–398
Acridine, use as extractant, 800
Activity coefficients, and concentration scales, 6, 49, 922–926
  Debye–Hückel expression, 36–39, 44–50, 53, 74, 84, 153, 156, 190, 197
  empirical formulas, 54–56
  Guggenheim's expression, 55, 64–65, 203
  hydration numbers and, 30, 50, 51
  of alkali halides, 232, 924, 925
  of alkylamine salts, 770, 776
  of components in mixtures, 6, 106, 110, 117–119
  of electrolytes, measurement of, 42–45
  of electrolytes, in organic solvents, 156–161

Activity coefficients—contd.
  of free ions, 196, 197, 224
  of invading electrolyte, in ion exchangers, 269–273
  —, abnormally low, 274
  of ion pairs, 197
  of neutral phosphorus esters, 659
  of nonelectrolyte in electrolyte solutions, 74–82
  of quaternary ammonium salts, 753
  of resinates, 273, 289–293, 374
  of single ions, 38–41
  of trace in bulk electrolytes, 60, 65 70–72, 899, 902
  quotients of, 61, 62, 203, 204, 455–463, 468–470
  ratio of, ion exchanger to external solution, 271, 274, 275, 372, 375, 380
  relation to salting power, 630
  stoichiometric, 6, 38, 42, 44, 157, 196, 379, 380
Affinity of a species to aqueous phase, 428, 429
Aggregates, mixed, of simple and complex amine salts, 476
Aggregation, degree of, 103, 168–171, 176, 177, 451, 452, 487
  of alkylamine salts, 757–765, 770
  of carboxylic acids, 554
  of dialkylphosphorus esters, 528, 531, 545
  of monoalkylphosphorus esters, 528, 530, 550
  of quaternary ammonium salts, 752, 754
  of solvents, 103, 104, 128, 129, 487
  of sulphonic acids, 554
Agitation, effect on rate of extraction or exchange, 254, 432, 433
Alcoholates, of inorganic salts, 627

Hyamine-1622, *see* quaternary ammonium salts
Hydration, of acidic phosphorus esters, 530
of aliphatic amines, 741
of neutral phosphorus esters, 653
of solutes, in ethers, 579, 591, 598, 610, 617, 632, 642
—, in neutral phosphorus esters, 669, 686, 701
Hydration of ions, absolute thermodynamic functions of, 13, 14, 18, 19
Born equation for, 15, 17, 21, 24, 27, 29
conventional thermodynamic functions for, 13, 14
enthalpy of, 11, 13, 14, 17, 18
entropy of, 11, 13, 14, 21, 27
free energy of, 11, 13, 16, 21, 27
permanent, 21, 29, 30
primary, 17, 21, 27, 29, 30
secondary, 17, 21, 29, 30
Hydration numbers, of ions, 10, 22, 25–33, 50–53, 58, 76, 82, 143, 145, 147
in ion exchangers, 264, 265, 269
from nuclear magnetic resonance, 31, 33
Hydriodic acid, activity coefficients of, 922
association of, 224
extraction of, by alkylamines, 770
—, by ether, 605
—, by neutral phosphorus esters, 670
Hydrobromic acid, activity coefficients of, 922
association of, 224–228
extraction of, by alkylamines, 770
—, by ethers, 605
—, by neutral phosphorus esters, 690
H_0 of, 59
Hydrocarbons, properties of, 928, 932
Hydrochloric acid, activity coefficients of, 60, 61, 922
—, in chloride solutions, 69–71
—, in ion exchangers, 272, 275
—, in organic solvents, 157, 158

Hydrochloric acid—*contd.*
association of, 224–228
extraction of, by acidic phosphorus esters, 546
—, by alkylamines, 770–776
—, by ethers and ketones, 605–610
—, by neutral phosphorus esters, 680
H_0 of, 59
solubility in aqueous, of ethers, 585
—, of tributylphosphate, 654
synergistic extraction of, 849
see also hydrogen chloride
Hydrofluoric acid, activity coefficients of, 922
extraction of, by alkylamines, 770
—, by ethers, 605
—, by neutral phosphorus esters, 690
Hydrogen bonding, in acidic phosphorus esters, 528, 529
in alkylamines, 742
in alkylamine salts, 748, 769, 790
in carboxylic acids, 554
in ether extracts, 602, 603
in neutral phosphorus ester extracts, 658, 663, 667, 676, 685
in water, 7, 8
of aqueous solvents, 123
of solvents, 103, 104
Hydrogen chloride, solubility in inert solvents, 132
Hydrogen dihalides, in ion exchangers, 397
in extraction systems, 474, 475, 774, 905
Hydrogen dinitrate, in ion exchangers, 288, 397
Hydrogen ions, hydration of, 57, 867
solvation of, 467
Hydrogen peroxide, extraction of, 615, 692
Hydrolysis, localized, in electrolyte solutions, 232
in ion exchangers, 344, 345
of metal ions, 229–231, 386, 458, 879–882
Hydrophobic bonding, in aqueous solutions, 20, 80, 123, 125, 170–174